Subject	Symbol	Meaning	Page
SEQUENCES	$\ldots$	and so forth	180
	$\displaystyle\sum_{k=m}^{n} a_k$	the summation from k equals m to n of a_k	183
	$\displaystyle\prod_{k=m}^{n} a_k$	the product from k equals m to n of a_k	187
	$n!$	n factorial	188
SET THEORY	$a \in A$	a is an element of A	76
	$a \notin A$	a is not an element of A	76
	$\{a_1, a_2, \ldots, a_n\}$	the set with elements $a_1, a_2, \ldots, a_n$	76
	$\{x \in D \,|\, P(x)\}$	the set of all x in D for which $P(x)$ is true	77
	$\mathbf{R}, \mathbf{R}^-, \mathbf{R}^+, \mathbf{R}^{nonneg}$	the sets of all real numbers, negative real numbers, positive real numbers, and nonnegative real numbers	76
	$\mathbf{Z}, \mathbf{Z}^-, \mathbf{Z}^+, \mathbf{Z}^{nonneg}$	the sets of all integers, negative integers, positive integers, and nonnegative integers	76
	$\mathbf{Q}, \mathbf{Q}^-, \mathbf{Q}^+, \mathbf{Q}^{nonneg}$	the sets of all rational numbers, negative rational numbers, positive rational numbers, and nonnegative rational numbers	76
	$A \subseteq B$	A is a subset of B	232
	$A \nsubseteq B$	A is not a subset of B	233
	$A = B$	A equals B	234
	$A \cup B$	A union B	236
	$A \cap B$	A intersect B	236
	$B - A$	the difference of B minus A	236
	A^c	the complement of A	236
	(x, y)	ordered pair	238
	$(x_1, x_2, \ldots, x_n)$	ordered n-tuple	238
	$A \times B$	the Cartesian product of A and B	238
	$A_1 \times A_2 \times \ldots \times A_n$	the Cartesian product of $A_1, A_2, \ldots, A_n$	238
	$\varnothing$	the empty set	260
	$\mathcal{P}(A)$	the power set of A	264

Continued on the back of this page.

Subject	Symbol	Meaning	Page
COUNTING AND PROBABILITY	$n(A)$	the number of elements in a set A	276
	$P(A)$	the probability of a set A	275
	$P(n, r)$	the number of r-permutations of a set of n elements	289
	$\binom{n}{r}$	n choose r, or the number of r-combinations of a set of n elements, or the number of r-element subsets of a set of n elements	306
	$[x_{i_1}, x_{i_2}, \ldots, x_{i_r}]$	multiset of size r	323
FUNCTIONS	$f : X \to Y$	f is a function from X to Y	345
	$f(x)$	the value of f at x	345
	$x \xrightarrow{f} y$	f sends x to y	345
	i_X	the identity function on X	349
	b^x	b raised to the power x	350
	$\exp_b(x)$	b raised to the power x	378
	$\log_b(x)$	logarithm with base b of x	350
	f^{-1}	the inverse function of f	382
	$f \circ g$	the composition of g and f	401
ALGORITHM EFFICIENCY	$x \cong y$	x is approximately equal to y	189
	$O(g(x))$	big-oh of g of x	485
	$O(g(n))$	big-oh of g of n	490
RELATIONS	$x \, R \, y$	x is related to y by R	534
	R^{-1}	the inverse relation of R	540
	$m \equiv n \pmod{d}$	m is congruent to n modulo d	559
	$[a]$	the equivalence class of a	561
	$x \preceq y$	x is related to y by a partial order relation $\preceq$	588

Continued on first page of back endpapers.

DISCRETE MATHEMATICS

with Applications

To Jayne and Ernest

DISCRETE MATHEMATICS

with Applications

Second Edition

Susanna S. Epp

DePaul University

Brooks/Cole Publishing Company

I(T)P An International Thomson Publishing Company

Pacific Grove • Albany • Belmont • Bonn • Boston • Cincinnati • Detroit • Johannesburg • London
Madrid • Melbourne • Mexico City • New York • Paris • Singapore • Tokyo • Toronto • Washington

For more information, contact:

BROOKS/COLE PUBLISHING COMPANY
511 Forest Lodge Road
Pacific Grove, CA 93950
USA

International Thomson Publishing Europe
Berkshire House 168-173
High Holborn
London WC1V 7AA
England

Thomas Nelson Australia
102 Dodds Street
South Melbourne, 3205
Victoria, Australia

Nelson Canada
1120 Birchmount Road
Scarborough, Ontario
Canada M1K 5G4

International Thomson Editores
Campos Eliseos 385, Piso 7
Col. Polanco
11560 Mexico C.F., Mexico

International Thomson Publishing GmbH
Konigswinterer Strasse 418
53227 Bonn, Germany

International Thomson Publishing Asia
211 Henderson Road
#05-10 Henderson Building
Singapore 0315

International Thomson Publishing Japan
Hirakawacho Kyyowa Building, 31
2-2-1 Hirakawacho
Chiyoda-ku, Tokyo 102
Japan

Sponsoring Editor: Steve Quigley
Editorial Assitant: John Ward
Production and Design: Pamela Rockwell
Marketing Manager: Marianne Rutter
Manufacturing Coordinator: Lisa Flanagan
Compositor: Interactive Composition Corporation
Cover Printer: John Pow Company
Text Printer and Binder: Courier Westford, Inc.

Library of Congress Cataloging-in-Publication Data
Epp, Susanna S.
   Discrete mathematics with applications / Susanna S. Epp.— 2nd
ed.
      p.   cm.
   Includes index.
   ISBN 0–534–94446–9 (acid-free)
   1. Mathematics.   I. Title.
QA39.2.E65   1995   94-31184
511—dc20   CIP

 This book is printed on recycled, acid-free paper.

Printed and bound in the United States of America
   96 97 98 99 — 10 9 8 7 6 5

Contents

Preface

Discrete mathematics concerns processes that consist of a sequence of individual steps. This distinguishes it from calculus, which studies continuously changing processes. While the ideas of calculus were fundamental to the science and technology of the industrial revolution, the ideas of discrete mathematics underlie the science and technology specific to the computer age.

Logic and Proof An important goal of a first course in discrete mathematics is to develop students' ability to think abstractly. This requires that students learn to use logically valid forms of argument, to avoid common logical errors, to understand what it means to reason from definitions, and to know how to use both direct and indirect argument to derive new results from those already known to be true.

Induction and Recursion An exciting development of recent years has been an increased appreciation for the power and beauty of "recursive thinking": using the assumption that a given problem has been solved for smaller cases, to solve it for a given case. Such thinking often leads to recurrence relations, which can be "solved" by various techniques, and to verification of solutions by mathematical induction.

Combinatorics Combinatorics is the mathematics of counting and arranging objects. Skill in using combinatorial techniques is needed in almost every discipline where mathematics is applied, from economics to biology, to computer science, to chemistry, to business management.

Algorithms and Their Analysis The word *algorithm* was largely unknown three decades ago. Yet now it is one of the first words encountered in the study of computer science. To solve a problem on a computer, it is necessary to find an algorithm or step-by-step sequence of instructions for the computer to follow. Designing an algorithm requires an understanding of the mathematics underlying the problem to be solved. Determining whether or not an algorithm is correct requires a sophisticated use of mathematical induction. Calculating the amount of time or memory space the algorithm will need requires knowledge of combinatorics, recurrence relations, functions, and O-notation.

Discrete Structures Discrete mathematical structures are made of finite or countably infinite collections of objects that satisfy certain properties. Those studied in this book are sets, Boolean algebras, functions, finite-start automata, relations, graphs, and trees. The concept of isomorphism is used to describe the state of affairs when two distinct structures are the same in their essentials and differ only in the labeling of the underlying objects.

Applications and Modeling   Mathematical topics are best understood when they are seen in a variety of contexts and used to solve problems in a broad range of applied situations. One of the profound lessons of mathematics is that the same mathematical model can be used to solve problems in situations that appear superficially to be totally dissimilar.

FEATURES OF THE SECOND EDITION

The most significant changes in the second edition are listed below.

Topic Reorganization   The chapters on logic have been tightened up to enable users to cover them more expeditiously. Set notation is now introduced in Chapter 2 rather than Chapter 5, and the barber problem, Russell's paradox, and the halting problem have been brought together into a single section. The introduction to algorithmic notation has been streamlined and is now part of the first applications section on algorithms. Proof by contradiction and proof by contraposition are introduced in a single section, and the classical proofs of the irrationality of $\sqrt{2}$ and the infinitude of the primes have been put next to each other. The introductory section on mathematical induction has been split and enlarged; proofs of inequalities and divisibility properties as well as more subtle induction problems are now in a section of their own. In Chapter 9 the first section, which contained some unnecessary review material, has been shortened to include only those topics needed to understand O-notation. The second and third sections now introduce the difficult ideas of O-notation and the analysis of algorithm efficiency in a simpler way without logarithmic or exponential functions. In the fourth and fifth sections those aspects of exponential and logarithmic functions that are most important for computer science are treated and algorithm efficiencies involving these functions are discussed. Chapters 7 and 10 have been revised so that instructors wishing to introduce functions as binary relations can cover the first section of Chapter 10 alongside the first section of Chapter 7.

New Material   New material has been added on two's complements, NAND and NOR gates, hash functions, the sieve of Eratosthenes, the characteristic function of a set, the Euler phi function, properties of Boolean algebras, and Stirling numbers of the second kind. Probability in the equally likely case has been incorporated throughout Chapter 6, and an entirely new section on r-combinations with repetition has been added. The discussion of algorithm efficiency in Chapter 9 has been expanded to include selection sort, insertion sort, and merge sort. Appendix B now contains an even greater number of complete solutions than the first edition to serve as an improved guide for students working independently.

SUPPORT FOR THE STUDENT

The prerequisites for this book are the same as for calculus except that trigonometry is not needed.

Worked Examples   To students, the most important part of a mathematics text is the collection of worked examples. This book contains many examples that are written in problem-solution form and are keyed in type and in difficulty to the exercises.

Exercises   In each section the mixture of exercises has been designed so that students with widely varying backgrounds and ability levels will find some exercises they can be sure to do successfully and also some exercises that will challenge them. Appendix B contains either full or partial solutions to a set of exercises whose selection was guided by a desire to give the most useful feedback to the student. Thus exercises with solutions are representative of all the exercises at the end of each section, the solutions are unusually complete, and when exercises are paired the one with a solution is stated first.

Spiral Approach to Concept Development   A number of concepts in this book appear in increasingly more sophisticated forms in successive chapters. Thus, for example, by the time students encounter the full-fledged concept of congruence relation and congruence class in Chapter 10, they have been introduced to *mod* and *div* in Chapter 3, studied partitions of the integers in Chapter 5, and considered *mod* and *div* as functions in Chapter 7. This approach builds in useful review and develops mathematical maturity in natural stages.

Inclusion of Applications   How to motivate students is one of the most important issues in mathematics education. Every concept in this book is applied in at least one and often in many different ways. Eight sections are explicitly devoted to applications, most of them to computer science, and several more sections are heavily weighted toward applications. Although these may best be appreciated by students with some prior or concurrent exposure to computer science *per se,* the treatment in this book is entirely self-contained.

Figures and Tables   I have tried to include figures and tables in every case where it seemed that doing so would help readers to a better understanding. In most tables a second color is used to add meaning.

Endpapers   The meanings of all special symbols used in this book and a list of reference formulas are given on the front and back endpapers.

SUPPORT FOR THE INSTRUCTOR

Exercises   The large variety of exercises at all levels of difficulty allows instructors great freedom to tailor a course to the abilities of their students and to assign whatever mixture they most prefer of exercises with and without answers. There is also an ample number of exercises to use for review assignments and exams. Many exercises are stated as questions rather than in "prove that" form. Instructors can use these exercises to stimulate class discussion on the role of proof and counterexample in problem solving.

Flexible Sections   Most sections are divided into subsections so that an instructor who is pressed for time can choose to cover certain subsections only and either omit the rest or leave them for the students to learn on their own. The division into subsections also makes it easier for instructors to break up sections if they wish to spend more than one day on them.

Presentation of Proof Methods   It is inevitable that the proofs and disproofs in this book will seem easy to instructors. Students, however, find many of them quite difficult. In showing students how to discover and construct proofs and disproofs, I tried to describe the kind of approaches that mathematicians use when confronting challenging problems in their own research.

Instructor's Manual   An instructor's manual is available. It contains suggestions about how to approach the material of each chapter and solutions for most exercises not fully solved in Appendix B.

ORGANIZATION

This book may be used effectively for a one- or two-semester course. Each chapter contains core sections, sections covering optional mathematical material, and sections covering optional applications. Instructors have the flexibility to choose whatever mixture will best serve the needs of their students. The following table shows a division of the sections into categories.

Chapter	Core Sections	Sections Containing Optional Mathematical Material	Sections Containing Optional Computer Science Applications
1	1.1–1.3		1.4, 1.5
2	2.1–2.3	2.2	
3	3.1–3.4, 3.6	3.5, 3.7	3.8
4	4.1–4.2	4.3, 4.4	4.5
5	5.1–5.3	5.4	5.4
6	6.1–6.4	6.5–6.7	
7	7.1, 7.3	7.5, 7.6	7.2, 7.4
8	8.1, 8.2	8.3, 8.4	
9	9.1, 9.2	9.4	9.3, 9.5
10	10.1–10.3	10.5	10.4, 10.5
11	11.1	11.2–11.5	11.5, 11.5

The tree diagram below shows, approximately, how the chapters of this book depend on each other. Chapters on different branches of the tree are sufficiently independent that instructors would need to make at most minor adjustments if they skipped chapters but followed paths along branches of the tree.

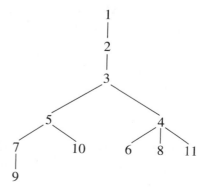

ACKNOWLEDGMENTS

I owe a debt of gratitude to many people at De Paul University for their support and encouragement throughout the years I worked on this book. A number of my colleagues used versions of this book in their courses and provided many excellent suggestions for improvement. For this, I am thankful to J. Marshall Ash, Allan Berele, William Chin, Barbara Cortzen, Constantine Georgakis, Sigrun Goes, Jerry Goldman, Lawrence Gluck, Leonid Krop, Lynn Narasimhan, Walter Pranger, Eric Rieders, Yuen-Fat Wong, and, most especially, Jeanne LaDuke. The hundreds of students to whom I have taught discrete mathematics had a profound influence on the book's form. By sharing their thoughts and thought processes with me, they taught me how better to teach them, and I am very grateful to them. I owe the DePaul University administration a special word of thanks for considering the writing of this book a worthwhile scholarly endeavor.

A number of reviewers read parts or all of my manuscript and offered valuable comments, suggestions, and corrections. I am particularly grateful to John Carroll of San Diego State University, Dr. Joseph S. Fulda, and Porter G. Webster of the University of Southern Mississippi for their unusual thoroughness and their encouragement. For their help with the first edition of this book, I would also like to thank Itshak Borosh, Texas A & M University; Douglas Campbell, Brigham Young University; David G. Cantor, University of California at Los Angeles; C. Patrick Collier, University of Wisconsin-Oshkosh; Kevan H. Croteau, University of Dubuque; Henry Etlinger, Rochester Institute of Technology; Melvin J. Friske, Wisconsin Lutheran College; Jerrold R. Griggs, University of South Carolina; Lillian E. Hupert, Loyola University of Chicago; John F. Morrison, Towson State University; Paul Pedersen, University of Denver; George Peck, Arizona State University; Roxy Peck, California Polytechnic State University, San Luis Obispo; Dix Pettey, University of Missouri; Anthony Ralston, State University of New York at Buffalo; and George Schultz, St. Petersburg Junior College, Clearwater. And for their assistance with the second edition, I am grateful to Nancy Baxter, Dickinson College; Irinel Drogan, University of Texas at Arlington; Ladnor Geissinger, University of North Carolina; Leonard T. Malinowski, Finger Lakes Community College; and John Roberts, University of Louisville. I am also very thankful for the suggestions of over fifty instructors from around the country who generously shared with me their experiences with the first edition and their ideas for improvement.

To my family, I owe thanks beyond measure. I am grateful to my mother, whose keen interest in the workings of the human intellect started me many years ago on the track that led ultimately to this book, and to my father, whose devotion to the written word has been a constant source of inspiration. I thank my children for their affection and cheerful acceptance of the demands this book placed on my life. And I am most grateful to my husband, Helmut, who for many years encouraged me with his faith in the value of this project and supported me with his advice and with extensive technical assistance through several generations of computer word processing systems.

1 The Logic of Compound Statements

Aristotle
(384 B.C.–322 B.C.)

The first great treatises on logic were written by the Greek philosopher Aristotle. They were a collection of rules for deductive reasoning that were intended to serve as a basis for the study of every branch of knowledge. In the seventeenth century, the German philosopher and mathematician Gottfried Leibniz conceived the idea of using symbols to mechanize the process of deductive reasoning in much the same way that algebraic notation had mechanized the process of reasoning about numbers and their relationships. Leibniz's idea was realized in the nineteenth century by the English mathematicians George Boole and Augustus De Morgan, who founded the modern subject of symbolic logic. With research continuing to the present day, symbolic logic has provided, among other things, the theoretical basis for many areas of computer science such as digital logic circuit design (see Sections 1.4 and 1.5), relational database theory (see Section 10.1), automata theory and computability (see Sections 7.2, 7.4, 7.6, and 10.4), and artificial intelligence (see Sections 2.3, 11.1, and 11.5).

1.1 LOGICAL FORM AND LOGICAL EQUIVALENCE

Logic is a science of the necessary laws of thought, without which no employment of the understanding and the reason takes place.

(Immanuel Kant, 1785)

The central concept of deductive logic is the concept of argument form. An argument is a sequence of statements aimed at demonstrating the truth of an assertion. To have confidence in the conclusion that you draw from an argument, you must be sure that the statements composing it either are acceptable on their own merits or follow from preceding statements. In logic, the form of an argument is distinguished from its content. Logical analysis won't help you determine the intrinsic merit of an argument's content, but it will help you analyze an argument's form to determine whether the truth of the conclusion follows *necessarily* from the truth of the preceding statements. For this reason logic is sometimes defined as the science of necessary inference or the science of reasoning.

Consider the following two arguments:

If the program syntax is faulty or if program execution results in division by zero, then the computer will generate an error message. Therefore, if the computer does not generate an error message, then the program syntax is correct and program execution does not result in division by zero.

If x is a real number such that $x < -2$ or $x > 2$, then $x^2 > 4$. Therefore, if $x^2 \leq 4$, then $x \geq -2$ and $x \leq 2$.

The content of these arguments is very different. Nevertheless, their *logical form* is the same. To illustrate the logical form we use letters of the alphabet (such as *p*, *q*, and *r*) to represent the component sentences and the expression "not *p*" to refer to the sentence "It is not the case that *p*." Then the common form of both the arguments above is as follows:

> If *p* or *q*, then *r*.
> Therefore, if not *r*, then not *p* and not *q*.

EXAMPLE 1.1.1   Identifying Logical Form

Fill in the blanks below so that argument (b) has the same form as argument (a). Then represent the common form of the arguments using letters to stand for component sentences.

a. If Jane is a math major or Jane is a computer science major, then Jane will take Math 150.
   Jane is a computer science major.
   Therefore, Jane will take Math 150.

b. If logic is easy or ___(1)___ , then ___(2)___ .
   I will study hard.
   Therefore, I will get an A in this course.

Solution   1. I (will) study hard.

2. I will get an A in this course.

> *common form*: If *p* or *q*, then *r*.
>   *q*.
>   Therefore, *r*.   ■

STATEMENTS

Most of the definitions of formal logic have been developed so that they agree with the natural or intuitive logic used by people who have been educated to think clearly and use language carefully. The differences that exist between formal and intuitive logic are necessary to avoid ambiguity and obtain consistency.

In any mathematical theory, new terms are defined by using those that have been previously defined. However, this process has to start somewhere. A few initial terms necessarily remain undefined. In logic, the words *sentence, true,* and *false* are the initial undefined terms.

DEFINITION

A **statement** (or **proposition**) is a sentence that is true or false but not both.

For example, "Two plus two equals four" and "Two plus two equals five" are both statements, the first because it is true and the second because it is false. On the other hand, the truth or falsity of "He is a college student" depends on the reference for the pronoun *he*. For some values of *he* the sentence is true; for others it is false. If the sentence were preceded by other sentences that made the pronoun's reference clear, then the sentence would be a statement. Considered on its own, however, the sentence is neither true nor false, and so it is not a statement. We will discuss ways of transforming sentences of this form into statements in Section 2.1.

Similarly "$x + y > 0$" is not a statement because for some values of x and y the sentence is true, whereas for others it is false. For instance, if $x = 1$ and $y = 2$, the sentence is true; if $x = -1$ and $y = 0$, the sentence is false.

COMPOUND STATEMENTS

We now introduce three symbols that are used to build more complicated logical expressions out of simpler ones. The symbol $\sim$ denotes *not,* $\wedge$ denotes *and,* and $\vee$ denotes *or.* Given a statement p, the sentence "$\sim p$" is read "not p" or "It is not the case that p" and is called the **negation of p**. In some computer languages the symbol $\neg$ is used in place of $\sim$. Given another statement q, the sentence "$p \wedge q$" is read "p and q" and is called the **conjunction of p and q**. The sentence "$p \vee q$" is read "p or q" and is called the **disjunction of p and q**.

In expressions that include the symbol $\sim$ as well as $\wedge$ or $\vee$, the **order of operation** is that $\sim$ is performed first. For instance, $\sim p \wedge q = (\sim p) \wedge q$. In logical expressions, as in ordinary algebraic expressions, the order of operations can be overridden through the use of parentheses. Thus $\sim(p \wedge q)$ represents the negation of the conjunction of p and q. In this, as in most treatments of logic, the symbols $\wedge$ and $\vee$ are considered coequal in order of operation, and an expression such as $p \wedge q \vee r$ is considered ambiguous. This expression must be written as either $(p \wedge q) \vee r$ or $p \wedge (q \vee r)$ to have meaning.

EXAMPLE 1.1.2   Translating from English to Symbols:
But and *Neither-Nor*

Write each of the following sentences symbolically, letting $p =$ "It is hot" and $q =$ "It is sunny."

a. It is not hot but it is sunny.
b. It is neither hot nor sunny.

Solution   a. The convention in logic is that the words *but* and *and* mean the same thing. Generally, *but* is used in place of *and* when the part of the sentence that follows is in some way unexpected. The given sentence is equivalent to "It is not hot and it is sunny," which can be written symbolically as $\sim p \wedge q$.
b. The phrase *neither A nor B* means the same as *not A and not B*. To say it is neither hot nor sunny means that it is not hot and it is not sunny. Therefore, the given sentence can be written symbolically as $\sim p \wedge \sim q$.

The notation for inequalities involves *and* and *or* statements. For instance, if x, a, and b are particular real numbers, then

$x \leq a$	means	$x < a$ or $x = a$
$a \leq x \leq b$	means	$a \leq x$ and $x \leq b$.

Note that

$$2 \leq x \leq 1 \qquad \text{means} \qquad 2 \leq x \quad \text{and} \quad x \leq 1,$$

which is false no matter what number x happens to be. By the way, the point of specifying x, a, and b to be *particular* real numbers is to ensure that sentences such as "$x < a$" and "$x \geq b$" are either true or false and hence that such sentences are statements. ∎

EXAMPLE 1.1.3   *And, Or, and Inequalities*

Suppose x is a particular real number. Let p, q, and r symbolize "$0 < x$," "$x < 3$," and "$x = 3$," respectively. Write the following inequalities symbolically:

a. $x \leq 3$ b. $0 < x < 3$ c. $0 < x \leq 3$

Solution   a. $q \vee r$ b. $p \wedge q$ c. $p \wedge (q \vee r)$ ∎

In Example 1.1.3 we built compound sentences out of component statements and the terms *not, and,* and *or.* If such sentences are to be statements, however, they must have well-defined **truth values**—they must be either true or false. We now define such compound sentences as statements by specifying their truth values in terms of the statements that compose them.

The negation of a statement is a statement that exactly expresses what it would mean for the given statement to be false. Therefore, the negation of a statement has opposite truth value from the statement.

DEFINITION

If p is a statement variable, the **negation** of p is "not p" or "It is not the case that p" and is denoted $\sim p$. It has opposite truth value from p: if p is true, $\sim p$ is false; if p is false, $\sim p$ is true.

The truth values for negation are summarized in a *truth table.*

Truth Table for $\sim p$

p	$\sim p$
T	F
F	T

In ordinary language the sentence "It is hot and it is sunny" is understood to be true when both conditions—being hot and being sunny—are satisfied. If it is hot but not sunny, or sunny but not hot, or neither hot nor sunny, the sentence is understood to be false. The formal definition of truth values for an *and* statement agrees with this general understanding.

> **DEFINITION**
>
> If p and q are statement variables, the **conjunction** of p and q is "p and q," denoted $p \wedge q$. It is true when, and only when, both p and q are true. If either p or q is false, or if both are false, $p \wedge q$ is false.

The truth values for conjunction can also be summarized in a truth table. The table is obtained by considering the four possible combinations of truth values for p and q. Each combination is displayed in one row of the table; the corresponding truth value for the whole statement is placed in the right-most column of that row. Note that the only row containing a T is the first one since the only way for an *and* statement to be true is for both component statements to be true.

Truth Table for $p \wedge q$

p	q	$p \wedge q$
T	T	T
T	F	F
F	T	F
F	F	F

By the way, the order of truth values for p and q in the table above is TT, TF, FT, FF. It is not necessary to write the truth values in this order, although it is customary to do so. We will use this order for all the truth tables in this book.

In the case of disjunction—statements of the form "p or q"—intuitive logic offers two alternative interpretations. In ordinary language *or* is sometimes used in an exclusive sense (p or q but not both) and sometimes in an inclusive sense (p or q or both). A waiter who says you may have "coffee, tea, or milk" uses the word *or* in an exclusive sense: extra payment is generally required if you want more than one beverage. On the other hand, a waiter who offers "cream or sugar" uses the word *or* in an inclusive sense: you are entitled to both cream and sugar if you wish to have them.

Mathematicians and logicians avoid possible ambiguity about the meaning of the word *or* by understanding it to mean the inclusive "and/or." The symbol $\vee$ comes from the Latin word *vel* which means *or* in its inclusive sense. To express the exclusive *or* the phrase *p or q but not both* is used.

> **DEFINITION**
>
> If p and q are statement variables, the **disjunction** of p and q is "p or q," denoted $p \vee q$. It is true when at least one of p or q is true and is false only when both p and q are false.

Here is the truth table for disjunction:

Truth Table for $p \lor q$

p	q	$p \lor q$
T	T	T
T	F	T
F	T	T
F	F	F

Note that the statement "$2 \leq 2$" ("2 is less than 2 or 2 equals 2") is true because $2 = 2$.

EVALUATING THE TRUTH OF MORE GENERAL COMPOUND STATEMENTS

Now that truth values have been assigned to $\sim p$, $p \land q$, and $p \lor q$, consider the question of assigning truth values to more complicated expressions such as $\sim p \lor q$, $(p \lor q) \land \sim(p \land q)$, and $(p \land q) \lor r$. Such expressions are called *statement forms* (or *propositional forms*). There is a close relationship between statement forms and *Boolean expressions,* which is discussed in Section 1.4.

DEFINITION

A **statement form** (or **propositional form**) is an expression made up of statement variables (such as p, q, and r) and logical connectives (such as $\sim$, $\land$, and $\lor$) that becomes a statement when actual statements are substituted for the component statement variable. The **truth table** for a given statement form displays the truth values that correspond to the different combinations of truth values for the variables.

To compute the truth values for a statement form, follow rules similar to those used to evaluate algebraic expressions. For each combination of truth values for the statement variables, first evaluate the expressions within the innermost parentheses, then evaluate the expressions within the next innermost set of parentheses, and so forth until you have the truth values for the complete expression.

EXAMPLE 1.1.4   Truth Table for *Exclusive Or*

Construct the truth table for the statement form $(p \lor q) \land \sim(p \land q)$. Note that when *or* is used in its exclusive sense, the statement "p or q" means "p or q but not both" or "p or q and not both p and q," which translates into symbols as $(p \lor q) \land \sim(p \land q)$. This is sometimes abbreviated $p \oplus q$ or p XOR q.

Solution   Set up columns labeled p, q, $p \lor q$, $p \land q$, $\sim(p \land q)$, and $(p \lor q) \land \sim(p \land q)$. Fill in the p and q columns with all the logically possible combinations of T's and F's. Then use the truth tables for $\lor$ and $\land$ to fill in the $p \lor q$ and $p \land q$ columns with the appropriate truth values. Next fill in the $\sim(p \land q)$ column by taking the opposites of the truth values for $p \land q$. For example, the entry for $\sim(p \land q)$ in the first row is F because in the first row the truth value of $p \land q$ is T. Finally, fill in the $(p \lor q) \land \sim(p \land q)$ column by considering the truth table for an *and* statement together with the computed truth values for $p \lor q$ and $\sim(p \land q)$. For example, the entry in the first row is F because the entry for $p \lor q$ is T, the entry for $\sim(p \land q)$ is F, and an *and* statement is false unless both components are true. The entry in the second row is T because both components are true.

Truth Table for *Exclusive Or:* $(p \lor q) \land \sim(p \land q)$

p	q	$p \lor q$	$p \land q$	$\sim(p \land q)$	$(p \lor q) \land \sim(p \land q)$
T	T	T	T	F	F
T	F	T	F	T	T
F	T	T	F	T	T
F	F	F	F	T	F

EXAMPLE 1.1.5   **Truth Table for $(p \land q) \lor \sim r$**

Construct a truth table for the statement form $(p \land q) \lor \sim r$.

Solution   Make columns headed p, q, r, $p \land q$, $\sim r$, and $(p \land q) \lor \sim r$. Since there are eight logically possible combinations of truth values for p, q, and r, enter these in the three left-most columns. Then fill in the truth values for $p \land q$ and for $\sim r$. Complete the table by considering the truth values for $(p \land q)$ and for $\sim r$ and the definition of an *or* statement. Since an *or* statement is false only when both components are false, the only rows in which the entry is F are the third, fifth, and seventh rows because those are the only rows in which the expressions $p \land q$ and $\sim r$ are both false. The entry for all the other rows is T.

p	q	r	$(p \land q)$	$\sim r$	$(p \land q) \lor \sim r$
T	T	T	T	F	T
T	T	F	T	T	T
T	F	T	F	F	F
T	F	F	F	T	T
F	T	T	F	F	F
F	T	F	F	T	T
F	F	T	F	F	F
F	F	F	F	T	T

The essential point about assigning truth values to compound statements is that it allows you—using logic alone—to judge the truth of a compound statement based on your knowledge of the truth of its component parts. Logic does not help you determine the truth or falsity of the component statements. Rather, logic helps link these separate pieces of information together into a coherent whole.

LOGICAL EQUIVALENCE

The statements

$$6 > 2 \quad \text{and} \quad 2 < 6$$

are two different ways of saying the same thing. Why? Because of the definition of the symbols $<$ and $>$. By contrast, the statements

Dogs bark and cats meow   and   Cats meow and dogs bark

are also two different ways of saying the same thing, but the reason has nothing to do with the definition of the words. It has to do with the logical form of the statements. Any two statements having the same form as these statements would either both be true or both be false. You can see this by examining the following truth table. The table shows that for each combination of truth values for p and q, $p \wedge q$ is true when, and only when, $q \wedge p$ is true. In such a case, the statement forms are called *logically equivalent*.

p	q	$p \wedge q$	$q \wedge p$
T	T	T	T
T	F	F	F
F	T	F	F
F	F	F	F

same truth values, so
$p \wedge q$ and $q \wedge p$ are
logically equivalent

DEFINITION

Two *statement forms* are called **logically equivalent** if, and only if, they have identical truth values for each possible substitution of statements for their statement variable. The logical equivalence of statement forms P and Q is denoted by writing $P \equiv Q$.

Two *statements* are called **logically equivalent** if, and only if, when the same statement variables are used to represent identical component statements, their forms are logically equivalent.

To test whether two statement forms P and Q are logically equivalent:

1. Construct the truth table for P.

2. Construct the truth table for Q using the same statement variables for identical component statements.

3. Check each combination of truth values of the statement variables to see whether the truth value of P is the same as the truth value of Q.

   a. If in each row the truth value of P is the same as the truth value of Q, then P and Q are logically equivalent.

   b. If in some row P has a different truth value from Q, then P and Q are not logically equivalent.

Alternatively, you can show that statement forms P and Q are not logically equivalent by finding concrete statements of each form, one of which is true and the other of which is false. This method is illustrated in part (b) of Example 1.1.7.

EXAMPLE 1.1.6   Double Negative Property: $\sim(\sim p) \equiv p$

Check that the negation of the negation of a statement is logically equivalent to the statement.

Solution

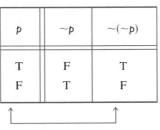

p	$\sim p$	$\sim(\sim p)$
T	F	T
F	T	F

same truth values, so p and $\sim(\sim p)$ are logically equivalent

EXAMPLE 1.1.7   Showing Nonequivalence

Show that the statement forms $\sim(p \land q)$ and $\sim p \land \sim q$ are not logically equivalent.

Solution   a. One way to show this is to use truth tables.

p	q	$\sim p$	$\sim q$	$p \land q$	$\sim(p \land q)$		$\sim p \land \sim q$
T	T	F	F	T	F		F
T	F	F	T	F	T	≠	F
F	T	T	F	F	T	≠	F
F	F	T	T	F	T		T

different truth values in rows 2 and 3, so $\sim(p \land q)$ and $\sim p \land \sim q$ are not logically equivalent

b. A second way to show that $\sim(p \wedge q)$ and $\sim p \wedge \sim q$ are not logically equivalent is by example. Let p be the statement "$0 < 1$" and let q be the statement "$1 < 0$." Then

$$\sim(p \wedge q) \quad \text{is} \quad \text{"It is not the case that both } 0 < 1 \text{ and } 1 < 0\text{,"}$$

which is true. On the other hand,

$$\sim p \wedge \sim q \quad \text{is} \quad \text{"}0 \not< 1 \quad \text{and} \quad 1 \not< 0\text{,"}$$

which is false. This example shows that there are concrete statements that when substituted for p and q make one of the statement forms true and the other false. Therefore, the statement forms are not logically equivalent. ∎

EXAMPLE 1.1.8   Negations of *And* and *Or:* De Morgan's Laws

For the statement "John is tall and Jim is redheaded" to be true, both components must be true. It follows that for the statement to be false, one or both components must be false. Thus the negation can be written as "John is not tall or Jim is not redheaded." In general, the negation of the conjunction of two statements is logically equivalent to the disjunction of their negations. That is, statements of the forms $\sim(p \wedge q)$ and $\sim p \vee \sim q$ are logically equivalent. Check this using truth tables.

Solution

p	q	$\sim p$	$\sim q$	$p \wedge q$	$\sim(p \wedge q)$	$\sim p \vee \sim q$
T	T	F	F	T	F	F
T	F	F	T	F	T	T
F	T	T	F	F	T	T
F	F	T	T	F	T	T

same truth values, so $\sim(p \wedge q)$ and $\sim p \vee \sim q$ are logically equivalent

Symbolically,

$$\boxed{\sim(p \wedge q) \equiv \sim p \vee \sim q.}$$

In the exercises at the end of this section you are asked to show the analogous law that the negation of the disjunction of two statements is logically equivalent to the conjunction of their negations:

$$\boxed{\sim(p \vee q) \equiv \sim p \wedge \sim q.}$$ ∎

The two logical equivalences of Example 1.1.8 are known as *De Morgan's laws* of logic in honor of Augustus De Morgan, who was the first to state them in formal mathematical terms.

Augustus De Morgan (1806–1871)

> **De Morgan's Laws**
>
> The negation of an *and* statement is logically equivalent to the *or* statement in which each component is negated.
>   The negation of an *or* statement is logically equivalent to the *and* statement in which each component is negated.

EXAMPLE 1.1.9   Applying De Morgan's Laws

Write negations for each of the following statements:

a. John is six feet tall and he weighs at least 200 pounds.
b. The bus was late or Tom's watch was slow.

Solution   a. John is not six feet tall or he weighs less than 200 pounds.
b. The bus was not late and Tom's watch was not slow.

The statement "neither p nor q" means the same as "$\sim p$ and $\sim q$." Thus an alternative answer is "Neither was the bus late nor was Tom's watch slow." ∎

   If x is a particular real number, saying that x is not less than 2 ($x \not< 2$) means that x does not lie to the left of 2 on the number line. This is equivalent to saying that either $x = 2$ or x lies to the right of 2 on the number line ($x = 2$ or $x > 2$). Hence,

$$x \not< 2 \quad \text{is equivalent to} \quad x \geq 2.$$

Similarly,

$$x \not> 2 \quad \text{is equivalent to} \quad x \leq 2,$$
$$x \not\leq 2 \quad \text{is equivalent to} \quad x > 2, \text{ and}$$
$$x \not\geq 2 \quad \text{is equivalent to} \quad x < 2.$$

EXAMPLE 1.1.10   Inequalities and De Morgan's Laws

Use De Morgan's laws to write the negation of $-1 < x \leq 4$.

Solution   The given statement is equivalent to

$$-1 < x \quad \text{and} \quad x \leq 4.$$

By De Morgan's laws, the negation is

$$-1 \not< x \quad \text{or} \quad x \not\leq 4,$$

which is equivalent to

$$-1 \geq x \quad \text{or} \quad x > 4.$$ ∎

EXAMPLE 1.1.11   A Cautionary Example

According to De Morgan's laws, the negation of

$$p: \text{Jim is tall and Jim is thin}$$

is

$$\sim p: \text{Jim is not tall or Jim is not thin}$$

because the negation of an *and* statement is the *or* statement in which the two components are negated.

Unfortunately, a potentially confusing aspect of the English language can arise when you are taking negations of this kind. Note that statement *p* can be written more compactly as

$$p': \text{Jim is tall and thin.}$$

When so written, another way to negate it is

$$\sim(p'): \text{Jim is not tall and thin.}$$

But in this form the negation looks like an *and* statement. Doesn't that violate De Morgan's laws?

Actually no violation occurs. The reason is that in formal logic the words *and* and *or* are allowed only between complete statements, not between sentence fragments.

One lesson to be learned from this example is that when you apply De Morgan's laws, you must have complete statements on either side of each *and* and on either side of each *or*. A deeper lesson is this: Although the laws of logic are extremely useful, they should be used as an *aid* to thinking, not as a mechanical substitute for it. ∎

TAUTOLOGIES AND CONTRADICTIONS

It has been said that all of mathematics reduces to tautologies. Although this is formally true, most working mathematicians think of their subject as having substance as well as form. Nonetheless, an intuitive grasp of basic logical tautologies is part of the equipment of anyone who reasons with mathematics.

DEFINITION

A **tautology** is a statement form that is always true regardless of the truth values of the individual statements substituted for its statement variables. A statement whose form is a tautology is called a **tautological statement.**

A **contradiction** is a statement form that is always false regardless of the truth values of the individual statements substituted for its statement variables. A statement whose form is a contradiction is called a **contradictory statement.**

According to this definition, the truth of a tautological statement and the falsity of a contradictory statement are due to the logical structure of the statements themselves and are independent of the meanings of the statements.

EXAMPLE 1.1.12   Tautologies and Contradictions

Show that the statement form $p \lor \sim p$ is a tautology and that the statement form $p \land \sim p$ is a contradiction.

Solution

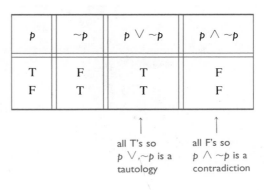

all T's so
$p \vee \sim p$ is a
tautology

all F's so
$p \wedge \sim p$ is a
contradiction

EXAMPLE 1.1.13   Logical Equivalence Involving Tautologies
and Contradictions

If t is a tautology and c is a contradiction, show that $p \wedge t \equiv p$ and $p \wedge c \equiv c$.

Solution

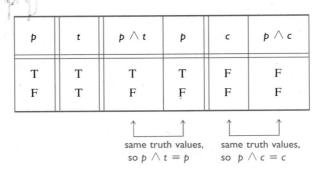

same truth values,
so $p \wedge t \equiv p$

same truth values,
so $p \wedge c \equiv c$

SUMMARY OF LOGICAL EQUIVALENCES

Knowledge of logically equivalent statements is very useful for constructing arguments. It often happens that it is difficult to see how a conclusion follows from one
form of a statement, whereas it is easy to see how it follows from a logically equivalent
form of the statement. A number of logical equivalences are summarized in
Theorem 1.1.1 for future reference.

     The proofs of laws 4 and 6, the first parts of laws 1 and 5, and the second part of
law 9 have already been given as examples in the text. Proofs of the other parts of the
theorem are left as exercises. In fact it can be shown that the first five laws of
Theorem 1.1.1 form a core from which the other laws can be derived. The first five
laws are the axioms for a mathematical structure known as a Boolean algebra, which
is discussed in Section 5.3.

     The equivalences of Theorem 1.1.1 are general laws of thought that occur in all
areas of human endeavor. They can also be used in a formal way to rewrite complicated statement forms more simply.

THEOREM 1.1.1   Logical Equivalences

Given any statement variables p, q, and r, a tautology t and a contradiction c, the following logical equivalences hold:

1. Commutative laws: $p \wedge q \equiv q \wedge p$ $p \vee q \equiv q \vee p$
2. Associative laws: $(p \wedge q) \wedge r \equiv p \wedge (q \wedge r)$ $(p \vee q) \vee r \equiv p \vee (q \vee r)$
3. Distributive laws: $p \wedge (q \vee r) \equiv (p \wedge q) \vee (p \wedge r)$ $p \vee (q \wedge r) \equiv (p \vee q) \wedge (p \vee r)$
4. Identity laws: $p \wedge t \equiv p$ $p \vee c \equiv p$
5. Negation laws: $p \vee \sim p \equiv t$ $p \wedge \sim p \equiv c$
6. Double negative law: $\sim(\sim p) \equiv p$
7. Idempotent laws: $p \wedge p \equiv p$ $p \vee p \equiv p$
8. De Morgan's laws: $\sim(p \wedge q) \equiv \sim p \vee \sim q$ $\sim(p \vee q) \equiv \sim p \wedge \sim q$
9. Universal bound laws: $p \vee t \equiv t$ $p \wedge c \equiv c$
10. Absorption laws: $p \vee (p \wedge q) \equiv p$ $p \wedge (p \vee q) \equiv p$
11. Negations of t and c: $\sim t \equiv c$ $\sim c \equiv t$

$$(p \wedge p) \vee (p \wedge q) \vee (\sim q \wedge p) \vee (\sim q \wedge q)$$
$$p$$

EXAMPLE 1.1.14   Simplifying Statement Forms

Use Theorem 1.1.1 to verify the logical equivalence

$$\sim(\sim p \wedge q) \wedge (p \vee q) \equiv p.$$

Solution   Use the laws of Theorem 1.1.1 to replace sections of the statement form on the left by logically equivalent expressions. Each time you do this, you obtain a logically equivalent statement form. Continue making replacements until you obtain the statement form on the right.

$$\sim(\sim p \wedge q) \wedge (p \vee q) \equiv (\sim(\sim p) \vee \sim q) \wedge (p \vee q) \qquad \text{by De Morgan's laws}$$
$$\equiv (p \vee \sim q) \wedge (p \vee q) \qquad \text{by the double negative law}$$
$$\equiv p \vee (\sim q \wedge q) \qquad \text{by the distributive law}$$
$$\equiv p \vee (q \wedge \sim q) \qquad \text{by the commutative law for } \wedge$$
$$\equiv p \vee c \qquad \text{by the negation law}$$
$$\equiv p \qquad \text{by the identity law} \qquad \blacksquare$$

Skill in simplifying statement forms is useful in constructing logically efficient computer programs and in designing digital logic circuits.

EXERCISE SET 1.1

Appendix B contains either full or partial solutions to all exercises with blue numbers. When the solution is not complete, the exercise number has an *H* next to it. A ◆ next to an exercise number signals that the exercise is more challenging than usual. Be careful not to get into the habit of turning too quickly to the solutions. Make every effort to work exercises on your own before checking your answers.

793 3297

$$\begin{array}{cc} P & c \\ T & F \\ F & F \end{array} \qquad \begin{array}{c} P \lor c \\ T \\ F \end{array}$$

In each of 1–4 represent the common form of each argument using letters to stand for component sentences, and fill in the blanks so that the argument in part (b) has the same logical form as the argument in part (a).

1. a. If all integers are rational, then the number 1 is rational.
   All integers are rational.
   Therefore, the number 1 is rational.

   b. If all algebraic expressions can be written in prefix notation, then ————————————.
   ————————————————————.
   Therefore, $(a + 2b) \cdot (a^2 - b)$ can be written in prefix notation.

2. a. If all computer programs contain errors, then this program contains an error.
   This program does not contain an error.
   Therefore, it is not the case that all computer programs contain errors.

   b. If ————, then ————.
   2 is not odd.
   Therefore, it is not the case that all prime numbers are odd.

3. a. This number is even or this number is odd.
   This number is not even.
   Therefore, this number is odd.

   b. ———— or logic is confusing.
   My mind is not shot.
   Therefore, ————.

4. a. If n is divisible by 6, then n is divisible by 3.
   If n is divisible by 3, then the sum of the digits of n is divisible by 3.
   Therefore, if n is divisible by 6, then the sum of the digits of n is divisible by 3.
   (Assume that n is a particular, fixed integer.)

   b. If ————————————————————,
   then the guard condition for the **while** loop is false.
   If ————————————————————,
   then program execution moves to the next instruction following the loop.
   Therefore, if x equals 0, then ——————————.
   (Assume that x is a particular variable in a particular computer program.)

5. Indicate which of the following sentences are statements.
   a. 1,024 is the smallest four-digit number that is a perfect square.
   b. She is a mathematics major.
   c. $128 = 2^6$ d. $x = 2^6$

6. Let $s =$ "stocks are increasing" and $i =$ "interest rates are steady".
   a. Stocks are increasing but interest rates are steady.
   b. Neither are stocks increasing nor are interest rates steady.

7. Juan is a math major but not a computer science major. ($m =$ "Juan is a math major," $c =$ "Juan is a computer science major")

8. Let $h =$ "John is healthy," $w =$ "John is wealthy," and $s =$ "John is wise".
   a. John is healthy and wealthy but not wise.
   b. John is not wealthy but he is healthy and wise.
   c. John is neither healthy, wealthy, nor wise.

9. Either Olga will go out for tennis or she will go out for track but not both. ($n =$ "Olga will go out for tennis," $k =$ "Olga will go out for track")

10. Let p be the statement "DATAENDFLAG is off," q the statement "ERROR equals 0," and r the statement "SUM is less than 1,000." Express the following sentences in symbolic notation.
    a. DATAENDFLAG is off, ERROR equals 0, and SUM is less than 1,000.
    b. DATAENDFLAG is off but ERROR is not equal to 0.
    c. DATAENDFLAG is off; however ERROR is not 0 or SUM is greater than or equal to 1,000.
    d. DATAENDFLAG is on and ERROR equals 0 but SUM is greater than or equal to 1,000.
    e. Either DATAENDFLAG is on or it is the case that both ERROR equals 0 and SUM is less than 1,000.

11. In the following sentence is the word *or* used in its inclusive or exclusive sense? A team wins the playoffs if it wins two games in a row or a total of three games.

Write truth tables for the statement forms in 12–16.

12. $\sim p \land q$

13. $(p \land q) \lor \sim(p \lor q)$

14. $p \land (q \land r)$

15. $\sim p \land (q \lor \sim r)$

16. $(p \lor (\sim p \lor q)) \land \sim(q \land \sim r)$

Determine which of the pairs of statement forms in 17–26 are logically equivalent. Justify your answers using truth tables. Read t to be a tautology and c to be a contradiction.

17. $p \lor (p \land q)$ and p

18. $\sim(p \lor q)$ and $\sim p \land \sim q$

19. $p \lor t$ and t

20. $p \vee c$ and p

21. $(p \wedge q) \wedge r$ and $p \wedge (q \wedge r)$

22. $p \wedge (q \vee r)$ and $(p \wedge q) \vee (p \wedge r)$

23. $(p \wedge q) \vee r$ and $p \wedge (q \vee r)$

24. $(p \vee q) \vee (p \wedge r)$ and $(p \vee q) \wedge r$

25. $((\sim p \vee q) \wedge (p \vee \sim r)) \wedge (\sim p \vee \sim q)$ and $\sim (p \vee r)$

26. $(r \vee p) \wedge ((\sim r \vee (p \wedge q)) \wedge (r \vee q))$ and $p \wedge q$

Use De Morgan's laws to write negations for the statements in 27–32.

27. Hal is a math major and Hal's sister is a computer science major.

28. Sam swims on Thursdays and Kate plays tennis on Saturdays.

29. The connector is loose or the machine is unplugged.

30. This computer program has a logical error in the first ten lines or it is being run with an incomplete data set.

31. The dollar is at an all-time high and the stock market is at a record low.

32. The train is late or my watch is fast.

Assume x is a particular real number and use De Morgan's laws to write negations for the statements in 33–36.

33. $-2 < x < 7$ 34. $-4 < x < -1$

35. $1 > x \geq 3$ 36. $0 \geq x > -5$

Use truth tables to establish which of the statement forms in 37–40 are tautologies and which are contradictions.

37. $(p \wedge q) \vee (\sim p \vee (p \wedge \sim q))$

38. $(p \wedge \sim q) \wedge (\sim p \vee q)$

39. $((\sim p \wedge q) \wedge (q \wedge r)) \wedge \sim q$

40. $(\sim p \vee q) \vee (p \wedge \sim q)$

In 41 and 42 below, a logical equivalence is derived from Theorem 1.1.1. Supply a reason for each step.

41. $(p \wedge \sim q) \vee (p \wedge q) \equiv p \wedge (\sim q \vee q)$ by (a) ___

$\equiv p \wedge (q \vee \sim q)$ by (b) ___

$\equiv p \wedge t$ by (c) ___

$\equiv p$ by (d) ___

Therefore, $(p \wedge \sim q) \vee (p \wedge q) \equiv p$.

42. $(p \vee \sim q) \wedge (\sim p \vee \sim q)$

$\equiv (\sim q \vee p) \wedge (\sim q \vee \sim p)$ by (a) ___

$\equiv \sim q \vee (p \wedge \sim p)$ by (b) ___

$\equiv \sim q \vee c$ by (c) ___

$\equiv \sim q$ by (d) ___

Therefore, $(p \vee \sim q) \wedge (\sim p \vee \sim q) \equiv \sim q$.

Use Theorem 1.1.1 to verify the logical equivalences in 43–47.

43. $(p \wedge \sim q) \vee p \equiv p$

44. $p \wedge (\sim q \vee p) \equiv p$

45. $\sim (p \vee \sim q) \vee (\sim p \wedge \sim q) \equiv \sim p$

46. $\sim ((\sim p \wedge q) \vee (\sim p \wedge \sim q)) \vee (p \wedge q) \equiv p$

47. $(p \wedge (\sim (\sim p \vee q))) \vee (p \wedge q) \equiv p$

♦48. In Example 1.1.4 the symbol $\oplus$ was introduced to denote *exclusive or*; so $p \oplus q \equiv (p \vee q) \wedge \sim (p \wedge q)$. Hence the truth table for *exclusive or* is as follows:

p	q	$p \oplus q$
T	T	F
T	F	T
F	T	T
F	F	F

a. Find simpler statement forms that are logically equivalent to $p \oplus p$ and $(p \oplus p) \oplus p$.

b. Is $(p \oplus q) \oplus r \equiv p \oplus (q \oplus r)$? Justify your answer.

c. Is $(p \oplus q) \wedge r \equiv (p \wedge r) \oplus (q \wedge r)$? Justify your answer.

♦49. In logic and in standard English, a double negative is equivalent to a positive. Is there any English usage in which a double positive is equivalent to a negative? Explain.

♦50. The rules for a certain frequent flyer club include the following statements: "Any member who fails to earn any mileage during the first twelve months after enrollment in the program may be removed from the program. Except as otherwise provided, any member who fails at any time to earn mileage for a period of three consecutive years is subject to termination of his or her membership and forfeiture of all accrued mileage. Notwithstanding this provision, no pre-July 1, 1993, member who has earned mileage (other than enrollment bonus) prior to July 1, 1994, shall be subject under this provision to the termination of his or her membership and to the cancellation of mileage accrued prior to July 1, 1994, until the amount of such mileage falls below 10,000 miles (the amount of mileage necessary for the lowest available award under the structure in place as of June 30, 1993), or until December 15, 1999, whichever comes first."

Let x be a particular member of this club, and let

p = "x fails to earn mileage during the first twelve months after enrollment,"

q = "x fails to earn mileage for a period of three consecutive years,"

r = "x became a member prior to July 1, 1993,"

s = "x currently has at least 10,000 miles for pre-July 1, 1994, mileage (not including enrollment bonus miles),"

t = "the current date is prior to December 15, 1999."

Use symbols to write the complete condition under which x's membership may be terminated.

1.2 CONDITIONAL STATEMENTS

. . . hypothetical reasoning implies the subordination of the real to the realm of the possible. . .
(Jean Piaget, 1972)

When you make a logical inference or deduction, you reason *from* a hypothesis *to* a conclusion. Your aim is to be able to say: "*If* such and such is known, *then* something or other must be the case."

Let p and q be statements. A sentence of the form "If p then q" is denoted symbolically by "$p \rightarrow q$"; p is called the *hypothesis* and q is called the *conclusion*. For instance, in

If 4,686 is divisible by 6, then 4,686 is divisible by 3

the hypothesis is "4,686 is divisible by 6" and the conclusion is "4,686 is divisible by 3." Such a sentence is called *conditional* because the truth of statement q is conditioned on the truth of statement p.

The notation $p \rightarrow q$ indicates that $\rightarrow$ is a connective, like $\wedge$ or $\vee$, that can be used to join statements to create new statements. To define $p \rightarrow q$ as a statement, therefore, we must specify the truth values for $p \rightarrow q$ as we specified truth values for $p \wedge q$ and for $p \vee q$. As is the case with the other connectives, the formal definition of truth values for $\rightarrow$ (if–then) is based on its everyday, intuitive meaning. Consider an example.

Suppose you go to interview for a job at a store and the owner of the store makes you the following promise:

If you show up for work Monday morning, then you will get the job.

Under what circumstances are you justified in saying the owner spoke falsely? That is, under what circumstances is the above sentence false? The answer is: You *do* show up for work Monday morning and you do *not* get the job. After all, the owner's promise only says you will get the job *if* a certain condition (showing up for work Monday morning) is met; it says nothing about what will happen if the condition is *not* met. So if the condition is not met, you cannot in fairness say the promise is false, regardless of whether or not you get the job.

The above example was intended to convince you that *the only combination of circumstances in which you would call a conditional sentence false occurs when the hypothesis is true and the conclusion is false.* In all other cases, you would not call the sentence false. This implies that the only row of the truth table for $p \rightarrow q$ that should be filled in with an F is the row where p is T and q is F. No other row should contain an F. But each row of a truth table must be filled in with either a T or an F. Thus all other rows of the truth table for $p \rightarrow q$ must be filled in with T's.

Truth Table for $p \rightarrow q$

p	q	$p \rightarrow q$
T	T	T
T	F	F
F	T	T
F	F	T

DEFINITION

If p and q are statement variables, the **conditional** of q by p is "If p then q" or "p implies q" and is denoted $p \rightarrow q$. It is false when p is true and q is false; otherwise it is true.

A conditional statement that is true by virtue of the fact that its hypothesis is false is often called **vacuously true** or **true by default**. Thus the statement "If you show up for work Monday morning, then you will get the job" is vacuously true if you do not show up for work Monday morning.

In expressions that include $\rightarrow$ as well as other logical operators such as $\wedge$, $\vee$, and $\sim$, the **order of operations** is that $\rightarrow$ is performed last. Thus, according to the specification of order of operations in Section 1.1, $\sim$ is performed first, then $\wedge$ and $\vee$, and finally $\rightarrow$.

EXAMPLE 1.2.1   Truth Table for $p \vee \sim q \rightarrow \sim p$

Construct a truth table for the statement form $p \vee \sim q \rightarrow \sim p$.

Solution   According to the order of operations given above, $p \vee \sim q \rightarrow \sim p$ means $(p \vee (\sim q)) \rightarrow (\sim p)$, and this order governs the construction of the truth table. First fill in the four possible combinations of truth values for p and q, and then enter the truth values for $\sim p$ and $\sim q$ using the definition of negation. Then fill in the $p \vee \sim q$ column using the definition of $\vee$. Finally, fill in the $p \vee \sim q \rightarrow \sim p$ column using the definition of $\rightarrow$. The only rows in which the hypothesis $p \vee \sim q$ is true and the conclusion $\sim p$ is false are the first and second rows. So you put F's in those two rows and T's in the other two rows.

p	q	$\sim p$	$\sim q$	$p \vee \sim q$	$p \vee \sim q \rightarrow \sim p$
T	T	F	F	T	F
T	F	F	T	T	F
F	T	T	F	F	T
F	F	T	T	T	T

LOGICAL EQUIVALENCES INVOLVING →

If at a certain stage of solving a problem you know that a statement p is true *or* that a statement q is true, you can deduce the truth of a statement r by showing two things: that the truth of r follows from the truth of p *and also* that the truth of r follows from the truth of q. Then no matter whether p or q is the case, the truth of r must follow. The division into cases method of analysis is based on the following logical equivalence.

EXAMPLE 1.2.2   **Showing that $p \vee q \to r \equiv (p \to r) \wedge (q \to r)$**

Use truth tables to show the logical equivalence of the statement forms $p \vee q \to r$ and $(p \to r) \wedge (q \to r)$.

Solution   First fill in the eight possible combinations of truth values for p, q, and r. Then fill in the columns for $p \vee q$, $p \to r$, and $q \to r$ using the definitions of *or* and *if–then*. For instance, the $p \to r$ column has F's in the second and fourth rows because these are the rows in which p is true and q is false. Next fill in the $p \vee q \to r$ column using the definition of *if–then*. The rows in which the hypothesis $p \vee q$ is true and the conclusion r is false are the second, fourth, and sixth. So F's go in these rows and T's in all the others. The complete table shows that $p \vee q \to r$ and $(p \to r) \wedge (q \to r)$ have the same truth values for each combination of truth values of p, q, and r. Hence the two statement forms are logically equivalent.

p	q	r	$p \vee q$	$p \to r$	$q \to r$	$p \vee q \to r$	$(p \to r) \wedge (q \to r)$
T	T	T	T	T	T	T	T
T	T	F	T	F	F	F	F
T	F	T	T	T	T	T	T
T	F	F	T	F	T	F	F
F	T	T	T	T	T	T	T
F	T	F	T	T	F	F	F
F	F	T	F	T	T	T	T
F	F	F	F	T	T	T	T

same truth values and so
$p \vee q \to r \equiv (p \to r) \wedge (q \to r)$ ∎

REPRESENTATION OF *IF-THEN* AS *OR*

In exercise 13a at the end of this section you are asked to use truth tables to show that

$$p \to q \equiv \sim p \vee q.$$

The logical equivalence of "if p then q" and "not p or q" is used only occasionally in everyday speech. Here is one instance.

EXAMPLE 1.2.3   **Application of Equivalence Between $\sim p \vee q$ and $p \to q$**

Rewrite the following statement in if–then form.

Either you get to work on time or you are fired.

Solution   Let $\sim p$ be

You get to work on time.

and q be

You are fired.

Then the given statement is $\sim p \vee q$. Also p is

You do not get to work on time.

So the equivalent if–then version, $p \rightarrow q$, is

If you do not get to work on time, then you are fired.   ■

THE NEGATION OF A CONDITIONAL STATEMENT

By definition, $p \rightarrow q$ is false if, and only if, its hypothesis, p, is true and its conclusion, q, is false. It follows that

> The negation of "if p then q" is logically equivalent to "p and not q."

This can be restated symbolically as follows:

> $$\sim(p \rightarrow q) \equiv p \wedge \sim q$$

You can also obtain this result by starting from the logical equivalence $p \rightarrow q \equiv \sim p \vee q$. Take the negation of both sides to obtain

$$\sim(p \rightarrow q) \equiv \sim(\sim p \vee q)$$
$$\equiv \sim(\sim p) \wedge (\sim q) \quad \text{by De Morgan's laws}$$
$$\equiv p \wedge \sim q \quad \text{by the double negative law.}$$

Yet another way to derive this result is to construct truth tables for $\sim(p \rightarrow q)$ and for $p \wedge \sim q$ and to check that they have the same truth values. (See exercise 13(b) at the end of this section.)

EXAMPLE 1.2.4   Negations of *If–Then* Statements

Write negations for each of the following statements:

a. If my car is in the repair shop, then I cannot get to class.
b. If Sara lives in Athens, then she lives in Greece.

Solution   a. My car is in the repair shop and I can get to class.
b. Sara lives in Athens and she does not live in Greece. (Sara might live in Athens, Georgia; Athens, Ohio; or Athens, Wisconsin.)   ■

▲ *CAUTION!*   *It is tempting to write the negation of an if–then statement as another if–then statement. Please resist that temptation! Remember that the negation of an if–then statement does not start with the word if.*

THE CONTRAPOSITIVE OF A CONDITIONAL STATEMENT

One of the most fundamental laws of logic is the equivalence between a conditional statement and its contrapositive.

DEFINITION

The **contrapositive** of a conditional statement of the form "If p then q" is

If $\sim q$ then $\sim p$.

Symbolically,

The contrapositive of $p \rightarrow q$ is $\sim q \rightarrow \sim p$.

The fact is that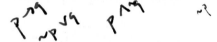

> A conditional statement is logically equivalent to its contrapositive.

You are asked to establish this equivalence in exercise 22.

EXAMPLE 1.2.5   Writing the Contrapositive

Write each of the following statements in its equivalent contrapositive form:

a. If Howard can swim across the lake, then Howard can swim to the island.
b. If today is Easter, then tomorrow is Monday.

Solution   a. If Howard cannot swim to the island, then Howard cannot swim across the lake.
b. If tomorrow is not Monday, then today is not Easter. ∎

When you are trying to solve certain problems, you may find that the contrapositive form of a conditional statement is easier to work with than the original statement. Replacing a statement by its contrapositive may give the extra push that helps you over the top in your search for a solution. This logical equivalence is also the basis for one of the most important laws of deduction, modus tollens (to be explained in Section 1.3), and for the contrapositive method of proof (to be explained in Section 3.6).

THE CONVERSE AND INVERSE OF A CONDITIONAL STATEMENT

The fact that a conditional statement and its contrapositive are logically equivalent is very important and has wide application. Two other variants of a conditional statement are *not* logically equivalent to the statement.

> **DEFINITION**
>
> Suppose a conditional statement of the form "If p then q" is given.
>
> **1.** The **converse** is "If q then p."
> **2.** The **inverse** is "If $\sim p$ then $\sim q$."
>
> Symbolically,
>
> $$\text{The converse of } p \rightarrow q \text{ is } q \rightarrow p,$$
>
> and
>
> $$\text{The inverse of } p \rightarrow q \text{ is } \sim p \rightarrow \sim q.$$

EXAMPLE 1.2.6   Writing the Converse and the Inverse

Write the converse and inverse of each of the following statements:

a. If Howard can swim across the lake, then Howard can swim to the island.
b. If today is Easter, then tomorrow is Monday.

Solution   a. *converse:* If Howard can swim to the island, then Howard can swim across the lake.

        inverse: If Howard cannot swim across the lake, then Howard cannot swim to the island.

b. *converse:* If tomorrow is Monday, then today is Easter.

        inverse: If today is not Easter, then tomorrow is not Monday. ■

▲ *CAUTION!*   *Many people mistakenly believe that if a conditional statement is true, then its converse and inverse are also true. This is not so. If a conditional statement is true, then its converse and inverse may or may not be true. For instance, on any Sunday except Easter, the conditional statement in Example 1.2.6(b) is true; yet both its converse and inverse are false.*

> 1. A conditional statement and its converse are *not* logically equivalent.
> 2. A conditional statement and its inverse are *not* logically equivalent.
> 3. The converse and the inverse of a conditional statement are logically equivalent to each other.

You are asked to use truth tables to verify the statements in the box above in exercises 20, 21, and 23 at the end of this section. Note that the truth of statement (3) also follows from the observation that the inverse of a conditional statement is the contrapositive of its converse.

ONLY IF AND THE BICONDITIONAL

To say "p only if q" means that p can take place *only* if q takes place also. That is, if q does not take place, then p cannot take place. Another way to say this is that if p occurs, then q must also occur (by the logical equivalence between a statement and its contrapositive).

> **DEFINITION**
>
> If p and q are statements,
>
> $$p \textbf{ only if } q \quad \text{means} \quad \text{"if not } q \text{ then not } p\text{,"}$$
>
> or, equivalently,
>
> $$\textit{"if } p \textit{ then } q\textit{."}$$

EXAMPLE 1.2.7   Converting *Only If* to *If–Then*

Use the contrapositive to rewrite the following statement in if–then form in two ways:

> John will break the world's record for the mile run only if
> he runs the mile in under four minutes.

Solution   *Version 1:* If John does not run the mile in under four minutes, then he will not break the world's record.

Version 2: If John breaks the world's record, then he will have run the mile in under four minutes. ∎

To say that John will break the world's record only if he runs the mile in under four minutes does not mean that John will break the world's record if he runs the mile in under four minutes. His time could be under four minutes but still not be fast enough to break the record.

▲ *CAUTION!*   "p *only if* q" *does not mean* "p *if* q."

> **DEFINITION**
>
> Given statement variables p and q, the **biconditional of p and q** is "p if, and only if, q" and is denoted $p \leftrightarrow q$. It is true if both p and q have the same truth values and is false if p and q have opposite truth values. The words *if and only if* are sometimes abbreviated **iff.**

The biconditional has the following truth table:

Truth Table for $p \leftrightarrow q$

p	q	$p \leftrightarrow q$
T	T	T
T	F	F
F	T	F
F	F	T

In order of operations $\leftrightarrow$ is coequal with $\rightarrow$. As with $\wedge$ and $\vee$, to indicate precedence between them parentheses must be used. Thus the full hierarchy of operations for the five logical connectives can be summarized as follows.

Order of Operations

1. $\sim$
2. $\wedge$, $\vee$
3. $\rightarrow$, $\leftrightarrow$

According to the separate definitions of *if* and *only if,* saying "*p* if, and only if, *q*" should mean the same as saying both "*p* if *q*" and "*p* only if *q*." The following truth table shows that this is the case:

Truth Table showing that $p \leftrightarrow q \equiv (p \rightarrow q) \wedge (q \rightarrow p)$

p	q	$p \rightarrow q$	$q \rightarrow p$	$p \leftrightarrow q$	$(p \rightarrow q) \wedge (q \rightarrow p)$
T	T	T	T	T	T
T	F	F	T	F	F
F	T	T	F	F	F
F	F	T	T	T	T

same truth values and so
$p \leftrightarrow q \equiv (p \rightarrow q) \wedge (q \rightarrow p)$

EXAMPLE 1.2.8   *If* and *Only If*

Rewrite the following statement as a conjunction of two if–then statements:

This computer program is correct if, and only if, it produces
the correct answer for all possible sets of input data.

Solution   If this program is correct, then it produces the correct answers for all possible sets of input data; and if this program produces the correct answers for all possible sets of input data, then it is correct.   ∎

Earlier it was noted that $p \rightarrow q \equiv \sim p \vee q$. Since $p \leftrightarrow q \equiv (p \rightarrow q) \wedge (q \rightarrow p)$, it follows that

$$p \leftrightarrow q \equiv (\sim p \vee q) \wedge (\sim q \vee p).$$

Consequently, any statement form containing $\rightarrow$ or $\leftrightarrow$ is logically equivalent to one containing only $\sim$, $\wedge$, and $\vee$. (See exercises 29–32.)

NECESSARY AND SUFFICIENT CONDITIONS

The phrases *necessary condition* and *sufficient condition,* as used in formal English, correspond exactly to their definitions in logic.

DEFINITION

If *r* and *s* are statements:

   r is a **sufficient condition** for *s*   means   "if *r* then *s*."

   r is a **necessary condition** for *s*   means   "if not *r* then not *s*."

In other words, to say "*r* is a sufficient condition for *s*" means that the occurrence of *r* is *sufficient* to guarantee the occurrence of *s*. On the other hand, to say "*r* is a necessary condition for *s*" means that if *r* does not occur, then *s* cannot occur either: The occurrence of *r* is *necessary* to obtain the occurrence of *s*. Note that because of the equivalence between a statement and its contrapositive,

---

   r is a necessary condition for *s*   also means   "if *s* then *r*."

Consequently,

---

   r is a necessary and sufficient condition for *s*   means   "*r* if, and only if, *s*."

EXAMPLE 1.2.9   Interpreting Necessary and Sufficient Conditions

Consider the statement "If John is eligible to vote, then he is at least 18 years old." The truth of the condition "John is eligible to vote" is *sufficient* to ensure the truth of the condition "John is at least 18 years old." In addition, the condition "John is at least 18 years old" is *necessary* for the condition "John is eligible to vote" to be true. If John were younger than 18, then he would not be eligible to vote.   ■

EXAMPLE 1.2.10   Converting a Sufficient Condition to If–Then Form

Rewrite the following statement in the form "If *A* then *B*":

> Pia's birth on U.S. soil is a sufficient condition
> for her to be a U.S. citizen.

Solution   If Pia was born on U.S. soil, then she is a U.S. citizen.   ■

EXAMPLE 1.2.11   Converting a Necessary Condition to If–Then Form

Use the contrapositive to rewrite the following statement in two ways:

> George's attaining age 35 is a necessary condition
> for his being president of the United States.

Solution   *Version 1:* If George has not attained the age of 35, then he cannot be president of the United States.

Version 2: If George can be president of the United States, then he has attained the age of 35.   ■

REMARKS

1. *In logic, a hypothesis and conclusion are not required to have related subject matters.*

    In ordinary speech we never say things like "If computers are machines, then Babe Ruth was a baseball player" or "If $2 + 2 = 5$, then Sting is president of the United States." We only formulate a sentence like "If p then q" if there is some connection of content between p and q.

    In logic, however, the two parts of a conditional statement need not have related meanings. The reason? If there were such a requirement, who would enforce it? What one person perceives as two unrelated clauses may seem related to someone else. There would have to be a central arbiter to check each conditional sentence before anyone could use it, to be sure its clauses were in proper relation. This is impractical, to say the least!

    Thus a statement like "if computers are machines, then Babe Ruth was a baseball player" is allowed, and it is even called true because both its hypothesis and its conclusion are true. Similarly, the statement "If $2 + 2 = 5$, then Sting is president of the United States" is allowed and is called true because its hypothesis is false, even though doing so may seem ridiculous.

    In mathematics it often happens that a carefully formulated definition that successfully covers the situations for which it was primarily intended is later seen to be satisfied by some extreme cases that the formulator did not have in mind. But those are the breaks, and it is important to get into the habit of exploring definitions fully to seek out and understand *all* their instances, even the unusual ones.

2. *In informal language, simple conditionals are often used to mean biconditionals.*

    The formal statement "p if, and only if, q" is seldom used in ordinary language. Frequently, when people intend the biconditional they leave out either the *and only if* or the *if and*. That is, they say either "p if q" or "p only if q" when they really mean "p if, and only if, q." For example, consider the statement "You will get dessert if, and only if, you eat your dinner." Logically, this is equivalent to the conjunction of the following two statements.

statement 1: If you eat your dinner, then you will get dessert.

statement 2: You will get dessert only if you eat your dinner.
        or
            If you do not eat your dinner, then you will not get dessert.

    Now how many parents in the history of the world have said to their children "You will get dessert if, and only if, you eat your dinner"? Not many! Most say either "If you eat your dinner you will get dessert" (these take the positive approach—they emphasize the reward) or "You will get dessert only if you eat your dinner" (these take the negative approach—they emphasize the punishment). Yet the parents who promise the reward intend to suggest the punishment as well, and those who threaten the punishment will certainly give the reward if earned. Both sets of parents expect that their conditional statements will be interpreted as biconditionals.

Since we often (correctly) interpret conditional statements as biconditionals, it is not surprising that we may come to believe (mistakenly) that conditional statements are always logically equivalent to their inverses and converses. In formal settings, however, statements must have unambiguous interpretations. If–then statements can't sometimes mean "if–then" and other times mean "if and only if." When using language in mathematics, science, or other situations where precision is important, it is essential to interpret if–then statements according to the formal definition and not to confuse them with their converses and inverses.

EXERCISE SET 1.2

Rewrite the statements in 1–4 in if–then form.

1. This loop will repeat exactly N times if it does not contain a **stop** or a **go to.**

2. I am on time for work if I catch the 8:05 bus.

3. Freeze or I'll shoot.

4. Fix my ceiling or I won't pay my rent.

Construct truth tables for the statement forms in 5–11.

5. $\sim p \vee q \to \sim q$ 6. $p \vee (\sim p \wedge q) \to q$

7. $p \wedge \sim q \to r$ 8. $\sim p \vee q \to r$

9. $p \wedge \sim r \leftrightarrow q \vee r$ 10. $(p \to r) \leftrightarrow (q \to r)$

11. $(p \to (q \to r)) \leftrightarrow ((p \wedge q) \to r)$

12. Use the logical equivalence established in Example 1.2.3, $p \vee q \to r \equiv (p \to r) \wedge (q \to r)$, to rewrite the following statement. (Assume x represents a fixed real number.)

$$\text{If } x > 2 \quad \text{or} \quad x < -2, \quad \text{then } x^2 > 4.$$

13. Use truth tables to verify that
   a. $p \to q \equiv \sim p \vee q$
   b. $\sim (p \to q) \equiv p \wedge \sim q$.

14. a. Show that the following statement forms are all logically equivalent.

   $$p \to q \vee r, \quad p \wedge \sim q \to r, \quad \text{and} \quad p \wedge \sim r \to q$$

   b. Use the logical equivalences established in part (a) to rewrite the following sentence in two different ways. (Assume n represents a fixed integer.)

   If n is prime, then n is odd or n is 2.

15. True or false? The negation of "If Sue is Luiz's mother, then Deana is his cousin" is "If Sue is Luiz's mother, then Deana is not his cousin."

16. Write negations for each of the following statements. (Assume that all variables represent fixed quantities or entities, as appropriate.)

a. If P is a square, then P is a rectangle.
b. If today is Thanksgiving, then tomorrow is Friday.
c. If r is rational, then the decimal expansion of r is repeating.
d. If n is prime, then n is odd or n is 2.
e. If x is nonnegative, then x is positive or x is 0.
f. If Tom is Ann's father, then Jim is her uncle and Sue is her aunt.
g. If n is divisible by 6, then n is divisible by 2 and n is divisible by 3.

17. Suppose that p and q are statements so that $p \to q$ is false. Find the truth values of each of the following:
   a. $\sim p \to q$ b. $p \vee q$ c. $q \to p$

H18. Write contrapositives for the statements of exercise 16.

H19. Write the converse and inverse for each statement of exercise 16.

Use truth tables to establish the truth of each statement in 20–23.

20. A conditional statement is not logically equivalent to its converse.

21. A conditional statement is not logically equivalent to its inverse.

22. A conditional statement and its contrapositive are logically equivalent to each other.

23. The converse and inverse of a conditional statement are logically equivalent to each other.

24. "Do you mean that you think you can find out the answer to it?" said the March Hare.
   "Exactly so," said Alice.
   "Then you should say what you mean," the March Hare went on.
   "I do," Alice hastily replied; "at least—at least I mean what I say—that's the same thing, you know."
   "Not the same thing a bit!" said the Hatter. "Why, you might just as well say that 'I see what I eat' is the same thing as 'I eat what I see'!"
   —from "A Mad Tea-Party" in
   Alice in Wonderland, by Lewis Carroll

The Hatter is right. "I say what I mean" is not the same thing as "I mean what I say." Rewrite each of these two sentences in if–then form and explain the logical relation between them. (This exercise is referenced in the introduction to Chapter 3.)

Use the contrapositive to rewrite the statements in 25 and 26 in if–then form in two ways. Assume that *only if* has its formal, logical meaning.

25. The Cubs will win the pennant only if they win tomorrow's game.

26. Sam will be allowed on Signe's racing boat only if he is an expert sailor.

27. Taking the long view on your edcuation, you go to the Prestige Corporation and ask what you should do in college to be hired when you graduate. The Personnel Director replies that you will be hired *only if* you major in mathematics or computer science, get a B average or better, and take accounting. You do, in fact, become a math major, get a B$^+$ average, and take accounting. You return to Prestige Corporation, make a formal application, and are turned down. Did the Personnel Director lie to you?

28. In formal contexts "*r* unless *s*" means "if not *s* then *r* ". Rewrite the following in if–then form:

    a. Payment will be made on the fifth unless a new hearing is granted.

    b. This door will not open unless a security code is entered.

In 29–32, (a) use the logical equivalences $p \rightarrow q \equiv {\sim}p \lor q$ and $p \leftrightarrow q \equiv ({\sim}p \lor q) \land ({\sim}q \lor p)$ to rewrite the given statement forms without using the symbols $\rightarrow$ or $\leftrightarrow$, and (b) use the logical equivalence $p \lor q \equiv {\sim}({\sim}p \land {\sim}q)$ to rewrite each statement form using only $\land$ and ${\sim}$.

29. $p \land {\sim}q \rightarrow r$ 30. ${\sim}p \lor q \rightarrow r \lor {\sim}q$

31. $(p \rightarrow r) \leftrightarrow (q \rightarrow r)$

32. $(p \rightarrow (q \rightarrow r)) \leftrightarrow ((p \land q) \rightarrow r)$

33. Given any statement form, is it possible to find a logically equivalent form that uses only ${\sim}$ and $\land$? Justify your answer.

Rewrite the statements in 34 and 35 in if–then form.

34. Catching the 8:05 bus is a sufficient condition for my being on time for work.

35. Having two 45° angles is a sufficient condition for this triangle to be a right triangle.

Use the contrapositive to rewrite the statements in 36 and 37 in if–then form in two ways.

36. Being divisible by 3 is a necessary condition for this number to be divisible by 9.

37. Doing his homework regularly is a necessary condition for Jim to pass the course.

Note that "a sufficient condition for *s* is *r*" means *r* is a sufficient condition for *s* and "a necessary condition for *s* is *r*" means *r* is a necessary condition for *s*. Rewrite the statements in 38 and 39 in if–then form.

38. A sufficient condition for Hal's team to win the championship is that it win the rest of its games.

39. A necessary condition for this computer program to be correct is that it not produce error messages during translation.

40. "If compound *X* is boiling, then its temperature must be at least 250°F." Assuming that this statement is true, which of the following must also be true?
    a. If the temperature of compound *X* is at least 250°F, then compound *X* is boiling.
    b. If the temperature of compound *X* is less than 250°F, then compound *X* is not boiling.
    c. Compound *X* will boil only if its temperature is at least 250°F.
    d. If compound *X* is not boiling, then its temperature is less than 250°F.
    e. A necessary condition for compound *X* to boil is that its temperature be at least 250°F.
    f. A sufficient condition for compound *X* to boil is that its temperature be at least 250°F.

"Contrariwise," continued Tweedledee, "if it was so, it might be; and if it were so, it would be; but as it isn't, it ain't. That's logic."
(Lewis Carroll, *Through the Looking Glass*)

1.3   VALID AND INVALID ARGUMENTS

In mathematics and logic an argument is not a dispute. It is a sequence of statements ending in a conclusion. In this section we show how to determine whether an argument is valid, that is, to determine whether the conclusion follows *necessarily* from the preceding statements.

DEFINITION

An **argument** is a sequence of statements. All statements but the final one are called **premises** (or **assumptions** or **hypotheses**). The final statement is called the **conclusion**. The symbol $\therefore$, read "therefore," is normally placed just before the conclusion.

It was shown in Section 1.1 that the logical form of an argument can be abstracted from the content of the argument. For example, the argument

> If Socrates is a human being, then Socrates is mortal;
>
> Socrates is a human being;
>
> $\therefore$ Socrates is mortal;

has the abstract form

> If p then q;
>
> p;
>
> $\therefore q$.

When considering the abstract form of an argument, think of p and q as variables for which statements may be substituted. An argument form is called *valid* if, and only if, whenever statements are substituted that make all the premises true, then the conclusion is also true.

DEFINITION

To say that an *argument form* is **valid** means that no matter what particular statements are substituted for the statement variables in its premises, if the resulting premises are all true, then the conclusion is also true.

To say that an *argument* is **valid** means that its form is valid.

The crucial fact about a valid argument is that the truth of its conclusion follows *necessarily* or *inescapably* or *by logical form alone* from the truth of its premises. It is impossible to have a valid argument with true premises and a false conclusion. When an argument is valid and its premises are true, the truth of the conclusion is said to be *inferred* or *deduced* from the truth of the premises. If a conclusion "ain't necessarily so," then it isn't a valid deduction.

Practically speaking, to test an argument form for validity:

1. Identify the premises and conclusion of the argument.
2. Construct a truth table showing the truth values of all the premises and the conclusion.
3. Find the rows (called **critical rows**) in which all the premises are true.
4. In each critical row, determine whether the conclusion of the argument is also true.
   a. If in each critical row the conclusion is also true, then the argument form is valid.
   b. If there is at least one critical row in which the conclusion is false, the argument form is invalid.

EXAMPLE 1.3.1   A Valid Argument Form

Show that the following argument form is valid:

$$p \lor (q \lor r)$$
$$\sim r$$
$$\therefore p \lor q$$

Solution

				premises		conclusion
p	q	r	$q \lor r$	$p \lor (q \lor r)$	$\sim r$	$p \lor q$
T	T	T	T	T	F	T
T	T	F	T	T	T	T
T	F	T	T	T	F	T
T	F	F	F	T	T	T
F	T	T	T	T	F	T
F	T	F	T	T	T	T
F	F	T	T	T	F	F
F	F	F	F	F	T	F

critical rows

In each row where the premises are both true the conclusion is also true, so the argument is valid.

Note that if you are in a hurry to check the validity of an argument, you need not fill in truth values for the conclusion except in the critical rows. The truth values in the other rows are irrelevant to the validity or invalidity of the argument. This is illustrated in the following example.

EXAMPLE 1.3.2   An Invalid Argument Form

Show that the following argument form is invalid.

$$p \rightarrow q \lor \sim r$$
$$q \rightarrow p \land r$$
$$\therefore p \rightarrow r$$

Solution   The truth table below shows that it is possible for an argument of this form to have true premises and a false conclusion.

						premises		conclusion
p	q	r	$\sim r$	$q \lor \sim r$	$p \land r$	$p \rightarrow q \lor \sim r$	$q \rightarrow p \land r$	$p \rightarrow r$
T	T	T	F	T	T	T	T	T
T	T	F	T	T	F	T	F	
T	F	T	F	F	T	F	T	
T	F	F	T	T	F	T	T	F
F	T	T	F	T	F	T	F	
F	T	F	T	T	F	T	F	
F	F	T	F	F	F	T	T	T
F	F	F	T	T	F	T	T	T

In this row the premises are true and the conclusion is false; hence the argument form is invalid.

MODUS PONENS AND MODUS TOLLENS

Consider the following argument form:

If *p* then *q*.

p

$\therefore$ *q*

Here is an argument of this form.

If the last digit of this number is a 0, then this number
is divisible by 10.

The last digit of this number is a 0.

$\therefore$ This number is divisible by 10.

The fact that this argument form is valid is called **modus ponens**. The term *modus ponens* is Latin meaning "method of affirming" (since the conclusion is an affirmation). Long before you saw your first truth table, you were undoubtedly being convinced by arguments of this form. Nevertheless, it is instructive to prove modus ponens, if for no other reason than to confirm the agreement between the formal definition of validity and the intuitive concept. To do so, we construct a truth table for the premises and conclusion.

		premises		conclusion	
p	*q*	$p \to q$	*p*	*q*	
T	T	T	T	T	← critical row
T	F	F	T	F	
F	T	T	F	T	
F	F	T	F	F	

The first row is the only one in which both premises are true, and the conclusion in that row is also true. Hence the argument form is valid.

Now consider this argument form:

If *p* then *q*.

~*q*

$\therefore$ ~*p*

The following is an example of an argument of this form:

If Zeus is human, then Zeus is mortal.

Zeus is not mortal.

$\therefore$ Zeus is not human.

An intuitive explanation for the validity of this argument form uses proof by contradiction. It goes like this:

Suppose

(1) If Zeus is human, then Zeus is mortal;

and

(2) Zeus is not mortal.

Must Zeus necessarily be nonhuman?

Yes!

Because, if Zeus were human, then by (1) he would be mortal.

But by (2) he is not mortal.

Hence, Zeus cannot be human.

The fact that this argument form is valid is called **modus tollens**. *Modus tollens* is Latin meaning "method of denying" (since the conclusion is a denial). The validity of modus tollens can be shown to follow from modus ponens together with the fact that a conditional statement is logically equivalent to its contrapositive. Or it can be established formally by using a truth table. (See exercise 11 on p. 40.)

Studies by cognitive psychologists have shown that while nearly 100% of college students have a solid, intuitive understanding of modus ponens, less than 60% are able to apply modus tollens correctly.* Yet in mathematical reasoning modus tollens is used almost as often as modus ponens. Thus it is important to study the form of modus tollens carefully to learn to use it effectively.

EXAMPLE 1.3.3   Recognizing Modus Ponens and Modus Tollens

Use modus ponens or modus tollens to fill in the blanks of the following arguments so that they become valid inferences.

a. If there are more pigeons than there are pigeonholes, then two pigeons roost in the same hole.
   There are more pigeons than there are pigeonholes.
   ∴ _____ .

b. If this number is divisible by 6, then it is divisible by 2.
   This number is not divisible by 2.
   ∴ _____ .

Solution   a. Two pigeons roost in the same hole.   by modus ponens
            b. This number is not divisible by 6.   by modus tollens   ■

ADDITIONAL VALID ARGUMENT FORMS

The following are additional examples of rules of inference, which state that certain forms of argument are valid. Verification of validity is left to the exercises at the end of this section.

EXAMPLE 1.3.4   Disjunctive Addition

The following argument forms are valid:

$$\begin{array}{ll} \text{a.} \quad p & \qquad \text{b.} \quad q \\ \therefore p \vee q & \qquad \therefore p \vee q \end{array}$$

These argument forms are used for making generalizations. For instance, according to the first, if p is true, then, more generally, "p or q" is true for *any* other statement q. As an example, suppose you are given the job of counting the number of

---

* *Cognitive Psychology and Its Implications*, 3d Edition, by John R. Anderson, New York: Freeman, 1990, pp. 292–297.

upperclassmen at your school. You ask what class Anton is in and are told he is a junior. Knowing that upperclassman means junior *or* senior, you generalize that he is an upperclassman and add him to your list.   ■

EXAMPLE 1.3.5   Conjunctive Simplification

The following argument forms are valid:

a. $p \wedge q$ b. $p \wedge q$
     $\therefore p$ $\therefore q$

These argument forms are used for particularizing. For instance, the first says that if both p and q are true, then, in particular, p is true.

When classifying objects according to some property, you often know much more about them than whether they do or do not have that property. When that happens, you discard extraneous information as you concentrate on the particular property of interest.   ■

Both generalization and particularization are used frequently in mathematics to tailor facts to fit into hypotheses of known theorems in order to draw further conclusions. Disjunctive syllogism, hypothetical syllogism, and proof by division into cases are also widely used tools.

EXAMPLE 1.3.6   Disjunctive Syllogism

The following argument forms are valid:

a. $p \vee q$ b. $p \vee q$
        $\sim q$ $\sim p$
     $\therefore p$ $\therefore q$

These argument forms say that when you have only two possibilities and you can rule one out, the other must be the case. For instance, suppose you know that for a particular number x, $x - 3 = 0$ or $x + 2 = 0$. If you also known that x is not negative, then $x \neq -2$ and so $x + 2 \neq 0$. By disjunctive syllogism you can conclude that $x - 3 = 0$.   ■

EXAMPLE 1.3.7   Hypothetical Syllogism

The following argument form is valid:

$$p \rightarrow q$$
$$q \rightarrow r$$
$$\therefore p \rightarrow r$$

Many arguments in mathematics contain chains of if–then statements. From the fact that one statement implies a second and the second implies a third, you can conclude that the first statement implies the third. Here is an example:

If 18,486 is divisible by 18, then 18,486 is divisible by 9.

If 18,486 is divisible by 9, then the sum of the digits of 18,486 is divisible by 9.

∴ If 18,486 is divisible by 18, then the sum of the digits of 18,486 is divisible by 9.   ■

EXAMPLE 1.3.8   Dilemma: Proof by Division into Cases

The following argument form is valid:

$$p \lor q$$
$$p \to r$$
$$q \to r$$
$$\therefore r$$

It often happens that you know one thing or another is true. If you can show that in either case a certain conclusion follows, then this conclusion must also be true. For instance, suppose you know that x is a particular nonzero real number. The trichotomy property of the real numbers says that any number is positive, negative, or zero. Thus (by disjunctive syllogism) you know that x is positive or x is negative. You can deduce that $x^2 > 0$ by arguing as follows:

x is positive or x is negative.

If x is positive, then $x^2 > 0$.

If x is negative, then $x^2 > 0$.

$\therefore x^2 > 0$.   ■

The rules of valid inference are used constantly in problem solving. Here is an example from everyday life.

EXAMPLE 1.3.9   Application: A More Complex Deduction

You are about to leave for school in the morning and discover you don't have your glasses. You know the following statements are true:

a. If my glasses are on the kitchen table, then I saw them at breakfast.
b. I was reading the newspaper in the living room or I was reading the newspaper in the kitchen.
c. If I was reading the newspaper in the living room, then my glasses are on the coffee table.
d. I did not see my glasses at breakfast.
e. If I was reading my book in bed, then my glasses are on the bed table.
f. If I was reading the newspaper in the kitchen, then my glasses are on the kitchen table.

Where are the glasses?

Solution   The glasses are on the coffee table. Here is a sequence of steps you might use to reach this answer, together with the rules of inference that allow you to draw the conclusion of each step:

1. The glasses are not on the kitchen table.   by (a), (d), and modus tollens
2. I did not read the newspaper in the kitchen.   by (f), (1), and modus tollens
3. I read the newspaper in the living room.   by (b), (2), and disjunctive syllogism
4. My glasses are on the coffee table.   by (c), (3), and modus ponens

Note that (e) was not needed to derive the conclusion. In mathematics as in real life, we frequently deduce a conclusion from just a part of the information available to us.

■

The preceding example shows how to use rules of inferential logic to solve an ordinary problem that could occur in real life. Normally, of course, you use these rules unconsciously. Occasionally, however, problems are so complex that it is helpful to use symbolic logic explicitly. The next example shows how you could do this for the situation described in Example 1.3.9.

EXAMPLE 1.3.10   Symbolizing a Situation to Find a Solution

Solve the problem of Example 1.3.9 symbolically.

Solution   Let p = My glasses are on the kitchen table.

q = I saw my glasses at breakfast.

r = I was reading the newspaper in the living room.

s = I was reading the newspaper in the kitchen.

t = My glasses are on the coffee table.

u = I was reading my book in bed.

v = My glasses are on the bed table.

Then the statements of Example 1.3.9 translate as follows:

a. $p \rightarrow q$	b. $r \vee s$	c. $r \rightarrow t$
d. $\sim q$	e. $u \rightarrow v$	f. $s \rightarrow p$

The following deductions can be made:

1.　　$p \rightarrow q$　　by (a)
　　　　$\sim q$　　by (d)
　　∴ $\sim p$　　by modus tollens

2.　　$s \rightarrow p$　　by (f)
　　　　$\sim p$　　by the conclusion of (1)
　　∴ $\sim s$　　by modus tollens

3.　　$r \vee s$　　by (b)
　　　　$\sim s$　　by the conclusion of (2)
　　∴ r　　by disjunctive syllogism

4.　　$r \rightarrow t$　　by (c)
　　　　r　　by the conclusion of (3)
　　∴ t　　by modus ponens

Hence t is true and the glasses are on the coffee table. ∎

FALLACIES

A **fallacy** is an error in reasoning that results in an invalid argument. Three common fallacies are using vague or ambiguous premises, begging the question (assuming what is to be proved), and jumping to conclusions without adequate grounds. In this section we discuss two other fallacies, called *converse error* and *inverse error,* which give rise to arguments that superficially resemble those that are valid by modus ponens and modus tollens but are not, in fact, valid.

As in previous examples, you can show that an argument is invalid by constructing a truth table for the argument form and finding at least one critical row in which all the premises are true but the conclusion is false. Another way is to find an argument of the same form with true premises and a false conclusion. The reason is that for an argument to be valid, *any* argument of the same form that has true premises must have a true conclusion.

EXAMPLE 1.3.11   Converse Error

Show that the following argument is invalid:

> If Zeke is a cheater, then Zeke sits in the back row.
>
> Zeke sits in the back row.
>
> ∴ Zeke is a cheater.

Solution   Many people recognize the invalidity of the above argument intuitively, reasoning something like this: The first premise gives information about Zeke *if* it is known he is a cheater. It doesn't give any information about him if it is not already known that he is a cheater. One can certainly imagine a person who is not a cheater but happens to sit in the back row. Then if that person's name is substituted for Zeke, the first premise is true by default and the second premise is also true but the conclusion is false.

The general form of the above argument is as follows:

$$p \rightarrow q$$
$$q \qquad \text{converse}$$
$$\therefore p$$

In exercise 12(a) at the end of this section you are asked to use a truth table to show that this form of argument is invalid. ∎

The fallacy underlying this invalid argument form is called the **converse error** because the conclusion of the argument would follow from the premises if the premise $p \rightarrow q$ were replaced by its converse. Such a replacement is not allowed, however, because a conditional statement is not logically equivalent to its converse.

Another common error in reasoning is called the *inverse error*. Consider the following argument:

> If interest rates are going up, stock market prices will go down.
>
> Interest rates are not going up.
>
> ∴ Stock market prices will not go down.

Note that this argument has the following form:

$$p \rightarrow q$$
$$\sim p \qquad \text{inverse}$$
$$\therefore \sim q$$

You are asked to give a truth table verification of the invalidity of this argument form in exercise 12(b) at the end of this section.

The fallacy underlying this invalid argument form is called the **inverse error** because the conclusion of the argument would follow from the premises if the premise $p \rightarrow q$ were replaced by its inverse. Such a replacement is not allowed, however, because a conditional statement is not logically equivalent to its inverse.

▲ *CAUTION!*   *It is possible for a valid argument to have a false conclusion, and for an invalid argument to have a true conclusion.*

Sometimes people lump together the ideas of validity and truth. If an argument seems valid, they accept the conclusion as true. And if an argument seems fishy (really a slang expression for invalid), they think the conclusion must be false.

This is not correct. Validity is a property of argument forms: If an argument is valid, then so is every other argument that has the same form. Similarly, if an argument is invalid, then so is every other argument that has the same form. What characterizes a valid argument is that no argument whose form is valid can have all true premises and a false conclusion. For each valid argument, there are arguments of that form with all true premises and a true conclusion, at least one false premise and a true conclusion, and at least one false premise and a false conclusion. On the other hand, for each invalid argument, there are arguments of that form with every combination of truth values for the premises and conclusion, including all true premises and a false conclusion.

EXAMPLE 1.3.12 ┐   **A Valid Argument with a False Conclusion**

The argument below is valid by modus ponens. But its major premise is false and so is its conclusion.

> If John Lennon was a rock star, then John Lennon had red hair.
> John Lennon was a rock star.
> ∴ John Lennon had red hair.   ■

EXAMPLE 1.3.13 ┐   **An Invalid Argument with a True Conclusion**

The argument below is invalid by the converse error but it has a true conclusion.

> If New York is a big city, then New York has tall buildings.
> New York has tall buildings.
> ∴ New York is a big city.   ■

CONTRADICTIONS AND VALID ARGUMENTS

The concept of logical contradiction can be used to make inferences through a technique of reasoning called the *contradiction rule*. Suppose p is some statement whose truth you wish to deduce.

Contradiction Rule

If you can show that the supposition that statement p is false leads logically to a contradiction, then you can conclude that p is true.

EXAMPLE 1.3.14 ┐   Contradiction Rule

Show that the following argument form is valid:

$$\sim p \rightarrow c, \text{ where } c \text{ is a contradiction}$$
$$\therefore p$$

Solution   Construct a truth table for the premise and the conclusion of this argument.

			premise	conclusion
p	$\sim p$	c	$\sim p \rightarrow c$	p
T	F	F	T	T
F	T	F	F	F

← There is only one critical row in which the premise is true, and in this row the conclusion is also true. Hence the argument is valid.

∎

The contradiction rule is the logical heart of the method of proof by contradiction. A slight variation also provides the basis for solving many logical puzzles by eliminating contradictory answers: If an assumption leads to a contradiction, then that assumption must be false.

EXAMPLE 1.3.15   Knights and Knaves

The logician Raymond Smullyan describes an island containing two types of people: knights who always tell the truth and knaves who always lie.* You visit the island and are approached by two natives who speak to you as follows:

> A says: B is a knight.
> B says: A and I are of opposite type.

What are A and B?

Solution   A and B are both knaves. To see this, reason as follows:

Suppose A is a knight.

∴ What A says is true.   by definition of knight

∴ B is a knight also   That's what A said.

∴ What B says is true.   by definition of knight

∴ A and B are of opposite types.   That's what B said.

∴ We have arrived at the following contradiction: A and B are both knights and A and B are of opposite type.

∴ The supposition is false.   by the contradiction rule

∴ A is not a knight.   negation of supposition

∴ A is a knave.   by disjunctive syllogism: It's given that all inhabitants are knights or knaves, so since A is not a knight, A is a knave.

∴ What A says is false.

∴ B is not a knight.

∴ B is a knave also.   by disjunctive syllogism

---

*Raymond Smullyan has written a delightful series of whimsical yet profound books of logical puzzles starting with *What Is the Name of This Book?* (Englewood Cliffs, New Jersey: Prentice-Hall, 1978). Other good sources of logical puzzles are the many excellent books of Martin Gardner, such as *Aha! Insight* and *Aha! Gotcha* (New York: W. H. Freeman, 1978, 1982).

This reasoning shows that if the problem has a solution at all, then A and B must both be knaves. It is conceivable, however, that the problem has no solution. The problem statement could be inherently contradictory. If you look back at the problem, though, you can see that it does work out for both A and B to be knaves. ∎

SUMMARY OF RULES OF INFERENCE

Table 1.3.1 summarizes some of the most important rules of inference.

TABLE 1.3.1   Valid Argument Forms

Modus ponens	$p \to q$ p $\therefore q$	Disjunctive syllogism	a. $p \lor q$ b. $p \lor q$ $\sim q$ $\sim p$ $\therefore p$ $\therefore q$
Modus tollens	$p \to q$ $\sim q$ $\therefore \sim p$	Hypothetical syllogism	$p \to q$ $q \to r$ $\therefore p \to r$
Disjunctive addition	a. p b. q $\therefore p \lor q$ $\therefore p \lor q$	Dilemma: proof by division into cases	$p \lor q$ $p \to r$ $q \to r$ $\therefore r$
Conjunctive simplification	a. $p \land q$ b. $p \land q$ $\therefore p$ $\therefore q$	Rule of contradiction	$\sim p \to c$ $\therefore p$
Conjunctive addition	p q $\therefore p \land q$		

(handwritten annotations: "F" "F" over modus tollens $p \to q$; "No truth table necessary"; and a truth table near Rule of contradiction: p $\sim p$ c $\sim p \to c$ p / T F F T T / F T F F F, "premise", "valid"*)*

EXERCISE SET 1.3

Use modus ponens or modus tollens to fill in the blanks in the arguments of 1–5 so as to produce valid inferences.

1. If $\sqrt{2}$ is rational, then $\sqrt{2} = a/b$ for some integers a and b.
    It is not true that $\sqrt{2} = a/b$ for some integers a and b.
    $\therefore$ _____.

2. If this is a **while** loop, then the body of the loop may never be executed.
    _____.
    $\therefore$ The body of the loop may never be executed.

3. If logic is easy, then I am a monkey's uncle.
    I am not a monkey's uncle.
    $\therefore$ _____.

4. If this polygon is a triangle, then the sum of its interior angles is 180°.

The sum of the interior angles of this polygon is not 180°.
$\therefore$ _____.

5. If they were unsure of the address, then they would have telephoned.
    _____.
    $\therefore$ They were sure of the address.

Use truth tables to determine whether the argument forms in 6–10 are valid.

6. $p \to q$
    $q \to p$
    $\therefore p \lor q$

7. p
    $p \to q$
    $\sim q \lor r$
    $\therefore r$

8. $p \lor q$
    $p \to \sim q$
    $p \to r$
    $\therefore r$

9. $p \to q$
    $p \to r$
    $\therefore p \to q \land r$

10. $p \wedge \sim q \to r$
$p \vee q$
$q \to p$
$\therefore r$

11. Prove modus tollens. In other words, prove that the following argument form is valid.

$$p \to q$$
$$\sim q$$
$$\therefore \sim p$$

12. Use truth tables to show that the following forms of argument are invalid:

a. $p \to q$
q
$\therefore p$
(converse error)

b. $p \to q$
$\sim p$
$\therefore \sim q$
(inverse error)

Use truth tables to show that the argument forms referred to in 13–20 are valid.

13. Example 1.3.4(a)

14. Example 1.3.4(b).

15. Example 1.3.5(a).

16. Example 1.3.5(b).

17. Example 1.3.6(a).

18. Example 1.3.6(b).

19. Example 1.3.7.

20. Example 1.3.8.

Use symbols to write the logical form of each argument in 21 and 22, and then use a truth table to test the argument for validity.

21. If Tom is not on team A, then Hua is on team B.
If Hua is not on team B, then Tom is on team A.
$\therefore$ Tom is not on team A or Hua is not on team B.

22. Oleg is a math major or Oleg is an economics major.
If Oleg is a math major, then Oleg is required to take Math 362.
$\therefore$ Oleg is an economics major or Oleg is not required to take Math 362.

Some of the arguments in 23–31 are valid while others exhibit the converse or the inverse error. Use symbols to write the logical form of each argument. If the argument is valid, identify the rule of inference that guarantees its validity. Otherwise state whether the converse or the inverse error is made.

23. If Jules solved this problem correctly, then Jules obtained the answer 2.
Jules obtained the answer 2.
$\therefore$ Jules solved this problem correctly.

24. This real number is rational or it is irrational.
This real number is not rational.
$\therefore$ This real number is irrational.

25. If I go to the movies, I won't finish my homework.
If I don't finish my homework, I won't do well on the exam tomorrow.

$\therefore$ If I go to the movies, I won't do well on the exam tomorrow.

26. If this number is larger than 2, then its square is larger than 4.
This number is not larger than 2.
$\therefore$ The square of this number is not larger than 4.

27. If there are as many rational numbers as there are irrational numbers, then the set of all irrational numbers is infinite.
The set of all irrational numbers is infinite.
$\therefore$ There are as many rational numbers as there are irrational numbers.

28. If at least one of these two numbers is divisible by 6, then the product of these two numbers is divisible by 6.
Neither of these two numbers is divisible by 6.
$\therefore$ The product of these two numbers is not divisible by 6.

29. If this computer program is correct, then it produces the correct output when run with the test data my teacher gave me.
This computer program produces the correct output when run with the test data my teacher gave me.
$\therefore$ This computer program is correct.

30. Sandra knows COBOL and Sandra knows C.
$\therefore$ Sandra knows C.

31. If I get a Christmas bonus, I'll buy a stereo.
If I sell my motorcycle, I'll buy a stereo.
$\therefore$ If I get a Christmas bonus or I sell my motorcycle, then I'll buy a stereo.

32. Give an example (other than Example 1.3.12) of a valid argument with a false conclusion.

33. Give an example (other than Example 1.3.13) of an invalid argument with a true conclusion.

34. Explain in your own words what distinguishes a valid form of argument from an invalid one.

35. Given the following information about a computer program, find the mistake in the program:
a. There is an undeclared variable or there is a syntax error in the first five lines.
b. If there is a syntax error in the first five lines, then there is a missing semicolon or a variable name is misspelled.
c. There is not a missing semicolon.
d. There is not a misspelled variable name.

36. In the back of an old cupboard you discover a note signed by a pirate famous for his bizarre sense of humor and love of logical puzzles. In the note he wrote that he had hidden treasure somewhere on the property. He listed five true statements (a–e below) and challenged the reader to use them to figure out the location of the treasure.

a. If this house is next to a lake, then the treasure is not in the kitchen.
b. If the tree in the front yard is an elm, then the treasure is in the kitchen.
c. This house is next to a lake.
d. The tree in the front yard is an elm or the treasure is buried under the flagpole.
e. If the tree in the back yard is an oak, then the treasure is in the garage.

Where is the treasure hidden?

37. You are visiting the island described in Example 1.3.15 and have the following encounters with natives.
   a. Two natives A and B address you as follows:
      A says: Both of us are knights.
      B says: A is a knave.
      What are A and B?
   b. Another two natives C and D approach you but only C speaks.
      C says: Both of us are knaves.
      What are C and D?
   c. You then encounter natives E and F.
      E says: F is a knave.
      F says: E is a knave.
      How many knaves are there?
   H d. Finally, you meet a group of six natives, U, V, W, X, Y, and Z, who speak to you as follows:
      U says: None of us is a knight.
      V says: At least three of us are knights.
      W says: At most three of us are knights.
      X says: Exactly five of us are knights.
      Y says: Exactly two of us are knights.
      Z says: Exactly one of us is a knight.
      Which are knights and which are knaves?

38. The famous detective Percule Hoirot was called in to solve a baffling murder mystery. He determined the following facts:
   a. Lord Hazelton, the murdered man, was killed by a blow on the head with a brass candlestick.
   b. Either Lady Hazelton or a maid, Sara, was in the dining room at the time of the murder.
   c. If the cook was in the kitchen at the time of the murder, then the butler killed Lord Hazelton with a fatal dose of strychnine.

d. If Lady Hazelton was in the dining room at the time of the murder, then the chauffeur killed Lord Hazelton.
e. If the cook was not in the kitchen at the time of the murder, then Sara was not in the dining room when the murder was committed.
f. If Sara was in the dining room at the time the murder was committed, then the wine steward killed Lord Hazelton.

Is it possible for the detective to deduce the identity of the murderer from the above facts? If so, who did murder Lord Hazelton? (Assume there was only one cause of death.)

39. Sharky, a leader of the underworld, was killed by one of his own band of four henchmen. Detective Sharp interviewed the men and determined that all were lying except for one. He deduced who killed Sharky on the basis of the following statements:
   a. Socko: Lefty killed Sharky.
   b. Fats: Muscles didn't kill Sharky.
   c. Lefty: Muscles was shooting craps with Socko when Sharky was knocked off.
   d. Muscles: Lefty didn't kill Sharky.

   Who did kill Sharky?

In 40–43 a set of premises and a conclusion are given. Use the valid argument forms listed in Table 1.3.1 to deduce the conclusion from the premises, giving a reason for each step as in Example 1.3.10.

40.
$$\sim p \lor q \to r$$
$$s \lor \sim q$$
$$\sim t$$
$$p \to t$$
$$\sim p \land r \to \sim s$$
$$\therefore \sim q$$

41.
$$p \lor q$$
$$q \to r$$
$$p \land s \to t$$
$$\sim r$$
$$\sim q \to u \land s$$
$$\therefore t$$

42.
$$\sim p \to r \land \sim s$$
$$t \to s$$
$$u \to \sim p$$
$$\sim w$$
$$u \lor w$$
$$\therefore \sim t \lor w$$

43.
$$p \to q$$
$$r \lor s$$
$$\sim s \to \sim t$$
$$\sim q \lor s$$
$$\sim s$$
$$\sim p \land r \to u$$
$$w \lor t$$
$$\therefore u \land w$$

1.4   APPLICATION: DIGITAL LOGIC CIRCUITS

Only connect!
(E. M. Forster,
Howards End)

In the late 1930s, a young M.I.T. graduate student named Claude Shannon noticed an analogy between the operations of switching devices, such as telephone switching circuits, and the operations of logical connectives. He used this analogy with striking success to solve problems of circuit design and wrote up his results in his master's thesis, which was published in 1938.

Claude Shannon

The drawing in Figure 1.4.1(a) shows the appearance of the two positions of a simple switch. When the switch is closed, current can flow from one terminal to the other; when it is open, current cannot flow. Imagine that such a switch is part of the circuit shown in Figure 1.4.1(b). The light bulb turns on if, and only if, current flows through it. And this happens if, and only if, the switch is closed.

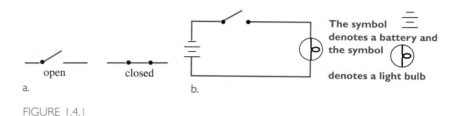

FIGURE 1.4.1

Now consider the more complicated circuits of Figures 1.4.2(a) and 1.4.2(b).

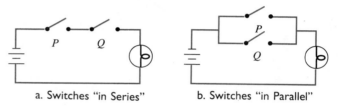

a. Switches "in Series"   b. Switches "in Parallel"

FIGURE 1.4.2

In the circuit of Figure 1.4.2(a) current flows and the light bulb turns on if, and only if, *both* switches P and Q are closed. The switches in this circuit are said to be **in series**. In the circuit of Figure 1.4.2(b) current flows and the light bulb turns on if, and only if, *at least one* of the switches P or Q is closed. The switches in this circuit are said to be **in parallel**. All possible behaviors of these circuits are described by Table 1.4.1.

TABLE 1.4.1

Switches		Light Bulb
P	Q	state
closed	closed	on
closed	open	off
open	closed	off
open	open	off

a. Switches in Series

Switches		Light Bulb
P	Q	state
closed	closed	on
closed	open	on
open	closed	on
open	open	off

b. Switches in Parallel

The INTEL Pentium integrated circuit, shown enlarged at the right, can function as the central processing unit of a powerful personal computer. It is a triumph of miniaturization, containing more than a million transistors that make up hundreds of thousands of digital logic circuits, all on an area not much larger than a fingernail.

Observe that if the words *closed* and *on* are replaced by T and *open* and *off* are replaced by F, Table (a) becomes the truth table for *and* and Table (b) becomes the truth table for *or*. Consequently, the switching circuit of Figure 1.4.2(a) is said to correspond to the logical expression $P \wedge Q$, and that of Figure 1.4.2(b) is said to correspond to $P \vee Q$.

More complicated circuits correspond to more complicated logical expressions. This correspondence has been used extensively in the design and study of circuits.

In the 1940s and 1950s, switches were replaced by electronic devices, with the physical states of closed and open corresponding to electronic states such as high and low voltages. The new electronic technology led to the development of modern digital systems such as electronic computers, electronic telephone switching systems, traffic light controls, electronic calculators, and the control mechanisms used in hundreds of other types of electronic equipment. The basic electronic components of a digital system are called *digital logic circuits*. The word *logic* indicates the important role of logic in the design of such circuits and the word *digital* indicates that the circuits process discrete, or separate, signals as opposed to continuous ones.

Electronic engineers continue to use the language of logic when they refer to values of signals produced by an electronic switch as being "true" or "false." But they generally use the symbols 1 and 0 rather than T and F to denote these values. The symbols 0 and 1 are called **bits**, short for *binary digits*. This terminology was introduced in 1946 by the statistician John Tukey.

John Tukey

BLACK BOXES AND GATES

Combinations of signal bits (1's and 0's) can be transformed into other combinations of signal bits (1's and 0's) by means of various circuits. Because a variety of different technologies are used in circuit construction, computer engineers and digital system designers find it useful to think of certain basic circuits as black boxes. The inside of a black box contains the detailed implementation of the circuit and is often ignored while attention is focused on the relation between the **input** and the **output** signals.

The operation of a black box is completely specified by constructing an **input/output table** that lists all its possible input signals together with their corresponding output signals. For example, the black box pictured above has three input signals. Since each of these signals can take the value 1 or 0, there are eight possible combinations of input signals. One possible correspondence of input to output signals is as follows:

An Input/Output Table

Input			Output
P	Q	R	S
1	1	1	1
1	1	0	0
1	0	1	0
1	0	0	1
0	1	1	0
0	1	0	1
0	0	1	1
0	0	0	0

The third row, for instance, indicates that for inputs $P = 1$, $Q = 0$, and $R = 1$, the output S equals 0.

An efficient method for designing more complicated circuits is to build them by connecting less complicated black box circuits. Three such circuits are known as NOT- , AND- , and OR-gates.

A **NOT-gate** (or **inverter**) is a circuit with one input signal and one output signal. If the input signal is 1, the output signal is 0. Conversely, if the input signal is 0, then the output signal is 1. An **AND-gate** is a circuit with two input signals and one output signal. If both input signals are 1, then the output signal is 1. Otherwise, the output signal is 0. An **OR-gate** also has two input signals and one output signal. If both input signals are 0, then the output signal is 0. Otherwise, the output signal is 1.

The actions of NOT- , AND- , and OR-gates are summarized in Figure 1.4.3, where P and Q represent input signals and R represents the output signal. It should be clear from Figure 1.4.3 that the actions of the NOT-, AND-, and OR-gates on signals correspond exactly to those of the logical connectives $\sim$, $\wedge$, and $\vee$ on statements, if the symbol 1 is identified with T and the symbol 0 is identified with F.

Gates can be combined into circuits in a variety of ways. If the rules on page 45 are obeyed, the result is a **combinational circuit**, one whose output at any time is determined entirely by its input at that time without regard to previous inputs.

Type of Gate	Symbolic Representation	Action
NOT	 P —[NOT]—o— R	<table><tr><td>Input P</td><td>Output R</td></tr><tr><td>1 0</td><td>0 1</td></tr></table>
AND	P — Q —[AND]— R	<table><tr><td>Input P Q</td><td>Output R</td></tr><tr><td>1 1 1 0 0 1 0 0</td><td>1 0 0 0</td></tr></table>
OR	P — Q —[OR]— R	<table><tr><td>Input P Q</td><td>Output R</td></tr><tr><td>1 1 1 0 0 1 0 0</td><td>1 1 1 0</td></tr></table>

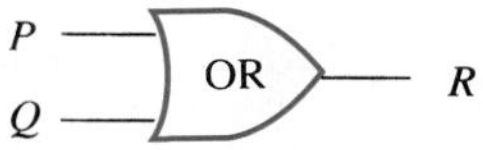

FIGURE 1.4.3

Never combine two input wires.   1.4.1

A single input wire can be split halfway and used as input for two separate gates.

1.4.2

An output wire can be used as input.   1.4.3

No output of a gate can eventually feed back into that gate.   1.4.4

Rule (1.4.4) is violated in more complex circuits, called **sequential circuits**, whose output at any given time depends both on the input at that time and also on previous inputs. These circuits are discussed in Section 7.2.

THE INPUT/OUTPUT TABLE FOR A CIRCUIT

If you are given a set of input signals for a circuit, you can find its output by tracing through the circuit gate by gate.

EXAMPLE 1.4.1   Determining Output for a Given Input

Indicate the output of the circuits in Figure 1.4.4 for the input signals shown.

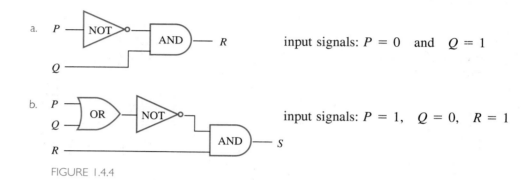

a. input signals: $P = 0$ and $Q = 1$

b. input signals: $P = 1$, $Q = 0$, $R = 1$

FIGURE 1.4.4

Solution   a. Move from left to right through the diagram, tracing the action of each gate on the input signals. The NOT-gate changes $P = 0$ to a 1, so both inputs to the AND-gate are 1; hence the output R is 1. This is illustrated by annotating the diagram as shown in Figure 1.4.5.

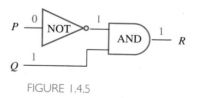

FIGURE 1.4.5

b. The output of the OR-gate is 1 since one of the input signals, P, is 1. The NOT-gate changes this 1 into a 0, so the two inputs to the AND-gate are 0 and $R = 1$. Hence the output S is 0. The trace is shown in Figure 1.4.6.

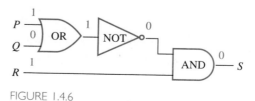

FIGURE 1.4.6

To construct the entire input/output table for a circuit, trace through the circuit to find the corresponding output signals for each possible combination of input signals.

EXAMPLE 1.4.2   Constructing the Input/Output Table
for a Circuit

Construct the input/output table for the circuit in Figure 1.4.7.

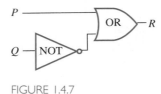

FIGURE 1.4.7

Solution   List the four possible combinations of input signals and find the output for each by tracing through the circuit.

Input		Output
P	Q	R
1	1	1
1	0	1
0	1	0
0	0	1

THE BOOLEAN EXPRESSION CORRESPONDING TO A CIRCUIT

In logic, variables such as p, q, and r represent statements, and a statement can have one of only two truth values: T (true) or F (false). A statement form is an expression, such as $p \wedge (\sim q \vee r)$, composed of statement variables and logical connectives.

As noted earlier, one of the founders of symbolic logic was the English mathematician George Boole. In his honor, any variable, such as a statement variable or an input signal, that can take one of only two values is called a *Boolean variable*. An expression composed of Boolean variables and the connectives $\sim$, $\wedge$, and $\vee$ is called a *Boolean expression*.*

Given a circuit consisting of combined NOT- , AND- , and OR-gates, a corresponding Boolean expression can be obtained by tracing the actions of the gates of the input variables.

George Boole (1815–1864)

---

*Strictly speaking, only meaningful expressions such as $(\sim p \wedge q) \vee (p \wedge r)$ or $\sim(\sim(p \wedge q) \vee r)$ are allowed, not meaningless ones like $p \sim q((rs \vee \wedge q \sim$. We use recursion to give a careful definition of Boolean expressions in Section 8.4.

EXAMPLE 1.4.3   Finding a Boolean Expression for a Circuit

Find the Boolean expressions that correspond to the circuits shown in Figure 1.4.8. A dot indicates a soldering of two wires; wires that cross without a dot are assumed not to touch.

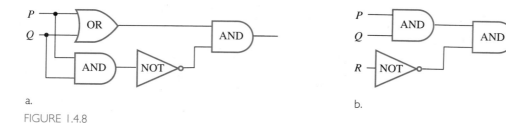

a.

b.

FIGURE 1.4.8

Solution   a. Trace through the circuit from left to right, indicating the output of each gate symbolically. See Figure 1.4.9.

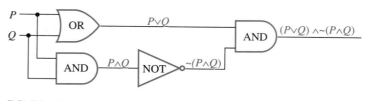

FIGURE 1.4.9

The final expression obtained, $(P \lor Q) \land \sim(P \land Q)$, is the expression for exclusive or: P or Q but not both.

b. See Figure 1.4.10. The Boolean expression corresponding to the circuit is $(P \land Q) \land \sim R$.

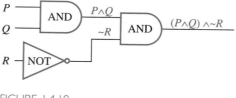

FIGURE 1.4.10   ■

Observe that the output of the circuit shown in Figure 1.4.8(b) is 1 for exactly one combination of inputs ($P = 1$, $Q = 1$, and $R = 0$) and is 0 for all other combinations of inputs. For this reason, the circuit can be said to "recognize" one particular combination of inputs. The output column of the input/output table has a 1 in exactly one row and 0's in all other rows.

DEFINITION

A **recognizer** is a circuit that outputs a 1 for exactly one particular combination of input signals and outputs 0's for all other combinations.

Input/Output Table for a Recognizer

P	Q	R	$(P \wedge Q) \wedge \sim R$
1	1	1	0
1	1	0	1
1	0	1	0
1	0	0	0
0	1	1	0
0	1	0	0
0	0	1	0
0	0	0	0

THE CIRCUIT CORRESPONDING TO A BOOLEAN EXPRESSION

The preceding examples showed how to find a Boolean expression corresponding to a circuit. The following examples show how to construct a circuit corresponding to a Boolean expression.

EXAMPLE 1.4.4 Constructing Circuits for Boolean Expressions

Construct circuits for the following Boolean expressions.

$$\text{a. } (\sim P \wedge Q) \vee \sim Q \qquad \text{b. } ((P \wedge Q) \wedge (R \wedge S)) \wedge T$$

Solution a. Write the input variables in a column on the left side of the diagram. Then go from the right side of the diagram to the left, working from the outermost part of the expression to the innermost part. Since the last operation executed when evaluating $(\sim P \wedge Q) \vee \sim Q$ is $\vee$, put an OR-gate at the extreme right of the diagram. One input to this gate is $\sim P \wedge Q$; so draw an AND-gate to the left of the OR-gate and show its output coming into the OR-gate. Since one input to the AND-gate is $\sim P$, draw a line from P to a NOT-gate and from there to the AND-gate. Since the other input to the AND-gate is Q, draw a line from Q directly to the AND-gate. The other input to the OR-gate is $\sim Q$, so draw a line from Q to a NOT-gate and from the NOT-gate to the OR-gate. The circuit you obtain is shown in Figure 1.4.11.

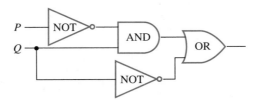

FIGURE 1.4.11

b. To start, put one AND-gate at the extreme right for the $\wedge$ between $((P \wedge Q) \wedge (R \wedge S))$ and T. To the left of that put the AND-gate corresponding to the $\wedge$ between $P \wedge Q$ and $R \wedge S$. To the left of that put the AND-gates corresponding

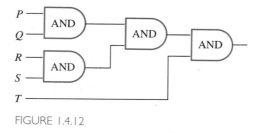

FIGURE 1.4.12

to the $\wedge$'s between P and Q and between R and S. The circuit is shown in Figure 1.4.12. ∎

It follows from Theorem 1.1.1 that all ways of adding parentheses to $P \wedge Q \wedge R \wedge S \wedge T$ are logically equivalent. Thus, for example,

$$((P \wedge Q) \wedge (R \wedge S)) \wedge T \equiv (P \wedge (Q \wedge R)) \wedge (S \wedge T).$$

It also follows that the circuit in Figure 1.4.13, which corresponds to $(P \wedge (Q \wedge R)) \wedge (S \wedge T)$, has the same input/output table as the circuit in Figure 1.4.12, which corresponds to $((P \wedge Q) \wedge (R \wedge S)) \wedge T$.

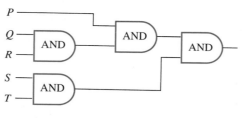

FIGURE 1.4.13

Each of the circuits in Figures 1.4.12 and 1.4.13 is, therefore, an implementation of the expression $P \wedge Q \wedge R \wedge S \wedge T$. Such a circuit is called a **multiple-input AND-gate** and is represented by the diagram shown in Figure 1.4.14. **Multiple-input OR-gates** can be constructed similarly.

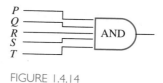

FIGURE 1.4.14

FINDING A CIRCUIT THAT CORRESPONDS TO A GIVEN INPUT/OUTPUT TABLE

To this point, we have discussed how to construct the input/output table for a circuit, how to find the Boolean expression corresponding to a given circuit, and how to construct the circuit corresponding to a given Boolean expression. Now we address the question of how to design a circuit (or find a Boolean expression) corresponding to a given input/output table. The way to do this is to put several recognizers together in parallel.

EXAMPLE 1.4.5   Designing a Circuit for a Given Input/Output Table

Design a circuit for the following input/output table:

Inputs			Outputs
P	Q	R	S
1	1	1	1
1	1	0	0
1	0	1	1
1	0	0	1
0	1	1	0
0	1	0	0
0	0	1	0
0	0	0	0

Solution   First construct a Boolean expression with this table as its truth table. To do this, identify each row for which the output is 1—in this case, the first, third, and fourth rows. For each such row, construct an *and* expression that produces a 1 (or true) for the exact combination of input values for that row and a 0 (or false) for all other combinations of input values. For example, the expression for the first row is $P \wedge Q \wedge R$ because $P \wedge Q \wedge R$ is 1 if $P = 1$ and $Q = 1$ and $R = 1$, and it is 0 for all other values of P, Q, and R. The expression for the third row is $P \wedge {\sim}Q \wedge R$ because $P \wedge {\sim}Q \wedge R$ is 1 if $P = 1$ and $Q = 0$ and $R = 1$, and it is 0 for all other values of P, Q, and R. Similarly, the expression for the fourth row is $P \wedge {\sim}Q \wedge {\sim}R$.

Now any Boolean expression with the given table as its truth table has the value 1 in case $P \wedge Q \wedge R = 1$, or in case $P \wedge {\sim}Q \wedge R = 1$, or in case $P \wedge {\sim}Q \wedge {\sim}R = 1$, and in no other cases. It follows that a Boolean expression with the given truth table is

$$(P \wedge Q \wedge R) \vee (P \wedge {\sim}Q \wedge R) \vee (P \wedge {\sim}Q \wedge {\sim}R). \qquad \text{1.4.5}$$

The circuit corresponding to this expression has the diagram shown in Figure 1.4.15. Observe that expression (1.4.5) is a disjunction of terms that are themselves conjunctions in which one of P or ${\sim}P$, one of Q or ${\sim}Q$, and one of R or ${\sim}R$ all appear. Such expressions are said to be in **disjunctive normal form** or **sum-of-products form**.

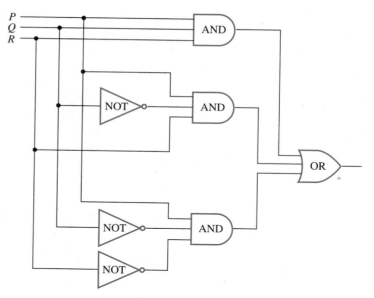

FIGURE 1.4.15

SIMPLIFYING COMBINATIONAL CIRCUITS

Consider the two combinational circuits shown in Figure 1.4.16.

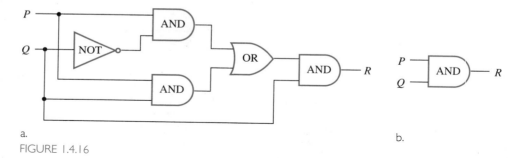

a.

FIGURE 1.4.16

b.

If you trace through circuit (a), you will find that its input/output table is

Input		Output
P	Q	R
1	1	1
1	0	0
0	1	0
0	0	0

which is the same as the input/output table for circuit (b). Thus these two circuits do the same job in the sense that they transform the same combinations of input signals into the same output signals. Yet circuit (b) is simpler than circuit (a) in that it contains many fewer logic gates. Thus, for one thing, it would be less expensive to construct.

DEFINITION

Two digital logic circuits are **equivalent** if, and only if, their input/output tables are identical.

Since logically equivalent statement forms have identical truth tables, you can determine that two circuits are equivalent by finding the Boolean expressions corresponding to the circuits and showing that these expressions, regarded as statement forms, are logically equivalent. Example 1.4.6 shows how this procedure works for circuits (a) and (b) above.

EXAMPLE 1.4.6   Showing That Two Circuits Are Equivalent

Find the Boolean expressions for each circuit in Figure 1.4.16. Use Theorem 1.1.1 to show that these expressions are logically equivalent when regarded as statement forms.

Solution   The Boolean expressions corresponding to circuits (a) and (b) are $((P \wedge \sim Q) \vee (P \wedge Q)) \wedge Q$ and $P \wedge Q$, respectively. By Theorem 1.1.1,

$$((P \wedge \sim Q) \vee (P \wedge Q)) \wedge Q$$
$$\equiv (P \wedge (\sim Q \vee Q)) \wedge Q \quad \text{by the distributive law (Theorem 1.1.1(3))}$$
$$\equiv (P \wedge (Q \vee \sim Q)) \wedge Q \quad \text{by the commutative law for } \vee \text{ (Theorem 1.1.1(1))}$$
$$\equiv (P \wedge t) \wedge Q \quad \text{by the negation law (Theorem 1.1.1(5))}$$
$$\equiv P \wedge Q \quad \text{by the identity law (Theorem 1.1.1(4)).}$$

It follows that the truth tables for $((P \wedge \sim Q) \vee (P \wedge Q)) \wedge Q$ and $P \wedge Q$ are the same. Hence the input/output tables for the circuits corresponding to these expressions are also the same, and so the circuits are equivalent. ∎

In general, you can simplify a combinational circuit by finding the corresponding Boolean expression, using the properties listed in Theorem 1.1.1 to find a Boolean expression that is simpler and logically equivalent to it (when both are regarded as statement forms), and constructing the circuit corresponding to this simpler Boolean expression.

Another way to simplify a circuit is to find an equivalent circuit that uses the fewest number of different kinds of logic gates. Two gates not previously introduced are particularly useful for this: NAND-gates and NOR-gates. A NAND-gate is a single gate that acts like an AND-gate followed by a NOT-gate. A NOR-gate acts like an OR-gate followed by a NOT-gate. Thus the output signal of a NAND-gate is 0 when, and only when, both input signals are 1, and the output signal for a NOR-gate is 1 when, and only when, both input signals are 0. The logical symbols corresponding to these gates are | (for NAND) and ↓ (for NOR), where | is called a **Scheffer stroke** and ↓ is called a **Peirce arrow**. The table below summarizes the action of these gates.

Type of Gate	Symbolic Representation	Action
NAND	P —⊐ NAND o— R Q —⊐	**Input** **Output** P Q $R = P \mid Q$ 1 1 0 1 0 1 0 1 1 0 0 1
NOR	P —⊐ NOR o— R Q —⊐	**Input** **Output** P Q $R = P \downarrow Q$ 1 1 0 1 0 0 0 1 0 0 0 1

It can be shown that any Boolean expression is equivalent to one written entirely with Scheffer strokes or entirely with Peirce arrows. Thus any digital logic circuit is equivalent to one that uses only NAND-gates or only NOR-gates. Example 1.4.7 develops part of the derivation of this result; the rest is left to the exercises.

EXAMPLE 1.4.7   Rewriting Expressions Using the Scheffer Stroke

Show that

a. $\sim p \equiv p \mid p$ and   b. $p \vee q \equiv (p \mid p) \mid (q \mid q)$.

Solution   a. $\sim p \quad \equiv \sim(p \wedge p)$ by the idempotent law for $\wedge$ (Theorem 1.1.1(7))
$\qquad\qquad \equiv p \mid p$ by definition of $\mid$.

b. $p \vee q \equiv \sim(\sim(p \vee q))$ by the double negative law (Theorem 1.1.1(6))
$\qquad\qquad \equiv \sim(\sim p \wedge \sim q)$ by De Morgan's laws (Theorem 1.1.1(8))
$\qquad\qquad \equiv \sim((p \mid p) \wedge (q \mid q))$ by part (a)
$\qquad\qquad \equiv (p \mid p) \mid (q \mid q)$ by definition of $\mid$.

EXERCISE SET 1.4

Give the output signals for the circuts in 1–4 if the input signals are as indicated.

1.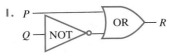

   input signals:
   $P = 1$ and $Q = 1$

2.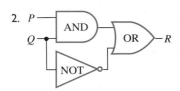

   input signals:
   $P = 1$ and $Q = 0$

3.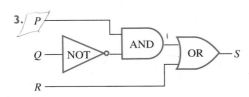

   input signals:
   $P = 1$, $Q = 0$, $R = 0$

4.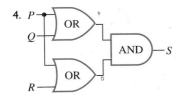

   input signals:
   $P = 0$, $Q = 1$, $R = 0$

In 5–8, write an input/output table for the circuit in the referenced exercise.

5. Exercise 1   6. Exercise 2

7. Exercise 3   8. Exercise 4

In 9–12, find the Boolean expression that corresponds to the circuit in the referenced exercise.

9. Exercise 1   10. Exercise 2

11. Exercise 3   12. Exercise 4

Construct circuits for the Boolean expressions in 13–17.

13. $\sim P \vee Q$

14. $\sim(P \vee Q)$

15. $P \vee (\sim P \wedge Q)$

16. $(P \wedge Q) \vee \sim R$

17. $(P \wedge \sim Q) \vee (\sim P \wedge R)$

For each of the tables in 18–21, construct (a) a Boolean expression having the given table as its truth table and (b) a circuit having the given table as its input/output table.

18.

P	Q	R	S
1	1	1	0
1	1	0	1
1	0	1	0
1	0	0	0
0	1	1	1
0	1	0	0
0	0	1	0
0	0	0	0

19.

P	Q	R	S
1	1	1	0
1	1	0	0
1	0	1	1
1	0	0	1
0	1	1	0
0	1	0	1
0	0	1	0
0	0	0	0

20.

P	Q	R	S
1	1	1	1
1	1	0	0
1	0	1	1
1	0	0	0
0	1	1	0
0	1	0	0
0	0	1	0
0	0	0	1

21.

P	Q	R	S
1	1	1	0
1	1	0	1
1	0	1	0
1	0	0	0
0	1	1	1
0	1	0	0
0	0	1	1
0	0	0	0

22. Design a circuit to take input signals P, Q, and R and output a 1 if, and only if, P and Q have the same value and Q and R have opposite values.

23. Design a circuit to take input signals P, Q, and R and output a 1 if, and only if, all three of P, Q, and R have the same value.

24. The lights in a classroom are controlled by two switches: one at the back and one at the front of the room. Moving either switch to the opposite position turns the lights off if they are on and on if they are off. Assume the lights have been installed so that when both switches are in the down position, the lights are off. Design a circuit to control the switches.

25. The ceiling light in an automobile is controlled by three switches: an automatic one in the driver's side door, another in the passenger's side door, and a manual one

in the ceiling. Moving any switch to the opposite position turns the light off if it is on and on if it is off. Assume the light has been installed so that when the doors are closed and the ceiling switch is in the back position, the light is off. Design a circuit to control the switches.

Use the properties listed in Theorem 1.1.1 to show that each pair of circuits in 26–29 have the same input/output table. (Find the Boolean expressions for the circuits and show that they are logically equivalent, when regarded as statement forms.)

26. a.

b.

27. a.

b.

28. a.

b.

29. a.

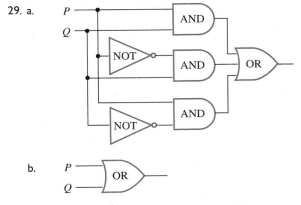

b.

P ———\
Q ———> OR >

For the circuits corresponding to the Boolean expressions in each of 30 and 31 there is an equivalent circuit with at most two logic gates. Find such a circuit.

30. $(P \wedge Q) \vee (\sim P \wedge Q) \vee (\sim P \wedge \sim Q)$

31. $(\sim P \wedge \sim Q) \vee (\sim P \wedge Q) \vee (P \wedge \sim Q)$

32. The Boolean expression for the circuit in Example 1.4.5 is

$$(P \wedge Q \wedge R) \vee (P \wedge \sim Q \wedge R) \vee (P \wedge \sim Q \wedge \sim R)$$

(a disjunctive normal form). Find a circuit with at most three logic gates that is equivalent to this circuit.

33. a. Show that for the Scheffer stroke $|$,

$$p \wedge q \equiv (p \mid q) \mid (p \mid q).$$

   b. Use the results of Example 1.4.7 and part (a) above to write $p \wedge (\sim q \vee r)$ using only Scheffer strokes.

34. Show that the following logical equivalences hold for the Peirce arrow $\downarrow$, where $p \downarrow q \equiv \sim (p \vee q)$.
    a. $\sim p \equiv p \downarrow p$ b. $p \vee q \equiv (p \downarrow q) \downarrow (p \downarrow q)$
    c. $p \wedge q \equiv (p \downarrow p) \downarrow (q \downarrow q)$
    H d. Write $p \rightarrow q$ using Peirce arrows only.
    e. Write $p \leftrightarrow q$ using Peirce arrows only.

1.5 APPLICATION: NUMBER SYSTEMS AND CIRCUITS FOR ADDITION

Counting in binary is just like counting in decimal if you are all thumbs.
(Glaser and Way)

In elementary school, you learned the meaning of decimal notation: that to interpret a string of decimal digits as a number, you mentally multiply each digit by its place value. For instance, 5,049 has a 5 in the thousand's place, a 0 in the hundred's place, a 4 in the ten's place, and a 9 in the one's place. Thus,

$$5,049 = 5 \cdot (1,000) + 0 \cdot (100) + 4 \cdot (10) + 9 \cdot (1).$$

Using exponential notation, this equation can be rewritten as

$$5,049 = 5 \cdot (10^3) + 0 \cdot (10^2) + 4 \cdot (10^1) + 9 \cdot (10^0).$$

More generally, decimal notation is based on the fact that any positive integer can be written uniquely as a sum of products of the form

$$d \cdot (10^n),$$

where each n is a nonnegative integer and each d is one of the decimal digits 0, 1, 2, 3, 4, 5, 6, 7, 8, or 9. The word *decimal* comes from the Latin root *deci*, meaning "ten." Decimal (or base 10) notation expresses a number as a string of digits in which each digit's position indicates the power of 10 by which it is multiplied. The right-most position is the one's place (or 10^0 place), to the left of that is the ten's place (or 10^1 place), to the left of that is the hundred's place (or 10^2 place), and so forth, as illustrated below.

place	10^3 thousands	10^2 hundreds	10^1 tens	10^0 ones
decimal digit	5	0	4	9

BINARY REPRESENTATION OF NUMBERS

There is nothing sacred about the number 10; we use 10 as a base for our usual number system because we happen to have ten fingers. In fact, any integer greater than 1 can serve as a base for a number system. In computer science, **base 2 notation**, or **binary notation**, is of special importance because the signals used in modern electronics are always in one of only two states. (The Latin root *bi* means "two.")

In Section 4.3, we show that any integer can be represented uniquely as a sum of powers of the form

$$d \cdot 2^n,$$

where each n is an integer and each d is one of the binary digits (or bits) 0 or 1. For example,

$$27 = 16 + 8 + 2 + 1$$
$$= 1 \cdot (2^4) + 1 \cdot (2^3) + 0 \cdot (2^2) + 1 \cdot (2^1) + 1 \cdot (2^0).$$

In binary notation, as in decimal notation, we write just the binary digits, and not the powers of the base. In binary notation, then,

$$1 \cdot (2^4) + 1 \cdot (2^3) + 0 \cdot (2^2) + 1 \cdot (2^1) + 1 \cdot (2^0)$$

$$27_{10} = 11011_2,$$

where the subscripts indicate the base, whether 10 or 2, in which the number is written. The places in binary notation correspond to the various powers of two. The right-most position is the one's place (or 2^0 place), to the left of that is the two's place (or 2^1 place), to the left of that is the four's place (or 2^2 place), and so forth as illustrated below.

place	2^4 sixteens	2^3 eights	2^2 fours	2^1 twos	2^0 ones
binary digit	1	1	0	1	1

As in the decimal notation, leading zeros may be added or dropped as desired. For example,

$$003_{10} = 3_{10} = 1 \cdot (2^1) + 1 \cdot (2^0) = 11_2 = 011_2.$$

EXAMPLE 1.5.1 Binary Notation for Integers from 1 to 9

Derive the binary notation for the integers from 1 to 9.

Solution

$$1_{10} = 1 \cdot (2^0) = 1_2$$
$$2_{10} = 1 \cdot (2^1) + 0 \cdot (2^0) = 10_2$$
$$3_{10} = 1 \cdot (2^1) + 1 \cdot (2^0) = 11_2$$
$$4_{10} = 1 \cdot (2^2) + 0 \cdot (2^1) + 0 \cdot (2^0) = 100_2$$
$$5_{10} = 1 \cdot (2^2) + 0 \cdot (2^1) + 1 \cdot (2^0) = 101_2$$
$$6_{10} = 1 \cdot (2^2) + 1 \cdot (2^1) + 0 \cdot (2^0) = 110_2$$

$$7_{10} = \qquad\qquad\qquad 1 \cdot (2^2) + 1 \cdot (2^1) + 1 \cdot (2^0) = \;\; 111_2$$

$$8_{10} = 1 \cdot (2^3) + 0 \cdot (2^2) + 0 \cdot (2^1) + 0 \cdot (2^0) = 1000_2$$

$$9_{10} = 1 \cdot (2^3) + 0 \cdot (2^2) + 0 \cdot (2^1) + 1 \cdot (2^0) = 1001_2$$ ∎

A list of powers of 2 is useful for doing binary to decimal conversions and the reverse. See Table 1.5.1.

TABLE 1.5.1   Powers of 2

power of 2	2^{10}	2^9	2^8	2^7	2^6	2^5	2^4	2^3	2^2	2^1	2^0
decimal form	1,024	512	256	128	64	32	16	8	4	2	1

EXAMPLE 1.5.2   Converting a Binary to a Decimal Number

Represent 110101_2 in decimal notation.

Solution
$$110101_2 = 1 \cdot (2^5) + 1 \cdot (2^4) + 0 \cdot (2^3) + 1 \cdot (2^2) + 0 \cdot (2^1) + 1 \cdot (2^0)$$
$$= 32 + 16 + 4 + 1$$
$$= 53_{10}$$

Alternatively, the schema below may be used.

2^5	2^4	2^3	2^2	2^1	2^0
32	16	8	4	2	1

$$
\begin{array}{cccccc}
1 & 1 & 0 & 1 & 0 & 1_2
\end{array}
$$

$$
\begin{aligned}
&\rightarrow 1\\
&\rightarrow 0\\
&\rightarrow 4\\
&\rightarrow 0\\
&\rightarrow 16\\
&\rightarrow \underline{32}\\
&\quad\; 53_{10}
\end{aligned}
$$ ∎

EXAMPLE 1.5.3   Converting a Decimal to a Binary Number

Represent 209 in binary notation.

Solution   Use Table 1.5.1 to write 209 as a sum of powers of 2, starting with the highest power of 2 that is less than 209 and continuing to lower powers.

Since 209 is between 128 and 256, the highest power of 2 that is less than 209 is 128. Hence

$$209_{10} = 128 + \text{a smaller number.}$$

Now $209 - 128 = 81$, and 81 is between 64 and 128. So the highest power of 2 that is less than 81 is 64. Hence

$$209_{10} = 128 + 64 + \text{a smaller number.}$$

Continuing in this way, you obtain

$$209_{10} = 128 + 64 + 16 + 1$$
$$= 1 \cdot (2^7) + 1 \cdot (2^6) + 0 \cdot (2^5) + 1 \cdot (2^4) + 0 \cdot (2^3)$$
$$+ 0 \cdot (2^2) + 0 \cdot (2^1) + 1 \cdot (2^0).$$

For each power of 2 that occurs in the sum, there is a 1 in the corresponding position of the binary number. For each power of 2 that is missing from the sum, there is a 0 in the corresponding position of the binary number. Thus

$$209_{10} = 11010001_2$$ ■

Another procedure for converting from decimal to binary notation is discussed in Section 4.1.

BINARY ADDITION AND SUBTRACTION

The computational methods of binary arithmetic are analogous to those of decimal arithmetic. In binary arithmetic the number 2 (10_2 in binary notation) plays a role similar to that of the number 10 in decimal arithmetic.

EXAMPLE 1.5.4   Addition in Binary Notation

Add 1101_2 and 111_2 using binary notation.

Solution   In binary notation $2_{10} = 10_2$. It is helpful to read 10_2 as "one oh base two." Also $1_{10} = 1_2$. Hence the translation of $1_{10} + 1_{10} = 2_{10}$ to binary notation is as follows:

$$\begin{array}{r} 1_2 \\ +\ 1_2 \\ \hline 10_2 \end{array}$$

It follows that adding two 1's together results in a carry of 1 when binary notation is used. Adding three 1's together also results in a carry of 1 since $3_{10} = 11_2$ ("one one base two").

$$\begin{array}{r} 1_2 \\ +\ 1_2 \\ +\ 1_2 \\ \hline 11_2 \end{array}$$

Thus the addition can be performed as follows:

$$\begin{array}{r} 1\ \ 1\ \ 1\quad \leftarrow \text{carry row} \\ 1\ 1\ 0\ 1_2 \\ +\quad\ 1\ 1\ 1_2 \\ \hline 1\ 0\ 1\ 0\ 0_2 \end{array}$$

■

EXAMPLE 1.5.5   Subtraction in Binary Notation

Subtract 1011_2 from 11000_2 using binary notation.

Solution   In decimal subtraction the fact that $10_{10} - 1_{10} = 9_{10}$ is used to borrow across several columns. For example, consider the following.

$$\begin{array}{r} 9\ \ 9 \\ 1\quad \leftarrow \text{borrow row} \\ 1\ 0\ 0\ 0_{10} \\ -\quad\ \ 5\ 8_{10} \\ \hline 9\ 4\ 2_{10} \end{array}$$

In binary subtraction it may also be necessary to borrow across more than one column. But when you borrow a 1_2 from 10_2, what remains is 1_2.

$$
\begin{array}{r}
10_2 \\
-\ 1_2 \\
\hline
1_2
\end{array}
$$

Thus the subtraction can be performed as follows:

$$
\begin{array}{r}
0\ \ 1\ \ 1 \qquad\qquad\ \\
\diagdown\ \diagdown\ 1 \quad \leftarrow \text{ borrow row} \\
1\ \diagdown\ 0\ 0\ 0_2 \qquad\qquad\quad\ \\
-\quad 1\ 0\ 1\ 1_2 \qquad\qquad\ \\
\hline
1\ 1\ 0\ 1_2 \qquad\qquad\ \\
\end{array}
$$

CIRCUITS FOR COMPUTER ADDITION

Consider the question of designing a circuit to produce the sum of two binary digits P and Q. Both P and Q can be either 0 or 1. And the following facts are known:

$$
\begin{aligned}
1_2 + 1_2 \quad &= 10_2, \\
1_2 + 0_2 = 1_2 &= 01_2, \\
0_2 + 1_2 = 1_2 &= 01_2, \\
0_2 + 0_2 = 0_2 &= 00_2.
\end{aligned}
$$

It follows that the circuit to be designed must have two outputs—one for the left binary digit (this is called the **carry**) and one for the right binary digit (this is called the **sum**). The carry output is 1 if both P and Q are 1; it is 0 otherwise. So the carry can be produced using the AND-gate circuit that corresponds to the Boolean expression $P \wedge Q$. The sum output is 1 if either P or Q, but not both, is 1. The sum can, therefore, be produced using a circuit that corresponds to the Boolean expression for the *exclusive or*: $(P \vee Q) \wedge {\sim}(P \wedge Q)$. (See Example 1.4.3(a).) Hence, a circuit to add two binary digits P and Q can be constructed as in Figure 1.5.1. This circuit is called a **half-adder**.

HALF-ADDER

Circuit

Input/Output Table

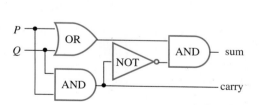

P	Q	Carry	Sum
1	1	1	0
1	0	0	1
0	1	0	1
0	0	0	0

FIGURE 1.5.1   Circuit to Add $P + Q$ Where P and Q Are Binary Digits

Now consider the question of how to construct a circuit to add two binary inte; each with more than one digit. Because the addition of two binary digits may r in a carry to the next column to the left, it may be necessary to add three binary at certain points. In the following example, the sum in the right column is the s two binary digits, and, because of the carry, the sum in the left column is the sum of three binary digits.

$$
\begin{array}{r}
1 \quad\quad \leftarrow \text{ carry row}\\
1 \; 1_2\\
+ \quad 1 \; 1_2\\
\hline
1 \; 1 \; 0_2
\end{array}
$$

So in order to construct a circuit that will add multidigit binary numbers, it is necessary to incorporate a circuit that will compute the sum of three binary digits. Such a circuit is called a **full-adder**. Consider a general addition of three binary digits P, Q, and R that results in a carry (or left-most digit) C and a sum (or right-most digit) S.

$$
\begin{array}{r}
P\\
+ \quad Q\\
+ \quad R\\
\hline
CS
\end{array}
$$

The operation of the full-adder is based on the fact that addition is a binary operation: Only two numbers can be added at one time. Thus P is first added to Q and then the result is added to R. For instance, consider the following addition:

$$
\left.
\begin{array}{r}
1_2\\
+ \quad 0_2\\
\end{array}
\right\} 1_2 \left.\begin{array}{r}\\ \\ \end{array}\right\} 1_2 + 1_2 = 10_2
$$
$$
\begin{array}{r}
+ \quad 1_2\\
\hline
10_2
\end{array}
$$

The process illustrated here can be broken down into steps that use half-adder circuits.

Step 1: Add P and Q using a half-adder to obtain a binary number with two digits.

$$
\begin{array}{r}
P\\
+ \quad Q\\
\hline
C_1 \, S_1
\end{array}
$$

Step 2: Add R to the sum $C_1 \, S_1$ of P and Q.

$$
\begin{array}{r}
C_1 \, S_1\\
+ \quad\quad R\\
\end{array}
$$

To do this, proceed as follows:

Step 2a: Add R to S_1 using a half-adder to obtain the two-digit number $C_2 \, S$.

$$
\begin{array}{r}
S_1\\
+ \quad R\\
\hline
C_2 \, S
\end{array}
$$

Then S is the right-most digit of the entire sum.

Step 2b: Determine the left-most digit, C, of the entire sum as follows: First note that it is impossible for both C_1 and C_2 to be 1's. For if $C_1 = 1$, then P and Q are both 1, and so $S_1 = 0$. Consequently, the addition of C_1 and C_2 gives a binary number C that has only one digit. Next observe that C will be a 1 in the case that the addition of P and Q gives a carry of 1 or in the case that the addition of S_1 (the right-most digit of $P + Q$) and R gives a carry of 1. In other words, $C = 1$ if, and only if, $C_1 = 1$ or $C_2 = 1$. It follows that the circuit shown in Figure 1.5.2 will compute the sum of three binary digits.

FULL-ADDER
Circuit

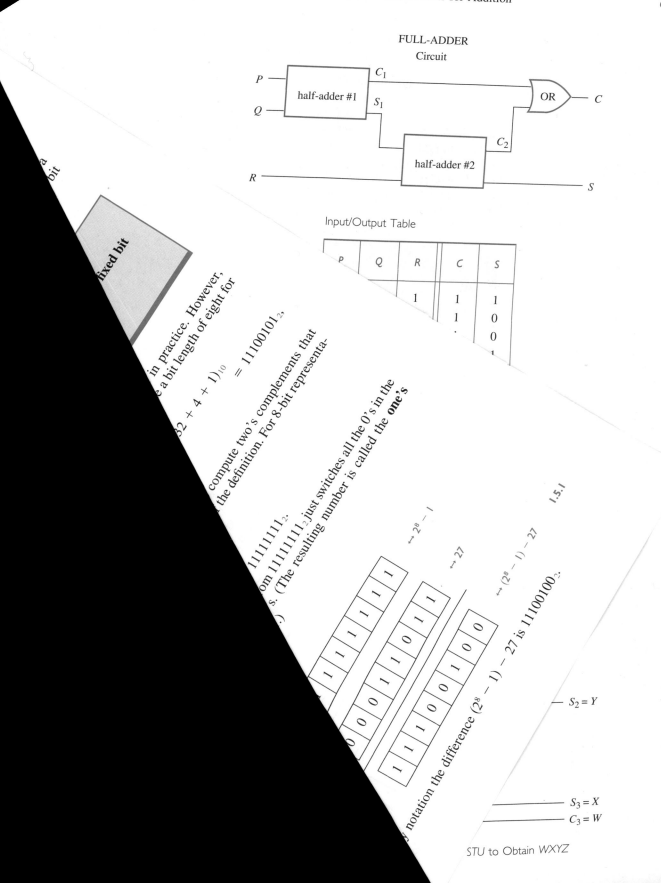

Input/Output Table

P	Q	R	C	S
		1	1	1
		1	0	0
		.	0	1
				1

in practice. However,
a bit length of eight for

$32 + 4 + 1)_{10}$ $= 11100101_2$

compute two's complements that
the definition. For 8-bit representa-

11111111_2.
om 11111111_2, just switches all the 0's in the
s. (The resulting number is called the **one's**

$\leftarrow 2^8 - 1$

$\leftarrow 27$

$\leftarrow (2^8 - 1) - 27$ **1.5.1**

notation the difference $(2^8 - 1) - 27$ is 11100100_2:

— $S_2 = Y$

— $S_3 = X$
— $C_3 = W$

STU to Obtain WXYZ

TWO'S COMPLEMENTS AND THE COMPUTER REPRESENTATION OF NEGATIVE INTEGERS

Typically, a fixed number of bits is used to represent integers on a computer, and these are required to represent negative as well as nonnegative integers. Sometimes a particular bit, normally the left-most, is used as a sign indicator, and the remaining bits are taken to be the absolute value of the number in binary notation. The problem with this approach is that the procedures for adding the resulting numbers are somewhat complicated and the representation of 0 is not unique. Another approach, using *two's complements*, makes it possible to add integers quite easily and results in unique representation for 0. The two's complement of an integer relative to a fixed length is defined as follows.

> **DEFINITION**
>
> Given a positive integer a, the **two's complement of a relative to a** **length** n is the n-bit binary representation of
>
> $$2^n - a.$$

Bit lengths of 16 and 32 are the most commonly used because the principles are the same for all bit lengths, we us simplicity in this discussion. Thus, for instance, since

$$(2^8 - 27)_{10} = (256 - 27)_{10} = 229_{10} = (128 + 64 + $$

the 8-bit two's complement of 27 is 11100101_2.

It turns out that there is a convenient way to involves less arithmetic than direct application of tions, it is based on three facts:

1. $2^8 - 27 = [(2^8 - 1) - 27] + 1.$
2. The binary representation of $2^8 - 1$ is
3. Subtracting an 8-bit binary number from number to 1's and all the 1's to 0 **complement** of the given number

By (2) and (3), for instance,

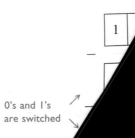

0's and 1's are switched

and so in binar

Now since $2^8 - 27 = [(2^8 - 1) - 27] + 1$, if we add 1 to (1.5.1), we obtain the 8-bit binary representation of $2^8 - 27$, which is the 8-bit two's complement of 27:

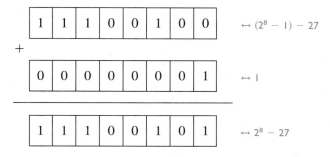

In general, for any integer a, $2^8 - a = (2^8 - 1) - a + 1$, and therefore:

To find the 8-bit two's complement of a positive integer a that is at most 255,

1. write the 8-bit binary representation for a,
2. switch all the 1's to 0's and 0's to 1's,
3. add 1 in binary notation.

EXAMPLE 1.5.6   **Finding a Two's Complement**

Find the 8-bit two's complement of 19.

Solution   Write the 8-bit binary representation for 19, switch all the 0's to 1's and 1's to 0's, and add 1.

$$19_{10} = (16 + 2 + 1)_{10} = 00010011_2 \xrightarrow{\text{switch 0's to 1's and 1's to 0's}}$$
$$11101100 \xrightarrow{\text{add 1}} 11101101$$

To check this result, note that

$$11101101_2 = (128 + 64 + 32 + 8 + 4 + 1)_{10} = 237_{10} = (256 - 19)_{10}$$
$$= (2^8 - 19)_{10},$$

which *is* the two's complement of 19.   ■

Observe that because

$$2^8 - (2^8 - a) = a$$

the two's complement of the two's complement of a number is the number itself, and therefore:

To find the decimal representation of the integer with a given 8-bit two's complement,

1. find the two's complement of the given two's complement,
2. write the decimal equivalent of the result.

EXAMPLE 1.5.7   Finding a Number with a Given Two's Complement

What is the decimal representation for the integer with two's complement 10101001?

Solution 10101001_2 $\xrightarrow{\text{switch 0's to 1's and 1's to 0's}}$ 01010110

$\xrightarrow{\text{add 1}}$ $01010111_2 = (64 + 16 + 4 + 2 + 1)_{10} = 87_{10}$

To check this result, note that the given number is

$$101000_2 = (128 + 32 + 8 + 1)_{10} = 169_{10} = (256 - 87)_{10} = (2^8 - 87)_{10},$$

which is the two's complement of 87.   ■

Now consider the two's complement of an integer n that satisfies the inequality $1 \leq n \leq 128$. Then

$$-1 \geq -n \geq -128 \qquad \text{because multiplying by } -1 \text{ reverses the direction of the inequality}$$

and

$$2^8 - 1 \geq 2^8 - n \geq 2^8 - 128 \qquad \text{by adding } 2^8 \text{ to all parts of the inequality.}$$

But $2^8 - 128 = 256 - 128 = 128 = 2^7$. Hence

$$2^7 \leq \text{the decimal form of the two's complement of } n < 2^8.$$

It follows that the 8-bit two's complement of an integer from 1 through 128 has a leading bit of 1. Note also that the ordinary 8-bit representation of an integer from 0 through 127 has a leading bit of 0. Consequently, eight bits can be used to represent both nonnegative and negative integers by representing each nonnegative integer up

TABLE 1.5.2

Integer	8-Bit Representation (ordinary 8-bit binary notation if nonnegative or 8-bit two's complement of absolute value if negative)	Decimal Form of Two's Complement for Negative Integers
127	01111111	
126	01111110	
.	.	
.	.	
.	.	
2	00000010	
1	00000001	
0	00000000	
−1	11111111	$2^8 - 1$
−2	11111110	$2^8 - 2$
−3	11111101	$2^8 - 3$
.	.	.
.	.	.
−127	10000001	$2^8 - 127$
−128	10000000	$2^8 - 128$

through 127 using ordinary 8-bit binary notation and representing each negative integer from -1 through -128 as the two's complement of its absolute value. That is, for any integer a from -128 through 127,

$$\text{the 8-bit representation of } a = \begin{cases} \text{the 8-bit binary representation of } a & \text{if } a \geq 0 \\ \text{the 8-bit binary representation of } 2^8 - |a| & \text{if } a < 0. \end{cases}$$

The representations are illustrated in Table 1.5.2.

COMPUTER ADDITION WITH NEGATIVE INTEGERS IN TWO'S COMPLEMENT FORM

The following statement summarizes how to use 8-bit representations and ordinary binary addition to add any two integers in the range -128 through 127. When modified for 16-bit and 32-bit representations, it describes an important method for adding integers in the ranges $-32,768$ through 32,767 and $-2,147,483,648$ through 2,147,483,647.

> To add two integers in the range -128 through 127 whose sum is also in the range -128 through 127,
>
> 1. convert both integers to their 8-bit representations (representing negative integers by using the two's complements of their absolute values),
> 2. add the resulting integers using ordinary binary addition,
> 3. truncate any leading 1 (overflow) that occurs in the 2^8th position,
> 4. convert the result back to decimal form (interpreting 8-bit integers with leading 0's as nonnegative and 8-bit integers with leading 1's as negative).

To see why this result is true, consider four cases: (1) both integers are nonnegative, (2) one integer is nonnegative and the other is negative and the absolute value of the negative integer is greater than that of the nonnegative one, (3) one integer is nonnegative and the other is negative and the absolute value of the negative integer is less than or equal to that of the nonnegative one, and (4) both integers are negative.

Case (1), where both integers are nonnegative, is easy because if two nonnegative integers from 0 through 127 are written in their 8-bit representation and if their sum is also in the range from 0 through 127, then the 8-bit representation of their sum has a leading 0 and is therefore interpreted correctly as a nonnegative integer. The example below illustrates what happens when 38 and 69 are added.

0	0	1	0	0	1	1	0	$\leftrightarrow$ 38

+

0	1	0	0	0	1	0	1	$\leftrightarrow$ 69

0	1	1	0	1	0	1	1	$\leftrightarrow$ 107

Both cases (2) and (3) involve adding a negative and a nonnegative integer. To be concrete, let the nonnegative integer be a and the negative integer be $-b$ and suppose both a and $-b$ are in the range -128 through 127. The crucial observation is that adding the 8-bit representations of a and $-b$ is equivalent to computing

$$a + (2^8 - b)$$

because the 8-bit representation of $-b$ is the binary representation of $2^8 - b$.

In case $|a| < |b|$, observe that

$$a + (2^8 - b) = 2^8 - (b - a),$$

and the binary representation of this number is the 8-bit representation of $-(b - a) = a + (-b)$. We must be careful to check that $2^8 - (b - a)$ is between 2^7 and 2^8. But it is because

$$2^7 = 2^8 - 2^7 \leq 2^8 - (b - a) < 2^8 \qquad \text{since } 0 < b - a \leq 128 = 2^7.$$

Hence in case $|a| < |b|$, adding the 8-bit representations of a and $-b$ gives the 8-bit representation of $a + (-b)$.

In case $|a| \geq |b|$, observe that

$$a + (2^8 - b) = 2^8 + (a - b).$$

Also

$$2^8 \leq 2^8 + (a - b) < 2^8 + 2^7 \qquad \text{because } 0 \leq a - b < 128.$$

So the binary representation of $a + (2^8 - b) = 2^8 + (a - b)$ has a leading 1 in the ninth (2^8th) position. This leading 1 is often called "overflow" because it does not fit in the 8-bit integer format. Now subtracting 2^8 from $2^8 + (a - b)$ is equivalent to truncating the leading 1 in the 2^8th position of the binary representation of the number. But

$$[a + (2^8 - b)] - 2^8 = 2^8 + (a - b) - 2^8 = a - b = a + (-b).$$

Hence in case $|a| \geq |b|$, adding the 8-bit representations of a and $-b$ and truncating the leading 1 (which is sure to be present) gives the 8-bit representation of $a + (-b)$.

EXAMPLE 1.5.8   Computing $a + (-b)$ Where $0 \leq a < b \leq 128$

Use 8-bit representations to compute $39 + (-89)$.

Solution   *Step 1*: Change from decimal to 8-bit representations using the two's complement to represent -89.

Since $39_{10} = (32 + 4 + 2 + 1)_{10} = 100111_2$, the 8-bit representation of 39 is 00100111. Now the 8-bit representation of -89 is the two's complement of 89. This is obtained as follows:

$$89_{10} = (64 + 16 + 8 + 1)_{10} = 01011001_2 \xrightarrow{\text{switch 0's to 1's and 1's to 0's}}$$

$$10100110 \xrightarrow{\text{add 1}} 10100111$$

So the 8-bit representation of -89 is 10100111.

Step 2: Add the 8-bit representations in binary notation and truncate the 1 in the 2^8th position if there is one:

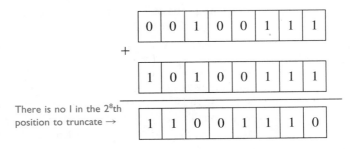

There is no 1 in the 2^8th
position to truncate →

Step 3: Find the decimal equivalent of the result. Since its leading bit is 1, this number is the 8-bit representation of a negative integer.

$$11001110 \xrightarrow{\text{switch 0's to 1's and 1's to 0's}} 00110001 \xrightarrow{\text{add 1}} 00110010$$
$$\leftrightarrow -(32 + 16 + 2)_{10} = -50_{10}$$

Note that since $39 - 89 = -50$, this procedure gives the correct answer. ■

EXAMPLE 1.5.9

Computing $a + (-b)$ Where $1 \le b \le a \le 127$

Use 8-bit representations to compute $39 + (-25)$.

Solution   *Step 1*: Change from decimal to 8-bit representations using the two's complement to represent -25.

As in Example 1.5.8, the 8-bit representation of 39 is 00100111. Now the 8-bit representation of -25 is the two's complement of 25, which is obtained as follows:

$$25_{10} = (16 + 8 + 1)_{10} = 00011001_2$$
$$\xrightarrow{\text{switch 0's to 1's and 1's to 0's}} 11100110 \xrightarrow{\text{add 1}} 11100111$$

So the 8-bit representation of -25 is 11100111.

Step 2: Add the 8-bit representations in binary notation and truncate the 1 in the 2^8th position if there is one:

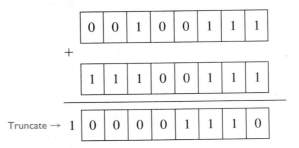

Truncate →

Step 3: Find the decimal equivalent of the result:

$$00001110_2 = (8 + 4 + 2)_{10} = 14_{10}.$$

Since $39 - 25 = 14$, this is the correct answer. ■

Case (4) involves adding two negative integers in the range from -1 through -128 whose sum is also in this range. To be specific, consider the sum $(-a) + (-b)$ where a, b, and $a + b$ are all in the range 1 through 128. In this case, the 8-bit representations of $-a$ and $-b$ are the 8-bit representations of $2^8 - a$ and $2^8 - b$. So if the 8-bit representations of $-a$ and $-b$ are added, the result is

$$(2^8 - a) + (2^8 - b) = [2^8 - (a + b)] + 2^8.$$

Recall that truncating a leading 1 in the ninth (2^8th) position of a binary number is equivalent to subtracting 2^8. So when the leading 1 is truncated from the 8-bit representation of $(2^8 - a) + (2^8 - b)$, the result is $2^8 - (a + b)$, which is the 8-bit representation of $-(a + b) = (-a) + (-b)$. (In exercise 34 you are asked to show that the sum $(2^8 - a) + (2^8 - b)$ *has* a leading 1 in the ninth (2^8th) position.)

EXAMPLE 1.5.10   Computing $(-a) + (-b)$ Where $1 \le a, b \le 128$ and $1 \le a + b \le 128$

Use 8-bit representations to compute $(-89) + (-25)$.

Solution   *Step 1*:   Change from decimal to 8-bit representations using the two's complements to represent -89 and -25.
The 8-bit representations of -89 and -25 were shown in Examples 1.5.8 and 1.5.9 to be 10100111 and 11100111, respectively.

Step 2:   Add the 8-bit representations in binary notation and truncate the 1 in the 2^8th position if there is one:

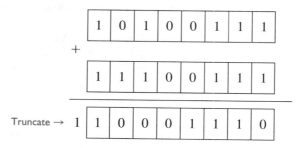

Step 3:   Find the decimal equivalent of the result. Because its leading bit is 1, this number is the 8-bit representation of a negative integer.

$$10001110 \xrightarrow{\text{switch 0's to 1's and 1's to 0's}} 01110001 \xrightarrow{\text{add 1}} 01110010_2$$
$$\leftrightarrow -(64 + 32 + 16 + 2)_{10} = -114_{10}$$

Since $(-89) + (-25) = -114$, that is the correct answer.   ∎

HEXADECIMAL NOTATION

It should now be obvious that numbers written in binary notation take up much more space than numbers written in decimal notation. Yet many aspects of computer operation can best be analyzed using binary numbers. **Hexadecimal notation** is even more compact than decimal notation, and it is much easier to convert back and forth between hexadecimal and binary notation than it is between binary and decimal notation. The word *hexadecimal* comes from the Greek root *hex-*, meaning "six," and the Latin root *deci-*, meaning "ten." Hence *hexadecimal* refers to "sixteen," and hexadecimal notation is also called **base 16 notation**. Hexadecimal notation is based on the fact that any integer can be uniquely expressed as a sum of numbers of the form

$$d \cdot (16^n),$$

where each n is a nonnegative integer and each d is one of the integers from 0 to 15. In order to avoid ambiguity, each hexadecimal digit must be represented by a single symbol. So digits 10 through 15 are represented by the first six letters of the alphabet. The sixteen hexadecimal digits are shown in Table 1.5.3, together with their decimal equivalents and, for future reference, their 4-bit binary equivalents.

TABLE 1.5.3

Decimal	Hexadecimal	4-Bit Binary Equivalent
0	0	0000
1	1	0001
2	2	0010
3	3	0011
4	4	0100
5	5	0101
6	6	0110
7	7	0111
8	8	1000
9	9	1001
10	A	1010
11	B	1011
12	C	1100
13	D	1101
14	E	1110
15	F	1111

EXAMPLE 1.5.11   Converting from Hexadecimal to Decimal Notation

Convert $3CF_{16}$ to decimal notation.

Solution   A schema similar to the one introduced in Example 1.5.2 can be used here.

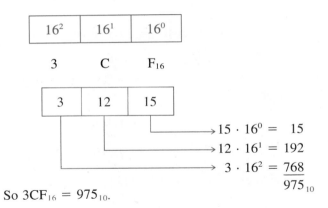

So $3CF_{16} = 975_{10}$.

Now consider how to convert from hexadecimal to binary notation. In the example below the numbers are rewritten using powers of two, and the laws of exponents are applied. The result suggests a general procedure.

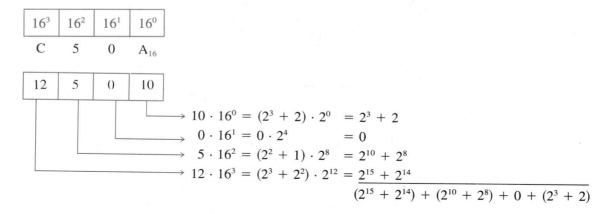

16^3	16^2	16^1	16^0

$$C \quad 5 \quad 0 \quad A_{16}$$

12	5	0	10

$$10 \cdot 16^0 = (2^3 + 2) \cdot 2^0 \quad = 2^3 + 2$$
$$0 \cdot 16^1 = 0 \cdot 2^4 \quad\quad\quad = 0$$
$$5 \cdot 16^2 = (2^2 + 1) \cdot 2^8 \quad = 2^{10} + 2^8$$
$$12 \cdot 16^3 = (2^3 + 2^2) \cdot 2^{12} = \underline{2^{15} + 2^{14}}$$
$$(2^{15} + 2^{14}) + (2^{10} + 2^8) + 0 + (2^3 + 2)$$

But

$$(2^{15} + 2^{14}) + (2^{10} + 2^8) + (2^3 + 2)$$
$$= 1100000000000000_2 + 010100000000_2$$ by the rules
$$+ 00000000_2 + 1010_2$$ for writing binary numbers

So

$$C50A_{16} = \underbrace{1100}_{C_{16}}\underbrace{0101}_{5_{16}}\underbrace{0000}_{0_{16}}\underbrace{1010}_{A_{16}}{}_2$$ by the rules for adding binary numbers

The procedure illustrated in this example can be generalized. In fact, the following sequence of steps will always give the correct answer:

To convert an integer from hexadecimal to binary notation:

1. Write each hexadecimal digit of the integer in fixed 4-bit binary notation.
2. Juxtapose the results.

EXAMPLE 1.5.12   Converting from Hexadecimal to Binary Notation

Convert $B09F_{16}$ to binary notation.

Solution $B_{16} = 11_{10} = 1011_2$, $0_{16} = 0_{10} = 0000_2$, $9_{16} = 9_{10} = 1001_2$, and $F_{16} = 15_{10} = 1111_2$. Consequently,

$$
\begin{array}{cccc}
B & 0 & 9 & F \\
\updownarrow & \updownarrow & \updownarrow & \updownarrow \\
1011 & 0000 & 1001 & 1111
\end{array}
$$

and the answer is 1011000010011111_2.   ∎

To convert integers written in binary notation into hexadecimal notation, reverse the steps of the previous procedure.

To convert an integer from binary to hexadecimal notation:

1. Group the digits of the binary number into sets of four, starting from the right and adding leading zeros as needed.
2. Convert the binary numbers in each set of four into hexadecimal digits.
3. Juxtapose those hexadecimal digits.

EXAMPLE 1.5.13   Converting from Binary to Hexadecimal Notation

Convert 100110110101001_2 to hexadecimal notation.

Solution   First group the binary digits in sets of four, working from right to left and adding leading 0's if necessary.

$$0100\ 1101\ 1010\ 1001.$$

Convert each group of four binary digits into a hexadecimal digit.

0100	1101	1010	1001
↕	↕	↕	↕
4	D	A	9

Then juxtapose the hexadecimal digits.

$$4DA9_{16}$$ ■

EXAMPLE 1.5.14   Reading a Memory Dump

The smallest addressable memory unit on most computers is one byte, or eight bits. In some debugging operations a dump is made of memory contents; that is, the contents of each memory location are displayed or printed out in order. Normally, the hexadecimal versions of the memory contents are given rather than the binary versions. Suppose, for example, that a segment of the memory dump looks like

$$A3\ BB\ 59\ 2E.$$

What is the actual content of the four memory locations?

Solution $A3 = 10100011$
                                   $BB = 10111011$
                                   $59 = 01011001$
                                   $2E = 00101110$ ■

EXERCISE SET 1.5

Represent the decimal integers in 1–5 in binary notation.

1. 19   2. 43   3. 287

4. 458   5. 1297

Represent the integers in 6–10 in decimal notation.

6. 1110_2 7. 10101_2 8. 110110_2

9. 1100101_2 10. 1000111_2

Perfom the arithmetic in 11–18 using binary notation.

11. $\begin{array}{r} 1011_2 \\ +\ \ 101_2 \\ \hline \end{array}$

12. $\begin{array}{r} 1001_2 \\ +\ 1111_2 \\ \hline \end{array}$

13. $\begin{array}{r} 101101_2 \\ +\ \ 11101_2 \\ \hline \end{array}$

14. $\begin{array}{r} 110111011_2 \\ +\ 100101 1010_2 \\ \hline \end{array}$

15. $\begin{array}{r} 10100_2 \\ -\ \ 1101_2 \\ \hline \end{array}$

16. $\begin{array}{r} 11010_2 \\ -\ \ 1101_2 \\ \hline \end{array}$

17. $\begin{array}{r} 101101_2 \\ -\ \ 10011_2 \\ \hline \end{array}$

18. $\begin{array}{r} 1110100_2 \\ -\ \ 10111_2 \\ \hline \end{array}$

19. Give the output signals S and T for the circuit below if the input signals P, Q, and R are as specified. Note that this is *not* the circuit for a full-adder.

 a. $P = 1, Q = 1, R = 1$

 b. $P = 0, Q = 1, R = 0$

 c. $P = 1, Q = 0, R = 1$

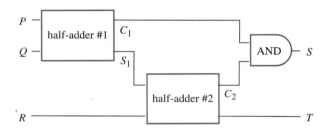

Find the 8-bit two's complements for the integers in 20–23.

20. 23   21. 67   22. 4   23. 108

Find the decimal representations for the integers with two's complements given in 24–27.

24. 11010011   25. 10001001   26. 11110010

27. 10111010

Use 8-bit representations to compute the sums in 28–33.

28. $57 + (-118)$ 29. $62 + (-18)$

30. $(-6) + (-73)$ 31. $79 + (-43)$

32. $(-15) + (-46)$ 33. $123 + (-94)$

◆34. Show that if a and b are integers in the range 1 through 128, and the sum of a and b is also in this range, then $2^8 \le (2^8 - a) + (2^8 - b) < 2^9$. Explain why it follows that the binary representation of $(2^8 - a) + (2^8 - b)$ has a leading 1 in the 2^8th position.

Convert the integers in 35–37 from hexadecimal to decimal notation.

35. A2BC_{16} 36. E0D_{16} 37. 29FB_{16}

Convert the integers in 38–40 from hexadecimal to binary notation.

38. 1C0ABE_{16} 39. B53DF8_{16} 40. 2ACF93_{16}

Convert the integers in 41–43 from binary to hexadecimal notation.

41. 00101110_2

42. 1011011111000101_2

43. 11101000111100_2

44. In addition to binary and hexadecimal, computer scientists also use octal notation (base 8) to represent numbers. Octal notation is based on the fact that any integer can be uniquely represented as a sum of numbers of the form $d \cdot (8^n)$, where each n is a nonnegative integer and each d is one of the integers from 0 to 7. Thus, for example, $5073_8 = 5 \cdot (8^3) + 0 \cdot (8^2) + 7 \cdot (8^1) + 3 \cdot (8^0) = 2619_{10}$.

 a. Convert 61502_8 to decimal notation.

 b. Describe methods for converting integers from octal to binary notation and the reverse that are similar to the methods used in Examples 1.5.12 and 1.5.13 for converting back and forth from hexadecimal to binary notation. Give examples showing that these methods result in correct answers.

The Logic of Quantified Statements

In Chapter 1 we discussed the logical analysis of compound statements—those made of simple statements joined by the connectives $\sim$, $\wedge$, $\vee$, $\rightarrow$, and $\leftrightarrow$. Such analysis casts light on many aspects of human reasoning, but it cannot be used to determine validity in the majority of everyday and mathematical situations. For example, the argument

> All human beings are mortal
>
> Socrates is a human being
>
> $\therefore$ Socrates is mortal

is intuitively perceived as correct. Yet its validity cannot be derived using the methods outlined in Section 1.3. To determine validity in examples like this, it is necessary to separate the statements into parts in much the same way that you separate declarative sentences into subjects and predicates. And you must analyze and understand the special role played by words that denote quantities such as "all" or "some." The symbolic analysis of predicates and quantified statements is called the **predicate calculus**. The symbolic analysis of ordinary compound statements (as outlined in Sections 1.1–1.3) is called the **statement calculus** (or the **propositional calculus**).

2.1 PREDICATES AND QUANTIFIED STATEMENTS I

People who call this "instinct" are merely giving the phenomenon a name, not explaining anything.
(Douglas Adams, *Dirk Gently's Holistic Detective Agency,* 1987)

As noted in Section 1.1, the sentence "He is a college student" is not a statement because it may be either true or false depending on the value of the pronoun *he*. Similarly, the sentence "$x + y$ is greater than 0" is not a statement because its truth value depends on the values of the variables x and y.

In grammar, the word *predicate* refers to the part of a sentence that gives information about the subject. In the sentence "James is a student at Bedford College," the word *James* is the subject and the phrase *is a student at Bedford College* is the predicate. The predicate is the part of the sentence from which the subject has been removed.

In logic, predicates can be obtained by removing any nouns from a statement. For instance, let P stand for the words "is a student at Bedford College" and let Q stand for the words "is a student at." Then both P and Q are *predicate symbols*. The sentences "x is a student at Bedford College" and "x is a student at y" are symbolized as $P(x)$ and as $Q(x, y)$ respectively, where x and y are *predicate variables* that take values in appropriate sets. When concrete values are substituted in place of predicate variables, a statement results. For simplicity, we define a *predicate* to be a predicate symbol together with suitable predicate variables. In some other treatments of logic, such objects are referred to as **propositional functions** or **open sentences**.

> ### DEFINITION
>
> A **predicate** is a sentence that contains a finite number of variables and becomes a statement when specific values are substituted for the variables. The **domain** of a predicate variable is the set of all values that may be substituted in place of the variable.

The sets in which predicate variables take their values may be described either in words or in symbols. When symbols are used, sets are normally denoted by upper-case letters and elements of sets by lower-case letters. The notation $x \in A$ indicates that x is an element of the set A, or, briefly, x is in A. Then $x \notin A$ means that x is not in A. One way to define a set is simply to indicate its elements between a pair of braces. For instance, $\{1, 2, 3\}$ refers to the set whose elements are 1, 2, and 3, and $\{1, 2, 3, \dots\}$ indicates the set of all positive integers. (The symbol ". . ." is called an **ellipsis** and is read "and so forth.") Two sets are equal if, and only if, they have exactly the same elements.

Certain sets of numbers are so frequently referred to that they are given special symbolic names. These are summarized in the table below.

Symbol	Set
R	set of all real numbers
Z	set of all integers*
Q	set of all rational numbers or quotients of integers

*The **Z** stands for the first letter of the German word for integers: *Zahlen.*

Addition of a superscript $+$ or $-$ or the letters *nonneg* indicates that only the positive or negative or nonnegative elements of the set are to be included. Thus $\mathbf{R}^+$ denotes the set of positive real numbers and $\mathbf{Z}^{nonneg}$ refers to the set of nonnegative integers: 0, 1, 2, 3, 4, and so forth. The set of nonnegative integers is also known as the set of *natural numbers* and is sometimes denoted $\mathbf{N}$.

The set of real numbers is ordinarily pictured as the set of all points on a line. This line is called *continuous* because it is imagined to have no holes. The set of integers is then pictured as a collection of points located at intervals one unit apart along the line. These points are called *discrete* because each is separated from the others. The name *discrete mathematics* comes from the distinction between continuous and discrete mathematical objects.

When an element in the domain of the variable of a one-variable predicate is substituted for the variable, the resulting statement is either true or false. The set of all such elements that make the predicate true is called the *truth set* of the predicate.

DEFINITION

If $P(x)$ is a predicate and x has domain D, the **truth set** of $P(x)$ is the set of all elements of D that make $P(x)$ true when substituted for x. The truth set of $P(x)$ is denoted

$$\{x \in D \mid P(x)\}$$

the set of all   such that

which is read "the set of all x in D such that $P(x)$."

As an example, let $P(x)$ be "x is a factor of 8" and suppose the domain of x is the set of all positive integers. Then the truth set $P(x)$ is $\{1, 2, 4, 8\}$ since 1, 2, 4, and **8** are exactly the positive integers that divide 8 evenly.

NOTATION

Let $P(x)$ and $Q(x)$ be predicates and suppose the common domain of x is D. The notation $P(x) \Rightarrow Q(x)$ means that every element in the truth set of $P(x)$ is in the truth set of $Q(x)$. The notation $P(x) \Leftrightarrow Q(x)$ means that $P(x)$ and $Q(x)$ have identical truth sets.

EXAMPLE 2.1.1   **Using $\Rightarrow$ and $\Leftrightarrow$**

Let $P(x)$ be "x is a factor of 8," $Q(x)$ be "x is a factor of 4," and $R(x)$ be "$x < 5$ and $x \neq 3$," and suppose the domain of x is $\mathbf{Z}^{+}$, the set of positive integers. Use the $\Rightarrow$ and $\Leftrightarrow$ symbols to indicate true relationships among $P(x)$, $Q(x)$, and $R(x)$.

Solution   1. As noted above, the truth set of $P(x)$ is $\{1, 2, 4, 8\}$. By similar reasoning, the truth set of $Q(x)$ is $\{1, 2, 4\}$. Since every element in the truth set of $Q(x)$ is in the truth set of $P(x)$, $Q(x) \Rightarrow P(x)$.

2. The truth set of $R(x)$ is $\{1, 2, 4\}$, which is identical to the truth set of $Q(x)$. Hence $R(x) \Leftrightarrow Q(x)$.

■

QUANTIFIERS: $\forall$ and $\exists$

One obvious way to change predicates into statements is to assign specific values to all their variables. For example, if x represents the number 35, the sentence "x is (evenly) divisible by 5" is a true statement since $35 = 5 \cdot 7$. Another way to obtain statements from predicates is to add **quantifiers**. Quantifiers are words that refer to quantities such as "some" or "all" and tell for how many elements a given predicate is true. The formal concept of quantifier was introduced into symbolic logic in the late nineteenth century by the American philosopher, logician, and engineer Charles Sanders Peirce and, independently, by the German logician Gottlob Frege.

The symbol $\forall$ denotes "for all" and is called the **universal quantifier**. For example, another way to express the sentence "All human beings are mortal" is to write

$$\forall \text{ human beings } x, x \text{ is mortal},$$

*Charles Sanders Peirce
(1839–1914)*

or, more formally,

$$\forall x \in S, \; x \text{ is mortal,}$$

where S denotes the set of all human beings. (Think "for all" when you see the symbol $\forall$.) The domain of the predicate variable is generally indicated between the $\forall$ symbol and the variable name (as in $\forall$ human beings x) or immediately following the variable name (as in $\forall x \in S$). Some other expressions that can be used instead of *for all* are *for every, for arbitrary, for any, for each,* and *given any.* In a sentence such as "$\forall$ real numbers x and y, $x + y = y + x$," the $\forall$ symbol is understood to refer to both x and y.*

Sentences that are quantified universally are defined as statements by giving them the truth values specified in the following definition:

> **DEFINITION**
>
> Let $Q(x)$ be a predicate and D the domain of x. A **universal statement** is a statement of the form "$\forall x \in D, Q(x)$." It is defined to be true if, and only if, $Q(x)$ is true for every x in D. It is defined to be false if, and only if, $Q(x)$ is false for at least one x in D. A value for x for which $Q(x)$ is false is called a **counterexample** to the universal statement.

EXAMPLE 2.1.2   **Truth and Falsity of Universal Statements**

a. Let $D = \{1, 2, 3, 4, 5\}$, and consider the statement

$$\forall x \in D, \; x^2 \geq x.$$

Show that this statement is true.

b. Consider the statement

$$\forall x \in \mathbf{R}, \; x^2 \geq x.$$

Find a counterexample to show that this statement is false.

Solution   a. Check that "$x^2 \geq x$" is true for each individual x in D.

$$1^2 \geq 1, \qquad 2^2 \geq 2, \qquad 3^2 \geq 3, \qquad 4^2 \geq 4, \qquad 5^2 \geq 5.$$

Hence "$\forall x \in D, x^2 \geq x$" is true.

b. *Counterexample*: Take $x = \frac{1}{2}$. Then x is in $\mathbf{R}$ (since $\frac{1}{2}$ is a real number) and

$$\left(\frac{1}{2}\right)^2 = \frac{1}{4} \neq \frac{1}{2}.$$

Hence "$\forall x \in \mathbf{R}, x^2 \geq x$" is false. ∎

The technique used to show the truth of the universal statement in Example 2.1.2(a) is called the **method of exhaustion**. It consists of showing the truth of the predicate separately for each individual element of the domain. (The idea is to exhaust the possibilities before you exhaust yourself!) This method can, in theory, be used

---

*More formal versions of symbolic logic would require writing a separate $\forall$ for each variable: "$\forall x \in \mathbf{R}$ $\forall y \in \mathbf{R}(x + y = y + x)$."

whenever the domain of the predicate variable is finite. In recent years the prevalence of digital computers has greatly increased the convenience of using the method of exhaustion. Computer expert systems, or knowledge-based systems, use this method to arrive at answers to many of the questions posed to them.

The symbol $\exists$ denotes "there exists" and is called the **existential quantifier**. For example, the sentence "There is a student in Math 140" can be written as

$$\exists \text{ a person } s \text{ such that } s \text{ is a student in Math 140,}$$

or, more formally,

$$\exists s \in S \text{ such that } s \text{ is a student in Math 140,}$$

where S is the set of all people. (Think "there exists" when you see the symbol $\exists$.) The domain of the predicate variable is generally indicated either between the $\exists$ symbol and the variable name or immediately following the variable name. The words *such that* are inserted just before the predicate. Some other expressions that can be used in place of *there exists* are *there is a, we can find a, there is at least one, for some,* and *for at least one.* In a sentence such as "$\exists$ integers m and n such that $m + n = m \cdot n$," the $\exists$ symbol is understood to refer to both m and n.*

Sentences that are quantified existentially are defined as statements by giving them the truth values specified in the following definition.

DEFINITION

Let $Q(x)$ be a predicate and D the domain of x. An **existential statement** is a statement of the form "$\exists x \in D$ such that $Q(x)$." It is defined to be true if, and only if, $Q(x)$ is true for at least one x in D. It is false if, and only if, $Q(x)$ is false for all x in D.

EXAMPLE 2.1.3   **Truth and Falsity of Existential Statements**

a. Consider the statement

$$\exists m \in \mathbf{Z} \text{ such that } m^2 = m.$$

Show that this statement is true.

b. Let $E = \{5, 6, 7, 8, 9, 10\}$ and consider the statement

$$\exists m \in E \text{ such that } m^2 = m.$$

Show that this statement is false.

Solution   a. Observe that $1^2 = 1$. Thus "$m^2 = m$" is true for at least one integer m. Hence "$\exists m \in \mathbf{Z}$ such that $m^2 = m$" is true.

b. Note that $m^2 = m$ is not true for any integers m from 5 to 10:

$$5^2 = 25 \neq 5, \qquad 6^2 = 36 \neq 6, \qquad 7^2 = 49 \neq 7, \qquad 8^2 = 64 \neq 8,$$
$$9^2 = 81 \neq 9, \qquad 10^2 = 100 \neq 10.$$

Thus "$\exists m \in E$ such that $m^2 = m$" is false.   ∎

---

*In more formal versions of symbolic logic the words *such that* are not written out (although they are understood) and a separate $\exists$ symbol is used for each variable: "$\exists m \in \mathbf{Z} \, \exists n \in \mathbf{Z}(m + n = m \cdot n)$."

It is important to be able to translate from formal into informal language when trying to make sense of mathematical concepts that are new to you. It is equally important to be able to translate from informal into formal language when thinking out a complicated problem.

EXAMPLE 2.1.4   Translating from Formal to Informal Language

Rewrite the following formal statements in a variety of equivalent but more informal ways. Do not use the symbol $\forall$ or $\exists$.

a. $\forall x \in \mathbf{R}, x^2 \geq 0$.
b. $\forall x \in \mathbf{R}, x^2 \neq -1$.
c. $\exists m \in \mathbf{Z}$ such that $m^2 = m$.

Solution   a. All real numbers have nonnegative squares.
Every real number has a nonnegative square.
Any real number has a nonnegative square.
x has a nonnegative square, for each real number x.
The square of any real number is nonnegative.
(Note that the singular noun is used to refer to the domain when the $\forall$ symbol is translated as *every, any,* or *each.*)
b. All real numbers have squares not equal to -1.
No real numbers have squares equal to -1.
(The words *none are* or *no . . . are* are equivalent to the words *all are not.*)
c. There is an integer whose square is equal to itself.
We can find at least one integer equal to its own square.
$m^2 = m$, for some integer m.
Some integer equals its own square.
Some integers equal their own squares.
(In ordinary English, this last statement might be taken to be true only if there are at least two integers equal to their own squares. In mathematics, we understand the last two statements to mean the same thing.)   ■

EXAMPLE 2.1.5   Translating from Informal to Formal Language

Rewrite each of the following statements formally. Use quantifiers and variables.

a. All triangles have three sides.
b. No dogs have wings.
c. Some programs are structured.

Solution   a. $\forall$ triangles t, t has three sides, or
$\forall t \in T$, t has three sides (where T is the set of all triangles).
b. $\forall$ dogs d, d does not have wings, or
$\forall d \in D$, d does not have wings (where D is the set of all dogs).
c. $\exists$ a program p such that p is structured, or
$\exists p \in P$ such that p is structured (where P is the set of all programs).   ■

UNIVERSAL CONDITIONAL STATEMENTS

A reasonable argument can be made that the most important form of statement in mathematics is the **universal conditional statement**:

$$\forall x, \text{ if } P(x) \text{ then } Q(x).$$

Familiarity with statements of this form is essential if you are to learn to speak mathematics.

EXAMPLE 2.1.6 Writing Universal Conditional Statements Informally

Rewrite the following formal statement in a variety of informal ways. Do not use quantifiers or variables.

$$\forall x \in \mathbf{R}, \text{ if } x > 2 \text{ then } x^2 > 4.$$

Solution   If a real number is greater than 2 then its square is greater than 4.

Whenever a real number is greater than 2, its square is greater than 4.

The square of any real number that is greater than 2 is greater than 4.

The squares of all real numbers greater than 2 are greater than 4.   ■

EXAMPLE 2.1.7 Writing Universal Conditional Statements Formally

Rewrite each of the following statements in the form

$$\forall \underline{\hspace{1cm}}, \text{ if } \underline{\hspace{1cm}} \text{ then } \underline{\hspace{1cm}}.$$

a. If a real number is an integer, then it is a rational number.
b. All bytes have eight bits.
c. No fire trucks are green.

Solution   a. $\forall$ real numbers x, if x is an integer, then x is a rational number, or
$\forall x \in \mathbf{R}$, if $x \in \mathbf{Z}$ then $x \in \mathbf{Q}$.
b. $\forall x$, if x is a byte, then x has eight bits.
c. $\forall x$, if x is a fire truck, then x is not green.

It is common, as in (b) and (c) above, to omit explicit identification of the domain of predicate variables in universal conditional statements.   ■

Careful thought about the meaning of universal conditional statements leads to another level of understanding for why the truth table for an if–then statement must be defined as it is. Consider again the statement

$$\forall \text{ real numbers } x, \text{ if } x > 2 \text{ then } x^2 > 4.$$

Your experience and intuition tell you that this statement is true. But that means that

$$\text{If } x > 2 \text{ then } x^2 > 4$$

must be true for every single real number x. Consequently, it must be true even for x that make its hypothesis "$x > 2$" false. In particular, both statements

$$\text{If } 1 > 2 \text{ then } 1^2 > 4 \quad \text{and} \quad \text{If } -3 > 2 \text{ then } (-3)^2 > 4$$

must be true. In both cases the hypothesis is false, but in the first case the conclusion "$1^2 > 4$" is false and in the second case the conclusion "$(-3)^2 > 4$" is true. Hence, regardless of whether its conclusion is true or false, an if–then statement with a false hypothesis must be true.

Note also that the definition of valid argument is a universal conditional statement:

$\forall$ combinations of truth values for the component statements,
if the premises are all true then the conclusion is also true.

EQUIVALENT FORMS OF UNIVERSAL AND EXISTENTIAL STATEMENTS

Observe that the two statements "$\forall$ real numbers x, if x is an integer then x is rational" and "$\forall$ integers x, x is rational" mean the same thing. Both have informal translations "All integers are rational." In fact, a statement of the form

$$\forall x \in U, \text{ if } P(x) \text{ then } Q(x)$$

can always be rewritten in the form

$$\forall x \in D, Q(x)$$

by narrowing U to be the domain D consisting of all values of the variable x that make $P(x)$ true. Conversely, a statement of the form

$$\forall x \in D, Q(x)$$

can be rewritten as

$$\forall x, \text{ if } x \text{ is in } D \text{ then } Q(x).$$

EXAMPLE 2.1.8   **Equivalent Forms for Universal Statements**

The following statements are equivalent:

$$\forall \text{ polygons } p, \text{ if } p \text{ is a square, then } p \text{ is a rectangle}$$

and

$$\forall \text{ squares } p, p \text{ is a rectangle.} \qquad \blacksquare$$

The existential statements

$$\exists x \in U \text{ such that } P(x) \text{ and } Q(x)$$

and

$$\exists x \in D \text{ such that } Q(x)$$

are also equivalent provided D is taken to consist of all elements in U that make $P(x)$ true.

EXAMPLE 2.1.9   **Equivalent Forms for Existential Statements**

The following statements are equivalent:

$$\exists \text{ a number } n \text{ such that } n \text{ is prime and } n \text{ is even}$$

and

$$\exists \text{ a prime number } n \text{ such that } n \text{ is even.} \qquad \blacksquare$$

IMPLICIT QUANTIFICATION

Consider the statement

> If a number is an integer, then it is a rational number.

As shown earlier, this statement is equivalent to a universal statement. However, it does not contain the telltale words *all* or *every* or *any* or *each*. The only clue to indicate its universal quantification comes from the presence of the indefinite article *a*. This is an example of *implicit* universal quantification.

Existential quantification can also be implicit. For instance, the statement "The number 24 can be written as a sum of two even integers" can be expressed formally as "∃ even integers m and n such that $24 = m + n$."

Mathematical writing contains many examples of implicitly quantified statements. Some occur, as in the first example above, through the presence of the word *a* or *an*. Others occur in cases where the general context of a sentence supplies part of its meaning. For example, in an algebra course in which the letter x is always used to indicate a real number, the predicate

$$\text{If } x > 2 \text{ then } x^2 > 4$$

is interpreted to mean the same as the statement

$$\forall \text{ real numbers } x, \text{ if } x > 2 \text{ then } x^2 > 4.$$

Some questions of quantification can be quite subtle. For instance, a mathematics text might contain the following:

a. $(x + 1)^2 = x^2 + 2x + 1$.
b. Solve $(x + 2)^2 = 25$.

Although neither (a) nor (b) contains explicit quantification, the reader is supposed to understand that the x in (a) is universally quantified whereas the x in (b) is existentially quantified. When the quantification is made explicit, (a) and (b) become the following:

a. $\forall$ real numbers x, $(x + 1)^2 = x^2 + 2x + 1$.
b. Show (by finding a value) that $\exists$ a real number x such that $(x + 2)^2 = 25$.

The quantification of a statement—whether universal or existential—crucially determines both how the statement can be applied and what method must be used to establish its truth. Thus it is important to be alert to the presence of hidden quantifiers when you read mathematics so that you will interpret statements in a logically correct way.

NEGATIONS OF QUANTIFIED STATEMENTS

Consider the statement "All mathematicians wear glasses." Many people would say that its negation is "No mathematicians wear glasses." In fact, the negation is "One or more mathematicians do not wear glasses" or "Some mathematicians do not wear glasses." After all, if even one mathematician does not wear glasses, the sweeping statement that all mathematicians wear glasses must be false.

The general form of the negation of a universal statement follows immediately from the definitions of negation and the truth values for universal and existential statements.

THEOREM 2.1.1

The negation of a statement of the form

$$\forall x \text{ in } D, Q(x)$$

is logically equivalent to a statement of the form

$$\exists x \text{ in } D \text{ such that } \sim Q(x).$$

Symbolically:

$$\sim(\forall x \in D, Q(x)) \equiv \exists x \in D \text{ such that } \sim Q(x).$$

Thus:

The negation of a universal statement ("all are") is logically equivalent to an existential statement ("some are not").

Now consider the statement "Some fish breathe air." What is its negation? Many people would answer incorrectly that it is "Some fish do not breathe air." Actually, the negation is "*No* fish breathe air." After all, if it is not true that some fish breathe air, then not a single fish breathes air. That is, no fish breathe air, or all fish are non–air-breathers.

The general form for the negation of an existential statement follows immediately from the definitions of negation and of the truth values for existential and universal statements.

THEOREM 2.1.2

The negation of a statement of the form

$$\exists x \text{ in } D \text{ such that } Q(x)$$

is logically equivalent to a statement of the form

$$\forall x \text{ in } D, \sim Q(x).$$

Symbolically:

$$\sim(\exists x \in D \text{ such that } Q(x)) \equiv \forall x \in D, \sim Q(x).$$

Thus:

The negation of an existential statement ("some are") is logically equivalent to a universal statement ("all are not").

EXAMPLE 2.1.10   Negating Quantified Statements

Write formal negations for each of the following statements:

a. $\forall$ primes p, p is odd.
b. $\exists$ a triangle T such that the sum of the angles of T equals $200°$.

Solution   a. By applying the rule for the negation of a $\forall$ statement, you can see that the answer is

$$\exists \text{ a prime } p \text{ such that } p \text{ is not odd.}$$

b. By applying the rule for the negation of a $\exists$ statement, you can see that the answer is

$$\forall \text{ triangles } T, \text{ the sum of the angles of } T \text{ does not equal } 200°. \quad ■$$

You need to exercise special care to avoid mistakes when writing negations of statements that are given informally. One way to avoid error is to rewrite the statement formally and take the negation using the formal rule.

EXAMPLE 2.1.11 More Negations

Rewrite the following statement formally. Then write formal and informal negations.

No politicians are honest.

Solution

formal version: $\forall$ politicians x, x is not honest.

formal negation: $\exists$ a politician x such that x is honest.

informal negation: Some politicians are honest. ∎

Another way to avoid error when taking negations of statements that are given in informal language is to ask yourself, "What *exactly* would it mean for the given statement to be false? What statement, if true, would be equivalent to saying that the given statement is false?"

EXAMPLE 2.1.12 Still More Negations

Write informal negations for each of the following statements:

a. All computer programs are finite.
b. Some computer hackers are over 40.

Solution

a. What exactly would it mean for this statement to be false? The statement asserts a property for all computer programs. So for it to be false would mean that there would be some computer program that does not have the property. Thus the answer is

Some computer programs are not finite.

b. This statement is equivalent to saying that there is at least one computer hacker with a certain property. So for it to be false would mean that not a single computer hacker has that property. Thus the negation is

No computer hackers are over 40.

Or:

All computer hackers are 40 or under. ∎

▲ CAUTION! *Informal negations of many universal statements can be constructed simply by inserting the word not or the words do not at an appropriate place. However, the resulting statements may be ambiguous. For example, a possible negation of "All mathematicians wear glasses" is "All mathematicians do not wear glasses." The problem is that this sentence has two meanings. With the proper verbal stress on the word not, it could be interpreted as the logical negation. (What! You say that all mathematicians wear glasses? Nonsense! All mathematicians do not wear glasses.) On the other hand, stated in a flat tone of voice (try it!), it would mean that all mathematicians are nonwearers of glasses; that is, not a single mathematician wears glasses. This is a much stronger statement than the logical negation: It implies the negation but is not equivalent to it.*

NEGATIONS OF UNIVERSAL CONDITIONAL STATEMENTS

Negations of universal conditional statements are of special importance in mathematics. The form of such negations can be derived from facts that have already been established.

By definition of the negation of a *for all* statement,

$$\sim(\forall x, P(x) \rightarrow Q(x)) \equiv \exists x \text{ such that } \sim(P(x) \rightarrow Q(x)). \qquad \textbf{2.1.1}$$

But the negation of an if–then statement is logically equivalent to an *and* statement. More precisely,

$$\sim(P(x) \rightarrow Q(x)) \equiv P(x) \wedge \sim Q(x). \qquad \textbf{2.1.2}$$

Substituting (2.1.2) into (2.1.1) gives

$$\sim(\forall x, P(x) \rightarrow Q(x)) \equiv \exists x \text{ such that } (P(x) \wedge \sim Q(x)).$$

Written less symbolically, this becomes the following.

$$\boxed{\sim(\forall x, \text{ if } P(x) \text{ then } Q(x)) \equiv \exists x \text{ such that } P(x) \text{ and } \sim Q(x)}$$

EXAMPLE 2.1.13   Negating Universal Conditional Statements

Write a formal negation for statement (a) and an informal negation for statement (b).

a. $\forall$ people p, if p is blond then p has blue eyes.
b. If a computer program has more than 100,000 lines, then it contains a bug.

Solution a. $\exists$ a person p such that p is blond and p does not have blue eyes.
b. There is at least one computer program that has more than 100,000 lines and does not contain a bug. ■

VACUOUS TRUTH OF UNIVERSAL STATEMENTS

Suppose a bowl sits on a table and next to the bowl is a pile of five blue and five gray balls, any of which may be placed in the bowl. If three blue balls and one gray ball are placed in the bowl, as shown in Figure 2.1.1(a), the statement "All the balls in the bowl are blue" would be false (since one of the balls in the bowl is gray).

Now suppose that no balls at all are placed in the bowl, as shown in Figure 2.1.1(b). Consider the statement

All the balls in the bowl are blue.

Is this statement true or false? The statement is false if, and only if, its negation is true. And its negation is

There exists a ball in the bowl that is not blue.

But the only way this negation can be true is for there actually to be a nonblue ball in the bowl. And there is not! Hence, the negation is false, and so the statement is true "by default."

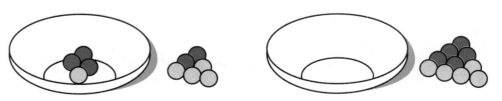

a.   b.

FIGURE 2.1.1

In general, a statement of the form

$$\forall x \text{ in } D, \text{ if } P(x) \text{ then } Q(x)$$

is called **vacuously true** or **true by default** if, and only if, $P(x)$ is false for every x in D.

By the way, in ordinary language the words *in general* mean that something is usually, but not always, the case. (In general, I take the bus home, but today I drove.) In mathematics, the words *in general* are used quite differently. When they occur just after discussion of a particular example (as in the paragraph above), they are a signal that what is to follow is a generalization of some aspect of the example that always holds true.

EXERCISE SET 2.1*

1. A menagerie consists of seven brown dogs, two black dogs, six gray cats, ten black cats, five blue birds, six yellow birds, and one black bird. Determine which of the following statements are true and which are false.
   a. There is an animal in the menagerie that is red.
   b. Every animal in the menagerie is a bird or a mammal.
   c. Every animal in the menagerie is brown or gray or black.
   d. There is an animal in the menagerie that is neither a cat nor a dog.
   e. No animal in the menagerie is blue.
   f. There are a dog, a cat and a bird in the menagerie that all have the same color.

2. Find the truth set of each predicate below.
   a. predicate: $x > 1/x$, domain: $\mathbf{R}$
   b. predicate: $n^2 \leq 30$, domain: $\mathbf{Z}$

3. Let $\mathbf{R}$ be the domain of the predicates "$x > 1$," "$x > 2$," "$|x| > 2$," and "$x^2 > 4$." Which of the following are true and which are false?
   a. $x > 2 \Rightarrow x > 1$ b. $x > 2 \Rightarrow x^2 > 4$
   c. $x^2 > 4 \Rightarrow x > 2$ d. $x^2 > 4 \Leftrightarrow |x| > 2$

Find counterexamples to show that the statements in 4–7 are false.

4. $\forall x \in \mathbf{R}, x > 1/x$.

5. $\forall a \in \mathbf{Z}, (a - 1)/a$ is not an integer.

6. $\forall$ positive integers m and n, $m \cdot n \geq m + n$.

7. $\forall$ real numbers x and y, $\sqrt{x + y} = \sqrt{x} + \sqrt{y}$.

8. Consider the following statement:

$$\forall \text{ basketball players } x, x \text{ is tall.}$$

Which of the following are equivalent ways of expressing this statement?
   a. Every basketball player is tall.
   b. Among all the basketball players, some are tall.
   c. Some of all the tall people are basketball players.
   d. Anyone who is tall is a basketball player.
   e. All people who are basketball players are tall.
   f. Anyone who is a basketball player is a tall person.

9. Consider the following statement:

$$\exists x \in \mathbf{R} \text{ such that } x^2 = 2.$$

Which of the following are equivalent ways of expressing this statement?
   a. The square of each real number is 2.
   b. Some real numbers have square 2.
   c. The number x has square 2, for some real number x.
   d. If x is a real number, then $x^2 = 2$.
   e. Some real number has square 2.
   f. There is at least one real number whose square is 2.

10. Rewrite the following statements informally in at least two different ways without using variables or the symbols $\forall$ or $\exists$:
   a. $\forall$ squares x, x is a rectangle.
   b. $\exists$ a set A such that A has 16 subsets.

11. Rewrite each of the following statements in the form "$\forall \underline{\quad} x, \underline{\quad}.$"
   a. All dinosaurs are extinct.
   b. Every real number is positive, negative, or zero.
   c. No irrational numbers are integers.
   d. No logicians are lazy.

12. Rewrite each of the following in the form "$\exists \underline{\quad} x$ such that $\underline{\quad}$":
   a. Some exercises have answers.
   b. Some real numbers are rational.

*Exercises with blue numbers or letters have solutions in Appendix B. The symbol H indicates that only a hint or a partial solution is given. The symbol ◆ signals that an exercise is more challenging than usual.

13. Consider the following statement:

$\forall$ integers n, if n^2 is even then n is even.

Which of the following are equivalent ways of expressing this statement?
a. All integers have even squares and are even.
b. Given any integer whose square is even, that integer is itself even.
c. For all integers, there are some whose square is even.
d. Any integer with an even square is even.
e. If the square of an integer is even, then that integer is even.
f. All even integers have even squares.

14. Rewrite the following statement informally in at least two different ways without using variables or the symbols $\forall$ or $\exists$:

$\forall$ students S, if S is in CSC 310
then S has taken MAT 140.

15. Rewrite each of the following statements in the form "$\forall$ _____, if _____ then _____."
a. All COBOL programs have at least 20 lines.
b. Any valid argument with true premises has a true conclusion.
c. The sum of any two even integers is even.
d. The product of any two odd integers is odd.

16. Rewrite each of the following statements in the two forms "$\forall x$, if _____ then _____" and "$\forall$ _____ x, _____" (without an if–then).
a. The square of any even integer is even.
b. Every computer science student needs to take assembly language programming.

17. Rewrite the following statements in the two forms "$\exists$ _____ x such that _____" and "$\exists x$ such that _____ and _____ ."
a. Some hatters are mad.
b. Some questions are easy.

18. Find an example in any mathematics or computer science text of a statement that is universal but is implicitly quantified. Copy the statement as it appears and rewrite it making the quantification explicit. Give a complete citation for your example including title, author, publisher, year, and page number.

19. Which of the following is a negation for "Every polynomial function is continuous"? More than one answer may be correct.
a. No polynomial function is continuous.
b. Some polynomial functions are not continuous.
c. Every polynomial function fails to be continuous.
d. There is a noncontinuous polynomial function.

In 20–23 write negations for each statement in the referenced exercise.

H20. exercise 11   H21. exercise 12

H22. exercise 15   H23. exercise 16

In each of 24–27 determine whether the proposed negation is correct. If it is not, write a correct negation.

24.   statement: The sum of any two irrational numbers is irrational.
proposed negation: The sum of any two irrational numbers is rational.

25.   statement: The product of any irrational number and any rational number is irrational.
proposed negation: The product of any irrational number and any rational number is rational.

26.   statement: For all integers n, if n^2 is even then n is even.
proposed negation: For all integers n, if n^2 is even then n is not even.

27.   statement: For all real numbers x_1 and x_2, if $x_1^2 = x_2^2$ then $x_1 = x_2$.
proposed negation: For all real numbers x_1 and x_2, if $x_1^2 = x_2^2$ then $x_1 \neq x_2$.

28. Let $D = \{-48, -14, -8, 0, 1, 3, 16, 23, 26, 32, 36\}$. Determine which of the following statements are true and which are false. Provide counterexamples for those statements that are false.
a. $\forall x \in D$, if x is odd then $x > 0$.
b. $\forall x \in D$, if x is less than 0 then x is even.
c. $\forall x \in D$, if x is even then $x \leq 0$.
d. $\forall x \in D$, if the ones digit of x is 2, then the tens digit is 3 or 4.
e. $\forall x \in D$, if the ones digit of x is 6, then the tens digit is 1 or 2.

Write negations for each of the statements in 29–36.

29. $\forall$ real numbers x, if $x > 3$ then $x^2 > 9$.

30. $\forall$ computer programs P, if P is correct then P compiles without error messages.

31. $\forall x \in \mathbf{R}$, if $x(x + 1) > 0$ then $x > 0$ or $x < -1$.

32. $\forall n \in \mathbf{Z}$, if n is prime then n is odd or $n = 2$.

33. $\forall$ integers a, b, and c, if $a - b$ is even and $b - c$ is even, then $a - c$ is even.

34. $\forall$ animals x, if x is a cat then x has whiskers and x has claws.

35. If an integer is divisible by 2, then it is even.

36. If the square of an integer is even, then the integer is even.

37. The statement "There are no easy questions on the exam" contains the words "there are." Is the statement existential? Write an informal negation for the statement, and then write the statement formally using quantifiers and variables.

◆ **38.** If $P(x)$ is a predicate that is defined for all real numbers x, let r be "$\forall x \in \mathbf{Z}, P(x)$," let s be "$\forall x \in \mathbf{Q}, P(x)$," and let t be "$\forall x \in \mathbf{R}, P(x)$."
  a. Find $P(x)$ (but not "$x \in \mathbf{Z}$") so that r is true and both s and t are false.
  b. Find $P(x)$ so that both r and s are true and t is false.

2.2   PREDICATES AND QUANTIFIED STATEMENTS II

It is not enough to have a good mind. The main thing is to use it well.
(René Descartes)

This section continues the discussion of predicates and quantified statements begun in Section 2.1. It contains an analysis of statements with more than one quantifier; an examination of the relation among $\forall$, $\exists$, $\wedge$, and $\vee$; a discussion of variants of universal conditional statements; an extension of the meaning of *necessary and sufficient* and *only if* to quantified statements; and an indication of some ways predicates are used in the computer language Prolog.

STATEMENTS CONTAINING MULTIPLE QUANTIFIERS

Many statements in mathematics contain more than one quantifier.

EXAMPLE 2.2.1   Writing Multiply Quantified Statements Informally

Rewrite each of the following statements without using variables or the symbols $\forall$ or $\exists$:

a. $\forall$ positive numbers x, $\exists$ a positive number y such that $y < x$.
b. $\exists$ a positive number x such that $\forall$ positive numbers y, $y < x$.

Solution **a.** Given any positive number, there is another positive number that is smaller than the given number.
  Given any positive number, we can find a smaller positive number.
  There is no smallest positive number.
  b. There is a positive number with the property that all positive numbers are smaller than this number.
  There is a positive number that is larger than all positive numbers. ∎

EXAMPLE 2.2.2   Writing Multiply Quantified Statements Formally

Rewrite the following statements formally using quantifiers and variables:

a. Everybody loves somebody.
b. Somebody loves everybody.

Solution **a.** $\forall$ people x, $\exists$ a person y such that x loves y.
  b. $\exists$ a person x such that $\forall$ people y, x loves y. ∎

In both Examples 2.2.1 and 2.2.2, (a) and (b) are the same except for the order of the quantifiers $\forall$ and $\exists$. Yet in both cases the two sentences have very different meanings. It is usually the case that if the order in which quantifiers are written is reversed, the meaning of the statement is changed.

EXAMPLE 2.2.3   The Definition of Limit

The definition of limit of a sequence, studied in calculus, uses both quantifiers $\forall$ and $\exists$ and also if–then. We say that the limit of the sequence a_n as n goes to infinity equals L and write

$$\lim_{n \to \infty} a_n = L$$

if, and only if, the values of a_n become *arbitrarily* close to L as n gets larger and larger without bound. More precisely, this means that given any positive number ε, we can find an integer N such that whenever n is larger than N, then the number a_n sits between $L - \varepsilon$ and $L + \varepsilon$ on the number line.

a_n must lie in here

Symbolically:

$\forall \varepsilon > 0, \exists$ an integer N such that $\forall$ integers n,
if $n > N$ then $L - \varepsilon < a_n < L + \varepsilon$.

Considering the logical complexity of this definition, it is no wonder that many students find it hard to understand. ∎

EXAMPLE 2.2.4   Finding Truth Values of Multiply Quantified Statements

A college cafeteria line has four stations: salads, main courses, desserts, and beverages. The salad station offers a choice of green salad or fruit salad; the main course station offers spaghetti or fish; the dessert station offers pie or cake; and the beverage station offers milk, soda, or coffee. Three students, Uta, Tim, and Yuen, go through the line and make the following choices:

Uta: green salad, spaghetti, pie, milk

Tim: fruit salad, fish, pie, cake, milk, coffee

Yuen: spaghetti, fish, pie, soda

Write each of the following statements informally and find its truth value:

a. $\exists$ an item I such that $\forall$ students S, S chose I.
b. $\exists$ a student S such that $\forall$ items I, S chose I.
c. $\exists$ a student S such that $\forall$ stations Z, $\exists$ an item I in Z such that S chose I.
d. $\forall$ students S and $\forall$ stations Z, $\exists$ an item I in Z such that S chose I.

Solution   a. There is an item that was chosen by every student. This is true; every student chose pie.
b. There is a student who chose every available item. This is false; no student chose all nine items.
c. There is a student who chose at least one item from every station. This is true; both Uta and Tim chose at least one item from every station.
d. Every student chose at least one item from every station. This is false; Yuen did not choose a salad. ∎

NEGATIONS OF MULTIPLY QUANTIFIED STATEMENTS

What is the negation of the following statement?

$\forall$ people $x, \exists$ a person y such that x loves y.

To derive the answer, ask yourself what it would mean for this statement to be false. The statement asserts that a certain property holds for all people x. So for it to be false would mean that there is a person for whom the property does not hold. Thus the negation of

$$\forall \text{ people } x, \exists \text{ a person } y \text{ such that } x \text{ loves } y$$

can be written

$$\exists \text{ a person } x \text{ such that } \sim(\exists \text{ a person } y \text{ such that } x \text{ loves } y).$$

Now rewrite the second part of the statement. What does it mean for it not to be the case that there exists a person y such that x loves y? It means that no matter what person y is selected, x does not love y. More formally,

$$\sim(\exists \text{ a person } y \text{ such that } x \text{ loves } y)$$

is logically equivalent to

$$\forall \text{ people } y, x \text{ does not love } y.$$

Thus the negation of

$$\forall \text{ people } x, \exists \text{ a person } y \text{ such that } x \text{ loves } y$$

can be written

$$\exists \text{ a person } x \text{ such that } \forall \text{ people } y, x \text{ does not love } y.$$

In less formal language, this says that the negation of

$$\text{Everybody loves somebody}$$

can be written

$$\text{There is somebody who does not love anybody.}$$

The procedure used to derive this negation can be generalized to arrive at the following fact:

The negation of
$$\forall x, \exists y \text{ such that } P(x, y)$$
is logically equivalent to
$$\exists x \text{ such that } \forall y, \sim P(x, y).$$

A similar sequence of reasoning can be used to derive the following:

The negation of
$$\exists x \text{ such that } \forall y, P(x, y)$$
is logically equivalent to
$$\forall x, \exists y \text{ such that } \sim P(x, y).$$

EXAMPLE 2.2.5   **Negating Multiply Quantified Statements**

Negate each of the following statements:

a. $\forall$ integers n, $\exists$ an integer k such that $n = 2k$. (Less formally: All integers are even.)
b. $\exists$ a person x such that $\forall$ people y, x loves y. (Less formally: Somebody loves everybody.)

Solution   a. *first version of negation:* $\exists$ an integer n such that $\sim(\exists$ an integer k such that $n = 2k)$.

final version of negation: $\exists$ an integer n such that $\forall$ integers k, $n \neq 2k$.
(A less formal version of the negation is: "There is some integer that is not equal to twice any other integer"; that is, "there is some integer that is not even.")

b. *first version of negation:* $\forall$ people x, $\sim(\forall$ people y, x loves $y)$.
final version of negation: $\forall$ people x, $\exists$ a person y such that x does not love y.
(A less formal version of the negation is "Nobody loves everybody.")   ∎

THE RELATION AMONG $\forall$, $\exists$, $\wedge$, AND $\vee$

The negation of a *for all* statement is a *there exists* statement, and the negation of a *there exists* statement is a *for all* statement. These facts are analogous to De Morgan's laws, which state that the negation of an *and* statement is an *or* statement and that the negation of an *or* statement is an *and* statement. This similarity is not accidental. In a sense, universal statements are generalizations of *and* statements and existential statements are generalizations of *or* statements.

If $Q(x)$ is a predicate and the domain D of x is the set $\{x_1, x_2, \ldots, x_n\}$, then the statements

$$\forall x \in D, Q(x)$$

and

$$Q(x_1) \wedge Q(x_2) \wedge \cdots \wedge Q(x_n)$$

are logically equivalent. For example, let $Q(x)$ be "$x \cdot x = x$" and suppose $D = \{0, 1\}$. Then

$$\forall x \in D, Q(x)$$

can be rewritten as

$$\forall \text{ binary digits } x, x \cdot x = x.$$

This is equivalent to

$$0 \cdot 0 = 0 \quad \text{and} \quad 1 \cdot 1 = 1,$$

which can be rewritten symbolically as

$$Q(0) \wedge Q(1).$$

Similarly, if $Q(x)$ is a predicate and $D = \{x_1, x_2, \ldots, x_n\}$, then the statements

$$\exists x \in D \text{ such that } Q(x)$$

and

$$Q(x_1) \vee Q(x_2) \vee \cdots \vee Q(x_n)$$

are logically equivalent. For example, let $Q(x)$ be "$x + x = x$" and suppose $D = \{0, 1\}$. Then

$$\exists x \in D \text{ such that } Q(x)$$

can be rewritten as

$$\exists \text{ a binary digit } x \text{ such that } x + x = x.$$

This is equivalent to

$$0 + 0 = 0 \quad \text{or} \quad 1 + 1 = 1,$$

which can be rewritten symbolically as

$$Q(0) \vee Q(1).$$

VARIANTS OF UNIVERSAL CONDITIONAL STATEMENTS

Recall from Section 1.2 that a conditional statement has a contrapositive, a converse, and an inverse. The definitions of these terms can be extended to universal conditional statements.

DEFINITION

Consider a statement of the form

$$\forall x \in D, \text{ if } P(x) \text{ then } Q(x).$$

1. Its **contrapositive** is the statement

$$\forall x \in D, \text{ if } \sim Q(x) \text{ then } \sim P(x).$$

2. Its **converse** is the statement

$$\forall x \in D, \text{ if } Q(x) \text{ then } P(x).$$

3. Its **inverse** is the statement

$$\forall x \in D, \text{ if } \sim P(x) \text{ then } \sim Q(x).$$

EXAMPLE 2.2.6   Contrapositive, Converse, and Inverse of a Universal Conditional Statement

Write the contrapositive, converse, and inverse for the following statement:

If a real number is greater than 2, then its square is greater than 4.

Solution   The formal version of this statement is $\forall x \in \mathbf{R}$, if $x > 2$ then $x^2 > 4$.

contrapositive: $\forall x \in \mathbf{R}$, if $x^2 \leq 4$ then $x \leq 2$; or,
If the square of a real number is less than or equal to 4, then the number is less than or equal to 2.

converse: $\forall x \in \mathbf{R}$, if $x^2 > 4$ then $x > 2$; or,
If the square of a real number is greater than 4, then the number is greater than 2.

inverse: $\forall x \in \mathbf{R}$, if $x \leq 2$ then $x^2 \leq 4$; or,
If a real number is less than or equal to 2, then the square of the number is less than or equal to 4.

Note that in solving this example we have used the equivalence of "$x \not> a$" and "$x \leq a$" for all real numbers x and a. (See page 11.) ■

In Section 1.2 we showed that a conditional statement is logically equivalent to its contrapositive and that it is not logically equivalent to either its converse or its inverse. The following discussion shows that these facts generalize to the case of universal conditional statements and their contrapositives, converses, and inverses.

Let $P(x)$ and $Q(x)$ be any predicates, let D be the domain of x, and consider the statement

$$\forall x \in D, \text{ if } P(x) \text{ then } Q(x)$$

and its contrapositive

$$\forall x \in D, \text{ if } \sim Q(x) \text{ then } \sim P(x).$$

Any particular x in D that makes "if $P(x)$ then $Q(x)$" true also makes "if $\sim Q(x)$ then $\sim P(x)$" true (by the logical equivalence between $p \rightarrow q$ and $\sim q \rightarrow \sim p$). It follows that the sentence "If $P(x)$ then $Q(x)$" is true for all x in D if, and only if, the sentence "If $\sim Q(x)$ then $\sim P(x)$" is true for all x in D. This is what is meant, in the predicate calculus, for the statements

$$\forall x \in D, \text{ if } P(x) \text{ then } Q(x) \quad \text{and} \quad \forall x \in D, \text{ if } \sim Q(x) \text{ then } \sim P(x)$$

to be logically equivalent to each other. Thus we write the following and say that a universal conditional statement is logically equivalent to its contrapositive:

$$\boxed{\forall x \in D, \text{ if } P(x) \text{ then } Q(x) \equiv \forall x \in D, \text{ if } \sim Q(x) \text{ then } \sim P(x)}$$

In Example 2.2.6 we noted that the statement

$$\forall x \in \mathbf{R}, \text{ if } x > 2 \text{ then } x^2 > 4$$

has the converse

$$\forall x \in \mathbf{R}, \text{ if } x^2 > 4 \text{ then } x > 2.$$

Observe that the statement is true whereas its converse is false (since, for instance, $(-3)^2 = 9 > 4$ but $-3 \not> 2$). This shows that a universal conditional statement may have a different truth value from its converse. Hence a universal conditional statement is not logically equivalent to its converse. This is written symbolically as follows:

$$\boxed{\forall x \in D, \text{ if } P(x) \text{ then } Q(x) \not\equiv \forall x \in D, \text{ if } Q(x) \text{ then } P(x).}$$

In the exercises at the end of this section you are asked to show similarly that a universal conditional statement is not logically equivalent to its inverse.

$$\boxed{\forall x \in D, \text{ if } P(x) \text{ then } Q(x) \not\equiv \forall x \in D, \text{ if } \sim P(x) \text{ then } \sim Q(x).}$$

NECESSARY AND SUFFICIENT CONDITIONS, ONLY IF

The definitions of *necessary, sufficient,* and *only if* can also be extended to apply to universal conditional statements.

DEFINITION

1. "$\forall x$, $r(x)$ is a **sufficient condition** for $s(x)$" means "$\forall x$, if $r(x)$ then $s(x)$."
2. "$\forall x$, $r(x)$ is a **necessary condition** for $s(x)$" means "$\forall x$, if $\sim r(x)$ then $\sim s(x)$" or, equivalently, "$\forall x$, if $s(x)$ then $r(x)$."
3. "$\forall x$, $r(x)$ **only if** $s(x)$" means "$\forall x$, if $\sim s(x)$ then $\sim r(x)$" or, equivalently, "$\forall x$, if $r(x)$ then $s(x)$."

EXAMPLE 2.2.7 Necessary and Sufficient Conditions

Rewrite the following statements as quantified conditional statements. Do not use the words *necessary* or *sufficient*.

a. Squareness is a sufficient condition for rectangularity.
b. Being at least 35 years old is a necessary condition for being President of the United States.

Solution a. $\forall x$, if x is a square, then x is a rectangle.
Or, in informal language:

> If a figure is a square, then it is a rectangle.

b. Using formal language, you could write the answer as

> $\forall$ people x, if x is younger than 35, then x cannot be President of the United States.

Or, by the equivalence between a statement and its contrapositive:

> $\forall$ people x, if x is President of the United States, then x is at least 35 years old. ∎

EXAMPLE 2.2.8 Only If

Rewrite the following as a universal conditional statement:

> A product of two numbers is 0 only if one of the numbers is 0.

Solution Using informal language, you could write the answer as

> If neither of two numbers is 0, then the product of the numbers is not 0.

Or, by the equivalence between a statement and its contrapositive,

> If a product of two numbers is 0, then one of the numbers is 0. ∎

PROLOG

The programming language Prolog (short for *pro*gramming in *log*ic) was developed in France in the 1970s by A. Colmerauer and P. Roussel to serve the needs of programmers working in the field of artificial intelligence. A simple Prolog program consists of a set of statements describing some situation together with questions about the situation. Built into the language are search and inference techniques needed to answer the questions by deriving the answers from the given statements. This frees the programmer from the necessity of having to write separate programs to answer each type of question. Example 2.2.9 gives a very simple example of a Prolog program.

EXAMPLE 2.2.9   A Prolog Program

Consider the following picture, which shows colored blocks stacked on a table.

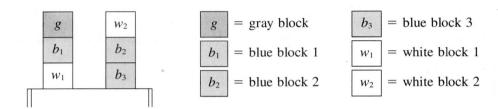

The following are statements in Prolog that describe this picture and ask two questions about it.*

isabove(g, b_1)   color(g, gray)   color(b_3, blue)
isabove(b_1, w_1)   color(b_1, blue)   color(w_1, white)
isabove(w_2, b_2)   color(b_2, blue)   color(w_2, white)
isabove(b_2, b_3)   isabove(X, Z) if isabove(X, Y) and isabove(Y, Z)
?color(b_1, blue)   ?isabove(X, w_1)

The statements "isabove(g, b_1)" and "color(g, gray)" are to be interpreted as "g is above b_1" and "g is colored gray." The statement "isabove(X, Z) if isabove(X, Y) and isabove(Y, Z)" is to be interpreted as "For all X, Y, and Z, if X is above Y and Y is above Z, then X is above Z." The program statement

$$?color(b_1, blue)$$

is a question asking whether block b_1 is colored blue. Prolog answers this by writing

$$Yes.$$

The statement

$$?isabove(X, w_1)$$

is a question asking for which blocks X is the predicate "X is above w_1" true. Prolog answers by giving a list of all such blocks. In this case, the answer is

$$X = b_1, X = g.$$

---

* Different Prolog implementations follow different conventions as to how to represent constant, variable, and predicate names and forms of questions and answers. The conventions used here are similar to those of Edinburgh Prolog.

Note that Prolog can find the solution $X = b_1$ by merely searching the original set of given facts. However, Prolog must *infer* the solution $X = g$ from the following statements:

$$\text{isabove}(g, b_1),$$
$$\text{isabove}(b_1, w_1),$$
$$\text{isabove}(X, Z) \text{ if isabove}(X, Y) \text{ and isabove}(Y, Z).$$

Write the answers Prolog would give if the following questions were added to the program above.

a. ?isabove(b_2, w_1)   b. ?color(w_1, X)   c. ?color(X, blue)

Solution   a. The question means "Is b_2 *above* w_1?"; so the answer is "No."
   b. The question means "For what colors X is the predicate 'w_1 is colored X' true?"; so the answer is "X = white."
   c. The question means "For what blocks is the predicate 'X is colored blue' true?"; so the answer is "$X = b_1$," "$X = b_2$," and "$X = b_3$."   ∎

EXERCISE SET 2.2

1. The following statement is true; "∀ nonzero real numbers x, ∃ a real number y such that $x \cdot y = 1$." For each x given below, find a y to make the predicate "$x \cdot y = 1$" true.
   a. $x = 2$ b. $x = -1$ c. $x = 3/4$

2. The following statement is true: "∀ real numbers x, ∃ an integer n such that $n > x$."* For each x given below, find an n to make the predicate "$n > x$" true.
   a. $x = 15.83$ b. $x = 10^8$ c. $x = 10^{10^{10}}$

In each of 3–8, (a) rewrite the statement in English without using the symbols ∀ or ∃ and expressing your answer as simply as possible, and (b) write a negation for the statement.

3. ∀ colors C, ∃ an animal A such that A is colored C.

4. ∃ a book b such that ∀ people p, p has read b.

5. ∀ odd integers n, ∃ an integer k such that $n = 2k + 1$.

6. ∀$r \in \mathbf{Q}$, ∃ integers a and b such that $r = a/b$.

7. ∀$x \in \mathbf{R}$, ∃ a real number y such that $x + y = 0$.

8. ∃$x \in \mathbf{R}$ such that for all real numbers y, $x + y = 0$.

9. Consider the statement "Everybody is older than somebody." Rewrite this statement in the form "∀ people x, ∃ ____ ."

10. Consider the statement "Somebody is older than everybody." Rewrite this statement in the form "∃ a person x such that ∀ ____ ."

In 11–17, (a) rewrite the statement formally using quantifiers and variables, and (b) write a negation for the statement.

11. Everybody trusts somebody.

12. Somebody trusts everybody.

13. Any even integer equals twice some other integer.

14. The number of rows in any truth table equals 2^n for some integer n.

15. Every action has an equal and opposite reaction.

16. There is a program that gives the correct answer to every question that is posed to it.

17. There is a prime number between every integer and its double.

For each of the statements in 18 and 19, (a) write a new statement by interchanging the symbols ∀ and ∃, and (b) state which is true: the given statement, the version with interchanged quantifiers, neither, or both.

18. ∀$x \in \mathbf{R}$, ∃$y \in \mathbf{R}$ such that $x < y$.

19. ∃$x \in \mathbf{R}$ such that ∀$y \in \mathbf{R}^-$ (the set of negative real numbers), $x > y$.

20. This exercise refers to Example 2.2.4. Determine whether each of the following statements is true or false.
   a. ∀ students S, ∃ a dessert D such that S chose D.
   b. ∀ students S, ∃ a salad T such that S chose T.

*This is called the Archimedean principle because it was first formulated (in geometric terms) by the great Greek mathematician Archimedes of Syracuse, who lived from about 287 to 212 B.C.

c. $\exists$ a dessert D such that $\forall$ students S, S chose D.
d. $\exists$ a beverage B such that $\forall$ students D, D chose B.
e. $\exists$ an item I such that $\forall$ students S, S did not choose I.
f. $\exists$ a station Z such that $\forall$ students S, $\exists$ an item I such that S chose I from Z.

21. How could you determine the truth or falsity of the following statements for the students in your discrete mathematics class? Assume that students will respond truthfully to questions that are asked of them.

   a. There is a student in this class who has dated at least one person from every residence hall at this school.
   b. There is a residence hall at this school with the property that every student in this class has dated at least one person from that residence hall.
   c. Every residence hall at this school has the property that if a student from this class has dated at least one person from that hall, then that student has dated at least two people from that hall.

Give the contrapositive, converse, and inverse of each statement in 22–29.

22. $\forall x \in \mathbf{R}$, if $x > 3$ then $x^2 > 9$.

23. $\forall$ computer programs P, if P is correct then P compiles without error messages.

24. If an integer is divisible by 6, then it is divisible by 3.

25. If the square of an integer is even, then the integer is even.

26. $\forall x \in \mathbf{R}$, if $x(x + 1) > 0$ then $x > 0$ or $x < -1$.

27. $\forall n \in \mathbf{Z}$, if n is prime then n is odd or $n = 2$.

28. $\forall$ integers a, b, and c, if $a - b$ is even and $b - c$ is even, then $a - c$ is even.

29. $\forall$ animals A, if A is a cat then A has whiskers and A has claws.

30. Give an example to show that a universal conditional statement is not logically equivalent to its inverse.

Rewrite each statement of 31–34 in if–then form.

31. Earning a grade of C– in this course is a sufficient condition for it to count toward graduation.

32. Being divisible by 6 is a sufficient condition for being divisible by 3.

33. Being on time each day is a necessary condition for keeping this job.

34. A grade-point average of at least 3.7 is a necessary condition for graduating with honors.

Use the facts that the negation of a $\forall$ statement is a $\exists$ statement and that the negation of an if–then statement is an *and* statement to rewrite each of statements 35–38 without using the words *sufficient* or *necessary*.

35. Divisibility by 4 is not a necessary condition for divisibility by 2.

36. Having a large income is not a necessary condition for a person to be happy.

37. Having a large income is not a sufficient condition for a person to be happy.

38. Being continuous is not a sufficient condition for a function to be differentiable.

39. The following statement is from *An Introduction to Programming*.* Rewrite it without using the words *necessary* or *sufficient*.

   The absence of error messages during translation of a computer program is only a necessary and not a sufficient condition for reasonable [program] correctness.

40. Find the answers Prolog would give if the following questions were added to the program given in Example 2.2.9:
   a. ?isabove(b_1, w_1)   b. ?isabove(w_1, g)
   c. ?color(w_2, blue)   d. ?color(X, white)
   e. ?isabove(X, b_1)   f. ?isabove(X, b_3)
   g. ?isabove(g, X)

41. Write the negation of the definition of limit of a sequence given in Example 2.2.3.

42. The notation $\exists!$ stands for the words "there exists a unique." Thus, for instance, "$\exists! \, x$ such that x is prime and x is even" means that there is one and only one even prime number. Which of the following statements are true and which are false? Explain.
   a. $\exists!$ real number x such that $\forall$ real numbers y, $xy = y$.
   b. $\exists!$ integer x such that $1/x$ is an integer.
   c. $\forall$ real numbers x, $\exists!$ real number y such that $x + y = 0$.

◆ 43. Suppose that $P(x)$ is a predicate and D is the domain of x. Rewrite the statement "$\exists! \, x \in D$ such that $P(x)$" without using the symbol $\exists!$. (See exercise 42 for the meaning of $\exists!$.)

*Richard Conway and David Gries, *An Introduction to Programming*, 2d ed. (Cambridge, Massachusetts: Winthrop, 1975), p. 224.

◆ **44.** Let $P(x)$ and $Q(x)$ be predicates and suppose D is the domain of x. For each pair of statements below, determine whether the statements have the same truth values. Justify your answers.

a. $\forall x \in D, (P(x) \wedge Q(x))$, and
   $(\forall x \in D, P(x)) \wedge (\forall x \in D, Q(x))$
b. $\exists x \in D, (P(x) \wedge Q(x))$, and
   $(\exists x \in D, P(x)) \wedge (\exists x \in D, Q(x))$
c. $\forall x \in D, (P(x) \vee Q(x))$, and
   $(\forall x \in D, P(x)) \vee (\forall x \in D, Q(x))$
d. $\exists x \in D, (P(x) \vee Q(x))$, and
   $(\exists x \in D, P(x)) \vee (\exists x \in D, Q(x))$

2.3 ARGUMENTS WITH QUANTIFIED STATEMENTS

The only complete safe-guard against reasoning ill, is the habit of reasoning well; familiarity with the principles of correct reasoning; and practice in applying those principles. (John Stuart Mill)

The rule of **universal instantiation** (in-stan-she-AY-shun) says that

> If some property is true of *everything* in a domain, then it is true of *any particular* thing in the domain.

Use of the words *universal instantiation* indicates that the truth of a property in a particular case follows as a special instance of its more general or universal truth. The validity of this argument form follows immediately from the definition of truth values for a universal statement. One of the most famous examples of universal instantiation is the following:

> All human beings are mortal.
> Socrates is a human being.
> ∴ Socrates is mortal.

Universal instantiation is *the* fundamental tool of deductive reasoning. Mathematical formulas, definitions, and theorems are like general templates that are used over and over in a wide variety of particular situations. A given theorem says that such and such is true for all things of a certain type. If, in a given situation, you have a particular object of that type, then by universal instantiation, you conclude that such and such is true for that particular object. You may repeat this process 10, 20, or more times in a single proof or problem solution.

As an example of universal instantiation, suppose you are doing a problem that requires you to simplify

$$r^{k+1} \cdot r,$$

where r is a particular real number and k is a particular integer. You know from your study of algebra that the following universal statements are true:

1. For all real numbers x and all integers m and n, $x^m \cdot x^n = x^{m+n}$.
2. For all real numbers x, $x^1 = x$.

So you proceed as follows:

$$r^{k+1} \cdot r = r^{k+1} \cdot r^1 \qquad \text{step 1}$$
$$= r^{(k+1)+1} \qquad \text{step 2}$$
$$= r^{k+2} \qquad \text{by basic algebra.}$$

The reasoning behind step 1 and step 2 is outlined below.

step 1:   For all real numbers x, $x^1 = x$.   universal truth
            r is a particular real number.   particular instance
            $\therefore r^1 = r$.   conclusion

step 2:   For all real numbers x and all integers
            m and n, $x^m \cdot x^n = x^{m+n}$.   universal truth
            r is a particular real number and $k + 1$
            and 1 are particular integers.   particular instance
            $\therefore r^{k+1} \cdot r^1 = r^{(k+1)+1}$.   conclusion

Both arguments are examples of universal instantiation.

UNIVERSAL MODUS PONENS

The rule of universal instantiation can be combined with modus ponens to obtain the rule called *universal modus ponens*.

Universal Modus Ponens

The following argument form is valid:

Formal Version	Informal Version
$\forall x$, if $P(x)$ then $Q(x)$	If x makes $P(x)$ true, then x makes $Q(x)$ true.
$P(a)$ for a particular a	a makes $P(x)$ true.
$\therefore Q(a)$	$\therefore a$ makes $Q(x)$ true.

The argument form above consists of two premises and a conclusion and at least one premise is quantified. An argument of this form is called a **syllogism**. The first and second premises are called its **major** and **minor premises**, respectively.

Note that the major premise of the argument form above could be written "All things that make $P(x)$ true make $Q(x)$ true," in which case the conclusion would follow by universal instantiation alone. However, the if–then form is more natural to use in the majority of mathematical situations.

EXAMPLE 2.3.1   Recognizing Universal Modus Ponens

Rewrite the following argument using quantifiers, variables, and predicate symbols. Is this argument valid? Why?

> If a number is even, then its square is even.
>
> k is a particular number that is even.
>
> $\therefore k^2$ is even.

Solution   The major premise of this argument can be rewritten as

$$\forall x, \text{ if } x \text{ is even then } x^2 \text{ is even.}$$

Let $E(x)$ be "x is even," let $S(x)$ be "x^2 is even," and let k stand for a particular number that is even. Then the argument has the following form:

$$\forall x, \text{ if } E(x) \text{ then } S(x).$$
$$E(k), \text{ for a particular } k.$$
$$\therefore\ S(k).$$

This form of argument is valid by universal modus ponens.   ■

EXAMPLE 2.3.2   Drawing Conclusions Using Universal Modus Ponens

Write the conclusion that can be inferred using universal modus ponens.

> If T is any right triangle with hypotenuse   Pythagorean
> c and legs a and b, then $c^2 = a^2 + b^2$.   theorem
>
> The triangle shown below is a right triangle with both legs equal to 1 and hypotenuse c.

$\therefore$ _____ .

Solution $c^2 = 1^2 + 1^2 = 2$ ■

USE OF UNIVERSAL MODUS PONENS IN A PROOF

In Chapter 3 we discuss methods of proving quantified statements. Here is a proof that the sum of any two even integers is even.

Suppose m and n are particular but arbitrarily chosen even integers. Then $m = 2r$ for some integer r,[1] and $n = 2s$ for some integer s.[2] Hence

$$m + n = 2r + 2s \qquad \text{by substitution}$$
$$= 2(r + s) \text{ [3]} \qquad \text{by factoring out the 2.}$$

Now $r + s$ is an integer,[4] and so $2(r + s)$ is even.[5] Thus $m + n$ is even.

The following expansion of the proof shows how each of the numbered steps is justified by arguments that are valid by universal modus ponens.

(1)   If an integer is even, then it equals twice some integer.
    m is a particular even integer.
$\therefore\ m$ equals twice some integer r.

(2)   If an integer is even, then it equals twice some integer.
    n is a particular even integer.
$\therefore\ n$ equals twice some integer s.

(3)   If a quantity is an integer, then it is a real number.
       r and s are particular integers.
     ∴ r and s are real numbers.
       For all a, b, and c, if a, b, and c are real numbers, then $ab + ac = a(b + c)$.
       $a = 2$, $b = r$, and $c = s$ are particular real numbers.
     ∴ $2r + 2s = 2(r + s)$.

(4)   For all m and n, if m and n are integers then $m + n$ is an integer.
       $m = r$ and $n = s$ are two particular integers.
     ∴ $r + s$ is an integer.

(5)   If a number equals twice some integer, then that number is even.
       $2(r + s)$ equals twice the integer $r + s$.
     ∴ $2(r + s)$ is even.

Of course, the actual proof that the sum of even integers is even does not explicitly contain the sequence of arguments given above. (Heaven forbid!) And, in fact, people who are good at analytical thinking are normally not even conscious that they are reasoning in this way. But that is because they have mastered the method so completely that it has become almost as automatic as breathing.

UNIVERSAL MODUS TOLLENS

Another crucially important rule of inference is *universal modus tollens*. The validity of this argument form results from combining universal instantiation with modus tollens. Universal modus tollens is the heart of proof of contradiction, which is one of the most important methods of mathematical argument.

Universal Modus Tollens

The following argument form is valid:

Formal Version	Informal Version
$\forall x$, if $P(x)$ then $Q(x)$.	If x makes $P(x)$ true, then x makes $Q(x)$ true.
$\sim Q(a)$, for a particular a.	a does not make $Q(x)$ true.
∴ $\sim P(a)$.	∴ a does not make $P(x)$ true.

EXAMPLE 2.3.3   Recognizing the Form of Universal Modus Tollens

Rewrite the following argument using quantifiers, variables, and predicate symbols. Write the major premise in conditional form. Is this argument valid? Why?

> All human beings are mortal.
> Zeus is not mortal.
> ∴ Zeus is not human.

Solution   The major premise can be rewritten as

$$\forall x, \text{ if } x \text{ is human then } x \text{ is mortal.}$$

Let $H(x)$ be "x is human," let $M(x)$ be "x is mortal," and let Z stand for Zeus. The argument becomes

$$\forall x, \text{ if } H(x) \text{ then } M(x)$$
$$\sim M(Z)$$
$$\therefore \sim H(Z).$$

This is valid by universal modus tollens.   ■

EXAMPLE 2.3.4   Drawing Conclusions Using Universal Modus Tollens

Write the conclusion that can be inferred using universal modus tollens.

All professors are absent-minded.

Tom Hutchins is not absent-minded.

$\therefore$ _____ .

Solution   Tom Hutchins is not a professor.   ■

PROVING VALIDITY OF ARGUMENTS WITH QUANTIFIED STATEMENTS

The intuitive definition of validity for arguments with quantified statements is the same as for arguments with compound statements. An argument is valid if, and only if, the truth of its conclusion follows *necessarily* from the truth of its premises. The formal definition is as follows:

DEFINITION

To say that an *argument form* is **valid** means the following: No matter what particular predicates are substituted for the predicate symbols in its premises, if the resulting premise statements are all true, then the conclusion is also true.

An *argument* is called **valid** if, and only if, its form is valid.

As already noted, the validity of universal instantiation follows immediately from the definition of truth values of a universal statement. General formal proofs of validity of arguments in the predicate calculus are beyond the scope of this book. We give the proof of the validity of universal modus ponens as an example to show that such proofs are possible and to give an idea of how they look.

The rule of universal modus ponens says that the following argument form is valid:

$$\forall x, \text{ if } P(x) \text{ then } Q(x)$$
$$P(a) \text{ for a particular } a$$
$$\therefore Q(a)$$

To prove that this is so, suppose the major and minor premises are both true. *[We must show that the conclusion "$Q(a)$" is also true.]* By the minor premise, $P(a)$ is true for a particular value of a. By the major premise and the rule of universal instantiation, the statement "If $P(a)$ then $Q(a)$" is true for that particular a. But by modus ponens, since the statements "If $P(a)$ then $Q(a)$" and "$P(a)$" are both true, it follows that $Q(a)$ is **true** also. *[This is what was to be shown.]*

The proof of validity given above is abstract and somewhat subtle. The proof is not given with the expectation that you will be able to make up such proofs yourself at this stage of your study. Rather it is intended as a glimpse of a more advanced treatment of the subject. One of the paradoxes of the formal study of logic is that the laws of logic are used to prove that the laws of logic are valid!

In the next part of this section we show how you can use diagrams to analyze the validity or invalidity of arguments that contain quantified statements. Diagrams do not provide totally rigorous proofs of validity and invalidity and in some complex settings they may even be confusing, but in many situations they are helpful and convincing.

USING DIAGRAMS TO TEST FOR VALIDITY

Consider the statement

All integers are rational numbers.

Or, formally,

$\forall$ integers n, n is a rational number.

Picture the set of all integers and the set of all rational numbers as disks. The truth of the given statement is represented by placing the integers disk entirely inside the rationals disk, as shown in Figure 2.3.1.

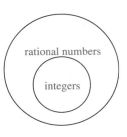

FIGURE 2.3.1

Since the two statements "$\forall x \in D, Q(x)$" and "$\forall x$, if x is in D then $Q(x)$" are logically equivalent, both can be represented by diagrams like the one above.

Perhaps the first person to use diagrams like these to analyze arguments was the German mathematician and philosopher Gottfried Wilhelm Leibniz. Leibniz (LIPE-nits) was far ahead of his time in anticipating modern symbolic logic. He also developed the main ideas of the differential and integral calculus at approximately the same time as (and independently of) Isaac Newton (1642–1727).

To test the validity of an argument diagrammatically, represent the truth of both premises with diagrams. Then analyze the diagrams to see whether they necessarily represent the truth of the conclusion as well.

G. W. Leibniz (1646–1716)

EXAMPLE 2.3.5 Using a Diagram to Show Validity

Use diagrams to show the validity of the following syllogism:

All human beings are mortal.

Zeus is not mortal.

∴ Zeus is not a human being.

Solution The major premise is pictured on the left in Figure 2.3.2 by placing a disk labeled "human beings" inside a disk labeled "mortals." The minor premise is pictured on the right in Figure 2.3.2 by placing a dot labeled "Zeus" outside the disk labeled "mortals."

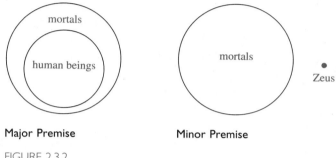

Major Premise   Minor Premise

FIGURE 2.3.2

The two diagrams fit together in only one way, as shown in Figure 2.3.3.

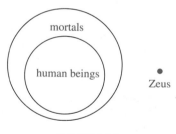

FIGURE 2.3.3

Since the Zeus dot is outside the mortals disk, it is necessarily outside the human beings disk. Thus the truth of the conclusion follows necessarily from the truth of the premises. It is impossible for the premises of this argument to be true and the conclusion false; hence the argument is valid. ∎

EXAMPLE 2.3.6 Using Diagrams to Show *Invalidity*

Use a diagram to show the invalidity of the following argument:

All human beings are mortal.

Felix is mortal.

∴ Felix is a human being.

Solution   The major and minor premises are represented diagrammatically in Figure 2.3.4.

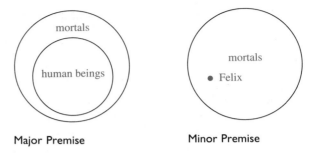

Major Premise   Minor Premise

FIGURE 2.3.4

All that is known is that the Felix dot is located *somewhere* inside the mortals disk. Where it is located with respect to the human beings disk cannot be determined. Either one of the situations shown in Figure 2.3.5 might be the case.

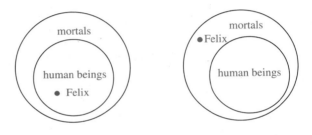

FIGURE 2.3.5

The conclusion "Felix is a human being" is true in the first case but not in the second (Felix might, for example, be a cat). Because the conclusion does not necessarily follow from the premises, the argument is invalid.   ■

The argument of Example 2.3.6 would be valid if the major premise were replaced by its converse. But since a universal conditional statement is not logically equivalent to its converse, such a replacement cannot, in general, be made. This argument exhibits the converse error.

Converse Error (Quantified Form)

The following argument form is *invalid:*

Formal Version	Informal Version
$\forall x$, if $P(x)$ then $Q(x)$.	If x makes $P(x)$ true, then x makes $Q(x)$ true.
$Q(a)$ for a particular a.	a makes $Q(x)$ true.
$\therefore P(a).$ ← invalid conclusion	$\therefore$ a makes $P(x)$ true. ← invalid conclusion

---

The following form of argument would be valid if a conditional statement were logically equivalent to its inverse. But it is not, and the argument form is invalid. It exhibits the inverse error. You are asked to show the invalidity of this argument form in the exercises at the end of this section.

Inverse Error (Quantified Form)

The following argument form is *invalid:*

Formal Version	Informal Version
$\forall x$, if $P(x)$ then $Q(x)$.	If x makes $P(x)$ true, then x makes $Q(x)$ true.
$\sim P(a)$, for a particular a.	a does not make $P(x)$ true.
$\therefore \sim Q(a)$. $\leftarrow$ invalid conclusion	$\therefore a$ does not make $Q(x)$ true. $\leftarrow$ invalid conclusion

---

EXAMPLE 2.3.7   An Argument with "No"

Use diagrams to test the following argument for validity:

> No polynomial functions have horizontal asymptotes.
> This function has a horizontal asymptote.
> $\therefore$ This function is not a polynomial.

Solution   A good way to represent the major premise diagrammatically is shown in Figure 2.3.6, two disks—a disk for polynomial functions and a disk for functions with horizontal asymptotes—that do not overlap at all. The minor premise is represented by placing a dot labeled "this function" inside the disk for functions with horizontal asymptotes.

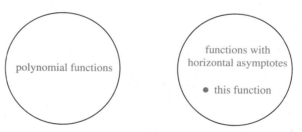

FIGURE 2.3.6

The diagram shows that "this function" must lie outside the polynomial functions disk, and so the truth of the conclusion necessarily follows from the truth of the premises. Hence the argument is valid. ∎

An alternative approach to this example is to transform the statement "No polynomial functions have horizontal asymptotes" into the equivalent form "$\forall x$, if x is a polynomial function, then x does not have a horizontal asymptote." If this is done, the argument can be seen to have the form

$$\forall x, \text{ if } P(x) \text{ then } Q(x);$$
$$\sim Q(a), \text{ for a particular } a;$$
$$\therefore \sim P(a);$$

where $P(x)$ is "x is a polynomial function" and $Q(x)$ is "x does not have a horizontal asymptote." This is valid by universal modus tollens.

REMARK ON THE CONVERSE AND INVERSE ERRORS

One reason so many people make converse and inverse errors is that the forms of the resulting arguments would be valid if the major premise were a biconditional rather than a simple conditional. And, as remarked in Section 1.2, many people tend to confuse biconditionals and conditionals.

Consider, for example, the following argument:

> All the town criminals frequent the Den of Iniquity bar.
>
> John frequents the Den of Iniquity bar.
>
> ∴ John is one of the town criminals.

The conclusion of this argument is invalid—it results from making the converse error. Therefore, it may be false even when the premises of the argument are true. This type of argument attempts unfairly to estabish guilt by association.

The closer, however, the major premise comes to being a biconditional, the more likely the conclusion is to be true. If hardly anyone but criminals frequents the bar and John also frequents the bar, then it is likely (though not certain) that John is a criminal. On the basis of the given premises, it might be sensible to be suspicious of John, but it would be wrong to convict him.

A variation of the converse error is, in fact, a very useful reasoning tool provided it is used with caution. It is the type of reasoning that is used by doctors to make medical diagnoses and by auto mechanics to repair cars. It is the type of reasoning used to generate explanations for phenomena. It goes like this: If a statement of the form

> For all x, if $P(x)$ then $Q(x)$

is true, and if

> $Q(a)$ is true, for a particular a,

then check out the statement $P(a)$; it just might be true. For instance, suppose a doctor knows that

> For all x, if x has pneumonia, then x has a fever and chills, coughs deeply, and feels exceptionally tired and miserable.

And suppose the doctor also knows that

> John has a fever and chills, coughs deeply, and feels exceptionally tired and miserable.

On the basis of this data, the doctor concludes that a diagnosis of pneumonia is a strong possibility, though not a certainty. The doctor will probably attempt to gain further support for this diagnosis through laboratory testing that is specifically designed to detect pneumonia. Note that the closer a set of symptoms comes to being a necessary and sufficient condition for an illness, the more certain the doctor can be of his or her diagnosis.

This form of reasoning has been named **abduction** by researchers working in artificial intelligence. It is used in certain computer programs called expert systems that attempt to duplicate the functioning of an expert in some field of knowledge.

EXERCISE SET 2.3

1. Let the following law of algebra be the first statement of an argument:

$$\text{For all real numbers } a \text{ and } b,$$
$$(a + b)^2 = a^2 + 2ab + b^2.$$

Suppose each of the following statements is, in turn, the second statement of the argument. Use universal instantiation or universal modus ponens to write the conclusion that follows.

a. $a = x$ and $b = y$ are particular real numbers.
b. $a = f_i$ and $b = f_j$ are particular real numbers.
c. $a = 3u$ and $b = 5v$ are particular real numbers.
d. $a = g(r)$ and $b = g(s)$ are particular real numbers.
e. $a = \log(t_1)$ and $b = \log(t_2)$ are particular real numbers.

Use universal instantiation or universal modus ponens to fill in valid conclusions for the arguments in 2–4.

2. If an integer n equals $2 \cdot k$ and k is an integer, then n is even.

    0 equals $2 \cdot 0$ and 0 is an integer.

    ∴ _____ .

3. For all real numbers a, b, c, and d, if $b \neq 0$ and $d \neq 0$, then $a/b + c/d = (ad + bc)/bd$.

    $a = 2$, $b = 3$, $c = 4$, and $d = 5$ are particular real numbers such that $b \neq 0$ and $d \neq 0$.

    ∴ _____ .

4. $\forall$ real numbers r, a, and b, if r is positive, then $(r^a)^b = r^{ab}$.

    $r = 5$, $a = 1/2$, and $b = 4$ are particular real numbers such that r is positive.

    ∴ _____ .

Use universal modus tollens to fill in valid conclusions for the arguments in 5 and 6.

5. All healthy people eat an apple a day.
    Harry does not eat an apple a day.

    ∴ _____ .

6. If a computer program is correct, then compilation of the program does not produce error messages.
    Compilation of this program produces error messages.

    ∴ _____ .

Some of the arguments in 7–18 are valid by universal modus ponens or universal modus tollens; others are invalid and exhibit the converse or the inverse error. State which are valid and which are invalid. Justify your answers.

7. All healthy people eat an apple a day.
    Helen eats an apple a day.
    ∴ Helen is a healthy person.

8. All freshmen must take writing.
    Caroline is a freshman.
    ∴ Caroline must take writing.

9. All healthy people eat an apple a day.
    Herbert is not a healthy person.
    ∴ Herbert does not eat an apple a day.

10. If a product of two numbers is 0, then at least one of the numbers is 0.
    For a particular number x, neither $(x - 1)$ nor $(x + 1)$ equals 0.
    ∴ The product $(x - 1)(x + 1)$ is not 0.

11. All cheaters sit in the back row.
    George sits in the back row.
    ∴ George is a cheater.

12. All honest people pay their taxes.
    Darth is not honest.
    ∴ Darth does not pay his taxes.

13. For all students x, if x studies discrete mathematics, then x is good at logic.
    Dawn studies discrete mathematics.
    ∴ Dawn is good at logic.

14. If compilation of a computer program produces error messages, then the program is not correct.
    Compilation of this program does not produce error messages.
    ∴ This program is correct.

15. Any sum of two rational numbers is rational.
    The sum $r + s$ is rational.
    ∴ The numbers r and s are both rational.

16. If a number is even, then twice that number is even.
    The number $2n$ is even, for a particular number n.
    ∴ n is even.

17.   If an infinite series converges, then the terms go to 0.

The terms of the infinite series $\sum_{n=1}^{\infty} \dfrac{1}{n}$ go to 0.

∴ The infinite series $\sum_{n=1}^{\infty} \dfrac{1}{n}$ converges.

18.   If an infinite series converges, then its terms go to 0.

The terms of the infinite series $\sum_{n=1}^{\infty} \dfrac{n}{n+1}$ do not go to 0.

∴ The infinite series $\sum_{n=1}^{\infty} \dfrac{n}{n+1}$ does not converge.

19. Rewrite the statement "No good cars are cheap" in the form "$\forall x$, if $P(x)$ then $\sim Q(x)$." Indicate whether each of the following arguments is valid or invalid, and justify your answers.

a.   No good car is cheap.
      A Rimbaud is a good car.
   ∴ A Rimbaud is not cheap.

b.   No good car is cheap.
      A Simbaru is not cheap.
   ∴ A Simbaru is a good car.

c.   No good car is cheap.
      A VX Roadster is cheap.
   ∴ A VX Roadster is not good.

d.   No good car is cheap.
      An Omnex is not a good car.
   ∴ An Omnex is cheap.

20. a. Use a diagram to show that the following argument can have true premises and a false conclusion.

All dogs are carnivorous.
Felix is not a dog.
∴ Felix is not carnivorous.

b. What can you conclude about the validity or invalidity of the following argument form?

$\forall x$, if $P(x)$ then $Q(x)$
$\sim P(a)$ for a particular a
∴ $\sim Q(a)$

Indicate whether the arguments in 21–26 are valid or invalid. Support your answers by drawing diagrams.

21.   All people are mice.
      All mice are mortal.
   ∴ All people are mortal.

22.   All discrete mathematics students can tell a valid argument from an invalid one.
      All thoughtful people can tell a valid argument from an invalid one.
   ∴ All discrete mathematics students are thoughtful.

23.   All teachers occasionally make mistakes.
      No gods ever make mistakes.
   ∴ No teachers are gods.

24.   No college cafeteria food is good.
      No good food is wasted.
   ∴ No college cafeteria food is wasted.

25.   All polynomial functions are differentiable.
      All differentiable functions are continuous.
   ∴ All polynomial functions are continuous.

26.   [Adapted from Lewis Carroll.]
      Nothing intelligible ever puzzles *me*.
      Logic puzzles me.
   ∴ Logic is unintelligible.

Exercises 27–28 are adapted from *Symbolic Logic* by Lewis Carroll.* Reorder the premises in each of the arguments to make it clear that the conclusion follows logically. (It may be helpful to rewrite some of the statements in if–then form and to replace some statements by their contrapositives.)

27.   1. I trust every animal that belongs to me.
      2. Dogs gnaw bones.
      3. I admit no animals into my study unless they will beg when told to do so.
      4. All the animals in the yard are mine.
      5. I admit every animal that I trust into my study.
      6. The only animals that are really willing to beg when told to do so are dogs.
   ∴ All the animals in the yard gnaw bones.

28.   1. When I work a logic example without grumbling, you may be sure it is one I understand.
      2. The arguments in these examples are not arranged in regular order like the ones I am used to.
      3. No easy examples make my head ache.
      4. I can't understand examples if the arguments are not arranged in regular order like the ones I am used to.
      5. I never grumble at an example unless it gives me a headache.
   ∴ These examples are not easy.

*Lewis Carroll, *Symbolic Logic* (New York: Dover, 1958), pp. 118, 120, 123.

In Exercises 29 and 30, a conclusion follows from the given premises but it is difficult to see because the premises are jumbled up. Reorder the premises to make it clear that a conclusion follows logically, and state the valid conclusion that can be drawn. (It may be helpful to rewrite some of the statements in if–then form and to replace some statements by their contrapositives.)

29. 1. No birds, except ostriches, are nine feet high.
    2. There are no birds in this aviary that belong to anyone but me.
    3. No ostrich lives on mince pies.
    4. I have no birds less than nine feet high.

30. 1. All writers who understand human nature are clever.
    2. No one is a true poet unless he can stir the hearts of men.
    3. Shakespeare wrote *Hamlet*.
    4. No writer who does not understand human nature can stir the hearts of men.
    5. None but a true poet could have written *Hamlet*.

◆ 31. Derive the rule of universal modus tollens from the rule of universal instantiation and the rule of modus tollens.

3 Elementary Number Theory and Methods of Proof

The underlying content of this chapter is likely to be familiar to you. It consists of properties of integers (whole numbers), rational numbers (integer fractions), and real numbers. The underlying theme of this chapter is the question of how to determine the truth or falsity of a mathematical statement.

Here is an example involving a concept used frequently in computer science. Given any real number x, the floor of x, or greatest integer in x, denoted $\lfloor x \rfloor$, is the largest integer that is less than or equal to x. On the number line, $\lfloor x \rfloor$ is the integer immediately to the left of x (or equal to x if x is, itself, an integer). Thus $\lfloor 2.3 \rfloor = 2$, $\lfloor 12.99999 \rfloor = 12$, and $\lfloor -1.5 \rfloor = -2$. Consider the following two questions:

1. For any real number x, is $\lfloor x - 1 \rfloor = \lfloor x \rfloor - 1$?
2. For any real numbers x and y, is $\lfloor x - y \rfloor = \lfloor x \rfloor - \lfloor y \rfloor$?

Take a few minutes to try to answer these questions for yourself.

It turns out that the answer to (1) is yes, whereas the answer to (2) is no. Are these the answers you got? If not, don't worry. In Section 3.5 you will learn the techniques you need to answer these questions and more. If you did get the correct answers, congratulations! You have excellent mathematical intuition. Now ask yourself, "How sure am I of my answers? Were they plausible guesses or absolute certainties? Was there any difference in certainty between my answers to (1) and (2)? Would 1 have been willing to bet a large sum of money on the correctness of my answers?"

One of the best ways to think of a mathematical proof is as a carefully reasoned argument to convince a skeptical listener (often yourself) that a given statement is true. Imagine the listener challenging your reasoning every step of the way, constantly asking, "Why is that so?" If you can counter every possible challenge, then your proof as a whole will be correct.

As an example, imagine proving to someone not very familiar with mathematical notation that if x is a number with $5x + 3 = 33$, then $x = 6$. You could argue as follows:

If $5x + 3 = 33$, then $5x + 3$ minus 3 will equal $33 - 3$ since subtracting the same number from two equal quantities gives equal results. But $5x + 3$ minus 3 equals $5x$ because adding 3 to $5x$ and then subtracting 3 just leaves $5x$. Also

$33 - 3 = 30$. Hence, $5x = 30$. This means that x is a number which when multiplied by 5 equals 30. But the only number with this property is 6. Therefore, $x = 6$.

There are, of course, other ways of phrasing this proof depending on the level of mathematical sophistication of the intended reader. In practice, mathematicians often omit reasons for certain steps of an argument when they are confident the reader can easily supply them. When you are first learning to write proofs, however, it is better to err on the side of supplying too many reasons rather than too few. All too frequently, when even the best mathematicians carefully examine some "details" in their arguments, they discover that those details are actually false. Probably the most important reason for requiring proof in mathematics is that writing a proof forces us to become aware of weaknesses in our arguments and in the unconscious assumptions we have made.

Sometimes correctness of a mathematical argument can be a matter of life or death. Suppose, for example, that a mathematician is part of a team charged with designing a new type of airplane engine, and suppose that the mathematician is given the job of determining whether the thrust delivered by various engine types is adequate. If you knew that the mathematician was only fairly sure but not positive of the correctness of his analysis, you would probably not want to ride in the resulting aircraft.

At a certain point in Lewis Carroll's *Alice in Wonderland* (see exercise 24 in Section 1.2), the March Hare tells Alice to "say what you mean." In other words, she should be precise in her use of language: If she means a thing, then that is exactly what she should say. In this chapter, perhaps more than in any other mathematics course you have ever taken, you will find it necessary to say what you mean. Precision of thought and language is essential to achieve the mathematical certainty that is needed if you are to have complete confidence in your solutions to mathematical problems.

3.1 DIRECT PROOF AND COUNTEREXAMPLE I: INTRODUCTION

Mathematics, as a science, commenced when first someone, probably a Greek, proved propositions about "any" things or about "some" things without specification of definite particular things. (Alfred North Whitehead, 1861–1947)

Both discovery and proof are integral parts of problem solving. When you think you have discovered a certain statement is true, try to figure out why it is true. If you succeed, you will know that your discovery is genuine. Even if you fail, the process of trying will give you insight into the nature of the problem and may lead to the discovery that the statement is false. For complex problems, the interplay between discovery and proof is not reserved to the end of the problem-solving process but is an important part of each step.

In this text we assume a familiarity with the laws of basic algebra, which are listed in Appendix A. We also use the fact that the set of integers is closed under addition, subtraction, and multiplication. This means that sums, differences, and products of integers are integers. Of course, most quotients of integers are not integers. For example, $3 \div 2$, which equals $3/2$, is not an integer, and $3 \div 0$ is not even a number.

The mathematical content of this section primarily concerns even and odd integers and prime and composite numbers.

DEFINITIONS

In order to evaluate the truth or falsity of a statement, you must understand what the statement is about. In other words, you must know the meanings of all terms that occur in the statement. Mathematicians define terms very carefully and precisely and consider it important to learn definitions virtually word for word.

DEFINITION

An integer n is **even** if, and only if, $n = 2k$ for some integer k. An integer n is **odd** if, and only if, $n = 2k + 1$ for some integer k.

Symbolically, if n is an integer, then

$$n \text{ is even} \iff \exists \text{ an integer } k \text{ such that } n = 2k.$$
$$n \text{ is odd} \iff \exists \text{ an integer } k \text{ such that } n = 2k + 1.$$

It follows from the definition that if you are doing a problem in which you happen to know a certain integer is even, you can deduce that it has the form $2k$ for some integer k. Conversely, if you know in some situation that a particular integer equals $2 \cdot$ (some integer), then you can deduce that the integer is even.

Know a particular integer n is even. $\xrightarrow{\text{deduce}}$ n has the form $2k$ for some integer k.

Know n has the form $2k$ for some integer k. $\xrightarrow{\text{deduce}}$ n is even.

EXAMPLE 3.1.1   Even and Odd Integers

a. Is 0 even?
b. Is -301 odd? If so, write it as $2k + 1$ for some integer k.
c. If a and b are integers, is $6a^2b$ even? Why?
d. If a and b are integers, is $10a + 8b + 1$ odd? Why?
e. Is every integer either even or odd?

Solution

a. Yes, $0 = 2 \cdot 0$.
b. Yes, $-301 = 2(-151) + 1$.
c. Yes, $6a^2b = 2(3a^2b)$, and since a and b are integers, so is $3a^2b$ (being a product of integers).
d. Yes, $10a + 8b + 1 = 2(5a + 4b) + 1$, and since a and b are integers, so is $5a + 4b$ (being a sum of products of integers).
e. The answer is yes although the proof is not obvious. (Try giving a reason yourself.) We will show in Section 3.4 that this fact results from another fact known as the quotient-remainder theorem. ∎

The integer 6, which equals $2 \cdot 3$, is a product of two smaller positive integers. On the other hand, 7 cannot be written as a product of two smaller positive integers; its only positive factors are 1 and 7. A positive integer, such as 7, that cannot be written as a product of two smaller positive integers is called *prime*.

> ### DEFINITION
>
> An integer n is **prime** if, and only if, $n > 1$ and for all positive integers r and s, if $n = r \cdot s$, then $r = 1$ or $s = 1$. An integer n is **composite** if, and only if, $n = r \cdot s$ for some positive integers r and s with $r \neq 1$ and $s \neq 1$.
>   Symbolically, if n is an integer that is greater than 1, then
>
> $$n \text{ is prime} \quad \Leftrightarrow \quad \forall \text{ positive integers } r \text{ and } s, \text{ if } n = r \cdot s$$
> $$\text{then } r = 1 \text{ or } s = 1.$$
>
> $$n \text{ is composite} \quad \Leftrightarrow \quad \exists \text{ positive integers } r \text{ and } s \text{ such that } n = r \cdot s$$
> $$\text{and } r \neq 1 \text{ and } s \neq 1.$$

EXAMPLE 3.1.2   Prime and Composite Numbers

a. Is 1 prime?
b. Is it true that every integer greater than 1 is either prime or composite?
c. Write the first six prime numbers.
d. Write the first six composite numbers.

Solution

a. No, a prime number is required to be greater than 1.
b. Yes, the two definitions are negations of each other.
c. 2, 3, 5, 7, 11, 13
d. 4, 6, 8, 9, 10, 12   ■

PROVING EXISTENTIAL STATEMENTS

According to the definition given in Section 2.1, a statement in the form

$$\exists x \in D \text{ such that } Q(x)$$

is true if, and only if,

$$Q(x) \text{ is true for at least one } x \text{ in } D.$$

One way to prove this is to find an x in D that makes $Q(x)$ true. Another way is to give a set of directions for finding such an x. Both of these methods are called **constructive proofs of existence**.

EXAMPLE 3.1.3   Constructive Proofs of Existence

a. Prove the following: $\exists$ an even integer n that can be written in two ways as a sum of two prime numbers.
b. Suppose that r and s are integers. Prove the following: $\exists$ an integer k such that $22r + 18s = 2k$.

Solution

a. Let $n = 10$. Then $10 = 5 + 5 = 3 + 7$ and 3, 5, and 7 are all prime numbers.
b. Let $k = 11r + 9s$. Then k is an integer because it is a sum of products of integers; and by substitution, $2k = 2(11r + 9s)$, which equals $22r + 18s$ by the distributive law of algebra.   ■

A **nonconstructive proof of existence** involves showing either (a) that the existence of a value of x that makes $Q(x)$ true is guaranteed by an axiom or a previously proved theorem or (b) that the assumption that there is no such x leads to a contradiction. The disadvantage of a nonconstructive proof is that it may give virtually no clue about where or how x may be found. The widespread use of digital computers in recent years has led to some dissatisfaction with this aspect of nonconstructive proofs and to increased efforts to produce constructive proofs containing directions for computer calculation of the quantity in question.

PROVING UNIVERSAL STATEMENTS

The vast majority of mathematical statements to be proved are universal. In discussing how to prove such statements, it is helpful to imagine them in a standard form:

$$\forall x \in D, \text{ if } P(x) \text{ then } Q(x).$$

In Section 2.1 we showed that any universal statement can be written in this form, and that when D is finite such a statement can be proved by the method of exhaustion. This method can also be used when there are only a finite number of elements that satisfy the condition $P(x)$.

EXAMPLE 3.1.4   **The Method of Exhaustion**

Use the method of exhaustion to prove the following statement:

$\forall n \in \mathbf{Z}$, if n is even and $4 \leq n \leq 30$, then n can be written as a sum of two prime numbers.

Solution

$$
\begin{array}{llll}
4 = 2 + 2 & 6 = 3 + 3 & 8 = 3 + 5 & 10 = 5 + 5 \\
12 = 5 + 7 & 14 = 11 + 3 & 16 = 5 + 11 & 18 = 7 + 11 \\
20 = 7 + 13 & 22 = 5 + 17 & 24 = 5 + 19 & 26 = 7 + 19 \\
28 = 11 + 17 & 30 = 11 + 19 & &
\end{array}
$$

∎

In most cases in mathematics, however, the method of exhaustion cannot be used. For instance, can you prove by exhaustion that every even integer greater than 4 can be written as a sum of two prime numbers? No. To do that you would have to check every even integer, and since there are infinitely many such numbers, this is an impossible task.

Even when the domain is finite it may be infeasible to use the method of exhaustion. Imagine, for example, trying to check by exhaustion that the multiplication circuitry of a particular computer gives the correct result for every pair of numbers in the computer's range. Since a typical computer would require thousands of years just to compute all possible products of all numbers in its range (not to mention the time it would take to check the accuracy of the answers), checking correctness by the method of exhaustion is obviously impractical.

The most powerful technique for proving a universal statement is one that works regardless of the size of the domain over which the statement is quantified. It is called the **method of generalizing from the generic particular**. This is the idea underlying the method:

> To show that every element of a domain satisfies a certain property, suppose x is a *particular* but *arbitrarily chosen* element of the domain and show that x satisfies the property.

Now suppose the property has the form "If $P(x)$ then $Q(x)$." How can you show that x satisfies the property? Recall that the only way "If $P(x)$ then $Q(x)$" can be false is for $P(x)$ to be true and $Q(x)$ to be false. Thus to show that "If $P(x)$ then $Q(x)$" is true, suppose $P(x)$ is true and show that $Q(x)$ must also be true. It follows that to prove a statement of the form "$\forall x \in D$, if $P(x)$ then $Q(x)$," you suppose x is a particular but arbitrarily chosen element of D that satisfies $P(x)$, and then you show that x satisfies $Q(x)$. This is called the method of *direct proof*.

Method of Direct Proof

1. Express the statement to be proved in the form "$\forall x \in D$, if $P(x)$ then $Q(x)$." (This step is often done mentally.)

2. Start the proof by supposing x is a particular but arbitrarily chosen element of D for which the hypothesis $P(x)$ is true. (This step is often abbreviated "Suppose $x \in D$ and $P(x)$.")

3. Show that the conclusion $Q(x)$ is true by using definitions, previously established results, and the rules for logical inference.

The point of having x be arbitrarily chosen (or generic) is to make a proof that can be generalized to all elements of D. By choosing x arbitrarily, you are making no special assumptions about x that are not also true of all other elements of D. The word *generic* means "sharing all the common characteristics of a group or class." Thus everything you deduce about a generic element x of D is equally true of any other element of D.

EXAMPLE 3.1.5   Proving a Theorem

Prove that if the sum of any two integers is even then so is their difference.

▲ CAUTION!   *The word two in this statement does not necessarily refer to two distinct integers. If a choice of integers is made arbitrarily, the integers will very likely be distinct but they might be the same.*

Solution   Whenever you are presented with a statement to be proved, explore it a bit to see whether you believe it to be true. In this case, you might imagine some pairs of integers whose sum is even and then check that their differences are also even. For instance, $8 + 4 = 12$, which is even, and $8 - 4 = 4$ is also even; $21 + 13 = 34$ is even and $21 - 13 = 8$ is also even. However, since you cannot possibly check all such pairs, you cannot know for sure that the statement is true in general by checking its truth in these particular instances. Many properties hold for a large number of examples and yet fail to be true in general. After all, the formal version of the statement to be proved is universally quantified over an infinite domain:

Formal Restatement: $\forall$ integers m and n, if $m + n$ is even then $m - n$ is even.

So to prove this statement in general, you need to show that no matter what two integers you start with, if their sum is even then so is their difference. Ask yourself, "Where am I starting from?" or "What am I supposing?"

Starting Point: Suppose m and n are particular but arbitrarily chosen integers such that $m + n$ is even. Or, in abbreviated form,

> Suppose m and n are any integers and $m + n$ is even.

Then ask yourself, "What conclusion do I need to show?"

To Show: $m - n$ is even.

Now ask yourself, "How do I get from the starting point to the conclusion?" Since both involve the term *even integer,* you must know what it means for an integer to be even. According to the definition, any even integer can be written in the form $2 \cdot$ (some integer). So

$$m + n = 2k \text{ for some integer } k.$$

Consider $m - n$. Can you conclude that $m - n$ is even? In other words, can $m - n$ be written as

$$2 \cdot \text{(some integer)?}$$

To see whether this is true, be prepared to play around with the hypothesis

$$m + n = 2k \text{ for some integer } k.$$

If you subtract n from both sides, you obtain

$$m = 2k - n,$$

which you can then substitute into the expression $m - n$:

$$m - n = (2k - n) - n) = 2k - 2n = 2(k - n).$$

But the right side of the string of equalities is $2 \cdot$ (something). Is that something an integer? Of course! Because both k and n are integers and the difference of two integers is an integer. ∎

This discussion is summarized by rewriting the statement as a theorem and giving a formal proof of it. (In mathematics, the word *theorem* refers to a statement that is known to be true.) The formal proof, as well as many others in this text, includes explanatory notes to make its logical flow apparent. Such comments are purely a convenience for the reader and could be omitted entirely. For this reason they are in italic and enclosed in square brackets: [ ].

Donald Knuth, one of the pioneers of the science of computing, has compared constructing a computer program from a set of specifications to writing a mathematical proof based on a set of axioms.* In keeping with this analogy, the bracketed comments can be thought of as similar to the explanatory documentation provided by a good programmer. Documentation is not necessary for a program to run, but it helps a human reader understand what is going on.

Most theorems, like the one above, can be analyzed to a point where you realize

---

*Donald E. Knuth, *The Art of Computer Programming,* 2d ed., vol. I (Reading, Massachusetts: Addison-Wesley, 1973), p. ix.

THEOREM 3.1.1

If the sum of any two integers is even, then so is their difference.

Proof:

Suppose m and n are *[particular but arbitrarily chosen]* integers so that $m + n$ is even. *[We must show that $m - n$ is even.]* By definition of even, $m + n = 2k$ for some integer k. Subtracting n from both sides gives $m = 2k - n$. So

$$m - n = (2k - n) - n \quad \text{by substitution}$$
$$= 2k - 2n \quad \text{by combining like terms (basic algebra)}$$
$$= 2(k - n) \quad \text{by factoring out a 2 (basic algebra).}$$

But $k - n$ is an integer because it is a difference of integers. Hence $m - n$ equals 2 times an integer, and so by definition of even, $m - n$ is even. *[This is what we needed to show.]*

---

that as soon as a certain thing is shown, the theorem will be proved. When that thing has been shown, it is natural to end the proof with the words "this is what we needed to show." The Latin words are *quod erat demonstrandum,* or Q.E.D. for short. Proofs in older mathematics books end with these initials.

Note that both the *if* and the *only if* parts of the definition of even were used in the proof of Theorem 3.1.1. Since $m + n$ was known to be even, the *only if* ($\Rightarrow$) part of the definition was used to deduce that $m + n$ had a certain general form. Then after some algebraic substitution and manipulation, the *if* ($\Leftarrow$) part of the definition was used to deduce that $m - n$ was even.

An alternate proof of the theorem in Example 3.1.5 uses properties of even and odd integers that you are asked to establish in the exercises, namely that a sum or difference of two even integers is even, a sum or difference of two odd integers is also even, and a sum or difference of one odd and one even integer is odd. It follows that if a sum of two integers is even, then either both integers are odd or both are even, and in either case their difference is even.

DIRECTIONS FOR WRITING PROOFS OF UNIVERSAL STATEMENTS

Over the years the following rules of style have become fairly standard for writing the final versions of proofs:

1. **Write the theorem to be proved.**
2. **Clearly mark the beginning of your proof with the word *Proof.***
3. **Make your proof self-contained.**

   This means that you should identify each variable used in your proof in the body of the proof. Thus you will begin proofs by introducing the initial variables and stating what kind of objects they are. The first sentence of your proof would be something like "Suppose m and n are integers" or "Let x be a real number that is greater than 2." This is similar to declaring variables and their data types at the beginning of a computer program.

At a later point in your proof you may introduce a new variable to help explain something. For example, knowing that a particular integer n is even, you may want to write $n = 2s$. But when you do this you must specify that s is an *integer*. Thus you will say, "Since n is even, $n = 2s$ for some integer s."

4. **Write proofs in complete English sentences.**

This does not mean that you should avoid using symbols and shorthand abbreviations, just that you should incorporate them into sentences. For example, the proof of Theorem 3.1.1 contains the sentence

$$\begin{aligned} \text{Then } m + n &= (2k - n) - n \\ &= 2k - 2n \\ &= 2(k - n) \end{aligned}$$

To read this as an English sentence, read the first equal sign as "equals" and each subsequent equal sign as "which equals."

It is rare that two proofs of a given statement, written by two different people, are identical. Even when the basic mathematical steps are the same, the two people may use different notation or may give differing amounts of explanation for their steps, or may choose different words to link the steps together into paragraph form. An important question is how detailed to make the explanations for the steps of a proof. This must ultimately be worked out between the writer of a proof and the intended reader, whether they be student and teacher, teacher and student, student and fellow student, or mathematician and colleague. Your teacher may provide explicit guidelines for you to use in your course. Or you may follow the example of the proofs in this book (which are generally explained rather fully in order to be understood by students at various stages of mathematical development). Remember that the phrases written inside brackets [ ] are intended to elucidate the logical flow or underlying assumptions of the proof and need not be written down at all. It is entirely your decision whether or not to include such phrases in your own proofs.

COMMON MISTAKES

The following are some of the most common mistakes people make when writing mathematical proofs.

1. **Arguing from examples.**

Looking at examples is one of the most helpful practices a problem solver can engage in and is encouraged by all good mathematics teachers. However, it is a mistake to think that a general statement can be proved by showing it to be true for some special cases. A universal statement may be true in many instances without being true in general.

Here is an example of this mistake. It is an incorrect "proof" of the fact that if the sum of any two integers is even, then so is their difference (Theorem 3.1.1).

This is true because if $m = 14$ and $n = 6$, then $m + n = 20$, which is even, and $m - n = 8$, which is also even.

Some people find this kind of argument convincing because it does, after all, consist of evidence in support of a true conclusion. But remember that when we discussed valid arguments we pointed out that an argument may be invalid and yet have a true conclusion. In the same way, an argument from examples may be mistakenly used to "prove" a true statement. In the example above, it is not sufficient to show that the conclusion "$m - n$ is even" is true for $m = 14$ and $n = 6$. You must give an argument to show that the conclusion is true for any integers m and n.

2. **Using the same letter to mean two different things.**

   Some beginning theorem provers give a new variable quantity the same letter name as a previously introduced variable. Consider the following proof fragment:

   > Suppose m and n are odd numbers. Then by definition of odd, $m = 2k + 1$ and $n = 2k + 1$ for some integer k.

   This is incorrect. Using the same symbol, k, in the expression for both m and n implies that $m = 2k + 1 = n$. This is inconsistent with the supposition that m and n are arbitrarily chosen odd integers. They are not necessarily equal to each other.

3. **Jumping to a conclusion.**

   To jump to a conclusion means to allege the truth of something without giving an adequate reason. Consider the following "proof" that if the sum of any two integers is even, then so is their difference.

   > Suppose m and n are integers and $m + n$ is even. By definition of even, $m + n = 2k$ for some integer k. Then $m = 2k - n$, and so $m - n$ is even.

   The problem with this "proof" is that the crucial calculation

   $$m - n = (2k - n) - n) = 2k - 2n = 2(k - n)$$

   is missing. The author of the proof has jumped prematurely to a conclusion.

4. **Begging the question.**

   To beg the question means to assume what is to be proved; it is a variation of jumping to a conclusion. As an example, consider the following "proof" of the fact that the product of any two odd integers is odd:

   > Suppose m and n are odd integers. If $m \cdot n$ is odd, then $m \cdot n = 2k + 1$ for some integer k. Also by definition of odd, $m = 2a + 1$ and $n = 2b + 1$ for some integers a and b. Then $m \cdot n = (2a + 1)(2b + 1) = 2k + 1$, which is odd by definition of odd. This is what was to be shown.

   The problem with this "proof" is that the author first states what it means for the conclusion to be true (that $m \cdot n$ can be expressed as $2k + 1$) and later just assumes this to be true (by setting $(2a + 1) \cdot (2b + 1)$ equal to $2k + 1$). Thus the author of the "proof" begs the question.

5. **Misuse of the word *if*.**

   Another common error is not serious in itself, but it reflects imprecise thinking that sometimes leads to problems later in a proof. This error involves using the word *if* when the word *because* or *since* is really meant. Consider the following proof fragment:

   > Suppose p is a prime number. If p is prime, then p cannot be written as a product of two smaller positive integers.

   The use of the word *if* in the second sentence is inappropriate. It suggests that the primeness of p is in doubt. But p is known to be prime by the first sentence. It cannot be written as a product of two smaller positive integers because it is prime. The correct word to use in the second sentence is *because* or *since*.

GETTING PROOFS STARTED

Believe it or not, once you understand the idea of generalizing from the generic particular and the method of direct proof, you can write the beginnings of proofs even for theorems you do not understand. The reason is that the starting point and what is to be shown in a proof depend only on the linguistic form of the statement to be proved and not on the content of the statement.

EXAMPLE 3.1.6   Proof Beginnings

Write the beginning of a proof for the following statement. Include the starting point (first sentence of the proof) and what is to be shown (the conclusion).

Every complete, bipartite graph is connected.   You are not expected to understand this statement.

Solution   Rewriting the statement formally makes it easier to identify the starting point and what is to be shown.

Formal Restatement: $\forall G$, if G is a complete, bipartite graph then G is connected.

The starting point includes the names of the variables and states what kind of objects they are. It also includes the hypothesis of the if–then part of the statement.

Starting Point: Suppose G is a particular but arbitrarily chosen complete, bipartite graph.

To Show: G is connected.

Thus the beginning of a proof looks as follows:

Proof: Suppose G is a particular but arbitrarily chosen complete, bipartite graph. *[We must show that G is connected.]*   ■

DISPROOF BY COUNTEREXAMPLE

Consider the question of disproving a statement of the form

$$\forall x \text{ in } D, \text{ if } P(x) \text{ then } Q(x).$$

Showing that this statement is false is equivalent to showing that its negation is true. The negation of the statement is

$$\exists x \text{ in } D \text{ such that } P(x) \text{ and not } Q(x).$$

Thus the method of disproof by counterexample can be written as follows:

Disproof by Counterexample

To disprove a statement of the form "$\forall x \in D$, if $P(x)$ then $Q(x)$," find a value of x in D for which $P(x)$ is true and $Q(x)$ is false. Such an x is called a **counterexample**.

EXAMPLE 3.1.7   Disproof by Counterexample

Disprove the following statement by finding a counterexample:

$$\forall \text{ real numbers } a \text{ and } b, \text{ if } a^2 = b^2 \text{ then } a = b.$$

Solution   To disprove this statement, you need to find real numbers a and b such that $a^2 = b^2$ and $a \neq b$. The fact that both positive and negative integers have positive squares helps the search. If you flip through some possibilities in your mind, you will quickly see that 1 and -1 will work (or 2 and -2, or 0.5 and -0.5, and so forth).

Counterexample: Let $a = 1$ and $b = -1$. Then $a^2 = 1^2 = 1$ and $b^2 = (-1)^2 = 1$, and so $a^2 = b^2$. But $a \neq b$ since $1 \neq -1$.   ■

It is a sign of intelligence to make generalizations. Frequently, after observing a property to hold in a large number of cases, you may guess that it holds in all cases. You may, however, run into difficulty when you try to prove your guess. Perhaps you just have not figured out the key to the proof. But perhaps your guess is false. Consequently, when you are having serious difficulty proving a general statement, you should interrupt your efforts to look for a counterexample. Analyzing the kinds of problems you are encountering in your proof efforts may help the search. It may even happen that if you find a counterexample and therefore prove the statement false, your understanding may be sufficiently clarified that you can formulate a more limited but true version of the statement. For instance, Example 3.1.7 shows that it is not always true that if the squares of two numbers are equal then the numbers are equal. However, it is true that if the squares of two *positive* numbers are equal then the numbers are equal.

Pierre de Fermat (1601–1665)

More than 350 years ago the French mathematician Pierre de Fermat claimed that it is impossible to find positive integers x, y, and z with $x^n + y^n = z^n$ if n is an integer that is at least 3. (For $n = 2$, the equation has many integer solutions, such as $3^2 + 4^2 = 5^2$ and $5^2 + 12^2 = 13^2$.) Fermat wrote his claim in the margin of a book, along with the comment "I have discovered a truly remarkable proof of this theorem which this margin is too small to contain." No proof, however, was found among his papers, and over the intervening years some of the greatest mathematical minds tried and failed to discover a proof or a counterexample, developing whole new branches of mathematics in the attempt.

In 1986 Kenneth Ribet of the University of California at Berkeley showed that if a certain other statement, the Taniyama conjecture, could be proved, Fermat's theorem would follow. Andrew Wiles, an English mathematician and faculty member at Princeton University, immediately set to work to prove the Taniyama conjecture, and in June of 1993 he presented the outline of a proof to worldwide acclaim. As this book is going to press, however, one important step of the proof remains to be filled in, and even using the latest and most sophisticated mathematical techniques, it is not at all clear that this can be done. Could Fermat's "proof" have been correct? Is there a simple solution that does not use modern techniques? Given the powerful minds that have worked on the problem, that seems highly unlikely, but further inquiry may lead to an even deeper understanding.

Besides the Fermat problem, one of the most famous remaining unsolved problems in mathematics is the Goldbach conjecture. In Example 3.1.4 it was shown that every even integer from 4 to 30 can be represented as a sum of two prime numbers. More than 250 years ago, Christian Goldbach (1690–1764) conjectured that every even integer greater than 2 can be so represented. Explicit computer-aided calculations have shown the conjecture to be true up to at least 100,000,000. But there is a huge chasm between 100,000,000 and infinity. As pointed out by James Gleick of *The New York Times,* many other plausible conjectures in number theory have proved false. Leonhard Euler (1707–1783), for example, proposed in the eighteenth century that $a^4 + b^4 + c^4 = d^4$ had no nontrivial whole number solutions. In other words, no three perfect fourth powers add up to another perfect fourth power. For small numbers, Euler's conjecture looked good. But in 1987, a Harvard mathematician, Noam Elkies, proved it wrong. One counterexample, found by Roger Frye of Thinking Machines Corporation in a long computer search, is $95,800^4 + 217,519^4 + 414,560^4 = 422,481^4$.*

EXERCISE SET 3.1†

1. Assume that m and n are particular integers. Justify your answers to each of the following questions:
   a. Is $6m + 8n$ even?
   b. Is $10mn + 7$ odd?
   c. If $m > n > 0$, is $m^2 - n^2$ composite?

2. Assume that r and s are particular integers. Justify your answers to each of the following questions:
   a. Is $4rs$ even?
   b. Is $6r + 4s^2 + 3$ odd?
   c. If r and s are both positive, is $r^2 + 2rs + s^2$ composite?

Prove the statements in 3–6.

3. There is an integer $n > 5$ such that $2^n - 1$ is prime.

4. There are real numbers a and b such that $\sqrt{a + b} = \sqrt{a} + \sqrt{b}$.

5. There are distinct positive integers the sum of whose reciprocals is an integer.

6. There is a real number x so that $2^x > x^{10}$.

Definition: An integer n is called a **perfect square** if, and only if, $n = k^2$ for some integer k.

7. There is a perfect square that can be written as a sum of two other perfect squares.

Prove the statements in 8 and 9 by the method of exhaustion.

8. Every positive even integer less than 26 can be expressed as a sum of three or fewer perfect squares. (For instance, $10 = 1^2 + 3^2$ and $16 = 4^2$.)

9. For each integer n such that $1 \le n \le 10$, $n^2 - n + 11$ is a prime number.

10. Fill in the blanks in the following proof that the sum of any two even integers is even:

   Proof:

   Suppose m and n are ___(a)___. By definition of even, $m = 2r$ and $n = 2s$ for some ___(b)___. By substitution,

$m + n = $ ___(c)___ $= 2(r + s)$. Since r and s are both integers, so is their sum $r + s$. Hence $m + n$ has the form $2 \cdot$ (some integer), and so ___(d)___ by definition of even.

Prove the statements in 11–14. Follow the directions for writing proofs of universal statements given in this section.

11. The negative of any even integer is even.

12. The sum of any two odd integers is even.

13. If n is any even integer, then $(-1)^n = 1$.

14. If n is any odd integer, then $(-1)^n = -1$.

Disprove the statements in 15 and 16 by giving a counterexample.

15. For all positive integers n, if n is prime then n is odd.

16. For all real numbers a and b, if $a < b$ then $a^2 < b^2$.

Each of the statements in 17–20 is true. For each, write the beginning of a proof (just the beginning, not the whole proof). Include the starting point and what is to be shown.

17. For all integers m, if $m > 1$ then $0 < \dfrac{1}{m} < 1$.

18. For all real numbers x, if $x > 1$ then $x^2 > x$.

19. For all integers m and n, if $m \cdot n = 1$ then $m = n = 1$ or $m = n = -1$.

20. For all real numbers x, if $0 < x < 1$ then $x^2 < x$.

Find the mistakes in the "proofs" shown in 21–24.

21. Theorem: For all integers k, if $k > 0$ then $k^2 + 2k + 1$ is composite.
   "Proof: For $k = 2$, $k^2 + 2k + 1 = 2^2 + 2 \cdot 2 + 1 = 9$. But $9 = 3 \cdot 3$, and so 9 is composite. Hence the theorem is true."

22. "Theorem": The sum of any two even integers equals $4k$ for some integer k.

---

*James Gleick, "Fermat's Last Theorem Still Has 0 Solutions," *New York Times,* 17 April 1988.

†Exercises with blue numbers have solutions in Appendix B. The symbol H indicates that only a hint or partial solution is given. The symbol ◆ signals that an exercise is more challenging than usual.

"Proof: Suppose m and n are any two even integers. By definition of even, $m = 2k$ for some integer k and $n = 2k$ for some integer k. By substitution, $m + n = 2k + 2k = 4k$. This is what was to be shown."

23. Theorem: For all integers k, if $k > 0$ then $k^2 + 2k + 1$ is composite.
    "Proof: Suppose k is any integer such that $k > 0$. If $k^2 + 2k + 1$ is composite, then $k^2 + 2k + 1 = r \cdot s$ for some integers r and s such that $1 < r < (k^2 + 2k + 1)$ and $1 < s < (k^2 + 2k + 1)$. Since $k^2 + 2k + 1 = r \cdot s$ and both r and s are strictly between 1 and $k^2 + 2k + 1$, then $k^2 + 2k + 1$ is not prime. Hence $k^2 + 2k + 1$ is composite as was to be shown."

24. Theorem: The product of an even integer and an odd integer is even.
    "Proof: Suppose m is an even integer and n is an odd integer. If $m \cdot n$ is even, then by definition of even there exists an integer r such that $m \cdot n = 2r$. Also since m is even, there exists an integer p such that $m = 2p$, and since n is odd there exists an integer q such that $n = 2q + 1$. Thus

$$m \cdot n = (2p) \cdot (2q + 1) = 2r,$$

where r is an integer. By definition of even, then, $m \cdot n$ is even, as was to be shown."

In 25–41 prove the statements that are true and give counterexamples to disprove those that are false.

25. The product of any two odd integers is odd.

26. The sum of any even and any odd integer is odd.

27. The difference of any two odd integers is odd.

28. The product of any even integer and any integer is even.

29. If a sum of two integers is even, then one of the summands is even. (In the expression $a + b$, a and b are called **summands**.)

30. The difference of any two even integers is even.

31. The difference of any two odd integers is even.

32. For all integers n and m, if $n - m$ is even then $n^3 - m^3$ is even.

33. For all integers n, if n is prime then $(-1)^n = -1$.

34. For all integers m, if $m > 2$ then $m^2 - 4$ is composite.

35. For all integers n, $n^2 - n + 11$ is a prime number.

36. For all integers n, $4(n^2 + n + 1) - 3n^2$ is a perfect square.

37. Every positive integer can be expressed as a sum of three or fewer perfect squares.

H◆38. Any product of four consecutive integers is one less than a perfect square.

39. For all nonnegative real numbers a and b, $\sqrt{ab} = \sqrt{a}\sqrt{b}$. (Note that if x is a nonnegative real number, then there is a unique nonnegative real number y, denoted $\sqrt{x}$, such that $y^2 = x$.)

40. For all nonnegative real numbers a and b, $\sqrt{a + b} = \sqrt{a} + \sqrt{b}$.

41. If m and n are positive integers and mn is a perfect square, then m and n are perfect squares.

42. If m and n are perfect squares, then $m + n + 2\sqrt{mn}$ is also a perfect square. Why?

◆43. To check whether an integer n greater than 1 is prime, it suffices to check that n is not divisible by any prime number less than or equal to $\sqrt{n}$. Explain why this is so.

H◆44. If p is a prime number, must $2^p - 1$ also be prime? Prove or give a counterexample.

◆45. If n is a nonnegative integer, must $2^{2^n} + 1$ be prime? Prove or give a counterexample.

46. When expressions of the form $(x - r)(x - s)$ are multiplied out, a quadratic polynomial is obtained. For instance, $(x - 2)(x - (-7)) = (x - 2)(x + 7) = x^2 + 5x - 14$.
    H a. What can be said about the coefficients of the polynomial obtained by multiplying out $(x - r)(x - s)$ when both r and s are odd integers? both r and s are even integers? one of r and s is even and the other is odd?
    b. Can $x^2 - 1253x + 255$ be factored over the integers? Use the result of part (a) to explain your answer.

◆47. Can $15x^3 + 7x^2 - 8x - 27$ be factored over the integers?

Such, then, is the whole art of convincing. It is contained in two principles: to define all notations used, and to prove everything by replacing mentally the defined terms by their definitions.
(Blaise Pascal, 1623–1662)

3.2 DIRECT PROOF AND COUNTEREXAMPLE II: RATIONAL NUMBERS

Sums, differences, and products of integers are integers. But most quotients of integers are not integers. Quotients of integers are, however, important; they are known as *rational numbers*.

> **DEFINITION**
>
> A real number r is **rational** if, and only if, $r = a/b$ for some integers a and b with $b \neq 0$. A real number that is not rational is **irrational**. More formally, if r is a real number then
>
> $$r \text{ is rational} \iff \exists \text{ integers } a \text{ and } b \text{ such that } r = a/b \text{ and } b \neq 0.$$

The word *rational* contains the word *ratio,* which is another word for quotient. A rational number is a fraction or ratio of integers.

EXAMPLE 3.2.1   Determining Whether Numbers Are Rational

a. Is 10/3 a rational number?
b. Is $-(5/39)$ a rational number?
c. Is 0.281 a rational number?
d. Is 7 a rational number?
e. Is 0 a rational number?
f. Is 2/0 a rational number?
g. Is 0.12121212 . . . a rational number (where the digits 12 are assumed to repeat forever)?
h. If m and n are integers and neither m nor n is zero, is $(m + n)/mn$ a rational number?

Solution

a. Yes, 10/3 is a quotient of the integers 10 and 3 and hence is rational.
b. Yes, $-(5/39) = -5/39$, which is a quotient of the integers -5 and 39 and hence is rational.
c. Yes, $0.281 = 281/1000$. Note that the real numbers represented on a typical calculator display are all finite decimals. An explanation similar to the one in this example shows that any such number is rational. It follows that a calculator with such a display can only represent rational numbers.
d. Yes, $7 = 7/1$.
e. Yes, $0 = 0/1$.
f. No, 2/0 is not a number (division by 0 is not allowed).
g. Yes. Let $x = 0.12121212.\ldots$ Then $100x = 12.12121212.\ldots$ Hence

$$100x - x = 12.12121212 \ldots - 0.12121212 \ldots = 12.$$

But also

$$100x - x = 99x \quad \text{by basic algebra.}$$

Hence

$$99x = 12,$$

and so

$$x = \frac{12}{99}.$$

Therefore $0.12121212 \ldots = 12/99$, which is a ratio of two nonzero integers and thus is a rational number.

Note that you can use an argument similar to this one to show that any repeating decimal is a rational number. In Section 7.4 we show that any rational number can be written as a repeating or terminating decimal.

h. Yes, since m and n are integers, so are $m + n$ and mn (because sums and products of integers are integers). Also $mn \neq 0$ by the **zero product property**. (One version of this property says that if neither of two real numbers is 0 then their product is also not 0. See exercise 8 at the end of this section.) It follows that $(m + n)/mn$ is a quotient of two integers with a nonzero denominator and hence is a rational number. ∎

MORE ON GENERALIZING FROM THE GENERIC PARTICULAR

Some people like to think of the method of generalizing from the generic particular as a challenge process. If you claim a property holds for all elements in a domain, then someone can challenge your claim by picking any element in the domain whatsoever and asking you to prove that that element satisifies the property. To prove your claim, you must be able to meet all such challenges. That is, you must have a way to convince the person that the property is true for an *arbitrarily chosen* element in the domain.

For example, suppose A claims that every integer is a rational number. B challenges this claim by asking A to prove it for $n = 7$. A observes that

$$7 = \frac{7}{1} \quad \text{which is a quotient of integers and hence rational.}$$

B accepts this explanation but challenges again with $n = -12$. A responds that

$$-12 = \frac{-12}{1} \quad \text{which is a quotient of integers and hence rational.}$$

Next B tries to trip up A by challenging with $n = 0$, but A answers

$$0 = \frac{0}{1} \quad \text{which is a quotient of integers and hence rational.}$$

As you can see, A is able to respond effectively to all B's challenges because A has a general procedure for putting integers into the form of rational numbers: A just divides whatever integer B gives by 1. That is, no matter what integer n B gives A, A writes

$$n = \frac{n}{1} \quad \text{which is a quotient of integers and hence rational.}$$

This discussion proves the following theorem.

THEOREM 3.2.1

Every integer is a rational number.

In exercise 11 at the end of this section you are asked to condense the above discussion into a formal proof.

PROVING PROPERTIES OF RATIONAL NUMBERS

The next example shows how to use the method of generalizing from the generic particular to prove a property of rational numbers.

EXAMPLE 3.2.2   A Sum of Rationals Is Rational

Prove that the sum of any two rational numbers is rational.

Solution   Begin by mentally or explicitly rewriting the statement to be proved in the form "∀_____, if _____ then _____."

Formal Restatement: ∀ real numbers r and s, if r and s are rational then $r + s$ is rational.

Next ask yourself, "Where am I starting from?" or "What am I supposing?" The answer gives you the starting point, or first sentence, of the proof.

Starting Point: Suppose r and s are particular but arbitrarily chosen real numbers such that r and s are rational; or, more simply,

Suppose r and s are rational numbers.

Then ask yourself, "What must I show to complete the proof?"

To Show: $r + s$ is rational.

Finally, of course, you ask, "How do I get from the starting point to the conclusion?" or "Why must $r + s$ be rational if both r and s are rational?" The answer depends in an essential way on the definition of rational.

Rational numbers are quotients of integers. So to say that r and s are rational means that

$$r = \frac{a}{b} \quad \text{and} \quad s = \frac{c}{d} \quad \text{for some integers } a, b, c, \text{ and } d \\ \text{where } b \neq 0 \text{ and } d \neq 0.$$

It follows by substitution that

$$r + s = \frac{a}{b} + \frac{c}{d}.$$

Hence you must show that the right-hand sum can be written as a single fraction or ratio of two integers with a nonzero denominator. But

$$\frac{a}{b} + \frac{c}{d} = \frac{ad}{bd} + \frac{bc}{bd} \qquad \text{rewriting the fraction with a common denominator}$$

$$= \frac{ad + bc}{bd} \qquad \text{adding fractions with a common denominator.}$$

Is this fraction a ratio of integers? Yes. Since products and sums of integers are integers, $ad + bc$ and bd are both integers. Is the denominator $bd \neq 0$? Yes, by the zero product property (since $b \neq 0$ and $d \neq 0$). Thus $r + s$ is a rational number.

This discussion is summarized as follows.

THEOREM 3.2.2

The sum of any two rational numbers is rational.

Proof:

Suppose r and s are rational numbers. *[We must show that $r + s$ is rational.]* Then by definition of rational, $r = a/b$ and $s = c/d$ for some integers a, b, c, and d with $b \neq 0$ and $d \neq 0$. So

$$r + s = \frac{a}{b} + \frac{c}{d} \qquad \text{by substitution}$$

$$= \frac{ad + bc}{bd} \qquad \text{by basic algebra.}$$

Let $p = ad + bc$ and $q = bd$. Then p and q are integers because products and sums of integers are integers and because a, b, c, and d are all integers. Also $q \neq 0$ by the zero product property. Thus

$$r + s = \frac{p}{q} \qquad \text{where } p \text{ and } q \text{ are integers and } q \neq 0.$$

So $r + s$ is rational by definition of a rational number. *[This is what was to be shown.]*

PROVING COROLLARIES

Mathematicians distinguish among several types of mathematical statements. Although the word *theorem* can be used to refer to any statement that is known to be true, mathematicians generally like to reserve this word for very important statements that have many and varied consequences. Then they use the word **proposition** to refer to true statements that are somewhat less consequential but are nonetheless worth writing down. They use the word **corollary** to refer to statements whose truth can be deduced almost immediately from theorems or propositions already proved but that are worth stating because of their own applications. And they use the word **lemma** to refer to true statements that do not seem to have much intrinsic interest but are needed to help prove other theorems or propositions.

EXAMPLE 3.2.3   The Double of a Rational Number

Derive the following as a corollary of Theorem 3.2.2.

COROLLARY 3.2.3

The double of a rational number is rational.

Solution   The double of a number is just its sum with itself. But since the sum of any two rational numbers is rational (Theorem 3.2.2), the sum of a rational number with itself is rational. Hence the double of a rational number is rational. Here is a formal version of this argument:

Proof:

Suppose r is any rational number. Then $2r = r + r$ is a sum of two rational numbers. So by Theorem 3.2.2, $2r$ is rational. ∎

EXERCISE SET 3.2

The numbers in 1–7 are all rational. Write each number as a ratio of two integers.

1. $-(26/7)$

2. 3.9602

3. $5/6 + 3/7$

4. $0.58585858\ldots$

5. $0.30303030\ldots$

6. $20.492492492492\ldots$

7. $6.3215215215\ldots$

8. The zero product property says that if a product of two real numbers is 0, then one of the numbers must be 0.
   a. Write this property formally using quantifiers and variables.
   b. Write the contrapositive of your answer to part (a).
   c. Write an informal version (without quantifier symbols or variables) for your answer to part (b).

9. Assume that a and b are both integers and that $a \neq 0$ and $b \neq 0$. Explain why $(b - a)/ab^2$ must be a rational number.

10. Assume that p and q are both integers and that $q \neq 0$. Explain why $(2p + 3q)/5q$ must be a rational number.

11. Prove that every integer is a rational number.

12. Fill in the blanks in the following proof that the square of any rational number is rational:
    Proof: Suppose that r is __(a)__ . By definition of rational, $r = a/b$ for some __(b)__ with $b \neq 0$. By substitution, $r^2 = $ __(c)__ $= a^2/b^2$. Since a and b are both integers, so are the products a^2 and __(d)__ . Also $b^2 \neq 0$ by the __(e)__ . Hence r^2 is a ratio of two integers with a nonzero denominator, and so __(f)__ by definition of rational.

Determine which of the statements in 13–17 are true and which are false. Prove each true statement and give a counterexample for each false statement. In case the statement is false, determine whether a small change would make it true. If so, make the change and prove the new statement.

13. The product of any two rational numbers is a rational number.

H 14. The quotient of any two rational numbers is a rational number.

15. The difference of any two rational numbers is a rational number.

16. Given any rational number r, $-r$ is also a rational number.

17. Given any two distinct rational numbers r and s with $r < s$, there is a rational number x such that $r < x < s$.

Derive the statements in 18–22 as corollaries of other theorems from the text or of statements you have proved true in the exercises.

18. The square of any rational number is rational.

19. The square of any odd integer is odd.

20. The square of any even integer is even.

21. If n is an odd integer, then $n^2 + n$ is even.

22. If r is any rational number, then $2r^2 - r + 1$ is rational.

23. It is a fact that if n is any nonnegative integer, then

$$1 + \frac{1}{2} + \frac{1}{2^2} + \frac{1}{2^3} + \cdots + \frac{1}{2^n} = \frac{1 - (1/2^{n+1})}{1 - (1/2)}.$$

(A more general form of this statement is proved in Section 4.2.) Is a number of this form rational? If so, express it as a ratio of two integers.

24. Suppose a, b, c, and d are integers and $a \neq c$. Suppose also that x is a real number that satisfies the equation

$$\frac{ax + b}{cx + d} = 1.$$

Must x be rational? If so, express x as a ratio of two integers.

◆ 25. Suppose a, b, and c are integers and x, y, and z are nonzero real numbers that satisfy the following equations:

$$\frac{xy}{x + y} = a \quad \text{and} \quad \frac{xz}{x + z} = b \quad \text{and} \quad \frac{yz}{y + z} = c.$$

Is x rational? If so, express it as a ratio of two integers.

26. Prove that if one solution to a quadratic equation of the form $x^2 + bx + c = 0$ is rational (where b and c are rational), then the other solution is also rational. (Use the fact that if the solutions of the equation are r and s, then $x^2 + bx + c = (x - r)(x - s)$.)

27. Prove that if a real number c satisfies a polynomial equation of the form

$$r_3 x^3 + r_2 x^2 + r_1 x + r_0 = 0,$$

where r_0, r_1, r_2, and r_3 are rational numbers, then c satisfies an equation of the form

$$n_3 x^3 + n_2 x^2 + n_1 x + n_0 = 0,$$

where n_0, n_1, n_2, and n_3 are integers.

◆ 28. Prove that for all real numbers c, if c is a root of a polynomial with rational coefficients, then c is a root of a polynomial with integer coefficients.

In 29–32 find the mistakes in the "proofs" that the sum of any two rational numbers is a rational number.

29. "Proof: Let rational numbers $r = \frac{1}{4}$ and $s = \frac{1}{2}$ be given. Then $r + s = \frac{1}{4} + \frac{1}{2} = \frac{3}{4}$, which is a rational number. This is what was to be shown."

30. "Proof: Suppose r and s are rational numbers. By definition of rational, $r = a/b$ for some integers a and b with $b \neq 0$, and $s = a/b$ for some integers a and b with $b \neq 0$. Then $r + s = a/b + a/b = 2a/b$. Let $p = 2a$. Then p is an integer since it is a product of integers. Hence $r + s = p/b$ where p and b are integers and $b \neq 0$. Thus $r + s$ is a rational number by definition of rational. This is what was to be shown."

31. "Proof: Suppose r and s are rational numbers. Then $r = a/b$ and $s = c/d$ for some integers a, b, c, and d with $b \neq 0$ and $d \neq 0$ (by definition of rational). Then $r + s = a/b + c/d$. But this is a sum of two fractions, which is a fraction. So $r + s$ is a rational number since a rational number is a fraction."

32. "Proof: Suppose r and s are rational numbers. If $r + s$ is rational, then by definition of rational $r + s = a/b$ for some integers a and b with $b \neq 0$. Also since r and s are rational, then $r = i/j$ and $s = m/n$ for some integers i, j, m, and n with $j \neq 0$ and $n \neq 0$. It follows that $r + s = i/j + m/n = a/b$, which is a quotient of two integers with a nonzero denominator. Hence it is a rational number. This is what was to be shown."

3.3   DIRECT PROOF AND COUNTEREXAMPLE III: DIVISIBILITY

The essential quality of a proof is to compel belief.
(Pierre de Fermat)

When you were first introduced to the concept of division in elementary school, you were probably taught that 12 divided by 3 is 4 because if you separate 12 objects into groups of 3, you get 4 groups with nothing left over.

You may also have been taught to describe this fact by saying that "12 is evenly divisible by 3" or "3 divides 12 evenly."

The notion of divisibility is the central concept of one of the most beautiful subjects in advanced mathematics: **number theory**, the study of properties of integers.

> **DEFINITION**
>
> If n and d are integers and $d \neq 0$, then
>
> n is **divisible by** d if, and only if, $n = d \cdot k$ for some integer k.
>
> Alternatively, we say that
>
> n **is a multiple of** d, or
>
> d **is a factor of** n, or
>
> d **is a divisor of** n, or
>
> d **divides** n.
>
> The notation $d \mid n$ is read "d divides n." Symbolically if n and d are integers and $d \neq 0$,
>
> $$d \mid n \quad \Leftrightarrow \quad \exists \text{ an integer } k \text{ such that } n = d \cdot k.$$

EXAMPLE 3.3.1   Divisibility

a. Is 21 divisible by 3?   b. Does 5 divide 40?   c. Does $7 \mid 42$?
d. Is 32 a multiple of -16?   e. Is 6 a factor of 54?   f. Is 7 a factor of -7?

Solution   a. Yes, $21 = 3 \cdot 7$.   b. Yes, $40 = 5 \cdot 8$.   c. Yes, $42 = 7 \cdot 6$.
d. Yes, $32 = (-16) \cdot (-2)$.   e. Yes, $54 = 6 \cdot 9$.   f. Yes, $-7 = 7 \cdot (-1)$. ∎

EXAMPLE 3.3.2   Divisors of Zero

If k is any nonzero integer, does k divide 0?

Solution   Yes, because $0 = k \cdot 0$. ∎

EXAMPLE 3.3.3   The Positive Divisors of a Positive Number

Suppose a and b are positive integers and $a \mid b$. Is $a \leq b$?

Solution   Yes. To say that $a \mid b$ means that $b = k \cdot a$ for some integer k. Now k must be a positive integer because both a and b are positive. It follows that

$$1 \leq k$$

because every positive integer is greater than or equal to 1. Multiplying both sides by a gives

$$a \leq k \cdot a = b$$

(since multiplying both sides of an inequality by a positive number preserves the inequality—property T19 of Appendix A). ∎

EXAMPLE 3.3.4   Divisors of 1

Which integers divide 1?

Solution   By Example 3.3.3 any positive integer that divides 1 is less than or equal to 1. Since $1 = 1 \cdot 1$, 1 divides 1, and there are no positive integers that are less than 1. So the only positive divisor of 1 is 1. On the other hand, if d is a negative integer that divides 1, then $1 = d \cdot k$, and so $1 = |d| \cdot |k|$. Hence $|d|$ is a positive integer that divides 1.

Thus $|d| = 1$, and so $d = -1$. It follows that the only divisors of 1 are 1 and -1. ∎

EXAMPLE 3.3.5 Divisibility of Algebraic Expressions

a. If a and b are integers, is $3a + 3b$ divisible by 3?

b. If k and m are integers, is $10km$ divisible by 5?

Solution

a. Yes. By the distributive law of algebra, $3a + 3b = 3(a + b)$; $a + b$ is an integer because it is a sum of two integers.

b. Yes. By the associative law of algebra, $10km = 5 \cdot (2km)$; $2km$ is an integer because it is a product of three integers. ∎

When the definition of divides is rewritten formally using the existential quantifier, the result is

$$d \mid n \quad \Leftrightarrow \quad \exists \text{ an integer } k \text{ such that } n = d \cdot k.$$

Since the negation of an existential statement is universal, it follows that d does not divide n (denoted $d \nmid n$) if, and only if, $\forall$ integers k, $n \neq d \cdot k$, or, in other words, the quotient n/d is not an integer.

> For all integers n and d with $d \neq 0$, $d \nmid n \quad \Leftrightarrow \quad \dfrac{n}{d}$ is not an integer.

EXAMPLE 3.3.6 Checking Nondivisibility

Does $4 \mid 15$?

Solution No, $\frac{15}{4} = 3.75$, which is not an integer. ∎

▲ CAUTION! *Be careful to distinguish between the notation $a \mid b$ and the notation a/b. The notation $a \mid b$ stands for the sentence "a divides b," which means that there is an integer k such that $b = a \cdot k$. Dividing both sides by a gives $b/a = k$, an integer. Thus $a \mid b$ if, and only if, b/a is an integer. On the other hand, the notation a/b stands for the fractional number a/b (the inverse fraction!), which may or may not be an integer.*

EXAMPLE 3.3.7 Prime Numbers and Divisibility

An alternative way to define a prime number is to say that an integer $n > 1$ is prime if, and only if, its only positive integer divisors are 1 and itself. ∎

PROVING PROPERTIES OF DIVISIBILITY

One of the most useful properties of divisibility is that it is transitive. If one number divides a second and the second number divides a third, then the first number divides the third.

EXAMPLE 3.3.8 Transitivity of Divisibility

Prove that for all integers a, b, and c, if $a \mid b$ and $b \mid c$, then $a \mid c$.

Solution Since the statement to be proved is already written formally, you can immediately pick out the starting point, or first sentence of the proof, and the conclusion that must be shown.

Starting Point: Suppose a, b, and c are particular but arbitrarily chosen integers such that $a \mid b$ and $b \mid c$.

To Show: $a \mid c$.

You need to show that $a \mid c$, or, in other words, that
$$c = a \cdot \text{(some integer)}.$$
But since $a \mid b$,
$$b = a \cdot r \quad \text{for some integer } r. \tag{3.3.1}$$
And since $b \mid c$,
$$c = b \cdot s \quad \text{for some integer } s. \tag{3.3.2}$$

Equation 3.3.2 expresses c in terms of b, and equation 3.3.1 expresses b in terms of a. Thus if you substitute 3.3.1 into 3.3.2, you will have an equation that expresses c in terms of a.
$$
\begin{aligned}
c &= b \cdot s \qquad \text{by equation 3.3.2}\\
&= (a \cdot r) \cdot s \qquad \text{by equation 3.3.1.}
\end{aligned}
$$
But $(a \cdot r) \cdot s = a \cdot (r \cdot s)$ by the associative law for multiplication. Hence
$$c = a \cdot (r \cdot s).$$

Now you are almost finished. You have expressed c as $a \cdot$ (something). It remains only to show that that something is an integer. But of course it is because it is a product of two integers.

This discussion is summarized as follows.

THEOREM 3.3.1   Transitivity of Divisibility

For all integers a, b, and c, if a divides b and b divides c, then a divides c.

Proof:

Suppose a, b, and c are *[particular but arbitrarily chosen]* integers such that a divides b and b divides c. *[We must show that a divides c.]* By definition of divisibility,
$$b = a \cdot r \quad \text{and} \quad c = b \cdot s \quad \text{for some integers } r \text{ and } s.$$
By substitution
$$
\begin{aligned}
c &= b \cdot s\\
&= (a \cdot r) \cdot s\\
&= a \cdot (r \cdot s) \qquad \text{by basic algebra.}
\end{aligned}
$$
Let $k = r \cdot s$. Then k is an integer since it is a product of integers, and therefore
$$c = a \cdot k \quad \text{where } k \text{ is an integer.}$$
Thus a divides c by definition of divisibility. *[This is what was to be shown.]*

It would appear from the definition of prime that to show an integer is prime you would need to show that it is not divisible by any integer greater than 1 and less than itself. In fact, you need only check divisibility by prime numbers. This follows from Theorem 3.3.1, Example 3.3.3, and the following theorem, which says that any integer greater than 1 is divisible by a prime number. The idea of the proof is quite simple. You start with a positive integer. If it is prime, you are done; if not, it is a product of two smaller positive factors. If one of these is prime, you are done; if not, you can pick one of the factors and write it as a product of still smaller positive factors. You can continue in this way, factoring the factors of the number you started with, until one of them turns out to be prime. This must happen eventually because all the factors can be chosen to be positive and each is smaller than the preceding one.

THEOREM 3.3.2   Divisibility by a Prime

Any integer $n > 1$ is divisible by a prime number.

Proof:

Suppose n is a *[particular but arbitrarily chosen]* integer that is greater than 1. *[We must show that there is a prime number that divides n.]* If n is prime, then n is divisible by a prime number (namely itself), and we are done. If n is not prime, then n is composite, and by definition of composite,

$$n = r_0 \cdot s_0 \quad \text{where } r_0 \text{ and } s_0 \text{ are integers and}$$
$$1 < r_0 < n \text{ and } 1 < s_0 < n \quad \text{by Example 3.3.3.}$$

It follows by definition of divisibility that $r_0 \mid n$.

If r_0 is prime, then r_0 is a prime number that divides n, and we are done. If r_0 is not prime, then r_0 is composite, and by definition of composite,

$$r_0 = r_1 \cdot s_1 \quad \text{where } r_1 \text{ and } s_1 \text{ are integers and}$$
$$1 < r_1 < r_0 \text{ and } 1 < s_1 < r_0 \quad \text{by Example 3.3.3.}$$

It follows by the definition of divisibility that $r_1 \mid r_0$. But we already know that $r_0 \mid n$. Consequently, by transitivity of divisibility, $r_1 \mid n$.

If r_1 is prime, then r_1 is a prime number that divides n, and we are done. If r_1 is not prime, then r_1 is composite, and by definition of composite,

$$r_1 = r_2 \cdot s_2 \quad \text{where } r_2 \text{ and } s_2 \text{ are integers and}$$
$$1 < r_2 < r_1 \text{ and } 1 < s_2 < r_1 \quad \text{by Example 3.3.3.}$$

It follows by definition of divisibility that $r_2 \mid r_1$. But we already know that $r_1 \mid n$. Consequently, by transitivity of divisibility, $r_2 \mid n$.

If r_2 is prime, then r_2 is a prime number that divides n, and we are done. If r_2 is not prime, then we may repeat the above process by factoring r_2 as $r_3 \cdot s_3$.

We may continue in this way, factoring successive factors of n until we find a prime factor. We must succeed in a finite number of steps because by Example 3.3.3 each new factor is both less than the previous one (which is less than n) and greater than 1, and there are fewer than n integers strictly between 1 and n.* Thus we obtain a sequence

$$r_0, r_1, r_2, \ldots r_k,$$

where $k \geq 0$, $1 < r_k < r_{k-1} < \cdots < r_2 < r_1 < r_0 < n$, and $r_i \mid n$ for each $i = 0, 1, 2, \ldots, k$. The condition for termination is that r_k should be prime. Hence r_k is a prime number that divides n. *[This is what we were to show.]*

---

*Strictly speaking, this statement is justified by an axiom for the integers called the well-ordering principle, which is discussed in Section 4.4. Theorem 3.3.2 can also be proved using strong mathematical induction, as shown in Example 4.4.1.

COUNTEREXAMPLES AND DIVISIBILITY

To show that a proposed divisibility property is not universally true, you need to find some integers for which it is false.

EXAMPLE 3.3.9   Checking a Proposed Divisibility Property

Is it true or false that for all integers a and b, if $a \mid b$ and $b \mid a$ then $a = b$?

Solution   This proposed property is false. Can you think of a counterexample just by concentrating for a minute or so?

The following discussion describes a mental process that may take just a few seconds. It is helpful to be able to use it consciously, however, to solve more difficult problems.

To discover the truth or falsity of a statement such as the one given above, start off much as you would if you were trying to prove it.

Starting Point: Suppose a and b are integers such that $a \mid b$ and $b \mid a$.

Ask yourself, "*Must* it follow that $a = b$, or *could* it happen that $a \neq b$ for some a and b?" Focus on the supposition. What does it mean? By definition of divisibility, the conditions $a \mid b$ and $b \mid a$ mean that

$$b = k \cdot a \quad \text{and} \quad a = l \cdot b \quad \text{for some integers } k \text{ and } l.$$

Must it follow that $a = b$, or can you find integers a and b that satisfy these equations for which $a \neq b$? The equations imply that

$$b = k \cdot a = k \cdot (l \cdot b) = (k \cdot l) \cdot b$$

Since $b \mid a$, $b \neq 0$, and so you can cancel b from the extreme left and right sides to obtain

$$1 = k \cdot l.$$

In other words, k and l are divisors of 1. But the only divisors of 1 are 1 and -1 (see Example 3.3.4). Thus k and l are both 1 or -1. If $k = l = 1$, then $b = a$. But if $k = l = -1$, then $b = -a$ and so $a \neq b$. This analysis suggests that you can find a counterexample by taking $b = -a$. For instance, try $a = 2$ and $b = -2$. This works because $a \mid b$ *[since $2 \mid (-2)$]* and $b \mid a$ *[since $(-2) \mid 2$]* but $a \neq b$ *[since $2 \neq -2$]*. Therefore the proposed divisibility property is false. ■

The search for a proof will frequently help you discover a counterexample (provided the statement you are trying to prove is, in fact, false). Conversely, in trying to find a counterexample for a statement, you may come to realize the reason that the statement is true (if it is, in fact, true). The important thing is to keep an open mind until you are convinced by the evidence of your own careful reasoning.

THE UNIQUE FACTORIZATION THEOREM

The most comprehensive statement about divisibility of integers is contained in a theorem known as the *unique factorization theorem* for the integers. Because of its importance, this theorem is also called the *fundamental theorem of arithmetic.* Although Euclid, who lived about 300 B.C., seems to have been acquainted with the theorem, it was first stated precisely by the great German mathematician Carl Friedrich Gauss (rhymes with *house*) in 1801.

The unique factorization theorem says that any integer greater than 1 is either prime or can be written as a product of prime numbers in a way that is unique except, perhaps, for the order in which the primes are written. For example,

$$72 = 2 \cdot 2 \cdot 2 \cdot 3 \cdot 3 = 2 \cdot 3 \cdot 3 \cdot 2 \cdot 2 = 3 \cdot 2 \cdot 2 \cdot 3 \cdot 2$$

and so forth. The three 2's and two 3's may be written in any order, but any factorization of 72 as a product of primes must contain exactly three 2's and two 3's—no other collection of prime numbers besides three 2's and two 3's multiplies out to 72.

THEOREM 3.3.3 Unique Factorization Theorem

Given any integer $n > 1$, there exist a positive integer k, distinct prime numbers $p_1, p_2, \ldots, p_k$, and positive integers $e_1, e_2, \ldots, e_k$ such that

$$n = p_1^{e_1} \cdot p_2^{e_2} \cdot p_3^{e_3} \cdot \cdots \cdot p_k^{e_k},$$

and any other expression of n as a product of prime numbers is identical to this except, perhaps, for the order in which the factors are written.

The proof of the unique factorization theorem is somewhat beyond the level of this book.

Because of the unique factorization theorem, any integer $n > 1$ can be put into a *standard factored form* in which the prime factors are written in ascending order from left to right.

DEFINITION

Given any integer $n > 1$, the **standard factored form** of n is an expression of the form

$$n = p_1^{e_1} \cdot p_2^{e_2} \cdot p_3^{e_3} \cdot \cdots \cdot p_k^{e_k},$$

where k is a positive integer; $p_1, p_2, \ldots, p_k$ are prime numbers; $e_1, e_2, \ldots, e_k$ are positive integers; and $p_1 < p_2 < \cdots < p_k$.

EXAMPLE 3.3.10 Writing Integers in Standard Factored Form

Write 3,300 in standard factored form.

Solution First find all the factors of 3,300. Then write them in ascending order:

$$3,300 = 100 \cdot 33 = 4 \cdot 25 \cdot 3 \cdot 11$$
$$= 2 \cdot 2 \cdot 5 \cdot 5 \cdot 3 \cdot 11 = 2^2 \cdot 3^1 \cdot 5^2 \cdot 11^1.$$

EXAMPLE 3.3.11 Using Unique Factorization to Solve a Problem

Suppose m is an integer such that

$$8 \cdot 7 \cdot 6 \cdot 5 \cdot 4 \cdot 3 \cdot 2 \cdot m = 17 \cdot 16 \cdot 15 \cdot 14 \cdot 13 \cdot 12 \cdot 11 \cdot 10.$$

Does $17 \mid m$?

Solution   Since 17 is one of the prime factors of the right-hand side of the equation, it is also a prime factor of the left-hand side (by the unique factorization theorem). But 17 does not equal any prime factor of 8, 7, 6, 5, 4, 3, or 2 (because it is too large). Hence 17 must occur as one of the prime factors of m, and so $17 \mid m$.   ■

EXERCISE SET 3.3

Give a reason for your answer in each of 1–12. Assume that all variables represent integers.

1. Is 52 divisible by 13?

2. Is 51 divisible by 17?

3. Is $(3k + 1) \cdot (3k + 2) \cdot (3k + 3)$ divisible by 3?

4. Is $2m(2m + 2)$ divisible by 4?

5. Is 29 a multiple of 3?

6. Is -3 a factor of 66?

7. Is $6a(a + b)$ a multiple of $3a$?

8. Is 4 a factor of $6a \cdot 10b$?

9. Does $7 \mid 34$?

10. Does $11 \mid 61$?

11. If $n = 4k + 1$, does 8 divide $n^2 - 1$?

12. If $n = 4k + 3$, does 8 divide $n^2 - 1$?

13. Fill in the blanks in the following proof that for all integers a and b, if $a \mid b$ then $a \mid (-b)$.

   Proof:

   Suppose a and b are any integers such that __(a)__ . By definition of divisibility, $b =$ __(b)__ for some __(c)__ k. By substitution, $-b =$ __(d)__ $= a \cdot (-k)$. But $-k = (-1) \cdot k$ is an integer since -1 and k are. Hence by definition of divisibility __(e)__ , as was to be shown.

Prove statements 14 and 15.

14. For all integers a, b, and c, if $a \mid b$ and $a \mid c$ then $a \mid (b + c)$.

15. For all integers a, b, and c, if $a \mid b$ and $a \mid c$ then $a \mid (b - c)$.

For each statement in 16–26, determine whether the statement is true or false. Prove the statement if it is true and give a counterexample if it is false. (Do not use the unique factorization theorem to help prove any of the statements.)

16. The sum of any three consecutive integers is divisible by 3. (Two integers are **consecutive** if, and only if, one is one more than the other.)

17. The product of any two even integers is a multiple of 4.

18. A necessary condition for an integer to be divisible by 6 is that it be divisible by 2.

19. A sufficient condition for an integer to be divisible by 8 is that it be divisible by 16.

20. For all integers n, $n(6n + 3)$ is divisible by 3.

H21. For all integers a, b, and c, if $a \mid b$ then $a \mid bc$.

22. For all integers a, b, and c, if a is a factor of c then ab is a factor of c.

H23. For all integers a, b, and c, if $a \mid (b + c)$ then $a \mid b$ and $a \mid c$.

24. For all integers a, b, and c, if $a \mid bc$ then $a \mid b$ or $a \mid c$.

25. For all integers a and b, if $a \mid b$ then $a^2 \mid b^2$.

26. For all integers a and b, if $a \mid 10b$ then $a \mid 10$ or $a \mid b$.

27. A fast-food chain has a contest in which a card with numbers on it is given to each customer who makes a purchase. If some of the numbers on the card add up to 100, then the customer wins \$100. A certain customer receives a card containing the numbers

   $$72, 21, 15, 36, 69, 81, 9, 27, 42, \text{ and } 63.$$

   Will the customer win \$100? Why or why not?

28. Two athletes run a circular track at a steady pace so that the first completes one round in eight minutes and the second in ten minutes. If they both start from the same spot at 4 P.M., when will be the first time they return to the start together?

29. The sieve of Eratosthenes, named after its inventor, the Greek scholar Eratosthenes (276–194 B.C.), provides a way to find all prime numbers less than or equal to some fixed number n. To construct it, write out all the integers from 2 to n. Cross out all multiples of 2 except 2 itself, then all multiples of 3 except 3 itself, then all multiples of 5 except 5 itself, and so forth. Continue crossing out the multiples of each successive prime number up to $\sqrt{n}$. The numbers that are not crossed out are all the prime numbers from 2 to n. Here is a sieve of Eratosthenes that includes the numbers from 2 to 27.

The multiples of 2 are crossed out with a /, the multiples of 3 with a \, and the multiples of 5 with a —.

2 3 A̶ 5 6̶ 7 8̶ 9̶ 1̶0̶ 11 1̶2̶ 13 1̶4̶
1̶5̶ 1̶6̶ 17 1̶8̶ 19 2̶0̶ 2̶1̶ 2̶2̶ 23 2̶4̶ 2̶5̶ 2̶6̶ 2̶7̶

Use the sieve of Eratosthenes to find all prime numbers less than 100.

30. It can be shown (see exercises 37–41) that an integer is divisible by 3 if, and only if, the sum of its digits is divisible by 3. An integer is divisible by 9 if, and only if, the sum of its digits is divisible by 9. An integer is divisible by 5 if, and only if, its right-most digit is a 5 or a 0. And an integer is divisible by 4 if, and only if, the number formed by its right-most two digits is divisible by 4. Check the following integers for divisibility by 3, 4, 5, and 9.
   a. 637,425,403,425   b. 12,858,306,120,312
   c. 517,924,440,926,512   d. 14,328,083,360,232

31. Use the unique factorization theorem to write the following integers in standard factored form.
   a. 702   b. 4,851   c. 8,925

H 32. Suppose m is an integer such that
$2 \cdot 3 \cdot 4 \cdot 5 \cdot 6 \cdot 7 \cdot 8 \cdot 9 \cdot m =$
$151 \cdot 150 \cdot 149 \cdot 148 \cdot 147 \cdot 146 \cdot 145 \cdot 144 \cdot 143.$
Does $151 \mid m$? Why?

33. Suppose n is an integer such that
$2 \cdot 3 \cdot 4 \cdot 5 \cdot n = 29 \cdot 28 \cdot 27 \cdot 26 \cdot 25.$
Does $29 \mid n$? Why?

34. Suppose that in standard factored form $a = p_1^{e_1} \cdot p_2^{e_2} \cdot \ldots \cdot p_k^{e_k}$, where k is a positive integer; $p_1, p_2, \ldots, p_k$ are prime numbers; and $e_1, e_2, \ldots, e_k$ are positive integers. What is the strandard factored form for a^2?

35. Suppose that in standard factored form $a = p_1^{e_1} \cdot p_2^{e_2} \cdot \ldots \cdot p_k^{e_k}$, where k is a positive integer; $p_1, p_2, \ldots, p_k$ are prime numbers; and $e_1, e_2, \ldots, e_k$ are positive integers. What is the standard factored form for a^3?

36. If n is an integer and $n > 1$, then $n!$ is the product of n and every other positive integer that is less than n. For example, $5! = 5 \cdot 4 \cdot 3 \cdot 2 \cdot 1$.
   a. Write $(20!)^2$ in standard factored form.
   b. How many zeros are at the end of $(20!)^2$ when it is written in decimal form?

DEFINITION:

Given any nonnegative integer n, the **decimal representation** of n is an expression of the form
$$d_k d_{k-1} \cdots d_2 d_1 d_0,$$
where k is a nonnegative integer; $d_0, d_1, d_2, \ldots, d_k$ (called the **decimal digits** of n) are integers from 0 to 9 inclusive;

$d_k \neq 0$ unless $n = 0$ and $k = 0$; and
$$n = d_k \cdot 10^k + d_{k-1} \cdot 10^{k-1} + \cdots$$
$$+ d_2 \cdot 10^2 + d_1 \cdot 10 + d_0.$$

(For example, $2{,}503 = 2 \cdot 10^3 + 5 \cdot 10^2 + 0 \cdot 10 + 3$.)

37. Prove that if n is any nonnegative integer whose decimal representation ends in 0, then $5 \mid n$. (*Hint:* If the decimal representation of a nonnegative integer n ends in d_0, then $n = 10m + d_0$ for some integer m.)

38. Prove that if n is any nonnegative integer whose decimal representation ends in 5, then $5 \mid n$.

39. Prove that if the decimal representation of a nonnegative integer n ends in $d_1 d_0$ and if $4 \mid (10d_1 + d_0)$ then $4 \mid n$. (*Hint:* If the decimal representation of a nonnegative integer n ends in $d_1 d_0$, then there is an integer s such that $n = 100 \cdot s + 10d_1 + d_0$.)

H◆ 40. Observe that
$$7{,}524 = 7 \cdot 1{,}000 + 5 \cdot 100 + 2 \cdot 10 + 4$$
$$= 7(999 + 1) + 5(99 + 1)$$
$$+ 2(9 + 1) + 4$$
$$= (7 \cdot 999 + 7) + (5 \cdot 99 + 5)$$
$$+ (2 \cdot 9 + 2) + 4$$
$$= (7 \cdot 999 + 5 \cdot 99 + 2 \cdot 9)$$
$$+ (7 + 5 + 2 + 4)$$
$$= (7 \cdot 111 \cdot 9 + 5 \cdot 11 \cdot 9 + 2 \cdot 9)$$
$$+ (7 + 5 + 2 + 4)$$
$$= (7 \cdot 111 + 5 \cdot 11 + 2) \cdot 9$$
$$+ (7 + 5 + 2 + 4)$$
$$= (\text{an integer divisible by 9})$$
$$+ (\text{the sum of the digits of } 7{,}524).$$

Since the sum of the digits of 7,524 is divisible by 9, 7,524 can be written as a sum of two integers each of which is divisible by 9. It follows from exercise 14 that 7,524 is divisible by 9.

Generalize the argument given in this example to all nonnegative integers n. In other words, prove that for any nonnegative integer n, if the sum of the digits of n is divisible by 9, then n is divisible by 9.

◆ 41. Prove that for any nonnegative integer n, if the sum of the digits of n is divisible by 3, then n is divisible by 3.

◆ 42. Given a positive integer n written in decimal form, the alternating sum of the digits of n is obtained by starting with the right-most digit, subtracting the digit immediately to its left, adding the next digit to the left, subtracting the next digit, and so forth. For example, the alternating sum of the digits of 180,928 is $8 - 2 + 9 - 0 + 8 - 1 = 22$. Justify the fact that for any nonnegative integer n, if the alternating sum of the digits of n is divisible by 11, then n is divisible by 11.

3.4   DIRECT PROOF AND COUNTEREXAMPLE IV: DIVISION INTO CASES AND THE QUOTIENT-REMAINDER THEOREM

Be especially critical of any statement following the word "obviously."
(Anna Pell Wheeler 1883–1966)

When you divide 11 by 4, you get a quotient of 2 and a remainder of 3.

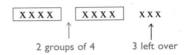

Another way to say this is that 11 equals 2 groups of 4 with 3 left over:

$$\boxed{\text{X X X X}} \quad \boxed{\text{X X X X}} \quad \text{X X X}$$

2 groups of 4   3 left over

Or,

$$11 = 2 \cdot 4 + 3.$$

2 groups of 4   3 left over

Of course, the number left over (3) is less than the size of the groups (4) because if more than 4 were left over, another group of 4 could be separated off.

The quotient-remainder theorem says that when any integer n is divided by any positive integer d, the result is a quotient q and a nonnegative remainder r that is smaller than d.

THEOREM 3.4.1   The Quotient-Remainder Theorem

Given any integer n and positive integer d, there exist unique integers q and r such that

$$n = d \cdot q + r \quad \text{and} \quad 0 \le r < d.$$

We give a proof of the quotient-remainder theorem in Section 4.4.

If n is positive, the quotient-remainder theorem can be illustrated on the number line as follows:

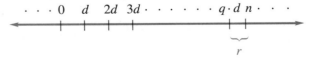

If n is negative, the picture changes. Since $n = d \cdot q + r$, where r is nonnegative, d must be multiplied by a negative integer q to go below n. Then the nonnegative integer r is added to come back up to n. This is illustrated below.

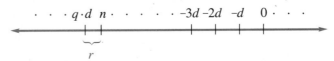

EXAMPLE 3.4.1   The Quotient-Remainder Theorem

For each of the following values of n and d, find integers q and r such that $n = d \cdot q + r$ and $0 \le r < d$:

a. $n = 54, d = 4$ b. $n = -54, d = 4$ c. $n = 54, d = 70$

Solution   a. $54 = 4 \cdot 13 + 2$; hence $q = 13$ and $r = 2$.
b. $-54 = (-14) \cdot 4 + 2$; hence $q = -14$ and $r = 2$.
c. $54 = 70 \cdot 0 + 54$; hence $q = 0$ and $r = 54$. ∎

div AND mod

A number of computer languages have built-in functions (called **div** and **mod** in Pascal, / and % in C, / and **rem** in Ada, and / and **mod** in PL/I) that compute the quotient and remainder of the division of a nonnegative integer by a positive integer.* The notation *div* is short for "divided by" and *mod* is short for "modulo."†

> **DEFINITION**
>
> Given a nonnegative integer n and a positive integer d,
>
> $$n \ div \ d = \text{the integer quotient obtained when } n \text{ is divided by } d, \text{ and}$$
> $$n \ mod \ d = \text{the integer remainder obtained when } n \text{ is divided by } d.$$
>
> Symbolically, if n and d are positive integers:
>
> $$n \ div \ d = q \quad \text{and} \quad n \ mod \ d = r \iff n = d \cdot q + r$$
> $$\text{where } q \text{ and } r \text{ are integers and } 0 \le r < d.$$

Note that it follows from the quotient-remainder theorem that $n \ mod \ d$ equals one of the integers from 0 through $d - 1$ (since the remainder of the division of n by d must be one of these integers). Note also that a necessary and sufficient condition for an integer n to be divisible by an integer d is that $n \ mod \ d = 0$.

EXAMPLE 3.4.2   *div and mod*

Compute 32 *div* 9 and 32 *mod* 9.

Solution

$$
\begin{array}{r}
3 \\
9\overline{)\,32} \\
27 \\
\hline
5
\end{array}
$$

$3 \leftarrow$ 32 div 9

$5 \leftarrow$ 32 mod 9

So 32 *div* 9 = 3 and 32 *mod* 9 = 5. ∎

---

*When n is negative, the values of *div* and *mod* vary dramatically from one language to another. We, therefore, define the functions for nonnegative values of n only.

†The modulo concept is discussed in greater detail in Section 10.3.

EXAMPLE 3.4.3

Computing the Day of the Week

Suppose today is Tuesday, and neither this year nor next year is a leap year. What day of the week will it be one year from today?

Solution   There are 365 days in a year that is not a leap year, and each week has seven days. Now

$$365 \; div \; 7 = 52 \quad \text{and} \quad 365 \; mod \; 7 = 1$$

because $365 = 52 \cdot 7 + 1$. Thus 52 weeks, or 364 days, from today will be a Tuesday, and so 365 days from today will be one day later, namely Wednesday.

More generally, if $DayT$ is the day of the week today and $DayN$ is the day of the week in N days, then

$$DayN = (DayT + N) \; mod \; 7,$$

where Sunday $= 0$, Monday $= 1, \ldots,$ Saturday $= 6$. ∎

REPRESENTATIONS OF INTEGERS

In Section 3.1 we defined an even integer to have the form $2k$ for some integer k. At that time we could have defined an odd integer to be one that was not even. Instead, because it was more useful for proving theorems, we specified that an odd integer has the form $2k + 1$ for some integer k. The quotient-remainder theorem brings these two ways of describing odd integers together by guaranteeing that any integer is either even or odd. To see why, let n be any integer, and consider what happens when n is divided by 2. By the quotient-remainder theorem (with $d = 2$), there exist integers q and r such that

$$n = 2 \cdot q + r \quad \text{and} \quad 0 \le r < 2.$$

But the only integers that satisfy $0 \le r < 2$ are $r = 0$ and $r = 1$. It follows that given any integer n, there exists an integer q with

$$n = 2q + 0 \quad \text{or} \quad n = 2q + 1.$$

In the case that $n = 2q + 0 = 2q$, n is even. In the case that $n = 2q + 1$, n is odd. Hence n is either even or odd.

The *parity* of an integer refers to whether the integer is even or odd. For instance, 5 has odd parity and 28 has even parity. We call the fact that any integer is either even or odd the **parity property**.

EXAMPLE 3.4.4

Consecutive Integers Have Opposite Parity

Prove that given any two consecutive integers, one is even and the other is odd.

Solution   Two integers are called *consecutive* if, and only if, one is one more than the other. So if one integer is m, the next consecutive integer is $m + 1$.

To prove the given statement, start by supposing that you have two particular but arbitrarily chosen consecutive integers. If the smaller is m, then the larger will be $m + 1$. How do you know for sure that one of these is even and the other is odd? You might imagine some examples: 4, 5; 12, 13; 1,073, 1,074. In the first two examples, the smaller of the two integers is even and the larger is odd; in the last example, it is the reverse. These observations suggest dividing the analysis into two cases.

Case 1: The smaller of the two integers is even.

Case 2: The smaller of the two integers is odd.

In the first case, when m is even, it appears that the next consecutive integer is odd. Is this always true? If an integer m is even, must $m + 1$ necessarily be odd? Of course the answer is yes. Because if m is even, then $m = 2k$ for some integer k, and so $m + 1 = 2k + 1$, which is odd.

In the second case, when m is odd, it appears that the next consecutive integer is even. Is this always true? If an integer m is odd, must $m + 1$ necessarily be even? Again, the answer is yes. For if m is odd, then $m = 2k + 1$ for some integer k, and so $m + 1 = (2k + 1) + 1 = 2k + 2 = 2(k + 1)$, which is even.

This discussion is summarized as follows.

THEOREM 3.4.2

Any two consecutive integers have opposite parity.

Proof:

Suppose that two *[particular but arbitrarily chosen]* consecutive integers are given; call them m and $m + 1$. *[We must show that one of m and m + 1 is even and that the other is odd.]* By the parity property, either m is even or m is odd. *[We break the proof into two cases depending on whether m is even or odd.]*

Case 1 (m is even):   In this case, $m = 2k$ for some integer k, and so $m + 1 = 2k + 1$, which is odd *[by definition of odd]*. Hence in this case one of m and $m + 1$ is even and the other is odd.

Case 2 (m is odd):   In this case, $m = 2k + 1$ for some integer k, and so $m + 1 = (2k + 1) + 1 = 2k + 2 = 2(k + 1)$. But $k + 1$ is an integer because it is a sum of two integers. Therefore, $m + 1$ equals twice some integer, and thus $m + 1$ is even. Hence in this case also one of m and $m + 1$ is even and the other is odd.

It follows that regardless of which case actually occurs for the particular m and $m + 1$ that are chosen, one of m and $m + 1$ is even and the other is odd. *[This is what was to be shown.]*

■

The division into cases in a proof is like the transfer of control for an **if–then–else** statement in a computer program. If m is even, control transfers to case 1; if not, control transfers to case 2. For any given integer, only one of the cases will apply. You must consider both cases, however, to obtain a proof that is valid for an arbitrarily given integer whether even or not.

There are times when division into more than two cases is called for. Suppose that at some stage of developing a proof, you know a statement of the form

$$A_1 \text{ or } A_2 \text{ or } A_3 \text{ or } \ldots \text{ or } A_n$$

is true, and suppose you want to deduce a conclusion C. By definition of *or*, you know that at least one of the statements A_i is true (although you may not know which). In this situation, you should use the method of division into cases. First assume A_1 is true and deduce C; next assume A_2 is true and deduce C; and so forth until you have assumed A_n is true and deduced C. At that point, you can conclude that regardless of which statement A_i happens to be true, the truth of C follows. Symbolically:

Given that A_1 or A_2 or A_3 or . . . or A_n, to show that
$(A_1$ or A_2 or A_3 or . . . or $A_n) \rightarrow C$, show all the implications

$$A_1 \rightarrow C,$$
$$A_2 \rightarrow C,$$
$$A_3 \rightarrow C,$$
$$\vdots$$
$$A_n \rightarrow C.$$

The procedure used to derive the parity property can be applied with other values of d to obtain a variety of alternate representations of integers.

EXAMPLE 3.4.5   Representations of Integers Modulo 4

Show that any integer can be written in one of the four forms

$$n = 4q \quad \text{or} \quad n = 4q + 1 \quad \text{or} \quad n = 4q + 2 \quad \text{or} \quad n = 4q + 3$$

for some integer q.

Solution   Given any integer n, apply the quotient-remainder theorem to n with $d = 4$. This implies that there exist an integer quotient q and a remainder r such that

$$n = 4 \cdot q + r \quad \text{and} \quad 0 \le r < 4.$$

But the only nonnegative remainders r that are less than 4 are 0, 1, 2, and 3. Hence

$$n = 4q \quad \text{or} \quad n = 4q + 1 \quad \text{or} \quad n = 4q + 2 \quad \text{or} \quad n = 4q + 3$$

for some integer q.   ∎

The next example illustrates how alternate representations for integers can help establish results in number theory. The solution is broken into two parts: a discussion and a formal proof. These correspond to the stages of actual proof development. Very few people, when asked to prove an unfamiliar theorem, immediately write down the kind of formal proof you find in a mathematics text. Most need to experiment with several possible approaches before they find one that works. A formal proof is much like the ending of a mystery story—the part in which the action of the story is systematically reviewed and all the loose ends are carefully tied together.

EXAMPLE 3.4.6   The Square of an Odd Integer

Prove that the square of any odd integer has the form $8m + 1$ for some integer m.

Solution   Begin by asking yourself, "Where am I starting from?" and "What do I need to show?" To help answer these questions, introduce variables to represent the quantities in the statement to be proved.

Formal Restatement: $\forall$ odd integers n, $\exists$ an integer m such that $n^2 = 8m + 1$.

From this, you can immediately identify the starting point and what is to be shown.

Starting Point: Suppose n is a particular but arbitrarily chosen odd integer.

To Show: $\exists$ an integer m such that $n^2 = 8m + 1$.

This looks tough. Why should there be an integer m with the property that $n^2 = 8m + 1$? That would say that $(n^2 - 1)/8$ is an integer, or that 8 divides $n^2 - 1$. Perhaps you could make use of the fact that $n^2 - 1 = (n - 1)(n + 1)$. Does 8 divide $(n - 1)(n + 1)$? Since n is odd, both $(n - 1)$ and $(n + 1)$ are even. That means that their product is divisible by 4. But that's not enough. You need to show that the product is divisible by 8. This seems to be blind alley.

You could try another tack. Since n is odd, you could represent n as $2q + 1$ for some integer q. Then $n^2 = (2q + 1)^2 = 4q^2 + 4q + 1 = 4(q^2 + q) + 1$. It is clear from this analysis that n^2 can be written in the form $4m + 1$, but it may not be clear that it can be written as $8m + 1$. This also seems to be a blind alley.

Yet another possibility is to use the result of Example 3.4.5. That example showed that any integer can be written in one of the four forms $4q$, $4q + 1$, $4q + 2$, or $4q + 3$. Two of these, $4q + 1$ and $4q + 3$, are odd. Thus any odd integer can be written in the form $4q + 1$ or $4q + 3$ for some integer q. You could try breaking into cases based on these two different forms.*

It turns out that this last possibility works. In each of the two cases, the conclusion follows readily by direct calculation. The details are shown in the following formal proof:

THEOREM 3.4.3

The square of any odd integer has the form $8m + 1$ for some integer m.

Proof:

Suppose n is a *[particular but arbitrarily chosen]* odd integer. By the quotient-remainder theorem, n can be written in one of the forms

$$4q \quad \text{or} \quad 4q + 1 \quad \text{or} \quad 4q + 2 \quad \text{or} \quad 4q + 3$$

for some integer q. In fact, since n is odd and $4q$ and $4q + 2$ are even, n must have one of the forms

$$4q + 1 \quad \text{or} \quad 4q + 3.$$

Case 1 $(n = 4q + 1$ for some integer $q)$: *[We must find an integer m such that $n^2 = 8m + 1$.]* Since $n = 4q + 1$,

$$
\begin{aligned}
n^2 &= (4q + 1)^2 && \text{by substitution} \\
&= (4q + 1)(4q + 1) && \text{by definition of square} \\
&= 16q^2 + 8q + 1 && \\
&= 8(2q^2 + q) + 1 && \text{by the laws of algebra.}
\end{aligned}
$$

[The choice of algebra steps was motivated by a desire to write the expression in the form $8 \cdot$ (some integer) $+ 1$.]

Let $m = 2q^2 + q$. Then m is an integer since 2 and q are integers and sums and products of integers are integers. Thus, substituting,

$$n^2 = 8m + 1 \quad \text{where } m \text{ is an integer.}$$

---

* Desperation can spur creativity. When you have tried all the obvious approaches without success and you really care about solving a problem, you reach into the odd corners of your memory for *anything* that may help.

Case 2 ($n = 4q + 3$ for some integer q): *[We must find an integer m such that* $n^2 = 8m + 1$.*]* Since $n = 4q + 3$,

$$
\begin{aligned}
n^2 &= (4q + 3)^2 && \text{by substitution} \\
&= (4q + 3)(4q + 3) && \text{by definition of square} \\
&= 16q^2 + 24q + 9 \\
&= 16q^2 + 24q + (8 + 1) \\
&= 8(2q^2 + 3q + 1) + 1 && \text{by the laws of algebra.}
\end{aligned}
$$

[Again, the motivation for the choice of algebra steps was the desire to write the expression in the form $8 \cdot$ *(some integer)* $+ 1.$*]*
Let $m = 2q^2 + 3q + 1$. Then m is an integer since 2, 3, and q are integers and sums and products of integers are integers. Thus, substituting,

$$ n^2 = 8m + 1 \quad \text{where } m \text{ is an integer.} $$

Cases 1 and 2 show that given any odd integer, whether of the form $4q + 1$ or $4q + 3$, $n^2 = 8m + 1$ for some integer m. *[This is what we needed to show.]*

■

Note that the result of Theorem 3.4.3 can also be written: For any odd integer n, $n^2 \bmod 8 = 1$.

EXERCISE SET 3.4

For each of the values of n and d given in 1–6, find integers q and r such that $n = dq + r$ and $0 \le r < d$.

1. $n = 70,\ d = 9$ 2. $n = 56,\ d = 5$

3. $n = 36,\ d = 40$ 4. $n = 5,\ d = 14$

5. $n = -45,\ d = 11$ 6. $n = -37,\ d = 9$

Evaluate the expressions in 7–10.

7. a. $43\ div\ 9$ b. $43\ mod\ 9$

8. a. $37\ div\ 10$ b. $37\ mod\ 10$

9. a. $61\ div\ 2$ b. $61\ mod\ 2$

10. a. $207\ div\ 4$ b. $207\ mod\ 4$

11. Check the correctness of formula (3.4.1) given in Example 3.4.3 for the following values of *DayT* and N.
   a. *DayT* $= 6$ (Saturday) and $N = 15$
   b. *DayT* $= 0$ (Sunday) and $N = 7$
   c. *DayT* $= 4$ (Thursday) and $N = 10$

◆ 12. Justify formula (3.4.1) for general values of *DayT* and N.

13. On a Monday a friend says he will meet you again in 30 days. What day of the week will that be?

14. January 1, 1990, was a Monday. What day of the week will January 1, 2000, be?

15. The following algorithm segment makes change; given an amount of money A between 1¢ and 99¢, it determines how many quarters (q), dimes (d), nickels (n), and pennies (p) are equal to A.

$$
\begin{aligned}
q &:= A\ div\ 25 \\
A &:= A\ mod\ 25 \\
d &:= A\ div\ 10 \\
A &:= A\ mod\ 10 \\
n &:= A\ div\ 5 \\
p &:= A\ mod\ 5
\end{aligned}
$$

   a. Trace this algorithm segment for $A = 69$.
   b. Trace this algorithm segment for $A = 83$.

16. A matrix **M** has 3 rows and 4 columns.

$$\begin{bmatrix} a_{11} & a_{12} & a_{13} & a_{14} \\ a_{21} & a_{22} & a_{23} & a_{24} \\ a_{31} & a_{32} & a_{33} & a_{34} \end{bmatrix}$$

The 12 entries in the matrix are to be stored in *row major* form in locations 8,304 to 8,315 in a computer's memory. This means that the entries in the first row (reading left to right) are stored first, then the entries in the second row, and finally the entries in the third row.
   a. Which location will a_{22} be stored in?
   b. Write a formula (in i and j) that gives the integer n so that a_{ij} is stored in location $8,304 + n$.
   c. Find formulas (in n) for r and s so that a_{rs} is stored in location $8,304 + n$.

17. Let **M** be a matrix with m rows and n columns, and suppose that the entries of **M** are stored in a computer's memory in row major form (see exercise 16) in locations $N, N + 1, N + 2, \ldots, N + mn - 1$. Find formulas in k for r and s so that a_{rs} is stored in location $N + k$.

18. Prove that the product of any two consecutive integers is even.

19. The result of exercise 18 suggests that the second apparent (blind alley) in the discussion of Example 3.4.6 might not be a blind alley after all. Write a new proof of Theorem 3.4.3 based on this observation.

20. Show that any integer n can be written in one of the three forms

$$n = 3q \quad \text{or} \quad n = 3q + 1 \quad \text{or} \quad n = 3q + 2$$

for some integer q.

21. a. Use the result of exercise 20 to prove that the product of any three consecutive integers is divisible by 3.
    b. Use the *mod* notation to rewrite the result of part (a).

22. Use the result of exercise 20 to prove that the square of any integer has the form $3k$ or $3k + 1$ for some integer k.

23. a. Prove that for all integers m and n, $m + n$ and $m - n$ are either both odd or both even.
    b. Find all solutions to the equation $m^2 - n^2 = 56$ for which both m and n are positive integers.
    c. Find all solutions to the equation $m^2 - n^2 = 40$ for which both m and n are positive integers.

Prove each of the statements in 24–31.

H 24. The product of any four consecutive integers is divisible by 8.

25. The square of any integer has the form $4k$ or $4k + 1$ for some integer k.

26. For any integer $n \geq 2$, $n^2 - 3$ is never divisible by 4.

H 27. The sum of any four consecutive integers is never divisible by 4.

28. For any integer n, $n(n^2 - 1)(n + 2)$ is divisible by 4.

H 29. Every prime number except 2 and 3 has the form $6q + 1$ or $6q + 5$ for some integer q.

30. If n is an odd integer, then $n^4 \bmod 16 = 1$.

31. If n is any integer that is not divisible by 2 or 3, then $n^2 \bmod 12 = 1$.

◆ 32. If m, n, and d are integers and $m \bmod d = n \bmod d$, does it necessarily follow that $m = n$? That $m - n$ is divisible by d? Prove your answers.

◆ 33. If m, n, and d are integers and $d \mid (m - n)$, what is the relation between $m \bmod d$ and $n \bmod d$? Prove your answer.

◆ 34. If m, n, a, b, and d are integers and $m \bmod d = a$ and $n \bmod d = b$, is $(m + n) \bmod d = a + b$? Is $(m + n) \bmod d = (a + b) \bmod d$? Prove your answers.

◆ 35. If m, n, a, b, and d are integers and $m \bmod d = a$ and $n \bmod d = b$, is $(m \cdot n) \bmod d = ab$? Is $(m \cdot n) \bmod d = ab \bmod d$? Prove your answers.

36. For any real number x, the absolute value of x, denoted $|x|$, is defined as follows:

$$|x| = \begin{cases} x & \text{if} \quad x \geq 0 \\ -x & \text{if} \quad x < 0 \end{cases}.$$

Prove (a)–(d) below.

a. For all real numbers x and y, $|x| \cdot |y| = |xy|$.
b. For all real numbers x, $-|x| \leq x \leq |x|$.
c. If c is a positive real number and x is any real number, then $-c \leq x \leq c$ if, and only if, $|x| \leq c$. (To prove a statement of the form "A if, and only if, B," you must prove "if A then B" and "if B then A.")
d. For all real numbers x and y, $|x + y| \leq |x| + |y|$. This result is called the **triangle inequality**. (*Hint:* Use (b) and (c) above.)

3.5 DIRECT PROOF AND COUNTEREXAMPLE V: FLOOR AND CEILING

Proof serves many purposes simultaneously. In being exposed to the scrutiny and judgment of a new audience, [a] proof is subject to a constant process of criticism and revalidation. Errors, ambiguities, and misunderstandings are cleared up by constant exposure. Proof is respectability. Proof is the seal of authority.

Proof, in its best instances, increases understanding by revealing the heart of the matter. Proof suggests new mathematics. The novice who studies proofs gets closer to the creation of new mathematics. Proof is mathematical power, the electric voltage of the subject which vitalizes the static assertions of the theorems.

Finally, proof is ritual, and a celebration of the power of pure reason.
(Philip J. Davis and Reuben Hersh, *The Mathematical Experience*, 1981)

Imagine a real number sitting on a number line. The *floor* and *ceiling* of the number are the integers to the immediate left and to the immediate right of the number (unless the number is, itself, an integer, in which case its floor and ceiling both equal the number itself). Many computer languages have built-in functions that compute floor and ceiling automatically. These functions are very convenient to use when writing certain kinds of computer programs. In addition, the concepts of floor and ceiling are important in analyzing the efficiency of many computer algorithms.

DEFINITION

Given any real number x, the **floor of x,** denoted $\lfloor x \rfloor$, is defined as follows:

$$\lfloor x \rfloor = \text{that unique integer } n \text{ such that } n \leq x < n + 1.$$

Symbolically, if x is a real number and n is an integer, then

$$\lfloor x \rfloor = n \quad \Leftrightarrow \quad n \leq x < n + 1.$$

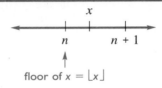

floor of $x = \lfloor x \rfloor$

DEFINITION

Given any real number x, the **ceiling of x,** denoted $\lceil x \rceil$, is defined as follows:

$$\lceil x \rceil = \text{that unique integer } n \text{ such that } n - 1 < x \leq n.$$

Symbolically, if x is a real number and n is an integer, then

$$\lceil x \rceil = n \quad \Leftrightarrow \quad n - 1 < x \leq n.$$

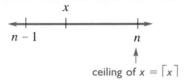

ceiling of $x = \lceil x \rceil$

EXAMPLE 3.5.1   Computing Floors and Ceilings

Compute $\lfloor x \rfloor$ and $\lceil x \rceil$ for each of the following values of x:

a. $25/4$ b. 0.999 c. -2.01

Solution   a. $25/4 = 6\frac{1}{4}$ and $6 < 6\frac{1}{4} < 7$; hence $\lfloor 25/4 \rfloor = 6$ and $\lceil 25/4 \rceil = 7$.
b. $0 < 0.999 < 1$; hence $\lfloor 0.999 \rfloor = 0$ and $\lceil 0.999 \rceil = 1$.
c. $-3 < -2.01 < -2$; hence $\lfloor -2.01 \rfloor = -3$ and $\lceil -2.01 \rceil = -2$.
Note that on some calculators $\lfloor x \rfloor$ is denoted INT (x).   ■

EXAMPLE 3.5.2   An Application

The 1,370 soldiers at a military base are given the opportunity to take buses into town for an evening out. Each bus holds a maximum of 40 passengers.

a. For reasons of economy, the base commander will only send full buses. What is the maximum number of buses the base commander will send?
b. If the base commander is willing to send a partially filled bus, how many buses will the commander need to allow all the soldiers to take the trip?

Solution   a. $\lfloor 1370/40 \rfloor = \lfloor 34.25 \rfloor = 34$ b. $\lceil 1370/40 \rceil = \lceil 34.25 \rceil = 35$ ■

EXAMPLE 3.5.3   Some General Values of Floor

If k is an integer, what are $\lfloor k \rfloor$ and $\lfloor k + 1/2 \rfloor$? Why?

Solution   Suppose k is an integer. Then

$$\lfloor k \rfloor = k \text{ because } k \text{ is an integer and } k \le k < k + 1,$$

and

$$\left\lfloor k + \frac{1}{2} \right\rfloor = k \text{ because } k \text{ is an integer and } k \le k + \frac{1}{2} < k + 1.$$ ■

EXAMPLE 3.5.4   Disproving an Alleged Property of Floor

Is the following statement true or false?

For all real numbers x and y, $\lfloor x + y \rfloor = \lfloor x \rfloor + \lfloor y \rfloor$.

Solution   The statement is false. As a counterexample, take $x = y = \frac{1}{2}$. Then

$$\lfloor x \rfloor + \lfloor y \rfloor = \left\lfloor \frac{1}{2} \right\rfloor + \left\lfloor \frac{1}{2} \right\rfloor = 0 + 0 = 0,$$

whereas

$$\lfloor x + y \rfloor = \left\lfloor \frac{1}{2} + \frac{1}{2} \right\rfloor = \lfloor 1 \rfloor = 1.$$

Hence $\lfloor x + y \rfloor \ne \lfloor x \rfloor + \lfloor y \rfloor$.

To arrive at this counterexample, you could have reasoned as follows: Suppose x and y are real numbers. Must it necessarily be the case that $\lfloor x + y \rfloor = \lfloor x \rfloor + \lfloor y \rfloor$, or could x and y be such that $\lfloor x + y \rfloor \ne \lfloor x \rfloor + \lfloor y \rfloor$? Imagine values that the various quantities could take. For instance, if both x and y are positive, then $\lfloor x \rfloor$ and $\lfloor y \rfloor$ are the integer parts of $\lfloor x \rfloor$ and $\lfloor y \rfloor$ respectively; just as

$$2\frac{3}{5} = 2 + \frac{3}{5}$$

integer part   fractional part

so is

$$x = \lfloor x \rfloor + \text{ fractional part of } x$$

and

$$y = \lfloor y \rfloor + \text{ fractional part of } y.$$

Thus if x and y are positive,

$$x + y = \lfloor x \rfloor + \lfloor y \rfloor + \textit{the sum of the fractional parts of x and y.}$$

But also

$$x + y = \lfloor x + y \rfloor + \textit{the fractional part of } (x + y).$$

These equations show that if there exist numbers x and y so that the sum of the fractional parts of x and y is at least 1, then a counterexample can be found. But there do exist such x and y; for instance, $x = \frac{1}{2}$ and $y = \frac{1}{2}$ as before.   ■

The analysis of Example 3.5.4 indicates that if x and y are positive and the sum of their fractional parts is less than 1, then $\lfloor x + y \rfloor = \lfloor x \rfloor + \lfloor y \rfloor$. In particular, if x is positive and m is a positive integer, then $\lfloor x + m \rfloor = \lfloor x \rfloor + \lfloor m \rfloor = \lfloor x \rfloor + m$. (The fractional part of m is 0; hence the sum of the fractional parts of x and m equals the fractional part of x, which is less than 1.) It turns out that you can use the definition of floor to show that this equation holds for all real numbers x and for all integers m.

EXAMPLE 3.5.5   Proving a Property of Floor

Prove that for all real numbers x and for all integers m, $\lfloor x + m \rfloor = \lfloor x \rfloor + m$.

Solution   Begin by supposing that x is a particular but arbitrarily chosen real number and that m is a particular but arbitrarily chosen integer. You must show that $\lfloor x + m \rfloor = \lfloor x \rfloor + m$. Since this is an equation involving $\lfloor x \rfloor$ and $\lfloor x + m \rfloor$, it is reasonable to give one of these quantities a name: Let $n = \lfloor x \rfloor$. By definition of floor,

$$n \text{ is an integer} \quad \text{and} \quad n \le x < n + 1.$$

This double inequality allows you to compute the value of $\lfloor x + m \rfloor$ in terms of n by adding m to all sides:

$$n + m \le x + m < n + m + 1.$$

Thus

$$\lfloor x + m \rfloor = n + m.$$

Now just substitute $\lfloor x \rfloor$ in place of n to obtain the equation that was to be shown. This discussion is summarized below.

THEOREM 3.5.1

For all real numbers x and all integers m, $\lfloor x + m \rfloor = \lfloor x \rfloor + m$.

Proof:

Suppose a real number x and an integer m are given. *[We must show that $\lfloor x + m \rfloor = \lfloor x \rfloor + m$.]* Let $n = \lfloor x \rfloor$. By definition of floor, n is an integer and

$$n \le x < n + 1.$$

Add m to all sides to obtain

$$n + m \le x + m < n + m + 1$$

[since adding a number to both sides of an inequality does not change the direction of the inequality]. Now $n + m$ is an integer *[since n and m are integers and a sum of integers is an integer],* and so by definition of floor

$$\lfloor x + m \rfloor = n + m.$$

But $n = \lfloor x \rfloor$. Hence by substitution

$$\lfloor x + m \rfloor = \lfloor x \rfloor + m$$

[as was to be shown].

The analysis of a number of computer algorithms, such as the binary search and merge sort algorithms, requires that you know the value of $\lfloor n/2 \rfloor$, where n is an integer. The formula for computing this value depends on whether n is even or odd.

THEOREM 3.5.2 The Floor of $n/2$

For any integer n,

$$\left\lfloor \frac{n}{2} \right\rfloor = \begin{cases} \dfrac{n}{2} & \text{if } n \text{ is even} \\[2mm] \dfrac{n-1}{2} & \text{if } n \text{ is odd.} \end{cases}$$

Proof:

Suppose n is a *[particular but arbitrarily chosen]* integer. By the quotient-remainder theorem, n is odd or n is even.

Case 1 (n is odd): In this case, $n = 2k + 1$ for some integer k. *[We must show that $\lfloor n/2 \rfloor = (n-1)/2$.]* By substitution,

$$\left\lfloor \frac{n}{2} \right\rfloor = \left\lfloor \frac{2k+1}{2} \right\rfloor = \left\lfloor \frac{2k}{2} + \frac{1}{2} \right\rfloor = \left\lfloor k + \frac{1}{2} \right\rfloor = k$$

because k is an integer and $k \le k + 1/2 < k + 1$. But since

$$n = 2k + 1,$$

then

$$n - 1 = 2k,$$

and so

$$k = \frac{n-1}{2}.$$

Thus on the one hand, $\lfloor n/2 \rfloor = k$, and on the other hand, $k = (n-1)/2$. It follows that $\lfloor n/2 \rfloor = \dfrac{n-1}{2}$ *[as was to be shown]*.

Case 2 (n is even): In this case, $n = 2k$ for some integer k. *[We must show that $\lfloor n/2 \rfloor = n/2$.]* The rest of the proof of this case is left as an exercise.

Given a nonnegative integer n and a positive integer d, the quotient-remainder theorem guarantees the existence of unique integers q and r such that

$$n = dq + r \quad \text{and} \quad 0 \le r < d.$$

The following theorem states that the floor notation can be used to describe q and r as follows:

$$q = \left\lfloor \frac{n}{d} \right\rfloor \quad \text{and} \quad r = n - d \cdot \left\lfloor \frac{n}{d} \right\rfloor.$$

Thus if, on a calculator or in a computer language, floor is built in but *div* and *mod* are not, *div* and *mod* can be defined as follows:

$$n \ div \ d = \left\lfloor \frac{n}{d} \right\rfloor \quad \text{and} \quad n \ mod \ d = n - d \cdot \left\lfloor \frac{n}{d} \right\rfloor.$$

Note that d divides n if, and only if, $n \ mod \ d = 0$, or, in other words, $n = d \cdot \lfloor n/d \rfloor$.

THEOREM 3.5.3

If n is a nonnegative integer and d is a positive integer, and if $q = \lfloor n/d \rfloor$ and $r = n - d \cdot \lfloor n/d \rfloor$, then

$$n = dq + r \quad \text{and} \quad 0 \leq r < d.$$

Proof:

Suppose n is a nonnegative integer, d is a positive integer, $q = \lfloor n/d \rfloor$, and $r = n - d \cdot \lfloor n/d \rfloor$. *[We must show that $n = dq + r$ and $0 \leq r < d$.]* By substitution

$$dq + r = d \cdot \left\lfloor \frac{n}{d} \right\rfloor + \left(n - d \cdot \left\lfloor \frac{n}{d} \right\rfloor \right) = n.$$

So it remains only to show that $0 \leq r < d$. But $q = \lfloor n/d \rfloor$. Thus by definition of floor,

$$q \leq \frac{n}{d} < q + 1.$$

Then

$$dq \leq n < dq + d \quad \text{by multiplying all sides by } d,$$

and so

$$0 \leq n - dq < d \quad \text{by subtracting } dq \text{ from all sides}$$

But

$$r = n - d\left\lfloor \frac{n}{d} \right\rfloor = n - dq.$$

Hence

$$0 \leq r < d \quad \text{by substitution.}$$

[This is what was to be shown.]

EXAMPLE 3.5.6   Computing *div* and *mod*

Use the floor notation to compute 3850 *div* 17 and 3850 *mod* 17.

Solution
$$3850 \ div \ 17 = \lfloor 3850/17 \rfloor = \lfloor 226.47 \rfloor = 226$$
$$3850 \ mod \ 17 = 3850 - 17 \cdot \lfloor 3850/17 \rfloor$$
$$= 3850 - 17 \cdot 226$$
$$= 3850 - 3842 = 8$$ ∎

EXERCISE SET 3.5

Compute $\lfloor x \rfloor$ and $\lceil x \rceil$ for each of the values of x in 1–4.

1. 37.999

2. 10/3

3. −14.00001

4. −57/2

5. Use the floor notation to express 589 *div* 12 and 589 *mod* 12.

6. If k is an integer, what is $\lceil k \rceil$? Why?

7. If k is an integer, what is $\lceil k + \frac{1}{2} \rceil$? Why?

8. Seven pounds of raw material are needed to manufacture each unit of a certain product. Express the number of units that can be produced from n pounds of raw material using either the floor or the ceiling notation. Which notation is more appropriate?

9. Boxes, each capable of holding 24 units, are used to ship a product from the manufacturer to a wholesaler. Express the number of boxes that would be required to ship n units of the product using either the floor or the ceiling notation. Which notation is more appropriate?

10. If 0 = Sunday, 1 = Monday, 2 = Tuesday, . . . , 6 = Saturday, then January 1 of year n occurs on the day of the week given by the formula below:

$$\left(n + \left\lfloor \frac{n-1}{4} \right\rfloor - \left\lfloor \frac{n-1}{100} \right\rfloor + \left\lfloor \frac{n-1}{400} \right\rfloor \right) mod \ 7.$$

   a. Use this formula to find January 1 of
   i. 2001   ii. 2025   iii. the year of your birth.
   Hb. Interpret the different components of this formula.

11. State a necessary and sufficient condition for the floor of a real number to equal that number.

12. Prove that if n is any even integer, then $\lfloor n/2 \rfloor = n/2$.

13. Suppose n and d are integers and $d \neq 0$. Prove each of the following.
   a. If $d \mid n$, then $n = \lfloor n/d \rfloor \cdot d$.
   b. If $n = \lfloor n/d \rfloor \cdot d$, then $d \mid n$.

   c. Use the floor notation to state a necessary and sufficient condition for an integer n to be divisible by an integer d.

Some of the statements in 14–22 are true and some are false. Prove each true statement and find a counterexample for each false statement.

H14. For all real numbers x and y, $\lfloor x - y \rfloor = \lfloor x \rfloor - \lfloor y \rfloor$.

H15. For all real numbers x, $\lfloor x - 1 \rfloor = \lfloor x \rfloor - 1$.

16. For all real numbers x, $\lfloor x^2 \rfloor = \lfloor x \rfloor^2$.

H17. For all integers n,
$$\lfloor n/3 \rfloor = \begin{cases} n/3 & \text{if } n \ mod \ 3 = 0 \\ (n-1)/3 & \text{if } n \ mod \ 3 = 1 \\ (n-2)/3 & \text{if } n \ mod \ 3 = 2. \end{cases}$$

H18. For all real numbers x and y, $\lceil x + y \rceil = \lceil x \rceil + \lceil y \rceil$.

H19. For all real numbers x, $\lceil x + 1 \rceil = \lceil x \rceil + 1$.

20. For all real numbers x and y, $\lceil xy \rceil = \lceil x \rceil \cdot \lceil y \rceil$.

21. For all odd integers n, $\lceil n/2 \rceil = (n + 1)/2$.

22. For all real numbers x and y, $\lceil xy \rceil = \lceil x \rceil \cdot \lfloor y \rfloor$.

Prove each of the statements in 23–29.

23. For any real number x, if x is not an integer, then $\lfloor x \rfloor + \lfloor -x \rfloor = -1$.

24. For any integer m and any real number x, if x is not an integer, then $\lfloor x \rfloor + \lfloor m - x \rfloor = m - 1$.

H25. For all real numbers x, $\lfloor \lfloor x/2 \rfloor /2 \rfloor = \lfloor x/4 \rfloor$.

26. For all real numbers x, if $x - \lfloor x \rfloor < 1/2$ then $\lfloor 2x \rfloor = 2 \cdot \lfloor x \rfloor$.

27. For all real numbers x, if $x - \lfloor x \rfloor \geq 1/2$ then $\lfloor 2x \rfloor = 2 \cdot \lfloor x \rfloor + 1$.

28. For any odd integer n,
$$\left\lfloor \frac{n^2}{4} \right\rfloor = \left(\frac{n-1}{2} \right) \left(\frac{n+1}{2} \right).$$

29. For any odd integer n,

$$\left\lceil \frac{n^2}{4} \right\rceil = \frac{n^2 + 3}{4}.$$

30. Find the mistake in the following "proof" that $\lfloor n/2 \rfloor = (n-1)/2$ if n is an odd integer. "Proof: Suppose n is any odd integer. Then $n = 2k + 1$ for some integer k.

Consequently,

$$\left\lfloor \frac{2k+1}{2} \right\rfloor = \frac{(2k+1)-1}{2} = \frac{2k}{2} = k.$$

But $n = 2k + 1$. Solving for k gives $k = (n-1)/2$. Hence by substitution, $\lfloor n/2 \rfloor = (n-1)/2$."

3.6   INDIRECT ARGUMENT: CONTRADICTION AND CONTRAPOSITION

Reductio ad absurdum is one of a mathematician's finest weapons. It is a far finer gambit than any chess gambit: a chess player may offer the sacrifice of a pawn or even a piece, but the mathematician offers the game.

(G. H. Hardy, 1877–1947)

In a direct proof you start with the hypothesis of a statement and make one deduction after another until you reach the conclusion. Indirect proofs are more roundabout. One kind of indirect proof, *argument by contradiction*, is based on the fact that either a statement is true or it is false but not both. Suppose you can show that the assumption that a given statement is not true leads logically to a contradiction, impossibility, or absurdity. Then that assumption must be false; hence, the given statement must be true. This method of proof is also known as *reductio ad impossibile* or *reductio ad absurdum* because it relies on reducing a given assumption to an impossibility or absurdity.

Argument by contradiction occurs in many different settings. For example, if a man accused of holding up a bank can prove that he was some place else at the time the crime was committed, he will certainly be acquitted. The logic of his defense is as follows:

Suppose I did commit the crime. Then at the time of the crime, I would have had to be at the scene of the crime. In fact, I was in a meeting with 20 people at that time, as they will testify. This contradicts the assumption that I committed the crime. Hence that assumption is false.

Another example occurs in debate. One technique of debate is to say, "Suppose for a moment that what my opponent says is correct." Starting from this supposition, the debater then deduces one statement after another until finally arriving at a statement that is completely ridiculous and unacceptable to the audience. By this means the debater shows the opponent's statement to be false.

The point of departure for a proof by contradiction is the supposition that the statement to be proved is false. The goal is to reason to a contradiction. Thus proof by contradiction has the following outline.

Method of Proof by Contradiction

1. Suppose the statement to be proved is false.
2. Show that this supposition leads logically to a contradiction.
3. Conclude that the statement to be proved is true.

Note that supposing a statement is false is the same thing as supposing the negation of the statement is true. When you begin a proof by contradiction, therefore, you must negate the statement to be proved.

There are no clear-cut rules for when to try a direct proof and when to try a proof by contradiction. There are some general guidelines, however. Proof by contradiction is indicated if you want to show that there is no object with a certain property, or if you want to show that a certain object does not have a certain property. The next two examples illustrate these situations.

EXAMPLE 3.6.1

There Is No Greatest Integer

Use proof by contradiction to show that there is no greatest integer.

Solution   In this example, the certain property is the property of being the greatest integer. You are to prove that there is no object with this property. Begin by supposing the negation: that there is an object with the property.

Starting Point: Suppose not. Suppose there is a greatest integer N.

This means that $N \geq n$ for all integers n.

To Show: This supposition leads logically to a contradiction.

Most small children believe there is a greatest integer—they often call it a "zillion." But with age and experience, they change their belief. At some point they realize that if there were a greatest integer, they could add 1 to it to obtain an integer that was greater still. Since that is a contradiction, no greatest integer can exist. This line of reasoning is the heart of the formal proof.

THEOREM 3.6.1

There is no greatest integer.

Proof:

Suppose not. *[We take the negation of the theorem and suppose it to be true.]* Suppose there is a greatest integer N. *[We must deduce a contradiction.]* Then $N \geq n$ for every integer n. Let $M = N + 1$. Now M is an integer since it is a sum of integers. Also $M > N$ since $M = N + 1$.

Thus M is an integer that is greater than the greatest integer, which is a contradiction. *[This contradiction shows that the supposition is false and, hence, that the theorem is true.]*

After a contradiction has been reached, the logic of the argument is always the same: "This is a contradiction. Hence the supposition is false and the theorem is true." Because of this, most mathematics texts end proofs by contradiction at the point at which the contradiction has been obtained.

The next example asks you to show that the sum of any rational number and any irrational number is irrational. One way to think of this is that a certain object (the sum of a rational and an irrational) does not have a certain property (the property of being rational). This suggests trying a proof by contradiction.

EXAMPLE 3.6.2   The Sum of a Rational Number and
                  an Irrational Number

Use proof by contradiction to show that the sum of any rational number and any irrational number is irrational.

Solution   Begin by supposing the negation of what you are to prove. Be very careful when writing down what this means. If you take the negation incorrectly, the entire rest of the proof will be flawed. In this example, the statement to be proved can be written formally as

$\forall$ real numbers r and s, if r is rational and
s is irrational, then $r + s$ is irrational.

From this you can see that the negation is

$\exists$ a rational number r and an irrational
number s such that $r + s$ is rational.

▲ CAUTION!   *The negation of "The sum of any irrational number and any rational number is irrational" is NOT "The sum of any irrational number and any rational number is rational."*

It follows that the starting point and what is to be shown are as follows:

Starting Point: Suppose not. That is, suppose there is a rational number r and an irrational number s such that $r + s$ is rational.

To Show: This supposition leads to a contradiction.

To derive a contradiction, you need to understand what you are supposing: There are numbers r and s such that r is rational, s is irrational, and $r + s$ is rational. By definition of rational and irrational, this means that s cannot be written as a quotient of any two integers but that r and $r + s$ can:

$$r = \frac{a}{b} \quad \text{for some integers } a \text{ and } b \text{ with } b \neq 0, \text{ and} \qquad \textbf{3.6.1}$$

$$r + s = \frac{c}{d} \quad \text{for some integers } c \text{ and } d \text{ with } d \neq 0. \qquad \textbf{3.6.2}$$

If you substitute (3.6.1) into (3.6.2), you obtain

$$\frac{a}{b} + s = \frac{c}{d}.$$

Subtracting a/b from both sides gives

$$s = \frac{c}{d} - \frac{a}{b}$$

$$= \frac{bc}{bd} - \frac{ad}{bd} \qquad \text{by rewriting } c/d \text{ and } a/b \text{ as equivalent fractions}$$

$$= \frac{bc - ad}{bd} \qquad \text{by the rule for subtracting fractions with the same denominator.}$$

But both $bc - ad$ and bd are integers because products and differences of integers are integers, and $bd \neq 0$ by the zero product property. Hence s can be expressed as a quotient of two integers. This is a contradiction.

This discussion is summarized in a formal proof.

THEOREM 3.6.2

The sum of any rational number and any irrational number is irrational.

Proof:

Suppose not. *[We take the negation of the theorem and suppose it to be true.]* Suppose there is a rational number r and an irrational number s such that $r + s$ is rational. *[We must deduce a contradiction.]* By definition of rational, $r = a/b$ and $r + s = c/d$ for some integers a, b, c, and d with $b \neq 0$ and $d \neq 0$. By substitution,

$$\frac{a}{b} + s = \frac{c}{d},$$

and so

$$s = \frac{c}{d} - \frac{a}{b} \qquad \text{by subtracting } a/b \text{ from both sides}$$

$$= \frac{bc - ad}{bd} \qquad \text{by the laws of algebra.}$$

Now $bc - ad$ and bd are both integers *[since a, b, c, and d are, and since products and differences of integers are integers]*, and $bd \neq 0$ *[by the zero product property]*. Hence s is a quotient of the two integers $bc - ad$ and bd with $bd \neq 0$. So by definition of rational, s is rational. This contradicts the supposition that s is irrational. *[Hence the supposition is false and the theorem is true.]*

■

ARGUMENT BY CONTRAPOSITION

A second form of indirect argument, *argument by contraposition,* is based on the logical equivalence between a statement and its contrapositive. To prove a statement by contraposition, you take the contrapositive of the statement, prove the contrapositive by a direct proof, and conclude that the original statement is true. The underlying reasoning is that since a conditional statement is logically equivalent to its contrapositive, if the contrapositive is true then the statement must also be true.

Method of Proof by Contraposition

1. Express the statement to be proved in the form

$$\forall x \text{ in } D, \text{ if } P(x) \text{ then } Q(x).$$

(This step may be done mentally.)

2. Rewrite this statement in the contrapositive form

$$\forall x \text{ in } D, \text{ if } Q(x) \text{ is false then } P(x) \text{ is false.}$$

(This step may also be done mentally.)
3. Prove the contrapositive by a direct proof.
   a. Suppose x is a (particular but arbitrarily chosen) element of D such that $Q(x)$ is false.
   b. Show that $P(x)$ is false.

EXAMPLE 3.6.3   **If the Square of an Integer Is Even, the Integer Is Even**

Prove that for all integers n, if n^2 is even then n is even.

Solution   First form the contrapositive of the statement to be proved.

Contrapositive: For all integers n, if n is not even then n^2 is not even.

But an integer is not even if, and only if, it is odd (by the parity property). So the contrapositive may be restated as follows:

Contrapositive: For all integers n, if n is odd then n^2 is odd.

Now prove the contrapositive using a direct proof. This is easy! Just suppose n is a *[particular but arbitrarily chosen]* integer that is odd and show that n^2 is odd. But by exercise 25 of Section 3.1, the product of any two odd integers is odd, and so $n^2 = n \cdot n$ is odd. This completes the proof of the contrapositive; hence the given statement is true by the logical equivalence between a statement and its contrapositive.

The following formal proof summarizes this discussion.

PROPOSITION 3.6.3

Given any integer n, if n^2 is even then n is even.

Proof (by contraposition):
Suppose n is any integer that is odd. *[We must show that n^2 is odd.]* By exercise 25 of Section 3.1, a product of two odd integers is odd. Hence $n^2 = n \cdot n$ is odd *[as was to be shown]*.

RELATION BETWEEN PROOF BY CONTRADICTION AND PROOF BY CONTRAPOSITION

Observe that any proof by contraposition can be recast in the language of proof by contradiction. In a proof by contraposition the statement

$$\forall x \text{ in } D, \text{ if } P(x) \text{ then } Q(x)$$

is proved by giving a direct proof of the equivalent statement

$$\forall x \text{ in } D, \text{ if } \sim Q(x) \text{ then } \sim P(x).$$

To do this, you suppose you are given an arbitrary element x of D such that $\sim Q(x)$. You then show that $\sim P(x)$. This is illustrated in Figure 3.6.1.

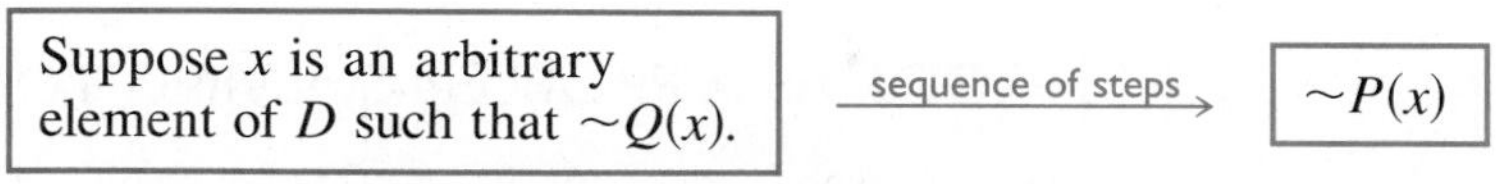

Suppose x is an arbitrary element of D such that $\sim Q(x)$. $\xrightarrow{\text{sequence of steps}}$ $\sim P(x)$

FIGURE 3.6.1   Proof by contraposition

Exactly the same sequence of steps can be used as the heart of a proof by contradiction for the given statement. The only thing that changes is the context in which the steps are written down.

To rewrite the proof as a proof by contradiction, you suppose there is an x in D such that $P(x)$ and $\sim Q(x)$. You then follow the steps of the proof by contraposition to deduce the statement $\sim P(x)$. But $\sim P(x)$ is a contradiction to the supposition that $P(x)$ and $\sim Q(x)$. (Because to contradict a conjunction of statements, it is only necessary to contradict one component.) This process is illustrated in Figure 3.6.2.

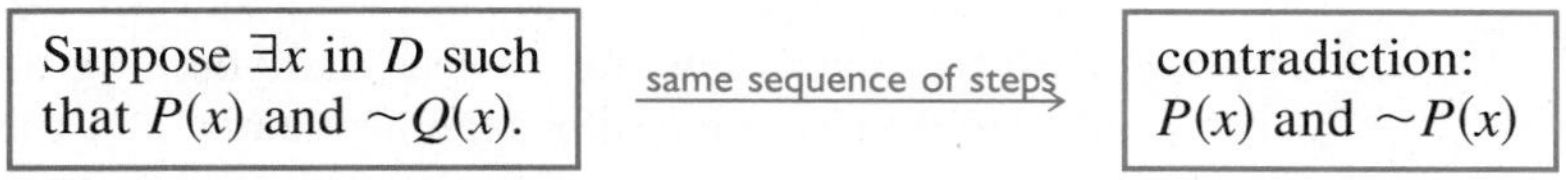

Suppose $\exists x$ in D such that $P(x)$ and $\sim Q(x)$. $\xrightarrow{\text{same sequence of steps}}$ contradiction: $P(x)$ and $\sim P(x)$

FIGURE 3.6.2   Proof by contradiction

As an example, here is a proof by contradiction that for any integer n, if n^2 is even then n is even.

Alternate Proof of Proposition 3.6.3 (by contradiction):

Suppose not. Suppose there exists an integer n such that n^2 is even and n is odd. *[We must deduce a contradiction.]* Since n is odd, n^2, which is the product of n with itself, is also odd (see exercise 25, Section 3.1). This contradicts the supposition that n^2 is even. *[Hence the supposition is false and the proposition is true.]*

Note that when you use proof by contraposition, you know exactly what conclusion you need to show, namely the negation of the hypothesis; whereas in proof by contradiction, it may be difficult to know what contradiction to head for. On the other hand, when you use proof by contradiction, once you have deduced any contradiction whatsoever, you are done. The main advantage of contraposition over contradiction is that you avoid having to take the negation (possibly incorrectly) of a complicated statement. The disadvantage of contraposition as compared with contradiction is that you can only use contraposition for a specific class of statements—those that are universal and conditional. The discussion above shows that any statement that can be proved by contraposition can be proved by contradiction. But the converse is not true. Statements such as "$\sqrt{2}$ is irrational" (discussed in the next section) can be proved by contradiction but not by contraposition.

PROOF AS A PROBLEM-SOLVING TOOL

Direct proof, disproof by counterexample, proof by contradiction, and proof by contraposition are all tools that may be used to help determine whether statements are true or false. Given a statement of the form

For all elements in a domain, if (hypothesis) then (conclusion),

imagine elements in the domain that satisfy the hypothesis. Ask yourself: Must they satisfy the conclusion? If you can see that the answer is "yes" in all cases, then the statement is true and your insight will form the basis for a direct proof. If after some thought it is not clear that the answer is "yes," try to think whether there are elements of the domain that satisfy the hypothesis and *not* the conclusion. If you are successful in finding some, then the statement is false and you have a counterexample. On the other hand, if you are not successful in finding such elements, perhaps none exist. Perhaps you can show that assuming the existence of elements in the domain that satisfy the hypothesis and not the conclusion leads logically to a contradiction. If so, then the given statement is true and you have the basis for a proof by contradiction. Alternatively, you could imagine elements of the domain for which the conclusion is false and ask whether such elements also fail to satisfy the hypothesis. If the answer in all cases is "yes," then you have a basis for a proof by contraposition.

Solving problems, especially difficult problems, is rarely a straightforward process. At any stage of following the guidelines above, you might want to try the method of a previous stage again. If, for example, you fail to find a counterexample for a certain statement, your experience in trying to find it might help you decide to reattempt a direct argument rather than trying an indirect one. Psychologists who have studied problem solving have found that the most successful problem solvers are those who are flexible and willing to use a variety of approaches without getting stuck in any one of them for very long. Sometimes mathematicians work for months (or longer) on difficult problems. Don't be discouraged if some problems in this book take you quite a while to solve.

Learning the skills of proof and disproof is much like learning other skills, such as those used in swimming, tennis, or playing a musical instrument. When you first start out, you may feel bewildered by all the rules, and you may not feel confident as you attempt new things. But with practice the rules become internalized and you can use them in conjunction with all your other powers—of balance, coordination, judgment, aesthetic sense—to concentrate on winning a meet, winning a match, or playing a concert successfully.

Now that you have worked through the first six sections of this chapter, return to the idea that, above all, a proof or disproof should be a convincing argument. You need to know how direct and indirect proofs and counterexamples are structured. But to use this knowledge effectively, you must use it in conjunction with your imaginative powers, your intuition, and especially your common sense.

EXERCISE SET 3.6

1. Fill in the blanks in the following proof that there is no least positive real number.
   Proof: Suppose not. Suppose that there is a real number x such that x is positive and __(a)__ for all positive real numbers y. Consider the number $x/2$. Then __(b)__ because x is positive and $x/2 < x$ because __(c)__ . Hence __(d)__ , which is a contradiction. *[Thus the supposition is false, and so there is no least positive real number.]*

Carefully formulate the negations of each of the statements in 2–4. Then prove each statement by contradiction.

2. There is no greatest even integer.

3. There is no greatest negative real number.

4. There is no least positive rational number.

5. When asked to prove that the difference of any rational number and any irrational number is irrational, a student begins: "Suppose not. Suppose the difference of any rational number and any irrational number is rational." Comment.

6. Prove by contradiction that the difference of any rational number and any irrational number is irrational.

Prove the statements in 7 and 8 by contraposition.

7. If a product of two positive real numbers is greater than 100, then at least one of the numbers is greater than 10.

8. If a sum of two real numbers is less than 50, then at least one of the numbers is less than 25.

Prove each of the statements in 9–14 in two ways: (a) by contraposition and (b) by contradiction.

9. The negative of any irrational number is irrational.

H 10. For all integers n, if n^2 is odd then n is odd.

H 11. For all integers n and all prime numbers p, if n^2 is divisible by p, then n is divisible by p.

12. For all integers a, b, and c, if $a \nmid bc$ then $a \nmid b$. (Recall that the symbol $\nmid$ means "does not divide.")

H 13. For all integers m and n, if $m + n$ is even then m and n are both even or m and n are both odd.

14. For all integers a, b, and c, if $a \mid b$ and $a \nmid c$, then $a \nmid (b + c)$. (*Hint for part (a):* To prove $p \to q \vee r$, it suffices to prove either $p \wedge \sim q \to r$ or $p \wedge \sim r \to q$. See exercise 14, Section 1.2.)

15. True or false? For all integers a and n, if $a \mid n^2$ then $a \mid n$. Justify your answer.

16. The following "proof" that every integer is rational is incorrect. Find the mistake.
    "Proof (by contradiction):
    Suppose not. Suppose every integer is irrational. Then the integer 1 is irrational. But $1 = 1/1$, which is rational. This is a contradiction. *[Hence the supposition is false and the theorem is true.]*"

Some of the statements in 17–23 are true and some are false. Use proof by contradiction to prove those that are true and give counterexamples to disprove those that are false. You may use the fact that $\sqrt{2}$ is irrational. This is proved in Section 3.7.

17. The sum of any two irrational numbers is irrational.

H 18. The product of any nonzero rational number and any irrational number is irrational.

H 19. The sum of any two positive irrational numbers is irrational.

20. The difference of any two irrational numbers is irrational.

21. The product of any two irrational numbers is irrational.

22. If a and b are rational numbers, $b \neq 0$, and r is an irrational number, then $a + br$ is irrational.

23. If r is any rational number and s is any irrational number, then r/s is irrational.

◆24. If a, b, and c are integers and $a^2 + b^2 = c^2$, must at least one of a and b be even? Justify your answer.

How flat and dead would be a mind that saw nothing in a negation but an opaque barrier! A live mind can see a window onto a world of possibilities.
(Douglas Hofstadter, *Gödel, Escher, Bach*, 1979)

3.7 TWO CLASSICAL THEOREMS

This section contains proofs of two of the most famous theorems in mathematics: that $\sqrt{2}$ is irrational and that there are infinitely many prime numbers. Both proofs are examples of indirect arguments and were well known more than 2,000 years ago, but they remain exemplary models of mathematical argument to this day.

THE IRRATIONALITY OF $\sqrt{2}$

When mathematics flourished at the time of the ancient Greeks, mathematicians believed that given any two line segments, say A: _____ and B: _____ , two integers, say a and b, could be found so that the ratio of the lengths of A and B would be in the same proportion as the ratio of a and b. Symbolically:

$$\frac{\text{length } A}{\text{length } B} = \frac{a}{b}.$$

Now it is easy to find a line segment of length $\sqrt{2}$; just take the diagonal of the unit square:

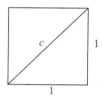

By the Pythagorean theorem, $c^2 = 1^2 + 1^2 = 2$, and so $c = \sqrt{2}$. If the belief of the ancient Greeks were correct, there would be integers a and b, such that

$$\frac{\text{length (diagonal)}}{\text{length (side)}} = \frac{a}{b}.$$

And this would imply that

$$\frac{c}{1} = \frac{\sqrt{2}}{1} = \sqrt{2} = \frac{a}{b}.$$

But then $\sqrt{2}$ would be a ratio of two integers, or, in other words, $\sqrt{2}$ would be rational.

In the fourth or fifth century B.C., the followers of the Greek mathematician and philosopher Pythagoras discovered that $\sqrt{2}$ was not rational. This discovery was very upsetting to them, for it undermined their deep, quasi-religious belief in the power of whole numbers to describe phenomena.

The following proof of the irrationality of $\sqrt{2}$ was known to Aristotle and is similar to that in the tenth book of Euclid's *Elements of Geometry*. The Greek mathematician Euclid is best known as a geometer. In fact, knowledge of the geometry in the first six books of his *Elements* has been considered an essential part of a liberal education for more than 2,000 years. Books 7–10 of his *Elements,* however, contain much that we would now call number theory.

The proof begins by supposing the negation: $\sqrt{2}$ is rational. This means that there exist integers m and n such that $\sqrt{2} = m/n$. Now if m and n have any common factors, these may be factored out to obtain a new fraction, equal to m/n, in which the numerator and denominator have no common factors. (For example, $18/12 = (6 \cdot 3)/(6 \cdot 2) = 3/2$, which is a fraction whose numerator and denominator have no common factors.) Thus without loss of generality we may assume that m and n had no common factors in the first place.* We will then derive the contradiction that m and

Euclid (fl. 300 B.C.)

---

*Strictly speaking, this deduction is a consequence of an axiom, called the "well-ordering principle," which is discussed in Section 4.4.

n do have a common factor of 2. The argument makes use of Proposition 3.6.3: If the square of an integer is even, then that integer is even.

THEOREM 3.7.1 Irrationality of $\sqrt{2}$

$\sqrt{2}$ is irrational.

Proof:

Suppose not. *[We take the negation and suppose it to be true.]* Suppose $\sqrt{2}$ is rational. Then there are integers m and n with no common factors so that

$$\sqrt{2} = \frac{m}{n} \qquad\qquad 3.7.1$$

[by dividing m and n by any common factors if necessary]. *[We must derive a contradiction.]* Squaring both sides of (3.7.1) gives

$$2 = \frac{m^2}{n^2}.$$

Or, equivalently,

$$m^2 = 2n^2. \qquad\qquad 3.7.2$$

Note that (3.7.2) implies that m^2 is even (by definition of even). It follows that m is even (by Proposition 3.6.3). We file this fact away for future reference and also deduce (by definition of even) that

$$m = 2k \quad \text{for some integer } k. \qquad\qquad 3.7.3$$

Substituting (3.7.3) into (3.7.2), we see that

$$m^2 = (2k)^2 = 4k^2 = 2n^2.$$

Dividing both sides of the right-most equation by 2 gives

$$n^2 = 2k^2.$$

Consequently, n^2 is even, and so n is even (by Proposition 3.6.3). But we also know that m is even. *[This is the fact we filed away.]* Hence both m and n have a common factor of 2. But this contradicts the supposition that m and n have no common factors. *[Hence the supposition is false and so the theorem is true.]*

Now that you have seen the proof that $\sqrt{2}$ is irrational, you can easily derive the irrationality of certain other real numbers.

EXAMPLE 3.7.1 Irrationality of $1 + 3\sqrt{2}$

Prove by contradiction that $1 + 3\sqrt{2}$ is irrational.

Solution The essence of the argument is the observation that if $1 + 3\sqrt{2}$ could be written as a fraction, then so could $\sqrt{2}$. But by Theorem 3.7.1, we know that to be a contradiction.

PROPOSITION 3.7.2

$1 + 3\sqrt{2}$ is irrational.

Proof:

Suppose not. Suppose $1 + 3\sqrt{2}$ is rational. *[We must derive a contradiction.]* Then by definition of rational,

$$1 + 3\sqrt{2} = \frac{a}{b} \quad \text{for some integers } a \text{ and } b \text{ with } b \neq 0.$$

It follows that

$$3\sqrt{2} = \frac{a}{b} - 1 \qquad \text{by subtracting I from both sides}$$

$$= \frac{a}{b} - \frac{b}{b} \qquad \text{by substitution}$$

$$= \frac{a - b}{b} \qquad \begin{array}{l}\text{by the rule for subtracting fractions} \\ \text{with a common denominator.}\end{array}$$

Hence

$$\sqrt{2} = \frac{a - b}{3b} \qquad \text{by dividing both sides by 3.}$$

But $a - b$ and $3b$ are integers (since a and b are integers and differences and products of integers are integers), and $3b \neq 0$ by the zero product property. Hence $\sqrt{2}$ is a quotient of the two integers $a - b$ and $3b$ with $3b \neq 0$, and so $\sqrt{2}$ is rational (by definition of rational.) This contradicts the fact that $\sqrt{2}$ is irrational. *[This contradiction shows that the supposition is false.]* Hence $1 + 3\sqrt{2}$ is irrational.

■

THE INFINITUDE OF THE SET OF PRIME NUMBERS

You know that a prime number is a positive integer that cannot be factored as a product of two smaller positive integers. Is the set of all such numbers infinite, or is there a largest prime number? The answer was known to Euclid, and a proof that the set of all prime numbers is infinite appears in Book 9 of his *Elements of Geometry*.

Euclid's proof requires one additional fact we have not yet established: If a prime number divides an integer a, then it does not divide $a + 1$.

PROPOSITION 3.7.3

For any integer a and any prime number p, if $p \mid a$, then $p \nmid (a + 1)$.

Proof:

Suppose not. Suppose there exists an integer a and a prime number p such that $p \mid a$ and $p \mid (a + 1)$. Then by definition of divisibility there exist integers r and s so that $a = pr$ and $a + 1 = ps$. It follows that $1 = (a + 1) - a = $

$ps - pr = p(s - r)$, and so (since $s - r$ is an integer) $p \mid 1$. But the only integer divisors of 1 are 1 and -1 (see Example 3.3.4), and since p is prime $p > 1$. This is a contradiction. *[Hence the supposition is false, and the proposition is true.]*

The idea of Euclid's proof is this: Suppose the set of prime numbers were finite. Then you could take the product of all the prime numbers and add one. By Theorem 3.3.2 this number must be divisible by some prime number. But by Proposition 3.7.3, this number is not divisible by any of the prime numbers in the set. Hence there must be a prime number that is not in the set of all prime numbers. This is impossible.

The following formal proof fills in the details of his outline.

THEOREM 3.7.4   Infinitude of the Primes

The set of prime numbers is infinite.

Proof (by contradiction):

Suppose not. Suppose the set of prime numbers is finite. *[We must deduce a contradiction.]* Then all the prime numbers can be listed, say, in ascending order:

$$p_1 = 2, p_2 = 3, p_3 = 5, p_4 = 7, p_5 = 11, \ldots, p_n.$$

Consider the integer

$$N = p_1 \cdot p_2 \cdot p_3 \cdot \ldots \cdot p_n + 1.$$

Then $N > 1$, and so by Theorem 3.3.2, N is divisible by some prime number p: $p \mid N$. Also since p is prime, p must equal one of the prime numbers $p_1, p_2, p_3, \ldots, p_n$. Thus $p \mid (p_1 \cdot p_2 \cdot p_3 \cdot \ldots \cdot p_n)$. By Proposition 3.7.3, then, $p \nmid (p_1 \cdot p_2 \cdot p_3 \cdot \ldots \cdot p_n + 1)$. So $p \nmid N$. Thus $p \mid N$ and $p \nmid N$, which is a contradiction. *[Hence the supposition is false and the theorem is true.]*

The proof of Theorem 3.7.3 shows that if you form the product of all prime numbers up to a certain point and add one, the result, N, is divisible by a prime number not on the list. The proof does not show that N is, itself, prime. In the exercises at the end of this section you are asked to find an example of an integer N constructed in this way that is not prime.

WHEN TO USE INDIRECT PROOF

The examples in this section and Section 3.6 have not provided a definitive answer to the question of when to prove a statement directly and when to prove it indirectly. Many theorems can be proved either way. Usually, however, when both types of proof are possible, indirect proof is clumsier than direct proof. In the absence of obvious clues suggesting indirect argument, try first to prove a statement directly. Then if that does not succeed, look for a counterexample. If the search for a counterexample is unsuccessful, look for a proof by contradiction or contraposition.

EXERCISE SET 3.7

1. Suppose a is an integer and p is a prime number such that $p \mid a$ and $p \mid (a + 3)$. What can you deduce about p? Why?

Some of the statements in 2–12 are true and some are false. Prove those that are true and disprove those that are false.

2. $6 - 7\sqrt{2}$ is irrational.

3. $4\sqrt{2} - 9$ is irrational.

H 4. $\sqrt{3}$ is irrational.

5. $\sqrt{4}$ is irrational.

6. $\sqrt{2}/4$ is rational.

H◆ 7. For any integer a, $4 \nmid (a^2 - 2)$.

H◆ 8. If n is any integer that is not a perfect square, then $\sqrt{n}$ is irrational.

◆ 9. $\sqrt{2} + \sqrt{3}$ is irrational.

10. $\sqrt[3]{2}$ is irrational.

H 11. If an integer is a perfect square, then its cube root is irrational.

12. The square root of an irrational number is irrational.

13. An alternative proof of the irrationality of $\sqrt{2}$ uses the unique factorization theorem to examine both sides of the equation $2n^2 = m^2$ and derive a contradiction. Write a proof that uses this approach.

14. If a, b, and c are odd integers, can $ax^2 + bx + c$ have a rational solution? Justify your answer.

◆ 15. Prove that $\log_2(3)$ is irrational.

H◆ 16. Prove that every integer greater than 11 is a sum of two composite numbers.

H 17. Fermat's Last Theorem says that for all integers $n > 2$, the equation $x^n + y^n = z^n$ has no positive integer solution (solution for which x, y, and z are positive integers). Prove the following: If for all prime numbers $p > 2$, $x^p + y^p = z^p$ has no positive integer solution, then for any integer $n > 2$ that is not a power of 2, $x^n + y^n = z^n$ has no positive integer solution.

18. Find the smallest nonprime integer of the form $p_1 \cdot p_2 \cdot \ldots \cdot p_n + 1$, where $p_1, p_2, \ldots,$ and p_n are consecutive prime numbers and $p_1 = 2$.

H◆ 19. Prove that if $p_1, p_2, \ldots,$ and p_n are distinct prime numbers with $p_1 = 2$ and $n > 1$, then $p_1 \cdot p_2 \cdot \ldots \cdot p_n + 1$ can be written in the form $4k + 3$ for some integer k.

H◆ 20. Prove that for all integers n, if $n > 2$ then there is a prime number p such that $n < p < n!$ where $n! = n(n - 1) \ldots 3 \cdot 2 \cdot 1$.

For exercises 21–22 note that to show there is a unique object with a certain property, first show that there is an object with the property and then show that if objects A and B have the property then A = B.

21. Prove that there exists a unique prime number of the form $n^2 - 1$, where n is an integer that is greater than or equal to 2.

22. Prove that there exists a unique prime number of the form $n^2 + 2n - 3$, where n is a positive integer.

23. Prove that there is at most one real number a with the property that $a + r = r$ for all real numbers r. (Such a number is called an *additive identity*.)

24. Prove that there is at most one real number b with the property that $br = r$ for all real numbers r. (Such a number is called a *multiplicative identity*.)

3.8   APPLICATION: ALGORITHMS

Begin at the beginning . . . and go on till you come to the end: then stop.
(Lewis Carroll, *Alice's Adventures in Wonderland,* 1865)

In this section we will show how the number theory facts developed in this chapter form the basis for some useful computer algorithms.

The word *algorithm* refers to a step-by-step method for performing some action. Some examples of algorithms in everyday life are food preparation recipes, directions for assembling equipment or hobby kits, sewing pattern instructions, and instructions for filling out income tax forms. Much of elementary school mathematics is devoted to learning algorithms for doing arithmetic such as multidigit addition and subtraction, multidigit (or long) multiplication, and long division.

The idea of a computer algorithm is credited to Ada Augusta, Countess of Lovelace. Trained as a mathematician, she became very interested in Charles Babbage's design for an "Analytical Engine," a machine similar in concept to a modern computer. Lady Lovelace extended Babbage's explorations of how such a machine would operate, recognizing that its importance lay "in the possibility of using a given sequence of instructions repeatedly, the number of times being either preassigned or dependent on the results of the computation." This is the essence of a modern computer algorithm.

AN ALGORITHMIC LANGUAGE

Lady Lovelace (1815–1852)

The algorithmic language used in this book is a kind of pseudocode, combining elements of Pascal, FORTRAN, BASIC, PL/I, C, Ada (named after Lady Lovelace), and ordinary, but fairly precise, English. We will use some of the formal constructs of computer languages—such as assignment statements, loops, and so forth—but we will ignore the more technical details such as the requirement for explicit end-of-statement delimiters, the range of integer values available on a particular installation, and so forth. The algorithms presented in this text are intended to be precise enough to be easily translated into virtually any high-level computer language.

In high-level computer languages, the term **variable** is used to refer to a specific storage location in a computer's memory. To say that the variable x has the value 3 means that the memory location corresponding to x contains the number 3. A given storage location can hold only one value at a time. So if a variable is given a new value during program execution, then the old value is erased. The **data type** of a variable indicates the set in which the variable takes its values, whether the set of integers, or real numbers, or character strings, or the set $\{0,1\}$ (for a Boolean variable), and so forth.

An **assignment statement** gives a value to a variable. It has the form

$$x := e,$$

where x is a variable and e is an expression. This is read "x is assigned the value e" or "let x be e." When an assignment statement is executed, the expression e is evaluated (using the current values of all the variables in the expression), and then its value is placed in the memory location corresponding to x (replacing any previous contents of this location).

Ordinarily, algorithm statements are executed one after another in the order in which they are written. **Conditional statements** allow this natural order to be overridden by using the current values of program variables to determine which algorithm statement will be executed next. Conditional statements are denoted either

a. **if** (*condition*)   or   b. **if** (*condition*) **then** s_1

    then s_1

    else s_2

where *condition* is a predicate involving algorithm variables and where s_1 and s_2 are algorithm statements or groups of algorithm statements. We generally use indentation to indicate that statements belong together as a unit. When ambiguity is possible, however, we may explicitly bind a group of statements together into a unit by preceding the group with the word **do** and following it with the words **end do.**

Execution of an **if–then–else** statement occurs as follows:

1. The *condition* is evaluated by substituting the current values of all algorithm variables appearing in it and evaluating the truth or falsity of the resulting statement.

2. If *condition* is true, then s_1 is executed and execution moves to the next algorithm statement following the **if–then–else** statement.

3. If *condition* is false, then s_2 is executed and moves to the next algorithm statement following the **if–then–else** statement.

Execution of an **if–then** statement is similar to execution of an **if–then–else** statement, except that if *condition* is false, execution passes immediately to the next algorithm statement following the **if–then** statement.

Often *condition* is called a **guard** because it is stationed before s_1 and s_2 and restricts access to them.

EXAMPLE 3.8.1   Execution of **if–then–else** and **if-then** Statements

Consider the following statements:

a. **if** $x > 2$ b. $y := 0$

    then $y := x + 1$ **if** $x > 2$ **then** $y := 2^x$

    else do $x := x - 1$

      $y := 3 \cdot x$ **end do**

What is the value of y after execution of these statements for the following values of x?

i. $x = 5$ ii. $x = 2$

Solution   a. (i) Because the value of x is 5 before execution, the guard condition $x > 2$ is true at the time it is evaluated. Hence the statement following **then** is executed, and so the value of $x + 1 = 5 + 1$ is computed and placed in the storage location corresponding to y. So after execution, $y = 6$.

(ii) Because the value of x is 2 before execution, the guard condition $x > 2$ is false at the time it is evaluated. Hence the statement following **else** is executed. The value of $x - 1 = 2 - 1$ is computed and placed in the storage location corresponding to x, and the value of $3 \cdot x = 3 \cdot 1$ is computed and placed in the storage location corresponding to y. So after execution, $y = 3$.

b. (i) Since $x = 5$ initially, the condition $x > 2$ is true at the time it is evaluated. So the statement following **then** is executed, and y obtains the value $2^5 = 32$.

(ii) Since $x = 2$ initially, the condition $x > 2$ is false at the time it is evaluated. Execution, therefore, moves to the next statement following the if–then statement, and the value of y does not change from its initial value of 0. ■

Iterative statements are used when a sequence of algorithm statements is to be executed over and over again. We will use two types of iterative statements: **while** loops and **for–next** loops.

A **while** loop has the form

          while *(condition)*

             [statements that make up
             the body of the loop]

          end while

where *condition* is a predicate involving algorithm variables. The word **while** marks the beginning of the loop and the words **end while** mark its end. Execution of a **while** loop occurs as follows:

1. The *condition* is evaluated by substituting the current values of all the algorithm variables and evaluating the truth or falsity of the resulting statement.

2. If *condition* is true, all the statements in the body of the loop are executed in order. Then execution moves back to the beginning of the loop and the process repeats.

3. If *condition* is false, execution passes to the next algorithm statement following the loop.

The loop is said to be **iterated** (IT-a-rate-ed) each time the statements in the body of the loop are executed. Each execution of the body of the loop is called an **iteration** (it-er-AY-shun) of the loop.

EXAMPLE 3.8.2   Tracing Execution of a **while** Loop

Trace the execution of the following algorithm segment by finding the values of all the algorithm variables each time they are changed during execution:

$$i := 1, s := 0$$
$$\textbf{while } (i \leq 2)$$
$$s := s + i$$
$$i := i + 1$$
$$\textbf{end while}$$

Solution   Since i is given an initial value of 1, the condition $i \leq 2$ is true when the **while** loop is entered. So the statements within the loop are executed in order:

$$s = 0 + 1 = 1 \quad \text{and} \quad i = 1 + 1 = 2.$$

Then execution passes back to the beginning of the loop.

   The condition $i \leq 2$ is evaluated using the current value of i, which is 2. The condition is true, and so the statements within the loop are executed again:

$$s = 1 + 2 = 3 \quad \text{and} \quad i = 2 + 1 = 3.$$

Then execution passes back to the beginning of the loop.

   The condition $i \leq 2$ is evaluated using the current value of i, which is 3. This time the condition is false, and so execution passes beyond the loop to the next statement of the algorithm.   ■

   The discussion above can be summarized in a table, called a **trace table**, that shows the current values of algorithm variables at various points during execution. The trace table for a **while** loop generally gives all values immediately following each iteration of the loop. ("After the zeroth iteration" means the same as "before the first iteration.")

Trace Table

Variable Name	Iteration Number		
	0	1	2
i	1	2	3
s	0	1	3

The second form of iteration we will use is a **for–next** loop. A **for–next** loop has the following form:

> **for** *variable* := *initial expression* **to** *final expression*
>   *[statements that make up*
>   *the body of the loop]*
> **next** *(same) variable*

A **for–next** loop is executed as follows:

1. The **for–next** loop *variable* is set equal to the value of *initial expression.*
2. A check is made to determine whether the value of *variable* is less than or equal to the value of *final expression.*
3. If the value of *variable* is less than or equal to the value of *final expression,* the statements in the body of the loop are executed, *variable* is increased by 1, and execution returns back to step 2.
4. If the value of *variable* is greater than the value of *final expression,* execution passes to the next algorithm statement following the loop.

EXAMPLE 3.8.3   Trace Table for a **for–next** Loop

Convert the **for–next** loop shown below to a **while** loop. Construct a trace table for the loop.

$$\textbf{for } i := 1 \textbf{ to } 4$$
$$x := i^2$$
$$\textbf{next } i$$

Solution   The given **for–next** loop is equivalent to the following:

$$i := 1$$
$$\textbf{while } (i \leq 4)$$
$$x := i^2$$
$$i := i + 1$$
$$\textbf{end while}$$

Its trace table is as follows:

Trace Table

| | | \multicolumn{5}{c}{Iteration Number} |
		0	1	2	3	4
Algorithm Variables	x		1	4	9	16
	i	1	2	3	4	5

A NOTATION FOR ALGORITHMS

We will express algorithms as subroutines that can be called upon by other algorithms as needed and used to transform a set of input variables with given values into a set of output variables with specific values. The output variables and their values are assumed to be returned to the calling algorithm. For example, the division algorithm specifies a procedure for taking any two positive integers as input and producing the quotient and remainder of the division of one number by the other as output. Whenever an algorithm requires such a computation, the algorithm can just "call" the division algorithm to do the job.

We generally include the following information when describing algorithms formally:

1. The name of the algorithm, together with a list of input and output variables.
2. A brief description of how the algorithm works.
3. The input variable names, labeled by data type (whether integer, real number, and so forth).
4. The statements that make up the body of the algorithm, possibly with explanatory comments.
5. The output variable names, labeled by data type.
6. An end statement.

al-Khowârizmî (c.780–c.850)

You may wonder where the word *algorithm* came from. It evolved from the last part of the name of the Persian mathematician Abu Ja'far Mohammed ibn Mûsâ al-Khowârizmî. During Europe's Dark Ages, the Arabic world enjoyed a period of intense intellectual activity. One of the great mathematical works of that period was a book written by al-Khowârizmî that contained foundational ideas for the subject of algebra. The translation of this book into Latin in the thirteenth century had a profound influence on the development of mathematics during the European Renaissance.

THE DIVISION ALGORITHM

For an integer a and a positive integer d, the quotient-remainder theorem guarantees the existence of integers q and r such that

$$a = d \cdot q + r \quad \text{and} \quad 0 \le r < d.$$

In this section, we give an algorithm to calculate q and r for given a and d where a is nonnegative. (The extension to negative a is left to the exercises at the end of this section.) The following example illustrates the idea behind the algorithm. Consider trying to find the quotient and the remainder of the division of 32 by 9, but suppose that you do not remember your multiplication table and have to figure out the answer from basic principles. The quotient represents that number of 9's that are contained in 32. The remainder is the number left over when all possible groups of 9 are subtracted. Thus you can calculate the quotient and remainder by repeatedly subtracting 9 from 32 until you obtain a number less than 9:

$$32 - 9 = 23 \ge 9, \text{ and}$$
$$32 - 9 - 9 = 14 \ge 9, \text{ and}$$
$$32 - 9 - 9 - 9 = 5 < 9.$$

This shows that 3 groups of 9 can be subtracted from 32 with 5 left over. Thus the quotient is 3 and the remainder is 5.

ALGORITHM 3.8.1   Division Algorithm

[Given a nonnegative integer a and a positive integer d, the aim of the algorithm is to find integers q and r that satisfy the conditions $a = d \cdot q + r$ and $0 \leq r < d$. This is done by subtracting d repeatedly from a until the result is less than d but is still nonnegative.

$$0 \leq a - d - d - d \ldots - d = a - d \cdot q < d.$$

The total number of d's that are subtracted is the quotient q. The quantity $a - d \cdot q$ equals the remainder r.]

Input: *a [a nonnegative integer], d [a positive integer]*

Algorithm Body:
   $r := a, q := 0$
   [Repeatedly subtract d from r until a number less than d is obtained. Add 1 to q each time d is subtracted.]

   while $(r \geq d)$
      $r := r - d$
      $q := q + 1$
   end while
   *[After execution of the **while** loop, $a = d \cdot q + r$.]*

Output: *q, r [nonnegative integers]*
end Algorithm 3.8.1

Note that the values of q and r obtained from the division algorithm are the same as those computed by the *div* and *mod* functions built into a number of computer languages. That is, if q and r are the quotient and remainder obtained from the division algorithm with input a and d, then the output variables q and r satisfy

$$q = a \ div \ d \quad \text{and} \quad r = a \ mod \ d.$$

The next example asks for a trace of the division algorithm.

EXAMPLE 3.8.4   Tracing the Division Algorithm

Trace the action of the Algorithm 3.8.1 on the input variables $a = 19$ and $d = 4$.

Solution   Make a trace table as shown below. The column under the kth iteration gives the states of the variables after the kth iteration of the loop.

		Iteration Number				
		0	1	2	3	4
	a	19				
	d	4				
Variable Names	r	19	15	11	7	3
	q	0	1	2	3	4

THE EUCLIDEAN ALGORITHM

The greatest common divisor of two integers a and b is the largest integer that divides both a and b. For example, the greatest common divisor of 12 and 30 is 6. The Euclidean algorithm provides a very efficient way to compute the greatest common divisor of two integers.

DEFINITION

Let a and b be integers that are not both zero. The **greatest common divisor** of a and b, denoted **gcd(a, b)**, is that integer d with the following properties:

1. d is a common divisor of both a and b. In other words,

$$d \mid a \quad \text{and} \quad d \mid b.$$

2. For all integers c, if c is a common divisor of both a and b, then c is less than or equal to d. In other words,

$$\text{for all integers } c, \text{ if } c \mid a \text{ and } c \mid b, \text{ then } c \le d.$$

EXAMPLE 3.8.5 Calculating Some gcd's

a. Find gcd(72, 63).
b. Find gcd(10^{20}, 6^{30}).
c. In the definition of greatest common divisor, gcd(0, 0) is not allowed. Why not? What would gcd(0, 0) equal if it were found in the same way as the greatest common divisors for other pairs of numbers?

Solution

a. $72 = 9 \cdot 8$ and $63 = 9 \cdot 7$. So $9 \mid 72$ and $9 \mid 63$, and no integer larger than 9 divides both 72 and 63. Hence, gcd(72, 63) = 9.

b. By the laws of exponents, $10^{20} = 2^{20} \cdot 5^{20}$ and $6^{30} = 2^{30} \cdot 3^{30} = 2^{20} \cdot 2^{10} \cdot 3^{30}$. It follows that

$$2^{20} \mid 10^{20} \quad \text{and} \quad 2^{20} \mid 6^{30},$$

and by the unique factorization theorem, no integer larger than 2^{20} divides both 10^{20} and 6^{30} (because no more than twenty 2's divide 10^{20}, no 3's divide 10^{20}, and no 5's divide 6^{30}). Hence gcd(10^{20}, 6^{30}) = 2^{20}.

c. Suppose gcd(0, 0) were defined to be the largest common factor that divides 0 and 0. The problem is that *every* positive integer divides 0 and there is no largest integer. So there is no largest common factor! ∎

Calculating gcd's using the approach illustrated in Example 3.8.5 only works when the numbers can be factored completely. By the unique factorization theorem, all numbers can, in principle, be factored completely. But, in fact, even using the highest-speed computers, the process is unfeasibly long for very large integers. Over 2,000 years ago, Euclid devised a method for finding greatest common divisors that is easy to use and is much more efficient than either factoring the numbers or repeatedly testing both numbers for divisibility by successively larger integers.

The Euclidean algorithm is based on the following two facts, which are stated as lemmas:

LEMMA 3.8.1

If r is a positive integer, then $\gcd(r, 0) = r$.

Proof:

Suppose r is a positive integer. *[We must show that the greatest common divisor of both r and 0 is r.]* Certainly, r is a common divisor of both r and 0 because r divides itself and also r divides 0 (since every positive integer divides 0). Also no integer larger than r can be a common divisor of r and 0 (since no integer larger than r can divide r). Hence r is the greatest common divisor of r and 0.

The proof of the second lemma is based on a clever pattern of argument that is used in many different areas of mathematics: To prove that $A = B$, prove that $A \leq B$ and that $B \leq A$.

LEMMA 3.8.2

If a and b are any integers with $b \neq 0$ and q and r are nonnegative integers such that

$$a = b \cdot q + r,$$

then

$$\gcd(a, b) = \gcd(b, r).$$

Proof:

[The proof is divided into two sections: (1) proof that $\gcd(a, b) \leq \gcd(b, r)$, and (2) proof that $\gcd(b, r) \leq \gcd(a, b)$. Since each gcd is less than or equal to the other, the two must be equal.]

1. $\gcd(a, b) \leq \gcd(b, r)$:

    a. We *will first show that any common divisor of a and b is also a common divisor of b and r.*

        Let c be a common divisor of a and b. Then $c \mid a$ and $c \mid b$, and so by definition of divisibility, $a = n \cdot c$ and $b = m \cdot c$, for some integers n and m. Now substitute into the equation

$$a = b \cdot q + r$$

to obtain

$$n \cdot c = (m \cdot c) \cdot q + r.$$

Then solve for r:

$$r = n \cdot c - (m \cdot c) \cdot q = (n - m \cdot q) \cdot c.$$

But $n - m \cdot q$ is an integer, and so by definition of divisibility, $c \mid r$. Now we already know that $c \mid b$; hence c is a common divisor of b and r [as was to be shown].

b. *Next we show that* $\gcd(a, b) \leq \gcd(b, r)$.

By part (a), every common divisor of a and b is a common divisor of b and r. It follows that the greatest common divisor of a and b is a common divisor of b and r. But then $\gcd(a, b)$ (being one of the common divisors of b and r) is less than or equal to the greatest common divisor of b and r:

$$\gcd(a, b) \leq \gcd(b, r).$$

2. $\gcd(b, r) \leq \gcd(a, b)$:

The second part of the proof is very similar to the first part. It is left as an exercise.

The Euclidean algorithm can be described as follows:

1. Let A and B be integers with $A > B \geq 0$.
2. To find the greatest common divisor of A and B, first check whether $B = 0$. If it is, then $\gcd(A, B) = A$ by Lemma 3.8.1. If it isn't, then $B > 0$ and the quotient-remainder theorem can be used to divide A by B to obtain a quotient q and a remainder r:

$$A = B \cdot q + r \quad \text{where } 0 \leq r < B.$$

By Lemma 3.8.2, $\gcd(A, B) = \gcd(B, r)$. Thus the problem of finding the greatest common divisor of A and B is reduced to the problem of finding the greatest common divisor of B and r.

What makes this piece of information useful is that B and r are smaller numbers than A and B. To see this, recall that we assumed

$$A > B \geq 0.$$

Also the r found by the quotient-remainder theorem satisfies

$$0 \leq r < B.$$

Putting these two inequalities together gives

$$0 \leq r < B < A.$$

So the largest number of the pair (B, r) is smaller than the largest number of the pair (A, B).

3. Now just repeat the process, starting again at (2), but use B instead of A and r instead of B. The repetitions are guaranteed to terminate eventually with $r = 0$ because each new remainder is less than the preceding one and all are nonnegative.

By the way, it is always the case that the number of steps required in the Euclidean algorithm is at most five times the number of digits in the smaller integer. This was proved by the French mathematician Gabriel Lamé (1795–1870).

The following example illustrates how to use the Euclidean algorithm.

EXAMPLE 3.8.6   Hand Calculation of gcd's Using the
Euclidean Algorithm

Use the Euclidean algorithm to find gcd(330, 156).

Solution   1. Divide 330 by 156:

$$\begin{array}{r} 2 \leftarrow \text{quotient} \\ 156 \overline{\smash{)}330} \\ \underline{312} \\ 18 \leftarrow \text{remainder} \end{array}$$

Thus $330 = 156 \cdot 2 + 18$ and hence $\gcd(330, 156) = \gcd(156, 18)$ by Lemma 3.8.2.

2. Divide 156 by 18:

$$\begin{array}{r} 8 \leftarrow \text{quotient} \\ 18 \overline{\smash{)}156} \\ \underline{144} \\ 12 \leftarrow \text{remainder} \end{array}$$

Thus $156 = 18 \cdot 8 + 12$ and hence $\gcd(156, 18) = \gcd(18, 12)$ by Lemma 3.8.2.

3. Divide 18 by 12:

$$\begin{array}{r} 1 \leftarrow \text{quotient} \\ 12 \overline{\smash{)}18} \\ \underline{12} \\ 6 \leftarrow \text{remainder} \end{array}$$

Thus $18 = 12 \cdot 1 + 6$ and hence $\gcd(18, 12) = \gcd(12, 6)$ by Lemma 3.8.2.

4. Divide 12 by 6:

$$\begin{array}{r} 2 \leftarrow \text{quotient} \\ 6 \overline{\smash{)}12} \\ \underline{12} \\ 0 \leftarrow \text{remainder} \end{array}$$

Thus $12 = 6 \cdot 2 + 0$ and hence $\gcd(12, 6) = \gcd(6, 0)$ by Lemma 3.8.2.

Putting all the equations above together gives

$$\begin{aligned} \gcd(330, 156) &= \gcd(156, 18) \\ &= \gcd(18, 12) \\ &= \gcd(12, 6) \\ &= \gcd(6, 0) \\ &= 6 \qquad \text{by Lemma 3.8.1.} \end{aligned}$$

Therefore, $\gcd(330, 156) = 6$. ∎

The following is a version of the Euclidean algorithm written using formal algorithm notation.

ALGORITHM 3.8.2   Euclidean Algorithm

[Given two integers A and B with $A > B \geq 0$, this algorithm computes gcd (A, B). It is based on two facts:

1. gcd $(a, b) = $ gcd (b, r) if a, b, q, and r are integers with $a = b \cdot q + r$ and $0 \leq r < b$;

2. gcd $(a, 0) = a.]$

Input: A, B *[integers with $A > B \geq 0]$*

Algorithm Body:
  $a := A, b := B$

  [If $b \neq 0$, compute a mod b, the remainder of the integer division of a by b, and set r equal to this value. Then repeat the process using b in place of a and r in place of b.]

  while $(b \neq 0)$
      $r := a \bmod b$

  [The value of a mod b can be obtained by calling the division algorithm.]
      $a := b$
      $b := r$

  end while

  *[After execution of the **while** loop, gcd $(A, B) = a.]$*
  gcd $:= a$

Output gcd *[a positive integer]*
end Algorithm 3.8.2

EXERCISE SET 3.8

Find the value of z when each of the algorithm segments in 1–2 is executed.

1. $i := 2$
   if $(i > 3$ or $i \leq 0)$
       then $z := 1$
       else $z := 0$

2. $i := 2$
   if $(i \leq 2$ or $i > 5)$
       then $z := 1$
       else $z := 0$

3. Consider the following algorithm segment:

   if $x \cdot y > 0$ **then do** $y := 2 \cdot x$
                          $x := x + 1$ **end do**
   $z := x \cdot y$

Find the value of z if prior to execution x and y have the values given below.

a. $x = 2, y = 3$ b. $x = 1, y = 1$

Find the values of a and e after execution of the loops in 4 and 5:

4. $a := 2$
   for $i := 1$ **to** 2
       $a := \dfrac{a}{2} + \dfrac{1}{a}$
   next i

5. $e := 0, f := 1$
   for $j := 1$ **to** 4
       $f := f \cdot j$
       $e := e + \dfrac{1}{f}$
   next j

Make a trace table to trace the action of Algorithm 3.8.1 for the input variables given in 6 and 7.

6. $a = 26, d = 7$ 7. $a = 54, d = 11$

Find the greatest common divisor of each of the pairs of integers in 8–11. (Use any method you wish.)

8. 27 and 72   9. 5 and 9

10. 5 and 10   11. 42 and 63

Use the Euclidean algorithm to hand calculate the greatest common divisors of each of the pairs of integers in 12–14.

12. 1,188 and 385   13. 544 and 1,001

14. 3,510 and 672

Make a trace table to trace the action of Algorithm 3.8.2 for the input variables given in 15 and 16.

15. 1,001 and 871   16. 2,628 and 738

17. Prove that for all positive integers a and b, $a \mid b$ if, and only if, $\gcd(a, b) = a$. (Note that to prove "A if, and only if, B," you need to prove "if A then B" and "if B then A.")

18. Write an algorithm that accepts the numerator and denominator of a fraction as input and produces as output the numerator and denominator of that fraction written in lowest terms. (The algorithm may call upon the Euclidean algorithm as needed.)

19. Complete the proof of Lemma 3.8.2 by proving the following: If a and b are any positive integers and q and r are any integers such that

$$a = bq + r \quad \text{and} \quad 0 \le r < b,$$

then

$$\gcd(b, r) \le \gcd(a, b).$$

20. The quotient-remainder theorem says not only that there exist quotients and remainders but also that the quotient and remainder of a division are unique. Prove the uniqueness. That is, prove that if a and d are integers with $d > 0$ and if q_1, r_1, q_2, and r_2 are integers such that

$$a = d \cdot q_1 + r_1 \quad \text{where } 0 \le r_1 < d$$

and

$$a = d \cdot q_2 + r_2 \quad \text{where } 0 \le r_2 < d,$$

then

$$q_1 = q_2 \quad \text{and} \quad r_1 = r_2.$$

H 21. a. Prove: If a and d are positive integers and q and r are integers such that $a = d \cdot q + r$ and $0 < r < d$, then

$$-a = d \cdot (-(q + 1)) + (d - r)$$

and

$$0 < d - r < d.$$

b. Indicate how to modify Algorithm 3.8.1 to allow for the input a to be negative.

22. a. Prove that if a, d, q, and r are integers such that $a = d \cdot q + r$ and $0 \le r < d$, then

$$q = \lfloor a/d \rfloor \quad \text{and} \quad r = a - \lfloor a/d \rfloor \cdot d.$$

b. In a computer language with a built-in floor function, div and mod can be calculated as follows:

$$a \ div \ d = \lfloor a/d \rfloor \quad \text{and} \quad a \ mod \ d = a - \lfloor a/d \rfloor \cdot d.$$

Rewrite the steps of Algorithm 3.8.2 for a computer language with a built-in floor function but without div and mod.

23. An alternative to the Euclidean algorithm uses subtraction rather than division to compute greatest common divisors. (After all, division is repeated subtraction.) It is based on the following lemma:

LEMMA 3.8.3

If $a \ge b > 0$, then $\gcd(a, b) = \gcd(b, a - b)$.

ALGORITHM 3.8.3   Computing gcd's by Subtraction

[Given two positive integers A and B, variables a and b are set equal to A and B. Then a repetitive process begins. If $a \ne 0$ and $b \ne 0$, then the larger of a and b is set equal to $a - b$ (if $a \ge b$) or to $b - a$ (if $a < b$), and the smaller of a and b is left unchanged. This process is repeated over and over until eventually a or b becomes 0. By Lemma 3.8.3, after each repetition of the process,

$$\gcd(A, B) = \gcd(a, b).$$

After the last repetition,

$$\gcd(A, B) = \gcd(a, 0) \quad or \quad \gcd(A, B) = \gcd(0, b)$$

depending on whether a or b is nonzero. But by Lemma 3.8.1,

$$\gcd(a, 0) = a \quad and \quad \gcd(0, b) = b.$$

Hence, after the last repetition,

$$\gcd(A, B) = a \ if \ a \ne 0 \quad or \quad \gcd(A, B) = b \ if \ b \ne 0.]$$

Input: A, B *[positive integers]*
Algorithm Body:
    $a := A$, $b := B$
    while $(a \ne 0 \text{ and } b \ne 0)$
        if $a \ge b$ **then** $a := a - b$
            else $b := b - a$
    end while
    if $a = 0$ **then** gcd $:= b$
        else gcd $:= a$
*[After execution of the **if–then–else** statement, gcd = gcd(A, B).]*

Output: gcd [*a positive integer*]

end Algorithm 3.8.3

a. Prove Lemma 3.8.3.
b. Trace the execution of Algorithm 3.8.3 for $A = 630$ and $B = 336$.

Exercises 24–28 refer to the following definition.

DEFINITION:

The **least common multiple** of two nonzero integers a and b, denoted **lcm(a, b)**, is the positive integer c such that

1. $a \mid c$ and $b \mid c$
2. for all integers m, if $a \mid m$ and $b \mid m$, then $c \mid m$.

24. Find

    a. lcm(12, 18)

    b. lcm($2 \cdot 3^2 \cdot 5$, $2^3 \cdot 3$)

    c. lcm(3500, 1960)

25. Prove that for all positive integers a and b, gcd(a, b) = lcm(a, b) if, and only if, $a = b$.

26. Prove that for all positive integers a and b, $a \mid b$ if, and only if, lcm(a, b) = b.

27. Prove that for all integers a and b, gcd(a, b) $\mid$ lcm(a, b).

28. Prove that for all positive integers a and b, gcd(a, b) $\cdot$ lcm(a, b) = ab.

4 Sequences and Mathematical Induction

One of the most important tasks of mathematics is to discover and characterize regular patterns, such as those associated with processes that are repeated. The main mathematical structure used to study repeated processes is the *sequence*. The main mathematical tool used to verify conjectures about patterns governing the arrangement of terms in sequences is *mathematical induction*. In this chapter we introduce the notation and terminology of sequences, show how to use both the ordinary and the strong forms of mathematical induction, and give an application showing how to prove the correctness of computer algorithms.

4.1 SEQUENCES

Imagine that a person decides to count his ancestors. He has two parents, four grandparents, eight great-grandparents, and so forth. These numbers can be written in a row as

$$2, 4, 8, 16, 32, 64, 128, \ldots$$

The symbol ". . ." is called an *ellipsis*. It is shorthand for "and so forth."

To express the pattern of the numbers, suppose that each is labeled by an integer giving its position in the row.

position in the row	1	2	3	4	5	6	7 . . .
number of ancestors	2	4	8	16	32	64	128 . . .

The number corresponding to position 1 is 2, which equals 2^1. The number corresponding to position 2 is 4, which equals 2^2. For positions 3, 4, 5, 6, and 7, the corresponding numbers are 8, 16, 32, 64, and 128, which equal 2^3, 2^4, 2^5, 2^6, and 2^7, respectively. For a general value of k, let A_k be the number of ancestors in the kth generation back. The pattern of computed values strongly suggests the following for each k:

$$A_k = 2^k. *$$

In this section we define the term **sequence** informally as a set of elements written in a row. (We give a more formal definition of sequence in terms of functions in Section 7.1.) In the sequence denoted

$$a_m, a_{m+1}, a_{m+2}, \ldots, a_n,$$

each individual element a_k (read "*a* sub *k*") is called a **term**. The k in a_k is called a **subscript** or **index**, m (which may be any integer) is the subscript of the **initial term**, and n (which must be greater than or equal to m) is the subscript of the **final term**. The notation

$$a_m, a_{m+1}, a_{m+2}, \ldots$$

denotes an **infinite sequence**. An **explicit formula** or **general formula** for a sequence is a rule that shows how the values of a_k depend on k.

The following example shows that it is possible for two different formulas to give sequences with the same terms.

EXAMPLE 4.1.1   **Finding Terms of Sequences Given by Explicit Formulas**

Define sequences $a_1, a_2, a_3, \ldots$ and $b_2, b_3, b_4, \ldots$ by the following explicit formulas:

$$a_k = \frac{k}{k+1} \quad \text{for all integers } k \geq 1,$$

$$b_i = \frac{i-1}{i} \quad \text{for all integers } i \geq 2.$$

Compute the first five terms of both sequences.

Solution

$$a_1 = \frac{1}{1+1} = \frac{1}{2} \qquad b_2 = \frac{2-1}{2} = \frac{1}{2}$$

$$a_2 = \frac{2}{2+1} = \frac{2}{3} \qquad b_3 = \frac{3-1}{3} = \frac{2}{3}$$

$$a_3 = \frac{3}{3+1} = \frac{3}{4} \qquad b_4 = \frac{4-1}{4} = \frac{3}{4}$$

$$a_4 = \frac{4}{4+1} = \frac{4}{5} \qquad b_5 = \frac{5-1}{5} = \frac{4}{5}$$

$$a_5 = \frac{5}{5+1} = \frac{5}{6} \qquad b_6 = \frac{6-1}{6} = \frac{5}{6}$$

As you can see, the first terms of both sequences are $\frac{1}{2}, \frac{2}{3}, \frac{3}{4}, \frac{4}{5}, \frac{5}{6}$; in fact, it can be shown that all terms of both sequences are identical. ∎

The next example shows that an infinite sequence may have only a finite number of values.

*Strictly speaking, the true value of A_k is probably less than 2^k when k is large because ancestors from one branch of the family tree may also appear on other branches of the tree.

EXAMPLE 4.1.2   An Alternating Sequence

Compute the first six terms of the sequence $c_0, c_1, c_2, \ldots$ defined as follows:

$$c_j = (-1)^j \quad \text{for all integers } j \geq 0.$$

Solution
$$c_0 = (-1)^0 = 1$$
$$c_1 = (-1)^1 = -1$$
$$c_2 = (-1)^2 = 1$$
$$c_3 = (-1)^3 = -1$$
$$c_4 = (-1)^4 = 1$$
$$c_5 = (-1)^5 = -1$$

Thus the first six terms are $1, -1, 1, -1, 1, -1$. By exercises 13 and 14 of Section 3.1, even powers of -1 equal 1 and odd powers equal -1. It follows that the sequence oscillates endlessly between 1 and -1.   ■

In Examples 4.1.1 and 4.1.2 the task was to compute initial values of a sequence given by an explicit formula. The next example treats the question of how to find an explicit formula for a sequence with given initial terms. Any such formula is a guess, but it is very useful to be able to make such guesses.

EXAMPLE 4.1.3   Finding an Explicit Formula to Fit
Given Initial Terms

Find an explicit formula for a sequence that has the following initial terms:

$$1, -\frac{1}{4}, \frac{1}{9}, -\frac{1}{16}, \frac{1}{25}, -\frac{1}{36}, \ldots$$

Solution   Denote the general term of the sequence by a_k and suppose the first term is a_1. Then observe that the denominator of each term is a perfect square. Thus the terms can be rewritten as

$$\frac{1}{1^2}, \frac{(-1)}{2^2}, \frac{1}{3^2}, \frac{(-1)}{4^2}, \frac{1}{5^2}, \frac{(-1)}{6^2}.$$
$$\updownarrow \quad \updownarrow \quad \updownarrow \quad \updownarrow \quad \updownarrow \quad \updownarrow$$
$$a_1 \quad a_2 \quad a_3 \quad a_4 \quad a_5 \quad a_6$$

Note that the denominator of each term equals the square of the subscript of that term, and that the numerator equals ± 1. Hence

$$a_k = \frac{\pm 1}{k^2}.$$

Also the numerator oscillates back and forth between $+1$ and -1; it is $+1$ when k is odd and -1 when k is even. To achieve this oscillation, insert a factor of $(-1)^{k+1}$ (or $(-1)^{k-1}$) into the formula for a_k. *[For when k is odd, $k + 1$ is even and thus $(-1)^{k+1} = +1$; and when k is even, $k + 1$ is odd and thus $(-1)^{k+1} = -1$.]* Consequently, an explicit formula that gives the correct first six terms is

$$a_k = \frac{(-1)^{k+1}}{k^2} \quad \text{for all integers } k \geq 1.$$

Note that making the first term a_0 would have led to the alternative formula

$$a_k = \frac{(-1)^k}{(k+1)^2} \quad \text{for all integers } k \geq 0.$$

You should check that this formula also gives the correct first six terms. ∎

▲ CAUTION!   *Two sequences may start off with the same initial values but diverge later on. See exercise 7 at the end of this section.*

SUMMATION NOTATION

Consider again the example in which $A_k = 2^k$ represented the number of ancestors a person has in the kth generation back. What is the total number of ancestors for the past six generations? The answer is

$$A_1 + A_2 + A_3 + A_4 + A_5 + A_6 = 2^1 + 2^2 + 2^3 + 2^4 + 2^5 + 2^6 = 126.$$

It is convenient to use a shorthand notation to write such sums. In 1772 the French mathematician Joseph Louis Lagrange introduced the capital Greek letter sigma, Σ, to denote the word *sum* (or *summation*), and the notation

$$\sum_{k=1}^{n} a_k$$

to represent the sum given in **expanded form** by

$$a_1 + a_2 + a_3 + \cdots + a_n.$$

More generally, if m and n are integers and $m \leq n$, then the **summation from k equals m to n of a_k** is the sum of all the terms $a_m, a_{m+1}, a_{m+2}, \ldots, a_n$. We write

$$\sum_{k=m}^{n} a_k = a_m + a_{m+1} + a_{m+2} + \cdots + a_n$$

and call k the **index** of the summation, m the **lower limit** of the summation, and n the **upper limit** of the summation.

Joseph Louis Lagrange
(1736–1813)

EXAMPLE 4.1.4   Computing Summations

Let $a_1 = -2$, $a_2 = -1$, $a_3 = 0$, $a_4 = 1$, and $a_5 = 2$. Compute the following:

a. $\displaystyle\sum_{k=1}^{5} a_k$ b. $\displaystyle\sum_{k=2}^{2} a_k$ c. $\displaystyle\sum_{k=1}^{2} a_{2k}$

Solution   a. $\displaystyle\sum_{k=1}^{5} a_k = a_1 + a_2 + a_3 + a_4 + a_5 = (-2) + (-1) + 0 + 1 + 2 = 0$

b. $\displaystyle\sum_{k=2}^{2} a_k = a_2 = -1$

c. $\displaystyle\sum_{k=1}^{2} a_{2k} = a_{2 \cdot 1} + a_{2 \cdot 2} = a_2 + a_4 = -1 + 1 = 0$ ∎

Oftentimes, the terms of a summation are expressed using an explicit formula. For instance, it is common to see summations such as the following:

$$\sum_{k=1}^{5} k^2 \quad \text{or} \quad \sum_{i=0}^{8} \frac{(-1)^i}{i+1}.$$

EXAMPLE 4.1.5 **When the Terms of a Summation Are Given by a Formula**

Compute the following summation:

$$\sum_{k=1}^{5} k^2.$$

Solution

$$\sum_{k=1}^{5} k^2 = 1^2 + 2^2 + 3^2 + 4^2 + 5^2 = 55.$$ ∎

The upper limit of a summation may be a variable, in which case the summation may be written in expanded form using an ellipsis.

EXAMPLE 4.1.6 **Changing from Summation Notation to Expanded Form**

Write the following summation in expanded form:

$$\sum_{i=0}^{n} \frac{(-1)^i}{i + 1}.$$

Solution

$$\sum_{i=0}^{n} \frac{(-1)^i}{i + 1} = \frac{(-1)^0}{0 + 1} + \frac{(-1)^1}{1 + 1} + \frac{(-1)^2}{2 + 1} + \frac{(-1)^3}{3 + 1} + \cdots + \frac{(-1)^n}{n + 1}$$

$$= \frac{1}{1} + \frac{(-1)}{2} + \frac{1}{3} + \frac{(-1)}{4} + \cdots + \frac{(-1)^n}{n + 1}$$

$$= 1 - \frac{1}{2} + \frac{1}{3} - \frac{1}{4} + \cdots + \frac{(-1)^n}{n + 1}$$ ∎

▲ *CAUTION!* *The expanded form of a sum may appear ambiguous for small values of n. For instance, consider*

$$1^2 + 2^2 + 3^2 + \cdots + n^2.$$

This expression is intended to represent the sum of squares of consecutive integers starting with 1^2 and ending with n^2. Thus, if $n = 1$ the sum is just 1^2, if $n = 2$ the sum is $1^2 + 2^2$, and if $n = 3$ the sum is $1^2 + 2^2 + 3^2$.

EXAMPLE 4.1.7 **Changing from Expanded Form to Summation Notation**

Express the following using summation notation:

$$\frac{1}{n} + \frac{2}{n + 1} + \frac{3}{n + 2} + \cdots + \frac{n + 1}{2n}.$$

Solution   The general term of this summation can be expressed as $(k + 1)/(n + k)$ for integers k from 0 to n. Hence

$$\frac{1}{n} + \frac{2}{n + 1} + \frac{3}{n + 2} + \cdots + \frac{n + 1}{2n} = \sum_{k=0}^{n} \frac{k + 1}{n + k}.$$ ∎

Certain sums are made up of terms, each of which is a difference. When you write such sums in expanded form, you sometimes see that successive cancellation of terms collapses the sum like a telescope.

EXAMPLE 4.1.8   **A Telescoping Sum**

Compute the following summation:

$$\sum_{k=1}^{n} \left(\frac{k}{k+1} - \frac{k+1}{k+2} \right).$$

Solution

$$\sum_{k=1}^{n} \left(\frac{k}{k+1} - \frac{k+1}{k+2} \right)$$

$$= \left(\frac{1}{1+1} - \frac{1+1}{1+2} \right) + \left(\frac{2}{2+1} - \frac{2+1}{2+2} \right) + \left(\frac{3}{3+1} - \frac{3+1}{3+2} \right) + \cdots$$

$$+ \left(\frac{n-1}{(n-1)+1} - \frac{(n-1)+1}{(n-1)+2} \right) + \left(\frac{n}{n+1} - \frac{n+1}{n+2} \right)$$

$$= \left(\frac{1}{2} - \frac{2}{3} \right) + \left(\frac{2}{3} - \frac{3}{4} \right) + \left(\frac{3}{4} - \frac{4}{5} \right) + \cdots$$

$$+ \left(\frac{n-1}{n} - \frac{n}{n+1} \right) + \left(\frac{n}{n+1} - \frac{n+1}{n+2} \right)$$

$$= \frac{1}{2} - \frac{n+1}{n+2}.$$

CHANGE OF VARIABLE

Observe that

$$\sum_{k=1}^{3} k^2 = 1^2 + 2^2 + 3^2$$

and also that

$$\sum_{i=1}^{3} i^2 = 1^2 + 2^2 + 3^2.$$

Hence

$$\sum_{k=1}^{3} k^2 = \sum_{i=1}^{3} i^2.$$

This equation illustrates the fact that the symbol used to represent the index of a summation can be replaced by any other symbol as long as the replacement is made in each location where the symbol occurs. As a consequence, the index of a summation is called a **dummy variable**.

The appearance of a summation can be altered by more complicated changes of variable as well. For example, observe that

$$\sum_{j=2}^{4} (j-1)^2 = (2-1)^2 + (3-1)^2 + (4-1)^2$$

$$= 1^2 + 2^2 + 3^2$$

$$= \sum_{k=1}^{3} k^2.$$

A general procedure to transform the first summation into the second is illustrated in Example 4.1.9 below.

EXAMPLE 4.1.9   Transforming a Sum by a Change of Variable

Transform the following summation by making the specified change of variable:

$$\text{summation: } \sum_{k=0}^{6} \frac{1}{k+1} \qquad \text{change of variable: } j = k+1$$

Solution   First calculate the lower and upper limits of the new summation:

$$\text{When } k = 0, \quad j = k+1 = 0+1 = 1.$$
$$\text{When } k = 6, \quad j = k+1 = 6+1 = 7.$$

Thus the new sum goes from $j = 1$ to $j = 7$.

Next calculate the general term of the new summation. You will need to replace each occurrence of k by an expression in j:

$$\text{Since } j = k+1, \text{ then } k = j-1.$$

$$\text{Hence } \frac{1}{k+1} = \frac{1}{(j-1)+1} = \frac{1}{j}.$$

Finally, put the steps together to obtain:

$$\sum_{k=0}^{6} \frac{1}{k+1} = \sum_{j=1}^{7} \frac{1}{j}. \qquad \qquad \text{4.1.1}$$

■

Equation 4.1.1 can be given an additional twist by noting that because the j in the right-hand summation is a dummy variable, it may be replaced by any other variable name, as long as the substitution is made in every location where j occurs. In particular, it is legal to substitute k in place of j to obtain

$$\sum_{j=1}^{7} \frac{1}{j} = \sum_{k=1}^{7} \frac{1}{k}. \qquad \qquad \text{4.1.2}$$

Putting equations (4.1.1) and (4.1.2) together gives

$$\sum_{k=0}^{6} \frac{1}{k+1} = \sum_{k=1}^{7} \frac{1}{k}.$$

EXAMPLE 4.1.10   When the Upper Limit Appears in the Expression to Be Summed

a. Transform the following summation by making the specified change of variable:

$$\text{summation: } \sum_{k=1}^{n+1} (n-k+1) \qquad \text{change of variable: } j = k-1$$

b. Transform the summation obtained in part (a) by changing all j's to k's.

Solution   a. When $k = 1$, then $j = k-1 = 1-1 = 0$. (So the new lower limit is 0.) When $k = n+1$, then $j = k-1 = (n+1)-1 = n$. (So the new upper limit is n.)

Since $j = k-1$, then $k = j+1$. Also note that n is a constant as far as the terms of the sum are concerned. It follows that

$$n-k+1 = n-(j+1)+1 = n-j-1+1 = n-j,$$

and so the general term of the new summation is $n-j$. Therefore,

$$\sum_{k=1}^{n+1} (n-k+1) = \sum_{j=0}^{n} (n-j). \qquad \qquad \text{4.1.3}$$

b. Changing all the j's to k's in the right-hand side of equation (4.1.3) gives

$$\sum_{j=0}^{n} (n - j) = \sum_{k=0}^{n} (n - k).$$ 4.1.4

Combining equations (4.1.3) and (4.1.4) results in

$$\sum_{k=1}^{n+1} (n - k + 1) = \sum_{k=0}^{n} (n - k).$$ ■

PRODUCT NOTATION

The notation for the product of a sequence of numbers is analogous to the notation for their sum. The Greek capital letter pi, Π, denotes a product. For example,

$$\prod_{k=1}^{5} a_k = a_1 \cdot a_2 \cdot a_3 \cdot a_4 \cdot a_5.$$

More generally, the **product from k equals m to n of a_k** is the product of all the terms $a_m, a_{m+1}, a_{m+2}, \ldots, a_n$. That is,

$$\prod_{k=m}^{n} a_k = a_m \cdot a_{m+1} \cdot a_{m+2} \cdot \ldots \cdot a_n.$$

EXAMPLE 4.1.11   Computing Products

Compute the following products:

a. $\displaystyle\prod_{k=1}^{5} k$ b. $\displaystyle\prod_{k=1}^{1} \frac{k}{k + 1}$

Solution   a. $\displaystyle\prod_{k=1}^{5} k = 1 \cdot 2 \cdot 3 \cdot 4 \cdot 5 = 120$ b. $\displaystyle\prod_{k=1}^{1} \frac{k}{k + 1} = \frac{1}{1 + 1} = \frac{1}{2}$ ■

PROPERTIES OF SUMMATIONS AND PRODUCTS

The following theorem states general properties of summations and products. The proof of the theorem is discussed in Section 8.4.

THEOREM 4.1.1

If $a_m, a_{m+1}, a_{m+2}, \ldots$ and $b_m, b_{m+1}, b_{m+2}, \ldots$ are sequences of real numbers and c is any real number, then the following equations hold for any integer $n \geq m$:

1. $\displaystyle\sum_{k=m}^{n} a_k + \sum_{k=m}^{n} b_k = \sum_{k=m}^{n} (a_k + b_k)$

2. $\displaystyle c \cdot \sum_{k=m}^{n} a_k = \sum_{k=m}^{n} c \cdot a_k$ generalized distributive law

3. $\displaystyle\left(\prod_{k=m}^{n} a_k\right) \cdot \left(\prod_{k=m}^{n} b_k\right) = \prod_{k=m}^{n} (a_k \cdot b_k).$

EXAMPLE 4.1.12   Using Properties of Summation and Product

Let $a_k = k + 1$ and $b_k = k - 1$ for all integers k. Write each of the following expressions as a single summation or product:

a. $\displaystyle\sum_{k=m}^{n} a_k + 2 \cdot \sum_{k=m}^{n} b_k$ b. $\displaystyle\left(\prod_{k=m}^{n} a_k\right) \cdot \left(\prod_{k=m}^{n} b_k\right)$

Solution   a. $\displaystyle\sum_{k=m}^{n} a_k + 2 \cdot \sum_{k=m}^{n} b_k = \sum_{k=m}^{n} (k + 1) + 2 \cdot \sum_{k=m}^{n} (k - 1)$ by substitution

$\displaystyle\qquad\qquad\qquad\qquad\quad = \sum_{k=m}^{n} (k + 1) + \sum_{k=m}^{n} 2 \cdot (k - 1)$ by Theorem 4.1.1 (2)

$\displaystyle\qquad\qquad\qquad\qquad\quad = \sum_{k=m}^{n} ((k + 1) + 2 \cdot (k - 1))$ by Theorem 4.1.1 (1)

$\displaystyle\qquad\qquad\qquad\qquad\quad = \sum_{k=m}^{n} (3k - 1)$ by algebraic simplification

b. $\displaystyle\left(\prod_{k=m}^{n} a_k\right) \cdot \left(\prod_{k=m}^{n} b_k\right) = \left(\prod_{k=m}^{n} (k + 1)\right) \cdot \left(\prod_{k=m}^{n} (k - 1)\right)$ by substitution

$\displaystyle\qquad\qquad\qquad\qquad\quad = \prod_{k=m}^{n} (k + 1) \cdot (k - 1)$ by Theorem 4.1.1 (3)

$\displaystyle\qquad\qquad\qquad\qquad\quad = \prod_{k=m}^{n} (k^2 - 1)$ by algebraic simplification   ∎

FACTORIAL NOTATION

The product of all consecutive integers up to a given integer occurs so often in mathematics that it is given a special notation—*factorial* notation.

DEFINITION

For each positive integer n, the quantity **n factorial**, denoted **$n!$**, is defined to be the product of all the integers from 1 to n:

$$n! = n \cdot (n - 1) \cdot \ldots \cdot 3 \cdot 2 \cdot 1.$$

Zero factorial is defined to be 1:

$$0! = 1.$$

The definition of zero factorial as 1 may seem odd, but, as you will see when you read Chapter 6, it is convenient for many mathematical formulas.

EXAMPLE 4.1.13   The First Ten Factorials

$0! = 1$ $1! = 1$

$2! = 2 \cdot 1 = 2$ $3! = 3 \cdot 2 \cdot 1 = 6$

$4! = 4 \cdot 3 \cdot 2 \cdot 1 = 24$ $5! = 5 \cdot 4 \cdot 3 \cdot 2 \cdot 1 = 120$

$6! = 6 \cdot 5 \cdot 4 \cdot 3 \cdot 2 \cdot 1 = 720$ $7! = 7 \cdot 6 \cdot 5 \cdot 4 \cdot 3 \cdot 2 \cdot 1 = 5{,}040$

$8! = 8 \cdot 7 \cdot 6 \cdot 5 \cdot 4 \cdot 3 \cdot 2 \cdot 1$ $9! = 9 \cdot 8 \cdot 7 \cdot 6 \cdot 5 \cdot 4 \cdot 3 \cdot 2 \cdot 1$

$\qquad = 40{,}320$ $\qquad = 362{,}880$ ∎

As you can see from the example above, the values of $n!$ grow very rapidly. For instance, $40! \cong 8.16 \times 10^{47}$, which is a number that is too large to be computed exactly using the standard integer arithmetic of the machine-specific implementations of many computer languages. (The symbol $\cong$ means "approximately equal.")

Note that the following formula holds for each positive integer n:

$$n! = n \cdot (n - 1)!$$

This formula provides a way to calculate successively higher values of the factorial in terms of lower values. It is an example of a recurrence relation, which is a type of relation of great importance in computer science and many areas of mathematics. (Recurrence relations are discussed in Chapter 8.) Example 4.1.14 illustrates the usefulness of this formula for making computations.

EXAMPLE 4.1.14   **Computing with Factorials**

Simplify the following expressions:

a. $\dfrac{8!}{7!}$ b. $\dfrac{5!}{2! \cdot 3!}$ c. $\dfrac{1}{2! \cdot 4!} + \dfrac{1}{3! \cdot 3!}$ d. $\dfrac{(n + 1)!}{n!}$ e. $\dfrac{n!}{(n - 3)!}$

Solution   a. $\dfrac{8!}{7!} = \dfrac{8 \cdot \cancel{7!}}{\cancel{7!}} = 8$

b. $\dfrac{5!}{2! \cdot 3!} = \dfrac{5 \cdot 4 \cdot \cancel{3!}}{2! \cdot \cancel{3!}} = \dfrac{5 \cdot 4}{2 \cdot 1} = 10$

c. $\dfrac{1}{2! \cdot 4!} + \dfrac{1}{3! \cdot 3!} = \dfrac{1}{2! \cdot 4!} \cdot \dfrac{3}{3} + \dfrac{1}{3! \cdot 3!} \cdot \dfrac{4}{4}$ by multiplying each numerator and denominator by just what is necessary to obtain a common denominator

$= \dfrac{3}{3 \cdot 2! \cdot 4!} + \dfrac{4}{3! \cdot 4 \cdot 3!}$ by rearranging factors

$= \dfrac{3}{3! \cdot 4!} + \dfrac{4}{3! \cdot 4!}$ because $3 \cdot 2! = 3!$ and $4 \cdot 3! = 4!$

$= \dfrac{7}{3! \cdot 4!}$ by the rule for adding fractions with a common denominator

$= \dfrac{7}{144}$

d. $\dfrac{(n + 1)!}{n!} = \dfrac{(n + 1) \cdot \cancel{n!}}{\cancel{n!}} = n + 1$

e. $\dfrac{n!}{(n - 3)!} = \dfrac{n \cdot (n - 1) \cdot (n - 2) \cdot \cancel{(n - 3)!}}{\cancel{(n - 3)!}} = n \cdot (n - 1) \cdot (n - 2)$

$= n^3 - 3n^2 + 2n$ ∎

SEQUENCES IN COMPUTER PROGRAMMING

An important data type in computer programming consists of finite sequences. In computer programming contexts, these are usually referred to as *one-dimensional arrays*. For example, consider a program that analyzes the wages paid to a sample of 50 workers. Such a program might compute the average wage and the difference between each individual wage and the average. This would require that each wage be stored in memory for retrieval later in the calculation. To avoid the use of entirely separate variable names for each of the 50 wages, each is written as a term of a one-dimensional array:

$$W[1], W[2], W[3], \ldots, W[50].$$

Note that the subscript labels are written inside square brackets. The reason is that until very recently it was impossible to type actual dropped subscripts on most computer keyboards.

The main difficulty programmers have when using one-dimensional arrays is keeping the labels straight.

EXAMPLE 4.1.15   Dummy Variable in a Loop

The index variable for a **for-next** loop is a dummy variable. For example, the following three algorithm segments all produce the same output:

1. **for** $i := 1$ **to** n 2. **for** $j := 0$ **to** $n - 1$ 3. **for** $k := 2$ **to** $n + 1$
    print $a[i]$ **print** $a[j + 1]$ **print** $a[k - 1]$

   next i **next** j **next** k ■

APPLICATION: ALGORITHM TO CONVERT FROM BASE 10 TO BASE 2 USING REPEATED DIVISION BY 2

Section 1.5 contains some examples of converting integers from decimal to binary notation. The method shown there, however, is only convenient to use with small numbers. A systematic algorithm to convert any nonnegative integer to binary notation uses repeated division by 2.

Suppose a is a nonnegative integer. Divide a by 2 using the quotient-remainder theorem to obtain a quotient $q[0]$ and a remainder $r[0]$. If the quotient is nonzero, divide by 2 again to obtain a quotient $q[1]$ and a remainder $r[1]$. Continue this process until a quotient of 0 is obtained. At each stage, the remainder must be less than the divisor, which is 2. Thus each remainder is either 0 or 1. The process is illustrated below for $a = 38$. (Read the divisions from the bottom up.)

0	remainder $= 1 = r[5]$
2 ⟌ 1	remainder $= 0 = r[4]$
2 ⟌ 2	remainder $= 0 = r[3]$
2 ⟌ 4	remainder $= 1 = r[2]$
2 ⟌ 9	remainder $= 1 = r[1]$
2 ⟌ 19	remainder $= 0 = r[0]$
2 ⟌ 38	

The results of all these divisions can be written as a sequence of equations:

$$38 = 19 \cdot 2 + 0,$$
$$19 = 9 \cdot 2 + 1,$$
$$9 = 4 \cdot 2 + 1,$$

$$4 = 2 \cdot 2 + 0,$$
$$2 = 1 \cdot 2 + 0,$$
$$1 = 0 \cdot 2 + 1.$$

By repeated substitution, then,

$$38 = 19 \cdot 2 + 0$$
$$= (9 \cdot 2 + 1) \cdot 2 + 0 = 9 \cdot 2^2 + 1 \cdot 2 + 0$$
$$= (4 \cdot 2 + 1) \cdot 2^2 + 1 \cdot 2 + 0 = 4 \cdot 2^3 + 1 \cdot 2^2 + 1 \cdot 2 + 0$$
$$= (2 \cdot 2 + 0) \cdot 2^3 + 1 \cdot 2^2 + 1 \cdot 2 + 0$$
$$= 2 \cdot 2^4 + 0 \cdot 2^3 + 1 \cdot 2^2 + 1 \cdot 2 + 0$$
$$= (1 \cdot 2 + 0) \cdot 2^4 + 0 \cdot 2^3 + 1 \cdot 2^2 + 1 \cdot 2 + 0$$
$$= 1 \cdot 2^5 + 0 \cdot 2^4 + 0 \cdot 2^3 + 1 \cdot 2^2 + 1 \cdot 2 + 0.$$

Note that each coefficient of a power of 2 on the right-hand side above is one of the remainders obtained in the repeated division of 38 by 2. This is true for the left-most 1 as well because $1 = 0 \cdot 2 + 1$. Thus

$$38_{10} = 100110_2 = (r[5]r[4]r[3]r[2]r[1]r[0]) .$$

In general, if a nonnegative integer a is repeatedly divided by 2 until a quotient of zero is obtained and the remainders are found to be $r[0], r[1], \ldots, r[k]$, then by the quotient-remainder theorem each $r[i]$ equals 0 or 1 and by repeated substitution from the theorem,

$$a = 2^k \cdot r[k] + 2^{k-1} \cdot r[k-1] + \cdots + 2^2 \cdot r[2]$$
$$+ 2^1 \cdot r[1] + 2^0 \cdot r[0].$$

4.1.5

Thus the binary representation for a can be read from equation (4.1.5) above:

$$a_{10} = (r[k]r[k-1] \cdots r[2]r[1]r[0])_2.$$

EXAMPLE 4.1.16

Converting from Decimal to Binary Notation Using Repeated Division by 2

Use repeated division by 2 to write the number 29_{10} in binary notation.

Solution

$$
\begin{array}{rl}
0 & \text{remainder} = r[4] = 1 \\
2 \,|\, 1 & \text{remainder} = r[3] = 1 \\
2 \,|\, 3 & \text{remainder} = r[2] = 1 \\
2 \,|\, 7 & \text{remainder} = r[1] = 0 \\
2 \,|\, 14 & \text{remainder} = r[0] = 1 \\
2 \,|\, 29 &
\end{array}
$$

Hence $29_{10} = (r[4]r[3]r[2]r[1]r[0])_2 = 11101_2.$ ∎

The procedure we have described for converting from base 10 to base 2 is formalized in the following algorithm:

ALGORITHM 4.1.1   Decimal to Binary Conversion Using Repeated
Division by 2

*[In Algorithm 4.1.1 the input is a nonnegative integer a. The aim of the algorithm is to
produce a sequence of binary digits $r[0], r[1], r[2], \ldots, r[k]$ so that the binary
representation of a is*

$$(r[k]r[k-1] \cdots r[2]r[1]r[0])_2.$$

That is,

$$a = 2^k \cdot r[k] + 2^{k-1} \cdot r[k-1] + \cdots + 2^2 \cdot r[2] + 2^1 \cdot r[1] + 2^0 \cdot r[0].]$$

Input: *a [a nonnegative integer]*

Algorithm Body:

$q := a, i := 0$
*[Repeatedly perform the integer division of q by 2 until q becomes 0. Store successive
remainders in a one-dimensional array $r[0], r[1], r[2], \ldots, r[k]$. Even if the initial
value of q equals 0, the loop should execute one time (so that $r[0]$ is computed). Thus
the guard condition for the **while** loop is $i = 0$ or $q \neq 0$.]*
while $(i = 0$ or $q \neq 0)$

    $r[i] := q \bmod 2$

    $q := q \operatorname{div} 2$
[$r[i]$ and q can be obtained by calling the division algorithm.]

      $i := i + 1$

end while

*[After execution of this step, the values of $r[0], r[1], \ldots, r[i-1]$ are all 0's
and 1's, and $a = (r[i-1]r[i-2] \cdots r[2]r[1]r[0])_2.]$*

Output: $r[0], r[1], r[2], \ldots, r[i-1]$ *[a sequence of integers]*

end Algorithm 4.1.1

EXERCISE SET 4.1*

**Write the first four terms of the sequences defined by
the formulas in 1–6.**

1. $a_k = \dfrac{k}{10 - k}$, for all integers $k \geq 1$.

2. $b_j = 1 + 2^j$, for all integers $j \geq 0$.

3. $c_i = \dfrac{(-1)^i}{3^i}$, for all integers $i \geq 0$.

4. $d_m = 1 - \left(\dfrac{1}{10}\right)^m$, for all integers $m \geq 1$.

5. $e_n = \left\lfloor \dfrac{n}{2} \right\rfloor \cdot 2$, for all integers $n \geq 0$.

6. $f_n = \left\lfloor \dfrac{n}{3} \right\rfloor \cdot 3$, for all integers $n \geq 0$.

7. Let $a_k = 2k + 1$ and $b_k = (k - 1)^3 + k + 2$ for all
integers $k \geq 0$. Show that the first three terms of these
sequences are identical but that their fourth terms differ.

**Compute the first fifteen terms of each of the sequences
in 8 and 9, and describe the general behavior of these
sequences in words. (A definition of logarithm is given in
Section 7.1.)**

8. $g_n = \lfloor \log_2 n \rfloor$ for all integers $n \geq 1$.

9. $h_n = n \cdot \lfloor \log_2 n \rfloor$ for all integers $n \geq 1$.

**Find explicit formulas for sequences with the initial
terms given in 10–16.**

10. $-1, 1, -1, 1, -1, 1$

11. $0, 1, -2, 3, -4, 5$

*Exercises with blue numbers have solutions in Appendix B. The symbol H indicates that only a hint or a partial solution is given. The symbol
◆ signals that an exercise is more challenging than usual.

12. $\frac{1}{3}, \frac{2}{4}, \frac{3}{5}, \frac{4}{6}, \frac{5}{7}, \frac{6}{8}$

13. $1 - \frac{1}{2}, \frac{1}{2} - \frac{1}{3}, \frac{1}{3} - \frac{1}{4}, \frac{1}{4} - \frac{1}{5}, \frac{1}{5} - \frac{1}{6}, \frac{1}{6} - \frac{1}{7}$

14. $\frac{1}{4}, \frac{2}{9}, \frac{3}{16}, \frac{4}{25}, \frac{5}{36}, \frac{6}{49}$

15. $\frac{1}{2}, -\frac{2}{3}, \frac{3}{4}, -\frac{4}{5}, \frac{5}{6}, -\frac{6}{7}$

16. $2, 6, 12, 20, 30, 42, 56$

◆ 17. Consider the sequence defined by $a_n = \dfrac{2n + (-1)^n - 1}{4}$ for all integers $n \geq 0$. Find an alternative explicit formula for a_n that uses the floor notation.

18. Let $a_0 = 2$, $a_1 = 3$, $a_2 = -2$, $a_3 = 1$, $a_4 = 0$, $a_5 = -1$, and $a_6 = -2$. Compute each of the summations and products below.

a. $\displaystyle\sum_{i=0}^{6} a_i$ b. $\displaystyle\sum_{i=0}^{0} a_i$ c. $\displaystyle\sum_{j=1}^{3} a_{2j}$ d. $\displaystyle\prod_{k=0}^{6} a_k$ e. $\displaystyle\prod_{k=2}^{2} a_k$

Compute the summations and products in 19–25.

19. $\displaystyle\sum_{k=1}^{5} (k + 1)$

20. $\displaystyle\prod_{k=2}^{4} k^2$

21. $\displaystyle\sum_{m=0}^{4} \frac{1}{2^m}$

22. $\displaystyle\prod_{j=1}^{5} (-1)^j$

23. $\displaystyle\sum_{k=-1}^{1} (k^3 + 2)$

24. $\displaystyle\sum_{n=1}^{10} \left(\frac{1}{n} - \frac{1}{n+1}\right)$

25. $\displaystyle\prod_{i=2}^{5} \frac{(i - 1) \cdot i}{(i + 1) \cdot (i + 2)}$

Write the summations in 26–28 in expanded form.

26. $\displaystyle\sum_{i=1}^{n} (-2)^i$ 27. $\displaystyle\sum_{j=1}^{n} j(j + 1)$ 28. $\displaystyle\sum_{k=0}^{n} \frac{1}{k!}$

Write each of 29–38 using summation or product notation.

29. $1^2 - 2^2 + 3^2 - 4^2 + 5^2 - 6^2 + 7^2$

30. $(1^3 - 1) + (2^3 - 1) + (3^3 - 1) + (4^3 - 1)$

31. $(2^2 + 1) \cdot (3^2 + 1) \cdot (4^2 + 1)$

32. $\dfrac{2}{3 \cdot 4} + \dfrac{3}{4 \cdot 5} + \dfrac{4}{5 \cdot 6} + \dfrac{5}{6 \cdot 7} + \dfrac{6}{7 \cdot 8}$

33. $1 + r + r^2 + r^3 + r^4 + r^5$

34. $(1 - r) \cdot (1 - r^2) \cdot (1 - r^3) \cdot (1 - r^4)$

35. $1^3 + 2^3 + 3^3 + \cdots + n^3$

36. $\dfrac{1}{2!} + \dfrac{2}{3!} + \dfrac{3}{4!} + \cdots + \dfrac{n}{(n + 1)!}$

37. $n + (n - 1) + (n - 2) + \cdots + 1$

38. $n + \dfrac{n - 1}{2!} + \dfrac{n - 2}{3!} + \dfrac{n - 3}{4!} + \cdots + \dfrac{1}{n!}$

Transform each of 39 and 40 by making the change of variable $i = k + 1$.

39. $\displaystyle\sum_{k=0}^{5} k \cdot (k - 1)$ 40. $\displaystyle\prod_{k=1}^{n} \frac{k^2}{k + 1}$

Transform each of 41–44 by making the change of variable $j = i - 1$.

41. $\displaystyle\sum_{i=1}^{n+1} \frac{(i - 1)^2}{i}$ 42. $\displaystyle\sum_{i=3}^{n+1} \frac{i}{i + n - 1}$

43. $\displaystyle\sum_{i=1}^{n-1} \frac{i}{(n - i)^2}$ 44. $\displaystyle\prod_{i=n}^{2n} \frac{n - i + 1}{i}$

Write each of 45 and 46 as a single summation or product.

45. $3 \cdot \displaystyle\sum_{k=1}^{n} (2k - 3) + \sum_{k=1}^{n} (4 - 5k)$

46. $\left(\displaystyle\prod_{k=1}^{n} \frac{k}{k + 1}\right) \cdot \left(\prod_{k=1}^{n} \frac{k + 1}{k + 2}\right)$

47. Check Theorem 4.1.1 for $m = 1$ and $n = 4$ by writing out the left- and right-hand sides of the equations in expanded form. The two sides are equal by repeated application of certain laws. What are these laws?

Compute each of 48–56.

48. $\dfrac{4!}{3!}$ 49. $\dfrac{5!}{7!}$ 50. $\dfrac{6!}{0!}$

51. $\dfrac{n!}{(n - 1)!}$ 52. $\dfrac{(n - 1)!}{(n + 1)!}$ 53. $\dfrac{n!}{(n - 2)!}$

54. $\dfrac{((n + 1)!)^2}{(n!)^2}$ 55. $\dfrac{n!}{(n - k)!}$ 56. $\dfrac{n!}{(n - k - 1)!}$

57. a. Prove that $n! + 2$ is divisible by 2, for all integers $n \geq 2$.

b. Prove that $n! + k$ is divisible k, for all integers $n \geq 2$ and $k = 2, 3, \ldots, n$.

H c. Given any integer $m \geq 2$, is it possible to find a sequence of $m - 1$ consecutive positive integers none of which is prime? Explain your answer.

58. Suppose $a[1], a[2], a[3], \ldots, a[m]$ is a one-dimensional array and consider the following algorithm segment:

sum := 0
for k := 1 **to** m
    sum := *sum* + $a[k]$
next k

Fill in the blanks below so that each algorithm segment performs the same job as the one given above.

a. *sum* := 0
    for *i* := 0 **to** _____
      sum := _____
    next *i*

b. *sum* := 0
    for *j* := 2 **to** _____
      sum := _____
    next *j*

Use repeated division by 2 to convert (by hand) the integers in 59–61 from base 10 to base 2.

59. 90   60. 196   61. 101

Make a trace table to trace the action of Algorithm 4.1.1 on the input in 62–64.

62. 23   63. 26   64. 37

65. Write an informal description of an algorithm (using repeated division by 16) to convert a nonnegative integer from decimal notation to hexadecimal notation (base 16).

Use the algorithm you developed for exercise 65 to convert the integers in 66–68 to hexadecimal notation.

66. 287   67. 684   68. 2622

69. Write a formal version of the algorithm you developed for exercise 65.

4.2 MATHEMATICAL INDUCTION I

*[Mathematical induction is]
the standard proof technique in computer science.*
(Anthony Ralston, 1984)

Mathematical induction is one of the more recently developed techniques of proof in the history of mathematics. It is used to check conjectures about the outcomes of processes that occur repeatedly and according to definite patterns. We introduce the technique with an example.

Some people claim that the United States penny is such a small coin that it should be abolished. They point out that frequently a person who drops a penny on the ground does not even bother to pick it up. Other people argue that abolishing the penny would not give enough flexibility for pricing merchandise. What prices could still be paid with exact change if the penny were abolished and another coin worth 2¢ were introduced? The answer is that the only prices that could not be paid with exact change would be 1¢ and 3¢. In other words:

Any whole number of cents of at least 4¢ can be obtained using 2¢ and 5¢ coins.

More formally:

For all integers $n \geq 4$, n cents can be obtained using 2¢ and 5¢ coins.

Even more formally:

For all integers $n \geq 4$, $P(n)$ is true where $P(n)$ is the sentence "n cents can be obtained using 2¢ and 5¢ coins."

You could check that $P(n)$ is true for a few particular values of n, as is done in the table below.

Number of Cents	How to Obtain It
4¢	2¢ + 2¢
5¢	5¢
6¢	2¢ + 2¢ + 2¢
7¢	5¢ + 2¢
8¢	2¢ + 2¢ + 2¢ + 2¢
9¢	5¢ + 2¢ + 2¢
10¢	5¢ + 5¢
11¢	5¢ + 2¢ + 2¢ + 2¢
12¢	5¢ + 5¢ + 2¢

The cases shown in the table give inductive evidence supporting the claim that $P(n)$ is true for general n. Indeed, $P(n)$ *is true for all* $n \geq 4$ *if, and only if, it is possible to continue filling in the table for arbitrarily large values of* n.

The kth line of the table gives information about how to obtain $k\phi$ using 2ϕ and 5ϕ coins. To continue the table to the next row, directions must be given for how to obtain $(k + 1)\phi$ using 2ϕ and 5ϕ coins. The secret is to observe first that if $k\phi$ can be obtained using at least one 5ϕ coin, then $(k + 1)\phi$ can be obtained by replacing the 5ϕ coin by three 2ϕ coins, as shown in Figure 4.2.1.

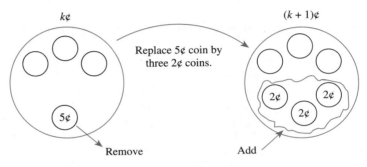

FIGURE 4.2.1

If, on the other hand, $k\phi$ is obtained without using a 5ϕ coin, then 2ϕ coins are used exclusively. And since the total is 4ϕ or more, at least two 2ϕ coins must be used. These two 2ϕ coins can be replaced by one 5ϕ coin to obtain a total of $(k + 1)\phi$, as shown in Figure 4.2.2.

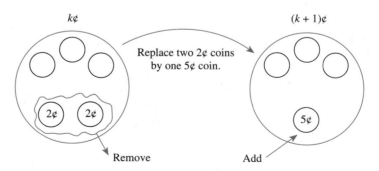

FIGURE 4.2.2

The structure of the argument above can be summarized as follows: To show that $P(n)$ is true for all integers $n \geq 4$, (1) show that $P(4)$ is true, and (2) show that the truth of $P(k + 1)$ follows necessarily from the truth of $P(k)$ for each $k \geq 4$. Any argument of this form is called an argument by *mathematical induction*.

PRINCIPLE OF MATHEMATICAL INDUCTION

Let $P(n)$ be a predicate that is defined for integers n, and let a be a fixed integer. Suppose the following two statements are true:

1. $P(a)$ is true.
2. For all integers $k \geq a$, if $P(k)$ is true then $P(k + 1)$ is true.

Then the statement

$$\text{for all integers } n \geq a, P(n)$$

is true.

The first known use of mathematical induction occúrs in the work of the Italian scientist Francesco Maurolico in 1575. In the seventeenth century both Pierre de Fermat and Blaise Pascal used the technique, Fermat calling it the "method of infinite descent." In 1883 Augustus De Morgan (best known for De Morgan's laws) described the process carefully and gave it the name mathematical induction.

To visualize the idea of mathematical induction, imagine a collection of dominoes positioned one behind the other in such a way that if any given domino falls backward, it makes the one behind it fall backward also. (See Figure 4.2.3.) Then imagine that the first domino falls backward. What happens? . . . They all fall down!

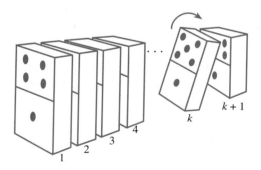

FIGURE 4.2.3   If the kth domino falls backward, it pushes the $(k + 1)$st domino backward also.

To see the connection between this image and the principle of mathematical induction, let $P(n)$ be the sentence "The nth domino falls backward." It is given that for each $k \geq 1$, if $P(k)$ is true (the kth domino falls backward), then $P(k + 1)$ is also true (the $(k + 1)$st domino falls backward). It is also given that $P(1)$ is true (the first domino falls backward). Thus by the principle of mathematical induction, $P(n)$ (the nth domino falls backward) is true for every integer $n \geq 1$.

Strictly speaking, the validity of proof by mathematical induction is an axiom. That is why it is referred to as the *principle* of mathematical induction rather than as a theorem. It is equivalent to the following property of the integers, which is easy to accept on intuitive grounds: Suppose S is any set of integers satisfying (1) $a \in S$, and (2) for all integers k, if $k \in S$ then $k + 1 \in S$. Then S must contain every integer greater than or equal to a. To understand the equivalence of this formulation and the one given earlier, just let S be the set of all integers for which $P(n)$ is true.

Proving a statement by mathematical induction is a two-step process. In step 1, the **basis step**, you prove that $P(a)$ is true for a particular integer a. In step 2, the

inductive step, you prove that for all integers $k \geq a$, *if $P(k)$ is true then $P(k + 1)$ is true*. Note that the inductive step can be written formally as

$$\forall \text{ integers } k \geq a, \text{ if } P(k) \text{ then } P(k + 1).$$

By the method of generalizing from the generic particular, to prove the inductive step you

> **suppose** that $P(k)$ is true, where k is a particular but arbitrarily chosen integer greater than or equal to a.

Then you

> **show** that $P(k + 1)$ is true.

The supposition that $P(k)$ is true is called the **inductive hypothesis**.
   Here is a formal version of the proof about coins developed informally above.

PROPOSITION 4.2.1

Let $P(n)$ be the property "$n\cent$ can be obtained using $2\cent$ and $5\cent$ coins." Then $P(n)$ is true for all integers $n \geq 4$.

Proof:

The property is true for $n = 4$:   The reason is that $4\cent = 2\cent + 2\cent$.

If the property is true for $n = k$, then it is true for $n = k + 1$:   Suppose $k\cent$ can be obtained using $2\cent$ and $5\cent$ coins for some integer $k \geq 4$. *[This is the inductive hypothesis.]* We must show that $(k + 1)\cent$ can be obtained using $2\cent$ and $5\cent$ coins. In case there is a $5\cent$ coin among those used to make up the $k\cent$, replace it by three $2\cent$ coins; the result will be $(k + 1)\cent$. In case no $5\cent$ coin is used to make up the $k\cent$, then at least two $2\cent$ coins must be used because $k \geq 4$. Remove two $2\cent$ coins and replace them by a $5\cent$ coin; the result will be $(k + 1)\cent$. Thus in either case $(k + 1)\cent$ can be obtained using $2\cent$ and $5\cent$ coins *[as was to be shown]*.

The following example shows how to use mathematical induction to prove a formula for the sum of the first n integers.

EXAMPLE 4.2.1   Sum of the First n Integers

Use mathematical induction to prove that

$$1 + 2 + \cdots + n = \frac{n(n + 1)}{2} \quad \text{for all integers } n \geq 1.$$

Solution   To construct a proof by induction, you must first identify $P(n)$. In this case,

$$P(n): \ 1 + 2 + \cdots + n = \frac{n(n+ 1)}{2}.$$

[To see that $P(n)$ is a sentence, note that its subject is "the sum of the integers from 1 to n" and its verb is "equals."]

In the basis step of the proof, you must show that $P(1)$ is true. Now $P(1)$ is obtained by substituting 1 in place of n in $P(n)$. The left-hand side of $P(1)$ is the sum of all the successive integers starting at 1 and ending at 1. This is just 1. Thus

$$P(1): \quad 1 = \frac{1(1+1)}{2}.$$

Of course, this equation is true because

$$\frac{1(1+1)}{2} = \frac{2}{2} = 1.$$

In the inductive step, you assume that $P(k)$ is true, for some integer k with $k \geq 1$. *[This assumption is the inductive hypothesis.]* You must then show that $P(k+1)$ is true. What are $P(k)$ and $P(k+1)$? $P(k)$ is obtained by substituting k for every n in $P(n)$. Thus

$$P(k): \quad 1 + 2 + \cdots + k = \frac{k(k+1)}{2}. \qquad \leftarrow \text{inductive hypothesis}$$

Similarly $P(k+1)$ is obtained by substituting the quantity $(k+1)$ for every n that appears in $P(n)$. Thus

$$P(k+1): \quad 1 + 2 + \cdots + (k+1) = \frac{(k+1)((k+1)+1)}{2},$$

or, equivalently,

$$P(k+1): \quad 1 + 2 + \cdots + (k+1) = \frac{(k+1)(k+2)}{2}. \qquad \leftarrow \text{to show}$$

Now the inductive hypothesis is the supposition that $P(k)$ is true. How can this supposition be used to show that $P(k+1)$ is true? $P(k+1)$ is an equation, and the truth of an equation can be shown in a variety of ways. One of the most straightforward is to transform the left-hand side into the right-hand side using algebra and other known facts and legal assumptions (such as the inductive hypothesis). In this case, the left-hand side of $P(k+1)$ is

$$1 + 2 + \cdots + (k+1),$$

which equals

$$(1 + 2 + \cdots + k) + (k+1) \qquad \text{by explicitly identifying the next-to-last term and regrouping.}$$

But by substitution from the inductive hypothesis,

$$(1 + 2 + \cdots + k) + (k+1)$$
$$= \frac{k(k+1)}{2} + (k+1) \qquad \text{since the inductive hypothesis says that } 1 + 2 + \cdots + k = \frac{k(k+1)}{2}.$$

Now use algebra to show that this expression equals the right-hand side of $P(k + 1)$:

$$\frac{k(k + 1)}{2} + (k + 1)$$

$$= \frac{k(k + 1)}{2} + \frac{2(k + 1)}{2} \qquad \text{multiply numerator and denominator of the second term by 2 to obtain a common denominator}$$

$$= \frac{k(k + 1) + 2(k + 1)}{2} \qquad \text{by adding fractions}$$

$$= \frac{(k + 2)(k + 1)}{2} \qquad \text{by factoring out } (k + 1)$$

$$= \frac{(k + 1)(k + 2)}{2} \qquad \text{by commuting the factors } (k + 1) \text{ and } (k + 2)$$

which equals the right-hand side of $P(k + 1)$.

This discussion is summarized as follows.

THEOREM 4.2.2   Sum of the First n Integers

For all integers $n \geq 1$,

$$1 + 2 + \cdots + n = \frac{n(n + 1)}{2}.$$

Proof (by mathematical induction):

The formula is true for $n = 1$:   To establish the formula for $n = 1$, we must show that $1 = \dfrac{1(1 + 1)}{2}$. But $\dfrac{1(1 + 1)}{2} = \dfrac{2}{2} = 1$, and so the formula is true for $n = 1$.

If the formula is true for $n = k$ then it is true for $n = k + 1$:

   [Suppose the formula $1 + 2 + \cdots + n = \dfrac{n(n + 1)}{2}$ is true when an integer $k \geq 1$ is substituted for n.]

   Suppose $1 + 2 + \cdots + k = \dfrac{k(k + 1)}{2}$, for some integer $k \geq 1$. *[This is the inductive hypothesis.]*

   [We must show that the formula $1 + 2 + \cdots + n = \dfrac{n(n + 1)}{2}$ is true when $k + 1$ is substituted for n.]

   We must show that $1 + 2 + \cdots + (k + 1) = \dfrac{(k + 1)((k + 1) + 1)}{2}$,

or equivalently, that $1 + 2 + \cdots + (k + 1) = \dfrac{(k + 1)(k + 2)}{2}$. *[We will show that the left-hand side of this equation equals the right-hand side.]*

$$1 + 2 + \cdots + (k + 1)$$

$$= 1 + 2 + \cdots + k + (k + 1) \qquad \text{The next-to-last term is } k \text{ because the terms are successive integers and the last term is } k + 1.$$

$$= \frac{k(k+1)}{2} + (k+1) \qquad \text{by substitution from the inductive hypothesis}$$

$$= \frac{k(k+1)}{2} + \frac{(k+1) \cdot 2}{2}$$

$$= \frac{(k+1)(k+2)}{2}$$

[This is what we needed to show.]

    [Since we have proved both the basis and inductive steps, we conclude that the theorem is true.]

■

The story is told that one of the greatest mathematicians of all time, Carl Friedrich Gauss (1777–1855), was given the problem of adding the numbers from 1 to 100 by his teacher when he was a young child. The teacher had asked his students to compute the sum, supposedly to gain himself some time to grade papers. But after just a few moments, Gauss produced the correct answer. Needless to say, the teacher was dumbfounded. How could young Gauss have calculated the quantity so rapidly? In his later years, Gauss explained that he had imagined the numbers paired according to the following schema.

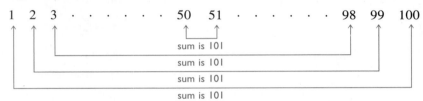

The sum of the numbers in each pair is 101, and there are 50 pairs in all; hence the total sum is $50 \cdot 101 = 5{,}050$.

EXAMPLE 4.2.2   **Applying the Formula for the Sum of the First *n* Integers**

a. Find $2 + 4 + 6 + \cdots + 500$.
b. For an integer $h \geq 2$, find $1 + 2 + 3 + \cdots + (h - 1)$.

Solution   a. $2 + 4 + 6 + \cdots + 500$

$$= 2 \cdot (1 + 2 + 3 + \cdots + 250)$$

$$= 2 \cdot \left(\frac{250 \cdot 251}{2} \right) \qquad \text{by applying the formula for the sum of the first } n \text{ integers with } n = 250$$

$$= 62{,}750.$$

b. $1 + 2 + 3 + \cdots + (h - 1)$

$$= \frac{(h - 1) \cdot [(h - 1) + 1]}{2} \qquad \text{by applying the formula for the sum of the first } n \text{ integers with } n = h - 1$$

$$= \frac{(h - 1) \cdot h}{2} \qquad \text{since } (h - 1) + 1 = h.$$

■

The next example asks for a proof of another famous and important formula in mathematics—the formula for the sum of a geometric sequence. In a **geometric sequence**, each term is obtained from the preceding one by multiplying by a constant factor. If the first term is 1 and the constant factor is r, then the sequence is $1, r, r^2, r^3, \ldots, r^n, \ldots$. The sum of the first n terms of this sequence is given by the formula

$$\sum_{i=0}^{n} r^i = \frac{r^{n+1} - 1}{r - 1}$$

for all integers $n \geq 0$ and real numbers r not equal to 1. The expanded form of this formula is

$$r^0 + r^1 + r^2 + \cdots + r^n = \frac{r^{n+1} - 1}{r - 1},$$

and because $r^0 = 1$ and $r^1 = r$, this can be rewritten as

$$1 + r + r^2 + \cdots + r^n = \frac{r^{n+1} - 1}{r - 1}$$

for $n \geq 1$.

EXAMPLE 4.2.3

Sum of a Geometric Sequence

Prove that $\sum_{i=0}^{n} r^i = \frac{r^{n+1} - 1}{r - 1}$, for all integers $n \geq 0$ and all real numbers r except 1.

Solution
In this example $P(n)$ is again an equation, although in this case it contains a real variable r:

$$P(n): \sum_{i=0}^{n} r^i = \frac{r^{n+1} - 1}{r - 1}.$$

Because r can be any real number other than 1, the proof begins by supposing that r is a particular but arbitrarily chosen real number not equal to 1. Then the proof continues by mathematical induction on n, starting with $n = 0$. In the basis step, you must show that $P(0)$ is true:

$$P(0): \sum_{i=0}^{0} r^i = \frac{r^{0+1} - 1}{r - 1}.$$

In the inductive step, you suppose $P(k)$ is true:

$$P(k): \sum_{i=0}^{k} r^i = \frac{r^{k+1} - 1}{r - 1}.$$

Then you show that $P(k + 1)$ is true:

$$P(k + 1): \sum_{i=0}^{k+1} r^i = \frac{r^{(k+1)+1} - 1}{r - 1},$$

or, equivalently,

$$P(k + 1): \sum_{i=0}^{k+1} r^i = \frac{r^{k+2} - 1}{r - 1}.$$

THEOREM 4.2.3   Sum of a Geometric Sequence

For any real number r except 1, and any integer $n \geq 0$,

$$\sum_{i=0}^{n} r^i = \frac{r^{n+1} - 1}{r - 1}.$$

Proof:

Suppose r is a particular but arbitrarily chosen real number that is not equal to 1. We must show that for all integers $n \geq 0$,

$$\sum_{i=0}^{n} r^i = \frac{r^{n+1} - 1}{r - 1}.$$

We show this by mathematical induction on n.

The formula is true for $n = 0$:   For $n = 0$ we must show that

$$\sum_{i=0}^{0} r^i = \frac{r^{0+1} - 1}{r - 1}.$$

The left-hand side of this equation is $r^0 = 1$. The right-hand side is

$$\frac{r^1 - 1}{r - 1} = \frac{r - 1}{r - 1} = 1$$

because $r^1 = r$ and $r \neq 1$. *[So the formula is true for $n = 0$.]*

If the formula is true for $n = k$ then it is true for $n = k + 1$:

[Suppose the formula $\sum_{i=0}^{n} r^i = \frac{r^{n+1} - 1}{r - 1}$ is true when an integer $k \geq 0$ is substituted in place of n.]

Suppose $\sum_{i=0}^{k} r^i = \frac{r^{k+1} - 1}{r - 1}$, for $k \geq 0$. *[This is the inductive hypothesis.]*

[We must show that the formula is true when $k + 1$ is substituted in place of n.]

We must show that $\sum_{i=0}^{k+1} r^i = \frac{r^{(k+1)+1} - 1}{r - 1}$. *[We will show that the left-hand side of this equation equals the right-hand side.]*

But

$$\sum_{i=0}^{k+1} r^i = \sum_{i=0}^{k} r^i + r^{k+1} \qquad \text{by writing the } (k + 1)\text{st term separately from the first } k \text{ terms}$$

$$= \frac{r^{k+1} - 1}{r - 1} + r^{k+1} \qquad \text{by substitution from the inductive hypothesis}$$

$$= \frac{r^{k+1} - 1}{r - 1} + \frac{r^{k+1}(r - 1)}{r - 1} \qquad \text{by multiplying the numerator and denominator of the second term by } (r - 1) \text{ to obtain a common denominator}$$

$$= \frac{(r^{k+1} - 1) + r^{k+1}(r - 1)}{r - 1} \qquad \text{by adding fractions}$$

$$= \frac{r^{k+1} - 1 + r^{k+1} \cdot r - r^{k+1}}{r - 1} \qquad \text{by multiplying out}$$

$$= \frac{r^{k+2} - 1}{r - 1} \qquad \begin{array}{l}\text{by cancelling the } r^{k+1}\text{'s and by the fact that}\\ r^{k+1} \cdot r = r^{k+1} \cdot r^1 = r^{k+2}.\end{array}$$

[This is what we needed to show.]
   [Since we have proved the basis and inductive steps, we conclude that the theorem is true.]

Note that the formula for the sum of a geometric sequence can be thought of as a family of different formulas in r, one for each real number r except 1.

EXAMPLE 4.2.4   Applying the Formula for the Sum
of a Geometric Sequence

In each of (a) and (b) below, assume that m is an integer that is greater than or equal to 3.
a. Find $1 + 3 + 3^2 + \cdots + 3^{m-2}$.
b. Find $3^2 + 3^3 + 3^4 + \cdots + 3^m$.

Solution   a. $1 + 3 + 3^2 + \cdots + 3^{m-2} = \dfrac{3^{(m-2)+1} - 1}{3 - 1} \qquad \begin{array}{l}\text{by applying the formula for the sum of a}\\ \text{geometric sequence with } r = 3 \text{ and}\\ n = m - 2\end{array}$

$$= \frac{3^{m-1} - 1}{2}.$$

b. $3^2 + 3^3 + 3^4 + \cdots + 3^m$

$$= 3^2 \cdot (1 + 3 + 3^2 + \cdots + 3^{m-2}) \qquad \text{by factoring out } 3^2$$

$$= 9 \cdot \left(\frac{3^{m-1} - 1}{2}\right) \qquad \text{by part (a).}$$

As with the formula for the sum of the first n integers, there is a way to think of the formula for the sum of the terms of a geometric sequence that makes it seem simple and intuitive. Let

$$S_n = 1 + r + r^2 + \cdots + r^n.$$

Then

$$rS_n = r + r^2 + r^3 + \cdots + r^{n+1},$$

and so

$$rS_n - S_n = (r + r^2 + r^3 + \cdots + r^{n+1}) - (1 + r + r^2 + \cdots + r^n)$$
$$= r^{n+1} - 1.$$

But

$$rS_n - S_n = (r - 1)S_n.$$

Hence

$$S_n = \frac{r^{n+1} - 1}{r - 1}.$$

This derivation of the formula is attractive and is quite convincing. However it is not as logically airtight as the proof by mathematical induction. To go from one step to another in the calculations above, the argument is made that each term among those indicated by the ellipsis (. . .) has such-and-such an appearance and when these are cancelled such-and-such occurs. But it is impossible actually to see each such term and each such calculation, and so the accuracy of these claims cannot be fully checked. With mathematical induction it is possible to focus exactly on what happens in the middle of the ellipsis and verify without doubt that the calculations are correct.

EXERCISE SET 4.2

1. Use mathematical induction (and the proof of Proposition 4.2.1 as a model) to show that any amount of money of at least 15¢ can be made up using 3¢ and 8¢ coins.

2. Use mathematical induction to show that any postage of at least 8¢ can be obtained using 3¢ and 5¢ stamps.

3. For each positive integer n, let $P(n)$ be the formula
$$1^2 + 2^2 + \cdots + n^2 = \frac{n(n + 1)(2n + 1)}{6}.$$
   a. Write $P(1)$. Is $P(1)$ true?
   b. Write $P(k)$.
   c. Write $P(k + 1)$.
   d. In a proof by mathematical induction that the formula holds for all integers $n \geq 1$, what must be shown in the inductive step?

4. For each integer n with $n \geq 2$, let $P(n)$ be the formula
$$\sum_{i=1}^{n-1} i(i + 1) = \frac{n(n - 1)(n + 1)}{3}.$$
   a. Write $P(2)$. Is $P(2)$ true?
   b. Write $P(k)$.
   c. Write $P(k + 1)$.
   d. In a proof by mathematical induction that the formula holds for all integers $n \geq 2$, what must be shown in the inductive step?

5. Fill in the missing pieces in the following proof that
$$1 + 3 + 5 + \cdots + (2n - 1) = n^2$$
for all integers $n \geq 1$.

Proof:

The formula is true for $n = 1$: To establish the formula for $n = 1$, we must show that when 1 is substituted in place of n, the left-hand side equals the right-hand side. But when $n = 1$, the left-hand side is the sum of all the odd integers from 1 to $2 \cdot 1 - 1$, which is the sum of the odd integers from 1 to 1, which is just 1. The right-hand side is __(a)__ , which also equals 1. So the formula is true for $n = 1$.

If the formula is true for $n = k$, *then it is true for* $n = k + 1$:
[*Suppose the formula* $1 + 3 + 5 + \cdots + (2n - 1) = n^2$ *is true when k is substituted in place of n.*] Suppose $1 + 3 + 5 + \cdots + (2k - 1) = $ __(b)__ . [*This is the inductive hypothesis.*] [*We must show that the formula* $1 + 3 + 5 + \cdots + (2n - 1) = n^2$ *is true when $k + 1$ is substituted for n.*] We must show that __(c)__ = __(d)__ . But

$$1 + 3 + 5 + \cdots + [2(k + 1) - 1]$$
$$= 1 + 3 + 5 + \cdots + (2k + 1) \qquad \text{by algebra}$$
$$= 1 + 3 + 5 + \cdots \qquad \text{the next-to-last term is}$$
$$\quad + (2k - 1) + (2k + 1) \qquad 2k - 1 \text{ because } \underline{(e)}$$
$$= \underline{(f)} + (2k + 1) \qquad \text{by inductive hypothesis}$$
$$= (k + 1)^2 \qquad \text{by algebra.}$$
[*This is what we needed to show.*]
   [*Since we have proved the basis and inductive steps, we conclude that the given statement is true.*]
   The proof above was heavily annotated to help make its logical flow more obvious. Normally such annotation is omitted.

6. Without using Theorem 4.2.2, use mathematical induction to prove that
$$2 + 4 + 6 + \cdots + 2n = n^2 + n,$$
for all integers $n \geq 1$.

7. Without using Theorem 4.2.2, use mathematical induction to prove that
$$1 + 5 + 9 + \cdots + (4n - 3) = n(2n - 1),$$
for all integers $n \geq 1$.

8. Without using Theorem 4.2.3, use mathematical induction to prove that
$$1 + 2 + 2^2 + \cdots + 2^n = 2^{n+1} - 1,$$
for all integers $n \geq 0$.

Prove each of the statements in 9–16 by mathematical induction.

9. $1^2 + 2^2 + \cdots + n^2 = \dfrac{n(n + 1)(2n + 1)}{6}$, for all integers $n \geq 1$.

10. $1^3 + 2^3 + \cdots + n^3 = \left[\dfrac{n(n+1)}{2}\right]^2$, for all integers $n \geq 1$.

11. $\dfrac{1}{1 \cdot 2} + \dfrac{1}{2 \cdot 3} + \cdots + \dfrac{1}{n(n+1)} = \dfrac{n}{n+1}$, for all integers $n \geq 1$.

12. $\displaystyle\sum_{i=1}^{n-1} i(i+1) = \dfrac{n(n-1)(n+1)}{3}$, for all integers $n \geq 2$.

13. $\displaystyle\sum_{i=1}^{n+1} i \cdot 2^i = n \cdot 2^{n+2} + 2$, for all integers $n \geq 0$.

H14. $\displaystyle\sum_{i=1}^{n} i(i!) = (n+1)! - 1$, for all integers $n \geq 1$.

15. $\left(1 - \dfrac{1}{2^2}\right) \cdot \left(1 - \dfrac{1}{3^2}\right) \cdot \cdots \cdot \left(1 - \dfrac{1}{n^2}\right) = \dfrac{n+1}{2n}$, for all integers $n \geq 2$.

16. $\displaystyle\prod_{i=0}^{n}\left(\dfrac{1}{2i+1} \cdot \dfrac{1}{2i+2}\right) = \dfrac{1}{(2n+2)!}$, for all integers $n \geq 0$.

17. The distributive law from algebra says that for all real numbers c, a_1, and a_2, $c(a_1 + a_2) = ca_1 + ca_2$. Use this law and mathematical induction to prove the generalized distributive law: For all integers $n \geq 2$, if c, a_1, a_2, $\ldots$, a_n are any real numbers, then $c(a_1 + a_2 + \cdots + a_n) = ca_1 + ca_2 + \cdots + ca_n$. (Use the definition $x_1 + x_2 + \cdots + x_m = (x_1 + x_2 + \cdots + x_{m-1}) + x_m$ for all integers $m \geq 2$ and all real numbers $x_1, x_2, \ldots, x_m$.)

◆18. If x is a real number not divisible by π, then for all integers $n \geq 1$,

$$\sin x + \sin 3x + \sin 5x + \cdots + \sin (2n-1)x = \dfrac{1 - \cos 2nx}{2 \sin x}.$$

Use the formula for the sum of the first n integers and/or the formula for the sum of a geometric sequence to find the sums in 19–25.

19. $3 + 4 + 5 + \cdots + 1{,}000$

20. $5 + 10 + 15 + 20 + \cdots + 300$

21. $1 + 2 + 3 + \cdots + (k - 1)$, where k is a positive integer

22. a. $1 + 2 + 2^2 + \cdots + 2^{25}$
   b. $2 + 2^2 + 2^3 + \cdots + 2^{26}$

23. $3 + 3^2 + 3^3 + \cdots + 3^n$, where n is an integer and $n \geq 2$

24. $1 + \dfrac{1}{2} + \dfrac{1}{2^2} + \cdots + \dfrac{1}{2^n}$, where n is a positive integer

25. $1 - 2 + 2^2 - 2^3 + \cdots + (-1)^n 2^n$, where n is a positive integer

H26. Find a formula in n, a, and d for the sum $(a + md) + (a + (m + 1)d) + (a + (m + 2)d) + \cdots + (a + (m + n)d)$, where m and n are integers, $n \geq 0$, and a and d are real numbers. Justify your answer.

27. Find a formula in a, r, m, and n for the sum $ar^m + ar^{m+1} + ar^{m+2} + \cdots + ar^{m+n}$, where m and n are integers, $n \geq 0$, and a and r are real numbers. Justify your answer.

28. You have two parents, four grandparents, eight great-grandparents, and so forth.
  a. If all your ancestors were distinct, what would be the total number of your ancestors for the past 40 generations (counting your parents' generation as number one)? (*Hint:* Use the formula for the sum of a geometric sequence.)
  b. Assuming that each generation represents 30 years, how long is 40 generations?
  c. The total number of people who have ever lived is approximately 10 billion, which equals 10^{10} people. Compare this fact with the answer to part (a). What do you deduce?

29. Find the mistake in the following proof fragment. Theorem: For any integer $n \geq 1$,

$$1^2 + 2^2 + \cdots + n^2 = \dfrac{n(n+1)(2n+1)}{6}.$$

"Proof (by mathematical induction): Certainly the theorem is true for $n = 1$ because $1^2 = 1$ and $\dfrac{1(1+1)(2 \cdot 1 + 1)}{6} = 1$. So the basis step is true. For the inductive step, suppose that for some $k \geq 1$, $k^2 = \dfrac{k(k+1)(2k+1)}{6}$. We must show that $(k+1)^2 = \dfrac{(k+1)((k+1)+1)(2(k+1)+1)}{6} \ldots$"

4.3 MATHEMATICAL INDUCTION II

In natural science courses, deduction and induction are presented as alternative modes of thought—deduction being to infer a conclusion from general principles using the laws of logical reasoning and induction being to enunciate a general principle after observing it to hold in a large number of specific instances. In this sense, then, *mathematical* induction is not inductive but deductive. Once proved by mathematical induction, a theorem is known just as certainly as if it were proved by any other mathematical method. However, inductive reasoning, in the natural sciences sense, is used in mathematics, but only to make conjectures not to prove them. For example, observe that

$$1 - \frac{1}{2} = \frac{1}{2}$$

$$\left(1 - \frac{1}{2}\right)\left(1 - \frac{1}{3}\right) = \frac{1}{3}$$

$$\left(1 - \frac{1}{2}\right)\left(1 - \frac{1}{3}\right)\left(1 - \frac{1}{4}\right) = \frac{1}{4}$$

This pattern seems so unlikely to occur by pure chance that it is reasonable to conjecture (though it is by no means certain) that the pattern holds true in general. In a case like this, a proof by mathematical induction (which you are asked to write in exercise 1 at the end of this section) gets to the essence of why the pattern holds in general. It reveals the mathematical mechanism that necessitates the truth of each successive case from the previous one. For instance, in this example observe that if

$$\left(1 - \frac{1}{2}\right)\left(1 - \frac{1}{3}\right) \cdots \left(1 - \frac{1}{k}\right) = \frac{1}{k},$$

then

$$\left(1 - \frac{1}{2}\right)\left(1 - \frac{1}{3}\right) \cdots \left(1 - \frac{1}{k}\right)\left(1 - \frac{1}{k+1}\right) = \frac{1}{k}\left(1 - \frac{1}{k+1}\right) = \frac{1}{k}\left(\frac{k+1-1}{k+1}\right)$$

$$= \frac{1}{k}\left(\frac{k}{k+1}\right) = \frac{1}{k+1}.$$

Thus mathematical induction makes knowledge of the general pattern a matter of mathematical certainty rather than vague conjecture.

In the remainder of this section we show how to use mathematical induction to prove additional kinds of statements such as divisibility properties of the integers and inequalities. The basic outlines of the proofs are the same in all cases, but the details of the basis and inductive steps differ from one to another.

In the example below, mathematical induction is used to establish a divisibility property.

EXAMPLE 4.3.1   Proving a Divisibility Property

Use mathematical induction to prove that for all integers $n \geq 1$, $2^{2n} - 1$ is divisible by 3.

Solution   As in the previous proofs by mathematical induction, you need to identify $P(n)$. In this example,

$P(n)$: $2^{2n} - 1$ is divisible by 3.

▲ CAUTION!   *Since P(n) is a sentence, not a number, we write P(n): . . . not P(n) = ...*

By substitution:

$$P(1): \quad 2^{2 \cdot 1} - 1 \text{ is divisible by 3.}$$
$$P(k): \quad 2^{2k} - 1 \text{ is divisible by 3.}$$
$$P(k + 1): \quad 2^{2(k+1)} - 1 \text{ is divisible by 3.}$$

Recall that an integer m is divisible by 3 if, and only if, $m = 3r$ for some integer r. Now the statement $P(1)$ is true because $2^{2 \cdot 1} - 1 = 2^2 - 1 = 4 - 1 = 3$, which is divisible by 3.

To prove the inductive step, you suppose that k is an integer greater than or equal to 1 such that $P(k)$ is true. This means that $2^{2k} - 1$ is divisible by 3. You must then prove the truth of $P(k + 1)$. Or, in other words, you must show that $2^{2(k+1)} - 1$ is divisible by 3. But

$$2^{2(k+1)} - 1 = 2^{2k+2} - 1$$
$$= 2^{2k} \cdot 2^2 - 1 \quad \text{by the laws of exponents}$$
$$= 2^{2k} \cdot 4 - 1.$$

The aim is to show that this quantity, $2^{2k} \cdot 4 - 1$, is divisible by 3. Why should that be so? By the inductive hypothesis, $2^{2k} - 1$ is divisible by 3, and $2^{2k} \cdot 4 - 1$ resembles $2^{2k} - 1$. Indeed, if you subtract $2^{2k} - 1$ from $2^{2k} \cdot 4 - 1$, you obtain $2^{2k} \cdot 3$, which is divisible by 3:

$$\underbrace{2^{2k} \cdot 4 - 1}_{\text{divisible by 3?}} - \underbrace{(2^{2k} - 1)}_{\text{divisible by 3}} = \underbrace{2^{2k} \cdot 3}_{\text{divisible by 3}}.$$

Adding $2^{2k} - 1$ to both sides gives

$$\underbrace{2^{2k} \cdot 4 - 1}_{\text{divisible by 3?}} = \underbrace{2^{2k} \cdot 3}_{\text{divisible by 3}} + \underbrace{2^{2k} - 1}_{\text{divisible by 3}}.$$

Both terms of the sum on the right-hand side of this equation are divisible by 3; hence the sum is divisible by 3. (See exercise 14 of Section 3.3.) Therefore, the left-hand side of the equation is also divisible by 3, which is what was to be shown.

This discussion is summarized as follows.

PROPOSITION 4.3.1

For all integers $n \geq 1$, $2^{2n} - 1$ is divisible by 3.

Proof (by mathematical induction):

The divisibility property is true for $n = 1$: To show the statement is true for $n = 1$, we must show that $2^{2 \cdot 1} - 1 = 2^2 - 1 = 3$, which is divisible by 3 *[since $3 = 3 \cdot 1$]*.

If the divisibility property is true for n = k then it is true for n = k + 1:

[Suppose the sentence "$2^{2n} - 1$ is divisible by 3" is true when the integer $k \geq 1$ is substituted for n.]

Suppose $2^{2k} - 1$ is divisible by 3, for some integer $k \geq 1$. *[Inductive hypothesis]*

[We must show that the sentence "$2^{2n} - 1$ is divisible by 3" is true when $k + 1$ is substituted for n.]

We must show that $2^{2(k+1)} - 1$ is divisible by 3.

But

$$
\begin{aligned}
2^{2(k+1)} - 1 &= 2^{2k+2} - 1 \\
&= 2^{2k} \cdot 2^2 - 1 &&\text{by the laws of exponents} \\
&= 2^{2k} \cdot 4 - 1 \\
&= 2^{2k}(3 + 1) - 1 \\
&= 2^{2k} \cdot 3 + (2^{2k} - 1) &&\text{by the laws of algebra}
\end{aligned}
$$

Clearly $2^{2k} \cdot 3$ is divisible by 3, and by the inductive hypothesis, so is $2^{2k} - 1$. Hence the sum of these quantities is divisible by 3, and so $2^{2(k+1)} - 1$ is divisible by 3 *[as was to be shown].*

[Since we have proved the basis and inductive steps, we conclude that the proposition is true.]

■

The next example illustrates the use of mathematical induction to prove an inequality.

EXAMPLE 4.3.2   **Proving an Inequality**

Use mathematical induction to prove that for all integers $n \geq 3$,

$$2n + 1 < 2^n.$$

Solution   In this example $P(n)$ is an inequality.

$$\boxed{P(n): \quad 2n + 1 < 2^n}$$

By substitution:

$$\boxed{\begin{aligned}
P(3): &\quad 2 \cdot 3 + 1 < 2^3 \\
P(k): &\quad 2k + 1 < 2^k \\
P(k + 1): &\quad 2(k + 1) + 1 < 2^{k+1}
\end{aligned}}$$

To prove the basis step, observe that the statement $P(3)$ is true because $2 \cdot 3 + 1 = 7$, $2^3 = 8$, and $7 < 8$.

To prove the inductive step, suppose $P(k)$ is true for an integer $k \geq 3$. *[This is the inductive hypothesis.]* This means that $2k + 1 < 2^k$ is assumed to be true for an integer $k \geq 3$. Then derive the truth of $P(k + 1)$. Or, in other words, show that the inequality $2(k + 1) + 1 < 2^{k+1}$ is true. But by multiplying out and regrouping

$$2(k + 1) + 1 = 2k + 3 = (2k + 1) + 2, \qquad \text{4.3.1}$$

and by substitution from the inductive hypothesis

$$(2k + 1) + 2 < 2^k + 2.$$

4.3.2

Hence

$$2(k + 1) + 1 < 2^k + 2 \qquad \text{The left-most part of equation (4.3.1)}$$

The left-most part of equation (4.3.1) is less than the right-most part of inequality (4.3.2).

If it can be shown that $2^k + 2$ is less than 2^{k+1}, then the desired inequality would be proved. But since the quantity 2^k can be added to or subtracted from an inequality without changing its direction,

$$2^k + 2 < 2^{k+1} \quad \text{if, and only if,} \quad 2 < 2^{k+1} - 2^k = 2^k(2 - 1) = 2^k.$$

And since multiplying or dividing an inequality by 2 does not change its direction,

$$2 < 2^k \quad \text{if, and only if,} \quad 1 = \frac{2}{2} < \frac{2^k}{2} = 2^{k-1} \qquad \text{by the laws of exponents.}$$

This last inequality is clearly true for all $k \geq 2$. Hence it is true that $2(k + 1) + 1 < 2^{k+1}$.

This discussion is made more flowing (but less intuitive) in the following formal proof:

PROPOSITION 4.3.2

For all integers $n \geq 3$, $2n + 1 < 2^n$.

Proof (by mathematical induction):

The inequality is true for $n = 3$: To prove the statement for $n = 3$, we must show that $2 \cdot 3 + 1 < 2^3$. But $2 \cdot 3 + 1 = 7$, $2^3 = 8$, and $7 < 8$. Hence the statement is true for $n = 3$.

If the inequality is true for $n = k$ then it is true for $n = k + 1$:
  [Suppose "$2n + 1 < 2^n$" is true when an integer $k \geq 3$ is substituted for n.]
  Suppose $2k + 1 < 2^k$, for some integer k such that $k \geq 3$. *[This is the inductive hypothesis.]*
  [We must show that "$2n + 1 < 2^n$" is true when $k + 1$ is substituted for n.]
  We must show that $2(k + 1) + 1 < 2^{k+1}$, or, equivalently, $2k + 3 < 2^{k+1}$.
But

$$2k + 3 = (2k + 1) + 2 \qquad \text{by algebra}$$
$$\Rightarrow \quad 2k + 3 < 2^k + 2^k \qquad \text{because } 2k + 1 < 2^k \text{ by the inductive hypothesis and because } 2 < 2^k \text{ for all integers } k \geq 2.$$

$$\therefore \ 2k + 3 < 2 \cdot 2^k = 2^{k+1} \qquad \text{by the laws of exponents.}$$

[This is what we needed to show.]
  [Since we have proved the basis and inductive steps, we conclude that the proposition is true.]

The last example of this section demonstrates how to use mathematical induction to show that the terms of a sequence satisfy a certain explicit formula.

EXAMPLE 4.3.3   Proving a Property of a Sequence

Define a sequence $a_1, a_2, a_3, \ldots$ as follows:*

$$a_1 = 2$$
$$a_k = 5a_{k-1} \quad \text{for all integers } k \geq 2.$$

a. Write the first four terms of the sequence.
b. Use mathematical induction to show that the terms of the sequence satisfy the formula

$$a_n = 2 \cdot 5^{n-1} \quad \text{for all integers } n \geq 1.$$

Solution   a.
$$a_1 = 2.$$
$$a_2 = 5a_{2-1} = 5a_1 = 5 \cdot 2 = 10$$
$$a_3 = 5a_{3-1} = 5a_2 = 5 \cdot 10 = 50$$
$$a_4 = 5a_{4-1} = 5a_3 = 5 \cdot 50 = 250.$$

b. To use mathematical induction to show that the formula holds in general, you begin by showing that the first term of the sequence satisfies the formula. Then you suppose that the kth term of the sequence (for some integer $k \geq 1$) satisfies the formula and show that the $(k + 1)$st term also satisfies the formula.

The formula holds for $n = 1$:   For $n = 1$, the formula gives $2 \cdot 5^{1-1} = 2 \cdot 5^0 = 2 \cdot 1 = 2$. But $a_1 = 2$ by definition of the sequence. Hence the formula holds for $n = 1$.

If the formula holds for $n = k$ then it holds for $n = k + 1$:   Let k be an integer with $k \geq 1$ and suppose that $a_k = 2 \cdot 5^{k-1}$. *[This is the inductive hypothesis.]* We must show that $a_{k+1} = 2 \cdot 5^{(k+1)-1} = 2 \cdot 5^k$. But

$$
\begin{aligned}
a_{k+1} &= 5a_{(k+1)-1} & &\text{by definition of } a_1, a_2, a_3, \ldots \\
&= 5a_k & &\text{since } (k + 1) - 1 = k \\
&= 5 \cdot (2 \cdot 5^{k-1}) & &\text{by inductive hypothesis} \\
&= 2 \cdot (5 \cdot 5^{k-1}) & &\text{by regrouping} \\
&= 2 \cdot 5^k & &\text{by the laws of exponents.}
\end{aligned}
$$

[This is what was to be shown.]
     [Since we have proved the basis and inductive steps, we conclude the formula holds for all terms of the sequence.]   ■

EXERCISE SET 4.3

1. Based on the discussion of the product $(1 - \frac{1}{2})(1 - \frac{1}{3})$ $(1 - \frac{1}{4}) \cdots (1 - \frac{1}{n})$ at the beginning of this section, conjecture a formula for general n. Prove your conjecture by mathematical induction.

2. Experiment with computing values of $(1 + \frac{1}{1})(1 + \frac{1}{2})$ $(1 + \frac{1}{3}) \cdots (1 + \frac{1}{n})$ for small values of n to conjecture a formula for this product for general n. Prove your conjecture by mathematical induction.

* This is an example of a recursive definition. The general subject of recursion is discussed in Chapter 8.

3. Observe that

$$\frac{1}{1 \cdot 3} = \frac{1}{3}$$

$$\frac{1}{1 \cdot 3} + \frac{1}{3 \cdot 5} = \frac{2}{5}$$

$$\frac{1}{1 \cdot 3} + \frac{1}{3 \cdot 5} + \frac{1}{5 \cdot 7} = \frac{3}{7}$$

$$\frac{1}{1 \cdot 3} + \frac{1}{3 \cdot 5} + \frac{1}{5 \cdot 7} + \frac{1}{7 \cdot 9} = \frac{4}{9}$$

Guess a general formula and prove it by mathematical induction.

H 4. Observe that

$$1 = 1,$$

$$1 - 4 = -(1 + 2),$$

$$1 - 4 + 9 = 1 + 2 + 3,$$

$$1 - 4 + 9 - 16 = -(1 + 2 + 3 + 4),$$

$$1 - 4 + 9 - 16 + 25 = 1 + 2 + 3 + 4 + 5.$$

Guess a general formula and prove it by mathematical induction.

5. Evaluate the sum $\displaystyle\sum_{k=1}^{n} \frac{k}{(k + 1)!}$ for $n = 1, 2, 3, 4,$ and 5. Make a conjecture about a formula for this sum for general n, and prove your conjecture by mathematical induction.

6. For each positive integer n, let $P(n)$ be the sentence

$$4^n - 1 \text{ is divisible by } 3.$$

a. Write $P(1)$. Is $P(1)$ true?
b. Write $P(k)$.
c. Write $P(k + 1)$.
d. In a proof by mathematical induction that this divisibility property holds for all integers $n \geq 1$, what must be shown in the inductive step?

7. For each positive integer n, let $P(n)$ be the inequality

$$n^2 < 2^n.$$

a. Write $P(5)$. Is $P(5)$ true?
b. Write $P(k)$.
c. Write $P(k + 1)$.
d. In a proof by mathematical induction that this inequality holds for all integers $n \geq 5$, what must be shown in the inductive step?

Prove each statement in 8–20 by mathematical induction.

8. $4^n - 1$ is divisible by 3, for each integer $n \geq 1$.

9. $2^{3n} - 1$ is divisible by 7, for each integer $n \geq 1$.

10. $n^3 - 7n + 3$ is divisible by 3, for each integer $n \geq 0$.

11. $3^{2n} - 1$ is divisible by 8, for each integer $n \geq 0$.

12. For any integer $n \geq 1$, $7^n - 2^n$ is divisible by 5.

H 13. For any integer $n \geq 1$, $x^n - y^n$ is divisible by $x - y$, where x and y are any integers with $x \neq y$.

H 14. $n^3 - n$ is divisible by 6, for each integer $n \geq 2$.

15. $n(n^2 + 5)$ is divisible by 6, for each integer $n \geq 1$.

16. $n^2 < 2^n$, for all integers $n \geq 5$.

17. $2^n < (n + 2)!$, for all integers $n \geq 0$.

18. $\sqrt{n} < \frac{1}{\sqrt{1}} + \frac{1}{\sqrt{2}} + \cdots + \frac{1}{\sqrt{n}}$, for all integers $n \geq 2$.

19. $1 + nx \leq (1 + x)^n$, for all real numbers $x > -1$ and integers $n \geq 2$.

20. a. $n^3 > 2n + 1$, for all integers $n \geq 2$.
b. $n! > n^2$, for all integers $n \geq 4$.

21. A sequence $a_1, a_2, a_3, \ldots$ is defined by letting $a_1 = 3$ and $a_k = 7a_{k-1}$ for all integers $k \geq 2$. Show that $a_n = 3 \cdot 7^{n-1}$ for all integers $n \geq 1$.

22. A sequence $b_0, b_1, b_2, \ldots$ is defined by letting $b_0 = 5$ and $b_k = 4 + b_{k-1}$ for all integers $k \geq 1$. Show that $b_n = 5 + 4n$ for all integers $n \geq 0$.

23. A sequence $c_0, c_1, c_2, \ldots$ is defined by letting $c_0 = 3$ and $c_k = (c_{k-1})^2$ for all integers $k \geq 1$. Show that $c_n = 3^{2^n}$ for all integers $n \geq 0$.

24. A sequence $d_1, d_2, d_3, \ldots$ is defined by letting $d_1 = 2$ and $d_k = \dfrac{d_{k-1}}{k}$ for all integers $k \geq 2$. Show that

$$d_n = \frac{2}{n!} \text{ for all integers } n \geq 1.$$

25. Prove that for all integers $n \geq 1$,

$$\frac{1}{3} = \frac{1 + 3}{5 + 7} = \frac{1 + 3 + 5}{7 + 9 + 11} = \cdots$$

$$= \frac{1 + 3 + \cdots + (2n - 1)}{(2n + 1) + \cdots + (4n - 1)}.$$

26. As each of a group of business people arrives at a meeting, each shakes hands with all the other people present. Use mathematical induction to show that if n people come to the meeting, then $[n(n - 1)]/2$ handshakes occur.

In order for a proof by mathematical induction to be valid, the basis statement must be true for $n = a$ and the argument of the inductive step must be correct for every integer $k \geq a$. In 27 and 28 find the mistakes in the "proofs" by mathematical induction.

27. "Theorem": For any integer $n \geq 1$, all the numbers in a set of n numbers are equal to each other.
"Proof (by mathematical induction): It is obviously true

that all the numbers in a set consisting of just one number are equal to each other, so the basis step is true. For the inductive step, let $A = \{a_1, a_2, \ldots, a_k, a_{k+1}\}$ be any set of $k + 1$ numbers. Form two subsets each of size k:

$$B = \{a_1, a_2, a_3, \ldots, a_k\} \quad \text{and}$$
$$C = \{a_1, a_3, a_4, \ldots, a_{k+1}\}.$$

(B consists of all the numbers in A except a_{k+1}, and C consists of all the numbers in A except a_2.) By inductive hypothesis all the numbers in B equal a_1 and all the numbers in C equal a_1 (since both sets have only k numbers). But every number in A is in B or C, so all the numbers in A equal a_1; hence all are equal to each other."

28. "Theorem": For all integers $n \geq 1$, $3^n - 2$ is even. "Proof (by mathematical induction): Suppose the theorem is true for an integer k, where $k \geq 1$. That is, suppose that $3^k - 2$ is even. We must show that $3^{k+1} - 2$ is even. But

$$3^{k+1} - 2 = 3^k \cdot 3 - 2 = 3^k(1 + 2) - 2$$
$$= (3^k - 2) + 3^k \cdot 2.$$

Now $3^k - 2$ is even by inductive hypothesis and $3^k \cdot 2$ is even by inspection. Hence the sum of the two quantities is even (by exercise 10, Section 3.1). It follows that $3^{k+1} - 2$ is even, which is what we needed to show."

29. (This problem requires knowledge of matrix multiplication.) Prove that

$$\begin{bmatrix} 2 & 1 \\ 0 & 2 \end{bmatrix}^n = \begin{bmatrix} 2^n & n \cdot 2^{n-1} \\ 0 & 2^n \end{bmatrix} \quad \text{for all integers } n \geq 1.$$

30. In a round-robin tournament each team plays every other team exactly once. If the teams are labeled T_1, $T_2, \ldots, T_n$, then the outcome of such a tournament can be represented by a drawing, called a *directed graph,* in which the teams are represented as dots and an arrow is drawn from one dot to another if, and only if, the

team represented by the first dot beats the team represented by the second dot. For example, the directed graph below shows one outcome of a round-robin tournament involving five teams, A, B, C, D, and E.

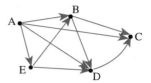

Use mathematical induction to show that in any round-robin tournament involving n teams, where $n \geq 2$, it is possible to label the teams $T_1, T_2, \ldots, T_n$ so that T_i beats T_{i+1} for all $i = 1, 2, \ldots, n - 1$. (For instance, one such labeling in the example above is $T_1 = A$, $T_2 = B$, $T_3 = C$, $T_4 = E$, $T_5 = D$. (*Hint:* Given $k + 1$ teams, pick one—say T'—and apply the inductive hypothesis to the remaining teams to obtain an ordering $T_1, T_2, \ldots, T_k$. Consider three cases: T' beats T_1, T' loses to the first m teams (where $1 \leq m \leq k - 1$) and beats the $(m + 1)$st team, and T' loses to all the other teams.)

$H \blacklozenge$ **31.** On the outside rim of a circular disk the integers from 1 through 30 are painted in random order. Show that no matter what this order is, there must be three successive integers whose sum is at least 45.

$\blacklozenge$ **32.** John and Sara gave a party that was attended by n married couples. As the guests arrived, some shook hands with their host or hostess or with other guests who had arrived earlier. When all the guests were assembled, John asked them and also Sara how many hands each had shaken. He discovered that everyone gave a different answer. Assuming that no guests shook their own hands or the hands of their spouses, how many hands did Sara shake? Prove your answer by mathematical induction.

4.4   STRONG MATHEMATICAL INDUCTION AND THE WELL-ORDERING PRINCIPLE

Mathematics takes us still further from what is human into the region of absolute necessity, to which not only the actual world, but every possible world, must conform.

(Bertrand Russell, 1902)

Strong mathematical induction is similar to ordinary mathematical induction in that it is a technique for establishing the truth of a sequence of statements about integers. Also, a proof by strong mathematical induction consists of a basis step and an inductive step. However, the basis step may contain proofs for several initial values, and in the inductive step the truth of the predicate $P(n)$ is assumed not just for one value of n but for *all* values through $k - 1$, and then the truth of $P(k)$ is proved.

PRINCIPLE OF STRONG MATHEMATICAL INDUCTION

Let $P(n)$ be a predicate that is defined for integers n, and let a and b be fixed integers with $a \leq b$. Suppose the following two statements are true:

1. $P(a)$, $P(a + 1)$, . . . , and $P(b)$ are all true. **(basis step)**
2. For any integer $k > b$, if $P(i)$ is true for all integers i with $a \leq i < k$, then $P(k)$ is true. **(inductive step)**

Then the statement

$$\text{for all integers } n \geq a, \; P(n)$$

is true. (The supposition that $P(i)$ is true for all integers i with $a \leq i < k$ is called the **inductive hypothesis.**)

It is apparent that if the principle of strong mathematical induction is true, then so is the principle of ordinary mathematical induction. For if the truth of $P(k)$ alone implies the truth of $P(k + 1)$ for all $k \geq a$, then the truth of $P(i)$ for all $a \leq i \leq k$ certainly implies the truth of $P(k + 1)$ for all $k \geq a$. Or, equivalently, the truth of $P(i)$ for all $a \leq i < k$ implies the truth of $P(k)$ for all $k > a$.

It can also be shown that if the principle of ordinary mathematical induction is true, then so is the principle of strong mathematical induction. A proof of this fact is sketched in the exercises at the end of this section.

The divisibility by a prime theorem (Theorem 3.3.2) states that any integer greater than 1 is divisible by a prime number. We prove this theorem below using strong mathematical induction.

EXAMPLE 4.4.1   Divisibility by a Prime

Prove that any integer greater than 1 is divisible by a prime number.

Solution   Let $P(n)$ be the divisibility property "n is divisible by a prime number." Use strong mathematical induction to prove that this property holds for every integer $n \geq 2$.

The divisibility property holds for $n = 2$:   The property holds because 2 is a prime number and $2 \mid 2$.

If the divisibility property holds for all i with $2 \leq i < k$, then it holds for k: Let k be an integer with $k > 2$. Suppose that:

For all integers i with $2 \leq i < k$, i is divisible by a prime number.

[This is the inductive hypothesis.]

[We must show that k is divisible by a prime number.] Either k is prime or k is not prime. If k is prime, then k is divisible by a prime number, namely itself. If k is not prime, then $k = a \cdot b$, where a and b are integers with $2 \leq a < k$ and $2 \leq b < k$. By the inductive hypothesis, a is divisible by a prime number p, and so by transitivity of divisibility, k is also divisible by p. Hence, regardless of whether k is prime or not, k is divisible by a prime number *[as was to be shown]*.

[Since we have proved the basis and the inductive steps of the strong mathematical induction, we conclude that the given statement is true.] ■

Sometimes strong mathematical induction must be used to show that the terms of certain sequences satisfy a certain property.

EXAMPLE 4.4.2   **Proving a Property of a Sequence**

Define a sequence $a_1, a_2, a_3, \ldots$ as follows:

$$a_1 = 0,$$
$$a_2 = 2,$$
$$a_k = 3 \cdot a_{\lfloor k/2 \rfloor} + 2 \quad \text{for all integers } k \geq 3.$$

a. Find the first seven terms of the sequence.
b. Prove that a_n is even for each integer $n \geq 1$.

Solution   a. $a_1 = 0,$
$a_2 = 2,$
$a_3 = 3 \cdot a_{\lfloor 3/2 \rfloor} + 2 = 3 \cdot a_1 + 2 = 3 \cdot 0 + 2 = 2,$
$a_4 = 3 \cdot a_{\lfloor 4/2 \rfloor} + 2 = 3 \cdot a_2 + 2 = 3 \cdot 2 + 2 = 8,$
$a_5 = 3 \cdot a_{\lfloor 5/2 \rfloor} + 2 = 3 \cdot a_2 + 2 = 3 \cdot 2 + 2 = 8,$
$a_6 = 3 \cdot a_{\lfloor 6/2 \rfloor} + 2 = 3 \cdot a_3 + 2 = 3 \cdot 2 + 2 = 8,$
$a_7 = 3 \cdot a_{\lfloor 7/2 \rfloor} + 2 = 3 \cdot a_3 + 2 = 3 \cdot 2 + 2 = 8.$

b. Let $P(n)$ be the property "a_n is even." Use strong mathematical induction to show that this property holds for all integers $n \geq 1$.

The property holds for n = 1 and n = 2:

$a_1 = 0$ and $a_2 = 2$ and both 0 and 2 are even integers.

If the property holds for all i with $1 \leq i < k$, then it holds for k:

Let k be an integer with $k > 2$ and suppose that

a_i is even for all integers i with $0 \leq i < k$.   *[This is the inductive hypothesis.]*

[We must show that a_k is even.] By definition of $a_1, a_2, a_3, \ldots$

$$a_k = 3 \cdot a_{\lfloor k/2 \rfloor} + 2 \quad \text{for all integers } k \geq 3.$$

Now $a_{\lfloor k/2 \rfloor}$ is even by inductive hypothesis *[because $k > 2$ and so $0 \leq \lfloor k/2 \rfloor < k$].* Thus $3 \cdot a_{\lfloor k/2 \rfloor}$ is even *[because odd · even = even],* and hence $3 \cdot a_{\lfloor k/2 \rfloor} + 2$ is even *[because even + even = even—see Section 3.1].* Consequently, a_k, which equals $3 \cdot a_{\lfloor k/2 \rfloor} + 2$, is even *[as was to be shown].*

[Since we have proved the basis and the inductive steps of the strong mathematical induction, we conclude that the given statement is true.] ■

A product of four numbers may be computed in a variety of different ways as indicated by parentheses in the product. For instance,

$((x_1x_2)x_3)x_4$ means multiply x_1 and x_2; multiply the result by x_3; then multiply that number by x_4.

And

$(x_1x_2)(x_3x_4)$ means multiply x_1 and x_2; multiply x_3 and x_4; then take the product of the two.

Note that in both examples above, although the factors are multiplied in a different order, the number of multiplications—three—is the same. Strong mathematical induction is used to prove a generalization of this fact.

▲ CONVENTION   *Let us agree to say that a single number x_1 is a product with one factor that can be computed with zero multiplications.*

EXAMPLE 4.4.3

The Number of Multiplications Needed to Multiply *n* Distinct Numbers

Prove that for any integer $n \geq 1$, if $x_1, x_2, \ldots, x_n$ are n distinct real numbers, then no matter how the parentheses are inserted into their product, the number of multiplications used to compute the product is $n - 1$.

Solution   Let $P(n)$ be the property "If $x_1, x_2, \ldots, x_n$ are n distinct real numbers, then no matter how the parentheses are inserted into their product, the number of multiplications used to compute the product is $n - 1$." Use strong mathematical induction to show that this property is true for all integers $n \geq 1$.

The property is true for n = 1:   By agreement, x_1 is a product with one factor and is computed using $1 - 1$, or 0, multiplications.

If the property is true for all i with $1 \leq i < k$, then it is true for k: Let $k > 1$ be an integer and suppose that

> For all i with $1 \leq i < k$, if $x_1, x_2, \ldots, x_i$ are i distinct real numbers, then no matter how the parentheses are inserted into their product, the number of multiplications used to compute the product is $i - 1$. *[This is the inductive hypothesis.]*

Consider a product of k distinct factors: $x_1, x_2, \ldots, x_k$. *[We must show that no matter how parentheses are inserted into the product of these factors, the number of multiplications is $k - 1$.]* When parentheses are inserted in order to compute the product of the factors $x_1, x_2, \ldots, x_n$, some multiplication must be the final one. (For instance, in the product $((x_1x_2)x_3)((x_4(x_5x_6))x_7)$, the final multiplication is between $((x_1x_2)x_3)$ and $((x_4(x_5x_6))x_7)$.) Consider the two factors in this final multiplication. Each is itself a product of fewer than k factors. Say the left-hand product consists of r_k and the right-hand product of s_k factors. Then $1 \leq r_k < k$ and $1 \leq s_k < k$, and so by the inductive hypothesis, the number of multiplications for the left-hand product is $r_k - 1$ and the number of multiplications for the right-hand product is $s_k - 1$. It follows that the number of multiplications to compute the product of all the factors $x_1, x_2, \ldots, x_k$ is

$$(r_k - 1) + (s_k - 1) + 1,$$

where the $+1$ at the end represents the final multiplication between the left-hand and right-hand products. But the sum of the factors in the left-hand product plus those in the right-hand product is the total number of factors in the product. Hence $r_k + s_k = k$, and the number of multiplications equals

$$(r_k - 1) + (s_k - 1) + 1 = (r_k + s_k) - 1 = k - 1.$$

[This is what was to be shown.]
    [Since we have proved the basis and the inductive steps of the strong mathematical induction, we conclude that the given statement is true.] ∎

BINARY REPRESENTATION OF INTEGERS

Strong mathematical induction makes possible a proof of the frequently used fact that every positive integer n has a unique binary integer representation. The proof looks complicated because of all the notation needed to write down the various steps. But the idea of the proof is simple. It is that if smaller integers than n have unique representations as sums of powers of 2, then the unique representation for n as a sum of powers of 2 can be found by taking the representation for $n/2$ (or for $(n - 1)/2$ if n is odd) and multiplying it by 2.

THEOREM 4.4.1   Existence and Uniqueness of
                     Binary Integer Representations

Given any positive integer n, n has a unique representation in the form

$$n = c_r \cdot 2^r + c_{r-1} \cdot 2^{r-1} + \cdots + c_2 \cdot 2^2 + c_1 \cdot 2 + c_0,$$

where r is a nonnegative integer, $c_r = 1$, and $c_j = 1$ or 0 for all $j = 0, 1, 2, \ldots, r - 1$.

Proof:

We give separate proofs by strong mathematical induction to show first the existence and second the uniqueness of the binary representation.

Existence (proof by strong mathematical induction):   Consider the formula

$$n = c_r \cdot 2^r + c_{r-1} \cdot 2^{r-1} + \cdots + c_2 \cdot 2^2 + c_1 \cdot 2 + c_0,$$

where r is a nonnegative integer, $c_r = 1$, and $c_j = 1$ or 0 for all $j = 0, 1, 2, \ldots, r - 1$.

[This is $P(n)$.]

The formula is true for $n = 1$:   Let $r = 0$ and $c_0 = 1$. Then $1 = c_r \cdot 2^r$, and so $n = 1$ can be written in the required form.

If the formula is true for all i with $1 \le i < k$, then it is true for k: Let k be an integer with $k > 1$. Suppose that for all integers i with $1 \le i < k$, i can be written in the required form

$$i = c_r \cdot 2^r + c_{r-1} \cdot 2^{r-1} + \cdots + c_2 \cdot 2^2 + c_1 \cdot 2 + c_0,$$

where r is a nonnegative integer, $c_r = 1$, and $c_j = 1$ or 0 for all $j = 0, 1, 2, \ldots, r - 1$.

[This is the inductive hypothesis.]

We must show that k can be written as a sum of powers of 2 in the required form:

Case 1 (k is even):   In this case $k/2$ is an integer, and since $1 \le k/2 < k$, then by inductive hypothesis

$$\frac{k}{2} = c_r \cdot 2^r + c_{r-1} \cdot 2^{r-1} + \cdots + c_2 \cdot 2^2 + c_1 \cdot 2 + c_0,$$

where r is a nonnegative integer, $c_r = 1$, and $c_j = 1$ or 0 for all $j = 0, 1, 2, \ldots, r - 1$.

Multiplying both sides of the equation by 2 gives

$$k = c_r \cdot 2^{r+1} + c_{r-1} \cdot 2^r + \cdots + c_2 \cdot 2^3 + c_1 \cdot 2^2 + c_0 \cdot 2,$$

which is a sum of powers of 2 of the required form.

Case 2 (k is odd):   In this case $(k - 1)/2$ is an integer, and since $1 \le (k - 1)/2 < k$, then by inductive hypothesis,

$$\frac{k - 1}{2} = c_r \cdot 2^r + c_{r-1} \cdot 2^{r-1} + \cdots + c_2 \cdot 2^2 + c_1 \cdot 2 + c_0,$$

where r is a nonnegative integer, $c_r = 1$, and $c_j = 1$ or 0 for all $j = 0, 1, 2, \ldots, r - 1$.

Multiplying both sides of the equation by 2 and adding 1 gives

$$k = c_r \cdot 2^{r+1} + c_{r-1} \cdot 2^r + \cdots + c_2 \cdot 2^3 + c_1 \cdot 2^2 + c_0 \cdot 2 + 1,$$

which is the sum of powers of 2 of the required form.

The arguments above show that regardless of whether k is even or odd, k has a representation of the required form. *[Or, in other words, $P(k)$ is true as was to be shown].*

[Since we have proved the basis step and the inductive step of the strong mathematical induction, the existence half of the theorem is true.]

Uniqueness: To prove uniqueness, suppose that there is an integer n with two different representations as a sum of nonnegative integer powers of two. Equating the two representations and canceling all identical terms gives

$$2^r + c_{r-1} \cdot 2^{r-1} + \cdots + c_1 \cdot 2 + c_0$$
$$= 2^s + d_{s-1} \cdot 2^{s-1} + \cdots + d_1 \cdot 2 + d_0 \qquad \text{4.4.1}$$

where r and s are nonnegative integers, $r < s$, and each c_i and each d_i equals 0 or 1. But by the formula for the sum of a geometric sequence (Theorem 4.2.3),

$$2^r + c_{r-1} \cdot 2^{r-1}$$
$$+ \cdots + c_1 \cdot 2 + c_0 \le 2^r + 2^{r-1} + \cdots + 2 + 1 = 2^{r+1} - 1$$
$$< 2^s$$
$$< 2^s + d_{s-1} \cdot 2^{s-1} + \cdots + d_1 \cdot 2 + d_0,$$

which contradicts (4.4.1). Hence the supposition is false, and so any integer n has only one representation as a sum of nonnegative integer powers of two.

THE WELL-ORDERING PRINCIPLE FOR THE INTEGERS

The well-ordering principle looks very different from both the ordinary and strong principles of mathematical induction, but it can be shown that all three principles are equivalent. That is, if any one of the three is true, then so are both of the others.

WELL-ORDERING PRINCIPLE FOR THE INTEGERS

Let S be a set containing one or more integers all of which are greater than some fixed integer. Then S has a least element.

EXAMPLE 4.4.4   Finding Least Elements

In each case, if the set has a least element, state what it is. If not, explain why the well-ordering principle is not violated.

a. The set of all positive real numbers.
b. The set of all nonnegative integers n such that $n^2 < n$.
c. The set of all nonnegative integers of the form $46 - 7k$, where k is an integer.

Solution

a. There is no least positive real number. For if x is any positive real number, then $x/2$ is a positive real number that is less than x. No violation of the well-ordering principle occurs because the well-ordering principle refers only to sets of integers and this set is not a set of integers.

b. There is no least nonnegative integer n such that $n^2 < n$ because there is *no* nonnegative integer that satisfies this inequality. The well-ordering principle is not violated because the well-ordering principle refers to sets that contain at least one or more elements.

c. The following table shows values of $46 - 7k$ for various values of k.

k	0	1	2	3	4	5	6	7	...	-1	-2	-3	...
$46 - 7k$	46	39	32	25	18	11	4	-3	...	53	60	67	...

The table suggests, and you can easily confirm, that $46 - 7k < 0$ for $k \geq 7$ and that $46 - 7k \geq 46$ for $k \leq 0$. Therefore, from the other values in the table it is clear that 4 is the least nonnegative integer of the form $46 - 7k$. This corresponds to $k = 6$. ∎

Another way to look at the analysis of Example 4.4.4(c) is to observe that subtracting six 7's from 46 leaves 4 left over and this is the least nonnegative integer obtained by repeated subtraction of 7's from 46. In other words, 6 is the quotient and 4 is the remainder for the division of 46 by 7. More generally, in the division of any integer n by any positive integer d, the remainder r is the least nonnegative integer of the form $n - dk$. This is the heart of the following proof of the existence part of the quotient-remainder theorem (the part that guarantees the existence of a quotient and a remainder of the division of an integer by a positive integer). For a proof of the uniqueness of the quotient and remainder, see exercise 19 of Section 3.8.

QUOTIENT-REMAINDER THEOREM (EXISTENCE PART)

Given any integer n and any positive integer d, there exist integers q and r such that

$$n = d \cdot q + r \quad \text{and} \quad 0 \leq r < d.$$

Proof:

Let S be the set of all nonnegative integers of the form

$$n - d \cdot k,$$

where k is an integer. This set has at least one element. *[For if n is nonnegative, then*

$$n - 0 \cdot d = n \geq 0,$$

and so $n - 0 \cdot d$ is in S. And if n is negative, then

$$n - n \cdot d = n \cdot \underbrace{(1 - d)}_{} \geq 0,$$

$$\uparrow \qquad \curvearrowright$$
$$\leq 0 \quad \leq 0 \text{ since d is a positive integer}$$

and so $n - n \cdot d$ is in S.] It follows by the well-ordering principle that S contains a least element r. Then for some specific integer $k = q$,

$$n - d \cdot q = r$$

[because every integer in S can be written in this form]. Adding $d \cdot q$ to both sides gives

$$n = d \cdot q + r.$$

Furthermore, $r < d$. *[For suppose $r \geq d$. Then*

$$n - d \cdot (q + 1) = n - d \cdot q - d = r - d \geq 0,$$

and so $n - d \cdot (q + 1)$ would be a nonnegative integer in S that would be smaller than r. But r is the smallest integer in S. This contradiction shows that the supposition $r \geq d$ must be false.] The preceding arguments prove that there exist integers r and q for which

$$n = d \cdot q + r \quad \text{and} \quad 0 \leq r < d.$$

[This is what was to be shown.]

Another consequence of the well-ordering principle is the fact that any strictly decreasing sequence of nonnegative integers is finite. That is, if $r_1, r_2, r_3, \ldots$ is a sequence of nonnegative integers satisfying

$$r_i > r_{i+1}$$

for all $i \geq 1$, then $r_1, r_2, r_3, \ldots$ is a finite sequence. *[For by the well-ordering principle such a sequence would have to have a least element r_k. It follows that r_k must be the final term of the sequence because if there were a term r_{k+1}, then since the sequence is strictly decreasing, $r_{k+1} < r_k$, which would be a contradiction.]* This fact is frequently used in computer science to prove that algorithms terminate after a finite number of steps and to prove that the guard conditions for loops eventually become false. It was also used implicitly in the proof of Theorem 3.3.2.

EXERCISE SET 4.4

1. Suppose $a_1, a_2, a_3, \ldots$ is a sequence defined as follows:

$$a_1 = 1, a_2 = 3,$$
$$a_k = a_{k-2} + 2a_{k-1} \quad \text{for all integers } k \geq 3.$$

Prove that a_n is odd for all integers $n \geq 1$.

2. Suppose $b_1, b_2, b_3, \ldots$ is a sequence defined as follows:

$$b_1 = 3, b_2 = 6,$$
$$b_k = b_{k-2} + b_{k-1} \quad \text{for all integers } k \geq 3.$$

Prove that b_n is divisible by 3 for all integers $n \geq 1$.

3. Suppose that $c_0, c_1, c_2, \ldots$ is a sequence defined as follows:

$$c_0 = 2, c_1 = 4, c_2 = 6,$$
$$c_k = 5c_{k-3} \quad \text{for all integers } k \geq 3.$$

Prove that c_n is even for all integers $n \geq 0$.

4. Suppose that $d_1, d_2, d_3, \ldots$ is a sequence defined as follows:

$$d_1 = \frac{9}{10}, d_2 = \frac{10}{11},$$
$$d_k = d_{k-1} \cdot d_{k-2} \quad \text{for all integers } k \geq 3.$$

Prove that $d_n \leq 1$ for all integers $n \geq 0$.

5. Suppose that $e_0, e_1, e_2, \ldots$ is a sequence defined as follows:

$$e_0 = 1, e_1 = 2, e_2 = 3,$$
$$e_k = e_{k-1} + e_{k-2} + e_{k-3} \quad \text{for all integers } k \geq 3.$$

Prove that $e_n \leq 3^n$ for all integers $n \geq 0$.

6. Suppose that $f_1, f_2, f_3, \ldots$ is a sequence defined as follows:

$$f_1 = 1, f_k = 2 \cdot f_{\lfloor k/2 \rfloor} \quad \text{for all integers } k \geq 2.$$

Prove that $f_n \leq n$ for all integers $n \geq 1$.

7. Suppose that $g_0, g_1, g_2, \ldots$ is a sequence defined as follows:

$$g_0 = 12, g_1 = 29,$$
$$g_k = 5g_{k-1} - 6g_{k-2} \quad \text{for all integers } k \geq 2.$$

Prove that $g_n = 5 \cdot 3^n + 7 \cdot 2^n$ for all integers $n \geq 0$.

8. Suppose that $h_0, h_1, h_2, \ldots$ is a sequence defined as follows:

$$h_0 = 1, h_1 = 2, h_2 = 3,$$
$$h_k = h_{k-1} + h_{k-2} + h_{k-3} \quad \text{for all integers } k \geq 3.$$

a. Prove that $h_n \leq 3^n$ for all integers $n \geq 0$.
b. Suppose that s is any real number such that $s^3 \geq s^2 + s + 1$. (This implies that $s > 1.83$.) Prove that $h_n \leq s^n$ for all $n \geq 2$.

H9. You begin solving a jigsaw puzzle by finding two pieces that match and fitting them together. Each subsequent step of the solution consists of fitting together two blocks made up of one or more pieces that have previously been assembled. Use strong mathematical induction to prove that the number of steps required to put together all n pieces of a jigsaw puzzle is $n - 1$.

10. Find the mistake in the following "proof" that purports to show that every nonnegative integer power of every nonzero real number is 1.

"Proof

Let r be any nonnegative real number and consider the formula '$r^n = 1$.'

The formula is true for n = 0:

$$r^0 = 1 \text{ by definition of zeroth power}$$

If the formula is true for all integers i with $0 \leq i < k$, then it is true for k: Let $k > 0$ be an integer, and suppose that $r^i = 1$ for all integers i with $0 \leq i < k$. *[We must show that $r^k = 1$.]* Now

$$r^k = r^{(k-1)+(k-1)-(k-2)} \quad \text{because } (k-1) + (k-1) - (k-2) = k$$

$$= \frac{r^{k-1} \cdot r^{k-1}}{r^{k-2}} \quad \text{by the laws of exponents}$$

$$= \frac{1 \cdot 1}{1} \quad \text{by inductive hypothesis}$$

$$= 1.$$

Thus $r^k = 1$ *[as was to be shown]*.
[Since we have proved the basis step and the inductive step, we conclude that $r^n = 1$ for all integers $n \geq 0$.]"

11. Use the well-ordering principle to prove that given any integer $n \geq 1$, there exists an odd integer m and a nonnegative integer k such that $n = 2^k \cdot m$. (*Hint:* Let $S =$ the set of all integers r such that $n = 2^i \cdot r$ for some integer $i \geq 0$.)

◆ 12. Use the well-ordering principle to prove that if a and b are any integers not both zero, then there exist integers u and v so that $\gcd(a,b) = ua + vb$. (*Hint:* Let S be the set of all positive integers of the form $ua + vb$ for some integers u and v.)

13. Suppose $P(n)$ is a predicate such that

1. $P(0), P(1), P(2)$ are all true,
2. for all integers $k \geq 0$, if $P(k)$ is true, then $P(3k)$ is true.

Must it follow that $P(n)$ is true for all integers $n \geq 0$? If yes, explain why; if no, give a counterexample.

◆ 14. Prove that if the principle of ordinary mathematical induction is true, then the principle of strong mathematical induction is also true. To do this, suppose the principle of ordinary mathematical induction is true. To derive the principle of strong mathematical induction, let $P(n)$ be a predicate that is defined for integers n, and suppose the following two statements are true:

1. $P(a), P(a + 1), \ldots, P(b)$ are all true.
2. For any integer $k > b$, if $P(i)$ is true for all integers i with $a \leq i < k$, then $P(k)$ is true.

You have to show that $P(n)$ is true for all integers $n \geq a$. To do so, let $Q(n)$ be the following predicate:

$Q(n)$: $P(j)$ is true for all integers j with $a \leq j \leq n$.

Then use ordinary mathematical induction to show that $Q(n)$ is true for all integers $n \geq b$. To do this, prove the following:

1. $Q(b)$ is true.
2. For any integer $k \geq b$, if $Q(k)$ is true then $Q(k + 1)$ is true.

15. Give examples to illustrate the proof of Theorem 4.4.1.

H 16. It is a fact that every integer $n \geq 1$ can be written in the form

$$c_r \cdot 3^r + c_{r-1} \cdot 3^{r-1} + \cdots + c_2 \cdot 3^2 + c_1 \cdot 3 + c_0,$$

where $c_r = 1$ or 2 and $c_i = 0, 1,$ or 2 for all integers $i = 0, 1, 2, \ldots, r - 1$. Sketch a proof of this fact.

H ◆ 17. Use mathematical induction to prove the existence part of the quotient-remainder theorem for integers $n \geq 0$.

◆ 18. Use the well-ordering principle to prove that every integer greater than 1 is divisible by a prime number.

H ◆ 19. Prove that the well-ordering principle implies the principle of mathematical induction.

4.5 APPLICATION: CORRECTNESS OF ALGORITHMS

What does it mean for a computer program to be correct? Each program is designed to do a specific task—calculate the mean or median of a set of numbers, compute the size of the paychecks for a company payroll, rearrange names in alphabetical order, and so forth. We will say that a program is correct if it produces the output specified in its accompanying documentation for each set of input data of the type specified in the documentation.*

Most computer programmers write their programs using a combination of logical analysis and trial and error. In order to get a program to run at all, the programmer must first fix all syntax errors (such as writing **ik** instead of **if**, failing to declare a variable, or using a restricted keyword for a variable name). When the syntax errors have been removed, however, the program may still contain logical errors that prevent it from producing correct output. Frequently, programs are tested using sets of sample data for which the correct output is known in advance. And often the sample data is deliberately chosen to test the correctness of the program under extreme circumstances. But for most programs the number of possible sets of input data is either infinite or unmanageably large, and so no amount of program testing can give perfect confidence that the program will be correct for all possible sets of legal input data.

Edsger W. Dijkstra

Since 1967, with the publication of a paper by Robert W. Lloyd,[†] considerable effort has gone into developing methods for proving programs correct at the same time they are composed. One of the pioneers in this effort, Edsger W. Dijkstra, asserts that "we now take the position that it is not only the programmer's task to produce a correct program but also to demonstrate its correctness in a convincing manner."[‡] Another leader in the field, David Gries, goes so far as to say that "a program and its proof should be developed hand-in-hand, with the *proof* usually leading the way."[‡] If such methods can eventually be used to write large scientific and commercial programs, the benefits to society will be enormous.

As with most techniques that are still in the process of development, methods for proving program correctness are somewhat awkward and unwieldy. In this section we give an overview of the general format of correctness proofs and the details of one crucial technique, the *loop invariant procedure*. At this point, we switch from using the term *program,* which refers to a particular programming language, to the more general term *algorithm*.

---

*Consumers of computer programs want an even more stringent definition of correctness. If a user puts in data of the wrong type, the user wants a decent error message, not a system crash.

[†]R. W. Lloyd, "Assigning meanings to programs," *Proc. Symp. Appl. Math.,* Amer. Math. Soc. **19** (1967), 19–32.

[‡]Edsger Dijkstra in O. J. Dahl, E. W. Dijkstra, and C. A. R. Hoare, *Structured Programming* (London: 1972), p. 5.
[‡]David Gries, *The Science of Programming* (New York: Springer-Verlag, 1981), p. 164.

ASSERTIONS

Consider an algorithm that is designed to produce a certain final state from a certain initial state. Both the initial and final states can be expressed as predicates involving the input and output variables. Often the predicate describing the initial state is called the **pre-condition of the algorithm** and the predicate describing the final state is called the **post-condition of the algorithm.**

EXAMPLE 4.5.1 Algorithm Pre-Conditions and Post-Conditions

Here are pre- and post-conditions for some typical algorithms.

a. Algorithm to compute a product of nonnegative integers

    pre-condition: The input variables m and n are nonnegative integers.

    post-condition: The output variable p equals $m \cdot n$.

b. Algorithm to find the quotient and remainder of the division of one positive integer by another

    pre-condition: The input variables a and b are positive integers.

    post-condition: The output variables q and r are integers such that $a = b \cdot q + r$ and $0 \le r < b$.

c. Algorithm to sort a one-dimensional array of real numbers

    pre-condition: The input variable $A[1], A[2], \ldots, A[n]$ is a one-dimensional array of real numbers.

    post-condition: The output variable $B[1], B[2], \ldots, B[n]$ is a one-dimensional array of real numbers with same elements as $A[1], A[2], \ldots, A[n]$ but with the property that $B[i] \le B[j]$ whenever $i \le j$. ∎

A proof of algorithm correctness consists of showing that if the pre-condition for the algorithm is true for a collection of values for the input variables and if the statements of the algorithms are executed, then the post-condition is also true.

The divide-and-conquer principle has been useful in many aspects of computer programming, and proving algorithm correctness is no exception. The steps of an algorithm are divided into sections with assertions about the current state of algorithm variables inserted at strategically chosen points:

    [Assertion 1: pre-condition of algorithm]

    {Algorithm statements}

    [Assertion 2]

    {Algorithm statements}

       ⋮

    [Assertion k − 1]

    {Algorithm statements}

    [Assertion k: post-condition of algorithm]

Successive pairs of assertions are then treated as pre- and post-conditions for the algorithm statements between them. For each $i = 1, 2, \ldots, k - 1$, one proves that if Assertion i is true and all the algorithm statements between Assertion i and Assertion $(i + 1)$ are executed, then Assertion $(i + 1)$ is true. Once all these individual proofs have been completed, one knows that Assertion k is true. And since Assertion 1 is the same as the pre-condition for the algorithm and Assertion k is the same as the post-condition for the algorithm, one concludes that the entire algorithm is correct with respect to its pre- and post-conditions.

LOOP INVARIANTS

The method of loop invariants is used to prove correctness of a loop with respect to certain pre- and post-conditions. It is based on the principle of mathematical induction. Suppose an algorithm contains a **while** loop and that entry to this loop is restricted by a condition G, called the **guard**. Suppose also that assertions describing the current states of algorithm variables have been placed immediately preceding and immediately following the loop. The assertion just preceding the loop is called the **pre-condition for the loop** and the one just following is called the **post-condition for the loop**. The annotated loop has the following appearance:

> *[pre-condition for loop]*
>
> **while** (G)
>
> > *[Statements in body of loop. None contain branching statements that lead outside the loop.]*
>
> **end while**

DEFINITION

A loop is defined as **correct with respect to its pre- and post-conditions** if, and only if, whenever the algorithm variables satisfy the pre-condition for the loop and the loop is executed, then the algorithm variables satisfy the post-condition for the loop.

Loop correctness is proved by the *loop invariant theorem,* stated as follows:

THEOREM 4.5.1   Loop Invariant Theorem

Let a **while** loop with guard G be given, together with pre- and post-conditions that are predicates in the algorithm variables. Also let a predicate $I(n)$, called the **loop invariant**, be given. If the following four properties are true, then the loop is correct with respect to its pre- and post-conditions.

I. *Basis Property:* The pre-condition for the loop implies that $I(0)$ is true before the first iteration of the loop.

II. *Inductive Property:* If the guard G and the loop invariant $I(k)$ are both true for an integer $k \geq 0$ before an iteration of the loop, then $I(k + 1)$ is true after iteration of the loop.

III. *Eventual Falsity of Guard:* After a finite number of iterations of the loop, the guard G becomes false.

IV. *Correctness of the Post-Condition:* If N is the least number of iterations after which G is false and $I(N)$ is true, then the values of the algorithm variables will be as specified in the post-condition of the loop.

The loop invariant theorem follows easily from the principle of mathematical induction.

Proof:

Assume that $I(n)$ is a predicate that satisfies properties I–IV of the loop invariant theorem. *[We will prove that the loop is correct with respect to its pre- and post-conditions.]* Properties I and II are the basis and inductive steps needed to prove the truth of the following statement:

For all integers $n \geq 0$, if the **while** loop iterates n times, then $I(n)$ is true.   4.5.1

So by the principle of mathematical induction, since both I and II are true, statement (4.5.1) is also true.

Property III says that the guard G eventually becomes false. At that point the loop will have been iterated some number, say N, of times. Since $I(n)$ is true after the nth iteration for every $n \geq 0$, then $I(n)$ is true after the Nth iteration. That is, after the Nth iteration the guard is false and $I(N)$ is true. But this is the hypothesis of property IV, which is an if–then statement. Since statement IV is true (by assumption) and its hypothesis is true (by the argument just given), it follows (by modus ponens) that its conclusion is also true. That is, the values of all algorithm variables after execution of the loop are as specified in the post-condition for the loop.

You may have noticed that the procedure outlined for proving loop correctness contains a rabbit in the hat—namely, the *loop invariant*. Where does it come from? The fact is that developing a good loop invariant is a tricky process. Although learning how to do it is beyond the scope of this book, it is worth pursuing in a more advanced course. Many people who have become good at the process claim it has significantly altered their outlook on programming and has greatly improved their ability to write good code.

Another tricky aspect of handling correctness proofs arises from the fact that execution of an algorithm is a dynamic process—it takes place in time. As execution progresses, the values of variables keep changing, yet often their names stay the same. In the following discussion, when we need to make a distinction between the values of a variable just before execution of an algorithm statement and just after execution of the statement, we will attach the subscripts *old* and *new* to the variable name.

EXAMPLE 4.5.2   Correctness of a Loop to Compute a Product

The following loop is designed to compute the product $m \cdot x$ for a nonnegative integer m and a real number x, without using a built-in multiplication operation. Prior to the loop, variables i and *product* have been introduced and given initial values $i = 0$ and *product* $= 0$.

[pre-condition: m is a nonnegative integer,
x is a real number, i = 0, and product = 0.]

    while $(i \neq m)$

        1. *product* := *product* + *x*

        2. $i := i + 1$

    end while

[post-condition: product = m · x]

Let the loop invariant be

$$\boxed{I(n): \quad i = n \text{ and } product = n \cdot x}$$

The guard condition G of the **while** loop is

$$\boxed{G: \quad i \neq m}$$

Use the loop invariant theorem to prove that the **while** loop is correct with respect to the given pre- and post-conditions.

Solution

I. **Basis Property** *[I(0) is true before the first iteration of the loop.]*
$I(0)$ is "$i = 0$ and *product* $= 0 \cdot x$", which is true before the first iteration of the loop because $0 \cdot x = 0$.

II. **Inductive Property** *[If $G \wedge I(k)$ is true before a loop iteration (where $k \geq 0$), then $I(k + 1)$ is true after the loop iteration.]*
    Suppose k is a nonnegative integer such that $G \wedge I(k)$ is true before an iteration of the loop. Then as execution reaches the top of the loop, $i \neq m$, *product* $= k \cdot x$, and $i = k$. Since $i \neq m$, the guard is passed and statement 1 is executed. Before execution of statement 1,

$$product_{\text{old}} = k \cdot x.$$

Thus execution of statement 1 has the following effect:

$$product_{\text{new}} = product_{\text{old}} + x = k \cdot x + x = (k + 1) \cdot x.$$

Similarly, before statement 2 is executed,

$$i_{\text{old}} = k.$$

So after execution of statement 2,

$$i_{\text{new}} = i_{\text{old}} + 1 = k + 1.$$

    Hence after the loop iteration, the statement $I(k + 1)$ ($i = k + 1$ and *product* $= (k + 1) \cdot x$) is true. This is what we needed to show.

III. **Eventual Falsity of Guard** *[After a finite number of iterations of the loop, G becomes false.]*
    The guard G is the condition $i \neq m$, and m is a nonnegative integer. By I and II, it is known that

for all integers $n \geq 0$, if the loop is iterated
n times, then $i = n$ and *product* $= n \cdot x$.

So after m iterations of the loop, $i = m$. Thus G becomes false after m iterations of the loop.

IV. Correctness of the Post-Condition *[If N is the least number of iterations after which G is false and I(N) is true, then the value of the algorithm variables will be as specified in the post-condition of the loop.]*

According to the post-condition, the value of *product* after execution of the loop should be $m \cdot x$. But if G becomes false after N iterations, $i = m$. And if $I(N)$ is true, $i = N$ and $product = N \cdot x$. Since both conditions (G false and $I(N)$ true) are satisfied, $m = i = N$ and $product = m \cdot x$ as required. ∎

In the remainder of this section, we present proofs of the correctness of the crucial loops in the division algorithm and the Euclidean algorithm. (These algorithms were given in Section 3.8.)

CORRECTNESS OF THE DIVISION ALGORITHM

The division algorithm is supposed to take a nonnegative integer a and a positive integer d and compute nonnegative integers q and r such that $a = d \cdot q + r$ and $0 \leq r < d$. Initially, the variables r and q are introduced and given the values $r = a$ and $q = 0$. The crucial loop, annotated with pre- and post-conditions, is the following:

[pre-condition: a is a nonnegative integer and d is a positive integer, r = a, and q = 0.]

while $(r \geq d)$

    1. $r := r - d$

    2. $q := q + 1$

end while

[post-condition: q and r are nonnegative integers with the property that a = q \cdot d + r and 0 \leq r < d.]

Proof:

To prove the correctness of the loop, let the loop invariant be

$$\boxed{I(n):\ r = a - n \cdot d \geq 0 \quad \text{and} \quad n = q}$$

The guard of the **while** loop is

$$\boxed{G:\ r \geq d}$$

I. Basis Property *[I(0) is true before the first iteration of the loop.]*

$I(0)$ is "$r = a - 0 \cdot d$ and $q = 0$." But by the pre-condition, $r = a$. So since $a = a - 0 \cdot d$, then $r = a - 0 \cdot d$. Also $q = 0$ by the pre-condition. Hence $I(0)$ is true before the first iteration of the loop.

II. Inductive Property *[If $G \wedge I(k)$ is true before an interation of the loop (where $k \geq 0$), then I(k + 1) is true after iteration of the loop.]*

Suppose k is a nonnegative integer such that $G \wedge I(k)$ is true before an iteration of the loop. Since G is true, $r \geq d$ and the loop is entered. Also since $I(k)$ is true, $r = a - k \cdot d \geq 0$ and $k = q$. Hence before execution of statements 1 and 2,

$$r_{\text{old}} \geq d \quad \text{and} \quad r_{\text{old}} = a - k \cdot d \quad \text{and} \quad q_{\text{old}} = k.$$

When statements 1 and 2 are executed, then,

$$r_{\text{new}} = r_{\text{old}} - d = (a - k \cdot d) - d = a - (k + 1) \cdot d \qquad \text{4.5.2}$$

and

$$q_{new} = q_{old} + 1 = k + 1.$$ 　　　4.5.3

In addition, since $r_{old} \geq d$ before execution of statements 1 and 2, after execution of these statements,

$$r_{new} = r_{old} - d \geq d - d \geq 0.$$ 　　　4.5.4

Putting equations (4.5.2), (4.5.3), and (4.5.4) together shows that after iteration of the loop,

$$r_{new} \geq 0 \quad \text{and} \quad r_{new} = a - (k + 1) \cdot d \quad \text{and} \quad q_{new} = k + 1.$$

Hence $I(k + 1)$ is true.

III. Eventual Falsity of the Guard *[After a finite number of iterations of the loop, G becomes false.]*

The guard G is the condition $r \geq d$. Each iteration of the loop reduces the value of r by d and yet leaves r nonnegative. Thus the values of r form a decreasing sequence of nonnegative integers, and so (by the well-ordering principle) there must be a smallest such r, say r_{min}. Then $r_{min} < d$. *[For if r_{min} were greater than d, the loop would iterate another time, and a new value of r equal to $r_{min} - d$ would be obtained. But this new value would be smaller than r_{min}, which would contradict the fact that r_{min} is the smallest remainder obtained by repeated iteration of the loop.]* Hence as soon as the value $r = r_{min}$ is computed, the value of r becomes less than d, and so the guard G is false.

IV. Correctness of the Post-Condition *[If N is the least number of iterations after which G is false and I(N) is true, then the values of the algorithm variables will be as specified in the post-condition of the loop.]*

Suppose that for some nonnegative integer N, G is false and $I(N)$ is true. Then $r < d$, $r = a - N \cdot d$, $r \geq 0$, and $q = N$. Since $q = N$, by substitution,

$$r = a - q \cdot d.$$

Or, adding $q \cdot d$ to both sides,

$$a = q \cdot d + r.$$

Combining the two inequalities involving r gives

$$0 \leq r < d.$$

But these are the values of q and r specified in the post-condition. So the proof is complete.

CORRECTNESS OF THE EUCLIDEAN THEOREM

The Euclidean algorithm is supposed to take integers A and B with $A > B \geq 0$ and compute their greatest common divisor. Just before the crucial loop, variables a, b, and r have been introduced with $a = A$, $b = B$, and $r = B$. The crucial loop, annotated with pre- and post-conditions, is the following:

[pre-condition: A and B are integers with
$A > B \geq 0$, $a = A$, $b = B$, $r = B$.]

    while $(b \neq 0)$

        1. $r := a \bmod b$

        2. $a := b$

        3. $b := r$

    end while

[post-condition: a = gcd(A, B)]

Proof:

To prove the correctness of the loop, let the invariant be

$$I(n): \qquad \gcd(a, b) = \gcd(A, B) \quad \text{and} \quad 0 \le b < a.$$

The guard of the **while** loop is

$$G: \qquad b \ne 0.$$

I. **Basis Property** *[I(0) is true before the first iteration of the loop.]*
   $I(0)$ is

$$\gcd(A, B) = \gcd(a, b) \quad \text{and} \quad 0 \le b < a.$$

According to the pre-condition,

$$a = A, \quad b = B, \quad r = B, \quad \text{and} \quad 0 \le B < A.$$

Hence $\gcd(A, B) = \gcd(a, b)$. Since $0 \le B < A$, $b = B$, and $a = A$ then $0 \le b < a$. Hence $I(0)$ is true.

II. **Inductive Property** *[If $G \wedge I(k)$ is true before an iteration of the loop (where $k \ge 0$), then $I(k + 1)$ is true after iteration of the loop.]*
   Suppose k is a nonnegative integer such that $G \wedge I(k)$ is true before an iteration of the loop. *[We must show that $I(k + 1)$ is true after iteration of the loop.]* Since G is true, $b_{\text{old}} \ne 0$ and the loop is entered. And since $I(k)$ is true, immediately before statement 1 is executed,

$$\gcd(a_{\text{old}}, b_{\text{old}}) = \gcd(A, B) \quad \text{and} \quad 0 \le b_{\text{old}} < a_{\text{old}}. \qquad \text{4.5.5}$$

After execution of statement 1,

$$r_{\text{new}} = a_{\text{old}} \bmod b_{\text{old}}.$$

Thus by the quotient-remainder theorem,

$$a_{\text{old}} = b_{\text{old}} \cdot q + r_{\text{new}} \quad \text{for some integer } q$$

and r_{new} has the property that

$$0 \le r_{\text{new}} < b_{\text{old}}. \qquad \text{4.5.6}$$

By Lemma 3.8.2,

$$\gcd(a_{\text{old}}, b_{\text{old}}) = \gcd(b_{\text{old}}, r_{\text{new}}),$$

and by the equation of (4.5.5),

$$\gcd(a_{\text{old}}, b_{\text{old}}) = \gcd(A, B).$$

Hence

$$\gcd(b_{\text{old}}, r_{\text{new}}) = \gcd(A, B). \qquad \text{4.5.7}$$

When statements 2 and 3 are executed,

$$a_{\text{new}} = b_{\text{old}} \quad \text{and} \quad b_{\text{new}} = r_{\text{new}}. \qquad \text{4.5.8}$$

Substituting equations (4.5.8) into equation (4.5.7),

$$\gcd(a_{\text{new}}, b_{\text{new}}) = \gcd(A, B). \qquad \text{4.5.9}$$

By inequality (4.5.6),

$$0 \le r_{\text{new}} < b_{\text{old}}.$$

So substituting the values from equations (4.5.8),

$$0 \le b_{\text{new}} < a_{\text{new}}. \qquad \text{4.5.10}$$

Hence after the iteration of the loop, by equation (4.5.9) and inequality (4.5.10),

$$\gcd(a, b) = \gcd(A, B) \quad \text{and} \quad 0 \le b < a,$$

which is $I(k + 1)$. *[This is what we needed to show.]*

III. Eventual Falsity of the Guard *[After a finite number of iterations of the loop, G becomes false.]*

Each value of b obtained by repeated iteration of the loop is nonnegative and less than the previous value of b. Thus by the well-ordering principle there is a least value b_{min}. The fact is that $b_{\text{min}} = 0$. *[For if b_{min} were not 0, then since r is given the value of b_{min} in statement 3, r would not be 0 either. But $r \ne 0$ means that the guard is true, and so the loop is iterated another time. In this iteration a value of r is calculated that is less than the previous value of b, b_{min}. Then the value of b is changed to r, which is less than b_{min}. This contradicts the fact that b_{min} is the least value of b obtained by repeated iteration of the loop. Hence $b_{\text{min}} = 0.]$* Since $b_{\text{min}} = 0$, the guard is false immediately following the loop iteration in which b_{min} is calculated.

IV. Correctness of the Post-Condition *[If N is the least number of iterations after which G is false and I(N) is true, then the values of the algorithm variables will be as specified in the post-condition.]*

Suppose that for some nonnegative integer N, G is false and $I(N)$ is true. *[We must show the truth of the post-condition: $a = \gcd(A, B)$.]* Since G is false, $b = 0$, and since $I(N)$ is true,

$$\gcd(a, b) = \gcd(A, B). \qquad \text{4.5.11}$$

Substituting $b = 0$ into equation (4.5.11) gives

$$\gcd(a, 0) = \gcd(A, B).$$

But by Lemma 3.8.1,

$$\gcd(a, 0) = a.$$

Hence $a = \gcd(A, B)$ *[as was to be shown].*

EXERCISE SET 4.5

Exercises 1–5 contain a while loop and a potential loop invariant condition. In each case show that the condition really is a loop invariant—in other words, if the condition is true before entry to the loop then it is also true after exit from the loop.

1. loop: **while** ($m \ge 0$ and $m \le 100$)

    $m := m + 1$

    $n := n - 1$

  end while

loop invariant: $m + n = 100$

2. loop: **while** ($m \ge 0$ and $m \le 100$)

    $m := m + 4$

    $n := n - 2$

  end while

loop invariant: $m + n$ is odd

3. loop:

$$\textbf{while } (m \geq 0 \text{ and } m \leq 100)$$
$$m := 3m$$
$$n := 5n$$
$$\textbf{end while}$$

Loop Invariant: $m^3 > n^2$

4. loop:

$$\textbf{while } (n \geq 0 \text{ and } n \leq 100)$$
$$n := n + 1$$
$$\textbf{end while}$$

Loop Invariant: $2^n < (n + 2)!$

5. loop:

$$\textbf{while } (n \geq 3 \text{ and } n \leq 100)$$
$$n := n + 1$$
$$\textbf{end while}$$

Loop Invariant: $2n + 1 \leq 2^n$

Exercises 6–9 each contain a while loop annotated with a pre- and a post-condition and also a loop invariant. In each case, use the loop invariant theorem to prove the correctness of the loop with respect to the pre- and post-conditions.

6. *[pre-condition: m is a nonnegative integer, x is a real number, i = 0, and exp = 1.]*

$$\textbf{while } (i \neq m)$$
$$1.\ exp := exp \cdot x$$
$$2.\ i := i + 1$$
$$\textbf{end while}$$

[post-condition: exp = exp · x^n and i = n]

Loop Invariant: $I(n)$ is "$exp = x^n$ and $i = n$".

7. *[pre-condition: largest = A[1] and i = 1]*

$$\textbf{while } (i \neq m)$$
$$1.\ i := i + 1$$
$$2.\ \textbf{if } A[i] > \text{largest } \textbf{then } \text{largest} := A[i]$$
$$\textbf{end while}$$

[post-condition: largest = maximum value of A[1], A[2], . . . , A[m]]

Loop Invariant: $I(n)$ is "*largest* = maximum value of $A[1], A[2], . . . , A[n + 1]$ and $i = n + 1$".

[pre-condition: sum = A[1] and i = 1]

$$\textbf{while } (i \neq m)$$
$$1.\ i := i + 1$$

2. $sum := sum + A[i]$

$$\textbf{end while}$$

[post-condition: sum = A[1] + A[2] + · · · + A[m]]

Loop Invariant: $I(n)$ is "$i = n + 1$ and $sum = A[1] + A[2] + · · · + A[n + 1]$".

9. *[pre-condition: a = A and A is a positive integer.]*

$$\textbf{while } (a > 0)$$
$$1.\ a := a - 2$$
$$\textbf{end while}$$

[post-condition: a = 0 if A is even and a = −1 if A is odd.]

Loop Invariant: $I(n)$ is "both a and A are even integers or both are odd integers, and $a \geq -1$".

◆H10. Prove correctness of the **while** loop of Algorithm 3.8.3 (in exercise 22 of Exercise Set 3.8) with respect to the following pre- and post-conditions:

pre-condition: A and B are positive integers, $a = A$, and $b = B$.

post-condition: One of a or b is zero and the other is nonzero. Whichever is nonzero equals gcd(A, B).

Use the loop invariant

$I(n)$ "(1) a and b are nonnegative integers with gcd$(a, b) = $ gcd(A, B).

(2) at most one of a and b equals 0,

(3) $0 \leq a + b \leq A + B - n$."

◆ 11. The following sentence could be added to the loop invariant for the Euclidean algorithm:

There exist integers u, v, s, and t such that
$$a = uA + vB \quad \text{and} \quad b = sA + tB. \qquad \textbf{4.5.12}$$

a. Show that this sentence is a loop invariant for

$$\textbf{while } (b \neq 0)$$
$$r := a \bmod b$$
$$a := b$$
$$b := r$$
$$\textbf{end while}$$

b. Show that if initially $a = A$ and $b = B$, then sentence (4.5.12) is true before the first iteration of the loop.

c. Explain how the correctness proof for the Euclidean algorithm together with the results of (a) and (b) above allow you to conclude that given any integers A and B with $A > B \geq 0$, there exist integers u and v so that gcd$(A, B) = uA + vB$.

d. By actually calculating u, v, s, and t at each stage of execution of the Euclidean algorithm, find integers u and v so that gcd$(330, 156) = 330u + 156v$.

5 Set Theory

Georg Cantor (1845–1918)

In the late nineteenth century Georg Cantor was the first to realize the potential usefulness of investigating properties of sets in general as distinct from properties of the elements that comprise them. Many mathematicians of his time resisted accepting the validity of Cantor's work. Now, however, abstract set theory is regarded as the foundation of mathematical thought. All mathematical objects (even numbers!) can be defined in terms of sets, and the language of set theory is used in every mathematical subject.

In this chapter we introduce the basic definitions and notation of set theory and show how to establish properties of sets through the use of proofs and counterexamples. Because properties of the empty set are proved differently from properties of other sets, we devote a separate section to the empty set. The chapter ends with a discussion of a famous "paradox" of set theory and its relation to computer science.

5.1 BASIC DEFINITIONS OF SET THEORY

The introduction of suitable abstractions is our only mental aid to organize and master complexity.
(E. W. Dijkstra)

The words *set* and *element* are undefined terms of set theory just as *sentence*, *true*, and *false* are undefined terms of logic. The founder of set theory, Georg Cantor, suggested imagining a set as a "collection into a whole M of definite and separate objects of our intuition or our thought. These objects are called the elements of M." Cantor used the letter M because it is the first letter of the German word for set: *Menge*. Following the spirit of his notation (though not the letter), let S denote a set and a an element of S. Then as indicated in Section 2.1, $a \in S$ means that a is an element of S, $a \notin S$ means that a is not an element of S, $\{1, 2, 3\}$ refers to the set whose elements are 1, 2, and 3, and $\{1, 2, 3, \ldots\}$ refers to the set of all positive integers. The **axiom of extension** says that a set is completely determined by its elements; the order in which the elements are listed is irrelevant, as is the fact that some elements may be listed more than once.

EXAMPLE 5.1.1   The { } Notation for Sets

a. Suppose that Ann, Bob, and Cal are three students in a discrete mathematics class. Since the following sets all have the same elements—namely Ann, Bob, and Cal—they all represent the same set:

$$\{Ann, Bob, Cal\}, \{Bob, Cal, Ann\}, \{Bob, Bob, Ann, Cal, Ann\}$$

b. {Ann} denotes the set whose only element is Ann, whereas the word *Ann* denotes Ann herself. Since these are different {Ann} ≠ Ann.

c. Sets can themselves be elements of other sets. For example, {1, {1}} has two elements: the number 1 and the set {1}.

d. Sometimes a set may appear to have more elements than it really has. For every nonnegative integer n, let $U_n = \{-n, n\}$. Then $U_2 = \{-2, 2\}$ and $U_1 = \{-1, 1\}$ each have two elements, but

$$U_0 = \{-0, 0\} = \{0\}$$

has only one element since $-0 = 0$.   ■

As noted in Section 2.1, we may also define a set by writing $A = \{x \in S \mid P(x)\}$, where the left-hand brace is read "the set of all" and the vertical bar is read "such that." Note that an element x is in A if, and only if, x is in S and $P(x)$ is true.

Occasionally we will write $\{x \mid P(x)\}$ without being specific about where the elements x come from. It turns out that unrestricted use of this notation can lead to genuine contradictions in set theory. We will discuss one of these in Section 5.4 and will be careful to use this notation purely as a convenience in cases where the set S could be specified if necessary.

EXAMPLE 5.1.2   **Sets Given by a Defining Property**

Recall that **R** denotes the set of all real numbers, **Z** the set of all integers, and $\mathbf{Z}^+$ the set of all positive integers. Describe

a. $\{x \in \mathbf{R} \mid -2 < x < 5\}$
b. $\{x \in \mathbf{Z} \mid -2 < x < 5\}$
c. $\{x \in \mathbf{Z}^+ \mid -2 < x < 5\}$.

Solution   a. $\{x \in \mathbf{R} \mid -2 < x < 5\}$ is the open interval of real numbers strictly between -2 and 5. It is pictured as follows:

b. $\{x \in \mathbf{Z} \mid -2 < x < 5\}$ is the set of all integers between -2 and 5. It is equal to the set $\{-1, 0, 1, 2, 3, 4\}$.

c. Since all the integers in $\mathbf{Z}^+$ are positive, $\{x \in \mathbf{Z}^+ \mid -2 < x < 5\} = \{1, 2, 3, 4\}$.   ■

SUBSETS

A basic relation between sets is that of subset.

DEFINITION

If A and B are sets, A is called a **subset** of B, written $A \subseteq B$, if, and only if, every element of A is also an element of B.

Symbolically:

$$A \subseteq B \quad \Leftrightarrow \quad \forall x, \text{ if } x \in A \text{ then } x \in B.$$

The phrases *A is contained in B* and *B contains A* are alternative ways of saying that A is a subset of B.

It follows from the definition of subset that a set A is not a subset of a set B, written $A \not\subseteq B$, if, and only if, there is at least one element of A that is not an element of B. Symbolically:

$$A \not\subseteq B \iff \exists x \text{ such that } x \in A \text{ and } x \notin B.$$

EXAMPLE 5.1.3   Subsets

Suppose B76, XR3, D54, ES2, and XL5 are the model numbers of certain pieces of equipment. Let $A = \{\text{B76, XR3, D54, XL5}\}$, $B = \{\text{B76, D54}\}$, and $C = \{\text{ES2, XL5}\}$.

a. Is $B \subseteq A$?   b. Is $C \subseteq A$?   c. Is $B \subseteq B$?

Solution   a. Yes. Both elements of B are in A.
b. No; ES2 is in C but not in A.
c. Yes. Both elements of B are in B. (The definition of subset implies that any set is a subset of itself.) ■

> **DEFINITION**
>
> Let A and B be sets. A is a **proper subset** of B if, and only if, every element of A is in B but there is at least one element of B that is not in A.

If sets A and B are represented as regions in the plane, relationships between A and B can be represented by pictures, called **Venn diagrams**, that were introduced by the British mathematician John Venn in 1881. For instance, the relationship $A \subseteq B$ can be pictured in one of two ways, as shown in Figure 5.1.1.

John Venn (1834–1923)

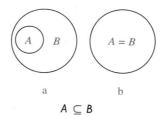

FIGURE 5.1.1 $A \subseteq B$

The relationship $A \not\subseteq B$ can be represented in three different ways with Venn diagrams as shown in Figure 5.1.2. If we allow the possibility that some subregions

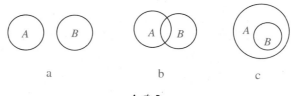

FIGURE 5.1.2
$A \not\subseteq B$

of Venn diagrams do not contain any points, then in Figure 5.1.1 diagram (b) can be viewed as a special case of diagram (a) by imagining that the part of B outside A does not contain any points. Similarly, diagrams (a) and (c) of Figure 5.1.2 can be viewed as special cases of diagram (b). To obtain (a) from (b), imagine that the region of overlap beween A and B does not contain any points. To obtain (c), imagine that the part of B that lies outside A does not contain any points. However, in all three diagrams it would be necessary to specify that there is a point in A that is not in B.

EXAMPLE 5.1.4   Relations Among Sets of Numbers

FIGURE 5.1.3

Since $\mathbf{Z}$, $\mathbf{Q}$, and $\mathbf{R}$ denote the sets of integers, rational numbers, and real numbers respectively, $\mathbf{Z}$ is a subset of $\mathbf{Q}$ because every integer is rational (any integer n can be written in the form $n/1$), and $\mathbf{Q}$ is a subset of $\mathbf{R}$ because every rational number is real (any rational number can be represented as a length on the number line). $\mathbf{Z}$ is a proper subset of $\mathbf{Q}$ because there are rational numbers that are not integers (for example, $1/2$), and $\mathbf{Q}$ is a proper subset of $\mathbf{R}$ because there are real numbers that are not rational (for example, $\sqrt{2}$). This is shown diagrammatically in Figure 5.1.3. ∎

It is important to distinguish clearly between the concepts of set membership ($\in$) and set containment ($\subseteq$). The following example illustrates some distinctions between them.

EXAMPLE 5.1.5   Distinction Between $\in$ and $\subseteq$

Which of the following are true statements?

a. $2 \in \{1, 2, 3\}$ b. $\{2\} \in \{1, 2, 3\}$ c. $2 \subseteq \{1, 2, 3\}$
d. $\{2\} \subseteq \{1, 2, 3\}$ e. $\{2\} \subseteq \{\{1\}, \{2\}\}$ f. $\{2\} \in \{\{1\}, \{2\}\}$

Solution   Only (a), (d), and (f) are true.

For (b) to be true, the set $\{1, 2, 3\}$ would have to contain the element $\{2\}$. But the only elements of $\{1, 2, 3\}$ are 1, 2, and 3, and 2 is not equal to $\{2\}$. Hence (b) is false.

For (c) to be true, the number 2 would have to be a set and every element in the set 2 would have to be an element of $\{1, 2, 3\}$. This is not the case, so (c) is false.

For (e) to be true, every element in the set containing only the number 2 would have to be an element of the set whose elements are $\{1\}$ and $\{2\}$. But 2 is not equal to $\{1\}$ or $\{2\}$, and so (e) is false. ∎

SET EQUALITY

Recall that by the principle of extension sets A and B are equal if, and only if, they have exactly the same elements. We restate this as a definition using the language of subsets.

DEFINITION

Given sets A and B, A **equals** B, written $A = B$, if, and only if, every element of A is in B and every element of B is in A.

Symbolically:

$$A = B \quad \Leftrightarrow \quad A \subseteq B \quad \text{and} \quad B \subseteq A.$$

This version of the definition of equality implies the following:

To know that a set A equals a set B, you must know
that $A \subseteq B$ and you must also know that $B \subseteq A$.

EXAMPLE 5.1.6   Set Equality

Let sets A, B, C, and D be defined as follows:

$$A = \{n \in \mathbf{Z} \mid n = 2p, \text{ for some integer } p\},$$
$$B = \text{the set of all even integers},$$
$$C = \{m \in \mathbf{Z} \mid m = 2q - 2, \text{ for some integer } q\},$$
$$D = \{k \in \mathbf{Z} \mid k = 3r + 1, \text{ for some integer } r\}.$$

a. Is $A = B$?   b. Is $A = C$?   c. Is $A = D$?

Solution   a. Yes. $A = B$ because every integer of the form $2p$, for some integer p, is even (so $A \subseteq B$), and every integer that is even can be written in the form $2p$, for some integer p (so $B \subseteq A$).

b. Yes. $A = C$ if, and only if, every element of A is in C and every element of C is in A. Considering the definitions of A and C, deciding whether $A = C$ involves deciding whether both of the following questions can be answered yes:

1. Can any integer that can be written in the form $2p$, for some integer p, also be written in the form $2q - 2$, for some integer q?
2. Can any integer that can be written in the form $2q - 2$, for some integer q, also be written in the form $2p$, for some integer p?

   To answer question (1), suppose an integer n equals $2p$, for some integer p. Can you find an integer q so that n equals $2q - 2$? If so, then

$$2q - 2 = 2p$$
$$2q = 2p + 2 = 2(p + 1)$$

and thus

$$q = p + 1.$$

So, if $n = 2p$, where p is an integer, let $q = p + 1$. Then q is an integer (since it is a sum of integers) and

$$2q - 2 = 2(p + 1) - 2 = 2p - 2 + 2 = 2p.$$

Hence the answer to question (1) is yes: $A \subseteq C$.

   To answer question (2), suppose an integer m equals $2q - 2$, for some integer q. Can you find an integer p such that m equals $2p$? If so, then

$$2p = 2q - 2 = 2(q - 1)$$

and thus

$$p = q - 1.$$

So, if $m = 2q - 2$, where q is an integer, let $p = q - 1$. Then p is an integer (since it is a difference of two integers), and

$$2p = 2(q - 1) = 2q - 2.$$

Hence the answer to (2) is yes: $C \subseteq A$.

   Since $A \subseteq C$ and $C \subseteq A$, then $A = C$ by definition of set equality.

c. No. $A \neq D$ for the following reason: $2 \in A$ since $2 = 2 \cdot 1$; but $2 \notin D$. For if 2 were an element of D, then 2 would equal $3r + 1$, for some integer r. Solving for r would give

$$3r + 1 = 2$$
$$3r = 2 - 1$$
$$3r = 1$$
$$r = \frac{1}{3}.$$

This argument shows that if 2 were an element of D, then there would be an integer r such that $r = 1/3$. But $1/3$ is not an integer, and so $2 \notin D$. Since there is an element in A that is not in D, $A \neq D$. ∎

OPERATIONS ON SETS

Most mathematical discussions are carried on within some context. For example, in a certain situation all sets being considered might be sets of real numbers. In such a situation, the set of real numbers would be called a **universal set** or a **universe of discourse** for the discussion.

DEFINITION

Let A and B be subsets of a universal set U.

1. The **union** of A and B, denoted $A \cup B$, is the set of all elements x in U such that x is in A or x is in B.

2. The **intersection** of A and B, denoted $A \cap B$, is the set of all elements x in U such that x is in A and x is in B.

3. The **difference** of B minus A (or **relative complement** of A in B), denoted $B - A$, is the set of all elements x in U such that x is in B and x is not in A.

4. The **complement** of A, denoted A^c, is the set of all elements x in U such that x is not in A.

Symbolically:

$$A \cup B = \{x \in U \mid x \in A \text{ or } x \in B\},$$
$$A \cap B = \{x \in U \mid x \in A \text{ and } x \in B\},$$
$$B - A = \{x \in U \mid x \in B \text{ and } x \notin A\},$$
$$A^c = \{x \in U \mid x \notin A\}.$$

Giuseppe Peano (1858–1932)

Thus the union of A and B is the set of elements in U that are in at least one of the sets A and B. The intersection of A and B is the set of elements common to both sets A and B. The difference of B minus A is the set of elements in B that are not in A. And the complement of A is the set of elements in the universal set U that are not in A. The symbols $\in$, $\cup$, and $\cap$ were introduced in 1889 by the Italian mathematician Giuseppe Peano.

EXAMPLE 5.1.7   Unions, Intersections, Differences, and Complements

Let the universal set be the set $\{a, b, c, d, e, f, g\}$ and let $A = \{a, c, e, g\}$ and $B = \{d, e, f, g\}$. Find $A \cup B$, $A \cap B$, $B - A$, and A^c.

Solution

$$A \cup B = \{a, c, d, e, f, g\} \qquad A \cap B = \{e, g\}$$
$$B - A = \{d, f\} \qquad A^c = \{b, d, f\}$$ ∎

EXAMPLE 5.1.8   An Example with Intervals

Let the universal set be the set **R** of all real numbers and let $A = \{x \in \mathbf{R} \mid -1 < x \le 0\}$ and $B = \{x \in \mathbf{R} \mid 0 \le x < 1\}$. These sets are shown on the number lines below.

Find $A \cup B$, $A \cap B$, and A^c.

Solution $A \cup B = \{x \in \mathbf{R} \mid -1 < x \le 0 \text{ or } 0 \le x < 1\} = \{x \in \mathbf{R} \mid -1 < x < 1\}$.

$A \cap B = \{x \in \mathbf{R} \mid -1 < x \le 0 \text{ and } 0 \le x < 1\} = \{0\}$.

$$A^c = \{x \in \mathbf{R} \mid \text{it is not the case that } -1 < x \le 0\}$$
$$= \{x \in \mathbf{R} \mid \text{it is not the case that } (-1 < x \text{ and } x \le 0)\} \quad \text{by definition of the double inequality}$$
$$= \{x \in \mathbf{R} \mid x \le -1 \text{ or } x > 0\} \quad \text{by De Morgan's law}$$

The Venn diagram representations for union, intersection, difference, and complement are shown in Figure 5.1.4.

Shaded region represents $A \cup B$.   Shaded region represents $A \cap B$.   Shaded region represents $B - A$.   Shaded region represents A^c.

FIGURE 5.1.4

CARTESIAN PRODUCTS

Recall that the definition of a set is unaffected by the order in which its elements are listed or the fact that some elements may be listed more than once. Thus $\{a, b\}$, $\{b, a\}$, and $\{a, a, b\}$ all represent the same set. The notation for an *ordered n-tuple* takes both order and multiplicity into account.

DEFINITION

Let n be a positive integer and let $x_1, x_2, \ldots, x_n$ be (not necessarily distinct) elements. The **ordered n-tuple, $(x_1, x_2, \ldots, x_n)$,** consists of $x_1, x_2, \ldots, x_n$ together with the ordering: first x_1, then x_2, and so forth up to x_n. An ordered 2-tuple is called an **ordered pair**, and an ordered 3-tuple is called an **ordered triple**.

Two ordered n-tuples $(x_1, x_2, \ldots, x_n)$ and $(y_1, y_2, \ldots, y_n)$ are **equal** if, and only if, $x_1 = y_1, x_2 = y_2, \ldots, x_n = y_n$.

Symbolically:

$$(x_1, x_2, \ldots, x_n) = (y_1, y_2, \ldots, y_n) \iff x_1 = y_1, x_2 = y_2, \ldots, x_n = y_n.$$

In particular:

$$(a, b) = (c, d) \iff a = c \text{ and } b = d.$$

EXAMPLE 5.1.9   Ordered n-tuples

a. Is $(1, 2) = (2, 1)$?
b. Is $(3, (-2)^2, \frac{1}{2}) = (\sqrt{9}, 4, \frac{3}{6})$?

Solution   a. No. By definition of equality of ordered pairs,

$$(1, 2) = (2, 1) \iff 1 = 2 \text{ and } 2 = 1.$$

But $1 \neq 2$, and so the ordered pairs are not equal.

b. Yes. By definition of equality of ordered triples,

$$(3, (-2)^2, \tfrac{1}{2}) = (\sqrt{9}, 4, \tfrac{3}{6}) \iff 3 = \sqrt{9} \text{ and } (-2)^2 = 4 \text{ and } \tfrac{1}{2} = \tfrac{3}{6}.$$

Because these equations are all true, the two ordered triples are equal. ■

DEFINITION

Given two sets A and B, the **Cartesian product** of A and B, denoted $A \times B$ (read "A cross B"), is the set of all ordered pairs (a, b), where a is in A and b is in B.

Given sets $A_1, A_2, \ldots, A_n$, the **Cartesian product** of $A_1, A_2, \ldots, A_n$, denoted $A_1 \times A_2 \times \cdots \times A_n$, is the set of all ordered n-tuples $(a_1, a_2, \ldots, a_n)$ where $a_1 \in A_1, a_2 \in A_2, \ldots, a_n \in A_n$.

Symbolically:

$$A \times B = \{(a, b) \mid a \in A \text{ and } b \in B\},$$

$$A_1 \times A_2 \times \cdots \times A_n = \{(a_1, a_2, \ldots, a_n) \mid a_1 \in A_1, a_2 \in A_2, \ldots, a_n \in A_n\}.$$

EXAMPLE 5.1.10   Cartesian Products

Let $A = \{x, y\}$, $B = \{1, 2, 3\}$, and $C = \{a, b\}$.

a. Find $A \times B$.   b. Find $(A \times B) \times C$.   c. Find $A \times B \times C$.

Solution
a. $A \times B = \{(x, 1), (x, 2), (x, 3), (y, 1), (y, 2), (y, 3)\}$

b. The Cartesian product of A and B is a set, so it may be used as one of the sets making up another Cartesian product. This is the case for $(A \times B) \times C$:

$(A \times B) \times C = \{(u, v) \mid u \in A \times B \text{ and } v \in C\}$ by definition of Cartesian product

$= \{((x, 1), a), ((x, 2), a), ((x, 3), a), ((y, 1), a),$
$((y, 2), a), ((y, 3), a), ((x, 1), b), ((x, 2), b), ((x, 3), b),$
$((y, 1), b), ((y, 2), b), ((y, 3), b)\}$

c. The Cartesian product $A \times B \times C$ is superficially similar to, but is not quite the same mathematical object as, $(A \times B) \times C$. $(A \times B) \times C$ is a set of ordered pairs of which one element is itself an ordered pair, whereas $A \times B \times C$ is a set of ordered triples. By definition of Cartesian product,

$A \times B \times C = \{(u, v, w) \mid u \in A, v \in B, \text{ and } w \in C\}$
$= \{(x, 1, a), (x, 2, a), (x, 3, a), (y, 1, a), (y, 2, a),$
$(y, 3, a), (x, 1, b), (x, 2, b), (x, 3, b), (y, 1, b),$
$(y, 2, b), (y, 3, b)\}.$ ■

FORMAL LANGUAGES

An English sentence can be regarded as a string of words, and an English word can be regarded as a string of letters. Not every string of letters is a legitimate word, and not every string of words is a grammatical sentence. We could say that a word is legitimate if it can be found in an unabridged English dictionary and that a sentence is grammatical if it satisfies the rules in a standard English grammar book.

Computer languages are similar to English in the sense that certain strings of characters are legitimate words of the language and certain strings of words can be put together according to certain rules to form syntactically correct programs. A compiler for a computer language analyzes the stream of characters in a program—first to recognize individual word and sentence units (this part of a compiler is called a lexical scanner), then to analyze the syntax, or grammar, of the sentences (this part is called a syntactic analyzer), and finally to translate the sentences into machine code (this part is called a code generator).

In computer science it has proved useful to look at languages from a very abstract point of view as strings of certain fundamental units. In computer science any finite set of symbols can be used as an **alphabet**. It is common to denote an alphabet by a capital Greek sigma: Σ. (This just happens to be the same notation as the symbol for summation. The two concepts have no other connection.)

A **string of characters of an alphabet Σ** (or a **string over Σ**) is either (1) an ordered n-tuple of elements of Σ written without parentheses or commas, or (2) the **null string** ε, which has no characters. For example, if the alphabet Σ consists of the two characters a and b, then aab, a, bb, $aabbbab$, and ε are all strings over Σ.

The **length of a string** of characters of an alphabet is the number of characters that make up the string. Thus if a and b are in the alphabet, the length of the string $aaba$ is 4. The null string has length 0.

A **formal language over an alphabet** is any set of strings of characters of the alphabet. The definitions above are summarized below.

alphabet Σ:	a finite set of characters
string over Σ:	an ordered n-tuple of characters of Σ, written without parentheses or commas, or the null string ε
formal language over Σ:	a set of strings over Σ

EXAMPLE 5.1.11   Examples of Formal Languages

Let the alphabet $\Sigma = \{a, b\}$.

a. Define a language L_1 over Σ to be the set of all strings that begin with the character a and have length of at most three characters. Find L_1.
b. A palindrome is a string that looks the same if the order of its characters is reversed. For instance, *baab* is a palindrome. Define a language L_2 over Σ to be the set of all palindromes obtained using the characters of Σ. Write ten elements of L_2.

Solution a. $L_1 = \{a, aa, ab, aaa, aab, aba, abb\}$
b. L_2 contains the following ten strings (among infinitely many others):

$$L_2: \varepsilon, a, b, aa, bb, aaa, bab, abba, baab, baabbbbbaab.$$

NOTATION

Let Σ be an alphabet. For each nonnegative integer n, let

$$\Sigma^n = \text{the set of all strings over } \Sigma \text{ that have length } n, \text{ and}$$
$$\Sigma^* = \text{the set of } all \text{ strings of finite length over } \Sigma.$$

Note that Σ^n is the Cartesian product of n copies of Σ.

EXAMPLE 5.1.12   The Languages Σ^n and Σ^*

Let $\Sigma = \{a, b\}$.

a. Find $\Sigma^0, \Sigma^1, \Sigma^2,$ and Σ^3.
b. Let $A = \Sigma^0 \cup \Sigma^1$ and $B = \Sigma^2 \cup \Sigma^3$. Describe A, B, and $A \cup B$.
c. Describe a systematic way of writing elements of Σ^*.

Solution a. $\Sigma^0 = \{\varepsilon\}, \Sigma^1 = \{a, b\}, \Sigma^2 = \{aa, ab, ba, bb\},$
$\Sigma^3 = \{aaa, aab, aba, abb, baa, bab, bba, bbb\}$
b. A is the set of all strings over Σ of length at most 1.
B is the set of all strings over Σ of length 2 or 3.
$A \cup B$ is the set of all strings over Σ of length at most 3.
c. Elements of Σ^* can be written systematically by starting with the null string ε, then writing all strings of length 1, then all strings of length 2, and so forth.

$$\Sigma^*: \varepsilon, a, b, aa, ab, ba, bb, aaa, aab, aba, abb, baa, bab, bba, bbb, aaaa, \ldots$$

Of course, the process of writing the strings in Σ^* would continue forever since Σ^* is an infinite set.

EXAMPLE 5.1.13   **A Language Consisting of Infix Expressions**

An expression such as $a + b$ in which a binary operator such as $+$ sits between the two quantities on which it acts is said to be written in **infix notation**. Alternate notations are called **prefix notation** (in which the binary operator precedes the quantities on which it acts) and **postfix notation** (in which the binary operator follows the quantities on which it acts). In prefix notation $a + b$ is written $+ab$. In postfix notation $a + b$ is written $ab+$.

Let $\Sigma = \{0, 1, +, -\}$ and let $L =$ the set of all strings over Σ obtained by writing either a 0 or a 1 first, then either a $+$ or a $-$, and finally either a 0 or a 1. List all elements of L.

Solution $$L = \{0 + 0, 0 + 1, 0 - 0, 0 - 1, 1 + 0, 1 + 1, 1 - 0, 1 - 1\}$$ ■

AN ALGORITHM TO CHECK WHETHER ONE SET IS A SUBSET OF ANOTHER (OPTIONAL)

You may get some additional insight into the concept of subset by considering an algorithm for checking whether one finite set is a subset of another. Order the elements of both sets and successively compare each element of the first set with each element of the second set. If some element of the first set is not found to equal any element of the second, then the first set is not a subset of the second. But if each element of the first set is found to equal an element of the second set, then the first set is a subset of the second. The following algorithm formalizes this reasoning.

ALGORITHM 5.1.1   Testing Whether $A \subseteq B$

[Input sets A and B are represented as one-dimensional arrays $a[1], a[2], \ldots, a[m]$ and $b[1], b[2], \ldots, b[n]$, respectively. Starting with $a[1]$ and for each successive $a[i]$ in A, a check is made to see whether $a[i]$ is in B. To do this, $a[i]$ is compared to successive elements of B. If $a[i]$ is not equal to any element of B, then answer is given the value "$A \nsubseteq B$." If $a[i]$ equals some element of B, the next successive element in A is checked to see whether it is in B. If every successive element of A is found to be in B, then answer never changes from its initial value "$A \subseteq B$."]

Input: m *[a positive integer]*, $a[1], a[2], \ldots, a[m]$ *[a one-dimensional array representing the set A]*, n *[a positive integer]*, $b[1], b[2], \ldots, b[n]$ *[a one-dimensional array representing the set B]*

Algorithm Body:

$$i := 1, \; answer := \text{“} A \subseteq B \text{”}$$

while ($i \leq m$ and $answer = \text{“} A \subseteq B \text{”}$)

$j := 1, found := \text{“no”}$

while ($j \leq n$ and $found = \text{“no”}$)

if $a[i] = b[j]$ **then** $found := \text{“yes”}$

$j := j + 1$

end while

[If found has not been given the value "yes" when execution reaches this point, then $a[i] \notin B$.]

> **if** *found* = "no" **then** *answer* := "$A \not\subseteq B$"
>
> $i := i + 1$
>
> **end while**
>
> *Output:*   *answer [a string]*
>
> *end Algorithm 5.1.1*

EXAMPLE 5.1.14   Tracing Algorithm 5.1.1

Trace the action of Algorithm 5.1.1 on the variables $i, j, found,$ and *answer* for $m = 3$, $n = 4$, and sets A and B represented as the arrays $a[1] = u$, $a[2] = v$, $a[3] = w$, $b[1] = w$, $b[2] = x$, $b[3] = y$, and $b[4] = u$.

Solution

i	1					2					3
j	1	2	3	4	5	1	2	3	4	5	
found	no			yes		no					
answer	$A \subseteq B$									$A \not\subseteq B$	

In the exercises at the end of this section, you are asked to write an algorithm to check whether a given element is in a given set. To do this, you can represent the set as a one-dimensional array and compare the given element with successive elements of the array to determine whether the two elements are equal. If they are, then the element is in the set; if the given element does not equal any element of the array, then the element is not in the set.

EXERCISE SET 5.1†

1. Which of the following sets are equal?
   a. $\{a, b, c, d\}$ b. $\{d, e, a, c\}$
   c. $\{d, b, a, c\}$ d. $\{a, a, d, e, c, e\}$

2. Is $4 = \{4\}$? Explain.

3. Which of the following sets are equal?
   a. $A = \{0, 1, 2\}$
   b. $B = \{x \in \mathbf{R} \mid -1 \le x < 3\}$
   c. $C = \{x \in \mathbf{R} \mid -1 < x < 3\}$
   d. $D = \{x \in \mathbf{Z} \mid -1 < x < 3\}$
   e. $E = \{x \in \mathbf{Z}^+ \mid -1 < x < 3\}$

4. a. Let $S = \{n \in \mathbf{Z} \mid n = (-1)^k,$ for some integer $k\}$. Describe S.
   b. Let
   $$T = \{m \in \mathbf{Z} \mid m = 1 + (-1)^i, \text{ for some integer } i\}.$$
   Describe T.

5. Let $A = \{c, d, f, g\}$, $B = \{f, j\}$, and $C = \{d, g\}$. Answer each of the following questions. Give reasons for your answers.
   a. Is $B \subseteq A$?   b. Is $C \subseteq A$?
   c. Is $C \subseteq C$?   d. Is C a proper subset of A?

6. a. Is $3 \in \{1, 2, 3\}$?   b. Is $1 \subseteq \{1\}$?
   c. Is $\{2\} \in \{1, 2\}$?   d. Is $\{3\} \in \{1, \{2\}, \{3\}\}$?
   e. Is $1 \in \{1\}$?   f. Is $\{2\} \subseteq \{1, \{2\}, \{3\}\}$?
   g. Is $\{1\} \subseteq \{1, 2\}$?   h. Is $1 \in \{\{1\}, 2\}$?
   i. Is $\{1\} \subseteq \{1, \{2\}\}$?   j. Is $\{1\} \subseteq \{1\}$?

7. Let $A = \{b, c, d, f, g\}$ and $B = \{a, b, c\}$. Find each of the following:
   a. $A \cup B$ b. $A \cap B$
   c. $A - B$ d. $B - A$

8. Let the universal set be the set $\mathbf{R}$ of all real numbers and let $A = \{x \in \mathbf{R} \mid 0 < x \le 2\}$ and

$B = \{x \in \mathbf{R} \mid 1 \le x < 4\}$. Find each of the following:

a. $A \cup B$ b. $A \cap B$ c. A^c
d. B^c e. $A^c \cap B^c$ f. $A^c \cup B^c$
g. $(A \cap B)^c$ h. $(A \cup B)^c$

9. Let the universal set be the set $\mathbf{R}$ of all real numbers and let $A = \{x \in \mathbf{R} \mid -2 \le x \le 1\}$ and $B = \{x \in \mathbf{R} \mid -1 < x < 3\}$. Find each of the following:

a. $A \cup B$ b. $A \cap B$ c. A^c
d. B^c e. $A^c \cap B^c$ f. $A^c \cup B^c$
g. $(A \cap B)^c$ h. $(A \cup B)^c$

10. Indicate which of the following relationships are true and which are false:

a. $\mathbf{Z}^+ \subseteq \mathbf{Q}$ b. $\mathbf{R}^- \subseteq \mathbf{Q}$
c. $\mathbf{Q} \subseteq \mathbf{Z}$ d. $\mathbf{Z}^- \cup \mathbf{Z}^+ = \mathbf{Z}$
e. $\mathbf{Q} \cap \mathbf{R} = \mathbf{Q}$ f. $\mathbf{Q} \cup \mathbf{Z} = \mathbf{Q}$
g. $\mathbf{Z}^+ \cap \mathbf{R} = \mathbf{Z}^+$ h. $\mathbf{Z} \cup \mathbf{Q} = \mathbf{Z}$

11. a. Write a negation for the following statement: $\forall$ sets A, if $A \subseteq \mathbf{R}$ then $A \subseteq \mathbf{Z}$. Which is true, the statement or its negation? Explain.
   b. Write a negation for the following statement: $\forall$ sets S, if $S \subseteq \mathbf{Q}^+$ then $S \subseteq \mathbf{Q}^-$. Which is true, the statement or its negation? Explain.

12. Let $A = \{m \in \mathbf{Z} \mid m = 2i - 1, \text{ for some integer } i\}$, $B = \{n \in \mathbf{Z} \mid n = 3j + 2, \text{ for some integer } j\}$, $C = \{p \in \mathbf{Z} \mid p = 2r + 1, \text{ for some integer } r\}$, and $D = \{q \in \mathbf{Z} \mid q = 3s - 1, \text{ for some integer } s\}$.
   a. Is $A = B$? Explain. b. Is $A = C$? Explain.
   c. Is $A = D$? Explain. d. Is $B = D$? Explain.

13. Let sets R, S, and T be defined as follows:

$$R = \{x \in \mathbf{Z} \mid x \text{ is divisible by } 2\},$$
$$S = \{y \in \mathbf{Z} \mid y \text{ is divisible by } 3\},$$
$$T = \{z \in \mathbf{Z} \mid z \text{ is divisible by } 6\}.$$

a. Is $R \subseteq T$? Explain. b. Is $T \subseteq R$? Explain.
c. Is $T \subseteq S$? Explain. d. Find $R \cap S$. Explain.

14. Let $A = \{a, b, c\}$, $B = \{b, c, d\}$, and $C = \{b, c, e\}$.
   a. Find $A \cup (B \cap C)$, $(A \cup B) \cap C$, and $(A \cup B) \cap (A \cup C)$. Which of these sets are equal?
   b. Find $A \cap (B \cup C)$, $(A \cap B) \cup C$, and $(A \cap B) \cup (A \cap C)$. Which of these sets are equal?
   c. Find $(A - B) - C$ and $A - (B - C)$. Are these sets equal?

15. Consider the Venn diagram at the top of the next column. For each of a–f, copy this diagram and shade the region corresponding to the indicated set.

a. $A \cap B$ b. $B \cup C$ c. A^c
d. $A - (B \cup C)$ e. $(A \cup B)^c$ f. $A^c \cap B^c$

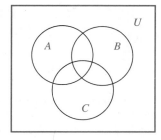

16. In each of the following draw a Venn diagram for sets A, B, and C that satisfies the given conditions:
   a. $A \subseteq B$; $C \subseteq B$; A and C have no elements in common
   b. $C \subseteq A$; B and C have no elements in common

17. Let $A = \{x, y, z, w\}$ and $B = \{a, b\}$. List the elements of each of the following sets:
   a. $A \times B$ b. $B \times A$
   c. $A \times A$ d. $B \times B$

18. Let $A = \{1, 2, 3\}$, $B = \{u, v\}$, and $C = \{m, n\}$. List the elements of each of the following sets:
   a. $A \times (B \times C)$ b. $(A \times B) \times C$ c. $A \times B \times C$

19. Let $\Sigma = \{x, y\}$ be an alphabet.
   a. Let L_1 be the language consisting of all strings over Σ that are palindromes and have length ≤ 4. List the elements of L_1 between braces.
   b. Let L_2 be the language consisting of all strings over Σ that begin with an x and have length ≤ 3. List the elements of L_2.
   c. Let L_3 be the language consisting of all strings over Σ of length ≤ 3 in which all the x's appear to the left of all the y's. List the elements of L_3 between braces.
   d. List between braces the elements of Σ^4, the set of strings of length 4 over Σ.
   e. Let $A = \Sigma^1 \cup \Sigma^2$ and $B = \Sigma^3 \cup \Sigma^4$. Describe A, B, and $A \cup B$ in words.

20. Let $\Sigma = \{1, 2, *, /\}$ and let L be the set of all strings over Σ obtained by writing first a number (1 or 2), then a second number (1 or 2), which can be the same as the first one, and finally an operation (* or /). List the elements of L between braces. (L is a set of postfix expressions.)

21. Trace the action of Algorithm 5.1.1 on the variables i, j, *found*, and *answer* for $m = 3$, $n = 3$, and sets A and B represented as the arrays $a[1] = u$, $a[2] = v$, $a[3] = w$, $b[1] = w$, $b[2] = u$, and $b[3] = v$.

22. Trace the action of Algorithm 5.1.1 on the variables i, j, *found*, and *answer* for $m = 4$, $n = 4$, and sets A and B represented as the arrays $a[1] = u$, $a[2] = v$, $a[3] = w$, $a[4] = x$, $b[1] = y$, $b[2] = u$, $b[3] = v$, $b[4] = z$.

23. Write an algorithm to determine whether a given element x belongs to a given set, which is represented as an array $a[1], a[2], \ldots, a[n]$.

5.2 PROPERTIES OF SETS

It is possible to list many relations involving unions, intersections, complements, and differences of sets. Some of these are true for all sets, whereas others fail to hold in some cases. In this section we show how to establish basic set properties using *element arguments* and discuss how to disprove properties by constructing counterexamples. We also show how to use algebraic techniques to derive additional set properties from ones already known to be true.*

We begin by listing some set properties that involve subset relations. As you read them, keep in mind that the operations of union, intersection, and difference take precedence over set inclusion. Thus, for example, $A \cap B \subseteq C$ means $(A \cap B) \subseteq C$.

THEOREM 5.2.1 Some Subset Relations

1. Inclusion of Intersection: For all sets A and B,

(a) $A \cap B \subseteq A$ and (b) $A \cap B \subseteq B$.

2. Inclusion in Union: For all sets A and B,

(a) $A \subseteq A \cup B$ and (b) $B \subseteq A \cup B$.

3. Transitive Property of Subsets: For all sets A, B, and C,

if $A \subseteq B$ and $B \subseteq C$, then $A \subseteq C$.

The conclusion of each part of Theorem 5.2.1 states that one set is a subset of another. Recall that by definition of subset, if X and Y are sets, then:

$$X \subseteq Y \iff \forall x, \text{ if } x \in X \text{ then } x \in Y.$$

Since the definition of subset is a universal conditional statement, the most basic way to prove that one set is a subset of another is as follows:

Basic (Element) Method for Proving That One Set Is a Subset of Another

Let sets X and Y be given. To prove that $X \subseteq Y$,

1. suppose that x is a particular but arbitrarily chosen element of X,

2. show that x is an element of Y.

In most set theoretic proofs the secret of getting from the assumption that x is in X to the conclusion that x is in Y is to think of the definitions of basic set operations in procedural terms. For example, the union of sets X and Y, $X \cup Y$, is defined as

$$X \cup Y = \{x \mid x \in X \text{ or } x \in Y\}.$$

This means that any time you know an element x is in $X \cup Y$, you can conclude that x must be in X or x must be in Y. Conversely, any time you know that a particular x is in some set X or is in some set Y, you can conclude that x is in $X \cup Y$. Thus, for any sets X and Y and any element x,

$$x \in X \cup Y \quad \text{if, and only if,} \quad x \in X \text{ or } x \in Y.$$

---

*An algebraic technique is one that involves using identities to transform expressions.

Procedural versions of the definitions of the other set operations are derived similarly and are summarized below.

Procedural Versions of Set Definitions

Let X and Y be subsets of a universal set U and suppose x and y are elements of U.

1. $x \in X \cup Y \iff x \in X$ or $x \in Y$
2. $x \in X \cap Y \iff x \in X$ and $x \in Y$
3. $x \in X - Y \iff x \in X$ and $x \notin Y$
4. $x \in X^c \iff x \notin X$
5. $(x, y) \in X \times Y \iff x \in X$ and $y \in Y$

EXAMPLE 5.2.1   Proof of a Subset Relation

Prove Theorem 5.2.1(1)(a): For all sets A and B, $A \cap B \subseteq A$.

Solution   We start by giving a proof of the statement and then explain how you can obtain such a proof yourself.

Proof:

Suppose A and B are any sets and suppose x is any element of $A \cap B$. Then $x \in A$ and $x \in B$ by definition of intersection. In particular, $x \in A$.

The underlying structure of this proof is not difficult, but it is more complicated than the brief length of the proof suggests. The first important thing to realize is that the statement to be proved is universal (it says that for *all* sets A and B, $A \cap B \subseteq A$). The proof, therefore, has the following outline:

Starting Point: Suppose A and B are any (particular but arbitrarily chosen) sets.
To Show: $A \cap B \subseteq A$

Now to prove that $A \cap B \subseteq A$, you must show that

$$\forall x, \text{ if } x \in A \cap B \text{ then } x \in A.$$

But this statement also is universal. So to prove it, you

suppose x is an element in $A \cap B$

and then you

show that x is in A.

Filling in the gap between the "suppose" and the "show" is easy if you use the procedural version of the definition of intersection: to say that x is in $A \cap B$ means that

x is in A and x is in B.

This allows you to complete the proof by deducing that, in particular,

x is in A

as was to be shown. Note that this deduction is just a special case of the valid argument form

$$p \wedge q$$
$$\therefore p.$$

■

In his book *Gödel, Escher, Bach*, Douglas Hofstadter introduces the fantasy rule for mathematical proof. Hofstadter points out that when you start a mathematical argument with *if*, *let*, or *suppose*, you are stepping into a fantasy world where not only are all the facts of the real world true but whatever you are supposing is also true. Once you are in that world, you can suppose something else. That sends you into a subfantasy world where not only is everything in the fantasy world true but also the new thing you are supposing. Of course you can continue stepping into new subfantasy worlds in this way indefinitely. You return one level closer to the real world each time you derive a conclusion that makes a whole if–then or universal statement true. Your aim in a proof is to continue deriving such conclusions until you return to the world from which you made your first supposition.

Occasionally, mathematical problems are stated in the following form:

> Suppose (*statement 1*). Prove that (*statement 2*).

When this phrasing is used, the author intends the reader to add statement 1 to his or her general mathematical knowledge and not to make explicit reference to it in the proof. In Hofstadter's terms, the author invites the reader to enter a fantasy world where statement 1 is known to be true and to prove statement 2 in this fantasy world. Thus the solver of such a problem would begin a proof with the starting point for a proof of statement 2. Consider, for instance, the following restatement of Example 5.2.1:

> Suppose A and B are arbitrarily chosen sets.
> Prove that $A \cap B \subseteq A$.

The proof would begin "Suppose $x \in A \cap B$," it being *understood* that sets A and B have already been chosen arbitrarily.

The proof of Example 5.2.1 is called an **element argument** because it shows one set to be a subset of another by demonstrating that every element in the one set is also an element in the other. In higher mathematics, element arguments are the standard method of establishing relations among sets. High school students are often allowed to justify set properties by using Venn diagrams. This method is appealing, but for it to be mathematically rigorous may be more complicated than you might expect. For instance, it is impossible to draw a single Venn diagram in which four circular disks represent sets in such a way that all 16 subsets appear as regions of the diagram. (Though it is possible to represent all the subsets if noncircular regions are used.) Even when appropriate Venn diagrams can be drawn, the verbal explanations needed to justify conclusions inferred from them are normally as long as a straightforward element proof.

SET IDENTITIES

An **identity** is an equation that is universally true for all elements in some set. For example, the equation $a + b = b + a$ is an identity for real numbers because it is true for all real numbers a and b. The collection of set properties in the next theorem consists entirely of set identities. That is, they are equations that are true for all sets in some universal set.

> **THEOREM 5.2.2**   Set Identities
>
> Let all sets referred to below be subsets of a universal set U.
>
> **1.** Commutative Laws: For all sets A and B,
>
> $\quad$ (a) $A \cap B = B \cap A$ and   (b) $A \cup B = B \cup A$.
>
> **2.** Associative Laws: For all sets A, B, and C,
>
> $\quad$ (a) $(A \cap B) \cap C = A \cap (B \cap C)$ and
> $\quad$ (b) $(A \cup B) \cup C = A \cup (B \cup C)$.
>
> **3.** Distributive Laws: For all sets, A, B, and C,
>
> $\quad$ (a) $A \cup (B \cap C) = (A \cup B) \cap (A \cup C)$ and
> $\quad$ (b) $A \cap (B \cup C) = (A \cap B) \cup (A \cap C)$.
>
> **4.** Intersection with U (U Acts as an Identity for $\cap$): For all sets A,
>
> $\quad A \cap U = A$.
>
> **5.** Double Complement Law: For all sets A,
>
> $\quad (A^c)^c = A$.
>
> **6.** Idempotent Laws: For all sets A,
>
> $\quad$ (a) $A \cap A = A$ and   (b) $A \cup A = A$.
>
> **7.** De Morgan's Laws: For all sets A and B,
>
> $\quad$ (a) $(A \cup B)^c = A^c \cap B^c$ and   (b) $(A \cap B)^c = A^c \cup B^c$.
>
> **8.** Union with U (U Acts as a Universal Bound for $\cup$):
>
> $\quad A \cup U = U$.
>
> **9.** Absorption Laws: For all sets A and B,
>
> $\quad$ (a) $A \cup (A \cap B) = A$ and   (b) $A \cap (A \cup B) = A$.
>
> **10.** Alternate Representation for Set Difference: For all sets A and B,
>
> $\quad A - B = A \cap B^c$.

The conclusion of each part of Theorem 5.2.2 is that one set equals another set. As was stated in Section 5.1:

$\quad$ Two sets are equal $\Leftrightarrow$ each is a subset of the other.

The method derived from this fact is the most basic way to prove equality of sets.

> **Basic Method for Proving That Sets Are Equal**
>
> Let sets X and Y be given. To prove that $X = Y$:
>
> **1.** prove that $X \subseteq Y$.
> **2.** prove that $Y \subseteq X$.

Later in this section we shall show a second, more algebraic method for proving set equality, and in Section 5.3 we shall show a third method that can be used in certain special circumstances.

EXAMPLE 5.2.2 **Proof of a Distributive Law**

Prove that for all sets A, B, and C,

$$A \cup (B \cap C) = (A \cup B) \cap (A \cup C).$$

Solution   The proof of this fact is somewhat more complicated than the proof in Example 5.2.1, so we first derive its logical structure, then find the core arguments, and end with a formal proof as a summary. As in Example 5.2.1, the statement to be proved is universal and so, by the method of generalizing from the generic particular, the proof has the following outline:

Starting Point: Suppose A, B, and C are arbitrarily chosen sets.

To Show: $A \cup (B \cap C) = (A \cup B) \cap (A \cup C)$.

Now two sets are equal if, and only if, each is a subset of the other. Hence, the following two statements must be proved:

$$A \cup (B \cap C) \subseteq (A \cup B) \cap (A \cup C)$$

and

$$(A \cup B) \cap (A \cup C) \subseteq A \cup (B \cap C).$$

Showing the first containment requires showing that

$$\forall x, \text{ if } x \in A \cup (B \cap C) \text{ then } x \in (A \cup B) \cap (A \cup C).$$

Showing the second containment requires showing that

$$\forall x, \text{ if } x \in (A \cup B) \cap (A \cup C) \text{ then } x \in A \cup (B \cap C).$$

Note that both of these statements are universal. So to prove the first containment, you

suppose you have any element x in $A \cup (B \cap C)$,

and then you

show that $x \in (A \cup B) \cap (A \cup C)$.

And to prove the second containment, you

suppose you have any element x in $(A \cup B) \cap (A \cup C)$,

and then you

show that $x \in A \cup (B \cap C)$.

In Figure 5.2.1, the structure of the proof is illustrated by the kind of diagram that is often used in connection with structured programs. The analysis in the diagram reduces the proof to two concrete tasks: filling in the steps indicated by dots in the two center boxes of Figure 5.2.1.

The top inner box goes from the supposition that $x \in A \cup (B \cap C)$ to the conclusion that $x \in (A \cup B) \cap (A \cup C)$. But when $x \in A \cup (B \cap C)$, then by definition of union $x \in A$ or $x \in B \cap C$. Now either of these possibilities might be

Suppose A, B, and C are sets. *[Show $A \cup (B \cap C) = (A \cup B) \cap (A \cup C)$. That is, show $A \cup (B \cap C) \subseteq (A \cup B) \cap (A \cup C)$ and $(A \cup B) \cap (A \cup C) \subseteq A \cup (B \cap C)$.]*

> Show $A \cup (B \cap C) \subseteq (A \cup B) \cap (A \cup C)$.   *[That is, show $\forall x$, if $x \in A \cup (B \cap C)$ then $x \in (A \cup B) \cap (A \cup C)$.]*
>
> > Suppose $x \in A \cup (B \cap C)$.   *[Show $x \in (A \cup B) \cap (A \cup C)$.]*
> > $$\vdots$$
> > Thus $x \in (A \cup B) \cap (A \cup C)$.
>
> Hence $A \cup (B \cap C) \subseteq (A \cup B) \cap (A \cup C)$.

> Show $(A \cup B) \cap (A \cup C) \subseteq A \cup (B \cap C)$.   *[That is, show $\forall x$, if $x \in (A \cup B) \cap (A \cup C)$ then $x \in A \cup (B \cap C)$.]*
>
> > Suppose $x \in (A \cup B) \cap (A \cup C)$.   *[Show $x \in A \cup (B \cap C)$.]*
> > $$\vdots$$
> > Thus $x \in A \cup (B \cap C)$.
>
> Hence $(A \cup B) \cap (A \cup C) \subseteq A \cup (B \cap C)$.

Thus $(A \cup B) \cap (A \cup C) = A \cup (B \cap C)$.

FIGURE 5.2.1

the case because x is assumed to be chosen arbitrarily from the set $A \cup (B \cap C)$. So you have to show you can reach the conclusion that $x \in (A \cup B) \cap (A \cup C)$ regardless of whether x happens to be in A or x happens to be in $B \cap C$. This leads you to break your analysis into two cases: $x \in A$ and $x \in B \cap C$. In case $x \in A$, your goal is to show that $x \in (A \cup B) \cap (A \cup C)$, which means that $x \in A \cup B$ and $x \in A \cup C$ (by definition of intersection). But when $x \in A$, both statements $x \in A \cup B$ and $x \in A \cup C$ are true by virtue of x's being in A. In case $x \in B \cap C$, your goal is also to show that $x \in (A \cup B) \cap (A \cup C)$, which means that $x \in A \cup B$ and $x \in A \cup C$. But when $x \in B \cap C$ then $x \in B$ and $x \in C$ (by definition of intersection), and so $x \in A \cup B$ (by virtue of x's being in B) and $x \in A \cup C$ (by virtue of x's being in C). This analysis shows that regardless of whether $x \in A$ or $x \in B \cap C$, the conclusion $x \in (A \cup B) \cap (A \cup C)$ follows. So you can fill in the steps in the top inner box.

To fill in the steps of the bottom inner box, you need to go from the supposition that $x \in (A \cup B) \cap (A \cup C)$ to the conclusion that $x \in A \cup (B \cap C)$. Now when $x \in (A \cup B) \cap (A \cup C)$ and when x happens to be in A, then the statement "$x \in A$ or $x \in B \cap C$" is certainly true, and so x is in $A \cup (B \cap C)$ by definition of union. But either x is in A or x is not in A. So it remains only to be shown that when $x \in (A \cup B) \cap (A \cup C)$ then $x \in A \cup (B \cap C)$, even in the case when x is not in A. Now to say that $x \in (A \cup B) \cap (A \cup C)$ means that $x \in A \cup B$ and $x \in A \cup C$ (by definition of union). But if $x \in A \cup B$, then x is in at least one of A or B, so if x is not in A then x must be in B. Similarly, if $x \in A \cup C$, then x is in at least one of A or C, so if x is not in A then x must be in C. Thus, when x is not in A and $x \in (A \cup B) \cap (A \cup C)$, then x is in both B and C, which means that $x \in B \cap C$. It follows that the statement "$x \in A$ or $x \in B \cap C$" is true, and so $x \in A \cup (B \cap C)$ by definition of union. This analysis shows that if $x \in (A \cup B) \cap (A \cup C)$, then regardless of whether $x \in A$ or $x \notin A$, you can conclude that $x \in A \cup (B \cap C)$. Hence you can fill in the steps of the bottom inner box.

A formal proof is shown below.

THEOREM 5.2.2(3)(a)

For all sets A, B, and C,

$$A \cup (B \cap C) = (A \cup B) \cap (A \cup C).$$

Proof: Suppose A and B are sets.

$A \cup (B \cap C) \subseteq (A \cup B) \cap (A \cup C)$:

Suppose $x \in A \cup (B \cap C)$. By definition of union, $x \in A$ or $x \in B \cap C$.

Case 1 ($x \in A$):   Since $x \in A$, $x \in A \cup B$ by definition of union and also $x \in A \cup C$ by definition of union. Hence $x \in (A \cup B) \cap (A \cup C)$ by definition of intersection.

Case 2 ($x \in B \cap C$):   Since $x \in B \cap C$, then $x \in B$ and $x \in C$ by definition of intersection. Since $x \in B$, $x \in A \cup B$ and since $x \in C$, $x \in A \cup C$ by definition of union. Hence $x \in (A \cup B) \cap (A \cup C)$ by definition of intersection.

In both cases $x \in (A \cup B) \cap (A \cup C)$. Hence $A \cup (B \cap C) \subseteq (A \cup B) \cap (A \cup C)$ by definition of subset.

$(A \cup B) \cap (A \cup C) \subseteq A \cup (B \cap C)$:

Suppose $x \in (A \cup B) \cap (A \cup C)$. By definition of intersection, $x \in A \cup B$ and $x \in A \cup C$. Consider the two cases $x \in A$ and $x \notin A$.

Case 1 ($x \in A$):   Since $x \in A$, we can immediately conclude that $x \in A \cup (B \cap C)$ by definition of union.

Case 2 ($x \notin A$):   Since $x \in A \cup B$, x is in at least one of A or B. But x is not in A; hence x is in B. Similarly, since $x \in A \cup C$, x is in at least one of A or C. But x is not in A; hence x is in C. We have shown that both $x \in B$ and $x \in C$, and so by definition of intersection, $x \in B \cap C$. It follows by definition of union that $x \in A \cup (B \cap C)$.

In both cases $x \in A \cup (B \cap C)$. Hence by definition of subset, $(A \cup B) \cap (A \cup C) \subseteq A \cup (B \cap C)$.

Since both subset relations have been proved, it follows by definition of set equality that $A \cup (B \cap C) = (A \cup B) \cap (A \cup C)$.

In the study of artificial intelligence, the types of reasoning used above to derive the proof are called *forward chaining* and *backward chaining*. First what is to be shown is viewed as a goal to be reached starting from a certain initial position: the starting point. Analysis of this goal leads to the realization that if a certain job is accomplished, then the goal will be reached. Call this job subgoal 1: SG_1. (For instance, if the goal is to show that $A \cup (B \cap C) = (A \cup B) \cap (A \cup C)$ then SG_1 would be to show that each set is a subset of the other.) Analysis of SG_1 shows that when yet another job is completed, then SG_1 will be reached. Call this job subgoal 2: SG_2. Continuing in this way, a chain of argument leading backward from the goal is constructed.

$$\boxed{\text{starting point}} \qquad\qquad \rightarrow SG_3 \rightarrow SG_2 \rightarrow SG_1 \rightarrow \boxed{\text{goal}}$$

At a certain point, backward chaining becomes difficult, but analysis of the current subgoal suggests it may be reachable by a direct line of argument, called forward chaining, beginning at the starting point. Using the information contained in the starting point, another piece of information, I_1, is deduced; from that another piece of information, I_2, is deduced; and so forth until finally one of the subgoals is reached. This completes the chain and proves the theorem. A completed chain is illustrated below.

$$\boxed{\text{starting point}} \rightarrow I_1 \rightarrow I_2 \rightarrow I_3 \rightarrow I_4 \rightarrow SG_3 \rightarrow SG_2 \rightarrow SG_1 \rightarrow \boxed{\text{goal}}$$

EXAMPLE 5.2.3 Proof of One of De Morgan's Laws

Prove that for all sets A and B, $(A \cup B)^c = A^c \cap B^c$.

Solution As in previous examples, the statement to be proved is universal, and so the starting point of the proof and the conclusion to be shown are as follows:

Starting Point: Suppose A and B are arbitrarily chosen sets.

To Show: $(A \cup B)^c = A^c \cap B^c$

To do this you must show that $(A \cup B)^c \subseteq A^c \cap B^c$ and that $A^c \cap B^c \subseteq (A \cup B)^c$. To show the first containment means to show that

$$\forall x, \text{ if } x \in (A \cup B)^c \text{ then } x \in A^c \cap B^c.$$

And to show the second containment means to show that

$$\forall x, \text{ if } x \in A^c \cap B^c \text{ then } x \in (A \cup B)^c.$$

Since each of these statements is universal and conditional, for the first containment, you

$$\textbf{suppose } x \in (A \cup B)^c,$$

and then you

$$\textbf{show that } x \in A^c \cap B^c.$$

And for the second containment, you

$$\textbf{suppose } x \in A^c \cap B^c,$$

and then you

$$\textbf{show that } x \in (A \cup B)^c.$$

To fill in the steps of these arguments, you use the procedural versions of the definitions of complement, union, and intersection, and at crucial points you use De Morgan's laws of logic.

Theorem 5.2.2(7)(a)

For all sets A and B, $\quad (A \cup B)^c = A^c \cap B^c$.

Proof:

Suppose A and B are sets.

$(A \cup B)^c \subseteq A^c \cap B^c$:   *[We must show that $\forall x$, if $x \in (A \cup B)^c$ then $x \in A^c \cap B^c$.]*

    Suppose $x \in (A \cup B)^c$. *[We must show that $x \in A^c \cap B^c$.]* By definition of complement,

$$x \notin A \cup B.$$

But to say that $x \notin A \cup B$ means that

$$\text{it is false that } (x \text{ is in } A \text{ or } x \text{ is in } B).$$

By De Morgan's laws of logic, this implies that

$$x \text{ is not in } A \text{ and } x \text{ is not in } B,$$

which can be written

$$x \notin A \quad \text{and} \quad x \notin B.$$

Hence $x \in A^c$ and $x \in B^c$ by definition of complement. It follows by definition of intersection that $x \in A^c \cap B^c$ *[as was to be shown]*. So $(A \cup B)^c \subseteq A^c \cap B^c$ by definition of subset.

$A^c \cap B^c \subseteq (A \cup B)^c$:   *[We must show that $\forall x$, if $x \in A^c \cap B^c$ then $x \in (A \cup B)^c$.]*

    Suppose $x \in A^c \cap B^c$. *[We must show that $x \in (A \cup B)^c$.]* By definition of intersection, $x \in A^c$ and $x \in B^c$, and by definition of complement,

$$x \notin A \quad \text{and} \quad x \notin B.$$

In other words,

$$x \text{ is not in } A \text{ and } x \text{ is not in } B.$$

By De Morgan's laws of logic this implies that

$$\text{it is false that } (x \text{ is in } A \text{ or } x \text{ is in } B),$$

which can be written

$$x \notin A \cup B$$

by definition of union. Hence by definition of complement, $x \in (A \cup B)^c$ *[as was to be shown]*. It follows that $A^c \cap B^c \subseteq (A \cup B)^c$ by definition of subset.

    Since both set containments have been proved, $(A \cup B)^c = A^c \cap B^c$ by definition of set equality.

The set property given in the next theorem says that if one set is a subset of another, then their intersection is the smaller of the two sets and their union is the larger of the two sets.

THEOREM 5.2.3   Intersection and Union with a Subset

For any sets A and B, if $A \subseteq B$, then

$$\text{(a) } A \cap B = A \quad \text{and} \quad \text{(b) } A \cup B = B.$$

Proof of part (a):

Suppose A and B are sets with $A \subseteq B$. To show part (a) we must show that $A \cap B \subseteq A$ and $A \subseteq A \cap B$. We already know that $A \cap B \subseteq A$ by the inclusion of intersection property. To show that $A \subseteq A \cap B$, let $x \in A$. *[We must show that $x \in A \cap B$.]* Since $A \subseteq B$, then $x \in B$ also. Hence

$$x \in A \quad \text{and} \quad x \in B,$$

and thus

$$x \in A \cap B$$

by definition of intersection *[as was to be shown]*.

The proof of part (b) is left as an exercise.

SHOWING THAT AN ALLEGED SET PROPERTY IS FALSE

Recall that to show a universal statement is false, it suffices to find one example (called a counterexample) for which it is false.

EXAMPLE 5.2.4   Finding a Counterexample for a Set Identity

Is the following set property true?

For all sets A, B, and C, $(A - B) \cup (B - C) = A - C$.

Solution   Observe that the property is true if, and only if,

the given equality holds for *all* sets A, B, and C.

So it is false if, and only if,

there are sets A, B, and C for which the equality does *not* hold.

You can picture sets A, B, and C by drawing a Venn diagram such as that shown in Figure 5.2.2 (p. 254). If you assume that any of the eight regions of the diagram may be empty of points, then the diagram is quite general.

Find and shade the region corresponding to $(A - B) \cup (B - C)$. Then shade the region corresponding to $A - C$. These are shown in Figure 5.2.3. When you compare the shaded regions you can see that there may be points in $(A - B) \cup (B - C)$ that are not in $A - C$. The property is therefore false, and a concrete counterexample consists of any sets A, B, and C with points inside regions shaded in one diagram but

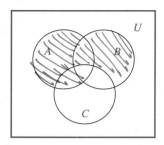

FIGURE 5.2.2

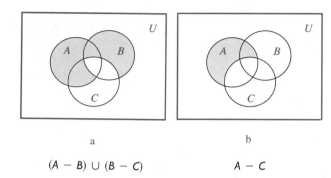

a	b

FIGURE 5.2.3 $(A - B) \cup (B - C)$ $A - C$

not the other. For example, A, B, and C could be taken to be the sets of *all* points inside each of the disks shown.

Alternatively, you can use the diagrams to help construct a discrete counterexample. The shading of the diagrams shows that for sets A, B, and C to be a counterexample, B must contain points that are not in either A or C, or there must be points in both A and C that are not in B. For example, you could take

$$A = \{a, b\}, \quad B = \{b, c\}, \quad \text{and} \quad C = \{a, d\}.$$

Then

$$A - B = \{a\}, \quad B - C = \{b, c\}, \quad \text{and} \quad A - C = \{b\}.$$

Hence

$$(A - B) \cup (B - C) = \{a, b, c\} \quad \text{whereas} \quad A - C = \{b\}.$$

So $(A - B) \cup (B - C) \neq A - C$. ■

DERIVING NEW SET PROPERTIES FROM OLD ONES ALGEBRAICALLY

Once a certain number of set properties have been established, new properties can be derived from them algebraically. To do this successfully, you need to use the fact that properties such as those of Theorems 5.2.1–5.2.3 are universal statements. Like the laws of algebra for real numbers, they apply to a wide variety of different situations. For example, one of the distributive laws states that

for all sets A, B, and C, $A \cap (B \cup C) = (A \cap B) \cup (A \cap C)$.

This law can be viewed as a general template into which *any* three particular sets can be placed.

For example, if A_1, A_2, and A_3 represent particular sets, then

$$A_1 \cap (A_2 \cup A_3) = (A_1 \cap A_2) \cup (A_1 \cap A_3),$$

$$A \cap (B \cup C) = (A \cap B) \cup (A \cap C)$$

where A_1 plays the role of A, A_2 plays the role of B, and A_3 plays the role of C. Similarly, if W, X, Y, and Z are any particular sets, then $W \cap X$ is also a set. Thus by the distributive law (with $W \cap X$ playing the role of A, Y playing the role of B, and Z playing the role of C),

$$(W \cap X) \cap (Y \cup Z) = ((W \cap X) \cap Y) \cup ((W \cap X) \cap Z),$$

$$A \cap (B \cup C) = (A \cap B) \cup (A \cap C)$$

where $W \cap X$ plays the role of A, Y plays the role of B, and Z plays the role of C.

EXAMPLE 5.2.5   Deriving a Set Difference Property

Prove that for all sets, A, B, and C

$$(A \cup B) - C = (A - C) \cup (B - C).$$

Solution   Let sets A, B, and C be given. Then

$$(A \cup B) - C = (A \cup B) \cap C^c \qquad \text{by the alternate representation of set difference law}$$

$$= C^c \cap (A \cup B) \qquad \text{by the commutative law for } \cap$$

$$= (C^c \cap A) \cup (C^c \cap B) \qquad \text{by the distributive law}$$

$$= (A \cap C^c) \cup (B \cap C^c) \qquad \text{by the commutative law for } \cap$$

$$= (A - C) \cup (B - C) \qquad \text{by the alternate representation of set difference law.}$$

■

EXAMPLE 5.2.6   Deriving a Generalized Associative Law

Prove that for any sets A_1, A_2, A_3, and A_4,

$$((A_1 \cup A_2) \cup A_3) \cup A_4 = A_1 \cup ((A_2 \cup A_3) \cup A_4).$$

Solution   Let sets A_1, A_2, A_3, and A_4 be given. Then

$$((A_1 \cup A_2) \cup A_3) \cup A_4 = (A_1 \cup (A_2 \cup A_3)) \cup A_4 \qquad \text{by the associative law for } \cup \text{ with } A_1 \text{ playing the role of } A, A_2 \text{ playing the role of } B, \text{ and } A_3 \text{ playing the role of } C$$

$$= A_1 \cup ((A_2 \cup A_3) \cup A_4) \qquad \text{by the associative law for } \cup \text{ with } A_1 \text{ playing the role of } A, A_2 \cup A_3 \text{ playing the role of } B, \text{ and } A_4 \text{ playing the role of } C.$$

■

▲ CAUTION!   *When doing problems similar to those of Examples 5.2.5 and 5.2.6, be sure to use the set properties exactly as they are stated.*

PROBLEM-SOLVING STRATEGY

How can you discover whether a given universal statement about sets is true or false? There are two basic approaches: the optimistic and the pessimistic. In the optimistic approach, you simple plunge in and start trying to prove the statement, asking yourself, "What do I need to show?" and "How do I show it?" In the pessimistic approach, you start by searching your mind for a set of conditions that must be fulfilled to construct a counterexample. With either approach you may have clear sailing and be immediately successful or you may run into difficulty. The trick is to be ready to switch to the other approach if the one you are trying does not look promising. For more difficult questions you may alternate several times between the two approaches before arriving at the correct answer.

EXERCISE SET 5.2

1. **a.** To say that an element is in $A \cap (B \cup C)$ means that it is in __(1)__ and in __(2)__ .
   b. To say that an element is in $(A \cap B) \cup C$ means that it is in __(1)__ or in __(2)__ .
   c. To say that an element is in $A - (B \cap C)$ means that it is in __(1)__ and not in __(2)__ .

2. The following are two proofs that for all sets A and B, $A - B \subseteq A$. The first is less formal, and the second is more formal. Fill in the blanks.

   a. Proof:

   Suppose A and B are any sets. To show that $A - B \subseteq A$, we must show that every element in __(1)__ is in __(2)__ . But any element in $A - B$ is in __(3)__ and not in __(4)__ (by definition of $A - B$). In particular, such an element is in A.

   b. Proof:

   Suppose A and B are any sets and $x \in A - B$. *[We must show that __(1)__ .]* By definition of set difference, $x \in$ __(2)__ and $x \notin$ __(3)__ . In particular, $x \in$ __(4)__ . *[which is what was to be shown].*

3. The following is a proof that for all sets A, B, and C, if $A \subseteq B$ and $B \subseteq C$, then $A \subseteq C$. Fill in the blanks.

   Proof:

   Suppose A, B, and C are sets and $A \subseteq B$ and $B \subseteq C$. To show that $A \subseteq C$, we must show that every element in __(1)__ is in __(2)__ . But given any element in A, that element is in __(3)__ (because $A \subseteq B$) and so that element is also in __(4)__ (because __(5)__). Hence $A \subseteq C$.

4. The following is a proof that for all sets A and B, if $A \subseteq B$ then $A \cup B \subseteq B$. Fill in the blanks.

   Proof:

   Suppose A and B are any sets and $A \subseteq B$. *[We must show that __(a)__ .]* Let $x \in$ __(b)__ . *[We must show that __(c)__ .]* By definition of union, $x \in$ __(d)__ __(e)__ $x \in$ __(f)__ . In case $x \in$ __(g)__ , then since $A \subseteq B$, $x \in$ __(h)__ . In case $x \in B$, then clearly $x \in B$. So in either case, $x \in$ __(i)__ *[as was to be shown].*

5. Prove that for all sets A and B, $B - A = B \cap A^c$.

6. The following is a proof that for any sets A, B, and C, $A \cap (B \cup C) = (A \cap B) \cup (A \cap C)$. Fill in the blanks.

   Proof:

   Suppose A, B, and C are any sets.

   (1) $A \cap (B \cup C) \subseteq (A \cap B) \cup (A \cap C)$: Let $x \in A \cap (B \cup C)$. *[We must show that $x \in$ __(a)__ .]* By definition of intersection, $x \in$ __(b)__ and $x \in$ __(c)__ . Thus $x \in A$ and by definition of union, $x \in B$ or __(d)__ .

   Case 1 $(x \in A$ and $x \in B)$: In this case, by definition of intersection $x \in$ __(e)__ , and so by definition of union, $x \in (A \cap B) \cup (A \cap C)$.

   Case 2 $(x \in A$ and $x \in C)$: In this case, __(f)__ .

   Hence in either case, $x \in (A \cap B) \cup (A \cap C)$ *[as was to be shown].*
   [So $A \cap (B \cup C) \subseteq (A \cap B) \cup (A \cap C)$ by definition of subset.]

   (2) $(A \cap B) \cup (A \cap C) \subseteq A \cap (B \cup C)$: Let $x \in (A \cap B) \cup (A \cap C)$. *[We must show that __(a)__ .]* By definition of union, $x \in$ __(b)__ or $x \in$ __(c)__ .

   Case 1 $(x \in A \cap B)$: In this case, by definition of intersection __(d)__ and __(e)__ . Since $x \in B$, by definition of union, $x \in B \cup C$. Hence $x \in A$ and $x \in B \cup C$, and so by definition of intersection, $x \in$ __(f)__ .

   Case 2 $(x \in A \cap C)$: In this case, __(g)__ .

   In either case $x \in A \cap (B \cup C)$ *[as was to be shown].*
   [Thus $(A \cap B) \cup (A \cap C) \subseteq A \cap (B \cup C)$ by definition of subset.]
   [Since both subset relations have been proved, it follows by definition of set equality that __(3)__ .]

H7. Prove that for all sets A and B, $(A \cap B)^c = A^c \cup B^c$.

8. Find a counterexample to show that the following statement is false: For all sets A and B, $(A \cap B) \cup C = A \cap (B \cup C)$.

Determine whether each of the statements in 9–24 is true or false. Use an element argument to prove each statement that is true directly from the definitions of the set operations. Find a counterexample for each statement that is false.

H9. For all sets A and B, $(A - B) \cup (A \cap B) = A$.

H10. For all sets A, B, and C.
$A - (B - C) = (A - B) - C$.

11. For all sets A, B, and C,
$(A - B) \cap (C - B) = (A \cap C) - B$.

12. For all sets A, B, and C,
$(A - B) \cap (C - B) = (A - (B \cup C)$.

13. For all sets A, B, and C, if $A \subseteq B$ then $A \cap C \subseteq B \cap C$.

14. For all sets A, B, and C, if $A \subseteq B$ then $A \cup C \subseteq B \cup C$.

15. For all sets A, B, and C, if $A \cap C = B \cap C$ then $A = B$.

16. For all sets A, B, and C, if $A \cup C = B \cup C$ then $A = B$.

17. For all sets A, B, and C, if $A \cap C \subseteq B \cap C$ and $A \cup C \subseteq B \cup C$, then $A = B$.

18. For all sets, A, B, and C,
$(A \cup B) \cap C = A \cup (B \cap C)$.

19. For all sets A and B, if $A \subseteq B$ then $B^c \subseteq A^c$.

20. For all sets A, B, and C, if $A \not\subseteq B$ and $B \not\subseteq C$ then $A \not\subseteq C$.

21. For all sets A, B, and C, if $A \subseteq B$ and $A \subseteq C$ then $A \subseteq B \cap C$.

22. For all sets A, B, and C, if $A \subseteq C$ and $B \subseteq C$ then $A \cup B \subseteq C$.

H23. For all sets A, B, and C,
$A \times (B \cup C) = (A \times B) \cup (A \times C)$.

24. For all sets A, B, and C,
$A \times (B \cap C) = (A \times B) \cap (A \times C)$.

25. Consider the Venn diagram below.

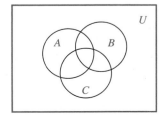

a. Illustrate one of the distributive laws by shading in the region corresponding to $A \cup (B \cap C)$ on one copy of the diagram and $(A \cup B) \cap (A \cup C)$ on another.

b. Illustrate the other distributive law by shading in the region corresponding to $A \cap (B \cup C)$ on one copy of the diagram and $(A \cap B) \cup (A \cap C)$ on another.

c. Illustrate one of De Morgan's laws by shading in the region corresponding to $(A \cup B)^c$ on one copy of the diagram and $A^c \cap B^c$ on the other. (Leave the set C out of your diagrams.)

d. Illustrate the other De Morgan's law by shading in the region corresponding to $(A \cap B)^c$ on one copy of the diagram and $A^c \cup B^c$ on the other. (Leave the set C out of your diagrams.)

26. Prove by mathematical induction that for any integer $n > 1$ and all sets $A_1, A_2, \ldots, A_n$ and B,

$$(A_1 - B) \cup (A_2 - B) \cup \cdots \cup (A_n - B)$$
$$= (A_1 \cup A_2 \cup \cdots \cup A_n) - B.$$

(Assume that if $n \geq 3$ and $C_1, C_2, \ldots, C_n$ are any sets, $C_1 \cup C_2 \cup \cdots \cup C_n$ is defined to be $(C_1 \cup C_2 \cup \cdots \cup C_{n-1}) \cup C_n$.)

27. Prove by mathematical induction that for any integer $n \geq 1$ and all sets $A_1, A_2, \ldots, A_n$ and B,

$$(A_1 - B) \cap (A_2 - B) \cap \cdots \cap (A_n - B)$$
$$= (A_1 \cap A_2 \cap \cdots \cap A_n) - B.$$

(Assume that if $n \geq 3$ and $C_1, C_2, \ldots C_n$ are any sets, $C_1 \cap C_2 \cap \cdots \cap C_n$ is defined to be $(C_1 \cap C_2 \cap \cdots \cap C_{n-1}) \cap C_n$.)

In 28–29 supply a reason for each of the steps in the derivation.

28. For all sets A, B, and C,
$(A \cup B) \cap C = (A \cap C) \cup (B \cap C)$.

Proof: Suppose A, B, and C are any sets. Then

$(A \cup B) \cap C$

$= C \cap (A \cup B)$ by ___(a)___

$= (C \cap A) \cup (C \cap B)$ by ___(b)___

$= (A \cap C) \cup (B \cap C)$ by ___(c)___.

29. For all sets A, B, and C, $(A \cup B) - (C - A) = A \cup (B - C)$.

Proof: Suppose A, B, and C are any sets. Then

$(A \cup B) - (C - A)$

$= (A \cup B) \cap (C - A)^c$ by ___(a)___

$= (A \cup B) \cap (C \cap A^c)^c$ by ___(b)___

$= (A \cup B) \cap (A^c \cap C)^c$ by ___(c)___

$= (A \cup B) \cap ((A^c)^c \cup C^c)$ by ___(d)___

$= (A \cup B) \cap (A \cup C^c)$ by ___(e)___

$= A \cup (B \cap C^c)$ by ___(f)___

$= A \cup (B - C)$ by ___(g)___

Derive each of the set properties in 30–36 from those given in Theorems 5.2.1–5.2.3. Be sure to use each property exactly as stated in the theorem.

H30. For all sets A, B, and C,
$(A \cap B) \cup C = (A \cup C) \cap (B \cup C)$.

31. For all sets A, B, and C,
$(A - B) - C = (A - C) - B$.

32. For all sets A, B, and C,
$(A - B) - (B - C) = A - B$.

33. For all sets A, B, and C,
$(A - B) \cup (B - A) = (A \cup B) - (A \cap B)$,

34. For all sets A, B, and C,
$(A - B) - C = A - (B \cup C)$,

35. For all sets A and B, $((A^c \cup B^c) - A)^c = A$.

36. For all sets A and B, $(B^c \cup (B^c - A))^c = B$.

H37. Find the mistake in the following "proof" that for all sets A, B, and C, if $A \subseteq B$ and $B \subseteq C$ then $A \subseteq C$.

"Proof: Suppose A, B, and C are sets such that $A \subseteq B$ and $B \subseteq C$. Since $A \subseteq B$, there is an element x so that $x \in A$ and $x \in B$. Since $B \subseteq C$, there is an element x so that $x \in B$ and $x \in C$. Hence there is an element x such that $x \in A$ and $x \in C$ and so $A \subseteq C$."

H38. Find the mistake in the following "proof."
"Theorem: For all sets A and B, $A^c \cup B^c \subseteq (A \cup B)^c$."
"Proof: Suppose A and B are sets, and suppose $x \in A^c \cup B^c$. Then $x \in A^c$ or $x \in B^c$ by definition of union. It follows that $x \notin A$ or $x \notin B$ by definition of complement, and so $x \notin A \cup B$ by definition of union. Thus $x \in (A \cup B)^c$ by definition of complement, and hence $A^c \cup B^c \subseteq (A \cup B)^c$."

39. Find the mistake in the following "proof" that for all sets A and B, $(A - B) \cup (A \cap B) \subseteq A$.
"Proof: Suppose A and B are sets, and suppose $x \in (A - B) \cup (A \cap B)$. If $x \in A$ then $x \in A - B$. Then by definition of difference, $x \in A$ and $x \notin B$. Hence $x \in A$, and so $(A - B) \cup (A \cap B) \subseteq A$ by definition of subset."

5.3   THE EMPTY SET, PARTITIONS, POWER SETS, AND BOOLEAN ALGEBRAS

If a fact goes against common sense, and we are nevertheless compelled to accept and deal with this fact, we learn to alter our notion of common sense.
(Phillip J. Davis and Reuben Hersh, *The Mathematical Experience*, 1981)

The discussion of sets so far has focused on the notion of set as defined by the elements that compose it. This being so, can there be a set that does not have any elements? It turns out that it is convenient to allow such a set. Otherwise, every time we wanted to define a set by specifying a property, it would be necessary to check that the result had elements and hence qualified for "sethood."

EXAMPLE 5.3.1   Sets with No Elements

a. Let A be the set of all animals and let $P = \{x \in A \mid x$ is a pink elephant$\}$. Then P does not have any elements.
b. Let S be the set of all real numbers x whose square is -1: $S = \{x \in \mathbf{R} \mid x^2 = -1\}$. Since the square of every real number is nonnegative, S does not have any elements.
c. Let $X = \{1, 3\}$, $Y = \{2, 4\}$, and $C = X \cap Y$. Then C does not have any elements.

∎

It is somewhat unsettling to talk about a set that does not have any elements, but it often happens in mathematics that the definitions formulated to fit one set of circumstances are satisfied by some extreme cases not originally anticipated. Yet changing the definitions to exclude those cases would seriously undermine the simplicity and elegance of the theory taken as a whole.

In this section, we investigate some properties of sets without elements. Many proofs involving such sets use the technique of proof by contradiction.

The first theorem states that a set with no elements is a subset of *every* set. Why is this true? Just ask yourself, "Could it possibly be false? Could there be a set without elements that is *not* a subset of some given set?" The crucial fact is that the negation of a universal statement is existential: If a set B is not a subset of a set A, then there exists x in B such that x is not in A. But if B has no elements, then no such x can exist.

THEOREM 5.3.1   A Set with No Elements Is a Subset
                    of Every Set

If $\emptyset$ is a set with no elements and A is any set, then $\emptyset \subseteq A$.

Proof (by contradiction):

Suppose not. *[We take the negation of the theorem and suppose it to be true.]* Suppose there exists a set $\emptyset$ with no elements and a set A such that $\emptyset \nsubseteq A$. *[We must deduce a contradiction.]* Then there would be an element of $\emptyset$ which is not an element of A *[by definition of subset]*. But there can be no such element since $\emptyset$ has no elements. This is a contradiction. *[Hence the supposition that there are sets $\emptyset$ and A , where $\emptyset$ has no elements and $\emptyset \nsubseteq A$, is false, and so the theorem is true.]*

The truth of Theorem 5.3.1 can also be understood by appeal to the notion of vacuous truth. If $\emptyset$ is a set with no elements and A is any set, then to say that $\emptyset \subseteq A$ is the same as saying that

$$\forall x \in \emptyset, x \in A.$$

But since $\emptyset$ has no elements, this statement is vacuously true.

How many sets with no elements are there? The answer is, only one.

COROLLARY 5.3.2   Uniqueness of the Empty Set

There is only one set with no elements.

Proof:

Suppose $\emptyset_1$ and $\emptyset_2$ are each sets with no elements. By Theorem 5.3.1 above, $\emptyset_1 \subseteq \emptyset_2$ since $\emptyset_1$ has no elements. Also $\emptyset_2 \subseteq \emptyset_1$ since $\emptyset_2$ has no elements. Thus $\emptyset_1 = \emptyset_2$ by definition of set equality.

It follows from Corollary 5.3.2 that the set of pink elephants is equal to the set of all real numbers whose square is -1 because each set has no elements! Since there is only one set with no elements, it has a special name.

> **DEFINITION**
>
> The unique set with no elements is called the **empty set**. It is denoted by the symbol Ø.

Note that while Ø is the set with no elements, the set {Ø} has one element, namely the empty set. This is similar to the convention in the computer programming language LISP in which () denotes the empty list and (()) denotes the list whose one element is the empty list.

Suppose you need to show that a certain set equals the empty set. By Corollary 5.3.2 it suffices to show that the set has no elements. For since there is only one set with no elements (namely Ø), if the given set has no elements it must equal Ø.

> **Element Method for Proving a Set Equals the Empty Set**
>
> To prove that a set X is equal to the empty set Ø, prove that X has no elements. To do this, suppose X has an element and derive a contradiction.

EXAMPLE 5.3.2   Proving That a Set Is Empty

Prove that for any set A, $A \cap \emptyset = \emptyset$.

Solution   Let A be a *[particular, but arbitrarily chosen]* set. To show that $A \cap \emptyset = \emptyset$, it suffices to show that $A \cap \emptyset$ has no elements *[by the element method for proving a set equals the empty set]*. Suppose not. That is, suppose there is an element x such that $x \in A \cap \emptyset$. Then by definition of intersection, $x \in A$ and $x \in \emptyset$. In particular, $x \in \emptyset$. But this is impossible since Ø has no elements. *[This contradiction shows that the supposition that there is an element x in $A \cap \emptyset$ is false. So $A \cap \emptyset$ has no elements as was to be shown.]* Thus $A \cap \emptyset = \emptyset$. ■

The following theorem gives a list of set properties that involve the empty set:

> **THEOREM 5.3.3**   Set Properties That Involve Ø
>
> Let all sets referred to below be subsets of a universal set U.
>
> **1.** Union with Ø (Ø Acts as an Identity for ∪): For all sets A,
> $$A \cup \emptyset = A.$$
>
> **2.** Intersection and Union with the Complement: For all sets A,
> (a) $A \cap A^c = \emptyset$ and   (b) $A \cup A^c = U$.
>
> **3.** Intersection with Ø (Ø Acts as a Universal Bound for ∩): For all sets A,
> $$A \cap \emptyset = \emptyset.$$
>
> **4.** Complements of U and Ø:
> (a) $U^c = \emptyset$ and   (b) $\emptyset^c = U$.

The proof of Theorem 5.3.3(3) was given in Example 5.3.2. Proofs of the other properties are left as exercises.

As shown in Section 5.2, algebraic methods can also be used to derive new set properties from old. These methods also work for derivations that involve the empty set.

EXAMPLE 5.3.3   Deriving a Set Identity from Known Properties

Use the properties in Theorems 5.2.1–5.2.3 and 5.3.3 to prove that for all sets A and B,

$$A - (A \cap B) = A - B.$$

Solution   Suppose A and B are sets. Then

$$A - (A \cap B) = A \cap (A \cap B)^c \qquad \text{by the alternate representation for set difference law}$$

$$= A \cap (A^c \cup B^c) \qquad \text{by De Morgan's laws}$$
$$= (A \cap A^c) \cup (A \cap B^c) \qquad \text{by the distributive law}$$
$$= \varnothing \cup (A \cap B^c) \qquad \text{by the intersection with the complement law}$$

$$= (A \cap B^c) \cup \varnothing \qquad \text{by the commutative law for } \cup$$
$$= A \cap B^c \qquad \text{by the union with } \varnothing \text{ law}$$
$$= A - B \qquad \text{by the alternate representation for set difference law.} \quad \blacksquare$$

To many people an algebraic proof seems more attractive than an element proof. But often an element proof is actually simpler. For instance, in Example 5.3.3 above, you could see immediately that $A - (A \cap B) = A - B$ because for an element to be in $A - (A \cap B)$ means that it is in A and not in both A and B, and this is the same as saying that it is in A and not in B.

PARTITIONS OF SETS

In many applications of set theory sets are divided up into nonoverlapping (or *disjoint*) pieces. Such a division is called a *partition*.

DEFINITION

Two sets are called **disjoint** if, and only if, they have no elements in common. Symbolically:

$$A \text{ and } B \text{ are disjoint} \quad \Leftrightarrow \quad A \cap B = \varnothing.$$

EXAMPLE 5.3.4   Disjoint Sets

a. Let $A = \{1, 3, 5\}$ and $B = \{2, 4, 6\}$. Are A and B disjoint?
b. Given any sets A and B, are $A - B$ and B disjoint?

Solution   a. Yes. By inspection A and B have no elements in common, or, in other words, $\{1, 3, 5\} \cap \{2, 4, 6\} = \varnothing$.

b. Yes. To prove that $A - B$ and B are disjoint, it is necessary to show that $(A - B) \cap B = \varnothing$. By the element method for proving a set equals the empty set, it suffices to show that $(A - B) \cap B$ has no elements. But by definition of intersection, any element in $(A - B) \cap B$ would be both in $A - B$ and in B. It follows by definition of difference that such an element would be both in B and not in B, which is impossible. This answer is formalized as follows:

PROPOSITION 5.3.4

Given any sets A and B, $(A - B)$ and B are disjoint.

Proof (by contradiction):

Suppose not. *[We take the negation of the theorem and suppose it to be true.]* Suppose there exist sets A and B such that $A - B$ and B are not disjoint. *[We must derive a contradiction.]* Then $(A - B) \cap B \neq \varnothing$, and so there is an element x in $(A - B) \cap B$. By definition of intersection, $x \in A - B$ and $x \in B$, and since $x \in A - B$, by definition of difference, $x \in A$ and $x \notin B$. Hence $x \in B$ and also $x \notin B$, which is a contradiction. *[Thus the supposition that there exist sets A and B such that $A - B$ and B are not disjoint is false, and hence the proposition is true.]*

DEFINITION

Sets $A_1, A_2, \ldots, A_n$ are **mutually disjoint** (or **pairwise disjoint or nonoverlapping**) if, and only if, no two sets A_i and A_j with distinct subscripts have any elements in common. More precisely, for all $i, j = 1, 2, \ldots, n$,

$$A_i \cap A_j = \varnothing \quad \text{whenever } i \neq j.$$

EXAMPLE 5.3.5   Mutually Disjoint Sets

a. Let $A_1 = \{3, 5\}$, $A_2 = \{1, 4, 6\}$, and $A_3 = \{2\}$. Are A_1, A_2, and A_3 mutually disjoint?

b. Let $B_1 = \{2, 4, 6\}$, $B_2 = \{3, 7\}$, and $B_3 = \{4, 5\}$. Are B_1, B_2, and B_3 mutually disjoint?

Solution   a. Yes. A_1 and A_2 have no elements in common, A_1 and A_3 have no elements in common, and A_2 and A_3 have no elements in common.

b. No. B_1 and B_3 both contain 4.   ∎

Suppose $A, A_1, A_2, A_3,$ and A_4 are the sets of points represented by the regions shown in Figure 5.3.1. Then A_1, A_2, A_3, and A_4 are subsets of A, and $A = A_1 \cup A_2 \cup A_3 \cup A_4$. Suppose boundaries are assigned to the regions representing

A_2, A_3, and A_4 in such a way that these sets are mutually disjoint. Then A is called a *union of mutually disjoint subsets,* and the collection of sets $\{A_1, A_2, A_3, A_4\}$ is said to be a *partition* of A.

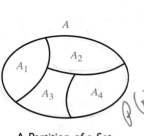

FIGURE 5.3.1   A Partition of a Set

DEFINITION

A collection of nonempty sets $\{A_1, A_2, \ldots, A_n\}$ is a **partition** of a set A if, and only if,

1. $A = A_1 \cup A_2 \cup \ldots \cup A_n$;
2. $A_1, A_2, \ldots, A_n$ are mutually disjoint.

EXAMPLE 5.3.6   Partitions of Sets

a. Let $A = \{1, 2, 3, 4, 5, 6\}$, $A_1 = \{1, 2\}$, $A_2 = \{3, 4\}$, and $A_3 = \{5, 6\}$. Is $\{A_1, A_2, A_3\}$ a partition of A?

b. Let $\mathbf{Z}$ be the set of all integers and let

$$T_0 = \{n \in \mathbf{Z} \mid n = 3k, \text{ for some integer } k\},$$
$$T_1 = \{n \in \mathbf{Z} \mid n = 3k + 1, \text{ for some integer } k\}, \text{ and}$$
$$T_2 = \{n \in \mathbf{Z} \mid n = 3k + 2, \text{ for some integer } k\}.$$

Is $\{T_0, T_1, T_2\}$ a partition of $\mathbf{Z}$?

Solution   a. Yes. By inspection, $A = A_1 \cup A_2 \cup A_3$ and the sets A_1, A_2, and A_3 are mutually disjoint.

b. Yes. By the quotient-remainder theorem every integer n can be represented in exactly one of the three forms

$$n = 3k \quad \text{or} \quad n = 3k + 1 \quad \text{or} \quad n = 3k + 2,$$

for some integer k. This implies that no integer can be in any two of the sets T_0, T_1, or T_2. So T_0, T_1, and T_2 are mutually disjoint. It also implies that every integer is in one of the sets T_0, T_1, or T_2. So $\mathbf{Z} = T_1 \cup T_2 \cup T_3$. ■

POWER SETS

There is a variety of situations in which it is useful to consider the set of all subsets of a particular set. The **power set axiom** guarantees that this is a set.

> **DEFINITION**
>
> Given a set A, the **power set** of A, denoted $\mathcal{P}(A)$, is the set of all subsets of A.

EXAMPLE 5.3.7   Power Set of a Set

Find the power set of the set $\{x, y\}$. That is, find $\mathcal{P}(\{x, y\})$.

Solution $\mathcal{P}(\{x, y\})$ is the set of all subsets of $\{x, y\}$. Now since $\varnothing$ is a subset of every set, $\varnothing \in \mathcal{P}(\{x, y\})$. Also any set is a subset of itself, so $\{x, y\} \in \mathcal{P}(\{x, y\})$. The only other subsets of $\{x, y\}$ are $\{x\}$ and $\{y\}$, so

$$\mathcal{P}(\{x, y\}) = \{\varnothing, \{x\}, \{y\}, \{x, y\}\}. \qquad\blacksquare$$

EXAMPLE 5.3.8   A Theorem About Power Sets

Prove the following theorem:

> **THEOREM 5.3.5**
>
> For all sets A and B, if $A \subseteq B$ then $\mathcal{P}(A) \subseteq \mathcal{P}(B)$.

Solution   Theorem 5.3.5 is equivalent to the statement that if a set A is a subset of a set B, then every subset of A is also a subset of B. But this is true by the transitive property for subsets! (See exercise 3 in Section 5.2.) This reasoning is the heart of the following formal proof:

> **Proof:**
>
> Suppose A and B are sets such that $A \subseteq B$. *[We must show that $\mathcal{P}(A) \subseteq \mathcal{P}(B)$.]*
>
> Suppose $X \in \mathcal{P}(A)$. *[We must show that $X \in \mathcal{P}(B)$.]* Since $X \in \mathcal{P}(A)$, then $X \subseteq A$ by definition of power set. But $A \subseteq B$. Hence $X \subseteq B$ by the transitive property for subsets. It follows that $X \in \mathcal{P}(B)$ by definition of power set *[as was to be shown]*.
>
> Thus $\mathcal{P}(A) \subseteq \mathcal{P}(B)$ by definition of subset *[as was to be shown]*.

$\blacksquare$

The following theorem states the important fact that if a set has n elements, then its power set has 2^n elements. The proof uses mathematical induction and is based on the following observations. Suppose X is a set and z is an element of X.

1. The subsets of X can be split into two groups: those that do not contain z and those that do contain z.
2. The subsets of X that do not contain z are the same as the subsets of $X - \{z\}$.

3. The subsets of X that do not contain z can be matched up one for one with the subsets of X that do contain z by matching each subset A that does not contain z to the subset $A \cup \{z\}$ that contains z. Thus there are as many subsets of X that contain z as there are subsets of X that do not contain z. For instance, if $X = \{x, y, z\}$, the following table shows the correspondence between subsets of X that do not contain z and subsets of X that contain z:

Subsets of X That Do Not Contain z		Subsets of X That Contain z
$\varnothing$	$\longleftrightarrow$	$\varnothing \cup \{z\} = \{z\}$
$\{x\}$	$\longleftrightarrow$	$\{x\} \cup \{z\} = \{x, z\}$
$\{y\}$	$\longleftrightarrow$	$\{y\} \cup \{z\} = \{y, z\}$
$\{x, y\}$	$\longleftrightarrow$	$\{x, y\} \cup \{z\} = \{x, y, z\}$

THEOREM 5.3.6

For all integers $n \geq 0$, if a set X has n elements then $\mathscr{P}(X)$ has 2^n elements.

Proof (by mathematical induction):

Consider the property "Any set with n elements has 2^n subsets."

The property is true for $n = 0$:   We must show that a set with zero elements has 2^0 subsets. But the only set with zero elements is the empty set, and the only subset of the empty set is itself. Thus a set with zero elements has one subset. Since $1 = 2^0$, the theorem is true for $n = 0$.

If the property is true for $n = k$, then it is true for $n = k + 1$:   Let k be any integer with $k \geq 0$ and suppose that any set with k elements has 2^k subsets. *[This is the inductive hypothesis.]* We must show that any set with $k + 1$ elements has 2^{k+1} subsets. Let X be a set with $k + 1$ elements and pick an element z in X. Observe that any subset of X either contains z or it does not. Furthermore, any subset of X that does not contain z is a subset of $X - \{z\}$. And any subset A of $X - \{z\}$ can be matched up with a subset B, equal to $A \cup \{z\}$, of X that contains z. Consequently, there are as many subsets of X that contain z as do not, and thus there are twice as many subsets of X as there are subsets of $X - \{z\}$. But $X - \{z\}$ has k elements, and so

$$\text{the number of subsets of } X - \{z\} = 2^k \quad \text{by inductive hypothesis.}$$

Therefore,

$$\text{the number of subsets of } X = 2 \cdot (\text{the number of subsets of } X - \{z\})$$
$$= 2 \cdot (2^k) \quad \text{by substitution}$$
$$= 2^{k+1} \quad \text{by basic algebra.}$$

[This is what was to be shown.]
    [Since we have proved both the basis step and the inductive step, we conclude that the theorem is true.]

BOOLEAN ALGEBRAS

If you look back at the logical equivalences of Theorem 1.1.1 and compare them to the set identities of Theorems 5.2.2 and 5.2.3, you will notice many similarities. These reflect a similarity of underlying structure between the set of all statement forms in a finite number of variables together with the operations of $\vee$ and $\wedge$ and the set of all subsets of a set together with the operations of $\cup$ and $\cap$. Both are special cases of a general algebraic structure known as a Boolean algebra. A **Boolean algebra** is a set S together with two operations generally denoted $+$ and $\cdot$ such that for all a and b in S, both $a + b$ and $a \cdot b$ are in S and such that the following axioms hold:

1. For all a and b in S,

$$a + b = b + a \qquad \text{Commutative Law of } +$$
$$a \cdot b = b \cdot a \qquad \text{Commutative Law of } \cdot.$$

2. For all a, b, and c in S,

$$(a + b) + c = a + (b + c) \qquad \text{Associative Law of } +$$
$$(a \cdot b) \cdot c = a \cdot (b \cdot c) \qquad \text{Associative Law of } \cdot.$$

3. For all a, b, and c in S,

$$a + (b \cdot c) = (a + b) \cdot (a + c) \qquad \text{Distributive Law of } + \text{ over } \cdot$$
$$a \cdot (b + c) = (a \cdot b) + (a \cdot c) \qquad \text{Distributive Law of } \cdot \text{ over } +.$$

4. There exist distinct elements 0 and 1 in S such that for all a in S,

$$a + 0 = a \qquad \text{0 Is an Identity for } +$$
$$a \cdot 1 = a \qquad \text{I Is an Identity for } \cdot.$$

5. For each a in S, there exists an element denoted $\bar{a}$ and called the *complement* or *negation of a in S* such that

$$a + \bar{a} = 1 \quad \text{and} \quad a \cdot \bar{a} = 0. \qquad \text{Complement Laws.}$$

For the set of statement forms in a finite number of variables, $\vee$ and $\wedge$ play the roles of $+$ and $\cdot$, the tautology t and contradiction c play the roles of 1 and 0, and $\sim$ plays the role of $^-$. For a set of subsets of a nonempty set U, $\cup$ and $\cap$ play the roles of $+$ and $\cdot$, U and $\emptyset$ play the roles of 1 and 0, and complementation ^{c} plays the role of $^-$.

EXERCISE SET 5.3

1. a. Is the number 0 in $\emptyset$? Why?
   b. Is $\emptyset = \{\emptyset\}$? Why?
   c. Is $\emptyset \in \{\emptyset\}$? Why?

2. Prove that for all sets A, $A \cup \emptyset = A$.

3. Fill in the blanks in the following proof that for all sets A and B, $(A - B) \cap (B - A) = \emptyset$.

   Proof:

   Let A and B be any sets and suppose $(A - B) \cap (B - A) \neq \emptyset$. That is, suppose there were an element

   x in __(a)__ . By definition of __(b)__ , $x \in A - B$ and $x \in$ __(c)__ . Then by definition of set difference, $x \in A$ and $x \notin B$ and $x \in$ __(d)__ and $x \notin$ __(e)__ . In particular $x \in A$ and $x \notin$ __(f)__ , which is a contradiction. Hence *[the supposition that $(A - B) \cap (B - A) \neq \emptyset$ is false, and so]* __(g)__ .

4. Prove that for all subsets A of a universal set U, $A \cap A^c = \emptyset$ and $A \cup A^c = U$.

5. Prove that if U denotes a universal set, then $U^c = \emptyset$ and $\emptyset^c = U$.

6. Draw Venn diagrams to describe sets A, B, and C that satisfy the given conditions.
   a. $A \cap B = \emptyset$, $A \subseteq C$, $C \cap B \neq \emptyset$
   b. $A \subseteq B$, $C \subseteq B$, $A \cap C \neq \emptyset$
   c. $A \cap B \neq \emptyset$, $B \cap C \neq \emptyset$, $A \cap C = \emptyset$

For each of 7–19 prove each statement that is true and find a counterexample for each statement that is false. Illustrate each statement of 7–17 by drawing a Venn diagram. Assume all sets are subsets of a universal set U.

7. For all sets A and B, $(A - B) \cap (A \cap B) = \emptyset$.

8. For all sets A, B, and C,
   $(A - C) \cap (B - C) \cap (A - B) = \emptyset$.

9. For all sets A and B, if $A \subseteq B$ then $A \cap B^c = \emptyset$.

10. For all sets A, B, and C, if $A \subseteq B$ then $A \cap (B \cap C)^c = \emptyset$.

11. For all sets A and B, if $B \subseteq A^c$ then $A \cap B = \emptyset$.

12. For all sets A and B, if $A^c \subseteq B$ then $A \cup B = U$.

13. For all sets A, B, and C, if $A \subseteq B$ and $B \cap C = \emptyset$ then $A \cap C = \emptyset$.

14. For all sets A, B, and C, if $B \cap C \subseteq A$, then $(A - B) \cap (A - C) = \emptyset$.

15. For all sets A, B, and C, if $B \subseteq C$ and $A \cap C = \emptyset$, then $A \cap B = \emptyset$.

16. For all sets A, B, and C, if $C \subseteq B - A$, then $A \cap C = \emptyset$.

17. For all sets A, B, and C, if $B \cap C \subseteq A$, then $(C - A) \cap (B - A) = \emptyset$.

18. For all sets A and B, if $A \cap B = \emptyset$ then $A \times B = \emptyset$.

19. For all sets A, $A \times \emptyset = \emptyset$.

20. a. Write a negation for the following statement: $\forall$ sets S, $\exists$ a set T such that $S \cap T = \emptyset$. Which is true, the statement or its negation? Explain.
    b. Write a negation for the following statement: $\exists$ a set S such that $\forall$ sets T, $S \cup T = \emptyset$. Which is true, the statement or its negation? Explain.

Derive the set properties of 21–25 from those listed in Theorems 5.2.1–5.2.3 and 5.3.3.

21. For all sets A and B, $A \cup (B - A) = A \cup B$.

22. For all sets A and B, $A - (A - B) = A \cap B$.

23. For all sets A and B, $A - (A \cap B) = A - B$.

24. For all sets A and B, $(A \cup B) - B = A - B$.

25. For all sets A and B, $(A - B) \cup (A \cap B) = A$.

Simplify the expressions in 26–28 using Theorems 5.2.1– 5.2.3 and 5.3.3.

H26. $A \cap ((B \cup A^c) \cap B^c)$

27. $(A - (A \cap B)) \cap (B - (A \cap B))$

28. $((A \cap (B \cup C)) \cap (A - B)) \cap (B \cup C^c)$

29. In Example 5.3.4 an element proof was given that for all sets A and B, $A - B$ and B are disjoint. Give an algebraic proof of this result.

30. In Example 5.3.2 an element proof was given that for all sets A, $A \cap \emptyset = \emptyset$. Use Theorem 5.2.1(1)(b) to write an alternative proof of this result.

31. Derive the set identity $A \cup (A \cap B) = A$ from the properties listed in Theorem 5.2.2(1)–(5). Start by showing that for all subsets B of a universal set U, $U \cup B = U$. Then intersect both sides with A and deduce the identity.

32. Derive the set identity $A \cap (A \cup B) = A$ from the properties listed in Theorem 5.2.2(1)–(5) and Theorem 5.3.3. Start by showing that for all subsets B of a universal set U, $\emptyset = \emptyset \cap B$. Then take the union of both sides with A and deduce the identity.

33. Given any sets A and B, define the **symmetric difference** of A and B, denoted $A \oplus B$, as follows:

$$A \oplus B = (A - B) \cup (B - A).$$

Prove each of the following for all sets A, B, and C in a universal set U.
   a. $A \oplus B = B \oplus A$
   ◆b. $A \oplus (B \oplus C) = (A \oplus B) \oplus C$
   c. $A \oplus \emptyset = A$
   d. $A \oplus A^c = U$
   e. $A \oplus A = \emptyset$
   f. If $A \oplus C = B \oplus C$, then $A = B$.

34. Suppose A, B, and C are sets.
   Ha. Are $A - B$ and $B - C$ necessarily disjoint? Explain.
   Hb. Are $A - B$ and $C - B$ necessarily disjoint? Explain.
   c. Are $A - (B \cup C)$ and $B - (A \cup C)$ necessarily disjoint? Explain.
   d. Are $A - (B \cap C)$ and $B - (A \cap C)$ necessarily disjoint? Explain.

35. a. Is $\{\{a, d, e\}, \{b, c\}, \{d, f\}\}$ a partition of $\{a, b, c, d, e, f\}$?
    b. Is $\{\{w, x, v\}, \{u, y, q\}, \{p, z\}\}$ a partition of $\{p, q, u, v, w, x, y, z\}$?
    c. Is $\{\{5, 4\}, \{7, 2\}, \{1, 3, 4\}, \{6, 8\}\}$ a partition of $\{1, 2, 3, 4, 5, 6, 7, 8\}$?
    d. Is $\{\{3, 7, 8\}, \{2, 9\}, \{1, 4, 5\}\}$ a partition of $\{1, 2, 3, 4, 5, 6, 7, 8, 9\}$?

e. Is $\{\{1, 5\}, \{4, 7\}, \{2, 8, 6, 3\}\}$ a partition of $\{1, 2, 3, 4, 5, 6, 7, 8\}$?

36. Let E be the set of all even integers and O the set of all odd integers. Is $\{E, O\}$ a partition of $\mathbf{Z}$, the set of all integers? Explain your answer.

37. Let $\mathbf{R}$ be the set of all real numbers. Is $\{\mathbf{R}^+, \mathbf{R}^-, \{0\}\}$ a partition of $\mathbf{R}$? Explain your answer.

38. Let $\mathbf{Z}$ be the set of all integers and let

$A_0 = \{n \in \mathbf{Z} \mid n = 4k, \text{ for some integer } k\},$
$A_1 = \{n \in \mathbf{Z} \mid n = 4k + 1, \text{ for some integer } k\},$
$A_2 = \{n \in \mathbf{Z} \mid n = 4k + 2, \text{ for some integer } k\}, \text{ and}$
$A_3 = \{n \in \mathbf{Z} \mid n = 4k + 3, \text{ for some integer } k\}.$

Is $\{A_0, A_1, A_2, A_3\}$ a partition of $\mathbf{Z}$? Explain your answer.

39. Let $\Sigma = \{0, 1\}$. Recall that Σ^* is the set of all strings over Σ and that Σ^n is the set of all strings over Σ of length n, where n is a nonnegative integer. Let m be a positive integer and let $A = \{s \in \Sigma^* \mid s \text{ has length no greater than } m\}$. Is $\{\Sigma^0, \Sigma^1, \Sigma^2, \ldots, \Sigma^m\}$ a partition of A? Explain your answer.

40. Suppose $A = \{1, 2\}$ and $B = \{2, 3\}$. Find each of the following.
   a. $\mathcal{P}(A \cap B)$　　b. $\mathcal{P}(A)$
   c. $\mathcal{P}(A \cup B)$　　d. $\mathcal{P}(A \times B)$

41. a. Suppose $A = \{1\}$ and $B = \{u, v\}$. Find $\mathcal{P}(A \times B)$.
   b. Suppose $X = \{a, b\}$ and $Y = \{x, y\}$. Find $\mathcal{P}(X \times Y)$.

42. a. Find $\mathcal{P}(\emptyset)$.
   b. Find $\mathcal{P}(\mathcal{P}(\emptyset))$.
   c. Find $\mathcal{P}(\mathcal{P}(\mathcal{P}(\emptyset)))$.

43. Determine which of the following statements are true and which are false. Prove each statement that is true and give a counterexample for each statement that is false.
   H a. For all sets A and B, $\mathcal{P}(A \cup B) = \mathcal{P}(A) \cup \mathcal{P}(B)$.
   b. For all sets A and B, $\mathcal{P}(A \cap B) = \mathcal{P}(A) \cap \mathcal{P}(B)$.
   c. For all sets A and B, $\mathcal{P}(A) \cup \mathcal{P}(B) \subseteq \mathcal{P}(A \cup B)$.
   d. For all sets A and B, $\mathcal{P}(A \times B) = \mathcal{P}(A) \times \mathcal{P}(B)$.

44. Let $S = \{a, b, c\}$ and for each integer $i = 0, 1, 2, 3$, let S_i be the set of all subsets of S that have i elements. List the elements in S_0, S_1, S_2, and S_3. Is $\{S_0, S_1, S_2, S_3\}$ a partition of $\mathcal{P}(S)$?

45. Let $S = \{a, b, c\}$ and let S_a be the set of all subsets of S that contain a, S_b the set of all subsets of S that contain b, S_c the set of all subsets of S that contain c, and $S_\emptyset$ the set whose only element is $\emptyset$. Is $\{S_a, S_b, S_c, S_\emptyset\}$ a partition of $\mathcal{P}(S)$?

46. Let $A = \{t, u, v, w\}$ and let S_1 be the set of all subsets of A that do not contain w and S_2 the set of all subsets of A that contain w.
   a. Find S_1.
   b. Find S_2.
   c. Are S_1 and S_2 disjoint?
   d. Compare the sizes of S_1 and S_2.
   e. How many elements are in $S_1 \cup S_2$?
   f. What is the relation between $S_1 \cup S_2$ and $\mathcal{P}(A)$?

◆H 47. The following problem, devised by Ginger Bolton, appeared in the January 1989 issue of the *College Mathematics Journal* (Vol. 20, No. 1, p. 68): Given a positive integer $n \geq 2$, let S be the set of all nonempty subsets of $\{2, 3, \ldots, n\}$. For each $S_i \in S$, let P_i be the product of the elements of S_i. Prove or disprove that

$$\sum_{i=1}^{2^{n-1}-1} P_i = \frac{(n + 1)!}{2} - 1.$$

◆ 48. Suppose B is a Boolean algebra and prove each of the properties below. Proofs of some of the properties depend on the establishment of some of the previous properties. For instance, the easiest proof of the property in (b) uses the property in (a).
   a. For all $x \in B$, (i) $x \cdot x = x$, and (ii) $x + x = x$.
   b. For all x and y in B, if $x \cdot y = 1$ then $x = y = 1$.
   c. For all $x \in B$, (i) $x + 1 = 1$, and (ii) $x \cdot 0 = 0$.
   d. For all x and y in B, (i) $(x + y) \cdot x = x$, and (ii) $(x \cdot y) + x = x$.
   e. For all x, y, and z in B, if $x + y = x + z$ and $x \cdot y = x \cdot z$, then $y = z$.

5.4　RUSSELL'S PARADOX AND THE HALTING PROBLEM

From the paradise created for us by Cantor, no one will drive us out.
(David Hilbert 1862–1943)

By the beginning of the twentieth century abstract set theory had gained such wide acceptance that a number of mathematicians were working hard to show that all of mathematics could be built upon a foundation of set theory. In the midst of this activity the English mathematician and philosopher Bertrand Russell discovered a "paradox" (really a genuine contradiction) that seemed to shake the very core of the foundation. The paradox assumes Cantor's definition of set as "any collection into a whole of definite and separate objects of our intuition or our thought."

Bertrand Russell (1872–1970)

Russell's Paradox: Most sets are not elements of themselves. For instance, the set of all integers is not an integer and the set of all horses is not a horse. However, some sets *are* elements of themselves. For instance, the set of all abstract ideas is an abstract idea. Let S be the set of all sets that are not elements of themselves:

$$S = \{A \mid A \text{ is a set and } A \notin A\}.$$

Is S an element of itself?

The answer is neither yes nor no. For if $S \in S$, then S satisfies the defining property for S, and hence $S \notin S$. But if $S \notin S$, then S is a set such that $S \notin S$ and so S satisfies the defining property for S, which implies that $S \in S$. Thus neither is $S \in S$ nor is $S \notin S$, which is a contradiction.

To help explain his discovery to lay people, Russell devised a puzzle, the barber puzzle, whose solution exhibits the same logic as his paradox.

EXAMPLE 5.4.1   **The Barber Puzzle**

In a certain town there is a male barber who shaves all those men, and only those men, who do not shave themselves. *Question:* Does the barber shave himself?

Solution   Neither yes nor no. If the barber shaves himself, he is a member of the class of men who shave themselves. But no member of this class is shaved by the barber, and so the barber does *not* shave himself. On the other hand, if the barber does not shave himself, he belongs to the class of men who do not shave themselves. But the barber shaves every man in this class, so the barber *does* shave himself.   ■

But how can the answer be neither yes nor no? Surely any barber either does or does not shave himself. You might try to think of circumstances that would make the paradox disappear. For instance, maybe the barber happens to have no beard and never shaves. But a condition of the puzzle is that the barber is a man who shaves *all* those men who do not shave themselves. If he does not shave then he does not shave himself, in which case he is shaved by the barber and the contradiction is as present as ever. Similarly, other attempts at resolution of the paradox by considering details of the barber's situation are doomed to failure.

So let's accept the fact that the paradox has no easy resolution and see where that thought leads. Since the barber neither shaves himself nor doesn't shave himself, the sentence "The barber shaves himself" is neither true nor false. But the sentence arose in a natural way from a description of a situation. If the situation actually existed, then the sentence would have to be true or false. So we are forced to conclude that the situation described in the puzzle simply cannot exist in the world as we know it.

In a similar way the conclusion to be drawn from Russell's paradox itself is that the object S is not a set. Because if it actually were a set, in the sense of satisfying the general properties of sets we have been assuming, then it either would be an element of itself or not.

In the years following Russell's discovery, several ways were found to define the basic concepts of set theory so as to avoid his contradiction. The way used in this text requires that, except for the power set whose existence is guaranteed by an axiom, whenever a set is defined using a predicate as a defining property, the stipulation must also be made that the set is a subset of a known set. This method does not allow us to talk about "the set of all sets that are not elements of themselves." We can only speak of "the set of all sets that are subsets of some known set and that are not elements of themselves." When this restriction is made, Russell's paradox ceases to be contradictory. Here is what happens:

Let U be a set of sets and let $S = \{A \in U \mid A \notin A\}$. Is $S \in S$?

The answer is simply no. For if $S \in S$, then S satisfies the defining property for S, and hence $S \notin S$.

Comment: In Russell's paradox both implications

$$S \in S \to S \notin S \quad \text{and} \quad S \notin S \to S \in S$$

are proved and the contradictory conclusion

$$\text{neither } S \in S \quad \text{nor} \quad S \notin S$$

is therefore deduced. When the definition of S is changed as above, only the implication $S \in S \to S \notin S$ can be proved (as was done above). This leads to the conclusion that $S \notin U$, which is not contradictory.

Kurt Gödel (1906–1978)

Russell's discovery had a profound impact on mathematics because even though his contradiction could be made to disappear by more careful definitions, its existence caused people to wonder whether other contradictions remained. In 1931 Kurt Gödel showed that it is not possible to prove in a mathematically rigorous way that mathematics is free of contradictions. You might think that Gödel's result would have caused mathematicians to give up their work in despair, but that has not happened. On the contrary, there has been more mathematical activity since 1931 than in any other period in history.

THE HALTING PROBLEM

Alan M. Turing (1912–1954)

Well before the actual construction of an electronic computer, Alan M. Turing deduced a profound theorem about how such computers would have to work. The argument he used is similar to that in Russell's paradox. It is also related to those used by Gödel to prove his theorem and by Cantor to prove that it is impossible to write all the real numbers in an infinitely long list, even given an infinitely long period of time (see Section 7.6).

If you have some experience programming computers, you know how badly an infinite loop can tie up a computer system. It would be useful to be able to preprocess a program and its data set by running it through a checking program that determines whether execution of the given program with the given data set would result in an infinite loop. Can an algorithm for such a program be written? In other words, can an algorithm be written that will accept any algorithm X and any data set D as input and will then print "halts" or "loops forever" to indicate whether X terminates in a finite number of steps or loops forever when run with data set D? In the 1930s Turing proved that the answer to this question is no.

THEOREM 5.4.1

There is no computer algorithm that will accept any algorithm X and data set D as input and then will output "halts" or "loops forever" to indicate whether X terminates in a finite number of steps when X is run with data set D.

Proof (by contradiction):

Suppose there is an algorithm, CheckHalt, such that if an algorithm X and a data set D are input,

CheckHalt(X, D) prints

"halts"
if X terminates in a finite number of steps when run with data set D

or

"loops forever"
if X does not terminate in a finite number of steps when run with data set D.

[To show that no algorithm such as CheckHalt can exist, we will deduce a contradiction.]

Observe that the sequence of characters making up an algorithm X can be regarded as a data set itself. Thus it is possible to consider running Check-Halt with input (X, X). Define a new algorithm, Test, as follows: For any input algorithm X,

Test(X)

loops forever if CheckHalt(X, X) prints "halts"

or

stops if CheckHalt(X, X) prints "loops forever".

Now run algorithm Test with input Test. If Test(Test) terminates after a finite number of steps, then the value of CheckHalt(Test, Test) is "halts", and so Test(Test) loops forever.

On the other hand, if Test(Test) does not terminate after a finite number of steps, then CheckHalt(Test, Test) prints "loops forever", and so Test(Test) terminates.

The two paragraphs above show that Test(Test) loops forever and also that it terminates. This is a contradiction. But the existence of Test follows logically from the supposition of the existence of an algorithm CheckHalt that can check any algorithm and data set for termination. *[Hence the supposition must be false, and there is no such algorithm.]*

In recent years the axioms for set theory which guarantee that Russell's paradox will not arise have been found inadequate to deal with the full range of recursively defined objects in computer science, and a new theory of "non-well-founded" sets has been developed. In addition, computer scientists and logicians working on programs to enable computers to process natural language have seen the importance of exploring further the kind of semantic issues raised by the barber puzzle and are developing new theories of logic to deal with them.

EXERCISE SET 5.4

In 1–5 determine whether each sentence is a statement. Explain your answers.

1. This sentence is false.

2. If $1 + 1 = 3$, then $1 = 0$.

3. | The sentence in this box is a lie. |

4. All real numbers with negative squares are prime.

5. This sentence is false or $1 + 1 = 3$.

6. a. Assuming that the following sentence is a statement, prove that $1 + 1 = 3$:

   If this sentence is true, then $1 + 1 = 3$.

   b. What can you deduce from part (a) about the status of "This sentence is true"? Why? (This example is known as **Löb's paradox**.)

H 7. The following two sentences were devised by the logician Saul Kripke. While not intrinsically paradoxical, they could be paradoxical under certain circumstances. Describe such circumstances.

   (i) Most of Nixon's assertions about Watergate are false.

   (ii) Everything Jones says about Watergate is true.

   (*Hint:* Suppose Nixon says (ii) and the only utterance Jones makes about Watergate is (i).)

8. Can there exist a computer program that has as output a list of all the computer programs that do not list themselves in their output? Explain your answer.

9. Can there exist a book that refers to all those books and only those books that do not refer to themselves? Explain your answer.

10. Some English adjectives are descriptive of themselves (for instance, the world *polysyllabic* is polysyllabic) whereas others are not (for instance, the world *monosyllabic* is not monosyllabic). The word *heterological* refers to an adjective that does not describe itself. Is *heterological* heterological? Explain your answer.

11. As strange as it may seem, it is possible to give a precise-looking verbal definition of an integer that, in fact, is not a definition at all. The following was devised by an English librarian, G. G. Berry, and reported by Bertrand Russell. Explain how it leads to a contradiction. Let n be "the smallest integer not describable in fewer than 12 English words." (Note that the total number of strings consisting of 11 or fewer English words is finite.)

H 12. Is there an algorithm which, for a fixed quantity a and any input algorithm X and data set D, can determine whether X prints a when run with data set D? Explain. (This problem is called the **printing problem**.)

6 Counting

"It's as easy as 1–2–3."

That's the saying. And in certain ways, it's true. But other aspects of counting aren't so simple. Have you ever agreed to meet a friend "in three days" and then realized that you and your friend might mean different things? For example, on the European continent, to meet in eight days means to meet on the same day as today one week hence; on the other hand, in English-speaking countries, to meet in seven days means to meet one week hence. The difference is that on the continent, all days including the first and the last are counted. In the English-speaking world, it's the number of 24-hour periods that are counted.

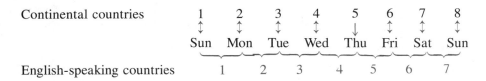

Continental countries	1	2	3	4	5	6	7	8
	Sun	Mon	Tue	Wed	Thu	Fri	Sat	Sun
English-speaking countries		1	2	3	4	5	6	7

The English convention for counting days follows the almost universal convention for counting hours. If it is 9 A.M. and two people anywhere in the world agree to meet in three hours, they mean that they will get back together again at 12 noon.

Musical intervals, on the other hand, are universally reckoned the way the Continentals count the days of a week. An interval of a third consists of two tones with a single tone in between, and an interval of a second consists of two adjacent tones. (See Figure 6.1.1.)

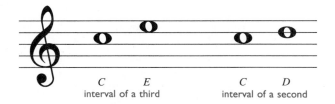

C   E   C   D
interval of a third   interval of a second

FIGURE 6.1.1

Of course, the complicating factor in all these examples is not how to count but rather what to count. And, indeed, in the more complex mathematical counting problems discussed in this chapter, it is what to count that is the central issue. Once one knows exactly what to count, the counting itself is as easy as $1-2-3$.

Reprinted by permission of UFS, Inc.

6.1 COUNTING AND PROBABILITY

Imagine tossing two coins and observing whether 0, 1, or 2 heads are obtained. It would be natural to guess that each of these events occurs about one-third of the time, but in fact this is not the case. Table 6.1.1 below shows actual data obtained from tossing two quarters 50 times.

TABLE 6.1.1   Experimental Data Obtained from Tossing Two Quarters 50 Times

Event	Tally	Frequency (Number of times the event occurred)	Relative Frequency (Fraction of times the event occurred)
2 heads obtained	⊞⊞⊞ ⊞⊞⊞ I	11	22%
1 head obtained	⊞⊞⊞ ⊞⊞⊞ ⊞⊞⊞ ⊞⊞⊞ ⊞⊞⊞ II	27	54%
0 heads obtained	⊞⊞⊞ ⊞⊞⊞ II	12	24%

As you can see, the relative frequency of obtaining exactly 1 head was roughly twice as great as that of obtaining either 2 heads or 0 heads. It turns out that the mathematical theory of probability can be used to predict that a result like this will almost always occur. To see how, call the two coins A and B, and suppose that each is perfectly balanced. Then each has an equal chance of coming up heads or tails, and when the two are tossed together the four outcomes pictured in Figure 6.1.2 are all equally likely.

Figure 6.1.2 shows that there is a 1 in 4 chance of obtaining two heads and a 1 in 4 chance of obtaining no heads. The chance of obtaining one head, however, is 2 in 4 because either A could come up heads and B tails or B could come up heads and A tails. So if you repeatedly toss two balanced coins and record the number of heads, you should expect relative frequencies similar to those shown in Table 6.1.1.

To formalize this analysis and extend it to more complex situations, we introduce the notions of random process, sample space, event, and probability. To say that a

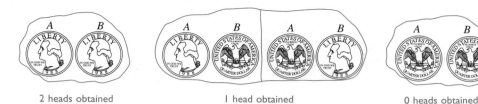

| 2 heads obtained | 1 head obtained | 0 heads obtained |

FIGURE 6.1.2   Equally Likely Outcomes from Tossing Two Balanced Coins

process is **random** means that when it takes place, one outcome from some set of outcomes is sure to occur but it is impossible to predict with certainty which outcome that will be. For instance, if an ordinary person performs the experiment of tossing an ordinary coin into the air and allowing it to fall flat on the ground, it can be predicted with certainty that the coin will land either heads up or tails up (so the set of outcomes can be denoted {heads, tails}), but it is not known for sure whether heads or tails will occur. We restricted this experiment to ordinary people because a skilled magician can toss a coin in a way that appears random but is not, and a physicist equipped with first-rate measuring devices may be able to analyze all the forces on the coin and correctly predict its landing position. Just a few of many examples of random processes or experiments are choosing winners in state lotteries, respondents in public opinion polls, and subjects to receive treatments or serve as controls in medical experiments. The set of outcomes that can result from a random process or experiment is called a *sample space.*

Andrei Nikolaevich Kolmogorov
(1903–1987)

DEFINITION

A **sample space** is the set of all possible outcomes of a random process or experiment. An **event** is a subset of a sample space.

In case an experiment has finitely many* outcomes and all outcomes are equally likely to occur, the *probability* of an event (set of outcomes) is just the ratio of the number of outcomes in the event to the total number of outcomes. Strictly speaking, this result can be deduced from a set of axioms for probability formulated in 1933 by the Russian mathematician Andrei Nikolaevich Kolmogorov. However, taking a naïve approach to probability, we simply state it as a principle.

EQUALLY LIKELY PROBABILITY FORMULA

If S is a finite sample space in which all outcomes are equally likely and E is an event in S, then the **probability of E**, denoted $P(E)$, is

$$P(E) = \frac{\text{the number of outcomes in } E}{\text{the total number of outcomes in } S}.$$

*In Section 7.3 the concepts of finite and infinite are defined formally.

NOTATION

For any finite set, $n(A)$ denotes the number of elements in A.

With this notation, the equally likely probability formula becomes

$$P(E) = \frac{n(E)}{n(S)}.$$

EXAMPLE 6.1.1   Probabilities for a Deck of Cards

An ordinary deck of cards contains 52 cards divided into four *suits*. The *red suits* are diamonds (♦) and hearts (♥) and the *black suits* are clubs (♣) and spades (♠). Each suit contains 13 cards of the following *denominations*: 2, 3, 4, 5, 6, 7, 8, 9, 10, J (jack), Q (queen), K (king), and A (ace). The cards J, Q, K, and A are called *face cards*. The American mathematician Persi Diaconis recently proved that seven shuffles are needed to thoroughly mix up the cards in a sorted deck. Imagine such a shuffled deck with the cards turned over so their values are hidden, and suppose you pick one card from the deck at random. (This means that you are as likely to pick any one card as any other.)

a. What is the sample space of outcomes?
b. What is the event that the chosen card is a black face card?
c. What is the probability that the chosen card is a black face card?

Solution   a. The outcomes in the sample space S are the 52 cards in the deck.
b. Let E be the event that a black face card is chosen. The outcomes in E are the jack, queen, king, and ace of spades and jack, queen, king, and ace of clubs. Symbolically,

$$E = \{J\clubsuit, Q\clubsuit, K\clubsuit, A\clubsuit, J\spadesuit, Q\spadesuit, K\spadesuit, A\spadesuit\}.$$

c. By part (b), $n(E) = 8$, and according to the description of the situation, all 52 outcomes in the sample space are equally likely. Therefore, by the equally likely probability formula, the probability that the chosen card is a black face card is

$$P(E) = \frac{n(E)}{n(S)} = \frac{8}{52} \cong 15.4\%. \qquad \blacksquare$$

EXAMPLE 6.1.2   Rolling a Pair of Dice

A die is one of a pair of dice. It is a cube with six sides, each containing from one to six dots, called *pips*. Suppose a blue die and a grey die are tossed together, and the numbers of dots that occur face up on each are recorded. The possible outcomes can be listed as follows where in each case the die on the left is blue and the one on the right is grey. A more compact notation identifies, say, [·] [··] with the string 24, [··] [··] with 53, and so forth.

a. Use the compact notation to write the sample space S of possible outcomes.
b. Use set notation to write the event E that the numbers showing face up have a sum of 6.
c. What is the probability that the numbers showing face up have a sum of 6?

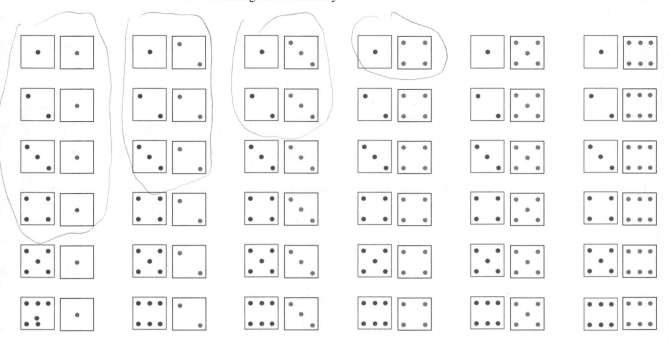

Solution
a. $S = \{11, 12, 13, 14, 15, 16, 21, 22, 23, 24, 25, 26, 31, 32, 33, 34, 35, 36,$
$41, 42, 43, 44, 45, 46, 51, 52, 53, 54, 55, 56, 61, 62, 63, 64, 65, 66\}.$

b. $E = \{15, 24, 33, 42, 51\}.$

c. The probability that the sum of the numbers is $6 = P(E) = \dfrac{n(E)}{n(S)} = \dfrac{5}{36}.$ ∎

COUNTING THE ELEMENTS OF A LIST

Some counting problems are as simple as counting the elements of a list. For instance, how many integers are there from 5 through 12? To answer this question, imagine going along the list of integers from 5 to 12, counting each in turn.

list:	5	6	7	8	9	10	11	12
	↕	↕	↕	↕	↕	↕	↕	↕
count:	1	2	3	4	5	6	7	8

So the answer is 8.

More generally, if m and n are integers and $m \leq n$, how many integers are there from m through n? To answer this question, note that $n = m + (n - m)$, where $n - m \geq 0$ [since $n \geq m$]. Note, also, that the element $m + 0$ is the first element of the list, the element $m + 1$ is the second element, the element $m + 2$ is the third, and so forth. In general, the element $m + i$ is the $(i + 1)$st element of the list.

list:	$m (= m + 0)$	$m + 1$	$m + 2$	$\ldots,$	$n (= m + (n - m))$
	↕	↕	↕		↕
count:	1	2	3	$\ldots$	$(n - m) + 1$

And so the number of elements in the list is $n - m + 1$.

This general result is important enough to be restated as a theorem, the formal proof of which uses mathematical induction. (See exercise 23 at the end of this section.) The heart of the proof is the observation that if the list $m, m + 1, \ldots, k$ has $k - m + 1$ numbers, then the list $m, m + 1, \ldots, k, k + 1$ has $(k - m + 1) + 1 = (k + 1) - m + 1$ numbers.

THEOREM 6.1.1   The Number of Elements in a List

If m and n are integers and $m \leq n$, then there are $n - m + 1$ integers from m to n inclusive.

EXAMPLE 6.1.3   Counting the Elements of a Sublist

a. How many three-digit integers (integers from 100 to 999 inclusive) are divisible by 5?
b. What is the probability that a randomly chosen three-digit integer is divisible by 5?

Solution
a. Imagine writing the three-digit integers in a row, noting those that are multiples of 5 and drawing arrows between each such integer and its corresponding multiple of 5.

100 101 102 103 104 105 106 107 108 109 110 $\cdots$ 994 995 996 997 998 999
 $\updownarrow$ $\updownarrow$ $\updownarrow$ $\updownarrow$
5 · 20   5 · 21   5 · 22   5 · 199

From the sketch it is clear that there are as many three-digit integers that are multiples of 5 as there are integers from 20 to 199 inclusive. By Theorem 6.1.1, there are $199 - 20 + 1$, or 180 such integers. Hence there are 180 integers that are divisible by 5.

b. By Theorem 6.1.1 the total number of integers from 100 through 999 is $999 - 100 + 1 = 900$. By part (a), 180 of these are divisible by 5. Hence the probability that a randomly chosen three-digit integer is divisible by 5 is $180/900 = 1/5$. ∎

EXAMPLE 6.1.4   Counting Elements of a One-Dimensional Array

Analysis of many computer algorithms requires skill at counting the elements of a one-dimensional array. Let $A[1], A[2], \ldots, A[n]$ be a one-dimensional array, where n is a positive integer.

a. Suppose the array is cut at a middle value $A[m]$ so that two subarrays are formed:

(1) $A[1], A[2], \ldots, A[m]$ and   (2) $A[m + 1], A[m + 2], \ldots, A[n]$.

How many elements does each subarray have?
b. What is the probability that a randomly chosen element of the array has an even subscript

(i) if n is even?   (ii) if n is odd?

Solution
a. Array (1) has the same number of elements as the list of integers from 1 through m. So by Theorem 6.1.1, it has m, or $m - 1 + 1$, elements. Array (2) has the same number of elements as the list of integers from $m + 1$ through n. So by Theorem 6.1.1, it has $n - m$, or $n - (m + 1) + 1$, elements.

b. (i) If n is even, each even subscript starting with 2 and ending with n can be matched up with an integer from 1 to $n/2$.

$$
\begin{array}{cccccccccccc}
1 & 2 & 3 & 4 & 5 & 6 & 7 & 8 & 9 & 10 & \cdots & n \\
& \updownarrow & & \updownarrow & & \updownarrow & & \updownarrow & & \updownarrow & & \updownarrow \\
& 2 \cdot 1 & & 2 \cdot 2 & & 2 \cdot 3 & & 2 \cdot 4 & & 2 \cdot 5 & & 2 \cdot n/2
\end{array}
$$

So there are $n/2$ array elements with even subscripts. Since the entire array has n elements, the probability that a randomly chosen element has an even subscript is $\dfrac{n/2}{n} = \dfrac{1}{2}$.

(ii) If n is odd, then the greatest even subscript of the array is $n - 1$. So there are as many even subscripts between 1 and n as there are from 2 through $n - 1$. Then the reasoning of (i) can be used to conclude that there are $(n - 1)/2$ array elements with even subscripts.

$$
\begin{array}{ccccccccc}
1 & 2 & 3 & 4 & 5 & 6 & \cdots & n - 1 & n \\
& \updownarrow & & \updownarrow & & \updownarrow & & \updownarrow & \\
& 2 \cdot 1 & & 2 \cdot 2 & & 2 \cdot 3 & \cdots & 2 \cdot (n - 1)/2 &
\end{array}
$$

Since the entire array has n elements, the probability that a randomly chosen element has an even subscript is $\dfrac{(n - 1)/2}{n} = \dfrac{n - 1}{2n}$. Observe that as n gets larger and larger, this probability gets closer and closer to $1/2$.

Note that the answers to (i) and (ii) can be combined using the floor notation. By Theorem 3.5.1, the number of array elements with even subscripts is $\lfloor n/2 \rfloor$, and so the probability that a randomly chosen element has an even subscript is $\dfrac{\lfloor n/2 \rfloor}{n}$. ■

EXERCISE SET 6.1*

1. Toss two coins 30 times and make a table showing the relative frequencies of 0, 1, and 2 heads. How do your values compare with those shown in Table 6.1.1?

2. In the example of tossing two quarters, what is the probability that at least one head is obtained? that coin A is a head? that coins A and B are either both heads or both tails?

3. For the sample space given in Example 6.1.1, write each of the following events as a set and compute its probability:
   a. The event that the chosen card is red and is not a face card.
   b. The event that the denomination of the chosen card is at least 10 (counting aces high).

4. For the sample space given in Example 6.1.2, write each of the following events as a set and compute its probability:
   a. The event that the sum of the numbers showing face up is 8.
   b. The event that the numbers showing face up are the same.

c. The event that the sum of the numbers showing face up is at most 5.

5. Suppose that a coin is tossed three times and the side showing face up on each toss is noted. Suppose also that on each toss heads and tails are equally likely. Let HHT indicate the outcome heads on the first two tosses and tails on the third, THT the outcome tails on the first and third tosses and heads on the second, and so forth.
   a. List the eight elements in the sample space whose outcomes are all the possible head–tail sequences obtained in the three tosses.
   b. Write each of the following events as a set and find its probability:
      (i) The event that exactly one toss results in a head.
      (ii) The event that at least two tosses result in a head.
      (iii) The event that no head is obtained.

6. Suppose that each child born is equally likely to be a boy or a girl. Consider a family with exactly three chil-

*Exercises with blue numbers have solutions in Appendix B. The symbol H indicates that only a hint or partial solution is given. The symbol ◆ indicates that an exercise is more challenging than usual.

dren. Let *BBG* indicate that the first two children born are boys and the third child is a girl, let *GBG* indicate that the first and third children born are girls and the second is a boy, and so forth.

a. List the eight elements in the sample space whose outcomes are all possible genders of the three children.

b. Write each of the following events as a set and find its probability:

(i) The event that exactly one child is a girl.
(ii) The event that at least two children are girls.
(iii) The event that no child is a girl.

7. Suppose that on a true/false exam you have no idea at all about the answers to three questions. You choose answers randomly and therefore have a 50–50 chance of being correct on any one question. Let *CCW* indicate that you were correct on the first two questions and wrong on the third, let *WCW* indicate that you were wrong on the first and third questions and correct on the second, and so forth.

a. List the elements in the sample space whose outcomes are all possible sequences of correct and incorrect responses on your part.

b. Write each of the following events as a set and find its probability:

(i) The event that exactly one answer is correct.
(ii) The event that at least two answers are correct.
(iii) The event that no answer is correct.

8. Three people have been exposed to a certain illness. Once exposed, a person has a 50–50 chance of actually becoming ill.

a. What is the probability that exactly one of the people becomes ill?

b. What is the probability that at least two of the people become ill?

c. What is the probability that none of the three people becomes ill?

9. When discussing counting and probability, we often consider situations that may appear frivolous or of little practical value, such as tossing coins, choosing cards, or rolling dice. The reason is that these relatively simple examples serve as models for a wide variety of more complex situations in the real world. In light of this remark, comment on the relationship between your answer to exercise 5 and your answers to exercises 6–8.

10. a. How many positive two-digit integers are multiples of 3?

b. What is the probability that a randomly chosen positive two-digit integer is a multiple of 3?

11. a. How many positive three-digit integers are multiples of 6?

b. What is the probability that a randomly chosen positive three-digit integer is a multiple of 6?

12. Suppose $A[1]$, $A[2]$, $A[3]$, . . . , $A[n]$ is a one-dimensional array and $n \geq 20$.

a. How many elements are in the array?

b. How many elements are in the subarray

$$A[4], A[5], \ldots , A[19]?$$

c. If $3 \leq m \leq n$, what is the probability that a randomly chosen array element is in the subarray

$$A[3], A[4], \ldots , A[m]?$$

d. What is the probability that a randomly chosen array element is in the subarray

$$A[\lfloor 19/2 \rfloor], A[\lfloor 19/2 \rfloor + 1], \ldots A[19]?$$

13. Suppose $A[1]$, $A[2]$, . . . , $A[n]$ is a one-dimensional array and $n \geq 2$. What is the probability that a randomly chosen array element is in the subarray

$$A[1], A[2], \ldots , A[\lfloor n/2 \rfloor]$$

a. if n is even?   b. if n is odd?

14. Suppose $A[1]$, $A[2]$, . . . , $A[n]$ is a one-dimensional array and $n \geq 2$. What is the probability that a randomly chosen array element is in the subarray

$$A[\lfloor n/2 \rfloor], A[\lfloor n/2 \rfloor + 1], \ldots , A[n]$$

a. if n is even?   b. if n is odd?

15. What is the 27th element in the one-dimensional array $A[42]$, $A[43]$, . . . , $A[100]$?

16. What is the 62nd element in the one-dimensional array $B[29]$, $B[30]$, . . . , $B[100]$?

17. If the largest of 56 consecutive integers is 279, what is the smallest?

18. If the largest of 87 consecutive integers is 326, what is the smallest?

19. How many even integers are there between 1 and 1,001?

20. How many integers are there between 1 and 1,001 that are multiples of 3?

21. A non-leap year has 365 days. Assuming January 1 is a Monday:

a. How many Sundays are there in the year?
b. How many Mondays are there in the year?

◆ 22. What is the largest number of elements that a set of integers from 1 through 100 can have so that no one element in the set is divisible by another? (*Hint:* Imagine writing all the numbers from 1 through 100 in the form $2^k \cdot m$, where $k \geq 0$ and m is odd.)

◆ 23. Prove Theorem 6.1.1. (Let m be any integer and prove the theorem by mathematical induction on n.)

6.2 POSSIBILITY TREES AND THE MULTIPLICATION RULE

Don't believe anything unless you have thought it through for yourself.
(Anna Pell Wheeler, 1883–1966)

A tree structure is a useful tool for keeping systematic track of all possibilities in situations in which events happen in order. The following example shows how to use such a structure to count the number of different outcomes of a tournament.

EXAMPLE 6.2.1 Possibilities for Tournament Play

Teams A and B are to play each other repeatedly until one wins two games in a row or a total of three games. One way in which this tournament can be played is for A to win the first game, B to win the second, and A to win the third and fourth games. Denote this by writing A–B–A–A.

a. How many ways can the tournament be played?
b. Assuming that all the ways of playing the tournament are equally likely, what is the probability that five games are needed to determine the tournament winner?

Solution a. The possible ways for the tournament to be played are represented by the distinct paths from "root" (the start) to "leaf" (a terminal point) in the tree shown sideways in Figure 6.2.1. The label on each branching point indicates the winner of the game. The notations in parentheses indicate the winner of the tournament.

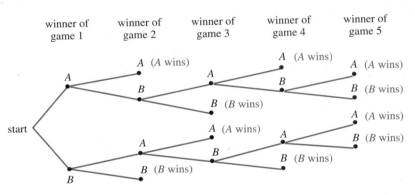

FIGURE 6.2.1   The Outcomes of a Tournament

The fact that there are ten paths from the root of the tree to its leaves shows that there are ten possible ways for the tournament to be played. They are (moving from the top down): A–A, A–B–A–A, A–B–A–B–A, A–B–A–B–B, B–A–A, A–B–B, B–A–B–A–A, B–A–B–A–B, B–A–B–B. and B–B. In five cases A wins, and in the other five B wins. The least number of games that must be played to determine a winner is two, and the most that will need to be played is five.

b. Since all the possible ways of playing the tournament listed in part (a) are assumed to be equally likely and the listing shows that five games are needed in four different cases (A–B–A–B–A, A–B–A–B–B, B–A–B–A–B, and B–A–B–A–A), the probability that five games are needed is $4/10 = 2/5$. ■

THE MULTIPLICATION RULE

Consider the following example. Suppose a computer installation has four input/output units (A, B, C, and D) and three central processing units (X, Y, and Z). Any input/output unit can be paired with any central processing unit. How many ways are there to pair an input/output unit with a central processing unit?

To answer this question, imagine the pairing of the two types of units as a two-step operation:

Step 1 is to choose the input/output unit.

Step 2 is to choose the central processing unit.

The possible outcomes of this operation are illustrated in the possibility tree of Figure 6.2.2.

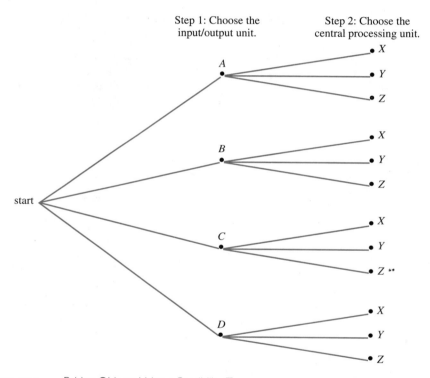

FIGURE 6.2.2   Pairing Objects Using a Possibility Tree

The top-most path from "root" to "leaf" indicates that input/output unit A is to be paired with central processing unit X. The next lower branch indicates that input/output unit A is to be paired with central processing unit Y. And so forth.

Thus the total number of ways to pair the two types of units is the same as the number of branches of the tree, which is

$$3 + 3 + 3 + 3 = 4 \cdot 3 = 12.$$

The idea behind this example can be used to prove the following rule. A formal proof uses mathematical induction and is left to the exercises.

THEOREM 6.2.1   The Multiplication Rule

If an operation consists of k steps and

the first step can be performed in n_1 ways,

the second step can be performed in n_2 ways (regardless of how the first step was performed),

$$\vdots$$

the kth step can be performed in n_k ways (regardless of how the preceding steps were performed),

then the entire operation can be performed in $n_1 \cdot n_2 \cdot \ldots \cdot n_k$ ways.

To apply the multiplication rule, think of the objects you are trying to count as the output of a multistep operation. The possible ways to perform a step may depend on how preceding steps were performed, but the *number* of ways to perform each step must be constant regardless of the action taken in prior steps.

EXAMPLE 6.2.2   Number of Personal Identification Numbers (PINs)

A typical PIN (personal identification number) is a sequence of any four symbols chosen from the 26 letters in the alphabet and the ten digits, with repetition allowed. How many different PINs are possible?

Solution   Typical PINs are CARE, 3387, B32B, and so forth. You can think of forming a PIN as a four-step operation.

Step 1 is to choose the first symbol.

Step 2 is to choose the second symbol.

Step 3 is to choose the third symbol.

Step 4 is to choose the fourth symbol.

There is a fixed number of ways to perform each step, namely 36, regardless of how preceding steps were performed. And so by the multiplication rule, there are $36 \cdot 36 \cdot 36 \cdot 36 = 36^4 = 1{,}679{,}616$ PINs in all.   ∎

Another way to look at the PINs of Example 6.2.2 is as ordered 4-tuples. For example, you can think of the PIN M2ZM as the ordered 4-tuple (M, 2, Z, M). Therefore, the total number of PINs is the same as the total number of ordered 4-tuples whose elements are either letters of the alphabet or digits. One of the most important uses of the multiplication rule is to derive a general formula for the number of elements in any Cartesian product of a finite number of finite sets. In Example 6.2.3, this is done for a Cartesian product of four sets.

EXAMPLE 6.2.3   The Number of Elements in a Cartesian Product

Suppose $A_1, A_2, A_3,$ and A_4 are sets with $n_1, n_2, n_3,$ and n_4 elements respectively. Show that the set $A_1 \times A_2 \times A_3 \times A_4$ has $n_1 \cdot n_2 \cdot n_3 \cdot n_4$ elements.

Solution   Each element in $A_1 \times A_2 \times A_3 \times A_4$ is an ordered 4-tuple of the form (a_1, a_2, a_3, a_4) where $a_1 \in A_1, a_2 \in A_2, a_3 \in A_3,$ and $a_4 \in A_4$. Imagine the process of constructing these ordered tuples as a four-step operation:

> Step 1 is to choose the first element of the 4-tuple.
>
> Step 2 is to choose the second element of the 4-tuple.
>
> Step 3 is to choose the third element of the 4-tuple.
>
> Step 4 is to choose the fourth element of the 4-tuple.

There are n_1 ways to perform step 1, n_2 ways to perform step 2, n_3 ways to perform step 3, and n_4 ways to perform step 4. Hence by the multiplication rule, there are $n_1 \cdot n_2 \cdot n_3 \cdot n_4$ ways to perform the entire operation. Therefore there are $n_1 \cdot n_2 \cdot n_3 \cdot n_4$ distinct 4-tuples in $A_1 \times A_2 \times A_3 \times A_4$.   ∎

EXAMPLE 6.2.4   Number of PINs Without Repetition

In Example 6.2.2 PINs were formed using four symbols, either letters of the alphabet or digits, and supposing that letters could be repeated. Now suppose that repetition is not allowed.

a. How many different PINs are there?
b. If all PINs are equally likely, what is the probability that a PIN chosen at random contains no repeated symbol?

Solution   a. Again think of forming a PIN as a four-step operation: Choose the first symbol, then the second, then the third, and then the fourth. There are 36 ways to choose the first symbol, 35 ways to choose the second (since the first symbol cannot be used again), 34 ways to choose the third (since the first two symbols cannot be reused), and 33 ways to choose the fourth (since the first three symbols cannot be reused). Thus, the multiplication rule can be applied to conclude that there are $36 \cdot 35 \cdot 34 \cdot 33 = 1,413,720$ different PINs with no repeated symbol.

b. By part (a) there are 1,413,720 PINs with no repeated symbol and by Example 6.2.2 there are 1,679,616 PINs in all. Thus the probability that a PIN chosen at random contains no repeated symbol is $\frac{1,413,720}{1,679,616} \cong .8417$. In other words, approximately 84% of PINs have no repeated symbol.   ∎

Any circuit with two input signals P and Q has an input/output table consisting of four rows corresponding to the four possible assignments of values to P and Q: 11, 10, 01, and 00. The next example shows that there are only 16 distinct ways in which such a circuit can function.

EXAMPLE 6.2.5   Number of Input/Output Tables for a Circuit with Two Input Signals

Consider the set of all circuits with two input signals P and Q. For each such circuit an input/output table can be constructed, but, as shown in Section 1.4, two such input/output tables may have the same values. How many distinct input/output tables can be constructed for circuits with input/output signals P and Q?

Solution   Fix the order of the input values for P and Q. Then two input/output tables are distinct if their output values differ in at least one row. For example, the input/output tables shown below are distinct since their output values differ in the first row.

P	Q	Output		P	Q	Output
1	1	1		1	1	0
1	0	0		1	0	0
0	1	1		0	1	1
0	0	0		0	0	0

For a fixed ordering of input values, you can obtain a complete input/output table by filling in the entries in the output column. You can think of this as a four-step operation:

Step 1 is to fill in the output value for the first row.
Step 2 is to fill in the output value for the second row.
Step 3 is to fill in the output value for the third row.
Step 4 is to fill in the output value for the fourth row.

Each step can be performed in exactly two ways: either a 1 or a 0 can be filled in. Hence by the multiplication rule, there are

$$2 \cdot 2 \cdot 2 \cdot 2 = 16$$

ways to perform the entire operation. It follows that there are $2^4 = 16$ distinct input/output tables for a circuit with two input signals P and Q. This means that such a circuit can function in only 16 distinct ways. ■

EXAMPLE 6.2.6   Counting the Number of Iterations of a Nested Loop

Consider the following nested loop:

```
for i := 1 to 4
    for j := 1 to 3
        [Statements in body of inner loop.
         None contain branching statements
         that lead out of the inner loop.]
    next j
next i
```

How many times will the inner loop be iterated when the algorithm is implemented and run?

Solution   The outer loop is iterated four times, and during each iteration of the outer loop, there are three iterations of the inner loop. Hence by the multiplication rule, the total number of iterations of the inner loop is $4 \cdot 3 = 12$. This is illustrated by the trace table below.

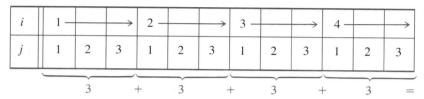

WHEN THE MULTIPLICATION RULE IS DIFFICULT OR IMPOSSIBLE TO APPLY

Consider the following problem:

> Three officers—a president, a treasurer, and a secretary—are to be chosen from among four people: Ann, Bob, Cyd, and Dan. Suppose that, for various reasons, Ann cannot be president and either Cyd or Dan must be secretary. How many ways can the officers be chosen?

It is natural to try to solve this problem using the multiplication rule. A person might answer as follows:

> There are three choices for president (all except Ann), three choices for treasurer (all except the one chosen as president), and two choices for secretary (Cyd or Dan). Therefore, by the multiplication rule, there are $3 \cdot 3 \cdot 2 = 18$ choices in all.

Unfortunately, this analysis is incorrect. The number of ways to choose the secretary varies depending on who is chosen for president and treasurer. For instance, if Bob is chosen for president and Ann for treasurer, then there are two choices for secretary: Cyd and Dan. But if Bob is chosen for president and Cyd for treasurer, then there is just one choice for secretary: Dan. The clearest way to see all the possible choices is to construct the possibility tree, as is shown in Figure 6.2.3.

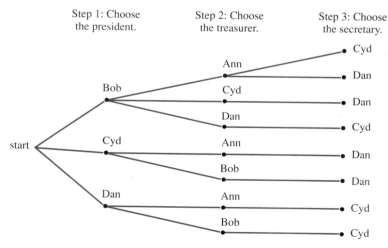

FIGURE 6.2.3

From the tree it is easy to see that there are only eight ways to choose a president, treasurer, and secretary so as to satisfy the given conditions.

Another way to solve this problem is somewhat surprising. It turns out that the steps can be reordered in a slightly different way so that the number of ways to perform each step is constant regardless of the way previous steps were performed.

EXAMPLE 6.2.7 A More Subtle Use of the Multiplication Rule

Reorder the steps for choosing the officers in the example above so that the total number of ways to choose officers can be computed using the multiplication rule.

Solution

Step 1: Choose the secretary.

Step 2: Choose the president.

Step 3: Choose the treasurer.

There are exactly two ways to perform step 1 (either Cyd or Dan may be chosen), two ways to perform step 2 (neither Ann nor the person chosen in step 1 may be chosen but either of the other two may), and two ways to perform step 3 (either of the two people not chosen as secretary or president may be chosen as treasurer). Thus by the multiplication rule the total number of ways to choose officers is $2 \cdot 2 \cdot 2 = 8$. A possibility tree illustrating this sequence of choices is shown in Figure 6.2.4. Note how balanced this tree is compared with the one in Figure 6.2.3.

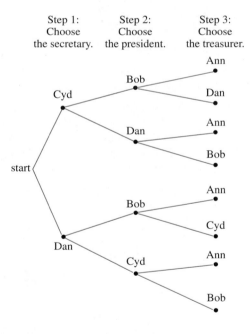

FIGURE 6.2.4

PERMUTATIONS

A **permutation** of a set of objects is an ordering of the objects in a row. For example, the set of elements *a*, *b*, and *c* has six permutations.

$$abc \quad acb \quad cba \quad bac \quad bca \quad cab$$

In general, given a set of *n* objects, how many permutations does the set have?

Imagine forming a permutation as an n-step operation:

Step 1 is to choose an element to write first.

Step 2 is to choose an element to write second.

$$\vdots \qquad \vdots$$

Step n is to choose an element to write nth.

Any element of the set can be chosen in step 1, so there are n ways to perform step 1. Any element except that chosen in step 1 can be chosen in step 2, so there are $n - 1$ ways to perform step 2. In general, the number of ways to perform each successive step is one less than the number of ways to perform the preceding step. At the point when the nth element is chosen, there is only one element left, so there is only one way to perform step n. Hence by the multiplication rule, there are

$$n \cdot (n - 1) \cdot (n - 2) \cdot \ldots \cdot 2 \cdot 1 = n!$$

ways to perform the entire operation. In other words, there are $n!$ permutations of a set of n elements. This reasoning is summarized in the following theorem. A formal proof uses mathematical induction and is left as an exercise.

THEOREM 6.2.2

For any integer n with $n \geq 1$, the number of permutations of a set with n elements is $n!$.

EXAMPLE 6.2.8   Permutations of the Letters in a Word

a. How many ways can the letters in the word *COMPUTER* be arranged in a row?
b. How many ways can the letters in the word *COMPUTER* be arranged if the letters *CO* must remain next to each other (in order) as a unit?
c. If letters of the word *COMPUTER* are randomly arranged in a row, what is the probability that the letters *CO* appear together in order?

Solution a. All the eight letters in the word *COMPUTER* are distinct, so the number of ways to arrange the letters equals the number of permutations of a set of eight elements. This equals $8! = 40{,}320$.

b. If the letter group *CO* is treated as a unit, then there are effectively only seven objects that are to be arranged in a row.

$$\boxed{\text{CO}}\ \boxed{\text{M}}\ \boxed{\text{P}}\ \boxed{\text{U}}\ \boxed{\text{T}}\ \boxed{\text{E}}\ \boxed{\text{R}}$$

So there are as many ways to write the letters as there are permutations of a set of seven elements, namely $7! = 5{,}040$.

c. $\dfrac{5{,}040}{40{,}320} = \dfrac{1}{8}$ ∎

EXAMPLE 6.2.9   Permutations of Objects Around a Circle

At a meeting of diplomats, the six participants are to be seated around a circular table. Since the table has no ends to confer particular status, it doesn't matter who sits in which chair. But it does matter how the diplomats are seated relative to each other. In other words, two seatings are considered the same if one is a rotation of the other. How many different ways can the diplomats be seated?

Solution Call the diplomats by the letters *A*, *B*, *C*, *D*, *E*, and *F*. Since only relative position matters, you can start with any diplomat, say *A*, place that diplomat anywhere, say in the top seat of the diagram shown in Figure 6.2.5, and then consider all arrangements of the other diplomats around that one. *B* through *F* can be arranged in the seats around diplomat *A* in all possible orders. So there are 5! = 120 ways to seat the group.

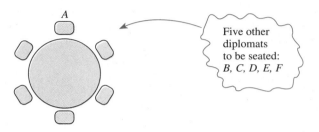

FIGURE 6.2.5

PERMUTATIONS OF SELECTED ELEMENTS

Given the set {*a*, *b*, *c*}, there are six ways to select two letters from the set and write them in order.

$$ab \quad ac \quad ba \quad bc \quad ca \quad cb$$

Each such ordering of two elements of {*a*, *b*, *c*} is called a 2-*permutation* of {*a*, *b*, *c*}.

DEFINITION

An *r*-**permutation** of a set of *n* elements is an ordered selection of *r* elements taken from the set of *n* elements. The number of *r*-permutations of a set of *n* elements is denoted $P(n, r)$.

THEOREM 6.2.3

If *n* and *r* are integers and $1 \leq r \leq n$, then the number of *r* permutations of a set of *n* elements is given by the formula

$$P(n, r) = n(n - 1)(n - 2) \cdots (n - r + 1) \qquad \text{first version}$$

or, equivalently,

$$P(n, r) = \frac{n!}{(n - r)!} \qquad \text{second version.}$$

A formal proof of this theorem uses mathematical induction and is based on the multiplication rule. The idea of the proof is the following.

Suppose a set of n elements is given. Formation of an r-permutation can be thought of as an r-step process. Step 1 is to choose the element to be first. Since the set has n elements, there are n ways to perform step 1. Step 2 is to choose the element to be second. Since the element chosen in step 1 is no longer available, there are $n - 1$ ways to perform step 2. Step 3 is to choose the element to be third. Since neither of the two elements chosen in the first two steps is available, there are $n - 2$ choices for step 3. This process is repeated r times, as shown below.

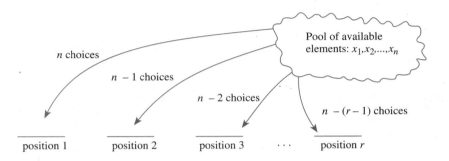

The number of ways to perform each successive step is one less than the number of ways to perform the preceding step. Step r is to choose the element to be rth. At the point just before step r is performed, $r - 1$ elements have already been chosen, and so there are

$$n - (r - 1) = n - r + 1$$

left to choose from. Hence there are $n - r + 1$ ways to perform step r. It follows by the mutiplication rule that the number of ways to form an r-permutation is $n \cdot (n - 1) \cdot (n - 2) \cdot \ldots \cdot (n - r + 1)$, and so

$$P(n, r) = n \cdot (n - 1) \cdot (n - 2) \cdot \ldots \cdot (n - r + 1).$$

Note that

$$\frac{n!}{(n - r)!} = \frac{n \cdot (n - 1) \cdot (n - 2) \cdot \ldots \cdot (n - r + 1) \cdot \cancel{(n - r)} \cdot \cancel{(n - r - 1)} \cdot \ldots \cdot \cancel{3} \cdot \cancel{2} \cdot \cancel{1}}{\cancel{(n - r)} \cdot \cancel{(n - r - 1)} \cdot \ldots \cdot \cancel{3} \cdot \cancel{2} \cdot \cancel{1}}$$
$$= n \cdot (n - 1) \cdot (n - 2) \cdot \ldots \cdot (n - r + 1).$$

Thus the formula can be written as

$$P(n, r) = \frac{n!}{(n - r)!}.$$

The second version of the formula is easier to remember. When you actually use it, however, first substitute the values of n and r and then immediately cancel the numerical value of $(n - r)!$ from numerator and denominator. Because factorials become so large so fast, direct use of the first version of the formula without cancellation can overload your calculator's capacity for exact arithmetic even when n and r are quite small. For instance, if $n = 15$ and $r = 2$, then

$$\frac{n!}{(n - r)!} = \frac{15!}{13!} = \frac{1{,}307{,}674{,}368{,}000}{6{,}227{,}020{,}800}.$$

But if you cancel $(n - r)! = 13!$ from numerator and denominator before multiplying out, you obtain

$$\frac{n!}{(n - r)!} = \frac{15!}{13!} = \frac{15 \cdot 14 \cdot 13!}{13!} = 15 \cdot 14 = 210.$$

In fact, many scientific calculators allow you to compute $P(n, r)$ simply by entering the values of n and r and pressing a key or making a menu choice. Alternate notations for $P(n, r)$ which you may see in your calculator manual are $_nP_r$, $P_{n,r}$ and ^nP_r.

EXAMPLE 6.2.10 Evaluating r-Permutations

a. Evaluate $P(5, 2)$.
b. How many 4-permutations are there of a set of seven objects?
c. How many 5-permutations are there of a set of five objects?

Solution   a. $P(5, 2) = \dfrac{5!}{(5 - 2)!} = \dfrac{5 \cdot 4 \cdot \cancel{3} \cdot \cancel{2} \cdot \cancel{1}}{\cancel{3} \cdot \cancel{2} \cdot \cancel{1}} = 20$

b. The number of 4-permutations of a set of seven objects is

$$P(7, 4) = \frac{7!}{(7 - 4)!} = \frac{7 \cdot 6 \cdot 5 \cdot 4 \cdot \cancel{3} \cdot \cancel{2} \cdot \cancel{1}}{\cancel{3} \cdot \cancel{2} \cdot \cancel{1}} = 7 \cdot 6 \cdot 5 \cdot 4 = 840.$$

c. The number of 5-permutations of a set of five objects is

$$P(5, 5) = \frac{5!}{(5 - 5)!} = \frac{5!}{0!} = \frac{5!}{1} = 5! = 120.$$

Note that the definition of 0! as 1 makes this calculation come out as it should, for the number of 5-permutations of a set of five objects is certainly equal to the number of permutations of the set. ∎

EXAMPLE 6.2.11 Permutations of Selected Letters of a Word

a. How many different ways can three of the letters of the word *BYTES* be chosen and written in a row?
b. How many different ways can this be done if the first letter must be *B*?

Solution   a. The answer equals the number of 3-permutations of a set of five elements. This equals

$$P(5, 3) = \frac{5!}{(5 - 3)!} = \frac{5 \cdot 4 \cdot 3 \cdot \cancel{2} \cdot \cancel{1}}{\cancel{2} \cdot \cancel{1}} = 5 \cdot 4 \cdot 3 = 60.$$

b. Since the first letter must be *B*, there are effectively only two letters to be chosen and placed in the other two positions. And since the *B* is used in the first position, there are four letters available to fill the remaining two positions.

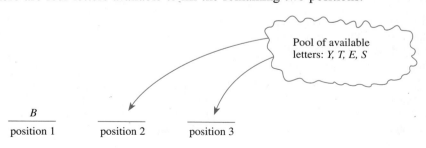

Pool of available letters: *Y, T, E, S*

B		
position 1	position 2	position 3

Hence the answer is the number of 2-permutations of a set of four elements, which is

$$P(4, 2) = \frac{4!}{(4 - 2)!} = \frac{4 \cdot 3 \cdot \cancel{2} \cdot \cancel{1}}{\cancel{2} \cdot \cancel{1}} = 4 \cdot 3 = 12.$$ ■

In many applications of the mathematics of counting, it is necessary to be skillful in working algebraically with quantities of the form $P(n, r)$. The next example shows a kind of problem that gives practice in developing such skill.

EXAMPLE 6.2.12   **Proving a Property of $P(n, r)$**

Prove that for all integers $n \geq 2$,

$$P(n, 2) + P(n, 1) = n^2.$$

Solution   Suppose n is an integer that is greater than or equal to 2. By Theorem 6.2.3,

$$P(n, 2) = \frac{n!}{(n - 2)!} = \frac{n \cdot (n - 1) \cdot \cancel{(n - 2)!}}{\cancel{(n - 2)!}} = n \cdot (n - 1)$$

and

$$P(n, 1) = \frac{n!}{(n - 1)!} = \frac{n \cdot \cancel{(n - 1)!}}{\cancel{(n - 1)!}} = n.$$

Hence

$$P(n, 2) + P(n, 1) = n \cdot (n - 1) + n = n^2 - n + n = n^2,$$

which is what we needed to show. ■

EXERCISE SET 6.2

In 1–4, use the fact that in baseball's World Series the first team to win four games wins the series.

1. Suppose team A wins the first three games. How many ways can the series be completed? (Draw a tree.)

2. Suppose team A wins the first two games. How many ways can the series be completed? (Draw a tree.)

3. How many ways can a World Series be played if team A wins four games in a row?

4. How many ways can a World Series be played if no team wins two games in a row?

5. In a competition between players X and Y, the first player to win three games in a row or a total of four games wins. How many ways can the competition be played if X wins the first two games? (Draw a tree.)

6. One urn contains two black balls (labeled $B1$ and $B2$) and one white ball. A second urn contains one black ball and two white balls (labeled $W1$ and $W2$). Suppose the following experiment is performed: One of the two

urns is chosen at random. Next a ball is randomly chosen from the urn. Then a second ball is chosen at random from the same urn without replacing the first ball.
a. Construct the possibility tree showing all possible outcomes of this experiment.
b. What is the total number of outcomes of this experiment?
c. What is the probability that two black balls are chosen?
d. What is the probability that two balls of opposite color are chosen?

7. One urn contains one blue ball and three red balls (labeled $R1$, $R2$, and $R3$). A second urn contains two red balls ($R4$ and $R5$) and two blue balls ($B1$ and $B2$). An experiment is performed in which one of the two urns is chosen at random and then two balls are randomly chosen from it, one after the other wihout replacement.
a. Construct the possibility tree showing all possible outcomes of this experiment.

b. What is the total number of outcomes of this experiment?

c. What is the probability that two red balls are chosen?

8. A person buying a personal computer system is offered a choice of three models of the basic unit, two models of keyboard, and two models of printer. How many distinct systems can be purchased?

9. Suppose there are three roads from city *A* to city *B* and five roads from city *B* to city *C*.
a. How many ways is it possible to travel from city *C* to city *C* via city *B*?
b. How many different round trip routes are there from city *A* to *B* to *C* to *B* and back to *A*?
c. How many different routes are there from city *A* to *B* to *C* to *B* and back to *A* in which no road is traversed twice?

10. Suppose there are two routes from Mill Creek to High Point, three routes from High Point to Grand Junction, two routes from Grand Junction to Devil's Fork, and four routes from High Point to Devil's Fork that bypass Grand Junction. (Draw a sketch.)
a. How many routes from Mill Creek to Devil's Fork pass through Grand Junction?
b. How many routes from Mill Creek to Devil's Fork bypass Grand Junction?

11. a. A bit string is a sequence of 0's and 1's. How many bit strings have length 8?
b. How many bit strings of length 8 begin with a 1?
c. How many bit strings of length 8 begin and end with a 1?
d. A fixed length 8-bit code (such as EBCDIC) represents symbols as bit strings of length 8. How many distinct symbols can be represented by such a code?

12. Hexadecimal numbers are made using the sixteen digits 0, 1, 2, 3, 4, 5, 6, 7, 8, 9, A, B, C, D, E, F. They are denoted by the subscript 16.
a. How many hexadecimal numbers are there from 10_{16} through FF_{16}?
b. How many hexadecimal numbers are there from 50_{16} through FF_{16}?
c. How many hexadecimal numbers are there from 30_{16} through AF_{16}?

13. A coin is tossed four times. Each time the result *H* for heads or *T* for tails is recorded. An outcome of *HHTT* means that heads were obtained on the first two tosses and tails on the second two. Assume that heads and tails are equally likely on each toss.
a. How many distinct outcomes are possible?
b. What is the probability that exactly two heads occur?
c. What is the probability that exactly one head occurs?

14. Suppose that in a certain state automobile license plates each have three letters followed by three digits.
a. How many different license plates are possible?
b. How many license plates could begin with *A* and end in 0?
c. How many license plates could begin with *PDQ*?
d. How many license plates are possible in which all the letters and digits are distinct?
e. How many license plates could begin with *AB* and have all letters and digits distinct?

15. A combination lock requires three selections of numbers, each from 1 through 20.
a. How many different combinations are possible?
b. Suppose the locks are constructed in such a way that no number can be used twice. How many different combinations are possible?

16. The diagram below shows the keypad for an automatic teller machine. As you can see, the same sequence of keys represents a variety of different PINs. For instance, 2133, AZDE, and BQ3F are all keyed in exactly the same way.

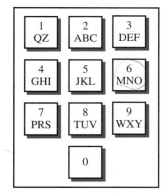

a. How many different PINs are represented by the same sequence of keys as 2133?
b. How many different PINs are represented by the same sequence of keys as 6809?
c. At an automatic teller machine, each PIN corresponds to a four-digit numeric sequence. For instance, TWJM corresponds to 8956. How many such numeric sequences contain no repeated digit?

17. Three officers—a president, a treasurer, and a secretary—are to be chosen from among four people: Ann, Bob, Cyd, and Dan. Suppose that Bob is not qualified to be treasurer and Cyd's other commitments make it impossible for her to be secretary. How many ways can the officers be chosen? Can the multiplication rule be used to solve this problem?

18. Modify Example 6.2.4 by supposing that a PIN must not begin with any of the letters A–M and must end

with a digit. Continue to assume that no symbol may be used more than once and that the total number of PINs is to be determined.

a. Find the error in the following "solution."

"Constructing a PIN is a four-step process.

Step 1 is to choose the left-most symbol.

Step 2 is to choose the second symbol from the left.

Step 3 is to choose the third symbol from the left.

Step 4 is to choose the right-most symbol.

Because none of the thirteen letters from A through M may be chosen in step 1, there are $36 - 13 = 23$ ways to perform step 1. There are 35 ways to perform step 2 and 34 ways to perform step 3 because previously used symbols may not be used. Since the symbol chosen in step 4 must be a previously unused digit, there are $10 - 3 = 7$ ways to perform step 4. Thus there are $23 \cdot 35 \cdot 34 \cdot 7 = 191{,}590$ different PINs that satisfy the given conditions."

b. Reorder steps 1–4 above so that

Step 1 is to choose the right-most symbol.

Step 2 is to choose the left-most symbol.

Step 3 is to choose the second symbol from the left.

Step 4 is to choose the third symbol from the left.

Use the multiplication rule to find the number of PINs that satisfy the given conditions.

19. a. How many integers are there from 10 through 99?
   b. How many odd integers are there from 10 through 99?
   c. How many integers from 10 through 99 have distinct digits?
   ◆ d. How many odd integers from 10 through 99 have distinct digits?
   e. What is the probability that a randomly chosen two-digit integer has distinct digits? has distinct digits and is odd?

20. a. How many integers are there from 100 through 999?
   b. How many odd integers are there from 100 through 999?
   c. How many integers from 100 through 999 have distinct digits?
   ◆ d. How many odd integers from 100 through 999 have distinct digits?
   e. What is the probability that a randomly chosen three-digit integer has distinct digits? has distinct digits and is odd?

In 21–25, determine how many times the innermost loops will be iterated when the algorithm segments are implemented and run. (Assume that m, n, p, a, b, c, and d are all positive integers.)

21. **for** $i := 1$ **to** 30
        for $j := 1$ **to** 15
                [Statements in body of inner loop.
                None contain branching statements that
                lead outside the loop.]
            next j
        next i

22. **for** $j := 1$ **to** m
        for $k := 1$ **to** n
                [Statements in body of inner loop.
                None contain branching statements that
                lead outside the loop.]
            next k
        next j

23. **for** $i := 1$ **to** m
        for $j := 1$ **to** n
            for $k := 1$ **to** p
                    [Statements in body of inner loop.
                    None contain branching statements that
                    lead outside the loop.]
                next k
            next j
        next i

24. **for** $i := 5$ **to** 50
        for $j := 10$ **to** 20
                [Statements in body of inner loop.
                None contain branching statements that
                lead outside the loop.]
            next j
        next i

25. Assume $a \leq b$ and $c \leq d$.
    for $i := a$ **to** b
        for $j := c$ **to** d
                [Statements in body of inner loop.
                None contain branching statements that
                lead outside the loop.]
            next j
        next i

H ◆ 26. Consider the decimal representations of numbers from 1 through 99,999. How many contain exactly one each of the digits 2, 3, 4, and 5?

◆ 27. Let $n = p_1^{k_1} p_2^{k_2} \cdots p_m^{k_m}$ where $p_1, p_2, \ldots,$ and p_m are distinct prime numbers and $k_1, k_2, \ldots,$ and k_m are positive integers. How many ways can n be written as a product of two positive integers which have no common factors
   a. assuming that order matters (i.e., $8 \cdot 15$ and $15 \cdot 8$ are regarded as different)?

b. assuming that order does not matter (i.e., $8 \cdot 15$ and $15 \cdot 8$ are regarded as the same)?

◆28. a. If p is a prime number and a is a positive integer, how many divisors does p^a have?

b. If p and q are prime numbers and a and b are positive integers, how many possible divisors does $p^a q^b$ have?

c. If p, q, and r are prime numbers and a, b, and c are positive integers, how many possible divisors does $p^a q^b r^c$ have?

d. If $p_1, p_2, \ldots, p_m$ are prime numbers and $a_1, a_2, \ldots, a_m$ are positive integers, how many possible divisors does $p_1{}^{a_1} p_2{}^{a_2} \cdots p_m{}^{a_m}$ have?

e. What is the smallest positive integer with exactly 12 divisors?

29. a. How many ways can the letters of the word *ALGORITHM* be arranged in a row?

b. How many ways can the letters of the word *ALGORITHM* be arranged in a row if A and L must remain together (in order) as a unit?

c. How many ways can the letters of the word *ALGORITHM* be arranged in a row if the letters *GOR* must remain together (in order) as a unit?

30. Six people attend the theater together.

a. How many ways can they be seated in a row?

b. Suppose one of the six is a doctor who must sit on the aisle in case she is paged. How many ways can the people be seated in a row of seats if exactly one of the seats is on the aisle and the doctor is in the aisle seat?

c. Suppose the six people consist of three married couples and each couple wants to sit together with the husband on the left. How many ways can the six be seated in a row?

31. Five people are to be seated around a circular table. Two seatings are considered the same if one is a rotation of the other. How many different seatings are possible?

32. Write all the 2-permutations of $\{W, X, Y, Z\}$.

33. Write all the 3-permutations of $\{a, b, c, d\}$.

34. Evaluate the following quantities.
   a. $P(6, 4)$ b. $P(6, 6)$ c. $P(6, 2)$ d. $P(6, 1)$

35. a. How many 3-permutations are there of a set of five objects?

b. How many 2-permutations are there of a set of seven objects?

36. a. How many ways can three of the letters of the word *ALGORITHM* be selected and written in a row?

b. How many ways can five of the letters of the word *ALGORITHM* be selected and written in a row?

c. How many ways can five of the letters of the word *ALGORITHM* be selected and written in a row if the first letter must be A?

d. How many ways can five of the letters of the word *ALGORITHM* be selected and written in a row if the first two letters must be *TH*?

37. Prove that for all integers $n \geq 2$,
$$P(n + 1, 3) = n^3 - n.$$

38. Prove that for all integers $n \geq 2$,
$$P(n + 1, 2) - P(n, 2) = 2P(n, 1).$$

39. Prove that for all integers $n \geq 3$,
$$P(n + 1, 3) - P(n, 3) = 3P(n, 2).$$

40. Prove that for all integers $n \geq 2$,
$$P(n, 1) = P(n, n - 1).$$

41. Prove Theorem 6.2.1 by mathematical induction.

H42. Prove Theorem 6.2.2 by mathematical induction.

◆43. Prove Theorem 6.2.3 by mathematical induction.

6.3   COUNTING ELEMENTS OF DISJOINT SETS: THE ADDITION RULE

The whole of science is nothing more than a refinement of everyday thinking.
(Albert Einstein, 1879–1955)

In the last section we discussed counting problems that can be solved using possibility trees. In this section we look at counting problems that can be solved by counting the number of elements in the union of two sets, the difference of two sets, or the intersection of two sets.

The basic rule underlying the calculation of the number of elements in a union or difference or intersection is the addition rule. This rule states that the number of elements in a union of mutually disjoint finite sets equals the sum of the number of elements in each of the component sets.

> **THEOREM 6.3.1** The Addition Rule
>
> Suppose a finite set A equals the union of k distinct mutually disjoint subsets $A_1, A_2, \ldots, A_k$. Then
> $$n(A) = n(A_1) + n(A_2) + \cdots + n(A_k).$$

A formal proof of this theorem uses mathematical induction and is left to the exercises.

EXAMPLE 6.3.1

Counting Code Words with Three or Fewer Letters

A computer access code word consists of from one to three letters chosen from the 26 in the alphabet with repetitions allowed. How many different code words are possible?

Solution   The set of all code words can be partitioned into subsets consisting of those of length 1, those of length 2, and those of length 3 as shown in Figure 6.3.1.

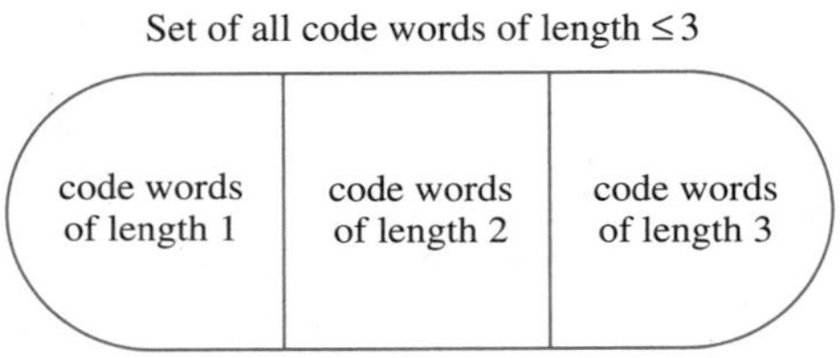

Set of all code words of length ≤ 3

| code words of length 1 | code words of length 2 | code words of length 3 |

FIGURE 6.3.1

By the addition rule, the total number of code words equals the number of code words of length 1, plus the number of code words of length 2, plus the number of code words of length 3. Now the

number of code words of length $1 = 26$ because there are 26 letters in the alphabet

number of code words of length $2 = 26^2$ because forming such a word can be thought of as a two-step process in which there are 26 ways to perform each step

number of code words of length $3 = 26^3$ because forming such a word can be thought of as a three-step process in which there are 26 ways to perform each step.

Hence

the total number of code words $= 26 + 26^2 + 26^3 = 18{,}278.$ ∎

EXAMPLE 6.3.2

Counting the Number of Integers Divisible by 5

How many three-digit integers (integers from 100 to 999 inclusive) are divisible by 5?

Solution   One solution to this problem was discussed in Example 6.1.3. Another approach uses the addition rule. Integers that are divisible by 5 end either in 5 or in 0. Thus the set of all three-digit integers that are divisible by 5 can be split into two mutually disjoint subsets A_1 and A_2 as shown in Figure 6.3.2.

Three-digit integers that are divisible by 5

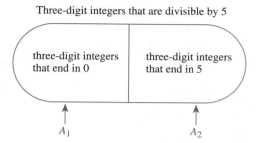

FIGURE 6.3.2

Now there are as many three-digit integers that end in 0 as there are possible choices for the left-most and middle digits (because the right-most digit must be a 0). As illustrated below, there are nine choices for the left-most digit (the digits 1 through 9) and ten choices for the middle digit (the digits 0 through 9). Hence $n(A_1) = 9 \cdot 10 = 90$.

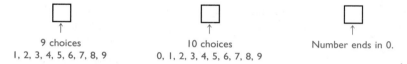

Similar reasoning (using 5 instead of 0) shows that $n(A_2) = 90$ also. So

$$\begin{bmatrix} \text{The number of} \\ \text{three-digit integers} \\ \text{that are divisible by 5} \end{bmatrix} = n(A_1) + n(A_2) = 90 + 90 = 180. \quad \blacksquare$$

THE DIFFERENCE RULE

An important consequence of the addition rule is the fact that if the number of elements in a set A and in a subset B of A are both known, then the number of elements that are in A and not in B can be computed.

THEOREM 6.3.2   The Difference Rule

If A is a finite set and B is a subset of A, then

$$n(A - B) = n(A) - n(B).$$

The difference rule is illustrated in Figure 6.3.3.

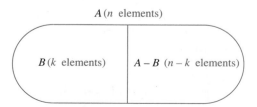

FIGURE 6.3.3   The Difference Rule

The difference rule holds for the following reason: If B is a subset of A, then $B \cup (A - B) = A$ and the two sets B and $A - B$ have no elements in common. Hence by the addition rule,

$$n(B) + n(A - B) = n(A).$$

Subtracting $n(B)$ from both sides gives the equation

$$n(A - B) = n(A) - n(B).$$

EXAMPLE 6.3.3   Counting PINs with Repeated Letters

As discussed in Examples 6.2.2 and 6.2.4, a PIN is made from exactly four symbols chosen from the 26 letters of the alphabet and the ten digits, with repetitions allowed.

a. How many PINs contain repeated symbols?
b. If all PINs are equally likely, what is the probability that a randomly chosen PIN contains a repeated symbol?

Solution a. According to Example 6.2.2, there are $36^4 = 1,679,616$ PINs when repetition is allowed, and by Example 6.2.4, there are 1,413,720 PINs when repetition is not allowed. Thus by the difference rule, there are

$$1,679,616 - 1,413,720 = 265,896$$

PINs that contain at least one repeated symbol.

b. By Example 6.2.2 there are 1,679,616 PINs in all, and by part (a) 265,896 of these contain at least one repeated symbol. Thus by the equally likely probability formula, the probability that a randomly chosen PIN contains a repeated symbol is

$$\frac{265,896}{1,679,616} \cong .1583.$$

An alternative solution to Example 6.3.3(b) is based on the observation that if S is the set of all PINs and A is the set of all PINs with no repeated symbol, then $S - A$ is the set of all PINs with at least one repeated symbol. If follows that

$$P(S - A) = \frac{n(S - A)}{n(S)} \qquad \text{by definition of probability in the equally likely case}$$

$$= \frac{n(S) - n(A)}{n(S)} \qquad \text{by the difference rule}$$

$$= \frac{n(S)}{n(S)} - \frac{n(A)}{n(S)} \qquad \text{by the laws of fractions}$$

$$= 1 - P(A) \qquad \text{by definition of probability in the equally likely case}$$

$$\cong 1 - .8417 \qquad \text{by Example 6.2.4}$$

$$\cong .1583.$$

This solution illustrates a more general property of probabilities: that the probability of the complement of an event is obtained by subtracting the probability of the event from the number 1.

Formula for the Probability of the Complement of an Event

If S is a finite sample space and A is an event in S, then

$$P(A^c) = 1 - P(A).$$

EXAMPLE 6.3.4

Number of Pascal Identifiers of Eight or Fewer Characters

In certain implementations of the computer language Pascal, identifiers must start with one of the 26 letters of the alphabet. The initial letter may be followed by other symbols chosen from the 26 letters of the alphabet plus the ten digits. Certain keywords, however, are reserved for commands and may not be used as identifiers. There are 35 such reserved words none of which has more than eight characters. How many Pascal identifiers are there that are less than or equal to eight characters in length?

Solution   The set of all Pascal identifiers with eight or fewer characters can be partitioned into eight subsets—identifiers of length 1, identifiers of length 2, and so on—as shown in Figure 6.3.4. The reserved words have various lengths (all less than or equal to eight), so the set of reserved words is shown overlapping the various subsets.

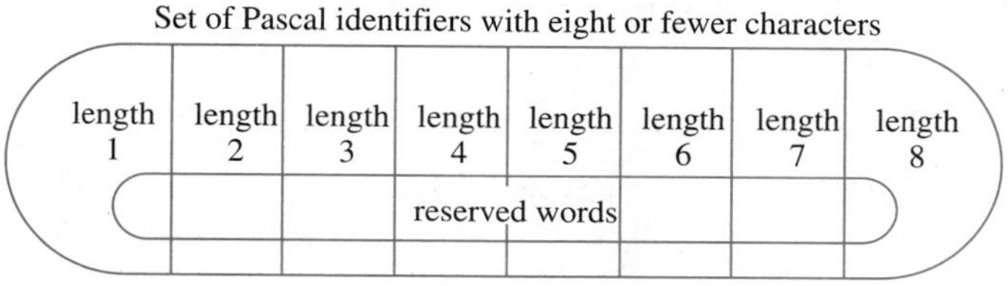

Set of Pascal identifiers with eight or fewer characters

length 1	length 2	length 3	length 4	length 5	length 6	length 7	length 8
			reserved words				

FIGURE 6.3.4

According to the rules for creating Pascal identifiers, there are

26 potential identifiers of length 1   *because there are 26 letters in the alphabet,*

$26 \cdot 36$ potential identifiers of length 2   *because the second symbol can be either a letter or a digit,*

$26 \cdot 36^2$ potential identifiers of length 3   *because the second and third symbols can be either letters or digits,*

$$\vdots$$

$26 \cdot 36^7$ potential identifiers of length 8   *because the second through eighth symbols can be either letters or digits.*

Thus by the addition rule, the number of potential Pascal identifiers with eight or fewer characters is

$$26 + 26 \cdot 36 + 26 \cdot 36^2 + 26 \cdot 36^3 + 26 \cdot 36^4 + 26 \cdot 36^5$$
$$+ 26 \cdot 36^6 + 26 \cdot 36^7 = 2,095,681,645,538.$$

Now 35 of these potential identifiers are reserved, so by the difference rule, the actual number of Pascal identifiers with eight or fewer characters is

$$2,095,681,645,538 - 35 = 2,095,681,645,503. \qquad \blacksquare$$

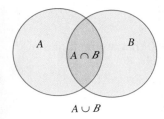

$A \cup B$

FIGURE 6.3.5

THE INCLUSION/EXCLUSION RULE

The addition rule says how many elements are in a union of sets if the sets are mutually disjoint. Now consider the question of how to determine the number of elements in a union of sets when some of the sets overlap. For simplicity, begin by looking at a union of two sets A and B, as shown in Figure 6.3.5.

First observe that the number of elements in $A \cup B$ varies according to the number of elements the two sets have in common. If A and B have no elements in common, then $n(A \cup B) = n(A) + n(B)$. If A and B coincide, then $n(A \cup B) = n(A)$. Thus any general formula for $n(A \cup B)$ must contain a reference to the number of elements the two sets have in common, $n(A \cap B)$, as well as to $n(A)$ and $n(B)$.

The simplest way to derive a formula for $n(A \cup B)$ is to reason as follows:* The number $n(A)$ counts the elements that are in A and not in B and also the elements that are in both A and B. Similarly, the number $n(B)$ counts the elements that are in B and not in A and also the elements that are in both A and B. Hence when the two numbers $n(A)$ and $n(B)$ are added, the elements that are in both A and B are counted twice. To get an accurate count of the elements in $A \cup B$, it is necessary to subtract the number of elements that are in both A and B. Since these are the elements in $A \cap B$,

$$n(A \cup B) = n(A) + n(B) - n(A \cap B).$$

A similar analysis gives a formula for the number of elements in a union of three sets, as stated in the Theorem 6.3.3.

THEOREM 6.3.3   The Inclusion/Exclusion Rule for
Two or Three Sets

If A, B, and C are any finite sets, then

$$n(A \cup B) = n(A) + n(B) - n(A \cap B)$$

and

$$n(A \cup B \cup C) = n(A) + n(B) + n(C) - n(A \cap B) - n(A \cap C)$$
$$- n(B \cap C) + n(A \cap B \cap C).$$

It can be shown using mathematical induction (see exercise 33 at the end of this section) that formulas analogous to those of Theorem 6.3.3 hold for unions of any finite number of sets.

EXAMPLE 6.3.5   **Counting Elements of a General Union**

a. How many integers from 1 through 1,000 are multiples of 3 or multiples of 5?
b. How many integers from 1 through 1,000 are neither multiples of 3 nor multiples of 5?

Solution   a. Let A = the set of all integers from 1 through 1,000 that are multiples of 3.
Let B = the set of all integers from 1 through 1,000 that are multiples of 5.

Then

$A \cup B$ = the set of all integers from 1 through 1,000 that are multiples of 3 or multiples of 5

---

*An alternative proof is outlined in exercise 31 at the end of this section.

and

$A \cap B$ = the set of all integers from 1 through 1,000 that are multiples of both 3 and 5

  = the set of all integers from 1 through 1,000 that are multiples of 15.

[Now calculate n(A), n(B), and n(A ∩ B) and use the inclusion/exclusion rule to solve for n(A ∪ B).]

Because every third integer from 3 through 999 is a multiple of 3, each can be represented in the form $3k$, for some integer k from 1 through 333. Hence, there are 333 multiples of 3 from 1 through 1,000, and so $n(A) = 333$.

$$
\begin{array}{ccccccccccc}
1 & 2 & 3 & 4 & 5 & 6 & \cdots & 996 & 997 & 998 & 999 \\
 & & \updownarrow & & & \updownarrow & & \updownarrow & & & \updownarrow \\
 & & 3 \cdot 1 & & & 3 \cdot 2 & & 3 \cdot 332 & & & 3 \cdot 333
\end{array}
$$

Similarly, each multiple of 5 from 1 through 1,000 has the form $5k$, for some integer k from 1 through 200.

$$
\begin{array}{cccccccccccccc}
1 & 2 & 3 & 4 & 5 & 6 & 7 & 8 & 9 & 10 & \cdots & 995 & 996 & 997 & 998 & 999 & 1{,}000 \\
 & & & & \updownarrow & & & & & \updownarrow & & \updownarrow & & & & & \updownarrow \\
 & & & & 5 \cdot 1 & & & & & 5 \cdot 2 & & 5 \cdot 199 & & & & & 5 \cdot 200
\end{array}
$$

So there are 200 multiples of 5 from 1 through 1,000 and $n(B) = 200$.

Finally, each multiple of 15 from 1 through 1,000 has the form $15k$, for some integer k from 1 through 66 (since $990 = 66 \cdot 15$).

$$
\begin{array}{ccccccccc}
1 & 2 \ldots & 15 \ldots & 30 \ldots & 975 \ldots & 990 \ldots & 999 & 1{,}000 \\
 & & \updownarrow & \updownarrow & \updownarrow & \updownarrow & & \\
 & & 15 \cdot 1 & 15 \cdot 2 & 15 \cdot 65 & 15 \cdot 66 & &
\end{array}
$$

So there are 66 multiples of 15 from 1 through 1,000 and $n(A \cap B) = 66$.

It follows by the inclusion/exclusion rule that

$$
\begin{aligned}
n(A \cup B) &= n(A) + n(B) - n(A \cap B) \\
&= 333 + 200 - 66 \\
&= 467.
\end{aligned}
$$

Thus there are 467 integers from 1 through 1,000 that are multiples of 3 or multiples of 5.

b. There are 1,000 integers from 1 through 1,000 and, by part (a), 467 of these are multiples of 3 or multiples of 5. Thus, by the set difference rule, there are $1{,}000 - 467 = 533$ that are neither multiples of 3 nor multiples of 5. ■

Note that in part (b) above the number of elements in the intersection of the complements of two sets, $n(A^c \cap B^c)$, was obtained using De Morgan's law, $A^c \cap B^c = (A \cup B)^c$. Then $n((A \cup B)^c)$ was calculated using the set difference rule $n((A \cup B)^c) = n(U) - n(A \cup B)$, where the universe U was the set of all integers from 1 through 1,000. Exercises 27–29 at the end of this section explore this technique further.

EXAMPLE 6.3.6 Counting the Number of Elements in an Intersection

A professor in an advanced computer course takes a survey on the first day of class to determine how many students know certain computer languages. The finding is that out of a total of 50 students in the class,

> 30 know Pascal;
>
> 18 know FORTRAN;
>
> 26 know COBOL;
>
> 9 know both Pascal and FORTRAN;
>
> 16 know both Pascal and COBOL;
>
> 8 know both FORTRAN and COBOL;
>
> 47 know at least one of the three languages.

Note that when we write "30 students know Pascal," we mean that the total number of students who know Pascal is 30 and we allow for the possibility that some of these students may know one or both of the other languages. If we want to say that 30 students know Pascal *only* (and not either of the other languages), we will say so explicitly.

   a. How many students know none of the three languages?
   b. How many students know all three languages?
   c. How many students know Pascal and FORTRAN but not COBOL? How many students know Pascal but neither FORTRAN nor COBOL?

Solution   a. By the difference rule, the number of students who know none of the three languages equals the number in the class minus the number who know at least one language. Thus the number of students who know none of the three languages is

$$50 - 47 = 3.$$

   b. Let

$$P = \text{the set of students who know Pascal,}$$
$$C = \text{the set of students who know COBOL, and}$$
$$F = \text{the set of students who know FORTRAN.}$$

Then by the inclusion/exclusion rule,

$$n(P \cup C \cup F) = n(P) + n(C) + n(F) - n(P \cap C) - n(P \cap F) \\ - n(C \cap F) + n(P \cap C \cap F).$$

So substituting known values,

$$47 = 30 + 26 + 18 - 9 - 16 - 8 + n(P \cap C \cap F).$$

Solving for $n(P \cap C \cap F)$ gives

$$n(P \cap C \cap F) = 6.$$

Hence there are six students who know all three languages. In general, if you know any seven of the eight terms in the inclusion/exclusion formula for three sets, you can solve for the eighth term.

   c. To answer the questions of part (c), look at the diagram in Figure 6.3.6. Since $n(P \cap C \cap F) = 6$, put the number 6 inside the innermost region. Then work outward to find the numbers of students represented by the other regions of the diagram. For example, since nine students know both Pascal and FORTRAN, and six know all three languages, $9 - 6 = 3$ students know Pascal and FORTRAN but not COBOL. Similarly, since 16 students know Pascal and COBOL and six know Pascal and COBOL and FORTRAN, then $16 - 6 = 10$ students know Pascal and COBOL but not FORTRAN. Now the total number of students who know Pascal

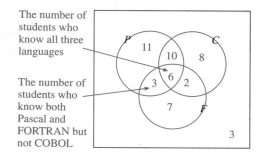

The number of students who know all three languages

The number of students who know both Pascal and FORTRAN but not COBOL

FIGURE 6.3.6

is 30. Of these 30, three also know FORTRAN but not COBOL, ten know COBOL but not FORTRAN, and six know both FORTRAN and COBOL. That leaves 11 students who know Pascal but neither of the other two languages.

A similar analysis can be used to fill in the numbers for the other regions of the diagram.   ■

EXERCISE SET 6.3

1. a. How many bit strings consist of from one through four digits? (Strings of different lengths are considered distinct. Thus 10 and 0010 are distinct strings.)
   b. How many bit strings consist of from five through eight digits?

2. a. How many strings of hexadecimal digits consist of from one through four digits? (Recall that hexadecimal numbers are constructed using the 16 digits 0, 1, 2, 3, 4, 5, 6, 7, 8, 9, A, B, C, D, E, F.)
   b. How many strings of hexadecimal digits consist of from three through five digits?

3. a. How many integers from 1 through 999 do not have any repeated digits?
   b. What is the probability that an integer chosen at random from 1 through 999 has at least one repeated digit?

4. How many arrangements of no more than three letters can be formed using the letters of the word *NETWORK* (with no repetitions allowed)?

5. a. How many four-digit integers (integers from 1,000 through 9,999) are divisible by 5?
   b. What is the probability that a four-digit integer chosen at random is divisible by 5?

6. In a certain state, license plates consist of from zero to three letters followed by from zero to four digits, with the provision, however, that a blank plate is not allowed.
   a. How many different license plates can the state produce?

   b. Suppose 85 letter combinations are not allowed because of their potential for giving offense. How many different license plates can the state produce?

◆7. A calculator has an eight-digit display and a decimal point that is located at the extreme right of the number displayed, at the extreme left, or between any pair of digits. The calculator can also display a minus sign at the extreme left of the number. How many distinct numbers can the calculator display?

8. a. Consider the following algorithm segment:

   for $i := 1$ **to** 4
       for $j := 1$ **to** i
           *[Statements in body of inner loop.
           None contain branching statements that
           lead outside the loop.]*
       next j
   next i

   How many times will the inner loop be iterated when the algorithm is implemented and run?
   b. Let n be a positive integer, and consider the following algorithm segment:

   for $i := 1$ **to** n
       for $j := 1$ **to** i
           *[Statements in body of inner loop.
           None contain branching statements that
           lead outside the loop.]*
       next j
   next i

   How many times will the inner loop be iterated when the algorithm is implemented and run?

9. a. How many ways can the letters of the word *QUICK* be arranged in a row?

b. How many ways can the letters of the word *QUICK* be arranged in a row if the *Q* and the *U* must remain next to each other in the order *QU*?

c. How many ways can the letters of the word *QUICK* be arranged in a row if the letters *QU* must remain together but may be in either the order *QU* or the order *UQ*?

10. a. How many ways can the letters of the word *DESIGN* be arranged in a row?

b. How many ways can the letters of the word *DESIGN* be arranged in a row if *G* and *N* must remain next to each other as either *GN* or *NG*?

11. A group of eight people are attending the movies together.

a. Two of the eight insist on sitting together. In how many ways can the eight be seated in a row?

b. Two of the people do not like each other and do not want to sit side-by-side. Now how many ways can the eight be seated in a row?

12. An early BASIC compiler recognized variable names according to the following rules: Numeric variable names had to begin with a letter and then the letter could be followed by another letter or a digit or by nothing at all. String variable names had to begin with the symbol $ followed by a letter, which could then be followed by another letter or a digit or by nothing at all. How many distinct variable names were recognized by this BASIC compiler?

H 13. Identifiers in a certain database language must begin with a letter and then the letter may be followed by other characters, which can be letters, digits, or under₋ scores (_). However, 82 keywords (all of 15 or fewer characters) are reserved and cannot be used as identifiers. How many identifiers with 30 or fewer characters are possible? (*Write the answer using the summation notation and evaluate it using a formula from Section 4.2.*)

14. a. Assuming that any seven digits can be used to form a telephone number, how many seven-digit telephone numbers do not have any repeated digits?

b. How many seven-digit telephone numbers have at least one repeated digit?

c. What is the probability that a randomly chosen seven-digit telephone number has at least one repeated digit?

15. a. How many strings of four hexadecimal digits do not have any repeated digits?

b. How many strings of four hexadecimal digits have at least one repeated digit?

c. What is the probability that a randomly chosen string of four hexadecimal digits has at least one repeated digit?

16. Just as the difference rule gives rise to a formula for the probability of the complement of an event, so the addition and inclusion/exclusion rules give rise to formulas for the probability of the union of mutually disjoint events and for a general union of (not necessarily mutually exclusive) events.

a. Prove that for mutually disjoint events *A* and *B*, $P(A \cup B) = P(A) + P(B)$.

b. Prove that for any events *A* and *B*,

$$P(A \cup B) = P(A) + P(B) - P(A \cap B).$$

H◆17. A combination lock requires three selections of numbers, each from 1 through 39. Suppose the lock is constructed in such a way that no number can be used twice in a row but the same number may occur both first and third. How many different combinations are possible?

◆18. a. How many integers from 1 through 100,000 contain the digit 6 exactly once?

b. How many integers from 1 through 100,000 contain the digit 6 at least once?

c. If an integer is chosen at random from 1 through 100,000, what is the probability that it contains two or more occurrences of the digit 6?

H◆19. Six new employees, two of whom are married to each other, are to be assigned six desks that are lined up in a row. If the assignment of employees to desks is made randomly, what is the probability that the married couple will have nonadjacent desks? (*Hint:* First find the probability that the couple has adjacent desks, and then subtract this number from 1.)

◆20. Consider strings of length *n* over the set $\{a, b, c, d\}$.

a. How many such strings contain at least one pair of consecutive characters that are the same?

b. If a string of length ten over $\{a, b, c, d\}$ is chosen at random, what is the probability that it contains at least one pair of consecutive characters that are the same?

21. a. How many integers from 1 through 1,000 are multiples of 4 or multiples of 7?

b. Suppose an integer from 1 through 1,000 is chosen at random. Use the result of part (a) to find the probability that the integer is a multiple of 4 or a multiple of 7.

c. How many integers from 1 through 1,000 are neither multiples of 4 nor multiples of 7?

22. a. How many integers from 1 through 1,000 are multiples of 2 or multiples of 9?

b. Suppose an integer from 1 through 100,000 is chosen at random. Use the result of part (a) to find the probability that the integer is a multiple of 2 or a multiple of 9.

c. How many integers from 1 through 100,000 are neither multiples of 2 nor multiples of 9?

23. A market research project studied student readership of certain news magazines by asking students to place checks underneath the names of all news magazines they read occasionally. Out of a sample of 100 students, it was found that 28 checked *Time,* 26 checked *Newsweek,* 14 checked *U.S. News and World Report,* 8 checked both *Time* and *Newsweek,* 4 checked both *Time* and *U.S. News,* 3 checked both *Newsweek* and *U.S. News,* and 2 checked all three. Note that some students who checked *Time* may also have checked one or both of the other magazines. A similar occurrence may be true for the other data.

   a. How many students checked at least one of the magazines?

   b. How many students checked none of the magazines?

   c. Let *T* be the set of students who checked *Time, N* the set of students who checked *Newsweek,* and *U* the set of students who checked *U.S. News.* Fill in the numbers for all eight regions of the diagram below.

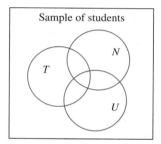

Sample of students

   d. How many students read *Time* and *Newsweek* but not *U.S. News*?

   e. How many students read *Newsweek* and *U.S. News* but not *Time*?

   f. How many students read *Newsweek* but neither of the other two?

24. A study was done to determine the efficacy of three different drugs—*A, B,* and *C*—in relieving headache pain. Over the period covered by the study, 40 subjects were given the chance to use all three drugs. The following results were obtained:

   23 reported relief from drug *A.*

   18 reported relief from drug *B.*

   31 reported relief from drug *C.*

   11 reported relief from both drugs *A* and *B.*

   19 reported relief from both drugs *A* and *C.*

   14 reported relief from both drugs *B* and *C.*

   37 reported relief from at least one of the drugs.

   Note that some of the 23 subjects who reported relief from drug *A* may also have reported relief from drugs

B or *C. A* similar occurrence may be true for the other data.

   a. How many people got relief from none of the drugs?

   b. How many people got relief from all three drugs?

   c. Let *A* be the set of all subjects who got relief from drug *A, B* the set of all subjects who got relief from drug *B,* and *C* the set of all subjects who got relief from drug *C.* Fill in the numbers for all eight regions of the diagram below.

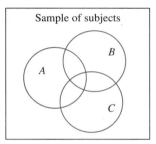

Sample of subjects

   d. How many subjects got relief from *A* only?

25. An interesting use of the inclusion/exclusion rule is to check survey numbers for consistency. For example, suppose a public opinion polltaker reports that out of a national sample of 1,200 adults, 675 are married, 682 are from 20 to 30 years old, 684 are female, 195 are married and are from 20 to 30 years old, 467 are married females, 318 are females from 20 to 30 years old, and 165 are married females from 20 to 30 years old. Are the polltaker's figures consistent? Could they have occurred as a result of an actual sample survey?

26. Fill in the reasons for each step below. If *A* and *B* are sets in a finite universe *U,* then

$$n(A \cap B)$$
$$= n(U) - n((A \cap B)^c) \qquad \underline{\text{(a)}}$$
$$= n(U) - n(A^c \cup B^c) \qquad \underline{\text{(b)}}$$
$$= n(U) - (n(A^c) + n(B^c) - n(A^c \cap B^c)) \qquad \underline{\text{(c)}}.$$

For each of exercises 27–29 below, the number of elements in a certain set can be found by computing the number in some larger universe that are not in the set and subtracting this from the total. In each case, as indicated by exercise 26, De Morgan's laws and the inclusion/exclusion rule can be used to compute the number that are not in the set.

27. How many positive integers less than 1,000 have no common factors with 1,000?

◆ **28.** How many permutations of *abcde* are there in which the first character is *a*, *b*, or *c* and the last character is *c*, *d*, or *e*?

◆ **29.** How many integers from 1 through 999,999 contain each of the digits 1, 2, and 3 at least once? (*Hint:* For each *i* = 1, 2, and 3, let A_i be the set of all integers from 1 through 999,999 that do not contain the digit *i*.)

30. Use mathematical induction to prove Theorem 6.3.1.

31. Prove the inclusion/exclusion rule for two sets *A* and *B* by showing that $A \cup B$ can be partitioned into $A - (A \cap B)$, $B - (A \cap B)$, and $A \cap B$ and then using the addition and difference rules.

32. Prove the inclusion/exclusion rule for three sets.

◆ **33.** Use mathematical induction to prove the general inclusion/exclusion rule:

If $A_1, A_2, \ldots, A_n$ are finite sets, then

$$n(A_1 \cup A_2 \cup \cdots \cup A_n)$$

$$= \sum_{1 \le i \le n} n(A_i) - \sum_{1 \le i < j \le n} n(A_i \cap A_j)$$

$$+ \sum_{1 \le i < j < k \le n} n(A_i \cap A_j \cap A_k) - \cdots$$

$$+ (-1)^{n+1} n(A_1 \cap A_2 \cap \cdots \cap A_n).$$

(The notation $\sum_{1 \le i < j \le n} n(A_i \cap A_j)$ means that quantities of the form $n(A_i \cap A_j)$ are to be added together for all integers *i* and *j* with $1 \le i < j \le n$.)

6.4   COUNTING SUBSETS OF A SET: COMBINATIONS

"But 'glory' doesn't mean 'a nice knock-down argument,'" Alice objected. "When I use a word," Humpty Dumpty said, in rather a scornful tone, "it means just what I choose it to mean—neither more nor less."

(Lewis Carroll,
Through the Looking Glass,
1872)

Consider the following question:

> Suppose five members of a group of twelve are to be chosen to work as a team on a special project. How many distinct five-person teams can be selected?

This question is answered in Example 6.4.5. It is a special case of the following more general question:

> Given a set *S* with *n* elements, how many subsets of size *r* can be chosen from *S*?

The number of subsets of size *r* that can be chosen from *S* equals the number of subsets of size *r* that *S* has. Each individual subset of size *r* is called an *r-combination* of the set.

DEFINITION

Let *n* and *r* be nonnegative integers with $r \le n$. An ***r*-combination** of a set of *n* elements is a subset of *r* of the *n* elements. The symbol $\binom{n}{r}$, read "*n* choose *r*," denotes the number of subsets of size *r* (*r*-combinations) that can be chosen from a set of *n* elements.

Note that on calculators the symbols $C(n, r)$, $_nC_r$, $C_{n,r}$, or ^nC_r are sometimes used instead of $\binom{n}{r}$.

EXAMPLE 6.4.1   **3-Combinations**

Let $S = \{\text{Ann, Bob, Cyd, Dan}\}$. Each committee consisting of three of the four people in *S* is a 3-combination of *S*.

a. List all such 3-combinations of *S*.   b. What is $\binom{4}{3}$?

Solution   a. Each 3-combination of S is a subset of S of size 3. But each subset of size 3 can be obtained by leaving out one of the elements of S. The 3-combinations are

$$\{Bob, Cyd, Dan\} \quad \text{leave out Ann}$$
$$\{Ann, Cyd, Dan\} \quad \text{leave out Bob}$$
$$\{Ann, Bob, Dan\} \quad \text{leave out Cyd}$$
$$\{Ann, Bob, Cyd\} \quad \text{leave out Dan.}$$

b. Because $\binom{4}{3}$ is the number of 3-combinations of a set with four elements, by

part (a), $\binom{4}{3} = 4$.   ∎

There are two distinct methods that can be used to select r objects from a set of n elements. In an **ordered selection**, it is not only what elements are chosen but also the order in which they are chosen that matters. Two ordered selections are said to be the same if the elements chosen are the same and also if the elements are chosen in the same order. An ordered selection of r elements from a set of n elements is an r-permutation of the set.

In an **unordered selection**, on the other hand, it is only the identity of the chosen elements that matters. Two unordered selections are said to be the same if they consist of the same elements, regardless of the order in which the elements are chosen. An unordered selection of r elements from a set of n elements is the same as a subset of size r or an r-combination of the set.

EXAMPLE 6.4.2   Unordered Selections

How many unordered selections of two elements can be made from the set $\{0, 1, 2, 3\}$?

Solution   An unordered selection of two elements from $\{0, 1, 2, 3\}$ is the same as a 2-combination, or subset of size 2, taken from the set. These can be listed systematically as follows:

$$\{0, 1\}, \{0, 2\}, \{0, 3\} \quad \text{subsets containing 0}$$
$$\{1, 2\}, \{1, 3\} \quad\quad\quad \text{subsets containing 1 but not already listed}$$
$$\{2, 3\} \quad\quad\quad\quad\quad \text{subsets containing 2 but not already listed.}$$

Since this listing exhausts all possibilities, there are six subsets in all. Thus $\binom{4}{2} = 6$, which is the number of unordered selections of two elements from a set of four.   ∎

When the values of n and r are small, it is reasonable to calculate values of $\binom{n}{r}$ using the method of **complete enumeration** (listing all possibilities) illustrated in Examples 6.4.1 and 6.4.2. But when n and r are large, it is not feasible to compute these numbers by listing and counting all possibilities.

The general values of $\binom{n}{r}$ can be found by a somewhat indirect but simple method. An equation is derived that contains $\binom{n}{r}$ as a factor. Then this equation is solved to obtain a formula for $\binom{n}{r}$. The method is illustrated by Example 6.4.3.

EXAMPLE 6.4.3   **Relation Between Permutations and Combinations**

Write all 2-permutations of the set {0, 1, 2, 3}. Find an equation relating the number of 2-permutations, $P(4, 2)$, and the number of 2-combinations, $\binom{4}{2}$, and solve this equation for $\binom{4}{2}$.

Solution   According to Theorem 6.2.3, the number of 2-permutations of the set {0, 1, 2, 3} is $P(4, 2)$, which equals

$$\frac{4!}{(4-2)!} = \frac{4 \cdot 3 \cdot \cancel{2} \cdot \cancel{1}}{\cancel{2} \cdot \cancel{1}} = 12.$$

Now the act of constructing a 2-permutation of {0, 1, 2, 3} can be thought of as a two-step process:

Step 1 is to choose a subset of two elements from {0, 1, 2, 3}.

Step 2 is to choose an ordering for the two-element subset.

This process can be illustrated by the possibility tree shown in Figure 6.4.1.

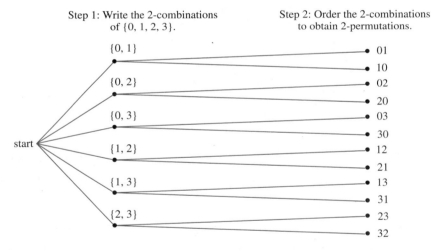

FIGURE 6.4.1   Relation Between Permutations and Combinations

The number of ways to perform step 1 is $\binom{4}{2}$, the same as the number of subsets of size 2 that can be chosen from {0, 1, 2, 3}. The number of ways to perform step 2 is 2!, the number of ways to order the elements in a subset of size 2. Because the number of ways of performing the whole process is the number of 2-permutations of the set {0, 1, 2, 3}, which equals $P(4, 2)$, it follows from the product rule that

$$P(4, 2) = \binom{4}{2} \cdot 2!.$$ This is an equation that relates $P(4, 2)$ and $\binom{4}{2}$.

Solving the equation for $\binom{4}{2}$ gives

$$\binom{4}{2} = \frac{P(4,\ 2)}{2!}$$

Recall that $P(4,\ 2) = \dfrac{4!}{(4-2)!}$. Hence, substituting,

$$\binom{4}{2} = \frac{\dfrac{4!}{(4-2)!}}{2!} = \frac{4!}{2! \cdot (4-2)!} = 6. \qquad \blacksquare$$

The reasoning used in Example 6.4.3 applies in the general case as well. To form an r-permutation of a set of n elements, first choose a subset of r of the n elements (there are $\binom{n}{r}$ ways to perform this step), and then choose an ordering for the r elements (there are $r!$ ways to perform this step). Thus the number of r-permutations is

$$P(n,\ r) = \binom{n}{r} \cdot r!.$$

Now solve for $\binom{n}{r}$ to obtain the formula

$$\binom{n}{r} = \frac{P(n,\ r)}{r!}.$$

Since $P(n,\ r) = \dfrac{n!}{(n-r)!}$, substitution gives

$$\binom{n}{r} = \frac{\dfrac{n!}{(n-r)!}}{r!} = \frac{n!}{r! \cdot (n-r)!}.$$

The result of this discussion is summarized and extended in Theorem 6.4.1.

THEOREM 6.4.1

The number of subsets of size r (or r-combinations) that can be chosen from a set of n elements, $\binom{n}{r}$, is given by the formula

$$\binom{n}{r} = \frac{P(n,\ r)}{r!} \qquad \text{first version}$$

or, equivalently,

$$\binom{n}{r} = \frac{n!}{r! \cdot (n-r)!} \qquad \text{second version}$$

where n and r are nonnegative integers with $r \leq n$.

Note that the analysis presented before the theorem proves the theorem in all cases where n and r are positive. If r is zero and n is any nonnegative integer, then $\binom{n}{0}$ is the number of subsets of size zero of a set with n elements. But you know from Section 5.3 that there is only one set that does not have any elements. Consequently, $\binom{n}{0} = 1$. Also

$$\frac{n!}{0! \cdot (n - 0)!} = \frac{\cancel{n!}}{1 \cdot \cancel{n!}} = 1$$

since $0! = 1$ by definition. (Remember we said that definition would turn out to be convenient!) Hence the formula

$$\binom{n}{0} = \frac{n!}{0! \cdot (n - 0)!}$$

holds for all integers $n \geq 0$, and so the theorem is true for all nonnegative integers n and r with $r \leq n$.

Many electronic calculators have keys for computing values of $\binom{n}{r}$. Theorem 6.4.1 enables you to compute these by hand as well.

EXAMPLE 6.4.4   Computing $\binom{n}{r}$ by Hand

Compute $\binom{8}{5}$.

Solution   By Theorem 6.4.1,

$$\begin{aligned}\binom{8}{5} &= \frac{8!}{5! \cdot (8 - 5)!} \\ &= \frac{8 \cdot 7 \cdot \cancel{6} \cdot \cancel{5} \cdot \cancel{4} \cdot \cancel{3 \cdot 2 \cdot 1}}{(\cancel{5} \cdot \cancel{4} \cdot \cancel{3} \cdot \cancel{2} \cdot 1) \cdot \cancel{(3 \cdot 2 \cdot 1)}} \qquad \text{always cancel common factors before multiplying} \\ &= 56.\end{aligned}$$ ∎

EXAMPLE 6.4.5   Calculating the Number of Teams

Consider again the problem of choosing five members from a group of twelve to work as a team on a special project. How many distinct five-person teams can be chosen?

Solution   The number of distinct five-person teams is the same as the number of subsets of size 5 (or 5-combinations) that can be chosen from the set of twelve. This number is $\binom{12}{5}$. By Theorem 6.4.1,

$$\binom{12}{5} = \frac{12!}{(5!) \cdot (12 - 5)!} = \frac{12 \cdot 11 \cdot \cancel{10} \cdot 9 \cdot 8 \cdot \cancel{7!}}{(\cancel{5} \cdot \cancel{4} \cdot \cancel{3} \cdot \cancel{2} \cdot 1) \cdot \cancel{7!}} = 11 \cdot 9 \cdot 8 = 792.$$

So there are 792 distinct five-person teams. ∎

The formula for the number of r-combinations of a set can be applied in a wide variety of situations. Some of these are illustrated in the following examples.

EXAMPLE 6.4.6   Teams That Contain Both or Neither

Suppose two members of the group of twelve insist on working as a pair—any team must either contain both or neither. How many five-person teams can be formed?

Solution   Call the two members of the group that insist on working as a pair A and B. Then any team formed must contain both A and B or neither A nor B. The set of all possible teams can be partitioned into two subsets as shown in Figure 6.4.2.

Because a team that contains both A and B contains exactly three other people from the remaining ten in the group, there are as many such teams as there are subsets of three people that can be chosen from the remaining ten. By Theorem 6.4.1, this number is

$$\binom{10}{3} = \frac{10!}{3! \cdot 7!} = \frac{10 \cdot \overset{3}{\cancel{9}} \cdot \overset{4}{\cancel{8}} \cdot \cancel{7!}}{\cancel{3} \cdot \cancel{2} \cdot 1 \cdot \cancel{7!}} = 120.$$

Because a team that contains neither A nor B contains exactly five people from the remaining ten, there are as many such teams as there are subsets of five people that can be chosen from the remaining ten. By Theorem 6.4.1, this number is

$$\binom{10}{5} = \frac{10!}{5! \cdot 5!} = \frac{\overset{2}{\cancel{10}} \cdot 9 \cdot \overset{2}{\cancel{8}} \cdot 7 \cdot \cancel{6} \cdot \cancel{5!}}{\cancel{5} \cdot \cancel{4} \cdot \cancel{3} \cancel{-2} \cdot 1 \cdot \cancel{5!}} = 252.$$

Because the set of teams that contain both A and B is disjoint from the set of teams that contain neither A nor B, by the addition rule,

$$\begin{bmatrix} \text{number of teams containing} \\ \text{both } A \text{ and } B \text{ or} \\ \text{neither } A \text{ nor } B \end{bmatrix} = \begin{bmatrix} \text{number of teams} \\ \text{containing} \\ \text{both } A \text{ and } B \end{bmatrix} + \begin{bmatrix} \text{number of teams} \\ \text{containing} \\ \text{neither } A \text{ nor } B \end{bmatrix}$$

$$= 120 + 252 = 372.$$

The reasoning in this example is summarized in Figure 6.4.2.

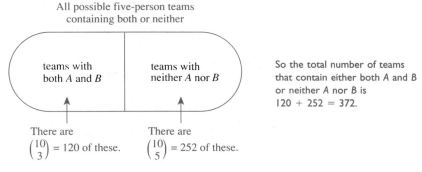

FIGURE 6.4.2

EXAMPLE 6.4.7   Teams That Do Not Contain Both

Suppose two members of the group don't get along and refuse to work together on a team. How many five-person teams can be formed?

Solution   Call the two people who refuse to work together C and D. There are two different ways to answer the given question: One uses the addition rule and the other uses the difference rule.

To use the addition rule, partition the set of all teams that don't contain both C and D into three subsets as shown in Figure 6.4.3.

Because any team that contains C but not D contains exactly four other people from the remaining ten in the group, by Theorem 6.4.1 the number of such teams is

$$\binom{10}{4} = \frac{10!}{4!(10-4)!} = \frac{10 \cdot \overset{3}{\cancel{9}} \cdot \cancel{8} \cdot 7 \cdot \cancel{6!}}{\cancel{4} \cdot \cancel{3} \cdot \cancel{2} \cdot 1 \cdot \cancel{6!}} = 210.$$

Similarly, there are $\binom{10}{4} = 210$ teams that contain D but not C. Finally, by the same reasoning as in Example 6.4.6, there are 252 teams that contain neither C nor D. Thus by the addition rule,

$$\begin{bmatrix} \text{number of teams that do} \\ \text{not contain both } C \text{ and } D \end{bmatrix} = 210 + 210 + 252 = 672.$$

This reasoning is summarized in Figure 6.4.3.

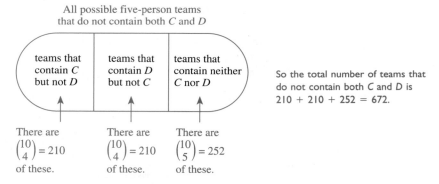

FIGURE 6.4.3

The alternative solution by the difference rule is based on the following observation: The set of all five-person teams that don't contain both C and D equals the set difference between the set of all five-person teams and the set of all five-person teams that contain both C and D. By Example 6.4.5, the total number of five-person teams is $\binom{12}{5} = 792$. Thus by the difference rule,

$$\begin{bmatrix} \text{number of teams that don't} \\ \text{contain both } C \text{ and } D \end{bmatrix} = \begin{bmatrix} \text{total number of} \\ \text{teams of five} \end{bmatrix} - \begin{bmatrix} \text{number of teams that} \\ \text{contain both } C \text{ and } D \end{bmatrix}$$

$$= \binom{12}{5} - \binom{10}{3} = 792 - 120 = 672.$$

This reasoning is summarized in Figure 6.4.4.

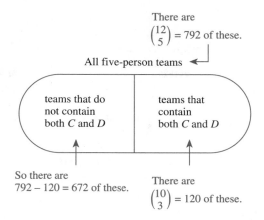

FIGURE 6.4.4

Before beginning the next example, a remark on the phrases *at least* and *at most* is in order:

At least n means "n or more."

At most n means "n or fewer."

For instance, if a set consists of three elements and you are to choose at least two, you will choose two or three; if you are to choose at most two, you will choose none, or one, or two.

EXAMPLE 6.4.8   **Teams with Members of Two Types**

Suppose the group of twelve consists of five men and seven women.

a. How many five-person teams can be chosen that consist of three men and two women?
b. How many five-person teams contain at least one man?
c. How many five-person teams contain at most one man?

Solution   a. To answer this question, think of forming a team as a two-step process:

Step 1 is to choose the men.

Step 2 is to choose the women.

There are $\binom{5}{3}$ ways to choose the three men out of the five and $\binom{7}{2}$ ways to choose the two women out of the seven. Hence by the product rule,

$$\begin{bmatrix}\text{number of teams of five that} \\ \text{contain three men and two women}\end{bmatrix} = \binom{5}{3} \cdot \binom{7}{2} = \frac{5!}{3!2!} \cdot \frac{7!}{2!5!}$$

$$= \frac{7 \cdot 6 \cdot 5 \cdot \overset{2}{\cancel{4}} \cdot \cancel{3} \cdot \cancel{2} \cdot 1}{\cancel{3} \cdot \cancel{2} \cdot 1 \cdot \cancel{2} \cdot 1}$$

$$= 210.$$

b. This question can also be answered either by the addition rule or by the difference rule. The solution by the difference rule is shorter and is shown first.

Observe that the set of five-person teams containing at least one man equals the set difference between the set of all five-person teams and the set of five-person teams that do not contain any men. See Figure 6.4.5.

Now a team with no men consists entirely of five women chosen from the seven women in the group. So there are $\binom{7}{5}$ such teams. Also, by Example 6.4.5, the total number of five-person teams is $\binom{12}{5} = 792$. Hence by the difference rule,

$$\begin{bmatrix} \text{number of teams} \\ \text{with at least} \\ \text{one man} \end{bmatrix} = \begin{bmatrix} \text{total number} \\ \text{of teams} \\ \text{of five} \end{bmatrix} - \begin{bmatrix} \text{number of teams} \\ \text{of five that do not} \\ \text{contain any men} \end{bmatrix}$$

$$= \binom{12}{5} - \binom{7}{5} = 792 - \frac{7!}{5! \cdot 2!}$$

$$= 792 - \frac{7 \cdot \overset{3}{\cancel{6}} \cdot \cancel{5!}}{\cancel{5!} \cdot \cancel{2} \cdot 1} = 792 - 21 = 771.$$

This reasoning is summarized in Figure 6.4.5.

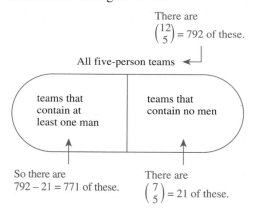

There are $\binom{12}{5} = 792$ of these.

All five-person teams

teams that contain at least one man

teams that contain no men

So there are $792 - 21 = 771$ of these.

There are $\binom{7}{5} = 21$ of these.

FIGURE 6.4.5

Alternatively, to use the addition rule observe that the set of teams containing at least one man can be partitioned as shown in Figure 6.4.6. The number of teams in each subset of the partition is calculated using the method illustrated in part (a). There are

$$\binom{5}{1} \cdot \binom{7}{4} \quad \text{teams with one man and four women;}$$

$$\binom{5}{2} \cdot \binom{7}{3} \quad \text{teams with two men and three women;}$$

$$\binom{5}{3} \cdot \binom{7}{2} \quad \text{teams with three men and two women;}$$

$$\binom{5}{4} \cdot \binom{7}{1} \quad \text{teams with four men and one woman;}$$

$$\binom{5}{5} \cdot \binom{7}{0} \quad \text{teams with five men and no women.}$$

Hence, by the addition rule,

$$\begin{bmatrix} \text{number of teams with} \\ \text{at least one man} \end{bmatrix}$$

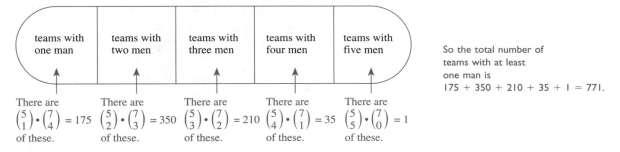

$$= 175 + 350 + 210 + 35 + 1 = 771.$$

This reasoning is summarized in Figure 6.4.6.

Teams with at least one man

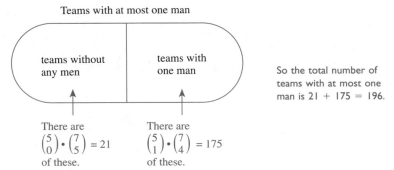

FIGURE 6.4.6

c. As shown in Figure 6.4.7, the set of teams containing at most one man can be partitioned into the set that does not contain any men and the set that contains exactly one man. Hence, by the addition rule,

$$\begin{bmatrix} \text{number of teams} \\ \text{with at} \\ \text{most one man} \end{bmatrix} = \begin{bmatrix} \text{number of} \\ \text{teams without} \\ \text{any men} \end{bmatrix} + \begin{bmatrix} \text{number of} \\ \text{teams with} \\ \text{one man} \end{bmatrix}$$

$$= \binom{5}{0} \cdot \binom{7}{5} + \binom{5}{1} \cdot \binom{7}{4} = 21 + 175 = 196.$$

This reasoning is summarized in Figure 6.4.7.

Teams with at most one man

FIGURE 6.4.7

EXAMPLE 6.4.9   Poker Hand Problems

The game of poker is played with an ordinary deck of cards (see Example 6.1.1). Various five-card holdings are given special names, and certain holdings beat certain other holdings. The named holdings are listed from highest to lowest below.

royal flush: 10, J, Q, K, A of the same suit

straight flush: five adjacent denominations of the same suit but not a royal flush— aces can be high or low so A, 2, 3, 4, 5 of the same suit is a straight flush.

four of a kind: four cards of one denomination, the fifth card can be any other in the deck

full house: three cards of one denomination, two cards of another denomination

flush: five cards of the same suit but not a straight or a royal flush

straight: five cards of adjacent denominations but not all of the same suit—aces can be high or low

three of a kind: three cards of the same denomination and two other cards of different denominations

two pairs: two cards of one denomination, two cards of a second denomination, and a fifth card of a third denomination

one pair: two cards of one denomination and three other cards all of different denominations

no pairs: all cards of different denominations but not a straight or straight flush

a. How many five-card poker hands contain two pairs?
b. If a five-card hand is dealt at random from an ordinary deck of cards, what is the probability that the hand contains two pairs?

Solution   a. Consider forming a hand with two pairs as a four-step process.

Step 1 is to choose the two denominations for the pairs.
Step 2 is to choose two cards from the smaller denomination.
Step 3 is to choose two cards from the larger denomination.
Step 4 is to choose one card from those remaining.

The number of ways to perform step 1 is $\binom{13}{2}$ because there are 13 denomina-

tions in all. The number of ways to perform steps 2 and 3 is $\binom{4}{2}$ because there are

four cards of each denomination, one in each suit. The number of ways to perform

step 4 is $\binom{44}{1}$ because removing the eight cards in the two chosen denominations

from the 52 in the deck leaves 44 from which to choose the fifth card. Thus

the total number of hands with two pairs $= \binom{13}{2}\binom{4}{2}\binom{4}{2}\binom{44}{1}$

$$= \frac{13!}{2!(13-2)!} \cdot \frac{4!}{2!(4-2)!} \cdot \frac{4!}{2!(4-2)!} \cdot \frac{44!}{1!(44-1)!}$$

$$= \frac{13 \cdot 12 \cdot 11!}{(2 \cdot 1) \cdot 11!} \cdot \frac{4 \cdot 3 \cdot 2!}{(2 \cdot 1) \cdot 2!} \cdot \frac{4 \cdot 3 \cdot 2!}{(2 \cdot 1) \cdot 2!} \cdot \frac{44 \cdot 43!}{1 \cdot 43!}$$

$$= 78 \cdot 6 \cdot 6 \cdot 44 = 123{,}552.$$

b. The total number of five-card hands from an ordinary deck of cards is $\binom{52}{5} =$ 2,598,960. Thus if all hands are equally likely, the probability of obtaining a hand with two pairs is $\dfrac{123,552}{2,598,960} \cong 4.75\%$.   ■

EXAMPLE 6.4.10   Number of Bit Strings with Fixed Number of 1's

How many eight-bit strings have exactly three 1's?

Solution   To solve this problem, imagine eight empty positions into which the 0's and 1's of the bit string will be placed.

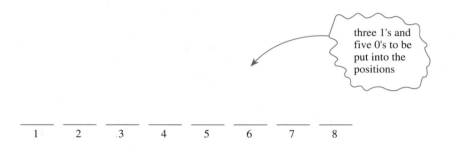

three 1's and five 0's to be put into the positions

$$\underline{\hspace{1cm}}\quad\underline{\hspace{1cm}}\quad\underline{\hspace{1cm}}\quad\underline{\hspace{1cm}}\quad\underline{\hspace{1cm}}\quad\underline{\hspace{1cm}}\quad\underline{\hspace{1cm}}\quad\underline{\hspace{1cm}}$$
$$\;\;1\qquad 2\qquad 3\qquad 4\qquad 5\qquad 6\qquad 7\qquad 8$$

Once a subset of three positions has been chosen from the eight to contain 1's, then the remaining five positions must all contain 0's (since the string is to have exactly three 1's). It follows that the number of ways to construct an eight-bit string with exactly three 1's is the same as the number of subsets of three positions that can be chosen from the eight into which to place the 1's. By Theorem 6.4.1, this equals

$$\binom{8}{3} = \frac{8!}{3! \cdot 5!} = \frac{8 \cdot 7 \cdot \cancel{6} \cdot \cancel{5!}}{\cancel{3} \cdot \cancel{2} \cdot \cancel{5!}} = 56. \qquad ■$$

EXAMPLE 6.4.11   Permutations of a Set with Repeated Elements

Consider various ways of ordering the letters in the word *MISSISSIPPI*:

$$\textit{IIMSSPISSIP,}\quad \textit{ISSSPMIIPIS,}\quad \textit{PIMISSSSIIP,}\quad \text{and so on.}$$

How many distinguishable orderings are there?

Solution   This example generalizes Example 6.4.10. Imagine placing the 11 letters of *MISSISSIPPI* one after another into 11 positions.

letters of MISSISSIPPI to be placed into the positions

$$\underline{\hspace{0.6cm}}\;\underline{\hspace{0.6cm}}\;\underline{\hspace{0.6cm}}\;\underline{\hspace{0.6cm}}\;\underline{\hspace{0.6cm}}\;\underline{\hspace{0.6cm}}\;\underline{\hspace{0.6cm}}\;\underline{\hspace{0.6cm}}\;\underline{\hspace{0.6cm}}\;\underline{\hspace{0.6cm}}\;\underline{\hspace{0.6cm}}$$
$$1\;\;2\;\;3\;\;4\;\;5\;\;6\;\;7\;\;8\;\;9\;\;10\;\;11$$

Because copies of the same letter cannot be distinguished from one another, once the positions for a certain letter are known, then all copies of the letter can go into the positions in any order. It follows that constructing an ordering for the letters can be thought of as a four-step process:

Step 1 is to choose a subset of four positions for the S's.

Step 2 is to choose a subset of four positions for the I's.

Step 3 is to choose a subset of two positions for the P's.

Step 4 is to choose a subset of one position for the M.

Since there are 11 positions in all, there are $\binom{11}{4}$ subsets of four positions for the S's. Once the four S's are in place, there are seven positions that remain empty, so there are $\binom{7}{4}$ subsets of four positions for the I's. After the I's are in place, there are three positions left empty, so there are $\binom{3}{2}$ subsets of two positions for the P's. That leaves just one position for the M. But $1 = \binom{1}{1}$. Hence by the multiplication rule,

$$
\begin{bmatrix} \text{number of ways to} \\ \text{position all the letters} \end{bmatrix} = \binom{11}{4} \cdot \binom{7}{4} \cdot \binom{3}{2} \cdot \binom{1}{1}
$$

$$
= \frac{11!}{4!\,7!} \cdot \frac{7!}{4!\,3!} \cdot \frac{3!}{2!\,1!} \cdot \frac{1!}{1!\,0!}
$$

$$
= \frac{11!}{4! \cdot 4! \cdot 2! \cdot 1!} = 34{,}650. \qquad \blacksquare
$$

In exercise 18 at the end of this section you are asked to show that changing the order in which the letters are placed into the positions does not change the answer to this example.

The same reasoning used in this example can be used to derive the following general theorem.

THEOREM 6.4.2

Suppose a collection consists of n objects of which:

n_1 are of type 1 and are indistinguishable from each other;

n_2 are of type 2 and are indistinguishable from each other;

$\vdots$

n_k are of type k and are indistinguishable from each other;

and suppose that $n_1 + n_2 + \cdots + n_k = n$. Then the number of distinct permutations of the n objects is

$$
\binom{n}{n_1} \cdot \binom{n - n_1}{n_2} \cdot \binom{n - n_1 - n_2}{n_3} \cdot \ldots \cdot \binom{n - n_1 - n_2 - \cdots - n_{k-1}}{n_k}
$$

$$
= \frac{n!}{n_1!\,n_2!\,n_3! \cdots n_k!}.
$$

SOME ADVICE ABOUT COUNTING

Students learning counting techniques often ask, "How do I know what to multiply and what to add? When do I use the multiplication rule and when do I use the addition rule?" Unfortunately, these questions have no easy answers. You need to imagine, as vividly as possible, the objects you are to count. You should then construct a model that would allow you to count the objects one by one if you had enough time. If you can imagine the elements to be counted as being obtained through a multistep process (in which each step is performed in a fixed number of ways regardless of how preceding steps were performed), then you can use the multiplication rule. The total number of elements is the product of the number of ways to perform each step. If, however, you can imagine the set of elements to be counted as being broken up into disjoint subsets, then you can use the addition rule. The total number of elements in the set is the sum of the number of elements in each subset.

One of the most common mistakes students make is to count certain possibilities more than once.

EXAMPLE 6.4.12   **Double Counting**

Consider again the problem of Example 6.4.8(b). A group consists of five men and seven women. How many teams of five contain at least one man?

▲ *CAUTION!*   *False Solution:*

Imagine constructing the team as a two-step process:

Step 1 is to choose a subset of one man from the five men.

Step 2 is to choose a subset of four others from the remaining eleven people.

Hence by the multiplication rule, there are $\binom{5}{1} \cdot \binom{11}{4} = 1{,}650$ five-person teams that contain at least one man.

Analysis of the False Solution:
The problem with the solution above is that some teams are counted more than once. Suppose the men are Anwar, Ben, Carlos, Dwayne, and Ed and the women are Fumiko, Gail, Hui-Fan, Inez, Jill, Kim, and Laura. According to the method described above, one possible outcome of the two-step process is as follows:

outcome of step 1: Anwar

outcome of step 2: Ben, Gail, Inez, and Jill.

So the team would be {Anwar, Ben, Gail, Inez, Jill}. But another possible outcome is

outcome of step 1: Ben

outcome of step 2: Anwar, Gail, Inez, and Jill,

which also gives the team {Anwar, Ben, Gail, Inez, Jill}. Thus this one team is given by two different branches of the possibility tree, and so it is counted twice.   ■

The best way to avoid mistakes such as the one described above is to mentally imagine the possibility tree corresponding to any use of the multiplication rule and the set partition corresponding to a use of the addition rule. Check how your division into steps works by applying it to some actual data—as was done in the analysis above—and try to pick data that are as typical or generic as possible.

It often helps to ask yourself: (1) "Am I counting everything?" and (2) "Am I counting anything twice?" When using the multiplication rule, these questions become (1) "Does every outcome appear as some branch of the tree?" and (2) "Does any outcome appear on more than one branch of the tree?" When using the addition rule, the questions become (1) "Does every outcome appear in some subset of the diagram?" and (2) "Do any two subsets in the diagram share common elements?"

EXERCISE SET 6.4

1. a. List all 2-combinations for the set $\{x_1, x_2, x_3\}$. Deduce the value of $\binom{3}{2}$.
   b. List all unordered selections of four elements from the set $\{a, b, c, d, e\}$. Deduce the value of $\binom{5}{4}$.

2. a. List all 3-combinations for the set $\{x_1, x_2, x_3, x_4, x_5\}$. Deduce the value of $\binom{5}{3}$.
   b. List all unordered selections of three elements from the set $\{x_1, x_2, x_3, x_4, x_5, x_6, x_7\}$. Deduce the value of $\binom{7}{3}$.

3. Write an equation relating $P(7, 2)$ and $\binom{7}{2}$.

4. Write an equation relating $P(8, 5)$ and $\binom{8}{5}$.

5. Compute each of the following.
   a. $\binom{5}{0}$ b. $\binom{5}{1}$ c. $\binom{5}{2}$ d. $\binom{5}{3}$
   e. $\binom{5}{4}$ f. $\binom{5}{5}$

6. A student council consists of 15 students.
   a. In how many ways can a committee of six be selected from the membership of the council?
   b. Two council members have the same major and are not permitted to serve together on a committee. How many ways can a committee of six be selected from the membership of the council?
   c. Two council members always insist on serving on committees together. If they can't serve together, they won't serve at all. How many ways can a committee of six be selected from the council membership?
   d. Suppose the council contains eight men and seven women.
      (i) How many committees of six contain three men and three women?
      (ii) How many committees of six contain at least one woman?
   e. Suppose the council consists of three freshmen, four sophomores, three juniors, and five seniors. How

many committees of eight contain two representatives from each class?

7. A computer programming team has 14 members.
   a. How many ways can a group of seven be chosen to work on a project?
   b. Suppose eight team members are women and six are men.
      (i) How many groups of seven can be chosen that contain four women and three men?
      (ii) How many groups of seven can be chosen that contain at least one man?
      (iii) How many groups of seven can be chosen that contain at most three women?
   c. Suppose two team members refuse to work together on projects. How many groups of seven can be chosen to work on a project?
   d. Suppose two team members insist on either working together or not at all on projects. How many groups of seven can be chosen to work on a project?

H8. An instructor gives an exam with fourteen questions. Students are allowed to choose any ten to answer.
   a. How many different choices of ten questions are there?
   b. Suppose six questions require proof and eight do not.
      (i) How many groups of ten questions contain four that require proof and six that do not?
      (ii) How many groups of ten questions contain at least one that requires proof?
      (iii) How many groups of ten questions contain at most three that require proof?
   c. Suppose the exam instructions specify that at most one of questions 1 and 2 may be included among the ten. How many different choices of ten questions are there?
   d. Suppose the exam instructions specify that either both questions 1 and 2 are to be included among the ten or neither is to be included. How many different choices of ten questions are there?

9. An all-male club is considering opening its membership to women. In a preliminary survey on the issue, 19 of the 30 members favored admitting women and 11 did not. A committee of six is to be chosen to give further study to the issue.

a. How many committees of six can be formed from the club membership?

b. How many of the committees will contain at least three men who, in the preliminary survey, favored opening the membership to women?

(If you do not have a calculator that computes values of $\binom{n}{r}$, write your answers as numerical expressions using the symbol $\binom{n}{r}$ for some particular values of n and r.)

10. Two new drugs are to be tested using a group of 40 laboratory mice, each tagged with a number for identification purposes. Drug A is to be given to 15 mice, drug B is to be given to another 15 mice, and the remaining 10 mice are to be used as controls. How many ways can the assignment of treatments to mice be made? (A single assignment involves specifying the treatment for each mouse—whether drug A, drug B, or no drug.)

◆11. For each poker holding below, (1) find the number of five-card poker hands with that holding; (2) find the probability that a randomly chosen set of five cards has that holding.

a. royal flush   b. straight flush   c. four of a kind
d. full house   e. flush   f. straight
g. three of a kind   h. one pair
i. no repeated denomination and not of five adjacent denominations

◆H 12. Assuming that all years have 365 days and all birthdays occur with equal probability, how large must n be so that in any randomly chosen group of n people, the probability that two or more have the same birthday is at least 1/2? (This is called the **birthday problem**.)

13. A coin is tossed ten times. In each case the outcome H (for heads) or T (for tails) is recorded. (One possible outcome of the ten tossings is denoted

THHTTTHTTH.)

a. What is the total number of possible outcomes of the coin-tossing experiment?

b. In how many of the possible outcomes are exactly five heads obtained?

c. In how many of the possible outcomes are at least nine heads obtained?

d. In how many of the possible outcomes is at least one head obtained?

e. In how many of the possible outcomes is at most one head obtained?

14. a. How many 16-bit strings contain exactly nine 1's?

b. How many 16-bit strings contain at least fourteen 1's?

c. How many 16-bit strings contain at least one 1?

d. How many 16-bit strings contain at most one 1?

15. a. How many even integers are in the set

$$\{1, 2, 3, \ldots, 100\}?$$

b. How many odd integers are in the set

$$\{1, 2, 3, \ldots, 100\}?$$

c. How many ways can two integers be selected from the set $\{1, 2, 3, \ldots, 100\}$ so that their sum is even?

d. How many ways can two integers be selected from the set $\{1, 2, 3, \ldots, 100\}$ so that their sum is odd?

16. Suppose that three computer boards in a production run of forty are defective. A sample of four is to be selected to be checked for defects.

a. How many different samples can be chosen?

b. How many samples will contain at least one defective board?

c. What is the probability that a randomly chosen sample of four contains at least one defective board?

17. Nine points labeled A, B, C, D, E, F, G, H, I are arranged in a plane in such a way that no three lie on the same straight line.

a. How many straight lines are determined by the nine points?

b. How many of these straight lines do not pass through point A?

c. How many triangles have three of the nine points as vertices?

d. How many of these triangles do not have A as a vertex?

18. Suppose that you placed the letters in Example 6.4.11 into positions in the following order: first the M, then the I's, then the S's, and then the P's. Show that you would obtain the same answer for the number of distinguishable orderings.

19. a. How many distinguishable ways can the letters of the word *HULLABALOO* be arranged?

b. How many distinguishable arrangements of the letters of *HULLABALOO* begin with U and end with L?

c. How many distinguishable arrangements of the letters of *HULLABALOO* contain the two letters *HU* next to each other in order?

20. a. How many distinguishable ways can the letters of the word *INTELLIGENCE* be arranged?

b. How many distinguishable arrangements of the letters of *INTELLIGENCE* begin with T and end with G?

c. How many distinguishable arrangements of the letters of *INTELLIGENCE* contain the letters *INT* next to each other in order and also the letters *IG* next to each other in order?

21. When the expression $(a + b)^4$ is multiplied out, terms of the form $aaaa$, $abaa$, $baba$, $bbba$, and so on are obtained. Let $\Sigma = \{a, b\}$. Then $\Sigma^4 = $ the set of all strings over Σ of length 4.

a. What is $n(\Sigma^4)$? In other words, how many strings of length 4 can be constructed using a's and b's?

b. How many elements of Σ^4 (strings of length 4) have three a's and one b?

c. How many elements of Σ^4 (strings of length 4) have two a's and two b's?

22. In Morse code, symbols are represented by variable-length sequences of dots and dashes. (For example, $A = \cdot -$, $1 = \cdot ----$, $? = \cdot \cdot -- \cdot \cdot$.) How many different symbols can be represented by sequences of six or fewer dots and dashes?

23. On an 8×8 chessboard, a rook is allowed to move any number of squares either horizontally or vertically. How many different paths can a rook follow from the bottom-left square of the board to the top-right square of the board if all moves are to the right or upward?

24. The number 42 has the prime factorization $2 \cdot 3 \cdot 7$. Thus 42 can be written in four ways as a product of two positive integer factors: $1 \cdot 42$, $6 \cdot 7$, $14 \cdot 3$, and $2 \cdot 21$.

a. How many distinct ways can the number 60 be written as a product of two positive integer factors?

b. If $n = p_1 \cdot p_2 \cdot p_3 \cdot p_4 \cdot p_5$, where the p_i are distinct prime numbers, how many ways can n be written as a product of two positive integer factors?

c. If $n = p_1 \cdot p_2 \cdot \ldots \cdot p_k$, where the p_i are all distinct prime numbers, how many ways can n be written as a product of two positive integer factors?

◆H25. A student council consists of three freshmen, four sophomores, three juniors, and five seniors. How many committees of eight members of the council contain at least one member from each class?

◆ 26. An alternative way to derive Theorem 6.4.1 uses the following *division rule*: Let n and k be integers so that k divides n. If a set consisting of n elements is divided into subsets that each contain k elements, then the number of such subsets is n/k. Explain how Theorem 6.4.1 can be derived using the division rule.

27. Find the error in the following reasoning: "Consider forming a poker hand with two pairs as a five-step process.

Step 1: Choose the denomination of one of the pairs.

Step 2: Choose the two cards of that denomination.

Step 3: Choose the denomination of the other of the pairs.

Step 4: Choose the two cards of that second denomination.

Step 5: Choose the fifth card from the remaining denominations.

There are $\binom{13}{1}$ ways to perform step 1, $\binom{4}{2}$ ways to perform step 2, $\binom{12}{1}$ ways to perform step 3, $\binom{4}{2}$ ways to perform step 4, and $\binom{44}{1}$ ways to perform step 5.

Therefore the total number of five-card poker hands with two pairs is $13 \cdot 6 \cdot 12 \cdot 6 \cdot 44 = 247,104$."

6.5 *r*- COMBINATIONS WITH REPETITION ALLOWED

The value of mathematics in any science lies more in disciplined analysis and abstract thinking than in particular theories and techniques.

(Alan Tucker, 1982)

In Section 6.4 we showed that there are $\binom{n}{r}$ r-combinations, or subsets of size r, of a set of n elements. In other words, there are $\binom{n}{r}$ ways to choose r distinct elements without regard to order from a set of n elements. For instance, there are $\binom{4}{3} = 4$ ways to choose three elements out of a set of four: $\{1, 2, 3\}$, $\{1, 2, 4\}$, $\{1, 3, 4\}$, $\{2, 3, 4\}$.

In this section we ask: How many ways are there to choose r elements without regard to order from a set of n elements *if repetition is allowed*? A good way to imagine this is to visualize the n elements as categories of objects from which multiple selections may be made. For instance, if the categories are labeled 1, 2, 3, and 4 and three elements are chosen, it is possible to choose two elements of type 3 and one of type 1, or all three of type 2, or one each of types 1, 2, and 4. We denote such choices by [3, 3, 1], [2, 2, 2], and [1, 2, 4] respectively. Note that because order does not matter, [3, 3, 1] = [3, 1, 3] = [1, 3, 3], for example.

DEFINITION

An **r-combination with repetition allowed**, or **multiset of size *r***, chosen from a set X of n elements is an unordered selection of elements taken from X with repetition allowed. If $X = \{x_1, x_2, \ldots, x_n\}$, we write an r-combination with repetition allowed, or multiset of size r, as $[x_{i_1}, x_{i_2}, \ldots, x_{i_r}]$ where each x_{i_j} is in X and some of the x_{i_j} may equal each other.

EXAMPLE 6.5.1   *r*-combinations with Repetition Allowed

Write a complete list to find the number of 3-combinations with repetition allowed, or multisets of size r, that can be selected from $\{1, 2, 3, 4\}$. Observe that because the order in which the elements are chosen does not matter, the elements of each selection may be written in increasing order, and writing the elements in increasing order will ensure that no combinations are overlooked.

Solution

$[1, 1, 1]; [1, 1, 2]; [1, 1, 3]; [1, 1, 4]$ — all combinations that include 1, 1
$[1, 2, 2]; [1, 2, 3]; [1, 2, 4];$ — all combinations that include 1, 2
$[1, 3, 3]; [1, 3, 4]; [1, 4, 4];$ — all combinations that include 1, 3 or 1, 4
$[2, 2, 2]; [2, 2, 3]; [2, 2, 4];$ — all combinations that include 2, 2
$[2, 3, 3]; [2, 3, 4]; [2, 4, 4];$ — all combinations that include 2, 3 or 2, 4
$[3, 3, 3]; [3, 3, 4]; [3, 4, 4];$ — all combinations that include 3, 3 or 3, 4
$[4, 4, 4]$ — the only combination that includes 4, 4.

Thus there are 20 3-combinations with repetition allowed. ∎

How could the number 20 have been predicted other than by making a complete list? Consider the numbers 1, 2, 3, and 4 as categories and imagine choosing a total of three numbers from the categories with multiple selections from any category allowed. The results of several such selections are represented by the table below.

Category 1	Category 2	Category 3	Category 4	Result of the Selection
	| × |	|	× ×	1 from category 2 2 from category 4
× |	| × |	×	1 each from categories 1, 3, and 4
× × × |	|	|	3 from category 1

As you can see, each selection of three numbers from the four categories can be represented by a string of vertical bars and crosses. Three vertical bars are used to separate the four categories and three crosses are used to indicate how many items from each category are chosen. Each distinct string of three vertical bars and three crosses represents a distinct selection. For instance, the string

$$\times \ \times \ | \ | \ \times \ |$$

represents the selection: two from category 1, none from category 2, one from category 3, and none from category 4. Thus the number of distinct selections of three elements that can be formed from the set $\{1, 2, 3, 4\}$ with repetition allowed equals the number of distinct strings of six symbols consisting of three $|$'s and three $\times$'s. But this equals the number of ways to select three positions out of six because once three positions have been chosen for the $\times$'s, the $|$'s are placed in the remaining three positions. So the answer is

$$\binom{6}{3} = \frac{6!}{3! \cdot (6 - 3)!} = \frac{6 \cdot 5 \cdot 4 \cdot 3!}{3 \cdot 2 \cdot 1 \cdot 3!} = 20,$$

as was obtained earlier by a careful listing.

The analysis of this example extends to the general case. To count the number of r-combinations with repetition allowed, or multisets of size r, that can be selected from a set of n elements, think of the elements of the set as categories. Then each r-combination with repetition allowed can be represented as a string of $n - 1$ vertical bars (to separate the n categories) and r crosses (to represent the r elements to be chosen). The number of $\times$'s in each category represents the number of times the element represented by that category is repeated.

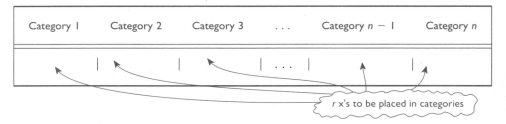

The number of strings of $n - 1$ vertical bars and r crosses is the number of ways to choose r positions, into which to place the r crosses, out of a total of $r + (n - 1)$ positions, leaving the remaining positions for the vertical bars. But by Theorem 6.4.1 this number is

$$\binom{r + n - 1}{r}.$$

This discussion proves the following.

THEOREM 6.5.1

The number of r-combinations with repetition allowed, or multisets of size r, that can be selected from a set of n elements is $\binom{r + n - 1}{r}$. This equals the number of ways r objects can be selected from n categories of objects with repetition allowed.

EXAMPLE 6.5.2 Selecting 15 Cans of Soft Drinks of Five Different Types

A person giving a party wants to set out 15 assorted cans of soft drinks for his guests. He shops at a store that sells five different types of soft drinks.

a. How many different selections of cans of 15 soft drinks can he make?
b. If root beer is one of the types of soft drink, how many different selections include at least six cans of root beer?
c. What is the probability that a randomly chosen selection of 15 soft drinks includes at least six cans of root beer?

Solution a. Think of the five different types of soft drinks as the n categories and the 15 cans of soft drinks to be chosen as the r objects. (So $n = 5$ and $r = 15$.) Each selection of cans of soft drinks is represented by a string of $5 - 1 = 4$ vertical bars (to separate the categories of soft drinks) and 15 crosses (to represent the cans selected). For instance, the string

$$\times \times \times \mid \times \times \times \times \times \times \times \mid \quad \mid \times \times \times \mid \times \times$$

represents a selection of three cans of soft drinks of type 1, seven of type 2, none of type 3, three of type 4, and two of type 5. The total number of selections of 15 cans of soft drinks of the five types is the number of strings of 19 symbols, $5 - 1 = 4$ of them $\mid$ and 15 of them $\times$:

$$\binom{15 + 5 - 1}{15} = \binom{19}{15} = \frac{19 \cdot \overset{6}{\cancel{18}} \cdot 17 \cdot \overset{2}{\cancel{16}} \cdot \cancel{15!}}{\cancel{15!} \cdot \cancel{4} \cdot \cancel{3} \cdot \cancel{2} \cdot 1} = 3{,}876.$$

b. If at least six cans of root beer are included, we can imagine choosing six such cans first and then choosing 9 additional cans. The choice of the nine additional cans can be represented as a string of 9 $\times$'s and 4 $\mid$'s. For example, if root beer is type 1, then the string $\times \times \times \mid \quad \mid \times \times \mid \times \times \times \times \mid$ represents a selection of three cans of root beer (in addition to the six chosen initially), none of type 2, two of type 3, four of type 4, and none of type 5. Thus the total number of selections of 15 cans of soft drinks of the five types, including at least six cans of root beer, is the number of strings of 13 symbols, $4 (= 5 - 1)$ of them $\mid$ and 9 of them $\times$:

$$\binom{9 + 4}{9} = \binom{13}{9} = \frac{13 \cdot \cancel{12} \cdot 11 \cdot \overset{5}{\cancel{10}} \cdot \cancel{9!}}{\cancel{9!} \cdot \cancel{4} \cdot \cancel{3} \cdot \cancel{2} \cdot 1} = 715.$$

c. The probability that a randomly chosen selection of cans will include at least six of root beer is the ratio of the number of selections that contain at least six cans of root beer (the answer to (b)) to the total number of selections (the answer to (a)). Therefore the probability is $715/3{,}876 \cong 18.45\%$. ∎

EXAMPLE 6.5.3 Counting Triples (i, j, k) with $1 \le i \le j \le k \le n$

If n is a positive integer, how many triples of integers from 1 through n can be formed in which the elements of the triple are written in increasing order but are not necessarily distinct? In other words, how many triples of integers (i, j, k) are there with $1 \le i \le j \le k \le n$?

Solution Any triple of integers (i, j, k) with $1 \le i \le j \le k \le n$ can be represented as a string of $n - 1$ vertical bars and three crosses, with the positions of the crosses indicating which three integers from 1 to n are included in the triple. The table below illustrates this for $n = 5$.

Category					Result of the Selection
1	2	3	4	5	
|	| ×× |		| ×		(3, 3, 5)
× |	× |		| × |		(1, 2, 4)

Thus the number of such triples is the same as the number of strings of $(n - 1)$ |'s and 3 ×'s, which is

$$\binom{3 + (n - 1)}{3} = \binom{n + 2}{3} = \frac{(n + 2)!}{3! \cdot (n + 2 - 3)!}$$
$$= \frac{(n + 2)(n + 1)n \cdot (n-1)!}{3! \cdot (n-1)!} = \frac{n(n + 1)(n + 2)}{6}. \qquad ■$$

Note that in Examples 6.5.2 and 6.5.3 the reasoning behind Theorem 6.5.1 was used rather than the statement of the theorem itself. Alternatively, in either example you could invoke Theorem 6.5.1 directly by recognizing that the items to be counted are either r-combinations with repetition allowed or are the same in number as such combinations. For instance, in Example 6.5.3 you might observe that there are exactly as many triples of integers (i, j, k) with $1 \le i \le j \le k \le n$ as there are 3-combinations of integers from 1 through n with repetition allowed because the elements of any such 3-combination can be written in increasing order in only one way.

EXAMPLE 6.5.4   Counting Iterations of a Loop

How many times will the innermost loop be iterated when the algorithm segment below is implemented and run? (Assume n is a positive integer.)

```
for k := 1 to n
    for j := 1 to k
        for i := 1 to j
            [Statements in the body of the inner loop,
            none containing branching statements that lead
            outside the loop]
        next i
    next j
next k
```

Solution   Construct a trace table for the values of k, j, and i for which the statements in the body of the innermost loop are executed. (See the table on the next page.) Because i goes from 1 to j, it is always the case that $i \le j$. Similarly because j goes from 1 to k, it is always the case that $j \le k$. To focus on the details of the table construction, consider what happens when $k = 3$. In this case, j takes each value 1, 2, and 3. When $j = 1$, i can only take the value 1 (because $i \le j$). When $j = 2$, i takes each value 1 and 2 (again because $i \leq j$). When $j = 3$, i takes each value 1, 2, and 3 (yet again because $i \leq j$).

k	1	2	→	3				→	...	n				→
j	1	1	2 →	1	2 →	3		→	...	1	2 →		...	n →
i	1	1	1	2	1	1	2	1	2	3	...	1	1	2	...	1	...	n

Observe that there is one iteration of the innermost loop for each column of this table, and there is one column of the table for each triple of integers (i, j, k) with $1 \leq i \leq j \leq k \leq n$. But in Example 6.5.3 the number of such triples was shown to be $[n(n + 1)(n + 2)]/6$. Thus there are $[n(n + 1)(n + 2)]/6$ iterations of the innermost loop.   ■

This solution in Example 6.5.4 is the most elegant and generalizable (see exercises 8 and 9) to the given problem. An alternative solution using summations is outlined in exercise 20.

EXAMPLE 6.5.5   The Number of Integral Solutions of an Equation

How many solutions are there to the equation $x_1 + x_2 + x_3 + x_4 = 10$ if x_1, x_2, x_3, and x_4 are nonnegative integers?

Solution   Think of the number 10 as divided into ten individual units and the variables x_1, x_2, x_3, and x_4 as four categories into which these units are placed. The number of units in each category x_i indicates the value of x_i in a solution of the equation. Each solution can, then, be represented by a string of three vertical bars (to separate the four categories) and ten crosses (to represent the ten individual units). For example, in the table below, the 2 crosses under x_1, 5 crosses under x_2, and 3 crosses under x_4 represent the solution $x_1 = 2$, $x_2 = 5$, $x_3 = 0$, and $x_4 = 3$.

Categories				Solution to the equation $x_1 + x_2 + x_3 + x_4 = 10$
x_1	x_2	x_3	x_4	
✕✕ | ✕✕✕✕✕ | | ✕✕✕				$x_1 = 2$, $x_2 = 5$, $x_3 = 0$, and $x_4 = 3$
✕✕✕✕ | ✕✕✕✕✕✕ | |				$x_1 = 4$, $x_2 = 6$, $x_3 = 0$, and $x_4 = 0$

Therefore, there are as many solutions to the equation as there are strings of ten crosses and 3 vertical bars, namely

$$\binom{10 + 3}{10} = \binom{13}{10} = \frac{13!}{10! \cdot (13 - 10)!} = \frac{13 \cdot 12 \cdot 11 \cdot \cancel{10!}}{\cancel{10!} \cdot 3 \cdot 2 \cdot 1} = 286.$$ ■

Example 6.5.6 illustrates a variation on Example 6.5.5.

328 CHAPTER 6 Counting

Additional Constraints on the Number of Solutions

How many integer solutions are there to the equation $x_1 + x_2 + x_3 + x_4 = 10$ if each $x_i \geq 1$?

Solution   In this case imagine starting by putting one cross in each of the four categories. Then distribute the remaining six crosses among the categories. Such a distribution can be represented by a string of three vertical bars and six crosses. For example, the string

$$\times \times \times \mid \cdot \mid \times \times \mid \times$$

indicates that there are three more crosses in category x_1 in addition to the one cross already there (so $x_1 = 4$), no more crosses in category x_2 in addition to the one already there (so $x_2 = 1$), two more crosses in category x_3 in addition to the one already there (so $x_3 = 3$), and one more cross in category x_4 in addition to the one already there (so $x_4 = 2$). It follows that the number of solutions to the equation that satisfy the given condition is the same as the number of strings of three vertical bars and six crosses, namely:

$$\binom{6+3}{6} = \binom{9}{6} = \frac{9!}{6! \cdot (9-6)!} = \frac{9 \cdot 8 \cdot 7 \cdot 6!}{6! \cdot 3 \cdot 2 \cdot 1} = 84.$$

An alternative solution to this example is based on the observation that since each $x_i \geq 1$, we may introduce new variables $y_i = x_i - 1$ for each $i = 1, 2, 3, 4$. Then each $y_i \geq 0$, and $y_1 + y_2 + y_3 + y_4 = 6$. So the number of solutions of $y_1 + y_2 + y_3 + y_4 = 6$ in nonnegative integers is the same as the number of solutions of $x_1 + x_2 + x_3 + x_4 = 10$ in positive integers. ∎

WHICH FORMULA TO USE?

Sections 6.2–6.5 have discussed four different ways of choosing k elements from n. The order in which the choices are made may or may not matter and repetition may or may not be allowed. The following table summarizes which formula to use in which situation.

	Order Matters	Order Does Not Matter
Repetition is allowed	n^k	$\binom{n+k-1}{k}$
Repetition is not allowed	$P(n, k)$	$\binom{n}{k}$

EXERCISE SET 6.5

1. a. According to Theorem 6.5.1, how many 5-combinations with repetition allowed can be chosen from a set of three elements?
   b. List all of the 5-combinations that can be chosen with repetition allowed from $\{1, 2, 3\}$.

2. a. According to Theorem 6.5.1, how many multisets of size four can be chosen from a set of three elements?
   b. List all of the multisets of size four that can be chosen from the set $\{a, b, c\}$.

3. A bakery produces six different kinds of pastry.
   a. How many different selections of twenty pastries are there?

b. Assuming that eclairs are one kind of pastry produced, how many different selections of twenty pastries are there if at least three must be eclairs?

c. If a selection of twenty pastries is chosen randomly, what is the probability that at least three are eclairs?

d. If a selection of twenty pastries is chosen randomly, what is the probability that exactly three are eclairs?

4. A camera shop stocks ten different types of batteries.
   a. How many ways can a total inventory of 30 batteries be distributed among the ten different types?
   b. Assuming that one of the types of batteries is A76, how many ways can a total inventory of 30 batteries be distributed among the ten different types if the inventory must include at least four A76 batteries?
   c. If an inventory of 30 batteries is selected at random from the ten different types, what is the probability that at least four A76 batteries will be included?
   d. If an inventory of 30 batteries is selected at random from the ten different types, what is the probability that exactly four A76 batteries will be included?

5. If n is a positive integer, how many 4-tuples of integers from 1 through n can be formed in which the elements of the 4-tuple are written in increasing order but are not necessarily distinct? In other words, how many 4-tuples of integers (i, j, k, m) are there with $1 \le i \le j \le k \le m \le n$?

6. If n is a positive integer, how many 5-tuples of integers from 1 through n can be formed in which the elements of the 5-tuple are written in decreasing order but are not necessarily distinct? In other words, how many 5-tuples of integers (h, i, j, k, m) are there with $n \ge h \ge i \ge j \ge k \ge m \ge 1$?

7. Another way to count the number of nonnegative integral solutions to an equation of the form $x_1 + x_2 + \cdots + x_n = m$ is to reduce the problem to one of finding the number of n-tuples $(y_1, y_2, \ldots, y_n)$ with $0 \le y_1 \le y_2 \le \cdots \le y_n \le m$. The reduction results from letting $y_i = x_1 + x_2 + \cdots + x_i$ for each $i = 1, 2, \ldots, n$. Use this approach to derive a general formula for the number of nonnegative integral solutions to $x_1 + x_2 + \cdots + x_n = m$.

In 8 and 9, how many times will the innermost loop be iterated when the algorithm segment is implemented and run? Assume n is a positive integer.

8. **for** $m := 1$ **to** n
     for $k := 1$ **to** m
       for $j := 1$ **to** k
         for $i := 1$ **to** j
           [Statements in the body of the inner loop, none containing branching statements that lead outside the loop]
         next i
       next j
     next k
   next m

9. **for** $k := 1$ **to** n
     for $j := k$ **to** n
       for $i := j$ **to** n
         [Statements in the body of the inner loop, none containing branching statements that lead outside the loop]
       next i
     next j
   next k

In 10–14, find how many solutions there are to the given equation that satisfy the given condition.

10. $x_1 + x_2 + x_3 = 20$, each x_i is a nonnegative integer.

11. $x_1 + x_2 + x_3 = 20$, each x_i is a positive integer.

12. $y_1 + y_2 + y_3 + y_4 = 32$, each y_i is a nonnegative integer.

13. $y_1 + y_2 + y_3 + y_4 = 32$, each y_i is an integer that is at least 2.

14. $a + b + c + d + e = 1000$, each of $a, b, c, d,$ and e is an integer that is at least 10.

15. a. A store sells 30 kinds of balloons. How many different combinations of 24 balloons can be chosen?
    b. What is the probability that a combination of 60 balloons chosen at random will contain at least one balloon of each kind?

16. A large pile of coins consists of pennies, nickles, dimes, and quarters (at least 20 of each).
    a. How many different collections of 20 coins can be chosen?
    b. What is the probability that a collection of 20 coins chosen at random will contain at least four coins of each type?

17. For how many integers from 1 to 10,000 is the sum of their digits equal to 10?

18. Suppose the bakery in exercise 3 has only ten eclairs but has at least twenty of each of the other kinds of pastry.
    a. How many different selections of twenty pastries are there?
    b. Suppose in addition to having only ten eclairs, the bakery has only eight napoleon slices. How many different selections of twenty pastries are there?

19. Suppose the camera shop in exercise 4 can obtain at most eight A76 batteries but can get at least thirty of each of the other types.
    a. How many ways can a total inventory of thirty batteries be distributed among the ten different types?
    b. Suppose that in addition to being able to obtain only eight A76 batteries, the store can only get six of type D303. How many ways can a total inventory of thirty batteries be distributed among the ten different types?

20. Observe that the number of columns in the trace table for Example 6.5.4 can be expressed as the sum

$$1 + (1 + 2) + (1 + 2 + 3) + \cdots + (1 + 2 + \cdots + n).$$

Explain why this is so, and show how this sum simplifies to the same expression given in the solution of Example 6.5.4.

Let us grant that the pursuit of mathematics is a divine madness of the human spirit, a refuge from the goading urgency of contingent happenings.
(Alfred North Whitehead, 1861–1947)

6.6   THE ALGEBRA OF COMBINATIONS

In this section we derive a number of useful formulas that give values of $\binom{n}{r}$ in special cases and explore relations among different values of $\binom{n}{r}$.

EXAMPLE 6.6.1   Values of $\binom{n}{n}$, $\binom{n}{n-1}$, $\binom{n}{n-2}$

Show that for all integers $n \geq 0$,

$$\binom{n}{n} = 1; \qquad\qquad 6.6.1$$

$$\binom{n}{n-1} = n, \quad \text{if } n \geq 1; \qquad\qquad 6.6.2$$

$$\binom{n}{n-2} = \frac{n(n-1)}{2}, \quad \text{if } n \geq 2. \qquad\qquad 6.6.3$$

Solution

$$\binom{n}{n} = \frac{n!}{n!(n-n)!} = \frac{1}{0!} = 1 \qquad \text{since } 0! = 1 \text{ by definition}$$

$$\binom{n}{n-1} = \frac{n!}{(n-1)! \cdot (n-(n-1))!}$$
$$= \frac{n \cdot (n-1)!}{(n-1)! \cdot (n-n+1)!} = \frac{n}{1} = n$$

$$\binom{n}{n-2} = \frac{n!}{(n-2)! \cdot (n-(n-2))!}$$
$$= \frac{n \cdot (n-1) \cdot (n-2)!}{(n-2)! \cdot 2!} = \frac{n(n-1)}{2} \qquad \blacksquare$$

Note that the result derived algebraically above that $\binom{n}{n}$ equals 1 agrees with the fact that a set with n elements has just one subset of size n, namely itself. Similarly, exercise 1 at the end of this section asks you to show algebraically that $\binom{n}{0} = 1$, which agrees with the fact that a set with n elements has one subset, the empty set, of size 0. In exercise 2 you are also asked to show algebraically that $\binom{n}{1} = n$. This result agrees with the fact that there are n subsets of size 1 that can be chosen from a set with n elements, namely the subsets consisting of each element taken alone.

EXAMPLE 6.6.2 $\binom{n}{r} = \binom{n}{n-r}$

In exercise 5 at the end of this section you are asked to verify algebraically that

$$\binom{n}{r} = \binom{n}{n-r}$$

for all nonnegative integers n and r with $r \leq n$.

An alternative way to deduce this formula is to interpret it as saying that a set A with n elements has exactly as many subsets of size r as it has subsets of size $n - r$. Derive the formula using this reasoning.

Solution   Observe that any subset of size r can be specified either by saying which r elements lie in the subset or by saying which $n - r$ elements lie outside the subset.

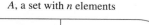

A, a set with n elements

| B, a subset with r elements | A − B, a subset with $n - r$ elements |

Any subset B with r elements completely determines a subset, A − B, with $n - r$ elements.

Suppose A has k subsets of size r: $B_1, B_2, \ldots, B_k$. Then each B_i can be paired up with exactly one set of size $n - r$, namely its complement $A - B_i$ as shown below.

$$
\begin{array}{ccc}
\text{subsets of size } r & & \text{subsets of size } n - r \\
B_1 & \longleftrightarrow & A - B_1 \\
B_2 & \longleftrightarrow & A - B_2 \\
\vdots & & \vdots \\
B_k & \longleftrightarrow & A - B_k
\end{array}
$$

All subsets of size r are listed in the left-hand column, and all subsets of size $n - r$ are listed in the right-hand column. Hence, the number of subsets of size r equals the number of subsets of size $n - r$, and so $\binom{n}{r} = \binom{n}{n-r}$.   ■

The type of reasoning used in this example is called *combinatorial,* which means that a result is obtained by counting things that are combined in different ways. A number of theorems have both combinatorial proofs and proofs that are purely algebraic.

EXAMPLE 6.6.3   New Formulas from Old by Substitution

The formulas established in Example 6.6.1 are true for all integers n in some specified range. For example, formula (6.6.3) states that

$$\binom{n}{n-2} = \frac{n(n-1)}{2} \qquad \text{for all integers } n \geq 2. \qquad \text{6.6.3}$$

The letter n in this formula is a dummy variable; it can be replaced by any other symbol or expression as long as each occurrence is replaced and the new symbol or expression represents an integer that is at least 2. Write the formulas obtained by substituting each of the following for n: $m + 1$, $s - 1$, and $n + 2$. Simplify the result, and give the range of values of each variable for which the formula holds.

Solution   a. $\dbinom{m + 1}{(m + 1) - 2} = \dfrac{(m + 1)((m + 1) - 1)}{2}$ for all integers $(m + 1) \geq 2$.

    So $\dbinom{m + 1}{m - 1} = \dfrac{m(m + 1)}{2}$ for all integers $m \geq 1$.

   b. $\dbinom{s - 1}{(s - 1) - 2} = \dfrac{(s - 1)((s - 1) - 1)}{2}$ for all integers $(s - 1) \geq 2$.

    So $\dbinom{s - 1}{s - 3} = \dfrac{(s - 1)(s - 2)}{2}$ for all integers $s \geq 3$.

   c. $\dbinom{n + 2}{(n + 2) - 2} = \dfrac{(n + 2)((n + 2) - 1)}{2}$ for all integers $n + 2 \geq 2$.

    So $\dbinom{n + 2}{n} = \dfrac{(n + 1)(n + 2)}{2}$ for all integers $n \geq 0$.   ∎

PASCAL'S FORMULA

Blaise Pascal (1623–1662)

Pascal's formula, named after the seventeenth-century French mathematician and philosopher Blaise Pascal, is one of the most famous and useful in combinatorics (which is the formal term for the study of counting and listing problems). It relates the value of $\dbinom{n + 1}{r}$ to the values of $\dbinom{n}{n - 1}$ and $\dbinom{n}{r}$. Specifically, it says that

$$\binom{n + 1}{r} = \binom{n}{r - 1} + \binom{n}{r}$$

whenever n and r are positive integers with $r \leq n$. This formula makes it easy to compute higher combinations in terms of lower ones: If the values of $\dbinom{n}{r}$ are known for all r, then the values of $\dbinom{n + 1}{r}$ can be computed for all r such that $0 < r \leq n$.

    Pascal's triangle, shown in Table 6.6.1, is a geometric version of Pascal's formula. Each entry in the triangle is a value of $\dbinom{n}{r}$. Pascal's formula translates into the fact that the entry in row $n + 1$, column r equals the sum of the entry in row n, column $r - 1$ plus the entry in row n, column r. That is, the entry in a given interior position equals the sum of the two entries directly above and to the above left. The left-most and right-most entries in each row are 1 because $\dbinom{n}{n} = 1$ by Example 6.6.1 and $\dbinom{n}{0} = 1$ by exercise 1 at the end of this section.

TABLE 6.6.1   Pascal's Triangle $\left(\text{Values of } \binom{n}{r}\right)$

$n \backslash r$	0	1	2	3	4	5	$\cdots$	$r-1$	r	$\cdots$
0	1									$\cdots$
1	1	1								$\cdots$
2	1	2	1							$\cdots$
3	1	3	3	1						$\cdots$
4	1	4	6 $+$	4	1					$\cdots$
5	1	5	10 $=$	10	5	1				$\cdots$
$\vdots$	$\vdots$	$\vdots$	$\vdots$	$\vdots$	$\vdots$	$\vdots$		$\vdots$	$\vdots$	$\vdots$
n	$\binom{n}{0}$	$\binom{n}{1}$	$\binom{n}{2}$	$\binom{n}{3}$	$\binom{n}{4}$	$\binom{n}{5}$	$\cdots$	$\binom{n}{r-1}+$	$\binom{n}{r}$	$\cdots$
$n+1$	$\binom{n+1}{0}$	$\binom{n+1}{1}$	$\binom{n+1}{2}$	$\binom{n+1}{3}$	$\binom{n+1}{4}$	$\binom{n+1}{5}$	$\cdots$		$=\binom{n+1}{r}$	$\cdots$
$\vdots$	$\vdots$	$\vdots$	$\vdots$	$\vdots$	$\vdots$	$\vdots$		$\vdots$	$\vdots$	$\vdots$

EXAMPLE 6.6.4   **Calculating $\binom{n}{r}$ Using Pascal's Triangle**

Use Pascal's triangle to compute the values of $\binom{6}{2}$ and $\binom{6}{3}$.

Solution   By construction, the value in row n, column r of Pascal's triangle is the value of $\binom{n}{r}$, for every pair of positive integers n and r with $r \leq n$. By Pascal's formula $\binom{n+1}{r}$ can be computed by adding together $\binom{n}{n-1}$ and $\binom{n}{r}$, which are located directly above and above left of $\binom{n+1}{r}$. Thus,

$$\binom{6}{2} = \binom{5}{1} + \binom{5}{2} = 5 + 10 = 15 \quad \text{and}$$

$$\binom{6}{3} = \binom{5}{2} + \binom{5}{3} = 10 + 10 = 20. \qquad \blacksquare$$

Pascal's formula can be derived by two entirely different arguments. One is algebraic; it uses the formula for the number of r-combinations obtained in Theorem 6.4.1. The other is combinatorial; it uses the definition of the number of r-combinations as the number of subsets of size r taken from a set with a certain number of elements. We give both proofs since both approaches have applications in many other situations.

THEOREM 6.6.1   Pascal's Formula

Let n and r be positive integers and suppose $r \leq n$. Then

$$\binom{n+1}{r} = \binom{n}{r-1} + \binom{n}{r}.$$

Proof (Algebraic Version):

Let n and r be positive integers with $r \leq n$. By Theorem 6.4.1,

$$\binom{n}{r-1} + \binom{n}{r} = \frac{n!}{(r-1)! \cdot (n-(r-1))!} + \frac{n!}{r! \cdot (n-r)!}$$

$$= \frac{n!}{(r-1)!(n-r+1)!} + \frac{n!}{r!(n-r)!}.$$

To add these fractions, a common denominator is needed, so multiply the numerator and denominator of the left-hand fraction by r and multiply the numerator and denominator of the right-hand fraction by $(n-r+1)$. Then

$$\binom{n}{r-1} + \binom{n}{r} = \frac{n!}{(r-1)!(n-r+1)!} \cdot \frac{r}{r} + \frac{n!}{r!(n-r)!} \cdot \frac{(n-r+1)}{(n-r+1)}$$

$$= \frac{r \cdot n!}{(n-r+1)! \cdot r \cdot (r-1)!} + \frac{n \cdot n! - r \cdot n! + n!}{(n-r+1) \cdot (n-r)! \cdot r!}$$

$$= \frac{n! \cdot r + n! \cdot n - n! \cdot r + n!}{(n-r+1)! \cdot r!} = \frac{n! \cdot (n+1)}{(n+1-r)! \cdot r!}$$

$$= \frac{(n+1)!}{((n+1)-r)! \cdot r!} = \binom{n+1}{r}.$$

Proof (Combinatorial Version):

Let n and r be positive integers with $r \leq n$. Suppose S is a set with $n+1$ elements. The number of subsets of S of size r can be calculated by thinking of S as consisting of two pieces: one with n elements $\{x_1, x_2, \ldots, x_n\}$ and the other with one element $\{x_{n+1}\}$.

Any subset of S with r elements either contains x_{n+1} or it does not. If it contains x_{n+1}, then it contains $r-1$ elements from the set $\{x_1, x_2, \ldots, x_n\}$. If it does not contain x_{n+1}, then it contains r elements from the set $\{x_1, x_2, \ldots, x_n\}$.

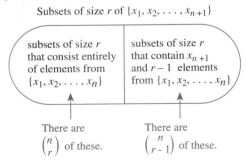

Subsets of size r of $\{x_1, x_2, \ldots, x_{n+1}\}$

subsets of size r that consist entirely of elements from $\{x_1, x_2, \ldots, x_n\}$ | subsets of size r that contain x_{n+1} and $r-1$ elements from $\{x_1, x_2, \ldots, x_n\}$

There are $\binom{n}{r}$ of these.

There are $\binom{n}{r-1}$ of these.

By Theorem 6.4.1, the set $\{x_1, x_2, \ldots, x_n, x_{n+1}\}$ has $\binom{n+1}{r}$ subsets of size r, the set $\{x_1, x_2, \ldots, x_n\}$ has $\binom{n}{r-1}$ subsets of size $r-1$, and the set

$\{x_1, x_2, \ldots, x_n\}$ has $\binom{n}{r}$ subsets of size r. By the addition rule,

$$\begin{bmatrix} \text{number of subsets of} \\ \{x_1, x_2, \ldots, x_n, x_{n+1}\} \\ \text{of size } r \end{bmatrix} = \begin{bmatrix} \text{number of subsets of} \\ \{x_1, x_2, \ldots, x_n\} \\ \text{of size } r-1 \end{bmatrix}$$

$$+ \begin{bmatrix} \text{number of subsets of} \\ \{x_1, x_2, \ldots, x_n\} \\ \text{of size } r \end{bmatrix}.$$

So,

$$\binom{n+1}{r} = \binom{n}{r-1} + \binom{n}{r}$$

as was to be shown.

EXAMPLE 6.6.5   **Deriving New Formulas from Pascal's Formula**

Use Pascal's formula to derive a formula for $\binom{n+2}{r}$ in terms of values of $\binom{n}{r}$, $\binom{n}{r-1}$, and $\binom{n}{r-2}$. Assume n and r are nonnegative integers and $2 \le r \le n$.

Solution   By Pascal's formula,

$$\binom{n+2}{r} = \binom{n+1}{r-1} + \binom{n+1}{r}.$$

Now apply Pascal's formula to $\binom{n+1}{r-1}$ and $\binom{n+1}{r}$ and substitute into the above to obtain

$$\binom{n+2}{r} = \left[\binom{n}{r-2} + \binom{n}{r-1}\right] + \left[\binom{n}{r-1} + \binom{n}{r}\right].$$

Combining the two middle terms gives

$$\binom{n+2}{r} = \binom{n}{r-2} + 2 \cdot \binom{n}{r-1} + \binom{n}{r}$$

for all nonnegative integers n and r such that $2 \le r \le n+2$.   ∎

EXERCISE SET 6.6

In 1–4, use Theorem 6.4.1 to compute the values of the indicated quantities. (Assume n is an integer.)

1. $\binom{n}{0}$, for $n \ge 0$

2. $\binom{n}{1}$, for $n \ge 1$

3. $\binom{n}{2}$, for $n \ge 2$

4. $\binom{n}{3}$, for $n \ge 3$

5. Use Theorem 6.4.1 to prove algebraically that $\binom{n}{r} = \binom{n}{n-r}$, for integers n and r with $0 \le r \le n$. (This can be done by direct calculation; it is not necessary to use mathematical induction.)

Apply substitution to the formulas of Example 6.6.1 to derive the formulas in 6–8. (Assume n, k, and r are integers.)

6. $\dbinom{n + k}{n + k - 1} = n + k$, for $n + k \geq 1$

7. $\dbinom{n + 3}{n + 1} = \dfrac{(n + 3)(n + 2)}{2}$, for $n \geq -1$

8. $\dbinom{k - r}{k - r} = 1$, for $k - r \geq 0$

9. Use Pascal's triangle given in Table 6.6.1 to compute the values of $\dbinom{6}{4}$ and $\dbinom{6}{5}$.

10. Complete the row of Pascal's triangle that corresponds to $n = 7$.

11. The row of Pascal's triangle that corresponds to $n = 8$ is as follows:

   1   8   28   56   70   56   28   8   1.

   What is the row that corresponds to $n = 9$?

12. Use Pascal's formula repeatedly to derive a formula for $\dbinom{n + 3}{r}$ in terms of values of $\dbinom{n}{k}$ with $k \leq r$. (Assume n and r are integers with $n + 3 \geq r \geq 0$.)

13. Prove that for all nonnegative integers n and r with $r + 1 \leq n$,
$$\dbinom{n}{r + 1} = \dfrac{n - r}{r + 1}\dbinom{n}{r}.$$

14. Prove by mathematical induction that if n is an integer and $n \geq 1$, then
$$\sum_{i=2}^{n+1}\dbinom{i}{2} = \dbinom{2}{2} + \dbinom{3}{2} + \cdots + \dbinom{n + 1}{2}$$
$$= \dbinom{n + 2}{3}.$$

H 15. Prove that if n is an integer and $n \geq 1$, then
$$1 \cdot 2 + 2 \cdot 3 + \cdots + n(n + 1) = 2\dbinom{n + 2}{3}.$$

16. Prove the following generalization of exercise 14: Let r be a fixed nonnegative integer. For all integers n with
$$n \geq r, \quad \sum_{i=r}^{n}\dbinom{i}{r} = \dbinom{n + 1}{r + 1}.$$

17. The sequence of **Catalan numbers**, named after the Belgian mathematician Eugène Catalan (1814–1894), arises in a variety of different contexts. It is defined as follows: For each integer $n \geq 1$,
$$C_n = \dfrac{1}{n + 1}\dbinom{2n}{n}.$$

   a. Find C_1, C_2, and C_3.

   b. Prove that $C_n = \dfrac{1}{4n + 2}\dbinom{2n + 2}{n + 1}$, for any integer $n \geq 1$.

18. Think of a set with $m + n$ elements as composed of two parts, one with m elements and the other with n elements. Give a combinatorial argument to show that
$$\dbinom{m + n}{r} = \dbinom{m}{0}\dbinom{n}{r} + \dbinom{m}{1}\dbinom{n}{r - 1} + \cdots + \dbinom{m}{r}\dbinom{n}{0},$$
where m and n are positive integers and r is an integer that is less than or equal to both m and n.

H 19. Prove that
$$\dbinom{n}{0}^2 + \dbinom{n}{1}^2 + \cdots + \dbinom{n}{n}^2 = \dbinom{2n}{n},$$
for all integers $n \geq 0$.

20. Let m be any nonnegative integer. Use mathematical induction and Pascal's formula to prove that
$$\dbinom{m}{0} + \dbinom{m + 1}{1} + \cdots + \dbinom{m + n}{n} = \dbinom{m + n + 1}{n},$$
for all integers $n \geq 0$.

21. Prove that if p is a prime number and r is an integer with $0 < r < p$, then $\dbinom{p}{r}$ is divisible by p.

6.7   THE BINOMIAL THEOREM

I'm very well acquainted, too, with matters mathematical, I understand equations both the simple and quadratical, About binomial theorem I am teaming with a lot of news, With many cheerful facts about the square of the hypotenuse.
(William S. Gilbert, *The Pirates of Penzance*, 1880)

In algebra a sum of two terms, such as $a + b$, is called a **binomial**. The *binomial theorem* gives an expression for the powers of a binomial $(a + b)^n$, for each positive integer n and all real numbers a and b.

Consider what happens when you calculate the first few powers of $a + b$. According to the distributive law of algebra, you take the sum of the products of all combinations of individual terms:

$$(a + b)^2 = (a + b) \cdot (a + b) = aa + ab + ba + bb,$$
$$(a + b)^3 = (a + b) \cdot (a + b) \cdot (a + b)$$
$$= aaa + aab + aba + abb + baa + bab + bba + bbb,$$
$$(a + b)^4 = (a + b) \cdot (a + b) \cdot (a + b) \cdot (a + b)$$

1st   2nd   3rd   4th
factor   factor   factor   factor

$$= aaaa + aaab + aaba + aabb + abaa + abab + abba + abbb$$
$$+ baaa + baab + baba + babb + bbaa + bbab + bbba + bbbb.$$

Now focus on the expansion of $(a + b)^4$. (It is concrete, and yet it has all the features of the general case.) A typical term of this expansion is obtained by multiplying one of the two terms from the first factor times one of the two terms from the second factor times one of the two terms from the third factor times one of the two terms from the fourth factor. For example, the term $abab$ is obtained by multiplying the a's and b's marked with arrows below.

$$\downarrow \qquad\qquad \downarrow \quad \downarrow \qquad\qquad \downarrow$$
$$(a + b) \cdot (a + b) \cdot (a + b) \cdot (a + b)$$

Since there are two possible values—a or b—for each term selected from one of the four factors, there are $2^4 = 16$ terms in the expansion of $(a + b)^4$.

Now some terms in the expansion are like terms and can be combined. Consider all possible orderings of three a's and one b, for example. By the techniques of Section 6.4, there are $\binom{4}{1} = 4$ of them. And each of the four occurs as a term in the expansion of $(a + b)^4$:

$$aaab \quad aaba \quad abaa \quad baaa.$$

By the commutative and associative laws of algebra, each such term equals a^3b, so all four are like terms. When like terms are combined, therefore, the coefficient of a^3b equals $\binom{4}{1}$.

Similarly, the expansion of $(a + b)^4$ contains the $\binom{4}{2} = 6$ different orderings of two a's and two b's,

$$aabb \quad abab \quad abba \quad baab \quad baba \quad bbaa,$$

all of which equal a^2b^2. So the coefficient of a^2b^2 equals $\binom{4}{2}$. By a similar analysis, the coefficient of ab^3 equals $\binom{4}{3}$. Also since there is only one way to order four a's, the coefficient of a^4 is 1 (which equals $\binom{4}{0}$), and since there is only one way to order four b's, the coefficient of b^4 is 1 (which equals $\binom{4}{4}$). Thus when all like terms are combined,

$$(a + b)^4 = \binom{4}{0}a^4 + \binom{4}{1}a^3b + \binom{4}{2}a^2b^2 + \binom{4}{3}ab^3 + \binom{4}{4}b^4$$
$$= a^4 + 4a^3b + 6a^2b^2 + 4ab^3 + b^4.$$

The binomial theorem generalizes this formula to an arbitrary nonnegative integer n.

THEOREM 6.7.1   Binomial Theorem

Given any real numbers a and b and any nonnegative integer n,

$$(a + b)^n = \sum_{k=0}^{n} \binom{n}{k} a^{n-k} b^k$$

$$= a^n + \binom{n}{1} a^{n-1} b^1 + \binom{n}{2} a^{n-2} b^2 + \cdots + \binom{n}{n-1} a^1 b^{n-1} + b^n.$$

Note that the second expression equals the first because $\binom{n}{0} = 1$ and $\binom{n}{n} = 1$, for all nonnegative integers n.

It is instructive to see two proofs of the binomial theorem: an algebraic and a combinatorial. Both require a precise definition of integer power.

DEFINITION

For any real number a and any nonnegative integer n, the **nonnegative integer powers of a** are defined as follows:

$$a^n = \begin{cases} 1 & \text{if } n = 0 \\ a \cdot a^{n-1} & \text{if } n > 0 \end{cases}$$

In some mathematical subjects, 0^0 is left undefined. Defining it to be 1, as is done here, makes it possible to write general formulas such as $\sum_{i=0}^{n} x^i = \dfrac{1}{1 - x}$ without having to exclude values of the variables that result in the expression 0^0.*

The algebraic version of the binomial theorem uses mathematical induction and calls upon Pascal's formula at a crucial point.

Proof of the Binomial Theorem (Algebraic Version):

Suppose a and b are real numbers. We prove that $(a + b)^n = \sum_{k=0}^{n} \binom{n}{k} a^{n-k} b^k$, for all integers $n \geq 0$, by induction on n.

True for $n = 0$ When $n = 0$, the binomial theorem states that

$$(a + b)^0 = \sum_{k=0}^{0} \binom{0}{k} a^{0-k} b^k.$$

---

*See *The Art of Computer Programming, Volume 1: Fundamental Algorithms,* Second Edition, by Donald E. Knuth (Reading, Massachusetts: Addison-Wesley Publishing Company, 1973), p. 56.

But the left-hand side is $(a + b)^0 = 1$ *[by definition of power]*, and the right-hand side is

$$\sum_{k=0}^{0} \binom{0}{k} a^{0-k} b^k = \binom{0}{0} a^{0-0} b^0$$

$$= \frac{0!}{0! \cdot (0 - 0)!} \cdot 1 \cdot 1 = \frac{1}{1 \cdot 1} = 1$$

also *[since $0! = 1$, $a^0 = 1$, and $b^0 = 1$]*. So the binomial theorem is true for $n = 0$.

If true for $n = m$ then true for $n = m + 1$ Let an integer $m \geq 1$ be given, and suppose the equality holds for m. That is, suppose

$$(a + b)^m = \sum_{k=0}^{m} \binom{m}{k} a^{m-k} b^k. \quad \text{[This is the inductive hypothesis.]}$$

We need to show that

$$(a + b)^{m+1} = \sum_{k=0}^{m+1} \binom{m + 1}{k} a^{(m+1)-k} b^k.$$

Now by definition of $(m + 1)$st power,

$$(a + b)^{m+1} = (a + b) \cdot (a + b)^m.$$

So by substitution from the inductive hypothesis,

$$(a + b)^{m+1} = (a + b) \cdot \sum_{k=0}^{m} \binom{m}{k} a^{m-k} b^k$$

$$= a \cdot \sum_{k=0}^{m} \binom{m}{k} a^{m-k} b^k + b \cdot \sum_{k=0}^{m} \binom{m}{k} a^{m-k} b^k$$

$$= \sum_{k=0}^{m} \binom{m}{k} a^{m+1-k} b^k + \sum_{k=0}^{m} \binom{m}{k} a^{m-k} b^{k+1}$$

by the generalized distributive law and the facts that $a \cdot a^{m-k} = a^{1+m-k} = a^{m+1-k}$ and $b \cdot b^k = b^{1+k} = b^{k+1}$.

We transform the second summation on the right-hand side by making the change of variable $j = k + 1$. When $k = 0$, then $j = 1$. When $k = m$, then $j = m + 1$. And since $k = j - 1$, the general term is

$$\binom{m}{k} a^{m-k} b^{k+1} = \binom{m}{j - 1} a^{m-(j-1)} b^j = \binom{m}{j - 1} a^{m+1-j} b^j.$$

Hence the second summation on the right-hand side above is

$$\sum_{j=1}^{m+1} \binom{m}{j - 1} a^{m+1-j} b^j.$$

But the j in this summation is a dummy variable; it can be replaced by the letter k, as long as the replacement is made everywhere the j occurs:

$$\sum_{j=1}^{m+1} \binom{m}{j - 1} a^{m+1-j} b^j = \sum_{k=1}^{m+1} \binom{m}{k - 1} a^{m+1-k} b^k.$$

Substituting back, we get

$$(a + b)^{m+1} = \sum_{k=0}^{m} \binom{m}{k} a^{m+1-k} b^k + \sum_{k=1}^{m+1} \binom{m}{k-1} a^{m+1-k} b^k.$$

[The reason for the above maneuvers was to make the powers of a and b agree so that we can add the summations together term by term, except for the first and the last terms, which we must write separately.]

Thus

$$(a + b)^{m+1} = \binom{m}{0} a^{m+1-0} b^0 + \sum_{k=1}^{m} \left[\binom{m}{k} + \binom{m}{k-1} \right] a^{m+1-k} b^k$$
$$+ \binom{m}{(m+1)-1} a^{m+1-(m+1)} b^{m+1}$$
$$= a^{m+1} + \sum_{k=1}^{m} \left[\binom{m}{k} + \binom{m}{k-1} \right] a^{m+1-k} b^k + b^{m+1}$$

since $a^0 = b^0 = 1$ and $\binom{m}{0} = \binom{m}{m} = 1.$

But

$$\left[\binom{m}{k} + \binom{m}{k-1} \right] = \binom{m+1}{k}$$ by Pascal's formula.

Hence

$$(a + b)^{m+1} = a^{m+1} + \sum_{k=1}^{m} \binom{m+1}{k} a^{(m+1)-k} b^k + b^{m+1}$$
$$= \sum_{k=0}^{m+1} \binom{m+1}{k} a^{(m+1)-k} b^k$$ because $\binom{m+1}{0} = \binom{m+1}{m+1} = 1,$

which is what we needed to show.

It is instructive to write out the product $(a + b) \cdot (a + b)^m$ without using the summation notation but using the inductive hypothesis about $(a + b)^m$:

$$(a + b)^{m+1} = (a + b) \cdot \left[a^m + \binom{m}{1} a^{m-1} b + \cdots + \binom{m}{k-1} a^{m-(k-1)} b^{k-1} \right.$$
$$\left. + \binom{m}{k} a^{m-k} b^k + \cdots + \binom{m}{m-1} ab^{m-1} + b^m \right].$$

You will see that the first and last coefficients are clearly 1, and that the term containing $a^{m+1-k} b^k$ is obtained from multiplying $a^{m-k} b^k$ by a and $a^{m-(k-1)} b^{k-1}$ by b *[since m + 1 − k = m − (k − 1)]*. Hence the coefficient of $a^{m+1-k} b^k$ equals the sum of $\binom{m}{k}$ and $\binom{m}{k-1}$. This is the crux of the algebraic proof.

If n and r are nonnegative integers and $r \le n$, then $\binom{n}{r}$ is called a **binomial coefficient** because it is one of the coefficients in the expansion of the binomial expression $(a + b)^n$.

The combinatorial proof of the binomial theorem goes as follows.

Proof of Binomial Theorem (Combinatorial Version):

[The combinatorial argument used here to prove the binomial theorem only works for $n \geq 1$. If we were only giving this combinatorial proof, we would have to prove the case $n = 0$ separately. Since we have already given a complete algebraic proof that includes the case $n = 0$, we do not reprove it here.]

Let a and b be real numbers and n an integer that is at least 1. The expression $(a + b)^n$ can be expanded (by the distributive law if $n \geq 2$) into products of n letters, where each letter is either a or b. For each $k = 0, 1, 2, \ldots, n$, the product

$$a^{n-k}b^k = \underbrace{a \cdot a \cdot a \cdot \ldots \cdot a}_{n-k \text{ factors}} \cdot \underbrace{b \cdot b \cdot b \cdot \ldots \cdot b}_{k \text{ factors}}$$

occurs as a term in the sum the same number of times as there are orderings of $(n - k)$ a's and k b's. But this number is $\binom{n}{k}$, the number of ways to choose k positions into which to place the b's. *[The other $n - k$ positions will be filled by a's.]* Hence when like terms are combined, the coefficient of $a^{n-k}b^k$ in the sum is $\binom{n}{k}$. Thus

$$(a + b)^n = \sum_{k=0}^{n} \binom{n}{k} a^{n-k}b^k.$$

This is what was to be proved.

EXAMPLE 6.7.1 **Substituting into the Binomial Theorem**

Expand the following expressions using the binomial theorem:

a. $(a + b)^5$ b. $(x - 4y)^4$

Solution a. $(a + b)^5 = \sum_{k=0}^{5} \binom{5}{k} a^{5-k}b^k$

$$= a^5 + \binom{5}{1}a^{5-1}b^1 + \binom{5}{2}a^{5-2}b^2 + \binom{5}{3}a^{5-3}b^3 + \binom{5}{4}a^{5-4}b^4 + b^5$$

$$= a^5 + 5a^4b + 10a^3b^2 + 10a^2b^3 + 5ab^4 + b^5.$$

b. Observe that $(x - 4y)^4 = (x + (-4y))^4$. So let $a = x$ and $b = (-4y)$, and substitute into the binomial theorem.

$$(x - 4y)^4 = \sum_{k=0}^{4} \binom{4}{k} x^{4-k}(-4y)^k$$

$$= x^4 + \binom{4}{1}x^{4-1}(-4y)^1 + \binom{4}{2}x^{4-2}(-4y)^2$$

$$+ \binom{4}{3}x^{4-3}(-4y)^3 + (-4y)^4$$

$$= x^4 + 4x^3(-4y) + 6x^2(16y^2) + 4x^1(-64y^3) + (256y^4)$$

$$= x^4 - 16x^3y + 96x^2y^2 - 256xy^3 + 256y^4.$$

■

EXAMPLE 6.7.2   **Estimating a Numerical Power Using the Binomial Theorem**

Which number is larger: $(1.01)^{1,000,000}$ or 10,000?

Solution   By the binomial theorem,

$$(1.01)^{1,000,000} = (1 + 0.01)^{1,000,000}$$

$$= 1 + \binom{1,000,000}{1} \cdot 1^{999,999}(0.01)^1 + \text{other positive terms}$$

$$= 1 + 1,000,000 \cdot 1 \cdot (0.01) + \text{other positive terms}$$

$$= 1 + 10,000 + \text{other positive terms}$$

$$= 10,001 + \text{other positive terms.}$$

Hence $(1.01)^{10,000} > 10,000$.   ■

EXAMPLE 6.7.3   **Deriving Another Combinatorial Identity
from the Binomial Theorem**

Use the binomial theorem to show that

$$2^n = \sum_{k=0}^{n}\binom{n}{k} = \binom{n}{0} + \binom{n}{1} + \binom{n}{2} + \cdots + \binom{n}{n}$$

for all integers $n \geq 0$.

Solution   Since $2 = 1 + 1$, $2^n = (1 + 1)^n$. Apply the binomial theorem to this expression by letting $a = 1$ and $b = 1$. Then

$$2^n = \sum_{k=0}^{n}\binom{n}{k} \cdot 1^{n-k} \cdot 1^k = \sum_{k=0}^{n}\binom{n}{k} \cdot 1 \cdot 1$$ ■

since $1^{n-k} = 1$ and $1^k = 1$. Consequently,

$$2^n = \sum_{k=0}^{n}\binom{n}{k} = \binom{n}{0} + \binom{n}{1} + \binom{n}{2} + \cdots + \binom{n}{n}.$$

NUMBER OF SUBSETS OF A SET

The formula derived in Example 6.7.3 makes possible the following alternative proof of the fact that a set with n elements has 2^n subsets (Theorem 5.3.5).

Proof of Theorem 5.3.5   Suppose S is a set with n elements. Then every subset of S has some number of elements k, where k is between 0 and n. It follows that the total number of subsets of S, $n(\mathcal{P}(S))$, can be expressed as the following sum:

$$\begin{bmatrix} \text{number of} \\ \text{subsets} \\ \text{of } S \end{bmatrix} = \begin{bmatrix} \text{number of} \\ \text{subsets of} \\ \text{size 0} \end{bmatrix} + \begin{bmatrix} \text{number of} \\ \text{subsets of} \\ \text{size 1} \end{bmatrix} + \cdots + \begin{bmatrix} \text{number of} \\ \text{subsets of} \\ \text{size } n \end{bmatrix}.$$

Now the number of subsets of size k of a set with n elements is $\binom{n}{k}$. Hence the

$$\text{number of subsets of } S = \binom{n}{0} + \binom{n}{1} + \binom{n}{2} + \cdots + \binom{n}{n}$$

$$= 2^n \quad \text{by Example 6.7.3.}$$

EXERCISE SET 6.7

Expand the expressions in 1–5 using the binomial theorem.

1. $(1 + x)^7$ 2. $(1 - x)^6$ 3. $(p + 3q)^4$

4. $(u^2 - 2v)^4$

5. $\left(\dfrac{2}{a} - \dfrac{a}{2}\right)^5$

6. In Example 6.7.1 it was shown that

$(a + b)^5 = a^5 + 5a^4b + 10a^3b^2 + 10a^2b^3 + 5ab^4 + b^5$.

Evaluate $(a + b)^6$ by substituting the expression above into the equation

$$(a + b)^6 = (a + b) \cdot (a + b)^5,$$

and then multiplying out and combining like terms.

In 7–10, find the coefficient of the given term when the expression is expanded by the binomial theorem.

7. x^6y^3 in $(x + y)^9$ 8. x^7 in $(2x + 3)^{10}$

9. a^5b^7 in $(a - 2b)^{12}$ 10. $u^{16}v^4$ in $(u^2 - v^2)^{10}$

In 11 and 12 indicate which number is larger. Use the binomial theorem to explain your answer.

11. $(1.1)^{10,000}$ or $1,000$ 12. $(1.2)^{4,000}$ or 800

13. Apply the binomial theorem to the expression $(1 - 1)^n$ to prove that for all integers $n \geq 1$,

$$\binom{n}{0} - \binom{n}{1} + \binom{n}{2} - \cdots + (-1)^n \binom{n}{n} = 0.$$

H 14. Use the binomial theorem to prove that for all integers $n \geq 0$,

$$3^n = \binom{n}{0} + 2\binom{n}{1} + 2^2\binom{n}{2} + \cdots + 2^n\binom{n}{n}.$$

Express each of the sums in 15 and 16 in closed form (without using a summation sign or $\cdots$).

H 15. $5^n\binom{n}{0} - 5^{n-1}\binom{n}{1} + 5^{n-2}\binom{n}{2} + \cdots$

$$+ (-1)^{n-1} \cdot 5 \cdot \binom{n}{n-1} + (-1)^n\binom{n}{n}$$

16. $\binom{n}{n}4^n - \binom{n}{n-1}4^{n-1} + \binom{n}{n-2}4^{n-2} + \cdots$

$$+ (-1)^{n-1}\binom{n}{1}4 + (-1)^n\binom{n}{0}$$

◆ 17. (For students who have studied calculus)

a. Explain how the equation below follows from the binomial theorem:

$$(1 + x)^n = \sum_{k=0}^{n} \binom{n}{k}x^k.$$

b. Write the formula obtained by taking the derivative of both sides of the equation in part (a) with respect to x.

c. Use the result of part (b) to derive the formulas below.

(i) 2^{n-1}

$$= \frac{1}{n}\left[\binom{n}{1} - 2\binom{n}{2} + 3\binom{n}{3} + \cdots + n\binom{n}{n}\right]$$

(ii) $\displaystyle\sum_{k=1}^{n} k\binom{n}{k}(-1)^k = 0$

d. Express $\displaystyle\sum_{k=1}^{n} k\binom{n}{k}3^k$ in closed form (without using a summation sign or $\cdots$).

7 Functions

Functions are ubiquitous in mathematics and computer science. That means you can hardly take two steps in these subjects without running into one. In this book we have already referred to truth tables and input/output tables (which are really Boolean functions), sequences (which are really functions defined on sets of integers), *mod* and *div* (which are really functions defined on Cartesian products of integers), and floor and ceiling (which are really functions from **R** to **Z**).

In this chapter we consider a wide variety of functions, focusing on those defined on discrete sets (such as finite sets or sets of integers) and discussing the functions that define finite-state automata. We then look at properties of functions such as one-to-one and onto, existence of inverse functions, and the interaction of composition of functions and the properties of one-to-one and onto. We end the chapter with a discussion of sizes of infinite sets and an application to computability.

7.1 FUNCTIONS DEFINED ON GENERAL SETS

The theory that has had the greatest development in recent times is without any doubt the theory of functions.
(Vito Volterra, 1888)

As used in ordinary language the word *function* indicates dependence of one varying quantity on another. If your teacher tells you that your grade in a course will be a function of your performance on the exams, you interpret this to mean that the teacher has some rule for translating exam scores into grades. To each collection of exam scores there corresponds a certain grade.

More generally, suppose two sets of objects are given—a first set and a second set—and suppose that with each element of the first set is associated a particular element of the second set. The relationship between the elements of the sets is called a *function*. Functions are generally denoted by single letters such as f, g, h, F, G, and so forth, although special functions are denoted by strings of letters or other symbols such as log, exp, and *mod*.

> **DEFINITION**
>
> A **function f from a set X to a set Y** is a relationship* between elements of X and elements of Y with the property that each element of X is related to a unique element of Y. The notation $f : X \rightarrow Y$ means that f is a function from X to Y. X is called the **domain of f** and Y is called the **co-domain of f**.
>
> Given an element x in X, there is a unique element y in Y that is related to x. We can think of x as *input* and y as the related *output*. We then say "f sends x to y" and write $x \xrightarrow{f} y$. The unique element y to which f sends x is denoted
>
> $$f(x) \quad \text{and is called} \quad \textbf{f of x, or}$$
> $$\text{the \textbf{value of } } f \textbf{ at } x \text{, or}$$
> $$\text{the \textbf{image of } } x \textbf{ under } f.$$
>
> The set of all values of f taken together is called the **range of f** or the **image of X under f**. Symbolically:
>
> $$\text{range of } f = \{ y \in Y \mid y = f(x), \text{ for some } x \text{ in } X \}.$$
>
> Given an element y in Y, there may exist elements in X with y as their image. The set of all such elements is called the **inverse image of y**. Symbolically:
>
> $$\text{inverse image of } y = \{ x \in X \mid f(x) = y \}.$$

The concept of function was developed over a period of centuries. The definition given above was first formulated for sets of numbers by the German mathematician Lejeune Dirichlet (DEER-ish-lay) in 1837.

ARROW DIAGRAMS

If X and Y are finite sets, you can define a function f from X to Y by making a list of elements in X and a list of elements in Y and drawing an arrow from each element in X to the corresponding element of Y. Such a drawing is called an **arrow diagram**. An example of an arrow diagram is shown in Figure 7.1.1.

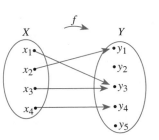

FIGURE 7.1.1

---

*In Chapter 10 we give a precise definition of the term *relation*.

The definition of function implies that the arrow diagram for a function f has the following two properties:

1. Every element of X has an arrow coming out of it;
2. No element of X has two arrows coming out of it that point to two different elements of Y.

Property (1) holds because the definition of function says that *each* element of X is sent to a unique element of Y. Property (2) holds because the definition of function says that each element of X is sent to a *unique* element of Y.

The range of f consists of all points in Y that have arrows pointing to them. The inverse image of an element y consists of all points in X that have arrows pointing from them to y.

Note that once X and Y have been given, the arrow diagram can also be specified by writing the set of all ordered pairs (x, y) for which there is an arrow from x to y. For instance, instead of drawing the arrows in Figure 7.1.1, we could write the set $\{(x_1, y_3), (x_2, y_1), (x_3, y_3), (x_4, y_4)\}$. In Chapter 10 we will discuss the formal definition of function, which specifies that a function from a set X to a set Y is a subset of $X \times Y$ satisfying certain properties.

EXAMPLE 7.1.1 Arrow Diagram of a Function

Let $X = \{a, b, c\}$ and $Y = \{1, 2, 3, 4\}$. Define a function f from X to Y by the arrow diagram in Figure 7.1.2.

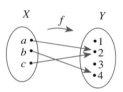

FIGURE 7.1.2

a. Write the domain and co-domain of f.
b. Find $f(a)$, $f(b)$, and $f(c)$.
c. What is the range of f?
d. Find the inverse images of 2, 4, and 1.
e. Represent f as a set of ordered pairs.

Solution
a. domain of $f = \{a, b, c\}$, co-domain of $f = \{1, 2, 3, 4\}$
b. $f(a) = 2, f(b) = 4, f(c) = 2$
c. range of $f = \{2, 4\}$
d. inverse image of 2 $= \{a, c\}$
  inverse image of 4 $= \{b\}$
  inverse image of 1 $= \emptyset$ *(since no arrows point to 1)*
e. $\{(a, 2), (b, 4), (c, 2)\}$ ∎

In Example 7.1.1 there are no arrows pointing to the 1 or the 3. This illustrates the fact that although each element of the domain of a function must have an arrow pointing out from it, there can be elements of the co-domain to which no arrows point. Note also that there are two arrows pointing to the 2—one coming from a and the other from c. This illustrates the fact that although no two arrows can start from the same element of the domain, there can be two or more arrows pointing to the same element of the co-domain.

EXAMPLE 7.1.2 Functions and Nonfunctions

Which of the arrow diagrams in Figure 7.1.3 define functions from $X = \{a, b, c\}$ to $Y = \{1, 2, 3, 4\}$?

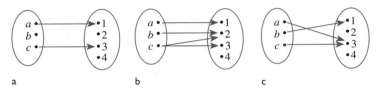

FIGURE 7.1.3

Solution   Only (c) defines a function. In (a) there is an element of X, namely b, that is not sent to any element of Y; that is, there is no arrow coming out of b. And in (b) the element c is not sent to a *unique* element of Y; that is, there are two arrows coming out of c, one pointing to 2 and the other to 3. ∎

FUNCTION MACHINES

Another useful way to think of a function is as a machine. Suppose f is a function from X to Y and an input x of X is given. Imagine f to be a machine that processes x in a certain way to produce the output $f(x)$. This is illustrated in Figure 7.1.4.

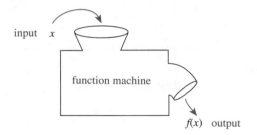

FIGURE 7.1.4

EXAMPLE 7.1.3 Function Machines

Define three functions f, g, and h as follows:

a. $f : \mathbf{R} \to \mathbf{R}$ is a **squaring function** : f sends each real number x to x^2. Symbolically,

$$f : x \to x^2.$$

b. $g : \mathbf{Z} \to \mathbf{Z}$ is the **successor function** : g sends each integer n to $n + 1$. Symbolically,

$$g : n \to n + 1.$$

c. $h : \mathbf{Q} \to \mathbf{Z}$ is a **constant function** : h sends each rational number r to 2. (No matter what the input, the output is always 2.) Symbolically,

$$h : r \to 2.$$

The functions f, g, and h are represented by the function machines in Figure 7.1.5.

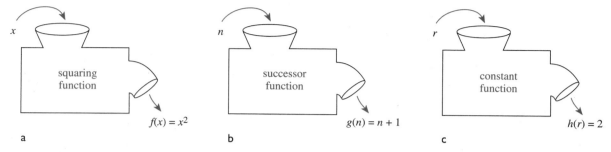

$f(x) = x^2$ $g(n) = n + 1$ $h(r) = 2$

a   b   c

FIGURE 7.1.5

A function is an entity in its own right. It can be thought of as a certain relationship between sets or as an input/output machine that operates according to a certain rule. This is the reason that a function is generally denoted by a single symbol or string of symbols, such as f, G, or log. In some mathematical contexts, however, the notation $f(x)$ is used to refer both to the value of f at x and to the function f itself. Because using the notation this way can lead to confusion, we avoid it whenever possible. In this book, unless explicitly stated otherwise, the symbol $f(x)$ always refers to the value of the function f at x and not to the function f itself.

> **DEFINITION**
>
> Suppose f and g are functions from X to Y. Then f **equals** g, written $f = g$, if, and only if,
>
> $$f(x) = g(x) \quad \text{for all } x \in X.$$

EXAMPLE 7.1.4   Equality of Functions

a. Define $f: \mathbf{R} \to \mathbf{R}$ and $g: \mathbf{R} \to \mathbf{R}$ by the following formulas:

$$f(x) = |x| \quad \text{for all } x \in \mathbf{R},$$
$$g(x) = \sqrt{x^2} \quad \text{for all } x \in \mathbf{R}.$$

Does $f = g$?

b. Suppose $F: \mathbf{R} \to \mathbf{R}$ and $G: \mathbf{R} \to \mathbf{R}$ are functions. Define new functions $F + G: \mathbf{R} \to \mathbf{R}$ and $G + F: \mathbf{R} \to \mathbf{R}$ as follows:

$$(F + G)(x) = F(x) + G(x) \quad \text{for all } x \in \mathbf{R},$$
$$(G + F)(x) = G(x) + F(x) \quad \text{for all } x \in \mathbf{R}.$$

Does $F + G = G + F$?

Solution   a. Yes. Since the absolute value of a number equals the square root of its square,

$$|x| = \sqrt{x^2} \quad \text{for all } x \in \mathbf{R}.$$

Hence $f = g$.

b. Again the answer is yes. For all real numbers x,

$$(F + G)(x) = F(x) + G(x) \quad \text{by definition of } F + G$$
$$= G(x) + F(x) \quad \text{by the commutative law for addition of real numbers}$$
$$= (G + F)(x) \quad \text{by definition of } G + F$$

Hence $F + G = G + F$. ■

EXAMPLES OF FUNCTIONS

The following examples illustrate some of the wide variety of different types of functions.

EXAMPLE 7.1.5   The Identity Function on a Set

Given a set X, define a function i_X from X to X by

$$i_X(x) = x \quad \text{for all } x \text{ in } X.$$

The function i_X is called the **identity function on X** because it sends each element of X to the element that is identical to it. Thus the identity function can be pictured as a machine that sends each piece of input directly to the output chute without changing it in any way.

Let X be any set and suppose that a_{ij}^k and $\phi(z)$ are elements of X. Find $i_X(a_{ij}^k)$ and $i_X(\phi(z))$.

Solution   Whatever is input to the identity function comes out unchanged. So $i_X(a_{ij}^k) = a_{ij}^k$ and $i_X(\phi(z)) = \phi(z)$. ■

EXAMPLE 7.1.6   Sequences

The formal definition of sequence specifies that a sequence is a function defined on the set of integers that are greater than or equal to a particular integer. For example, the sequence denoted

$$1, \; -\frac{1}{2}, \frac{1}{3}, \; -\frac{1}{4}, \frac{1}{5}, \; \ldots, \; \frac{(-1)^n}{n + 1}, \ldots$$

can be thought of as the function f from the nonnegative integers to the real numbers that associates $0 \to 1$, $1 \to -\frac{1}{2}$, $2 \to \frac{1}{3}$, $3 \to -\frac{1}{4}$, $4 \to \frac{1}{5}$, and, in general, $n \to (-1)^n/(n + 1)$. In other words, $f: \mathbf{Z}^{nonneg} \to \mathbf{R}$ is the function defined as follows:

$$\text{Send each integer } n \geq 0 \text{ to } f(n) = \frac{(-1)^n}{n + 1}.$$

In fact, there are many functions that can be used to define a given sequence. For instance, express the sequence above as a function from the set of *positive* integers to the set of real numbers.

Solution   Define $g: \mathbf{Z}^+ \to \mathbf{R}$ by $g(n) = \dfrac{(-1)^{n+1}}{n}$, for each $n \in \mathbf{Z}^+$. Then $g(0) = 1$, $g(1) = -\dfrac{1}{2}$, $g(2) = \dfrac{1}{3}$, and in general $g(n + 1) = \dfrac{(-1)^{n+2}}{n + 1} = \dfrac{(-1)^n}{n + 1} = f(n)$. ■

EXAMPLE 7.1.7   A Function Defined on a Power Set

Recall from Section 5.3 that $\mathcal{P}(A)$ denotes the set of all subsets of the set A. Define a function $F : \mathcal{P}(\{a,\ b,\ c\}) \to \mathbf{Z}^{nonneg}$ as follows: for each $X \in \mathcal{P}(\{a,\ b,\ c\})$,

$$F(X) = \text{the number of elements in } X.$$

Draw an arrow diagram for F.

Solution

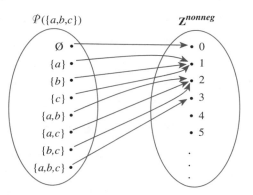

EXAMPLE 7.1.8   Function Defined on a Language

Recall from Section 5.1 that if Σ is an alphabet then Σ^* is the set of all strings over Σ. (The symbol ε represents the null string.) Let $\Sigma = \{a,\ b\}$ and define a function $g : \Sigma^* \to \mathbf{Z}$ as follows : for each string $s \in \Sigma^*$,

$$g(s) = \text{the number of } a\text{'s in } s.$$

Find the following.

a. $g(\varepsilon)$ b. $g(bb)$ c. $g(ababb)$ d. $g(bbbaa)$

Solution   a. 0   b. 0   c. 2   d. 2

EXAMPLE 7.1.9   The Logarithmic Function

Let b be a positive real number. For each positive real number x, the **logarithm with base b of x**, written $\log_b x$, is the exponent to which b must be raised to obtain x.* Symbolically,

$$\log_b x = y \iff b^y = x.$$

The **logarithmic function with base b** is the function from $\mathbf{R}^+$ to $\mathbf{R}$ that takes each positive real number x to $\log_b x$. Find the following:

a. $\log_3 9$ b. $\log_2(\tfrac{1}{2})$ c. $\log_{10}(1)$ d. $\log_2(2^m)$

Solution   a. $\log_3 9 = 2$ because $3^2 = 9$.
b. $\log_2(\tfrac{1}{2}) = -1$ because $2^{-1} = \tfrac{1}{2}$.
c. $\log_{10}(1) = 0$ because $10^0 = 1$.

*It is not obvious but it is true that for any positive real number x there is a unique real number y so that $b^y = x$. Most calculus books contain a derivation of this result.

d. $\log_2(2^m) = m$ because the exponent to which 2 must be raised to obtain 2^m is m. ∎

EXAMPLE 7.1.10   ## Encoding and Decoding Functions

When messages are communicated across a transmission channel, they are frequently coded in special ways to reduce the possibility that they will be garbled by interfering noise in the transmission lines. For example, suppose a message consists of a sequence of 0's and 1's. A simple way to encode the message is to write each bit three times. Thus the message

$$00101111$$

would be encoded as

$$000000111000111111111111.$$

The receiver of the message decodes it by replacing each section of three identical bits by the one bit to which all three are equal.

Let $\Sigma = \{0, 1\}$. Then Σ^* is the set of all strings in 0 and 1. Let L be the set of all strings over Σ that consist of consecutive triples of identical bits. The encoding and decoding processes described above are actually functions from Σ^* to L and from L to Σ^*. The encoding function E is the function from Σ^* to L defined as follows: for each string $s \in \Sigma^*$,

$$E(s) = \text{the string obtained from } s \text{ by replacing each bit of } s \text{ by the same bit written three times.}$$

The decoding function D is defined as follows: for each string $t \in L$,

$$D(t) = \text{the string obtained from } t \text{ by replacing each consecutive triple of three identical bits of } t \text{ by a single copy of that bit.}$$

The advantage of this particular coding scheme is that it makes it possible to do a certain amount of error correction when interference in the transmission channels has introduced errors into the stream of bits. If the receiver of the coded message observes that one of the sections of three consecutive bits that should be identical does not consist of identical bits, then one bit differs from the other two. In this case, if errors are rare, it is likely that the single bit that is different is the one in error, and this bit is changed to agree with the other two before decoding. ∎

EXAMPLE 7.1.11   ## The Hamming Distance Function

The Hamming distance function, named after the computer scientist Richard W. Hamming, is very important in coding theory. It gives a measure of the "difference" between two strings of 0's and 1's that have the same length. Let $\Sigma = \{0, 1\}$. Then Σ^n is the set of all strings of 0's and 1's of length n. Define a function $H: \Sigma^n \times \Sigma^n \to \mathbf{Z}^{nonneg}$ as follows: for each pair of strings $(s, t) \in \Sigma^n \times \Sigma^n$,

$$H(s, t) = \text{the number of positions in which } s \text{ and } t \text{ have different values.}$$

Thus, letting $n = 5$,

$$H(11111, 00000) = 5$$

Richard Hamming

because 11111 and 00000 differ in all five positions, whereas

$$H(11000, 00000) = 2$$

because 11000 and 00000 differ only in the first two positions.

a. Find $H(00101, 01110)$.   b. Find $H(10001, 01111)$.

Solution   a. 3   b. 4   ∎

BOOLEAN FUNCTIONS

In Section 1.4 we showed how to find input/output tables for certain digital logic circuits. Any such input/output table defines a function in the following way: The elements in the input column can be regarded as ordered tuples of 0's and 1's; the set of all such ordered tuples is the domain of the function. The elements in the output column are all either 0 or 1; thus $\{0, 1\}$ is taken to be co-domain of the function. The relationship is that which sends each input element to the output element in the same row. Thus, for instance, the input/output table of Figure 7.1.6(a) defines the function with the arrow diagram shown in Figure 7.1.6(b).

More generally, the input/output table corresponding to a circuit with n input wires has n input columns. Such a table defines a function from the set of all n-tuples of 0's and 1's to the set $\{0, 1\}$.

Input			Output
P	Q	R	S
1	1	1	1
1	1	0	1
1	0	1	0
1	0	0	1
0	1	1	0
0	1	0	0
0	0	1	0
0	0	0	1

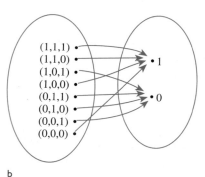

a   b

FIGURE 7.1.6   Two Representations of a Boolean Function.

DEFINITION

An (**n-place**) **Boolean function** f is a function whose domain is the set of all ordered n-tuples of 0's and 1's and whose co-domain is the set $\{0, 1\}$. More formally, the domain of a Boolean function can be described as the Cartesian product of n copies of the set $\{0, 1\}$, which is denoted $\{0, 1\}^n$. Thus $f : \{0, 1\}^n \to \{0, 1\}$.

It is customary to omit one set of parentheses when referring to functions defined on Cartesian products. For example, we write $f(1, 0, 1)$ rather than $f((1, 0, 1))$.

EXAMPLE 7.1.12 A Boolean Function

Consider the three-place Boolean function defined from the set of all 3-tuples of 0's and 1's to $\{0, 1\}$ as follows: For each triple (x_1, x_2, x_3) of 0's and 1's,

$$f(x_1, x_2, x_3) = (x_1 + x_2 + x_3) \bmod 2.$$

Describe f using an input/output table.

Solution
$$f(1, 1, 1) = (1 + 1 + 1) \bmod 2 = 3 \bmod 2 = 1$$
$$f(1, 1, 0) = (1 + 1 + 0) \bmod 2 = 2 \bmod 2 = 0$$

The rest of the values of f can be calculated similarly to obtain the following table.

Input			Output
x_1	x_2	x_3	$(x_1 + x_2 + x_3) \bmod 2$
1	1	1	1
1	1	0	0
1	0	1	0
1	0	0	1
0	1	1	0
0	1	0	1
0	0	1	1
0	0	0	0

■

CHECKING WHETHER A FUNCTION IS WELL DEFINED

It can sometimes happen that what appears to be a function defined by a rule is not really a function at all. To give an example, suppose we wrote, "Define a function $f: \mathbf{R} \to \mathbf{R}$ by the formula

$$f(x) = \sqrt{-x^2}   \text{ for all real numbers } x."$$

This definition is contradictory: on the one hand, f is supposed to be a function from the real numbers to the real numbers; but on the other hand, $\sqrt{-x^2}$ is only a real number when $x = 0$. In a situation like this we say that f is **not well defined** because the formula does not define a function.

EXAMPLE 7.1.13 A Function That Is Not Well Defined

Recall that $\mathbf{Q}$ represents the set of all rational numbers. Suppose you read that a function $f: \mathbf{Q} \to \mathbf{Z}$ is to be defined by the formula

$$f\left(\frac{m}{n}\right) = m   \text{ for all integers } m \text{ and } n \text{ with } n \neq 0.$$

That is, the integer associated by f to the number m/n is m. Is f well defined? Why?

Solution　The function f is not well defined. The reason is that fractions have more than one representation as quotients of integers. For instance, $\frac{1}{2} = \frac{3}{6}$. Now if f were a function, then the definition of a function would imply that $f(\frac{1}{2}) = f(\frac{3}{6})$ since $\frac{1}{2} = \frac{3}{6}$. But applying the formula for f, you find that

$$f\left(\frac{1}{2}\right) = 1 \quad \text{and} \quad f\left(\frac{3}{6}\right) = 3,$$

and so

$$f\left(\frac{1}{2}\right) \neq f\left(\frac{3}{6}\right).$$

This contradiction shows that f is not well defined and is, therefore, not a function. ∎

Note that the phrase *well-defined function* is actually redundant; for a function to be well defined really means that it is worthy of being called a function.

EXERCISE SET 7.1*

1. Let $X = \{1, 3, 5\}$ and $Y = \{s, t, u, v\}$. Define $f : X \rightarrow Y$ by the following arrow diagram.

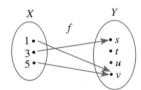

a. Write the domain of f and the co-domain of f.
b. Find $f(1)$, $f(3)$, and $f(5)$.
c. What is the range of f?
d. What is the inverse image of s? of u? of v?
e. Represent f as a set of ordered pairs.

2. Let $X = \{1, 3, 5\}$ and $Y = \{s, t, u, v\}$. Define $g : X \rightarrow Y$ by the following arrow diagram.

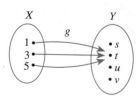

a. Write the domain of g and the co-domain of g.
b. Find $g(1)$, $g(3)$, and $g(5)$.
c. What is the range of g?

d. What is the inverse image of t? of u?
e. Represent g as a set of ordered pairs.

3. Let $X = \{2, 4, 5\}$ and $Y = \{1, 2, 4, 6\}$. Which of the following arrow diagrams determine functions from X to Y?

a.
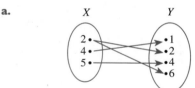

b.
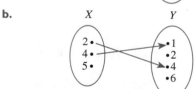

c.
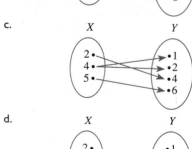

d.

e.

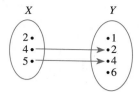

4. a. Find all functions from $X = \{a, b\}$ to $Y = \{u, v\}$.
  b. Find all functions from $X = \{a, b, c\}$ to $Y = \{u\}$.
  c. Find all functions from $X = \{a, b, c\}$ to $Y = \{u, v\}$.

5. a. How many functions are there from a set with three elements to a set with four elements?
  b. How many functions are there from a set with five elements to a set with two elements?
  c. How many functions are there from a set with m elements to a set with n elements, where m and n are positive integers?

6. Define functions f and g from $\mathbf{R}$ to $\mathbf{R}$ by the formulas: for all $x \in \mathbf{R}$,

$$f(x) = 2x \quad \text{and} \quad g(x) = \frac{2x^3 + 2x}{x^2 + 1}.$$

Show that $f = g$.

7. Define functions H and K from $\mathbf{R}$ to $\mathbf{R}$ by the formulas: for all $x \in \mathbf{R}$,

$$H(x) = \lfloor x \rfloor + 1 \quad \text{and} \quad K(x) = \lceil x \rceil.$$

Does $H = K$? Explain.

8. Let F and G be functions from the set of all real numbers to itself. Define the product functions $F \cdot G : \mathbf{R} \to \mathbf{R}$ and $G \cdot F : \mathbf{R} \to \mathbf{R}$ as follows:

$$(F \cdot G)(x) = F(x) \cdot G(x) \quad \text{for all } x \in \mathbf{R},$$
$$(G \cdot F)(x) = G(x) \cdot F(x) \quad \text{for all } x \in \mathbf{R}.$$

Does $F \cdot G = G \cdot F$? Explain.

9. Let F and G be functions from the set of all real numbers to itself. Define new functions $F - G : \mathbf{R} \to \mathbf{R}$ and $G - F : \mathbf{R} \to \mathbf{R}$ as follows:

$$(F - G)(x) = F(x) - G(x) \quad \text{for all } x \in \mathbf{R},$$
$$(G - F)(x) = G(x) - F(x) \quad \text{for all } x \in \mathbf{R}.$$

Does $F - G = G - F$? Explain.

10. Let $i_{\mathbf{Z}}$ be the identity function defined on the set of all integers, and suppose that e, b_i^{jk}, and $K(t)$ all represent integers. Find
  a. $i_{\mathbf{Z}}(e)$ **b.** $i_{\mathbf{Z}}(b_i^{jk})$ **c.** $i_{\mathbf{Z}}(K(t))$

11. Find functions defined on the set of nonnegative integers that define the sequences whose first six terms are given below.
  a. $1, -\dfrac{1}{3}, \dfrac{1}{5}, -\dfrac{1}{7}, \dfrac{1}{9}, -\dfrac{1}{11}$
  b. $0, -2, 4, -6, 8, -10$

12. Let $A = \{1, 2, 3, 4, 5\}$ and define a function $F : \mathscr{P}(A) \to \mathbf{Z}$ as follows: for all sets X in $\mathscr{P}(A)$,

$$F(X) = \begin{cases} 0 & \text{if } X \text{ has an even number of elements} \\ 1 & \text{if } X \text{ has an odd number of elements.} \end{cases}$$

Find the following.
  a. $F(\{1, 3, 4\})$ **b.** $F(\varnothing)$
  c. $F(\{2, 3\})$ **d.** $F(\{2, 3, 4, 5\})$

13. Let $\Sigma = \{a, b\}$ and $\Sigma^* =$ the set of all strings over Σ.
  a. Define $f : \Sigma^* \to \mathbf{Z}$ as follows: for each string s in Σ^*,

$$f(s) = \begin{cases} \text{the number of } b\text{'s to the left of the left-most } a \text{ in } s \\ 0 \quad \text{if } s \text{ contains no } a\text{'s.} \end{cases}$$

Find $f(aba)$, $f(bbab)$, and $f(b)$. What is the range of f?

  b. Define $g : \Sigma^* \to \Sigma^*$ as follows: for each string s in Σ^*,

$$g(s) = \text{the string obtained by writing the characters of } s \text{ in reverse order.}$$

Find $g(aba)$, $g(bbab)$, and $g(b)$. What is the range of g?

14. Fill in the blanks below.
  a. $\log_2 8 = 3$ because _____.
  b. $\log_3 \frac{1}{9} = -2$ because _____.
  c. $\log_4 4 = 1$ because _____.
  d. $\log_5 5^n = n$ because _____.
  e. $\log_4 1 = 0$ because _____.

15. Find exact values for each of the following quantities. Do not use a calculator.
  a. $\log_3 81$ **b.** $\log_2 1,024$ **c.** $\log_3 \frac{1}{27}$ **d.** $\log_2 1$
  e. $\log_4 \frac{1}{4}$ **f.** $\log_2 2$ **g.** $\log_2 2^k$

16. Use the definition of logarithm to prove that for any positive real number b with $b \neq 1$, $\log_b b = 1$.

17. Use the definition of logarithm to prove that for any positive real number b with $b \neq 1$, $\log_b 1 = 0$.

18. If b and y are positive real numbers such that $\log_b y = 3$, what is $\log_{1/b} (y)$? Why?

19. If b and y are positive real numbers such that $\log_b y = 2$, what is $\log_{b^2} (y)$? Why?

20. Let $A = \{2, 3, 5\}$ and $B = \{x, y\}$. Let p_1 and p_2 be the **projections of $A \times B$ onto the first and second coordinates**. That is, for each pair $(a, b) \in A \times B$, $p_1(a, b) = a$ and $p_2(a, b) = b$.
  a. Find $p_1(2, y)$ and $p_1(5, x)$. What is the range of p_1?
  b. Find $p_2(2, y)$ and $p_2(5, x)$. What is the range of p_2?

21. Observe that *mod* and *div* can be defined as functions from $\mathbf{Z}^{nonneg} \times \mathbf{Z}^+$ to $\mathbf{Z}$. For each ordered pair (n, d) consisting of a nonnegative integer n and a positive integer d, let

   $mod(n, d) = n \bmod d$ (the nonnegative remainder obtained when n is divided by d).

   $div(n, d) = n \, div \, d$ (the integer quotient obtained when n is divided by d).

   Find each of the following.
   a. $mod(67, 10)$ and $div(67, 10)$
   b. $mod(57, 8)$ and $div(57, 8)$
   c. $mod(32, 4)$ and $div(32, 4)$

22. Consider the coding and decoding functions E and D defined in Example 7.1.10.
   a. Find $E(0110)$ and $D(111111000111)$.
   b. Find $E(1010)$ and $D(000000111111)$.

23. Consider the Hamming distance function defined in Example 7.1.11.
   a. Find $H(10101, 00011)$.
   b. Find $H(00110, 10111)$.

24. A permutation on a set can be regarded as a function from the set to itself. For instance, one permutation of $\{1, 2, 3, 4\}$ is 2341. It can be identified with the function that sends each position number to the number occupying that position. Since position 1 is occupied by 2, 1 is sent to 2 or $1 \rightarrow 2$; since position 2 is occupied by 3, 2 is sent to 3 or $2 \rightarrow 3$; and so forth. The entire permutation can be written using arrows as follows.

$$1 \quad 2 \quad 3 \quad 4$$
$$\downarrow \quad \downarrow \quad \downarrow \quad \downarrow$$
$$2 \quad 3 \quad 4 \quad 1$$

   a. Use arrows to write each of the six permutations of $\{1, 2, 3\}$.
   b. Use arrows to write each of the permutations of $\{1, 2, 3, 4\}$ that keep 2 and 4 fixed.
   c. Which permutations of $\{1, 2, 3\}$ keep no elements fixed?
   d. Use arrows to write all permutations of $\{1, 2, 3, 4\}$ that keep no elements fixed.

25. Draw arrow diagrams for the Boolean functions defined by the following input/output tables.

a.

Input		Output
P	Q	R
1	1	0
1	0	1
0	1	0
0	0	1

b.

Input			Output
P	Q	R	S
1	1	1	1
1	1	0	1
1	0	1	0
1	0	0	1
0	1	1	0
0	1	0	0
0	0	1	0
0	0	0	1

26. Fill in the following table to show the values of all possible two-place Boolean functions.

Input	f_1	f_2	f_3	f_4	f_5	f_6	f_7	f_8	f_9	f_{10}	f_{11}	f_{12}	f_{13}	f_{14}	f_{15}	f_{16}
1 1																
1 0																
0 1																
0 0																

27. Consider the three-place Boolean function f defined by the following rule: for each triple (x_1, x_2, x_3) of 0's and 1's,

$$f(x_1, x_2, x_3) = (3x_1 + x_2 + 2x_3) \bmod 2.$$

   a. Find $f(1, 1, 1)$ and $f(0, 1, 1)$.
   b. Describe f using an input/output table.

28. Student A tries to define a function $g: \mathbf{Q} \rightarrow \mathbf{Z}$ by the rule

$$g\left(\frac{m}{n}\right) = m - n, \text{ for all integers } m \text{ and } n \text{ with } n \neq 0.$$

   Student B claims that g is not well defined. Justify student B's claim.

29. Student C tries to define a function $h: \mathbf{Q} \rightarrow \mathbf{Q}$ by the rule

$$h\left(\frac{m}{n}\right) = \frac{m^2}{n}, \text{ for all integers } m \text{ and } n \text{ with } n \neq 0.$$

   Student D claims that h is not well defined. Justify student D's claim.

30. On certain computers the integer data type goes from

−2,147,483,648 through 2,147,483,647. Let S be the set of all integers from −2,147,483,648 through 2,147,483,647. Try to define a function $f: S \to S$ by the rule $f(n) = n^2$ for each n in S. Is f well defined? Why?

31. Given a set S and a subset A, the **characteristic function of A**, denoted χ_A, is the function defined from S to $\mathbf{Z}$ with the property that for all $u \in S$,

$$\chi_A(u) = \begin{cases} 1 & \text{if } u \in A \\ 0 & \text{if } u \notin A. \end{cases}$$

Show that each of the following holds for all subsets A and B of S and all $u \in S$.

a. $\chi_{A \cap B}(u) = \chi_A(u) \cdot \chi_B(u)$
b. $\chi_{A \cup B}(u) = \chi_A(u) + \chi_B(u) - \chi_A(u) \cdot \chi_B(u)$

Each of exercises 32–36 refers to the Euler phi function, denoted ϕ, which is defined as follows: For each integer $n \geq 1$, $\phi(n)$ is the number of positive integers less than or equal to n that have no common factors with n except ± 1. For example, $\phi(10) = 4$ because there are four positive integers less than or equal to 10 that have no common factors with 10 except ± 1; namely, 1, 3, 7, and 9.

32. Find each of the following.
a. $\phi(15)$ b. $\phi(2)$ c. $\phi(5)$
d. $\phi(12)$ e. $\phi(11)$ f. $\phi(1)$

◆ 33. Prove that if p is a prime number and n is an integer with $n \geq 1$, then $\phi(p^n) = p^n - p^{n-1}$.

H 34. Prove that there are infinitely many integers n for which $\phi(n)$ is a perfect square.

H 35. Use the inclusion/exclusion principle to prove that if $n = pq$, where p and q are prime numbers, then $\phi(n) = (p - 1)(q - 1)$.

36. Use the inclusion/exclusion principle to prove that if

$n = pqr$, where p, q, and r are prime numbers, then $\phi(n) = (p - 1)(q - 1)(r - 1)$.

Exercises 37–44 refer to the following definition:

Definition: If $f: X \to Y$ is a function and $A \subseteq X$ and $C \subseteq Y$, then

$$f(A) = \{y \in Y \mid y = f(x) \text{ for some } x \text{ in } A\}$$

and

$$f^{-1}(C) = \{x \in X \mid f(x) \in C\}.$$

$f(A)$ is called the **image of A**, or **range of f** and $f^{-1}(C)$ is called the **inverse image of C**.

Determine which of the properties in 37–44 are true for all functions f from a set X to a set Y and which are false for some function f. Justify your answers.

◆ 37. For all subsets A and B of X, if $A \subseteq B$, then $f(A) \subseteq f(B)$.

◆ 38. For all subsets A and B of X, $f(A \cup B) = f(A) \cup f(B)$.

◆ 39. For all subsets A and B of X, $f(A \cap B) = f(A) \cap f(B)$.

◆ 40. For all subsets A and B of X, $f(A - B) = f(A) - f(B)$.

◆ 41. For all subsets C and D of Y, if $C \subseteq D$, then $f^{-1}(C) \subseteq f^{-1}(D)$.

◆ H 42. For all subsets C and D of Y, $f^{-1}(C \cup D) = f^{-1}(C) \cup f^{-1}(D)$.

◆ 43. For all subsets C and D of Y, $f^{-1}(C \cap D) = f^{-1}(C) \cap f^{-1}(D)$.

◆ 44. For all subsets C and D of Y, $f^{-1}(C - D) = f^{-1}(C) - f^{-1}(D)$.

7.2 APPLICATION: FINITE-STATE AUTOMATA

The world of the future will be an ever more demanding struggle against the limitations of our intelligence, not a comfortable hammock in which we can lie down to be waited upon by our robot slaves.

(Norbert Wiener, 1964)

The kind of circuit discussed in Section 1.4 is called a *combinational circuit*. Such a circuit is characterized by the fact that its output is completely determined by its input/output table, or in other words, by a Boolean function. Its output does not depend in any way on the history of previous inputs to the circuit. For this reason, a combinational circuit is said to have no memory.

Combinational circuits are very important in computer design, but they are not the only type of circuits used. Equally important are *sequential circuits*. For sequential circuits one cannot predict the output corresponding to a particular input unless one also knows something about the prior history of the circuit, or more technically, unless one knows the state the circuit was in before receiving the input. The behavior of a sequential circuit is a function not only of the input to the circuit but also of the state the circuit is in when the input is received. A computer memory circuit is a type of sequential circuit.

A **finite-state automaton** (aw-TAHM-uh-tahn) is an idealized machine that embodies the essential idea of a sequential circuit. Each piece of input to a finite-state automaton leads to a change in the state of the automaton, which in turn affects how subsequent input is processed. Imagine, for example, the act of dialing a telephone number. Dialing 1–800 puts the telephone circuit in a state of readiness to receive the final seven digits of a toll-free call, whereas dialing 328 leads to a state of expectation for the four digits of a local call. Vending machines operate similarly. Just knowing that you put a nickel into a vending machine is not enough for you to be able to predict what the behavior of the machine will be. You also have to know the state the machine was in when the nickel was inserted. If 55¢ had already been deposited, you might get a soft drink or some candy, but if the nickel was the first coin deposited, you would probably get nothing at all. The most important finite-state automata (this is the plural of automaton) in modern life are digital computers. The number of states of a digital computer is generally very large, but it *is* finite, and, like any finite-state automaton, a computer changes state in predictable ways in response to the sequence of inputs.

It is interesting to note that the basic theory of automata was developed to answer very theoretical questions about the foundations of mathematics posed by the great German mathematician David Hilbert in 1900. The pathbreaking work on automata was done in the mid-1930s by the English mathematician and logician Alan Turing. In the 1940s and 1950s, Turing's work played an important role in the development of real-world automatic computers.

David Hilbert (1862–1943)

EXAMPLE 7.2.1   A Simple Vending Machine

A vending machine dispenses pieces of candy that cost 20¢ each. The machine accepts nickels and dimes only and does not give change. As soon as the amount deposited equals or exceeds 20¢, the machine releases a piece of candy. The next coin deposited starts the process over again. The operation of the machine is represented by the diagram of Figure 7.2.1.

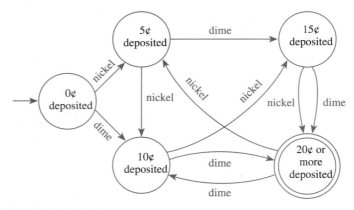

FIGURE 7.2.1   A Vending Machine

Each circle represents a state of the machine: the state in which 0¢ has been deposited, 5¢, 10¢, 15¢, and 20¢ or more. The unlabeled arrow pointing to "0¢ deposited" indicates that this is the initial state of the machine. The double circle around "20¢ or more deposited" indicates that candy is released when the machine has reached this state. (It is called an *accepting state* of the machine because when the machine is in this state it has accepted the input sequence of coins as payment for candy.) The arrows that link the states indicate what happens when a particular input is made to the machine in each of its various states. For instance, the arrow labeled "nickel" that goes from "0¢ deposited" to "5¢ deposited" indicates that when the machine is in the state "0¢ deposited" and a nickel is inserted, the machine goes to the state "5¢ deposited." The arrow labeled "dime" that goes from "15¢ deposited" to "20¢ or more deposited" indicates that when the machine is in the state "15¢ deposited" and a dime is inserted, the machine goes to the state "20¢ or more deposited" and candy is dispensed. (In this case the purchaser would pay 25¢ for the candy since the machine does not return change.) The arrow labeled "dime" that goes from "20¢ or more deposited" to "10¢ deposited" indicates that when the machine is in the state "20¢ or more deposited" and a dime is inserted, the machine goes back to the state "10¢ deposited." (This corresponds to the fact that after the machine has dispensed a piece of candy, it starts operation all over again.)

Equivalently, the operation of the vending machine can be represented by a *next-state table* as shown in Table 7.2.1.

TABLE 7.2.1   Next-state Table

			Input	
			Nickel	Dime
State	→	0¢ deposited	5¢ deposited	10¢ deposited
		5¢ deposited	10¢ deposited	15¢ deposited
		10¢ deposited	15¢ deposited	20¢ or more deposited
		15¢ deposited	20¢ or more deposited	20¢ or more deposited
	◎	20¢ or more deposited	5¢ deposited	10¢ deposited

The arrow pointing to "0¢ deposited" in the table indicates that the machine begins operation in this state. The double circle next to "20¢ or more deposited" indicates that candy is released when the machine has reached this state. Entries in the body of the table are interpreted in the obvious way. For instance, the entry in the third row of the column labeled *Dime* shows that when the machine is in state "10¢ deposited" and a dime is deposited, it goes to state "20¢ or more deposited."

Note that Table 7.2.1 conveys exactly the same information as the diagram of Figure 7.2.1. If the diagram is given, the table can be constructed; and if the table is given, the diagram can be drawn.   ∎

Observe that the vending machine described in Example 7.2.1 can be thought of as having a primitive memory: it "remembers" how much money has been deposited (within limits) by referring to the state it is in. This capability for storing and acting upon stored information is what gives finite-state automata their tremendous power.

DEFINITION OF A FINITE-STATE AUTOMATON

A general *finite-state automaton* is completely described by giving a set of states together with an indication of which is the initial state and which are the accepting states (when something special happens), a list of all input elements, and specification for a *next-state function* that defines which state is produced by each input in each state. This is formalized in the following definition:

DEFINITION

A **finite-state automaton** A consists of five objects:

1. a set I, called the **input alphabet**, of input symbols;
2. a set S of **states** the automaton can be in;
3. a designated state s_0, called the **initial state**;
4. a designated set of states called the set of **accepting states**;
5. a **next-state function** $N: S \times I \rightarrow S$ that associates a "next-state" to each ordered pair consisting of a "current state" and a "current input." For each state s in S and input symbol m in I, $N(s, m)$ is called the state to which A goes if m is input to A when A is in state s.

The operation of a finite-state automaton is commonly described by a diagram called a **(state-)transition diagram**, similar to that of Figure 7.2.1. It is called a *transition diagram* because it shows the transitions the machine makes from one state to another in response to various inputs. In a transition diagram states are represented by circles and accepting states by double circles. There is one arrow that points to the initial state and other arrows that are labeled with input symbols and point from each state to other states to indicate the action of the next-state function. Specifically, an arrow from state s to state t labeled m means that $N(s, m) = t$.

The **next-state table** for an automaton shows the values of the next-state function N for all possible states s and input symbols i. In the **annotated next-state table** the initial state is indicated by an arrow and the accepting states are marked by double circles.

EXAMPLE 7.2.2   A Finite-State Automaton Given by a Transition Diagram

Consider the finite-state automaton A defined by the transition diagram shown in Figure 7.2.2.

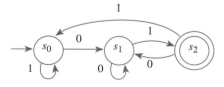

FIGURE 7.2.2

a. What are the states of A?
b. What are the input symbols of A?
c. What is the initial state of A?
d. What are the accepting states of A?
e. Find $N(s_1, 1)$.
f. Find the annotated next-state table for A.

Solution

a. The states of A are s_0, s_1, and s_2 *[since these are the labels of the circles]*.
b. The input symbols of A are 0 and 1 *[since these are the labels of the arrows]*.
c. The initial state of A is s_0 *[since the unlabeled arrow points to s_0]*.
d. The only accepting state of A is s_2 *[since this is the only state marked by a double circle]*.
e. $N(s_1, 1) = s_2$ *[since there is an arrow from s_1 to s_2 labeled 1]*.

f.

	Input	
	0	1
State $\rightarrow$ s_0	s_1	s_0
s_1	s_1	s_2
◎ s_2	s_1	s_0

EXAMPLE 7.2.3

A Finite-State Automaton Given by an Annotated Next-State Table

Consider the finite-state automaton A defined by the following annotated next-state table:

		Input		
		a	b	c
State	$\rightarrow$ U	Z	Y	Y
	◎ V	V	V	V
	Y	Z	V	Y
	◎ Z	Z	Z	Z

a. What are the states of A?
b. What are the input symbols of A?
c. What is the initial state of A?
d. What are the accepting states of A?
e. Find $N(U, c)$.
f. Draw the transition diagram for A.

Solution
a. The states of A are U, V, Y, and Z.
b. The input symbols of A are a, b, and c.
c. The initial state of A is U *[since the arrow points to U]*.
d. The accepting states of A are V and Z *[since these are marked with double circles.]*.
e. $N(U, c) = Y$ *[since the entry in the row labeled U and the column labeled c of the next-state table is Y]*.
f. The transition diagram for A is shown in Figure 7.2.3. It can be drawn more compactly by labeling arrows with multiple-input symbols where appropriate. This is illustrated in Figure 7.2.4.

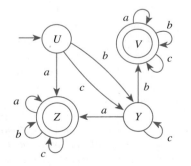

FIGURE 7.2.3

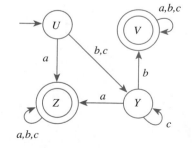

FIGURE 7.2.4

THE LANGUAGE ACCEPTED BY AN AUTOMATON

Now suppose a string of input symbols is fed into a finite-state automaton in sequence. After each successive input symbol has changed the state of the automaton, the automaton ends up in a certain state. This may be either an accepting state or a nonaccepting state. In this way, a finite-state automaton separates the set of all strings of input symbols into two subsets: those that send the automaton to an accepting state and those that do not. Those strings that send the automaton to an accepting state are said to be *accepted* by the automaton.

DEFINITION

Let A be a finite-state automaton with set of input symbols I. Let I^* be the set of all strings over I, and let w be a string in I^*. Then **w is accepted by A** if, and only if, A goes to an accepting state when the symbols of w are input to A in sequence starting when A is in its initial state. The **language accepted by A**, denoted $L(A)$, is the set of all strings that are accepted by A.

EXAMPLE 7.2.4   Finding the Language Accepted by an Automaton

Consider the finite-state automaton A defined in Example 7.2.2 and shown again below.

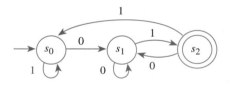

a. To what states does A go if the symbols of the following strings are input to A in sequence starting from the initial state?
   (i) 01   (ii) 0011   (iii) 0101100   (iv) 10101
b. Which of the strings in part (a) send A to an accepting state?
c. What is the language accepted by A?

Solution   a. (i) s_2 (ii) s_0 (iii) s_1 (iv) s_2
b. The strings 01 and 10101 send A to an accepting state.
c. Observe that if w is any string that ends in 01, then w is accepted by A. For if w is any string of length $n \geq 2$, then after the first $n - 2$ symbols of w have been input, A is in one of its three states: s_0, s_1, or s_2. But from any of these three states, input of the symbols 01 in sequence sends A first to s_1 and then to the accepting state s_2. Hence, any string that ends in 01 is accepted by A.

   Also note that the only strings accepted by A are those that end in 01. (That is, no other strings besides those ending in 01 are accepted by A.) The reason for this is that the only accepting state of A is s_2 and the only arrow pointing to s_2 comes from s_1 and is labeled 1. Thus in order for an input string w of length n to send A to an accepting state, the last symbol of w must be a 1 and the first $n - 1$ symbols of w must send A to state s_1. Now three arrows point to s_1, one from each of the three states of A, and all are labeled 0. Thus, the last of the first $n - 1$ symbols of

w must be 0, or, in other words, the next-to-the-last symbol of w must be 0. Hence the last two symbols of w must be 01. ∎

A finite-state automaton with multiple accepting states can have output devices attached to each one so that the automaton can classify input strings into a variety of different categories, one for each accepting state. This is how finite-state automata are used in the lexical scanner component of a computer compiler to group the symbols from a stream of input characters into identifiers, keywords, and so forth.

THE EVENTUAL-STATE FUNCTION

Now suppose a finite-state automaton is in one of its states (not necessarily the initial state) and a string of input symbols is fed into it in sequence. To what state will the automaton eventually go? The function that gives the answer to this question for every possible combination of input strings and states of the automaton is called the *eventual-state function.*

DEFINITION

Let A be a finite-state automaton with set of states S, set of input symbols I, and next-state function $N: S \times I \rightarrow S$. Let I^* be the set of all strings over I, and define the **eventual-state function** $N^*: S \times I^* \rightarrow S$ as follows:

For any state s and for any input string w,

$$N^*(s, w) = \begin{bmatrix} \text{the state to which } A \text{ goes if the} \\ \text{symbols of } w \text{ are input to } A \text{ in sequence} \\ \text{starting when } A \text{ is in state } s \end{bmatrix}.$$

EXAMPLE 7.2.5   **Computing Values of the Eventual-State Function**

Consider again the finite-state automaton of Example 7.2.2 shown below. Find $N^*(s_1, 10110)$.

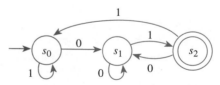

Solution   By definition of the eventual-state function,

$$N^*(s_1, 10110) = \begin{bmatrix} \text{the state to which } A \text{ goes if the} \\ \text{symbols of } 10110 \text{ are input to } A \text{ in} \\ \text{sequence starting when } A \text{ is in state } s_1 \end{bmatrix}.$$

By referring to the transition diagram for A, you can see that starting from s_1, when a 1 is input, A goes to s_2; then when a 0 is input, A goes back to s_1; after that, when a 1 is input, A goes to s_2; from there, when a 1 is input, A goes to s_0; and finally when a 0 is input, A goes back to s_1. This sequence of state transitions can be written as follows:

$$s_1 \xrightarrow{\;1\;} s_2 \xrightarrow{\;0\;} s_1 \xrightarrow{\;1\;} s_2 \xrightarrow{\;1\;} s_0 \xrightarrow{\;0\;} s_1.$$

Thus, after all the symbols of 10110 have been input in sequence, the eventual state of A is s_1. So

$$N^*(s_1, 10110) = s_1. \qquad \blacksquare$$

The definitions of string and language accepted by an automaton can be restated symbolically using the eventual-state function. Suppose A is a finite-state automaton with set of input symbols I and next-state function N, and suppose I^* is the set of all strings over I and w is a string in I^*.

> w is accepted by A $\iff$ $N^*(s_0, w)$ is an accepting state of A

> $L(A) = \{w \in I^* \mid N^*(s_0, w)$ is an accepting state of $A\}$

DESIGNING A FINITE-STATE AUTOMATON

Now consider the problem of starting with a description of a language and designing an automaton to accept exactly that language.

EXAMPLE 7.2.6

A Finite-State Automaton That Accepts the Set of Strings of 0's and 1's for Which the Number of 1's Is Divisible by 3

Design a finite-state automaton A that accepts the set of all strings of 0's and 1's such that the number of 1's in the string is divisible by 3.

Solution   Let s_0 be the initial state of A, s_1 its state after one 1 has been input, and s_2 its state after two 1's have been input. Note that s_0 is the state of A after zero 1's have been input, and since zero is divisible by 3 ($0 = 0 \cdot 3$), s_0 must be an accepting state. The states s_0, s_1, and s_2 must be different from one another because from state s_0 three 1's are needed to reach a new total divisible by 3, whereas from state s_1 two additional 1's are necessary and from state s_2 just one more is required.

Now the state of A after three 1's have been input can also be taken to be s_0 because after three 1's have been input, three more are needed to reach a new total divisible by 3. More generally, if $3k$ 1's have been input to A, where k is any nonnegative integer, then three more are needed for the total to again be divisible by 3 (since $3k + 3 = 3(k + 1)$). So the state in which $3k$ 1's have been input, for any nonnegative integer k, can be taken to be the initial state s_0.

By similar reasoning, the states in which $3k + 1$ 1's and $3k + 2$ 1's have been input, where k is a nonnegative integer, can be taken to be s_1 and s_2, respectively.

Now every nonnegative integer can be written in one of the three forms $3k$, $3k + 1$, or $3k + 2$ (see Section 3.4), so the three states s_0, s_1, and s_2 are all that is needed to create A. Thus the states of A can be drawn and labeled as shown at left.

Next consider the possible inputs to A in each of its states. No matter what state A is in, if a 0 is input the total number of 1's in the input string remains unchanged. Thus there is a loop at each state labeled 0.

Now suppose a 1 is input to A when it is in state s_0. Then A goes to state s_1 (since the total number of 1's in the input string has changed from $3k$ to $3k + 1$). Similarly, if a 1 is input to A when it is in state s_1, then A goes to state s_2 (since the total number of 1's in the input string has changed from $3k + 1$ to $3k + 2$). Finally, if a 1 is input to A when it is in state s_2, then it goes to state s_0 (since the total number of 1's in the input string becomes $(3k + 2) + 1 = 3k + 3 = 3(k + 1)$, which is a multiple of 3.)

It follows that the transition diagram has the appearance shown below.

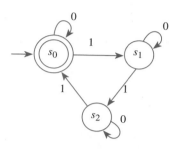

This automaton accepts the set of strings for which the number of 1's is divisible by 3.

EXAMPLE 7.2.7

A Finite-State Automaton That Accepts the Set of All Strings of 0's and 1's Containing Exactly One 1

Design a finite-state automaton A to accept the set of all strings of 0's and 1's that contain exactly one 1.

Solution   The automaton A must have at least two distinct states:

s_0: initial state;

s_1: state to which A goes when the input string contains exactly one 1.

If A is in state s_0 and a 0 is input, A may as well stay in state s_0 (since it still needs to wait for a 1 to move to state s_1), but as soon as a 1 is input, A moves to state s_1. Thus a partial drawing of the transition diagram is as shown below.

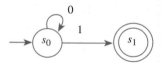

Now consider what happens when A is in state s_1. If a 0 is input, the input string still has a single 1, so A stays in state s_1. But if a 1 is input, then the input string contains more than one 1, so A must leave s_1 (since no string with more than one 1 is to be accepted by A). It cannot go back to state s_0 because there is a way to get from s_0 to s_1, and after input of the second 1 A can never return to state s_1. So A must go to a third state, s_2, from which there is no return to s_1. Thus from s_2 every input may as well leave A in state s_2. It follows that the completed transition diagram for A has the appearance shown below.

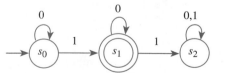

This automaton accepts a set of strings with exactly one 1.

SIMULATING A FINITE-STATE AUTOMATON USING SOFTWARE

Suppose items have been coded with strings of 0's and 1's. A program is to be written to govern the processing of items coded with strings that end 011; items coded any other way are to be ignored. This situation can be modeled by the finite-state automaton shown in Figure 7.2.5.

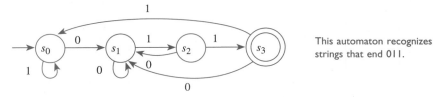

This automaton recognizes strings that end 011.

FIGURE 7.2.5

The symbols of the item code are fed into this automaton in sequence, and every string of item code symbols sends the automaton to one of the four states s_0, s_1, s_2, or s_3. If state s_3 is reached, the item is processed; if not, the item is ignored.

The action of this finite-state automaton can be simulated by a computer algorithm as given in Algorithm 7.2.1.

ALGORITHM 7.2.1   A Finite-State Automaton

[This algorithm simulates the action of the finite-state automaton of Figure 7.2.5 by mimicking the functioning of the transition diagram. The states are denoted 0, 1, 2, and 3.]

Input: *string*   *[a string of 0's and 1's plus an end marker e]*

Algorithm Body:

    state := 0

    symbol := first symbol in the input string

    while (*symbol* ≠ *e*)

        if *state* = 0   **then if** *symbol* = 0

                              then *state* := 1

                              else *state* := 0

        else if *state* = 1 **then if** *symbol* = 0

                              then *state* := 1

                              else *state* := 2

        else if *state* = 2 **then if** *symbol* = 0

                              then *state* := 1

                              else *state* := 3

        else if *state* = 3 **then if** *symbol* = 0

                              then *state* := 1

                              else *state* := 0

        symbol := next symbol in the input string

    end while

    *[After execution of the **while** loop, the value of state is 3 if, and only if, the input string ends in 011e.]*

end Algorithm 7.2.1

Note how use of the finite-state automaton allows the creator of the algorithm to focus on each step of the analysis of the input string independently of the other steps.

An alternative way to program this automaton is to enter the values of the next-state function directly as a two-dimensional array. This is done in Algorithm 7.2.2.

ALGORITHM 7.2.2   A Finite-State Automaton

[This algorithm simulates the action of the finite-state automaton of Figure 7.2.5 by repeated application of the next-state function. The states are denoted 0, 1, 2, and 3.]

Input: *string*   *[a string of 0's and 1's plus an end marker e]*

Algorithm Body:

$N(0, 0) := 1$, $N(0, 1) := 0$, $N(1, 0) := 1$, $N(1, 1) := 2$,
$N(2, 0) := 1$, $N(2, 1) := 3$, $N(3, 0) := 1$, $N(3, 1) := 0$

state := 0

symbol := first symbol in the input string

while (*symbol* ≠ *e*)

   state := *N*(*state*, *symbol*)

   symbol := next symbol in the input string

end while

*[After execution of the **while** loop, the value of state is 3 if, and only if, the input string ends in 011e.]*

end Algorithm 7.2.2

EXERCISE SET 7.2

1. Find the state of the vending machine in Example 7.2.1 after each of the following sequences of coins have been input.
   a. nickel, dime, nickel
   b. nickel, dime, dime
   c. dime, nickel, nickel, nickel, dime

In 2–7 a finite-state automaton is given by a transition diagram. For each automaton

a. find its states;
b. find its input symbols:
c. find its initial state;
d. find its accepting states;
e. write its annotated next-state table.

3.

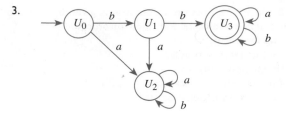

2.

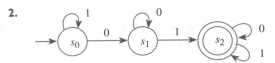

4.

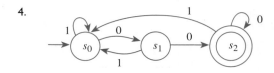

5.

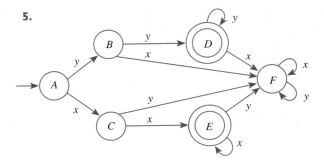

6.

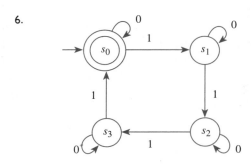

7.

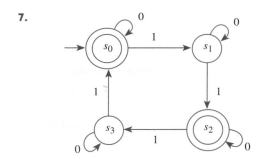

In 8 and 9 a finite-state automaton is given by an anno-tated next-state table. For each automaton
a. **find its states;**
b. **find its input symbols;**
c. **find its initial state;**
d. **find its accepting states;**
e. **draw its transition diagram.**

8. Next-State Table

		Input	
		0	1
→	s_0	s_1	s_2
State	s_1	s_1	s_2
◎	s_2	s_1	s_2

9. Next-State Table

		Input	
		0	1
→	s_0	s_0	s_1
◎	s_1	s_1	s_2
State	s_2	s_2	s_3
	s_3	s_3	s_0

10. A finite-state automaton A, given by the transition dia-gram below, has next-state function N and eventual-state function N^*.

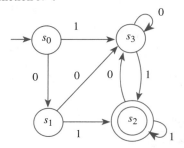

a. Find $N(s_1, 1)$ and $N(s_0, 1)$.
b. Find $N(s_2, 0)$ and $N(s_1, 0)$.
c. Find $N^*(s_0, 10011)$ and $N^*(s_1, 01001)$.
d. Find $N^*(s_2, 11010)$ and $N^*(s_0, 01000)$.

11. A finite-state automaton A, given by the transition dia-gram below, has next-state function N and eventual-state function N^*.

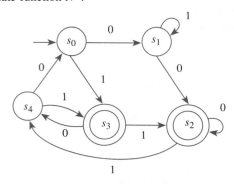

a. Find $N(s_3, 0)$ and $N(s_2, 1)$.
b. Find $N(s_0, 0)$ and $N(s_4, 1)$.
c. Find $N^*(s_0, 010011)$ and $N^*(s_3, 01101)$.
d. Find $N^*(s_0, 1111)$ and $N^*(s_2, 00111)$.

12. Consider again the finite-state automaton of exercise 2.
a. To what state does the automaton go when the sym-bols of the following strings are input to it in se-quence starting from the initial state?

(i) 1110001   (ii) 0001000   (iii) 11110000

b. Which of the strings in part (a) send the automaton to an accepting state?

c. What is the language accepted by the automaton?

13. Consider again the finite-state automaton of exercise 3.

a. To what state does the automaton go when the symbols of the following strings are input to it in sequence starting from the initial state?

(i) *bb* (ii) *aabbbaba* (iii) *babbbbbabaa*
(iv) *bbaaaabaa*

b. Which of the strings in part (a) send the automaton to an accepting state?

c. What is the language accepted by the automaton?

In each of 14–19 find the language accepted by the automaton in the referenced exercise.

14. exercise 4 **15.** exercise 5 16. exercise 6

17. exercise 7 **18.** exercise 8 19. exercise 9

20. Design a finite-state automaton with input alphabet equal to {0, 1} that accepts the set of all strings for which the final three input symbols are 1.

H **21.** Design a finite-state automaton with input alphabet equal to {*a, b*} that accepts the set of all strings of length at least 2 for which the final two input symbols are the same.

22. Design a finite-state automaton with input alphabet equal to {0, 1} that accepts the set of all strings that start with 01 or 10.

23. A string of 0's and 1's is said to have *even parity* if it contains an even number of 1's and is said to have *odd parity* if it contains an odd number of 1's. Design a finite-state automaton to accept the set of all strings of 0's and 1's that have even parity.

24. Design a finite-state automaton to accept the set of all strings of 0's and 1's that begin 01.

25. Design a finite-state automaton to accept the set of all strings of 0's and 1's that begin 101.

26. Design a finite-state automaton to accept the set of all strings of 0's and 1's that end 10.

27. Design a finite-state automaton to accept the set of all strings of *a*'s and *b*'s that contain exactly two *b*'s.

28. Design a finite-state automaton to accept the set of all strings of 0's and 1's that start with 0 and contain exactly one 1.

29. Design a finite-state automaton to accept the set of all strings of 0's and 1's that contain the pattern 010.

30. A simplified telephone switching system allows the following strings as legal telephone numbers:

a. a string of seven digits that does not start 00, 01, 10, or 11 (*a local call string*);

b. a 1 followed by a three-digit *area code string* (any digit except 0 or 1 followed by a 0 or 1 followed by any digit) followed by a seven-digit local call string;

c. a 0 alone or followed by a three-digit area code string plus a seven-digit local call string.

Design a finite-state automaton to recognize legal telephone numbers.

31. Write a computer algorithm that simulates the action of the finite-state automaton of exercise 2 by mimicking the action of the transition diagram.

32. Write a computer algorithm that simulates the action of the finite-state automaton of exercise 8 by repeated application of the next-state function.

7.3 ONE-TO-ONE AND ONTO, INVERSE FUNCTIONS

Don't accept a statement just because it is printed.
(Anna Pell Wheeler, 1883-1966)

In this section we discuss two important properties that functions may satisfy: the property of being *one-to-one* and the property of being *onto*. Functions that satisfy both properties are called *one-to-one correspondences* or *one-to-one onto functions*. When a function is a one-to-one correspondence, the elements of its domain and co-domain match up perfectly, and we can define an *inverse function* from the co-domain to the domain that "undoes" the action of the function.

ONE-TO-ONE FUNCTIONS

In Section 7.1 we noted that a function may send several elements of its domain to the same element of its co-domain. In terms of arrow diagrams, this means that two or more arrows that start in the domain can point to the same element in the co-domain. On the other hand, a function may associate a different element of its co-domain to each element of its domain, which would mean that no two arrows that start in the domain would point to the same element of its co-domain. A function with this property is called *one-to-one* or *injective*. For a one-to-one function, each element of the range is the image of at most one element of the domain.

DEFINITION

Let F be a function from a set X to a set Y. F is **one-to-one** (or **injective**) if, and only if, for all elements x_1 and x_2 in X,

$$\text{if } F(x_1) = F(x_2), \text{ then } x_1 = x_2.$$

Or, equivalently,

$$\text{if } x_1 \neq x_2, \text{ then } F(x_1) \neq F(x_2).$$

Symbolically:

$$F: X \to Y \text{ is one-to-one} \iff \forall x_1, x_2 \in X, \text{ if } F(x_1) = F(x_2) \text{ then } x_1 = x_2.$$

To obtain a precise statement of what it means for a function *not* to be one-to-one, take the negation of one of the equivalent versions of the definition above. Thus:

A function $F: X \to Y$ is *not* one-to-one $\iff$ $\exists$ elements x_1 and x_2 in X with $F(x_1) = F(x_2)$ and $x_1 \neq x_2$.

That is, if elements x_1 and x_2 can be found that have the same function value but are not equal, then F is not one-to-one.

In terms of arrow diagrams, a one-to-one function can be thought of as a function that separates points. That is, it takes distinct points of the domain to distinct points of the co-domain. A function that is not one-to-one fails to separate points. That is, at least two points of the domain are taken to the same point of the co-domain. This is illustrated in Figure 7.3.1.

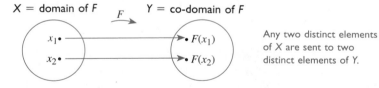

FIGURE 7.3.1 (a)   A One-to-One Function Separates Points

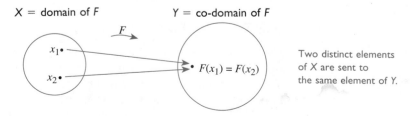

FIGURE 7.3.1 (b)   A Function That Is Not One-to-One Collapses Points Together

EXAMPLE 7.3.1 | ## Identifying One-to-One Functions Defined on Finite Sets

a. Which of the arrow diagrams in Figure 7.3.2 define one-to-one functions?

domain of F   co-domain of F   domain of G   co-domain of G

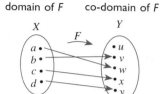   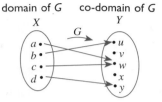

FIGURE 7.3.2

b. Let $X = \{1, 2, 3\}$ and $Y = \{a, b, c, d\}$. Define $H: X \to Y$ by specifying that $H(1) = c$, $H(2) = a$, and $H(3) = d$. Define $K: X \to Y$ by specifying that $K(1) = d$, $K(2) = b$, and $K(3) = d$. Are either H or K one-to-one?

Solution | a. F is one-to-one but G is not. F is one-to-one because no two different elements of X are sent by F to the same element of Y. G is not one-to-one because the elements a and c are both sent by G to the same element of Y: $G(a) = G(c) = w$ but $a \neq c$.

b. H is one-to-one but K is not. H is one-to-one because each of the three elements of the domain of H is sent by H to a different element of the co-domain: $H(1) \neq H(2)$, $H(1) \neq H(3)$, and $H(2) \neq H(3)$. K, however, is not one-to-one because $K(1) = K(2) = d$ but $1 \neq 3$. ■

Consider the problem of writing a computer algorithm to check whether a function F is one-to-one. If F is defined on a finite set and there is an independent algorithm to compute values of F, then an algorithm to check whether F is one-to-one can be written as follows: Represent the domain of F as a one-dimensional array $a[1], a[2], \ldots, a[n]$ and use a nested loop to examine all possible pairs $(a[i], a[j])$ where $i < j$. If there is a pair $(a[i], a[j])$ for which $F(a[i]) = F(a[j])$ and $a[i] \neq a[j]$, then F is not one-to-one. If, however, all pairs have been examined without finding such a pair, then F is one-to-one. You are asked to write such an algorithm in the exercises at the end of this section.

ONE-TO-ONE FUNCTIONS ON INFINITE SETS

Now suppose f is a function defined on an infinite set X. By definition, f is one-to-one if, and only if, the following universal statement is true:

$$\forall x_1, x_2 \in X, \text{ if } f(x_1) = f(x_2) \text{ then } x_1 = x_2.$$

Thus to prove f is one-to-one, you will generally use the method of direct proof:

suppose x_1 and x_2 are elements of X such that $f(x_1) = f(x_2)$

and

show that $x_1 = x_2$.

To show that f is *not* one-to-one, you will ordinarily

find elements x_1 and x_2 in X so that $f(x_1) = f(x_2)$ but $x_1 \neq x_2$.

EXAMPLE 7.3.2   **Proving or Disproving That Functions Are One-to-One**

Define $f: \mathbf{R} \to \mathbf{R}$ and $g: \mathbf{Z} \to \mathbf{Z}$ by the rules

$$f(x) = 4x - 1 \quad \text{for all} \quad x \in \mathbf{R}$$

and

$$g(n) = n^2 \quad \text{for all} \quad n \in \mathbf{Z}.$$

a. Is f one-to-one? Prove or give a counterexample.
b. Is g one-to-one? Prove or give a counterexample.

Solution   It is usually best to start by taking a positive approach to answering questions like these. Try to prove the given functions are one-to-one and see if you run into difficulty. If you finish without running into any problems, then you have a proof. If you do encounter a problem, then analyzing the problem may lead you to discover a counterexample.

a. The function $f: \mathbf{R} \to \mathbf{R}$ is defined by the rule

$$f(x) = 4x - 1 \quad \text{for all real numbers } x.$$

To prove that f is one-to-one it is necessary to prove that

$$\forall \text{ real numbers } x_1 \text{ and } x_2, \text{ if } f(x_1) = f(x_2) \text{ then } x_1 = x_2.$$

Substituting the definition of f into the outline of a direct proof, you

suppose x_1 and x_2 are any real numbers such that $4x_1 - 1 = 4x_2 - 1$,

and

show that $x_1 = x_2$.

Can you reach what is to be shown from the supposition? Of course. Just add 1 to both sides of the equation in the supposition and then divide both sides by 4. This discussion is summarized in the following formal answer.

Answer to (a)

If the function $f: \mathbf{R} \to \mathbf{R}$ is defined by the rule $f(x) = 4x - 1$, for all real numbers x, then f is one-to-one.

Proof:
Suppose x_1 and x_2 are real numbers such that $f(x_1) = f(x_2)$. *[We must show that $x_1 = x_2$.]*

$$4x_1 - 1 = 4x_2 - 1.$$

Adding 1 to both sides gives

$$4x_1 = 4x_2,$$

and dividing both sides by 4 gives

$$x_1 = x_2,$$

which is what was to be shown.

b. The function $g: \mathbf{Z} \to \mathbf{Z}$ is defined by the rule

$$g(n) = n^2 \quad \text{for all integers } n.$$

As above, you start as if you were going to prove that g is one-to-one. Substituting the definition of g into the outline of a direct proof, you

suppose n_1 and n_2 are integers such that $n_1^2 = n_2^2$,

and

try to show that $n_1 = n_2$.

Can you reach what is to be shown from the supposition? No! It is quite possible for two numbers to have the same squares and yet be different. For example, $2^2 = (-2)^2$ but $2 \neq -2$.

Thus trying to prove that g is one-to-one runs into a difficulty. But analyzing this difficulty leads to the discovery of a counterexample, which shows that g is not one-to-one.

This discussion is summarized as follows.

Answer to (b)

If the function $g : \mathbf{Z} \to \mathbf{Z}$ is defined by the rule $g(n) = n^2$, for all $n \in \mathbf{Z}$, then g is not one-to-one.

Counterexample:
Let $n_1 = 2$ and $n_2 = -2$. Then

$$g(n_1) = g(2) = 2^2 = 4 \quad \text{and also}$$
$$g(n_2) = g(-2) = (-2)^2 = 4.$$

Hence

$$g(n_1) = g(n_2) \quad \text{but} \quad n_1 \neq n_2,$$

and so g is not one-to-one.

APPLICATION: HASH FUNCTIONS

Imagine a set of student records, each of which includes the student's social security number, and suppose the records are to be stored in a table in which a record can be located if the social security number is known. One way to do this would be to place the record with social security number n into position n of the table. However, since social security numbers have nine digits, this method would require a table with 999,999,999 positions. The problem is that creating such a table for a small set of records would be very wasteful of computer memory space. **Hash functions** are functions defined from larger to smaller sets of integers, frequently using the *mod* function, which provide part of the solution to this problem. We illustrate how to define and use a hash function with a very simple example.

EXAMPLE 7.3.3 **A Hash Function**

Suppose there are no more than seven student records. Define a function h from the set of all social security numbers (ignoring hyphens) to the set $\{0, 1, 2, 3, 4, 5, 6\}$ as follows:

$$h(n) = n \bmod 7 \quad \text{for all social security numbers } n.$$

TABLE 7.3.1

0	356-63-3102
1	
2	513-40-8716
3	223-79-9061
4	
5	328-34-3419
6	

To use your calculator to find $n \bmod 7$, use the formula $n \bmod 7 = n - 7 \cdot (n \operatorname{div} 7)$. (See Section 3.4.) In other words, divide n by 7, multiply the integer part of the result by 7, and subtract that number from n. For instance, since $328343419/7 = 46906202.71 \ldots$

$$h(328\text{-}34\text{-}3419) = 328343419 - (7 \cdot 46906202) = 5.$$

As a first approximation to solving the problem of storing the records, try to place the record with social security number n in position $h(n)$. For instance, if the social security numbers are 328-34-3419, 356-63-3102, 223-79-9061, and 513-40-8716, the positions of the records are shown in Table 7.3.1.

The problem with this approach is that h may not be one-to-one; h might assign the same position in the table to records with different social security numbers. Such an assignment is called a **collision**. When collisions occur, various **collision resolution methods** are used. One of the simplest is the following: If, when the record with social security number n is to be placed, position $h(n)$ is already occupied, start from that position and search downward to place the record in the first empty position that occurs, going back up to the beginning of the table if necessary. To locate a record in the table from its social security number, n, you compute $h(n)$ and search downward from that position to find the record with social security number n. If there are not too many collisions, this is a very efficient way to store and locate records.

Suppose the social security number for another record to be stored is 908-37-1011. Find the position in Table 7.3.1 into which this record would be placed.

Solution   When you compute h you find that $h(908\text{-}37\text{-}1011) = 2$, which is already occupied by the record with social security number 513-40-8716. Searching downward from position 2, you find that position 3 is also occupied but position 4 is free.

$$908\text{-}37\text{-}1011 \;\overset{h}{\rightarrow}\; \underset{\substack{\uparrow \\ \text{occupied}}}{2} \;\rightarrow\; \underset{\substack{\uparrow \\ \text{occupied}}}{3} \;\rightarrow\; \underset{\substack{\uparrow \\ \text{free}}}{4}$$

So you place the record with social security number n into position 4.   ■

ONTO FUNCTIONS

It was noted in Section 7.1 that there may be an element of the co-domain of a function that is not the image of any element in the domain. On the other hand, a function may have the property that *every* element of its co-domain is the image of some element of its domain. Such a function is called *onto* or *surjective*. When a function is onto, its range is equal to its co-domain.

> **DEFINITION**
>
> Let F be a function from a set X to a set Y. F is **onto** (or **surjective**) if, and only if, given any element y in Y it is possible to find an element x in X with the property that $y = F(x)$.
>   Symbolically:
>
> $$F : X \rightarrow Y \text{ is onto} \quad \Leftrightarrow \quad \forall y \in Y, \exists x \in X \text{ such that } F(x) = y.$$

To obtain a precise statement of what it means for a function *not* to be onto, take the negation of the definition of onto:

$$F: X \to Y \text{ is } not \text{ onto} \quad \Leftrightarrow \quad \exists y \text{ in } Y \text{ such that } \forall x \in X, F(x) \neq y.$$

That is, there is some element in Y that is *not* the image of any element in X.

In terms of arrow diagrams, a function is onto if each element of the co-domain has an arrow pointing to it from some element of the domain. A function is not onto if at least one element in its co-domain does not have an arrow pointing to it. This is illustrated in Figure 7.3.3.

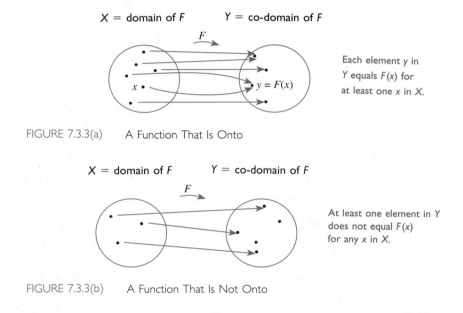

X = domain of F   Y = co-domain of F

Each element y in Y equals F(x) for at least one x in X.

FIGURE 7.3.3(a)   A Function That Is Onto

X = domain of F   Y = co-domain of F

At least one element in Y does not equal F(x) for any x in X.

FIGURE 7.3.3(b)   A Function That Is Not Onto

EXAMPLE 7.3.4   Identifying Onto Functions Defined on Finite Sets

a. Which of the arrow diagrams in Figure 7.3.4 define onto functions?

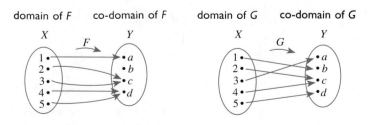

domain of F   co-domain of F   domain of G   co-domain of G

FIGURE 7.3.4

b. Let $X = \{1, 2, 3, 4\}$ and $Y = \{a, b, c\}$. Define $H: X \to Y$ by specifying that $H(1) = c$, $H(2) = a$, $H(3) = c$, and $H(4) = b$. Define $K: X \to Y$ by specifying that $K(1) = c$, $K(2) = b$, $K(3) = b$, and $K(4) = c$. Is either H or K onto?

Solution   a. F is not onto because $b \neq F(x)$ for any x in X. G is onto because each element of Y equals $G(x)$ for some x in X: $a = G(3)$, $b = G(1)$, $c = G(2) = G(4)$, and $d = G(5)$.

b. H is onto but K is not. H is onto because each of the three elements of the co-domain of H is the image of some element of the domain of H: $a = H(2)$, $b = H(4)$, and $c = H(1) = H(3)$. K, however, is not onto because $a \neq K(x)$ for any x in $\{1, 2, 3, 4\}$.   ■

It is possible to write a computer algorithm to check whether a function F is onto, provided F is defined from a finite set X to a finite set Y and there is an independent algorithm to compute values of F. Represent X and Y as one-dimensional arrays $a[1]$, $a[2], \ldots, a[n]$ and $b[1], b[2], \ldots, b[n]$, respectively, and use a nested loop to pick each element y of Y in turn and search through the elements of X to find an x such that y is the image of x. If any search is unsuccessful, then F is not onto. If each such search is successful, then F is onto. You are asked to write such an algorithm in the exercises at the end of this section.

ONTO FUNCTIONS ON INFINITE SETS

Now suppose F is a function from a set X to a set Y and suppose Y is infinite. By definition, F is onto if, and only if, the following universal statement is true:

$$\forall y \in Y, \exists x \in X \text{ such that } F(x) = y.$$

Thus to prove F is onto, you will ordinarily use the method of generalizing from the generic particular:

suppose that y is any element of Y

and

show that there is an element of X with $F(x) = y$.

To prove F is *not onto*, you will usually

find an element y of Y such that $y \neq F(x)$ for *any* x in X.

EXAMPLE 7.3.5   **Proving or Disproving That Functions Are Onto**

Define $f : \mathbf{R} \to \mathbf{R}$ and $h : \mathbf{Z} \to \mathbf{Z}$ by the rules

$$f(x) = 4x - 1 \quad \text{for all } x \in \mathbf{R}$$

and

$$h(n) = 4n - 1 \quad \text{for all } n \in \mathbf{Z}.$$

a. Is f onto? Prove or give a counterexample.
b. Is h onto? Prove or give a counterexample.

Solution   a. The best approach is to start trying to prove that f is onto and be alert for difficulties that might indicate that it is not. Now $f : \mathbf{R} \to \mathbf{R}$ is the function defined by the rule

$$f(x) = 4x - 1 \quad \textbf{for all real numbers } x.$$

To prove that f is onto, you must prove

$$\forall y \in Y, \exists x \in X \text{ such that } f(x) = y.$$

Substituting the definition of f into the outline of a proof by the method of generalizing from the generic particular, you

suppose y is a real number

and

show that there exists a real number x such that $y = 4x - 1$.

Scratch Work   If such a real number x exists, then

$$4x - 1 = y$$
$$4x = y + 1 \qquad \text{by adding 1 to both sides}$$
$$x = \frac{y + 1}{4} \qquad \text{by dividing both sides by 4.}$$

Thus *if* such a number x exists, it must equal $(y + 1)/4$. Does such a number exist? Yes. To show this, let $x = (y + 1)/4$, and then check that (1) x is a real number and (2) the steps above are valid if followed in reverse order to conclude $y = 4x - 1$. The formal answer below summarizes this process.

Answer to (a)

> If $f: \mathbf{R} \rightarrow \mathbf{R}$ is the function defined by the rule $f(x) = 4x - 1$ for all real numbers x, then f is onto.
>
> Proof:
>
> Let $y \in \mathbf{R}$. *[We must show that $\exists x$ in $\mathbf{R}$ such that $f(x) = y$.]* Let $x = (y + 1)/4$. Then x is a real number since sums and quotients (other than by 0) of real numbers are real numbers. It follows that
>
> $$f(x) = f\left(\frac{y + 1}{4}\right) \qquad \text{by substitution}$$
> $$= 4 \cdot \left(\frac{y + 1}{4}\right) - 1 \qquad \text{by definition of } f$$
> $$= (y + 1) - 1 = y \qquad \text{by basic algebra.}$$
>
> This is what was to be shown.

b. The function $h: \mathbf{Z} \rightarrow \mathbf{Z}$ is defined by the rule

$$h(n) = 4n - 1 \quad \text{for all integers } n.$$

To prove that h is onto, it would be necessary to prove that

$$\forall \text{ integers } m, \exists \text{ an integer } n \text{ such that } h(n) = m.$$

Substituting the definition of h into the outline of a proof by the method of generalizing from the generic particular, you

suppose m is any integer

and

try to show that there is an integer n with $4n - 1 = m$.

Can you reach what is to be shown from the supposition? No! If $4n - 1 = m$, then

$$n = \frac{m + 1}{4}$$ by adding 1 and dividing by 4.

But n must be an integer. And when, for example, $m = 0$, then

$$n = \frac{0 + 1}{4} = \frac{1}{4},$$

which is *not* an integer.

Thus, trying to prove that h is onto runs into a difficulty, and this difficulty reveals a counterexample that shows h is not onto.

This discussion is summarized in the following formal answer.

Answer to (b)

> If the function $h : \mathbf{Z} \to \mathbf{Z}$ is defined by the rule $h(n) = 4n - 1$ for all integers n, then h is not onto.
>
> **Counterexample:**
>
> The co-domain of h is $\mathbf{Z}$ and $0 \in \mathbf{Z}$. But $h(n) \neq 0$ for any integer n. For if $h(n) = 0$, then
>
> $$4n - 1 = 0$$ by definition of h
> $$4n = 1$$ by adding 1 to both sides
> $$n = \frac{1}{4}$$ by dividing both sides by 4.
>
> But $1/4$ is not an integer. Hence there is no integer n for which $f(n) = 0$, and so f is not onto.

THE EXPONENTIAL AND LOGARITHMIC FUNCTIONS

The **exponential function with base b**, denoted $\exp_b$, is the function from $\mathbf{R}$ to $\mathbf{R}^+$ defined as follows: for all real numbers x,

$$\exp_b(x) = b^x$$

where $b^0 = 1$ and $b^{-x} = 1/b^x$.*

When working with the exponential function, it is useful to recall the laws of exponents from elementary algebra.

---

* That the quantity b^x is a real number for any real number x follows from the least-upper-bound property of the real number system. (See Appendix A.)

> If b and c are any positive real numbers and u and v are any real numbers, the following laws of exponents hold true:
>
> $$b^u b^v = b^{u+v} \qquad \text{7.3.1}$$
>
> $$(b^u)^v = b^{uv} \qquad \text{7.3.2}$$
>
> $$(bc)^u = b^u c^u \qquad \text{7.3.3}$$

The logarithmic function with base b was defined in Example 7.1.9 to be the function from $\mathbf{R}^+$ to $\mathbf{R}$ with the property that for each positive real number x,

$$\log_b(x) = \text{the exponent to which } b \text{ must be raised to obtain } x.$$

Or, equivalently, for each positive real number x and real number y.

$$\log_b x = y \quad \Leftrightarrow \quad b^y = x.$$

It can be shown using calculus that both the exponential and logarithmic functions are one-to-one and onto. Therefore, by definition of one-to-one, the following properties hold true:

> For any positive real number b,
>
> $$\text{if } b^u = b^v \text{ then } u = v \quad \text{for all numbers } u \text{ and } v, \qquad \text{7.3.4}$$
>
> and
>
> $$\text{if } \log_b u = \log_b v \text{ then } u = v \quad \text{for all positive real numbers } u \text{ and } v. \quad \text{7.3.5}$$

These properties are used to derive many additional facts about exponents and logarithms. One example is given below.

EXAMPLE 7.3.6   Using the One-to-Oneness of the Exponential Function

Use the definition of logarithm, the laws of exponents, and the one-to-oneness of the exponential function (property (7.3.4.)) to show that for any positive real numbers b, c, and x,

> $$\log_c x = \frac{\log_b x}{\log_b c}. \qquad \text{7.3.6}$$

Solution   Suppose positive real numbers b, c, and x are given. Let

$$(1)\ u = \log_b c \qquad (2)\ v = \log_c x \qquad (3)\ w = \log_b x.$$

Then by definition of logarithm,

$$(1')\ c = b^u \qquad (2')\ x = c^v \qquad (3')\ x = b^w.$$

Substituting $(1')$ into $(2')$ gives

$$x = c^v = (b^u)^v = b^{uv} \quad \text{by 7.3.2}$$

But by (2), $x = b^w$ also. Hence

$$b^{uv} = b^w,$$

and so by the one-to-oneness of the exponential function (property 7.3.4),

$$uv = w.$$

Substituting from (1), (2), and (3) gives that

$$(\log_b c)(\log_c x) = \log_b x.$$

And dividing both sides by $\log_b c$ results in

$$\log_c x = \frac{\log_b x}{\log_b c}.$$ ∎

EXAMPLE 7.3.7   **Computing Logarithms with Base 2 on a Calculator**

In computer science it is often necessary to compute logarithms with base 2. Most calculators do not have keys to compute logarithms with base 2 but do have keys to compute logarithms with base 10 (called **common logarithms** and often denoted simply log) and logarithms with base e (called **natural logarithms** and usually denoted ln). Suppose your calculator shows that $\log_{10} 5 \cong 0.6989700043$ and $\log_{10} 2 \cong 0.3010299957$. Use formula 7.3.6 from Example 7.3.6 to find an approximate value for $\log_2 5$.

Solution   By formula 7.3.6,

$$\log_2 5 = \frac{\log_{10} 5}{\log_{10} 2} \cong \frac{0.6989700043}{0.3010299957} \cong 2.321928095.$$ ∎

ONE-TO-ONE CORRESPONDENCES

Consider a function $F: X \to Y$ that is both one-to-one and onto. Given any element x in X, there is a unique corresponding element $y = F(x)$ in Y (since F is a function). Also given any element y in Y, there is an element x in X such that $F(x) = y$ (since F is onto) and there is only one such x (since F is one-to-one). Thus, a function that is one-to-one and onto sets up a pairing between the elements of X and the elements of Y that pairs each element of X with exactly one element of Y and each element of Y with exactly one element of X. Such a pairing is called a *one-to-one correspondence* or *bijection* and is illustrated by the arrow diagram in Figure 7.3.5. In Chapter 6 we frequently used one-to-one correspondences to count the number of elements in a set. The pairing of Figure 7.3.5, for example, shows that there are five elements in the set X.

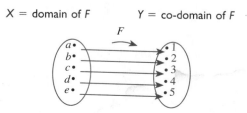

FIGURE 7.3.5   An Arrow Diagram for a One-to-One Correspondence

> **DEFINITION**
>
> A **one-to-one correspondence** (or **bijection**) from a set X to a set Y is a function $F: X \to Y$ that is both one-to-one and onto.

EXAMPLE 7.3.8   **A Function from a Power Set to a Set of Strings**

Let $\mathscr{P}(\{a, b\})$ be the set of all subsets of $\{a, b\}$ and let Σ^2 be the set of all strings of length 2 made up of 0's and 1's. Then $\mathscr{P}(\{a, b\}) = \{\varnothing, \{a\}, \{b\}, \{a, b\}\}$ and $\Sigma^2 = \{00, 01, 10, 11\}$. Define a function h from $\mathscr{P}(\{a, b\})$ to Σ^2 as follows: Given any subset A of $\{a, b\}$, a is either in A or not in A and b is either in A or not in A. If a is in A, write a 1 in the first position of the string $h(A)$. If a is not in A, write a 0 in the first position of the string $h(A)$. Similarly, if b is in A, write a 1 in the second position of the string $h(A)$. If b is not in A, write a 0 in the second position of the string $h(A)$. This definition is summarized in the following table:

h			
Subset of $\{a, b\}$	Status of a	Status of b	String in Σ^2
$\varnothing$	not in	not in	00
$\{a\}$	in	not in	10
$\{b\}$	not in	in	01
$\{a, b\}$	in	in	11

Is h a one-to-one correspondence?

Solution   The arrow diagram shown in Figure 7.3.6 shows clearly that h is a one-to-one correspondence. It is onto because each element of Σ^2 has an arrow pointing to it. It is one-to-one because each element of Σ^2 has no more than one arrow pointing to it.

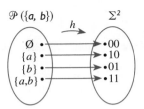

FIGURE 7.3.6

EXAMPLE 7.3.9   **A String-Reversing Function**

Let $\Sigma = \{x, y\}$ and recall that Σ^* is the set of all finite strings over Σ. Define $g: \Sigma^* \to \Sigma^*$ by the rule

For all strings $s \in \Sigma^*$,

$$g(s) = \text{the string obtained by writing the characters of } s \text{ in reverse order.}$$

Is g a one-to-one correspondence from Σ^* to itself?

Solution The answer is yes. To show that g is a one-to-one correspondence, it is necessary to show that g is one-to-one and onto.

To see that g is one-to-one, suppose that for some strings s_1 and s_2 in Σ^*, $g(s_1) = g(s_2)$. *[We must show that $s_1 = s_2$.]* Now to say that $g(s_1) = g(s_2)$ is the same as saying that the string obtained by writing the characters of s_1 in reverse order equals the string obtained by writing the characters of s_2 in reverse order. But if s_1 and s_2 are equal when written in reverse order, then they must be equal to start with. In other words, $s_1 = s_2$ *[as was to be shown].*

To show that g is onto, suppose t is a string in Σ^*. *[We must find a string s in Σ^* such that $g(s) = t$.]* Let $s = g(t)$. By definition of g, $s = g(t)$ is the string in Σ^* obtained by writing the characters of t in reverse order. But when the order of the characters of a string is reversed once and then reversed again, the original string is recovered. Thus

$$g(s) = g(g(t)) = \text{the string obtained by writing the characters}$$
$$\text{of } t \text{ in reverse order and then writing}$$
$$\text{those characters in reverse order again}$$
$$= t.$$

This is what was to be shown. ∎

INVERSE FUNCTIONS

If F is a one-to-one correspondence from a set X to a set Y, then there is a function from Y to X that "undoes" the action of F; that is, it sends each element of Y back to the element of X that it came from. This function is called the *inverse function* for F.

THEOREM 7.3.1

Suppose $F: X \to Y$ is a one-to-one correspondence; that is, suppose F is one-to-one and onto. Then there is a function $F^{-1}: Y \to X$ that is defined as follows:

Given any element y in Y,

$$F^{-1}(y) = \text{that unique element } x \text{ in } X \text{ such that } F(x) \text{ equals } y.$$

In other words,

$$F^{-1}(y) = x \iff y = F(x).$$

The proof of Theorem 7.3.1 follows immediately from the definition of one-to-one and onto. Given an element y in Y, there is an element x in X with $F(x) = y$ because F is onto; x is unique because F is one-to-one.

DEFINITION

The function F^{-1} of Theorem 7.3.1 is called the **inverse function** for F.

Note that according to this definition, the logarithmic function with base $b > 0$ is the inverse of the exponential function with base b.

The diagram on the next page illustrates the fact that an inverse function sends each element back to where it came from.

X = domain of F   Y = co-domain of F

$x = F^{-1}(y)$ •   • $F(x) = y$

F

F^{-1}

EXAMPLE 7.3.10   **Finding an Inverse Function for a Function Given by an Arrow Diagram**

Define the inverse function for the one-to-one correspondence h given in Example 7.3.8.

Solution   The arrow diagram for h^{-1} is obtained by tracing the h-arrows back from Σ^2 to $\mathcal{P}(\{a, b\})$ as shown below.

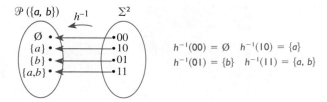

$\mathcal{P}(\{a, b\})$ h^{-1} Σ^2

Ø •   •00
$\{a\}$ •   •10
$\{b\}$ •   •01
$\{a,b\}$ •   •11

$h^{-1}(00) = \emptyset$ $h^{-1}(10) = \{a\}$
$h^{-1}(01) = \{b\}$ $h^{-1}(11) = \{a, b\}$

EXAMPLE 7.3.11   **Finding an Inverse Function for a Function Given in Words**

Define the inverse function for the one-to-one correspondence g given in Example 7.3.9.

Solution   The function $g: \Sigma^* \to \Sigma^*$ is defined by the rule

For all strings s in Σ^*,

$g(s)$ = the string obtained by writing the
characters of s in reverse order.

Now if the characters of Σ^* are written in reverse order and then written in reverse order again, the original string is recovered. Thus given any string t in Σ^*,

$g^{-1}(t)$ = the unique string that, when written
in reverse order, equals t

= the string obtained by writing the
characters of t in reverse order

= $g(t)$.

Hence $g^{-1}: \Sigma^* \to \Sigma^*$ is the same as g, or, in other words, $g^{-1} = g$.   ■

EXAMPLE 7.3.12   **Finding an Inverse Function for a Function Given by a Formula**

The function $f: \mathbf{R} \to \mathbf{R}$ defined by the formula

$f(x) = 4x - 1$ for all real numbers x

was shown to be one-to-one in Example 7.3.2 and onto in Example 7.3.5. Find its inverse function.

Solution   By definition of f^{-1},

$$f^{-1}(y) = \text{that unique real number } y \text{ such that } f(x) = y.$$

But

$$f(x) = y$$
$$\Leftrightarrow \quad 4x - 1 = y \qquad \text{by definition of } f$$
$$\Leftrightarrow \quad x = \frac{y + 1}{4} \qquad \text{by adding 1 and dividing both sides by 4.}$$

Hence $f^{-1}(y) = \dfrac{y + 1}{4}$. ∎

The following theorem follows easily from the definitions.

THEOREM 7.3.2

If X and Y are sets and $F : X \to Y$ is one-to-one and onto, then $F^{-1} : Y \to X$ is also one-to-one and onto.

Proof:

F^{-1} **is one-to-one:** Suppose y_1 and y_2 are elements of Y such that $F^{-1}(y_1) = F^{-1}(y_2)$. *[We must show that $y_1 = y_2$.]* Let $x = F^{-1}(y_1) = F^{-1}(y_2)$. Then $x \in X$, and by definition of F^{-1},

$$F(x) = y_1 \quad \text{since } x = F^{-1}(y_1)$$

and

$$F(x) = y_2 \quad \text{since } x = F^{-1}(y_2).$$

Consequently, $y_1 = y_2$ since each is equal to $F(x)$. This is what was to be shown.

F^{-1} **is onto:** Suppose $x \in X$. *[We must show that there exists an element y in Y such that $F^{-1}(y) = x$.]* Let $y = F(x)$. Then $y \in Y$, and by definition of F^{-1}, $F^{-1}(y) = x$. This is what was to be shown.

EXERCISE SET 7.3

1. a. The definition of one-to-one is stated in two ways:

$$\forall x_1, x_2 \in X, \text{ if } F(x_1) = F(x_2) \text{ then } x_1 = x_2$$

and

$$\forall x_1, x_2 \in X, \text{ if } x_1 \neq x_2 \text{ then } F(x_1) \neq F(x_2).$$

Why are these two statements logically equivalent?

b. Give a counterexample to show that the following statement is false: A function $f : X \to Y$ is one-to-one if, and only if, every element of X is sent to exactly one element of Y.

c. Let $f : X \to Y$ be a function. True or false? A sufficient condition for f to be one-to-one is that for all elements y in Y, there is at most one x in X with $f(x) = y$.

2. Let $X = \{1, 5, 9\}$ and $Y = \{3, 4, 7\}$.

a. Define $f : X \to Y$ by specifying that

$$f(1) = 4, \quad f(5) = 7, \quad f(9) = 4.$$

Is f one-to-one? Is f onto? Explain your answers.

b. Define $g : X \to Y$ by specifying that

$$g(1) = 7, \quad g(5) = 3, \quad g(9) = 4.$$

Is g one-to-one? Is g onto? Explain your answers.

3. Let $X = \{a, b, c, d\}$ and $Y = \{x, y, z\}$. Define functions F and G by the arrow diagrams below.

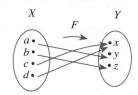

domain of F   co-domain of F

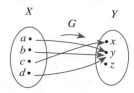

domain of G   co-domain of G

a. Is F one-to-one? Why or why not? Onto? Why or why not?

b. Is G one-to-one? Why or why not? Onto? Why or why not?

4. Let $X = \{a, b, c\}$ and $Y = \{w, x, y, z\}$. Define functions H and K by the arrow diagrams below.

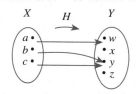

domain of H   co-domain of H

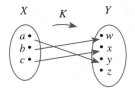

domain of K   co-domain of K

a. Is H one-to-one? Why or why not? Onto? Why or why not?

b. Is K one-to-one? Why or why not? Onto? Why or why not?

5. Let $X = \{1, 2, 3\}$, $Y = \{1, 2, 3, 4\}$, and $Z = \{1, 2\}$.
a. Define a function $f : X \to Y$ that is one-to-one but not onto.

b. Define a function $g : X \to Z$ that is onto but not one-to-one.

c. Define a function $h : X \to X$ that is neither one-to-one nor onto.

d. Define a function $k : X \to X$ that is one-to-one and onto but is not the identity function on X.

6. a. How many one-to-one functions are there from a set with three elements to a set with four elements?

b. How many one-to-one functions are there from a set with three elements to a set with two elements?

c. How many one-to-one functions are there from a set with three elements to a set with three elements?

d. How many one-to-one functions are there from a set with three elements to a set with five elements?

H e. How many one-to-one functions are there from a set with m elements to a set with n elements, where $m \le n$?

7. a. How many onto functions are there from a set with three elements to a set with two elements?

b. How many onto functions are there from a set with three elements to a set with five elements?

H c. How many onto functions are there from a set with four elements to a set with two elements?

d. How many onto functions are there from a set with five elements to a set with three elements?

H◆ e. Let $c_{m,n}$ be the number of onto functions from a set of m elements to a set of n elements, where $m \ge n \ge 1$. Find a formula relating $c_{m,n}$ to $c_{m-1,n}$ and $c_{m-1,n-1}$.

8. a. Define $f : \mathbf{Z} \to \mathbf{Z}$ by the rule $f(n) = 2n$, for all integers n.
   (i) Is f one-to-one? Prove or give a counterexample.
   (ii) Is f onto? Prove or give a counterexample.

b. Let $2\mathbf{Z}$ denote the set of all even integers. That is, $2\mathbf{Z} = \{n \in \mathbf{Z} \mid n = 2k, \text{ for some integer } k\}$. Define $h : \mathbf{Z} \to 2\mathbf{Z}$ by the rule $h(n) = 2n$, for all integers n. Is h onto? Prove or give a counterexample.

9. a. Define $g : \mathbf{Z} \to \mathbf{Z}$ by the rule $g(n) = 3n - 2$, for all integers n.
   (i) Is g one-to-one? Prove or give a counterexample.
   (ii) Is g onto? Prove or give a counterexample.

b. Define $G : \mathbf{R} \to \mathbf{R}$ by the rule $G(x) = 3x - 2$, for all real numbers x. Is G onto? Prove or give a counterexample.

10. a. Define $H : \mathbf{R} \to \mathbf{R}$ by the rule $H(x) = x^2$, for all real numbers x.
   (i) Is H one-to-one? Prove or give a counterexample.
   (ii) Is H onto? Prove or give a counterexample.

b. Define $K : \mathbf{R}^{nonneg} \to \mathbf{R}^{nonneg}$ by the rule $K(x) = x^2$, for all nonnegative real numbers x. Is K onto? Prove or give a counterexample.

H 11. Explain the mistake in the following "proof."

Theorem: The function $f : \mathbf{Z} \to \mathbf{Z}$ defined by the formula $f(n) = 4n + 3$, for all integers n, is one-to-one.

"**Proof:** Suppose any integer n is given. Then by definition of f, there is only one possible value for $f(n)$, namely, $4n + 3$. Hence f is one-to-one."

In each of 12–15 a function f is defined on a set of real numbers. Determine whether or not f is one-to-one and justify your answer.

12. $f(x) = \dfrac{x + 1}{x}$, for all real numbers $x \neq 0$

13. $f(x) = \dfrac{x}{x^2 + 1}$, for all real numbers x

14. $f(x) = \dfrac{2x + 1}{x}$, for all real numbers $x \neq 0$

15. $f(x) = \dfrac{x - 1}{x + 1}$, for all real numbers $x \neq -1$

16. Referring to Example 7.3.3, assume records with the following social security numbers are to be placed in sequence into Table 7.3.1. Find the position into which each record is placed.
  a. 417-30-2072   b. 364-98-1703   c. 283-09-0787

17. Define Floor : $\mathbf{R} \rightarrow \mathbf{Z}$ by the formula Floor$(x) = \lfloor x \rfloor$, for all real numbers x.
  a. Is Floor one-to-one? Prove or give a counterexample.
  b. Is Floor onto? Prove or give a counterexample.

18. Let $\Sigma = \{0, 1\}$ and define $l : \Sigma^* \rightarrow \mathbf{Z}^{nonneg}$ by

  $l(s) = $ the length of s,   for all strings s in Σ^*.

  a. Is l one-to-one? Prove or give a counterexample.
  b. Is l onto? Prove or give a counterexample.

19. Define $F : \mathcal{P}(\{a, \ b, \ c\}) \rightarrow \mathbf{Z}$ as follows: for all $A \in \mathcal{P}(\{a, b, c\})$,

  $F(A) = $ the number of elements in A.

  a. Is F one-to-one? Prove or give a counterexample.
  b. Is F onto? Prove or give a counterexample.

20. Let $\Sigma = \{a, b\}$ and define $N : \Sigma^* \rightarrow \mathbf{Z}$ by

  $N(s) = $ the number of a's in s,   for all $s \in \Sigma^*$.

  a. Is N one-to-one? Prove or give a counterexample.
  b. Is N onto? Prove or give a counterexample.

21. Let $\Sigma = \{a, b\}$ and define $C : \Sigma^* \rightarrow \Sigma^*$ by

  $C(s) = as$,   for all $s \in \Sigma^*$.

  ($C(s)$ is concatenation by a on the left.)
  a. Is C one-to-one? Prove or give a counterexample.
  b. Is C onto? Prove or give a counterexample.

◆22. Define $F : \mathbf{Z}^+ \times \mathbf{Z}^+ \rightarrow \mathbf{Z}^+ \times \mathbf{Z}^+$ and $G : \mathbf{Z}^+ \times \mathbf{Z}^+ \rightarrow \mathbf{Z}^+ \times \mathbf{Z}^+$ as follows: for all $(n, m) \in \mathbf{Z}^+ \times \mathbf{Z}^+$,

$$F(n, m) = 3^n 6^m \quad \text{and} \quad G(n, m) = 3^n 5^m.$$

  a. Is F one-to-one? Prove or give a counterexample.
  b. Is G one-to-one? Prove or give a counterexample.

23. a. Is $\log_8 27 = \log_2 3$? Why or why not?
  b. Is $\log_9 25 = \log_3 5$? Why or why not?

The properties of logarithms established in 24–26 are used in Sections 9.4 and 9.5.

24. Prove that for all positive real numbers b, x, and y with $b \neq 1$,

$$\log_b\left(\frac{x}{y}\right) = \log_b x - \log_b y.$$

25. Prove that for all positive real numbers b, x, and y with $b \neq 1$,

$$\log_b (xy) = \log_b x + \log_b y.$$

26. Prove that for all real numbers a, b, and x with b and x positive and $b \neq 1$,

$$\log_b (x^a) = a \cdot \log_b x.$$

Exercises 27 and 28 use the following definition: If $f : \mathbf{R} \rightarrow \mathbf{R}$ and $g : \mathbf{R} \rightarrow \mathbf{R}$ are functions, the function $(f + g) : \mathbf{R} \rightarrow \mathbf{R}$ is defined by the formula $(f + g)(x) = f(x) + g(x)$ for all real numbers x.

27. If $f : \mathbf{R} \rightarrow \mathbf{R}$ and $g : \mathbf{R} \rightarrow \mathbf{R}$ are both one-to-one, is $f + g$ also one-to-one? Justify your answer.

28. If $f : \mathbf{R} \rightarrow \mathbf{R}$ and $g : \mathbf{R} \rightarrow \mathbf{R}$ are both onto, is $f + g$ also onto? Justify your answer.

Exercises 29 and 30 use the following definition: If $f : \mathbf{R} \rightarrow \mathbf{R}$ is a function and c is a nonzero real number, the function $(c \cdot f) : \mathbf{R} \rightarrow \mathbf{R}$ is defined by the formula $(c \cdot f)(x) = c \cdot f(x)$ for all real numbers x.

29. Let $f : \mathbf{R} \rightarrow \mathbf{R}$ be a function and c a nonzero real number. If f is one-to-one, is $c \cdot f$ also one-to-one? Justify your answer.

30. Let $f : \mathbf{R} \rightarrow \mathbf{R}$ be a function and c a nonzero real number. If f is onto, is $c \cdot f$ also onto? Justify your answer.

Let $X = \{a, b, c, d, e\}$ and $Y = \{s, t, u, v, w\}$. In each of 31 and 32, a one-to-one correspondence $F : X \rightarrow Y$ is defined by an arrow diagram. In each case draw an arrow diagram for F^{-1}.

31.

32.

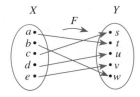

In 33–45 indicate which of the functions in the referenced exercise are one-to-one correspondences. For each function that is a one-to-one correspondence, find the inverse function.

33. exercise 8a

34. exercise 8b

35. exercise 9a

36. exercise 9b

37. exercise 10b

38. exercise 17

39. exercise 18

40. exercise 19

41. exercise 20

42. exercise 12, with the co-domain taken to be the set of all real numbers not equal to 1

H 43. exercise 13, with the co-domain taken to be the set of all real numbers

44. exercise 14, with the co-domain taken to be the set of all real numbers not equal to 2

45. exercise 15, with the co-domain taken to be the set of all real numbers not equal to 1

46. In Example 7.3.8 a one-to-one correspondence was defined from the power set of $\{a, b\}$ to the set of all strings of 0's and 1's that have length 2. Thus the elements of these two sets can be matched up exactly, and so the two sets have the same number of elements.
   a. Let $X = \{x_1, x_2, \ldots, x_n\}$ be a set with n elements. Use Example 7.3.8 as a model to define a one-to-one correspondence from $\mathcal{P}(X)$, the set of all subsets of X, to the set of all strings of 0's and 1's that have length n.
   b. Use the one-to-one correspondence of part (a) to deduce that a set with n elements has 2^n subsets. (This provides an alternative proof of Theorem 5.3.5.)

H 47. Write a computer algorithm to check whether a function from one finite set to another is one-to-one. Assume the existence of an independent alogrithm to compute values of the function.

H 48. Write a computer algorithm to check whether a function from one finite set to another is onto. Assume the existence of an independent algorithm to compute values of the function.

7.4 APPLICATION: THE PIGEONHOLE PRINCIPLE

The shrewd guess, the fertile hypothesis, the courageous leap to a tentative conclusion—these are the most valuable coin of the thinker at work.
(Jerome S. Bruner, 1960)

The pigeonhole principle states that if n pigeons fly into m pigeonholes and $n > m$, then at least one hole must contain two or more pigeons. This principle is illustrated in Figure 7.4.1 for $n = 5$ and $m = 4$. Illustration (a) shows the pigeons perched next to their holes, and (b) shows the correspondence from pigeons to pigeonholes. The pigeonhole principle is sometimes called the *Dirichlet box principle* because it was first stated formally by P.G.L. Dirichlet (1805–1859).

Illustration (b) suggests the following mathematical way to phrase the principle.

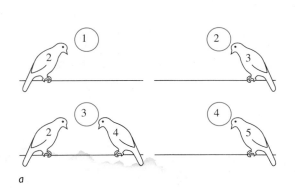

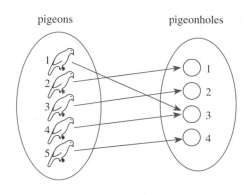

a

b

FIGURE 7.4.1

> **PIGEONHOLE PRINCIPLE**
>
> A function from one finite set to a smaller finite set cannot be one-to-one:
> There must be at least two elements in the domain that have the same image
> in the co-domain.

Thus an arrow diagram for a function from a finite set to a smaller finite set must
have at least two arrows from the domain that point to the same element of the
co-domain. In Figure 7.4.1(b), arrows from pigeons 1 and 4 both point to pigeonhole 3.

Since the truth of the pigeonhole principle is easy to accept on an intuitive basis,
we move immediately to applications, leaving a formal proof to the end of the section.
Applications of the pigeonhole principle range from the totally obvious to the extremely subtle. A representative sample is given in the examples and exercises that
follow.

EXAMPLE 7.4.1   **Applying the Pigeonhole Principle**

a. In a group of six people, must there be at least two who were born in the same
   month? In a group of thirteen people, must there be at least two who were born in
   the same month? Why?
b. Among the residents of New York City, must there be at least two people with the
   same number of hairs on their heads? Why?

Solution   a. A group of six people need not contain two who were born in the same month. For
   instance, the six people could have birthdays in each of the six months January
   through June.

   A group of thirteen people, however, must contain at least two who were born
   in the same month, for there are only twelve months in a year and $13 > 12$. To get
   at the essence of this reasoning, think of the thirteen people as the pigeons and the
   twelve months of the year as the pigeonholes. Denote the thirteen people by the
   symbols $x_1, x_2, \ldots, x_{13}$ and define a function B from the set of people to the set
   of twelve months as shown in the following arrow diagram.

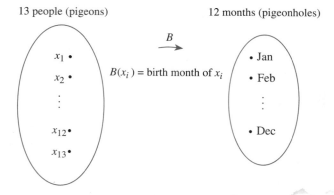

The pigeonhole principle says that no matter what the particular assignment of
months to people, there must be at least two arrows pointing to the same month.
Thus at least two people must have been born in the same month.

b. The answer is yes. In this example the pigeons are the people of New York City and the pigeonholes are all possible numbers of hairs on any individual's head. Call the population of New York City P. It is known that P is at least 5,000,000. Also the maximum number of hairs on any person's head is known to be no more than 300,000. Define a function H from the set of people in New York City $\{x_1, x_2, \ldots, x_p\}$ to the set $\{0, 1, 2, 3, \ldots, 300\,000\}$, as shown below.

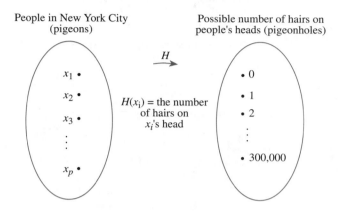

Since the number of people in New York City is larger than the number of possible hairs on their heads, the function H is not one-to-one; at least two arrows point to the same number. But that means that at least two people have the same number of hairs on their heads.   ∎

EXAMPLE 7.4.2   **Finding the Number to Pick to Ensure a Result**

A drawer contains ten black and ten white socks. You reach in and pull some out without looking at them. What is the *least* number of socks you must pull out to be sure to get a matched pair? Explain how the answer follows from the pigeonhole principle.

Solution   If you pick just two socks, they may have different colors. But when you pick a third sock, it must be the same color as one of the socks already chosen. Hence the answer is three.

   This answer could be phrased more formally as follows: Let the socks pulled out be denoted $s_1, s_2, s_3, \ldots, s_n$ and consider the function C that sends each sock to its color, as shown below.

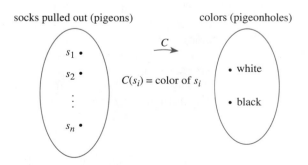

If $n = 2$, C could be a one-to-one correspondence (if the two socks pulled out were of different colors). But if $n > 2$, then the number of elements in the domain of C is larger than the number of elements in the co-domain of C. Thus, by the pigeonhole principle, C is not one-to-one: $C(s_i) = C(s_j)$ for some $s_i \neq s_j$. This means that if at least three socks are pulled out, then at least two of them have the same color. ∎

EXAMPLE 7.4.3 **Selecting a Pair of Integers with a Certain Sum**

Let $A = \{1, 2, 3, 4, 5, 6, 7, 8\}$.

a. If five integers are selected from A, must at least one pair of the integers have a sum of 9?

b. If four integers are selected from A, must at least one pair of the integers have a sum of 9?

Solution a. Yes, because the set A can be partitioned into four subsets:

$$\{1, 8\}, \quad \{2, 7\}, \quad \{3, 6\}, \quad \text{and} \quad \{4, 5\},$$

each consisting of two integers whose sum is 9. If five integers are selected from A, then by the pigeonhole priniciple at least two must be from the same subset. But then the sum of these two integers is 9.

To see precisely how the pigeonhole principle applies, let the pigeons be the five selected integers (call them a_1, a_2, a_3, a_4, and a_5) and let the pigeonholes be the subsets of the partition. The function P from pigeons to pigeonholes is defined by letting $P(a_i)$ be the subset that contains a_i.

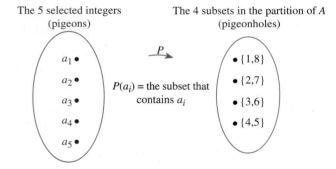

The function P is well defined because for each integer a_i in the domain, a_i belongs to one of the subsets (since the union of the subsets is A) and a_i does not belong to more than one subset (since the subsets are disjoint).

Because there are more pigeons than pigeonholes, at least two pigeons must go to the same hole. Thus two distinct integers are sent to the same set. But that implies that those two integers are the two distinct elements of the set, so their sum is 9. More formally, by the pigeonhole principle, since P is not one-to-one, there are integers a_i and a_j such that

$$P(a_i) = P(a_j) \quad \text{and} \quad a_i \neq a_j.$$

But then, by definition of P, a_i and a_j belong to the same subset. Since the elements in each subset add up to 9, $a_i + a_j = 9$.

b. The answer is no. This is a case where the piegonhole principle does not apply; the number of piegons is not larger than the number of piegonholes. For instance, if you select the numbers 1, 2, 3, and 4, then since the largest sum of any two of these numbers is 7, no two of them add up to 9.   ■

APPLICATION TO DECIMAL EXPANSIONS OF FRACTIONS

One important consequence of the piegonhole principle is the fact that the decimal expansion of any fraction either terminates or repeats. A terminating decimal is one like

$$3.625$$

and a repeating decimal is one like

$$2.38\overline{246},$$

where the bar over the digits 246 means that these digits are repeated forever.*

Recall that the decimal expansion of a fraction is obtained by dividing its numerator by its denominator using long division. For example, the decimal expansion of 4/33 is obtained as follows:

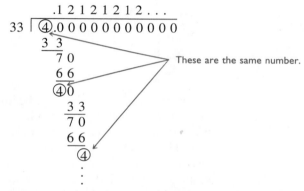

Because the number 4 reappears as a remainder in the long-division process, the sequence of quotients and remainders that give the digits of the decimal expansion repeats forever; hence the digits of the decimal expansion repeat forever.

In general, when one integer is divided by another, it is the pigeonhole principle (together with the quotient-remainder theorem) that guarantees that such a repetition of remainders and hence decimal digits must always occur. This is explained in the following example. The analysis in the example uses an obvious generalization of the pigeonhole principle, namely that a function from an infinite set to a finite set cannot be one-to-one.

EXAMPLE 7.4.4   The Decimal Expansion of a Fraction

Consider a fraction a/b, where for simplicity a and b are both assumed to be positive. The decimal expansion of a/b is obtained by dividing the a by the b as illustrated on page 392 for $a = 3$ and $b = 14$.

---

*Strictly speaking, a terminating decimal like 3.625 can be regarded as a repeating decimal by adding trailing zeros: $3.625 = 3.625\overline{0}$. This can also be written as $3.624\overline{9}$.

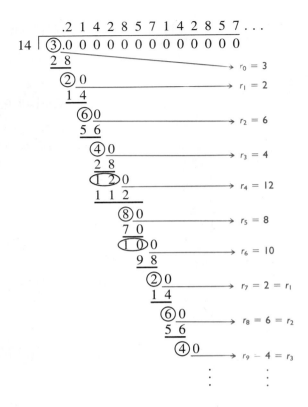

Let $r_0 = a$ and $r_1, r_2, r_3, \ldots$ be the successive remainders obtained in the long division of a by b. By the quotient-remainder theorem each remainder must be between 0 and $b - 1$. (In this example, a is 3 and b is 14, and so the remainders are from 0 to 13.) If some remainder $r_i = 0$, then the division terminates and a/b has a terminating decimal expansion. If no $r_i = 0$, then the division process and hence the sequence of remainders continues forever. By the pigeonhole principle, since there are more remainders than values the remainders can take, some remainder value must repeat: $r_j = r_k$, for some indices j and k with $j < k$. This is illustrated below for $b = 14$.

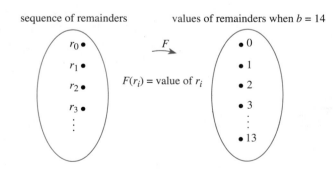

It follows that the decimal digits obtained from the divisions between r_j and r_{k-1} repeat forever. In the case of 3/14, the repetition begins with $r_7 = 2 = r_1$ and the decimal expansion repeats the quotients obtained from the divisions from r_1 through r_6 forever: $3/14 = 0.2\overline{142857}$. ■

APPLICATION TO FINITE-STATE AUTOMATA

The following example is an application of the pigeonhole principle to finite-state automata. Because a finite-state automaton can assume only a finite number of states and because there are infinitely many input sequences, by the pigeonhole principle there must be at least one state to which the automaton returns over and over again. It follows that it is possible to specify a language that is not the language accepted by any finite-state automaton. (Recall that the language accepted by a finite-state automaton is the set of all strings which, when input to the automaton, send it to one of its accepting states.)

EXAMPLE 7.4.5 **Showing That There Is No Finite-State Automaton That Accepts a Certain Language**

Let the language L consist of all strings of the form $a^k b^k$, where k is a positive integer. Symbolically, L is the language over the alphabet $\Sigma = \{a, b\}$ defined by

$$L = \{s \in \Sigma^* \mid s = a^k b^k, \text{ where } k \text{ is a positive integer}\}.$$

Use the pigeonhole principle to show that there is no finite-state automaton that accepts L.

Solution *[Use a proof by contradiction.]* Suppose not. Suppose there is a finite-state automaton A that accepts L. *[A contradiction will be derived.]* Since A has only a finite number of states, these states can be denoted $s_1, s_2, s_3, \ldots, s_n$, where n is a positive integer. Consider all input strings that consist entirely of a's: $a, a^2, a^3, a^4, \ldots$. Now there are infinitely many such strings and only finitely many states. Thus, by the pigeonhole principle, there must be a state s_m and two input strings a^p and a^q with $p \neq q$ such that when either a^p or a^q is input to A, A goes to state s_m. (See Figure 7.4.2.) *[The pigeons are the strings of a's, the pigeonholes are the states, and the correspondence associates each string with the state to which A goes when the string is input.]*

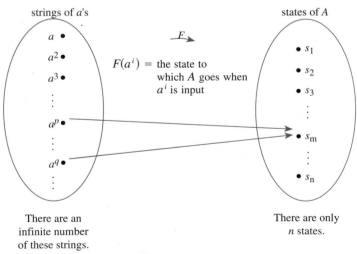

strings of a's

states of A

F

$F(a^i) =$ the state to which A goes when a^i is input

$a \bullet$

$a^2 \bullet$

$a^3 \bullet$

$\vdots$

$a^p \bullet$

$\vdots$

$a^q \bullet$

$\vdots$

$\bullet\ s_1$

$\bullet\ s_2$

$\bullet\ s_3$

$\vdots$

$\bullet\ s_m$

$\vdots$

$\bullet\ s_n$

There are an infinite number of these strings.

There are only n states.

Since F is not one-to-one, $\exists$ strings a^p and a^q with $p \neq q$ such that both a^p and a^q send A to the same state s_m.

FIGURE 7.4.2

Now by supposition A accepts L. Hence A accepts the string

$$a^p b^p.$$

This means that after p a's have been input, at which point A is in state s_m, inputting p additional b's sends A into an accepting state, say s_a. But that implies that

$$a^q b^p$$

also sends A into the accepting state s_a, and so $a^q b^p$ is accepted by A. The reason is that after q a's have been input, A is also in state s_m, and from that point inputting p additional b's sends A to state s_a, which is an accepting state. Pictorially, if $p < q$ then

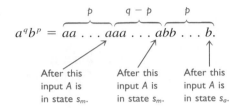

[*Remember the only factors that determine the next state of a finite-state automaton are the current state and the current input. The history of how the automaton came to be in its current state is irrelevant.*]

Now, by supposition, A accepts L. So since s is accepted by A, $s \in L$. But by definition of L, L consists only of those strings that have equal numbers of a's and b's, and so since $p \neq q$, $s \notin L$. Hence $s \in L$ and $s \notin L$, which is a contradiction.

It follows that the supposition is false, and so there is no finite-state automaton that accepts L. ∎

GENERALIZED PIGEONHOLE PRINCIPLE

A generalization of the pigeonhole principle states that if n pigeons fly into m pigeon-holes and, for some positive integer k, $n > k \cdot m$, then at least one pigeonhole contains $k + 1$ or more pigeons. This is illustrated in Figure 7.4.3 for $m = 4$, $n = 9$, and $k = 2$. Since $9 > 2 \cdot 4$, at least one pigeonhole contains three $(2 + 1)$ or more pigeons. (In this example, it is pigeonhole 3 that contains three pigeons.)

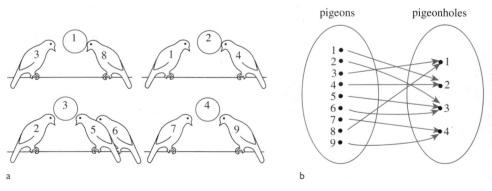

FIGURE 7.4.3

GENERALIZED PIGEONHOLE PRINCIPLE

For any function f from a finite set X to a finite set Y and for any positive integer k, if $n(X) > k \cdot n(Y)$, then there is some $y \in Y$ such that y is the image of at least $k + 1$ distinct elements of X.

EXAMPLE 7.4.6   Applying the Generalized Pigeonhole Principle

Show how the generalized pigeonhole principle implies that in a group of 85 people at least 4 must have the same last initial.

Solution   In this example the pigeons are the 85 people and the pigeonholes are the 26 possible last initials of their names. Note that

$$85 > 3 \cdot 26 = 78.$$

Consider the function I from people to initials defined by the following arrow diagram.

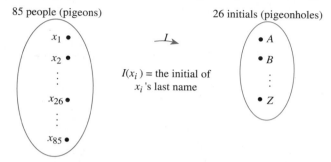

Since $85 > 3 \cdot 26$, the generalized pigeonhole principle states that some initial must be the image of at least four $(3 + 1)$ people. Thus at least four people have the same last initial. ∎

Consider the following contrapositive form of the generalized pigeonhole principle.

> **GENERALIZED PIGEONHOLE PRINCIPLE (CONTRAPOSITIVE FORM)**
>
> For any function f from a finite set X to a finite set Y and for any positive integer k, if for each $y \in Y$, $f^{-1}(y)$ has at most k elements, then X has at most $k \cdot n(Y)$ elements.

You may find it natural to use the contrapositive form of the generalized pigeonhole principle in certain situations. For instance, the result of Example 7.4.6 can be explained as follows:

Suppose no 4 people out of the 85 had the same last initial. Then at most 3 would share any particular one. By the contrapositive form of the generalized pigeonhole principle, this would imply that the total number of people is at most $3 \cdot 26 = 78$. But this contradicts the fact that there are 85 people in all. Hence at least 4 people share a last initial.

EXAMPLE 7.4.7   Using the Contrapositive Form of the Generalized Pigeonhole Principle

There are 42 students who are to share 12 computers. Each student uses exactly 1 computer and no computer is used by more than 6 students. Show that at least 5 computers are used by 3 or more students.

Solution a. **Using an Argument by Contradiction:** Suppose not. Suppose that 4 or fewer computers are used by 3 or more students. *[A contradiction will be derived.]* Then 8 or more computers are used by 2 or fewer students. Divide the set of computers into two subsets: C_1 and C_2. Into C_1 place 8 of the computers used by 2 or fewer students; into C_2 place the computers used by 3 or more students plus any remaining computers (to make a total of 4 computers in C_2). (See Figure 7.4.4.) Since at most 6 students are served by any one computer, by the contrapositive form of the generalized pigeonhole principle, the computers in set C_2 serve at most $6 \cdot 4 = 24$ students. Since at most 2 students are served by any one computer in C_1, by the contrapositive form of the generalized pigeonhole principle, the computers in set C_1 serve at most $2 \cdot 8 = 16$ students. Hence the total number of students served by the computers is $24 + 16 = 40$. But this contradicts the fact that each of the 42 students is served by a computer. So the supposition is false: At least 5 computers are used by 3 or more students.

The Set of 12 Computers

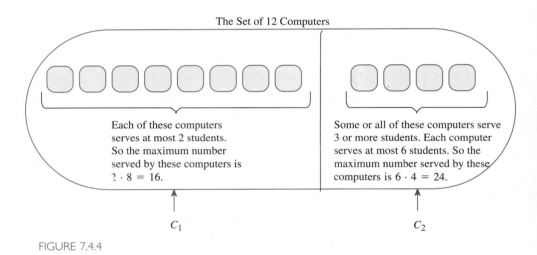

Each of these computers serve at most 2 students. So the maximum number served by these computers is $2 \cdot 8 = 16$.

Some or all of these computers serve 3 or more students. Each computer serves at most 6 students. So the maximum number served by these computers is $6 \cdot 4 = 24$.

C_1

C_2

FIGURE 7.4.4

b. **Using a Direct Argument:** Let k be the number of computers used by 3 or more students. *[We must show that $k \geq 5$.]* Because each computer is used by at most 6 students, these computers are used by at most $6k$ students (by the contrapositive form of the generalized pigeonhole principle). The remaining $12 - k$ computers are each used by at most 2 students. Hence, taken together, they are used by at most $2(12 - k) = 24 - 2k$ students (again, by the contrapositive form of the generalized pigeonhole principle). Thus the maximum number of students served by the computers is $6k + (24 - 2k) = 4k + 24$. Because 42 students are served by the computers, $4k + 24 \geq 42$. Solving for k gives that $k \geq 4.5$, and since k is an integer, this implies that $k \geq 5$ *[as was to be shown]*. ∎

PROOF OF THE PIGEONHOLE PRINCIPLE

The truth of the pigeonhole principle depends essentially on the sets involved being finite. Formal definitions of finite and infinite can be stated as follows:

DEFINITION

A set is called **finite** if, and only if, it is the empty set or there is a one-to-one correspondence from $\{1, 2, \ldots, n\}$ to it, where n is a positive integer. In the first case, the **number of elements** in the set is said to be 0, and in the second case it is said to be n. A set that is not finite is called **infinite**.

Note that it follows immediately from the definition that for a set to be finite means that it is empty or can be written in the form $\{x_1, x_2, \ldots, x_n\}$ where n is a positive integer.

THEOREM 7.4.1   The Pigeonhole Principle

For any function f from a finite set X to a finite set Y, if $n(X) > n(Y)$, then f is not one-to-one.

Proof:

Suppose f is any function from a finite set X to a finite set Y where $n(X) > n(Y)$. Let $n(Y) = m$, and denote the elements of Y by $y_1, y_2, \ldots, y_m$. Recall that for each y_i in Y, the inverse image set $f^{-1}(y_i) = \{x \in X \mid f(x) = y_i\}$. Now consider the collection of all the inverse image sets for all the elements of Y:

$$f^{-1}(y_1), f^{-1}(y_2), \ldots, f^{-1}(y_m).$$

By definition of function, each element of X is sent by f to some element of Y. Hence each element of X is in one of the inverse image sets, and so the union of all these sets equals X. But also by definition of function, no element of X is sent by f to more than one element of Y. Thus each element of X is in only one of the inverse image sets, and so the inverse image sets are mutually disjoint. By the addition rule, therefore,

$$n(X) = n(f^{-1}(y_1)) + n(f^{-1}(y_2)) + \cdots + n(f^{-1}(y_m)). \qquad \textbf{7.4.1}$$

Now suppose that f *is* one-to-one *[which is the opposite of what we want to prove]*. Then each set $f^{-1}(y_i)$ has at most one element, and so

$$n(f^{-1}(y_1)) + n(f^{-1}(y_2)) + \cdots + n(f^{-1}(y_m)) \qquad \textbf{7.4.2}$$
$$\leq \underbrace{1 + 1 + \cdots + 1}_{m \text{ terms}} = m$$

Putting (7.4.1) and (7.4.2) together gives that

$$n(X) \leq m = n(Y).$$

This contradicts the fact that $n(X) > n(Y)$, and so the supposition that f is one-to-one must be false. Hence f is not one-to-one *[as was to be shown]*.

An important theorem that follows from the pigeonhole principle states that a function from one finite set to another finite set of the same size is one-to-one if, and only if, it is onto. We will show in Section 7.6 that this result does not hold for infinite sets.

THEOREM 7.4.2

Let X and Y be finite sets with the same number of elements and suppose f is a function from X to Y. Then f is one-to-one if, and only if, f is onto.

Proof:

Suppose f is a function from X to Y, where X and Y are finite sets each with k elements.

If f is one-to-one, then f is onto: Suppose f is one-to-one. Then $f(x_1), f(x_2), \ldots, f(x_m)$ are all distinct. Consider the set S of all elements of Y that are not the image of any element of X:

Then the sets

$$\{f(x_1)\}, \{f(x_2)\}, \ldots, \{f(x_m)\} \quad \text{and} \quad S$$

are mutually disjoint. By the addition rule,

$$n(Y) = n(\{f(x_1)\}) + n(\{f(x_2)\}) + \cdots + n(\{f(x_m)\}) + n(S)$$

$$= \underbrace{1 + 1 + \cdots + 1}_{m \text{ terms}} + n(S) \qquad \text{because each } \{f(x_i)\} \text{ is a singleton set}$$

$$= m + n(S).$$

Thus

$$m = m + n(S) \qquad \text{because } n(Y) = m,$$

$$\Rightarrow \quad n(S) = 0 \qquad \text{by subtracting } m \text{ from both sides.}$$

Hence S is empty, and so there is no element of Y that is not the image of some element of X. Consequently, f is onto.

If f is onto, then f is one-to-one: Suppose f is onto. Then $f^{-1}(y_i) \neq \varnothing$ and so $n(f^{-1}(y_i)) \geq 1$ for all $i = 1, 2, \ldots, m$. As in the proof of the pigeonhole principle (Theorem 7.4.1), X is the union of the mutually disjoint sets $f^{-1}(y_1)$, $f^{-1}(y_2), \ldots, f^{-1}(y_m)$. By the addition principle,

$$n(X) = \underbrace{n(f^{-1}(y_1)) + n(f^{-1}(y_2)) + \cdots + n(f^{-1}(y_m))}_{m \text{ terms, each} \geq 1} \geq m. \qquad \textbf{7.4.3}$$

Now if any one of the sets $f^{-1}(y_i)$ has more than one element, then the sum in (7.4.3) is greater than m. But we know this is not the case because $n(X) = m$. Hence each set $f^{-1}(y_i)$ has exactly one element, and thus f is one-to-one *[as was to be shown]*.

Note that Theorem 7.4.2 applies in particular to the case $X = Y$. Thus a one-to-one function from a finite set to itself is onto, and an onto function from a finite set to itself is one-to-one. Such functions can be identified with permutations of the sets on which they are defined. For instance, the function defined by the diagram below can be identified with the permutation $cdba$ obtained by listing the images of $a, b, c,$ and d in order.

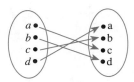

EXERCISE SET 7.4

1. a. If 4 cards are selected from a standard 52-card deck, must at least 2 be of the same suit? Why?
   b. If 5 cards are selected from a standard 52-card deck, must at least 2 be of the same suit? Why?

2. a. If 13 cards are selected from a standard 52-card deck, must at least 2 be of the same denomination? Why?
   b. If 20 cards are selected from a standard 52-card deck, must at least 2 be of the same denomination? Why?

3. A small town has only 500 residents. Must there be 2 residents who have the same birthday? Why?

4. In a group of 700 people, must there be 2 who have the same first and last initials? Why?

5. a. Given any set of four integers, must there be two that have the same remainder when divided by 3? Why?
   b. Given any set of three integers, must there be two that have the same remainder when divided by 3? Why?

6. a. Given any set of seven integers, must there be two that have the same remainder when divided by 6? Why?
   b. Given any set of seven integers, must there be two that have the same remainder when divided by 8? Why?

H 7. Let $S = \{3, 4, 5, 6, 7, 8, 9, 10, 11, 12\}$. Suppose six integers are chosen from S. Must there be two integers whose sum is 15? Why?

8. Let $T = \{1, 2, 3, 4, 5, 6, 7, 8, 9\}$. Suppose five integers are chosen from T. Must there be two integers whose sum is 10? Why?

9. a. If seven integers are chosen from between 1 and 12 inclusive, must at least one of them be odd? Why?
   b. If ten integers are chosen from between 1 and 20 inclusive, must at least one of them be even? Why?

10. If $n + 1$ integers are chosen from the set

$$\{1, 2, 3, \ldots, 2n\},$$

where n is a positive integer, must at least one of them be odd? Why?

11. If $n + 1$ integers are chosen from the set

$$\{1, 2, 3, \ldots, 2n\},$$

where n is a positive integer, must at least one of them be even? Why?

12. How many cards must you pick from a standard 52-card deck to be sure of getting at least 1 red card? Why?

13. Suppose six pairs of similar-looking boots are thrown together in a pile. How many individual boots must you pick to be sure of getting a matched pair? Why?

14. How many integers from 0 through 60 must you pick in order to be sure of getting at least one that is odd? at least one that is even?

15. If n is a positive integer, how many integers from 0 through $2n$ must you pick in order to be sure of getting at least one that is odd? at least one that is even?

16. How many integers from 1 through 100 must you pick in order to be sure of getting one that is divisible by 5?

17. How many integers must you pick in order to be sure that at least two of them have the same remainder when divided by 7?

18. How many integers must you pick in order to be sure that at least two of them have the same remainder when divided by 15?

19. How many integers from 100 through 999 must you pick in order to be sure that at least two of them have a digit in common? (For example, 256 and 530 have the common digit 5.)

20. If repeated divisions by 20,483 are performed, how many distinct remainders can be obtained?

21. Is $0.10100100010000100001\ldots$ rational or irrational? Why?

H 22. Show that within any set of thirteen integers chosen from 2 through 40, there are at least two integers with a common divisor greater than 1.

H 23. Let L be the language consisting of all strings of the form

$$a^m b^n, \text{ where } m \text{ and } n \text{ are} \\ \text{positive integers and } m \geq n.$$

Show that there is no finite-state automaton that accepts L.

24. Let L be the language consisting of all strings of the form

$$a^m b^n, \text{ where } m \text{ and } n \text{ are} \\ \text{positive integers and } m \leq n.$$

Show that there is no finite-state automaton that accepts L.

H 25. Let L be the language consisting of all strings of the form

$$a^n, \text{ where } n = m^2, \text{ for some positive integer } m.$$

Show that there is no finite-state automaton that accepts L.

26. In a group of 30 people, must at least 3 have been born in the same month? Why?

27. In a group of 30 people, must at least 4 have been born in the same month? Why?

28. In a group of 2,000 people, must at least 5 have the same birthday? Why?

29. A programmer writes 500 lines of computer code in 17 days. Must there have been at least 1 day when the programmer wrote 30 or more lines of code? Why?

30. A certain college class has 40 students. All the students in the class are known to be from 17 through 34 years of age. You want to make a bet that the class contains at least x students of the same age. How large can you make x and yet be sure to win your bet?

31. A penny collection contains twelve 1967 pennies, seven 1968 pennies, and eleven 1971 pennies. If you are to pick some pennies without looking at the dates, how many must you pick to be sure of getting at least five pennies from the same year?

H 32. A group of 15 executives are to share 5 secretaries. Each executive is assigned exactly 1 secretary and no secretary is assigned to more than 4 executives. Show that at least 3 secretaries are assigned to 3 or more executives.

H ◆ 33. Let A be a set of six positive integers each of which is less than 13. Show that there must be two distinct subsets of A whose elements when added up give the same sum. (For example, if $A = \{5, 12, 10, 1, 3, 4\}$ then the elements of the subsets $S_1 = \{1, 4, 10\}$ and $S_2 = \{5, 10\}$ both add up to 15.)

H ◆ 34. Given a set of 52 distinct integers, show that there must be 2 whose sum or difference is divisible by 100.

H ◆ 35. Show that if 101 integers are chosen from 1 to 200 inclusive, there must be 2 with the property that one is divisible by the other.

◆ 36. a. Suppose $a_1, a_2, \ldots, a_n$ is a sequence of n integers none of which is divisible by n. Show that at least one of the differences $a_i - a_j$ (for $i \neq j$) must be divisible by n.

H b. Show that every finite sequence $x_1, x_2, \ldots, x_n$ of n integers has a consecutive subsequence $x_{i+1}, x_{i+2}, \ldots, x_j$ whose sum is divisible by n. (For instance, the sequence 3, 4, 17, 7, 16 has the consecutive subsequence 17, 7, 16 whose sum is divisible by 5.)*

H ◆ 37. Observe that the sequence 12, 15, 8, 13, 7, 18, 19, 11, 14, 10 has three increasing subsequences of length four: 12, 15, 18, 19; 12, 13, 18, 19; and 8, 13, 18, 19. It also has one decreasing subsequence of length four: 15, 13, 11, 10. Show that in any sequence of $n^2 + 1$ distinct real numbers, there must be a sequence of length $n + 1$ that is either strictly increasing or strictly decreasing.

38. Suppose X and Y are finite sets, X has more elements than Y, and $F: X \to Y$ is a function. By the pigeonhole principle, there exist elements a and b in X such that $a \neq b$ and $F(a) = F(b)$. Write a computer algorithm to find such a pair of elements a and b.

---

*James E. Schultz and William F. Burger, "An Approach to Problem-Solving Using Equivalence Classes Modulo n," *College Mathematics Journal (15)*, No. 5, 1984, 401–405.

7.5   COMPOSITION OF FUNCTIONS

*It is no paradox to say that
in our most theoretical
moods we may be nearest
to our most practical
applications.*
(Alfred North Whitehead)

Consider two functions, the successor function and the squaring function, defined from **Z** (the set of integers) to **Z**, and imagine that each is represented by a machine. If the two machines are hooked up so that the output from the successor function is used as input to the squaring function, then they work together to operate as one larger machine. In this larger machine, an integer n is first increased by 1 to obtain $n + 1$; then the quantity $n + 1$ is squared to obtain $(n + 1)^2$. This is illustrated in the following drawing.

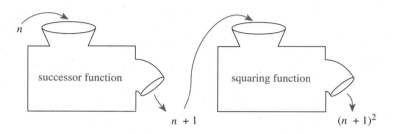

Combining functions in this way is called *composing* them; the resulting function is called the *composition* of the two functions. Note that the composition can only be formed if the output of the first function is acceptable input to the second function. That is, the range of the first function must be contained in the domain of the second function.

DEFINITION

Let $f: X \rightarrow Y'$ and $g: Y \rightarrow Z$ be functions with the property that the range of f is a subset of the domain of g. Define a new function $g \circ f: X \rightarrow Z$ as follows:

$$(g \circ f)(x) = g(f(x)) \quad \text{for all } x \in X,$$

where $g \circ f$ is read "g circle f" and $g(f(x))$ is read "g of f of x." The function $g \circ f$ is called the **composition of f and g**. (We put the f first when we say "the composition of f and g" because an element x is first acted upon by f and then by g.)

This definition is shown schematically below.

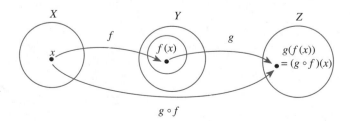

EXAMPLE 7.5.1   Composition of Functions Defined by Formulas

Let $f: \mathbf{Z} \to \mathbf{Z}$ be the successor function and let $g: \mathbf{Z} \to \mathbf{Z}$ be the squaring function. Then $f(n) = n + 1$ for all $n \in \mathbf{Z}$ and $g(n) = n^2$ for all $n \in \mathbf{Z}$.

a. Find the compositions $g \circ f$ and $f \circ g$.
b. Is $g \circ f = f \circ g$? Explain.

Solution   a. The functions $g \circ f$ and $f \circ g$ are defined as follows:

$$(g \circ f)(n) = g(f(n)) = g(n + 1) = (n + 1)^2 \quad \text{for all } n \in \mathbf{Z},$$

and

$$(f \circ g)(n) = f(g(n)) = f(n^2) = n^2 + 1 \quad \text{for all } n \in \mathbf{Z}.$$

Thus

$$(g \circ f)(n) = (n + 1)^2 \quad \text{and} \quad (f \circ g)(n) = n^2 + 1 \quad \text{for all } n \in \mathbf{Z}.$$

b. Two functions from one set to another are equal if, and only if, they take the same values. In this case,

$$(g \circ f)(1) = (1 + 1)^2 = 4, \text{ whereas } (f \circ g)(1) = 1^2 + 1 = 2.$$

Thus the two functions $g \circ f$ and $f \circ g$ are not equal:

$$g \circ f \neq f \circ g. \qquad \blacksquare$$

Example 7.5.1 illustrates the important fact that composition of functions is not a commutative operation: *for general functions F and G, F ∘ G need not necessarily equal G ∘ F* (although the two *may* be equal).

EXAMPLE 7.5.2   Composition of Functions Defined on Finite Sets

Let $X = \{1, 2, 3\}$, $Y' = \{a, b, c, d\}$, $Y = \{a, b, c, d, e\}$, and $Z = \{x, y, z\}$. Define functions $f: X \to Y'$ and $g: Y \to Z$ by the arrow diagrams below.

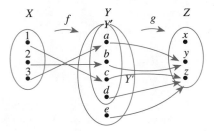

That is, $f(1) = c$, $f(2) = b$, $f(3) = a$, and $g(a) = y$, $g(b) = y$, $g(c) = z$, $g(d) = z$, and $g(e) = z$. Find the arrow diagram for $g \circ f$. What is the range of $g \circ f$?

Solution   To find the arrow diagram for $g \circ f$, just trace the arrows all the way across from X to Z through Y. The result is shown below.

$$(g \circ f)(1) = g(f(1)) = g(c) = z,$$
$$(g \circ f)(2) = g(f(2)) = g(b) = y,$$
$$(g \circ f)(3) = g(f(3)) = g(a) = y.$$

The range of $g \circ f$ is $\{y, z\}$. ■

Recall that the identity function on a set X, i_X, is the function from X to X defined by the formula

$$i_X(x) = x \quad \text{for all } x \in X.$$

That is, the identity function on X sends each element of X to itself. What happens when an identity function is composed with another function?

EXAMPLE 7.5.3   **Composition with the Identity Function**

Let $X = \{a, b, c, d\}$ and $Y = \{u, v, w\}$, and suppose $f: X \rightarrow Y$ is given by the arrow diagram shown below.

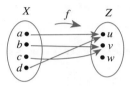

Find $f \circ i_X$ and $i_Y \circ f$.

Solution   The values of $f \circ i_X$ are obtained by tracing through the arrow diagram shown below.

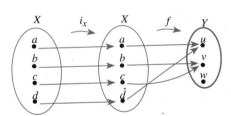

$$(f \circ i_X)(a) = f(i_X(a)) = f(a) = u$$
$$(f \circ i_X)(b) = f(i_X(b)) = f(b) = v$$
$$(f \circ i_X)(c) = f(i_X(c)) = f(c) = v$$
$$(f \circ i_X)(d) = f(i_X(d)) = f(d) = u$$

Note that for all elements x in X,

$$(f \circ i_X)(x) = f(x).$$

By definition of equality of functions, this means that $f \circ i_X = f$.

Similarly, the equality $i_Y \circ f = f$ can be verified by tracing through the arrow diagram below for each x in X and noting that in each case $(i_Y \circ f)(x) = f(x)$.

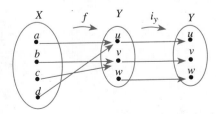

More generally, the composition of any function with an identity function equals the function. ■

THEOREM 7.5.1

If f is a function from a set X to a set Y, and i_X is the identity function on X, and i_Y is the identity function on Y, then

$$\text{(a) } f \circ i_X = f \quad \text{and} \quad \text{(b) } i_Y \circ f = f.$$

Proof of (a):

Suppose f is a function from a set X to a set Y and i_X is the identity function on X. Then for all x in X,

$$(f \circ i_X)(x) = f(i_X(x)) = f(x).$$

Hence by definition of equality of functions, $f \circ i_X = f$ as was to be shown.

Proof of (b):

This is exercise 12 at the end of this section.

Now let f be a function from a set X to a set Y, and suppose f has an inverse function f^{-1}. Recall that f^{-1} is the function from Y to X with the property that

$$f^{-1}(y) = x \quad \Leftrightarrow \quad f(x) = y.$$

What happens when f is composed with f^{-1}? Or when f^{-1} is composed with f?

EXAMPLE 7.5.4   **Composing a Function with Its Inverse**

Let $X = \{a, b, c\}$ and $Y = \{x, y, z\}$. Define $f : X \to Y$ by the following arrow diagram.

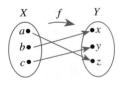

Then f is one-to-one and onto. So f^{-1} exists and is found by tracing the arrows backwards, as shown below.

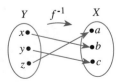

Now $f^{-1} \circ f$ is found by following the arrows from X to Y by f and back to X by f^{-1}. If you do this, you will see that

$$(f^{-1} \circ f)(a) = f^{-1}(f(a)) = f^{-1}(z) = a$$
$$(f^{-1} \circ f)(b) = f^{-1}(f(b)) = f^{-1}(x) = b$$

and
$$(f^{-1} \circ f)(c) = f^{-1}(f(c)) = f^{-1}(y) = c.$$

Thus the composition of f and f^{-1} sends each element to itself. So by definition of the identity function,
$$f^{-1} \circ f = i_X.$$

In a similar way, you can see that
$$f \circ f^{-1} = i_Y. \qquad \blacksquare$$

More generally, the composition of any function with its inverse (if it has one) is an identity function. Intuitively, the function sends an element in its domain to an element in its co-domain and the inverse function sends it back again, so the composition of the two sends each element to itself. This reasoning is formalized in Theorem 7.5.2.

THEOREM 7.5.2

If $f: X \rightarrow Y$ is a one-to-one and onto function with inverse function $f^{-1}: Y \rightarrow X$, then

(a) $f^{-1} \circ f = i_X$ and (b) $f \circ f^{-1} = i_Y$.

Proof of (a):

Suppose $f: X \rightarrow Y$ is a one-to-one and onto function with inverse function $f^{-1}: Y \rightarrow X$. *[To show that $f^{-1} \circ f = i_X$, we must show that for all $x \in X$, $(f^{-1} \circ f)(x) = x$.]* Let x be an element in X. Then
$$(f^{-1} \circ f)(x) = f^{-1}(f(x))$$

by definition of composition of functions. Now the inverse function f^{-1} satisfies the condition
$$f^{-1}(b) = a \quad \Leftrightarrow \quad f(a) = b \quad \text{for all } a \in X \text{ and } b \in Y. \qquad 7.5.1$$

Let
$$x' = f^{-1}(f(x)). \qquad 7.5.2$$

Apply property (7.5.1) with x' playing the role of a and $f(x)$ playing the role of b. Then
$$f(x') = f(x).$$

But since f is one-to-one, this implies that $x' = x$. Substituting x for x' in equation (7.5.2) gives
$$x = f^{-1}(f(x)).$$

Then by definition of composition of functions,
$$(f^{-1} \circ f)(x) = x,$$

as was to be shown.

Proof of (b):

This is exercise 13 at the end of this section.

COMPOSITION OF ONE-TO-ONE AND ONTO FUNCTIONS

The composition of functions interacts in interesting ways with the properties of being one-to-one and onto. What happens, for instance, when two one-to-one functions are composed? Must their composition be one-to-one? For example, let $X = \{a, b, c\}$, $Y = \{w, x, y, z\}$, and $Z = \{1, 2, 3, 4, 5\}$, and define one-to-one functions $f: X \to Y$ and $g: Y \to Z$ as shown in the arrow diagrams of Figure 7.5.1.

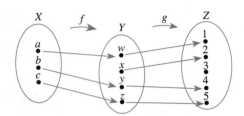

FIGURE 7.5.1

Then $g \circ f$ is the function with the arrow diagram shown in Figure 7.5.2.

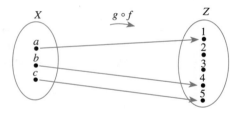

FIGURE 7.5.2

From the diagram it is clear that for these particular functions, the composition is one-to-one. This result is no accident. It turns out that the composition of two one-to-one functions is always one-to-one.

THEOREM 7.5.3

If $f: X \to Y$ and $g: Y \to Z$ are both one-to-one functions, then $g \circ f$ is one-to-one.

By the method of direct proof, the proof of Theorem 7.5.3 has the following starting point and conclusion to be shown.

Starting Point: Suppose f is a one-to-one function from X to Y and g is a one-to-one function from Y to Z.

To Show: $g \circ f$ is a one-to-one function from X to Z.

The conclusion to be shown says that a certain function is one-to-one. How do you show that? By substituting into the definition of one-to-one, you see that

$g \circ f$ is one-to-one $\Leftrightarrow$ $\forall x_1, x_2 \in X$, if $(g \circ f)(x_1) = (g \circ f)(x_2)$ then $x_1 = x_2$.

By the method of direct proof, then, to show $g \circ f$ is one-to-one, you

suppose x_1 and x_2 are elements of X such that $(g \circ f)(x_1) = (g \circ f)(x_2)$,

and you

show that $x_1 = x_2$.

Now the heart of the proof begins. To show that $x_1 = x_2$, you work forward from the supposition that $(g \circ f)(x_1) = (g \circ f)(x_2)$ using the fact that f and g are both one-to-one. By definition of composition,

$$(g \circ f)(x_1) = g(f(x_1)) \quad \text{and} \quad (g \circ f)(x_2) = g(f(x_2)).$$

Since the left-hand sides of the equations are equal, so are the right-hand sides. Thus

$$g(f(x_1)) = g(f(x_2)).$$

Now just stare at the above equation for a moment. It says that

$$g(\text{something}) = g(\text{something else}).$$

Because g is a one-to-one function, any time g of one thing equals g of another thing, those two things are equal. Hence

$$f(x_1) = f(x_2).$$

But f is also a one-to-one function. Any time f of one thing equals f of another thing, those two things are equal. Therefore

$$x_1 = x_2.$$

This is what was to be shown!

This discussion is summarized in the following formal proof.

Proof of Theorem 7.5.3:

Suppose $f: X \to Y$ and $g: Y \to Z$ are both one-to-one functions. *[We must show that $g \circ f$ is one-to-one.]* Suppose x_1 and x_2 are elements of X such that

$$(g \circ f)(x_1) = (g \circ f)(x_2).$$

[We must show that $x_1 = x_2$.] By definition of composition of functions,

$$g(f(x_1)) = g(f(x_2)).$$

Since g is one-to-one,

$$f(x_1) = f(x_2).$$

And since f is one-to-one,

$$x_1 = x_2.$$

This is what was to be shown.

Now consider what happens when two onto functions are composed. For example, let $X = \{a, b, c, d, e\}$, $Y = \{w, x, y, z\}$, and $Z = \{1, 2, 3\}$. Define onto functions $f: X \to Y$ and $g: Y \to Z$ by the following arrow diagrams.

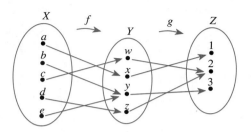

Then $g \circ f$ is the function with the arrow diagram shown below.

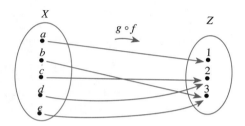

It is clear from the diagram that $g \circ f$ is onto.

It turns out that the composition of any two onto functions (that can be composed) is onto.

THEOREM 7.5.4

If $f: X \to Y$ and $g: Y \to Z$ are both onto functions, then $g \circ f$ is onto.

By the method of direct proof, the proof of Theorem 7.5.4 has the following starting point and conclusion to be shown:

Starting Point: Suppose f is an onto function from X to Y and g is an onto function from Y to Z.

To Show: $g \circ f$ is an onto function from X to Z.

The conclusion to be shown says that a certain function is onto. How do you show that? By substituting into the definition of onto, observe that

$g \circ f: X \to Z$ is onto $\Leftrightarrow$ given any element of Z, it is possible to find an element of X such that $(g \circ f)(x) = z$.

Since this statement is universal, to prove it you

suppose z is a *[particular but arbitrarily chosen]* element of Z

and

> **show** that there is an element x in X such that $(g \circ f)(x) = z$.

Now begins the heart of the proof. To find x, reason from the supposition that z is in Z using the fact that both g and f are onto. Imagine arrow diagrams for the functions f and g.

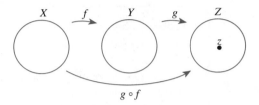

You have a particular element z in Z, and you need to find an element x in X such that if x is sent over to Z by $g \circ f$, its image will be z. Since g is onto, z is at the tip of some arrow coming from Y. That is, there is an element y in Y such that

$$g(y) = z. \qquad\qquad \textbf{7.5.3}$$

Now the arrow diagrams can be drawn as follows.

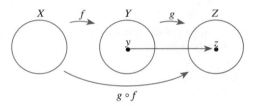

But f also is onto. So every element in Y is at the tip of an arrow coming from X. In particular, y is at the tip of some arrow, so there is an element x in X such that

$$f(x) = y. \qquad\qquad \textbf{7.5.4}$$

The diagram, therefore, can be drawn as shown below.

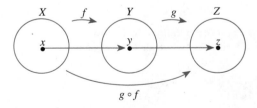

Now just substitute equation (7.5.4) into equation (7.5.3) to obtain

$$g(f(x)) = z.$$

But by definition of $g \circ f$,

$$g(f(x)) = (g \circ f)(x).$$

Hence

$$(g \circ f)(x) = z.$$

Thus x is an element of X that is sent by $g \circ f$ to z, and so x is the element you were supposed to find.

This discussion is summarized in the following formal proof:

Proof of Theorem 7.5.4:

Suppose $f: X \to Y$ and $g: Y \to Z$ are both onto functions. *[We must show that $g \circ f$ is onto.]* Let z be a *[particular but arbitrarily chosen]* element of Z. *[We must show the existence of an element x in X such that $(g \circ f)(x) = z$.]* Since g is onto, there is an element y in Y such that $g(y) = z$. And since f is onto, there is an element x in X such that $f(x) = y$. Hence there exists an element x in X such that

$$(g \circ f)(x) = g(f(x)) = g(y) = z.$$

If follows that $g \circ f$ is onto.

EXERCISE SET 7.5

In each of 1 and 2 functions f and g are defined by arrow diagrams. Find $g \circ f$ and $f \circ g$ and determine whether $g \circ f$ equals $f \circ g$.

1.

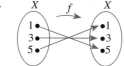 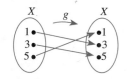

2.

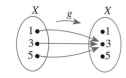

In each of 3–5 functions F and G are defined by formulas. Find $G \circ F$ and $F \circ G$ and determine whether $G \circ F$ equals $F \circ G$.

3. $F(x) = x^3$ and $G(x) = x - 1$, for all real numbers x.

4. $F(x) = x^5$ and $G(x) = x^4$, for all real numbers x.

5. $F(n) = 2n$ and $G(n) = \lfloor n/2 \rfloor$, for all integers n.

6. Let $\Sigma = \{a, b\}$ and let $l: \Sigma^* \to \mathbf{Z}$ be the length function:

for all strings $s \in \Sigma^*$,

$$l(s) = \text{the number of characters in } s.$$

Let $f: \mathbf{Z} \to \{0, 1, 2\}$ be the *mod* 3 function:

for all integers n, $f(n) = n \bmod 3.$

What is $(f \circ l)(abaa)$? $(f \circ l)(baaab)$? $(f \circ l)(aaa)$?

7. Let $F: \mathbf{R} \to \mathbf{R}$ be defined by the formula $F(x) = x^2/3$ for all $x \in \mathbf{R}$, and let $G: \mathbf{R} \to \mathbf{Z}$ be the floor function: $G(x) = \lfloor x \rfloor$ for all $x \in \mathbf{R}$. What is $(G \circ F)(2)$? $(G \circ F)(-3)$? $(G \circ F)(5)$?

The functions of each pair in 8–10 are inverse to each other. For each pair, check that both compositions give the identity function.

8. $F: \mathbf{R} \to \mathbf{R}$ and $F^{-1}: \mathbf{R} \to \mathbf{R}$ are defined by

$$F(x) = 3x + 2, \quad \text{for all } x \in \mathbf{R}$$

and

$$F^{-1}(y) = \frac{y - 2}{3}, \quad \text{for all } y \in \mathbf{R}.$$

9. $G: \mathbf{R}^+ \to \mathbf{R}^+$ and $G^{-1}: \mathbf{R}^+ \to \mathbf{R}^+$ are defined by

$$G(x) = x^2, \quad \text{for all } x \in \mathbf{R}^+$$

and

$$G^{-1}(x) = \sqrt{x}, \quad \text{for all } x \in \mathbf{R}^+.$$

10. H and H^{-1} are both defined from $\mathbf{R} - \{1\}$ to $\mathbf{R} - \{1\}$ by the formula

$$H(x) = H^{-1}(x) = \frac{x + 1}{x - 1}, \quad \text{for all } x \in \mathbf{R} - \{1\}.$$

11. Explain how it follows from the definition of logarithm that
   a. $\log_b (b^x) = x$,   for all real numbers x.
   b. $b^{\log_b x} = x$,   for all positive real numbers x.

12. Prove Theorem 7.5.1(b): If f is any function from a set X to a set Y, then $i_Y \circ f = f$, where i_Y is the identity function on Y.

13. Prove Theorem 7.5.2(b): If $f: X \rightarrow Y$ is a one-to-one and onto function with inverse function $f^{-1}: Y \rightarrow X$, then $f \circ f^{-1} = i_Y$, where i_Y is the identity function on Y.

14. Suppose Y and Z are sets and $g: Y \rightarrow Z$ is a one-to-one function. This means that if g takes the same value on any two elements of Y, then those elements are equal. Thus, for example, if a and b are elements of Y and $g(a) = g(b)$, then it can be inferred that $a = b$. What can be inferred in the following situations?
   a. s_k and s_m are elements of Y and $g(s_k) = g(s_m)$.
   b. $z/2$ and $t/2$ are elements of Y and $g(z/2) = g(t/2)$.
   c. $f(x_1)$ and $f(x_2)$ are elements of Y and
      $g(f(x_1)) = g(f(x_2))$.

15. If $f: X \rightarrow Y$ and $g: Y \rightarrow Z$ are functions and $g \circ f: X \rightarrow Z$ is one-to-one, must both f and g be one-to-one? Prove or give a counterexample.

16. If $f: X \rightarrow Y$ and $g: Y \rightarrow Z$ are functions and $g \circ f: X \rightarrow Z$ is onto, must both f and g be onto? Prove or give a counterexample.

H 17. If $f: X \rightarrow Y$ and $g: Y \rightarrow Z$ are functions and $g \circ f: X \rightarrow Z$ is one-to-one, must f be one-to-one? Prove or give a counterexample.

H 18. If $f: X \rightarrow Y$ and $g: Y \rightarrow Z$ are functions and $g \circ f: X \rightarrow Z$ is onto, must g be onto? Prove or give a counterexample.

19. Let $f: W \rightarrow X$, $g: X \rightarrow Y$, and $h: Y \rightarrow Z$ be functions. Must $h \circ (g \circ f) = (h \circ g) \circ f$? Prove or give a counterexample.

In each of 20 and 21 find $g \circ f$, $(g \circ f)^{-1}$, g^{-1}, f^{-1}, and $f^{-1} \circ g^{-1}$, and state how $(g \circ f)^{-1}$ and $f^{-1} \circ g^{-1}$ are related.

20. Let $X = \{a, c, b\}$, $Y = \{x, y, z\}$, and $Z = \{u, v, w\}$. Define $f: X \rightarrow Y$ and $g: Y \rightarrow Z$ as shown in the arrow diagram below.

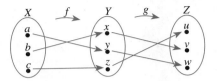

21. Define $f: \mathbf{R} \rightarrow \mathbf{R}$ and $g: \mathbf{R} \rightarrow \mathbf{R}$ by the formulas
   $$f(x) = x + 2 \quad \text{and} \quad g(x) = -x \quad \text{for all } x \in \mathbf{R}.$$

22. Prove or give a counterexample: If $f: X \rightarrow Y$ and $g: Y \rightarrow X$ are functions such that $g \circ f = i_X$ and $f \circ g = i_Y$, then f and g are both one-to-one and onto and $g = f^{-1}$.

H 23. Suppose $f: X \rightarrow Y$ and $g: Y \rightarrow Z$ are both one-to-one and onto. Prove that $(g \circ f)^{-1}$ exists and that $(g \circ f)^{-1} = f^{-1} \circ g^{-1}$.

Exercises 24–26 refer to exercises 37–44 of Section 7.1. Determine which of the properties in 24 and 25 are true for all functions $f: X \rightarrow Y$ and which are false for some function f. Also determine whether the property in 26 is true for all functions $f: X \rightarrow Y$ and $g: Y \rightarrow Z$ or false for some functions f and g. Justify your answers.

H ◆ 24. For all subsets A of X, $f^{-1}(f(A)) = A$.

◆ 25. For all subsets C of Y, $f(f^{-1}(C)) = C$.

◆ 26. For all subsets E of Z, $(g \circ f)^{-1}(E) = f^{-1}(g^{-1}(E))$.

7.6   CARDINALITY WITH APPLICATIONS TO COMPUTABILITY

There are as many squares as there are numbers because they are just as numerous as their roots.
(Galileo Galilei, 1632)

Historically, the term *cardinal number* was introduced to describe the size of a set ("This set has *eight* elements") as distinguished from an *ordinal number* that refers to the order of an element in a sequence ("This is the *eighth* element in the row"). The definition of cardinal number derives from the primitive technique of representing numbers by fingers or tally marks. Small children, when asked how old they are, will usually answer by holding up a certain number of fingers, each finger being paired with a year of their life. As was discussed in Section 7.3, a pairing of the elements of two sets is called a one-to-one correspondence. We say that two finite sets whose elements can be paired by a one-to-one correspondence have the *same size*. This is illustrated by the diagram at the top of the following page.

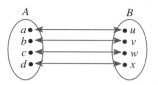

A B

The elements of set A can be put into one-to-one correspondence with the elements of set B.

Galileo Galilei (1564–1642)

Now a **finite set** is one that has no elements at all or that can be put into one-to-one correspondence with a set of the form $\{1, 2, \ldots, n\}$ for some positive integer n. By contrast, an **infinite set** is a nonempty set that cannot be put into one-to-one correspondence with $\{1, 2, \ldots, n\}$ for any positive integer n. Suppose that, as suggested by the quote from Galileo at the beginning of this section, we extend the concept of size to infinite sets by saying that one infinite set has the same size as another if, and only if, the first set can be put into one-to-one correspondence with the second. What consequences follow from such a definition? Do all infinite sets have the same size, or are some infinite sets larger than others? These are the questions we address in this section. The answers are sometimes surprising and have interesting applications to determining what can and cannot be computed on a computer.

DEFINITION

Let A and B be any sets. **A has the same cardinality as B** if, and only if, there is a one-to-one correspondence from A to B. In other words, A has the same cardinality as B if, and only if, there is a function f from A to B that is one-to-one and onto.

The following theorem gives some basic properties of cardinality, most of which follow from statements proved earlier about one-to-one and onto functions.

THEOREM 7.6.1

For all sets A, B, and C,

a. A has the same cardinality as A (*reflexive property of cardinality*).
b. If A has the same cardinality as B, then B has the same cardinality as A (*symmetric property of cardinality*).
c. If A has the same cardinality as B and B has the same cardinality as C, then A has the same cardinality as C (*transitive property of cardinality*).

Proof:

a. Suppose A is any set. *[To show that A has the same cardinality as A, we must show there is a one-to-one correspondence from A to A.]* Consider the identity function i_A from A to A. This function is one-to-one because if x_1 and x_2 are any elements in A with $i_A(x_1) = i_A(x_2)$, then by definition of i_A, $x_1 = x_2$. The identity function is also onto because if y is any element of A, then $y = i_A(y)$ by definition of i_A. Hence i_A is a one-to-one correspondence from A to A. *[So there exists a one-to-one correspondence from A to A as was to be shown.]*

b. Suppose A and B are any sets and A has the same cardinality as B. *[We must show that B has the same cardinality as A.]* Since A has the same cardinality as B, there is a function f from A to B that is one-to-one and onto. But then by Theorems 7.3.1 and 7.3.2, there is a function f^{-1} from B to A that is also one-to-one and onto. Hence B has the same cardinality as A *[as was to be shown]*.

c. Suppose A, B, and C are any sets and A has the same cardinality as B and B has the same cardinality as C. *[We must show that A has the same cardinality as C.]* Since A has the same cardinality as B, there is a function f from A to B that is one-to-one and onto, and since B has the same cardinality as C, there is a function g from B to C that is one-to-one and onto. But then by Theorems 7.5.3 and 7.5.4, $g \circ f$ is a function from A to C that is one-to-one and onto. Hence A has the same cardinality as C *[as was to be shown]*.

Note that Theorem 7.6.1(b) makes it possible to say simply that two sets have the same cardinality instead of always having to say that one set has the same cardinality as another. That is, the following definition can be made.

DEFINITION

A and B **have the same cardinality** if, and only if, A has the same cardinality as B or B has the same cardinality as A.

COUNTABLE SETS

The set $\mathbf{Z}^+$ of counting numbers $\{1, 2, 3, 4, \ldots\}$ is, in a sense, the most basic of all infinite sets. A set A having the same cardinality as this set is called *countably infinite*. The reason is that the one-to-one correspondence between the two sets can be used to "count" the elements of A: If f is a one-to-one and onto function from $\mathbf{Z}^+$ to A, then $f(1)$ can be designated as the first element of A, $f(2)$ as the second element of A, $f(3)$ as the third element of A, and so forth. This is illustrated graphically in Figure 7.6.1. Because f is one-to-one, no element is ever counted twice, and because it is onto, every element of A is counted eventually.

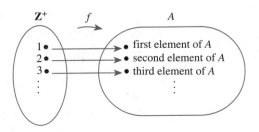

FIGURE 7.6.1　"Counting" a Countably Infinite Set

> **DEFINITION**
>
> A set is called **countably infinite** if, and only if, it has the same cardinality as the set of positive integers $\mathbf{Z}^+$. A set is called **countable** if, and only if, it is finite or countably infinite. A set that is not countable is called **uncountable**.

EXAMPLE 7.6.1   Countability of **Z**, the Set of All Integers

Show that the set **Z** of all integers is countable.

Solution   The set **Z** of all integers is certainly not finite. So if it is countable, it must be because it is countably infinite. To show **Z** is countably infinite, find a function from the positive integers $\mathbf{Z}^+$ to **Z** that is one-to-one and onto. Looked at in one light, this contradicts common sense; judging from the diagram below there appear to be more than twice as many integers as there are positive integers.

$$\underbrace{\cdots -5 \quad -4 \quad -3 \quad -2 \quad -1 \quad 0 \quad \overbrace{1 \quad 2 \quad 3 \quad 4 \quad 5}^{\text{positive integers}} \cdots}_{\text{all integers}}$$

But you were alerted that results in this section might be surprising. Try to think of a way to "count" the set of all integers anyway.

The trick is to start in the middle and work outward systematically. Let the first integer be 0, the second 1, the third -1, the fourth 2, the fifth -2, and so forth as shown in Figure 7.6.2, starting at 0 and swinging outward in back and forth arcs from positive to negative integers and back again, picking up one additional integer at each swing.

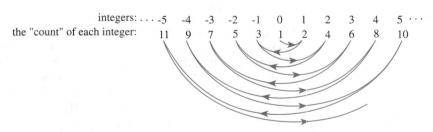

FIGURE 7.6.2   "Counting" the Set of All Integers

It is clear from the diagram that no integer is counted twice (so the function is one-to-one) and every integer is counted eventually (so the function is onto). Consequently, this diagram defines a function from $\mathbf{Z}^+$ to **Z** that is one-to-one and onto. Even though in one sense there seem to be more integers than positive integers, the elements of the two sets can be paired up one for one. It follows by definition of cardinality that $\mathbf{Z}^+$ has the same cardinality as **Z**. Thus **Z** is countably infinite and hence countable.

The diagrammatic description of the above function is acceptable as given. You can check, however, that the function can also be described by the explicit formula

$$f(n) = \begin{cases} \dfrac{n}{2} & \text{if } n \text{ is an even positive integer} \\[2ex] -\dfrac{n-1}{2} & \text{if } n \text{ is an odd positive integer.} \end{cases} \qquad \blacksquare$$

Theorem 7.4.2 states that a function from one finite set to another set of the same size is one-to-one if, and only if, it is onto. This result does not hold for infinite sets. Although it is true that for two infinite sets A and B to have the same size there must exist a function from A to B that is both one-to-one and onto, it is also always the case that there are other functions from A to B that are one-to-one but not onto and onto but not one-to-one. For instance, in Example 7.6.1, a function f was constructed from $\mathbf{Z}^+$ to $\mathbf{Z}$ that was one-to-one and onto. This showed that $\mathbf{Z}^+$ and $\mathbf{Z}$ have the same size. But another function from $\mathbf{Z}^+$ to $\mathbf{Z}$, the "inclusion" function g given by $g(n) = n$ for all positive integers n, is one-to-one and not onto. And a variation of the function defined in Example 7.6.1 also goes from $\mathbf{Z}^+$ to $\mathbf{Z}$ and is onto but not one-to-one. (See exercise 5 at the end of this section.)

Another important characteristic of an infinite set (often taken to be its defining property) is that it can have the same cardinality as a proper subset of itself. This characteristic was illustrated in Example 7.6.1, which showed that $\mathbf{Z}$ and its proper subset $\mathbf{Z}^+$ have the same cardinality. It is also exhibited in Example 7.6.2 below, which shows that even though it may seem reasonable to say that there are twice as many integers as there are even integers, the elements of $\mathbf{Z}$ and $2\mathbf{Z}$ can be matched up exactly, and so according to the definition, $\mathbf{Z}$ and $2\mathbf{Z}$ have the same cardinality. It follows that the set of all even integers is countable.

EXAMPLE 7.6.2 **Countability of 2Z, the Set of All Even Integers**

Show that the set $2\mathbf{Z}$ of all even integers is countable.

Solution Consider the function h from $\mathbf{Z}$ to $2\mathbf{Z}$ defined as follows:

$$h(n) = 2n \quad \text{for all } n \in \mathbf{Z}.$$

A (partial) arrow diagram for h is shown below.

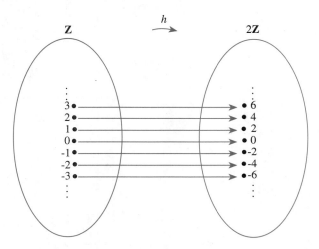

By exercises 8b and 34 of Section 7.3, h is one-to-one and onto. Hence $\mathbf{Z}$ has the same cardinality as $2\mathbf{Z}$. But Example 7.6.1 showed that $\mathbf{Z}^+$ has the same cardinality as $\mathbf{Z}$. So by the transitive property of cardinality, $\mathbf{Z}^+$ has the same cardinality as $2\mathbf{Z}$.

It follows by definition of countably infinite that $2\mathbf{Z}$ is countably infinite and thus countable. ∎

THE SEARCH FOR LARGER INFINITIES

Every infinite set we have discussed so far has been countably infinite. Do any larger infinities exist? Are there uncountable sets? Here is one candidate.

Imagine the number line as shown below.

$$\cdots \quad -4 \quad -3 \quad -2 \quad -1 \quad 0 \quad 1 \quad 2 \quad 3 \quad 4 \quad \cdots$$

Observe that the integers are spread along the number line at discrete intervals. The rational numbers, on the other hand, are *dense*: Between any two rational numbers (no matter how close) lies another rational number (the average of the two numbers, for instance). This suggests the conjecture that the infinity of the set of rational numbers is larger than the infinity of the set of integers.

Amazingly, this conjecture is false. Despite the fact that the rational numbers are crowded onto the number line whereas the integers are quite separated, the set of all rational numbers can be put into one-to-one correspondence with the set of integers. The next example gives a partial proof of this fact. It shows that the set of all positive rational numbers can be put into one-to-one correspondence with the set of all positive integers. In exercise 7 at the end of this section you are asked to use this result, together with a technique similar to that of Example 7.6.1, to show that the set of *all* rational numbers is countable.

EXAMPLE 7.6.3   **The Set of All Positive Rational Numbers Is Countable**

Show that the set $\mathbf{Q}^+$ of all positive rational numbers is countable.

Solution   Display the elements of the set $\mathbf{Q}^+$ of positive rational numbers in a grid as shown in Figure 7.6.3.

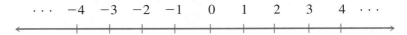

$$
\begin{array}{cccccc}
\frac{1}{1} & \frac{1}{2} & \frac{1}{3} & \frac{1}{4} & \frac{1}{5} & \frac{1}{6} \cdots \\
\frac{2}{1} & \frac{2}{2} & \frac{2}{3} & \frac{2}{4} & \frac{2}{5} & \frac{2}{6} \cdots \\
\frac{3}{1} & \frac{3}{2} & \frac{3}{3} & \frac{3}{4} & \frac{3}{5} & \frac{3}{6} \cdots \\
\frac{4}{1} & \frac{4}{2} & \frac{4}{3} & \frac{4}{4} & \frac{4}{5} & \frac{4}{6} \cdots \\
\frac{5}{1} & \frac{5}{2} & \frac{5}{3} & \frac{5}{4} & \frac{5}{5} & \frac{5}{6} \cdots \\
\frac{6}{1} & \frac{6}{2} & \frac{6}{3} & \frac{6}{4} & \frac{6}{5} & \frac{6}{6} \cdots \\
\end{array}
$$

FIGURE 7.6.3

Define a function F from $\mathbf{Z}^+$ to $\mathbf{Q}^+$ by starting to count at $\frac{1}{1}$ and following the arrows as indicated, skipping over any number that has already been counted.

To be specific: Set $F(1) = \frac{1}{1}$, $F(2) = \frac{1}{2}$, and $F(3) = \frac{2}{1}$. Then skip $\frac{2}{2}$ since $\frac{2}{2} = \frac{1}{1}$, which was counted first. After that, set $F(4) = \frac{3}{1}$, $F(5) = \frac{1}{3}$, $F(6) = \frac{1}{4}$, $F(7) = \frac{2}{3}$, $F(8) = \frac{3}{2}$, $F(9) = \frac{4}{1}$, and $F(10) = \frac{5}{1}$. Then skip $\frac{4}{2}$, $\frac{3}{3}$, and $\frac{2}{4}$ (since $\frac{4}{2} = \frac{2}{1}$, $\frac{3}{3} = \frac{1}{1}$, and $\frac{2}{4} = \frac{1}{2}$) and set $F(11) = \frac{1}{5}$. Continue in this way, defining $F(n)$ for each positive integer n.

Note that every positive rational number appears somewhere in the grid, and the counting procedure is set up so that every point in the grid is reached eventually. Thus the function F is onto. Also by skipping numbers that have already been counted, no number is counted twice. So F is one-to-one. Consequently, F is a function from $\mathbf{Z}^+$ to $\mathbf{Q}^+$ that is one-to-one and onto, and so $\mathbf{Q}^+$ is countably infinite and hence countable. ■

In 1874 the German mathematician Georg Cantor achieved success in the search for a larger infinity by showing that the set of all real numbers is uncountable. His method of proof was somewhat complicated, however. We give a proof of the uncountability of the set of all real numbers between 0 and 1 using a simpler technique introduced by Cantor in 1891 and now called the *Cantor diagonalization process*. Over the intervening years, this technique and variations on it have been used to establish a number of important results in logic and the theory of computation.

Before stating and proving Cantor's theorem, we note that every real number, which is a measure of location on a number line, can be represented by a decimal expansion of the form

$$a_0 \cdot a_1 a_2 a_3 \ldots,$$

where a_0 is an integer (positive, negative, or zero) and for each $i \geq 1$, a_i is an integer from 0 through 9. Although a formal proof of this result is somewhat advanced, it can be illustrated with a simple example.

Consider the point P in Figure 7.6.4. Figure 7.6.4(a) shows P located between 1 and 2. When the interval from 1 to 2 is divided into ten equal subintervals (see Figure 7.6.4(b)) P is seen to lie between 1.6 and 1.7. If the interval from 1.6 to 1.7 is itself divided into ten equal subintervals (see Figure 7.6.4(c)), then P is seen to lie between 1.62 and 1.63 but closer to 1.62 than to 1.63. So to two-decimal-place accuracy, the decimal expansion for P is 1.62.

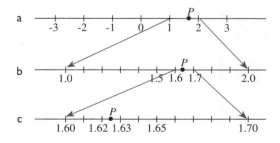

FIGURE 7.6.4

Assuming that any interval of real numbers, no matter how small, can be divided into ten equal subintervals, the process of obtaining additional digits in the decimal expansion for P can, in theory, be repeated indefinitely. If at any stage P is seen to be a subdivision point, then all further digits in the expansion may be taken to be 0. If not, then the process gives an expansion with an infinite number of digits.

The resulting decimal representation for P is unique except for numbers that end in infinitely repeating 9's or infinitely repeating 0's. For example (see exercise 16 at the end of this section),

$$0.199999 \ldots = 0.200000 \ldots .$$

Let us agree to express any such decimal in the form that ends in all 0's.

THEOREM 7.6.2   (Cantor)

The set of all real numbers between 0 and 1 is uncountable.

Proof (by contradiction):

Suppose the set of all real numbers between 0 and 1 is countable. Then the decimal representations of these numbers can be written in a list as follows:

$$0.a_{11}a_{12}a_{13} \cdots a_{1n} \cdots$$
$$0.a_{21}a_{22}a_{23} \cdots a_{2n} \cdots$$
$$0.a_{31}a_{32}a_{33} \cdots a_{3n} \cdots$$
$$\vdots$$
$$0.a_{n1}a_{n2}a_{n3} \cdots a_{nn} \cdots$$
$$\vdots$$

[We will derive a contradiction by showing that there is a number between 0 and 1 that does not appear on this list.]

For each pair of positive integers i and j, the jth decimal digit of the ith number on the list is a_{ij}. In particular, the first decimal digit of the first number on the list is a_{11}, the second decimal digit of the second number on the list is a_{22}, and so forth. As an example, suppose the list of real numbers between 0 and 1 starts out as follows:

$$0.\,②\;\;0\;\;1\;\;4\;\;8\;\;8\;\;0\;\;2\ldots$$
$$0.\,1\;\;①\;\;6\;\;6\;\;6\;\;0\;\;2\;\;1\ldots$$
$$0.\,0\;\;3\;\;③\;\;5\;\;3\;\;3\;\;2\;\;0\ldots$$
$$0.\,9\;\;6\;\;7\;\;⑦\;\;6\;\;8\;\;0\;\;9\ldots$$
$$0.\,0\;\;0\;\;0\;\;3\;\;①\;\;0\;\;0\;\;2\ldots$$
$$\vdots$$

The diagonal elements are circled: a_{11} is 2, a_{22} is 1, a_{33} is 3, a_{44} is 7, a_{55} is 1, and so forth.

Construct a new decimal number $d = 0.d_1d_2d_3 \cdots d_n \cdots$ as follows:

$$d_n = \begin{cases} 1 & \text{if } a_{nn} \neq 1 \\ 2 & \text{if } a_{nn} = 1. \end{cases}$$

In the above example,

$$d_1 \text{ is 1 because } a_{11} = 2 \neq 1,$$
$$d_2 \text{ is 2 because } a_{22} = 1,$$
$$d_3 \text{ is 1 because } a_{33} = 3 \neq 1,$$

$$d_4 \text{ is } 1 \text{ because } a_{44} = 7 \neq 1,$$
$$d_5 \text{ is } 2 \text{ because } a_{55} = 1,$$

and so forth. Hence, d would equal $0.12112 \ldots$.

The crucial observation is that *for each integer n, d differs in the nth decimal position from the nth number on the list.* But this implies that d is not on the list! In other words, d is a real number between 0 and 1 that is not on the list of *all* real numbers between 0 and 1. This contradiction shows the falseness of the supposition that the set of all numbers between 0 and 1 is countable. Hence the set of all real numbers between 0 and 1 is uncountable.

Along with demonstrating the existence of an uncountable set, Cantor developed a whole arithmetic theory of infinite sets of various sizes. One of the most basic theorems of the theory states that any subset of a countable set is countable.

THEOREM 7.6.3

Any subset of any countable set is countable.

Proof:

Let A be a particular but arbitrarily chosen countable set and let B be any subset of A. *[We must show that B is countable.]* Either B is finite or it is infinite. If B is finite, then B is countable by definition of countable, and we are done. So suppose B is infinite. Since A is countable, the distinct elements of A can be represented as a sequence

$$a_1, a_2, a_3, \ldots.$$

Define a function $g : \mathbf{Z}^+ \to B$ inductively as follows:

1. Search sequentially through elements of $a_1, a_2, a_3, \ldots$ until an element of B is found. *[This must happen eventually since $B \subseteq A$ and $B \neq \varnothing$.]* Call that element $g(1)$.

2. For each integer $k \geq 2$, suppose $g(k-1)$ has been defined. Then $g(k-1) = a_i$ for some a_i in $\{a_1, a_2, a_3, \ldots\}$. Starting with a_{i+1}, search sequentially through $a_{i+1}, a_{i+2}, a_{i+3}, \ldots$ trying to find an element of B. One must be found eventually because B is infinite, and $\{g(1), g(2), \ldots, g(k-1)\}$ is a finite set. When an element of B is found, define it to be $g(k)$.

By (1) and (2) above, the function g is defined for each positive integer.

Since the elements of $a_1, a_2, a_3, \ldots$ are all distinct, g is one-to-one. Furthermore, the searches for elements of B are sequential: Each picks up where the previous one left off. Thus every element of A is reached during some search. But all the elements of B are located somewhere in the sequence $a_1, a_2, a_3, \ldots$, and so every element of B is eventually found and made the image of some integer. Hence g is onto. These remarks show that g is a one-to-one correspondence from $\mathbf{Z}^+$ to B. So B is countably infinite and thus countable.

It follows from Theorem 7.6.3 that any set with an uncountable subset is uncountable. (For if such a set were countable, then every subset would have to be countable.) Consequently, the set of all real numbers is uncountable (since the subset of numbers between 0 and 1 is uncountable). In fact, the set of all real numbers has the same cardinality as the set of all real numbers between 0 and 1! (See exercises 20 and 21 at the end of this section.)

EXAMPLE 7.6.4   The Cardinality of the Set of All Real Numbers

Show that the set of all real numbers has the same cardinality as the set of real numbers between 0 and 1.

Solution   Let S be the open interval of real numbers between 0 and 1:

$$S = \{x \in \mathbf{R} \mid 0 < x < 1\}.$$

Imagine picking up S and bending it into a circle as shown below. Since S does not include either endpoint 0 or 1, the top-most point of the circle is omitted from the drawing.

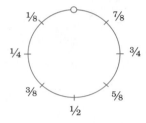

Define a function $F: S \rightarrow \mathbf{R}$ as follows:

Draw a number line and place the interval, S, bent into a circle, tangent to the line above the point 0. This is shown below.

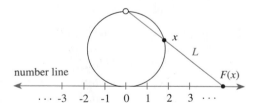

For each point x on the circle representing S, draw a straight line L through the top-most point of the circle and x. Let $F(x)$ be the point of intersection of L and the number line. ($F(x)$ is called the *projection* of x onto the number line.)

It is clear from the geometry of the situation that distinct points on the circle go to distinct points on the number line; so F is one-to-one. In addition, given any point y on the number line, a line can be drawn through y and the top-most point of the circle. This line must intersect the circle at some point x, and, by definition, $y = F(x)$. Thus F is onto. Hence F is a one-to-one correspondence from S to $\mathbf{R}$, and so S and $\mathbf{R}$ have the same cardinality. ∎

APPLICATION: CARDINALITY AND COMPUTABILITY

Knowledge of the countability and uncountability of certain sets can be used to answer a question of computability. We begin by showing that a certain set is countable.

EXAMPLE 7.6.5

Countability of the Set of Computer Programs in a Computer Language

Show that the set of all computer programs in a given computer language is countable.

Solution   This result is a consequence of the fact that any computer program in any language can be regarded as a finite string of symbols in the (finite) alphabet of the language.

Given any computer language, set up a binary code to translate the symbols of the alphabet of the language to strings of 0's and 1's. (For instance, you could use codes such as ASCII or EBCDIC to translate the symbols of present-day computer languages.) Let P be the set of all computer programs in the language. Given any computer program in P, translate all the symbols in the program to 0's and 1's. In this way any program can be represented as a (long) string of 0's and 1's. Order these strings by length, putting shorter before longer, and order all strings of a given length by regarding each string as a binary number and writing the numbers in ascending order. Either P is finite or P is infinite. If P is finite, then P is countable and we are done. So suppose P is infinite. Define a function $F: \mathbf{Z}^+ \rightarrow P$ by specifying that

$$F(n) = \text{the } n\text{th program in the list} \quad \text{for each } n \in \mathbf{Z}^+.$$

By construction, F is one-to-one and onto, and so P is countably infinite and hence countable. As a simple example, suppose the following are all the programs in P that translate into bit strings of length less than or equal to 5:

$$10111, 11, 0010, 1011, 01, 00100, 1010, 00010.$$

Ordering these by length gives

length 2: 11, 01
length 4: 0010, 1011, 1010
length 5: 10111, 00100, 00010

And ordering those of each given length by size of the binary number it represents gives

$$
\begin{aligned}
01 \quad &= F(1) \\
11 \quad &= F(2) \\
0010 \quad &= F(3) \\
1010 \quad &= F(4) \\
1011 \quad &= F(5) \\
00010 &= F(6) \\
00100 &= F(7) \\
10111 &= F(8)
\end{aligned}
$$

Note that when viewed purely as numbers, ignoring leading zeros, $0010 = 00010$. This shows the necessity of first ordering the strings by length before arranging them in ascending numerical order. ∎

The final example of this section shows that a certain set is uncountable and hence that there must exist a noncomputable function.

EXAMPLE 7.6.6   The Cardinality of a Set of Functions and Computability

a. Let T be the set of all functions from the positive integers to the set $\{0, 1, 2, 3, 4,$ $5, 6, 7, 8, 9\}$. Show that T is uncountable.
b. Derive the consequence that there are noncomputable functions. Specifically, show that for any computer language there must be a function F from $\mathbf{Z}^+$ to $\{0, 1, 2, 3, 4, 5, 6, 7, 8, 9\}$ with the property that no computer program can be written in the language to take arbitrary values as input and output the corresponding function values.

Solution   a. Let S be the set of all real numbers between 0 and 1. As noted before, any number in S can be represented in the form

$$0.a_1 a_2 a_3 \ldots a_n \ldots,$$

where each a_i is an integer from 0 to 9. This representation is unique if decimals that end in all 9's are omitted.

Define a function F from S to a subset of T (the set of all functions from $\mathbf{Z}^+$ to $\{0, 1, 2, 3, 4, 5, 6, 7, 8, 9\}$) as follows:

$$F(0.a_1 a_2 a_3 \ldots a_n \ldots) = \text{the function that sends each}$$
$$\text{positive integer } n \text{ to } a_n.$$

Choose the co-domain of F to be exactly that subset of T that makes F onto. That is, define the co-domain of F to equal the image of F. Note that F is one-to-one because if $F(x_1) = F(x_2)$, then each decimal digit of x_1 equals the corresponding decimal digit of x_2, and so $x_1 = x_2$. Thus F is a one-to-one correspondence from S to a subset of T. But S is uncountable by Theorem 7.6.2. Hence T has an uncountable subset, and so by Theorem 7.6.3, T is uncountable.
b. Part (a) shows that the set T of all functions from $\mathbf{Z}^+$ to $\{0, 1, 2, 3, 4, 5, 6, 7, 8, 9\}$ is uncountable. But Example 7.6.5 shows that given any computer language, the set of all programs in that language is countable. Consequently, in any computer language there are not enough programs to compute values of every function in T. There must exist functions that are not computable! ■

EXERCISE SET 7.6

1. Show that "there are as many squares as there are numbers" by exhibiting a one-to-one correspondence from the positive integers, $\mathbf{Z}^+$, to the set S of all squares of positive integers:

$$S = \{n \in \mathbf{Z}^+ \mid n = k^2, \text{ for some positive integer } k\}.$$

2. Let $3\mathbf{Z} = \{n \in \mathbf{Z} \mid n = 3k, \text{ for some integer } k\}$.

a. Prove that $\mathbf{Z}$ and $3\mathbf{Z}$ have the same cardinality.

b. Use the result of (a) to prove that $3\mathbf{Z}$ is countable.

3. Show that the set of all nonnegative integers is countable by exhibiting a one-to-one correspondence between $\mathbf{Z}^+$ and $\mathbf{Z}^{nonneg}$.

Exercises 4 and 5 refer to the function f described in Example 7.6.1.

4. a. Check that the formula for f given at the end of the example produces the correct values for $n = 1, 2, 3,$ and 4.

b. Use the floor function to write a formula for f as a single algebraic expression for all positive integers n.

H5. Show how to adapt the description of f to define a function h from $\mathbf{Z}^+$ to $\mathbf{Z}$ that "counts" every integer twice (so that, for instance, $h(1) = h(2) = 0$ and $h(3) = h(4) = 1$). The function h should be onto but not one-to-one.

6. Show that the set of all bit strings is countable.

7. Show that **Q**, the set of all rational numbers, is countable.

8. Show that the set **Q** of all rational numbers is dense along the number line by showing that given any two rational numbers r_1 and r_2 with $r_1 < r_2$, there exists a rational number x such that $r_1 < x < r_2$.

H 9. Must the average of two irrational numbers always be irrational? Prove or give a counterexample.

H ◆ 10. Show that the set of all irrational numbers is dense along the number line by showing that given any two real numbers, there is an irrational number in between.

11. Give two examples of functions from **Z** to **Z** that are one-to-one but not onto.

12. Give two examples of functions from **Z** to **Z** that are onto but not one-to-one.

13. Define a function $g : \mathbf{Z}^+ \times \mathbf{Z}^+ \to \mathbf{Z}^+$ by the formula $g(m, n) = 2^m 3^n$ for all $(m, n) \in \mathbf{Z}^+ \times \mathbf{Z}^+$. Show that g is one-to-one and use this result to prove that $\mathbf{Z}^+ \times \mathbf{Z}^+$ is countable.

14. **a.** Explain how to use the following diagram to show that $\mathbf{Z}^{nonneg} \times \mathbf{Z}^{nonneg}$ and $\mathbf{Z}^{nonneg}$ have the same cardinality.

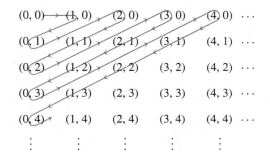

H ◆ **b.** Define a function $H : \mathbf{Z}^{nonneg} \times \mathbf{Z}^{nonneg} \to \mathbf{Z}^{nonneg}$ by the formula

$$H(m, n) = n + \frac{(m + n)(m + n + 1)}{2}$$

for all nonnegative integers m and n. Interpret the action of H geometrically using the diagram of part (a).

◆ 15. Prove that the function H defined in exercise 14 is a one-to-one correspondence.

H 16. Prove that $0.1999 \ldots = 0.2$.

In 17–21, let $S = \{x \in \mathbf{R} \mid 0 < x < 1\}$.

17. Let $U = \{x \in \mathbf{R} \mid 0 < x < 2\}$. Prove that S and U have the same cardinality.

H 18. Let $V = \{x \in \mathbf{R} \mid 2 < x < 5\}$. Prove that S and V have the same cardinality.

19. Let a and b be real numbers with $a < b$, and let $W = \{x \in \mathbf{R} \mid a < x < b\}$. Prove that S and W have the same cardinality.

20. Draw the graph of the function f defined by the following formula:

for all real numbers x with $0 < x < 1$,

$$f(x) = \tan\left(\pi x - \frac{\pi}{2}\right).$$

Use the graph to explain why S and **R** have the same cardinality.

◆ 21. Define a function g from the set of real numbers to S by the following formula:

for all real numbers x,

$$g(x) = \frac{1}{2} \cdot \left(\frac{x}{1 + |x|}\right) + \frac{1}{2}.$$

Prove that g is a one-to-one correspondence. What conclusion can you draw from this fact?

22. Prove that any infinite set contains a countably infinite subset.

23. Prove that any union of two countable sets is countable.

24. Use the result of exercise 23 to prove that the set of all irrational numbers is uncountable.

◆ 25. Let S be the set of all solutions to all equations of the form $x^2 + ax + b$, where a and b are integers. Use the result of exercise 23 to prove that S is countable.

H 26. Let $\mathcal{P}(S)$ be the set of all subsets of a set S, and let T be the set of all functions from S to $\{0, 1\}$. Show that $\mathcal{P}(S)$ and T have the same cardinality.

H 27. Let S be a set and let $\mathcal{P}(S)$ be the set of all subsets of S. Show that S and $\mathcal{P}(S)$ do not have the same cardinality.

◆ 28. Prove that there are as many functions from $\mathbf{Z}^+$ to $\{0, 1, 2, 3, 4, 5, 6, 7, 8, 9\}$ as there are functions from $\mathbf{Z}^+$ to $\{0, 1\}$.

H 29. Prove that if A and B are any countable sets, then $A \times B$ is countable.

◆ 30. Prove that a countable union of countable sets is countable.

8 Recursion

A sequence is said to be defined recursively if certain initial values are specified and later terms of the sequence are defined by relating them to a fixed number of earlier terms. In the first section of this chapter, we give a variety of examples that show how to analyze certain kinds of problems by thinking recursively to obtain a recursively defined sequence. In the next two sections we address the problem of finding an explicit formula for a sequence that is defined recursively. And in the final section we discuss more general recursive definitions, such as the one used for the careful formulation of the concept of Boolean expression, and the idea of recursive function.

8.1 RECURSIVELY DEFINED SEQUENCES

So, Nat'ralists observe, a Flea Hath smaller Fleas that on him prey, And these have smaller Fleas to bite 'em, And so proceed ad infinitum.
(Jonathan Swift, 1733)

A sequence can be defined in a variety of different ways. One informal way is to write the first few terms with the expectation that the general pattern will be obvious. We might say, for instance, "consider the sequence 3, 5, 7, ... " Unfortunately, misunderstandings can occur when this approach is used. The next term of the sequence could be 9 if we mean the sequence of odd integers or it could be 11 if we mean the sequence of odd prime numbers.

A second way to define a sequence is to give an explicit formula for its nth term. For example, a sequence $a_0, a_1, a_2, \ldots$ can be specified by writing

$$a_n = \frac{(-1)^n}{n + 1} \quad \text{for all integers } n \geq 0.$$

The advantage of defining a sequence by such an explicit formula is that each term of the sequence is uniquely determined and any term can be computed in a fixed, finite number of steps. In this case, for instance,

$$a_0 = \frac{(-1)^0}{0 + 1} = 1, \qquad a_1 = \frac{(-1)^1}{1 + 1} = -\frac{1}{2}, \qquad \text{and so forth.}$$

A third way to define a sequence is to use recursion. This requires giving both an equation, called a *recurrence relation,* that relates later terms in the sequence to earlier terms and a specification, called *initial conditions,* of the values of the first few terms of the sequence. The initial conditions are also called the *base* or *bottom* of the recursion. For instance, a sequence $b_0, b_1, b_2, \ldots$ can be defined recursively as follows: For all integers $k \geq 2$,

$$(1)\ b_k = b_{k-1} + b_{k-2} \quad \text{recurrence relation}$$
$$(2)\ b_0 = 1, \quad b_1 = 3 \quad \text{initial conditions.}$$

Since b_0 and b_1 are given, b_2 can be computed using the recurrence relation.

$$b_2 = b_1 + b_0 \quad \text{by substituting } k = 2 \text{ into (1)}$$
$$= 3 + 1 \quad \text{since } b_1 = 3 \text{ and } b_0 = 1 \text{ by (2)}$$
$$(3)\ \therefore b_2 = 4$$

Then, since both b_1 and b_2 are now known, b_3 can be computed using the recurrence relation.

$$b_3 = b_2 + b_1 \quad \text{by substituting } k = 3 \text{ into (1)}$$
$$= 4 + 3 \quad \text{since } b_2 = 4 \text{ by (3) and } b_1 = 3 \text{ by (2)}$$
$$(4)\ \therefore b_3 = 7$$

In general, the recurrence relation says that any term of the sequence after b is the sum of the two preceding terms. Thus

$$b_4 = b_3 + b_2 = 7 + 4 = 11,$$
$$b_5 = b_4 + b_3 = 11 + 7 = 18,$$

and so forth. It should be clear that any later term of the sequence can be computed from this point by continuing in a step-by-step fashion.

Sometimes it is very difficult or impossible to find an explicit formula for a sequence, but it *is* possible to define the sequence using recursion. Note that defining sequences recursively is similar to proving theorems by mathematical induction. The recurrence relation is like the inductive step and the initial conditions are like the basis step. Indeed, the fact that sequences can be defined recursively is equivalent to the fact that mathematical induction works as a method of proof.

DEFINITION

A **recurrence relation** for a sequence $a_0, a_1, a_2, \ldots$ is a formula that relates each term a_k to certain of its predecessors $a_{k-1}, a_{k-2}, \ldots, a_{k-i}$, where i is a fixed integer and k is any integer greater than or equal to i. The **initial conditions** for such a recurrence relation specify the values of $a_0, a_1, a_2, \ldots, a_{i-1}$.

EXAMPLE 8.1.1   Computing Terms of a Recursively Defined Sequence

Define a sequence $c_0, c_1, c_2, \ldots$ recursively as follows: For all integers $k \geq 2$,

$$(1)\ c_k = c_{k-1} + k \cdot c_{k-2} + 1 \quad \text{recurrence relation}$$
$$(2)\ c_0 = 1 \quad \text{and} \quad c_1 = 2 \quad \text{initial conditions.}$$

Find $c_2, c_3,$ and c_4.

Solution

$$c_2 = c_1 + 2 \cdot c_0 + 1 \quad \text{by substituting } k = 2 \text{ into (1)}$$
$$= 2 + 2 \cdot 1 + 1 \quad \text{since } c_1 = 2 \text{ and } c_0 = 1 \text{ by (2)}$$
$$(3)\ \therefore c_2 = 5$$
$$c_3 = c_2 + 3 \cdot c_1 + 1 \quad \text{by substituting } k = 3 \text{ into (1)}$$
$$= 5 + 3 \cdot 2 + 1 \quad \text{since } c_2 = 5 \text{ by (3) and } c_1 = 2 \text{ by (2)}$$

$(4) \therefore c_3 = 12$

$$c_4 = c_3 + 4 \cdot c_2 + 1 \quad \text{by substituting } k = 4 \text{ into (1)}$$
$$= 12 + 4 \cdot 5 + 1 \quad \text{since } c_3 = 12 \text{ by (4) and } c_2 = 5 \text{ by (3)}$$

$(5) \therefore c_4 = 33$ ■

A given recurrence relation may be expressed in several different ways.

EXAMPLE 8.1.2   Writing a Recurrence Relation in More Than One Way

Let $s_0, s_1, s_2, \ldots$ be a sequence that satisfies the following recurrence relation:

$$\text{for all integers } k \geq 1, \quad s_k = 3s_{k-1} - 1.$$

Explain why the following statement is true:

$$\text{for all integers } k \geq 0, \quad s_{k+1} = 3s_k - 1.$$

Solution   In informal language, the recurrence relation says that any term of the sequence equals 3 times the previous term minus 1. Now for any integer $k \geq 0$, the term previous to s_{k+1} is s_k. Thus for any integer $k \geq 0$, $s_{k+1} = 3s_k - 1$. ■

A sequence defined recursively need not start with a subscript of zero. Also, a given recurrence relation may be satisfied by many different sequences; the actual values of the sequence are determined by the initial conditions.

EXAMPLE 8.1.3   Sequences That Satisfy the Same Recurrence Relation

Let $a_1, a_2, a_3, \ldots$ and $b_1, b_2, b_3, \ldots$ satisfy the recurrence relation that the kth term equals 3 times the $(k - 1)$st term for all integers $k \geq 1$:

$$(1)\ a_k = 3a_{k-1} \quad \text{and} \quad b_k = 3b_{k-1}.$$

But suppose that the initial conditions for the sequences are different:

$$(2)\ a_1 = 2 \quad \text{and} \quad b_1 = 1.$$

Find (a) a_2, a_3, a_4 and (b) b_2, b_3, b_4.

Solution   a. $a_2 = 3a_1 = 3 \cdot 2 = 6$ b. $b_2 = 3b_1 = 3 \cdot 1 = 3$

$a_3 = 3a_2 = 3 \cdot 6 = 18$ $b_3 = 3b_2 = 3 \cdot 3 = 9$

$a_4 = 3a_3 = 3 \cdot 18 = 54$ $b_4 = 3b_3 = 3 \cdot 9 = 27$

Thus

$$a_1, a_2, a_3, \ldots \text{ begins } 2, 6, 18, 54, \ldots \text{ and}$$
$$b_1, b_2, b_3, \ldots \text{ begins } 1, 3, 9, 27, \ldots.$$ ■

EXAMPLE 8.1.4   Showing That a Sequence Given by an Explicit Formula Satisifies a Certain Recurrence Relation

Show that the sequence $1, -1!, 2!, -3!, 4!, \ldots, (-1)^n n!, \ldots$, for $n \geq 0$, satisfies the recurrence relation

$$s_k = -k \cdot s_{k-1} \quad \text{for all integers } k \geq 1.$$

Solution   The recurrence relation specifies that the kth term of the sequence equals $-k$ times the $(k-1)$st term. Call the general term of the sequence s_n starting with $s_0 = 1$. Then by definition of the sequence,

$$s_n = (-1)^n \cdot n! \quad \text{for each integer } n \geq 0.$$

Substitute k and $k - 1$ for n to get

$$s_k = (-1)^k \cdot k! \qquad \qquad \text{8.1.1}$$

$$s_{k-1} = (-1)^{k-1} \cdot (k-1)! \qquad \qquad \text{8.1.2}$$

It follows that

$$
\begin{aligned}
-k \cdot s_{k-1} &= -k \cdot [(-1)^{k-1} \cdot (k-1)!] && \text{by substitution from (8.1.2)} \\
&= (-1) \cdot k \cdot (-1)^{k-1} \cdot (k-1)! \\
&= (-1) \cdot (-1)^{k-1} \cdot k \cdot (k-1)! \\
&= (-1)^k \cdot k! && \text{by basic algebra} \\
&= s_k && \text{by substitution from (8.1.1).} \quad \blacksquare
\end{aligned}
$$

EXAMPLES OF RECURSIVELY DEFINED SEQUENCES

Recursion is one of the central ideas of computer science. To solve a problem recursively means to find a way to break it down into smaller subproblems each having the same form as the original problem, and to do this in such a way that when the process is repeated many times the last of the subproblems are small and easy to solve and the solutions of the subproblems can be woven together to form a solution to the original problem.

Probably the most difficult part of solving problems recursively is to figure out how knowing the solution to smaller subproblems of the same type as the original problem will give you a solution to the problem as a whole. You *suppose* you know the solutions to smaller subproblems and ask yourself how you would best make use of that knowledge to solve the larger problem. The supposition that the smaller subproblems have already been solved has been called the *recursive paradigm* or the *recursive leap of faith.* Once you take this leap, you are right in the middle of the most difficult part of the problem, but generally, though difficult, the path to a solution from this point is short. The recursive leap of faith is similar to the inductive hypothesis in a proof by mathematical induction.

EXAMPLE 8.1.5   The Tower of Hanoi

According to legend, a certain Hindu temple contains three thin diamond poles on one of which, at the time of creation, God placed 64 golden disks that decrease in size as they rise from the base. See Figure 8.1.1.

The priests of the temple work unceasingly to transfer all the disks one by one from the first pole to one of the others, but they must never place a larger disk on top of a smaller one. As soon as they have completed their task, "tower, temple, and Brahmins alike will crumble into dust, and with a thunderclap the world will vanish."*

The question is this: Assuming the priests work as efficiently as possible, how long will it be from the time of creation until the end of the world?

Edouard Lucas

*Actually, the Tower of Hanoi puzzle was invented in 1883 by the French mathematician Édouard Lucas, who made up the legend to accompany it. This quotation is from an account that appeared in *La Nature* by DeParville as translated by W. W. Rouse Ball in *Mathematical Recreations and Essays,* printed in 1905 by Macmillan Company Ltd.

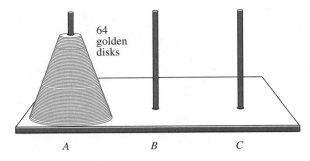

64
golden
disks

A   B   C

FIGURE 8.1.1

Solution   An elegant and efficient way to solve this problem is to think recursively. Suppose that you, somehow or other, have found the most efficient way possible to transfer a tower of $k - 1$ disks one by one from one pole to another, obeying the restriction that you never place a larger disk on top of a smaller one. What is the most efficient way to move a tower of k disks from one pole to another? The answer is sketched in Figure 8.1.2, where pole A is the initial pole and pole C is the target pole.

Step 1 is to move the top $k - 1$ disks from pole A to pole B. (If $k > 2$, execution of this step will require a number of moves of individual disks among the three poles. But the point of thinking recursively is not to get caught up in imagining the details of how those moves will occur.) Step 2 is to move the bottom disk from pole A to pole C. Step 3 is to move the top $k - 1$ disks from pole B to pole C. (Again, if $k > 2$ execution of this step will require more than one move.)

To see that this sequence of moves is most efficient, observe that to move the bottom disk of a stack of k disks from one pole to another, you must first move the top $k - 1$ disks to a third pole to get them out of the way. Thus moving the stack of k disks from pole A to pole C requires at least two transfers of the top $k - 1$ disks, one to move them off the bottom disk to free the disk so it can be moved and another to move them back on top of the bottom disk after the bottom disk has been moved to pole C. If the bottom disk were not moved directly from pole A to pole C but were moved to pole B first, at least two additional transfers of the top $k - 1$ disks would be necessary, one to move them from pole A to pole C so that the bottom disk could be moved from pole A to pole B and another to move them off pole C so that the bottom disk could be moved onto pole C. This would increase the total number of moves and result in a less efficient transfer.

Thus the minimum sequence of moves must include going from the initial position (a) to position (b) to position (c) to position (d). It follows that

$$
\begin{bmatrix} \text{the minimum} \\ \text{number of moves} \\ \text{needed to transfer} \\ \text{a tower of } k \text{ disks} \\ \text{from pole } A \text{ to} \\ \text{pole } C \end{bmatrix} = \begin{bmatrix} \text{the minimum} \\ \text{number of} \\ \text{moves needed} \\ \text{to go from} \\ \text{position (a)} \\ \text{to position (b)} \end{bmatrix} + \begin{bmatrix} \text{the minimum} \\ \text{number of} \\ \text{moves needed} \\ \text{to go from} \\ \text{position (b)} \\ \text{to position (c)} \end{bmatrix} + \begin{bmatrix} \text{the minimum} \\ \text{number of} \\ \text{moves needed} \\ \text{to go from} \\ \text{position (c)} \\ \text{to position (d)} \end{bmatrix} \qquad \textbf{8.1.3}
$$

For each integer $n \geq 1$, let

$$
m_n = \begin{bmatrix} \text{the minimum number of moves needed to move} \\ \text{a tower of } n \text{ disks from one pole to another} \end{bmatrix}
$$

Note that the numbers m_n are independent of the labeling of the poles; it takes the same minimum number of moves to transfer n disks from pole A to pole C as to transfer n

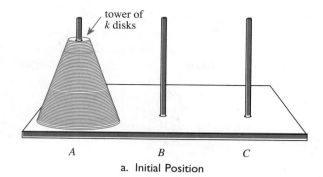

a. Initial Position

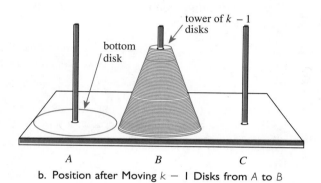

b. Position after Moving $k - 1$ Disks from A to B

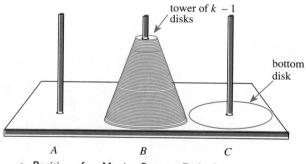

c. Position after Moving Bottom Disks from A to C

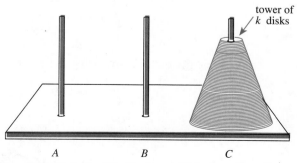

d. Position after Moving $k - 1$ Disks from B to C

FIGURE 8.1.2   Moves for the Tower of Hanoi

disks from pole A to pole B, for example. Also the values of m_n are independent of the number of larger disks that may lie below the top n, provided these remain stationary while the top n are moved. For the disks on the bottom are all larger than the ones on the top, and so the top disks can be moved from pole to pole as if the bottom disks were not present.

Now going from position (a) to position (b) requires m_{k-1} moves, going from position (b) to position (c) requires just one move, and going from position (c) to position (d) requires m_{k-1} moves. By substitution into equation (8.1.3), therefore,

$$m_k = m_{k-1} + 1 + m_{k-1}$$
$$= 2m_{k-1} + 1 \qquad \text{for all integers } k \geq 2.$$

The initial condition, or base, of this recursion is found by using the definition of the sequence.

$$m_1 = \left[\begin{array}{l} \text{the minimum number of moves needed to move} \\ \text{a tower of one disk from one pole to another} \end{array} \right]$$
$$= 1,$$

since just one move is needed to move one disk from one pole to another! Hence the complete recursive specification of the sequence $m_1, m_2, m_3, \ldots$ is as follows: For all integers $k \geq 2$,

$$\text{(1) } m_k = 2m_{k-1} + 1 \qquad \text{recurrence relation}$$
$$\text{(2) } m_1 = 1 \qquad \text{initial condition.}$$

Here is a computation of the next five terms of the sequence:

$$\text{(3) } m_2 = 2m_1 + 1 = 2 \cdot 1 + 1 = 3 \qquad \text{by (1) and (2),}$$
$$\text{(4) } m_3 = 2m_2 + 1 = 2 \cdot 3 + 1 = 7 \qquad \text{by (1) and (3),}$$
$$\text{(5) } m_4 = 2m_3 + 1 = 2 \cdot 7 + 1 = 15 \qquad \text{by (1) and (4),}$$
$$\text{(6) } m_5 = 2m_4 + 1 = 2 \cdot 15 + 1 = 31 \qquad \text{by (1) and (5),}$$
$$\text{(7) } m_6 = 2m_5 + 1 = 2 \cdot 31 + 1 = 63 \qquad \text{by (1) and (6).}$$

Going back to the legend, suppose the priests work rapidly and move one disk every second. Then the time from the beginning of creation to the end of the world would be m_{64} seconds. In the next section we derive an explicit formula for m_n. Meanwhile, we can compute m_{64} on a calculator or a computer by continuing the process started above (Try it!). The approximate result is

$$1.844674 \times 10^{19} \text{ seconds} \cong 5.84542 \times 10^{11} \text{ years}$$
$$\cong 584.5 \text{ billion years,}$$

which is obtained by using the estimate of

$$60 \cdot 60 \cdot 24 \cdot (365.25) = 31,557,600$$

seconds per minute	minutes per hour	hours per day	days per year	seconds per year
↑	↑	↖	↖	↑

seconds in a year (figuring 365.25 days in a year to take leap years into account). Surprisingly, this figure is close to scientific estimates of the life of the universe! ∎

EXAMPLE 8.1.6   **The Fibonacci Numbers**

One of the earliest examples of a recursively defined sequence arises in the writings of Leonardo of Pisa, commonly known as Fibonacci, who was the greatest European mathematician of the Middle Ages. In 1202 Fibonacci posed the following problem.

> A single pair of rabbits (male and female) is born at the beginning of a year. Assume the following conditions:
>
> 1. Rabbit pairs are not fertile during their first month of life but thereafter give birth to one new male/female pair at the end of every month;
> 2. No rabbits die.
>
> How many rabbits will there be at the end of the year?

Solution   One way to solve this problem is to plunge right into the middle of it using recursion. Suppose you know how many rabbit pairs there were at the ends of previous months. How many will there be at the end of the current month? The crucial observation is that the number of rabbit pairs born at the end of month k is the same as the number of pairs alive at the end of month $k - 2$. Why? Because it is exactly the rabbit pairs that were alive at the end of month $k - 2$ that were fertile during month k. The rabbits born at the end of month $k - 1$ were not.

Fibonacci (Leonardo of Pisa) (c. 1175–1250)

month $k - 2$ $k - 1$ k

Each pair alive here ↑   gives birth to a pair here ↑.

Now the number of rabbit pairs alive at the end of month k equals the ones alive at the end of month $k - 1$ plus the pairs newly born at the end of the month. Thus

$$
\begin{bmatrix} \text{the number} \\ \text{of rabbit} \\ \text{pairs alive} \\ \text{at the end} \\ \text{of month } k \end{bmatrix} = \begin{bmatrix} \text{the number} \\ \text{of rabbit} \\ \text{pairs alive} \\ \text{at the end of} \\ \text{month } k-1 \end{bmatrix} + \begin{bmatrix} \text{the number} \\ \text{of rabbit} \\ \text{pairs born} \\ \text{at the end} \\ \text{of month } k \end{bmatrix}
$$

$$
= \begin{bmatrix} \text{the number} \\ \text{of rabbit} \\ \text{pairs alive} \\ \text{at the end of} \\ \text{month } k-1 \end{bmatrix} + \begin{bmatrix} \text{the number} \\ \text{of rabbit} \\ \text{pairs alive} \\ \text{at the end of} \\ \text{month } k-2 \end{bmatrix} \qquad 8.1.4
$$

For each integer $n \geq 1$, let

$$
F_n = \begin{bmatrix} \text{the number of rabbit pairs} \\ \text{alive at the end of month } n \end{bmatrix}
$$

and let

$$
F_0 = \text{the initial number of rabbit pairs}
$$
$$
= 1.
$$

Then by substitution into equation (8.1.4), for all integers $k \geq 2$,

$$
F_k = F_{k-1} + F_{k-2}.
$$

Now $F_0 = 1$ as already noted, and $F_1 = 1$ also because the first pair of rabbits is not fertile until the second month. Hence the complete specification of the Fibonacci sequence is as follows: For all integers $k \geq 2$,

$$(1)\ F_k = F_{k-1} + F_{k-2} \quad \text{recurrence relation}$$
$$(2)\ F_0 = 1, \quad F_1 = 1 \quad \text{initial conditions.}$$

To answer Fibonacci's question, compute F_2, F_3, and so forth through F_{12}:

$$(3)\ F_2 = F_1 + F_0\ \ = 1 + 1\ \ \ = 2 \quad \text{by (1) and (2),}$$
$$(4)\ F_3 = F_2 + F_1\ \ = 2 + 1\ \ \ = 3 \quad \text{by (1), (2) and (3),}$$
$$(5)\ F_4 = F_3 + F_2\ \ = 3 + 2\ \ \ = 5 \quad \text{by (1), (3) and (4),}$$
$$(6)\ F_5 = F_4 + F_3\ \ = 5 + 3\ \ \ = 8 \quad \text{by (1), (4) and (5),}$$
$$(7)\ F_6 = F_5 + F_4\ \ = 8 + 5\ \ \ = 13 \quad \text{by (1), (5) and (6),}$$
$$(8)\ F_7 = F_6 + F_5\ \ = 13 + 8\ \ = 21 \quad \text{by (1), (6) and (7),}$$
$$(9)\ F_8 = F_7 + F_6\ \ = 21 + 13 = 34 \quad \text{by (1), (7) and (8),}$$
$$(10)\ F_9 = F_8 + F_7\ \ = 34 + 21 = 55 \quad \text{by (1), (8) and (9),}$$
$$(11)\ F_{10} = F_9 + F_8\ \ = 55 + 34 = 89 \quad \text{by (1), (9) and (10),}$$
$$(12)\ F_{11} = F_{10} + F_9 = 89 + 55 = 144 \quad \text{by (1), (10) and (11),}$$
$$(13)\ F_{12} = F_{11} + F_{10} = 144 + 89 = 233 \quad \text{by (1), (11) and (12).}$$

At the end of the twelfth month there are 233 rabbit pairs or 466 rabbits in all. ∎

EXAMPLE 8.1.7   Compound Interest

On your twenty-first birthday you get a letter informing you that on the day you were born an eccentric rich aunt deposited $1,000 in a bank account earning 5.5% interest compounded annually and she now intends to turn the account over to you provided you can figure out how much it is worth. What is the amount currently in the account?

Solution   To approach this problem recursively, observe that

$$\begin{bmatrix} \text{the amount in} \\ \text{the account at} \\ \text{the end of any} \\ \text{particular year} \end{bmatrix} = \begin{bmatrix} \text{the amount in} \\ \text{the account at} \\ \text{the end of the} \\ \text{previous year} \end{bmatrix} + \begin{bmatrix} \text{the interest} \\ \text{earned on the} \\ \text{account during} \\ \text{the year} \end{bmatrix}.$$

Now the interest earned during the year equals the interest rate, $5.5\% = 0.055$, times the amount in the account at the end of the previous year. Thus

$$\begin{bmatrix} \text{the amount in} \\ \text{the account at} \\ \text{the end of any} \\ \text{particular year} \end{bmatrix} = \begin{bmatrix} \text{the amount in} \\ \text{the account at} \\ \text{the end of the} \\ \text{previous year} \end{bmatrix} + (0.055) \cdot \begin{bmatrix} \text{the amount in} \\ \text{the account at} \\ \text{the end of the} \\ \text{previous year} \end{bmatrix}. \quad \textbf{8.1.5}$$

For each positive integer n, let

$$A_n = \begin{bmatrix} \text{the amount in the account} \\ \text{at the end of year } n \end{bmatrix}$$

and let

$$A_0 = \begin{bmatrix} \text{the initial amount} \\ \text{in the account} \end{bmatrix} = \$1,000.$$

Then for any particular year k, substitution into equation (8.1.5) gives

$$A_k = A_{k-1} + (0.055) \cdot A_{k-1}$$
$$= (1 + 0.055) \cdot A_{k-1} = (1.055) \cdot A_{k-1} \qquad \text{by factoring out } A_{k-1}.$$

Consequently, the values of the sequence $A_0, A_1, A_2, \ldots$ are completely specified as follows: for all integers $k \geq 1$,

(1) $A_k = (1.055) \cdot A_{k-1}$ recurrence relation

(2) $A_0 = \$1,000$ initial condition.

The number 1.055 is called the *growth factor* of the sequence.

In the next section we derive an explicit formula for the value of the account in any year n. The value on your twenty-first birthday can also be computed by repeated substitution as follows:

(3) $A_1 = 1.055 \cdot A_0 = (1.055) \cdot \$1,000 \quad = \$1,055.00$ by (1) and (2),

(4) $A_2 = 1.055 \cdot A_1 = (1.055) \cdot \$1,055 \quad \cong \$1,113.02$ by (1) and (3),

(5) $A_3 = 1.055 \cdot A_2 \cong (1.055) \cdot \$1,113.02 \cong \$1,174.24$ by (1) and (4),

$$\vdots \qquad \vdots \qquad \qquad \qquad \vdots$$

(22) $A_{20} = 1.055 \cdot A_{19} \cong (1.055) \cdot \$2,765.65 \cong \$2,917.76$ by (1) and (21),

(23) $A_{21} = 1.055 \cdot A_{20} \cong (1.055) \cdot \$2,917.76 \cong \$3,078.23$ by (1) and (22).

Thus the amount in the account is \$3,078.23 (to the nearest cent). Fill in the dots (to check the arithmetic) and collect your money! ∎

EXAMPLE 8.1.8 **Compound Interest with Compounding Several Times a Year**

When an annual interest rate of i is compounded m times per year, then the interest rate paid per period is i/m. For instance, if $6\% = 0.06$ annual interest is compounded quarterly, then the interest rate paid per quarter is $0.06/4 = 0.015$. For each integer $k \geq 1$, let P_k = the amount on deposit at the end of the kth period assuming no additional deposits or withdrawals. Then the interest earned during the kth period equals the amount on deposit at the end of the $(k-1)$st period times the interest rate for the period:

$$\text{interest earned during } k\text{th period} = P_{k-1} \cdot \left(\frac{i}{m}\right).$$

The amount on deposit at the end of the kth period, P_k, equals the amount at the end of the $(k-1)$st period, P_{k-1}, plus the interest earned during the kth period:

$$P_k = P_{k-1} + P_{k-1} \cdot \left(\frac{i}{m}\right) = P_{k-1}\left(1 + \frac{i}{m}\right). \qquad 8.1.6$$

Suppose \$100 is left on deposit at 6% compounded quarterly. How much will the account be worth at the end of one year assuming no additional deposits or withdrawals?

Solution For each integer $n \geq 1$, let P_n = the amount on deposit after n consecutive quarters assuming no additional deposits or withdrawals, and let P_0 be the initial \$100. Then by equation (8.1.6) with $i = 0.06$ and $m = 4$, a recurrence relation for the sequence $P_0, P_1, P_2, \ldots$ is

(1) $P_k = P_{k-1}(1 + 0.015) = (1.015) \cdot P_{k-1}$ for all integers $k \geq 1$.

The amount on deposit at the end of one year (four quarters), P_4, can be found by successive substitution:

$(2)\ P_0 = \$100,$
$(3)\ P_1 = 1.015 \cdot P_0 = (1.015) \cdot \$100 \quad = \$101.50 \quad$ by (1) and (2),
$(4)\ P_2 = 1.015 \cdot P_1 = (1.015) \cdot \$101.50 \cong \$103.02 \quad$ by (1) and (3),
$(5)\ P_3 = 1.015 \cdot P_2 \cong (1.015) \cdot \$103.02 \cong \$104.57 \quad$ by (1) and (4),
$(6)\ P_4 = 1.015 \cdot P_3 \cong (1.015) \cdot \$104.57 \cong \$106.14 \quad$ by (1) and (5).

Hence after one year there is $106.17 (to the nearest cent) in the account.   ■

EXAMPLE 8.1.9   Number of Bit Strings with a Certain Property

a. Make a list of all bit strings of lengths 0, 1, 2, and 3 that do not contain the bit pattern 11.
b. For each integer $n \geq 0$, let

$$s_n = \left[\begin{array}{l} \text{the number of bit strings of length } n \\ \text{that do not contain the pattern 11} \end{array} \right].$$

Find $s_0, s_1, s_2,$ and s_3.
c. Find the number of bit strings of length ten that do not contain the pattern 11.

Solution   a. One way to solve this problem is to make a list of all bit strings of lengths 0, 1, 2, and 3 and to cross off all those that contain the pattern 11:

length 0: ε
length 1: 0, 1
length 2: 00, 01, 10, 11̶
length 3: 000, 001, 010, 0̶1̶1̶, 100, 101, 1̶1̶0̶, 1̶1̶1̶

b. Counting the number of strings of each length listed in part (a) gives

$$s_0 = 1, \quad s_1 = 2, \quad s_2 = 3, \quad \text{and} \quad s_3 = 5.$$

c. To find the number of strings of length ten that do not contain the pattern 11, you could list all $2^{10} = 1,024$ strings of length ten and cross off those that contain the pattern 11, as was done in part (a). However, this approach would be very time-consuming. A more efficient solution uses recursion.

Suppose the number of bit strings of length less than some integer k that do not contain the pattern 11 is known. To use recursion to find the number of bit strings of length k that do not contain the pattern 11, you have to describe strings that do not contain the pattern 11 in terms of shorter strings that do not contain the pattern 11.

Consider the set of all bit strings of length k that do not contain the pattern 11. Any string in the set begins with either a 0 or a 1. If the string begins with a 0, the remaining $k - 1$ characters can be any sequence of 0's and 1's except that the pattern 11 cannot appear. If the string begins with a 1, then the second character must be a 0, for otherwise the string would contain the pattern 11; the remaining $k - 2$ characters can be any sequence of 0's and 1's that does not contain the pattern 11. Thus the set of all bit strings of length k that do not contain the pattern 11 can be partitioned into two mutually disjoint subsets as shown in Figure 8.1.3.

Set of all bit strings of length k that do not contain the pattern 11

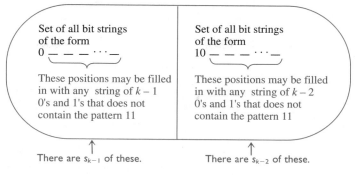

FIGURE 8.1.3   Partition of a Set of Bit Strings

By the addition rule, the number of elements in the entire set equals the sum of the numbers of elements in the two disjoint subsets:

$$\begin{bmatrix} \text{the number of} \\ \text{bit strings of} \\ \text{length } k \text{ that} \\ \text{do not contain} \\ \text{the pattern 11} \end{bmatrix} = \begin{bmatrix} \text{the number of} \\ \text{bit strings of} \\ \text{length } k - 1 \text{ that} \\ \text{do not contain} \\ \text{the pattern 11} \end{bmatrix} + \begin{bmatrix} \text{the number of} \\ \text{bit strings of} \\ \text{length } k - 2 \text{ that} \\ \text{do not contain} \\ \text{the pattern 11} \end{bmatrix}. \qquad \textbf{8.1.7}$$

So by definition of $s_0, s_1, s_2, \ldots$, for all integers $k \geq 2$,

$$(1)\ s_k = s_{k-1} + s_{k-2} \qquad \text{recurrence relation}$$

and by part (b)

$$(2)\ s_0 = 1 \quad s_1 = 2 \qquad \text{initial conditions.}$$

It follows that

$$(3)\ s_2 = s_1 + s_0 = 2 + 1 = 3 \qquad \text{by (1) and (2),}$$
$$(4)\ s_3 = s_2 + s_1 = 3 + 2 = 5 \qquad \text{by (1), (2), and (3),}$$
$$(5)\ s_4 = s_3 + s_2 = 5 + 3 = 8 \qquad \text{by (1), (3), and (4),}$$
$$\vdots \qquad \vdots$$
$$(10)\ s_{10} = s_9 + s_8 = 89 + 55 = 144 \qquad \text{by (1), (8), and (9).}$$

Hence there are 144 bit strings of length ten that do not contain the bit pattern 11.

Note that, because of the similarity in the defining relations, the sequence $s_0, s_1, s_2, \ldots$ has almost the same values as the Fibonacci sequence. ■

THE NUMBER OF PARTITIONS OF A SET INTO **r** SUBSETS

In an ordinary (or *singly indexed*) sequence, integers n are associated to numbers a_n. In a *doubly indexed* sequence, ordered pairs of integers (m, n) are associated to numbers $a_{m, n}$. For example, combinations can be thought of as terms of the doubly indexed sequence defined by $C_{n, r} = \binom{n}{r}$ for all integers n and r with $0 \leq r \leq n$.

An important example of a doubly indexed sequence is the sequence of *Stirling numbers of the second kind*. These numbers, named after the English mathematician James Stirling, arise in a surprisingly large variety of counting problems. They are defined recursively and can be interpreted in terms of partitions of a set.

Observe that if a set of three elements $\{x_1, x_2, x_3\}$ is partitioned into two subsets, then one of the subsets has one element and the other has two elements. Therefore, there are three ways the set can be partitioned:

$$\{x_1, x_2\}\{x_3\} \quad \text{put } x_3 \text{ by itself}$$
$$\{x_1, x_3\}\{x_2\} \quad \text{put } x_2 \text{ by itself}$$
$$\{x_2, x_3\}\{x_1\} \quad \text{put } x_1 \text{ by itself.}$$

In general, let

> $S_{n,r}$ = the number of ways a set of size n can be partitioned into r subsets

Then, by the above, $S_{3,2} = 3$. The numbers $S_{n,r}$ are called **Stirling numbers of the second kind.**

EXAMPLE 8.1.10   Values of Stirling Numbers

Find $S_{4,1}$, $S_{4,2}$, $S_{4,3}$, and $S_{4,4}$.

Solution   Given a set with four elements, denote it by $\{x_1, x_2, x_3, x_4\}$. The Stirling number $S_{4,1} = 1$ because a set of four elements can be partitioned into one subset in only one way:

$$\{x_1, x_2, x_3, x_4\}.$$

Similarly, $S_{4,4} = 1$ because there is only one way to partition a set of four elements into four subsets:

$$\{x_1\}\{x_2\}\{x_3\}\{x_4\}.$$

The number $S_{4,2} = 7$. The reason is that any partition of $\{x_1, x_2, x_3, x_4\}$ into two subsets must consist either of two subsets of size two or of one subset of size three and one subset of size one. The partitions for which both subsets have size two must pair x_1 with either x_2 or x_3 or x_4, which gives rise to the three partitions listed below:

$$\{x_1, x_2\}\{x_3, x_4\} \quad x_2 \text{ paired with } x_1$$
$$\{x_1, x_3\}\{x_2, x_4\} \quad x_3 \text{ paired with } x_1$$
$$\{x_1, x_4\}\{x_2, x_3\} \quad x_4 \text{ paired with } x_1.$$

The partitions for which one subset has size one and the other has size three can have any one of the four elements in the subset of size one, which leads to these four partitions:

$$\{x_1\}\{x_2, x_3, x_4\} \quad x_1 \text{ by itself}$$
$$\{x_2\}\{x_1, x_3, x_4\} \quad x_2 \text{ by itself}$$
$$\{x_3\}\{x_1, x_2, x_4\} \quad x_3 \text{ by itself}$$
$$\{x_4\}\{x_1, x_2, x_3\} \quad x_4 \text{ by itself}$$

It follows that the total number of ways the set $\{x_1, x_2, x_3, x_4\}$ can be partitioned into two subsets is $3 + 4 = 7$.

Finally, $S_{4,3} = 6$ because any partition of a set of four elements into three subsets must have two elements in one subset and the other two elements in subsets by themselves. There are $\binom{4}{2} = 6$ ways to choose the two elements to put together, which results in the following six possible partitions.

$$\{x_1, x_2\}\{x_3\}\{x_4\} \qquad \{x_2, x_3\}\{x_1\}\{x_4\}$$
$$\{x_1, x_3\}\{x_2\}\{x_4\} \qquad \{x_2, x_4\}\{x_1\}\{x_3\}$$
$$\{x_1, x_4\}\{x_2\}\{x_3\} \qquad \{x_3, x_4\}\{x_1\}\{x_2\}$$ ∎

EXAMPLE 8.1.11   Finding a Recurrence Relation for $S_{n,r}$

Find a recurrence relation relating $S_{n,r}$ to values of the sequence with lower indices than n and r and give initial conditions for the recursion.

Solution   To solve this problem recursively, suppose a procedure has been found to count both the number of ways to partition a set of $n - 1$ elements into $r - 1$ subsets and the number of ways to partition a set of $n - 1$ elements into r subsets. The partitions of a set of n elements $\{x_1, x_2, \ldots, x_n\}$ into r subsets can be divided, as shown in Figure 8.1.4, into those that contain the set $\{x_n\}$ and those that do not.

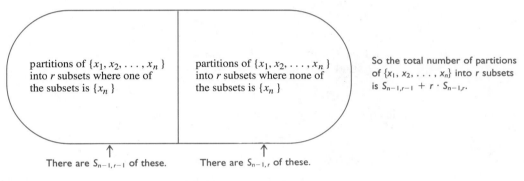

Partitions of $\{x_1, x_2, \ldots, x_n\}$ into r subsets

partitions of $\{x_1, x_2, \ldots, x_n\}$ into r subsets where one of the subsets is $\{x_n\}$

partitions of $\{x_1, x_2, \ldots, x_n\}$ into r subsets where none of the subsets is $\{x_n\}$

So the total number of partitions of $\{x_1, x_2, \ldots, x_n\}$ into r subsets is $S_{n-1,r-1} + r \cdot S_{n-1,r}$.

There are $S_{n-1,r-1}$ of these.   There are $S_{n-1,r}$ of these.

FIGURE 8.1.4

To obtain the result shown in Figure 8.1.4, first count the number of partitions of $\{x_1, x_2, \ldots, x_n\}$ into r subsets where one of the subsets is $\{x_n\}$. To do this, imagine taking any one of the $S_{n-1,r-1}$ partitions of $\{x_1, x_2, \ldots, x_{n-1}\}$ into $r - 1$ subsets and adding the subset $\{x_n\}$ to the partition. (For example, if $n = 4$ and $r = 3$, you would take one of the three partitions of $\{x_1, x_2, x_3\}$ into two subsets, namely

$$\{x_1, x_2\}\{x_3\}, \quad \{x_1, x_3\}\{x_2\}, \quad \text{or} \quad \{x_2, x_3\}\{x_1\},$$

and add $\{x_4\}$. The result would be one of the partitions

$$\{x_1, x_2\}\{x_3\}\{x_4\}, \quad \{x_1, x_3\}\{x_2\}\{x_4\}, \quad \text{or} \quad \{x_2, x_3\}\{x_1\}\{x_4\}.)$$

Clearly, any partition of $\{x_1, x_2, \ldots, x_n\}$ into r subsets with $\{x_n\}$ as one of the subsets can be obtained in this way. Hence $S_{n-1,r-1}$ is the number of partitions of $\{x_1, x_2, \ldots, x_n\}$ into r subsets of which one is $\{x_n\}$.

Next, count the number of partitions of $\{x_1, x_2, \ldots, x_n\}$ into r subsets where $\{x_n\}$ is *not* one of the subsets of the partition. Imagine taking any one of the $S_{n-1,r}$ partitions of $\{x_1, x_2, \ldots, x_{n-1}\}$ into r subsets. Now imagine choosing one of the r subsets of the partition and adding in the element x_n. The result is a partition of $\{x_1, x_2, \ldots, x_n\}$ into r subsets none of which is the singleton subset $\{x_n\}$. Since the element x_n could have been added to any one of the r subsets of the partition, it follows from the multiplication rule that there are $r \cdot S_{n-1,r}$ partitions of this type. (For instance, if $n = 4$ and $r = 3$, you would take the (unique) partition of $\{x_1, x_2, x_3\}$ into three subsets, namely $\{x_1\}\{x_2\}\{x_3\}$, and add x_4 to one of these sets. The result would be one of the partitions

$$\{x_1, x_4\}\{x_2\}\{x_3\}, \quad \{x_1\}\{x_2, x_4\}\{x_3\}, \quad or \quad \{x_1\}\{x_2\}\{x_3, x_4\}.)$$

x_4 is added to $\{x_1\}$ x_4 is added to $\{x_2\}$ x_4 is added to $\{x_3\}$

Clearly, any partition of $\{x_1, x_2, \ldots, x_n\}$ into r subsets, none of which is $\{x_n\}$, can be obtained in the way described above, for when x_n is removed from whatever subset contains it in such a partition, the result is a partition of $\{x_1, x_2, \ldots, x_{n-1}\}$ into r subsets. Hence $r \cdot S_{n-1, r}$ is the number of partitions of $\{x_1, x_2, \ldots, x_n\}$ that do not contain $\{x_n\}$.

Since any partition of $\{x_1, x_2, \ldots, x_n\}$ either contains $\{x_n\}$ or not,

$$\begin{bmatrix} \text{the number of partitions} \\ \text{of } \{x_1, x_2, \ldots, x_n\} \\ \text{into } r \text{ subsets} \end{bmatrix} = \begin{bmatrix} \text{the number of partitions of} \\ \{x_1, x_2, \ldots, x_n\} \text{ into } r \text{ subsets} \\ \text{of which } \{x_n\} \text{ is one} \end{bmatrix}$$

$$+ \begin{bmatrix} \text{the number of partitions of} \\ \{x_1, x_2, \ldots, x_n\} \text{ into } r \text{ subsets} \\ \text{none of which is } \{x_n\} \end{bmatrix}$$

Thus

$$S_{n, r} = S_{n-1, r-1} + r \cdot S_{n-1, r}$$

for all integers n and r with $1 \le r \le n$.

The initial conditions for the recurrence relation are

$$S_{n, 1} = 1 \quad and \quad S_{n, n} = 1 \quad \text{for all integers } n \ge 1$$

because there is only one way to partition $\{x_1, x_2, \ldots, x_n\}$ into one subset, namely

$$\{x_1, x_2, \ldots, x_n\},$$

and only one way to partition $\{x_1, x_2, \ldots, x_n\}$ into n subsets, namely

$$\{x_1\}\{x_2\}, \ldots, \{x_n\}. \quad ■$$

EXERCISE SET 8.1*

Find the first four terms of each of the recursively defined sequences in 1–8.

1. $a_k = 2 \cdot a_{k-1} + k$, for all integers $k \ge 2$
   $a_1 = 1$

2. $b_k = b_{k-1} + 2 \cdot k$, for all integers $k \ge 2$
   $b_1 = 2$

3. $c_k = k \cdot (c_{k-1})^2$, for all integers $k \ge 1$
   $c_0 = 1$

4. $d_k = k \cdot (d_{k-1})^2$, for all integers $k \ge 1$
   $d_0 = 2$

5. $s_k = s_{k-1} + 2s_{k-2}$, for all integers $k \ge 2$
   $s_0 = 1, s_1 = 1$

6. $t_k = t_{k-1} + 2t_{k-2}$, for all integers $k \ge 2$
   $t_0 = -1, t_1 = 1$

7. $u_k = k \cdot u_{k-1} - u_{k-2}$, for all integers $k \ge 3$
   $u_1 = 1, u_2 = 1$

8. $v_k = v_{k-1} + v_{k-2} + 1$, for all integers $k \ge 3$
   $v_1 = 1, v_2 = 2$

9. Define a sequence $a_0, a_1, a_2, \ldots$ by the formula $a_n = 3n + 1$, for all integers $n \ge 0$. Show that this sequence satisfies the recurrence relation $a_k = a_{k-1} + 3$, for all integers $k \ge 1$.

10. Define a sequence $b_0, b_1, b_2, \ldots$ by the formula $b_n = 5^n$, for all integers $n \ge 0$. Show that this sequence satisfies the recurrence relation $b_k = 5b_{k-1}$, for all integers $k \ge 1$.

11. Show that the sequence $0, 1, 3, 7, \ldots, 2^n - 1, \ldots,$ for $n \ge 0$, satisfies the recurrence relation

$$c_k = 2c_{k-1} + 1, \quad \text{for all integers } k \ge 1.$$

---

*Exercises with *blue* numbers have solutions in Appendix B. The symbol H indicates that only a hint or partial solution is given. The symbol ◆ signals that an exercise is more challenging than usual.

12. Show that the sequence $1, -1, \dfrac{1}{2}, \dfrac{-1}{3}, \ldots,$ $\dfrac{(-1)^n}{n!}, \ldots,$ for $n \geq 0$, satisfies the recurrence relation

$$s_k = \frac{-s_{k-1}}{k}, \text{ for all integers } k \geq 1.$$

13. Show that the sequence $2, 3, 4, 5, \ldots, 2 + n, \ldots,$ for $n \geq 0$, satisfies the recurrence relation

$$t_k = 2t_{k-1} - t_{k-2}, \text{ for all integers } k \geq 2.$$

14. Show that the sequence $0, 1, 3, 7, \ldots, 2^n - 1, \ldots,$ for $n \geq 0$, satisifies the recurrence relation

$$d_k = 3d_{k-1} - 2d_{k-2}, \text{ for all integers } k \geq 2.$$

15. Define a sequence $a_0, a_1, a_2, \ldots$ by the formula

$$a_n = (-2)^{\lfloor n/2 \rfloor} = \begin{cases} (-2)^{n/2} & \text{if } n \text{ is even} \\ (-2)^{(n-1)/2} & \text{if } n \text{ is odd} \end{cases}$$

for all integers $n \geq 0$. Show that this sequence satisfies the recurrence relation $a_k = -2a_{k-2}$, for all integers $k \geq 2$.

16. The sequence of Catalan numbers was defined in Exercise Set 6.6 by the formula $C_n = \dfrac{1}{n+1}\dbinom{2n}{n}$, for each integer $n \geq 1$. Show that this sequence satisfies the recurrence relation $C_k = \dfrac{4k-2}{k+1}C_{k-1}$, for all integers $k \geq 2$.

17. Use the recurrence relation and values for the Tower of Hanoi sequence $m_1, m_2, m_3, \ldots$ discussed in Example 8.1.5 to compute m_7 and m_8.

18. Suppose that in addition to the requirement that they never move a larger disk on top of a smaller one, the priests who move the disks of the Tower of Hanoi are also only allowed to move disks one by one from one pole to an *adjacent* pole. Let

$$a_n = \begin{bmatrix} \text{the minimum number of moves} \\ \text{needed to transfer a tower of } n \\ \text{disks from pole } A \text{ to pole } C \end{bmatrix}.$$

a. Find a_1, a_2, and a_3.
b. Find a_4.
c. Find a recurrence relation for $a_1, a_2, a_3, \ldots$.

19. Suppose the same situation as in exercise 18. Let

$$b_n = \begin{bmatrix} \text{the minimum number of moves} \\ \text{needed to transfer a tower of } n \\ \text{disks from pole } A \text{ to pole } B \end{bmatrix}.$$

a. Find b_1, b_2, and b_3.
b. Find b_4.

c. Show that $b_k = a_{k-1} + 1 + b_{k-1}$ for all integers $k \geq 2$, where $a_1, a_2, a_3, \ldots$ is the sequence defined in exercise 18.
d. Show that $b_k \leq 3b_{k-1} + 1$ for all integers $k \geq 2$.

20. Suppose that the Tower of Hanoi problem has four poles in a row instead of three. A disk can be transferred from one pole to any other pole, but at no time may a larger disk be placed on top of a smaller disk. Let s_n be the minimum number of moves needed to transfer the entire tower of n disks from the left-most to the right-most pole.
a. Find s_1, s_2, and s_3.
b. Find s_4.
c. Show that $s_k \leq 2s_{k-2} + 3$ for all integers $k \geq 3$.

21. In a Double Tower of Hanoi there are three poles in a row and $2n$ disks, two of each of n different sizes, where n is any positive integer. Initially one of the poles contains all the disks placed on top of each other in pairs of decreasing size. Disks are transferred one by one from one pole to another, but at no time may a larger disk be placed on top of a smaller disk. However, a disk may be placed on top of one of the same size. Let t_n be the minimum number of moves needed to transfer a tower of $2n$ disks from one pole to another.
a. Find t_1 and t_2.
b. Find t_3.
c. Find a recurrence relation for $t_1, t_2, t_3, \ldots$.

22. Use the recurrence relation and values for the Fibonacci sequence $F_0, F_1, F_2, \ldots$ given in Example 8.1.6 to compute F_{13} and F_{14}.

23. The Fibonacci sequence satisfies the recurrence relation $F_k = F_{k-1} + F_{k-2}$, for all integers $k \geq 2$.
a. Explain why the following is true:

$$F_{k+1} = F_k + F_{k-1}, \text{ for all integers } k \geq 1.$$

b. Write an equation expressing F_{k+2} in terms of F_{k+1} and F_k.
c. Write an equation expressing F_{k+3} in terms of F_{k+2} and F_{k+1}.

24. Prove each of the following for the Fibonacci sequence $F_0, F_1, F_2, \ldots$.
a. $F_k^2 - F_{k-1}^2 = F_k \cdot F_{k+1} - F_{k+1} \cdot F_{k-1}$, for all integers $k \geq 1$.
b. $F_{k+1}^2 - F_k^2 - F_{k-1}^2 = 2 \cdot F_k \cdot F_{k-1}$, for all integers $k \geq 1$.
c. $F_{k+1}^2 - F_k^2 = F_{k-1} \cdot F_{k+2}$, for all integers $k \geq 1$.
d. $F_{n+2}F_n - F_{n+1}^2 = (-1)^n$, for all integers $n \geq 0$. (Use mathematical induction.)

25. (For students who have studied calculus) Find $\lim_{n \to \infty} \dfrac{F_{n+1}}{F_n}$ where $F_1, F_2, F_3, \ldots$ is the Fibonacci sequence. (Assume that the limit exists.)

26. (For students who have studied calculus) Define x_0, x_1, $x_2, \ldots$ as follows:

$$x_k = \sqrt{2 + x_{k-1}}, \text{ for all integers } k \geq 1$$
$$x_0 = 0.$$

Find $\lim_{n \to \infty} x_n$. Assume that the limit exists.

27. A single pair of rabbits (male and female) is born at the beginning of a year. Assume the following conditions (which are more realistic than Fibonacci's):
    (1) Rabbit pairs are not fertile during their first month of life, but thereafter give birth to four new male/female pairs at the end of every month;
    (2) No rabbits die.
    a. Let r_n = the number of pairs of rabbits alive at the end of month n, for each integer $n \geq 1$, and let $r_0 = 1$. Find a recurrence relation for $r_0, r_1, r_2, \ldots$.
    b. Compute r_0, r_1, r_2, r_3, r_4, r_5, and r_6.
    c. How many rabbits will there be at the end of the year?

28. A single pair of rabbits (male and female) is born at the beginning of a year. Assume the following conditions:
    (1) Rabbit pairs are not fertile during their first *two* months of life, but thereafter give birth to three new male/female pairs at the end of every month;
    (2) No rabbits die.
    a. Let s_n = the number of pairs of rabbits alive at the end of month n, for each integer $n \geq 1$, and let $s_0 = 1$. Find a recurrence relation for $s_0, s_1, s_2, \ldots$.
    b. Compute s_0, s_1, s_2, s_3, s_4, and s_5.
    c. How many rabbits will there be at the end of the year?

29. Suppose a certain amount of money is deposited in an account paying 8% annual interest compounded quarterly. For each positive integer n, let R_n = the amount on deposit at the end of the nth quarter assuming no additional deposits or withdrawals, and let R_0 be the initial amount deposited.
    a. Find a recurrence relation for $R_0, R_1, R_2, \ldots$.
    b. If $R_0 =$ \$500, find the amount of money on deposit at the end of one year.

30. Suppose a certain amount of money is deposited in an account paying 6% annual interest compounded monthly. For each positive integer n, let S_n = the amount on deposit at the end of the nth month, and let S_0 be the initial amount deposited.
    a. Find a recurrence relation for $S_0, S_1, S_2, \ldots$, assuming no additional deposits or withdrawals during the year.
    b. If $S_0 =$ \$1,000, find the amount of money on deposit at the end of one year.

31. a. Make a list of all bit strings of lengths zero, one, two, three, and four that do not contain the bit pattern 111.

b. For each integer $n \geq 0$, let d_n = the number of bit strings of length n that do not contain the bit pattern 111. Find d_0, d_1, d_2, d_3, and d_4.
c. Find a reccurrence relation for $d_0, d_1, d_2, \ldots$.
d. Use the results of parts (b) and (c) to find the number of bit strings of length five that do not contain the pattern 111.

32. a. Let $\Sigma = \{a, b, c\}$. Then Σ^* is the set of all strings over Σ. Make a list of all strings over Σ of lengths zero, one, two, and three that do not contain the pattern aa.
    b. For each integer $n \geq 0$, let s_n = the number of strings over Σ of length n that do not contain the pattern aa. Find s_0, s_1, s_2, and s_3.
    H c. Find a recurrence relation for $s_0, s_1, s_2, \ldots$.
    d. Use the results of parts (b) and (c) to find the number of strings over Σ of length four that do not contain the pattern aa.

33. For each integer $n \geq 0$, let a_n be the number of bit strings of length n that do not contain the pattern 101.
    a. Show that $a_k = a_{k-1} + a_{k-3} + \cdots + a_1 + 3$, for all integers $k \geq 3$.
    b. Use the result of part (a) to show that if $k \geq 4$, then $a_k = 2a_{k-1} - a_{k-2} + a_{k-3}$.

34. With each step you take when climbing a staircase, you can move up either one stair or two stairs. As a result, you can climb the entire staircase taking one stair at a time, taking two at a time, or taking any combination of one- and two-stair increments. For each integer $n \geq 1$, if the staircase consists of n stairs, let c_n be the number of different ways to climb the staircase. Find a recurrence relation for $c_1, c_2, c_3, \ldots$.

35. A set of blocks contains blocks of heights 1, 2, and 4 inches. Imagine constructing towers by piling blocks of different heights directly on top of one another. (A tower of height 6 inches could be obtained using six 1-inch blocks, three 2-inch blocks, one 2-inch block with one 4-inch block on top, one 4-inch block with one 2-inch block on top, and so forth.) Let t_n be the number of ways to construct a tower of height n inches using blocks from the set. (Assume an infinite supply of blocks of each size.) Find a recurrence relation for $t_1, t_2, t_3 \ldots$.

36. For each integer $n \geq 2$ let a_n be the number of permutations of $\{1, 2, 3, \ldots, n\}$ in which no number is more than one place removed from its "natural" position. Thus $a_1 = 1$ since the one permutation of $\{1\}$, namely 1, does not move 1 from its natural position. Also $a_2 = 2$ since neither of the two permutations of $\{1, 2\}$, namely 12 and 21, moves either number more than one place from its natural position.
    a. Find a_3.
    b. Find a recurrence relation for $a_1, a_2, a_3, \ldots$.

◆37. A row in a classroom has n seats. Let s_n be the number of ways nonempty sets of students can sit in the row so that no student is seated directly adjacent to any other student. (For instance, a row of three seats could contain a single student in any of the seats or a pair of students in the two outer seats. So $s_3 = 4$.) Find a recurrence relation for $s_1, s_2, s_3, \ldots$.

◆38. Let P_n be the number of partitions of a set with n elements. Show that

$$P_n = \binom{n-1}{0}P_{n-1} + \binom{n-1}{1}P_{n-2} + \cdots + \binom{n-1}{n-1}P_0$$

for all integers $n \geq 1$.

Exercises 39–44 refer to the sequence of Stirling numbers of the second kind.

39. Find $S_{5,4}$ by exhibiting all the partitions of $\{x_1, x_2, x_3, x_4, x_5\}$ into four subsets.

40. Use the values computed in Example 8.1.10 and the recurrence relation and initial conditions found in Example 8.1.11 to compute $S_{5,2}$.

41. Use the values computed in Example 8.1.10 and the recurrence relation and initial conditions found in Example 8.1.11 to compute $S_{5,3}$.

42. Find the total number of different partitions of a set with five elements.

43. Use mathematical induction and the recurrence relation found in Example 8.1.11 to prove that for all integers $n \geq 2$, $S_{n,2} = 2^{n-1} - 1$.

44. Use mathematical induction and the recurrence relation found in Example 8.1.11 to prove that for all integers $n \geq 2$, $\sum_{k=1}^{n} 3^{n-k} S_{k,2} = S_{n+1,3}$.

H45. If X is a set with n elements and Y is a set with m elements, express the number of onto functions from X and Y using Stirling numbers of the second kind. Justify your answer.

In 46 and 47, assume that $F_0, F_1, F_2, \ldots$ is the Fibonacci sequence.

◆46. Use strong mathematical induction to prove that $F_n < 2^n$ for all integers $n \geq 1$.

H◆47. Prove that for all integers $n \geq 0$, $\gcd(F_{n+1}, F_n) = 1$.

48. A gambler decides to play successive games of black jack until he loses three times in a row. (Thus the gambler could play five games by losing the first, winning the second, and losing the final three or by winning the first two and losing the final three. These possibilities can be symbolized as $LWLLL$ and $WWLLL$.) Let g_n be the number of ways the gambler can play n games.
   a. Find g_3, g_4, and g_5.
   b. Find g_6.
   Hc. Find a recurrence relation for $g_3, g_4, g_5, \ldots$.

◆49. A *derangement* of the set $\{1, 2, \ldots, n\}$ is a permutation that moves every element of the set away from its "natural" position. Thus 21 is a derangement of $\{1, 2\}$ and 231 and 312 are derangements of $\{1, 2, 3\}$. For each positive integer n, let d_n be the number of derangements of the set $\{1, 2, \ldots, n\}$.
   a. Find d_1, d_2, and d_3.
   b. Find d_4.
   Hc. Find a recurrence relation for $d_1, d_2, d_3, \ldots$.

50. Note that a product $x_1 \cdot x_2 \cdot x_3$ may be parenthesized in two different ways: $(x_1 \cdot x_2) \cdot x_3$ and $x_1 \cdot (x_2 \cdot x_3)$. Similarly, there are several different ways to parenthesize $x_1 \cdot x_2 \cdot x_3 \cdot x_4$. Two such ways are $(x_1 \cdot x_2) \cdot (x_3 \cdot x_4)$ and $x_1 \cdot ((x_2 \cdot x_3) \cdot x_4)$. Let P_n be the number of different ways to parenthesize the product $x_1 \cdot x_2 \cdot \ldots \cdot x_n$. Show that if $P_1 = 1$, then

$$P_n = \sum_{k=1}^{n-1} P_k P_{n-k}, \text{ for all integers } n \geq 2.$$

(It turns out that the sequence $P_1, P_2, P_3, \ldots$, is the same as the sequence of Catalan numbers.)

8.2 SOLVING RECURRENCE RELATIONS BY ITERATION

The keener one's sense of logical deduction, the less often one makes hard and fast inferences.
(Bertrand Russell, 1872–1970)

Suppose you have a sequence that satisfies a certain recurrence relation and initial conditions. It is often helpful to know an explicit formula for the sequence, especially if you need to compute terms with very large subscripts or if you need to examine general properties of the sequence. Such an explicit formula is called a **solution** to the recurrence relation. In this section and the next, we discuss methods for solving recurrence relations. In the text and exercises of this section, we will show that the Tower of Hanoi sequence of Example 8.1.5 satisfies the formula

$$m_n = 2^n - 1,$$

and the compound interest sequence of Example 8.1.7 satisfies

$$A_n = (1.055)^n \cdot \$1,000.$$

In Section 8.3 we will show that the Fibonacci sequence of Example 8.1.6 satisfies the formula

$$F_n = \frac{1}{\sqrt{5}}\left[\left(\frac{1+\sqrt{5}}{2}\right)^{n+1} - \left(\frac{1-\sqrt{5}}{2}\right)^{n+1}\right].$$

THE METHOD OF ITERATION

The most basic method for finding an explicit formula for a recursively defined sequence is **iteration**. Iteration works as follows: Given a sequence $a_0, a_1, a_2, \ldots$ defined by a recurrence relation and initial conditions, you start from the initial conditions and calculate successive terms of the sequence until you see a pattern developing. At that point you guess an explicit formula.

EXAMPLE 8.2.1   Finding an Explicit Formula

Let $a_0, a_1, a_2, \ldots$ be the sequence defined recursively as follows: For all integers $k \geq 1$,

(1) $a_k = a_{k-1} + 2$ recurrence relation
(2) $a_0 = 1$ initial condition.

Use iteration to guess an explicit formula for the sequence.

Solution   Recall that to say

$$a_k = a_{k-1} + 2 \quad \text{for all integers } k \geq 1$$

means

$$a_\square = a_{\square-1} + 2 \quad \text{no matter what positive integer is placed into the box } \square.$$

In particular,

$$a_1 = a_0 + 2,$$
$$a_2 = a_1 + 2,$$
$$a_3 = a_2 + 2,$$

and so forth. Now use the initial condition to begin a process of successive substitutions into these equations, not just of numbers (as was done in Section 8.1) but of *numerical expressions*.

The reason for using numerical expressions rather than numbers is that in these problems you are seeking a numerical pattern that underlies a general formula. The secret of success is to leave most of the arithmetic undone. However, you do need to eliminate parentheses as you go from one step to the next. Otherwise, you will soon end up with a bewilderingly large nest of parentheses. Also, it is almost always helpful to use shorthand notations for regrouping additions, subtractions, and multiplications. Thus, for instance, you would write

$$5 \cdot 2 \quad \text{instead of} \quad 2 + 2 + 2 + 2 + 2$$

and

$$2^5 \quad \text{instead of} \quad 2 \cdot 2 \cdot 2 \cdot 2 \cdot 2.$$

Notice that you don't lose any information about the number patterns when you use these shorthand notations.

Here's how the process works for the given sequence:

$a_0 = 1$ the initial condition

$a_1 = a_0 + 2 = 1 + 2$ by substitution

$a_2 = a_1 + 2 = (1 + 2) + 2 \qquad\qquad = 1 + 2 + 2$ eliminate parentheses

$a_3 = a_2 + 2 = (1 + 2 + 2) + 2 \qquad = 1 + 2 + 2 + 2$ eliminate parentheses again; write $3 \cdot 2$ instead of $2 + 2 + 2$?

$a_4 = a_3 + 2 = (1 + 2 + 2 + 2) + 2 = 1 + 2 + 2 + 2 + 2$ eliminate parentheses again; definitely write $4 \cdot 2$ instead of $2 + 2 + 2 + 2$—the length of the string of 2's is getting out of hand.

Since it appears helpful to use the shorthand $k \cdot 2$ in place of $2 + 2 + \cdots + 2$ (k times), we do so, starting again from a_0.

$a_0 = 1 \qquad\qquad\qquad\qquad = 1 + 0 \cdot 2$ the initial condition

$a_1 = a_0 + 2 = 1 + 2 \qquad\qquad = 1 + 1 \cdot 2$ by substitution

$a_2 = a_1 + 2 = (1 + 2) + 2 \qquad = 1 + 2 \cdot 2$

$a_3 = a_2 + 2 = (1 + 2 \cdot 2) + 2 = 1 + 3 \cdot 2$

$a_4 = a_3 + 2 = (1 + 3 \cdot 2) + 2 = 1 + 4 \cdot 2$ At this point it certainly seems likely that the general pattern is $1 + n \cdot 2$; check whether the next calculation supports this.

$a_5 = a_4 + 2 = (1 + 4 \cdot 2) + 2 = 1 + 5 \cdot 2$ It does! So go ahead and write an answer. It's only a guess after all.

Guess: $a_n = 1 + n \cdot 2 = 1 + 2n$

The answer obtained for this problem is just a guess. To be sure of the correctness of this guess, you will need to check it by mathematical induction. Later in this section, we will show how to do this.   ■

A sequence like the one in Example 8.2.1, in which each term equals the previous term plus a fixed constant, is called an *arithmetic sequence*. In the exercises at the end of this section you are asked to show that the $(n + 1)$st term of an arithmetic sequence always equals the initial value of the sequence plus n times the fixed constant.

DEFINITION

A sequence $a_0, a_1, a_2, \ldots$ is called an **arithmetic sequence** if, and only if, there is a constant d such that

$$a_k = a_{k-1} + d \quad \text{for all integers } k \geq 1.$$

Or, equivalently,

$$a_n = a_0 + d \cdot n \quad \text{for all integers } n \geq 0.$$

EXAMPLE 8.2.2   An Arithmetic Sequence

Under the force of gravity, an object falling in a vacuum falls about 9.8 meters farther each second than it fell the second before. Thus, neglecting air resistance, a skydiver leaving an airplane falls approximately 9.8 meters between 0 and 1 seconds after departure, $9.8 + 9.8 = 19.6$ meters between 1 and 2 seconds after departure, and so forth. If air resistance is neglected, how many meters would the diver fall between 60 and 61 seconds after leaving the airplane?

Solution   Let d_n be the distance the skydiver would fall between n and $n + 1$ seconds after exiting the airplane if there were no air resistance. Thus d_0 is the distance fallen between 0 and 1 seconds after exiting, d_1 is the distance fallen between 1 and 2 seconds after exiting, and so forth. Then $d_0 = 9.8$, and since the diver would fall 9.8 meters farther each second than the second before,

$$d_k = d_{k-1} + 9.8 \text{ meters} \quad \text{for all integers } k \geq 1.$$

It follows that $d_0, d_1, d_2, \ldots$ is an arithmetic sequence with a constant adder of 9.8 and that

$$d_n = d_0 + n \cdot (9.8) \quad \text{for each integer } n \geq 0.$$

Hence between the 60th and the 61st seconds after exiting, the diver would fall

$$d_{60} = 9.8 + 60 \cdot (9.8) = 597.8 \text{ meters.}$$

Note that 597.8 meters is approximately equal to 1,961 feet or about three or four city blocks, which is a long way to fall in one second. Of course, this result was obtained by neglecting air resistance, which in fact cuts the diver's speed considerably. ∎

In an arithmetic sequence, each term equals the previous term plus a fixed constant. In a geometric sequence, each term equals the previous term *times* a fixed constant. Geometric sequences arise in a large variety of applications such as compound interest, certain models of population growth, radioactive decay, and the number of operations needed to execute certain computer algorithms.

EXAMPLE 8.2.3   The Explicit Formula for a Geometric Sequence

Let r be a fixed nonzero constant and suppose a sequence $a_0, a_1, a_2, \ldots$ is defined recursively as follows:

$$a_k = r \cdot a_{k-1} \quad \text{for all integers } k \geq 1,$$
$$a_0 = a.$$

Use iteration to guess an explicit formula for this sequence.

Solution
$$a_0 = a$$
$$a_1 = r \cdot a_0 = r \cdot \widehat{a}$$
$$a_2 = r \cdot a_1 = r \cdot (r \cdot a) = r^2 \cdot a$$
$$a_3 = r \cdot a_2 = r \cdot (r^2 \cdot a) = r^3 \cdot a$$
$$a_4 = r \cdot a_3 = r \cdot (r^3 \cdot a) = r^4 \cdot a$$
$$\vdots$$

Guess: $a_n = r^n \cdot a$ for any arbitrary integer $n \geq 0$

In the exercises at the end of this section, you are asked to prove that this formula is correct. ■

DEFINITION

A sequence $a_0, a_1, a_2, \ldots$ is called a **geometric sequence** if, and only if, there is a constant r such that

$$a_k = r \cdot a_{k-1} \quad \text{for all integers } k \geq 1.$$

Or, equivalently,

$$a_n = a_0 \cdot r^n \quad \text{for all integers } n \geq 0.$$

EXAMPLE 8.2.4   **A Geometric Sequence**

As shown in Example 8.1.7, if a bank pays interest at a rate of 5.5% per year compounded annually and A_n denotes the amount in the account at the end of year n, then $A_k = (1.055)A_{k-1}$, for all integers $k \geq 1$, assuming no deposits or withdrawals during the year. Suppose the initial amount deposited is $1,000. Assuming no additional deposits or withdrawals are made,

a. How much will the account be worth at the end of 21 years?
b. In how many years will the account be worth $10,000?

Solution
a. $A_0, A_1, A_2, \ldots$ is a geometric sequence with initial value 1,000 and constant multiplier 1.055. Hence,

$$A_n = \$1{,}000 \cdot (1.055)^n \quad \text{for all integers } n \geq 0.$$

After 21 years, the amount in the account will be

$$A_{21} = \$1{,}000 \cdot (1.055)^{21} \cong \$3{,}078.23.$$

This is the same answer as that obtained in Example 8.1.7 but computed much more easily (at least if a calculator with a powering key— $\boxed{x^y}$ —is used).

b. Let t be the number of years needed for the account to grow to $10,000. Then

$$\$10{,}000 = \$1{,}000 \cdot (1.055)^t.$$

Dividing both sides by 1,000 gives

$$10 = (1.055)^t,$$

and taking logarithms with base 10 of both sides results in

$$\log_{10} 10 = \log_{10}(1.055)^t.$$

Then

$$1 = t \log_{10}(1.055)$$

because $\log_{10} 10 = 1$ and
$\log_b(x^a) = a \log_b(x)$ (see exercise
16 of Section 7.1)

and so

$$t = \frac{1}{\log_{10} 1.055} \cong 43.$$

Hence the account will grow to $10,000 in approximately 43 years.   ■

An important property of a geometric sequence with constant multiplier greater than 1 is that its terms increase very rapidly in size as the subscripts get larger and larger. For instance, the first ten terms of a geometric sequence with a constant multiplier of 10 are

$$1, 10, 10^2, 10^3, 10^4, 10^5, 10^6, 10^7, 10^8, 10^9.$$

So by its tenth term the sequence already has the value $10^9 = 1,000,000,000 = 1$ billion. The following box indicates some quantities that are approximately equal to certain powers of 10.

$10^6 \cong$ number of bytes of memory in a personal computer

$10^7 \cong$ number of seconds in a year

$10^8 \cong$ number of bytes of memory in a large mainframe computer

$10^{11} \cong$ number of neurons in a human brain

$10^{17} \cong$ age of the universe in seconds

$10^{31} \cong$ number of seconds to process all possible positions of a checkers game if moves are processed at a rate of 1 per billionth of a second

$10^{81} \cong$ number of atoms in the universe

$10^{111} \cong$ number of seconds to process all possible positions of a chess game if moves are processed at a rate of 1 per billionth of a second

USING FORMULAS TO SIMPLIFY SOLUTIONS OBTAINED BY ITERATION

Explicit formulas obtained by iteration can often be simplified by using formulas such as those developed in Section 4.2. For instance, according to the formula for the sum of a geometric sequence with initial term 1 (Theorem 4.2.3), for each real number r except $r = 1$,

$$1 + r + r^2 + \cdots + r^n = \frac{r^{n+1} - 1}{r - 1} \quad \text{for all integers } n \geq 0.$$

And according to the formula for the sum of the first n integers (Theorem 4.2.2),

$$1 + 2 + 3 + \cdots + n = \frac{n(n + 1)}{2} \quad \text{for all integers } n \geq 1.$$

EXAMPLE 8.2.5

An Explicit Formula for the Tower of Hanoi Sequence

Recall that the Tower of Hanoi sequence $m_1, m_2, m_3, \ldots$ of Example 8.1.5 satisfies the recurrence relation

$$m_k = 2m_{k-1} + 1 \quad \text{for all integers } k \geq 2$$

and has the initial condition

$$m_1 = 1.$$

Use iteration to guess an explicit formula for this sequence, making use of formula from Section 4.2 to simplify the answer.

Solution   By iteration

$$m_1 = 1$$

$$m_② = 2m_1 + 1 = 2 \cdot 1 + 1 \qquad\qquad = 2^① + 1,$$

$$m_③ = 2m_2 + 1 = 2(2 + 1) + 1 \qquad\qquad = 2^② + 2 + 1,$$

$$m_④ = 2m_3 + 1 = 2(2^2 + 2 + 1) + 1 \qquad = 2^③ + 2^2 + 2 + 1.$$

$$m_⑤ = 2m_4 + 1 = 2(2^3 + 2^2 + 2 + 1) + 1 = 2^④ + 2^3 + 2^2 + 2 + 1.$$

These calculations show that each term up to m_5 is a sum of successive powers of 2, starting with $2^0 = 1$ and going up to 2^k, where k is 1 less than the subscript of the term. The pattern would seem to continue to higher terms because each term is obtained from the preceding one by multiplying by 2 and adding 1; multiplying by 2 raises the exponent of each component of the sum by 1, and adding 1 adds back the 1 that was lost when the previous 1 was multiplied by 2. For instance, for $n = 6$,

$$m_6 = 2m_5 + 1 = 2(2^4 + 2^3 + 2^2 + 2 + 1) + 1 = 2^5 + 2^4 + 2^3 + 2^2 + 2 + 1.$$

Thus its seems that in general

$$m_n = 2^{n-1} + 2^{n-2} + \cdots + 2^2 + 2 + 1.$$

By the formula for the sum of a geometric sequence (Theorem 4.2.3),

$$2^{n-1} + 2^{n-2} + \cdots + 2^2 + 2 + 1 = \frac{2^n - 1}{2 - 1} = 2^n - 1.$$

Hence the explicit formula seems to be

$$m_n = 2^n - 1 \quad \text{for all integers } n \geq 1. \qquad\blacksquare$$

▲ CAUTION!   *It is not true that*

$$2 \cdot (2 + 1) + 1 = 2^2 + 1 + 1 \quad \leftarrow \text{This is false.}$$

A common mistake people make when doing problems such as this is to misuse the laws of algebra. For instance, by the distributive law,

$$a \cdot (b + c) = a \cdot b + a \cdot c \quad \text{for all real numbers } a, b, \text{ and } c.$$

Thus, in particular, for a = 2, b = 2, and c = 1,

$$2 \cdot (2 + 1) = 2 \cdot 2 + 2 \cdot 1 = 2^2 + 2.$$

It follows that

$$2 \cdot (2 + 1) + 1 = (2^2 + 2) + 1 = 2^2 + 2 + 1.$$

EXAMPLE 8.2.6

Using the Formula for the Sum of the First *n* Integers

Let K_n be the picture obtained by drawing n dots (which we call *vertices*) and joining each pair of vertices by a line segment (which we call an *edge*). (In Chapter 11 we discuss these objects in a more general context.) Then K_1, K_2, K_3, and K_4 are as follows.

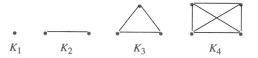

K_1 K_2 K_3 K_4

Observe that K_5 may be obtained from K_4 by adding one vertex and drawing edges between this new vertex and all the vertices of K_4 (the old vertices). The reason this procedure gives the correct result is that each pair of old vertices is already joined by an edge, and adding the new edges joins each pair of vertices consisting of an old and the new.

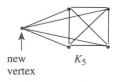

new
vertex K_5

Thus

the number of edges of $K_5 = 4 +$ the number of edges of K_4.

By the same reasoning, for all integers $k \geq 2$, the number of edges of K_k is $k - 1$ more than the number of edges of K_{k-1}. That is, if for each integer $n \geq 1$

$$s_n = \text{the number of edges of } K_n,$$

then

$$s_k = s_{k-1} + (k - 1) \quad \text{for all integers } k \geq 2.$$

Use iteration to find an explicit formula for $s_1, s_2, s_3, \ldots$.

Solution   Because

$$s_k = s_{k-1} + (k - 1) \quad \text{for } \textit{all} \text{ integers } k \geq 2$$

and

$$s_① = ⓪ \qquad | \; - \; |$$

then, in particular,

$$s_② = s_1 + 1 = \underbrace{0 + ①}; \qquad 2 \; - \; |$$

$$s_③ = s_2 + 2 = \overline{(0 + 1)} + 2 = \underbrace{0 + 1 + ②}; \qquad 3 \; - \; |$$

$$s_④ = s_3 + 3 = \overline{(0 + 1 + 2)} + 3 = \underbrace{0 + 1 + 2 + ③}, \qquad 4 \; - \; |$$

$$s_⑤ = s_4 + 4 = \overline{(0 + 1 + 2 + 3)} + 4 = 0 + 1 + 2 + 3 + ④, \qquad 5 \; - \; |$$

$$\vdots$$

Guess: $\quad s_ⓝ = 0 + 1 + 2 + \cdots + \boxed{(n - 1)}$

But by Theorem 4.2.2,

$$0 + 1 + 2 + 3 + \cdots + (n - 1) = \frac{(n - 1) \cdot n}{2} = \frac{n(n - 1)}{2}.$$

Hence it appears that

$$s_n = \frac{n(n - 1)}{2}. \qquad ■$$

CHECKING THE CORRECTNESS OF A FORMULA BY MATHEMATICAL INDUCTION

As you can see from some of the previous examples, the process of solving a recurrence relation by iteration can involve complicated calculations. It is all too easy to make a mistake and come up with the wrong formula. That is why it is important to confirm your calculations by checking the correctness of your formula. The most common way to do this is to use mathematical induction.

EXAMPLE 8.2.7 **Using Mathematical Induction to Verify the Correctness of a Solution to a Recurrence Relation**

In Example 8.2.5 we obtained a formula for the Tower of Hanoi sequence. Use mathematical induction to show that this formula is correct.

Solution What does it mean to show the correctness of a formula for a recursively defined sequence? You are given a sequence of numbers that satisfies a certain recurrence relation and initial condition. Your job is to show that each term of the sequence satisfies the proposed explicit formula. To do this, you need to prove the following statement:

If $m_1, m_2, m_3, \ldots$ is the sequence defined by

$$m_k = 2m_{k-1} + 1 \quad \text{for all integers } k \geq 2, \text{ and}$$
$$m_1 = 1,$$

then $\qquad m_n = 2^n - 1 \quad \text{for all integers } n \geq 1.$

To prove this by mathematical induction, you prove a basis step (that the formula holds for $n = 1$) and an inductive step (that if the formula holds for an integer $n = k$, then it holds for $n = k + 1$). In other words, you show that

1. $m_1 = 2^1 - 1$
2. If $m_k = 2^k - 1$, for some integer $k \geq 1$, then $m_{k+1} = 2^{k+1} - 1$.

Proof of Correctness:

The formula holds for $n = 1$:   Observe that $m_1 = 1$ by definition of the sequence $m_1, m_2, m_3, \ldots$. And $2^1 - 1 = 1$ by basic algebra. Hence $m_1 = 2^1 - 1$, and so the formula holds for $n = 1$.

If the formula holds for $n = k$ then it holds for $n = k + 1$:   Suppose that

$$m_k = 2^k - 1 \quad \text{for some integer } k \geq 1. \quad \text{This is the inductive hypothesis.}$$

We must show that $m_{k+1} = 2^{k+1} - 1$. But

$$
\begin{aligned}
m_{k+1} &= 2m_{(k+1)-1} + 1 &&\text{by definition of } m_1, m_2, m_3, \ldots \\
&= 2m_k + 1 \\
&= 2(2^k - 1) + 1 &&\text{by substitution from the inductive hypothesis} \\
&= 2^{k+1} - 2 + 1 &&\text{by the distributive law and the fact that } 2 \cdot 2^k = 2^{k+1} \\
&= 2^{k+1} - 1 &&\text{by basic algebra}
\end{aligned}
$$

[This is what was to be shown.]
[Since the basis and inductive steps have been proved, it follows by mathematical induction that the given formula holds for all integers $n \geq 1$.] ■

DISCOVERING THAT AN EXPLICIT FORMULA IS INCORRECT

The following example shows how the process of trying to verify a formula by mathematical induction may reveal a mistake.

EXAMPLE 8.2.8   Using Verification by Mathematical Induction to Find a Mistake

Let $c_0, c_1, c_2, \ldots$ be the sequence defined as follows:

$$c_k = 2c_{k-1} + k \quad \text{for all integers } k \geq 1,$$
$$c_0 = 1.$$

Suppose your calculations suggest that $c_0, c_1, c_2, \ldots$ satisfies the following explicit formula:

$$c_n = 2^n + n \quad \text{for all integers } n \geq 0.$$

Is this formula correct?

Solution   Start to prove the statement by mathematical induction and see what develops. The proposed formula passes the basis step of the inductive proof with no trouble, for on the one hand, $c_0 = 1$ by definition of $c_0, c_1, c_2, \ldots$, and on the other hand, $2^0 + 0 = 1 + 0 = 1$ also.

In the inductive step you suppose

$$c_k = 2^k + k \quad \text{for some integer } k \geq 0 \quad \text{This is the inductive hypothesis.}$$

and then you must show that

$$c_{k+1} = 2^{k+1} + (k + 1).$$

To do this, you start with c_{k+1}, substitute from the recurrence relation, and then use the inductive hypothesis as follows:

$$c_{k+1} = 2c_k + (k + 1) \qquad \text{by the recurrence relation}$$
$$= 2 \cdot (2^k + k) + (k + 1) \qquad \text{by substitution from the inductive hypothesis}$$
$$= 2^{k+1} + 3k + 1 \qquad \text{by basic algebra.}$$

To finish the verification, therefore, you need to show that

$$2^{k+1} + 3k + 1 = 2^{k+1} + (k + 1).$$

Now this equation is equivalent to

$$2k = 0 \qquad \text{by subtracting } 2^{k+1} + k + 1 \text{ from both sides,}$$

which is equivalent to

$$k = 0 \qquad \text{by dividing both sides by 2.}$$

But this is false since k may be *any* nonnegative integer. Hence the sequence $c_0, c_1, c_2, \ldots$ does not satisfy the proposed formula. ∎

Once you have found a proposed formula to be false, you should look back at your calculations to see where you made a mistake, correct it, and try again.

EXERCISE SET 8.2

1. The formula

$$1 + 2 + 3 + \cdots + n = \frac{n(n + 1)}{2}$$

is true for all integers $n \geq 1$. Use this fact to solve each of the following problems:

a. If k is an integer and $k \geq 2$, find a formula for $1 + 2 + 3 + \cdots + (k - 1)$.

b. If n is an integer and $n \geq 1$, find a formula for $3 + 2 + 4 + 6 + 8 + \cdots + 2n$.

c. If n is an integer and $n \geq 1$, find a formula for $3 + 3 \cdot 2 + 3 \cdot 3 + \cdots + 3 \cdot n + n$.

2. The formula

$$1 + r + r^2 + \cdots + r^n = \frac{r^{n+1} - 1}{r - 1}$$

is true for all real numbers r except $r = 1$ and for all integers $n \geq 0$. Use this fact to solve each of the following problems:

a. If i is an integer and $i \geq 1$, find a formula for $1 + 2 + 2^2 + \cdots + 2^{i-1}$.

b. If n is an integer and $n \geq 1$, find a formula for $3^{n-1} + 3^{n-2} + \cdots + 3^2 + 3 + 1$.

c. If n is an integer and $n \geq 2$, find a formula for $2^n + 2^{n-2} \cdot 3 + 2^{n-3} \cdot 3 + \cdots + 2^2 \cdot 3 + 2 \cdot 3 + 3$.

d. If n is an integer and $n \geq 1$, find a formula for $2^n - 2^{n-1} + 2^{n-2} - 2^{n-3} + \cdots + (-1)^{n-1} \cdot 2 + (-1)^n$.

In each of 3–12 a sequence is defined recursively. Use iteration to guess an explicit formula for the sequence. Use the formulas from Section 4.2 to simplify your answers whenever possible.

3. $a_k = k \cdot a_{k-1}$, for all integers $k \geq 1$
   $a_0 = 1$

4. $b_k = \dfrac{b_{k-1}}{1 + b_{k-1}}$, for all integers $k \geq 1$
   $b_0 = 1$

5. $c_k = 3c_{k-1} + 1$, for all integers $k \geq 2$
   $c_1 = 1$

6. $d_k = 2d_{k-1} + 3$, for all integers $k \geq 2$
   $d_1 = 2$

7. $e_k = e_{k-1} + 2k$, for all integers $k \geq 1$
   $e_0 = 3$

8. $f_k = f_{k-1} + 3k + 1$, for all integers $k \geq 1$
   $f_0 = 0$

9. $v_k = v_{k-1} + 2^k$, for all integers $k \geq 2$
   $v_1 = 1$

10. $w_k = 2^k - w_{k-1}$, for all integers $k \geq 1$
    $w_0 = 1$

◆11. $x_k = 3x_{k-1} + k$, for all integers $k \geq 2$
    $x_1 = 1$

12. $u_k = u_{k-1} + k^2$, for all integers $k \geq 2$
    $u_1 = 1$

13. Solve the recurrence relation obtained as the answer to exercise 18(c) of Section 8.1.

14. Solve the recurrence relation obtained as the answer to exercise 21(c) of Section 8.1.

15. Suppose d is a fixed constant and $a_0, a_1, a_2, \ldots$ is a sequence that satisfies the recurrence relation $a_k = a_{k-1} + d$, for all integers $k \geq 1$. Use mathematical induction to prove that $a_n = a_0 + n \cdot d$, for all integers $n \geq 0$.

16. A worker is promised a bonus if he can increase his productivity by 2 units a day for a period of 30 days. If on day 0 he produces 170 units, how many units must he produce on day 30 to qualify for the bonus?

17. A runner targets herself to improve her time on a certain course by 3 seconds a day. If on day 0 she runs the course in 3 minutes, how fast must she run it on day 14 to stay on target?

18. Suppose r is a fixed constant and $a_0, a_1, a_2, \ldots$ is a sequence that satisfies the recurrence relation $a_k = r \cdot a_{k-1}$, for all integers $k \geq 1$. Use mathematical induction to prove that $a_n = a_0 \cdot r^n$, for all integers $n \geq 0$.

19. As shown in Example 8.1.8, if a bank pays interest at a rate of i compounded m times a year, then the amount of money P_k at the end of k time periods (where one time period $= 1/m$th of a year) satisfies the recurrence relation $P_k = [1 + (i/m)]P_{k-1}$ with initial condition $P_0 = $ the initial amount deposited. Find an explicit formula for P_n.

20. Suppose the population of a country increases at a steady rate of 3% per year. If the population is 50 million at a certain time, what will it be 25 years later?

21. A chain letter works as follows: One person sends a copy of the letter to five friends, each of whom sends a copy to five friends, each of whom sends a copy to five friends, and so forth. How many people will have received copies of the letter after the twentieth repetition of this process, assuming no person receives more than one copy?

22. A certain computer algorithm executes twice as many operations when it is run with an input of size k as when it is run with an input of size $k - 1$ (where k is an integer that is greater than 1). When the algorithm is run with an input of size 1, it executes seven operations. How many operations does it execute when it is run with an input of size 25?

23. A person saving for retirement makes an initial deposit of $1,000 to a bank account earning interest at a rate of

6% per year compounded monthly, and each month she adds an additional $100 to the account.
  a. For each nonnegative integer n, let A_n be the amount in the account at the end of n months. Find a recurrence relation relating A_k to A_{k-1}.
  H b. Use iteration to find an explicit formula for A_n.
  c. Use mathematical induction to prove the correctness of the formula you obtained in part (b).
  d. How much will the account be worth at the end of 20 years? At the end of 40 years?
  H e. In how many years will the account be worth $10,000?

In 24–35 use mathematical induction to verify the correctness of the formula you obtained in the referenced exercise.

24. Exercise 3   25. Exercise 4

26. Exercise 5   27. Exercise 6

H 28. Exercise 7   29. Exercise 8

30. Exercise 9   H 31. Exercise 10

32. Exercise 11   33. Exercise 12

34. Exercise 13   35. Exercise 14

36. A sequence is defined recursively as follows:

$$v_k = v_{\lfloor k/2 \rfloor} + v_{\lfloor (k+1)/2 \rfloor} + 2 \quad \text{for all integers } k \geq 2,$$
$$v_1 = 1.$$

  a. Use iteration to guess an explicit formula for the sequence.
  b. Use strong mathematical induction to verify that the formula of part (a) is correct.

37. A sequence is defined recursively as follows:

$$s_k = 2s_{k-2}, \text{ for all integers } k \geq 2,$$
$$s_0 = 1, s_1 = 2.$$

  a. Use iteration to guess an explicit formula for the sequence.
  b. Use strong mathematical induction to check that the formula of part (a) is correct.

38. A sequence is defined recursively as follows:

$$t_k = k - t_{k-1} \text{ for all integers } k \geq 1,$$
$$t_0 = 0.$$

  a. Use iteration to guess an explicit formula for the sequence.
  b. Use strong mathematical induction to check that the formula of part (a) is correct.

◆ 39. A sequence is defined recursively as follows:

$$w_k = w_{k-2} + k, \text{ for all integers } k \geq 3,$$
$$w_1 = 1, \quad w_2 = 2.$$

H**a.** Use iteration to guess an explicit formula for the sequence.

b. Use strong mathematical induction to check that the formula of part (a) is correct.

◆ **40.** A sequence is defined recursively as follows:

$$u_k = u_{k-2} \cdot u_{k-1}, \text{ for all integers } k \geq 2,$$
$$u_0 = u_1 = 2.$$

H**a.** Use iteration to guess an explicit formula for the sequence.

b. Use strong mathematical induction to check that the formula of part (a) is correct.

In 41 and 42 determine whether the given recursively defined sequence satisfies the explicit formula

$$a_n = (n-1)^2, \text{ for all integers } n \geq 1.$$

41. $a_k = 2a_{k-1} + k - 1$, for all integers $k \geq 2$
$a_1 = 0$

42. $a_k = (a_{k-1} + 1)^2$, for all integers $k \geq 2$
$a_1 = 0$

43. A single line divides a plane into two regions. Two lines (by crossing) can divide a plane into four regions; three lines can divide it into seven regions (see the figure). Let P_n be the maximum number of regions into which n lines divide a plane, where n is a positive integer.

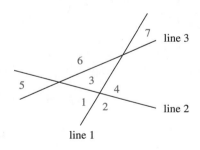

H**a.** Derive a recurrence relation for P_k in terms of P_{k-1}, for all integers $k \geq 2$.

b. Use iteration to guess an explicit formula for P_n.

H**44.** Compute $\begin{bmatrix} 1 & 1 \\ 1 & 0 \end{bmatrix}^n$ for small values of n (up to about 5 or 6). Conjecture explicit formulas for the entries in this matrix, and prove your conjecture using mathematical induction.

45. In economics the behavior of an economy from one period to another is often modeled by recurrence relations. Let Y_k be the income in period k and C_k the consumption in period k. In one economic model, income in any period is assumed to be the sum of consumption in that period plus investment and government expenditures (which are assumed to be constant from period to period), and consumption in each period is assumed to be a linear function of the income of the preceding period. That is,

$$Y_k = C_k + E \qquad \text{where } E \text{ is the sum of investment plus government expenditures}$$

$$C_k = c + mY_{k-1} \qquad \text{where } c \text{ and } m \text{ are constants.}$$

Substituting the second equation into the first gives $Y_k = E + c + mY_{k-1}$.

a. Use iteration on the above recurrence relation to obtain

$$Y_n = (E + c)\left(\frac{m^n - 1}{m - 1}\right) + m^n Y_0,$$

for all integers $n \geq 1$.

b. (For students who have studied calculus) Show that if $0 < m < 1$, then $\lim_{n \to \infty} Y_n = \dfrac{E + c}{1 - m}$.

8.3 SECOND-ORDER LINEAR HOMOGENEOUS RECURRENCE RELATIONS WITH CONSTANT COEFFICIENTS

Genius is 1% inspiration and 99% perspiration.
(Thomas Alva Edison, 1932)

In Section 8.2 we discussed finding explicit formulas for recursively defined sequences using iteration. This is a basic technique that does not require any special tools beyond the ability to discern patterns. In many cases, however, a pattern is not readily discernible and other methods must be used. A variety of techniques is available for finding explicit formulas for special classes of recursively defined sequences. The method explained in this section is one that works for the Fibonacci and other similarly defined sequences.

> **DEFINITION**
>
> A **second-order linear homogeneous recurrence relation with constant coefficients** is a recurrence relation of the form
>
> $$a_k = A \cdot a_{k-1} + B \cdot a_{k-2} \quad \text{for all integers } k \geq \text{some fixed integer,}$$
>
> where A and B are fixed real numbers with $B \neq 0$.

"Second-order" refers to the fact that the expression for a_k contains the two previous terms a_{k-1} and a_{k-2}, "linear" to the fact that a_{k-1} and a_{k-2} appear in separate terms and to the first power, "homogeneous" to the fact that the total degree of each term is the same (thus there is no constant term), and "constant coefficients" to the fact that A and B are fixed real numbers that do not depend on k.

EXAMPLE 8.3.1   Second-Order Linear Homogeneous Recurrence Relations with Constant Coefficients

State whether each of the following is a second-order linear homogeneous recurrence relation with constant coefficients:

a. $a_k = 3a_{k-1} + 2a_{k-2}$ b. $b_k = b_{k-1} + b_{k-2} + b_{k-3}$

c. $c_k = \dfrac{1}{2}c_{k-1} - \dfrac{3}{7}c_{k-2}$ d. $d_k = d_{k-1}^2 + d_{k-1} \cdot d_{k-2}$

e. $e_k = 2e_{k-2}$ f. $f_k = 2f_{k-1} + 1$

g. $g_k = g_{k-1} + g_{k-2}$ h. $h_k = (-1)h_{k-1} + (k-1)h_{k-2}$

Solution   a. Yes; $A = 3$ and $B = 2$ b. No; not second order

c. Yes; $A = \dfrac{1}{2}$ and $B = -\dfrac{3}{7}$ d. No; not linear

e. Yes; $A = 0$ and $B = 2$ f. No; not homogeneous

g. Yes; $A = 1$ and $B = 1$ h. No; nonconstant coefficients   ∎

THE DISTINCT ROOTS CASE

Consider a second-order linear homogeneous recurrence relation with constant coefficients:

$$a_k = A \cdot a_{k-1} + B \cdot a_{k-2} \quad \text{for all integers } k \geq 2, \qquad \textbf{8.3.1}$$

where A and B are fixed real numbers. *Suppose* that for some number t with $t \neq 0$, the sequence

$$1, t, t^2, t^3, \ldots, t^n, \ldots$$

satisfies relation (8.3.1). This means that each term of the sequence equals A times the previous term plus B times the term before that. So for all integers $k \geq 2$,

$$t^k = A \cdot t^{k-1} + B \cdot t^{k-2}.$$

Since $t \neq 0$, this equation may be divided by t^{k-2} to obtain

$$t^2 = A \cdot t + B.$$

Or, equivalently,

$$t^2 - At - B = 0. \qquad \textbf{8.3.2}$$

This is a quadratic equation and the values of t that make it true can be found either by factoring or by using the quadratic formula.

Now work backward. *Suppose t is any number that satisfies equation (8.3.2). Does the sequence $1, t, t^2, t^3, \ldots, t^n, \ldots$ satisfy relation (8.3.1)? To answer this question multiply equation (8.3.2) by t^{k-2} to obtain

$$t^{k-2} \cdot t^2 - t^{k-2} \cdot At - t^{k-2} \cdot B = 0.$$

This is equivalent to

$$t^k - A \cdot t^{k-1} - B \cdot t^{k-2} = 0$$

or

$$t^k = A \cdot t^{k-1} + B \cdot t^{k-2}.$$

Hence the answer is yes: $1, t, t^2, t^3, \ldots, t^n, \ldots$ satisfies relation (8.3.1).

This discussion proves the following lemma.

LEMMA 8.3.1

Let A and B be real numbers. A recurrence relation of the form

$$a_k = A \cdot a_{k-1} + B \cdot a_{k-2} \qquad \text{8.3.1}$$

is satisfied by the sequence

$$1, t, t^2, t^3, \ldots, t^n, \ldots,$$

where t is a nonzero real number, if, and only if, t satisfies the equation

$$t^2 - At - B = 0. \qquad \text{8.3.2}$$

Equation (8.3.2) is called the *characteristic equation* of the recurrence relation.

DEFINITION

Given a second-order linear homogeneous recurrence relation with constant coefficients:

$$a_k = A \cdot a_{k-1} + B \cdot a_{k-2} \quad \text{for all integers } k \geq 2, \qquad \text{8.3.1}$$

the **characteristic equation of the relation** is

$$t^2 - At - B = 0. \qquad \text{8.3.2}$$

EXAMPLE 8.3.2 **Using the Characteristic Equation to Find Solutions to a Recurrence Relation**

Consider the recurrence relation that specifies that the kth term of a sequence equals the sum of the $(k-1)$st term plus twice the $(k-2)$nd term. That is,

$$a_k = a_{k-1} + 2 \cdot a_{k-2} \quad \text{for all integers } k \geq 2. \qquad \text{8.3.3}$$

Find all sequences that satisfy relation (8.3.3) and have the form

$$1, t, t^2, t^3, \ldots, t^n, \ldots,$$

where t is nonzero.

Solution   By Lemma 8.3.1 relation (8.3.3) is satisfied by a sequence $1, t, t^2, t^3, \ldots, t^n, \ldots$ if, and only if, t satisfies the characteristic equation

$$t^2 - t - 2 = 0.$$

Since

$$t^2 - t - 2 = (t - 2)(t + 1),$$

the only possible values of t are 2 and -1. It follows that the sequences

$$1, 2, 2^2, 2^3, \ldots, 2^n, \ldots \quad \text{and} \quad 1, -1, (-1)^2, (-1)^3, \ldots, (-1)^n, \ldots$$

are both solutions for relation (8.3.3) and there are no other solutions of this form. Note that these sequences can be rewritten more simply as

$$1, 2, 2^2, 2^3, \ldots, 2^n, \ldots \quad \text{and} \quad 1, -1, 1, -1, \ldots, (-1)^n, \ldots \quad ■$$

The example above shows how to find two distinct sequences that satisfy a given second-order linear homogeneous recurrence relation with constant coefficients. It turns out that any linear combination of such sequences produces another sequence that also satisfies the relation.

LEMMA 8.3.2

If $r_0, r_1, r_2, \ldots$ and $s_0, s_1, s_2, \ldots$ are sequences that satisfy the same second-order linear homogeneous recurrence relation with constant coefficients, and if C and D are *any* numbers, then the sequence $a_0, a_1, a_2, \ldots$ defined by the formula

$$a_n = C \cdot r_n + D \cdot s_n \quad \text{for all integers } n \geq 0$$

also satisfies the same recurrence relation.

Proof:

Suppose $r_0, r_1, r_2, \ldots$ and $s_0, s_1, s_2, \ldots$ are sequences that satisfy the same second-order linear homogeneous recurrence relation with constant coefficients. In other words, suppose that for some real numbers A and B

$$r_k = A \cdot r_{k-1} + B \cdot r_{k-2} \quad \text{and} \quad s_k = A \cdot s_{k-1} + B \cdot s_{k-2} \qquad \text{8.3.4}$$

for all integers $k \geq 2$. Suppose also that C and D are any numbers. Let $a_0, a_1, a_2, \ldots$ be the sequence defined by

$$a_n = C \cdot r_n + D \cdot s_n \quad \text{for all integers } n \geq 0. \qquad \text{8.3.5}$$

[We must show that $a_0, a_1, a_2, \ldots$ satisfies the same recurrence relation as $r_0, r_1, r_2, \ldots$ and $s_0, s_1, s_2, \ldots$. That is, we must show that $a_k = A \cdot a_{k-1} + B \cdot a_{k-2}$, for all integers $k \geq 2$.]

For all integers $k \geq 2$,

$A \cdot a_{k-1} + B \cdot a_{k-2}$

$\quad = A \cdot (C \cdot r_{k-1} + D \cdot s_{k-1}) + B \cdot (C \cdot r_{k-2} + D \cdot s_{k-2})$ by substitution from (8.3.5)

$\quad = C \cdot (A \cdot r_{k-1} + B \cdot r_{k-2}) + D \cdot (A \cdot s_{k-1} + B \cdot s_{k-2})$ by basic algebra

$\quad = C \cdot r_k + D \cdot s_k$ by substitution from (8.3.4)

$\quad = a_k$ by substitution from (8.3.5).

Hence $a_0, a_1, a_2, \ldots$ satisfies the same recurrence relation as $r_0, r_1, r_2, \ldots$ and $s_0, s_1, s_2, \ldots$ *[as was to be shown]*.

Given a second-order linear homogeneous recurrence relation with constant coefficients, if the characteristic equation has two distinct roots, then Lemmas 8.3.1 and 8.3.2 can be used together to find a sequence that satisfies both the recurrence relation and the specific initial conditions.

EXAMPLE 8.3.3

Finding the Linear Combination That Satisfies the Initial Conditions

Find a sequence that satisfies the recurrence relation of Example 8.3.2

$$a_k = a_{k-1} + 2 \cdot a_{k-2} \quad \text{for all integers } k \geq 2, \qquad \text{8.3.3}$$

and that also satisfies the initial conditions

$$a_0 = 1 \quad \text{and} \quad a_1 = 8.$$

Solution By Example 8.3.2, the sequences

$$1, 2, 2^2, 2^3, \ldots, 2^n, \ldots \quad \text{and} \quad 1, -1, 1, -1, \ldots, (-1)^n, \ldots$$

both satisfy relation (8.3.3) (though neither satisfies the given initial conditions). By Lemma 8.3.2, therefore, any sequence $a_0, a_1, a_2, \ldots$ that satisfies an explicit formula of the form

$$a_n = C \cdot 2^n + D \cdot (-1)^n, \qquad \text{8.3.6}$$

where C and D are numbers, also satisfies relation (8.3.3). You can find C and D so that $a_0, a_1, a_2, \ldots$ satisfies the specified initial conditions by substituting $n = 0$ and $n = 1$ into equation (8.3.6) and solving for C and D:

$$a_0 = 1 = C \cdot 2^0 + D \cdot (-1)^0,$$
$$a_1 = 8 = C \cdot 2^1 + D \cdot (-1)^1.$$

When you simplify, you obtain the system

$$1 = C + D$$
$$8 = 2C - D,$$

which can be solved in various ways. For instance, if you add the two equations, you get

$$9 = 3C,$$

and so

$$C = 3.$$

Then by substituting into $1 = C + D$, you get

$$D = -2.$$

It follows that the sequence $a_0, a_1, a_2, \ldots$ given by

$$a_n = 3 \cdot 2^n + (-2) \cdot (-1)^n = 3 \cdot 2^n - 2 \cdot (-1)^n,$$

for integers $n \geq 0$, satisfies both the recurrence relation and the given initial conditions. ■

    The techniques of Examples 8.3.2 and 8.3.3 can be used to find an explicit formula for *any* sequence that satisfies a second-order linear homogeneous recurrence relation with constant coefficients for which the characteristic equation has distinct roots, provided the first two terms of the sequence are known. This is made precise in the next theorem.

THEOREM 8.3.3   Distincts Roots Theorem

Suppose a sequence $a_0, a_1, a_2, \ldots$ satisfies a recurrence relation

$$a_k = A \cdot a_{k-1} + B \cdot a_{k-2} \qquad\qquad\text{8.3.1}$$

for some real numbers A and B and all integers $k \geq 2$. If the characteristic equation

$$t^2 - At - B = 0 \qquad\qquad\text{8.3.2}$$

has two distinct roots r and s, then $a_0, a_1, a_2 \ldots$ satisfies the explicit formula

$$a_n = C \cdot r^n + D \cdot s^n,$$

where C and D are the numbers whose values are determined by the values a_0 and a_1.

Note: To say "C and D are determined by the values of a_0 and a_1" means that C and D are the solutions to the system of simultaneous equations

$$a_0 = C \cdot r^0 + D \cdot s^0 \quad \text{and} \quad a_1 = C \cdot r^1 + D \cdot s^1.$$

Or, equivalently,

$$a_0 = C + D \quad \text{and} \quad a_1 = C \cdot r + D \cdot s.$$

In exercise 19 at the end of this section you are asked to verify that this system always has a solution when $r \neq s$.

Proof:

Suppose a sequence a_0, a_1, a_2, ... satisfies the recurrence relation $a_k = A \cdot a_{k-1} + B \cdot a_{k-2}$ for some real numbers A and B and for all integers $k \geq 2$, and suppose the characteristic equation $t^2 - At - B = 0$ has two distinct roots r and s. We will show that

$$\text{for all integers } n \geq 0, \quad a_n = C \cdot r^n + D \cdot s^n,$$

where C and D are numbers such that

$$a_0 = C \cdot r^0 + D \cdot s^0 \quad \text{and} \quad a_1 = C \cdot r^1 + D \cdot s^1.$$

Consider the formula $a_n = C \cdot r^n + D \cdot s^n$. We use strong mathematical induction to prove that the formula holds for all integers $n \geq 0$. In the basis step, we prove not only that the formula holds for $n = 0$ but also that it holds for $n = 1$. The reason we do this is that in the inductive step we need the formula to hold for $n = 0$ and $n = 1$ in order to prove that it holds for $n = 2$.

The formula holds for $n = 0$ and $n = 1$:   The truth of the formula for $n = 0$ and $n = 1$ is automatic because C and D are exactly those numbers that make the following equations true:

$$a_0 = C \cdot r^0 + D \cdot s^0 \quad \text{and} \quad a_1 = C \cdot r^1 + D \cdot s^1.$$

If $k \geq 2$ and the formula holds for all integers i with $0 \leq i < k$, then it holds for k: Suppose that $k \geq 2$ and for all integers i with $0 \leq i < k$,

$$a_i = C \cdot r^i + D \cdot s^i. \quad \text{This is the inductive hypothesis.}$$

We must show that

$$a_k = C \cdot r^k + D \cdot s^k.$$

Now by the inductive hypothesis,

$$a_{k-1} = C \cdot r^{k-1} + D \cdot s^{k-1} \quad \text{and} \quad a_{k-2} = C \cdot r^{k-2} + D \cdot s^{k-2}.$$

So

$$
\begin{aligned}
a_k &= A \cdot a_{k-1} + B \cdot a_{k-2} && \text{by definition} \\
&&& \text{of } a_0, a_1, a_2, \ldots \\
&= A \cdot (C \cdot r^{k-1} + D \cdot s^{k-1}) + B \cdot (C \cdot r^{k-2} + D \cdot s^{k-2}) && \text{by inductive} \\
&&& \text{hypothesis} \\
&= C \cdot (A \cdot r^{k-1} + B \cdot r^{k-2}) + D \cdot (A \cdot s^{k-1} + B \cdot s^{k-2}) && \text{by combining} \\
&&& \text{terms involving} \\
&&& C \text{ and } D \\
&&& \text{together} \\
&= C \cdot r^k + D \cdot s^k && \text{by Lemma 8.3.1.}
\end{aligned}
$$

This is what was to be shown.
[The reason the last equality follows from Lemma 8.3.1 is that since r and s satisfy the characteristic equation (8.3.2), the sequences $r^0, r^1, r^2, \ldots$ and $s^0, s^1, s^2, \ldots$ satisfy the recurrence relation (8.3.1).]

Remark: The t of Lemma 8.3.1 and the C and D of Lemma 8.3.2 and Theorem 8.3.3 are referred to simply as numbers. This is to allow for the possibility of complex as well as real number values. If both roots of the characteristic equation of the recurrence relation are real numbers, then C and D will be real. But if the roots are nonreal complex numbers, then C and D will be nonreal complex numbers.

The next example shows how to use the distinct roots theorem to find an explicit formula for the Fibonacci sequence.

EXAMPLE 8.3.4   **A Formula for the Fibonacci Sequence**

The Fibonacci sequence $F_0, F_1, F_2, \ldots$ satisfies the recurrence relation

$$F_k = F_{k-1} + F_{k-2} \quad \text{for all integers } k \geq 2$$

with initial conditions

$$F_0 = F_1 = 1.$$

Find an explicit formula for this sequence.

Solution   The Fibonacci sequence satisfies part of the hypothesis of the distinct roots theorem since the Fibonacci relation is a second-order linear homogeneous recurrence relation with constant coefficients ($A = 1$ and $B = 1$). Is the second part of the hypothesis also satisfied? Does the characteristic equation

$$t^2 - t - 1 = 0$$

have distinct roots? By the quadratic formula, the roots are

$$t = \frac{1 \pm \sqrt{1 - 4(-1)}}{2} = \begin{cases} \dfrac{1 + \sqrt{5}}{2} \\ \dfrac{1 - \sqrt{5}}{2} \end{cases}$$

and so the answer is yes. It follows from the distinct roots theorem that the Fibonacci sequence satisfies the explicit formula

$$F_n = C \cdot \left(\frac{1 + \sqrt{5}}{2}\right)^n + D \cdot \left(\frac{1 - \sqrt{5}}{2}\right)^n \quad \text{for all integers } n \geq 0, \qquad \text{8.3.7}$$

where C and D are the numbers whose values are determined by the fact that $F_0 = F_1 = 1$. To find C and D, write

$$F_0 = 1 = C \cdot \left(\frac{1 + \sqrt{5}}{2}\right)^0 + D \cdot \left(\frac{1 - \sqrt{5}}{2}\right)^0 = C \cdot 1 + D \cdot 1 = C + D$$

and

$$F_1 = 1 = C \cdot \left(\frac{1 + \sqrt{5}}{2}\right)^1 + D \cdot \left(\frac{1 - \sqrt{5}}{2}\right)^1$$

$$= C \cdot \left(\frac{1 + \sqrt{5}}{2}\right) + D \cdot \left(\frac{1 - \sqrt{5}}{2}\right).$$

So the problem is to find numbers C and D such that

$$C + D = 1$$

and

$$C \cdot \left(\frac{1 + \sqrt{5}}{2} \right) + D \cdot \left(\frac{1 - \sqrt{5}}{2} \right) = 1.$$

This may look complicated, but in fact it is just a system of two equations in two unknowns. In exercise 7 at the end of this section, you are asked to show that

$$C = \frac{1 + \sqrt{5}}{2\sqrt{5}} \quad \text{and} \quad D = \frac{-(1 - \sqrt{5})}{2\sqrt{5}}.$$

Substituting these values for C and D into formula (8.3.7) gives

$$F_n = \left(\frac{1 + \sqrt{5}}{2\sqrt{5}} \right) \cdot \left(\frac{1 + \sqrt{5}}{2} \right)^n + \left(\frac{-(1 - \sqrt{5})}{2\sqrt{5}} \right) \cdot \left(\frac{1 - \sqrt{5}}{2} \right)^n.$$

Or, simplifying,

$$F_n = \frac{1}{\sqrt{5}} \cdot \left(\frac{1 + \sqrt{5}}{2} \right)^{n+1} - \frac{1}{\sqrt{5}} \cdot \left(\frac{1 - \sqrt{5}}{2} \right)^{n+1} \qquad \text{8.3.8}$$

for all integers $n \geq 0$. Remarkably, even though the formula for F_n involves $\sqrt{5}$, all of the values of the Fibonacci sequence are integers. It is also interesting to note that the numbers $(1 + \sqrt{5})/2$ and $(1 - \sqrt{5})/2$ are related to the golden ratio of Greek mathematics. (See exercise 24 at the end of this section.) ■

THE SINGLE-ROOT CASE

Consider again the recurrence relation

$$a_k = A \cdot a_{k-1} + B \cdot a_{k-2} \quad \text{for all integers } k \geq 2, \qquad \text{8.3.1}$$

where A and B are real numbers, but suppose now that the characteristic equation

$$t^2 - At - B = 0 \qquad \text{8.3.2}$$

has a single real root r. By Lemma 8.3.1, one sequence that satisfies the recurrence relation is

$$1, r, r^2, r^3, \ldots, r^n, \ldots .$$

But another sequence that also satisfies the relation is

$$0, r, 2r^2, 3r^3, \ldots, nr^n, \ldots .$$

To see why this is so, observe that since r is the unique root of $t^2 - At - B = 0$, then the left-hand side of the equation can be factored as $(t - r)^2$, and so

$$t^2 - At - B = (t - r)^2 = t^2 - 2rt + r^2. \qquad \text{8.3.9}$$

Equating coefficients in equation (8.3.9) gives

$$A = 2r \quad \text{and} \quad B = -r^2. \qquad \text{8.3.10}$$

Let $s_0, s_1, s_2, \ldots$ be the sequence defined by the formula

$$s_n = n \cdot r^n \quad \text{for all integers } n \geq 0.$$

Then

$$A \cdot s_{k-1} + B \cdot s_{k-2} = A \cdot (k-1) \cdot r^{k-1} + B \cdot (k-2) \cdot r^{k-2} \quad \text{by definition}$$

$$= 2r \cdot (k-1)r^{k-1} - r^2 \cdot (k-2)r^{k-2} \quad \begin{array}{l}\text{by substitution} \\ \text{from (8.3.10)}\end{array}$$

$$= 2(k-1)r^k - (k-2)r^k$$

$$= (2k-2-k+2) \cdot r^k$$

$$= k \cdot r^k \quad \text{by basic algebra}$$

$$= s_k \quad \text{by definition}$$

So $s_0, s_1, s_2, \ldots$ satisfies the recurrence relation. This argument proves the following lemma.

LEMMA 8.3.4

Let A and B be real numbers and suppose the characteristic equation

$$t^2 - At - B = 0$$

has a single root r. Then the sequences $1, r^1, r^2, r^3, \ldots, r^n, \ldots$ and $0, r, 2r^2, 3r^3, \ldots, nr^n, \ldots$ both satisfy the recurrence relation

$$a_k = A \cdot a_{k-1} + B \cdot a_{k-2}$$

for all integers $k \geq 2$.

Lemmas 8.3.2 and 8.3.4 can be used to establish the *single-root theorem,* which tells how to find an explicit formula for any recursively defined sequence satisfying a second-order linear homogeneous recurrence relation with constant coefficients for which the characteristic equation has just one root. Taken together, the distinct-roots and single-root theorems cover all second-order linear homogeneous recurrence relations with constant coefficients. The proof of the single-root theorem is very similar to that of the distinct-roots theorem and is left as an exercise.

THEOREM 8.3.5 Single-Root Theorem

Suppose a sequence $a_0, a_1, a_2, \ldots$ satisfies a recurrence relation

$$a_k = A \cdot a_{k-1} + B \cdot a_{k-2}$$

for some real numbers A and B with $B \neq 0$ and for all integers $k \geq 2$. If the characteristic equation $t^2 - At - B = 0$ has a single (real) root r, then $a_0, a_1, a_2, \ldots$ satisfies the explicit formula

$$a_n = C \cdot r^n + D \cdot n \cdot r^n,$$

where C and D are the real numbers whose values are determined by the values of a_0 and any other known value of the sequence.

EXAMPLE 8.3.5   **Single-Root Case**

Suppose a sequence $b_0, b_1, b_2, \ldots$ satisfies the recurrence relation

$$b_k = 4b_{k-1} - 4b_{k-2} \quad \text{for all integers } k \geq 2, \qquad \textbf{8.3.11}$$

with initial conditions

$$b_0 = 1 \quad \text{and} \quad b_1 = 3.$$

Find an explicit formula for $b_0, b_1, b_2, \ldots$.

Solution   This sequence satisfies part of the hypothesis of the single-root theorem since it satisfies a second-order linear homogeneous recurrence relation with constant coefficients ($A = 4$ and $B = -4$). The single-root condition is also met because the characteristic equation

$$t^2 - 4t + 4 = 0$$

has the unique root $r = 2$ *[since $t^2 - 4t + 4 = (t - 2)^2$]*.

It follows from the single-root theorem that $b_0, b_1, b_2, \ldots$ satisfies the explicit formula

$$b_n = C \cdot 2^n + D \cdot n \cdot 2^n \quad \text{for all integers } n \geq 0, \qquad \textbf{8.3.12}$$

where C and D are the real numbers whose values are determined by the fact that $b_0 = 1$ and $b_1 = 3$. To find C and D, write

$$b_0 = 1 = C \cdot 2^0 + D \cdot 0 \cdot 2^0 = C$$

and

$$b_1 = 3 = C \cdot 2^1 + D \cdot 1 \cdot 2^1 = 2 \cdot C + 2 \cdot D.$$

So the problem is to find numbers C and D such that

$$C = 1$$

and

$$2 \cdot C + 2 \cdot D = 3.$$

Substitute $C = 1$ into the second equation to obtain

$$2 + 2 \cdot D = 3,$$

and so

$$D = \frac{1}{2}.$$

Now substitute $C = 1$ and $D = \frac{1}{2}$ into formula (8.3.12) to conclude that

$$b_n = 2^n + \frac{1}{2} \cdot n \cdot 2^n = 2^n \cdot \left(1 + \frac{n}{2}\right) \quad \text{for all integers } n \geq 0. \qquad \blacksquare$$

EXAMPLE 8.3.6   **Gambler's Ruin**

A gambler repeatedly bets $1 that a coin will come up heads when tossed. Each time the coin comes up heads, the gambler wins $1; each time it comes up tails he loses $1. The gambler will quit playing either when he is ruined (loses all his money) or when he has M (where M is a value he has decided in advance). Let P_n be the probability that the gambler is ruined when he begins playing with $\$n$. Then if the coin is fair (has an equal chance of coming up heads or tails),

$$P_{k-1} = \frac{1}{2} \cdot P_k + \frac{1}{2} \cdot P_{k-2} \quad \text{for each integer } k \text{ with } 2 \le k \le M.$$

(This follows from the fact that if the gambler has $\$(k-1)$, then he has equal chance of winning \$1 or losing \$1, and if he wins \$1 then his chance of being ruined is P_k, while if he loses \$1 then his chance of being ruined is P_{k-2}.) Also $P_0 = 1$ (because if he has \$0 he is certain of being ruined) and $P_M = 0$ (because once he has $\$M$ he quits and so stands no chance of being ruined). Find an explicit formula for P_n. How should the gambler choose M to minimize his chance of being ruined?

Solution   Multiplying both sides of $P_{k-1} = \frac{1}{2} \cdot P_k + \frac{1}{2} \cdot P_{k-2}$ by 2 and subtracting P_{k-2} from both sides gives

$$P_k = 2 \cdot P_{k-1} - P_{k-2},$$

which is a second-order homogeneous recurrence relation with constant coefficients. Its characteristic equation is

$$t^2 - 2t + 1 = 0,$$

which has the single root $r = 1$. Thus by the single-root theorem,

$$P_n = C \cdot r^n + D \cdot n \cdot r^n = C + D \cdot n$$

(since $r = 1$), where C and D are determined by two values of the sequence. But $P_0 = 1$ and $P_M = 0$. Hence

$$1 = P_0 = C + D \cdot 0 = C,$$
$$0 = P_M = C + D \cdot M = 1 + D \cdot M.$$

It follows that $C = 1$ and $D = -\dfrac{1}{M}$, and so

$$P_n = 1 - \frac{1}{M} \cdot n = \frac{M - n}{M} \quad \text{for each integer } n \text{ with } 0 \le n \le M.$$

For instance, a gambler who starts with \$20 and decides to quit either if his total grows to \$100 or if he goes broke has the following chance of going broke:

$$P_{20} = \frac{100 - 20}{100} = \frac{80}{100} = 80\%.$$

Observe that the larger M is relative to n, the closer P_n is to 1. In other words, the larger the amount of money the gambler sets himself as a target, the more likely he is to go broke. Conversely, the more modest he is in his goal, the more likely he is to reach it. ∎

EXERCISE SET 8.3

1. Which of the following are second-order linear homogeneous recurrence relations with constant coefficients?
   a. $a_k = 2a_{k-1} - 5a_{k-2}$
   b. $b_k = kb_{k-1} + b_{k-2}$
   c. $c_k = 3c_{k-1} \cdot c_{k-2}^2$
   d. $d_k = 3d_{k-1} + d_{k-2}$
   e. $r_k = r_{k-1} - r_{k-2} - 2$
   f. $s_k = 10s_{k-2}$

2. Which of the following are second-order linear homogeneous recurrence relations with constant coefficients?
   a. $a_k = (k - 1)a_{k-1} + 2ka_{k-2}$
   b. $b_k = -b_{k-1} + 7b_{k-2}$
   c. $c_k = 3c_{k-1} + 1$
   d. $d_k = 3d_{k-1}^2 + d_{k-2}$
   e. $r_k = r_{k-1} - 6r_{k-3}$
   f. $s_k = s_{k-1} + 10s_{k-2}$

3. Let $a_0, a_1, a_2, \ldots$ be the sequence defined by the explicit formula

   $$a_n = C \cdot 2^n + D, \text{ for all integers } n \geq 0,$$

   where C and D are real numbers.
   a. Find C and D so that $a_0 = 1$ and $a_1 = 3$. What is a_2 in this case?
   b. Find C and D so that $a_0 = 0$ and $a_1 = 2$. What is a_2 in this case?

4. Let $b_0, b_1, b_2, \ldots$ be the sequence defined by the explicit formula

   $$b_n = C \cdot 3^n + D \cdot (-2)^n, \text{ for all integers } n \geq 0,$$

   where C and D are real numbers.
   a. Find C and D so that $b_0 = 0$ and $b_1 = 5$. What is b_2 in this case?
   b. Find C and D so that $b_0 = 3$ and $b_1 = 4$. What is b_2 in this case?

5. Let $a_0, a_1, a_2, \ldots$ be the sequence defined by the explicit formula

   $$a_n = C \cdot 2^n + D, \text{ for all integers } n \geq 0,$$

   where C and D are real numbers. Show that for any choice of C and D,

   $$a_k = 3a_{k-1} - 2a_{k-2}, \text{ for all integers } k \geq 2.$$

6. Let $b_0, b_1, b_2, \ldots$ be the sequence defined by the explicit formula

   $$b_n = C \cdot 3^n + D \cdot (-2)^n, \text{ for all integers } n \geq 0,$$

   where C and D are real numbers. Show that for any choice of C and D,

   $$b_k = b_{k-1} + 6b_{k-2}, \text{ for all integers } k \geq 2.$$

7. Solve the system of equations in Example 8.3.4 to obtain

   $$C = \frac{1 + \sqrt{5}}{2\sqrt{5}} \text{ and } D = \frac{-(1 - \sqrt{5})}{2\sqrt{5}}.$$

In each of 8–10: (a) suppose a sequence of the form 1, t, t², t³, ... , tⁿ, ... , where t ≠ 0, satisfies the given recurrence relation (but not necessarily the initial conditions), and find all possible values of t; (b) suppose a sequence satisfies the given initial conditions as well as the recurrence relation and find an explicit formula for the sequence.

8. $a_k = 2a_{k-1} + 3a_{k-2}$, for all integers $k \geq 2$
   $a_0 = 1, a_1 = 2$

9. $b_k = 7b_{k-1} - 10b_{k-2}$, for all integers $k \geq 2$
   $b_0 = 2, b_1 = 2$

10. $c_k = c_{k-1} + 6c_{k-2}$, for all integers $k \geq 2$
    $c_0 = 0, c_1 = 3$

In each of 11–15 suppose a sequence satisfies the given recurrence relation and initial conditions. Find an explicit formula for the sequence.

11. $d_k = 4d_{k-2}$, for all integers $k \geq 2$
    $d_0 = 1, d_1 = -1$

12. $e_k = 9e_{k-2}$, for all integers $k \geq 2$
    $e_0 = 0, e_1 = 2$

13. $r_k = 2r_{k-1} - r_{k-2}$, for all integers $k \geq 2$
    $r_0 = 1, r_1 = 4$

14. $s_k = -4s_{k-1} - 4s_{k-2}$, for all integers $k \geq 2$
    $s_0 = 0, s_1 = -1$

15. $t_k = 6t_{k-1} - 9t_{k-2}$, for all integers $k \geq 2$
    $t_0 = 1, t_1 = 3$

H16. Find an explicit formula for the sequence of exercise 32 in Section 8.1.

17. Find an explicit formula for the sequence of exercise 34 in Section 8.1.

18. Suppose that the sequences $s_0, s_1, s_2, \ldots$ and $t_0, t_1, t_2, \ldots$ both satisfy the same second-order linear homogeneous recurrence relation with constant coefficients:

    $$s_k = 5s_{k-1} - 4s_{k-2}, \text{ for all integers } k \geq 2,$$
    $$t_k = 5t_{k-1} - 4t_{k-2}, \text{ for all integers } k \geq 2.$$

    Show that the sequence $2s_0 + 3t_0$, $2s_1 + 3t_1$, $2s_2 + 3t_2, \ldots$ also satisfies the same relation. In other words, show that

    $$2s_k + 3t_k = 5 \cdot [2s_{k-1} + 3t_{k-1}] - 4 \cdot [2s_{k-2} + 3t_{k-2}]$$

    for all integers $k \geq 2$. Do *not* use Lemma 8.3.2.

19. Show that if r, s, a_0, and a_1 are numbers with $r \neq s$, then there exist unique numbers C and D so that

    $$C + D = a_0$$
    $$Cr + Ds = a_1.$$

20. Show that if r is a nonzero real number, k and l are distinct integers, and a_k and a_l are any real numbers, then there exist unique real numbers C and D so that

    $$Cr^k + kDr^k = a_k$$
    $$Cr^l + lDr^l = a_l.$$

H21. Prove Theorem 8.3.5 for the case where the values of C and D are determined by a_0 and a_1.

Exercises 22 and 23 are intended for students who are familiar with complex numbers.

22. Find an explicit formula for a sequence $a_0, a_1, a_2, \ldots$ that satisfies

$$a_k = 2a_{k-1} - 2a_{k-2}, \text{ for all integers } k \geq 2$$

with initial conditions $a_0 = 1$ and $a_1 = 2$.

23. Find an explicit formula for a sequence $b_0, b_1, b_2, \ldots$ that satisfies

$$b_k = 2b_{k-1} - 5b_{k-2}, \text{ for all integers } k \geq 2$$

with initial conditions $b_0 = 1$ and $b_1 = 1$.

24. The numbers $\dfrac{1 + \sqrt{5}}{2}$ and $\dfrac{1 - \sqrt{5}}{2}$ that appear in the explicit formula for the Fibonacci sequence are related to a quantity called the *golden ratio* in Greek mathematics. Consider a rectangle of length ϕ units and height 1, where $\phi > 1$.

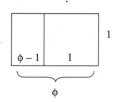

Divide the rectangle into a rectangle and a square as shown above: The square is 1 unit on each side and the rectangle has sides of lengths 1 and $\phi - 1$. The ancient Greeks considered the outer rectangle to be perfectly proportioned (saying that the lengths of its sides were in a *golden ratio* to each other) if the ratio of the length to the width of the outer rectangle equals the ratio of the length to the width of the inner rectangle. That is,

$$\frac{\phi}{1} = \frac{1}{\phi - 1}.$$

a. Show that ϕ satisfies the quadratic equation $t^2 - t - 1 = 0$.

b. Find the two solutions of $t^2 - t - 1 = 0$ and call them ϕ_1 and ϕ_2.

c. Express the explicit formula for the Fibonacci sequence in terms of ϕ_1 and ϕ_2.

H 25. A gambler repeatedly bets that a die will come up 6 when rolled. Each time the die comes up 6, the gambler wins \$1; each time it does not the gambler loses \$1. He will quit playing either when he is ruined or when he wins \$300. If P_n is the probability that the gambler is ruined when he begins play with \$$n$, then $P_{k-1} = \frac{1}{6} \cdot P_k + \frac{5}{6} \cdot P_{k-2}$ for all integers k with $2 \leq k \leq 300$. Also $P_0 = 1$ and $P_{300} = 0$. Find an explicit formula for P_n and use it to calculate P_{20}.

◆26. A circular disk is cut into n distinct sectors, each shaped like a piece of pie and all meeting at the center point of the disk. Each sector is to be painted either red, green, yellow, or blue in such a way that no two adjacent sectors are painted the same color. Let S_n be the number of ways to paint the disk.

H a. Find a recurrence relation for S_k in terms of S_{k-1} and S_{k-2} for each integer $k \geq 4$.

b. Find an explicit formula for S_n for $n \geq 2$.

Genie: Oh, aren't you acquainted with recursive acronyms? I thought everybody knew about them. You see, "GOD" stands for "GOD Over Djinn"—which can be expanded as "GOD Over Djinn, Over Djinn"—and that can, in turn, be expanded to "GOD Over Djinn, Over Djinn, Over Djinn"—which can, in its turn, be further expanded .... You can go as far as you like.
Achilles: But I'll never finish!
Genie: Of course not. You can never totally expand GOD.
(Douglas Hofstadter, *Gödel, Escher, Bach*, 1979)

8.4 GENERAL RECURSIVE DEFINITIONS

Sequences of numbers are not the only objects that can be defined recursively. In this section we discuss recursive definitions for sets, sums, products, unions, intersections, and functions.

RECURSIVELY DEFINED SETS

To define a set of objects recursively, you identify a few core objects as belonging to the set and give rules showing how to build new set elements from old. More formally, a recursive definition for a set consists of the following three components:

   I. BASE: A statement that certain objects belong to the set.

  II. RECURSION: A collection of rules indicating how to form new set objects from those already known to be in the set.

 III. RESTRICTION: A statement that no objects belong to the set other than those coming from I and II.

EXAMPLE 8.4.1   **Recursive Definition of Boolean Expressions**

The set of Boolean expressions was introduced in Section 1.4 as "legal" expressions involving letters from the alphabet such as p, q, and r, and the symbols $\wedge$, $\vee$, and $\sim$ *[a legal expression being, for instance, $p \wedge (q \vee \sim r)$ and an illegal one being $\wedge \sim pqr \vee$].* To make precise which expressions are legal, the set of Boolean expressions over a general alphabet is defined recursively.

   I. BASE: Each symbol of the alphabet is a Boolean expression.
   II. RECURSION: If P and Q are Boolean expressions, then so are

$$\text{(a) } (P \wedge Q) \quad \text{and} \quad \text{(b) } (P \vee Q) \quad \text{and} \quad \text{(c) } \sim P.$$

   III. RESTRICTION: There are no Boolean expressions over the alphabet other than those obtained from I and II.

Derive the fact that the following is a Boolean expression over the English alphabet $\{a, b, c, \ldots, x, y, z\}$:

$$(\sim(p \wedge q) \vee (\sim r \wedge p)).$$

Solution   1. By I, p, q, and r are Boolean expressions.
   2. By (1) and II(a) and (c), $(p \wedge q)$ and $\sim r$ are Boolean expressions.
   3. By (2) and II(c) and (a), $\sim(p \wedge q)$ and $(\sim r \wedge p)$ are Boolean expressions.
   4. By (3) and II(b), $(\sim(p \wedge q) \vee (\sim r \wedge p))$ is a Boolean expression.   ∎

EXAMPLE 8.4.2   **The Set of Strings over an Alphabet**

Consider a finite alphabet $\Sigma = \{a, b\}$. The set of all finite strings over Σ, denoted Σ^*, is defined recursively as follows:

   I. BASE: ε is in Σ^*, where ε is the null string.
   II. RECURSION: If $s \in \Sigma^*$, then

$$\text{(a) } sa \in \Sigma^* \quad \text{and} \quad \text{(b) } sb \in \Sigma^*,$$

   where sa and sb are the concatenations of s with a and b respectively.
   III. RESTRICTION: Nothing is in Σ^* other than objects defined in I and II above.

Derive the fact that $ab \in \Sigma^*$.

Solution   1. By I, $\varepsilon \in \Sigma^*$.
   2. By (1) and II(a), $\varepsilon a \in \Sigma^*$. But εa is the concatenation of the null string and a, which equals a. So $a \in \Sigma^*$.
   3. By (2) and II(b), $ab \in \Sigma^*$.   ∎

EXAMPLE 8.4.3   **Sets of Strings with Certain Properties**

In *Gödel, Escher, Bach*, Douglas Hofstadter introduces the following recursively defined set of strings of M's, I's, and U's, which he calls the *MIU*-system*:

   I. BASE: *MI* is in the *MIU*-system.

---

*Douglas Hofstadter, *Gödel, Escher, Bach* (New York: Basic Books), pp. 33–35.

    II. RECURSION:

      a. If *xI* is in the *MIU*-system, where *x* is a string, then *xIU* is in the *MIU*-system. (In other words, you can add a *U* to any string that ends in *I*. For example, since *MI* is in the system so is *MIU*.)

      b. If *Mx* is in the *MIU*-system, where *x* is a string, then *Mxx* is in the *MIU*-system. (In other words, you can repeat all the characters in a string that follow an initial *M*. For example, if *MUI* is in the system so is *MUIUI*.)

      c. If *xIIIy* is in the *MIU*-system, where *x* and *y* are strings (possibly null), then *xUy* is also in the *MIU*-system. (In other words, you can replace *III* by *U*. For example, if *MIIII* is in the system so are *MIU* and *MUI*.)

      d. If *xUUy* is in the *MIU*-system, where *x* and *y* are strings (possibly null), then *xUy* is also in the *MIU*-system. (In other words, you can replace *UU* by *U*. For example, if *MIIUU* is in the system so is *MIIU*.)

    III. RESTRICTION: No strings other than those derived from I and II are in the *MIU*-system.

Derive the fact that *MUIU* is in the *MIU*-system.

Solution
1. By I, *MI* is in the *MIU*-system.
2. By (1) and II(b), *MII* is in the *MIU*-system.
3. By (2) and II(b), *MIIII* is in the *MIU*-system.
4. By (3) and II(c), *MUI* is in the *MIU*-system.
5. By (3) and II(a), *MUIU* is in the *MIU*-system. ∎

EXAMPLE 8.4.4   **Parenthesis Structures**

Certain configurations of parentheses in algebraic expressions are "grammatical" *[such as (( ))( ) and ( )( )( )]* whereas others are not *[such as)( ))) and ( )))((( ]*. Here is a recursive definition to generate the set *P* of grammatical configurations of parentheses.

    I. BASE: ( ) is in *P*.

    II. RECURSION:
      a. If *E* is in *P*, so is (*E*).
      b. If *E* and *F* are in *P*, so is *EF*.

    III. RESTRICTION: No configurations of parentheses are in *P* other than those derived from I and II above.

Derive the fact that (( ))( ) is in *P*.

Solution
1. By I, ( ) is in *P*.
2. By (1) and II(a), (( )) is in *P*.
3. By (2), (1), and II(b), (( ))( ) is in *P*. ∎

RECURSIVE DEFINITIONS OF SUM, PRODUCT, UNION, AND INTERSECTION

Addition and multiplication are called *binary* operations because only two numbers can be added or multiplied at a time. Careful definitions of sums and products of more than two numbers use recursion.

> **DEFINITION**
>
> Given numbers $a_1, a_2, \ldots, a_n$, where n is a positive integer, the **summation from $i = 1$ to n of the a_i,** denoted $\sum_{i=1}^{n} a_i$, is defined as follows:
>
> $$\sum_{i=1}^{1} a_i = a_1 \quad \text{and} \quad \sum_{i=1}^{n} a_i = \left(\sum_{i=1}^{n-1} a_i \right) + a_n, \quad \text{if } n > 1.$$
>
> The **product from $i = 1$ to n of the a_i,** denoted $\prod_{i=1}^{n} a_i$, is defined by
>
> $$\prod_{i=1}^{1} a_i = a_1 \quad \text{and} \quad \prod_{i=1}^{n} a_i = \left(\prod_{i=1}^{n-1} a_i \right) \cdot a_n, \quad \text{if } n > 1.$$

The effect of these definitions is to specify an *order* in which sums and products of more than two numbers are computed. For example,

$$\sum_{i=1}^{4} a_i = \left(\sum_{i=1}^{3} a_i \right) + a_4 = \left(\left(\sum_{i=1}^{2} a_i \right) + a_3 \right) + a_4$$

$$= ((a_1 + a_2) + a_3) + a_4.$$

Sometimes these recursive definitions are started at $n = 0$ by decreeing that $\sum_{i=1}^{0} a_i = 0$ and $\prod_{i=1}^{0} a_i = 1$. Before rejecting these definitions as formalistic nonsense, observe that the usual computer algorithms to compute sums and products use them in a very natural way. For instance, to compute the sum of $a[1], a[2], \ldots, a[n]$, one normally writes

$$sum := 0$$
$$\textbf{for } k := 1 \textbf{ to } n$$
$$sum := sum + a[k]$$
$$\textbf{next } k.$$

The recursive definitions are used with mathematical induction to establish various properties of general finite sums and products.

EXAMPLE 8.4.5   A Sum of Sums

Prove that for any positive integer n, if $a_1, a_2, \ldots, a_n$ and $b_1, b_2, \ldots, b_n$ are real numbers, then

$$\sum_{i=1}^{n} (a_i + b_i) = \sum_{i=1}^{n} a_i + \sum_{i=1}^{n} b_i.$$

Solution   The proof is by mathematical induction.

The formula holds for $n = 1$:   Suppose a_1 and b_1 are real numbers. Then

$$\sum_{i=1}^{1} (a_i + b_i) = a_1 + b_1 \qquad \text{by definition of } \Sigma$$

$$= \sum_{i=1}^{1} a_i + \sum_{i=1}^{1} b_i \qquad \text{also by definition of } \Sigma.$$

If the formula holds for n = k, then it holds for n = k + 1:   Suppose $a_1, a_2, \ldots, a_k, a_{k+1}$ and $b_1, b_2, \ldots, b_k, b_{k+1}$ are real numbers and that for some $k \geq 1$

$$\sum_{i=1}^{k} (a_i + b_i) = \sum_{i=1}^{k} a_i + \sum_{i=1}^{k} b_i. \quad \text{This is the inductive hypothesis.}$$

We must show that

$$\sum_{i=1}^{k+1} (a_i + b_i) = \sum_{i=1}^{k+1} a_i + \sum_{i=1}^{k+1} b_i.$$

[We will show that the left-hand side of this equation equals the right-hand side.]

But

$$
\begin{aligned}
\sum_{i=1}^{k+1} (a_i + b_i) &= \sum_{i=1}^{k} (a_i + b_i) + (a_{k+1} + b_{k+1}) && \text{by definition of } \Sigma \\
&= \left(\sum_{i=1}^{k} a_i + \sum_{i=1}^{k} b_i \right) + (a_{k+1} + b_{k+1}) && \text{by inductive hypothesis} \\
&= \left(\sum_{i=1}^{k} a_i + a_{k+1} \right) + \left(\sum_{i=1}^{k} b_i + b_{k+1} \right) && \begin{array}{l}\text{by the associative and commutative} \\ \text{laws of algebra}\end{array} \\
&= \sum_{i=1}^{k+1} a_i + \sum_{i=1}^{k+1} b_i && \text{by definition of } \Sigma.
\end{aligned}
$$

This is what was to be shown.   ■

Like sum and product, union and intersection are also binary operations, and unions and intersections of more than two sets can be defined recursively.

DEFINITION

Given sets $A_1, A_2, \ldots, A_n$, where n is a positive integer, the **union of the A_i from $i = 1$ to n**, denoted $\bigcup_{i=1}^{n} A_i$, is defined by

$$\bigcup_{i=1}^{1} A_i = A_1 \quad \text{and} \quad \bigcup_{i=1}^{n} A_i = \left(\bigcup_{i=1}^{n-1} A_i \right) \cup A_n.$$

The **intersection of the A_i from $i = 1$ to n,** denoted $\bigcap_{i=1}^{n} A_i$, is defined by

$$\bigcap_{i=1}^{1} A_i = A_1 \quad \text{and} \quad \bigcap_{i=1}^{n} A_i = \left(\bigcap_{i=1}^{n-1} A_i \right) \cap A_n.$$

EXAMPLE 8.4.6   A Generalized De Morgan Law

Prove that for all integers $n \geq 1$, if $A_1, A_2, \ldots, A_n$ are sets, then

$$\left(\bigcup_{i=1}^{n} A_i \right)^c = \bigcap_{i=1}^{n} (A_i)^c.$$

Solution The proof is by mathematical induction.

The formula holds for n = 1: We must show that

$$\left(\bigcup_{i=1}^{1} A_i \right)^c = \left(\bigcap_{i=1}^{1} (A_i)^c \right).$$

But

$$\left(\bigcup_{i=1}^{1} A_i \right)^c = (A_1)^c = \left(\bigcap_{i=1}^{1} A_i \right)^c.$$

If the formula is true for n = k, then it is true for n = k + 1 Suppose that for some integer $k \geq 1$,

$$\left(\bigcup_{i=1}^{k} A_i \right)^c = \bigcap_{i=1}^{k} (A_i)^c. \qquad \text{This is the inductive hypothesis.}$$

We must show that

$$\left(\bigcup_{i=1}^{k+1} A_i \right)^c = \bigcap_{i=1}^{k+1} (A_i)^c.$$

But

$$\left(\bigcup_{i=1}^{k+1} A_i \right)^c = \left(\left(\bigcup_{i=1}^{k} A_i \right) \cup A_n \right)^c \qquad \text{by the recursive definition of union}$$

$$= \left(\bigcup_{i=1}^{k} A_i \right)^c \cap (A_n)^c \qquad \text{by De Morgan's law for two sets}$$

$$= \left(\bigcap_{i=1}^{k} (A_i)^c \right) \cap (A_n)^c \qquad \text{by inductive hypothesis}$$

$$= \left(\bigcap_{i=1}^{k+1} (A_i)^c \right) \qquad \text{by the recursive definition of intersection.}$$

This is what was to be shown. ■

RECURSIVE FUNCTIONS

A function is said to be **defined recursively** or to be a **recursive function** if its rule of definition refers to itself. Because of this self-reference, it is sometimes difficult to tell whether a given recursive function is well defined. Recursive functions are of great importance in the theory of computation in computer science.

EXAMPLE 8.4.7 **McCarthy's 91 Function**

The following function $M : \mathbf{Z}^+ \to \mathbf{Z}$ was defined by John McCarthy, a pioneer in the theory of computation and in the study of artificial intelligence:

$$M(n) = \begin{cases} n - 10 & \text{if } n > 100 \\ M(M(n + 11)) & \text{if } n \leq 100 \end{cases}$$

for all positive integers n. Find $M(99)$.

Solution   By repeated use of the definition of M,

$$
\begin{aligned}
M(99) &= M(M(110)) && \text{since } 99 \le 100 \\
&= M(100) && \text{since } 110 > 100 \\
&= M(M(111)) && \text{since } 100 \le 100 \\
&= M(101) && \text{since } 111 > 100 \\
&= 91 && \text{since } 101 > 100.
\end{aligned}
$$

John McCarthy

The remarkable thing about this function is that it takes the value 91 for all positive integers less than or equal to 101. (You are asked to show this in the exercises at the end of this section.) Of course, for $n > 101$, $M(n)$ is well defined because it equals $n - 10$. ∎

EXAMPLE 8.4.8   The Ackermann Function

In the 1920s the German logician and mathematician Wilhelm Ackermann first defined a version of the function that now bears his name. This function is important in computer science because it helps answer the question of what can and cannot be computed on a computer. It is defined on the set of all pairs of nonnegative integers as follows:

$$
\begin{aligned}
A(0, n) &= n + 1 && \text{for all nonnegative integers } n, && \textbf{8.4.1} \\
A(m, 0) &= A(m - 1, 1) && \text{for all positive integers } m, && \textbf{8.4.2} \\
A(m, n) &= A(m - 1, A(m, n - 1)) && \text{for all positive integers } m \text{ and } n. && \textbf{8.4.3}
\end{aligned}
$$

Find $A(1, 2)$.

Solution
$$
\begin{aligned}
A(1, 2) &= A(0, A(1, 1)) && \text{by (8.4.3) with } m = 1 \text{ and } n = 2 \\
&= A(0, A(0, A(1, 0))) && \text{by (8.4.3) with } m = 1 \text{ and } n = 1 \\
&= A(0, A(0, A(0, 1))) && \text{by (8.4.2) with } m = 1 \\
&= A(0, A(0, 2)) && \text{by (8.4.1) with } n = 1 \\
&= A(0, 3) && \text{by (8.4.1) with } n = 2 \\
&= 4 && \text{by (8.4.1) with } n = 3.
\end{aligned}
$$

Wilhelm Ackermann

The special properties of the Ackermann function are a consequence of its phenomenal rate of growth. While the values of $A(0, 0) = 1$, $A(1, 1) = 3$, $A(2, 2) = 7$, and $A(3, 3) = 61$ are not especially impressive,

$$
A(4, 4) \cong 2^{2^{2^{2^{65536}}}}
$$

and the values of $A(n, n)$ continue to increase with extraordinary rapidity thereafter. ∎

The argument is somewhat technical, but it is not difficult to show that the Ackermann function is well defined. The following is an example of a recursive "definition" that does not define a function.

EXAMPLE 8.4.9 A Recursive "Function" That Is Not Well Defined

Consider the following attempt to define a recursive function G from $\mathbf{Z}^+$ to $\mathbf{Z}$. For all integers $n \geq 1$,

$$G(n) = \begin{cases} 1 & \text{if } n \text{ is } 1 \\ 1 + G\left(\dfrac{n}{2}\right) & \text{if } n \text{ is even} \\ G(3n - 1) & \text{if } n \text{ is odd and } n > 1. \end{cases}$$

Is G well defined? Why?

Solution   Suppose G is a function. Then by definition of G,

$G(1) = 1,$
$G(2) = 1 + G(1) = 1 + 1 = 2,$
$G(3) = G(8) = 1 + G(4) = 1 + (1 + G(2)) = 1 + (1 + 2) = 4,$
$G(4) = 1 + G(2) = 1 + 2 = 3.$

However,

$$\begin{aligned} G(5) = G(14) = 1 + G(7) &= 1 + G(20) \\ &= 1 + (1 + G(10)) = 1 + (1 + (1 + G(5))) \\ &= 3 + G(5). \end{aligned}$$

Subtracting $G(5)$ from both sides gives

$$0 = 3,$$

which is false. Since the supposition that G is a function leads logically to a false statement, it follows that G is not a function. ∎

A slight modification of the formula of Example 8.4.9 produces a "function" whose status of definition is unknown. Consider the following formula:

$$H(n) = \begin{cases} 1 & \text{if } n \text{ is } 1 \\ 1 + H\left(\dfrac{n}{2}\right) & \text{if } n \text{ is even} \\ H(3n + 1) & \text{if } n \text{ is odd and } n > 1. \end{cases}$$

Lothar Collatz, who devised it in a slightly different form, has conjectured that the formula defines a function on the set of all positive integers. At the present time, however, it is only known for sure that $H(n)$ is computable for all integers n with $1 \leq n < 10^9$.

EXERCISE SET 8.4

1. Consider the set of Boolean expressions defined in Example 8.4.1. Give derivations showing that each of the following is a Boolean expression over the English alphabet $\{a, b, c, \ldots, x, y, z\}$.

   a. $(\sim p \lor (q \land (r \lor \sim s)))$
   b. $((p \lor q) \lor \sim((p \land \sim s) \land r))$

2. Let Σ^* be defined as in Example 8.4.2. Give derivations showing that each of the following is in Σ^*.

   a. aab b. bb

3. Consider the MIU-system discussed in Example 8.4.3. Give derivations showing that each of the following is in the MIU-system.

   a. $MIUI$ b. $MUIIU$

H◆4. Is the string MU in the MIU-system?

5. Consider the set P of parenthesis structures defined in Example 8.4.4. Give derivations showing that each of the following is in P.

   a. $()(())$ b. $(())(())$

◆6. Determine whether either of the following parenthesis structures is in the set P defined in Example 8.4.4.

   a. $()(()$ b. $(()()))(()$

7. The set of arithmetic expressions over the real numbers can be defined recursively as follows:

   I. BASE: Each real number r is an arithmetic expression.

   II. RECURSION: If u and v are arithmetic expressions, then the following are also arithmetic expressions:

      a. $(+u)$ b. $(-u)$ c. $(u + v)$

      d. $(u - v)$ e. $(u \cdot v)$ f. $\left(\dfrac{u}{v}\right)$

   III. RESTRICTION: There are no arithmetic expressions over the real numbers other than those obtained from I and II.

   (Note that the *expression* $\left(\dfrac{u}{v}\right)$ is legal even though the value of v may be 0.) Give derivations showing that each of the following is an arithmetic expression.

   a. $((2 \cdot (0.3 - 4.2)) + (-7))$

   b. $\left(\dfrac{(9 \cdot (6.1 + 2))}{((4 - 7) \cdot 6)}\right)$

8. Give a recursive definition for the set of all strings of 0's and 1's that have the same number of 0's as 1's.

9. Give a recursive definition for the set of all strings of 0's and 1's for which all the 0's precede all the 1's.

10. Give a recursive definition for the set of all strings of a's and b's that contain an even number of a's.

11. Give a recursive definition for the set of all strings of a's and b's that contain exactly one a.

12. Use the recursive definition of summation together with mathematical induction to prove that for all positive integers n, if $a_1, a_2, \ldots, a_n$ and c are real numbers, then

$$\sum_{i=1}^{n} c \cdot a_i = c \cdot \left(\sum_{i=1}^{n} a_i\right).$$

13. Use the recursive definition of product together with mathematical induction to prove that for all positive integers n, if $a_1, a_2, \ldots, a_n$ and $b_1, b_2, \ldots, b_n$ are real numbers, then

$$\prod_{i=1}^{n}(a_i \cdot b_i) = \left(\prod_{i=1}^{n} a_i\right) \cdot \left(\prod_{i=1}^{n} b_i\right).$$

14. Use the recursive definition of product together with mathematical induction to prove that for all positive integers n, if $a_1, a_2, \ldots, a_n$ and c are real numbers, then

$$\prod_{i=1}^{n}(c \cdot a_i) = c^n \cdot \left(\prod_{i=1}^{n} a_i\right).$$

15. The triangle inequality for absolute value states that for all real numbers a and b, $|a + b| \le |a| + |b|$. Use the recursive definition of summation, the triangle inequality, the definition of absolute value, and mathematical induction to prove that for all positive integers n, if $a_1, a_2, \ldots, a_n$ are real numbers, then

$$\left|\sum_{i=1}^{n} a_i\right| \le \sum_{i=1}^{n} |a_i|.$$

16. Use the recursive definitions of union and intersection to prove the following general distributive law: For all positive integers n, if A and $B_1, B_2, \ldots, B_n$ are sets, then

$$A \cap \left(\bigcup_{i=1}^{n} B_i\right) = \bigcup_{i=1}^{n}(A \cap B_i).$$

17. Use the recursive definitions of union and intersection to prove the following general distributive law: For all positive integers n, if A and $B_1, B_2, \ldots, B_n$ are sets, then

$$A \cup \left(\bigcap_{i=1}^{n} B_i\right) = \bigcap_{i=1}^{n}(A \cup B_i).$$

18. Use the recursive definitions of union and intersection to prove the following general De Morgan's law: For all positive integers n, if $A_1, A_2, \ldots, A_n$ are sets, then

$$\left(\bigcap_{i=1}^{n} A_i \right)^c = \bigcup_{i=1}^{n} (A_i)^c.$$

19. Use the definition of McCarthy's 91 function in Example 8.4.7 to show the following:

    a. $M(86) = M(91)$ b. $M(91) = 91$

◆ 20. Prove that McCarthy's 91 function equals 91 for all positive integers less than or equal to 101.

21. Use the definition of the Ackermann function in Example 8.4.8 to compute the following:

    a. $A(1, 1)$ b. $A(2, 1)$

22. Use the definition of the Ackermann function to show the following:

    a. $A(1, n) = n + 2$, for all nonnegative integers n.
    b. $A(2, n) = 3 + 2n$, for all nonnegative integers n.
    c. $A(3, n) = 8 \cdot 2^n - 3$, for all nonnegative integers n.

23. Compute $H(2), H(3), H(4), H(5), H(6),$ and $H(7)$ for the "function" H defined after Example 8.4.9.

24. Student A tries to define a function $F : \mathbf{Z}^+ \to \mathbf{Z}$ by the rule

$$F(n) = \begin{cases} 1 & \text{if } n \text{ is 1} \\ F\left(\dfrac{n}{2}\right) & \text{if } n \text{ is even} \\ 1 + F(5n - 9) & \text{if } n \text{ is odd and } n > 1 \end{cases}$$

for all integers $n \geq 1$. Student B claims that F is not well defined. Justify student B's claim.

25. Student C tries to define a function $G : \mathbf{Z}^+ \to \mathbf{Z}$ by the rule

$$G(n) = \begin{cases} 1 & \text{if } n \text{ is 1} \\ G\left(\dfrac{n}{2}\right) & \text{if } n \text{ is even} \\ 2 \cdot G(3n - 2) & \text{if } n \text{ is odd and } n > 1 \end{cases}$$

for all integers $n \geq 1$. Student D claims that G is not well defined. Justify student D's claim.

9 *O*-Notation and the Efficiency of Algorithms

René Descartes (1596–1650)

In 1637 the French mathematician and philosopher René Descartes published his great philosophical work *Discourse on Method.* An appendix to this work, called "Geometry," laid the foundation for the subject of analytic geometry, in which geometric methods are applied to the study of algebraic objects, such as functions, equations, and inequalities, and algebraic methods are used to study geometric objects, such as straight lines, circles, and half-planes.

The analytic geometry of Descartes provides the foundation for the main topic of this chapter: *O*-notation (read "big-oh notation") and its application to the analysis of algorithms. In Section 9.1 we briefly discuss certain properties of graphs of real-valued functions of a real variable that are needed to understand *O*-notation. In Section 9.2 we introduce the concept of *O*-notation for power and polynomial functions, and in Section 9.3 we show how *O*-notation is used to study the efficiency of algorithms. Because the analysis of algorithms often involves logarithmic and exponential functions, we develop the needed properties of these functions in Section 9.4 and apply them to several algorithms in Section 9.5.

9.1 REAL-VALUED FUNCTIONS OF A REAL VARIABLE AND THEIR GRAPHS

The first precept was never to accept a thing as true until I knew it as such without a single doubt.
(René Descartes, 1637)

A **Cartesian plane** or **two-dimensional Cartesian coordinate system** is a pictorial representation of $\mathbf{R} \times \mathbf{R}$ obtained by setting up a one-to-one correspondence between ordered pairs of real numbers and points in a Euclidean plane. To obtain it, two perpendicular lines, called the **horizontal** and **vertical axes**, are drawn in the plane. Their point of intersection is called the **origin**, and a unit of distance is chosen for each axis. An ordered pair (x, y) of real numbers corresponds to the point P that lies $|x|$ units to the right or left of the vertical axis and $|y|$ units above or below the horizontal axis. On each axis the positive direction is marked with an arrow.

A **real-valued function of a real variable** is a function from one set of real numbers to another. If f is such a function, then to each real number x in the domain of f, there is a unique corresponding real number $f(x)$. Thus it is possible to define the *graph of f* as follows.

DEFINITION

Let f be a real-valued function of a real variable. The **graph of** f is the set of all points (x, y) in the Cartesian coordinate plane with the property that x is in the domain of f and $y = f(x)$.

The definition of *graph* means that for all x in the domain of f:

$$y = f(x) \quad \Leftrightarrow \quad \text{the point } (x, y) \text{ lies on the graph of } f.$$

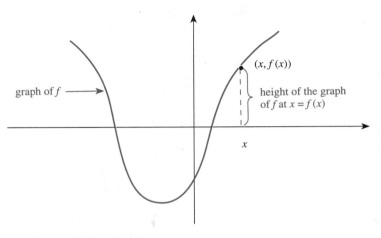

FIGURE 9.1.1   Graph of a Function f

Note that if $f(x)$ can be written as an algebraic expression in x, the graph of the function f is the same as the graph of the equation $y = f(x)$ where x is restricted to lie in the domain of f.

POWER FUNCTIONS

A function that sends a real number x to a particular power, x^a, is called a *power function*. For applications in computer science, we are almost invariably concerned with situations where x and a are nonnegative, and so we restrict our definition to these cases.

DEFINITION

Let a be any nonnegative number. Define p_a, the **power function with exponent** a, as follows:

$$p_a(x) = x^a \quad \text{for each nonnegative real number } x.$$

EXAMPLE 9.1.1   **Graphs of Power Functions**

Plot the graphs of the power functions $p_0, p_{1/2}, p_1$, and p_2 on the same coordinate axes.

Solution   Because the power function with exponent zero satisfies $p_0(x) = x^0 = 1$ for all non-negative numbers x,* all points of the form $(x, 1)$ lie on the graph of p_0 for all such x. So the graph is just a horizontal half-line of height 1 lying above the horizontal axis. Similarly, $p_1(x) = x$ for all nonnegative numbers x, and so the graph of p_1 consists of all points of the form (x, x) where x is nonnegative. The graph is, therefore, the half-line of slope 1 that emanates from $(0, 0)$.

Since for each nonnegative number x, $p_{1/2}(x) = x^{1/2} = \sqrt{x}$, any point with coordinates $(x, \sqrt{x})$, where x is nonnegative, is on the graph of $p_{1/2}$. For instance, the graph of $p_{1/2}$ contains the points $(0, 0), (1, 1), (4, 2)$, and $(9, 3)$. Similarly, since $p_2(x) = x^2$, any point with coordinates (x, x^2) lies on the graph of p_2. Thus, for instance, the graph of p_2 contains the points $(0, 0), (1, 1), (2, 4)$, and $(3, 9)$.

The graphs of all four functions are shown in Figure 9.1.2.

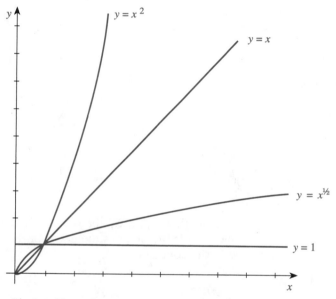

FIGURE 9.1.2   Graphs of Some Power Functions

THE FLOOR FUNCTION

The floor and ceiling functions arise in many computer science contexts. Example 9.1.2 illustrates the graph of the floor function. In exercise 5 at the end of this section you are asked to draw the graph of the ceiling function.

EXAMPLE 9.1.2   **Graph of the Floor Function**

Recall that each real number is either an integer itself or sits between two consecutive integers: For each real number x, there exists a unique integer n such that $n \le x < n + 1$. The floor of a number is the integer immediately to its left on the

---

*As in Section 6.7 (see page 338), we simply define $0^0 = 1$.

number line. More formally, the floor function F is defined by the rule

For each real number x,

$$F(x) = \lfloor x \rfloor$$

= the greatest integer that is less than or equal to x

= the unique integer n such that $n \le x < n + 1$.

Graph the floor function.

Solution   If n is any integer, then for each real number x in the interval $n \le x < n + 1$, the floor of x, $\lfloor x \rfloor$, equals n. Thus on each such interval the graph of the floor function is horizontal; for each x in the interval the height of the graph is n.

It follows that the graph of the floor function consists of horizontal line segments, like a staircase, as shown in Figure 9.1.3. The open circles at the right-hand edge of each step are used to show that those points are *not* on the graph.

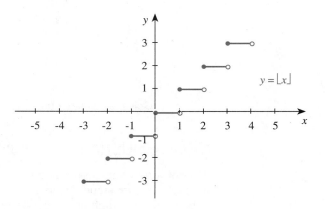

FIGURE 9.1.3   Graph of the Floor Function

GRAPHING FUNCTIONS DEFINED ON SETS OF INTEGERS

Many real-valued functions used in computer science are defined on sets of integers and not on intervals of real numbers. If you know what the graph of a function looks like when it is given by a certain formula on an interval of real numbers, you can obtain the graph of the function defined by the same formula on the set of integers in the interval by selecting out only those points on the known graph with integers as their first coordinates. For instance, if f is the function defined by the same formula as the power function p_1 but having as its domain the set of nonnegative integers, then $f(n) = n$ for all nonnegative integers n. The graphs of p_1, reproduced from Example 9.1.2, and f are shown side-by-side below.

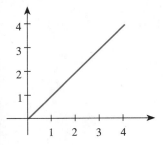

Graph of p_1 where $p_1(x) = x$
for all nonnegative
real numbers x

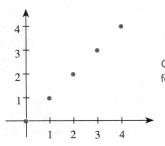

Graph of f where $f(n) = n$
for all nonnegative integers n

EXAMPLE 9.1.3   **Graph of a Function Defined on a Set of Integers**

Consider an integer version of the power function $p_{1/2}$. In other words, define a function g by the formula $g(n) = n^{1/2}$ for all nonnegative integers n. Draw the graph of g.

Solution   Look back at the graph of $p_{1/2}$ in Figure 9.1.2. Draw the graph of g by reproducing only those points on the graph of $p_{1/2}$ with integer first coordinates. Thus for each nonnegative integer n, the point $(n, n^{1/2})$ is on the graph of g.

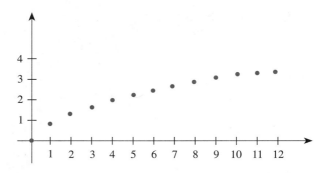

Graph of g where $g(n) = n^{1/2}$ for all nonnegative integers n

GRAPH OF A MULTIPLE OF A FUNCTION

A *multiple* of a function is obtained by multiplying every value of the function by a fixed number. To understand the concept of O-notation, it is helpful to understand the relation between the graph of a function and the graph of a multiple of the function.

> **DEFINITION**
>
> Let f be a real-valued function of a real variable and let M be any real number. The function $M \cdot f$, called the **multiple of f by M or M times f**, is the real-valued function with the same domain as f that is defined by the rule
>
> $$(M \cdot f)(x) = M \cdot f(x) \quad \text{for all } x \in \text{domain of } f.$$

If the graph of a function is known, the graph of any multiple can easily be deduced. Specifically, if f is a function and M is a real number, the height of the graph of $M \cdot f$ at any real number x is M times the quantity $f(x)$. To sketch the graph of $M \cdot f$ from the graph of f, you plot the heights $M \cdot f(x)$ based on knowledge of M and visual inspection of the heights $f(x)$.

EXAMPLE 9.1.4   **Graph of a Multiple of a Function**

Let f be the function whose graph is shown below. Sketch the graph of $2f$.

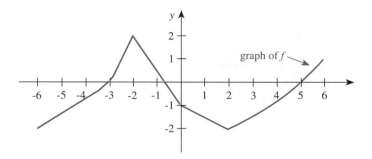

Solution   At each real number x, you obtain the height of the graph of $2f$ by measuring the height of the graph of f at x and multiplying that number by 2. The result is the graph shown below. Note that the general shapes of f and $2f$ are very similar, but the graph of $2f$ is "stretched out": the "highs" are twice as high and the "lows" are twice as low.

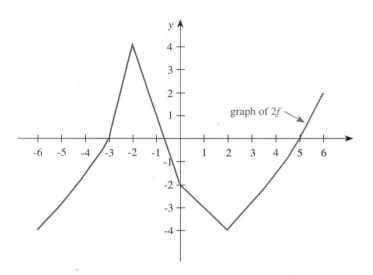

INCREASING AND DECREASING FUNCTIONS

Consider the *absolute value function, A* , which is defined as follows:

$$A(x) = |x| = \begin{cases} x & \text{if } x \geq 0 \\ -x & \text{if } x < 0 \end{cases} \qquad \text{for all real numbers } x.$$

When $x \geq 0$, the graph of A is the same as the graph of $y = x$, the straight line with slope 1 that passes through the origin $(0, 0)$. For $x < 0$, the graph of A is the same as the graph of $y = -x$, which is the straight line with slope -1 that passes through $(0, 0)$. (See Figure 9.1.4 on page 482.)

Note that as you trace from left to right along the graph to the left of the origin, the height of the graph continually *decreases*. For this reason, the absolute value function is said to be *decreasing* on the set of real numbers less than 0. On the other hand, as you trace from left to right along the graph to the right of the origin, the height of the graph continually *increases*. Consequently, the absolute value function is said to be *increasing* on the set of real numbers greater than 0.

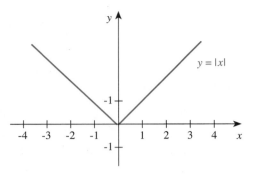

FIGURE 9.1.4   Graph of the Absolute Value Function

Since the height of the graph of a function f at a point x is $f(x)$, these geometric concepts translate to the following analytic definition.

> ### DEFINITION
>
> Let f be a real-valued function defined on a set of real numbers and suppose the domain of f contains a set S. We say that f is **increasing on the set S** if, and only if,
>
> for all real numbers x_1 and x_2 in S, if $x_1 < x_2$ then $f(x_1) < f(x_2)$.
>
> We say that f is **decreasing on the set S** if, and only if,
>
> for all real numbers x_1 and x_2 in S, if $x_1 < x_2$ then $f(x_1) > f(x_2)$.
>
> We say that f is an **increasing** (or **decreasing**) **function** if, and only if, f is increasing (or decreasing) on its entire domain.

Figure 9.1.5 illustrates the analytic definitions of increasing and decreasing.

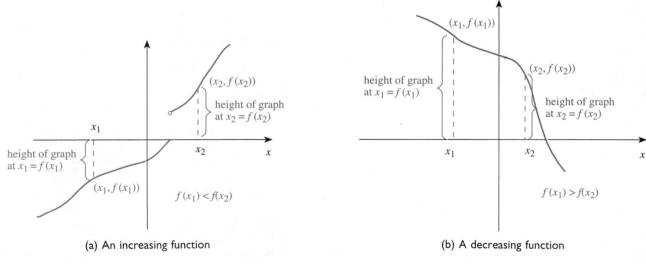

(a) An increasing function   (b) A decreasing function

FIGURE 9.1.5

It follows almost immediately from the definitions that both increasing functions and decreasing functions are one-to-one. You are asked to show this in the exercises.

EXAMPLE 9.1.5   **A Positive Multiple of an Increasing Function Is Increasing**

Suppose that f is a real-valued function of a real variable that is increasing on a set S of real numbers and suppose M is any positive real number. Show that $M \cdot f$ is also increasing on S.

Solution   Suppose x_1 and x_2 are particular but arbitrarily chosen elements of S such that

$$x_1 < x_2.$$

[We must show that $(M \cdot f)(x_1) < (M \cdot f)(x_2)$.] From the facts that $x_1 < x_2$ and f is increasing, it follows that

$$f(x_1) < f(x_2).$$

Then

$$M \cdot f(x_1) < M \cdot f(x_2),$$

since multiplying both sides of the inequality by a positive number does not change the direction of the inequality. Hence by definition of $M \cdot f$,

$$(M \cdot f)(x_1) < (M \cdot f)(x_2),$$

and, consequently, $M \cdot f$ is increasing on S.   ■

It is also true that a positive multiple of a decreasing function is decreasing, that a negative multiple of a increasing function is decreasing, and that a negative multiple of a decreasing function is increasing. The proofs of these facts are left to the exercises.

EXERCISE SET 9.1*

1. The graph of a function f is shown below.
   a. Is $f(0)$ positive or negative?
   b. For what values of x does $f(x) = 0$?
   c. As x increases from -3 to -1, do the values of f increase or decrease?
   d. As x increases from 0 to 4, do the values of f increase or decrease?

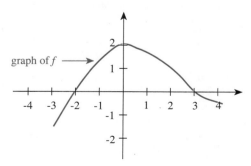

graph of f ⟶

2. Draw the graphs of the power functions $p_{1/3}$ and $p_{1/4}$ on the same set of axes. When $0 < x < 1$, which is greater: $x^{1/3}$ or $x^{1/4}$? When $x > 1$, which is greater: $x^{1/3}$ or $x^{1/4}$?

3. Draw the graphs of the power functions p_3 and p_4 on the same set of axes. When $0 < x < 1$, which is greater: x^3 or x^4? When $x > 1$, which is greater: x^3 or x^4?

4. Draw the graphs of $y = 2\lfloor x \rfloor$ and $y = \lfloor 2x \rfloor$ for all real numbers x. What can you conclude from these graphs?

Graph each of the functions defined in 5–8 below.

5. $g(x) = \lceil x \rceil$ for all real numbers x (Recall that the ceiling of x, $\lceil x \rceil$, is the least integer that is greater than or equal to x. That is, $\lceil x \rceil =$ the unique integer n such that $n - 1 < x \le n$.)

6. $h(x) = \lceil x \rceil - \lfloor x \rfloor$ for all real numbers x

7. $F(x) = \lfloor x^{1/2} \rfloor$ for all real numbers x

8. $G(x) = x - \lfloor x \rfloor$ for all real numbers x

*Exercises with blue numbers have solutions in Appendix B. The symbol H indicates that only a partial solution is given. The symbol ◆ signals that an exercise is more challenging than usual.

In each of 9–12 a function is defined on a set of integers. Graph each function.

9. $f(n) = |n|$ for each integer n

10. $g(n) = (n/2) + 1$ for each integer n

11. $h(n) = \lfloor n/2 \rfloor$ for each integer $n \geq 0$

12. $k(n) = \lfloor n^{1/2} \rfloor$ for each integer $n \geq 0$

13. The graph of a function f is shown below. Find the intervals on which f is increasing and on which f is decreasing.

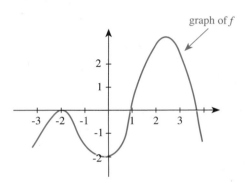

14. Show that the function $f : \mathbf{R} \to \mathbf{R}$ defined by the rule $f(x) = 2x - 3$ is increasing on the set of all real numbers.

15. Show that the function $g : \mathbf{R} \to \mathbf{R}$ defined by the rule $g(x) = -(x/3) + 1$ is decreasing on the set of all real numbers.

16. Let h be the function from $\mathbf{R}$ to $\mathbf{R}$ defined by the formula $h(x) = x^2$ for all real numbers x.
    a. Show that h is decreasing on the set of all real numbers less than zero.
    b. Show that h is increasing on the set of all real numbers greater than zero.

17. Let $k : \mathbf{R} \to \mathbf{R}$ be the function defined by the formula $k(x) = (x - 1)/x$ for all real numbers $x \neq 0$.
    a. Show that k is increasing for all real numbers $x > 0$.
    b. Is k increasing or decreasing for $x < 0$? Prove your answer.

18. Show that if a function $f : \mathbf{R} \to \mathbf{R}$ is increasing, then f is one-to-one.

19. Given real-valued functions f and g with the same domain D, the sum of f and g, denoted $f + g$, is defined as follows:

    For all real numbers x, $(f + g)(x) = f(x) + g(x)$.

    Show that if f and g are both increasing on a set S, then $f + g$ is also increasing on S.

20. a. Let m be any positive integer and define $f(x) = x^m$ for all nonnegative real numbers x. Use the binomial theorem to show that f is an increasing function.
    b. Let m and n be any positive integers and let $g(x) = x^{m/n}$ for all nonnegative real numbers x. Prove that g is an increasing function.
    The results of this exercise are used in the exercises for Sections 9.2 and 9.4.

21. Let f be the function whose graph is shown below. Draw the graph of $3f$.

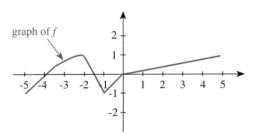

22. Let h be the function whose graph is shown below. Draw the graph of $2h$.

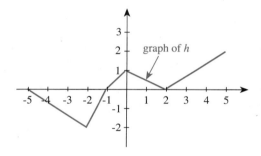

23. Let f be a real-valued function of a real variable. Show that if f is decreasing on a set S and if M is any positive real number, then $M \cdot f$ is decreasing on S.

24. Let f be a real-valued function of a real variable. Show that if f is increasing on a set S and if M is any negative real number, then $M \cdot f$ is decreasing on S.

25. Let f be a real-valued function of a real variable. Show that if f is decreasing on a set S and if M is any negative real number, then $M \cdot f$ is increasing on S.

In 26 and 27, functions f and g are defined. In each case draw the graphs of f and $2g$ on the same set of axes and find a number x_0 so that $f(x) \leq 2g(x)$ for all $x > x_0$. If you have a graphing calculator, you can find approximate values of x_0 by using the zoom and trace features. If not, you can find x_0 by solving a quadratic equation.

26. $f(x) = x^2 + 10x + 3$ and $g(x) = x^2$ for all real numbers $x \geq 0$

27. $f(x) = 2x^2 + 126x + 35$ and $g(x) = 3x^2$ for all real numbers $x \geq 0$

9.2   *O*-NOTATION

Although this may seem a
paradox, all exact science
is dominated by the idea of
approximation.
(Bertrand Russell,
1872–1970)

The *O*-notation (read "big-oh notation") provides a special way to compare relative sizes of functions that is very useful in the analysis of computer algorithms. It often happens that the time or memory space requirements for the algorithms available to do a certain job differ from each other on such a grand scale that differences of just a constant factor are completely overshadowed. The *O*-notation makes use of approximations that highlight these large-scale differences while ignoring differences of a constant factor and differences that only occur for small sets of input data. The notation was introduced in 1892 by the German mathematician Paul Bachman.

  The idea of the *O*-notation is this. Suppose that f and g are both real-valued functions of a real variable x. If, for large values of x, the graph of f lies closer to the horizontal axis than the graph of some positive multiple of g, then f is said to be *of order g*. This is written $f(x)$ *is* $O(g(x))$ and is illustrated in Figure 9.2.1.

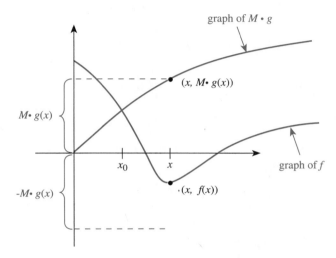

graph of $M \cdot g$

$(x, M \cdot g(x))$

$M \cdot g(x)$

x_0 x

$-M \cdot g(x)$ $\cdot (x, f(x))$

graph of f

The graph of f lies closer to the horizontal axis than the graph of $M \cdot g$ for $x > x_0$, where M is a positive constant.

FIGURE 9.2.1   The Graph of f Compared with the Graph of $M \cdot g$

  In analytic terms, saying that for large values of x the graph of f lies closer to the horizontal axis than the graph of $M \cdot g$ is the same as saying that there is a real number x_0 such that for all real numbers $x > x_0$,

$$-M \cdot g(x) \leq f(x) \leq M \cdot g(x).$$

Or, equivalently,

$$|f(x)| \leq M \cdot |g(x)|.$$

DEFINITION

Let f and g be real-valued functions that are defined on the same set of real numbers. Then **f is of order g**, written **$f(x)$ is $O(g(x))$**, if, and only if, there exists a positive real number M and a real number x_0 such that for all x in the common domain of f and g,

$$|f(x)| \leq M \cdot |g(x)|, \text{ whenever } x > x_0.$$

The sentence "$f(x)$ is $O(g(x))$" is also read "f of x is big-oh of g of x" or "g is a big-oh approximation for f."

Remark on Notation: In Section 7.1 we stated that we would generally make a careful distinction between a function f and its value $f(x)$. The traditional use of the order notation violates this general rule. In the statement "$f(x)$ is $O(g(x))$," the symbols $f(x)$ and $g(x)$ are understood to refer to the functions f and g defined by the expressions $f(x)$ and $g(x)$, respectively. For instance, the statement

$$3\sqrt{x} + 4 \quad \text{is} \quad O(x^{1/2})$$

means that f is of order g where f and g are defined by $f(x) = 3\sqrt{x} + 4$ and $g(x) = x^{1/2}$ with some common domain (usually the largest set of real numbers for which both function formulas are defined). In this case, the common domain could be the set of all nonnegative real numbers (since $\sqrt{x}$ is only defined for $x \geq 0$).

EXAMPLE 9.2.1   Translating to O-Notation

Use O-notation to express the following:

a. $|17x^6 - 3x^3 + 2x + 8| \leq 30|x^6|$, for all real numbers $x > 1$.

b. $\left|\dfrac{15\sqrt{x}(2x + 9)}{x + 1}\right| \leq 45|\sqrt{x}|$, for all real numbers $x > 6$.

Solution   a. Let $M = 30$ and $x_0 = 1$. By definition of O-notation, the given statement translates to

$$17x^6 - 3x^3 + 2x + 8 \quad \text{is} \quad O(x^6).$$

b. Let $M = 45$ and $x_0 = 6$. By definition of O-notation, the given statement translates to

$$\frac{15\sqrt{x}(2x + 9)}{x + 1} \quad \text{is} \quad O(\sqrt{x}). \qquad \blacksquare$$

ORDERS OF POWER FUNCTIONS

Observe that if

$$1 < x,$$

then

$$x < x^2 \qquad \text{multiplying both sides by } x \text{ (which is positive)}$$

and so

$$x^2 < x^3 \qquad \text{multiplying again by } x.$$

Thus if $1 < x$, then

$$1 < x < x^2 < x^3 \qquad \text{by transitivity of } <.$$

The following generalization of this result is developed in exercises 15 and 16 at the end of this section.

> For any rational numbers r and s,
>
> $$\text{if } x > 1 \text{ and } r < s, \text{ then } x^r < x^s. \qquad \qquad 9.2.1$$

Property (9.2.1) has the following consequence for orders.

For any rational numbers r and s,

$$\text{if} \quad r < s, \quad \text{then} \ x^r \ \text{is} \ O(x^s).$$

9.2.2

The relation among the graphs of various positive power functions of x is shown graphically in Figure 9.2.2.

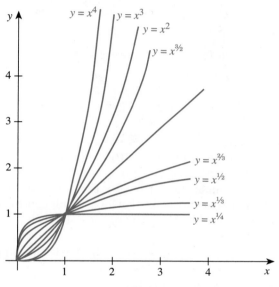

If $r < s$, the graph of $y = x^r$ lies underneath the graph of $y = x^s$ for $x > 1$.

FIGURE 9.2.2   Graphs of Powers of *x*

ORDERS OF POLYNOMIAL FUNCTIONS

The following example shows how to use property (9.2.1) to derive a polynomial inequality.

EXAMPLE 9.2.2   A Polynomial Inequality

Show that for any real number x:

$$\text{if} \ x > 1, \ \text{then} \ 3x^3 + 2x + 7 \le 12x^3.$$

Solution   Suppose x is a real number and $x > 1$. Then by property (9.2.1),

$$x < x^3 \quad \text{and} \quad 1 < x^3.$$

Multiply the left-hand inequality by 2 and the right-hand inequality by 7 to get

$$2x < 2x^3 \quad \text{and} \quad 7 < 7x^3.$$

Now add $3x^3 \le 3x^3$, $2x < 2x^3$, and $7 < 7x^3$ to obtain

$$3x^3 + 2x + 7 \le 3x^3 + 2x^3 + 7x^3 = 12x^3.$$ ■

The method of Example 9.2.2 is used in the next example (more compactly) to show that a polynomial function has a certain order.

EXAMPLE 9.2.3 **Using the Definition to Show That a Polynomial Function with Positive Coefficients Has a Certain Order**

Use the definition of order to show that $2x^4 + 3x^3 + 5$ is $O(x^4)$.

Solution   In this example the functions f and g referred to in the definition of O-notation are defined as follows: For all real numbers x,

$$f(x) = 2x^4 + 3x^3 + 5,$$
$$g(x) = x^4 \qquad \text{since both expressions } 2x^4 + 3x^3 + 5 \text{ and } x^4 \text{ are defined for all real numbers } x.$$

Observe that for all real numbers $x > 1$,

$$|2x^4 + 3x^3 + 5| = 2x^4 + 3x^3 + 5 \qquad 2x^4 + 3x^3 + 5 \text{ is positive because } x > 1$$
$$\Rightarrow |2x^4 + 3x^3 + 5| \le 2x^4 + 3x^4 + 5x^4 \qquad \text{by (9.2.1), } x^3 < x^4 \text{ and } 1 < x^4, \text{ and so } 3x^3 < 3x^4 \text{ and } 5 < 5x^4$$
$$\Rightarrow |2x^4 + 3x^3 + 5| \le 10x^4 \qquad \text{since } 2 + 3 + 5 = 10$$
$$\Rightarrow |2x^4 + 3x^3 + 5| \le 10 \cdot |x^4| \qquad \text{since } x^4 \text{ is nonnegative.}$$

(When the implication arrow, $\Rightarrow$, is placed at the beginning of a line, it means that the truth of the statement in that line is implied by the truth of the statement in the previous line.) Let $M = 10$ and $x_0 = 1$. Then

$$|2x^4 + 3x^3 + 5| \le M \cdot |x^4| \quad \text{for all real numbers } x > x_0.$$

Hence by definition of O-notation,

$$2x^4 + 3x^3 + 5 \quad \text{is} \quad O(x^4). \qquad \blacksquare$$

A general polynomial function can have negative as well as positive coefficients. To show that such a polynomial has a certain order, you need to use the triangle inequality for absolute value. This says that

$$|a + b| \le |a| + |b| \quad \text{for all real numbers } a \text{ and } b.$$

If $-b$ is substituted in place of b, then by the triangle inequality,

$$|a - b| = |a + (-b)| \le |a| + |-b| = |a| + |b| \qquad \text{since } |-b| = |b|.$$

Thus, also,

$$|a - b| \le |a| + |b| \quad \text{for all real numbers } a \text{ and } b.$$

EXAMPLE 9.2.4   Using the Definition to Show That a General Polynomial Function Has a Certain Order

Use the definition of order to show that $7x^3 - 2x + 3$ is $O(x^3)$.

Solution   Observe that for all real numbers $x > 1$,

$$|7x^3 - 2x + 3| \le |7x^3| + |2x| + |3|$$ by the triangle inequality

$$\Rightarrow \quad |7x^3 - 2x + 3| \le 7x^3 + 2x + 3$$ since when $x > 1$ then $7x^3$, $2x$, and 1 are all positive, and so the absolute value signs may be dropped

$$\Rightarrow \quad |7x^3 - 2x + 3| \le 7x^3 + 2x^3 + 3x^3$$ by (9.2.1), $x < x^3$ and $1 < x^3$, and so $2x < 2x^3$ and $3 < 3x^3$

$$\Rightarrow \quad |7x^3 - 2x + 3| \le 12x^3$$ since $7 + 2 + 3 = 12$

$$\Rightarrow \quad |7x^3 - 2x + 3| \le 12|x^3|$$ since when $x > 1$ then x^3 is positive so the absolute value sign may be added.

Let $M = 12$ and $x_0 = 1$. Then

$$|7x^3 - 2x + 3| \le M \cdot |x^3| \quad \text{for all real numbers } x > x_0.$$

Hence by definition of O-notation,

$$7x^3 - 2x + 3 \quad \text{is} \quad O(x^3).$$ ∎

The method illustrated in Example 9.2.4 can be generalized to show that any polynomial function is big-oh of the power function of its highest order term or of any larger power function. A proof of this fact is developed in exercises 15, 17, and 18 at the end of this section.

THEOREM 9.2.1   On Polynomial Orders

If $a_0, a_1, a_2, \ldots, a_n$ are real numbers and $a_n \ne 0$, then

$$a_n x^n + a_{n-1} x^{n-1} + \cdots + a_1 x + a_0 \quad \text{is} \quad O(x^m), \text{ for all } m \ge n.$$

EXAMPLE 9.2.5   Calculating Polynomial Orders Using the Theorem on Polynomial Orders

Use the theorem on polynomial orders to find orders for the functions given by the following formulas.

a. $f(x) = 7x^5 + 5x^3 - x + 4$, for all real numbers x.

b. $g(x) = \dfrac{(x - 1)(x + 1)}{4}$, for all real numbers x.

Solution   a. By direct application of the theorem on polynomial orders, $7x^5 + 5x^3 - x + 4$ is $O(x^5)$. Note that according to the theorem, it is also correct to write that $7x^5 + 5x^3 - x + 4$ is $O(x^6)$, or $O(x^7)$, or $O(x^n)$, for any integer $n \ge 5$, although the choice of $n = 5$ is "best" in a sense to be explained later in this section.

b. $g(x) = \dfrac{(x-1)(x+1)}{4}$

$\qquad = \dfrac{1}{4}(x^2 - 1)$

$\qquad = \dfrac{1}{4}x^2 - \dfrac{1}{4} \qquad$ by algebra.

Thus $g(x)$ is $O(x^2)$ by the theorem on polynomial orders. ∎

ORDERS FOR FUNCTIONS OF INTEGER VARIABLES

It is traditional to use the symbol x to denote a real number variable while n is used to represent an integer variable. Thus given a statement of the form

$$f(n) \quad \text{is} \quad O(g(n)),$$

it is assumed that f and g are functions defined on sets of *integers*. If it is true that

$$f(x) \quad \text{is} \quad O(g(x)),$$

where f and g are functions defined for *real numbers,* then it is certainly true that $f(n)$ is $O(g(n))$. The reason is that if $f(x)$ is $O(g(x))$, then the inequality $|f(x)| \leq M \cdot |g(x)|$ holds for all real numbers $x > x_0$ (for some number x_0). Hence, in particular, the inequality $|f(n)| \leq M \cdot |g(n)|$ is true for all integers $n > x_0$.

EXAMPLE 9.2.6   **An Order for the Sum of the First n Integers**

Sums of the form $1 + 2 + 3 + \cdots + n$ arise in the analysis of computer algorithms such as the selection sort. Show that for a positive integer variable n,

$$1 + 2 + 3 + \cdots + n \quad \text{is} \quad O(n^2).$$

Solution   Two different solutions to this problem are shown because each illustrates a useful technique.

1. By the formula for the sum of the first n integers (see Section 4.2), for all positive integers n,

$$1 + 2 + 3 + \cdots + n = \frac{n(n+1)}{2}.$$

But

$$\frac{n(n+1)}{2} = \frac{1}{2}n^2 + \frac{1}{2}n \qquad \text{by basic algebra.}$$

And by the theorem on polynomial orders,

$$\frac{1}{2}n^2 + \frac{1}{2}n \quad \text{is} \quad O(n^2).$$

Hence

$$1 + 2 + 3 + \cdots + n \quad \text{is} \quad O(n^2).$$

2. Each positive integer n is greater than every positive integer that precedes it. Hence for each positive integer n,

$$1 + 2 + \cdots + n \le \underbrace{n + n + \cdots + n}_{n \text{ terms}} = n \cdot n = n^2.$$

It follows that

$$1 + 2 + \cdots + n \quad \text{is} \quad O(n^2). \qquad \blacksquare$$

"BEST" BIG-OH APPROXIMATIONS

Now consider the problem of showing that one function f is *not* big-oh of another function g. Probably the most natural way to proceed is to suppose that f *is* big-oh of g and derive a contradiction.

EXAMPLE 9.2.7   Showing That a Function Is Not Big-Oh of Another

Show that x^2 is not $O(x)$.

Solution   *[Argue by contradiction.]* Suppose that x^2 is $O(x)$. *[Derive a contradiction.]* By the supposition that x^2 is $O(x)$, there exist a positive real number M and a real number x_0 such that

$$|x^2| \le M \cdot x \quad \text{for all real numbers } x > x_0. \qquad (*)$$

Let x be a positive real number that is greater than both M and x_0. Then

$$x \cdot x > M \cdot x \qquad \text{by multiplying both sides of } x > M \text{ by } x, \text{ which is positive}$$

$$\Rightarrow \quad |x^2| > M \cdot |x| \qquad \text{because } x \text{ is positive.}$$

Thus there is a real number $x > x_0$ such that

$$|x^2| > M \cdot |x|.$$

This contradicts $(*)$. Hence the supposition is false, and so x^2 is not $O(x)$. $\blacksquare$

The technique used in Example 9.2.7 can be extended and generalized to prove that any polynomial function in x of degree n is *not* big-oh of the mth power function if $m < n$.

If n is a positive integer, $a_0, a_2, \ldots, a_n$ are real numbers, and $a_n \ne 0$, then

$$a_n x^n + a_{n-1} x^{n-1} + \cdots + a_1 x + a_0 \text{ is not } O(x^m) \qquad 9.2.3$$

for any integer m with $m < n$.

Frequently, when the order notation is used in the analysis of computer algorithms, it is desired to find "a best big-oh approximation" for a given function from among a select set of functions. Roughly speaking, a best-oh approximation for a given function from a specified set of functions is a "smallest" function in the set that is an order of the given function.

> **DEFINITION**
>
> Suppose S is a set of functions from a subset of $\mathbf{R}$ to $\mathbf{R}$ and suppose f is a real-valued function of a real variable. We say that a function g in S is **a best big-oh approximation for f in S** if, and only if,
>
> 1. $f(x)$ is $O(g(x))$;
> 2. for any h in S, if $f(x)$ is $O(h(x))$, then $g(x)$ is $O(h(x))$.

If f is a polynomial function, then we almost always look for a best big-oh approximation for f within the set of integral power functions.

EXAMPLE 9.2.8   **Finding a Best Big-Oh Approximation for a Polynomial**

Let f be the function defined by the rule

$$f(x) = 5x^3 - 2x + 1 \quad \text{for all real numbers } x,$$

and let S be the set of all integral power functions. That is, S is the set of all functions g defined by a formula $g(x) = x^n$ for all real numbers x, where n is an integer. Show that the cubic power function (power function of degree 3) is a best big-oh approximation for f in S.

Solution   By the theorem on polynomial orders,

$$5x^3 - 2x + 1 \text{ is } O(x^n) \quad \text{for all integers } n \geq 3.$$

By property (9.2.3),

$$5x^3 - 2x + 1 \text{ is not } O(x^m) \quad \text{for all integers } m < 3.$$

Thus f is big-oh of all integral power functions with exponents greater than or equal to 3 and *is not* big-oh of any integral power function with exponent less than 3. It follows that x^3 is a best big-oh approximation for f from among the set of all integral power functions. ∎

The theorem on polynomial orders and property (9.2.3) both extend as follows to functions made up of rational power functions. (See exercises 16, 17, 27–29, and 34.)

> If $a_0, a_1, \ldots, a_n$ are real numbers with $a_n \neq 0$ and if $r_0, r_1, \ldots, r_n$ are rational numbers with $r_0 < r_1 < \cdots < r_n$, then
>
> 1. $a_n x^{r_n} + a_{n-1} x^{r_{n-1}} + \cdots + a_1 x^{r_1} + a_0 x^{r_0}$ is $O(x^r)$ for all rational numbers $r \geq r_n$.
> 2. $a_n x^{r_n} + a_{n-1} x^{r_{n-1}} + \cdots + a_1 x^{r_1} + a_0 x^{r_0}$ is not $O(x^s)$ for any rational number $s < r_n$.
>
> 9.2.4

EXAMPLE 9.2.9 **Finding a Best Big-Oh Approximation from the Set of Rational Power Functions**

Find a best big-oh approximation from the set of all rational power functions for the function f defined by

$$f(x) = \frac{(\sqrt{x} + 2)(x^2 - 2)}{4} \quad \text{for all real numbers } x > 0.$$

Solution Multiply out the expression for $F(x)$ to obtain

$$f(x) = \frac{(\sqrt{x} + 2)(x^2 - 2)}{4} = \frac{1}{4}x^{5/2} + \frac{1}{2}x^2 - \frac{1}{2}x^{1/2} - 1.$$

Then by property (9.2.3), $f(x)$ is $O(x^r)$ for all rational numbers $r \geq 5/2$ and $f(x)$ is not $O(x^s)$ for any rational number $s < 5/2$. So a best big-oh approximation for f is the power function that takes each positive real number x to $x^{5/2}$. ∎

EXERCISE SET 9.2

1. Draw graphs of functions f, g, and $3g$ with the following properties: f and g are defined and for all real numbers $x \geq 0$, $|f(x)| > |g(x)|$ for all x with $0 \leq x \leq 2$ and $|f(x)| \leq 3|g(x)|$ for all $x > 2$. Which of the following is true: $f(x)$ is $O(g(x))$ or $g(x)$ is $O(f(x))$?

2. The following is a formal definition of the O-notation, written using quantifiers and variables:

$f(x)$ is $O(g(x))$ if, and only if, $\exists$ a positive real number M and a real number x_0 such that $\forall$ real numbers $x > x_0$,

$$|f(x)| \leq M \cdot |g(x)|.$$

a. Write the formal negation for this definition using the symbols $\forall$ and $\exists$.
b. Restate the negation less formally without using the symbols $\forall$ and $\exists$.

Express each of the statements in 3–4 using O-notation.

3. $|5x^8 - 9x^7 + 2x^5 + 3x - 1| \leq 20 \cdot |x^8|$, for all real numbers $x > 1$.

4. $\left| \dfrac{(x^2 - 1)(12x + 25)}{3x^2 + 4} \right| \leq 4 \cdot |x|$, for all real numbers $x > 3$.

5. a. Show that for any real number x, if $x > 1$ then $|2x^2 + 15x + 4| \leq 21|x^2|$.
   b. Use O-notation to express the result of part (a).

6. a. Show that for any real number x, if $x > 1$ then $|23x^4 + 8x^3 + 4x| \leq 35|x^4|$.
   b. Use O-notation to express the result of part (a).

7. a. Show that $\dfrac{8x + 45}{x + 1} \leq 9$ for all real numbers $x > 36$.
   b. Use the result of part (a) to show that $\left| \dfrac{15x^4(8x + 1)}{(x + 1)} \right| \leq 135|x^4|$ for all real numbers $x > 36$.
   c. Use O-notation to express the result for part (b).

8. a. Show that $\dfrac{(12x^2 + 60)}{(5x^2 + 4)} \leq 3$ for all real numbers $x > 4$.
   b. Use the result of part (a) to show that $\left| \dfrac{(12x^{7/2} + 60x^{3/2})}{(5x^2 + 4)} \right| \leq 3|x^{3/2}|$ for all real numbers $x > 4$.
   c. Use O-notation to express the result of part (b).

Prove each of the statements in 9–14 directly from the definition of O-notation. Do not use the theorem on polynomial orders.

9. $7x^2 + 12x$ is $O(x^2)$

10. $6x^4 + x^2 + 13$ is $O(x^4)$

11. $100x^5 - 50x^3 - 18x^2 + 12x$ is $O(x^5)$

12. $10x^3 + x^2 - 5x + 6$ is $O(x^3)$

13. $\lceil x^2 \rceil$ is $O(x^2)$

14. $\lfloor \sqrt{n} \rfloor$ is $O(\sqrt{n})$

H 15. Prove that if $x > 1$ and m and n are integers with $m < n$, then $x^m < x^n$.

H 16. a. Let n be any positive integer. Use proof by contraposition to show that for all positive real numbers x, if $x > 1$ then $x^{1/n} > 1$.

b. Let p, q, r, and s be integers with q and s nonzero and $(p/q) > (r/s)$. Use part (a) and exercise 15 to show that for any real number x, if $x > 1$ then $x^{p/q} > x^{r/s}$.

17. a. Show that if f, g, and h are functions from $\mathbf{R}$ to $\mathbf{R}$ and $f(x)$ is $O(h(x))$ and $g(x)$ is $O(h(x))$ then $f(x) + g(x)$ is $O(h(x))$.

b. How does it follow from part (a) that $x^4 + x^2$ is $O(x^4)$?

c. Show that if f is a function from $\mathbf{R}$ to $\mathbf{R}$, $f(x)$ is $O(g(x))$, and c is any real number, then $cf(x)$ is $O(g(x))$.

d. How does it follow from parts (a) and (c) that $12x^5 - 34x^2 + 7$ is $O(x^5)$?

18. Show how the theorem on polynomial orders follows from exercise 15 and parts (a) and (c) of exercise 17.

Use the theorem on polynomial orders to prove each of the statements in 19–22.

19. $\dfrac{(x + 1)(x - 2)}{4}$ is $O(x^2)$.

20. $\dfrac{x}{3}(4x^2 - 1)$ is $O(x^3)$.

21. $\dfrac{n(n + 1)(2n + 1)}{6}$ is $O(n^3)$.

22. $\left[\dfrac{n(n + 1)}{2} \right]^2$ is $O(n^4)$.

Prove each of the statements in 23–26 assuming n is a variable that takes positive integer values. (Use formulas from the exercise set of Section 4.2 and the theorem on polynomial orders as appropriate.)

23. $1^2 + 2^2 + 3^2 + \cdots + n^2$ is $O(n^3)$.

24. $1^3 + 2^3 + 3^3 + \cdots + n^3$ is $O(n^4)$.

25. $\displaystyle\sum_{i=1}^{n} i(i + 1)$ is $O(n^3)$.

26. $2 + 4 + 6 + \cdots + 2n$ is $O(n^2)$.

Explain how each statement in 27 and 28 follows from exercise 16 and parts (a) and (c) of exercise 17.

27. $4x^{4/3} - 15x + 7$ is $O(x^{4/3})$.

28. $\sqrt{x}(38x^5 + 9)$ is $O(x^{11/2})$.

29. Show how part 1 of property (9.2.4) follows from part (b) of exercise 16 and parts (a) and (c) of exercise 17.

30. Show that if f, g, and h are functions from $\mathbf{R}$ to $\mathbf{R}$ and $f(x)$ is $O(g(x))$ and $g(x)$ is $O(h(x))$, then $f(x)$ is $O(h(x))$.

◆ 31. a. Use mathematical induction to prove that $\sqrt{1} + \sqrt{2} + \sqrt{3} + \cdots + \sqrt{n} \leq n^{3/2}$ for all integers $n \geq 1$.

b. What can you conclude from part (a) about the order of $\sqrt{1} + \sqrt{2} + \sqrt{3} + \cdots + \sqrt{n}$?

◆ 32. a. Use mathematical induction to prove that $1^{1/3} + 2^{1/3} + 3^{1/3} + \cdots + n^{1/3} \leq n^{4/3}$, for all integers $n \geq 1$.

b. What can you conclude from part (a) about the order of $1^{1/3} + 2^{1/3} + 3^{1/3} + \cdots + n^{1/3}$?

H ◆ 33. a. Show that $1 + \dfrac{1}{2} + \dfrac{1}{3} + \cdots + \dfrac{1}{\lfloor \sqrt{n} \rfloor} \leq \sqrt{n}$, for all integers $n \geq 1$.

b. Show that for all integers $n \geq 2$,

$$\dfrac{1}{\lfloor \sqrt{n} \rfloor + 1} + \dfrac{1}{\lfloor \sqrt{n} \rfloor + 2} + \cdots + \dfrac{1}{n} \leq \sqrt{n}.$$

c. Show how it follows from parts (a) and (b) that $1 + \dfrac{1}{2} + \dfrac{1}{3} + \cdots + \dfrac{1}{n}$ is $O(\sqrt{n})$.

H 34. Prove that if r and s are rational numbers with $r > s$, then x^r is not $O(x^s)$. (See exercise 20, Section 9.1.)

Find best big-oh approximations from the set of all rational power functions for the functions defined by the expressions in 35–38.

35. $(x + 1)(2x^2 - 5)$

36. $x(25x - 4)(x^2 + 3)$

37. $\dfrac{\sqrt{x}(3x + 5)}{2x}$

38. $\dfrac{(2x^{5/2} + 1)(x - 1)}{x^{1/2}}$

39. (Requires the concept of limit from calculus)
◆ a. Prove that if g is a best big-oh approximation for f in a set S of functions and if $\lim_{x \to \infty} \dfrac{h(x)}{k(x)} = c$, a nonzero constant, then $g(x)$ is a best big-oh approximation for $\dfrac{f(x)h(x)}{k(x)}$ in S.

b. Use the result of part (a) to find best big-oh approximations from the set of all rational power functions for the functions defined by the following expressions:

i. $\dfrac{(2x^2 + 5)(x + 4)}{(x + 1)}$

ii. $\dfrac{(25x^{3/2} + 4x^{1/2} + 7)(4x^3 + 5x + 2)}{7x^3 - 3x^2 + 1}$

Exercises 40–42 use the following definition, which requires the concept of limit from calculus.

> **DEFINITION**
>
> If f and g are real-valued functions of a real variable and $\lim_{x \to \infty} g(x) \neq 0$, then
>
> $$f(x) \text{ is } o(g(x)) \quad \Leftrightarrow \quad \lim_{x \to \infty} \frac{f(x)}{g(x)} = 0.$$
>
> The notation $f(x)$ is $o(g(x))$ is read "$f(x)$ is little-oh of $g(x)$."

40. Prove that if $f(x)$ is $o(g(x))$, then $f(x)$ is $O(g(x))$.

41. Prove that if $f(x)$ and $g(x)$ are both $o(h(x))$, then for all real numbers a and b, $af(x) + bg(x)$ is $o(h(x))$.

42. Prove that for any positive real numbers a and b, if $a < b$ then x^a is $o(x^b)$.

9.3 APPLICATION: EFFICIENCY OF ALGORITHMS I

As soon as an Analytical Engine exists, it will necessarily guide the future course of the science. Whenever any result is sought by its aid, the question will then arise—by what course of calculation can these results be arrived at by the machine in the shortest time?
(Charles Babbage, 1864)

Charles Babbage (1792–1871)

Charles Babbage's Analytical Engine was similar in concept to a modern computer. The above quotation suggests that Babbage anticipated the importance of analyzing the efficiencies of computer algorithms well over a hundred years ago. In the 1950s and 1960s, a number of mathematicians and computer scientists contributed to the development of algorithm analysis, especially Donald Knuth, whose three volumes called *The Art of Computer Programming* provide a foundation for the subject that is both elegant and mathematically rigorous.*

Understanding the relative efficiencies of algorithms designed to do the same job is of much more than academic interest. In industrial and scientific settings the choice of an efficient over an inefficient program may result in the saving of many thousands of dollars or may make the difference between being able or not being able to do a project at all.

Two aspects of algorithm efficiency are important: the amount of time required to execute the algorithm and the amount of memory space needed when it is run. In this chapter we introduce basic techniques for calculating time efficiency. Similar techniques exist for calculating space efficiency. Occasionally, one algorithm may make more efficient use of time but less efficient use of memory space than another, forcing a trade-off based on the resources available to the user.

TIME EFFICIENCY OF AN ALGORITHM

How can the time efficiency of an algorithm be calculated? The answer depends on several factors. One is the size of the set of data that is input to the algorithm; for example, it takes longer for a sort algorithm to process 1,000,000 items than

Donald Knuth

*Donald E. Knuth, *The Art of Computer Programming*—vol. 1: *Fundamental Algorithms* (1973), vol. 2: *Seminumerical Algorithms* (1981), vol. 3: *Searching and Sorting* (1973) (Reading, Massachusetts: Addison-Wesley).

100 items. Consequently, the execution time of an algorithm is generally expressed as a function of its input size. Another factor that may affect the run time of an algorithm is the nature of the input data. For instance, a program that searches sequentially through a list of length n to find a data item requires only one step if the item is first on the list, but it uses n steps if the item is last on the list. Thus algorithms are frequently analyzed in terms of their "best case," "worst case," and "average case" performances for an input of size n.

Roughly speaking, the analysis of an algorithm for time efficiency begins by trying to count the number of elementary operations that must be performed when the algorithm is executed with an input of size n (either in the best case, worst case, or average case). What is classified as an "elementary operation" may vary depending on the nature of the problem the algorithms being compared are designed to solve. For instance, to compare two algorithms designed to evaluate a polynomial, the crucial issue is the number of additions and multiplications that are needed; whereas to compare two algorithms designed to search a list for a particular element, the important distinction is the number of comparisons that are required. For simplicity, we will classify the following as **elementary operations**: addition, subtraction, multiplication, division, and comparison.

When algorithms are implemented in a particular programming language and run on a particular computer, some operations are executed faster than others and, of course, there are differences in execution times from one machine to another. In certain practical situations these factors are taken into account when deciding which algorithm or which machine to use to solve a particular problem. In other cases, however, the machine is fixed, and rough estimates are all that is need to determine the clear superiority of one algorithm over another. Since each elementary operation is executed in time no longer than the slowest, the time efficiency of an algorithm is approximately proportional to the number of elementary operations required to execute the algorithm.

Consider the example of two algorithms, A and B, designed to do a certain job. Suppose that for an input of size n the number of elementary operations needed to perform algorithm A is no greater than $20n$ (at least for large n) and the number of elementary operations needed to perform algorithm B is no greater than $4n^2$ (for large n). Note that $20n < 4n^2$ whenever $n > 5$, and, in fact, $20n$ is very much less than $4n^2$ when n is large. For instance if $n = 1,000$, then $20n = 20,000$ whereas $4n^2 = 4,000,000$. We say that algorithm A is $O(n)$ (or has order n) and algorithm B is $O(n^2)$ (or has order n^2).

DEFINITION

Let A be an algorithm.

1. Suppose the number of elementary operations performed when A is executed for an input of size n depends on n alone and not on the nature of the input data; say it equals $f(n)$. If $f(n)$ is $O(g(n))$, we say that **A is $O(g(n))$** or **A is of order $g(n)$**.

2. Suppose the number of elementary operations performed when A is executed for an input of size n depends on the nature of the input data as well as on n.

   a. Let $b(n)$ be the *minimum* number of elementary operations required to execute A for all possible input sets of size n. If $b(n)$ is $O(g(n))$, we say that **in the best case A is $O(g(n))$** or **A has a best case order of $g(n)$**.

> **b.** Let $w(n)$ be the *maximum* number of elementary operations required to execute A for all possible input sets of size n. If $w(n)$ is $O(g(n))$, we say that **in the worst case A is $O(g(n))$** or **A has a worst case order of $g(n)$.**

Some of the most common orders used to describe algorithm efficiencies are shown in Table 9.3.1. As you see from the table, differences between the orders of various types of algorithms are more than astronomical. The time required for an algorithm of order 2^n to operate on a data set of size 100,000 is approximately $10^{29,976}$ times the estimated 15 billion years since the universe began (according to one theory of cosmology). On the other hand, an algorithm of order $\log_2 n$ needs at most a fraction of a second to process the same data set.

TABLE 9.3.1   Time Comparisons of Some Algorithm Orders

Approximate Time to Execute $f(n)$ Operations Assuming One Operation per Microsecond*

$f(n)$	$n = 10$	$n = 100$	$n = 1,000$	$n = 10,000$	$n = 100,000$
$\log_2 n$	0.000003 second	0.000007 second	0.00001 second	0.000013 second	0.000017 second
n	0.00001 second	0.0001 second	0.001 second	0.01 second	0.1 second
$n \cdot \log_2 n$	0.000033 second	0.0007 second	0.01 second	0.13 second	1.7 second
n^2	0.0001 second	0.01 second	1 second	1.67 minutes	2.78 hours
n^3	0.001 second	1 second	16.7 minutes	11.6 days	31.7 years
2^n	0.001 second	40×10^{16} years	3.4×10^{287} years	$6.3 \times 10^{2,996}$ years	$3.2 \times 10^{30,089}$ years

*one microsecond = one millionth of a second

EXAMPLE 9.3.1   **Computing an Order of an Algorithm Segment**

Assume n is a positive integer and consider the following algorithm segment:

$$p := 0, x := 2$$
$$\textbf{for } i := 2 \textbf{ to } n$$
$$p := (p + i) \cdot x$$
$$\textbf{next } i$$

a. Compute the actual number of additions and multiplications that must be performed when this algorithm segment is executed.
b. Find an order for this algorithm segment from among the set of power functions.

Solution   a. There is one multiplication and one addition for each iteration of the loop, so there are twice as many multiplications and additions as there are iterations of the loop. Now the number of iterations of the **for–next** loop equals the top index of the loop minus the bottom index plus 1; that is, $n - 2 + 1 = n - 1$. Hence there are $2 \cdot (n - 1) = 2n - 2$ multiplications and additions.

b. By the theorem on polynomial orders,

$$2n - 2 \quad \text{is} \quad O(n),$$

and so this algorithm segment is $O(n)$. ∎

As noted in Section 9.2, any function that is $O(n)$ is $O(n^s)$ where s is any rational number greater than 1. So we could say that the algorithm of Example 9.3.1 is $O(n^{3/2})$ or $O(n^2)$ or even $O(n^{100})$. In general, however, the *best* big-oh approximation (as defined in Section 9.2) for a polynomial function is the power function whose degree is the same as the polynomial. This is the power function of smallest degree that is a big-oh approximation for the polynomial.

EXAMPLE 9.3.2   The Order of an Algorithm with a Nested Loop

Assume n is a positive and consider the following algorithm segment:

$$s := 0$$
$$\textbf{for } i := 1 \textbf{ to } n$$
$$\qquad \textbf{for } j := 1 \textbf{ to } i$$
$$\qquad\qquad s := s + j \cdot (i - j + 1)$$
$$\qquad \textbf{next } j$$
$$\textbf{next } i$$

a. Compute the actual number of additions, subtractions, and multiplications that must be performed when this algorithm segment is executed.

b. Find an order for this algorithm segment from among the set of power functions.

Solution   a. There are two additions, one multiplication, and one subtraction for each iteration of the inner loop. So the total number of additions, multiplications, and subtractions is four times the number of iterations of the inner loop. Now the inner loop is iterated

$$\text{one time when } i = 1,$$
$$\text{two times when } i = 2,$$
$$\text{three times when } i = 3,$$
$$\vdots$$
$$n \text{ times when } i = n.$$

You can see this easily if you construct a table that shows the values of i and j for which the statements in the inner loop are executed. There is one iteration for each column in the table.

i	1	2 →		3 →			4 →				$\cdots$	n →				
j	1	1	2	1	2	3	1	2	3	4	$\cdots$	1	2	3	$\cdots$	n

1   2   3   4 n

Hence the total number of iterations of the inner loop is

$$1 + 2 + 3 + \cdots + n = \frac{n(n+1)}{2} \quad \text{by Theorem 4.2.2,}$$

and so the number of additions, subtractions, and multiplications is

$$4 \cdot \frac{n(n+1)}{2} = 2n(n+1).$$

Note that an alternative method for computing the number of columns of the table uses an approach discussed in Section 6.5. Observe that the number of columns in the table is the same as the number of ways to place two ×'s in n categories, 1, 2, ..., n, where the location of the ×'s indicates the values of i and j. By Theorem 6.5.1, this number is

$$\binom{n-1+2}{2} = \binom{n+1}{2} = \frac{(n+1)!}{2!((n+1)-2)!} = \frac{(n+1)n(n-1)!}{2(n-1)!} = \frac{n(n+1)}{2}.$$

Although this alternative method appears more complicated than the preceding one, it is simpler when the number of loop nestings exceeds two. (See exercise 14.)

b. By the theorem on polynomial orders, $2n(n+1) = 2n^2 + 2n$ is $O(n^2)$, and so this algorithm segment is $O(n^2)$. ∎

EXAMPLE 9.3.3   When the Number of Iterations Depends on the Floor Function

Assume n is a positive integer and consider the following algorithm segment.

$$\textbf{for } i := \lfloor n/2 \rfloor \textbf{ to } n$$
$$a := n - i$$
$$\textbf{next } i$$

a. Compute the actual number of subtractions that must be performed when this algorithm segment is executed.
b. Find an order for this algorithm segment from among the set of power functions.

Solution a. There is one subtraction for each iteration of the loop, and the loop is iterated $n - \lfloor \frac{n}{2} \rfloor + 1$ times. If n is even, then $\lfloor \frac{n}{2} \rfloor = \frac{n}{2}$, and so the number of subtractions is

$$n - \left\lfloor \frac{n}{2} \right\rfloor + 1 = n - \frac{n}{2} + 1 = \frac{n+2}{2}.$$

If n is odd, then $\lfloor \frac{n}{2} \rfloor = \frac{n-1}{2}$, and so the number of subtractions is

$$n - \left\lfloor \frac{n}{2} \right\rfloor + 1 = n - \frac{n-1}{2} + 1 = \frac{2n - (n-1) + 2}{2} = \frac{n+3}{2}.$$

b. By the theorem on polynomial orders,

$$\frac{n+2}{2} \quad \text{is} \quad O(n) \quad \text{and} \quad \frac{n+3}{2} \quad \text{is} \quad O(n)$$

also. Hence, regardless of whether n is even or odd, this algorithm segment is $O(n)$. ∎

THE SEQUENTIAL SEARCH ALGORITHM

The object of a search algorithm is to hunt through an array of data in an attempt to find a particular item x. In a sequential search, x is compared to the first element in the array, then the second, then the third, and so forth. The search is stopped if a match is found at any stage. On the other hand, if the entire array is processed without finding a match, the x is not in the array. An example of a sequential search is shown diagrammatically in Figure 9.3.1.

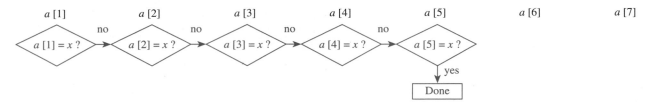

FIGURE 9.3.1   Sequential Search of $a[1]$, $a[2]$, . . . , $a[7]$ for x where $x = a[5]$

EXAMPLE 9.3.4   **Best- and Worst-Case Orders for Sequential Search**

Find best- and worst-case orders for the sequential search algorithm from among the set of power functions.

Solution   Suppose the sequential search algorithm is applied to an input array $a[1]$, $a[2]$, . . . , $a[n]$ to find an element x. In the best case, the algorithm requires only one comparison between x and the elements of $a[1]$, $a[2]$, . . . , $a[n]$. This occurs when x is the first element of the array. Thus in the best case, the sequential search algorithm is $O(1)$. (Note that $O(1) = O(n^0)$.) In the worst case, however, the algorithm requires n comparisons. This occurs when $x = a[n]$ or when x does not appear in the array at all. Thus in the worst case the sequential search algorithm is $O(n)$.   ∎

THE INSERTION SORT ALGORITHM

Insertion sort is an algorithm for arranging the elements of a one-dimensional array $a[1]$, $a[2]$, . . . , $a[n]$ into increasing order. In the first step, $a[2]$ is compared to $a[1]$. If $a[2] < a[1]$, the values of $a[1]$ and $a[2]$ are interchanged. Otherwise, the values of $a[1]$ and $a[2]$ are left as they were. Thus after the first step, the subarray $a[1]$, $a[2]$ is in increasing order. The idea of the algorithm is to gradually lengthen the section of the array that is known to be in increasing order by inserting each successive array value in its correct position relative to the preceding ones.

For instance, in the second step, $a[3]$ is inserted into the appropriate position relative to $a[1]$ and $a[2]$. The way step 2 proceeds is illustrated in Figure 9.3.2.

In general, at the beginning of the kth step of the algorithm, the subarray $a[1]$, $a[2]$, . . . , $a[k]$ is in increasing order. Then $a[k + 1]$ is inserted into $a[1]$, $a[2]$, . . . , $a[k]$ so that $a[1]$, $a[2]$, . . . , $a[k + 1]$ will be in increasing order. To achieve this, $a[k + 1]$ is compared to successive array elements $a[i]$ starting with $a[1]$. At each stage, if $a[k + 1] \leq a[i]$, then the value of $a[k + 1]$ is inserted into $a[i]$, and the values previously assigned to $a[i]$, $a[i + 1]$, . . . , $a[k]$ are shifted to $a[i + 1]$, $a[i + 2]$, . . . , $a[k + 1]$, respectively. After the insertion and shift, $a[1]$, $a[2]$, . . . , $a[k + 1]$ is in increasing order and the algorithm proceeds in increasing order

Beginning of step 2: $a[1], a[2], a[3], a[4], \ldots, a[n]$

Compare $a[3]$ to $a[1]$.

If $a[3] \leq a[1]$:

Give $a[1]$ the value of $a[3]$.

Shift the previous values of $a[1]$ and $a[2]$ to $a[2]$ and $a[3]$ $\left.\right\}$ $a[1] \rightarrow a[2] \rightarrow a[3]$

Now the array $a[1], a[2], a[3]$ is in increasing order.

If $a[3] > a[1]$, then the value of $a[1]$ is the smallest of the three values $a[1], a[2],$ and $a[3]$, and so the task is to put the values of $a[2]$ and $a[3]$ in order.

Compare $a[3]$ and $a[2]$.

If $a[3] \leq a[2]$:

Give $a[2]$ the value of $a[3]$.

Shift the previous value of $a[2]$ to $a[3]$. $\left.\right\}$ $a[1] \quad a[2] \rightarrow a[3]$

Now the array $a[1], a[2], a[3]$ is in increasing order.

If $a[3] > a[2]$, then the array $a[1], a[2], a[3]$ is already in increasing order.

So however step 2 is performed, at the end of it the subarray $a[1], a[2], a[3]$ is sorted.

in increasing order

End of step 2: $a[1], a[2], a[3], a[4], \ldots, a[n]$

FIGURE 9.3.2   Step 2 of Insertion Sort

to the $(k + 1)$st step. If it should happen that $a[k + 1]$ has been compared to each of $a[1], a[2], \ldots, a[k]$ and no value changes have occurred, then $a[1], a[2], \ldots, a[k + 1]$ was in increasing order to start with, and the algorithm also proceeds to the $(k + 1)$st step. In either case, at the end of the kth step, $a[1], a[2], \ldots, a[k + 1]$ is in increasing order. The overall action of the kth step is illustrated in Figure 9.3.3. Note that when the insertion sort is applied to an array of length n, the kth step is executed for each value of $k = 1, 2, \ldots, n - 1$.

sorted subarray

$a[1], a[2], a[3], \ldots, a[k], a[k + 1], a[k + 2], \ldots, a[n]$
$\uparrow$

Step k: Insert the value of $a[k + 1]$ into its proper position relative to $a[1], a[2], \ldots, a[k]$. At the end of this step $a[1], a[2], \ldots, a[k + 1]$ is sorted.

FIGURE 9.3.3   Step k of Insertion Sort

EXAMPLE 9.3.5   Implementing Insertion Sort

Construct a table showing the result of each step when insertion sort is applied to the array $a[1] = 6, a[2] = 3, a[3] = 5, a[4] = 4,$ and $a[5] = 2$.

Solution   The top row of the table below shows the initial values of the array and the bottom row shows the final values. The result of each step is shown in a separate row. For each step the sorted section of the array is shaded.

	a[1]	a[2]	a[3]	a[4]	a[5]
initial	6	3	5	4	2
result of step 1	3	6	5	4	2
result of step 2	3	5	6	4	2
result of step 3	3	4	5	6	2
result of step 4	2	3	4	5	6

EXAMPLE 9.3.6   **Finding an Order for Insertion Sort**

a. How many comparisons are performed when insertion sort is applied to the array $a[1], a[2], \ldots, a[n]$?
b. Find an order for insertion sort from among the set of power functions.

Solution   a. For each $k = 1, 2, \ldots, n - 1$, the maximum number of comparisons that occur when $a[k + 1]$ is inserted into $a[1], a[2], \ldots, a[k]$ is k, which happens either when $a[k + 1]$ is greater than every $a[1], a[2], \ldots, a[k]$ or when $a[k + 1]$ is greater than every $a[1], a[2], \ldots, a[k - 1]$ but is less than $a[k]$. Therefore, by the addition rule the total number of comparisons is

$$1 + 2 + \cdots + (n - 1) = \frac{(n - 1)((n - 1) + 1)}{2} \quad \text{by Theorem 4.2.2}$$

$$= \frac{n(n - 1)}{2}$$

$$= \frac{1}{2}n^2 - \frac{1}{2}n \quad \text{by algebra.}$$

b. By the theorem on polynomial orders, $\frac{1}{2}n^2 - \frac{1}{2}n$ is $O(n^2)$. So the insertion sort algorithm is $O(n^2)$. ∎

EXERCISE SET 9.3

1. Suppose a computer takes 1 microsecond (a millionth of a second) to execute each operation. Approximately how long will it take for the computer to execute the following numbers of operations? Convert your answers into seconds, minutes, hours, days, weeks, or years, as appropriate. For example, instead of 2^{50} microseconds, write 35.7 years.

a. $\log_2 200$ b. 200 c. $200 \cdot \log_2 200$
d. 200^2 e. 200^8 f. 2^{200}

2. If an algorithm requires cn^2 operations when performed with an input of size n (where c is a constant),
a. how many operations will be required when the input size is increased from m to $2m$ (where m is a positive integer)?

b. by what factor will the number of operations increase when the input size is doubled?

c. by what factor will the number of operations increase when the input size is increased by a factor of ten?

3. If an algorithm requires cn^3 operations when performed with an input of size n (where c is a constant),
a. how many operations will be required when the input size is increased from m to $2m$ (where m is a positive integer)?
b. by what factor will the number of operations increases when the input size is doubled?
c. by what factor will the number of operations increase when the input size is increased by a factor of ten?

Exercises 4–5 explore the fact that for relatively small values of n, algorithms with larger orders can be more efficient than algorithms with smaller orders.

4. Suppose that when run with an input of size n, algorithm A requires $2n^2$ operations and algorithm B requires $80n^{3/2}$ operations.
a. What are orders for algorithms A and B from among the set of power functions?
b. For what values of n is algorithm A more efficient than algorithm B?
c. For what values of n is algorithm B at least 100 times more efficient than algorithm A?

5. Suppose that when run with an input of size n, algorithm A requires $4n^2$ operations and algorithm B requires $200n^{5/3}$ operations.
a. What are orders for algorithms A and B from among the set of power functions?
b. For what values of n is algorithm A more efficient than algorithm B?
c. For what values of n is algorithm B at least 100 times more efficient than algorithm A?

For each of the algorithm segments in 6–14 assume that n is a positive integer and
a. **compute the actual number of additions, subtractions, multiplications, divisions, and comparisons that must be performed when the algorithm segment is executed;**
b. **find an order for the algorithm segment from among the set of power functions.**

6. **for** $i := 3$ **to** $n - 1$
 $\quad a := 3 \cdot n + 2 \cdot i - 1$
 next i

7. $max := a[1]$
 for $i := 2$ **to** n
 $\quad$ **if** $max < a[i]$ **then** $max := a[i]$
 next i

8. **for** $i := 1$ **to** $\left\lfloor \dfrac{n}{2} \right\rfloor$
 $\quad a := n - i$
 next i

9. **for** $i := 1$ **to** n
 $\quad$ **for** $j := 1$ **to** $2n$
 $\quad\quad a := 2 \cdot n + i \cdot j$
 $\quad$ **next** j
 next i

10. **for** $k := 1$ **to** $n - 1$
 $\quad max := a[k]$
 $\quad$ **for** $i := k + 1$ **to** n
 $\quad\quad$ **if** $max < a[i]$ **then** $max := a[i]$
 $\quad$ **next** i
 $a[k] := max$
 next k

11. **for** $i := 1$ **to** $n - 1$
 $\quad$ **for** $j := i + 1$ **to** n
 $\quad\quad$ **if** $a[j] < a[i]$ **then do**
 $\quad\quad\quad temp := a[i]$
 $\quad\quad\quad a[i] := a[j]$
 $\quad\quad\quad a[j] := temp$
 $\quad\quad$ **end do**
 $\quad$ **next** j
 next i

12. **for** $i := 1$ **to** n
 $\quad$ **for** $j := 1$ **to** $\left\lfloor \dfrac{i + 1}{2} \right\rfloor$
 $\quad\quad a := (n - i) \cdot (n - j)$
 $\quad$ **next** j
 next i

13. **for** $i := 1$ **to** n
 $\quad$ **for** $j := 1$ **to** $2n$
 $\quad\quad$ **for** $k := 1$ **to** n
 $\quad\quad\quad x := i \cdot j \cdot k$
 $\quad\quad$ **next** k
 $\quad$ **next** j
 next i

$H\blacklozenge$14. **for** $i := 1$ **to** n
 $\quad$ **for** $j := 1$ **to** i
 $\quad\quad$ **for** $k := 1$ **to** j
 $\quad\quad\quad x := i \cdot j \cdot k$
 $\quad\quad$ **next** k
 $\quad$ **next** j
 next i

15. Construct a table showing the result of each step when insertion sort is applied to the array $a[1] = 6$, $a[2] = 2$, $a[3] = 1$, $a[4] = 8$, and $a[5] = 4$.

16. Construct a table showing the result of each step when insertion sort is applied to the array $a[1] = 7$, $a[2] = 3$, $a[3] = 6$, $a[4] = 9$, and $a[5] = 5$.

17. How many comparisons actually occur when insertion sort is applied to the array of exercise 15?

18. How many comparisons actually occur when insertion sort is applied to the array of exercise 16?

19. According to Example 9.3.6, the maximum number of comparisons needed to perform insertion sort on an array of length five is $1 + 2 + 3 + 4 = 10$. Find an array of length five that requires this maximum number of comparisons when insertion sort is applied to it.

Exercises 20–26 and 28 refer to selection sort. *Selection sort* is another algorithm for arranging the elements of a one-dimensional array $a[1], a[2], \ldots, a[n]$ into increasing order. For each $i = 1, 2, \ldots, n - 1$, $a[i]$ is compared with each successive element $a[k]$ of the subarray $a[i + 1], a[i + 2], \ldots, a[n]$ starting with $a[i + 1]$. After each comparison, if it has been found that the value of $a[i]$ is greater than the value of $a[k]$, the two values are interchanged. This is illustrated below.

$$a[1] \; a[2] \ldots \boxed{a[i]} \; a[i + 1] \ldots a[n]$$

↑

*i*th step: Compare $a[i]$ to each of
$a[i + 1], \ldots, a[n]$ and
interchange values each
time $a[i]$ is greater.

Thus $a[1]$ is compared with each element of $a[2], a[3], \ldots, a[n]$ with interchanges of value made as appropriate. At the end of this step, $a[1]$ has the least (or minimum) value of the array. Then $a[2]$ is compared with each element of $a[3], a[4], \ldots, a[n]$, with interchanges made as appropriate. At the end of this step, $a[2]$ is greater than or equal to $a[1]$ and less than or equal to all the other $a[i]$. The process continues until finally $a[n - 1]$ is compared with $a[n]$. If the value of $a[n - 1]$ is greater, the two values are interchanged. At that point the entire array is in increasing order.

20. Construct a table showing the interchanges that occur when selection sort is applied to the array $a[1] = 5$, $a[2] = 3$, $a[3] = 4$, $a[4] = 6$, and $a[5] = 2$.

21. Construct a table showing the interchanges that occur when selection sort is applied to the array $a[1] = 6$, $a[2] = 4$, $a[3] = 5$, $a[4] = 8$, and $a[5] = 1$.

22. How many interchanges actually occur when selection sort is applied to the array of exercise 20?

23. How many interchanges actually occur when selection sort is applied to the array of exercise 21?

24. How many comparisons occur when selection sort is applied to array $a[1], a[2], a[3], a[4]$?

25. a. How many comparisons occur when $a[1]$ is compared to each of $a[2,] \, a[3], \ldots, a[n]$?

    b. How many comparisons occur when $a[2]$ is compared to each of $a[3], a[4], \ldots, a[n]$?

    c. How many comparisons occur when $a[i]$ is compared to each of $a[i + 1], a[i + 2], \ldots, a[n]$?

    H d. Find an order for selection sort from among the set of power functions.

26. Find an array of length five that requires the maximum number of interchanges when selection sort is applied to it.

27. Write a formal algorithm to implement insertion sort.

28. Write a formal algorithm to implement selection sort.

Exercises 29–32 refer to the following algorithm to compute the value of a real polynomial.

Algorithm 9.3.1   Term-by-term Polynomial Evaluation
[This algorithm computes the value of the real polynomial $a[n]x^n + a[n - 1]x^{n-1} + \cdots + a[2]x^2 + a[1]x + a[0]$ by computing each term separately, starting with $a[0]$, and adding it on to an accumulating sum.]

Input: n *[a nonnegative integer]*, $a[0], a[1], a[2], \ldots, a[n]$ *[an array of real numbers]*, x *[a real number]*

Algorithm Body:
$polyval := a[0]$
    for $i := 1$ **to** n
      $term := a[i]$
      for $j := 1$ **to** i
        $term := term \cdot x$
      next j
      $polyval := polyval + term$
    next i
[At this point
    $polyval = a[n]x^n + a[n - 1]x^{n-1} + \cdots + a[2]x^2 + a[1]x + a[0].]$

Output: *polyval [a real number]*
end Algorithm 9.3.1

29. Trace Algorithm 9.3.1 for the input $n = 3$, $a[0] = 2$, $a[1] = 1$, $a[2] = -1$, $a[3] = 3$, and $x = 2$.

30. Trace Algorithm 9.3.1 for the input $n = 2$, $a[0] = 5$, $a[1] = -1$, $a[2] = 2$, and $x = 3$.

31. Let s_n = the number of additions and multiplications that must be performed when Algorithm 9.3.1 is executed for a polynomial of degree n. Express s_n as a function of n.

32. Find an order for Algorithm 9.3.1 from among the set of power functions.

Exercises 33–36 refer to another algorithm, known as Horner's rule, for finding the value of a real polynomial.

Algorithm 9.3.2 Horner's Rule

[This algorithm computes the value of the real polynomial
$a[n]x^n + a[n - 1]x^{n-1} + \cdots + a[2]x^2 + a[1]x + a[0]$
by nesting successive additions and multiplications as indicated in the following parenthesization:

$$((\cdots ((a[n]x + a[n - 1])x + a[n - 2])x \\ + \cdots + a[2])x + a[1])x + a[0].$$

At each stage, starting with a[n], the current value of polyval is multiplied by x and the next lower coefficient of the polynomial is added on.]

Input: *n [a nonnegative integer], a[0], a[1], a[2], . . . , a[n]*
    [an array of real numbers], x [a real number]

Algorithm Body:

$polyval := a[n]$

for $i := 1$ **to** n

$polyval := polyval \cdot x + a[n - i]$

next i

[At this point
$polyval = a[n]x^n + a[n - 1]x^{n-1} \\ + \cdots + a[2]x^2 + a[1]x + a[0].]$

Output: *polyval [a real number]*

end Algorithm 9.3.2

33. Trace Algorithm 9.3.2 for the input $n = 3$, $a[0] = 2$, $a[1] = 1$, $a[2] = -1$, $a[3] = 3$, and $x = 2$.

34. Trace Algorithm 9.3.2 for the input $n = 2$, $a[0] = 5$, $a[1] = -1$, $a[2] = 2$, and $x = 3$.

H 35. Let $t_n =$ the number of additions and multiplications that must be performed when Algorithm 9.3.2 is executed for a polynomial of degree n. Express t_n as a function of n.

36. Find an order for Algorithm 9.3.2 from among the set of power functions. How does this order compare with that of Algorithm 9.3.1?

9.4 EXPONENTIAL AND LOGARITHMIC FUNCTIONS: GRAPHS AND ORDERS

We ought never to allow ourselves to be persuaded of the truth of anything unless on the evidence of our own reason.
(René Descartes, 1596–1650)

Exponential and logarithmic functions are of great importance in mathematics in general and in computer science in particular. Several important computer algorithms have execution times that involve logarithmic functions of the size of the input data (which means they are relatively efficient for large data sets), and some have execution times that are exponential functions of the size of the input data (which means they are quite inefficient for large data sets). In addition, since exponential and logarithmic functions arise naturally in the descriptions of many growth and decay processes and in the computation of many kinds of probabilities, these functions are used to analyze computer operating systems, in queuing theory, and in the theory of information.

GRAPHS OF EXPONENTIAL FUNCTIONS

As defined in Section 7.3, the exponential function with base $b > 0$ is the function that sends each real number x to b^x. The graph of the exponential function with base 2 (together with a partial table of its values) is shown in Figure 9.4.1. Note that the values of this function increase with extraordinary rapidity. If we tried to continue drawing the graph using the scale shown in Figure 9.4.1, we would have to plot the point $(10, 2^{10})$ more than 21 feet above the horizontal axis. And the point $(30, 2^{30})$ would be located more than 610,080 miles above the axis—well beyond the moon!

x	2^x
0	$2^0 = 1$
1	$2^1 = 2$
2	$2^2 = 4$
3	$2^3 = 8$
-1	$2^{-1} = 0.5$
-2	$2^{-2} = 0.25$
-3	$2^{-3} = 0.125$
0.5	$2^{0.5} \cong 1.414$
-0.5	$2^{-0.5} \cong 0.707$

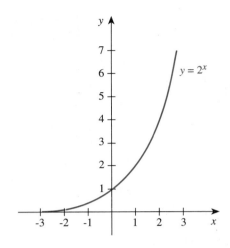

FIGURE 9.4.1   The Exponential Function with Base 2

The graph of any exponential function with base $b > 1$ has a shape that is similar to the graph of the exponential function with base 2. If $0 < b < 1$, then $1/b > 0$ and the graph of the exponential function with base b is the reflection across the vertical axis of the exponential function with base $1/b$. These facts are illustrated in Figure 9.4.2.

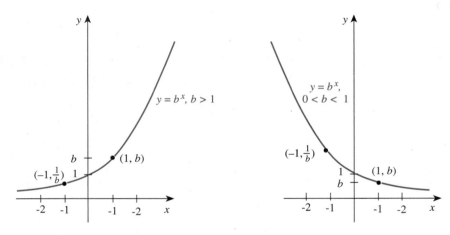

a. Graph of the exponential function with base $b > 1$

b. Graph of the exponential function with base b where $0 < b < 1$

FIGURE 9.4.2   Graphs of Exponential Functions

GRAPHS OF LOGARITHMIC FUNCTIONS

Logarithms were first introduced by the Scotsman John Napier. Astronomers and navigators found them so useful for reducing the time needed to do multiplication and division that they quickly gained wide acceptance and played a crucial role in the remarkable development of those areas in the seventeenth century. Nowadays, however, electronic calculators and computers are available to handle most computations quickly and conveniently, and logarithms and logarithmic functions are used primarily as conceptual tools.

John Napier (1550–1617)

Recall the definition of the logarithmic function with base b from Section 7.1. We state it formally below.

DEFINITION

The **logarithmic function with base b, $\log_b : \mathbf{R}^+ \to \mathbf{R}$**, is the function that sends each positive real number x to the number $\log_b x$, which is the exponent to which b must be raised to obtain x.

If b is a positive real number not equal to 1, then the logarithmic function with base b is, in fact, the inverse of the exponential function with base b. (See exercise 10 at the end of this section.)

The graph of the logarithmic function with base $b > 1$ is shown in Figure 9.4.3.

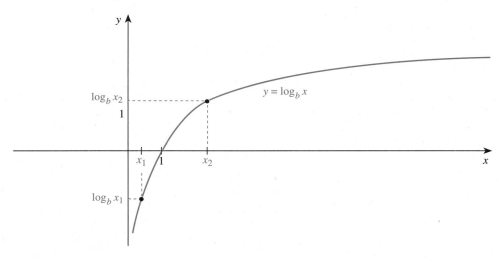

FIGURE 9.4.3   The Graph of the Logarithmic Function with Base $b > 1$

Observe that, if its base b is greater than 1, the logarithmic function is increasing. Analytically, this means that

---

if $b > 1$, then for all positive numbers x_1 and x_2

$$\text{if } x_1 < x_2, \text{ then } \log_b(x_1) < \log_b(x_2).$$ 9.4.1

Notice, however, that the logarithmic function "grows" very slowly; that is, you must go very far out on the horizontal axis to find points whose logarithms are large numbers. For instance, $\log_2(1{,}024)$ is only 10 and $\log_2(1{,}048{,}576)$ is just 20.

The following example shows how to make use of the increasing nature of the logarithmic function with base 2 to derive a remarkably useful property.

EXAMPLE 9.4.1   **Base 2 Logarithms of Numbers Between Two Consecutive Powers of Two**

Prove the following property:

a.

> If k is an integer and x is a real number with $2^k \leq x < 2^{k+1}$, then $\lfloor \log_2 x \rfloor = k$.
>
> **9.4.2**

b. Describe property (9.4.2) in words and give a graphical interpretation of the property for $x > 1$.

Solution   a. Suppose that k is an integer and x is a real number with

$$2^k \leq x < 2^{k+1}.$$

Because the logarithmic function with base 2 is increasing, this implies that

$$\log_2(2^k) \leq \log_2 x < \log_2(2^{k+1}).$$

But $\log_2(2^k) = k$ *[the exponent to which you must raise 2 to get 2^k is k]* and $\log_2(2^{k+1}) = k + 1$. Hence

$$k \leq \log_2 x < k + 1.$$

By definition of the floor function, then

$$k = \lfloor \log_2 x \rfloor.$$

b. Recall that the floor of a positive number is its integer part. For instance, $\lfloor 2.82 \rfloor = 2$. Hence property (9.4.2) can be described in words as follows:

> If x is a positive number that lies between two consecutive integer powers of two, the floor of the logarithm with base 2 of x is the exponent of the smaller power of two.

Graphical interpretation:

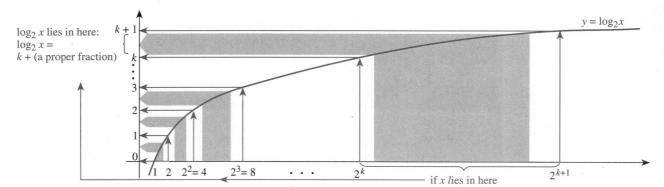

One consequence of property (9.4.2) does not appear particularly interesting in its own right but is frequently needed as a step in the analysis of algorithm efficiency.

EXAMPLE 9.4.2   When $\lfloor \log_2(n-1) \rfloor = \lfloor \log_2 n \rfloor$

Prove the following property:

For any odd integer $n > 1$, $\lfloor \log_2(n-1) \rfloor = \lfloor \log_2 n \rfloor$.   9.4.3

Solution   If n is an odd integer that is greater than one, then n lies strictly between two successive powers of 2:

$$2^k < n < 2^{k+1} \quad \text{for some integer } k > 0. \qquad 9.4.4$$

It follows that $2^k \le n - 1$ because $2^k < n$ and both 2^k and n are integers. Consequently,

$$2^k \le n - 1 < 2^{k+1}. \qquad 9.4.5$$

Applying property (9.4.2) to both (9.4.4) and (9.4.5) gives

$$\lfloor \log_2 n \rfloor = k \quad \text{and also} \quad \lfloor \log_2(n-1) \rfloor = k.$$

Hence $\lfloor \log_2 n \rfloor = \lfloor \log_2(n-1) \rfloor$. ∎

APPLICATION: NUMBER OF BITS NEEDED TO REPRESENT AN INTEGER IN BINARY NOTATION

Given a positive integer n, how many binary digits are needed to represent n? To answer this question, recall from Section 4.4 that any positive integer n can be written in a unique way as

$$n = 2^k + c_{k-1} \cdot 2^{k-1} + \cdots + c_2 \cdot 2^2 + c_1 \cdot 2 + c_0,$$

where k is a nonnegative integer and each $c_0, c_1, c_2, \ldots, c_{k-1}$ is either 0 or 1. Then the binary representation of n is

$$1c_{k-1}c_{k-2} \cdots c_2 c_1 c_0,$$

and so the number of binary digits needed to represent n is $k + 1$.

What is $k + 1$ as a function of n? Observe that since each $c_i \le 1$,

$$n = 2^k + c_{k-1} \cdot 2^{k-1} + \cdots + c_2 \cdot 2^2 + c_1 \cdot 2 + c_0 \le 2^k + 2^{k-1} + \cdots + 2^2 + 2 + 1.$$

But by the formula for the sum of a geometric sequence (Theorem 4.2.3),

$$2^k + 2^{k-1} + \cdots + 2^2 + 2 + 1 = \frac{2^{k+1} - 1}{2 - 1} = 2^{k+1} - 1.$$

So by transitivity of order,

$$n \leq 2^{k+1} - 1 < 2^{k+1} \qquad \textbf{9.4.6}$$

In addition, because each $c_i \geq 0$,

$$2^k \leq 2^k + c_{k-1} \cdot 2^{k-1} + \cdots + c_2 \cdot 2^2 + c_1 \cdot 2 + c_0 = n. \qquad \textbf{9.4.7}$$

Putting inequalities (9.4.6) and (9.4.7) together gives the double inequality

$$2^k \leq n < 2^{k+1}.$$

But then by property (9.4.2),

$$k = \lfloor \log_2 n \rfloor.$$

Thus the number of binary digits needed to represent n is $\lfloor \log_2 n \rfloor + 1$.

EXAMPLE 9.4.3   **Number of Bits in a Binary Representation**

How many binary digits are needed to represent 52,837 in binary notation?

Solution   If you compute the logarithm with base 2 using formula (7.3.6) in Example 7.3.6 and a calculator that gives you approximate values of logarithms with base 10, you find that

$$\log_2(52,837) \cong \frac{\log_{10}(52,837)}{\log_{10}(2)} \cong \frac{4.722938151}{0.3010299957} \cong 15.7.$$

Thus the binary representation of 52,837 has $\lfloor 15.7 \rfloor + 1 = 15 + 1 = 16$ binary digits.   ∎

APPLICATION: USING LOGARITHMS TO SOLVE RECURRENCE RELATIONS

In Chapter 8 we discussed methods for solving recurrence relations. One class of recurrence relations that is very important in computer science has solutions that can be expressed in terms of logarithms. One such recurrence relation is discussed in the next example.

EXAMPLE 9.4.4   **A Recurrence Relation with a Logarithmic Solution**

Define a sequence $a_1, a_2, a_3, \ldots$, recursively as follows:

$$a_1 = 1,$$
$$a_k = 2 \cdot a_{\lfloor k/2 \rfloor} \quad \text{for all integers } k \geq 2.$$

a. Use iteration to guess an explicit formula for this sequence.
b. Use strong mathematical induction to confirm the correctness of the formula obtained in part (a).

Solution   a. Begin by iterating to find the values of the first few terms of the sequence.

$$a_1 = 1 \qquad\qquad\qquad\qquad\qquad\qquad 1 = 2^0$$

$$a_2 = 2 \cdot a_{\lfloor 2/2 \rfloor} = 2 \cdot a_1 = 2 \cdot 1 = 2$$
$$a_3 = 2 \cdot a_{\lfloor 3/2 \rfloor} = 2 \cdot a_1 = 2 \cdot 1 = 2 \qquad 2 = 2^1$$

$$a_4 = 2 \cdot a_{\lfloor 4/2 \rfloor} = 2 \cdot a_2 = 2 \cdot 2 = 4 \qquad 4 = 2^2$$
$$a_5 = 2 \cdot a_{\lfloor 5/2 \rfloor} = 2 \cdot a_2 = 2 \cdot 2 = 4$$
$$a_6 = 2 \cdot a_{\lfloor 6/2 \rfloor} = 2 \cdot a_3 = 2 \cdot 2 = 4$$
$$a_7 = 2 \cdot a_{\lfloor 7/2 \rfloor} = 2 \cdot a_3 = 2 \cdot 2 = 4$$

$$a_8 = 2 \cdot a_{\lfloor 8/2 \rfloor} = 2 \cdot a_4 = 2 \cdot 4 = 8 \qquad 8 = 2^3$$
$$a_9 = 2 \cdot a_{\lfloor 9/2 \rfloor} = 2 \cdot a_4 = 2 \cdot 4 = 8$$
$$\vdots$$
$$a_{15} = 2 \cdot a_{\lfloor 15/2 \rfloor} = 2 \cdot a_7 = 2 \cdot 4 = 8$$
$$a_{16} = 2 \cdot a_{\lfloor 16/2 \rfloor} = 2 \cdot a_8 = 2 \cdot 8 = 16 \qquad 16 = 2^4$$

Note that in each case when the subscript n is between two powers of 2, a_n equals the smaller power of 2. More precisely,

If $2^i \le n < 2^{i+1}$, then $a_n = 2^i$.   9.4.8

But since n satisfies the inequality

$$2^i \le n < 2^{i+1},$$

then (by property 9.4.2 of Example 9.4.1)

$$i = \lfloor \log_2 n \rfloor.$$

Substituting into statement (9.4.8) gives

$$a_n = 2^{\lfloor \log_2 n \rfloor}.$$

b. The following proof shows that if $a_1, a_2, a_3, \ldots$ is a sequence of numbers that satisfies

$$a_1 = 1,$$

and

$$a_k = 2 \cdot a_{\lfloor k/2 \rfloor} \quad \text{for all } k \ge 2,$$

then

$$a_n = 2^{\lfloor \log_2 n \rfloor} \quad \text{for all integers } n \ge 1.$$

Proof:

The formula holds for $n = 1$:   By definition of the sequence, $a_1 = 1$. And $2^{\lfloor \log_2 1 \rfloor} = 2^0 = 1$ also. Hence $a_1 = 2^{\lfloor \log_2 1 \rfloor}$, and so the formula holds for $n = 1$.

If $k \ge 2$ and the formula holds for all integers i with $1 \le i < k$, then the formula holds for k: Let k be an integer that is greater than or equal to 2, and suppose $a_i = 2^{\lfloor \log_2 i \rfloor}$ for all integers i with $1 \le i < k$. Now either k is odd or k is even.

Case 1 (k is odd): In this case,

$$
\begin{aligned}
a_k &= 2 \cdot a_{\lfloor k/2 \rfloor} && \text{by definition of } a_1, a_2, a_3, \ldots \\
&= 2 \cdot a_{(k-1)/2} && \text{because } \lfloor k/2 \rfloor = (k-1)/2 \text{ since } k \text{ is odd} \\
&= 2 \cdot 2^{\lfloor \log_2((k-1)/2) \rfloor} && \text{by inductive hypothesis (since } k \geq 2, \ 1 \leq k/2 < k) \\
&= 2^{\lfloor \log_2((k-1)/2) \rfloor + 1} && \text{by the laws of exponents from algebra (7.3.1)} \\
&= 2^{\lfloor \log_2(k-1) - \log_2 2 \rfloor + 1} && \text{by the identity } \log_b(x/y) = \log_b x - \log_b y \text{ derived} \\
&&& \text{in exercise 24 of Section 7.3} \\
&= 2^{\lfloor \log_2(k-1) - 1 \rfloor + 1} && \text{since } \log_2 2 = 1 \\
&= 2^{\lfloor \log_2(k-1) \rfloor - 1 + 1} && \text{by substituting } x = \log_2 (k-1) \text{ into the identity} \\
&&& \lfloor x - 1 \rfloor = \lfloor x \rfloor - 1 \text{ derived in exercise 15 of} \\
&&& \text{Section 3.5} \\
&= 2^{\lfloor \log_2(k-1) \rfloor} \\
&= 2^{\lfloor \log_2 k \rfloor} && \text{by property (9.4.3)}
\end{aligned}
$$

Case 2 (k is even): The analysis of this case is very similar to that of case 1 and is left as an exercise. ■

EXPONENTIAL AND LOGARITHMIC ORDERS

Now consider the question, "How do graphs of logarithmic and exponential functions compare with graphs of power functions?" It turns out that for large enough values of *x*, the graph of the logarithmic function with any base $b > 1$ lies *below* the graph of any positive power function, and the graph of the exponential function with any base $b > 1$ lies *above* the graph of any positive power function. In analytic terms, this says the following:

For all real numbers *b* and *r* with $b > 1$ and $r > 0$,

$$\log_b x \leq x^r \quad \text{for all sufficiently large values of } x. \qquad \text{9.4.9}$$

and

$$x^r \leq b^x \quad \text{for all sufficiently large values of } x. \qquad \text{9.4.10}$$

These statements have the following implications for *O*-notation.

For all real numbers *b* and *r* with $b > 1$ and $r > 0$,

$$\log_b x \text{ is } O(x^r) \qquad \text{9.4.11}$$

and

$$x^r \text{ is } O(b^x) \qquad \text{9.4.12}$$

Another important function in the analysis of algorithms is the function *f* defined by the formula

$$f(x) = x \log_b x \quad \text{for all real numbers } x > 0.$$

For large values of x, the graph of this function fits in between the graphs of the identity function and the squaring function. More precisely:

> For all real numbers b with $b > 1$ and for all sufficiently large values of x,
>
> $$x \leq x \log_b x \leq x^2.$$
>
> 9.4.13

The O-notation versions of these facts are as follows:

> For all real numbers $b > 1$,
>
> $$x \quad \text{is} \quad O(x \log_b x) \quad \text{and} \quad x \log_b x \quad \text{is} \quad O(x^2).$$
>
> 9.4.14

Although proofs of some of these facts require calculus, proofs of some cases can be obtained using the algebra of inequalities. (See the exercises at the end of this section.) The drawing of Figure 9.4.4 illustrates the relationships among some power functions, the logarithmic function with base 2, the exponential function with base 2, and the function defined by the formula $x \rightarrow x \log_2 x$. Note that different scales are used on the horizontal and vertical axes.

The inequalities above can be used to derive additional orders involving the logarithmic function.

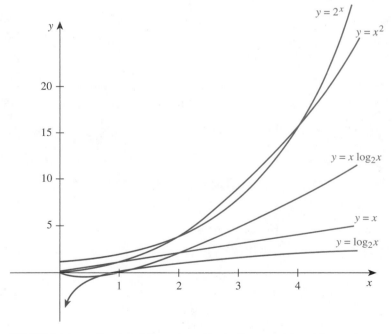

FIGURE 9.4.4   Graphs of Some Logarithmic, Exponential, and Power Functions

EXAMPLE 9.4.5 **Deriving an Order from Logarithmic Inequalities**

Show that $x + x \log_2 x$ is $O(x \log_2 x)$.

Solution   According to property (9.4.13) with $b = 2$, there is a number x_0 so that for all $x > x_0$,

$$x < x \log_2 x$$
$$\Rightarrow \quad x + x \log_2 x < 2x \log_2 x \qquad \text{by adding } x \log_2 x \text{ to both sides}$$

Therefore, if x_0 is taken to be greater than 2, then

$$\left| x + x \log_2 x \right| < 2 \cdot \left| x \log_2 x \right| \qquad \text{because when } x > 2, x \log_2 x > 0, \text{ and so } \left| x + x \log_2 x \right| = x + x \log_2 x \text{ and } x \log_2 x = \left| x \log_2 x \right|.$$

Let $M = 2$. Then $\left| x + x \log_2 x \right| \leq M \cdot \left| x \log_2 x \right|$ for all $x > x_0$.

Hence by definition of *O*-notation, $x + x \log_2 x$ is $O(x \log_2 x)$.   ■

Example 9.4.5 illustrates a special case of a useful general fact about *O*-notation: *If one function "dominates" another (in the sense of being larger for large values of the variable), then the sum of the two is big-oh of the dominating function.* (See exercise 17a at the end of Section 9.2.)

Example 9.4.6 shows that any two logarithmic functions with bases greater than 1 have the same order.

EXAMPLE 9.4.6 **Logarithm with Base *b* is Big-Oh of Logarithm with Base *c***

Show that if b and c are real numbers such that $b > 1$ and $c > 1$, then $\log_b x$ is $O(\log_c x)$.

Solution   Suppose b and c are real numbers and $b > 1$ and $c > 1$. To show that $\log_b x$ is $O(\log_c x)$, a positive real number M and a real number x_0 must be found such that

$$\left| \log_b x \right| \leq M \cdot \left| \log_c x \right| \qquad \text{for all real numbers } x > x_0.$$

By property (7.3.6) in Example 7.3.6,

$$\log_b x = \frac{\log_c x}{\log_c b} = \left(\frac{1}{\log_c b} \right) \cdot \log_c x. \tag{*}$$

Since $b > 1$ and the logarithmic function with base c is strictly increasing, then $\log_c b > \log_c 1 = 0$, and so $\dfrac{1}{\log_c b} > 0$ also. Furthermore, if $x > 1$, then $\log_b x > 0$ and $\log_c x > 0$, and so $\left| \log_b x \right| = \log_b x$ and $\left| \log_c x \right| = \log_c x$. Let $M = \dfrac{1}{\log_c b}$ and $x_0 = 1$. Then by substitution into (*),

$$\left| \log_b x \right| \leq M \cdot \left| \log_c x \right| \qquad \text{for all real numbers } x > x_0.$$

Hence by definition of *O*-notation,

$$\log_b x \quad \text{is} \quad O(\log_c x).$$ ■

Example 9.4.7 shows how a logarithmic order can arise from the computation of a certain kind of sum. It requires the following fact from calculus:

The area underneath the graph of $y = \dfrac{1}{x}$ between $x = 1$ and $x = n$ equals $\ln n$.

(Recall that $\ln n = \log_e n$.) This is illustrated in Figure 9.4.5.

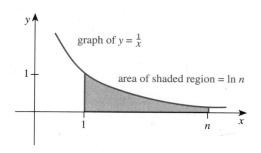

FIGURE 9.4.5   Area Under Graph of $y = \dfrac{1}{x}$ Between $x = 1$ and $x = n$

EXAMPLE 9.4.7   **Order of Harmonic Sum**

Sums of the form $1 + \dfrac{1}{2} + \cdots + \dfrac{1}{n}$ are called *harmonic sums*. They occur in the analysis of various computer algorithms such as quick sort. Show that $1 + \dfrac{1}{2} + \dfrac{1}{3} + \cdots + \dfrac{1}{n}$ is $O(\ln n)$ by executing the following steps:

a. Interpret Figure 9.4.5 to show that

$$\frac{1}{2} + \frac{1}{3} + \cdots + \frac{1}{n} \le \ln n.$$

b. Show that if n is an integer that is at least 3, then $1 \le \ln n$.
c. Deduce from (a) and (b) that if the integer n is greater than or equal to 3, then

$$1 + \frac{1}{2} + \frac{1}{3} + \cdots + \frac{1}{n} \le 2 \cdot \ln n.$$

d. Deduce from (c) that

$$1 + \frac{1}{2} + \frac{1}{3} + \cdots + \frac{1}{n} \quad \text{is} \quad O(\ln n).$$

Solution   a. In Figure 9.4.5 draw rectangles whose bases are the intervals between each pair of integers from 1 to n and whose heights are the heights of the graph of $y = \dfrac{1}{x}$ above the right-hand endpoints of the intervals. This is shown in Figure 9.4.6. Now the area of each rectangle is its base times its height. Since all the rectangles

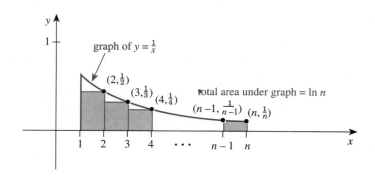

FIGURE 9.4.6

have base 1, the area of each rectangle equals its height. Thus

$$\text{the area of the rectangle from 1 to 2 is } \frac{1}{2};$$

$$\text{the area of the rectangle from 2 to 3 is } \frac{1}{3};$$

$$\vdots$$

$$\text{the area of the rectangle from } n - 1 \text{ to } n \text{ is } \frac{1}{n}.$$

So the sum of the areas of all the rectangles is $\frac{1}{2} + \frac{1}{3} + \cdots + \frac{1}{n}$. From the picture it is clear that this sum is no larger than the area underneath the graph of f between $x = 1$ and $x = n$, which is known to equal $\ln n$. Hence

$$\frac{1}{2} + \frac{1}{3} + \cdots + \frac{1}{n} \le \ln n.$$

b. Suppose n is an integer and $n \ge 3$. Since $e \cong 2.718$, then $n \ge e$. Now the logarithmic function with base e is strictly increasing. Thus

$$\text{since } e \le n, \text{ then } 1 = \ln e \le \ln n.$$

c. By part (a),

$$\frac{1}{2} + \frac{1}{3} + \cdots + \frac{1}{n} \le \ln n,$$

and by part (b),

$$1 \le \ln n.$$

Adding these two inequalities together gives

$$1 + \frac{1}{2} + \frac{1}{3} + \cdots + \frac{1}{n} \le 2 \cdot \ln n \quad \text{for any integer } n \ge 3.$$

d. By part (c), for all integers $n \ge 3$,

$$1 + \frac{1}{2} + \frac{1}{3} + \cdots + \frac{1}{n} \le 2 \cdot \ln n$$

$$\Rightarrow \left| 1 + \frac{1}{2} + \frac{1}{3} + \cdots + \frac{1}{n} \right| \le 2 \cdot \left| \ln n \right|$$
because for $n > 2$, $\ln n > 0$ (since the logarithmic function with base e is strictly increasing).

Let $M = 2$ and $x_0 = 2$. Then

$$\left| 1 + \frac{1}{2} + \frac{1}{3} + \cdots + \frac{1}{n} \right| \le M \cdot \left| \ln n \right| \quad \text{for all } n > x_0.$$

Hence by definition of O-notation,

$$1 + \frac{1}{2} + \frac{1}{3} + \cdots + \frac{1}{n} \quad \text{is} \quad O(\ln n). \qquad \blacksquare$$

EXERCISE SET 9.4

Graph each function defined in 1–8 below.

1. $f(x) = 3^x$ for all real numbers x

2. $g(x) = (\frac{1}{3})^x$ for all real numbers x

3. $h(x) = \log_{10} x$ for all positive real numbers x

4. $k(x) = \log_2 x$ for all positive real numbers x

5. $F(x) = \lfloor \log_2 x \rfloor$ for all positive real numbers x

6. $G(x) = \lceil \log_2 x \rceil$ for all positive real numbers x

7. $H(x) = x \log_2 x$ for all positive real numbers x

8. $K(x) = x \log_{10} x$ for all positive real numbers x

9. The scale of the graph shown in Figure 9.4.1 is one-fourth inch to each unit. If the point $(2, 2^{64})$ is plotted on the graph of $y = 2^x$, how many miles would it lie above the horizontal axis? What is the ratio of the height of the point to the distance of the earth from the sun? (There are 12 inches per foot and 5,280 feet per mile. The earth is approximately 93,000,000 miles from the sun on average.)

10. a. Use the definition of logarithm to show that $\log_b b^x = x$ for all real numbers x.
    b. Use the definition of logarithm to show that $b^{\log_b x} = x$ for all positive real numbers x.
    c. By the result of exercise 22 in Section 7.5, if $f : X \to Y$ and $g : Y \to X$ are functions and $g \circ f = i_X$ and $f \circ g = i_Y$, then f and g are inverse functions. Use this result to show that $\log_b$ and $\exp_b$ (the exponential function with base b) are inverse functions.

11. Let $b > 1$.
    a. Use the fact that $u = \log_b v \Leftrightarrow v = b^u$ to show that a point (u, v) lies on the graph of the logarithmic function with base b if, and only if, (v, u) lies on the graph of the exponential function with base b.
    b. Plot several pairs of points of the form (u, v) and (v, u) on a coordinate system. Describe the geometric relationship between the locations of the points in each pair.
    c. Draw the graphs of $y = \log_2 x$ and $y = 2^x$. Describe the geometric relationship between these graphs.

12. Give a graphical interpretation for property (9.4.2) in Example 9.4.1(a) for $0 < x < 1$.

H 13. Suppose a positive real number x satisfies the inequality $10^m \le x < 10^{m+1}$ where m is an integer. What can be inferred about $\lfloor \log_{10} x \rfloor$? Justify your answer.

14. Prove that if x is a positive real number and k is a nonnegative integer such that $2^k < x \le 2^{k+1}$, then $\lceil \log_2 x \rceil = k + 1$.

15. Describe in words the statement proved in exercise 14.

16. If n is an odd integer and $n > 1$, is $\lceil \log_2(n - 1) \rceil = \lceil \log_2(n) \rceil$? Justify your answer.

H 17. If n is an odd integer and $n > 1$, is $\lceil \log_2(n + 1) \rceil = \lceil \log_2(n) \rceil$? Justify your answer.

18. If n is an odd integer and $n > 1$, is $\lfloor \log_2(n + 1) \rfloor = \lfloor \log_2(n) \rfloor$? Justify your answer.

19. How many binary digits are needed to represent each of the following in binary notation?
    a. 148,206   b. 5,067,329

20. It was shown in the text that the number of binary digits needed to represent a positive integer n is $\lfloor \log_2 n \rfloor + 1$. Can this also be given as $\lceil \log_2 n \rceil$? Why or why not?

In each of 21 and 22, a sequence is specified by a recurrence relation and initial conditions. In each case,
    a. use iteration to guess an explicit formula for the sequence;
    b. use strong mathematical induction to confirm the correctness of the formula you obtained in part (a).

21. $a_k = a_{\lfloor k/2 \rfloor} + 2$ for all integers $k \ge 2$
    $a_1 = 1$

22. $b_k = 3b_{\lfloor k/3 \rfloor}$ for all integers $k \ge 3$
    $b_1 = 1, b_2 = 1$
    (See exercise 17, Section 3.5.)

H 23. Define a sequence $c_1, c_2, c_3, \ldots$ recursively as follows:

$$c_1 = 0,$$
$$c_k = 2 \cdot c_{\lfloor k/2 \rfloor} + k, \text{ for all integers } k \ge 2.$$

Use strong mathematical induction to show that $c_n \le n^2$ for all integers $n \ge 1$.

◆H 24. Use strong mathematical induction to show that for the sequence of exercise 23, $c_n \le n \cdot \log_2 n$, for all integers $n \ge 4$.

Exercises 25–28 refer to properties 9.4.9 and 9.4.10. To solve them, think big!

25. Find a real number $x > 1$ so that $\log_2 x < x^{1/10}$.

26. Find a real number $x > 1$ so that $x^{50} < 2^x$.

27. Find a real number $x > 1$ so that $x < 1.0001^x$.

28. Use a graphing calculator or computer graphing program to find two distinct approximate values of x so that $x = 1.0001^x$. On what approximate intervals is $x > 1.0001^x$? On what approximate intervals is $x < 1.0001^x$?

29. Use O-notation to express the following statement:

$$\left|\, 7x^2 + 3x \log_2 x \,\right| \leq 10 \cdot \left|\, x^2 \,\right|, \text{ for all real numbers}$$
$x > 2$.

30. Show that $2x + \log_2 x$ is $O(x)$.

31. Show that $x^2 + 5x \log_2 x$ is $O(x^2)$.

Prove each of the statements in 32–34 assuming n is an integer variable that takes positive integer values. Use identities from Section 4.2 as needed.

32. $1 + 2 + 2^2 + 2^3 + \cdots + 2^n$ is $O(2^{n+1})$.

H 33. $n + \dfrac{n}{2} + \dfrac{n}{4} + \cdots + \dfrac{n}{2^n}$ is $O(n)$.

34. $\dfrac{2n}{3} + \dfrac{2n}{3^2} + \dfrac{2n}{3^3} + \cdots + \dfrac{2n}{3^n}$ is $O(n)$.

35. Quantities of the form

$$kn + kn \log_2 n, \text{ for positive integers } k \text{ and } n$$

arise in the analysis of the merge sort algorithm in computer science. Show that for any positive integer k,

$$kn + kn \log_2 n \quad \text{is} \quad O(n \log_2 n).$$

36. Show that $n^2 + 2^n$ is $O(2^n)$.

37. Calculate the values of the harmonic sums

$$1 + \frac{1}{2} + \frac{1}{3} + \cdots + \frac{1}{n} \text{ for } n = 2, 3, 4, \text{ and } 5.$$

38. Use part (c) of Example 9.4.7 to show that

$$n + \frac{n}{2} + \frac{n}{3} + \cdots + \frac{n}{n} \quad \text{is} \quad O(n \ln n).$$

39. Use the fact that $\log_2 x = \left(\dfrac{1}{\log_e 2}\right) \cdot \log_e x$ and $\log_e x = \ln x$, for all positive numbers x, and part (c) of Example 9.4.7 to show that

$$1 + \frac{1}{2} + \frac{1}{3} + \cdots + \frac{1}{n} \quad \text{is} \quad O(\log_2 n).$$

40. a. Show that $\lfloor \log_2 n \rfloor$ is $O(\log_2 n)$.
  b. Show that $\lfloor \log_2 n \rfloor + 1$ is $O(\log_2 n)$.

H 41. a. Show that if n is a variable that takes positive integer values, then $n!$ is $O(n^n)$.
  b. Use part (a) to show that $\log_2(n!)$ is $O(n \log_2 n)$.

42. Show that if n is a variable that takes positive integer values, then 2^n is $O(n!)$.

43. Prove by mathematical induction that for all integers $n \geq 1$, $\log_2 n \leq n$.

44. Prove by mathematical induction that for all integers $n \geq 1$, $n \leq 10^n$.

H◆ 45. a. (Calculus needed.) Use the fact that

$$\lim_{k \to \infty} \left(1 + \frac{1}{k}\right)^k = e \text{ to prove by mathematical}$$
induction that

$$n \log_2 n \leq 2 \log_2 (n!), \text{ for all integers } n \geq 2.$$

  b. What can you deduce from part (a) about an order for $n \log_2 n$?

H◆ 46. Show that $n \log_2 n$ is $O(\log_2 (n!))$ without using calculus by showing that for all integers $n \geq 4$,

$$\log_2 (n!) \geq \frac{1}{4}(n \log_2 n).$$

H 47. Let r be a real number with $r \geq 1$. Use the binomial theorem to show that $(1 + r)^n > n$ for all integers $n \geq 1$.

◆ 48. a. For all positive real numbers u, $\log_2 u < u$. Use this fact to show that for any positive integer n, $\log_2 x < nx^{1/n}$ for all real numbers $x > 0$.
  b. Interpret the statement of part (a) using O-notation.

◆ 49. a. For all real numbers x, $x < 2^x$. Use this fact to show that for any positive integer n, $x^n < n^n 2^x$ for all real numbers $x > 0$.
  b. Interpret the statement of part (a) using O-notation.

◆ 50. For all positive real numbers u, $\log_2 u < u$. Use this fact and the result of exercise 20 in Section 9.1 to prove the following: For all integers $n \geq 1$, $\log_2 x < x^{1/n}$ for all real numbers $x > (2n)^{2n}$.

◆ 51. Use the result of exercise 50 above to prove the following: For all integers $n \geq 1$, $x^n < 2^x$ for all real numbers $x > (2n)^{2n}$.

Exercises 52 and 53 use L'Hôpital's rule from calculus.

52. a. Let b be any real number greater than 1. Use L'Hôpital's rule and mathematical induction to prove that for all integers $n \geq 1$,

$$\lim_{x \to \infty} \frac{x^n}{b^x} = 0.$$

  b. Use the result of part (a) and the definition of O-notation to prove that x^n is $O(b^x)$ for any integer $n \geq 1$.

53. a. Let b be any real number greater than 1. Use L'Hôpital's rule to prove that for all integers $n \geq 1$,

$$\lim_{x \to \infty} \frac{\log_b x}{x^{1/n}} = 0.$$

  b. Use the result of part (a) and the definition of O-notation to prove that $\log_b x$ is $O(x^{1/n})$ for any integer $n \geq 1$.

9.5 APPLICATION: EFFICIENCY OF ALGORITHMS II

Have you ever played the "guess my number" game? A person thinks of a number between two other numbers, say 1 and 10 or 1 and 100 for example, and you try to figure out what it is using the least number of guesses. Each time you guess a number, the person tells you whether you are correct, too low, or too high. If you have ever played this game, you have probably already hit upon the most efficient strategy: Begin by guessing a number as close to the middle of the two given numbers as possible. If your guess is too high, then the number is between the lower of the two given numbers and the one you first chose. If your guess is too low, then the number is between the number you first chose and the higher of the two given numbers. In either case, you take as your next guess a number as close as possible to the middle of the new range in which you now know the number lies. You repeat this process as many times as necessary until you have found the person's number.

The technique described above is an example of a general strategy called **divide and conquer**, which works as follows: To solve a problem, reduce it to a fixed number of smaller problems of the same kind, which can themselves be reduced to the same fixed number of smaller problems of the same kind, and so forth until easily resolved problems are obtained. In this case, the problem of finding a particular number in a given range of numbers is reduced at each stage to finding a particular number in a range of numbers approximately half as long.

It turns out that algorithms using a divide-and-conquer strategy are generally quite efficient and almost always have orders involving logarithmic functions. In this section we define the *binary search* algorithm, which is the formalization of the guess-my-number game described above, and we compare the efficiency of binary search to the sequential search discussed in Section 9.3. Then we develop a divide-and-conquer algorithm for sorting, *merge sort,* and compare its efficiency with that of insertion sort and selection sort, which were also discussed in Section 9.3.

BINARY SEARCH

While a sequential search can be performed on an array whose elements are in any order, a binary search can only be performed on an array whose elements are arranged in ascending (or descending) order. Given an array $a[1], a[2], \ldots, a[n]$ of distinct elements arranged in ascending order, consider the problem of trying to find a particular element x in the array. To use binary search, x is first compared to the "middle element" of the array. If the two are equal, the search is successful. If the two are not equal, then because the array elements are in ascending order, comparing the values of x and the middle array element narrows the search either to the lower subarray (consisting of all the array elements below the middle element) or to the upper subarray (consisting of all array elements above the middle element). The search continues by repeating this basic process over and over on successively smaller subarrays. The search terminates either when a match occurs or when the subarray to which the search has been narrowed contains no elements. The efficiency of the algorithm is a result of the fact that at each step the length of the subarray to be searched is roughly half the length of the array of the previous step. This process is illustrated in Figure 9.5.1.

To write down a formal algorithm for binary search, we introduce a variable *index* whose final value will tell us whether or not x is in the array and, if so, will indicate the location of x. Since the array goes from $a[1]$ to $a[n]$, we intialize *index* to be 0. If and when x is found, the value of *index* is changed to the subscript of the array element

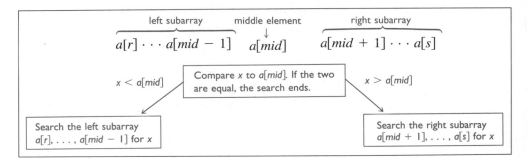

FIGURE 9.5.1   One Iteration of the Binary Search Process

equaling x. If index still has the value 0 when the algorithm is complete, then x is not one of the elements in the array. Figure 9.5.2 shows the action of a particular binary search.

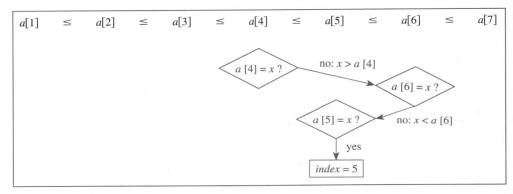

FIGURE 9.5.2   Binary Search of $a[1], a[2], \ldots, a[7]$ for x where $x = a[5]$

Formalizing a binary search algorithm also requires that we be more precise about the meaning of the "middle element" of an array. (This issue was side-stepped by careful choice of n in Figure 9.5.2.) If the array consists of an even number of elements, there are two elements in the middle. For instance, both $a[6]$ and $a[7]$ are equally in the middle of the following array.

$$\underbrace{a[3] \quad a[4] \quad a[5]}_{\text{three elements}} \quad \underbrace{a[6] \quad a[7]}_{\substack{\text{two middle} \\ \text{elements}}} \quad \underbrace{a[8] \quad a[9] \quad a[10]}_{\text{three elements}}$$

In a case such as this the algorithm must choose which of the two middle elements to take, the smaller or the larger. The choice is arbitrary—either would do. We will write the algorithm to choose the smaller. The index of the smaller of the two middle elements is the floor of the average of the top and bottom indices of the array. That is, if

bot = the bottom index of the array,

top = the top index of the array,

mid = the lower of the two middle indices of the array,

then

$$mid = \left\lfloor \frac{bot + top}{2} \right\rfloor.$$

In this case $bot = 3$ and $top = 10$, and so the index of the "middle element" is

$$mid = \left\lceil \frac{3 + 10}{2} \right\rceil = \left\lceil \frac{13}{2} \right\rceil = \lfloor 6.5 \rfloor = 6.$$

The following is a formal algorithm for a binary search.

ALGORITHM 9.5.1   Binary Search

[The aim of this algorithm is to search for an element x in an ascending array of elements a[1], a[2], . . . , a[n]. If x is found, the variable index is set equal to the index of the array element where x is located. If x is not found, index is not changed from its initial value, which is 0. The variables bot and top denote the bottom and top indices of the array currently being examined.]

Input: *n [a positive integer], a[1], a[2], . . . , a[n] [an array of data items given in ascending order], x [a data item of the same data type as the elements of the array]*

Algorithm Body:

$index := 0, bot := 1, top := n$

[Compute the middle index of the array, mid. Compare x to a[mid]. If the two are equal, the search is successful. If not, repeat the process either for the lower or the upper subarray, either giving top the new value mid − 1 or bot the new value mid + 1. Each iteration of the loop either decreases the value of top or increases the value of bot. Thus if the looping is not stopped by success in the search process, eventually the value of top will become less than the value of bot. This occurrence stops the looping process and indicates that x is not an element of the array.]

while ($top \geq bot$ and $index = 0$)

$$mid := \left\lfloor \frac{bot + top}{2} \right\rfloor$$

if $a[mid] = x$ **then** $index := mid$

if $a[mid] > x$

   then $top := mid - 1$

   else $bot := mid + 1$

end while

[If index has the value 0 at this point, then x is not in the array. Otherwise, index gives the index of the array where x is located.]

Output: *index [a nonnegative integer]*

end Algorithm 9.5.1

EXAMPLE 9.5.1   Tracing the Binary Search Algorithm

Trace the action of Algorithm 9.5.1 on the variables *index, bot, top, mid,* and the values of x given in (a) and (b) below for the input array

$$a[1] = \text{Ann}, a[2] = \text{Dawn}, a[3] = \text{Erik}, a[4] = \text{Gail}, a[5] = \text{Juan},$$
$$a[6] = \text{Matt}, a[7] = \text{Max}, a[8] = \text{Rita}, a[9] = \text{Tsuji}, a[10] = \text{Yuen}$$

where alphabetical ordering is used to compare elements of the array.

a. $x = \text{Max}$ b. $x = \text{Sara}$

Solution a.

index	0				7
bot	1	6		7	
top	10		7		
mid		5	8	6	7

b.

index	0			
bot	1	6	9	
top	10			8
mid		5	8	9

THE EFFICIENCY OF THE BINARY SEARCH ALGORITHM

The idea of the derivation of the efficiency of the binary search algorithm is not difficult. Here it is in brief. At each stage of the binary search process, the length of the new subarray to be searched is approximately half that of the previous one, and in the worst case every subarray down to a subarray with a single element must be searched. Consequently, in the worst case the maximum number of iterations of the **while** loop in the binary search algorithm is one more than the number of times the original input array can be cut approximately in half. If the length n of this array is a power of 2 ($n = 2^k$ for some integer k), then n can be halved exactly $k = \log_2 n = \lfloor \log_2 n \rfloor$ times before an array of length 1 is reached. If n is not a power of 2, then $n = 2^k + m$ for some integer k (where $m < 2^k$), and so n can be split approximately in half k times also. But in this case $k = \lfloor \log_2 n \rfloor$ also. Thus in the worst case, the number of iterations of the **while** loop in the binary search algorithm, which is proportional to the number of comparisons required to execute it, is $\lfloor \log_2 n \rfloor + 1$. The derivation is concluded by noting that $\lfloor \log_2 n \rfloor + 1$ is $O(\log_2 n)$.

  The details of the derivation are developed in Examples 9.5.2–9.5.6. Throughout the derivation, for each integer $n \geq 1$, let

> w_n = the number of iterations of the **while** loop
>   in a *worst-case* execution of the binary search
>   algorithm for an input array of length n.

  The first issue to consider is this. If the length of the input array for one iteration of the **while** loop is known, what is the greatest possible length of the array input to the next iteration?

EXAMPLE 9.5.2 The Length of the Input Array to the Next Iteration of the Loop

Prove that if an array of length k is input to the **while** loop of the binary search algorithm, then after one unsuccessful iteration of the loop, the input to the next iteration is an array of length at most $\lfloor k/2 \rfloor$.

Solution Consider what occurs when an array of length k is input to the **while** loop in the case where $x \neq a[mid]$:

$$\underbrace{a[bot], a[bot+1], \ldots, a[mid-1]}_{\text{new input to the while loop if } x < a[mid]},\ \underset{\uparrow\ \text{“middle element”}}{a[mid]},\ \underbrace{a[mid+1], \ldots, a[top-1], a[top]}_{\text{new input to the while loop if } x > a[mid]}.$$

Since the input array has length k, the value of mid depends on whether k is odd or even. In both cases we match up the array elements with the integers from 1 to k and analyze the lengths of the left and right subarrays. In case k is odd, both the left and the right subarrays have length $\lfloor k/2 \rfloor$. In case k is even, the left subarray has length $\lfloor k/2 \rfloor - 1$ and the right subarray has length $\lfloor k/2 \rfloor$. The reasoning behind these results is shown in Figure 9.5.3.

k odd: $a[bot] \;\cdots\; a[mid-1] \;\; a[mid] \;\; a[mid+1] \;\cdots\; a[top]$

$$1 \;\cdots\; \frac{k+1}{2} - 1 \qquad \frac{k+1}{2} \qquad \frac{k+1}{2} + 1 \;\cdots\; k$$

$$\text{length} = \frac{k+1}{2} - 1 = \frac{k-1}{2} = \left\lfloor \frac{k}{2} \right\rfloor \qquad \text{middle element} \qquad \text{length} = k - \left(\frac{k+1}{2} + 1\right) + 1 = \frac{k-1}{2} = \left\lfloor \frac{k}{2} \right\rfloor$$

k even: $a[bot] \;\cdots\; a[mid-1] \;\; a[mid] \;\; a[mid+1] \;\cdots\; a[top]$

$$1 \;\cdots\; \frac{k}{2} - 1 \qquad \frac{k}{2} \qquad \frac{k}{2} + 1 \;\cdots\; k$$

$$\text{length} = \frac{k}{2} - 1 = \left\lfloor \frac{k}{2} \right\rfloor - 1 \qquad \text{middle element} \qquad \text{length} = k - \left(\frac{k}{2} + 1\right) + 1 = \frac{k}{2} = \left\lfloor \frac{k}{2} \right\rfloor$$

FIGURE 9.5.3 Lengths of the Left and Right Subarrays

Because the maximum of the numbers $\lfloor k/2 \rfloor$ and $\lfloor k/2 \rfloor - 1$ is $\lfloor k/2 \rfloor$, in the worst case this will be the length of the array input to the next iteration of the loop. ■

To find the order of the algorithm, a formula for $w_1, w_2, w_3, \ldots$ is needed. The next example derives a recurrence relation for the sequence.

EXAMPLE 9.5.3 A Recurrence Relation for $w_1, w_2, w_3, \ldots$

Prove that the sequence $w_1, w_2, \ldots, w_n, \ldots$ satisfies the recurrence relation and initial condition

$$w_1 = 1,$$
$$w_k = 1 + w_{\lfloor k/2 \rfloor} \quad \text{for all integers } k > 1.$$

Solution   Example 9.5.2 showed that given an input array of length k to the **while** loop, the worst that can happen is that the next iteration of the loop will have to search an array of length $\lfloor k/2 \rfloor$. Hence the maximum number of iterations of the loop is 1 more than the maximum number necessary to execute it for an input array of length $\lfloor k/2 \rfloor$. In symbols,

$$w_k = 1 + w_{\lfloor k/2 \rfloor}.$$

Also

$$w_1 = 1$$

because for an input array of length 1 ($bot = top$), the **while** loop iterates only one time.   ∎

Now that a recurrence relation for $w_1, w_2, w_3, \ldots$ has been found, iteration can be used to come up with a good guess for an explicit formula.

EXAMPLE 9.5.4   **An Explicit Formula for $w_1, w_2, w_3, \ldots$**

Apply iteration to the recurrence relation found in Example 9.5.3 to conjecture an explicit formula for $w_1, w_2, w_3, \ldots$

Solution   Begin by iterating to find the values of the first few terms of the sequence.

$$w_① = ①$$
$$w_② = 1 + w_{\lfloor 2/2 \rfloor} = 1 + w_1 = 1 + 1 = ②$$
$$w_3 = 1 + w_{\lfloor 3/2 \rfloor} = 1 + w_1 = 1 + 1 = 2$$

$1 = 2^0;\ 1 = 0 + 1$
$2 = 2^1;\ 2 = 1 + 1$

$$w_④ = 1 + w_{\lfloor 4/2 \rfloor} = 1 + w_2 = 1 + 2 = ③$$
$$w_5 = 1 + w_{\lfloor 5/2 \rfloor} = 1 + w_2 = 1 + 2 = 3$$
$$w_6 = 1 + w_{\lfloor 6/2 \rfloor} = 1 + w_3 = 1 + 2 = 3$$
$$w_7 = 1 + w_{\lfloor 7/2 \rfloor} = 1 + w_3 = 1 + 2 = 3$$

$4 = 2^2;\ 3 = 2 + 1$

$$w_⑧ = 1 + w_{\lfloor 8/2 \rfloor} = 1 + w_4 = 1 + 3 = ④$$
$$w_9 = 1 + w_{\lfloor 9/2 \rfloor} = 1 + w_4 = 1 + 3 = 4$$
$$\vdots$$
$$w_{15} = 1 + w_{\lfloor 15/2 \rfloor} = 1 + w_7 = 1 + 3 = 4$$

$8 = 2^3;\ 4 = 3 + 1$

$$w_{⑯} = 1 + w_{\lfloor 16/2 \rfloor} = 1 + w_8 = 1 + 4 = ⑤$$

$16 = 2^4;\ 5 = 4 + 1$

Note that in each case when the subscript n is between two powers of 2, w_n is 1 more than the exponent of the lower power of 2. In other words:

$$\text{If } 2^i \leq n < 2^{i+1}, \text{ then } w_n = i + 1. \qquad \textbf{9.5.1}$$

But if

$$2^i \leq n < 2^{i+1},$$

then *[by property (9.4.2) of Example 9.4.1]*

$$i = \lfloor \log_2 n \rfloor.$$

Substitution into statement (9.5.1) gives the conjecture that

$$w_n = \lfloor \log_2 n \rfloor + 1.$$ ∎

Now mathematical induction can be used to verify the correctness of the formula found in Example 9.5.4.

EXAMPLE 9.5.5

Verifying the Correctness of the Formula

Use strong mathematical induction to show that if $w_1, w_2, w_3, \ldots$ is a sequence of numbers that satisfies the recurrence relation and initial condition

$$w_1 = 1,$$

$$w_k = 1 + w_{\lfloor k/2 \rfloor} \quad \text{for all integers } k > 1,$$

then

$$w_n = \lfloor \log_2 n \rfloor + 1 \quad \text{for all integers } n \geq 1.$$

Solution   The formula holds for $n = 1$:

For $n = 1$, $w_1 = 1$ and $\lfloor \log_2 1 \rfloor + 1 = \lfloor 0 \rfloor + 1 = 1$ also. Hence the statement is true for $n = 1$.

If $k \geq 2$ and the formula holds for all integers i with $1 \leq i < k$, then it holds for k:

Suppose that for some integer $k \geq 2$,

$$w_i = \lfloor \log_2 i \rfloor + 1 \quad \text{for all integers } i \text{ with } 1 \leq i < k. \qquad \text{This is the inductive hypothesis.}$$

We must show that

$$w_k = \lfloor \log_2 k \rfloor + 1.$$

Consider the two cases: k is odd and k is even.

Case 1 (k is odd): In this case $\lfloor k/2 \rfloor = \dfrac{k-1}{2}$, and so

$$
\begin{aligned}
w_k &= 1 + w_{\lfloor k/2 \rfloor} && \text{by the recurrence relation} \\
&= 1 + w_{(k-1)/2} && \text{because } \lfloor k/2 \rfloor = (k-1)/2 \text{ since } k \text{ is odd} \\
&= 1 + \left(\left\lfloor \log_2 \left(\frac{k-1}{2} \right) \right\rfloor + 1 \right) && \text{by inductive hypothesis} \\
&= \lfloor \log_2(k-1) - \log_2 2 \rfloor + 2 && \begin{array}{l}\text{by substituting into the identity} \\ \log_b (x/y) = \log_b x - \log_b y \text{ derived in} \\ \text{exercise 24 of Section 7.3}\end{array} \\
&= \lfloor \log_2(k-1) - 1 \rfloor + 2 && \text{since } \log_2 2 = 1 \\
&= (\lfloor \log_2(k-1) \rfloor - 1) + 2 && \begin{array}{l}\text{by substituting } x = \log_2(k-1) \text{ into the} \\ \text{identity } \lfloor x - 1 \rfloor = \lfloor x \rfloor - 1 \text{ derived in exercise} \\ \text{15 of Section 3.5}\end{array} \\
&= \lfloor \log_2 k \rfloor + 1 && \text{by property (9.4.3) in Example 9.4.2}
\end{aligned}
$$

Case 2 (k is even): In this case, it can also be shown that $w_k = \lfloor \log_2 k \rfloor + 1$. The analysis is very similar to that of case 1 and is left as an exercise.

Hence regardless of whether k is odd or k is even,

$$w_k = \lfloor \log_2 k \rfloor + 1,$$

as was to be shown.

[Since both the basis and the inductive steps have been demonstrated, the proof by strong mathematical induction is complete.] ■

The final example shows how to use the formula for $w_1, w_2, w_3, \ldots$ to find a worst-case order for the algorithm.

EXAMPLE 9.5.6   The Binary Search Algorithm Is Logarithmic

Given that by Example 9.5.5, for all positive integers n,

$$w_n = \lfloor \log_2 n \rfloor + 1,$$

show that the binary search algorithm is $O(\log_2 n)$.

Solution   For any integer $n > 2$,

$$w_n = \lfloor \log_2 n \rfloor + 1 \qquad \text{by Example 9.5.5}$$

$$\Rightarrow \quad w_n \leq \log_2 n + 1 \qquad \text{because } \lfloor x \rfloor \leq x \text{ for all real numbers } x$$

$$\Rightarrow \quad w_n \leq \log_2 n + \log_2 n \qquad \begin{array}{l} \text{since the logorithm with base 2 is increasing,} \\ \text{if } 2 < n \text{ then } 1 = \log_2 2 < \log_2 n \end{array}$$

$$\Rightarrow \quad w_n \leq 2 \cdot \log_2 n.$$

Both w_n and $\log_2 n$ are positive for $n > 2$. Therefore,

$$|w_n| \leq 2 \cdot |\log_2 n| \quad \text{for all integers } n > 2.$$

Let $M = 2$ and $x_0 = 2$. Then

$$|w_n| \leq M \cdot |\log_2 n| \quad \text{for all integers } n > x_0.$$

Hence by definition of O-notation,

$$w_n \quad \text{is} \quad O(\log_2 n).$$

But w_n, the number of iterations of the **while** loop, is proportional to the number of comparisons performed when the binary search algorithm is executed. Thus the binary search algorithm is $O(\log_2 n)$. ∎

Examples 9.5.2–9.5.6 show that in the worst case the binary search algorithm has order $\log_2 n$. As noted in Section 9.3, in the worst case the sequential search algorithm has the order n. This difference in efficiency becomes increasingly more important as n gets larger and larger. Assuming one loop iteration is performed each millionth of a second, then performing n iterations for $n = 100,000$ requires 0.1 second, whereas performing $\log_2 n$ iterations requires 0.000017 second. For $n = 100,000,000$ the times are 1.67 minutes and 0.000027 second, respectively. And for $n = 100,000,000,000$ the respective times are 2.78 hours and 0.000037 second.

MERGE SORT

Note that it is much easier to write a detailed algorithm for sequential search than for binary search. Yet binary search is much more efficient than sequential search. Such trade-offs often occur in computer science. Frequently, the straightforward "obvious" solution to a problem is less efficient than a clever solution that is more complicated to describe.

In the text and exercises for Section 9.3, we gave two methods for sorting, insertion sort and selection sort, both of which are formalizations of methods human beings often use in ordinary situations. Can a divide-and-conquer approach be used to find a sorting method more efficient than these? It turns out that the answer is an emphatic "yes." In fact, over the past few decades computer scientists have developed several divide-and-conquer sorting methods all of which are somewhat more complex to describe but are significantly more efficient than either insertion sort or selection sort.

One of these methods, **merge sort**, is obtained by thinking recursively. Imagine that an efficient way for sorting arrays of length less than k is already known. How can such knowledge be used to sort an array of length k? One way is to suppose the array of length k is split into two roughly equal parts and each part is sorted using the known method. Is there an efficient way to combine the parts into a sorted array? Sure. Just "merge" them.

Figure 9.5.4 illustrates how a merge works. Imagine that the elements of two ordered subarrays, 2, 5, 6, 8 and 3, 6, 7, 9, are written on slips of paper (to make them easy to move around). Place the slips for each subarray in two columns on a tabletop, one at the left and one at the right. Along the bottom of the tabletop, set up eight positions into which the slips will be moved. Then, one-by-one, bring down the slips from the bottoms of the columns. At each stage compare the numbers on the slips currently at the column bottoms, and move the slip containing the smaller number down into the next position in the array as a whole. If at any stage the two numbers are equal, take, say, the slip on the left to move into the next position. And if at any stage one of the columns is empty, just move the slips from the other column into position one-by-one in order.

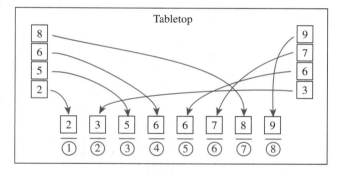

FIGURE 9.5.4   Merging Two Sorted Subarrays to Obtain a Sorted Array

One important observation about the merging algorithm described above: It requires memory space to move the array elements around. A second set of array positions as long as the original one is needed into which to place the elements of the two subarrays in order. In Figure 9.5.4 this second set of array positions is represented by the positions set up at the bottom of the tabletop. Of course, once the elements of the original array have been placed into this new array, they can be moved back in order into the original array positions.

In terms of time, however, merging is efficient because the total number of comparisons needed to merge two subarrays into an array of length k is just $k - 1$. You can see why by analyzing Figure 9.5.4. Observe that at each stage, the decision about which slip to move is made by comparing the numbers on the slips currently at the bottoms of the two columns except when one of the columns is empty, in which case no comparisons are made at all. Thus in the worst case there will be one comparison for each of the k positions in the final array except the very last one (because when the last slip is placed into position the other column is sure to be empty), or a total of $k - 1$ comparisons in all.

The merge sort algorithm is recursive: Its defining statements include references to itself. The algorithm is well defined, however, because at each stage the length of the array that is input to the algorithm is shorter than at the previous stage, so that, ultimately, the algorithm only has to deal with arrays of length one, which are already sorted. Specifically, merge sort works as follows:

> Given an array of elements that can be put into order, if the array consists of a single element leave it as it is. It is already sorted. Otherwise
>
> **1.** divide the array into the two subarrays of as equal lengths as possible;
> **2.** use merge sort to sort each subarray;
> **3.** merge the two subarrays together.

Figure 9.5.5 illustrates a merge sort in a particular case.

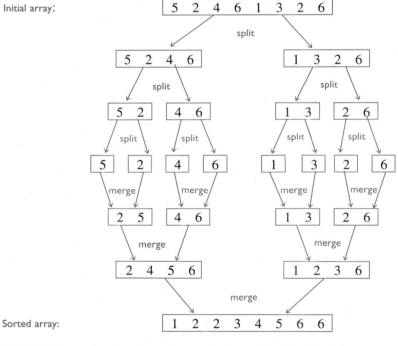

FIGURE 9.5.5   Applying Merge Sort to the Array 5, 2, 4, 6, 1, 3, 2, 6

As in the case of the binary search algorithm, in order to formalize merge sort we must decide at exactly what point to split each array. Given an array $a[bot], a[bot + 1], \ldots, a[top]$, let $mid = \lfloor (bot + top)/2 \rfloor$. Take the left subarray to be $a[bot]$, $a[bot + 1]$, . . . , $a[mid]$ and the right subarray to be $a[mid + 1]$, $a[mid + 2]$, . . . , $a[top]$. The following is a formal version of merge sort.

ALGORITHM 9.5.2 Merge Sort

[The aim of this algorithm is to take an array of elements $a[r]$, $a[r + 1]$, . . . , $a[s]$ (where $r \leq s$) and to order it. The output array is denoted $a[r]$, $a[r + 1]$, . . . , $a[s]$ also. It has the same values as the input array, but they are in ascending order. The input array is split into two nearly equal-length subarrays each of which is ordered using merge sort. Then the two subarrays are merged together.]

Input: r and s *[positive integers with $r < s$]*, $a[r]$, $a[r + 1]$, . . . , $a[s]$ *[an array of data items that can be ordered]*

Algorithm Body:

$bot := r$, $top := s$

while ($bot < top$)

$$mid := \left\lfloor \frac{bot + top}{2} \right\rfloor$$

call **merge sort** with input bot, mid, and
$a[bot]$, $a[bot + 1]$, . . . , $a[mid]$

call **merge sort** with input $mid + 1$, top and
$a[mid + 1]$, $a[mid + 2]$, . . . , $a[top]$

[After these steps, the arrays $a[bot]$, $a[bot + 1]$, . . . , $a[mid]$ and $a[mid + 1]$, $a[mid + 2]$, . . . , $a[top]$ are both in order.]

merge $a[bot]$, $a[bot + 1]$, . . . , $a[mid]$ and
$a[mid + 1]$, $a[mid + 2]$, . . . , $a[top]$

[This step can be done with a call to a merge algorithm. To put the final array in ascending order, the merge algorithm must be written so as to take two arrays in ascending order and merge them into an array in ascending order.]

end while

Output: $a[r]$, $a[r + 1]$, . . . , $a[s]$ *[an array with the same elements as the input array but in ascending order]*

To derive the efficiency of merge sort, let

$m_n =$ the maximum number of comparisons used when merge sort is applied to an array of length n.

Then $m_1 = 0$ because no comparisons are used when merge sort is applied to an array of length 1. Also for any integer $k > 1$, consider an array $a[bot]$, $a[bot + 1]$, . . . , $a[top]$ of length k that is split into two subarrays, $a[bot]$, $a[bot + 1]$, . . . , $a[mid]$ and $a[mid + 1]$, $a[mid + 2]$, . . . , $a[top]$, where $mid = \lfloor (bot + top)/2 \rfloor$. In exercise 24 you are asked to show that the right subarray has length $\lfloor k/2 \rfloor$ and the left subarray has length $\lceil k/2 \rceil$. From the previous discussion of the merge process, it is known that to merge two subarrays into an array of length k, at most $k - 1$ comparisons are needed.

Consequently,

$$
\left[\begin{array}{c} \text{the number of comparisons} \\ \text{when merge sort is applied} \\ \text{to an array of length } k \end{array}\right] = \left[\begin{array}{c} \text{the number of comparisons} \\ \text{when merge sort is applied} \\ \text{to an array of length } \lfloor k/2 \rfloor \end{array}\right]
$$

$$
+ \left[\begin{array}{c} \text{the number of comparisons} \\ \text{when merge sort is applied} \\ \text{to an array of length } \lceil k/2 \rceil \end{array}\right] + \left[\begin{array}{c} \text{the number of comparisons} \\ \text{used to merge two subarrays} \\ \text{into an array of length } k \end{array}\right].
$$

Or in other words,

$$
m_k = m_{\lfloor k/2 \rfloor} + m_{\lceil k/2 \rceil} + (k - 1) \quad \text{for all integers } k > 1.
$$

In exercise 25 you are asked to use this recurrence relation to show that

$$
m_n \leq 2n \log_2 n \quad \text{for all integers } n \geq 1.
$$

It follows that merge sort is $O(n \log_2 n)$.

In the text and exercises for Section 9.3, we showed that insertion sort and selection sort are both $O(n^2)$. How much difference can it make that merge sort is $O(n \log_2 n)$? If $n = 1,000,000$ and a computer is used that performs one operation each millionth of a second, the time needed to perform $n \log_2 n$ operations is about 20 seconds whereas the time needed to perform n^2 operations is approximately 11.6 days.

TRACTABLE AND INTRACTABLE PROBLEMS

At an opposite extreme from an algorithm such as binary search, which has logarithmic order, is an algorithm with exponential order. For example, consider an algorithm to direct the movement of each of the 64 disks in the Tower of Hanoi puzzle as they are transferred one by one from one pole to another. In Section 8.2 we showed that such a transfer requires $2^{64} - 1$ steps. If a computer took a millionth of a second to calculate each transfer step, the total time to calculate all the steps would be

$$
(2^{64} - 1) \cdot \left(\frac{1}{10^6}\right) \cdot \left(\frac{1}{60}\right) \cdot \left(\frac{1}{60}\right) \cdot \left(\frac{1}{24}\right) \cdot \left(\frac{1}{365.25}\right) \cong 584,542 \text{ years.}
$$

| number of moves | moves per second | seconds per minute | minutes per hour | hours per day | days per year |

Problems whose only solution algorithms have at least exponential order are called **intractable** because even for moderate input sizes, the amount of time required to execute them on a computer is impractically long. In general, a problem with a solution algorithm whose order can be expressed as a polynomial in the input size is called **tractable**. It is true that an algorithm with polynomial order may take a long time to execute if the degree of the polynomial is large. But compared to algorithms with exponential order, the time required is vastly less for large input sizes. Specifically, if you take the time required to execute an algorithm with polynomial order and divide it by the time needed for an algorithm with exponential order, the fraction approaches zero as the input size gets larger and larger.

In recent years computer scientists have identified a fairly large class of problems, called **NP-complete**, all of which appear to be intractable. At present this is only a conjecture; it has not been proved. However, one curious fact is known for sure about this class of problems: If any one of them can be solved with an algorithm whose order is a polynomial, then so can all the others. In Section 11.2 we discuss one of these NP-complete problems, the traveling salesperson problem.

A FINAL NOTE

This section and the previous one on algorithm efficiency have offered only a partial view of what is involved in analyzing a computer algorithm. For one thing, it is assumed that searches and sorts take place in the memory of the computer. Searches and sorts on disk-based files require different algorithms, though the methods for their analysis are similar. For another thing, as mentioned at the beginning of Section 9.3, time efficiency is not the only factor that matters in the decision about which algorithm to choose. The amount of memory space required is also important, and there are mathematical techniques to estimate space efficiency very similar to those used to estimate time efficiency. Furthermore, as parallel processing of data becomes increasingly prevalent, current methods of algorithm analysis are being modified and extended to apply to algorithms designed for this new technology.

EXERCISE SET 9.5

1. Use the facts that $\log_2 10 \cong 3.32$ and $\log_2(10^a) = a \cdot \log_2 10$, for all real numbers a, to find $\log_2(1,000)$, $\log_2(1,000,000)$, and $\log_2(1,000,000,000,000,000)$.

2. If an algorithm requires $c \lfloor \log_2 n \rfloor$ operations when performed with an input of size n (where c is a constant),
   a. how many operations will be required when the input size is increased from m to m^2 (where m is a positive integer power of 2)?
   b. by what factor will the number of operations increase when the input size is increased from m to m^{10} (where m is a positive integer power of 2)?
   c. When n increases from $128(=2^7)$ to $268,435,456$ $(=2^{28})$, by what factor is $c \lfloor \log_2 n \rfloor$ increased?

Exercises 3 and 4 illustrate that for relatively small values of n, algorithms with larger orders can be more efficient than algorithms with smaller orders. Use a graphing calculator or computer to answer these questions.

3. For what values of n is an algorithm that requires n operations more efficient than an algorithm that requires $\lfloor 50 \log_2 n \rfloor$ operations?

4. For what values of n is an algorithm that requires $\lfloor n^2/10 \rfloor$ operations more efficient than an algorithm that requires $\lfloor n \log_2 n \rfloor$ operations?

In 5 and 6, trace the action of the binary search algorithm (Algorithm 9.5.1) on the variables index, bot, top, mid, and the given values of x for the input array $a[1] = $ Chia, $a[2] = $ Doug, $a[3] = $ Jan, $a[4] = $ Jim, $a[5] = $ José, $a[6] = $

Mary, $a[7] = $ Rob, $a[8] = $ Roy, $a[9] = $ Sue, $a[10] = $ Usha, where alphabetical ordering is used to compare elements of the array.

5. a. $x = $ Chia
   b. $x = $ Max

6. a. $x = $ Amanda
   b. $x = $ Roy

7. Suppose bot and top are positive integers with $bot \le top$. Consider the array

$$a[bot], a[bot + 1], \ldots, a[top].$$

   a. How many elements are in this array?
   b. Show that if the number of elements in the array is odd, then the quantity $bot + top$ is even.
   c. Show that if the number of elements in the array is even, then the quantity $bot + top$ is odd.

Exercises 8–11 refer to the following algorithm segment. For each positive integer n, let a_n be the number of iterations of the while loop.

while $(n > 0)$
  $n := n$ div 2
end while

8. Trace the action of this algorithm segment on n when the initial value of n is 27.

9. Find a recurrence relation for a_n.

10. Find an explicit formula for a_n.

11. Find an order for this algorithm segment.

Exercises 12–15 refer to the following algorithm segment. For each positive integer n, let b_n be the number of iterations of the while loop.

$$\textbf{while } (n > 0)$$
$$n := n \text{ div } 3$$
$$\textbf{end while}$$

12. Trace the action of this algorithm segment on n when the initial value of n is 424.

13. Find a recurrence relation for b_n.

14. Find an explicit formula for b_n.

15. Find an order for this algorithm segment.

16. Complete the proof of case 2 of the strong induction argument in Example 9.5.5. In other words, show that if k is an even integer and $w_i = \lfloor \log_2 i \rfloor + 1$ for all integers i with $1 \le i < k$, then $w_k = \lfloor \log_2 k \rfloor + 1$.

For 17–19, modify the binary search algorithm (Algorithm 9.5.1) to take the upper of the two middle array elements in case the input array has even length. In other words, in Algorithm 9.5.1 replace

$$mid := \left\lfloor \frac{bot + top}{2} \right\rfloor \text{ with } mid := \left\lceil \frac{bot + top}{2} \right\rceil.$$

17. Trace the modified binary search algorithm for the same input as was used in Example 9.5.1.

18. Suppose an array of length k is input to the **while** loop of the modified binary search algorithm. Show that after one iteration of the loop, if $a[mid] \neq x$ the input to the next iteration is an array of length at most $\lfloor k/2 \rfloor$.

19. Let w_n be the number of iterations of the **while** loop in a worst-case execution of the modified binary search algorithm for an input array of length n. Show that $w_k = 1 + w_{\lfloor k/2 \rfloor}$ for $k \ge 2$.

In 20 and 21, draw a diagram like that of Figure 9.5.4 to show how to merge the given subarrays into a single array in ascending order.

20. 3, 5, 6, 9, 12 and 2, 4, 7, 9, 11

21. F, K, L, R, U and C, E, L, P, W (alphabetical order)

In 22 and 23, draw a diagram like that of Figure 9.5.5 to show how merge sort works for the given input arrays.

22. R, G, B, U, C, F, H, G (alphabetical order)

23. 5, 2, 3, 9, 7, 4, 3, 2

24. Show that given an array $a[bot], a[bot + 1], \ldots, a[top]$ of length k, if $mid = \lfloor (bot + top)/2 \rfloor$ then

a. the subarray $a[mid + 1], a[mid + 2], \ldots, a[top]$ has length $\lfloor k/2 \rfloor$.

b. the subarray $a[bot], a[bot + 1], \ldots, a[mid]$ has length $\lceil k/2 \rceil$.

H25. The recurrence relation for $m_1, m_2, m_3, \ldots,$ which arises in the calculation of the efficiency of merge sort, is

$$m_1 = 0$$
$$m_k = m_{\lfloor k/2 \rfloor} + m_{\lceil k/2 \rceil} + k - 1.$$

Show that for all integers $n \ge 1$, $m_n \le 2n \log_2 n$.

26. You might think that $n - 1$ multiplications are needed to compute x^n since

$$x^n = \underbrace{x \cdot x \cdot \ldots \cdot x.}_{n - 1 \text{ multiplications}}$$

But observe that, for instance, since $6 = 4 + 2$,

$$x^6 = x^4 \cdot x^2 = (x^2)^2 \cdot x^2.$$

Thus x^6 can be computed using three multiplications: one to compute x^2, one to compute $(x^2)^2$, and one to multiply $(x^2)^2$ times x^2. Similarly, since $11 = 8 + 2 + 1$,

$$x^{11} = x^8 \cdot x^2 \cdot x^1 = ((x^2)^2)^2 \cdot x^2 \cdot x$$

and so x^{11} can be computed using five multiplications: one to compute x^2, one to compute $(x^2)^2$, one to compute $((x^2)^2)^2$, one to multiply $((x^2)^2)^2$ times x^2, and one to multiply that product by x.

a. Write an algorithm to take a real number x and a positive integer n and compute x^n by

(i) calling Algorithm 4.1.1 to find the binary representation of n:

$$(r[k]\ r[k - 1] \cdots r[0])_2,$$

where each $r[i]$ is 0 or 1;

(ii) computing $x^2, x^{2^2}, x^{2^3}, \ldots, x^{2^k}$ by squaring, then squaring again, and so forth;

(iii) computing x^n using the fact that

$$x^n = x^{r[k]2^k + \cdots + r[2]2^2 + r[1]2^1 + r[0]2^0}$$
$$= x^{r[k]2^k} \cdot \ldots \cdot x^{r[2]2^2} \cdot x^{r[1]2^1} \cdot x^{r[0]2^0}$$

b. Show that the number of multiplications performed by the algorithm of part (a) is less than or equal to $2 \cdot \lfloor \log_2 n \rfloor$.

10 Relations

There are many kinds of relationships in the world. For instance, we say that two people are related by blood if they share a common ancestor and that they are related by marriage if one shares a common ancestor with the spouse of the other. We also speak of the relationship between boyfriend and girlfriend, between student and teacher, between people who work for the same employer, and between people who share a common ethnic background.

Similarly, the objects of mathematics and computer science may be related in various ways. Two digital logic circuits may be said to be related if they have the same input/output table. A set A may be said to be related to a set B if A is a subset of B, or if A is not a subset of B, or if A is the complement of B. A number x may be said to be related to a number y if $x < y$, or if x divides y, or if $x^2 + y^2 = 1$. Two identifiers in a computer program may be said to be related if they have the same first eight characters, or if the same memory location is used to store their values when the program is executed. And the list could go on!

In this chapter we discuss the mathematics of relations defined on sets, focusing on ways to represent relations and exploring various properties they may have. The concept of equivalence relation is introduced in Section 10.3 and applied in Section 10.4 to the question of determining whether two finite-state automata accept the same language. Partial order relations are discussed in Section 10.5 and an application is given showing how to use these relations to help coordinate and guide the flow of individual tasks that must be performed to accomplish a complex, large-scale project.

10.1 RELATIONS ON SETS

Strange as it may sound, the power of mathematics rests on its evasion of all unnecessary thought and on its wonderful saving of mental operations.
(Ernst Mach, 1838–1916)

Let $A = \{0, 1, 2\}$ and $B = \{1, 2, 3\}$. Let us say that an element x in A is related to an element y in B if, and only if, x is less than y. Let us use the notation $x \, R \, y$ as a shorthand for the sentence "x is related to y." Then

$$
\begin{aligned}
0 \, R \, 1 \quad &\text{since} \quad 0 < 1, \\
0 \, R \, 2 \quad &\text{since} \quad 0 < 2, \\
0 \, R \, 3 \quad &\text{since} \quad 0 < 3, \\
1 \, R \, 2 \quad &\text{since} \quad 1 < 2, \\
1 \, R \, 3 \quad &\text{since} \quad 1 < 3, \quad \text{and} \\
2 \, R \, 3 \quad &\text{since} \quad 2 < 3.
\end{aligned}
$$

On the other hand, if the notation $x \not{R} y$ represents the sentence "x is not related to y," then

$$0 \not{R} 0 \quad \text{since} \quad 0 \not< 0,$$
$$1 \not{R} 1 \quad \text{since} \quad 1 \not< 1,$$
$$2 \not{R} 1 \quad \text{since} \quad 2 \not< 1, \quad \text{and}$$
$$2 \not{R} 2 \quad \text{since} \quad 2 \not< 2.$$

Recall that the Cartesian product of A and B, $A \times B$, consists of all ordered pairs whose first element is in A and whose second element is in B:

$$A \times B = \{(x, y) \mid x \in A \text{ and } y \in B\}.$$

In this case,

$$A \times B = \{(0, 1), (0, 2), (0, 3), (1, 1), (1, 2), (1, 3), (2, 1), (2, 2), (2, 3)\}.$$

The elements of some ordered pairs in $A \times B$ are related while the elements of other ordered pairs are not. Consider the set of all ordered pairs in $A \times B$ whose elements are related:

$$\{(0, 1), (0, 2), (0, 3), (1, 2), (1, 3), (2, 3)\}.$$

Observe that knowing which ordered pairs lie in this set is equivalent to knowing which elements are related to which. The relation itself can, therefore, be thought of as the totality of ordered pairs whose elements are related by the given condition. The formal mathematical definition of relation, based on this idea, was introduced by the American mathematician and logician C. S. Peirce in the nineteenth century.

DEFINITION

Let A and B be sets. A **(binary) relation R from A to B** is a subset of $A \times B$. Given an ordered pair (x, y) in $A \times B$, **x is related to y by R**, written $x R y$, if, and only if, (x, y) is in R.

The notation for relations may be written symbolically as follows.

$$x R y \quad \Leftrightarrow \quad (x, y) \in R$$

The notation $x \not{R} y$ means that x is not related to y by R.

$$x \not{R} y \quad \Leftrightarrow \quad (x, y) \notin R$$

The term *binary* is used in the definition above to refer to the fact that the relation is a subset of the Cartesian product of two sets. Because we mostly discuss binary relations in this text, when we use the term *relation* by itself, we will mean binary relation. A more general type of relation, called an *n-ary relation*, is defined later in this section.

EXAMPLE 10.1.1

A Binary Relation as a Subset

Let $A = \{1, 2\}$ and $B = \{1, 2, 3\}$ and define a binary relation R from A to B as follows:

$$\text{given any } (x, y) \in A \times B, \quad (x, y) \in R \iff x - y \text{ is even.}$$

a. State explicitly which ordered pairs are in $A \times B$ and which are in R.
b. Is 1 R 3? Is 2 R 3? Is 2 R 2?

Solution   a. $A \times B = \{(1, 1), (1, 2), (1, 3), (2, 1), (2, 2), (2, 3)\}$. To determine explicitly the composition of R, examine each ordered pair in $A \times B$ to see whether its elements satisfy the defining condition for R.

$(1, 1) \in R$ because $1 - 1 = 0$ and 0 is even   since $0 = 2 \cdot 0$.

$(1, 2) \notin R$ because $1 - 2 = -1$ and -1 is not even   since $-1 \ne 2 \cdot k$, for any integer k.

$(1, 3) \in R$ because $1 - 3 = -2$ and -2 is even.
$(2, 1) \notin R$ because $2 - 1 = 1$ and 1 is not even.
$(2, 2) \in R$ because $2 - 2 = 0$ and 0 is even.
$(2, 3) \notin R$ because $2 - 3 = -1$ and -1 is not even.

Thus

$$R = \{(1, 1), (1, 3), (2, 2)\}.$$

b. Yes, 1 R 3 since $(1, 3) \in R$.
No, 2 $\not R$ 3 since $(2, 3) \notin R$.
Yes, 2 R 2 since $(2, 2) \in R$.   ■

EXAMPLE 10.1.2

The Congruence Modulo 2 Relation

Generalize the relation defined in Example 10.1.1 to the set of all integers $\mathbf{Z}$. That is, define a binary relation E from $\mathbf{Z}$ to $\mathbf{Z}$ as follows:

$$\text{for all } (m, n) \in \mathbf{Z} \times \mathbf{Z}, \quad m \, E \, n \iff m - n \text{ is even.}$$

a. Is 4 E 0? Is 2 E 6? Is 3 E (-3)? Is 5 E 2?
b. List five integers that are related by E to 1.
c. Prove that if n is any odd integer, then n E 1.

Solution   a. Yes, 4 E 0 because $4 - 0 = 4$ and 4 is even.
Yes, 2 E 6 because $2 - 6 = -4$ and -4 is even.
Yes, 3 E (-3) because $3 - (-3) = 6$ and 6 is even.
No, 5 $\not E$ 2 because $5 - 2 = 3$ and 3 is not even.

b. There are many such lists. One is

$\quad\quad\quad 1$ because $1 - 1 = 0$ is even,
$\quad\quad\quad 3$ because $3 - 1 = 2$ is even,
$\quad\quad\quad 5$ because $5 - 1 = 4$ is even,
$\quad\quad -1$ because $-1 - 1 = -2$ is even,
$\quad\quad -3$ because $-3 - 1 = -4$ is even.

c. **Proof:** Suppose n is any odd integer. Then $n = 2k + 1$ for some integer k. Now by definition of E, n E 1 if, and only if, $n - 1$ is even. But by substitution,

$$n - 1 = (2k + 1) - 1 = 2k,$$

and since k is an integer, $2k$ is even. Hence n E 1 *[as was to be shown]*.

It can be shown (see exercise 4 at the end of this section) that integers m and n are related by E if, and only if, $m \bmod 2 = n \bmod 2$ (that is, that both are even or both are odd). When this occurs m and n are said to be **congruent modulo 2**. ∎

EXAMPLE 10.1.3   The Circle Relation

Define a binary relation C from **R** to **R** as follows:

$$\text{for any } (x, y) \in \mathbf{R} \times \mathbf{R}, \quad (x, y) \in C \iff x^2 + y^2 = 1.$$

a. Is $(1, 0) \in C$? Is $(0, 0) \in C$? Is $(-\frac{1}{2}, \frac{\sqrt{3}}{2}) \in C$? Is $-2 \, C \, 0$? Is $0 \, C \, (-1)$? Is $1 \, C \, 1$?
b. Draw a graph for C by plotting the points of C in the Cartesian plane.

Solution   a.   Yes, $(1, 0) \in C$ because $1^2 + 0^2 = 1$.
No, $(0, 0) \notin C$ because $0^2 + 0^2 = 0 \neq 1$.
Yes, $(-\frac{1}{2}, \frac{\sqrt{3}}{2}) \in C$ because $(-\frac{1}{2})^2 + (\frac{\sqrt{3}}{2})^2 = \frac{1}{4} + \frac{3}{4} = 1$.
No, $-2 \, \cancel{C} \, 0$ because $(-2)^2 + 0^2 = 4 \neq 1$.
Yes, $0 \, C \, (-1)$ because $0^2 + (-1)^2 = 1$.
No, $1 \, \cancel{C} \, 1$ because $1^2 + 1^2 = 2 \neq 1$.

b.

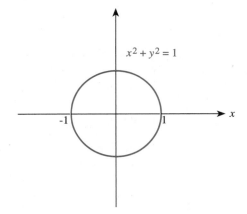

$x^2 + y^2 = 1$

∎

EXAMPLE 10.1.4   A Relation on a Set of Strings

Let A be the set of all strings of length 6 consisting of x's and y's. Then A is denoted Σ^6 where $\Sigma = \{x, y\}$. Define a binary relation R from A to A as follows: For all strings s and t in A,

$$s \, R \, t \iff \text{the first four characters of } s \text{ equal the first four characters of } t.$$

Is $xxyxyx \, R \, xxxyxy$? Is $yxyyyx \, R \, yxyyxy$? Is $xyxxxx \, R \, yxxxxx$?

Solution   No, $xxyxyx \, \cancel{R} \, xxxyxy$ because $xxyx \neq xxxy$.
Yes, $yxyyyx \, R \, yxyyxy$ because $yxyy = yxyy$.
No, $xyxxxx \, \cancel{R} \, yxxxxx$ because $xyxx \neq yxxx$. ∎

ARROW DIAGRAM OF A RELATION

Suppose R is a relation from a set A to a set B. The **arrow diagram for** R is obtained as follows:

1. Represent the elements of A as points in one region and the elements of B as points in another region.
2. For each x in A and y in B, draw an arrow from x to y if, and only if, x is related to y by R. Symbolically:

$$\textbf{Draw an arrow from } x \textbf{ to } y \quad \Leftrightarrow \quad x\, R\, y \quad \Leftrightarrow \quad (x, y) \in R.$$

EXAMPLE 10.1.5   Arrow Diagrams of Relations

Let $A = \{1, 2, 3\}$ and $B = \{1, 3, 5\}$ and define relations S and T from A to B as follows:

for all $(x, y) \in A \times B$, $\quad (x, y) \in S \quad \Leftrightarrow \quad x < y \qquad$ S is a "less than" relation.
$$T = \{(2, 1), (2, 5)\}.$$

Draw arrow diagrams for S and T.

Solution

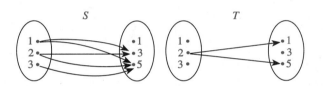

These example relations illustrate that it is quite possible to have an element of A that does not have an arrow coming out of it. Also, it is possible to have several arrows coming out of the same element of A pointing in different directions. ∎

RELATIONS AND FUNCTIONS

With the introduction of Georg Cantor's set theory in the late nineteenth century, it began to seem possible to put mathematics on a firm logical foundation by developing all the different branches of mathematics from logic and set theory alone. In 1914, a crucial breakthrough in using sets to specify mathematical structures was made by Norbert Wiener (1894–1964), a young American who had recently received his Ph.D. from Harvard. What Wiener showed was that an ordered pair can be defined as a certain type of set. Unfortunately, his definition was somewhat awkward. At about the same time, the German mathematician Felix Hausdorff (1868–1942) offered another definition, but it turned out to have a slight flaw. Finally, in 1921 the Polish mathematician Kazimierez Kuratowski (1896–1980) published the version of the definition that has since become standard. It specifies that

$$(a, b) = \{\{a\}, \{a, b\}\}.$$

Note that this definition implies the fundamental property of ordered pairs:

$$(a, b) = (c, d) \quad \Leftrightarrow \quad a = c \text{ and } b = d.$$

The importance of this definition is that it makes it possible to define binary relations using nothing other than set theory because Cartesian products are defined as sets of ordered pairs and binary relations are defined as subsets of Cartesian products. The concept of function is then defined as the following special kind of a binary relation.

> **DEFINITION**
>
> A **function F from a set A to a set B** is a relation from A to B that satisfies the following two properties:
>
> **1.** For every element x in A, there is an element y in B such that $(x, y) \in F$.
> **2.** For all elements x in A and y and z in B,
>
> $$\text{if } (x, y) \in F \text{ and } (x, z) \in F, \text{ then } y = z.$$
>
> If F is a function from A to B, we write
>
> $$y = F(x) \quad \Leftrightarrow \quad (x, y) \in F.$$

Note that $y = F(x)$ if, and only if, y is the second element of an ordered pair in F whose first element is x. Note also that properties (1) and (2) can be stated less formally as follows: A binary relation F from A to B is a function if, and only if:

1. every element of A is the first element of an ordered pair of F
2. no two distinct ordered pairs in F have the same first element.

EXAMPLE 10.1.6   Functions and Relations on Finite Sets

Let $A = \{2, 4, 6\}$ and $B = \{1, 3, 5\}$. Which of the relations R and S defined below are functions from A to B?

a. $R = \{(2, 5), (4, 1), (4, 3), (6, 5)\}$.
b. For all $(x, y) \in A \times B$, $(x, y) \in S \quad \Leftrightarrow \quad y = x + 1$.

Solution   a. R is not a function because it does not satisfy property (2). The ordered pairs $(4, 1)$ and $(4, 3)$ have the same first element but different second elements. You can see this graphically if you draw the arrow diagram for R.

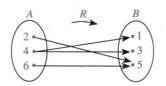

b. S is not a function because it does not satisfy property (1). It is not true that every element of A is the first element of an ordered pair in S. For example, $6 \in A$ but there is no y in B such that $y = 6 + 1 = 7$. You can also see this graphically by drawing the arrow diagram for S.

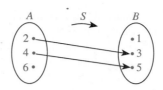

EXAMPLE 10.1.7 **Functions and Relations on Sets of Real Numbers**

a. In Example 10.1.3 the circle relation C was defined as follows:

$$\text{for all } (x, y) \in \mathbf{R} \times \mathbf{R}, \quad (x, y) \in C \;\Leftrightarrow\; x^2 + y^2 = 1.$$

Is C a function?

b. Define a relation from $\mathbf{R}$ to $\mathbf{R}$ as follows:

$$\text{for all } (x, y) \in \mathbf{R} \times \mathbf{R}, \quad (x, y) \in L \;\Leftrightarrow\; y = x - 1.$$

Is L a function?

Solution   a. The graph of C, shown below, indicates that C does not satisfy either function property. To see why C does not satisfy property (1), observe that there are many real numbers x such that $(x, y) \notin C$ for any y.

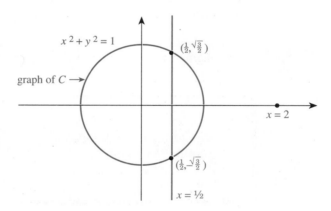

For instance, when $x = 2$, there is no real number y so that

$$x^2 + y^2 = 2^2 + y^2 = 4 + y^2 = 1$$

because if there were, then

$$y^2 = -3,$$

which is not the case for any real number y.

To see why C does not satisfy property (2), note that for some values of x there are two distinct values of y so that $(x, y) \in C$. One way to see this graphically is to observe that there are vertical lines, such as $x = \frac{1}{2}$, that intersect the graph of C at two separate points: $(\frac{1}{2}, \frac{\sqrt{3}}{2})$ and $(\frac{1}{2}, -\frac{\sqrt{3}}{2})$.

b. L is a function. For each real number x, $y = x - 1$ is a real number, and so there is a real number y with $(x, y) \in L$. Also if $(x, y) \in L$ and $(x, z) \in L$, then $y = x - 1$ and $z = x - 1$, and so $y = z$.

You can also check these results by inspecting the graph of L, shown below. Note that for every real number x, the vertical line through $(x, 0)$ passes through the graph of L exactly once. This indicates both that every real number x is the first element of an ordered pair in L and also that no two distinct ordered pairs in L have the same first element.

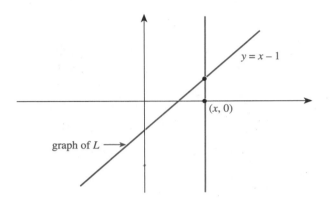

THE INVERSE OF A RELATION

If R is a relation from A to B, then a relation R^{-1} from B to A can be defined by interchanging the elements of all the ordered pairs of R.

DEFINITION

Let R be a relation from A to B. Define the inverse relation R^{-1} from B to A as follows:

$$R^{-1} = \{(y, x) \in B \times A \mid (x, y) \in R\}.$$

This definition can be written operationally as follows:

$$\text{for all } x \in X \text{ and } y \in Y, \quad (y, x) \in R^{-1} \iff (x, y) \in R.$$

EXAMPLE 10.1.8   **The Inverse of a Finite Relation**

Let $A = \{2, 3, 4\}$ and $B = \{2, 6, 8\}$ and let R be the "divides" relation from A to B:

$$\text{for all } (x, y) \in A \times B, \quad x \, R \, y \iff x \mid y \quad \text{\footnotesize x divides y.}$$

a. State explicitly which ordered pairs are in R and R^{-1}, and draw arrow diagrams for R and R^{-1}.

b. Describe R^{-1} in words.

Solution a. $R = \{(2, 2), (2, 6), (2, 8), (3, 6), (4, 8)\}$
$R^{-1} = \{(2, 2), (6, 2), (8, 2), (6, 3), (8, 4)\}$

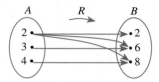

To draw the arrow diagram for R^{-1}, you can copy the arrow diagram for R but reverse the direction of the arrows.

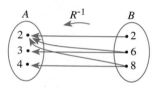

Or you can redraw the diagram so that B is on the left.

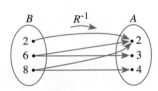

b. R^{-1} is defined in words as follows:

for all $(x, y) \in B \times A$, $\quad y\, R^{-1}\, x \quad \Leftrightarrow \quad y$ is a multiple of x. ■

EXAMPLE 10.1.9 **The Inverse of an Infinite Relation**

Define a relation R from $\mathbf{R}$ to $\mathbf{R}$ as follows:

for all $(x, y) \in \mathbf{R} \times \mathbf{R}$, $\quad x\, R\, y \quad \Leftrightarrow \quad y = 2 \cdot |x|$.

Draw the graphs of R and R^{-1} in the Cartesian plane. Is R^{-1} a function?

Solution A point (v, u) is on the graph of R^{-1} if, and only if, (u, v) is on the graph of R. Note that if $x \geq 0$, then the graph of $y = 2 \cdot |x| = 2x$ is a straight line with slope 2. And if $x < 0$, then the graph of $y = 2 \cdot |x| = 2 \cdot (-x) = -2x$ is a straight line with slope -2. Some sample values are tabulated and the graphs are shown on page 542.

$$R = \{(x, y) \mid y = 2 \cdot |x|\} \qquad\qquad R^{-1} = \{(y, x) \mid y = 2 \cdot |x|\}$$

x	y
0	0
1	2
-1	2
2	4
-2	4

1st coordinate   2nd coordinate

y	x
0	0
2	1
2	-1
4	2
4	-2

1st coordinate   2nd coordinate

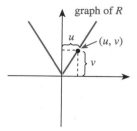

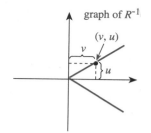

Note that R^{-1} is not a function because, for instance, both $(2, 1)$ and $(2, -1)$ are in R^{-1}. ∎

DIRECTED GRAPH OF A RELATION

In the remaining sections of this chapter, we discuss important properties of relations that are defined from a set to itself.

DEFINITION

A **binary relation on a set A** is a binary relation from A to A.

When a binary relation R is defined *on* a set A, the arrow diagram of the relation can be modified so that it becomes a **directed graph**. Instead of representing A as two separate sets of points, represent A only once, and draw an arrow from each point of A to each related point. As with an ordinary arrow diagram,

for all points x and y in A,

there is an arrow from x to y ⇔ $x\,R\,y$ ⇔ $(x, y) \in R$.

If a point is related to itself, a loop is drawn that extends out from the point and goes back to it.

EXAMPLE 10.1.10   Directed Graph of a Relation

Let $A = \{3, 4, 5, 6, 7, 8\}$ and define a binary relation R on A as follows:

for all $x, y \in A$, $\quad x\,R\,y$ ⇔ $2 \mid (x - y)$.

Draw the directed graph of R.

Solution   Note that $3\,R\,3$ because $3 - 3 = 0$ and $2\,|\,0$ since $0 = 2 \cdot 0$. Thus there is a loop from 3 to itself. Similarly, there is a loop from 4 to itself, from 5 to itself, and so forth, since the difference of each integer with itself is 0 and $2\,|\,0$.

Note also that $3\,R\,5$ because $3 - 5 = -2 = 2 \cdot (-1)$. And $5\,R\,3$ because $5 - 3 = 2 = 2 \cdot 1$. Hence there is an arrow from 3 to 5 and also an arrow from 5 to 3. The other arrows in the directed graph, as shown below, are obtained by similar reasoning.

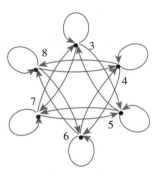

N-ARY RELATIONS AND RELATIONAL DATABASES

N-ary relations form the mathematical foundation for relational database theory. A binary relation is a subset of the Cartesian product of two sets; similarly, an *n-ary* relation is a subset of the Cartesian product of n sets.

> ### DEFINITION
>
> Given sets $A_1, A_2, \ldots, A_n$, an n-ary relation R on $A_1 \times A_2 \times \cdots \times A_n$ is a subset of $A_1 \times A_2 \times \cdots \times A_n$. The special cases of 2-ary, 3-ary, and 4-ary relations are called **binary**, **ternary**, and **quaternary relations**, respectively.

EXAMPLE 10.1.11   A Simple Database

The following is a radically simplified version of a database that might be used in a hospital. Let A_1 be a set of positive integers, A_2 a set of alphabetic character strings, A_3 a set of numeric character strings, and A_4 a set of alphabetic character strings. Define a quaternary relation R on $A_1 \times A_2 \times A_3 \times A_4$ as follows:

$(a_1, a_2, a_3, a_4) \in R \iff$ a patient with patient ID number a_1, named a_2, was admitted on date a_3, with primary diagnosis a_4.

At a particular hospital this relation might contain the following 4-tuples:

(011985, John Schmidt, 020795, asthma)

(574329, Tak Kurosawa, 011495, pneumonia)

(466581, Mary Lazars, 010395, appendicitis)

(008352, Joan Kapłan, 112494, gastritis)

(011985, John Schmidt, 021795, pneumonia)

(244388, Sarah Wu, 010395, broken leg)

(778400, Jamal Baskers, 122794, appendicitis)

In discussions of relational databases the tuples are normally thought of as being written in tables. Each row of the table corresponds to one tuple, and the header for each column gives the descriptive attribute for the elements in the column.

Operations within a database allow the data to be manipulated in many different ways. For example, in the database language SQL, if the above database is denoted S, the result of the query

SELECT Patient_ID# , Name FROM S WHERE

Admission_Date = 010395

would be a list of the ID numbers and names of all patients admitted on 01-03-95:

466581   Mary Lazars,

244388   Sarah Wu.

This is obtained by taking the intersection of the set $A_1 \times A_2 \times \{010395\} \times A_4$ with the database and then projecting onto the first two coordinates. (See exercise 20 of Section 7.1.) Similarly, SELECT can be used to obtain a list of all admission dates of a given patient. For John Schmidt this list is

02-07-95   and

02-17-95

Individual entries in a database can be added, deleted, or updated, and most databases can sort data entries in various ways. In addition, entire databases can be merged and the entries common to two databases can be moved to a new database.   ∎

EXERCISE SET 10.1*

1. Let $A = \{2, 3, 4\}$ and $B = \{6, 8, 10\}$ and define a binary relation R from A to B as follows:

   for all $(x, y) \in A \times B$, $(x, y) \in R \iff x \mid y$.

   a. Is 4 R 6? Is 4 R 8? Is $(3, 8) \in R$? Is $(2, 10) \in R$?
   b. Write R as a set of ordered pairs.

2. Let $C = \{2, 3, 4, 5\}$ and $D = \{3, 4\}$ and define a binary relation S from C to D as follows:

   for all $(x, y) \in C \times D$, $(x, y) \in S \iff x \geq y$.

   a. Is 2 S 4? Is 4 S 3? Is $(4, 4) \in S$? Is $(3, 2) \in S$?
   b. Write S as a set of ordered pairs.

3. As in Example 10.1.2, the **congruence modulo 2** relation E is defined from $\mathbf{Z}$ to $\mathbf{Z}$ as follows: for all integers m and n, $m \, E \, n \iff m - n$ is even.

   a. Is 0 E 0? Is 5 E 2? Is $(6, 6) \in E$? Is $(-1, 7) \in E$?
   b. Prove that for any even integer n, $n \, E \, 0$.

H 4. Prove that for all integers m and n, $m - n$ is even if,

and only if, both m and n are even or both m and n are odd.

5. The **congruence modulo 3** relation, T, is defined from $\mathbf{Z}$ to $\mathbf{Z}$ as follows: for all integers m and n, $m \, T \, n \iff 3 \mid (m - n)$.

   a. Is 10 T 1? Is 1 T 10? Is $(2, 2) \in T$? Is $(8, 1) \in T$?
   b. List five integers n such that $n \, T \, 0$.
   c. List five integers n such that $n \, T \, 1$.
   d. List five integers n such that $n \, T \, 2$.

H e. Make and prove a conjecture about which integers are related by T to 0, which integers are related by T to 1, and which integers are related by T to 2.

6. Define a binary relation S from $\mathbf{R}$ to $\mathbf{R}$ as follows:

   for all $(x, y) \in \mathbf{R} \times \mathbf{R}$, $x \, S \, y \iff x \geq y$.

   a. Is $(2, 1) \in S$? Is $(2, 2) \in S$? Is 2 S 3? Is $(-1) \, S \, (-2)$?
   b. Draw the graph of S in the Cartesian plane.

7. Define a binary relation R from $\mathbf{R}$ to $\mathbf{R}$ as follows:

for all $(x, y) \in \mathbf{R} \times \mathbf{R}$, $x\,R\,y \iff y = x^2$.

a. Is $(2, 4) \in R$? Is $(4, 2) \in R$? Is $(-3)\,R\,9$? Is $9\,R\,(-3)$?

b. Draw the graph of R in the Cartesian plane.

8. Define a binary relation P on $\mathbf{Z}$ as follows:

for all $m, n \in \mathbf{Z}$,

$m\,P\,n \iff m$ and n have a common prime factor.

a. Is $15\,P\,25$?   **b.** Is $22\,P\,27$?   **c.** Is $0\,P\,5$?
d. Is $8\,P\,8$?

9. Let $X = \{a, b, c\}$. Recall that $\mathcal{P}(X)$ is the power set of X. Define a binary relation $\mathcal{R}$ on $\mathcal{P}(X)$ as follows:

for all $A, B \in \mathcal{P}(X)$,

$A\,\mathcal{R}\,B \iff A$ has the same number of elements as B.

a. Is $\{a, b\}\,\mathcal{R}\,\{b, c\}$?   **b.** Is $\{a\}\,\mathcal{R}\,\{a, b\}$?
c. Is $\{c\}\,\mathcal{R}\,\{b\}$?

10. Let $X = \{a, b, c\}$. Define a binary relation $\mathcal{J}$ on $\mathcal{P}(X)$ as follows:

for all $A, B \in \mathcal{P}(X)$. $A\,\mathcal{J}\,B \iff A \cap B \neq \varnothing$.

a. Is $\{a\}\,\mathcal{J}\,\{c\}$?   **b.** Is $\{a, b\}\,\mathcal{J}\,\{b, c\}$?
c. Is $\{a, b\}\,\mathcal{J}\,\{a, b, c\}$?

11. Let $\Sigma = \{a, b\}$. Then Σ^4 is the set of all strings over Σ of length 4. Define a relation R on Σ^4 as follows:

for all $s, t \in \Sigma^4$, $s\,R\,t \iff s$ has the same first two characters as t.

a. Is *abaa R abba*?   **b.** Is *aabb R bbaa*?
c. Is *aaaa R aaab*?

H 12. Let $A = \{4, 5, 6\}$ and $B = \{5, 6, 7\}$ and define binary relations R, S, and T from A to B as follows:

for all $(x, y) \in A \times B$, $(x, y) \in R \iff x \geq y$.
for all $(x, y) \in A \times B$, $x\,S\,y \iff 2\,|\,(x - y)$.
$T = \{(4, 7), (6, 5), (6, 7)\}$.

a. Draw arrow diagrams for R, S, and T.
b. Indicate whether any of the relations R, S, or T are functions.

13. a. Find all binary relations from $\{0, 1\}$ to $\{1\}$.
b. Find all functions from $\{0, 1\}$ to $\{1\}$.
c. What fraction of the binary relations from $\{0, 1\}$ to $\{1\}$ are functions?

14. Find four binary relations from $\{a, b\}$ to $\{x, y\}$ that are not functions.

H 15. Suppose A is a set with m elements and B is a set with n elements.
a. How many binary relations are there from A to B? Explain.

b. How many functions are there from A to B? Explain.
c. What fraction of the binary relations from A to B are functions?

16. Define a binary relation P from $\mathbf{R}$ to $\mathbf{R}$ as follows:

for all real numbers x and y,
$$(x, y) \in P \iff x = y^2.$$

Is P a function? Explain.

17. Let $A = \{3, 4, 5\}$ and $B = \{4, 5, 6\}$ and let R be the "less than" relation. That is,

for all $(x, y) \in A \times B$, $x\,R\,y \iff x < y$.

State explicitly which ordered pairs are in R and R^{-1}.

18. Let $A = \{3, 4, 5\}$ and $B = \{4, 5, 6\}$ and let S be the "divides" relation. That is,

for all $(x, y) \in A \times B$, $x\,S\,y \iff x\,|\,y$.

State explicitly which ordered pairs are in S and S^{-1}.

19. Let $\Sigma = \{a, b\}$. Then Σ^* is the set of all strings over Σ. Define a relation T on Σ^* as follows:

for all $s, t \in \Sigma^*$, $s\,T\,t \iff t = as$

(that is, t is the concatenation of a with s)

a. Is *ab T aab*?   **b.** Is *aab T ab*?
c. Is *ba T aba*?   **d.** Is *aba T^{-1} ba*?
e. Is *abb T^{-1} bba*?   **f.** Is *abba T^{-1} bba*?

20. Define a relation R from $\mathbf{R}$ to $\mathbf{R}$ as follows:

for all $(x, y) \in \mathbf{R} \times \mathbf{R}$, $x\,R\,y \iff y = \lfloor x \rfloor$.

Draw the graphs of R and R^{-1} in the Cartesian plane.

21. a. Rewrite the definition of one-to-one function using the notation of the definition of a function as a relation.
b. Rewrite the definition of onto function using the notation of the definition of function as a relation.

22. a. Suppose a function $F : X \to Y$ is one-to-one but not onto. Is F^{-1} (the inverse relation for F) a function? Explain your answer.
b. Suppose a function $F : X \to Y$ is onto but not one-to-one. Is F^{-1} (the inverse relation for F) a function? Explain your answer.

Draw the directed graphs of the binary relations defined in 23–27 below.

23. Define a binary relation R on $A = \{0, 1, 2, 3\}$ by $R = \{(0, 0), (1, 2), (2, 2)\}$.

24. Define a binary relation S on $B = \{a, b, c, d\}$ by $S = \{(a, b), (a, c), (b, c), (d, d)\}$.

25. Let $A = \{2, 3, 4, 5, 6, 7, 8\}$ and define a binary relation R on A as follows:

for all $x, y \in A$, $x\,R\,y \iff x\,|\,y$.

H26. Let $A = \{5, 6, 7, 8, 9, 10\}$ and define a binary relation S on A as follows:

$$\text{for all } x, y \in A, \quad x\,S\,y \iff 2 \mid (x - y).$$

27. Let $A = \{2, 3, 4, 5, 6, 7, 8\}$ and define a binary relation T on A as follows:

$$\text{for all } x, y \in A, \quad x\,T\,y \iff 3 \mid (x - y).$$

28. In Example 10.1.11 the result of the query SELECT Patient_ID#, Name FROM S WHERE Primary_Diagnosis = X is the projection onto the first two coordinates of the intersection of the set $A_1 \times A_2 \times A_3 \times \{X\}$ with the database.
   a. Find the result of the query SELECT Patient_ID#, Name FROM S WHERE Primary_Diagnosis = pneumonia.
   b. Find the result of the query SELECT Patient_ID#, Name FROM S WHERE Primary_Diagnosis = appendicitis.

Exercises 29–33 refer to unions and intersections of relations. Since binary relations are subsets of Cartesian products, their unions and intersections can be calculated as for any subsets. Given two relations R and S from A to B,

$$R \cup S = \{(x, y) \in A \times B \mid (x, y) \in R \text{ or } (x, y) \in S\}$$
$$R \cap S = \{(x, y) \in A \times B \mid (x, y) \in R \text{ and } (x, y) \in S\}$$

29. Let $A = \{2, 4\}$ and $B = \{6, 8, 10\}$ and define binary relations R and S from A to B as follows:

$$\text{for all } (x, y) \in A \times B, \quad x\,R\,y \iff x \mid y.$$
$$\text{for all } (x, y) \in A \times B, \quad x\,S\,y \iff y - 4 = x.$$

State explicitly which ordered pairs are in $A \times B$, R, S, $R \cup S$, and $R \cap S$.

30. Let $A = \{-1, 1, 2, 4\}$ and $B = \{1, 2\}$ and define binary relations R and S from A to B as follows:

$$\text{for all } (x, y) \in A \times B, \quad x\,R\,y \iff |x| = |y|.$$
$$\text{for all } (x, y) \in A \times B, \quad x\,S\,y \iff x - y \text{ is even.}$$

State explicitly which ordered pairs are in $A \times B$, R, S, $R \cup S$, and $R \cap S$.

31. Define R and S from $\mathbf{R}$ to $\mathbf{R}$ as follows:

$$R = \{(x, y) \in \mathbf{R} \times \mathbf{R} \mid x < y\} \quad \text{and}$$
$$S = \{(x, y) \in \mathbf{R} \times \mathbf{R} \mid x = y\}.$$

That is, R is the "less than" relation and S is the "equals" relation from $\mathbf{R}$ to $\mathbf{R}$. Graph R, S, $R \cup S$, and $R \cap S$ in the Cartesian plane.

32. Define binary relations R and S from $\mathbf{R}$ to $\mathbf{R}$ as follows:

$$R = \{(x, y) \in \mathbf{R} \times \mathbf{R} \mid x^2 + y^2 = 4\} \quad \text{and}$$
$$S = \{(x, y) \in \mathbf{R} \times \mathbf{R} \mid x = y\}.$$

Graph R, S, $R \cup S$, and $R \cap S$ in the Cartesian plane.

33. Define binary relations R and S from $\mathbf{R}$ to $\mathbf{R}$ as follows:

$$R = \{(x, y) \in \mathbf{R} \times \mathbf{R} \mid y = |x|\} \quad \text{and}$$
$$S = \{(x, y) \in \mathbf{R} \times \mathbf{R} \mid y = 1\}.$$

Graph R, S, $R \cup S$, and $R \cap S$ in the Cartesian plane.

10.2   REFLEXIVITY, SYMMETRY, AND TRANSITIVITY

Mathematics is the tool specially suited for dealing with abstract concepts of any kind and there is no limit to its power in this field.
(P. A. M. Dirac, 1902–1984)

Let $A = \{2, 3, 4, 6, 7, 9\}$ and define a relation R on A as follows:

$$\text{for all } x, y \in A, \quad x\,R\,y \iff 3 \mid (x - y).$$

Then $2\,R\,2$ because $2 - 2 = 0$ and $3 \mid 0$. Similarly, $3\,R\,3$, $4\,R\,4$, $6\,R\,6$, $7\,R\,7$, and $9\,R\,9$. Also $6\,R\,3$ because $6 - 3 = 3$ and $3 \mid 3$. And $3\,R\,6$ because $3 - 6 = -(6 - 3) = -3$ and $3 \mid (-3)$. Similarly, $3\,R\,9$, $9\,R\,3$, $6\,R\,9$, $9\,R\,6$, $4\,R\,7$, and $7\,R\,4$. Thus the directed graph for R has the appearance shown in Figure 10.2.1.

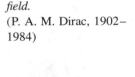

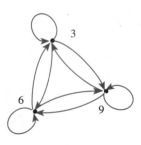

FIGURE 10.2.1

This graph has three important properties:

1. Each point of the graph has an arrow looping around from it back to itself.
2. In each case where there is an arrow going from one point to a second, there is an arrow going from the second point back to the first.
3. In each case where there is an arrow going from one point to a second and from a second point to a third, there is an arrow going from the first point to the third. That is, there are no "incomplete directed triangles" in the graph.

Properties (1), (2), and (3) correspond to properties of general binary relations called *reflexivity, symmetry,* and *transitivity*.

DEFINITION

Let R be a binary relation on a set A.

1. R is **reflexive** if, and only if, for all $x \in A$, $x \, R \, x$.
2. R is **symmetric** if, and only if, for all $x, y \in A$, **if** $x \, R \, y$ then $y \, R \, x$.
3. R is **transitive** if, and only if, for all $x, y, z \in A$, **if** $x \, R \, y$ and $y \, R \, z$ then $x \, R \, z$.

Because of the equivalence of the expressions $x \, R \, y$ and $(x, y) \in R$ for all x and y in A, the reflexive, symmetric, and transitive properties can also be written as follows:

---

1. R is reflexive $\Leftrightarrow$ for all x in A, $(x, x) \in R$.
2. R is symmetric $\Leftrightarrow$ for all x and y in A, **if** $(x, y) \in R$ then $(y, x) \in R$.
3. R is transitive $\Leftrightarrow$ for all $x, y,$ and z in A, **if** $(x, y) \in R$ and $(y, z) \in R$ then $(x, z) \in R$.

In informal terms, properties (1)–(3) say the following:

1. **Reflexive:** Each element is related to itself.
2. **Symmetric:** If any one element is related to any other element, then the second element is related to the first.
3. **Transitive:** If any one element is related to a second and that second element is related to a third, then the first element is related to the third.

▲ *CAUTION!*   *One caution about the informal phrasing: The first, second, and third elements referred to need not all be distinct. This is the disadvantage of informal phrasing; it sometimes masks some nuances of the full formal definition.*

Note that the definitions of reflexivity, symmetry, and transitivity are universal statements. This means that to prove a relation has one of the properties, you use either the method of exhaustion or the method of generalizing from the generic particular.

Now consider what it means for a relation *not* to have one of the properties defined above. Recall that the negation of a universal statement is existential. Hence if R is a binary relation on a set A, then

1. R is **not reflexive** $\Leftrightarrow$ there is an element x in A such that $x \not{R} x$ [*that is, such that* $(x, x) \notin R$];

2. R is **not symmetric** $\Leftrightarrow$ there are elements x and y in A such that $x R y$ but $y \not{R} x$ [*that is, such that* $(x, y) \in R$ *but* $(y, x) \notin R$];

3. R is **not transitive** $\Leftrightarrow$ there are elements x, y, and z in A such that $x R y$ and $y R z$ but $x \not{R} z$ [*that is, such that* $(x, y) \in R$ *and* $(y, z) \in R$ *but* $(x, z) \notin R$].

It follows that you can show a binary relation does *not* have one of the properties by finding a counterexample.

EXAMPLE 10.2.1   **Properties of Binary Relations on Finite Sets**

Let $A = \{0, 1, 2, 3\}$ and define relations R, S, and T on A as follows:

$$R = \{(0, 0), (0, 1), (0, 3), (1, 0), (1, 1), (2, 2), (3, 0), (3, 3)\},$$
$$S = \{(0, 0), (0, 2), (0, 3), (2, 3)\},$$
$$T = \{(0, 1), (2, 3)\}.$$

a. Is R reflexive? symmetric? transitive?
b. Is S reflexive? symmetric? transitive?
c. Is T reflexive? symmetric? transitive?

Solution   a. The directed graph of R has the appearance shown in Figure 10.2.2.

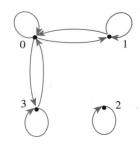

FIGURE 10.2.2

R is reflexive: There is a loop at each point of the directed graph. This means that each element of A is related to itself, so R is reflexive.

R is symmetric: In each case where there is an arrow going from one point of the graph to a second, there is an arrow going from the second point back to the first. This means that whenever one element of A is related by R to a second, then the second is related to the first. Hence R is symmetric.

R is not transitive: There is an arrow going from 1 to 0 and an arrow going from 0 to 3, but there is no arrow going from 1 to 3. This means that there are elements of A—0, 1, and 3—such that $1 R 0$ and $0 R 3$ but $1 \not{R} 3$. Hence R is not transitive.

b. The directed graph of S has the appearance shown in Figure 10.2.3.

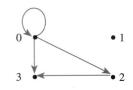

FIGURE 10.2.3

S is not reflexive: There is no loop at 1 for example. Thus $(1, 1) \notin S$, and so S is not reflexive.

S is not symmetric: There is an arrow from 0 to 2 but not from 2 to 0. Hence $(0, 2) \in S$ but $(2, 0) \notin S$, and so S is not symmetric.

S is transitive: There are three cases for which there is an arrow going from one point of the graph to a second and from the second point to a third: Namely, there are arrows going from 0 to 2 and from 2 to 3; there are arrows going from 0 to 0 and from 0 to 2; and there are arrows going from 0 to 0 and from 0 to 3. In each case there is an arrow going from the first point to the third. (Note again that the "first," "second," and "third" points need not be distinct.) This means that whenever $(x, y) \in S$ and $(y, z) \in S$ then $(x, z) \in S$, for all $x, y, z \in \{0, 1, 2, 3\}$, and so S is transitive.

FIGURE 10.2.4

c. The directed graph of T has the appearance shown in Figure 10.2.4.

T is not reflexive: There is no loop at 0 for example. Thus $(0, 0) \notin T$, so T is not reflexive.

T is not symmetric: There is an arrow from 0 to 1 but not from 1 to 0. Thus $(0, 1) \in T$ but $(1, 0) \notin T$, and so T is not symmetric.

T is transitive: The transitivity condition is vacuously true for T. That is, T is transitive by default because it is *not* not transitive! To see this, observe that the transitivity condition says that

for all $x, y, z \in A$,   if $(x, y) \in T$ and $(y, z) \in T$ then $(x, z) \in T$.

The only way for this to be false would be for there to exist elements of A that make the hypothesis true and the conclusion false. That is, there would have to be elements x, y, and z in A such that

$$(x, y) \in T \quad \text{and} \quad (y, z) \in T \quad \text{and} \quad (x, z) \notin T.$$

In other words, there would have to be two ordered pairs in T that have the potential to "link up" by having the *second* element of one pair be the *first* element of the other pair. But the only elements in T are $(0, 1)$ and $(2, 3)$ and these do not have the potential to link up. Hence the hypothesis is never true. It follows that it is impossible for T *not* to be transitive, and thus T is transitive.   ■

When a binary relation R is defined on a finite set A, it is possible to write computer algorithms to check whether R is reflexive, symmetric, and transitive. One way to do this is to represent A as a one-dimensional array, $(a[1], a[2], \ldots, a[n])$, and use the algorithm of exercise 23 in Section 5.1 to check whether an ordered pair in $A \times A$ is in R. Checking whether R is reflexive can be done with a loop that examines each element $a[i]$ of A in turn. If, for some i, $(a[i], a[i]) \notin R$, then R is not reflexive. Otherwise, R is reflexive. Checking for symmetry can be done with a nested loop that examines each pair $(a[i], a[j])$ of $A \times A$ in turn. If, for some i and j, $(a[i], a[j]) \in R$ and $(a[j], a[i]) \notin R$, then R is not symmetric. Otherwise, R is symmetric. Checking whether R is transitive can be done with a triply nested loop that examines each triple $(a[i], a[j], a[k])$ of $A \times A \times A$ in turn. If for some triple, $(a[i], a[j]) \in R$, $(a[j], a[k]) \in R$, and $(a[i], a[k]) \notin R$, then R is not transitive. Otherwise, R is transitive. In the exercises for this section, you are asked to formalize these algorithms.

THE TRANSITIVE CLOSURE OF A RELATION

Generally speaking, a relation fails to be transitive because it fails to contain certain ordered pairs. For example, if $(1, 3)$ and $(3, 4)$ are in a relation R, then the pair $(1, 4)$ *must* be in R if R is to be transitive. To obtain a transitive relation from one that is not transitive, it is necessary to add ordered pairs. Roughly speaking, the relation obtained by adding the least number of ordered pairs to ensure transitivity is called the *transitive closure* of the relation. In a sense made precise by the formal definition, the transitive closure of a relation is the smallest transitive relation that contains the relation.

DEFINITION

Let A be a set and R a binary relation on A. The **transitive closure** of R is the binary relation R^t on A that satisfies the following three properties:

1. R^t is transitive;
2. $R \subseteq R^t$;
3. If S is any other transitive relation that contains R, then $R^t \subseteq S$.

EXAMPLE 10.2.2   Transitive Closure of a Relation

Let $A = \{0, 1, 2, 3\}$ and consider the relation R defined on A as follows:

$$R = \{(0, 1), (1, 2), (2, 3)\}.$$

Find the transitive closure of R.

Solution   Every ordered pair in R is in R^t, so

$$\{(0, 1), (1, 2), (2, 3)\} \subseteq R^t.$$

Thus the directed graph of R contains the arrows shown in Figure 10.2.5.

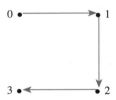

FIGURE 10.2.5

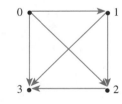

FIGURE 10.2.6

Since there are arrows going from 0 to 1 and from 1 to 2, R^t must have an arrow going from 0 to 2. Hence $(0, 2) \in R^t$. Then $(0, 2) \in R^t$ and $(2, 3) \in R^t$, so since R^t is transitive, $(0, 3) \in R^t$. Also since $(1, 2) \in R^t$ and $(2, 3) \in R^t$, then $(1, 3) \in R^t$. Thus R^t contains at least the following ordered pairs:

$$\{(0, 1), (0, 2), (0, 3), (1, 2), (1, 3), (2, 3)\}.$$

But this relation *is* transitive; hence it equals R^t. Note that the directed graph of R^t is as shown in Figure 10.2.6. ∎

PROPERTIES OF RELATIONS ON INFINITE SETS

Suppose a binary relation R is defined on an infinite set A. To prove the relation is reflexive, symmetric, or transitive, first write down what is to be proved. For instance, for symmetry you need to prove that

$$\forall x, y \in A, \text{ if } x\,R\,y \text{ then } y\,R\,x.$$

Then use the definitions of A and R to rewrite the statement for the particular case in question. For instance, for the "equality" relation on the set of real numbers, the rewritten statement is

$$\forall x, y \in \mathbf{R}, \text{ if } x = y \text{ then } y = x.$$

Sometimes the truth of the rewritten statement will be immediately obvious (as it is here). At other times you will need to prove it using the method of generalizing from the generic particular. We give examples of both cases in this section. We begin with the relation of equality, one of the simplest and yet most important binary relations.

EXAMPLE 10.2.3   **Properties of Equality**

Define a binary relation R on **R** (the set of all real numbers) as follows: for all real numbers x and y.

$$x \, R \, y \quad \Leftrightarrow \quad x = y.$$

a. Is R reflexive?   b. Is R symmetric?   c. Is R transitive?

Solution   a. *R is reflexive:* R is reflexive if, and only if, the following statement is true:

$$\text{for all } x \in \mathbf{R}, \quad x \, R \, x.$$

Since $x \, R \, x$ just means that $x = x$, this is the same as saying

$$\text{for all } x \in \mathbf{R}, \quad x = x.$$

But this statement is certainly true; every real number is equal to itself.

b. *R is symmetric:* R is symmetric if, and only if, the following statement is true:

$$\text{for all } x, y \in \mathbf{R}, \quad \text{if } x \, R \, y \text{ then } y \, R \, x.$$

By definition of R, $x \, R \, y$ means that $x = y$ and $y \, R \, x$ means that $y = x$. Hence R is symmetric if, and only if,

$$\text{for all } x, y \in \mathbf{R}, \quad \text{if } x = y \text{ then } y = x.$$

But this statement is certainly true; if one number is equal to a second, then the second is equal to the first.

c. *R is transitive:* R is transitive if, and only if, the following statement is true:

$$\text{for all } x, y, z \in \mathbf{R}, \quad \text{if } x \, R \, y \text{ and } y \, R \, z \text{ then } x \, R \, z.$$

By definition of R, $x \, R \, y$ means that $x = y$, $y \, R \, z$ means that $y = z$, and $x \, R \, z$ means that $x = z$. Hence R is transitive if, and only if, the following statement is true:

$$\text{for all } x, y, z \in \mathbf{R}, \quad \text{if } x = y \text{ and } y = z \text{ then } x = z.$$

But this statement is certainly true; if one real number equals a second and the second equals a third, then the first equals the third.   ■

EXAMPLE 10.2.4   **Properties of "Less Than"**

Define a relation R on **R** (the set of all real numbers) as follows: for all $x, y \in$ R,

$$x \, R \, y \quad \Leftrightarrow \quad x < y.$$

a. Is R reflexive?   b. Is R symmetric?   c. Is R transitive?

Solution   a. *R is not reflexive:* R is reflexive if, and only if, $\forall x \in \mathbf{R}, x \, R \, x$. By definition of R, this means that $\forall x \in \mathbf{R}, x < x$. But this is false: $\exists x \in \mathbf{R}$ such that $x \not< x$. As a counterexample, let $x = 0$ and note that $0 \not< 0$. Hence R is not reflexive.

b. *R is not symmetric:* R is symmetric if, and only if, $\forall x, y \in \mathbf{R}$, if $x \, R \, y$ then $y \, R \, x$. By definition of R, this means that $\forall x, y \in \mathbf{R}$, if $x < y$ then $y < x$. But this is false: $\exists x, y \in \mathbf{R}$ such that $x < y$ and $y \not< x$. As a counterexample, let $x = 0$ and $y = 1$ and note that $0 < 1$ but $1 \not< 0$. Hence R is not symmetric.

c. *R is transitive:* R is transtive if, and only if, for all $x, y, z \in \mathbf{R}$, if $x \, R \, y$ and $y \, R \, z$ then $x \, R \, z$. By definition of R this means that for all $x, y, z \in \mathbf{R}$, if $x < y$ and $y < z$ then $x < z$. But this statement is true by the transitive law of order for real numbers (Appendix A, T17). Hence R is transitive. ∎

Sometimes a property is "universally false" in the sense that it is false for *every* element of its domain. It follows immediately, of course, that the property is false for each particular element of the domain and hence counterexamples abound. In such a case, it may seem more natural to prove the universal falseness of the property rather than give a single counterexample. In the example above, for instance, you might find it natural to answer (a) and (b) as follows:

Alternative Answer to (a): R is not reflexive because $x \not< x$ for any real number x (by the trichotomy law—Appendix A, T16).

Alternative Answer to (b): R is not symmetric because for all x and y in A, if $x < y$, then $y \not< x$ (by the trichotomy law).

EXAMPLE 10.2.5   **Properties of Congruence Modulo 3**

Define a relation R on $\mathbf{Z}$ (the set of all integers) as follows: for all integers m and n,

$$m \, R \, n \quad \Leftrightarrow \quad 3 \mid (m - n).$$

This relation is called **congruence modulo 3**.

a. Is R reflexive?   b. Is R symmetric?   c. Is R transitive?

Solution   a. *R is reflexive:*   To show that R is reflexive, it is necessary to show that

$$\text{for all } m \in \mathbf{Z}, \quad m \, R \, m.$$

By definition of R, this means that

$$\text{for all } m \in \mathbf{Z}, \quad 3 \mid (m - m).$$

Or, since $m - m = 0$,

$$\text{for all } m \in \mathbf{Z}, \quad 3 \mid 0.$$

But this is true: $3 \mid 0$ since $0 = 3 \cdot 0$. Hence R is reflexive. This reasoning is formalized in the following proof.

Proof of Reflexivity: Suppose m is a particular but arbitrarily chosen integer. *[We must show that $m \, R \, m$.]* Now $m - m = 0$. But $3 \mid 0$ since $0 = 3 \cdot 0$. Hence $3 \mid (m - m)$. So by definition of R, $m \, R \, m$ *[as was to be shown]*.

b. *R is symmetric:*   To show that R is symmetric, it is necessary to show that

$$\text{for all } m, n \in \mathbf{Z}, \quad \text{if } m \, R \, n \text{ then } n \, R \, m.$$

By definition of R this means that

$$\text{for all } m, n \in \mathbf{Z}, \quad \text{if } 3 \mid (m - n) \text{ then } 3 \mid (n - m).$$

Is this true? Suppose m and n are particular but arbitrarily chosen intergers such that $3 \mid (m - n)$. Must it follow that $3 \mid (n - m)$? By definition of "divides," since

$$3 \mid (m - n),$$

then

$$m - n = 3k \quad \text{for some integer } k.$$

The crucial observation is that $n - m = -(m - n)$. Hence, you can multiply both sides of this equation by -1 to obtain

$$-(m - n) = -3k,$$

which is equivalent to

$$n - m = 3 \cdot (-k).$$

Since $-k$ is an integer, this equation shows that

$$3 \mid (n - m).$$

It follows that R is symmetric.

The reasoning above is formalized in the following proof.

Proof of Symmetry: Suppose m and n are particular but arbitrarily chosen integers that satisfy the condition $m \, R \, n$. *[We must show that $n \, R \, m$.]* By definition of R, since $m \, R \, n$ then $3 \mid (m - n)$. By definition of "divides," this means that $m - n = 3k$, for some integer k. Multiplying both sides by -1 gives $n - m = 3 \cdot (-k)$. Since $-k$ is an integer, this equation shows that $3 \mid (n - m)$. Hence, by definition of R, $n \, R \, m$ *[as was to be shown].*

c. *R is transitive:*   To show that R is transitive, it is necessary to show that

for all $m, n \in \mathbf{Z}$,   if $m \, R \, n$ and $n \, R \, p$ then $m \, R \, p$.

By definition of R this means that

for all $m, n \in \mathbf{Z}$,   if $3 \mid (m - n)$ and $3 \mid (n - p)$ then $3 \mid (m - p)$.

Is this true? Suppose m, n, and p are particular but arbitrarily chosen integers such that $3 \mid (m - n)$ and $3 \mid (n - p)$. Must it follow that $3 \mid (m - p)$? By definition of "divides," since

$$3 \mid (m - n) \quad \text{and} \quad 3 \mid (n - p),$$

then

$$m - n = 3r \quad \text{for some integer } r,$$

and

$$n - p = 3s \quad \text{for some integer } s.$$

The crucial observation is that $(m - n) + (n - p) = m - p$. Hence add these two equations together to obtain

$$(m - n) + (n - p) = 3r + 3s,$$

which is equivalent to

$$m - p = 3 \cdot (r + s).$$

Since r and s are integers, $r + s$ is an integer, and so this equation shows that

$$3 \mid (m - p).$$

It follows that R is transitive.

The reasoning above is formalized in the following proof.

Proof of Transitivity: Suppose m, n, and p are particular but arbitrarily chosen integers that satisfy the condition $m \, R \, n$ and $n \, R \, p$. *[We must show that $m \, R \, p$.]* By definition of R, since $m \, R \, n$ and $n \, R \, p$, then $3 \mid (m - n)$ and $3 \mid (n - p)$. By definition of "divides," this means that $m - n = 3r$ and $n - p = 3s$, for some integers r and s. Adding the two equations gives $(m - n) + (n - p) = 3r + 3s$, and simplifying gives that $m - p = 3 \cdot (r + s)$. Since $r + s$ is an integer, this equation shows that $3 \mid (m - p)$. Hence, by definition of R, $m \, R \, p$ *[as was to be shown].* ■

EXERCISE SET 10.2

In 1–8 a number of binary relations are defined on the set $A = \{0, 1, 2, 3\}$. For each relation

   a. draw the directed graph;
   b. determine whether the relation is reflexive;
   c. determine whether the relation is symmetric;
   d. determine whether the relation is transitive.

Give a counterexample in each case in which the relation does not satisfy one of the properties.

1. $R_1 = \{(0, 0), (0, 1), (0, 3), (1, 1), (1, 0), (2, 3), (3, 3)\}$

2. $R_2 = \{(0, 0), (0, 1), (1, 1), (1, 2), (2, 2), (2, 3)\}$

3. $R_3 = \{(2, 3), (3, 2)\}$

4. $R_4 = \{(1, 2), (2, 1), (1, 3), (3, 1)\}$

5. $R_5 = \{(0, 0), (0, 1), (0, 2), (1, 2)\}$

6. $R_6 = \{(0, 1), (0, 2)\}$

7. $R_7 = \{(0, 3), (2, 3)\}$

8. $R_8 = \{(0, 0), (1, 1)\}$

In 9–11, R, S, and T are binary relations defined on $A = \{0, 1, 2, 3\}$.

9. Let $R = \{(0, 1), (0, 2), (1, 1), (1, 3), (2, 2), (3, 0)\}$. Find R', the transitive closure of R.

10. Let $S = \{(0, 0), (0, 3), (1, 0), (1, 2), (2, 0), (3, 2)\}$. Find S', the transitive closure of S.

11. Let $T = \{(0, 2), (1, 0), (2, 3), (3, 1)\}$. Find T', the transitive closure of T.

In 12–36 determine whether or not the given binary relation is reflexive, symmetric, transitive, or none of these. Justify your answers.

12. R is the "greater than or equal to" relation on the set of real numbers: for all $x, y \in \mathbf{R}$, $x R y \Leftrightarrow x \geq y$.

13. C is the circle relation on the set of real numbers: for all $x, y \in \mathbf{R}$, $x C y \Leftrightarrow x^2 + y^2 = 1$.

14. D is the binary relation defined on $\mathbf{R}$ as follows: for all $x, y \in \mathbf{R}$, $x D y \Leftrightarrow xy \geq 0$.

15. E is the congruence modulo 2 relation on $\mathbf{Z}$: for all $m, n \in \mathbf{Z}$, $m E n \Leftrightarrow 2 \mid (m - n)$.

16. F is the congruence modulo 5 relation on $\mathbf{Z}$: for all $m, n \in \mathbf{Z}$, $m F n \Leftrightarrow 5 \mid (m - n)$.

17. O is the binary relation defined on $\mathbf{Z}$ as follows: for all $m, n \in \mathbf{Z}$, $m O n \Leftrightarrow m - n$ is odd.

18. D is the "divides" relation on $\mathbf{Z}$: for all integers m and n, $m D n \Leftrightarrow m \mid n$.

19. A is the "absolute value" relation on $\mathbf{R}$: for all real numbers x and y, $x A y \Leftrightarrow |x| = |y|$.

20. Recall that a prime number is an integer that is greater than 1 and has no positive integer divisors other than 1 and itself. (In particular, 1 is not prime.) A binary relation P is defined on $\mathbf{Z}$ as follows: for all $m, n \in \mathbf{Z}$, $m P n \Leftrightarrow \exists$ a prime number p such that $p \mid m$ and $p \mid n$.

21. Let $\Sigma = \{0, 1\}$ and $A = \Sigma^*$. A binary relation L is defined on Σ^* as follows: for all strings $s, t \in \Sigma^*$, $s L t \Leftrightarrow l(s) < l(t)$ where l is the length function (that is, the number of characters in s is less than the number of characters in t).

22. Let $\Sigma = \{0, 1\}$ and $A = \Sigma^*$. A binary relation G is defined on Σ^* as follows: for all $s, t \in \Sigma^*$, $s G t \Leftrightarrow$ the number of 0's in s is greater than the number of 0's in t.

23. Let $X = \{a, b, c\}$ and $\mathscr{P}(X)$ be the power set of X (the set of all subsets of X). A binary relation # is defined on $\mathscr{P}(X)$ as follows: for all $A, B \in \mathscr{P}(X)$, $A \# B \Leftrightarrow$ the number of elements in A equals the number of elements in B.

24. Let $X = \{a, b, c\}$ and $\mathscr{P}(X)$ be the power set of X. A binary relation $\mathscr{R}$ is defined on $\mathscr{P}(X)$ as follows: for all $A, B \in \mathscr{P}(X)$, $A \mathscr{R} B \Leftrightarrow n(A) < n(B)$ (that is, the number of elements in A is less than the number of elements in B).

25. Let $X = \{a, b, c\}$ and $\mathscr{P}(X)$ be the power set of X. A binary relation $\mathscr{N}$ is defined on $\mathscr{P}(X)$ as follows: for all $A, B \in \mathscr{P}(X)$, $A \mathscr{N} B \Leftrightarrow n(A) \neq n(B)$ (that is, the number of elements in A is not equal to the number of elements in B).

26. Let A be a nonempty set and $\mathscr{P}(A)$ the power set of A. Define the "subset" relation $\mathscr{S}$ on $\mathscr{P}(A)$ as follows: for all $X, Y \in \mathscr{P}(A)$, $X \mathscr{S} Y \Leftrightarrow X \subseteq Y$.

27. Let A be a nonempty set and $\mathscr{P}(A)$ the power set of A. Define the "not equal to" relation $\mathscr{R}$ on $\mathscr{P}(A)$ as follows: for all $X, Y \in \mathscr{P}(A)$, $X \mathscr{R} Y \Leftrightarrow X \neq Y$.

28. Let A be a nonempty set and $\mathscr{P}(A)$ the power set of A. Define the "relative complement" relation $\mathscr{C}$ on $\mathscr{P}(A)$ as follows: for all $X, Y \in \mathscr{P}(A)$, $X \mathscr{C} Y \Leftrightarrow Y = A - X$.

29. Let A be a set with at least two elements and $\mathscr{P}(A)$ the power set of A. Define a relation $\mathscr{R}$ on $\mathscr{P}(A)$ as follows: for all $X, Y \in \mathscr{P}(A)$, $X \mathscr{R} Y \Leftrightarrow X \subseteq Y$ or $Y \subseteq X$.

30. Let A be the set of all English statements. A binary relation I is defined on A as follows: for all $p, q \in A$, $p \, I \, q \iff p \to q$ is true.

31. Let $A = \mathbf{R} \times \mathbf{R}$. A binary relation $\mathscr{R}$ is defined on A as follows: for all (x_1, y_1) and (x_2, y_2) in A, $(x_1, y_1) \, \mathscr{R} \, (x_2, y_2) \iff x_1 = x_2$.

32. Let $A = \mathbf{R} \times \mathbf{R}$. A binary relation $\mathscr{R}$ is defined on A as follows: for all (x_1, y_1) and (x_2, y_2) in A, $(x_1, y_1) \, \mathscr{R} \, (x_2, y_2) \iff y_1 = y_2$.

33. Let A be the "punctured plane"; that is, A is the set of all points in the Cartesian plane except the origin $(0, 0)$. A binary relation R is defined on A as follows: for all p_1 and p_2 in A, $p_1 \, R \, p_2 \iff p_1$ and p_2 lie on the same half line emanating from the origin.

34. Let A be the set of people living in the world today. A binary relation R is defined on A as follows: for all $p, q \in A$, $p \, R \, q \iff p$ lives within 100 miles of q.

35. Let A be the set of all lines in the plane. A binary relation R is defined on A as follows: for all l_1 and l_2 in A, $l_1 \, R \, l_2 \iff l_1$ is parallel to l_2. (Assume that a line is parallel to itself.)

36. Let A be the set of all lines in the plane. A binary relation R is defined on A as follows: for all l_1 and l_2 in A, $l_1 \, R \, l_2 \iff l_1$ is perpendicular to l_2.

37. Let A be a set with eight elements.

a. How many binary relations are there on A?
b. How many binary relations on A are reflexive?
c. How many binary relations on A are symmetric?
d. How many binary relations on A are both reflexive and symmetric?

38. Write a computer algorithm to test whether a binary relation R defined on a finite set A is reflexive, where $A = \{a[1], a[2], \ldots, a[n]\}$.

39. Write a computer algorithm to test whether a binary relation R defined on a finite set A is symmetric, where $A = \{a[1], a[2], \ldots, a[n]\}$.

40. Write a computer algorithm to test whether a binary relation R defined on a finite set A is transitive, where $A = \{a[1], a[2], \ldots, a[n]\}$.

◆ **41.** Let R be a binary relation on a set A and let R^t be the transitive closure of R. Prove that for all x and y in A, $x \, R^t \, y$ if, and only if, there is a sequence of elements of A, $x_1, x_2, \ldots, x_n$, such that $x = x_1$, $x_1 \, R \, x_2$, $x_2 \, R \, x_3$, $\ldots, x_{n-1} \, R \, x_n$, and $x_n = y$.

◆ **42.** Write a computer program to find the transitive closure of a binary relation R defined on a finite set $A = \{a[1], a[2], \ldots, a[n]\}$.

43. Suppose R and S are binary relations on a set A.

a. If R and S are reflexive, is $R \cap S$ reflexive? Why?
H**b.** If R and S are symmetric, is $R \cap S$ symmetric? Why?
c. If R and S are transitive, is $R \cap S$ transitive? Why?

44. Suppose R and S are binary relations on a set A.

a. If R and S are reflexive, is $R \cup S$ reflexive? Why?
b. If R and S are symmetric, is $R \cup S$ symmetric? Why?
c. If R and S are transitive, is $R \cup S$ transitive? Why?

In 45–52 the following definitions are used: A binary relation on a set A is defined to be

irreflexive if, and only if, for all $x \in A$, $x \, \not{R} \, x$;

asymmetric if, and only if, for all $x, y \in A$, if $x \, R \, y$ then $y \, \not{R} \, x$;

intransitive if, and only if, for all $x, y, z \in A$, if $x \, R \, y$ and $y \, R \, z$ then $x \, \not{R} \, z$.

For each of the binary relations in the referenced exercise, determine whether the relation is irreflexive, asymmetric, intransitive, or none of these.

45. Exercise 1   **46.** Exercise 2

47. Exercise 3   **48.** Exercise 4

49. Exercise 5   **50.** Exercise 6

51. Exercise 7   **52.** Exercise 8

10.3 EQUIVALENCE RELATIONS

"You are sad," the Knight said in an anxious tone: "let me sing you a song to comfort you."

"Is it very long?" Alice asked, for she had heard a good deal of poetry that day.

"It's long," said the Knight, "but it's very, very beautiful. Everybody that hears me sing it—either it brings the tears into the eyes, or else—"

"Or else what?" said Alice, for the Knight had made a sudden pause.

"Or else it doesn't, you know. The name of the song is called 'Haddocks' Eyes.'"

"Oh, that's the name of the song, is it?" Alice said, trying to feel interested.

"No, you don't understand," the Knight said, looking a little vexed. "That's what the name is called. *The name really is 'The Aged Aged Man.'"*

"Then I ought to have said 'That's what the song is called'?" Alice corrected herself.

"No, you oughtn't: that's quite another thing! The song *is called 'Ways and Means': but that's only what it's* called, *you know!"*

"Well, what is *the song, then?" said Alice, who was by this time completely bewildered.*

"I was coming to that," the Knight said. "The song really is 'A-sitting on a Gate': and the tune's my own invention."

So saying, he stopped his horse and let the reins fall on its neck: then, slowly beating time with one hand, and with a faint smile lighting up his gentle foolish face, as if he enjoyed the music of his song, he began.

(Lewis Carroll, *Through the Looking Glass,* 1872)

You know from your early study of fractions that each fraction has many equivalent forms. For example,

$$\frac{1}{2}, \frac{2}{4}, \frac{3}{6}, \frac{-1}{-2}, \frac{-3}{-6}, \frac{15}{30}, \ldots, \text{ and so on}$$

are all different ways to represent the same number. They may look different; they may be called different names; but they are all equal. The idea of grouping together things that "look different but are really the same" is the central idea of equivalence relations.

THE RELATION INDUCED BY A PARTITION

Recall that a partition of a set A is a collection of nonempty mutually disjoint subsets whose union is A. The diagram of Figure 10.3.1 illustrates a partition of a set A by sets $A_1, A_2, \ldots, A_6$.

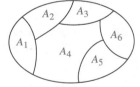

$A_i \cap A_j = \emptyset$, whenever $i \neq j$
$A_1 \cup A_2 \cup \cdots \cup A_6 = A$

FIGURE 10.3.1
A Partition of a Set

DEFINITION

Given a partition of a set A the **binary relation induced by the partition**, R, is defined on A as follows:

for all $x, y \in A$, $x R y$ $\Leftrightarrow$ there is a subset A of the partition
                                    such that both x and y are in A.

EXAMPLE 10.3.1   Relation Induced by a Partition

Let $A = \{0, 1, 2, 3, 4\}$ and consider the following partition of A:

$$\{0, 3, 4\}, \{1\}, \{2\}.$$

Find the relation R induced by this partition.

Solution   Since $\{0, 3, 4\}$ is a subset of the partition.

$\quad 0\ R\ 3$ because both 0 and 3 are in $\{0, 3, 4\}$,

$\quad 3\ R\ 0$ because both 3 and 0 are in $\{0, 3, 4\}$,

$\quad 0\ R\ 4$ because both 0 and 4 are in $\{0, 3, 4\}$,

$\quad 4\ R\ 0$ because both 4 and 0 are in $\{0, 3, 4\}$,

$$3\ R\ 4 \quad \text{because both 3 and 4 are in } \{0, 3, 4\}, \quad \text{and}$$

$$4\ R\ 3 \quad \text{because both 4 and 3 are in } \{0, 3, 4\}.$$

Also,

$$0\ R\ 0 \quad \text{because both 0 and 0 are in } \{0, 3, 4\}$$

[This statement may seem strange, but, after all, it is not false!],

$$3\ R\ 3 \quad \text{because both 3 and 3 are in } \{0, 3, 4\}, \quad \text{and}$$

$$4\ R\ 4 \quad \text{because both 4 and 4 are in } \{0, 3, 4\}.$$

Since $\{1\}$ is a subset of the partition,

$$1\ R\ 1 \quad \text{because both 1 and 1 are in } \{1\},$$

and since $\{2\}$ is a subset of the partition,

$$2\ R\ 2 \quad \text{because both 2 and 2 are in } \{2\}.$$

Hence

$$R = \{(0, 0), (0, 3), (0, 4), (1, 1), (2, 2), (3, 0), (3, 3), (3, 4), (4, 0), (4, 3), (4, 4)\}.$$

■

The fact is that a relation induced by a partition of a set satisfies all three properties studied in Section 10.2: reflexivity, symmetry, and transitivity.

THEOREM 10.3.1

Let A be a set with a partition and let R be the relation induced by the partition. Then R is reflexive, symmetric, and transitive.

Proof:

Suppose A is a set with a partition. In order to simplify notation, we assume that the partition consists of only a finite number of sets. The proof for an infinite partition is identical except for notation. Denote the partition subsets by

$$A_1, A_2, \ldots, A_n.$$

Then $A_i \cap A_j = \varnothing$ whenever $i \neq j$, and $A_1 \cup A_2 \cup \cdots \cup A_n = A$. The relation R induced by the partition is defined as follows: for all $x, y \in A$,

$$x\ R\ y \quad \Leftrightarrow \quad \text{there is a set } A_i \text{ of the partition}$$
$$\text{such that } x \in A_i \text{ and } y \in A_i.$$

R is reflexive: [For R to be reflexive means that each element of A is related by R to itself. But by definition of R , for an element x to be related to itself means that x is in the same subset of the partition as itself. Well, if x is in some subset of the partition, then it is certainly in the same subset as itself. But x is in some subset of the partition because the union of the subsets of the partition is all of A. This reasoning is formalized as follows.]

Suppose $x \in A$. Since $A_1, A_2, \ldots, A_n$ is a partition of A, it follows that $x \in A_i$ for some index i. But then the statement there is a set A_i of the partition such that $x \in A_i$ and $x \in A_i$

is true.* So by definition of R, $x \, R \, x$.

R is symmetric: *[For R to be symmetric means that any time one element is related to a second, then the second is related to the first. Now for one element x to be related to a second element y means that x and y are in the same subset of the partition. But if this is the case, then y is in the same subset of the partition as x. So y is related to x by definition of R. This reasoning is formalized as follows.]*

Suppose x and y are elements of A such that $x \, R \, y$. Then

there is a subset A_i of the partition such that $x \in A_i$ and $y \in A_i$

by definition of R. It follows that the statement

there is a subset A_i of the partition such that $y \in A_i$ and $x \in A_i$

is also true.† Hence by definition of R, $y \, R \, x$.

R is transitive: *[For R to be transitive means that any time one element of A is related by R to a second and that second is related to a third, then the first element is related to the third. But for one element to be related to another means that there is a subset of the partition that contains both. So suppose x, y, and z are elements such that x is in the same subset as y and y is in the same subset as z. Must x be in the same subset as z? Yes, because the subsets of the partition are mutually disjoint. Since the subset that contains x and y has an element in common with the subset that contains y and z (namely y), then the two subsets are equal. But that means that x, y, and z are all in the same subset, and so in particular x and z are in the same subset. Hence x is related by R to z. This reasoning is formalized as follows.]*

Suppose x, y, and z are in A and $x \, R \, y$ and $y \, R \, z$. By definition of R there are subsets A_i and A_j of the partition such that

x and y are in A_i and y and z are in A_j.

Suppose $A_i \neq A_j$. *[We will deduce a contradiction.]* Then $A_i \cap A_j = \emptyset$ since $\{A_1, A_2, A_3, \ldots, A_n\}$ is a partition of A. But y is in A_i and y is in A_j also. Hence $A_i \cap A_j \neq \emptyset$. *[This is a contradiction.]* Thus $A_i = A_j$. It follows that x, y, and z are all in A_i, and so in particular,

x and z are in A_i.

Thus by definition of R, $x \, R \, z$.

DEFINITION OF AN EQUIVALENCE RELATION

A binary relation that satisfies the three properties of reflexivity, symmetry, and transitivity is called an *equivalence relation*.

---

*Since the statement forms p and $p \wedge p$ are logically equivalent, if p is true then $p \wedge p$ is also true.
†This follows from the fact that the statement forms $p \wedge q$ and $q \wedge p$ are logically equivalent.

Carl Friedrich Gauss
(1777–1855)

> **DEFINITION**
>
> Let A be a nonempty set and R a binary relation on A. R is an **equivalence relation** if, and only if, R is reflexive, symmetric, and transitive.

Thus according to Theorem 10.3.1, the relation induced by a partition is an equivalence relation. Another example is congruence modulo 3. In Example 10.2.5 it was shown that this relation is reflexive, symmetric, and transitive. Hence it, also, is an equivalence relation.

The following notation is used frequently when referring to congruence relations. It was introduced by Carl Friedrich Gauss in the first chapter of his book *Disquisitiones Arithmeticae*. This work, which was published when Gauss was only 24, laid the foundation for modern number theory.

> **NOTATION**
>
> Let m and n be integers and let d be a positive integer. The notation
>
> $$m \equiv n\,(\textbf{\textit{mod d}})$$
>
> is read "m is congruent to n modulo d" and means that
>
> $$d \mid (m - n).$$
>
> Symbolically, $m \equiv n\,(mod\ d) \iff d \mid (m - n)$

Exercise 12 at the end of this section asks you to show that $m \equiv n\,(mod\ d)$ if, and only if, $m\ mod\ d = n\ mod\ d$, where m, n, and d are integers and d is positive.

EXAMPLE 10.3.2   **Evaluating Congruences**

Determine which of the following congruences are true and which are false.

a. $12 \equiv 7\,(mod\ 5)$ b. $6 \equiv -8\,(mod\ 4)$ c. $3 \equiv 3\,(mod\ 7)$

Solution
a. True. $12 - 7 = 5 = 5 \cdot 1$. Hence $5 \mid (12 - 7)$, and so $12 \equiv 7\,(mod\ 5)$.
b. False. $6 - (-8) = 14$, and $4 \nmid 14$ because $14 \neq 4 \cdot k$ for any integer k. Hence $6 \not\equiv -8\,(mod\ 4)$.
c. True. $3 - 3 = 0 = 7 \cdot 0$. Hence $7 \mid (3 - 3)$, and so $3 \equiv 3\,(mod\ 7)$. ∎

EXAMPLE 10.3.3   **Equivalence of Digital Logic Circuits Is
an Equivalence Relation**

Let S be the set of all digital logic circuits with a fixed number n of inputs. Define a binary relation $\mathscr{E}$ on S as follows: for all circuits C_1 and C_2 in S,

$$C_1\ \mathscr{E}\ C_2 \iff C_1 \text{ has the same input/output table as } C_2.$$

If $C_1\ \mathscr{E}\ C_2$, then circuit C_1 is said to be *equivalent* to circuit C_2. Prove that $\mathscr{E}$ is an equivalence relation on S.

Solution   *$\mathscr{E}$ is reflexive:*   Suppose C is a digital logic circuit in S. *[We must show that $C \mathscr{E} C$.]* Certainly C has the same input/output table as itself. So by definition of $\mathscr{E}$, $C \mathscr{E} C$ *[as was to be shown].*

$\mathscr{E}$ is symmetric: Suppose C_1 and C_2 are digital logic circuits in S such that $C_1 \mathscr{E} C_2$. *[We must show that $C_2 \mathscr{E} C_1$.]* By definition of $\mathscr{E}$, since $C_1 \mathscr{E} C_2$, then C_1 has the same input/output table as C_2. It follows that C_2 has the same input/output table as C_1. Hence by definition of $\mathscr{E}$, $C_2 \mathscr{E} C_1$ *[as was to be shown.]*

$\mathscr{E}$ is transitive: Suppose C_1, C_2, and C_3 are digital logic circuits in S such that $C_1 \mathscr{E} C_2$ and $C_2 \mathscr{E} C_3$. *[We must show that $C_1 \mathscr{E} C_3$.]* By definition of $\mathscr{E}$, since $C_1 \mathscr{E} C_2$ and $C_2 \mathscr{E} C_3$, then

$$C_1 \text{ has the same input/output table as } C_2$$

and

$$C_2 \text{ has the same input/output table as } C_3.$$

It follows that

$$C_1 \text{ has the same input/output table as } C_3.$$

Hence by definition of $\mathscr{E}$, $C_1 \mathscr{E} C_3$ *[as was to be shown].*

Since $\mathscr{E}$ is reflexive, symmetric, and transitive, then $\mathscr{E}$ is an equivalence relation on S. ∎

Certain implementations of computer languages (such as Pascal, BASIC, and C) do not place a limit on the allowable length of an identifier. This permits a programmer to be as precise as necessary in naming variables without having to worry about exceeding length limitations. However, compilers for such languages often ignore all but some specified number of initial characters: As far as the compiler is concerned, two identifiers are the same if they have the same initial characters even though they may look different to a human reader of the program. For example, to a compiler that ignores all but the first eight characters of an identifier, the following identifiers would be the same:

$$\text{NumberOfScrews} \qquad \text{NumberOfBolts.}$$

Obviously, in using such a language, the programmer has to be sure to avoid giving two distinct identifiers the same first eight characters. When a compiler lumps identifiers together in this way, it sets up an equivalence relation on the set of all possible identifiers in the language. Such a relation is described in the next example.

EXAMPLE 10.3.4   **A Binary Relation on a Set of Identifiers**

Let L be the set of all allowable identifiers in a certain computer language, and define a relation R on L as follows: for all strings s and t in L,

$s \; R \; t \;\;\; \Leftrightarrow \;\;\;$ **the first eight characters of s equal the first eight characters of t.**

Prove that R is an equivalence relation on L.

Solution   *R is reflexive:*   Let $s \in L$. *[We must show that $s \; R \; s$.]* Clearly s has the same first eight characters as itself. So by definition of R, $s \; R \; s$ *[as was to be shown].*

R is symmetric: Let s and t be in L and suppose that $s \; R \; t$. *[We must show that $t \; R \; s$.]* By definition of R, since $s \; R \; t$, the first eight characters of s equal the first eight characters of t. But then the first eight characters of t equal the first eight characters of s. And so by definition of R, $t \; R \; s$ *[as was to be shown].*

R is transitive: Let *s*, *t*, and *u* be in *L* and suppose that *s R t* and *t R u*. *[We must show that s R u.]* By definition of *R*, since *s R t* and *t R u*, the first eight characters of *s* equal the first eight characters of *t* and the first eight characters of *t* equal the first eight characters of *u*. Hence the first eight characters of *s* equal the first eight characters of *u*. So by definition of *R*, *s R u* *[as was to be shown]*.

Since *R* is reflexive, symmetric, and transitive, *R* is an equivalence relation on *L*. ∎

EQUIVALENCE CLASSES OF AN EQUIVALENCE RELATION

Suppose there is an equivalence relation on a certain set. If *a* is any particular element of the set, then one can ask: "What is the subset of all elements that are related to *a*?" This subset is called the *equivalence class* of *a*.

> **DEFINITION**
>
> Suppose *A* is a set and *R* is an equivalence relation on *A*. For each element *a* in *A*, the **equivalence class of *a***, denoted **[*a*]** and called the **class of *a*** for short, is the set of all elements *x* in *A* such that *x* is related to *a* by *R*.

Written symbolically, this definition becomes the following:

$$[a] = \{x \in A \mid x \, R \, a\}$$

When several equivalence relations on a set are under discussion, the notation $[a]_R$ is often used to denote the equivalence class of *a* under *R*.

The procedural version of this definition is as follows:

$$\text{for all } x \in A, \quad x \in [a] \iff x \, R \, a.$$

EXAMPLE 10.3.5   **Equivalence Classes of a Relation Defined on a Finite Set**

Let *A* = {0, 1, 2, 3, 4} and define a binary relation *R* on *A* as follows:

$$R = \{(0, 0), (0, 4), (1, 1), (1, 3), (2, 2), (4, 0), (3, 3), (3, 1), (4, 4)\}.$$

The directed graph for *R* is as shown in Figure 10.3.2. As can be seen by inspection, *R* is an equivalence relation on *A*. Find the distinct equivalence classes of *R*.

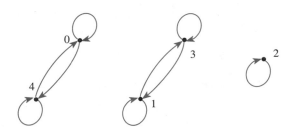

FIGURE 10.3.2

Solution   First find the equivalence class of every element of A.

$$[0] = \{x \in A \mid x\,R\,0\} = \{0, 4\}$$
$$[1] = \{x \in A \mid x\,R\,1\} = \{1, 3\}$$
$$[2] = \{x \in A \mid x\,R\,2\} = \{2\}$$
$$[3] = \{x \in A \mid x\,R\,3\} = \{1, 3\}$$
$$[4] = \{x \in A \mid x\,R\,4\} = \{0, 4\}$$

Note that $[0] = [4]$ and $[1] = [3]$. Thus the *distinct* equivalence classes of the relation are

$$\{0, 4\}, \{1, 3\}, \text{ and } \{2\}. \qquad \blacksquare$$

When a problem asks you to find the *distinct* equivalence classes of an equivalence relation you will generally solve the problem in two steps. In the first step you either explicitly construct (as in Example 10.3.5) or imagine constructing (as in infinite cases) the equivalence class for every element of the domain A of the relation. Usually several of the classes will contain exactly the same elements. So in the second step you must take a careful look at the classes to determine which are the same. You then indicate the distinct equivalence classes by describing them without duplication.

EXAMPLE 10.3.6   **Equivalence Classes of Identifiers**

In Example 10.3.4 it was shown that the relation R of having the same first eight characters is an equivalence relation on the set L of allowable identifiers in a computer language. Describe the distinct equivalence classes of R.

Solution   By definition of R, two strings in L are related by R if, and only if, they have the same first eight characters. Given any string s in L, therefore,

$$[s] = \{t \in L \mid t\,R\,s\}$$
$$= \{t \in L \mid \text{the first eight characters of } t \text{ equal the first}$$
$$\text{eight characters of } s\}.$$

Thus the distinct equivalence classes of R are sets of strings such that (1) each class consists entirely of strings all of which have the same first eight characters, and (2) any two distinct classes contain strings that differ somewhere in their first eight characters. $\qquad \blacksquare$

EXAMPLE 10.3.7   **Equivalence Classes of the Identity Relation**

Let A be any set and define a relation R on A as follows: for all x and y in A,

$$x\,R\,y \quad \Leftrightarrow \quad x = y.$$

Then R is an equivalence relation. *[To prove this, just generalize the argument used in Example 10.2.3.]* Describe the distinct equivalence classes of R.

Solution   Given any a in A, the class of a is

$$[a] = \{x \in A \mid x\,R\,a\}.$$

But by definition of R, $a\,R\,x$ if, and only if, $a = x$. So

$$[a] = \{x \in A \mid x = a\}$$
$$= \{a\} \qquad \text{since the only element of } A \text{ that equals } a \text{ is } a.$$

Hence, given any a in A,

$$[a] = \{a\}.$$

Now if $x \neq a$, then $\{x\} \neq \{a\}$. Consequently, all the classes of all the elements of A are distinct, and the distinct equivalence classes of R are all the single-element subsets of A.   ∎

In each of Examples 10.3.5, 10.3.6, and 10.3.7, the set of distinct equivalence classes of the relation consists of mutually disjoint subsets whose union is the entire domain A of the relation. This means that the set of equivalence classes of the relation forms a partition of the domain A. In fact, it is always the case that the equivalence classes of an equivalence relation partition the domain of the relation into a union of mutually disjoint subsets. We establish the truth of this statement in stages, first proving two lemmas and then the main theorem.

The first lemma says that if two elements of A are related by an equivalence relation R, then their equivalence classes are the same.

LEMMA 10.3.2

Suppose A is a set, R is an equivalence relation on A, and a and b are elements of A. If $a \, R \, b$, then $[a] = [b]$.

This lemma says that if a certain condition is satisfied, then $[a] = [b]$. Now $[a]$ and $[b]$ are *sets,* and two sets are equal if, and only if, each is a subset of the other. Hence the proof of the lemma consists of two parts: first, a proof that $[a] \subseteq [b]$ and, second, a proof that $[b] \subseteq [a]$. To show each subset relation, it is necessary to show that every element in the left-hand set is an element of the right-hand set.

Proof of Lemma 10.3.2:

Let A be a set, R an equivalence relation on A, and suppose

a and b are elements of A such that $a \, R \, b$.

[We must show that $[a] = [b]$.]

Proof that $[a] \subseteq [b]$:   Let $x \in [a]$. *[We must show that $x \in [b]$.]* Since

$$x \in [a]$$

then

$$x \, R \, a$$

by definition of class. But

$$a \, R \, b$$

by hypothesis. So by transitivity of R,

$$x \, R \, b.$$

Hence

$$x \in [b]$$

by definition of class. *[This is what was to be shown.]*

Proof that $[b] \subseteq [a]$:   Let $x \in [b]$. *[We must show that $x \in [a]$.]* Since

$$x \in [b]$$

then

$$x \, R \, b$$

by definition of class. Now

$$a \, R \, b$$

by hypothesis. So since R is symmetric,

$$b \, R \, a$$

also. Then since R is transitive and $x \, R \, b$ and $b \, R \, a$,

$$x \, R \, a.$$

Hence,

$$x \in [a]$$

by definition of class. *[This is what was to be shown.]*

Since $[a] \subseteq [b]$ and $[b] \subseteq [a]$, it follows that $[a] = [b]$ by definition of set equality.

The second lemma says that any two equivalence classes of an equivalence relation are either mutually disjoint or identical.

LEMMA 10.3.3

If A is a set, R is an equivalence relation on A, and a and b are elements of A, then

$$\text{either} \quad [a] \cap [b] = \varnothing \quad \text{or} \quad [a] = [b].$$

The statement of Lemma 10.3.3 has the form

$$\text{if } p \text{ then } q \text{ or } r,$$

where p is the statement "A is a set, R is an equivalence relation on A, and a and b are elements of A," q is the statement "$[a] \cap [b] = \varnothing$," and r is the statement "$[a] = [b]$." To prove the lemma, we will prove the logically equivalent statement*

---

*See exercise 14 in Section 1.2.

if p and not q then r.

That is, we will prove the following:

> If A is a set, R is an equivalence relation on A, a and b are elements of A, and $[a] \cap [b] \neq \emptyset$, then $[a] = [b]$.

Proof of Lemma 10.3.3:

Suppose A is a set, R is an equivalence relation on A, a and b are elements of A, and

$$[a] \cap [b] \neq \emptyset.$$

[We must show that $[a] = [b]$.] Since $[a] \cap [b] \neq \emptyset$, then there exists an element x in A such that $x \in [a] \cap [b]$. By definition of intersection,

$$x \in [a] \quad \text{and} \quad x \in [b]$$

and so

$$x \, R \, a \quad \text{and} \quad x \, R \, b$$

by definition of class. Since R is symmetric *[being an equivalence relation]* and $x \, R \, a$, then $a \, R \, x$. But R is also transitive *[since it is an equivalence relation]*, and so since $a \, R \, x$ and $x \, R \, b$ then

$$a \, R \, b.$$

Now a and b satisfy the hypothesis of Lemma 10.3.2. Hence by that lemma,

$$[a] = [b].$$

[This is what was to be shown.]

THEOREM 10.3.4

If A is a nonempty set and R is an equivalence relation on A, then the distinct equivalence classes of R form a partition of A; that is, the union of the equivalence classes is all of A and the intersection of any two distinct classes is empty.

The proof of Theorem 10.3.4 is divided into two parts: first, a proof that A is the union of the equivalence classes of R and, second, a proof that the intersection of any two distinct equivalence classes is empty. The proof of the first part follows from the fact that the relation is reflexive. The proof of the second part follows from Lemma 10.3.3.

Proof of Theorem 10.3.4:

Suppose A is a set and R is an equivalence relation on A. For notational simplicity, we assume that R has only a finite number of distinct equivalence classes, which we denote
$$A_1, A_2, \ldots, A_n,$$
where n is a positive integer. (When the number of classes is infinite, the proof is identical except for notation.)

Proof that $A = A_1 \cup A_2 \cup A_n$: *[We must show that $A \subseteq A_1 \cup A_2 \cup \cdots \cup A_n$ and that $A_1 \cup A_2 \cup \cdots \cup A_n \subseteq A.$]*

To show that $A \subseteq A_1 \cup A_2 \cup \cdots \cup A_n$, suppose x is any element of A. *[We must show that $x \in A_1 \cup A_2 \cup \cdots \cup A_n.$]* By reflexivity of R, $x\,R\,x$. But this implies that $x \in [x]$ by definition of class. Since x is in *some* equivalence class, it must be in one of the distinct equivalence classes $A_1, A_2, \ldots,$ or A_n. Thus $x \in A_i$ for some index i, and hence $x \in A_1 \cup A_2 \cup \cdots \cup A_n$ by definition of union *[as was to be shown]*.

To show that $A_1 \cup A_2 \cup \cdots A_n \subseteq A$, suppose x is any element of $A_1 \cup A_2 \cup \cdots \cup A_n$. *[We must show that $x \in A.$]* Then $x \in A_i$ for some $i = 1, 2, \ldots,$ or n, by definition of union. But each A_i is an equivalence class of R. And equivalence classes are subsets of A. Hence $A_i \subseteq A$ and so $x \in A$ *[as was to be shown]*.

Since $A \subseteq A_1 \cup A_2 \cup \cdots \cup A_n$ and $A_1 \cup A_2 \cup \cdots \cup A_n \subseteq A$, then $A = A_1 \cup A_2 \cup \cdots \cup A_n$ by definition of set equality.

Proof that any two distinct classes of R are mutually disjoint: Suppose that A_i and A_j are any two distinct equivalence classes of R. *[We must show that A_i and A_j are disjoint.]* Since A_i and A_j are distinct, then $A_i \neq A_j$. And since A_i and A_j are equivalence classes of R, there must exist elements a and b in A such that $A_i = [a]$ and $A_j = [b]$. By Lemma 10.3.3,
$$\text{either} \quad [a] \cap [b] = \emptyset \quad \text{or} \quad [a] = [b].$$
But $[a] \neq [b]$ because $A_i \neq A_j$. Hence $[a] \cap [b] = \emptyset$. Thus $A_i \cap A_j = \emptyset$, and so A_i and A_j are disjoint *[as was to be shown]*.

EXAMPLE 10.3.8   Equivalence Classes of Congruence Modulo 3

Let R be the relation of congruence modulo 3 on the set $\mathbf{Z}$ of all integers. That is, for all integers m and n,
$$m\,R\,n \quad \Leftrightarrow \quad 3 \mid (m - n) \quad \Leftrightarrow \quad m \equiv n \;(mod\; 3).$$
Describe the distinct equivalence classes of R.

Solution   For each integer a,
$$[a] = \{x \in \mathbf{Z} \mid x\,R\,a\}$$
$$= \{x \in \mathbf{Z} \mid 3 \mid (x - a)\}$$
$$= \{x \in \mathbf{Z} \mid x - a = 3 \cdot k, \text{ for some integer } k\}.$$
So
$$[a] = \{x \in \mathbf{Z} \mid x = 3 \cdot k + a, \text{ for some integer } k\}.$$

In particular,

$$[0] = \{x \in \mathbf{Z} \mid x = 3 \cdot k + 0, \text{ for some integer } k\}$$
$$= \{x \in \mathbf{Z} \mid x = 3 \cdot k, \text{ for some integer } k\}$$
$$= \{\ldots -9, -6, -3, 0, 3, 6, 9, \ldots\},$$
$$[1] = \{x \in \mathbf{Z} \mid x = 3 \cdot k + 1, \text{ for some integer } k\}$$
$$= \{\ldots -8, -5, -2, 1, 4, 7, 10, \ldots\},$$
$$[2] = \{x \in \mathbf{Z} \mid x = 3 \cdot k + 2, \text{ for some integer } k\}$$
$$= \{\ldots -7, -4, -1, 2, 5, 8, 11, \ldots\}.$$

Now since $3 \, R \, 0$, then by Lemma 10.3.2,

$$[3] = [0].$$

More generally, by the same reasoning,

$$[0] = [3] = [-3] = [6] = [-6] = \ldots, \text{ and so on.}$$

Similarly,

$$[1] = [4] = [-2] = [7] = [-5] = \ldots, \text{ and so on.}$$

And

$$[2] = [5] = [-1] = [8] = [-4] = \ldots, \text{ and so on.}$$

Notice that every integer is in one of the three classes $[0], [1]$, or $[2]$. Hence the distinct equivalence classes are

$$\{x \in \mathbf{Z} \mid x = 3 \cdot k, \text{ for some integer } k\},$$
$$\{x \in \mathbf{Z} \mid x = 3 \cdot k + 1, \text{ for some integer } k\}, \quad \text{and}$$
$$\{x \in \mathbf{Z} \mid x = 3 \cdot k + 2, \text{ for some integer } k\}.$$

In words, the three classes of congruence modulo 3 are (1) the set of all integers that are divisible by 3, (2) the set of all integers that leave a remainder of 1 when divided by 3, and (3) the set of all integers that leave a remainder of 2 when divided by 3. ∎

Example 10.3.8 illustrates a very important property of equivalence classes, namely that an equivalence class may have many different names. In Example 10.3.8, for instance, the class of 0, $[0]$, may also be *called* the class of 3, $[3]$, or the class of -6, $[-6]$. But what the class *is* is the set

$$\{x \in \mathbf{Z} \mid x = 3 \cdot k, \text{ for some integer } k\}.$$

(The quote at the beginning of this section refers in a humorous way to the philosophically interesting distinction between what things are *called* and what they *are*.)

DEFINITION

Suppose R is an equivalence relation on a set A and S is an equivalence class of R. A **representative** of the class S is any element a such that $[a] = S$.

In the exercises at the end of this section, you are asked to show that if x is any element of an equivalence class S, then $S = [x]$. Hence *any* element of an equivalence class is a representative of that class.

EXAMPLE 10.3.9   **Equivalence Classes of Digital Logic Circuits**

In Example 10.3.3 it was shown that the relation of equivalence among circuits is an equivalence relation. Let S be the set of all digital logic circuits with exactly two inputs and one output. The binary relation $\mathscr{E}$ is defined on S as follows: for all C_1 and C_2 in S,

$$C_1 \mathscr{E} C_2 \quad \Leftrightarrow \quad C_1 \text{ has the same input/output table as } C_2.$$

Describe the equivalence classes of this relation. How many distinct equivalence classes are there? Write representative circuits for two of the distinct classes.

Solution   Given a circuit C, the equivalence class of C is the set of all circuits with two input signals and one output signal that have the same input/output table as C. Now each input/output table has exactly four rows, corresponding to the four possible combinations of inputs: 11, 10, 01, and 00. A typical input/output table is the following:

Input		Output
P	Q	R
1	1	0
1	0	0
0	1	0
0	0	1

There are exactly as many such tables as there are binary strings of length 4. The reason is that distinct input/output tables can be formed by changing the pattern of the four 0's and 1's in the output column, and there are as many ways to do that as there are strings of four 0's and 1's. But the number of binary strings of length 4 is $2^4 = 16$. Hence there are sixteen distinct input/output tables.

This implies that there are exactly 16 equivalence classes of circuits, one for each distinct input/output table. However, there are infinitely many circuits that give rise to each table. For instance, two representative circuits for the above input/output table are shown in Figure 10.3.3.

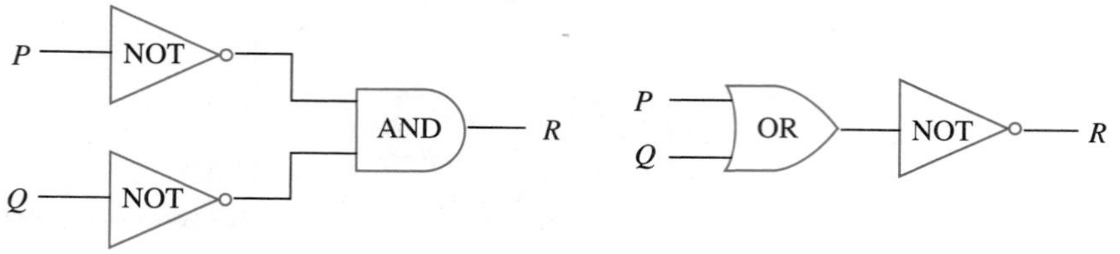

FIGURE 10.3.3

EXAMPLE 10.3.10 **Rational Numbers Are Really Equivalence Classes**

For a moment forget what you know about fractional arithmetic and look at the numbers

$$\frac{1}{3} \quad \text{and} \quad \frac{2}{6}$$

as *symbols*. Considered as symbolic expressions, these *appear* quite different. In fact, if they were written as ordered pairs

$$(1, 3) \quad \text{and} \quad (2, 6)$$

they would *be* different. The fact that we regard them as "the same" is a specific instance of our general agreement to regard any two numbers

$$\frac{a}{b} \quad \text{and} \quad \frac{c}{d}$$

as equal provided the *cross products* are equal: $ad = bc$. This can be formalized as follows using the language of equivalence relations.

Let A be the set of all ordered pairs of integers for which the second element of the pair is nonzero. Symbolically,

$$A = \mathbf{Z} \times (\mathbf{Z} - \{0\}).$$

Define a binary relation R on A as follows: for all $(a, b), (c, d) \in A$,

$$(a, b) \, R \, (c, d) \quad \Leftrightarrow \quad ad = bc.$$

The fact is that R is an equivalence relation.

a. Prove that R is transitive. (Proofs that R is reflexive and symmetric are left to the exercises.)

b. Describe the distinct equivalence classes of R.

Solution a. *[We must show that for all $(a, b), (c, d), (e, f) \in A$, if $(a, b) \, R \, (c, d)$ and $(c, d) \, R \, (e, f)$, then $(a, b) \, R \, (e, f)$.]* Suppose $(a, b), (c, d)$, and (e, f) are particular but arbitrarily chosen elements of A such that $(a, b) \, R \, (c, d)$ and $(c, d) \, R \, (e, f)$. *[We must show that $(a, b) \, R \, (e, f)$.]* By definition of R,

$$(1) \; ad = bd \quad \text{and} \quad (2) \; cf = de.$$

Since the second elements of all ordered pairs in A are nonzero, $b \neq 0, d \neq 0$, and $f \neq 0$. Dividing (1) by b gives

$$c = \frac{ad}{b},$$

and substituting this into (2) gives

$$\left(\frac{ad}{b}\right) \cdot f = de.$$

Since $d \neq 0$, both sides of this equation may be divided by d. We can also multiply both sides by b to obtain

$$af = be.$$

It follows by definition of R that $(a, b) \, R \, (e, f)$ *[as was to be shown]*.

b. There is one equivalence class for each distinct rational number. Each equivalence class consists of all ordered pairs (a, b) that if written as fractions a/b would equal each other. The reason for this is that the condition for two rational numbers to be equal is the same as the condition for two ordered pairs to be related. For instance, the class of $(1, 2)$ is

$$[(1, 2)] = \{(1, 2), (-1, -2), (2, 4), (-2, -4), (3, 6), (-3, -6), \ldots\}$$

since $\dfrac{1}{2} = \dfrac{-1}{-2} = \dfrac{2}{4} = \dfrac{-2}{-4} = \dfrac{3}{6} = \dfrac{-3}{-6}$ and so forth.   ∎

It is possible to expand the result of Example 10.3.10 to define operations of addition and multiplication on the equivalence classes of R that satisfy all the same properties as the addition and multiplication of rational numbers. (See exercise 34.) It follows that the rational numbers can be defined as equivalence classes of ordered pairs of integers. Similarly (see exercise 35), it can be shown that all integers, negative and zero included, can be defined as equivalence classes of ordered pairs of positive integers. But in the late nineteenth century, F. L. G. Frege and Giuseppe Peano showed that the positive integers can be defined entirely in terms of sets. And just a little earlier, Richard Dedekind (1848–1916) showed that all real numbers can be defined as sets of rational numbers. All together, these results show that the real numbers can be defined using logic and set theory alone.

EXERCISE SET 10.3

1. Each of the following partitions of $\{0, 1, 2, 3, 4\}$ induces a relation R on $\{0, 1, 2, 3, 4\}$. In each case, find the ordered pairs in R.
   a. $\{0, 2\}, \{1\}, \{3, 4\}$
   b. $\{0\}, \{1, 3, 4\}, \{2\}$
   c. $\{0\}, \{1, 2, 3, 4\}$

In 2–9 the relation R is an equivalence relation on the set A. Find the distinct equivalence classes of R.

2. $A = \{0, 1, 2, 3, 4\}$,
   $R = \{(0, 0), (0, 4), (1, 1), (1, 3), (2, 2), (3, 1), (3, 3), (4, 0), (4, 4)\}$
   $(b \; d)$

3. $A = \{a, b, c, d\}$,
   $R = \{(a, a), (b, b), (b, d), (c, c), (d, b), (d, d)\}$

4. $A = \{1, 2, 3, 4, \ldots, 20\}$, R is defined on A as follows:
   for all $x, y \in A$, $x \, R \, y \iff 4 \mid (x - y)$.

5. $A = \{-4, -3, -2, -1, 0, 1, 2, 3, 4, 5\}$, R is defined on A as follows:
   for all $x, y \in A$, $x \, R \, y \iff 3 \mid (x - y)$.

6. $A = \{(1, 3), (2, 4), (-4, -8), (3, 9), (1, 5), (3, 6)\}$, R is defined on A as follows: for all $(a, b), (c, d) \in A$,
   $(a, b) \, R \, (c, d) \iff ad = bc$.

7. $X = \{a, b, c\}, A = \mathcal{P}(X)$, R is defined on A as follows:
   for all $U, V \in \mathcal{P}(X)$, $U \, R \, V \iff n(U) = n(V)$.

(That is, the number of elements in U equals the number of elements in V.)

8. $\Sigma = \{a, b\}, A = \Sigma^4$, R is defined on A as follows: for all $s, t \in \Sigma^4$,
   $s \, R \, t \iff$ the first two characters of s equal the first two characters of t.

9. $A = \{-5, -4, -3, -2, -1, 0, 1, 2, 3, 4, 5\}$, R is defined on A as follows:
   for all $m, n \in \mathbf{Z}$, $m \, R \, n \iff 3 \mid (m^2 - n^2)$.

10. Determine which of the following congruence relations are true and which are false.
    a. $17 \equiv 2 \pmod{5}$ b. $4 \equiv -5 \pmod{7}$
    c. $-2 \equiv -8 \pmod{3}$ d. $-6 \equiv 22 \pmod{2}$

11. a. Let R be the relation of congruence modulo 3. Which of the following equivalence classes are equal?
    $$[7], [-4], [-6], [17], [4], [27], [19]$$
    b. Let R be the relation of congruence modulo 7. Which of the following equivalence classes are equal?
    $$[35], [3], [-7], [12], [0], [-2], [17]$$

12. a. Prove that for all integers m and n, $m \equiv n \pmod{3}$ if, and only if, $m \bmod 3 = n \bmod 3$.
    b. Prove that for all integers m and n and any positive integer d, $m \equiv n \pmod{d}$ if, and only if, $m \bmod d = n \bmod d$.

13. **a.** Give an example of two sets that are distinct but not disjoint.
   b. Find sets A_1 and A_2 and elements x, y, and z such that x and y are in A_1 and y and z are in A_2 but x and z are not both in either of the sets A_1 or A_2.

Each of the relations given in 14–26 is an equivalence relation. Describe the distinct equivalence classes of each relation.

14. A is the set of all students at your college.
   a. R is the relation defined on A as follows: for all $x, y \in A$,

   $x R y \iff x$ has the same major (or double major) as y.

   (Assume "undeclared" is a major.)
   b. S is the relation defined on A as follows: for all $x, y \in A$,

   $x S y \iff x$ is the same age as y.

15. E is the relation defined on $\mathbf{Z}$ as follows:

   for all $m, n \in \mathbf{Z}$, $m E n \iff 2 \mid (m - n)$.

16. F is the relation defined on $\mathbf{Z}$ as follows:

   for all $m, n \in \mathbf{Z}$, $m F n \iff 4 \mid (m - n)$.

17. Let A be the set of all statement forms in three variables p, q, and r. $\mathcal{R}$ is the relation defined on A as follows: for all P and Q in A,

   $P \mathcal{R} Q \iff P$ and Q have the same truth table.

18. Let P be a set of parts shipped to a company from various suppliers. S is the relation defined on P as follows: for all $x, y \in P$,

   $x S y \iff x$ has the same part number and is shipped from the same supplier as y.

19. A is the "absolute value" relation defined on $\mathbf{R}$ as follows:

   for all $x, y \in \mathbf{R}$, $x A y \iff |x| = |y|$.

20. I is the relation defined on $\mathbf{R}$ as follows:

   for all $x, y \in \mathbf{R}$, $x I y \iff x - y$ is an integer.

21. D is the relation defined on $\mathbf{Z}$ as follows:

   for all $m, n \in \mathbf{Z}$, $m D n \iff 3 \mid (m^2 - n^2)$.

22. Define P on the set $\mathbf{R} \times \mathbf{R}$ of ordered pairs of real numbers as follows: for all $(w, x), (y, z) \in \mathbf{R} \times \mathbf{R}$,

   $(w, x) P (y, z) \iff w = y$.

23. Let A be the set of identifiers in a computer program. It is common for identifiers to be used for only a short part of the execution time of a program and not to be used again to execute other parts of the program. In such cases, arranging for identifiers to share memory loca-

tions makes efficient use of a computer's memory capacity. Define R on A as follows: for all identifiers x and y,

   $x R y \iff$ the values of x and y are stored in the same memory location during execution of the program.

24. Let A be the set of all straight lines in the Cartesian plane. Define a relation $\|$ on A as follows:

   for all l_1 and l_2 in A, $l_1 \| l_2 \iff l_1$ is parallel to l_2.

25. Let P be the set of all points in the Cartesian plane except the origin. R is the relation defined on P as follows: for all p_1 and p_2 in P,

   $p_1 R p_2 \iff p_1$ and p_2 lie on the same half-line emanating from the origin.

26. Let A be the set of points in the rectangle with x and y coordinates between 0 and 1. That is,

   $$A = \{(x, y) \in \mathbf{R} \times \mathbf{R} \mid 0 \le x \le 1 \quad \text{and} \quad 0 \le y \le 1\}.$$

   Define a relation R on A as follows: for all (x_1, y_1) and (x_2, y_2) in A,

   $$\begin{aligned}
   (x_1, y_1) \, R \, (x_2, y_2) \iff & \; (x_1, y_1) = (x_2, y_2); \quad \text{or} \\
   & \; x_1 = 0 \quad \text{and} \quad x_2 = 1 \quad \text{and} \quad y_1 = y_2; \quad \text{or} \\
   & \; x_1 = 1 \quad \text{and} \quad x_2 = 0 \quad \text{and} \quad y_1 = y_2; \quad \text{or} \\
   & \; y_1 = 0 \quad \text{and} \quad y_2 = 1 \quad \text{and} \quad x_1 = x_2; \quad \text{or} \\
   & \; y_1 = 1 \quad \text{and} \quad y_2 = 0 \quad \text{and} \quad x_1 = x_2; \quad \text{or} \\
   & \; x_1 = x_2 = 0 \quad \text{and} \quad y_1 = y_2 = 1; \quad \text{or} \\
   & \; x_1 = x_2 = 1 \quad \text{and} \quad y_1 = y_2 = 0.
   \end{aligned}$$

   In other words, all points along the top edge of the rectangle are related to the points along the bottom edge directly beneath them, and all points directly opposite each other along the left and right edges are related to each other. The points in the interior of the rectangle are not related to anything other than themselves.

Let R be an equivalence relation on a set A. Prove each of the statements in 27–31 directly from the definitions of equivalence relation and equivalence class without using the results of Lemma 10.3.2, Lemma 10.3.3, or Theorem 10.3.4.

27. For all a in A, $a \in [a]$.

28. For all a and b in A, if $b \in [a]$ then $a R b$.

29. For all a, b, and c in A, if $b R c$ and $c \in [a]$ then $b \in [a]$.

30. For all a and b in A, if $[a] = [b]$ then $a R b$.

31. For all a and b in A, if $a \in [b]$ then $[a] = [b]$.

32. Find an additional representative circuit for the input/output table of Example 10.3.9.

33. Let R be the binary relation defined in Example 10.3.10.
    a. Prove that R is reflexive.
    b. Prove that R is symmetric.
    c. List four distinct elements in $[(1, 3)]$.
    d. List four distinct elements in $[(2, 5)]$.

◆ 34. In Example 10.3.10 define operations of addition $(+)$ and multiplication $(\cdot)$ as follows: for all (a, b), $(c, d) \in A$,
$$[(a, b)] + [(c, d)] = [(ad + bc, bd)]$$
$$[(a, b)] \cdot [(c, d)] = [(ac, bd)].$$

    a. Prove that this addition is well defined. That is, show that if $[(a, b)] = [a', b')]$ and $[(c, d)] = [(c', d')]$, then $[(ad + bc, bd)] = [(a'd' + b'c', b'd')]$.
    b. Prove that this multiplication is well defined. That is, show that if $[(a, b)] = [a', b')]$ and $[(c, d)] = [(c', d')]$, then $[(ac, bd)] = [(a'c', b'd')]$.
    c. Show that $[(0, 1)]$ is an identity element for addition. That is, show that for any $(a, b) \in A$,
$$[(a, b)] + [(0, 1)] = [(a, b)].$$
    d. Find an identity element for multiplication.
    e. For any $(a, b) \in A$, show that $[(-a, b)]$ is an inverse for $[(a, b)]$ for addition. That is, show that $[(-a, b)] + [(a, b)] = [(0, 1)]$.
    f. Given any $(a, b) \in A$ with $a \neq 0$, find an inverse for $[(a, b)]$ for multiplication.

35. Let $A = \mathbf{Z}^+ \times \mathbf{Z}^+$. Define a binary relation R on A as follows: for all (a, b) and (c, d) in A,
$$(a, b) \, R \, (c, d) \quad \Leftrightarrow \quad a + d = c + b.$$

    a. Prove that R is reflexive.
    b. Prove that R is symmetric.
    H c. Prove that R is transitive.
    d. List five elements in $[(1, 1)]$.
    e. List five elements in $[(3, 1)]$.
    f. List five elements in $[(1, 2)]$.
    g. Describe the distinct equivalence classes of R.

36. The following argument claims to prove that the requirement that an equivalence relation be reflexive is redundant. In other words, it claims to show that if a relation is symmetric and transitive, then it is reflexive. Find the mistake in the argument.
"Proof: Let R be a binary relation on a set A and suppose R is symmetric and transitive. For any two elements x and y in A, if $x \, R \, y$ then $y \, R \, x$ since R is symmetric. But then it follows by transitivity that $x \, R \, x$. Hence R is reflexive."

37. Let R be a binary relation on a set A and suppose R is symmetric and transitive. Prove the following: If for every x in A there is a y in A such that $x \, R \, y$, then R is an equivalence relation.

38. Refer to the quote at the beginning of this section to answer the following questions.
    a. What is the name of the Knight's song called?
    b. What is the name of the Knight's song?
    c. What is the Knight's song called?
    d. What *is* the Knight's song?
    e. What is your (full, legal) name?
    f. What are you called?
    g. What *are* you? (Do not answer this on paper; just think about it.)

10.4 APPLICATION: SIMPLIFYING FINITE-STATE AUTOMATA

Our life is frittered away by detail. . . . Simplify, simplify.
(Henry David Thoreau, *Walden,* 1854)

Recall that any string input to a finite-state automaton either sends the automaton to an accepting state or not, and that the set of all strings accepted by an automaton is the language accepted by the automaton. It often happens that in creating an automaton to do a certain job (as in compiler construction, for example), the automaton that emerges "naturally" from the development process is unnecessarily complicated; that is, there may be an automaton with fewer states that accepts exactly the same language. It is desirable to find such an automaton because the memory space required to store an automaton with n states is approximately proportional to n^2. Thus approximately 10,000 memory spaces are required to store an automaton with 100 states, whereas only about 100 memory spaces are needed to store an automaton with 10 states. In addition, the fewer the number of states that an automaton has, the easier it is to write a computer alogrithm based on it; and to see that two automata both accept the same language, it is easiest to simplify each to a minimal number of states and compare the simplified automata. In this section we show how to take a given automaton and simplify it in the sense of finding an automaton with fewer states that accepts the same language.

EXAMPLE 10.4.1 An Overview

Consider the finite-state automata A and A' in Figure 10.4.1. A moment's thought should convince you that A' accepts all those strings and only those strings that contain an even number of 1's. But A, although it appears more complicated, accepts exactly those strings also. Thus the two automata are "equivalent" in the sense that they accept the same language, even though A' has fewer states than A.

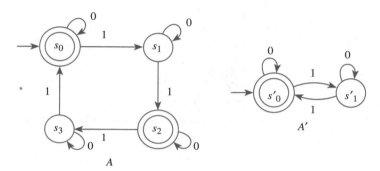

FIGURE 10.4.1

Roughly speaking, the reason for the equivalence of these automata is that some of the states of A can be combined without affecting the acceptance or nonacceptance of any input string. It turns out that s_2 can be combined with state s_0 and s_3 can be combined with state s_1. (How to figure out which states can be combined is explained later in this section.) The automaton with the two combined states $\{s_0, s_2\}$ and $\{s_1, s_3\}$ is called the *quotient automaton* of A and is denoted $\overline{A}$. Its transition diagram is obtained by combining the circles for s_0 and s_2 and for s_1 and s_3 and by replacing any arrow from a state s to a state t by an arrow from the combined state containing s to the combined state containing t. For instance, since there is an arrow labeled 1 from s_1 to s_2 in A, there is an arrow labeled 1 from $\{s_1, s_3\}$ to $\{s_0, s_2\}$ in $\overline{A}$. The complete transition diagram for $\overline{A}$ is shown in Figure 10.4.2. As you can see, except for labeling the names of the states, it is identical to the diagram for A'.

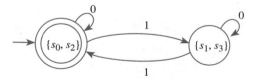

FIGURE 10.4.2 ∎

In general, simplification of a finite-state automaton involves identifying "equivalent states" that can be combined without affecting the action of the automaton on input strings. Mathematically speaking, this means defining an equivalence relation on the set of states of the automaton and forming a new automaton whose states are the equivalence classes of the relation. The rest of this section is devoted to developing an algorithm to carry out this process in a practical way.

*-EQUIVALENCE OF STATES

Two states of a finite-state automaton are said to be *-*equivalent* (read "star equivalent") if any string accepted by the automaton when it starts from one of the states is accepted by the automaton when it starts from the other state. Recall that the value of the eventual-state function, N^*, for a state s and input string w is the state to which the automaton goes if the characters of w are input in sequence when the automaton is in state s.

DEFINITION

Let A be a finite-state automaton with next-state function N, and eventual-state function N^*. Define a binary relation on the set of states of A as follows: Given any states s and t of A, we say that **s and t are *-equivalent** and write **$s\ R_* \ t$** if, and only if, for all input strings w,

either both $N^*(s, w)$ and $N^*(t, w)$ are accepting states or both are nonaccepting states.

In other words, states s and t are *-equivalent if, and only if, for all input strings w,

$$N^*(s, w) \text{ is an accepting state} \quad \Leftrightarrow \quad N^*(t, w) \text{ is an accepting state.}$$

Or, more simply, for all input strings w,

$$\begin{bmatrix} A \text{ goes to an accepting state if } w \\ \text{is input when } A \text{ is in state } s \end{bmatrix} \Leftrightarrow \begin{bmatrix} A \text{ goes to an accepting state if } w \\ \text{is input when } A \text{ is in state } t \end{bmatrix}.$$

It follows immediately by substitution into the definition that

---

R_* is an equivalence relation on S, the set of states of A. **10.4.1**

You are asked to prove this formally in the exercises at the end of this section.

k-EQUIVALENCE OF STATES

From a procedural point of view, it is difficult to determine the *-equivalence of two states using the definition directly. According to the definition, the action of the automaton starting in states s and t on *all* input strings must be known in order to tell whether s and t are equivalent. But since most languages have infinitely many input strings, you cannot check individually the effect of every string that is input to an automaton. As a practical matter, you can tell whether or not two states s and t are *-equivalent by using an iterative procedure based on a simpler kind of equivalence of states called *k-equivalence.* Two states are *k-equivalent* if any string *of length less than or equal to k* that is accepted by the automaton when it starts from one of the states is accepted by the automaton when it starts from the other state.

> DEFINITION
>
> Let A be a finite-state automaton with next-state function N and eventual-state function N^*. Define a binary relation on the set of states of A as follows: Given any states s and t of A and an integer $k \geq 0$, we say that s is **k-equivalent** to t and write $s\ R_k\ t$ if, and only if, for all input strings w *of length less than or equal to k,* either $N^*(s, w)$ and $N^*(t, w)$ are both accepting states or are both nonaccepting states.

Certain useful facts follow quickly from the definition of k-equivalence:

> For each integer $k \geq 0$, k-equivalence is an equivalence relation.   10.4.2
>
> For each integer $k \geq 0$, the k-equivalence classes partition the set of all states of the automaton into a union of mutually disjoint subsets.   10.4.3
>
> For each integer $k \geq 1$, if two states are k-equivalent, then they are also $(k - 1)$-equivalent.   10.4.4
>
> For each integer $k \geq 1$, each k-equivalence class is a subset of a $(k - 1)$-equivalence class.   10.4.5
>
> Any two states that are k-equivalent for all integers $k \geq 0$ are $*$-equivalent.   10.4.6

Proofs of these facts are left to the exercises.

The following theorem gives a recursive description of k-equivalence of states. It says, first, that any two states are 0-equivalent if, and only if, either both are accepting states or both are nonaccepting states and, second, that any two states are k-equivalent (for $k \geq 1$) if, and only if, they are $(k - 1)$-equivalent and for any input symbols their next-states are also $(k - 1)$-equivalent.

> THEOREM 10.4.1
>
> Let A be a finite-state automaton with next-state function N. Given any states s and t in A,
>
> **1.** s is 0-equivalent to t $\Leftrightarrow$ $\begin{bmatrix} \text{either } s \text{ and } t \text{ are both accepting states} \\ \text{or are both nonaccepting states} \end{bmatrix}$
>
> **2.** for every integer $k \geq 1$,
> $\quad\quad s$ is k-equivalent to t $\Leftrightarrow$ $\begin{bmatrix} s \text{ and } t \text{ are } (k - 1)\text{-equivalent, and} \\ \text{for any input symbol } m, N(s, m) \text{ and} \\ N(t, m) \text{ are also } (k - 1)\text{-equivalent.} \end{bmatrix}$

The truth of Theorem 10.4.1 follows from the fact that inputting a string w of length k has the same effect as inputting the first symbol of w and then the remaining $k - 1$ symbols of w. The detailed proof of this theorem is somewhat above the level of this book.

Theorem 10.4.1 implies that if you know which states are $(k-1)$-equivalent (where k is a positive integer) and if you know the action of the next-state function, then you can figure out which states are k-equivalent. Specifically, if s and t are $(k-1)$-equivalent states whose next-states are $(k-1)$-equivalent for any input symbol m, then s and t are k-equivalent. Thus the k-equivalence classes are obtained by subdividing the $(k-1)$-equivalence classes according to the action of the next-state function on the members of the classes. An example should make this procedure clear.

EXAMPLE 10.4.2   Finding k-Equivalence Classes

Find the 0-equivalence classes, the 1-equivalence classes, and the 2-equivalence classes for the states of the automaton shown below.

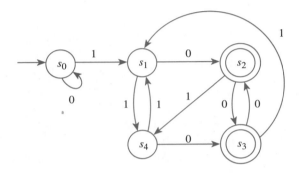

Solution
1. *0-equivalence classes:*   By Theorem 10.4.1, two states are 0-equivalent if, and only if, both are accepting states or both are nonaccepting states. Thus there are two sets of 0-equivalent states:

    $\{s_0, s_1, s_4\}$ (the nonaccepting states)   and $\{s_2, s_3\}$ (the accepting states),

    and so

    the 0-equivalence classes are $\{s_0, s_1, s_4\}$ and $\{s_2, s_3\}$.

2. *1-equivalence classes:*   By Theorem 10.4.1, two states are 1-equivalent if, and only if, they are 0-equivalent and after input of any input symbol, their next-states are 0-equivalent. Thus s_1 is not 1-equivalent to s_0 because when a 0 is input to the automaton in state s_1 it goes to state s_2, whereas when a 0 is input to the automaton in state s_0 it goes to state s_0, and s_2 and s_0 are not 0-equivalent. On the other hand, s_1 *is* 1-equivalent to s_4 because when a 0 is input to the automaton in states s_1 or s_4 the next-states are s_2 and s_3, which are 0-equivalent; and when a 1 is input to the automaton in states s_1 or s_4 the next-states are s_4 and s_1, which are 0-equivalent. By a similar argument, s_2 is 1-equivalent to s_3. Since 1-equivalent states must also be 0-equivalent *[by property (10.4.4)]*, no other pairs of states can be 1-equivalent. Hence

    the 1-equivalence classes are $\{s_0\}$, $\{s_1, s_4\}$, and $\{s_2, s_3\}$.

3. *2-equivalence classes:*   By Theorem 10.4.1, two states are 2-equivalent if, and only if, they are 1-equivalent and after input of any input symbol, their next-states are 1-equivalent. Now s_1 is 2-equivalent to s_4 because they are 1-equivalent and when a 1 is input to the automaton in states s_1 or s_4 the next-states are s_4 and s_1, which are 1-equivalent; and when a 0 is input to the automaton in states s_1 or s_4 the next-states are s_2 and s_3, which are 1-equivalent. Similarly, s_2 is 2-equivalent to s_3.

Since 2-equivalent states must also be 1-equivalent *[by property* (10.4.4)*]*, no other pairs of states can be 2-equivalent. Hence

the 2-equivalence classes are $\{s_0\}$, $\{s_1, s_4\}$, and $\{s_2, s_3\}$.

Note that the set of 2-equivalence classes equals the set of 1-equivalence classes.

∎

FINDING THE ∗-EQUIVALENCE CLASSES

Example 10.4.2 illustrates the relative ease with which the sets of k-equivalence classes of states can be found. But to simplify a finite-state automaton, you need to find the set of ∗-equivalence classes of states. The next theorem says that for some integer K, the set of ∗-equivalence classes equals the set of K-equivalence classes.

THEOREM 10.4.2

If A is a finite-state automaton, then for some integer $K \geq 0$, the set of K-equivalence classes of states of A equals the set of $(K + 1)$-equivalence classes of states of A, and for all such K these are both equal to the set of ∗-equivalence classes of states of A.

The detailed proof of Theorem 10.4.2 is somewhat technical, but the idea of the proof is not hard to understand. Theorem 10.4.2 follows from the fact that for each positive integer k, the k-equivalence classes are obtained by subdividing the $(k - 1)$-equivalence classes according to a certain rule that is the same for each k. Since the number of states of the automaton is finite, this subdivision process cannot continue forever and so for some integer $K \geq 0$, the set of K-equivalence classes equals the set of $(K + 1)$-equivalence classes. Moreover, the set of m-equivalence classes equals the set of K-equivalence classes for every integer $m \geq K$. But this implies that the set of ∗-equivalence classes equals the set of K-equivalence classes.

EXAMPLE 10.4.3   **Finding ∗-Equivalence Classes of R**

Let A be the finite-state automaton defined in Example 10.4.2. Find the ∗-equivalence classes of states of A.

Solution   According to Example 10.4.2, the set of 1-equivalence classes for A equals the set of 2-equivalence classes. By Theorem 10.4.2, then, the set of ∗-equivalence classes also equals the set of 1-equivalence classes. Hence

the ∗-equivalence classes are $\{s_0\}$, $\{s_1, s_4\}$, and $\{s_2, s_3\}$.

In the notation of Section 10.3, the equivalence classes are denoted

$$[s_0] = \{s_0\} \quad [s_1] = \{s_1, s_4\} = [s_4] \quad [s_2] = \{s_2, s_3\} = [s_3].$$

∎

THE QUOTIENT AUTOMATON

We next define the *quotient automaton* $\overline{A}$ of an automaton A. However, in order for all parts of the definition to make sense, we must point out two facts.

> No *-equivalence class of states of A can contain both accepting and nonaccepting states.   **10.4.7**

The reason this is true is that the 0-equivalence classes divide the set of states of A into accepting and nonaccepting states, and the *-equivalence classes are subsets of 0-equivalence classes.

> If two states are *-equivalent, then their next-states are also *-equivalent for any input symbol m.   **10.4.8**

This is true for the following reason. Suppose states s and t are *-equivalent. Then any input string that sends A to an accepting state when A is in state s sends A to an accepting state when A is in state t. Now suppose m is any input symbol and consider the next-states $N(s, m)$ and $N(t, m)$. Inputting a string of length k to A when A is in state $N(s, m)$ or $N(t, m)$ produces the same effect as inputting a certain string of length $k + 1$ to A when A is in state s or t (namely the concatenation of m with the string of length k). Hence any string that sends A to an accepting state when A is in state $N(s, m)$ also sends A to an accepting state when A is in state $N(t, m)$. It follows that $N(s, m)$ and $N(t, m)$ are *-equivalent. Complete proofs of properties (10.4.7) and (10.4.8) are left to the exercises.

Now we can define the quotient automaton $\overline{A}$ of A. It is the finite-state automaton whose states are the *-equivalence classes of states of A, whose initial state is the *-equivalence class containing the initial state of A, whose accepting states are of the form $[s]$ where s is an accepting state of A, whose input symbols are the same as the input symbols of A and whose next-state function is derived from the next-state function for A in the following way: To find the next-state of $\overline{A}$ for a state s and an input symbol m, pick any state t in $[s]$ and look to see what next-state A goes to if m is input when A is in state t; the equivalence class of this state is the next-state of $\overline{A}$.

DEFINITION

Let A be a finite-state automaton with set of states S, set of input symbols I, and next-state function N. The **quotient automaton $\overline{A}$** is defined as follows:

1. the set of states, $\overline{S}$, of $\overline{A}$ is the set of *-equivalence classes of states of A;
2. the set of input symbols, $\overline{I}$, of $\overline{A}$ equals I;
3. the initial state of $\overline{A}$ is $[s_0]$, where s_0 is the initial state of A;
4. the accepting states of $\overline{A}$ are the states of the form $[s]$, where s is an accepting state of A;
5. the next-state function $\overline{N} : \overline{S} \times I \rightarrow \overline{S}$ is defined as follows:

   for all states $[s]$ in $\overline{S}$ and input symbols m in I, $\overline{N}([s], m) = [N(s, m)]$.

   (That is, if m is input to $\overline{A}$ when $\overline{A}$ is in state $[s]$, then $\overline{A}$ goes to the state that is the *-equivalence class of $N(s, m)$.)

Note that since the states of $\overline{A}$ are *sets* of states of A, $\overline{A}$ generally has fewer states than A. (A and $\overline{A}$ have the same number of states only in the case where each $*$-equivalence class of states contains just one element.) Also, by property (10.4.7), each accepting state of $\overline{A}$ consists entirely of accepting states of A. Furthermore, property (10.4.8) guarantees that the next-state function $\overline{N}$ is well defined.

By construction, a quotient automaton $\overline{A}$ accepts exactly the same strings as A. We state this formally as Theorem 10.4.3. The details of the proof are somewhat above the level of this book.

THEOREM 10.4.3

If A is a finite-state automaton, then the quotient automaton $\overline{A}$ accepts exactly the same language as A. In other words, if $L(A)$ denotes the language accepted by A and $L(\overline{A})$ the language accepted by $\overline{A}$, then

$$L(A) = L(\overline{A}).$$

CONSTRUCTING THE QUOTIENT AUTOMATON

Let A be a finite-state automaton with set of states S, next-state function N, relation R_* of $*$-equivalence of states, and relations R_k of k-equivalence of states. It follows from Theorems 10.4.2 and 10.4.3 and from the definition of quotient automaton that to find the quotient automaton $\overline{A}$ of A, you can proceed as follows:

1. Find the set of 0-equivalence classes of S.
2. For each integer $k \geq 1$, subdivide the $(k-1)$-equivalence classes of S (as described earlier) to find the k-equivalence classes of S. Stop subdividing when you observe that for some integer K the set of $(K+1)$-equivalence classes equals the set of K-equivalence classes. At this point, conclude that the set of K-equivalence classes equals the set of $*$-equivalence classes.
3. Construct the quotient automaton $\overline{A}$ whose states are the $*$-equivalence classes of states of A and whose next-state function $\overline{N}$ is given by

$$\overline{N}([s], m) = [N(s, m)] \quad \text{for any state of } \overline{A} \text{ and any input symbol } m,$$

where s is any state in $[s]$. *[That is, to see where $\overline{A}$ goes if m is input to $\overline{A}$ when it is in state s, look to see where A goes if m is input to A when it is in state s. The $*$-equivalence class of that state is the answer.]*

EXAMPLE 10.4.4   Constructing a Quotient Automaton

Consider the automaton A of Examples 10.4.2 and 10.4.3 shown on page 580 in Figure 10.4.3. Find the quotient automaton of A.

Solution   According to Example 10.4.3, the $*$-equivalence classes of the states of A are

$$\{s_0\}, \{s_1, s_4\}, \text{ and } \{s_2, s_3\}.$$

Hence the states of the quotient automaton $\overline{A}$ are

$$[s_0] = \{s_0\}, \quad [s_1] = \{s_1, s_4\} = [s_4], \quad [s_2] = \{s_2, s_3\} = [s_3].$$

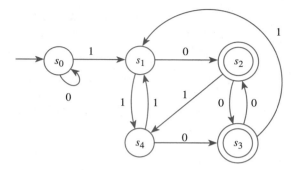

FIGURE 10.4.3

The accepting states of A are s_2 and s_3, so the accepting state of $\overline{A}$ is $[s_2] = [s_3]$. The next-state function $\overline{N}$ of $\overline{A}$ is defined as follows: for all states $[s]$ and input symbols m of $\overline{A}$,

$$\overline{N}([s], m) = [N(s, m)] = \text{the *-equivalence class of } N(s, m).$$

Thus,

$$\overline{N}([s_0], 0) = [N(s_0, 0)] = \text{the *-equivalence class of } N(s_0, 0).$$

But $N(s_0, 0) = s_0$; so

$$\overline{N}([s_0], 0) = \text{the *-equivalence class of } s_0 = [s_0].$$

Similarly,

$$\overline{N}([s_0], 1) = [N(s_0, 1)] = [s_1]$$
$$\overline{N}([s_1], 0) = [N(s_1, 0)] = [s_2]$$
$$\overline{N}([s_1], 1) = [N(s_1, 1)] = [s_4] = [s_1]$$
$$\overline{N}([s_2], 0) = [N(s_2, 0)] = [s_3] = [s_2]$$
$$\overline{N}([s_2], 1) = [N(s_2, 1)] = [s_4] = [s_1].$$

The transition diagram for $\overline{A}$ is, therefore, as shown in Figure 10.4.4.

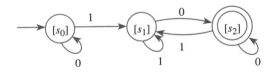

FIGURE 10.4.4

By Theorem 10.4.3 this automaton accepts the same language as the original automaton. ∎

EQUIVALENT AUTOMATA

Output devices may be attached to the states of finite-state automata to indicate whether they are accepting or nonaccepting states. For example, accepting states might produce an output of 1 and nonaccepting states an output of 0. Then a finite-state automaton can be thought of as an input/output device whose input consists of strings and whose output consists of 0's and 1's. Recall that a circuit can be thought of as a black box that transforms combinations of input signals into output signals. Two circuits that produce identical output signals for each combination of input signals are called *equivalent*. Similarly, a finite-state automaton can be regarded as a black box that processes input strings and produces output signals (indicating whether or not the strings are accepted). Two finite-state automata are called *equivalent* if they produce identical output signals for each input string. But this means that two finite-state automata are equivalent if, and only if, they accept the same language.

DEFINITION

Let A and A' be finite-state automata with the same set of input symbols I. Let $L(A)$ denote the language accepted by A and $L(A')$ the language accepted by A'. Then A is said to be **equivalent** to A' if, and only if, $L(A) = L(A')$.

EXAMPLE 10.4.5   **Showing That Two Automata Are Equivalent**

Show that A and A' in Figure 10.4.5 are equivalent.

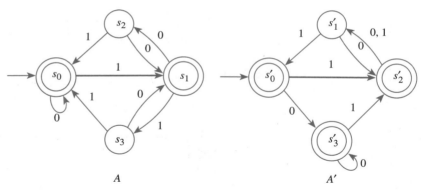

A A'

FIGURE 10.4.5

(The label 0, 1 on an arrow of a transition diagram means that for either input 0 or 1 the next-state of the automaton is the state to which the arrow points.)

Solution   *For the automaton A:*   The 0-equivalence classes are

$$\{s_0, s_1\} \quad \text{and} \quad \{s_2, s_3\} \qquad \text{since } s_0 \text{ and } s_1 \text{ are accepting states and } s_2 \text{ and } s_3 \text{ are nonaccepting states.}$$

The 1-equivalence classes are

$$\{s_0\}, \quad \{s_1\}, \quad \text{and} \quad \{s_2, s_3\} \qquad \text{since } s_0 \text{ and } s_1 \text{ are not 1-equivalent (because } N(s_0, 1) = s_1, \text{ whereas } N(s_1, 1) = s_3 \text{ and } s_1 \text{ is not 0-equivalent to } s_3) \text{ but } s_2 \text{ and } s_3 \text{ are 1-equivalent.}$$

The 2-equivalence classes are

$$\{s_0\}, \quad \{s_1\}, \quad \text{and} \quad \{s_2, s_3\} \qquad \text{since } s_2 \text{ and } s_3 \text{ are 1-equivalent.}$$

This discussion shows that the set of 1-equivalence classes equals the set of 2-equivalence classes, so by Theorem 10.4.2 this is equal to the set of *-equivalence classes. Hence the *-equivalence classes are

$$\{s_0\}, \quad \{s_1\}, \quad \text{and} \quad \{s_2, s_3\}.$$

For the automaton A′:   By similar reasoning as above, the 0-equivalence classes are

$$\{s'_0, s'_2, s'_3\} \quad \text{and} \quad \{s'_1\}.$$

The 1-equivalence classes are

$$\{s'_0, s'_3\}, \quad \{s'_2\}, \quad \text{and} \quad \{s'_1\}.$$

The 2-equivalence classes are the same as the 1-equivalence classes, which are therefore equal to the *-equivalence classes. Thus the *-equivalence classes are

$$\{s'_0, s'_3\}, \quad \{s'_2\}, \quad \text{and} \quad \{s'_1\}.$$

To calculate the next-state functions for $\overline{A}$ and $\overline{A'}$, you repeatedly use the fact that in the quotient automaton the next-state of $[s]$ and m is the class of the next-state of s and m. For instance,

$$\overline{N}([s_1], 1) = [N(s_1, 1)] = [s_3] = [s_2]$$

and

$$\overline{N'}([s'_0], 0) = [N(s'_0, 0)] = [s'_3] = [s'_0]$$

where N is the next-state function for A and N' is the next-state function for A'.

The complete transition diagrams for the quotient automata $\overline{A}$ and $\overline{A'}$ are shown in Figure 10.4.6.

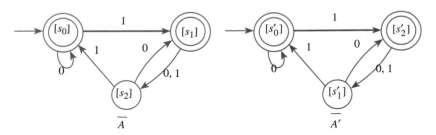

FIGURE 10.4.6

As you can see, except for the labeling of the names of the states, $\overline{A}$ and $\overline{A'}$ are identical and hence accept the same language. But by Theorem 10.4.3, each original automaton accepts the same language as its quotient automaton. Thus A and A' accept the same language, and so they are equivalent.   ■

In mathematics an object such as a finite-state automaton is called a *structure.* In general, when two mathematical structures are the same in all respects except for the labeling given to their elements, they are called **isomorphic,** which comes from the Greek words *isos,* meaning "same" or "equal," and *morphe,* meaning "form." It can be shown that two automata are equivalent if, and only if, their quotient automata are isomorphic, provided that "inaccessible states" have first been removed. (Inaccessible states are those that cannot be reached by inputting any string of symbols to the automaton when it is in its initial state.)

EXERCISE SET 10.4

1. Consider the finite-state automaton A given by the following transition diagram:

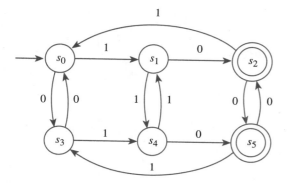

a. Find the 0- , 1- , and 2-equivalence classes of states of A.

b. Draw the transition diagram for $\bar{A}$, the quotient automaton of A.

2. Consider the finite-state automaton A given by the following transition diagram:

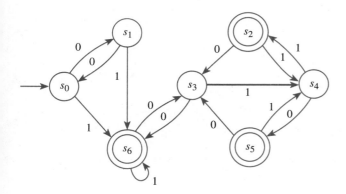

a. Find the 0- , 1- , and 2-equivalence classes of states of A.

b. Draw the transition diagram for $\bar{A}$, the quotient automaton of A.

3. Consider the finite-state automaton A discussed in Example 10.4.1:

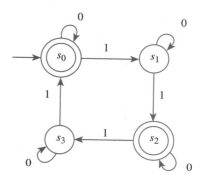

a. Find the 0- and 1-equivalence classes of states of A.

b. Draw the transition diagram of $\bar{A}$, the quotient automaton of A.

4. Consider the finite-state automaton given by the following transition diagram:

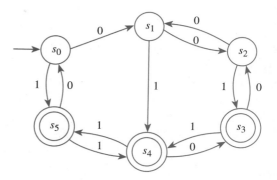

a. Find the 0- , 1- , 2- , and 3-equivalence classes of states of A.

b. Draw the transition diagram for $\bar{A}$, the quotient automaton of A.

5. Consider the finite-state automaton given by the following transition diagram:

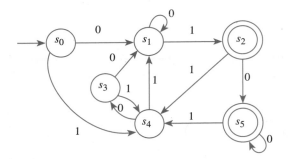

a. Find the 0-, 1-, 2-, and 3-equivalence classes of states of A.

b. Draw the transition diagram for $\bar{A}$, the quotient automaton of A.

6. Consider the finite-state automaton given by the following transition diagram:

Ha. Find the 0- , 1- , 2- , and 3-equivalence classes of states of A.

b. Draw the transition diagram for $\bar{A}$, the quotient automaton of A.

7. Are the two automata shown below equivalent?

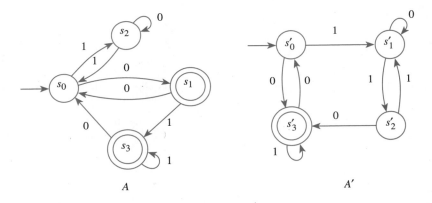

A A'

8. Are the two automata shown below equivalent?

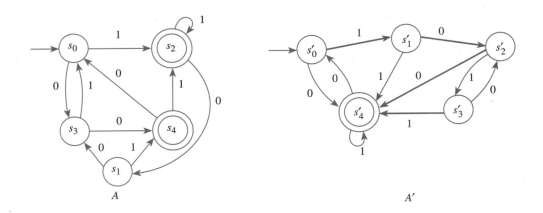

A A'

9. Are the two automata shown below equivalent?

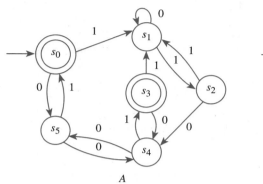

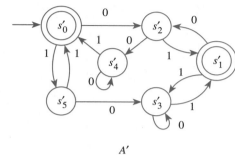

A   *A'*

10. Are the two automata shown below equivalent?

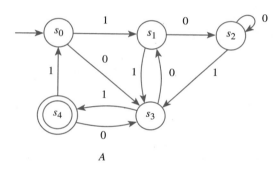

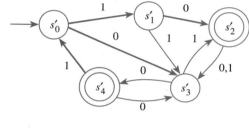

A   *A'*

H **11.** Prove property (10.4.1).

12. How should the proof of property (10.4.1) be modified to prove property (10.4.2)?

13. Prove property (10.4.3).

14. Prove property (10.4.4).

H **15.** Prove property (10.4.5).

16. Prove property (10.4.6).

H **17.** Prove that if two states of a finite-state automaton are k-equivalent for some integer k, then those states are m-equivalent for all nonnegative integers $m < k$.

18. Write a complete proof of property (10.4.7).

H **19.** Write a complete proof of property (10.4.8).

10.5 PARTIAL ORDER RELATIONS

There is no branch of mathematics, however abstract, which may not some day be applied to phenomena of the real world.
(Nicolai Ivanovitch Lobachevsky, 1793–1856)

In order to obtain a degree in computer science at a certain university, a student must take a specified set of required courses, some of which must be completed before others can be started. Given the prerequisite structure of the program, one might ask what is the least number of school terms needed to fulfill the degree requirements, or what is the maximum number of courses that can be taken in the same term, or whether there is a sequence in which a part-time student can take the courses one per term. Later in this section we will show how representing the prerequisite structure of the program as a partial order relation makes it relatively easy to answer such questions.

ANTISYMMETRY

In Section 10.2 we defined three properties of relations: reflexivity, symmetry, and transitivity. A fourth property of relations is called *antisymmetry*. In terms of the arrow diagram of a relation, saying that a relation is antisymmetric is the same as saying that whenever there is an arrow going from one element to another *distinct* element, there is *not* an arrow going back from the second to the first.

> **DEFINITION**
>
> Let R be a relation on a set A. R is **antisymmetric** if, and only if,
>
> for all a and b in A,   if $a\,R\,b$ and $b\,R\,a$ then $a = b$.

By taking the negation of the definition, you can see that a relation R is **not antisymmetric** if, and only if,

there are elements a and b in A such that $a\,R\,b$ and $b\,R\,a$ but $a \neq b$.

EXAMPLE 10.5.1   **Testing for Antisymmetry of Finite Relations**

Let R_1 and R_2 be the relations on $\{0, 1, 2\}$ defined as follows. Draw the directed graphs for R_1 and R_2 and indicate which relations are antisymmetric.

a. $R_1 = \{(0, 2), (1, 2), (2, 0)\}$
b. $R_2 = \{(0, 0), (0, 1), (0, 2), (1, 1), (1, 2)\}$

Solution   a. R_1 is not antisymmetric.

Since $0\,R_1\,2$ and $2\,R_1\,0$ but $0 \neq 2$, R_1 is not antisymmetric.

b. R_2 is antisymmetric.

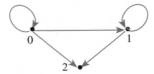

In order for R_2 not to be antisymmetric, there would have to exist a pair of distinct elements of A such that each is related to the other by R_2. But you can see by inspection that no such pair exists. ∎

EXAMPLE 10.5.2   **Testing for Antisymmetry of "Divides" Relations**

Let R_1 be the "divides" relation on the set of all positive integers and let R_2 be the "divides" relation on the set of all integers.

$$\text{for all } a, b \in Z^+, \quad a\,R_1\,b \;\Leftrightarrow\; a \mid b.$$
$$\text{for all } a, b \in Z, \quad a\,R_2\,b \;\Leftrightarrow\; a \mid b.$$

a. Is R_1 antisymmetric? Prove or give a counterexample.
b. Is R_2 antisymmetric? Prove or give a counterexample.

Solution   a. R_1 is antisymmetric.

Proof:

Suppose a and b are positive integers such that $a\, R_1\, b$ and $b\, R_1\, a$. *[We must show that $a = b$.]* By definition of R_1, $a \mid b$ and $b \mid a$. So by definition of divides there are integers k_1 and k_2 with $b = k_1 \cdot a$ and $a = k_2 \cdot b$. It follows that

$$b = k_1 \cdot a = k_1 \cdot (k_2 \cdot b) = (k_1 \cdot k_2) \cdot b.$$

Dividing both sides by b gives

$$k_1 \cdot k_2 = 1.$$

Now since a and b are both positive integers, k_1 and k_2 are both positive integers also. But the only product of two positive integers that equals 1 is $1 \cdot 1$. Thus

$$k_1 = k_2 = 1$$

and so

$$a = k_2 \cdot b = 1 \cdot b = b.$$

[This is what was to be shown.]

b. R_2 is not antisymmetric.

Counterexample:

Let $a = 2$ and $b = -2$. Then $a \mid b$ (since $-2 = (-1) \cdot 2$) and $b \mid a$ (since $2 = (-1) \cdot (-2)$). Hence $a\, R_2\, b$ and $b\, R_2\, a$ but $a \neq b$. ∎

One conclusion to be drawn from Example 10.5.2 is that the size of the set on which a relation is defined may affect whether or not it is antisymmetric.

PARTIAL ORDER RELATIONS

A relation that is reflexive, antisymmetric, and transitive is called a *partial order*.

DEFINITION

Let R be a binary relation defined on a set A. R is a **partial order relation** if, and only if, R is reflexive, antisymmetric, and transitive.

Two fundamental partial order relations are the "less than or equal to" relation on a set of real numbers and the "subset" relation on a set of sets. These can be thought of as models, or paradigms, for general partial order relations.

EXAMPLE 10.5.3   The "Subset" Relation

Let $\mathscr{A}$ be any collection of sets and define the "subset" relation, $\subseteq$, on $\mathscr{A}$ as follows: for all $U, V \in \mathscr{A}$,

$$U \subseteq V \quad \Leftrightarrow \quad \text{for all } x, \text{ if } x \in U \text{ then } x \in V.$$

By an argument almost identical to that of exercise 26 of Section 10.2, $\subseteq$ is reflexive and transitive. Finish the proof that $\subseteq$ is a partial order relation by proving that $\subseteq$ is antisymmetric.

Solution   For $\subseteq$ to be antisymmetric means that for all sets U and V in $\mathscr{A}$, if $U \subseteq V$ and $V \subseteq U$ then $U = V$. But this is true by definition of equality of sets.   ■

EXAMPLE 10.5.4   A "Divides" Relation on a Set of Positive Integers

Let $|$ be the "divides" relation on a set A of positive integers. That is, for all $a, b \in A$,

$$a \mid b \iff b = k \cdot a \text{ for some integer } k.$$

Prove that $|$ is a partial order relation on A.

Solution $|$ *is reflexive:*   *[We must show that for all $a \in A$, $a \mid a$.]* Suppose $a \in A$. Then $a = 1 \cdot a$ so $a \mid a$ by definition of divisibility.

$|$ *is antisymmetric:*   *[We must show that for all $a, b \in A$, if $a \mid b$ and $b \mid a$ then $a = b$.]* The proof of this is virtually identical to that of Example 10.5.2(a).

$|$ *is transitive:*   To show transitivity means to show that for all $a, b, c \in A$, if $a \mid b$ and $b \mid c$ then $a \mid c$. But this was proved as Theorem 3.3.1.

Since $|$ is reflexive, antisymmetric, and transitive, $|$ is a partial order relation on A.   ■

EXAMPLE 10.5.5   The "Less Than or Equal to" Relation

Let S be a set of real numbers and define the "less than or equal to" relation, $\leq$, on S as follows: for all real numbers x and y in S,

$$x \leq y \iff x < y \text{ or } x = y.$$

Show that $\leq$ is a partial order relation.

Solution $\leq$ *is reflexive:*   For $\leq$ to be reflexive means that $x \leq x$ for all real numbers x in S. But $x \leq x$ means that $x < x$ or $x = x$, and $x = x$ is always true.

$\leq$ *is antisymmetric:*   For $\leq$ to be antisymmetric means that for all real numbers x and y in S, if $x \leq y$ and $y \leq x$ then $x = y$. This follows immediately from the definition of $\leq$ and the trichotomy property (see Appendix A, T16), which says that given any real numbers x and y, exactly one of the following holds: $x < y$ or $x = y$ or $x > y$.

$\leq$ *is transitive:*   For $\leq$ to be transitive means that for all real numbers $x, y,$ and z in S, if $x \leq y$ and $y \leq z$ then $x \leq z$. This follows from the definition of $\leq$ and the transitivity property of order (see Appendix A, T17), which says that given any real numbers $x, y,$ and z, if $x < y$ and $y < z$ then $x < z$.   ■

Notation: Because of the special paradigmatic role played by the $\leq$ relation in the study of partial order relations, the symbol $\preceq$ is often used to refer to a general partial order relation, and the notation $x \preceq y$ is read "x is less than or equal to y" or "y is greater than or equal to x."

LEXICOGRAPHIC ORDER

To figure out which of two words comes first in an English dictionary, you compare their letters one by one from left to right. If all letters have been the same to a certain point and one word runs out of letters, that word comes first in the dictionary. For example, *play* comes before *playhouse*. If all letters up to a certain point are the same and the next letters differ, then the word whose next letter is located earlier in the alphabet comes first in the dictionary. For instance, *playhouse* comes before *playmate*.

More generally, if Σ is any set with a partial order relation, then a *dictionary* or *lexicographic* order can be defined on Σ^*, the set of strings over Σ, as follows.

THEOREM 10.5.1

Let Σ be a finite set and suppose R is a partial order relation defined on Σ. Define a relation $\preceq$ on Σ^*, the set of all strings over Σ, as follows:

For any positive integers m and n and $a_1a_2 \cdots a_m$ and $b_1b_2 \cdots b_n$ in Σ^*:

1. If $m \leq n$ and $a_i = b_i$ for all $i = 1, 2, \ldots, m$, then

$$a_1a_2 \cdots a_m \preceq b_1b_2 \cdots b_n.$$

2. If for some integer k with $k \leq m$, $k \leq n$, and $k \geq 1$, $a_i = b_i$ for all $i = 1, 2, \ldots, k-1$, and $a_k \, R \, b_k$ but $a_k \neq b_k$, then

$$a_1a_2 \cdots a_m \preceq b_1b_2 \cdots b_n.$$

3. If ϵ is the null string and s is any string in Σ^*, then $\epsilon \preceq s$.

If no strings are related other than by these three conditions, then $\preceq$ is a partial order relation.

The proof of Theorem 10.5.1 is technical but straightforward. It is left to the exercises.

DEFINITION

The partial order relation of Theorem 10.5.1 is called the **lexicographic order for Σ^*** (corresponding to the partial order R on Σ).

EXAMPLE 10.5.6   A Lexicographic Order

Let $\Sigma = \{x, y\}$ and let R be the following partial order relation on Σ:

$$R = \{(x, x), (x, y), (y, y)\}.$$

Denote the lexicographic order for Σ^* corresponding to R by $\preceq$.

a. Is $x \preceq xx$? $x \preceq xy$? $xx \preceq xxx$? $yxy \preceq yxyxxx$?
b. Is $x \preceq y$? $xx \preceq xyx$? $xxxy \preceq xy$? $yxyxxyy \preceq yxyxy$?
c. Is $\epsilon \preceq x$? $\epsilon \preceq xy$? $\epsilon \preceq yyxy$?

Solution   a. Yes in all cases, by property (1) of the definition of $\preceq$.
b. Yes in all cases, by property (2) of the definition of $\preceq$.
c. Yes in all cases, by property (3) of the definition of $\preceq$.

HASSE DIAGRAMS

Let $A = \{1, 2, 3, 9, 18\}$ and consider the "divides" relation on A:

$$\text{for all } a, b \in A, \quad a \mid b \iff b = k \cdot a \text{ for some integer } k.$$

The directed graph of this relation has the appearance shown in Figure 10.5.1.

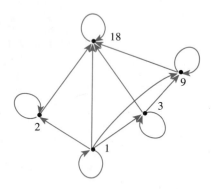

FIGURE 10.5.1

Note that there is a loop at every vertex, all other arrows point in the same direction (upward), and any time there is an arrow from one point to a second and from the second point to a third, there is an arrow from the first point to the third. Given any partial order relation defined on a finite set, it is possible to draw the directed graph so that all of these properties are satisfied. This makes it possible to associate a somewhat simpler graph, called a **Hasse diagram** (named after Helmut Hasse, a twentieth-century German number theorist), with a partial order relation defined on a finite set. To obtain a Hasse diagram proceed as follows:

Start with a directed graph of the relation in which all arrows point upward. Then eliminate

1. the loops at all the vertices,
2. all arrows whose existence is implied by the transitive property,
3. the direction indicators on the arrows.

For the relation given above, the Hasse diagram is as shown in Figure 10.5.2.

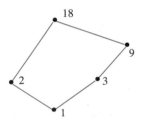

FIGURE 10.5.2

EXAMPLE 10.5.7   Constructing a Hasse Diagram

Consider the "subset" relation, $\subseteq$, on the set $\mathcal{P}$ ($\{a, b, c\}$). That is, for all sets U and V in $\mathcal{P}(\{a, b, c\})$,

$$U \subseteq V \quad \Leftrightarrow \quad \forall x, \text{ if } x \in U \text{ then } x \in V.$$

Construct the Hasse diagram for this relation.

Solution   Draw the directed graph of the relation in such a way that all arrows except loops point upward. This is shown in Figure 10.5.3.

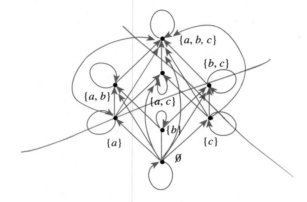

FIGURE 10.5.3

Then strip away all loops, unnecessary arrows, and direction indicators to obtain the Hasse diagram shown in Figure 10.5.4.

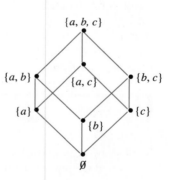

FIGURE 10.5.4

To recover the directed graph of a relation from the Hasse diagram, just reverse the instructions given above, using the knowledge that the original directed graph was sketched so that all arrows pointed upward:

1. reinsert the direction markers on the arrows making all arrows point upward;
2. add loops at each vertex;
3. for each sequence of arrows from one point to a second and from that second point to a third, add an arrow from the first point to the third.

EXAMPLE 10.5.8

Obtaining the Directed Graph of a Partial Order Relation from the Hasse Diagram of the Relation

A partial order relation has the Hasse diagram shown in Figure 10.5.5. Find the directed graph of *R*.

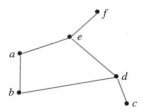

FIGURE 10.5.5

Solution

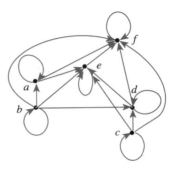

PARTIALLY AND TOTALLY ORDERED SETS

Given any two real numbers x and y, either $x \le y$ or $y \le x$. In a situation like this, the elements x and y are said to be *comparable*. On the other hand, given two subsets A and B of $\{a, b, c\}$, it may be the case that neither $A \subseteq B$ nor $B \subseteq A$. For instance, let $A = \{a, b\}$ and $B = \{b, c\}$. Then $A \not\subseteq B$ and $B \not\subseteq A$. In such a case, A and B are said to be *noncomparable*.

DEFINITION

Suppose R is a partial order relation on a set A. Elements a and b of A are said to be **comparable** if, and only if, either $a\,R\,b$ or $b\,R\,a$. Otherwise a and b are called **noncomparable**.

When all the elements of a partial order relation are comparable, the relation is called a *total order*.

DEFINITION

If R is a partial order relation on a set A, and for any two elements a and b in A either $a\,R\,b$ or $b\,R\,a$, then R is a **total order relation** on A.

Both the "less than or equal to" relation on sets of real numbers and the lexico-graphic order of the set of words in a dictionary are total order relations. Note that the Hasse diagram for a total order relation can be drawn as a single vertical "chain."

Many important partial order relations have elements that are not comparable and are, therefore, not total order relations. For instance, the subset relation on $\mathcal{P}(\{a, b, c\})$ is not a total order because as shown above the subsets $\{a, b\}$ and $\{a, c\}$ of $\{a, b, c\}$ are not comparable. In addition, a "divides" relation is not a total order relation unless the elements are all powers of a single integer. (See exercise 21 at the end of this section.)

A set A is called a **partially ordered set** (or **poset**) with respect to a relation $\preceq$ if, and only if, $\preceq$ is a partial order relation on A. For instance, the set of real numbers is a partially ordered set with respect to the "less than or equal to" relation $\leq$, and a set of sets is partially ordered with respect to the "subset" relation $\subseteq$. It is entirely straightforward to show that *any subset of a partially ordered set is partially ordered.* (See exercise 32 at the end of this section.) This, of course, assumes the "same definition" for the relation on the subset as for the set as a whole. A set A is called a **totally ordered set** with respect to a relation $\preceq$ if, and only if, A is partially ordered with respect to $\preceq$ and $\preceq$ is a total order.

A set that is partially ordered but not totally ordered may have totally ordered subsets. Such subsets are called *chains.*

DEFINITION

Let A be a set that is partially ordered with respect to a relation $\preceq$. A subset B of A is called a **chain** if, and only if, each pair of elements in B is comparable. In other words, $a \preceq b$ or $b \preceq a$ for all a and b in A. The **length of a chain** is one less than the number of elements in the chain.

Observe that if B is a chain in A, then B is a totally ordered set with respect to the "restriction" of $\preceq$ to B. (See exercise 32 at the end of this section.)

EXAMPLE 10.5.9   **A Chain of Subsets**

The set $\mathcal{P}(\{a, b, c\})$ is partially ordered with respect to the subset relation. Find a chain of length 3 in $\mathcal{P}(\{a, b, c\})$.

Solution   Since $\varnothing \subseteq \{a\} \subseteq \{a, b\} \subseteq \{a, b, c\}$, the set

$$S = \{\varnothing, \{a\}, \{a, b\}, \{a, b, c\}\}$$

is a chain of length 3 in $\mathcal{P}(\{a, b, c\})$.   ■

In exercise 36 at the end of this section you are asked to show that a set that is partially ordered with respect to a relation $\preceq$ is totally ordered with respect to $\preceq$ if, and only if, it is a chain.

A *maximal element* in a partially ordered set is an element that is greater than or equal to every element *to which it is comparable.* (There may be many elements to which it is *not* comparable.) A *greatest element* in a partially ordered set is an element that is greater than or equal to *every* element in the set. (So it is comparable to every element in the set.) Minimal and least elements are defined similarly.

DEFINITION

Let a set A be partially ordered with respect to a relation $\preceq$.

1. An element a in A is called a **maximal element of A** if, and only if, for all b in A, either $b \preceq a$ or b and a are not comparable.

2. An element a in A is called a **greatest element of A** if, and only if, for all b in A, $b \preceq a$.

3. An element a in A is called a **minimal element of A** if, and only if, for all b in A, either $a \preceq b$ or b and a are not comparable.

4. An element a in A is called a **least element of A** if, and only if, for all b in A, $a \preceq b$.

A greatest element is maximal but a maximal element need not to be a greatest element. Similarly, a least element is minimal but a minimal element need not be a least element. Furthermore, a set that is partially ordered with respect to a relation can have at most one greatest element and one least element (see exercise 38 at the end of this section), but it may have more than one maximal or minimal element. The next example illustrates some of these facts.

EXAMPLE 10.5.10   Maximal, Minimal, Greatest, and Least Elements

Let $A = \{a, b, c, d, e, f, g, h, i\}$ have the partial ordering $\preceq$ defined by the Hasse diagram in Figure 10.5.6. Find all maximal, minimal, greatest, and least elements of A.

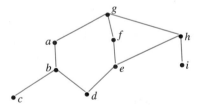

FIGURE 10.5.6

Solution   There is just one maximal element, g, which is also the greatest element. The minimal elements are c, d, and i, and there is no least element. ■

TOPOLOGICAL SORTING

Is it possible to input the sets of $\mathcal{P}(\{a, b, c\})$ into a computer in a way that is *compatible* with the subset relation $\subseteq$ in the sense that if set U is a subset of set V then U is input before V? The answer, as it turns out, is yes. For instance, the following input order satisfies the given condition:

$$\varnothing, \{a\}, \{b\}, \{c\}, \{a, b\}, \{a, c\}, \{b, c\}, \{a, b, c\}.$$

Another input order that also satisfies the condition is

$$\varnothing, \{a\}, \{b\}, \{a, b\}, \{c\}, \{b, c\}, \{a, c\}, \{a, b, c\}.$$

DEFINITION

Given partial order relations $\preceq$ and $\preceq'$ on a set A, $\preceq'$ is **compatible** with $\preceq$ if, and only if, for all a and b in A, if $a \preceq b$ then $a \preceq' b$.

Given an arbitrary partial order relation $\preceq$ on a set A, is there a total order $\preceq'$ on A that is compatible with $\preceq$? If the set on which the partial order is defined is finite, then the answer is yes. A total order that is compatible with a given order is called a *topological sorting*.

DEFINITION

Given partial order relations $\preceq$ and $\preceq'$ on a set A, $\preceq'$ is a **topological sorting** for $\preceq$ if, and only if, $\preceq'$ is a total order that is compatible with $\preceq$.

The construction of a topological sorting for a general finite partially ordered set is based on the fact that *any partially ordered set that is finite and nonempty has a minimal element*. (See exercise 37 at the end of this section.) To create a total order for a partially ordered set, simply pick any minimal element and make it number one. Then consider the set obtained when this element is removed. Since the new set is a subset of a partially ordered set, it is partially ordered. If it is empty, stop the process. If not, pick a minimal element from it and call that element number two. Then consider the set obtained when this element also is removed. If this set is empty, stop the process. If not, pick a minimal element and call it number three. Continue in this way until all the elements of the set have been used up.

Here is a somewhat more formal version of the algorithm described above to construct a topological sorting for a relation $\preceq$ defined on a nonempty finite set A:

1. Pick any minimal element x in A. *[Such an element exists since A is nonempty.]*
2. Set $A := A - \{x\}$.
3. Repeat steps a–c while $A \neq \emptyset$.
   a. Pick any minimal element y in A.
   b. Define $x \preceq' y$.
   c. Set $A := A - \{y\}$ and $x := y$.
   [Completion of steps 1–3 of this algorithm gives enough information to construct the Hasse diagram for the total ordering $\preceq'$. We have already shown how to use the Hasse diagram to obtain a complete directed graph for a relation.]

EXAMPLE 10.5.11   A Topological Sorting

Consider the set $A = \{2, 3, 4, 6, 18, 24\}$ ordered by the "divides" relation $|$. The Hasse diagram of this relation is shown in Figure 10.5.7.

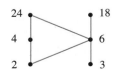

FIGURE 10.5.7

The ordinary "less than or equal to" relation $\le$ on this set is a topological sorting for it since for positive integers a and b, if $a \mid b$ then $a \le b$. Find another topological sorting for this set.

Solution   The set has two minimal elements: 2 and 3. Either one may be chosen; say you pick 3. The beginning of the total order is

total order: 3.

Set $A = A - \{3\}$. You can indicate this by removing 3 from the Hasse diagram, as shown in Figure 10.5.8.

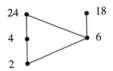

FIGURE 10.5.8

Next choose a minimal element from $A - \{3\}$. Only 2 is minimal, and so you must pick it. The total order thus far is

total order: $3 \le 2$.

Set $A = (A - \{3\}) - \{2\} = A - \{3, 2\}$. You can indicate this by removing 2 from the Hasse diagram, as is shown in Figure 10.5.9.

FIGURE 10.5.9

Choose a minimal element from $A - \{3, 2\}$. Again you have two choices: 4 and 6. Say you pick 6. The total order for the elements chosen thus far is

total order: $3 \le 2 \le 6$.

You continue in this way until every element of A has been picked. One possible sequence of choices gives

total order: $3 \le 2 \le 6 \le 18 \le 4 \le 24$.

You can verify that this order is compatible with the "divides" partial order by checking that for each pair of elements a and b in A such that $a \mid b$, then $a \le b$. Note that it is *not* the case that if $a \le b$ then $a \mid b$. ∎

AN APPLICATION

To return to the example that introduced this section, note that the following defines a partial order relation on the set of courses required for a computer science degree: for all required courses x and y,

$$x \le y \iff x = y \quad \text{or} \quad x \text{ is a prerequisite for } y$$

If the Hasse diagram for the relation is drawn, then the questions raised at the beginning of this section can be answered easily. For instance, suppose the Hasse diagram for the requirements at a particular university is as shown in Figure 10.5.10.

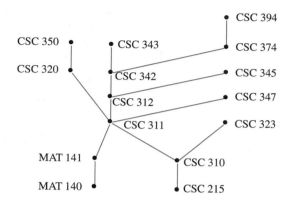

FIGURE 10.5.10

The minimum number of school terms needed to complete the requirements is the length of the longest chain, which is 7 (215, 310, 311, 312, 342, 374, 394, for example). The maximum number of courses that could be taken in the same term (assuming the university allows it) is the maximum number of noncomparable courses, which is 6 (320, 343, 374, 345, 347, 323, for example). A part-time student could take the courses in a sequence determined by constructing a topological sorting for the set. (One such sorting is 140, 215, 141, 310, 311, 323, 320, 312, 350, 347, 345, 342, 350, 343, 374, 394. There are many others.)

PERT AND CPM

Two important and widely used applications of partial order relations are **PERT** (Program Evaluation and Review Technique) and **CPM** (Critical Path Method). These techniques came into being in the 1950s to deal with the complexities of scheduling the individual activities needed to complete very large projects, and although they are very similar, their developments were independent. PERT was developed by the U.S. Navy to help organize the construction of the Polaris submarine, and CPM was developed by the E. I. Du Pont de Nemours company for scheduling chemical plant maintenance. Here is a somewhat simplified example of the way the techniques work.

EXAMPLE 10.5.12   A Job Scheduling Problem

At an automobile assembly plant, the job of assembling an autombile can be broken down into these tasks:

1. Build frame.
2. Install engine, power train components, gas tank.
3. Install brakes, wheels, tires.
4. Install dashboard, floor, seats.
5. Install electrical lines.
6. Install gas lines.

7. Install brake lines.

8. Attach body panels to frame.

9. Paint body.

Certain of these tasks can be carried out at the same time while some cannot be started until others are finished. Table 10.5.1 summarizes the order in which tasks can be performed and the time required to perform each task.

TABLE 10.5.1

Task	Immediately Preceding Tasks	Time Needed to Perform Task
1		7 hours
2	1	6 hours
3	1	3 hours
4	2	6 hours
5	2, 3	3 hours
6	4	1 hour
7	2, 3	1 hour
8	4, 5	2 hours
9	6, 7, 8	5 hours

Let T be the set of all tasks and consider the partial order relation $\leq$ defined on T as follows: for all tasks x and y in T,

$$x \leq y \iff x = y \text{ or } x \text{ precedes } y.$$

If the Hasse diagram of this relation is turned sideways (as is customary in PERT and CPM analysis), it has the appearance shown in Figure 10.5.11.

What is the minimum time required to assemble a car? You can determine this by working from left to right across the diagram, noting for each task (say, just above the box representing that task) the minimum time needed to complete that task starting from the beginning of the assembly process. For instance, you can put a 7 above the box for task 1 because task 1 requires 7 hours. Task 2 requires completion of task 1

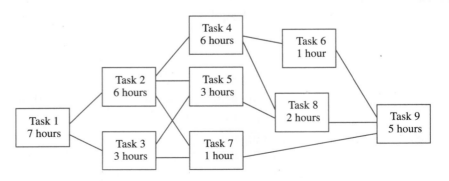

FIGURE 10.5.11

(7 hours) plus 6 hours for itself, so the minimum time required to complete task 2 starting at the beginning of the assembly process is $7 + 6 = 13$ hours. You can put a 13 above the box for task 2. Similarly, you can put a 10 above the box for task 3 because $7 + 3 = 10$. Now consider what number you should write above the box for task 5. The minimum time to complete tasks 2 and 3 starting from the beginning of the assembly process are 13 and 10 hours respectively. Since *both* tasks must be completed before task 5 can be started, the minimum time to complete task 5, starting from the beginning, is the time needed for task 5 itself (3 hours) plus the *maximum* of the times to complete tasks 2 and 3 (13 hours), and this equals $3 + 13 = 16$ hours. Thus you should place the number 16 above the box for task 5. The same reasoning leads you to place a 14 above the box for task 7. Similarly, you can place a 19 above the box for task 4, a 20 above the box for task 6, a 21 above the box for task 8, and a 26 above the box for task 9 as shown in Figure 10.5.12.

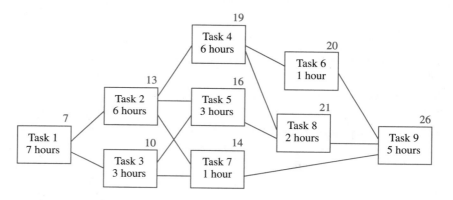

FIGURE 10.5.12

This analysis shows that at least 26 hours are required to complete task 9 starting from the beginning of the assembly process. But when task 9 is finished, the assembly is complete. So 26 hours is the minimum time needed to accomplish the whole process.

Note that the minimum time required to complete tasks 1, 2, 4, 8, and 9 in sequence is exactly 26 hours. This means that a delay in performing any one of these tasks causes a delay in the total time required for assembly of the car. For this reason the path through tasks 1, 2, 4, 8, and 9 is called a **critical path**. ■

EXERCISE SET 10.5

1. Each of the following is a relation on $\{0, 1, 2, 3\}$. Draw directed graphs for each relation and indicate which relations are antisymmetric.

   a. $R_1 = \{(0, 0), (0, 2), (1, 0), (1, 3), (2, 2), (3, 0), (3, 1)\}$

   b. $R_2 = \{(0, 1), (0, 2), (1, 1), (1, 2), (1, 3), (2, 2), (3, 2)\}$

   c. $R_3 = \{(0, 0), (0, 3), (1, 0), (1, 3), (2, 2), (3, 3), (3, 2)\}$

   d. $R_4 = \{(0, 0), (1, 0), (1, 2), (1, 3), (2, 0), (2, 1), (3, 2), (3, 0)\}$

2. Let P be the set of all people in the world and define a relation R on P as follows: for all $x, y \in P$,

$$x \, R \, y \quad \Leftrightarrow \quad x \text{ is no older than } y.$$

   Is R antisymmetric? Prove or give a counterexample.

3. Let $\Sigma = \{a, b\}$ and let Σ^* be the set of all strings over Σ. Define a relation R on Σ^* as follows: for all s, $t \in \Sigma^*$,

$$s \, R \, t \quad \Leftrightarrow \quad l(s) \leq l(t),$$

   where $l(x)$ denotes the length of a string x. Is R antisymmetric? Prove or give a counterexample.

4. Let R be the "less than" relation on the set $\mathbf{R}$ of all real numbers: for all $x, y \in \mathbf{R}$,

$$x\,R\,y \iff x < y.$$

Is R antisymmetric? Prove or give a counterexample.

5. Let $\mathbf{R}$ be the set of all real numbers and define a relation R on $\mathbf{R} \times \mathbf{R}$ as follows: for all (a, b) and (c, d) in $\mathbf{R} \times \mathbf{R}$,

$$(a, b)\,R\,(c, d) \iff \text{either } a < c \text{ or both } a = c \text{ and } b \le d.$$

Is R a partial order relation? Prove or give a counterexample.

6. Let P be the set of all people who have ever lived and define a relation R on P as follows: for all $r, s \in P$,

$$r\,R\,s \iff r \text{ is an ancestor of } s \text{ or } r = s.$$

Is R a partial order relation? Prove or give a counterexample.

7. Define a relation R on the set $\mathbf{Z}$ of all integers as follows: for all $m, n \in \mathbf{Z}$,

$$m\,R\,n \iff \text{every prime factor of } m \text{ is a prime factor of } n.$$

Is R a partial order relation? Prove or give a counterexample.

8. Define a relation R on the set $\mathbf{Z}$ of all integers as follows: for all $m, n \in \mathbf{Z}$,

$$m\,R\,n \iff m + n \text{ is even.}$$

Is R a partial order relation? Prove or give a counterexample.

9. Define a relation R on the set of all real numbers $\mathbf{R}$ as follows: for all $x, y \in \mathbf{R}$,

$$x\,R\,y \iff x^2 \le y^2.$$

Is R a partial order relation? Prove or give a counterexample.

10. Suppose R and S are antisymmetric relations on a set A. Must $R \cup S$ also be antisymmetric? Explain.

11. Let $\Sigma = \{a, b\}$ and suppose Σ has the partial order relation R where $R = \{(a, a), (a, b), (b, b)\}$. Let $\preceq$ be the corresponding lexicographic order on Σ^*. Indicate which of the following statements are true, and for each true statement cite either (1), (2), or (3) of the definition of lexicographic order given in Theorem 10.5.1 as a reason.

a. $aab \preceq aaba$ b. $bbab \preceq bba$
c. $\varepsilon \preceq aba$ d. $aba \preceq abb$
e. $bbab \preceq bbaa$ f. $ababa \preceq ababaa$
g. $bbaba \preceq bbabb$

12. Prove Theorem 10.5.1.

13. Let $A = \{a, b\}$. Describe all partial order relations on A.

14. Let $A = \{a, b, c\}$.
a. Describe all partial order relations on A for which a is a maximal element.
b. Describe all partial order relations on A for which a is a minimal element.

H 15. Suppose a relation R on a set A is reflexive, symmetric, transitive, and antisymmetric. What can you conclude about R? Prove your answer.

16. Consider the "divides" relation on each of the following sets A. Draw the Hasse diagram for each relation.
a. $A = \{1, 2, 4, 5, 10, 15, 20\}$
b. $A = \{2, 3, 4, 6, 8, 9, 12, 18\}$

17. Consider the "subset" relation on $\mathcal{P}(S)$ for each of the following sets S. Draw the Hasse diagram for each relation.
a. $S = \{0, 1\}$ b. $S = \{0, 1, 2\}$

18. Let $S = \{0, 1\}$ and consider the partial order relation R defined on $S \times S$ as follows: for all ordered pairs (a, b) and (c, d) in $S \times S$,

$$(a, b)\,R\,(c, d) \iff \text{either } a < c \text{ or } a = c \text{ and } b \le d,$$

where $<$ denotes the usual "less than" and $\le$ denotes the usual "less than or equal to" relation for real numbers. Draw the Hasse diagram for R.

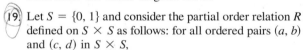
19. Let $S = \{0, 1\}$ and consider the partial order relation R defined on $S \times S$ as follows: for all ordered pairs (a, b) and (c, d) in $S \times S$,

$$(a, b)\,R\,(c, d) \iff a \le c \text{ and } b \le d,$$

where $\le$ denotes the usual "less than or equal to" relation for real numbers. Draw the Hasse diagram for R.

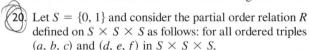
20. Let $S = \{0, 1\}$ and consider the partial order relation R defined on $S \times S \times S$ as follows: for all ordered triples (a, b, c) and (d, e, f) in $S \times S \times S$,

$$(a, b, c)\,R\,(d, e, f) \iff a \le d, b \le e, \text{ and } c \le f,$$

where $\le$ denotes the usual "less than or equal to" relation for real numbers. Draw the Hasse diagram for R.

21. Consider the "divides" relation defined on the set $A = \{1, 2, 2^2, 2^3, \ldots, 2^n\}$, where n is a nonnegative integer.
a. Prove that this relation is a total order on A.
b. Draw the Hasse diagram for this relation for $n = 4$.

In 22–29, find all greatest, least, maximal, and minimal elements for the relations in each of the referenced exercises.

22. exercise 16(a).
23. exercise 16(b).

24. exercise 17(a).
25. exercise 17(b).

26. exercise 18.   **28.** exercise 20.

27. exercise 19.   **29.** exercise 21.

30. Each of the following sets is partially ordered with respect to the "less than or equal to" relation, $\leq$, for real numbers. In each case, determine whether the set has a greatest or least element.
  a. $\mathbf{R}$
  b. $\{x \in \mathbf{R} \mid 0 \leq x \leq 1\}$
  c. $\{x \in \mathbf{R} \mid 0 < x < 1\}$
  d. $\{x \in \mathbf{Z} \mid 0 < x < 10\}$

31. How many total orderings are there on a set with n elements? Explain your answer.

H 32. Suppose R is a partial order relation on a set A and B is a subset of A. The **restriction of R to B** is defined as follows:

  the restriction of R to B
  $$= \{(x, y) \mid x \in B, y \in B, \text{ and } (x, y) \in R\}.$$

  In other words, two elements of B are related by the restriction of R to B if, and only if, they are related by R. Prove that the restriction of R to B is a partial order relation on B. (In less formal language this says that a subset of a partially ordered set is partially ordered.)

33. The set $\mathcal{P}(\{w, x, y, z\})$ is partially ordered with respect to the "subset" relation $\subseteq$. Find a chain of length 4 in $\mathcal{P}(\{w, x, y, z\})$.

34. The set $A = \{2, 4, 3, 6, 12, 18, 24\}$ is partially ordered with respect to the "divides" relation. Find a chain of length 3 in A.

35. Find a chain of length 2 for the relation defined in exercise 19.

36. Prove that a partially ordered set is totally ordered if, and only if, it is a chain.

37. Prove that a nonempty finite partially ordered set has
  a. at least one minimal element,
  b. at least one maximal element.

38. Prove that a finite partially ordered set has
  a. at most one greatest element,
  b. at most one least element.

39. Draw a Hasse diagram for a partially ordered set that has two maximal elements and two minimal elements and is such that each element is related to exactly two other elements.

40. Draw a Hasse diagram for a partially ordered set that has three maximal elements and three minimal elements and is such that each element is either greater than or less than exactly two other elements.

41. Use the algorithm given in the text to find a topological sorting for the relation of exercise 16(a) that is different from the "less than or equal to" relation $\leq$.

42. Use the algorithm given in the text to find a topological sorting for the relation of exercise 16(b) that is different from the "less than or equal to" relation $\leq$.

43. Use the algorithm given in the text to find a topological sorting for the relation of exercise 19.

44. Use the algorithm given in the text to find a topological sorting for the relation of exercise 20.

45. Use the algorithm given in the text to find a topological sorting for the "subset" relation on $\mathcal{P}(\{a, b, c, d\})$.

46. A set S of jobs can be ordered by writing $x \preceq y$ to mean that either $x = y$ or x must be done before y, for all x and y in S. The following is a Hasse diagram for this relation for a particular set S of jobs:

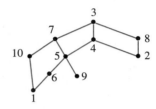

  a. If one person is to perform all the jobs, one after another, find an order in which the jobs can be done.
  b. Suppose enough people are available to perform any number of jobs simultaneously.
    (i) If each job requires one day to perform, what is the least number of days needed to perform all ten jobs?
    (ii) What is the maximum number of jobs that can be performed at the same time?

47. Suppose the tasks described in Example 10.5.12 require the following performance times:

Task	Time Needed to Perform Task
1	9 hours
2	7 hours
3	4 hours
4	5 hours
5	7 hours
6	3 hours
7	2 hours
8	4 hours
9	6 hours

  a. What is the minimum time required to assemble a car?
  b. Find a critical path for the assembly process.

11 Graphs and Trees

Graphs and trees have already appeared in this book as convenient visualizations to use in a variety of situations. For instance, a possibility tree shows all possible outcomes of a multistep operation with a finite number of outcomes for each step, a transition diagram for a finite-state automaton shows all the changes of state of the automaton that correspond to various input values, the directed graph of a relation on a set shows which elements of the set are related to which, and a PERT diagram shows which tasks must precede which in executing a project.

In this chapter we present some of the mathematics of graphs and trees, discussing concepts such as the degree of a vertex, connectedness, Euler and Hamiltonian circuits, representation of graphs by matrices, isomorphisms of graphs, the relation between the number of vertices and the number of edges of a tree, rooted trees, and the spanning tree of a graph. Applications include uses of graphs and trees in the study of artificial intelligence, chemistry, scheduling problems, and transportation systems.

11.1 GRAPHS: AN INTRODUCTION

The whole of mathematics consists in the organization of a series of aids to the imagination in the process of reasoning.
(Alfred North Whitehead, 1861–1947)

Imagine an organization that has acquired six different computers over the years. In an effort to upgrade computer services, the organization proposes to connect the computers to form an integrated system. It is not necessary that every computer be linked with every other computer, however. In fact, analysis shows that the following connections are optimal:

Connect computer A with B, C, D, and E;

connect computer B with A and C;

connect computer C with A, B, D, and E;

connect computer D with A and C;

connect computer E with A, C, and F;

connect computer F with E.

This information can be conveniently displayed in the diagram shown below.

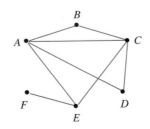

A drawing such as this is an illustration of a *graph*. The dots are called *vertices* (plural of *vertex*) and the line segments joining vertices are called *edges*. As you can see from the drawing, it is possible for two edges to cross at a point that is not a vertex. Note also that the type of graph described here is quite different from the "graph of an equation" or the "graph of a function."

In general, a graph consists of a set of vertices and a set of edges connecting various vertices. The edges may be straight or curved and should either connect one vertex to another or a vertex to itself as shown below.

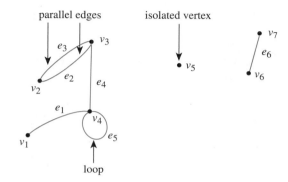

In this drawing, the vertices have been labeled with v's and the edges with e's. When an edge connects a vertex to itself (as e_5 does), it is called a *loop*. When two edges connect the same pair of vertices (as e_2 and e_3 do), they are said to be *parallel*. It is quite possible for a vertex to be unconnected by an edge to any other vertex in the graph (as v_5 is), and in that case the vertex is said to be *isolated*. The formal definition of a graph is as follows.

DEFINITION

A **graph** G consists of two finite sets: a set $V(G)$ of **vertices** and a set $E(G)$ of **edges**, where each edge is associated with a set consisting of either one or two vertices called its **endpoints**. The correspondence from edges to endpoints is called the **edge-endpoint function**. An edge with just one endpoint is called a **loop**, and two distinct edges with the same set of endpoints are said to be **parallel**. An edge is said to **connect** its endpoints; two vertices that are connected by an edge are called **adjacent**; and a vertex that is an endpoint of a loop is said to be **adjacent to itself**. An edge is said to be **incident on** each of its endpoints, and two edges incident on the same endpoint are called **adjacent**. A vertex on which no edges are incident is called **isolated**. A graph with no vertices is called **empty**, and one with at least one vertex is called **nonempty**.

Graphs have pictorial representations in which the vertices are represented by dots and the edges by line segments. A given pictorial representation uniquely determines a graph.

EXAMPLE 11.1.1   **Terminology**

For the graph pictured in Figure 11.1.1,

a. Write the vertex set and the edge set, and give a table showing the edge-endpoint function;
b. Find all edges that are incident on v_1, all vertices that are adjacent to v_1, all edges that are adjacent to e_1, all loops, all parallel edges, all vertices that are adjacent to themselves, and all isolated vertices.

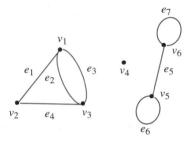

FIGURE 11.1.1

Solution   a. vertex set $= \{v_1, v_2, v_3, v_4, v_5, v_6\}$
edge set $= \{e_1, e_2, e_3, e_4, e_5, e_6, e_7\}$
edge-endpoint function:

Edge	Endpoints
e_1	$\{v_1, v_2\}$
e_2	$\{v_1, v_3\}$
e_3	$\{v_1, v_3\}$
e_4	$\{v_2, v_3\}$
e_5	$\{v_5, v_6\}$
e_6	$\{v_5\}$
e_7	$\{v_6\}$

Note that the isolated vertex v_4 does not appear in this table. Although each edge must have either one or two endpoints, a vertex need not be an endpoint of an edge.

b. e_1, e_2, and e_3 are incident on v_1.
v_2 and v_3 are adjacent to v_1.
e_2, e_3, and e_4 are adjacent to e_1.
e_6 and e_7 are loops.
e_2 and e_3 are parallel.
v_5 and v_6 are adjacent to themselves.
v_4 is an isolated vertex.   ∎

As noted earlier, a given pictorial representation uniquely determines a graph. However, a given graph may have more than one pictorial representation. Such things as the lengths or curvatures of the edges and the relative position of the vertices on the page may vary from one pictorial representation to another.

EXAMPLE 11.1.2   Drawing More Than One Picture for a Graph

Consider the graph specified as follows:

$$\text{vertex set} = \{v_1, v_2, v_3, v_4\}$$
$$\text{edge set} = \{e_1, e_2, e_3, e_4\}$$
edge-endpoint function:

Edge	Endpoints
e_1	$\{v_1, v_3\}$
e_2	$\{v_2, v_4\}$
e_3	$\{v_2, v_4\}$
e_4	$\{v_3\}$

Both drawings (1) and (2) shown below are pictorial representations of this graph.

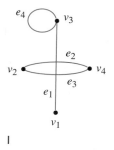   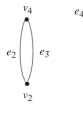

1   2

EXAMPLE 11.1.3   Labeling Drawings to Show They Represent
the Same Graph

Consider the two drawings shown below. Label vertices and edges in such a way that both drawings represent the same graph.

1   2

Solution   Imagine putting one end of a piece of string at the top vertex of drawing (1) (call this vertex v_1), then laying the string to the next adjacent vertex on the lower right (call this vertex v_2), then to the next adjacent vertex on the upper left (v_3), and so forth, returning finally to the top vertex v_1. Call the first edge e_1, the second e_2, and so forth as shown in Figure 11.1.2.

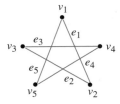

FIGURE 11.1.2

Now imagine picking up the piece of string, together with its labels and repositioning it as shown in Figure 11.1.3.

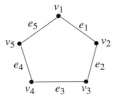

FIGURE 11.1.3

This is the same as drawing (2). So both drawings are representations of the graph with vertex set $\{v_1, v_2, v_3, v_4, v_5\}$, edge set $\{e_1, e_2, e_3, e_4, e_5\}$, and edge-endpoint function as follows:

Edge	Endpoints
e_1	$\{v_1, v_2\}$
e_2	$\{v_2, v_3\}$
e_3	$\{v_3, v_4\}$
e_4	$\{v_4, v_5\}$
e_5	$\{v_5, v_1\}$

■

In Chapter 10 we discussed the directed graph of a binary relation on a set. The general definition of directed graph is similar to the definition of graph, except that one associates an *ordered pair* of vertices with each edge instead of a *set* of vertices. Thus, each edge of a directed graph can be drawn as an arrow going from the first vertex to the second vertex of the ordered pair.

> **DEFINITION**
>
> A **directed graph**, or **digraph**, consists of two finite sets: a set $V(G)$ of vertices and a set $D(G)$ of directed edges, where each edge is associated with an ordered pair of vertices called its **endpoints**. If edge e is associated with the pair (v, w) of vertices, then e is said to be the (**directed**) **edge** from v to w.

Note that each directed graph has an associated ordinary (undirected) graph, which is obtained by ignoring the directions of the edges.

EXAMPLES OF GRAPHS

You have already seen a number of examples of directed and undirected graphs in this book. Flowcharts, possibility trees, transition diagrams for finite-state automata, and Hasse diagrams (including PERT diagrams) can all be viewed as graphs if additional structure is stripped away. The following examples illustrate some other situations in which graphs are used.

EXAMPLE 11.1.4   Communication System

Computer, telephone, electric power, gas pipeline, and air transport systems can all be represented by graphs. Questions that arise in the design of such systems involve choosing connecting edges to minimize cost, optimize a certain type of service, and so forth. A typical communication system is shown below.

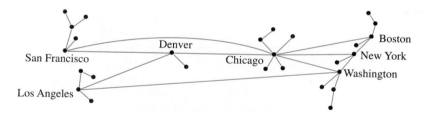

EXAMPLE 11.1.5   Using a Graph to Represent Knowledge

In many applications of artifical intelligence a knowledge base of information is collected and represented inside a computer. Because of the way the knowledge is represented and because of the properties that govern the artificial intelligence program, the computer is not limited to retrieving data in the same form as it was entered; it can also derive new facts from the knowledge base by using certain built-in rules of inference. For example, from the knowledge that the *Los Angeles Times* is a big-city daily and that a big-city daily contains national news, an artificial intelligence program could infer that the *Los Angeles Times* contains national news. The directed graph shown in Figure 11.1.4 is a pictorial representation for a simplified knowledge base about periodical publications.

According to this knowledge base, what paper finish does the *New York Times* use?

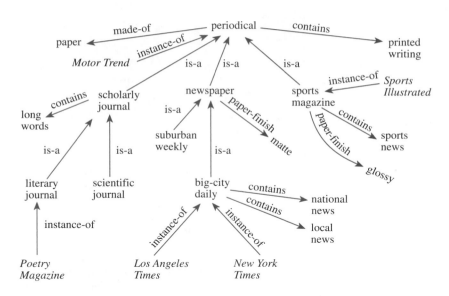

FIGURE 11.1.4

Solution   The arrow going from *New York Times* to big-city daily (labeled "instance-of") shows that the *New York Times* is a big-city daily. The arrow going from big-city daily to newspaper (labeled "is-a") shows that a big-city daily is a newspaper. The arrow going from newspaper to matte (labeled "paper-finish") indicates that the paper finish on a newspaper is matte. Hence it can be inferred that the paper finish on the *New York Times* is matte. ∎

EXAMPLE 11.1.6   **Vegetarians and Cannibals**

The following is a variation of a famous puzzle often used as an example in the study of artificial intelligence. It concerns an island on which all the people are of one of two types, either vegetarians or cannibals. Initially, two vegetarians and two cannibals are on the left bank of a river. With them is a boat that can hold a maximum of two people. The aim of the puzzle is to find a way to transport all the vegetarians and cannibals to the right bank of the river. What makes this difficult is that at no time can the number of cannibals on either bank outnumber the number of vegetarians. Otherwise, disaster befalls the vegetarians!

Solution   A systematic way to approach this problem is to introduce a notation that can indicate all possible arrangements of vegetarians, cannibals, and the boat on the banks of the river. For example, you could write (vvc/Bc) to indicate that there are two vegetarians and one cannibal on the left bank and one cannibal and the boat on the right bank. Then $(vvccB/)$ would indicate the initial position in which both vegetarians, both cannibals, and the boat are on the left bank of the river. The aim of the puzzle is to figure out a sequence of moves to reach the position $(/Bvvcc)$ in which both vegetarians, both cannibals, and the boat are on the right bank of the river.

Construct a graph whose vertices are the various arrangements that can be reached in a sequence of legal moves starting from the initial position. Connect vertex x to vertex y if it is possible to reach vertex y in one legal move from vertex x. For instance, from the initial position there are four legal moves: one vegetarian and one cannibal can take the boat to the right bank; two cannibals can take the boat to the right bank; one cannibal can take the boat to the right bank; or two vegetarians can take the boat to the right bank. You can show these by drawing edges connecting vertex $(vvccB/)$ to vertices (vc/Bvc), (vv/Bcc), $(vvcBc)$, and (cc/Bvv). (It might seem natural to draw directed edges rather than undirected edges from one vertex to another. The rationale for drawing undirected edges is that each legal move is reversible.) From the position (vc/Bvc), the only legal moves are to go back to $(vvccB/)$ or to go to $(vvcB/c)$. You can also show these by drawing in edges. Continue this process until finally you reach $(/Bvvcc)$. From Figure 11.1.5 it is apparent that one successful sequence of moves is $(vvccB/) \rightarrow (vc/Bvc) \rightarrow (vvcB/c) \rightarrow (c/Bvvc) \rightarrow (ccB/vv) \rightarrow (/Bvvcc)$.

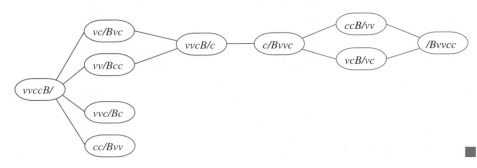

FIGURE 11.1.5

SPECIAL GRAPHS

One important class of graphs consists of graphs that do not have any loops or parallel edges. Such graphs are called *simple*. In a simple graph no two edges share the same set of endpoints, so specifying two endpoints completely determines an edge.

> ### DEFINITION AND NOTATION
>
> A **simple graph** is a graph that does not have any loops or parallel edges. In a simple graph, an edge with endpoints v and w is denoted $\{v, w\}$.

EXAMPLE 11.1.7   A Simple Graph

Draw all simple graphs with the four vertices $\{u, v, w, x\}$ and two edges, one of which is $\{u, v\}$.

Solution   Each possible edge of a simple graph corresponds to a subset of two vertices. Given four vertices, there are $\binom{4}{2} = 6$ such subsets in all: $\{u, v\}$, $\{u, w\}$, $\{u, x\}$, $\{v, w\}$, $\{v, x\}$, and $\{w, x\}$. Now one edge of the graph is specified to be $\{u, v\}$, so any of the remaining five from this list can be chosen to be the second edge. Thus the possibilities are as shown in Figure 11.1.6.

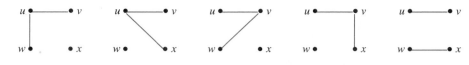

FIGURE 11.1.6

Another important class of graphs consists of those that are "complete" in the sense that all pairs of vertices are connected by edges.

DEFINITION

A **complete graph on *n* vertices**, denoted K_n, is a simple graph with n vertices $v_1, v_2, \ldots v_n$ whose set of edges contains exactly one edge for each pair of distinct vertices.*

EXAMPLE 11.1.8   **Complete Graphs on *n* Vertices: K_2, K_3, K_4, K_5**

The complete graphs K_2, K_3, K_4, and K_5 can be drawn as shown in Figure 11.1.7.

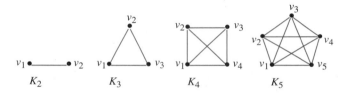

FIGURE 11.1.7

In yet another class of graphs the vertex set can be separated into two subsets: Each vertex in one of the subsets is connected by exactly one edge to each vertex in the other subset, but not to any vertices in its own subset. Such a graph is called *complete bipartite*.

DEFINITION

A **complete bipartite graph on (*m*, *n*) vertices**, denoted $K_{m,n}$, is a simple graph with vertices $v_1, v_2, \ldots, v_m$ and $w_1, w_2, \ldots, w_n$ that satisfies the following properties:

for all $i, k = 1, 2, \ldots, m$ and for all $j, l = 1, 2, \ldots, n$,

1. there is an edge from each vertex v_i to each vertex w_j;

2. there is not an edge from any vertex v_i to any other vertex v_k;

3. there is not an edge from any vertex w_j to any other vertex w_l.

---

*The K stands for the German word *komplett*, which means "complete."

EXAMPLE 11.1.9   **Complete Bipartite Graphs: $K_{3,2}$ and $K_{3,3}$**

The bipartite graphs $K_{3,2}$ and $K_{3,3}$ are illustrated below.

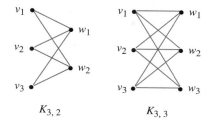

$K_{3,2}$ $K_{3,3}$

DEFINITION

A graph H is said to be a **subgraph** of a graph G if, and only if, every vertex in H is also a vertex in G, every edge in H is also an edge in G, and every edge in H has the same endpoints as in G.

EXAMPLE 11.1.10   **Subgraphs**

List all nonempty subgraphs of the grah G with vertex set $\{v_1, v_2\}$ edge set $\{e_1, e_2, e_3\}$, where the endpoints of e_1 are v_1 and v_2, the endpoints of e_2 are v_1 and v_2, and e_3 is a loop at v_1.

Solution G can be drawn as shown in Figure 11.1.8.

FIGURE 11.1.8

There are 11 nonempty subgraphs of G, which can be grouped according to those that do not have any edges, those that have one edge, those that have two edges, and those that have three edges. The 11 nonempty subgraphs are shown in Figure 11.1.9.

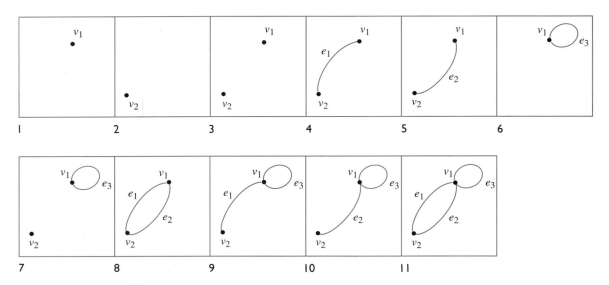

FIGURE 11.1.9

THE CONCEPT OF DEGREE

The *degree of a vertex* is the number of edges that are incident on (or stick out of) the vertex. We will show that the sum of the degrees of all the vertices in a graph is twice the number of edges of the graph.

> **DEFINITION**
>
> Let G be a graph and v a vertex of G. The **degree of v**, denoted **deg(v)**, equals the number of edges that are incident on v, with an edge that is a loop counted twice. The **total degree of G** is the sum of the degrees of all the vertices of G.

Since an edge that is a loop is counted twice, the degree of a vertex can be obtained from the drawing of a graph by counting how many end segments of edges are incident on the vertex. This is illustrated below.

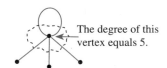

The degree of this vertex equals 5.

EXAMPLE 11.1.11   Degree of a Vertex and Total Degree of a Graph

Find the degree of each vertex of the graph G shown below. Then find the total degree of G.

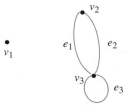

Solution

deg(v_1) = 0 since no edge is incident on v_1 (v_1 is isolated).

deg(v_2) = 2 since both e_1 and e_2 are incident on v_2.

deg(v_3) = 4 since e_1 and e_2 are incident on v_3 and the loop e_3 is also incident on v_3 (and contributes 2 to the degree of v_3).

total degree of G = deg(v_1) + deg(v_2) + deg(v_3) = 0 + 2 + 4 = 6.

Note that the total degree of G, which is 6, equals twice the number of edges of G, which is 3. Roughly speaking, this is true because each edge has two end segments, and each end segment is counted once toward the degree of some vertex. This result generalizes to any graph. ■

THEOREM 11.1.1

If G is any graph, then the sum of the degrees of all the vertices of G equals twice the number of edges of G. Specifically, if the vertices of G are $v_1, v_2, \ldots, v_n$, where n is a positive integer, then

the total degree of G = deg(v_1) + deg(v_2) + $\cdots$ + deg(v_n)

$$= 2 \cdot \text{(the number of edges of } G)$$

Proof:

Let G be a particular but arbitrarily chosen graph. If G does not have any vertices then it does not have any edges, and so its total degree, which is 0, is twice the number of its edges, which is also 0. If G has n vertices v_1, $v_2, \ldots, v_n$ and m edges, where n is a positive integer and m is a nonnegative integer, we claim that each edge of G contributes 2 to the total degree of G. For suppose e is an arbitrarily chosen edge with endpoints v_i and v_j. This edge contributes 1 to the degree of v_i and 1 to the degree of v_j. As shown below, this is true even if $i = j$ because an edge that is a loop is counted twice in computing the degree of the vertex on which it is incident.

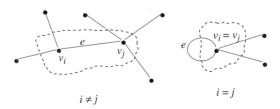

Therefore, e contributes 2 to the total degree of G. Since e was arbitrarily chosen, this shows that *each* edge of G contributes 2 to the total degree of G. Thus

the total degree of G = $2 \cdot$ (the number of edges of G).

The following corollary is an immediate consequence of Theorem 11.1.1.

> ### COROLLARY 11.1.2
>
> The total degree of a graph is even.
>
> Proof:
> Since the total degree of G equals 2 times the number of edges, which is an integer, the total degree of G is even.

EXAMPLE 11.1.12   **Determining Whether Certain Graphs Exist**

Draw a graph with the specified properties or show that no such graph exists.

a. Graph with four vertices of degrees 1, 1, 2, and 3.
b. Graph with four vertices of degrees 1, 1, 3, and 3.
c. Simple graph with four vertices of degrees 1, 1, 3 and 3.

Solution   a. No such graph is possible. By Corollary 11.1.2, the total degree of a graph is even. But a graph with four vertices of degrees 1, 1, 2, and 3 would have a total degree of $1 + 1 + 2 + 3 = 7$, which is odd.

b. Let G be any of the graphs shown in Figure 11.1.10.

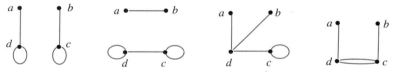

FIGURE 11.1.10

In each case, no matter how the edges are labeled, $\deg(a) = 1$, $\deg(b) = 1$, $\deg(c) = 3$, and $\deg(d) = 3$.

c. There is no simple graph with four vertices of degrees 1, 1, 3, and 3.

Proof (by contradiction):   Suppose there were a simple graph G with four vertices of degrees 1, 1, 3, and 3. Call a and b the vertices of degree 1 and c and d the vertices of degree 3. Since $\deg(c) = 3$ and G does not have any loops or parallel edges (because it is simple), there must be edges that connect c to a, b, and d. (See Figure 11.1.11.)

FIGURE 11.1.11

By the same reasoning, there must be edges connecting d to a, b, and c. (See Figure 11.1.12.)

FIGURE 11.1.12

But then $\deg(a) \geq 2$ and $\deg(b) \geq 2$, which contradicts the supposition that these vertices have degree 1. Hence the supposition is false, and consequently there is no simple graph with four vertices of degrees 1, 1, 3, and 3. ∎

EXAMPLE 11.1.13

Application to an Acquaintance Graph

Is it possible in a group of nine people for each to be friends with exactly five others?

Solution   The answer is no. Imagine constructing an "acquaintance graph" in which each of the nine people is represented by a dot and two dots are joined by an edge if, and only if, the people they represent are friends. Suppose each of the people were friends with exactly five others. Then the degree of each of the nine vertices of the graph would be five, and so the total degree of the graph would be 45. But this contradicts Corollary 11.1.2, which says that the total degree of a graph is even. This contradiction shows that the supposition is false, and hence it is impossible for each person in a group of nine people to be friends with exactly five others. ∎

The following proposition is easily deduced from Corollary 11.1.2 using properties of even and odd integers.

PROPOSITION 11.1.3

In any graph there are an even number of vertices of odd degree.

Proof:

Suppose G is any graph, and suppose G has n vertices of odd degree and m vertices of even degree, where n and m are nonnegative integers. *[We must show that n is even.]* If n is 0, then, since 0 is even, G has an even number of vertices of odd degree. So suppose that $n \geq 1$. Let E be the sum of the degrees of all the vertices of even degree, O the sum of the degrees of all the vertices of odd degree, and T the total degree of G. If $u_1, u_2, \ldots, u_m$ are the vertices of even degree and $v_1, v_2, \ldots, v_n$ are the vertices of odd degree, then

$$E = \deg(u_1) + \deg(u_2) + \cdots + \deg(u_m),$$
$$O = \deg(v_1) + \deg(v_2) + \cdots + \deg(v_n), \quad \text{and}$$
$$T = \deg(u_1) + \cdots + \deg(u_m) + \deg(v_1) + \cdots + \deg(v_n) = E + O.$$

Now T, the total degree of G, is an even integer by Corollary 11.1.2. Also E is even since either E is 0, which is even, or E is a sum of the numbers $\deg(u_i)$, each of which is even. But

$$T = E + O,$$

and, therefore,

$$O = T - E.$$

Hence O is a difference of two even integers, and so O is even.

By assumption, $\deg(v_i)$ is odd for all $i = 1, 2, \ldots, n$. Thus O, an even integer, is a sum of the n odd integers $\deg(v_1), \deg(v_2), \ldots, \deg(v_n)$. But if a sum of n odd integers is even, then n is even. (See exercise 32 at the end of this section.) Therefore, n is even *[as was to be shown].*

EXAMPLE 11.1.14   **Applying the Fact That the Number of Vertices with Odd Degree Is Even**

Is there a graph with ten vertices of degrees 1, 1, 2, 2, 2, 3, 4, 4, 4, and 6?

Solution   No. Such a graph would have three vertices of odd degree, which is impossible by Proposition 11.1.3.

Note that this same result could have been deduced directly from Corollary 11.1.2 by computing the total degree $(1 + 1 + 2 + 2 + 2 + 3 + 4 + 4 + 4 + 6 = 29)$ and noting that it is odd. However, use of Proposition 11.1.3 gives the result without the need to do any addition. ■

EXERCISE SET 11.1

In 1 and 2, graphs are represented by drawings. Define each graph formally by specifying its vertex set, its edge set, and a table giving the edge-endpoint function.

1.

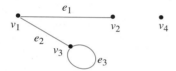

2.

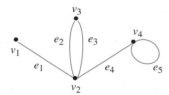

In 3 and 4, draw pictures of the specified graphs.

3. Graph G has vertex set $\{v_1, v_2, v_3, v_4, v_5\}$ and edge set $\{e_1, e_2, e_3, e_4\}$, with edge-endpoint function as follows:

Edge	Endpoints
e_1	$\{v_1, v_2\}$
e_2	$\{v_1, v_2\}$
e_3	$\{v_2, v_3\}$
e_4	$\{v_2\}$

4. Graph H has vertex set $\{v_1, v_2, v_3, v_4, v_5\}$ and edge set $\{e_1, e_2, e_3, e_4\}$ with edge-endpoint function as follows:

Edge	Endpoints
e_1	$\{v_1\}$
e_2	$\{v_2, v_3\}$
e_3	$\{v_2, v_3\}$
e_4	$\{v_1, v_5\}$

In 5–7, show that the two drawings represent the same graph by labeling the vertices and edges of the right-hand drawing to correspond to those of the left-hand drawing.

5.

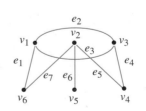

6.

*Exercises with blue numbers or letters have solutions in Appendix B. The symbol H indicates that only a hint or partial solution is given. The symbol ◆ signals that an exercise is more challenging than usual.

7.

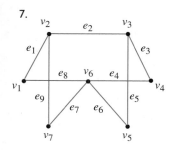

For each of the graphs in 8 and 9,

(i)   **find all edges that are incident on v_1;**

(ii) **find all vertices that are adjacent to v_3;**

(iii) **find all edges that are adjacent to e_1;**

(iv) **find all loops;**

(v)   **find all parallel edges;**

(vi) **find all isolated vertices;**

(vii) **find the degree of v_3;**

(viii) **find the total degree of the graph.**

8.

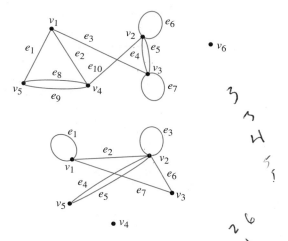

9.

10. Use the graph of Example 11.1.5 to determine
   a. whether *Sports Illustrated* contains printed writing;
   b. whether *Poetry Magazine* contains long words.

11. Find three other winning sequences of moves for the vegetarians and the cannibals problem of Example 11.1.6.

12. Another famous puzzle used as an example in the study of artificial intelligence seems first to have appeared in a collection of problems, *Problems for the Quickening of the Mind,* which was compiled about 775 A.D. It involves a wolf, a goat, a bag of cabbage, and a ferry-man. From an initial position on the left bank of a river, the ferryman is to transport the wolf, the goat, and the

cabbage to the right bank. The difficulty is that the ferryman's boat is only big enough for him to transport one object at a time, other than himself. Yet, for obvious reasons, the wolf cannot be left alone with the goat, and the goat cannot be left alone with the cabbage. How should the ferryman proceed?

13. Solve the vegetarians and cannibals puzzle for the case where there are three vegetarians and three cannibals to be transported from one side of a river to another.

H 14. Two jugs *A* and *B* have capacities of three quarts and five quarts, respectively. Can you use the jugs to measure out exactly one quart of water while obeying the following restrictions: You may fill either jug to capacity from a water tap; you may empty the contents of either jug into a drain; and you may pour water from either jug into the other.

In each of 15–23, either draw a graph with the specified properties or explain why no such graph exists.

15. Graph with five vertices of degrees 1, 2, 3, 3, and 5.

16. Graph with four vertices of degrees 1, 2, 3, and 3.

17. Graph with four vertices of degrees 1, 1, 1, and 4.

18. Graph with four vertices of degrees 1, 2, 3, and 4.

19. Simple graph with four vertices of degrees 1, 2, 3, and 4.

20. Simple graph with five vertices of degrees 2, 3, 3, 3, and 5.

21. Simple graph with five vertices of degrees 1, 1, 1, 2, and 3.

22. Simple graph with six edges and all vertices of degree 3.

23. Simple graph with nine edges and all vertices of degree 3.

24. Find all nonempty subgraphs of each of the following graphs.

a.

b.

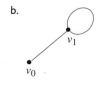

c.

25. a. In a group of 15 people, is it possible for each person to have exactly 3 friends? Explain. (Assume that friendship is a symmetric relationship: If x is a friend of y, then y is a friend of x.)
  b. In a group of 4 people, is it possible for each person to have exactly 3 friends? Why?

26. In a group of 25 people, is it possible for each to shake hands with exactly 1 other person? Explain.

27. Is there a simple graph, each of whose vertices has even degree? Explain.

28. Suppose a graph has vertices of degrees 0, 2, 2, 3, and 9. How many edges does the graph have?

29. Suppose a graph has vertices of degrees 1, 1, 4, 4, and 6. How many edges does the graph have?

30. Suppose that G is a graph with v vertices and e edges and that the degree of each vertex is at least $d_{\min}$ and at most $d_{\max}.$ Show that

$$\frac{1}{2}d_{\min} \cdot v \le e \le \frac{1}{2}d_{\max} \cdot v.$$

31. Prove that any sum of an odd number of odd integers is odd.

32. Deduce from exercise 31 that for any positive integer n, if a sum of n odd integers is even, then n is even.

H33. Recall that K_n denotes a complete graph on n vertices.
  a. Draw K_6.
  b. Show that for all integers $n \ge 1$, the number of edges of K_n is $\dfrac{n(n-1)}{2}.$

34. Use the result of exercise 33 to show that the number of edges of a simple graph with n vertices is less than or equal to $\dfrac{n(n-1)}{2}.$

35. Is there a nonempty simple graph with twice as many edges as vertices? Explain. (You may find it helpful to use the result of exercise 34.)

36. Recall that $K_{m,n}$ denotes a complete bipartite graph on (m, n) vertices.
  a. Draw $K_{4,2}$ **b.** Draw $K_{1,3}$ **c.** Draw $K_{3,4}$
  d. How many vertices of $K_{m,n}$ have degree m? degree n?
  e. What is the total degree of $K_{m,n}$?
  f. Find a formula in terms of m and n for the number of edges of $K_{m,n}.$ Explain.

37. A **bipartite graph** G is a simple graph whose vertex set can be partitioned into two mutually disjoint nonempty subsets V_1 and V_2 such that vertices in V_1 may be connected to vertices in V_2, but no vertices in V_1 are connected to other vertices in V_1 and no vertices in V_2 are connected to other vertices in V_2. For example, the graph G illustrated in (i) can be redrawn as shown in (ii). From the drawing in (ii), you can see that G is

bipartite with mutually disjoint vertex sets $V_1 = \{v_1, v_3, v_5\}$ and $V_2 = \{v_2, v_4, v_6\}$.

(i)

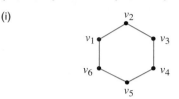

(ii)

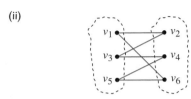

Find which of the following graphs are bipartite. Redraw the bipartite graphs so that their bipartite nature is evident.

a.

b.

c.

d.

e.

f.

38. Suppose r and s are any positive integers. Does there exist a graph G with the property that G has vertices of degrees r and s and of no other degrees? Explain.

Definition: If G is a simple graph, the **complement of G**, denoted **G'**, is obtained as follows: The vertex set of G' is identical to the vertex set of G. However, two distinct vertices v and w of G' are connnected by an edge if, and only if, v and w are not connected by an edge in G. For example, if G is the graph

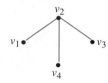

then G' is

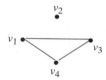

39. Find the complement of each of the following graphs.

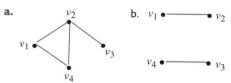

40. a. Find the complement of the graph K_4, the complete graph on four vertices. (See Example 11.1.8.)
    b. Find the complement of the graph $K_{3,2}$, the complete bipartite graph on (3, 2) vertices. (See Example 11.1.9.)

41. Suppose that in a group of five people A, B, C, D, and E the following pairs of people are acquainted with each other: A and C, A and D, B and C, C and D, C and E.
    a. Draw a graph to represent this situation.

    b. Draw a graph that illustrates who among these five people are *not* acquainted. That is, draw an edge between two people if, and only if, they are not acquainted.

H 42. Let G be a simple graph with n vertices. What is the relation between the number of edges of G and the number of edges of the complement G'?

43. Show that at a party with at least two people, there are at least two mutual acquaintances or at least two mutual strangers.

44. a. In a simple graph, must every vertex have degree that is less than the number of vertices in the graph? Why?
    b. Can there be a simple graph that has four vertices each with a different degree?
 H ◆ c. Can there be a simple graph that has n vertices each with a different degree?

H ◆ 45. In a group of two or more people, must there always be at least two people who are acquainted with the same number of people within the group? Why?

H 46. In this exercise a graph is used to help solve a scheduling problem. Eleven faculty members in a mathematics department serve on the following committees:

Undergraduate Education: Bergen, Jones, Wong, Cohen

Graduate Education: Gatto, Moussa, Cohen, Pranger

Colloquium: Vagi, Goldman, Ash

Library: Cortzen, Bergen, Vagi

Hiring: Gatto, Goldman, Moussa, Jones

Personnel: Moussa, Georgakis, Cortzen

The committees must all meet during the first week of classes, but there are only three time slots available. Find a schedule that will allow all faculty members to attend the meetings of all committees on which they serve. To do this, represent each committee as the vertex of a graph and draw an edge between two vertices if the two committees have a common member.

11.2 PATHS AND CIRCUITS

One can begin to reason only when a clear picture has been formed in the imagination.
(W. W. Sawyer,
Mathematician's Delight,
1943)

The subject of graph theory began in the year 1736 when the great mathematician Leonhard Euler published a paper that contained the solution to the following puzzle.

The town of Königsberg in Prussia (now Kaliningrad in Russia) was built at a point where two branches of the Pregel River came together. It consisted of an island and some land along the river banks. These were connected by seven bridges as shown in Figure 11.2.1.

The question is this: Is it possible for a person to take a walk around town, starting and ending at the same location and crossing each of the seven bridges exactly once?

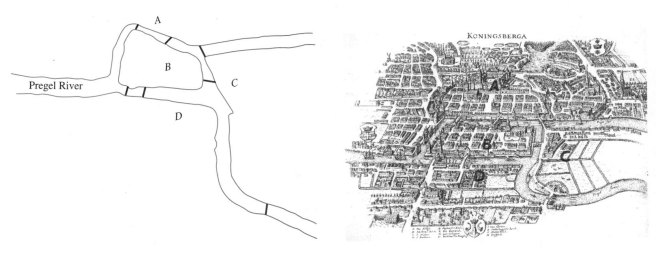

FIGURE 11.2.1   The Seven Bridges of Königsberg

Leonhard Euler (1707–1783)

To solve this puzzle, Euler translated it into a graph theory problem. He noticed that all points of a given land mass can be identified with each other since a person can travel from any one point to any other point of the same land mass without crossing a bridge. Thus for the purpose of solving the puzzle, the map of Königsberg can be identified with the graph shown in Figure 11.2.2 in which the vertices *A*, *B*, *C*, and *D* represent land masses and the seven edges represent the seven bridges.

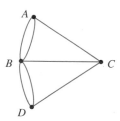

FIGURE 11.2.2 Graph Version of Königsberg Map

In terms of this graph, the question becomes the following:

Is it possible to find a route through the graph that starts and ends at some vertex, one of *A*, *B*, *C*, or *D*, and traverses each edge exactly once?

Equivalently:

Is it possible to trace this graph, starting and ending at the same point, without ever lifting your pencil from the paper?

Take a few minutes to think about the question yourself. Can you find a route that meets the requirements? Try it!

Looking for a route is frustrating because you continually find yourself at a vertex that does not have an unused edge on which to leave while elsewhere there are unused edges that must still be traversed. If you start at vertex A, for example, each time you pass through vertices B, C, or D, you use up two edges because you arrive on one edge and depart on a different one. So, if it is possible to find a route that uses all the edges of the graph and starts and ends at A, then the total number of arrivals and departures from each vertex B, C, and D must be a multiple of 2. Or, in other words, the degrees of the vertices B, C, and D must be even. But they are not: $\deg(B) = 5$, $\deg(C) = 3$, and $\deg(D) = 3$. Hence there is no route that solves the puzzle by starting and ending at A. Similar reasoning can be used to show that there are no routes that solve the puzzle by starting and ending at B, C, or D. Therefore, it is impossible to travel all around the city crossing each bridge exactly once.

DEFINITIONS

Travel in a graph is accomplished by moving from one vertex to another along a sequence of adjacent edges. In the graph of Figure 11.2.3, for instance, you can go from u_1 to u_4 by taking f_1 to u_2 and then f_7 to u_4. This is represented by writing

$$u_1 f_1 u_2 f_7 u_4.$$

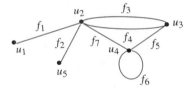

FIGURE 11.2.3

Or you could take the roundabout route

$$u_1 f_1 u_2 f_3 u_3 f_4 u_2 f_3 u_3 f_5 u_4 f_6 u_4 f_7 u_2 f_3 u_3 f_5 u_4.$$

Certain types of sequences of adjacent vertices and edges are of special importance in graph theory: those that do not have a repeated edge, those that do not have a repeated vertex, and those that start and end at the same vertex.

DEFINITION

Let G be a graph and let v and w by vertices in G.

A **walk from v to w** is a finite alternating sequence of adjacent vertices and edges of G. Thus a walk has the form

$$v_0 e_1 v_1 e_2 \ldots v_{n-1} e_n v_n,$$

where the v's represent vertices, the e's represent edges, $v_0 = v$, $v_n = w$, and for all $i = 1, 2, \ldots n$, v_{i-1} and v_i are the endpoints of e_i. The **trivial walk from v to v** consists of the single vertex v.

A **path from v to w** is a walk from v to w that does not contain a repeated edge. Thus a path from v to w is a walk of the form

$$v = v_0 e_1 v_1 e_2 \ldots v_{n-1} e_n v_n = w,$$

where all the e_i are distinct (that is, $e_i \neq e_k$ for any $i \neq k$).

A **simple path from v to w** is a path that does not contain a repeated vertex. Thus a simple path is a walk of the form

$$v = v_0 e_1 v_1 e_2 \ldots v_{n-1} e_n v_n = w,$$

where all the e_i are distinct and all the v_j are also distinct (that is, $v_j \neq v_m$ for any $j \neq m$).

A **closed walk** is a walk that starts and ends at the same vertex.

A **circuit** is a closed walk that does not contain a repeated edge. Thus a circuit is a walk of the form

$$v_0 e_1 v_1 e_2 \ldots v_{n-1} e_n v_n,$$

where $v_0 = v_n$ and all the e_i are distinct.

A **simple circuit** is a circuit that does not have any other repeated vertex except the first and last. Thus a simple circuit is a walk of the form

$$v_0 e_1 v_1 e_2 \ldots v_{n-1} e_n v_n,$$

where all the e_i are distinct and all the v_j are distinct except that $v_0 = v_n$.

For ease of reference, these definitions are summarized in the following table:

	Repeated Edge?	Repeated Vertex?	Starts and Ends at Same Point?
walk	allowed	allowed	allowed
path	no	allowed	allowed
simple path	no	no	no
closed walk	allowed	allowed	yes
circuit	no	allowed	yes
simple circuit	no	first and last only	yes

Often a walk can be specified unambiguously by giving either a sequence of edges or a sequence of vertices. The next two examples show how this is done.

EXAMPLE 11.2.1   Notation for Walks

a. In the graph of Figure 11.2.4, the notation $e_1 e_2 e_4 e_3$ refers unambiguously to the walk $v_1 e_1 v_2 e_2 v_3 e_4 v_3 e_3 v_2$. On the other hand, the notation e_1 is ambiguous if used to refer to a walk. It could mean either $v_1 e_1 v_2$ or $v_2 e_1 v_1$.

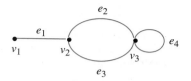

FIGURE 11.2.4

b. In the graph of Figure 11.2.4, the notation v_2v_3 is ambiguous if used to refer to a walk. It could mean $v_2e_2v_3$ or $v_2e_3v_3$. On the other hand, in the graph of Figure 11.2.5, the notation $v_1v_2v_2v_3$ refers unambiguously to the walk $v_1e_1v_2e_2v_2e_3v_3$.

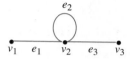

FIGURE 11.2.5   ∎

Note that if a graph G does not have any parallel edges, then any walk in G is uniquely determined by its sequence of vertices.

EXAMPLE 11.2.2   Walks, Paths, and Circuits

In the graph of Figure 11.2.6, determine which of the following walks are paths, simple paths, circuits, and simple circuits:

a. $v_1e_1v_2e_3v_3e_4v_3e_5v_4$ b. $e_1e_3e_5e_5e_6$ c. $v_2v_3v_4v_5v_3v_6v_2$
d. $v_2v_3v_4v_5v_6v_2$ e. $v_2v_3v_4v_5v_6v_3v_2$ f. v_1

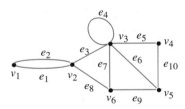

FIGURE 11.2.6

Solution   a. This walk has a repeated vertex but does not have a repeated edge, so it is a path from v_1 to v_4 but not a simple path.
b. This is just a walk from v_1 to v_5. It is not a path because it has a repeated edge.
c. This walk starts and ends at v_2 and does not have a repeated edge, so it is a circuit. Since the vertex v_3 is repeated in the middle, it is not a simple circuit.
d. This walk starts and ends at v_2, does not have a repeated edge, and does not have a repeated vertex. Thus it is a simple circuit.
e. This is just a closed walk starting and ending at v_2. It is neither a circuit nor a simple circuit because edge e_3 and vertex v_3 are repeated.
f. The first vertex of this walk is the same as its last vertex. (Try to disprove this statement if you are inclined not to believe it!) Also, this walk has neither a repeated vertex nor a repeated edge. Thus it is a simple circuit. (A circuit such as this is called a **trivial circuit**.)   ∎

CONNECTEDNESS

It is easy to understand the concept of connectedness on an intuitive level. Roughly speaking, a graph is connected if it is possible to travel from any vertex to any other vertex along a sequence of adjacent edges of the graph. The formal definition of connectedness is stated in terms of walks.

DEFINITION

Let G be a graph. Two **vertices v and w of G are connected** if, and only if, there is a walk from v to w. The **graph G is connected** if, and only if, given *any* two vertices v and w in G, there is a walk from v to w. Symbolically:

G is connected $\Leftrightarrow$ $\forall$ vertices $v, w \in V(G)$, $\exists$ a walk from v to w.

If you take the negation of this definition, you will see that a graph G is *not connected* if, and only if, there are two vertices of G that are not connected by any walk.

EXAMPLE 11.2.3   Connected and Disconnected Graphs

Which of the graphs below are connected?

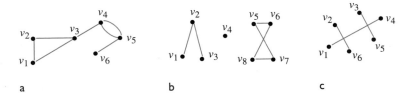

a   b   c

Solution   The graph represented in (a) is connected while those of (b) and (c) are not. To understand why (c) is not connected, recall that in a drawing of a graph two edges may cross at a point that is not a vertex. Thus the graph in (c) can be redrawn as shown below.

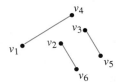

Some useful facts relating circuits and connectedness are collected in the following lemma. Proofs are left to the exercises.

LEMMA 11.2.1

Let G be a graph.

a. If G is connected, then any two distinct vertices of G can be connected by a simple path.

b. If vertices v and w are part of a circuit in G and one edge is removed from the circuit, then there still exists a path from v to w in G.

c. If G is connected and G contains a circuit, then an edge of the circuit can be removed without disconnecting G.

Look back at Example 11.2.3. The graphs in (b) and (c) are both made up of three pieces, each of which is itself a connected graph. A *connected component* of a graph is a connected subgraph of largest possible size.

DEFINITION

A graph H is a **connected component** of a graph G if, and only if,

1. H is a subgraph of G;

2. H is connected;

3. no connected subgraph of G has H as a subgraph and contains vertices or edges that are not in H.

The fact is that any graph is a kind of union of its connected components.

EXAMPLE 11.2.4   Connected Components

Find all connected components of the graph G shown below.

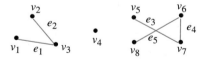

Solution G has three connected components: H_1, H_2, and H_3 with vertex sets V_1, V_2, and V_3 and edge sets E_1, E_2, and E_3, where

$$V_1 = \{v_1, v_2, v_3\}, \qquad E_1 = \{e_1, e_2\},$$
$$V_2 = \{v_4\}, \qquad E_2 = \varnothing,$$
$$V_3 = \{v_5, v_6, v_7, v_8\}, \qquad E_3 = \{e_3, e_4, e_5\}.$$ ■

EULER CIRCUITS

Now we return to consider general problems similar to the puzzle of the Königsberg bridges. The following definition is made in honor of Euler.

DEFINITION

Let G be a graph. An **Euler circuit** for G is a circuit that contains every vertex and every edge of G. That is, an Euler circuit for G is a sequence of adjacent vertices and edges in G that starts and ends at the same vertex, uses every vertex of G at least once, and uses every edge of G exactly once.

The analysis used earlier to solve the puzzle of the Königsberg bridges generalizes to prove the following theorem:

THEOREM 11.2.2

If a graph has an Euler circuit, then every vertex of the graph has even degree.

Proof:

Suppose G is a graph that has an Euler circuit. *[We must show that given any vertex v of G, the degree of v is even.]* Let v be any particular but arbitrarily chosen vertex of G. Since the Euler circuit contains every edge of G, it contains all edges incident on v. Now imagine taking a journey that begins in the middle of one of the edges adjacent to the start of the Euler circuit and continues around the Euler circuit to end in the middle of the starting edge. (See Figure 11.2.7.)

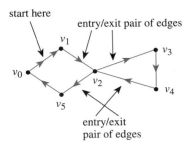

Euler circuit: $v_0v_1v_2v_3v_4v_2v_5v_0$
Let v be v_2.
Each time v_2 is entered by one edge it is exited by another edge.

FIGURE 11.2.7   Example for Proof of Theorem 11.2.2

Each time v is entered by traveling along one edge, it is immediately exited by traveling along another edge (since the journey ends in the *middle* of an edge). Because the Euler circuit uses every edge of G exactly once, every edge incident on v is traversed exactly once in this process. Hence the edges incident on v occur in entry/exit pairs, and consequently the degree of v must be a multiple of 2. But that means that the degree of v is even *[as was to be shown]*.

Recall that the contrapositive of a statement is logically equivalent to the statement. The contrapositive of Theorem 11.2.2 is as follows:

CONTRAPOSITIVE VERSION OF THEOREM 11.2.2

If some vertex of a graph has odd degree, then the graph does not have an Euler circuit.

This version of Theorem 11.2.2 is useful for showing that a given graph does *not* have an Euler circuit.

EXAMPLE 11.2.5   **Showing That a Graph Does Not Have an Euler Circuit**

Show that the graph below does not have an Euler circuit.

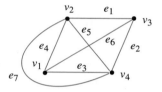

Solution   Vertices v_1 and v_3 both have degree 3, which is odd. Hence by (the contrapositive form of) Theorem 11.2.2, this graph does not have an Euler circuit.   ■

Now consider the converse of Theorem 11.2.2: If every vertex of a graph has even degree, then the graph has an Euler circuit. Is this true? The answer is no. There is a graph G such that every vertex of G has even degree but G does not have an Euler circuit. (In fact, there are many such graphs.) Figure 11.2.8 shows one example.

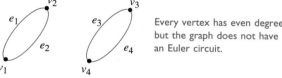

Every vertex has even degree, but the graph does not have an Euler circuit.

FIGURE 11.2.8

Note that the graph in Figure 11.2.8 is not connected. It turns out that although the converse of Theorem 11.2.2 is false, a modified converse is true: If every vertex of a graph has even degree *and* if the graph is connected, then the graph has an Euler circuit. The proof of this fact is constructive: It contains an algorithm to find an Euler circuit for any connected graph in which every vertex has even degree.

THEOREM 11.2.3

If every vertex of a nonempty graph has even degree and if the graph is connected, then the graph has an Euler circuit.

Proof:

Suppose that G is any nonempty connected graph and that every vertex of G has even degree. *[We must find an Euler circuit for G.]* Construct a circuit C by the following algorithm:

Step 1:   Pick any vertex v of G at which to start.
[This step can be accomplished because the vertex set of G is nonempty by assumption.]

Step 2:   Pick any sequence of adjacent vertices and edges, starting and ending at v and never repeating an edge. Call the resulting circuit C.
[This step can be performed for the following reasons: Since the degree of each vertex of G is even, as each vertex other than v is entered by traveling on one edge, it can be exited by traveling on another previously unused edge. Thus a sequence of distinct adjacent edges can be produced indefinitely as long as v is not reached. But since the number of edges of the graph is finite (by definition of graph), the sequence of distinct edges cannot go on forever. Thus the sequence must eventually return to the starting vertex v.]

Step 3:   Check whether C contains every edge and vertex of G. If so, C is an Euler circuit, and we are finished. If not, perform the following steps.

> *Step 3a* Remove all edges of C from G and also any vertices that become isolated when the edges of C are removed. Call the resulting subgraph G'.
> *[Note that G' may not be connected (as illustrated in Figure 11.2.9), but every vertex of G' has even degree (since removing the edges of C removes an even number of edges from each vertex and the difference of two even integers is even).]*

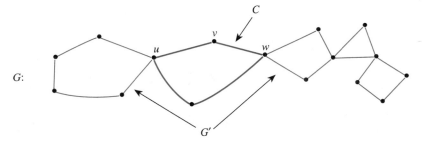

FIGURE 11.2.9

> *Step 3b* Pick any vertex u common to both C and G'.
> *[There must be at least one such vertex since G is connected. Exercise 44 at the end of this section asks for a proof of this fact. (In Figure 11.2.9 there are two such vertices: u and w.)]*

Step 3c Pick any sequence of adjacent vertices and edges of G', starting and ending at w and never repeating an edge. Call the resulting circuit C'.

[This can be done since the degree of each vertex of G' is even and G' is finite. See the justification for step 2.]

Step 3d Patch C and C' together to create a new circuit C'' as follows: Start at v and follow C all the way to w. Then follow C' all the way back to w. After that, continue along the untraveled portion of C to return to v.

[The effect of executing steps 3c and 3d for the graph of Figure 11.2.9 is shown in Figure 11.2.10.]

G:

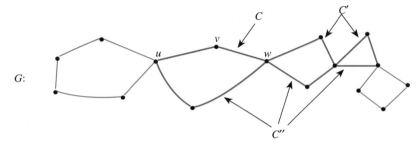

FIGURE 11.2.10

Step 3e Let $C = C''$ and go back to step 3.

Since the graph G is finite, execution of the steps outlined in this algorithm must eventually terminate. At that point an Euler circuit for G will have been constructed. (Note that because of the element of choice in steps 1, 2, 3b, and 3c, a variety of different Euler circuits can be produced by using this algorithm.)

EXAMPLE 11.2.6 Finding an Euler Circuit

Use Theorem 11.2.3 to check that the graph below has an Euler circuit. Then use the algorithm from the proof of the theorem to find an Euler circuit for the graph.

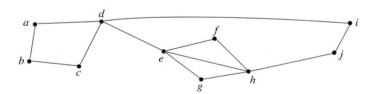

Solution Observe that $\deg(a) = \deg(b) = \deg(c) = \deg(f) = \deg(g) = \deg(i) = \deg(j) = 2$ and that $\deg(d) = \deg(e) = \deg(h) = 4$. Hence all vertices have even degree. Also, the graph is connected. Thus by Theorem 11.2.3 the graph has an Euler circuit.

To construct an Euler circuit using the algorithm of Theorem 11.2.3, let $v = a$ and let C be

$$C: abcda.$$

C is represented by the labeled edges shown below.

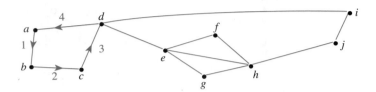

Observe that C is not an Euler circuit for the graph but that C intersects the rest of the graph at d. Let C' be

$$C': deghjid.$$

Path C' into C to obtain

$$C'': abcdeghjida.$$

Set $C = C''$. Then C is represented by the labeled edges shown below.

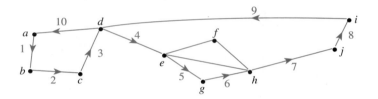

Observe that C is not an Euler circuit for the graph but that it intersects the rest of the graph at e. Let C' be

$$C': efhe.$$

Patch C' into C to obtain

$$C'': abcdefheghjida.$$

Set $C = C''$. Then C is represented by the labeled edges shown below.

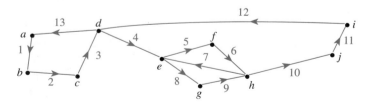

Since C includes every edge of the graph exactly once, C is an Euler circuit for the graph. ∎

In exercise 45 at the end of this section you are asked to show that any graph with an Euler circuit is connected. This result can be combined with Theorems 11.2.2 and

11.2.3 to give a complete characterization of graphs that have Euler circuits, as stated in Theorem 11.2.4.

THEOREM 11.2.4

A graph G has an Euler circuit if, and only if, G is connected and every vertex of G has even degree.

A corollary to Theorem 11.2.4 gives a criterion for determining when it is possible to find a walk from one vertex of a graph to another, passing through every vertex of the graph at least once and every edge of the graph exactly once.

DEFINITION

Let G be a graph and let v and w be two vertices of G. An **Euler path from v to w** is a sequence of adjacent edges and vertices that starts at v, ends at w, passes through every vertex of G at least once, and traverses every edge of G exactly once.

COROLLARY 11.2.5

Let G be a graph and let v and w be two vertices of G. There is an Euler path from v to w if, and only if, G is connected, v and w have odd degree, and all other vertices of G have even degree.

The proof of this corollary is left as an exercise.

EXAMPLE 11.2.7   Finding an Euler Path

The floor plan shown below is for a house that is open for public viewing. Is it possible to find a path that starts in room A, ends in room B, and passes through every interior doorway of the house exactly once? If so, find such a path.

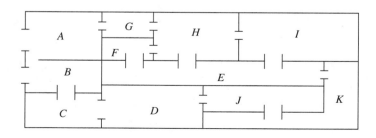

Solution   Let the floor plan of the house be represented by the graph below.

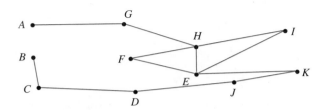

Each vertex of this graph has even degree except for *A* and *B*, which each have degree 1. Hence by Corollary 11.2.5, there is an Euler path from *A* to *B*. One such path is

$$AGHFEIHEKJDCB.$$ ■

HAMILTONIAN CIRCUITS

Sir Wm. Hamilton
(1805–1865)

Theorem 11.2.4 completely answers the following question: Given a graph *G*, is it possible to find a circuit for *G* in which all the *edges* of *G* appear exactly once? A related question is this: Given a graph *G*, is it possible to find a circuit for *G* in which all the *vertices* of *G* appear exactly once (except the first and the last)? In 1859 the Irish mathematician Sir William Rowan Hamilton introduced a puzzle in the shape of a dodecahedron (DOH-dek-a-HEE-dron). (Figure 11.2.11 contains a drawing of a dodecahedron, which is a solid figure with twelve identical pentagonal faces.)

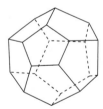

FIGURE 11.2.11   Dodecahedron

Each vertex was labeled with the name of a city—London, Paris, Hong Kong, New York, and so on. The problem Hamilton posed was to start at one city and tour the world by visiting each other city exactly once and returning to the starting city. The puzzle was quite easy to solve by imagining the surface of the dodecahedron stretched out and laid flat in the plane as shown in Figure 11.2.12.

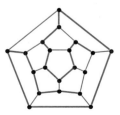

FIGURE 11.2.12

The circuit denoted with thick lines is one solution. Note that although every city is visited, many edges are omitted from the circuit. (More difficult versions of the puzzle required that certain cities be visited in a certain order.)

The following definition is made in honor of Hamilton.

DEFINITION

Given a graph G, a **Hamiltonian circuit** for G is a simple circuit that includes every vertex of G. That is, a Hamiltonian circuit for G is a sequence of adjacent vertices and distinct edges in which every vertex of G appears exactly once.

Note that while an Euler circuit for a graph G must include every vertex of G, it may visit some vertices more than once and hence may not be a Hamiltonian circuit. On the other hand, a Hamiltonian circuit for G does not need to include all the edges of G and hence may not be an Euler circuit.

Despite the analogous-sounding definitions of Euler and Hamiltonian circuits, the mathematics of the two are very different. Theorem 11.2.4 gives a simple criterion for determining whether or not a given graph has an Euler circuit. Unfortunately, there is no analogous criterion for determining whether or not a given graph has a Hamiltonian circuit, nor is there even an efficient algorithm for finding such a circuit. There is, however, a simple technique that can be used in many cases to show that a graph does *not* have a Hamiltonian circuit. This follows from the following considerations:

Suppose a graph G has a Hamiltonian circuit C given concretely as

$$C: v_0 e_1 v_1 e_2 \ldots v_{n-1} e_n v_n.$$

Since C is a simple circuit, all the e_i are distinct and all the v_j are distinct except that $v_0 = v_n$. Let H be the subgraph of G that is formed by taking the vertices and edges of C. An example of such an H is shown in Figure 11.2.13.

H is indicated by the thick lines.

FIGURE 11.2.13

Note that H has the same number of edges as it has vertices since all its n edges are distinct and so are its n vertices $v_1, v_2, \ldots, v_n$. Also by definition of Hamiltonian circuit, every vertex of G is a vertex of H, and H is connected since any two of its vertices lie on a circuit. In addition, every vertex of H has degree 2. The reason for this is that there are exactly two edges incident on any vertex, namely e_i and e_{i+1} for any vertex v_i except $v_0 = v_n$ and e_1 and e_n for $v_0(= v_n)$. These observations have established the truth of the following proposition:

PROPOSITION 11.2.6

If a graph G has a Hamiltonian circuit then G has a subgraph H with the following properties:

1. H contains every vertex of G;
2. H is connected;
3. H has the same number of edges as vertices;
4. every vertex of H has degree 2.

Recall that the contrapositive of a statement is logically equivalent to the statement. The contrapositive of Proposition 11.2.6 says that if a graph G does *not* have a subgraph H with properties (1)–(4), then G does *not* have a Hamiltonian circuit.

EXAMPLE 11.2.8   **Showing That a Graph Does Not Have a Hamiltonian Circuit**

Prove that the graph G shown in Figure 11.2.14 does not have a Hamiltonian circuit.

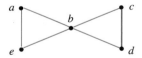

FIGURE 11.2.14

Solution   If G has a Hamiltonian circuit, then by Proposition 11.2.6, G has a subgraph H that (1) contains every vertex of G, (2) is connected, (3) has the same number of edges as vertices, and (4) is such that every vertex has degree 2. Suppose such a subgraph H exists. In other words, suppose there is a connected subgraph H of G such that H has five vertices (a, b, c, d, e) and five edges and such that every vertex of H has degree 2. Since the degree of b in G is 4 and every vertex of H has degree 2, two edges incident on b must be removed from G to create H. Edge $\{a, b\}$ cannot be removed because if it were, vertex a would have degree less than 2 in H. Similar reasoning shows that edges $\{e, b\}$, $\{b, a\}$, and $\{b, d\}$ cannot be removed either. It follows that the degree of b in H must be 4, which contradicts the condition that every vertex in H has degree 2 in H. Hence no such subgraph H exists, and so G does not have a Hamiltonian circuit. ∎

The traveling salesperson problem, discussed in the next example, is a variation of the problem of finding a Hamiltonian circuit for a graph.

EXAMPLE 11.2.9   A Traveling Salesperson Problem

Imagine that the drawing in Figure 11.2.15 is a map showing four cities and the distances in kilometers between them. Suppose that a salesperson must travel to each city exactly once, starting and ending in city A. Which route from city to city will minimize the total distance that must be traveled?

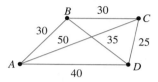

FIGURE 11.2.15

Solution   This problem can be solved by writing all possible Hamiltonian circuits starting and ending at A and calculating the total distance traveled for each.

Route	Total Distance (in kilometers)	
ABCDA	30 + 30 + 25 + 40 = 125	
ABDCA	30 + 35 + 25 + 50 = 140	
ACBDA	50 + 30 + 35 + 40 = 155	
ACDBA	140	[*ABDCA* backwards]
ADBCA	155	[*ACBDA* backwards]
ADCBA	125	[*ABCDA* backwards]

Thus either route *ABCDA* or *ADCBA* gives a minimum total distance of 125 kilometers.   ■

The general traveling salesperson problem involves finding a Hamiltonian circuit to minimize the total distance traveled for an arbitrary graph with n vertices in which each edge is marked with a distance. One way to solve the general problem is to use the method of Example 11.2.9: Write down all Hamiltonian circuits starting and ending at a particular vertex, compute the total distance for each, and pick one for which this total is minimal. However, even for medium-sized values of n this method is impractical. In the language of Chapter 9, any algorithm to implement this method has exponential order. Observe that for a complete graph with 30 vertices, there would be $29! \cong 8.84 \times 10^{30}$ different Hamiltonian circuits starting and ending at a particular vertex to check. Even if each circuit could be found and its total distance computed in just one microsecond, it would require approximately 2.8×10^{17} years to finish the computation. At present, there is no known algorithm for solving the general traveling salesperson problem that is more efficient. However, there are efficient algorithms that find "pretty good" solutions—that is, circuits that, while not necessarily having the least possible total distance, have smaller total distances than most other Hamiltonian circuits.

EXERCISE SET 11.2

1. In the graph below, determine whether the following walks are paths, simple paths, closed walks, circuits, simple circuits, or are just walks.

   a. $v_0 e_1 v_1 e_{10} v_5 e_9 v_2 e_2 v_1$

   b. $v_4 e_7 v_2 e_9 v_5 e_{10} v_1 e_3 v_2 e_9 v_5$

   c. v_2

   d. $v_5 v_2 v_3 v_4 v_4 v_5$

   e. $v_2 v_3 v_4 v_5 v_2 v_4 v_3 v_2$

   f. $e_5 e_8 e_{10} e_3$

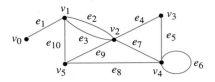

2. In the graph below, determine whether the following walks are paths, simple paths, closed walks, circuits, simple circuits, or are just walks.

   a. $v_1 e_2 v_2 e_3 v_3 e_4 v_4 e_5 v_2 e_2 v_1 e_1 v_0$

   b. $v_2 v_3 v_4 v_5 v_2$

   c. $v_4 v_2 v_3 v_4 v_5 v_2 v_4$

   d. $v_2 v_1 v_5 v_2 v_3 v_4 v_2$

   e. $v_0 v_5 v_2 v_3 v_4 v_2 v_1$

   f. $v_5 v_4 v_2 v_1$

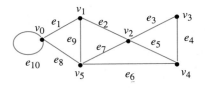

3. Let G be the graph

   and consider the walk $v_1 e_1 v_2 e_2 v_1$.

   a. Can this walk be written unambiguously as $v_1 v_2 v_1$? Why?

   b. Can this walk be written unambiguously as $e_1 e_2$? Why?

4. Consider the following graph.

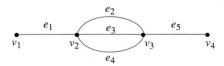

   a. How many simple paths are there from v_1 to v_4?
   b. How many paths are there from v_1 to v_4?
   c. How many walks are there from v_1 to v_4?

5. Consider the following graph.

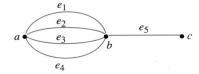

   a. How many simple paths are there from a to c?
   b. How many paths are there from a to c?
   c. How many walks are there from a to c?

6. An edge whose removal disconnects the graph of which it is a part is called a **bridge**. Find all bridges for each of the following graphs.

   a.

   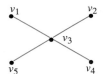

   b.

   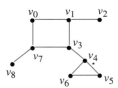

   c.

   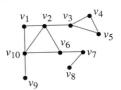

7. Given any positive integer n, (a) find a connected graph with n edges such that removal of just one edge disconnects the graph; (b) find a connected graph with n edges that cannot be disconnected by the removal of a single edge.

8. Find the number of connected components for each of
   the following graphs.

   a.

   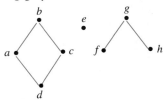

   b.

   c.

   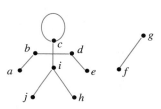

   d.

9. Each of (a)–(c) describes a graph. In each case answer
   yes, no, or not necessarily to the question: Does the
   graph have an Euler circuit? Justify your answers.
   a. G is a connected graph with five vertices of degrees
      2, 2, 3, 3, and 4.
   b. G is a connected graph with five vertices of degrees
      2, 2, 4, 4, and 6.
   c. G is a graph with five vertices of degrees 2, 2, 4, 4,
      and 6.

10. Figure 11.2.8 shows a graph for which every vertex has
    even degree but which does not have an Euler circuit.
    Give another example of a graph satisfying these prop-
    erties.

11. Is it possible for a citizen of Königsberg to make a tour
    of the city and cross each bridge exactly twice? (See
    Figure 11.2.1.) Why?

**Determine which of the graphs in 12–17 have Euler cir-
cuits. Find Euler circuits for those graphs that have them.**

12.

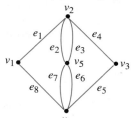

13.

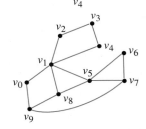

14.

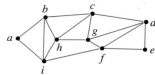

15.

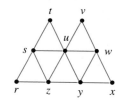

16.

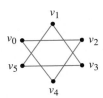

17.

18. Is it possible to take a walk around the city whose map is shown below, starting and ending at the same point and crossing each bridge exactly once? If so, how can this be done?

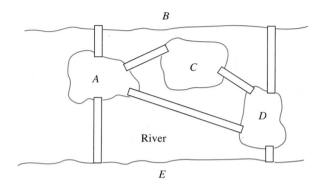

River

22. The following is a floor plan of a house. Is it possible to enter the house in room *A*, travel through every interior doorway of the house exactly once, and exit out of room *E*? If so, how can this be done?

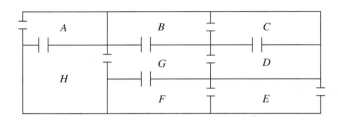

Find Hamiltonian circuits for each of the graphs in 23 and 24.

23.

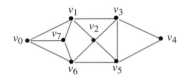

For each of the graphs in 19–21, determine whether there is an Euler path from *u* to *w*. If there is, find such a path.

19.

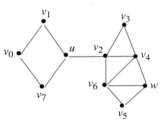

24.

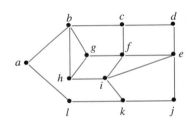

Show that none of the graphs in 25–27 has a Hamiltonian circuit.

20.

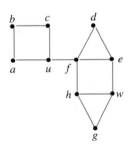

H 25.

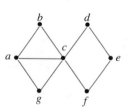

21.

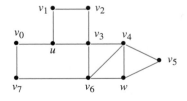

26.

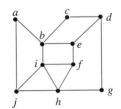

27.

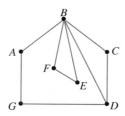

In 28–31 find Hamiltonian circuits for those graphs that have them. Explain why the other graphs do not.

H **28.**

29.

30.

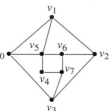

31.

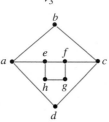

H **32.** Give two examples of graphs that have Euler circuits but not Hamiltonian circuits.

H **33.** Give two examples of graphs that have Hamiltonian circuits but not Euler circuits.

H **34.** Give two examples of graphs that have circuits that are both Euler circuits and Hamiltonian circuits.

H **35.** Give two examples of graphs that have Euler circuits and Hamiltonian circuits that are not the same.

36. A traveler in Europe wants to visit each of the cities shown on the map in the next column exactly once, starting and ending in Brussels. The distance (in kilo-

meters) between each pair of cities is given in the table. Find a Hamiltonian circuit that minimizes the total distance traveled. (Use the map to narrow the possible circuits down to just a few. Then use the table to find the total distance for each of those.)

	Berlin	Brussels	Düsseldorf	Luxembourg	Munich
Brussels	783				
Düsseldorf	564	223			
Luxembourg	764	219	224		
Munich	585	771	613	517	
Paris	1,057	308	497	375	832

37. a. Prove that if a walk in a graph contains a repeated edge, then the walk contains a repeated vertex.

b. Explain how it follows from part (a) that any walk with no repeated vertex has no repeated edge.

38. Prove Lemma 11.2.1(a): If G is a connected graph, then any two distinct vertices of G can be connected by a simple path.

39. Prove Lemma 11.2.1(b): If vertices v and w are part of a circuit in a graph G and one edge is removed from the circuit, then there still exists a path from v to w in G.

40. Prove Lemma 11.2.1(c): If a graph G is connected and G contains a circuit, then an edge of the circuit can be removed without disconnecting G.

41. Prove that if there is a path in a graph G from a vertex v to a vertex w, then there is a path from w to v.

42. If a graph contains a circuit that starts and ends at a vertex v, does the graph contain a simple circuit that starts and ends at v? Why?

43. Prove that if there is a circuit in a graph that starts and ends at a vertex v and if w is another vertex in the circuit, then there is a circuit in the graph that starts and ends at w.

44. Let G be a connected graph and let C be a circuit in G. Let G' be the subgraph obtained by removing all the edges of C from G and also any vertices that become isolated when the edges of C are removed. Prove that if G' is nonempty, then there exists a vertex v such that v is in both C and G'.

45. Prove that any graph with an Euler circuit is connected.

46. Prove Corollary 11.2.5.

47. For what values of n does the complete graph K_n with n vertices have (a) an Euler circuit? (b) a Hamiltonian circuit?

◆ 48. For what values of m and n does the complete bipartite graph on (m, n) vertices have (a) an Euler circuit? (b) a Hamiltonian circuit?

◆ 49. What is the maximum number of edges a simple disconnected graph with n vertices can have? Prove your answer.

◆ 50. Show that a graph is bipartite if, and only if, it does not have a circuit with an odd number of edges. (See exercise 37 of Section 11.1 for the definition of bipartite graph.)

11.3   MATRIX REPRESENTATIONS OF GRAPHS

Order and simplification are the first steps toward the mastery of a subject. (Thomas Mann, *The Magic Mountain,* 1924)

How can graphs be represented inside a computer? It happens that all the information needed to specify a graph can be conveyed by a structure called a *matrix,* and matrices (plural of matrix) are easy to represent inside computers. This section contains some basic definitions about matrices and matrix operations, a description of the relation between graphs and matrices, and some applications.

MATRICES

Matrices are two-dimensional analogues of sequences. They are also called two-dimensional arrays.

> **DEFINITION**
>
> An $m \times n$ (read "m by n") **matrix A over a set S** is a rectangular array of elements of S arranged into m rows and n columns:
>
> $$\mathbf{A} = \begin{bmatrix} a_{11} & a_{12} & \cdots & a_{1j} & \cdots & a_{1n} \\ a_{21} & a_{22} & \cdots & a_{2j} & \cdots & a_{2n} \\ \vdots & \vdots & & \vdots & & \vdots \\ a_{i1} & a_{i2} & \cdots & a_{ij} & \cdots & a_{in} \\ \vdots & \vdots & & \vdots & & \vdots \\ a_{m1} & a_{m2} & \cdots & a_{mj} & \cdots & a_{mn} \end{bmatrix} \quad \leftarrow i\text{th row of } \mathbf{A}$$
>
> $\uparrow$
> jth column of $\mathbf{A}$
>
> We write $\mathbf{A} = (a_{ij})$.

The **ith row of A** is

$$[a_{i1}\ a_{i2}\ \cdots\ a_{in}]$$

and the **jth column of A** is

$$\begin{bmatrix} a_{1j} \\ a_{2j} \\ \cdot \\ \cdot \\ \cdot \\ a_{mj} \end{bmatrix}.$$

The entry a_{ij} in the ith row and jth column of **A** is called the **ijth entry of A**. An $m \times n$ matrix is said to have **size $m \times n$**. If **A** and **B** are matrices, then $\mathbf{A} = \mathbf{B}$ if, and only if, **A** and **B** have the same size and the corresponding entries of **A** and **B** are all equal; that is,

$$a_{ij} = b_{ij} \quad \text{for all } i = 1, 2, \ldots, m \text{ and } j = 1, 2, \ldots, n.$$

A matrix for which the number of rows and columns are equal is called a **square matrix**. If **A** is a square matrix of size $n \times n$, then the **main diagonal of A** consists of all the entries $a_{11}, a_{22}, \ldots, a_{nn}$:

$$\begin{bmatrix} a_{11} & a_{12} & \cdots & a_{1i} & \cdots & a_{1n} \\ a_{21} & a_{22} & \cdots & a_{2i} & \cdots & a_{2n} \\ \cdot & \cdot & & \cdot & & \cdot \\ \cdot & \cdot & & \cdot & & \cdot \\ a_{i1} & a_{i2} & \cdots & a_{ii} & \cdots & a_{in} \\ \cdot & \cdot & & \cdot & & \cdot \\ a_{n1} & a_{n2} & \cdots & a_{ni} & \cdots & a_{nn} \end{bmatrix} \leftarrow \text{main diagonal of } \mathbf{A}$$

EXAMPLE 11.3.1 **Matrix Terminology**

The following is a 3×3 matrix over the set of integers.

$$\begin{bmatrix} 1 & 0 & -3 \\ 4 & -1 & 5 \\ -2 & 2 & 0 \end{bmatrix}$$

a. What is the entry in row 2, column 3?
b. What is the second column of **A**?
c. What are the entries in the main diagonal of **A**?

Solution   a. 5   b. $\begin{bmatrix} 0 \\ -1 \\ 2 \end{bmatrix}$ c. 1, -1, and 0

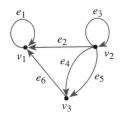

FIGURE 11.3.1

MATRICES AND DIRECTED GRAPHS

Consider the directed graph shown in Figure 11.3.1. This graph can be represented by the matrix $\mathbf{A} = (a_{ij})$ for which $a_{ij} = $ the number of arrows from v_i to v_j, for all $i = 1, 2, 3$ and $j = 1, 2, 3$. Thus $a_{11} = 1$ because there is one arrow from v_1 to v_1, $a_{12} = 0$ because there is no arrow from v_1 to v_2, $a_{23} = 2$ because there are two arrows from v_2 to v_3, and so forth. **A** is called the *adjacency matrix* of the directed graph. For convenient reference, the rows and columns of an adjacency matrix $\mathbf{A}$ are often labeled with the vertices of G as shown below.

$$\mathbf{A} = \begin{array}{c} \\ v_1 \\ v_2 \\ v_3 \end{array} \begin{array}{ccc} v_1 & v_2 & v_3 \\ \left[\begin{array}{ccc} 1 & 0 & 0 \\ 1 & 1 & 2 \\ 1 & 0 & 0 \end{array}\right] \end{array}$$

DEFINITION

Let G be a directed graph with ordered vertices $v_1, v_2, \ldots, v_n$. The **adjacency matrix of G** is the matrix $\mathbf{A} = (a_{ij})$ over the set of nonnegative integers such that

$a_{ij} =$ the number of arrows from v_i to v_j for all $i, j = 1, 2, \ldots, n$.

Note that nonzero entries along the main diagonal of an adjacency matrix indicate the presence of loops, and entries larger than 1 correspond to parallel edges. Moreover, if the vertices of a directed graph are reordered, then the entries in the rows and columns of the corresponding adjacency matrix are moved around.

EXAMPLE 11.3.2   **The Adjacency Matrix of a Graph**

The two directed graphs in Figure 11.3.2 differ only in the ordering of their vertices. Find their adjacency matrices.

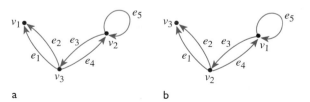

a   b

FIGURE 11.3.2

Solution   Since both graphs have three vertices, both adjacency matrices are 3×3 matrices. For (a) all entries in the first row are 0 since there are no arrows from v_1 to any other vertex. For (b) the first two entries in the first row are 1 and the third entry is 0 since from v_1 there are single arrows to v_1 and to v_2 and no arrows to v_3. Continuing the analysis in this way, you obtain the following two adjacency matrices:

a. $\begin{array}{c} \\ v_1 \\ v_2 \\ v_3 \end{array} \begin{array}{ccc} v_1 & v_2 & v_3 \\ \left[\begin{array}{ccc} 0 & 0 & 0 \\ 0 & 1 & 1 \\ 2 & 1 & 0 \end{array}\right] \end{array}$ b. $\begin{array}{c} \\ v_1 \\ v_2 \\ v_3 \end{array} \begin{array}{ccc} v_1 & v_2 & v_3 \\ \left[\begin{array}{ccc} 1 & 1 & 0 \\ 1 & 0 & 2 \\ 0 & 0 & 0 \end{array}\right] \end{array}$ ■

If you are given a square matrix with nonnegative integer entries, you can construct a directed graph with that matrix as its adjacency matrix. However, the matrix does not tell you how to label the edges, so the directed graph is not uniquely determined.

EXAMPLE 11.3.3   Obtaining a Directed Graph from a Matrix

Let

$$A = \begin{bmatrix} 0 & 1 & 1 & 0 \\ 1 & 1 & 0 & 2 \\ 0 & 0 & 1 & 1 \\ 2 & 1 & 0 & 0 \end{bmatrix}.$$

Draw a directed graph having **A** as its adjacency matrix.

Solution   Let G be the graph corresponding to **A** and let v_1, v_2, v_3, v_4 be the vertices of G. Label **A** across the top and down the left side with these vertex names as shown.

$$A = \begin{array}{c c} & \begin{array}{cccc} v_1 & v_2 & v_3 & v_4 \end{array} \\ \begin{array}{c} v_1 \\ v_2 \\ v_3 \\ v_4 \end{array} & \begin{bmatrix} 0 & 1 & 1 & 0 \\ 1 & 1 & 0 & 2 \\ 0 & 0 & 1 & 1 \\ 2 & 1 & 0 & 0 \end{bmatrix}. \end{array}$$

Then, for instance, the 2 in the fourth row and the first column means that there are two arrows from v_4 to v_1. The 0 in the first row and the fourth column means that there is no arrow from v_1 to v_4. A corresponding directed graph is shown in Figure 11.3.3 (without edge labels because the matrix does not determine those).

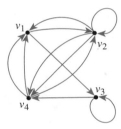

FIGURE 11.3.3

MATRICES AND (UNDIRECTED) GRAPHS

Once you know how to associate a matrix with a directed graph, the definition of the matrix corresponding to an undirected graph should seem natural to you. As before, you must order the vertices of the graph, but in this case you simply set the ijth entry of the adjacency matrix equal to the number of edges connecting the ith and jth vertices of the graph.

> **DEFINITION**
>
> Let G be an (undirected) graph with ordered vertices $v_1, v_2, \ldots, v_n$. The **adjacency matrix of G** is the matrix $\mathbf{A} = (a_{ij})$ over the set of nonnegative integers such that
>
> $$a_{ij} = \text{the number of edges connecting } v_i \text{ and } v_j$$
>
> for all $i, j = 1, 2, \ldots, n$.

EXAMPLE 11.3.4   Finding the Adjacency Matrix of a Graph

Find the adjacency matrix for the graph G shown below.

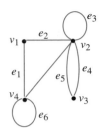

Solution

$$\mathbf{A} = \begin{array}{c} \\ v_1 \\ v_2 \\ v_3 \\ v_4 \end{array} \overset{\begin{array}{cccc} v_1 & v_2 & v_3 & v_4 \end{array}}{\begin{bmatrix} 0 & 1 & 0 & 1 \\ 1 & 1 & 2 & 1 \\ 0 & 2 & 0 & 0 \\ 1 & 1 & 0 & 1 \end{bmatrix}}$$

Note that if the matrix $\mathbf{A} = (a_{ij})$ in Example 11.3.4 is flipped across its main diagonal, it looks the same: $a_{ij} = a_{ji}$, for all $i, j = 1, 2, \ldots, n$. Such a matrix is said to be *symmetric*.

DEFINITION

An $n \times n$ square matrix $\mathbf{A} = (a_{ij})$ is called **symmetric** if, and only if, for all $i, j = 1, 2, \ldots, n$,

$$a_{ij} = a_{ji}.$$

EXAMPLE 11.3.5   Symmetric Matrices

Which of the following matrices are symmetric?

a. $\begin{bmatrix} 1 & 0 \\ 1 & 2 \end{bmatrix}$ b. $\begin{bmatrix} 0 & 1 & 2 \\ 1 & 1 & 0 \\ 2 & 0 & 3 \end{bmatrix}$ c. $\begin{bmatrix} 2 & 0 & 0 \\ 0 & 1 & 0 \end{bmatrix}$

Solution   Only (b) is symmetric. In (a) the entry in the first row and the second column differs from the entry in the second row and the first column; the matrix in (c) is not even square.

It is easy to see that the matrix of *any* undirected graph is symmetric since it is always the case that the number of edges joining v_i and v_j equals the number of edges joining v_j and v_i for all $i, j = 1, 2, \ldots, n$.

MATRICES AND CONNECTED COMPONENTS

Consider a graph G, as shown below, that consists of several connected components.

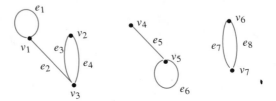

The adjacency matrix of G is

$$
\mathbf{A} = \left[
\begin{array}{ccc:cc:cc}
1 & 0 & 1 & 0 & 0 & 0 & 0 \\
0 & 0 & 2 & 0 & 0 & 0 & 0 \\
1 & 2 & 0 & 0 & 0 & 0 & 0 \\
\hdashline
0 & 0 & 0 & 0 & 1 & 0 & 0 \\
0 & 0 & 0 & 1 & 1 & 0 & 0 \\
\hdashline
0 & 0 & 0 & 0 & 0 & 0 & 2 \\
0 & 0 & 0 & 0 & 0 & 2 & 0
\end{array}
\right].
$$

As you can see, $\mathbf{A}$ consists of square matrix blocks (of different sizes) down its diagonal and blocks of 0's everywhere else. The reason is that vertices in each connected component share no edges with vertices in other connected components. For instance, since v_1, v_2, and v_3 share no edges with v_4, v_5, v_6, or v_7, all entries in the top three rows to the right of the third column are 0 and all entries in the left three columns below the third row are also 0. Sometimes matrices whose entries are all 0's are themselves denoted 0. If this convention is followed here, $\mathbf{A}$ is written as

$$
\mathbf{A} = \left[
\begin{array}{ccc|cc|cc}
1 & 0 & 1 & & \mathbf{0} & & \mathbf{0} \\
0 & 0 & 2 & & & & \\
1 & 2 & 0 & & & & \\
\hline
& \mathbf{0} & & 0 & 1 & & \mathbf{0} \\
& & & 1 & 1 & & \\
\hline
& \mathbf{0} & & & \mathbf{0} & 0 & 2 \\
& & & & & 2 & 0
\end{array}
\right].
$$

The above reasoning can be generalized to prove the following theorem:

THEOREM 11.3.1

Let G be a graph with connected components $G_1, G_2, \ldots, G_k$. If there are n_i vertices in each connected component G_i and these vertices are numbered consecutively, then the adjacency matrix of G has the form

$$
\begin{bmatrix}
A_1 & O & O & \ldots & O & O \\
O & A_2 & O & \ldots & O & O \\
O & O & A_3 & \ldots & O & O \\
\vdots & \vdots & \vdots & & \vdots & \vdots \\
O & O & O & \ldots & O & A_k
\end{bmatrix}
$$

where each A_i is the $n_i \times n_i$ adjacency matrix of G_i, for all $i = 1, 2, \ldots, k$, and the O's represent matrices whose entries are all 0.

MATRIX MULTIPLICATION

Matrix multiplication is an enormously useful operation that arises in many contexts, including the investigation of walks in graphs. Although matrix multiplication can be defined in quite abstract settings, the definition for matrices whose entries are real numbers will be sufficient for our applications. The product of two matrices is built up of *scalar* or *dot* products of their individual rows and columns.

DEFINITION

Suppose that all entries in matrices $\mathbf{A}$ and $\mathbf{B}$ are real numbers. If the number of elements, n, in the ith row of $\mathbf{A}$ equals the number of elements in the jth column of $\mathbf{B}$, then the **scalar product** or **dot product** of the ith row of $\mathbf{A}$ and the jth column of $\mathbf{B}$ is the real number obtained as follows:

$$
\begin{bmatrix} a_{i1} & a_{i2} & \cdots & a_{in} \end{bmatrix}
\begin{bmatrix} b_{1j} \\ b_{2j} \\ \vdots \\ b_{nj} \end{bmatrix}
= a_{i1}b_{1j} + a_{i2}b_{2j} + \cdots + a_{in}b_{nj}.
$$

EXAMPLE 11.3.6   Multiplying a Row and a Column

$$
\begin{bmatrix} 3 & 0 & -1 & 2 \end{bmatrix}
\begin{bmatrix} -1 \\ 2 \\ 3 \\ 0 \end{bmatrix}
= 3 \cdot (-1) + 0 \cdot 2 + (-1) \cdot 3 + 2 \cdot 0
$$

$$
= -3 + 0 - 3 + 0 = -6 \qquad \blacksquare
$$

More generally, if **A** and **B** are matrices whose entries are real numbers and if **A** and **B** have *compatible sizes* in the sense that the number of columns of **A** equals the number of rows of **B**, then the product **AB** is defined. It is the matrix whose ijth entry is the scalar product of the ith row of **A** times the jth column of **B**, for all possible values of i and j.

DEFINITION

Let $\mathbf{A} = (a_{ij})$ be an $m \times k$ matrix and $\mathbf{B} = (b_{ij})$ a $k \times n$ matrix with real entries. The (matrix) product of **A** times **B**, denoted **AB**, is that matrix (c_{ij}) defined as follows:

$$
\begin{bmatrix}
a_{11} & a_{12} & \cdots & a_{1k} \\
a_{21} & a_{22} & \cdots & a_{2k} \\
\vdots & \vdots & & \vdots \\
a_{i1} & a_{i2} & \cdots & a_{ik} \\
\vdots & \vdots & & \vdots \\
a_{m1} & a_{m2} & \cdots & a_{mk}
\end{bmatrix}
\begin{bmatrix}
b_{11} & b_{12} & \cdots & b_{1j} & \cdots & b_{1n} \\
b_{21} & b_{22} & \cdots & b_{2j} & \cdots & b_{2n} \\
\vdots & & & \vdots & & \vdots \\
 & & & \vdots & & \\
b_{k1} & b_{k2} & \cdots & b_{kj} & \cdots & b_{kn}
\end{bmatrix}
$$

$$
= \begin{bmatrix}
c_{11} & c_{12} & \cdots & c_{1j} & \cdots & c_{1n} \\
c_{21} & c_{22} & \cdots & c_{2j} & \cdots & c_{2n} \\
\vdots & \vdots & & \vdots & & \vdots \\
c_{i1} & c_{i2} & \cdots & c_{ij} & \cdots & c_{in} \\
\vdots & \vdots & & \vdots & & \vdots \\
c_{m1} & c_{m2} & \cdots & c_{mj} & \cdots & c_{mn}
\end{bmatrix}
$$

where

$$
c_{ij} = a_{i1}b_{1j} + a_{i2}b_{2j} + \cdots + a_{ik}b_{kj} = \sum_{r=1}^{k} a_{ir}b_{rj},
$$

for all $i = 1, 2, \ldots, m$ and $j = 1, 2, \ldots, n$.

EXAMPLE 11.3.7 Computing a Matrix Product

Let $\mathbf{A} = \begin{bmatrix} 2 & 0 & 3 \\ -1 & 1 & 0 \end{bmatrix}$ and $\mathbf{B} = \begin{bmatrix} 4 & 3 \\ 2 & 2 \\ -2 & -1 \end{bmatrix}$. Compute **AB**.

Solution **A** has size 2×3 and **B** has size 3×2, so the number of columns of **A** equals the number of rows of **B** and the matrix product of **A** and **B** can be computed. Then

$$
\begin{bmatrix} 2 & 0 & 3 \\ -1 & 1 & 0 \end{bmatrix}
\begin{bmatrix} 4 & 3 \\ 2 & 2 \\ -2 & -1 \end{bmatrix}
= \begin{bmatrix} c_{11} & c_{12} \\ c_{21} & c_{22} \end{bmatrix},
$$

where

$$c_{11} = 2 \cdot 4 + 0 \cdot 2 + 3 \cdot (-2) = 2 \qquad \begin{bmatrix} 2 & 0 & 3 \\ -1 & 1 & 0 \end{bmatrix}\begin{bmatrix} 4 & 3 \\ 2 & 2 \\ -2 & -1 \end{bmatrix},$$

$$c_{12} = 2 \cdot 3 + 0 \cdot 2 + 3 \cdot (-1) = 3 \qquad \begin{bmatrix} 2 & 0 & 3 \\ -1 & 1 & 0 \end{bmatrix}\begin{bmatrix} 4 & 3 \\ 2 & 2 \\ -2 & -1 \end{bmatrix},$$

$$c_{21} = (-1) \cdot 4 + 1 \cdot 2 + 0 \cdot (-2) = -2 \qquad \begin{bmatrix} 2 & 0 & 3 \\ -1 & 1 & 0 \end{bmatrix}\begin{bmatrix} 4 & 3 \\ 2 & 2 \\ -2 & -1 \end{bmatrix},$$

$$c_{22} = (-1) \cdot 3 + 1 \cdot 2 + 0 \cdot (-1) = -1 \qquad \begin{bmatrix} 2 & 0 & 3 \\ -1 & 1 & 0 \end{bmatrix}\begin{bmatrix} 4 & 3 \\ 2 & 2 \\ -2 & -1 \end{bmatrix}.$$

Hence

$$\mathbf{AB} = \begin{bmatrix} 2 & 3 \\ -2 & -1 \end{bmatrix}. \qquad \blacksquare$$

Matrix multiplication is both similar to and different from multiplication of real numbers. One difference is that although the product of any two numbers can be formed, only matrices with compatible sizes can be multiplied. Also multiplication of real numbers is commutative (for all real numbers a and b, $a \cdot b = b \cdot a$), whereas matrix multiplication is not. For instance,

$$\begin{bmatrix} 1 & 1 \\ 0 & 1 \end{bmatrix}\begin{bmatrix} 0 & 1 \\ 0 & 1 \end{bmatrix} = \begin{bmatrix} 0 & 2 \\ 0 & 1 \end{bmatrix}, \quad \text{but} \quad \begin{bmatrix} 0 & 1 \\ 0 & 1 \end{bmatrix}\begin{bmatrix} 1 & 1 \\ 0 & 1 \end{bmatrix} = \begin{bmatrix} 0 & 1 \\ 0 & 1 \end{bmatrix}.$$

On the other hand, both real number and matrix multiplications are associative $((ab)c = a(bc)$, for all elements a, b, and c for which the products are defined). This is proved in Example 11.3.8 for multiplications of 2×2 matrices. Additional exploration of matrix multiplication is contained in the exercises.

EXAMPLE 11.3.8   Associativity of Matrix Multiplication
for 2×2 Matrices

Prove that if $\mathbf{A}$, $\mathbf{B}$, and $\mathbf{C}$ are 2×2 matrices over the set of real numbers, then $(\mathbf{AB})\mathbf{C} = \mathbf{A}(\mathbf{BC})$.

Solution   Suppose $\mathbf{A} = (a_{ij})$, $\mathbf{B} = (b_{ij})$, and $\mathbf{C} = (c_{ij})$ are particular but arbitrarily chosen 2×2 matrices with real entries. Since the number of rows and columns are all the same, $\mathbf{AB}$, $\mathbf{BC}$, $(\mathbf{AB})\mathbf{C}$, and $\mathbf{A}(\mathbf{BC})$ are defined. Let $\mathbf{AB} = (d_{ij})$ and $\mathbf{BC} = (e_{ij})$. Then for all integers $i = 1, 2$ and $j = 1, 2$,

the ijth entry of $(\mathbf{AB})\mathbf{C} = \displaystyle\sum_{r=1}^{2} d_{ir}c_{rj}$ by definition of the product of **AB** and **C**

$$= d_{i1}c_{1j} + d_{i2}c_{2j}$$ by definition of Σ

$$= \left(\sum_{r=1}^{2} a_{ir}b_{r1} \right) \cdot c_{1j} + \left(\sum_{r=1}^{2} a_{ir}b_{r2} \right) \cdot c_{2j}$$ by definition of the product of **A** and **B**

$$= (a_{i1}b_{11} + a_{i2}b_{21}) \cdot c_{1j}$$ by definition of Σ
$$+ (a_{i1}b_{12} + a_{i2}b_{22}) \cdot c_{2j}$$

$$= a_{i1}b_{11}c_{1j} + a_{i2}b_{21}c_{1j} + a_{i1}b_{12}c_{2j} + a_{i2}b_{22}c_{2j}.$$

Similarly, the ijth entry of $\mathbf{A}(\mathbf{BC})$ is

$$(\mathbf{A}(\mathbf{BC}))_{ij} = \sum_{r=1}^{2} a_{ir}e_{rj}$$

$$= a_{i1}e_{1j} + a_{i2}e_{2j}$$

$$= a_{i1}\left(\sum_{r=1}^{2} b_{1r}c_{rj} \right) + a_{i2}\left(\sum_{r=1}^{2} b_{2r}c_{rj} \right)$$

$$= a_{i1}(b_{11}c_{1j} + b_{12}c_{2j}) + a_{i2}(b_{21}c_{1j} + b_{22}c_{2j})$$

$$= a_{i1}b_{11}c_{1j} + a_{i1}b_{12}c_{2j} + a_{i2}b_{21}c_{1j} + a_{i2}b_{22}c_{2j}$$

$$= a_{i1}b_{11}c_{1j} + a_{i2}b_{21}c_{1j} + a_{i1}b_{12}c_{2j} + a_{i2}b_{22}c_{2j}.$$

Comparing the results of the two computations shows that for all i and j,

the ijth entry of $(\mathbf{AB})\mathbf{C}$ = the ijth entry of $\mathbf{A}(\mathbf{BC})$.

Since all corresponding entries are equal, $(\mathbf{AB})\mathbf{C} = \mathbf{A}(\mathbf{BC})$ as was to be shown. ■

As far as multiplicative identities are concerned, there are both similarities and differences between real numbers and matrices. You know that the number 1 acts as a multiplicative identity for products of real numbers. It turns out that there are certain matrices, called *identity matrices,* that act as multiplicative identities for certain matrix products. For instance, mentally perform the following matrix multiplications to check that for any real numbers a, b, c, d, e, f, g, h and i,

$$\begin{bmatrix} 1 & 0 \\ 0 & 1 \end{bmatrix} \begin{bmatrix} a & b & c \\ d & e & f \end{bmatrix} = \begin{bmatrix} a & b & c \\ d & e & f \end{bmatrix}$$

and

$$\begin{bmatrix} a & b & c \\ d & e & f \\ g & h & i \end{bmatrix} \begin{bmatrix} 1 & 0 & 0 \\ 0 & 1 & 0 \\ 0 & 0 & 1 \end{bmatrix} = \begin{bmatrix} a & b & c \\ d & e & f \\ g & h & i \end{bmatrix}.$$

These computations show that $\begin{bmatrix} 1 & 0 \\ 0 & 1 \end{bmatrix}$ acts as an identity on the left side for

multiplication with 2×3 matrices and $\begin{bmatrix} 1 & 0 & 0 \\ 0 & 1 & 0 \\ 0 & 0 & 1 \end{bmatrix}$ acts as an identity on the right side for multiplication with 3×3 matrices. Note that $\begin{bmatrix} 1 & 0 \\ 0 & 1 \end{bmatrix}$ cannot act as an identity on the right side for multiplication with 2×3 matrices because the sizes are not compatible.

DEFINITION

For each positive integer n, the $n \times n$ **identity matrix**, denoted $\mathbf{I}_n = (\delta_{ij})$ or just $\mathbf{I}$ (if the size of the matrix is obvious from context), is the $n \times n$ matrix in which all the entries in the main diagonal are 1's and all other entries are 0's. In other words, $\delta_{ij} = 1$ if $i = j$, and $\delta_{ij} = 0$ if $i \neq j$, for all $i, j = 1, 2, 3, \ldots, n$.

*Leopold Kronecker
(1823–1891)*

The German mathematician Leopold Kronecker introduced the symbol δ_{ij} to make matrix computations more convenient. In his honor, this symbol is called the *Kronecker delta.*

EXAMPLE 11.3.9   **An Identity Matrix Acts As an Identity**

Prove that if $\mathbf{A}$ is any $m \times n$ matrix and $\mathbf{I}$ is the $n \times n$ identity matrix, then $\mathbf{AI} = \mathbf{A}$. (In exercise 14 at the end of this section you are asked to show that if $\mathbf{I}$ is the $m \times m$ identity matrix, then $\mathbf{IA} = \mathbf{A}$.)

Proof:

Let $\mathbf{A}$ be any $n \times n$ matrix and let A_{ij} be the *ij*th entry of $\mathbf{A}$ for all integers $i = 1, 2, \ldots, m$ and $j = 1, 2, \ldots, n$. Consider the product $\mathbf{AI}$, where $\mathbf{I}$ is the $n \times n$ identity matrix. Observe that

$$\begin{bmatrix} a_{11} & a_{12} & \cdots & a_{1n} \\ a_{21} & a_{22} & \cdots & a_{2n} \\ \cdot & \cdot & & \cdot \\ \cdot & \cdot & & \cdot \\ \cdot & \cdot & & \cdot \\ a_{m1} & a_{m2} & \cdots & a_{mn} \end{bmatrix} \begin{bmatrix} 1 & 0 & \cdots & 0 \\ 0 & 1 & \cdots & 0 \\ \cdot & \cdot & & \cdot \\ \cdot & \cdot & & \cdot \\ \cdot & \cdot & & \cdot \\ 0 & 0 & \cdots & 1 \end{bmatrix} = \begin{bmatrix} a_{11} & a_{12} & \cdots & a_{1n} \\ a_{21} & a_{22} & \cdots & a_{2n} \\ \cdot & \cdot & & \cdot \\ \cdot & \cdot & & \cdot \\ \cdot & \cdot & & \cdot \\ a_{m1} & a_{m2} & \cdots & a_{mn} \end{bmatrix}$$

because

$$\begin{aligned}
\text{the } ij\text{th entry of } \mathbf{AI} &= \sum_{r=1}^{n} a_{ir}\delta_{rj} && \text{by definition of } I \\
&= a_{i1}\delta_{1j} + a_{i2}\delta_{2j} + \cdots && \text{by definition of } \Sigma \\
&\quad + a_{ij}\delta_{jj} + \cdots + a_{in}\delta_{nj} \\
&= a_{ij}\delta_{jj} && \text{since } \delta_{kj} = 0 \\
&= a_{ij} && \text{whenever } k \neq j \text{ and } \delta_{jj} = 1 \\
&= \text{the } ij\text{th entry of } \mathbf{A}
\end{aligned}$$

Thus $\mathbf{AI} = \mathbf{A}$ as was to be shown.   ∎

There are also similarities and differences between real numbers and matrices with respect to the computation of powers. Any number can be raised to a nonnegative integer power, but a matrix can only be multiplied by itself if it has the same number of rows as columns. As for real numbers, however, the definition of matrix powers is recursive. Just as any number to the zero power is defined to be 1, so any $n \times n$ matrix to the zero power is defined to be the $n \times n$ identity matrix. The nth power of an $n \times n$ matrix $\mathbf{A}$ is defined to be the product of $\mathbf{A}$ with its $(n - 1)$st power.

DEFINITION

For any $n \times n$ matrix A, the **powers of A** are defined as follows:

$$\mathbf{A}^0 = \mathbf{I} \quad \text{where } \mathbf{I} \text{ is the } n \times n \text{ identity matrix,}$$
$$\mathbf{A}^n = \mathbf{A}\mathbf{A}^{n-1} \quad \text{for all integers } n \geq 1.$$

EXAMPLE 11.3.10   **Powers of a Matrix**

Let $\mathbf{A} = \begin{bmatrix} 1 & 2 \\ 2 & 0 \end{bmatrix}$. Compute $\mathbf{A}^0$, $\mathbf{A}^1$, $\mathbf{A}^2$, and $\mathbf{A}^3$.

Solution

$$\mathbf{A}^0 = \text{the } 2 \times 2 \text{ identity matrix} = \begin{bmatrix} 1 & 0 \\ 0 & 1 \end{bmatrix}$$

$$\mathbf{A}^1 = \mathbf{A}\mathbf{A}^0 = \mathbf{A}\mathbf{I} = \mathbf{A}$$

$$\mathbf{A}^2 = \mathbf{A}\mathbf{A}^1 = \mathbf{A}\mathbf{A} = \begin{bmatrix} 1 & 2 \\ 2 & 0 \end{bmatrix}\begin{bmatrix} 1 & 2 \\ 2 & 0 \end{bmatrix} = \begin{bmatrix} 5 & 2 \\ 2 & 4 \end{bmatrix}$$

$$\mathbf{A}^3 = \mathbf{A}\mathbf{A}^2 = \begin{bmatrix} 1 & 2 \\ 2 & 0 \end{bmatrix}\begin{bmatrix} 5 & 2 \\ 2 & 4 \end{bmatrix} = \begin{bmatrix} 9 & 10 \\ 10 & 4 \end{bmatrix} \quad ■$$

COUNTING WALKS OF LENGTH N

A walk in a graph consists of an alternating sequence of vertices and edges. If repeated edges are counted each time they occur, then the number of edges in the sequence is called the **length** of the walk. For instance, the walk $v_2 e_3 v_3 e_4 v_2 e_2 v_2 e_3 v_3$ has length 4 (counting e_3 twice). Consider the graph G of Figure 11.3.4.

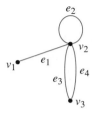

FIGURE 11.3.4

How many distinct walks of length 2 connect v_2 and v_2? You can list the possibilities systematically as follows: From v_1, the first edge of the walk must go to *some* vertex of G: v_1, v_2, or v_3. There is one walk of length 2 from v_2 to v_2 that starts by going from v_2 to v_1:

$$v_2 e_1 v_1 e_1 v_2.$$

There is one walk of length 2 from v_2 to v_2 that starts by going from v_2 to v_2:

$$v_2 e_2 v_2 e_2 v_2.$$

And there are four walks of length 2 from v_2 to v_2 that start by going from v_2 to v_3:

$$v_2 e_3 v_3 e_4 v_2,$$
$$v_2 e_4 v_3 e_3 v_2,$$
$$v_2 e_3 v_3 e_3 v_2,$$
$$v_2 e_4 v_3 e_4 v_2.$$

So the answer is six.

The general question of finding the number of walks that have a given length and connect two particular vertices of a graph can easily be answered using matrix multiplication. Consider the adjacency matrix $\mathbf{A}$ of the graph G of Figure 11.3.4:

$$\mathbf{A} = \begin{array}{c} \\ v_1 \\ v_2 \\ v_3 \end{array} \begin{array}{ccc} v_1 & v_2 & v_3 \\ \left[\begin{array}{ccc} 0 & 1 & 0 \\ 1 & 1 & 2 \\ 0 & 2 & 0 \end{array}\right] \end{array}.$$

Compute $\mathbf{A}^2$ as follows:

$$\begin{bmatrix} 0 & 1 & 0 \\ 1 & 1 & 2 \\ 0 & 2 & 0 \end{bmatrix} \begin{bmatrix} 0 & 1 & 0 \\ 1 & 1 & 2 \\ 0 & 2 & 0 \end{bmatrix} = \begin{bmatrix} 1 & 1 & 2 \\ 1 & 6 & 2 \\ 2 & 2 & 4 \end{bmatrix}.$$

Note that the entry in the second row and the second column is 6, which equals the number of walks of length 2 from v_2 to v_2. This is no accident! To compute a_{22}, you multiply the second row of $\mathbf{A}$ times the second column of $\mathbf{A}$ to obtain a sum of three terms:

$$\begin{bmatrix} 1 & 1 & 2 \end{bmatrix} \begin{bmatrix} 1 \\ 1 \\ 2 \end{bmatrix} = 1 \cdot 1 + 1 \cdot 1 + 2 \cdot 2.$$

Observe that

$$\begin{bmatrix} \text{the first term} \\ \text{of this sum} \end{bmatrix} = \begin{bmatrix} \text{number of} \\ \text{edges from} \\ v_2 \text{ to } v_1 \end{bmatrix} \cdot \begin{bmatrix} \text{number of} \\ \text{edges from} \\ v_1 \text{ to } v_2 \end{bmatrix} = \begin{bmatrix} \text{number of pairs} \\ \text{of edges from} \\ v_2 \text{ to } v_1 \text{ and } v_1 \text{ to } v_2 \end{bmatrix}.$$

Now consider the ith term of this sum, for each $i = 1, 2$, and 3. It equals the number of edges from v_2 to v_i times the number of edges from v_i to v_2. By the multiplication rule this equals the number of pairs of edges from v_2 to v_i and from v_i back to v_2. But this equals the number of walks of length 2 that start and end at v_2 and pass through v_i. Since this analysis holds for each term of the sum for $i = 1, 2$, and 3, the sum as a whole equals the total number of walks of length 2 that start and end at v_2:

$$1 \cdot 1 + 1 \cdot 1 + 2 \cdot 2 = 1 + 1 + 4 = 6.$$

More generally, if $\mathbf{A}$ is the adjacency matrix of a graph G, the ijth entry of $\mathbf{A}^2$ equals the number of walks of length 2 connecting the ith vertex to the jth vertex of G. Even more generally, if n is any positive integer, the ijth entry of $\mathbf{A}^n$ equals the number of walks of length n connecting the ith and the jth vertices of G.

THEOREM 11.3.2

If G is a graph with vertices $v_1, v_2, \ldots, v_m$ and **A** is the adjacency matrix of G, then for each positive integer n,

the ijth entry of $\mathbf{A}^n$ = the number of walks of length n from v_i to v_j
for all integers $i, j = 1, 2, \ldots, m$.

Proof:

Suppose G is a graph with vertices $v_1, v_2, \ldots, v_m$ and **A** is the adjacency matrix of G. We use mathematical induction to show that for each positive integer n,

$$\text{the } ij\text{th entry of } \mathbf{A}^n = \text{the number of walks of}$$
$$\text{length } n \text{ from } v_i \text{ to } v_j$$

for all integers $i, j = 1, 2, \ldots, m$.

The equality holds for n = 1:

The ijth entry of $\mathbf{A}^1$ = the ijth entry of **A**   because $\mathbf{A}^1 = \mathbf{A}$

= the number of edges   by definition of adjacency matrix
connecting v_i to v_j

= the number of walks of
length 1 from v_i to v_j.

If the equality holds for n = k, then it holds for n = k + 1:   Suppose that for some integer $k \geq 1$, the ijth entry of $\mathbf{A}^k$ = the number of walks of length k from v_i to v_j. *[This is the inductive hypothesis.]* We must show that the ijth entry of $\mathbf{A}^{k+1}$ = the number of walks of length $k + 1$ from v_i to v_j.

Let $\mathbf{A} = (a_{ij})$ and $\mathbf{A}^k = (b_{ij})$. Since $\mathbf{A}^{k+1} = \mathbf{A}\mathbf{A}^k = \mathbf{A}\mathbf{A}^k$, the ijth entry of $\mathbf{A}^{k+1}$ is obtained by multiplying the ith row of **A** by the jth column of $\mathbf{A}^k$:

$$\text{the } ij\text{th entry of } \mathbf{A}^{k+1} = a_{i1}b_{1j} + a_{i2}b_{2j} + \cdots + a_{im}b_{mj} \qquad 11.3.1$$

for all $i, j = 1, 2, \ldots, m$. Now consider the individual terms of this sum: a_{i1} is the number of edges from v_i to v_1; and, by inductive hypothesis, b_{1j} is the number of walks of length k from v_1 to v_j. But any edge from v_i to v_1 can be joined with any walk of length k from v_1 to v_j to create a walk of length $k + 1$ from v_i to v_j with v_1 as its second vertex. Thus by the multiplication rule

$$a_{i1}b_{1j} = \begin{bmatrix} \text{the number of walks of length } k + 1 \text{ from} \\ v_i \text{ to } v_j \text{ that have } v_1 \text{ as their second vertex} \end{bmatrix}.$$

More generally, for each integer $r = 1, 2, \ldots, m$,

$$a_{ir}b_{rj} = \begin{bmatrix} \text{the number of walks of length } k + 1 \text{ from} \\ v_i \text{ to } v_j \text{ that have } v_r \text{ as their second vertex} \end{bmatrix}.$$

Since any walk of length $k + 1$ from v_i to v_j must have *one* of the vertices $v_1, v_2, \ldots, v_m$ as its second vertex, the total number of walks of length $k + 1$ from v_i to v_j equals the sum in (11.3.1), which equals the ijth entry of $\mathbf{A}^{k+1}$. Hence

the ijth entry of $\mathbf{A}^{k+1}$ = the number of walks of length $k + 1$ from v_i to v_j

[as was to be shown].

Since both the basis and inductive steps have been proved, the given equality is true for all integers $n \geq 1$.

EXERCISE SET 11.3

1. Find real numbers a, b, and c so that the following are true.

   a. $\begin{bmatrix} a + b & a - c \\ c & b - a \end{bmatrix} = \begin{bmatrix} 1 & 0 \\ -1 & 3 \end{bmatrix}$

   b. $\begin{bmatrix} 2a & b + c \\ c - a & 2b - a \end{bmatrix} = \begin{bmatrix} 4 & 3 \\ 1 & -2 \end{bmatrix}$

2. Find the adjacency matrices for the following directed graphs.

   a. 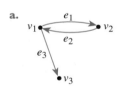   b.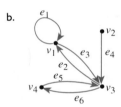

3. Find directed graphs that have the following adjacency matrices:

   a. $\begin{bmatrix} 1 & 0 & 1 & 2 \\ 0 & 0 & 1 & 0 \\ 0 & 2 & 1 & 1 \\ 0 & 1 & 1 & 0 \end{bmatrix}$ b. $\begin{bmatrix} 0 & 1 & 0 & 0 \\ 2 & 0 & 1 & 0 \\ 1 & 2 & 1 & 0 \\ 0 & 0 & 1 & 0 \end{bmatrix}$

4. Find adjacency matrices for the following (undirected) graphs.

   a. 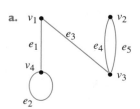   b.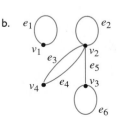

c. K_4, the complete graph on four vertices
d. $K_{2,3}$, the complete bipartite graph on (2, 3) vertices.

5. Find graphs that have the following adjacency matrices.

   a. $\begin{bmatrix} 1 & 0 & 1 \\ 0 & 1 & 2 \\ 1 & 2 & 0 \end{bmatrix}$ b. $\begin{bmatrix} 0 & 2 & 0 \\ 2 & 1 & 0 \\ 0 & 0 & 1 \end{bmatrix}$

6. The following are adjacency matrices for graphs. In each case determine whether the graph is connected by analyzing the matrix without drawing the graph.

   a. $\begin{bmatrix} 0 & 1 & 1 \\ 1 & 1 & 0 \\ 1 & 0 & 0 \end{bmatrix}$ b. $\begin{bmatrix} 0 & 2 & 0 & 0 \\ 2 & 0 & 0 & 0 \\ 0 & 0 & 1 & 1 \\ 0 & 0 & 1 & 1 \end{bmatrix}$

7. Suppose that for all i, all the entries in the ith row and ith column of the adjacency matrix of a graph are 0. What can you conclude about the graph?

8. Find each of the following products.

   a. $[2 \quad -1]\begin{bmatrix} 1 \\ 3 \end{bmatrix}$ b. $[4 \quad -1 \quad 7]\begin{bmatrix} 1 \\ 2 \\ 0 \end{bmatrix}$

9. Find each of the following products.

   a. $\begin{bmatrix} 3 & 0 \\ 1 & -2 \end{bmatrix}\begin{bmatrix} 1 & -1 & 4 \\ 0 & 2 & 1 \end{bmatrix}$

   b. $\begin{bmatrix} 2 & 0 & 1 \\ 0 & -1 & 0 \end{bmatrix}\begin{bmatrix} 1 & 3 \\ 5 & -4 \\ -2 & 2 \end{bmatrix}$

   c. $\begin{bmatrix} -1 \\ 2 \end{bmatrix}[2 \quad 3]$

10. Let $\mathbf{A} = \begin{bmatrix} 1 & 1 & -1 \\ 0 & -2 & 1 \end{bmatrix}$, $\mathbf{B} = \begin{bmatrix} -2 & 0 \\ 1 & 3 \end{bmatrix}$, and

$\mathbf{C} = \begin{bmatrix} 0 & -2 \\ 3 & 1 \\ 1 & 0 \end{bmatrix}$. For each of the following, determine

whether the indicated product exists and compute it if it does.

a. **AB** b. **BA** c. $\mathbf{A}^2$ d. **BC** e. **CB** f. $\mathbf{B}^2$

g. $\mathbf{B}^3$ h. $\mathbf{C}^2$ i. **AC** j. **CA**

11. Give an example different from that in the text to show that matrix multiplication is not commutative. That is, find 2×2 matrices **A** and **B** such that **AB** and **BA** both exist but $\mathbf{AB} \neq \mathbf{BA}$.

12. Let O denote the matrix $\begin{bmatrix} 0 & 0 \\ 0 & 0 \end{bmatrix}$. Find 2×2 matrices **A** and **B** such that $\mathbf{A} \neq \mathbf{O}$ and $\mathbf{B} \neq \mathbf{O}$, but $\mathbf{AB} = \mathbf{O}$.

13. Let O denote the matrix $\begin{bmatrix} 0 & 0 \\ 0 & 0 \end{bmatrix}$. Find 2×2 matrices **A** and **B** such that $\mathbf{A} \neq \mathbf{B}$, $\mathbf{B} \neq \mathbf{O}$, and $\mathbf{AB} \neq \mathbf{O}$, but $\mathbf{BA} = \mathbf{O}$.

In 14–18 assume that the entries of all matrices are real numbers.

H**14.** Prove that if **I** is the $m \times m$ identity matrix and **A** is any $m \times n$ matrix, then $\mathbf{IA} = \mathbf{A}$.

15. Prove that if **A** is an $m \times m$ symmetric matrix, then $\mathbf{A}^2$ is symmetric.

16. Prove that matrix multiplication is associative: If **A**, **B**, and **C** are any $m \times k$, $k \times r$, and $r \times n$ matrices, respectively, then $(\mathbf{AB})\mathbf{C} = \mathbf{A}(\mathbf{BC})$.

17. Use mathematical induction to prove that if **A** is any $m \times m$ matrix, then $\mathbf{A}^n \mathbf{A} = \mathbf{A}\mathbf{A}^n$ for all integers $n \geq 1$. (You will need to use the result of exercise 16.)

18. Use mathematical induction to prove that if **A** is an $m \times m$ symmetric matrix, then for any integer $n \geq 1$, $\mathbf{A}^n$ is also symmetric.

19. a. Let $\mathbf{A} = \begin{bmatrix} 1 & 1 & 2 \\ 1 & 0 & 1 \\ 2 & 1 & 0 \end{bmatrix}$. Find $\mathbf{A}^2$ and $\mathbf{A}^3$.

b. Let G be the graph with vertices v_1, v_2, and v_3 and with **A** as its adjacency matrix. Use the answers to part (a) to find the number of walks of length 2 from v_1 to v_3 and the number of walks of length 3 from v_1 to v_3. Do not draw G to solve this problem.

c. Examine the calculations you performed in answering part (a) to find five walks of length 2 from v_3 to v_3. Then draw G and find the walks by visual inspection.

20. The following is an adjacency matrix for a graph:

$$\begin{array}{c} \\ v_1 \\ v_2 \\ v_3 \\ v_4 \end{array} \begin{array}{cccc} v_1 & v_2 & v_3 & v_4 \\ \begin{bmatrix} 0 & 1 & 1 & 0 \\ 1 & 0 & 2 & 1 \\ 1 & 2 & 0 & 1 \\ 0 & 1 & 1 & 1 \end{bmatrix} \end{array}$$

Answer the following questions by examining the matrix and its powers only, not by drawing the graph:

a. How many walks of length 2 are there from v_2 to v_3?

b. How many walks of length 2 are there from v_3 to v_4?

c. How many walks of length 3 are there from v_1 to v_4?

d. How many walks of length 3 are there from v_2 to v_3?

21. Let **A** be the adjacency matrix for K_3, the complete graph on three vertices. Use mathematical induction to prove that for each positive integer n, all the entries along the main diagonal of $\mathbf{A}^n$ are equal to each other and all the entries that do not lie along the main diagonal are equal to each other.

22. a. Draw a graph that has

$$\begin{bmatrix} 0 & 0 & 0 & 1 & 2 \\ 0 & 0 & 0 & 1 & 1 \\ 0 & 0 & 0 & 2 & 1 \\ 1 & 1 & 2 & 0 & 0 \\ 2 & 1 & 1 & 0 & 0 \end{bmatrix}$$

as its adjacency matrix. Is this graph bipartite? (For a definition of bipartite, see exercise 37 in Section 11.1.)

Definition: Given an $m \times n$ matrix **A** whose ijth entry is denoted a_{ij}, the **transpose of A** is the matrix $\mathbf{A}^t$ whose ijth entry is a_{ji}, for all $i = 1, 2, \ldots, m$ and $j = 1, 2, \ldots, n$.

Note that the first row of **A** becomes the first column of $\mathbf{A}^t$, the second row of **A** becomes the second column of $\mathbf{A}^t$, and so forth. For instance,

$$\text{if } \mathbf{A} = \begin{bmatrix} 0 & 2 & 1 \\ 1 & 2 & 3 \end{bmatrix}, \text{ then } \mathbf{A}^t = \begin{bmatrix} 0 & 1 \\ 2 & 2 \\ 1 & 3 \end{bmatrix}.$$

H**b.** Show that a graph with n vertices is bipartite if, and only if, for some labeling of its vertices, its adjacency matrix has the form

$$\begin{bmatrix} \mathbf{O} & \mathbf{A} \\ \mathbf{A}^t & \mathbf{O} \end{bmatrix}$$

where **A** is a $k \times (n - k)$ matrix, for some integer k such that $0 < k < n$, the top left O represents a $k \times k$ matrix all of whose entries are 0, and the bottom right O represents an $(n - k) \times (n - k)$ matrix all of whose entries are 0.

23. a. Let G be a graph with n vertices and let v and w be distinct vertices of G. Prove that if there is a walk from v to w, then there is a walk from v to w that has length less than or equal to $n - 1$.

Hb. If $\mathbf{A} = (a_{ij})$ and $\mathbf{B} = (b_{ij})$ are any $m \times n$ matrices, the matrix $\mathbf{A} + \mathbf{B}$ is the $m \times n$ matrix whose ijth entry is $a_{ij} + b_{ij}$ for all $i = 1, 2, \ldots, m$ and $j = 1, 2, \ldots, n$. Let G be a graph with n vertices where $n > 1$, and let $\mathbf{A}$ be the adjacency matrix of G. Prove that G is connected if, and only if, every entry of $\mathbf{A} + \mathbf{A}^2 + \cdots + \mathbf{A}^{n-1}$ is positive.

11.4   ISOMORPHISMS OF GRAPHS

Thinking is a momentary dismissal of irrelevancies.
(R. Buckminster Fuller, 1969)

Recall from Section 11.1 that the two drawings shown in Figure 11.4.1 both represent the same graph: Their vertex and edge sets are identical and their edge-endpoint functions are the same.

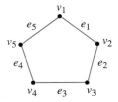
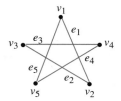

FIGURE 11.4.1

Call this graph G. Now consider the graph G' represented in Figure 11.4.2.

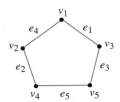

FIGURE 11.4.2

Observe that G' is a different graph from G (for instance, in G the endpoints of e_1 are v_1 and v_2 whereas in G' the endpoints of e_1 are v_1 and v_3). Yet G' is certainly very similar to G. In fact, if the vertices and edges of G' are relabeled by the functions shown in Figure 11.4.3, then G' becomes the same as G.

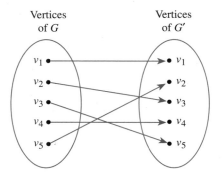
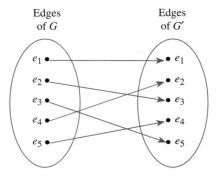

FIGURE 11.4.3

Note that these relabeling functions are one-to-one and onto.

Two graphs that are the same except for the labeling of their vertices and edges are called *isomorphic*. As noted in Section 10.4, the word *isomorphism* comes from the Greek meaning "same form." Isomorphic graphs are those that have essentially the same form.

DEFINITION

Let G and G' be graphs with vertex sets $V(G)$ and $V(G')$ and edge sets $E(G)$ and $E(G')$, respectively. **G is isomorphic to G'** if, and only if, there exist one-to-one correspondences $g: V(G) \to V(G')$ and $h: E(G) \to E(G')$ that preserve the edge-endpoint functions of G and G' in the sense that

for all $v \in V(G)$ and $e \in E(G)$,

$$v \text{ is an endpoint of } e \quad \Leftrightarrow \quad g(v) \text{ is an endpoint of } h(e). \qquad \textbf{11.4.1}$$

In words, G is isomorphic to G' if, and only if, the vertices and edges of G and G' can be matched up by one-to-one, onto functions so that the edges between corresponding vertices correspond to each other.

It is common in mathematics to identify objects that are isomorphic. For instance, if we are given a graph G with five vertices such that each pair of vertices is connected by an edge, then we may identify G with K_5, saying that G *is* K_5 rather than that G is isomorphic to K_5.

EXAMPLE 11.4.1   **Showing That Two Graphs Are Isomorphic**

Show that the two graphs in Figure 11.4.4 are isomorphic.

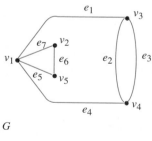

G G'

FIGURE 11.4.4

Solution   To solve this problem, you must find functions $g: V(G) \to V(G')$ and $h: E(G) \to E(G')$ such that for all $v \in V(G)$ and $e \in E(G)$, v is an endpoint of e if, and only if, $g(v)$ is an endpoint of $h(e)$. Setting up such functions is partly a matter of trial and error and partly a matter of deduction. For instance, since e_2 and e_3 are parallel (have the same endpoints), $h(e_2)$ and $h(e_3)$ must be parallel also. So $h(e_2) = f_1$ and $h(e_3) = f_2$ or $h(e_2) = f_2$ and $h(e_3) = f_1$. Also the endpoints of e_2 and e_3 must correspond to the endpoints of f_1 and f_2, and so $g(v_3) = w_1$ and $g(v_4) = w_5$ or $g(v_3) = w_5$ and $g(v_4) = w_1$.

Similarly, since v_1 is the endpoint of four distinct edges (e_1, e_7, e_5, and e_4), $g(v_1)$ must also be the endpoint of four distinct edges (because every edge incident on $g(v_1)$

is the image under h of an edge incident on v_1 and h is one-to-one and onto). But the only vertex in G' that has four edges coming out of it is w_2, and so $g(v_1) = w_2$. Now if $g(v_3) = w_1$, then since v_1 and v_3 are endpoints of e_1 in G, $g(v_1) = w_2$ and $g(v_3) = w_1$ must be endpoints of $h(e_1)$ in G'. This implies that $h(e_1) = f_3$.

By continuing in this way, possibly making some arbitrary choices as you go, you eventually can find functions g and h to define the isomorphism between G and G'. One pair of functions (there are several) is shown in Figure 11.4.5.

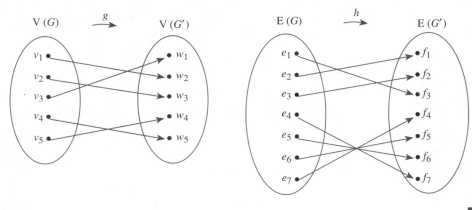

FIGURE 11.4.5

It is not hard to show that graph isomorphism is an equivalence relation on a set of graphs; in other words it is reflexive, symmetric, and transitive. To prove the reflexive property, it must be shown that any graph is isomorphic to itself. Such an isomorphism can be defined using the identity functions on the set of vertices and on the set of edges.

To prove that graph isomorphism is symmetric, it must be shown that if a graph G is isomorphic to a graph G' then G' is isomorphic to G. But this is true because if g and h are vertex and edge correspondences from G to G' that preserve the edge-endpoint functions, then g^{-1} and h^{-1} are vertex and edge correspondences from G' to G that preserve the edge-endpoint functions. Note that as a consequence of the symmetry property, you can simply say "G and G' are isomorphic" instead of "G is isomorphic to G'" or "G' is isomorphic to G."

Finally, to establish that graph isomorphism is transitive, it must be shown that if a graph G is isomorphic to a graph G' and if G' is isomorphic to G'', then G is isomorphic to G''. But this follows from the fact that if g_1 and h_1 are vertex and edge correspondences from G to G' that preserve the edge-endpoint functions of G and G' and g_2 and h_2 are vertex and edge correspondences from G' to G'' that preserve the edge-endpoint functions of G' and G'', then $g_2 \circ g_1$ and $h_2 \circ h_1$ are vertex and edge correspondences from G to G'' that preserve the edge-endpoint functions of G and G''.

EXAMPLE 11.4.2   Finding Representatives of Isomorphism Classes

Find all nonisomorphic graphs that have two vertices and two edges. In other words, find a collection of representative graphs with two vertices and two edges such that every such graph is isomorphic to one in the collection.

Solution There are four nonisomorphic graphs that have two vertices and two edges. These can be drawn without vertex and edge labels because any two labelings give isomorphic graphs. (See Figure 11.4.6.)

a   b   c   d

FIGURE 11.4.6

To see that these four drawings show all the nonisomorphic graphs that have two vertices and two edges, first note whether one of the edges joins the two vertices or not. If it does, there are two possibilities: The other edge can also join the two vertices (as in (a)) or it can be a loop incident on one of them (as in (b)—it makes no difference *which* vertex is chosen to have the loop because interchanging the two vertex labels gives isomorphic graphs). If neither edge joins the two vertices, then both edges are loops. In this case, there are only two possibilities: Either both loops are incident on the same vertex (as in (c)) or the two loops are incident on separate vertices (as in (d)). There are no other possibilities for placing the edges, so the listing is complete. ■

Now consider the question "Is there a general method to figure out whether graphs G and G' are isomorphic?" In other words, is there some algorithm that will accept graphs G and G' as input and produce a statement as to whether they are isomorphic? In fact, there is such an algorithm. It consists of generating all one-to-one, onto functions from the set of vertices of G to the set of vertices of G' and from the set of edges of G to the set of edges of G' and checking each pair to determine whether it preserves the edge-endpoint functions of G and G'. The problem with this algorithm is that it takes an unreasonably long time to perform, even on a high-speed computer. If G and G' each have n vertices and m edges, the number of one-to-one correspondences from vertices to vertices is $n!$ and the number of one-to-one correspondences from edges to edges is $m!$, so the total number of pairs of functions to check is $n! \cdot m!$. For instance, if $m = n = 20$, there would be $20! \cdot 20! \cong 5.9 \times 10^{36}$ pairs to check. Assuming that each check takes just 1 microsecond, the total time would be approximately 1.9×10^{23} years!

Unfortunately, there is no more efficient general method known for checking whether two graphs are isomorphic. However, there are some simple tests that can be used to show that certain pairs of graphs are *not* isomorphic. For instance, if two graphs are isomorphic, then they have the same number of vertices (because there is a one-to-one correspondence from the vertex set of one graph to the vertex set of the other). It follows that if you are given two graphs, one with 16 vertices and the other with 17, you can immediately conclude that the two are not isomorphic. More generally, a property that is preserved by graph isomorphism is called an *isomorphic invariant*. For instance, "having 16 vertices" is an isomorphic invariant: If one graph has 16 vertices, then so does any graph that is isomorphic to it.

DEFINITION

A property P is called an **isomorphic invariant** if, and only if, given any graphs G and G', if G has property P and G' is isomorphic to G, then G' has property P.

THEOREM 11.4.1

Each of the following properties is an invariant for graph isomorphism, where n, m, and k are all nonnegative integers:

1. has n vertices;
2. has m edges;
3. has a vertex of degree k;
4. has m vertices of degree k;
5. has a circuit of length k;
6. has a simple circuit of length k;
7. has m simple circuits of length k;
8. is connected;
9. has an Euler circuit;
10. has a Hamiltonian circuit.

EXAMPLE 11.4.3   Showing That Two Graphs Are Not Isomorphic

Show that the pairs of graphs in Figure 11.4.7 are not isomorphic by finding an isomorphic invariant that they do not share.

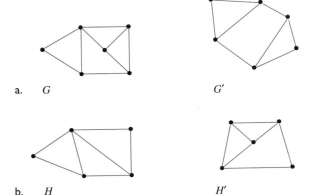

a. G G'

b. H H'

FIGURE 11.4.7

Solution   a. G has nine edges; G' has only eight.
b. H has a vertex of degree 4; H' does not.   ■

We prove part (3) of Theorem 11.4.1 below and leave the proofs of the other parts as exercises.

EXAMPLE 11.4.4   Proof of Theorem 11.4.1, Part (3)

Prove that if G is a graph that has a vertex of degree k and G' is isomorphic to G, then G' has a vertex of degree k.

Proof:

Suppose G and G' are isomorphic graphs and G has a vertex v of degree k, where k is a nonnegative integer. [*We must show that G' has a vertex of degree k.*] Since G and G' are isomorphic, there are one-to-one, onto functions g and h from the vertices of G to the vertices of G' and from the edges of G to the edges of G' that preserve the edge-endpoint functions in the sense that for all edges e and all vertices u of G, u is an endpoint of e if, and only if, $g(u)$ is an endpoint of $h(e)$. An example for a particular vertex v is shown below.

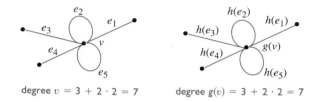

degree $v = 3 + 2 \cdot 2 = 7$ degree $g(v) = 3 + 2 \cdot 2 = 7$

Let $e_1, e_2, \ldots, e_m$ be the m distinct edges that are incident on a vertex v in G, where m is a nonnegative integer. Then $h(e_1), h(e_2), \ldots, h(e_m)$ are m distinct edges that are incident on $g(v)$ in G'. *[The reason that $h(e_1), h(e_2), \ldots, h(e_m)$ are distinct is that h is one-to-one and $e_1, e_2, \ldots, e_m$ are distinct. And the reason that $h(e_1), h(e_2), \ldots, h(e_m)$ are incident on $g(v)$ is that g and h preserve the edge-endpoint functions of G and G' and $e_1, e_2, \ldots, e_m$ are incident on v.]*

Also, there are no edges incident on $g(v)$ other than the ones that are images under g of edges incident on v *[because g is onto and g and h preserve the edge-endpoint functions of G and G']*. Thus the number of edges incident on v equals the number of edges incident on $g(v)$.

Finally, an edge e is a loop at v if, and only if, $h(e)$ is a loop at $g(v)$, so the number of loops incident on v equals the number of loops incident on $g(v)$. *[For since g and h preserve the edge-endpoint functions of G and G', a vertex w is an endpoint of e in G if, and only if, $g(w)$ is an endpoint of $h(e)$ in G'. It follows that v is the only endpoint of v in G if, and only if, $g(v)$ is the only endpoint of $h(e)$ in G'.]*

Now the degree of v, which is k, equals the number of edges incident on v plus the number of edges incident on v that are loops (since each loop contributes 2 to the degree of v). But we have already shown that the number of edges incident on v equals the number of edges incident on $g(v)$ and that the number of loops incident on v equals the number of loops incident on $g(v)$. Hence $g(v)$ also has degree k.   ∎

GRAPH ISOMORPHISM FOR SIMPLE GRAPHS

When graphs G and G' are both simple, the definition of G being isomorphic to G' can be written without referring to the correspondence between the edges of G and the edges of $G.'$

DEFINITION

If G and G' are simple graphs then **G is isomorphic to G'** if, and only if, there exists a one-to-one correspondence g from the vertex set $V(G)$ of G to the vertex set $V(G')$ of G' that preserves the edge-endpoint functions of G and G' in the sense that

for all vertices u and v of G,

$\{u, v\}$ is an edge in G $\iff$ $\{g(u), g(v)\}$ is an edge in G'.   11.4.2

EXAMPLE 11.4.5   Isomorphism of Simple Graphs

Are the two graphs shown below isomorphic? If so, define an isomorphism.

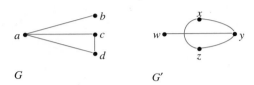

G G'

Solution   Yes. Define $f: V(G) \rightarrow V(G')$ by the arrow diagram shown below.

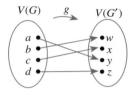

Then g is one-to-one and onto by inspection. The fact that g preserves the edge-endpoint functions of G and G' is shown by the following table:

Edges of G	Edges of G'
$\{a, b\}$	$\{y, w\} = \{g(a), g(b)\}$
$\{a, c\}$	$\{y, x\} = \{g(a), g(c)\}$
$\{a, d\}$	$\{y, z\} = \{g(a), g(d)\}$
$\{c, d\}$	$\{x, z\} = \{g(c), g(d)\}$

EXERCISE SET 11.4

For each pair of graphs G and G' in 1–5, determine whether G and G' are isomorphic. If they are, give functions $g: V(G) \rightarrow V(G')$ and $h: E(G) \rightarrow E(G')$ that define the isomorphism. If they are not, give an isomorphic invariant that they do not share.

1.
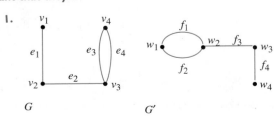

G G'

2.
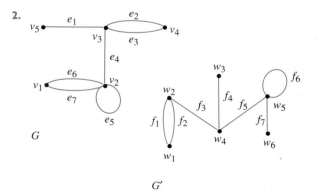

G G'

3.

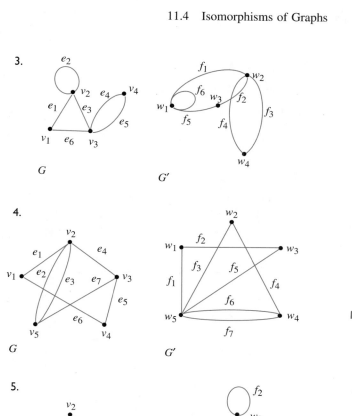

4.

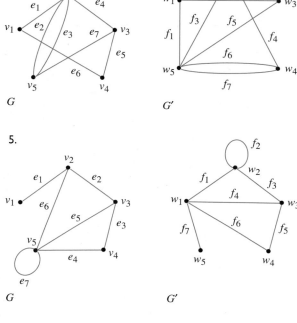

5.

8.

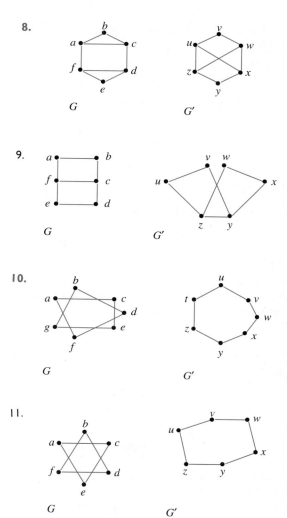

9.

10.

11.

For each pair of simple graphs G and G' in 6–13, determine whether G and G' are isomorphic. If they are, give a function $g: V(G) \to V(G')$ that defines the isomorphism. If they are not, give an isomorphic invariant that they do not share.

6.

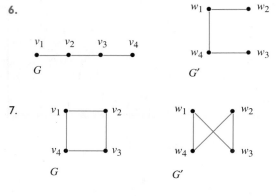

7.

12.

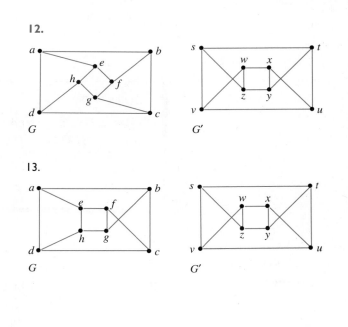

13.

14. Draw all nonisomorphic simple graphs with three vertices.

15. Draw all nonisomorphic simple graphs with four vertices.

16. Draw all nonisomorphic graphs with three vertices and no more than two edges.

17. Draw all nonisomorphic graphs with four vertices and no more than two edges.

H18. Draw all nonisomorphic graphs with four vertices and three edges.

19. Draw all nonisomorphic graphs with six vertices, all having degree 2.

20. Draw four nonisomorphic graphs with six vertices, two of degree 4 and four of degree 3.

Prove that each of the properties in 21–29 is an invariant for graph isomorphism. Assume that n, m, and k are all nonnegative integers.

21. Has n vertices

22. Has m edges

23. Has a circuit of length k

24. Has a simple circuit of length k

H25. Has m vertices of degree k

26. Has m simple circuits of length k

H27. Is connected

28. Has an Euler circuit

29. Has a Hamiltonian circuit

30. Show that the following two graphs are not isomorphic by supposing they are isomorphic and deriving a contradiction.

11.5 TREES

A fool sees not the same tree that a wise man sees.
(William Blake,
1757–1827)

If a friend asks what you are studying and you answer "trees," your friend is likely to infer you are taking a course in botany. But trees are also a subject for mathematical investigation. In mathematics, a tree is a connected graph that does not contain any but trivial circuits. (Recall that a trivial circuit is one that consists of a single vertex.) Despite the formality of the definition, mathematical trees are similar in certain ways to their botanical namesakes.

DEFINITION

A graph is said to be **circuit-free** if, and only if, it has no nontrivial circuits. A graph is called a **tree** if, and only if, it is circuit-free and connected. A **trivial tree** is a graph that consists of a single vertex, and an **empty tree** is a tree that does not have any vertices or edges. A graph is called a **forest** if, and only if, it is circuit-free.

EXAMPLE 11.5.1   Trees and Non-Trees

All the graphs shown in Figure 11.5.1 are trees, whereas those in Figure 11.5.2 are not.

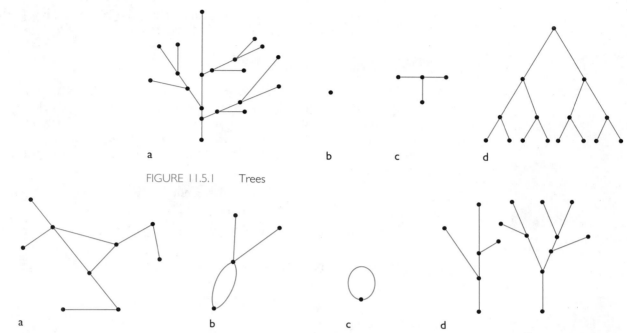

FIGURE 11.5.1   Trees

Note that the graphs in (a), (b), and (c) all have circuits and that the graph in (d) is not connected.

FIGURE 11.5.2   Non-Trees

EXAMPLES OF TREES

The following examples give just a few of the many and varied situations in which mathematical trees arise.

EXAMPLE 11.5.2   A Decision Tree

During orientation week a college administers an exam to all entering students to determine placement in the mathematics curriculum. The exam consists of two parts and placement recommendations are made as indicated by the tree shown in Figure 11.5.3. Read the tree from left to right to decide what course should be recommended for a student who scored 9 on part I and 7 on part II.

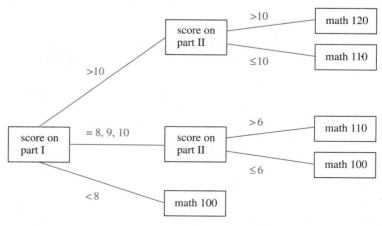

FIGURE 11.5.3

Solution
Since the student scored 9 on part I, the score on part II is checked. Since it is greater than 6, the student should be advised to take Math 110. ∎

EXAMPLE 11.5.3

A Syntactic Derivation Tree

In the last thirty years Noam Chomsky and others have developed new ways to describe the syntax (or grammatical structure) of natural languages such as English. This work has proved useful in constructing compilers for high-level computer languages. In this study trees are often used to show the derivation of grammatically correct sentences from certain basic rules. A very small subset of English grammar, for example, specifies that

Noam Chomsky

1. a sentence can be produced by writing first a noun phrase and then a verb phrase;
2. a noun phrase can be produced by writing an article and then a noun;
3. a noun phrase can also be produced by writing an article, then an adjective, and then a noun;
4. a verb phrase can be produced by writing a verb and then a noun phrase;
5. one article is "the";
6. one adjective is "young";
7. one verb is "caught";
8. one noun is "man";
9. one (other) noun is "ball."

John Backus

It is customary to express rules such as these (called **productions**) using the shorthand notation shown below. This notation, introduced by John Backus in 1959 and modified by Peter Naur in 1960, was used to describe the computer language Algol. It is called the **Backus-Naur notation**. In the notation, the symbol | represents the word *or*, and angle brackets ⟨ ⟩ are used to enclose terms to be defined (such as a sentence or noun phrase).

1. ⟨sentence⟩ → ⟨noun phrase⟩⟨verb phrase⟩
2, 3. ⟨noun phrase⟩ → ⟨article⟩⟨noun⟩ | ⟨article⟩⟨adjective⟩⟨noun⟩
4. ⟨verb phrase⟩ → ⟨verb⟩⟨noun phrase⟩
5. ⟨article⟩ → the
6. ⟨adjective⟩ → young
7, 8. ⟨noun⟩ → man | ball
9. ⟨verb⟩ → caught

Peter Naur

The derivation of the sentence "The young man caught the ball" from the above rules is described by the tree shown below.

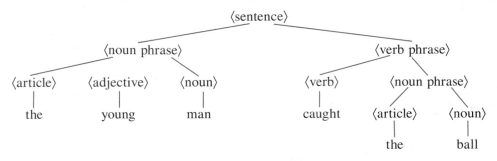

In the study of linguistics, **syntax** refers to the grammatical structure of sentences and **semantics** refers to the meanings of words and their interrelations. A sentence can be syntactically correct but semantically incorrect, as in the nonsensical sentence "The young ball caught the man," which can be derived from the rules given above. Or a sentence can contain syntactic errors but not semantic ones as, for instance, when a two-year-old child says "Me hungry!" ∎

EXAMPLE 11.5.4   ### Structure of Hydrocarbon Molecules

Arthur Cayley (1821–1895)

The German physicist Gustav Kirchoff (1824–1887) was the first to analyze the behavior of mathematical trees in connection with the investigation of electrical circuits. Soon after (and independently) the English mathematician Arthur Cayley used the mathematics of trees to enumerate all isomers for certain hydrocarbons. Hydrocarbon molecules are composed of carbon and hydrogen; each carbon atom can form up to four chemical bonds with other atoms and each hydrogen atom can form one bond with another atom. Thus the structure of hydrocarbon molecules can be represented by graphs such as those shown in Figure 11.5.4, in which the vertices represent atoms of hydrogen and carbon, denoted H and C, and the edges represent the chemical bonds between them.

Butane   Isobutane

FIGURE 11.5.4

Note that each of these graphs has four carbon atoms and ten hydrogen atoms, but the two graphs show different configurations of atoms. When two molecules have the same chemical formulae (in this case $C_4 H_{10}$) but different chemical bonds, they are called *isomers*.

Certain *saturated hydrocarbon* molecules contain the maximum number of hydrogen atoms for a given number of carbon atoms. Cayley showed that if such a saturated hydrocarbon molecule has k carbon atoms, then it has $2k + 2$ hydrogen atoms. The first step in doing so is to prove that the graph of such a saturated hydrocarbon molecule is a tree. Prove this using proof by contradiction. (You are asked to finish the derivation of Cayley's result in exercise 4 at the end of this section.)

Solution   Suppose there is a hydrocarbon molecule that contains the maximum number of hydrogen atoms for the number of its carbon atoms, and whose graph G is not a tree. *[We must derive a contradiction.]* Since G is not a tree, G is not connected or G has a nontrivial circuit. But the graph of any molecule is connected (all the atoms in a molecule must be connected to each other), and so G must have a nontrivial circuit. Now the edges of the circuit can only link carbon atoms because every vertex of a circuit has degree at least 2 and a hydrogen atom vertex has degree 1. Delete one edge of the circuit and add two new edges to join each of the newly disconnected carbon atom vertices to a hydrogen atom vertex as shown below.

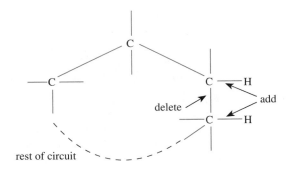

The resulting molecule has two more hydrogen atoms than the given molecule, but the number of carbon atoms is unchanged. This contradicts the supposition that the given molecule has the maximum number of hydrogen atoms for the given number of carbon atoms. Hence the supposition is false, and so G is a tree. ∎

CHARACTERIZING TREES

There is a somewhat surprising relation between the number of vertices and the number of edges of a tree. It turns out that if n is a positive integer, then any tree (no matter what the shape) with n vertices has $n - 1$ edges. Perhaps even more surprisingly, a partial converse to this fact is also true, namely, any *connected* graph with n vertices and $n - 1$ edges is a tree. It follows from these facts that if even one new edge (but no new vertex) is added to a tree, the resulting graph must contain a nontrivial circuit. Also, from the fact that removing an edge from a circuit does not disconnect a graph, it can be shown that every connected graph has a subgraph that is a tree. It follows that if n is a positive integer, any graph with n vertices and *fewer* than $n - 1$ edges is not connected.

A small but very important fact necessary to derive the first main theorem about trees is that any nontrivial tree must have at least one vertex of degree 1.

LEMMA 11.5.1

Any tree that has more than one vertex has at least one vertex of degree 1.

A constructive way to understand this lemma is to imagine being given a tree T with more than one vertex. You pick a vertex v at random and then search outward along a path from v looking for a vertex of degree 1. As you reach each new vertex, you check whether it has degree 1. If it does, you are finished. If it does not, you exit from the vertex along a different edge from the one you entered on. Because T is circuit-free, the vertices included in the path never repeat. And since the number of vertices of T is finite, the process of building a path must eventually terminate. When that happens, the final vertex v' of the path must have degree 1. This process is illustrated below.

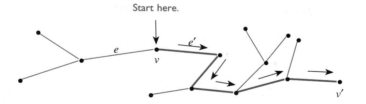

Search outward from v to find vertex v' of degree 1.

This discussion is made precise in the following proof.

Proof:

Let T be a particular but arbitrarily chosen tree that has more than one vertex, and consider the following algorithm:

Step 1:   Pick a vertex v of T and let e be an edge incident on v.
[If there were no edge incident on v, then v would be an isolated vertex. But this would contradict the assumption that T is connected (since it is a tree) and has at least two vertices.]

Step 2:   While $\deg(v) > 1$, repeat steps 2a, 2b, and 2c:

   Step 2a: Choose e' to be an edge incident on v such that $e' \neq e$.
   [Such an edge exists because deg (v) > 1 and so there are at least two edges incident on v.]

   Step 2b: Let v' be the vertex at the other end of e' from v. [Since T is a tree, e' cannot be a loop and therefore e' has two distinct endpoints.]

   Step 2c: Let $e = e'$ and $v = v'$. [This is just a renaming process in preparation for a repetition of step 2.]

   The algorithm just described must eventually terminate because the set of vertices of the tree T is finite. When it does, a vertex v of degree 1 will have been found.

Using Lemma 11.5.1 it is not difficult to show that, in fact, any tree that has more than one vertex has at least *two* vertices of degree 1. This extension of Lemma 11.5.1 is left to the exercises at the end of this section.

DEFINITION

Let T be a tree. A vertex of degree 1 in T is called a **terminal vertex** (or a **leaf**), and a vertex of degree greater than 1 in T is called an **internal vertex** (or a **branch vertex**).

EXAMPLE 11.5.5   Terminal and Internal Vertices

Find all terminal vertices and all internal vertices in the following tree:

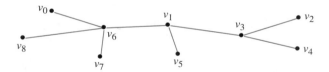

Solution   The terminal vertices are v_0, v_2, v_4, v_5, v_7, and v_8. The internal vertices are v_6, v_1, and v_3. ∎

The following is the first of the two main theorems about trees:

THEOREM 11.5.2

For any positive integer n, any tree with n vertices has $n - 1$ edges.

The proof is by mathematical induction. To do the inductive step, you assume the theorem is true for a positive integer k and then show it is true for $k + 1$. So you assume you have a tree T with $k + 1$ vertices, and you must show that T has $(k + 1) - 1 = k$ edges. As you do this, you are free to use the inductive hypothesis that any tree with k vertices has $k - 1$ edges. To make use of the inductive hypothesis, you need to reduce the tree T with $k + 1$ vertices to a tree with just k vertices. But by Lemma 11.5.1, T has a vertex v of degree 1, and since T is connected, v is attached to the rest of T by a single edge e as sketched below.

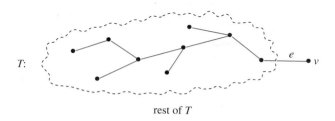

rest of T

Now if e and v are removed from T, what remains is a tree T' with $(k + 1) - 1 = k$ vertices. By inductive hypothesis, then, T' has $k - 1$ edges. But the original tree T has one more vertex and one more edge than T'. Hence T must have $(k - 1) + 1 = k$ edges, as was to be shown. A formal version of this argument is given on the next page.

Proof (by mathematical induction):

Let $P(n)$ be the property

$$\text{Any tree with } n \text{ vertices has } n - 1 \text{ edges.}$$

We use mathematical induction to show that this property is true for all integers $n \geq 1$.

The property is true for $n = 1$:   Let T be any tree with one vertex. Then T has zero edges (since it contains no loops). But $0 = 1 - 1$. So the property is true for $n = 1$.

If the property is true for some integer $k \geq 1$, then it is true for $k + 1$:   Suppose $k \geq 1$ is a positive integer and the property is true for k. In other words, suppose that any tree with k vertices has $k - 1$ edges. *[This is the inductive hypothesis.]* We must show that the property is true for $k + 1$. In other words, we must show that any tree with $k + 1$ vertices has $(k + 1) - 1 = k$ edges.

Let T be a particular but arbitrarily chosen tree with $k + 1$ vertices. *[We must show that T has k edges.]* Since k is a positive integer, $(k + 1) \geq 2$, and so T has more than one vertex. Hence by Lemma 11.5.1, T has a vertex v of degree 1. Also since T has more than one vertex, there is at least one other vertex in T besides v. Thus there is an edge e connecting v to the rest of T. Define a subgraph T' of T so that

$$V(T') = V(T) - \{v\}$$
$$E(T') = E(T) - \{e\}.$$

Then

1. The number of vertices of T' is $(k + 1) - 1 = k$;
2. T' is circuit-free (since T is circuit-free and removing an edge and a vertex cannot create a circuit);
3. T' is connected (see exercise 24 at the end of this section).

Hence, by the definition of tree, T' is a tree. Since T' has k vertices, by inductive hypothesis

$$\text{the number of edges of } T' = (\text{the number of vertices of } T') - 1$$
$$= k - 1.$$

But then

$$\text{the number of edges of } T = (\text{the number of edges of } T') + 1$$
$$= (k - 1) + 1$$
$$= k.$$

[This is what was to be shown.]

EXAMPLE 11.5.6   Determining Whether a Graph Is a Tree

A graph G has ten vertices and twelve edges. Is it a tree?

Solution   No. By Theorem 11.5.2, any tree with ten vertices has nine edges, not twelve. ■

EXAMPLE 11.5.7   Finding Nonisomorphic Trees

Find all nonisomorphic trees with four vertices.

Solution   By Theorem 11.5.2, any tree with four vertices has three edges. Thus the total degree of a tree with four vertices must be 6. Also, every tree with more than one vertex has at least two vertices of degree 1 (see the comment following Lemma 11.5.1 and exercise 29 at the end of this section). So the following combinations of degrees for the vertices are the only ones possible:

$$1, 1, 1, 3 \quad \text{and} \quad 1, 1, 2, 2.$$

There are nonisomorphic trees corresponding to both of these possibilities, as shown below.

To prove the second major theorem about trees, we need another lemma.

LEMMA 11.5.3

If G is any connected graph, C is any nontrivial circuit in G, and one of the edges of C is removed from G', then the graph that remains is connected.

Essentially, the reason Lemma 11.5.3 is true is that any two vertices in a circuit are connected by two distinct paths. It is possible to draw the graph so that one of these goes "clockwise" and the other goes "counterclockwise" around the circuit. For example, in the circuit shown in Figure 11.5.5 the clockwise path from v_2 to v_3 is

$$v_2 e_3 v_3$$

while the counterclockwise path from v_2 to v_3 is

$$v_2 e_2 v_1 e_1 v_0 e_6 v_5 e_5 v_4 e_4 v_3.$$

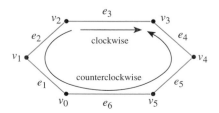

FIGURE 11.5.5

Proof:

Suppose G is a connected graph, C is a circuit in G, and e is an edge of C. Form a subgraph G' of G by removing e from G. Thus

$$V(G') = V(G)$$
$$E(G') = E(G) - \{e\}.$$

We must show that G' is connected. *[To show a graph is connected, we must show that if u and w are any vertices of the graph, then there exists a walk in G' from u to w.]* Suppose u and w are any two vertices of G'. *[We must find a walk from u to w.]* Since the vertex sets of G and G' are the same, u and w are both vertices of G, and since G is connected, there is a walk W in G from u to w.

Case 1 (e is not an edge of W): The only edge in G that is not in G' is e. So in this case W is also a walk in G'. Hence u is connected to w by a walk in G'.

Case 2 (e is an edge of W): In this case the walk W from u to w includes a section of the circuit C that contains e. Let C be denoted as follows:

$$C: v_0 e_1 v_1 e_2 v_2 \ldots e_n v_n (= v_0).$$

Now e equals one of the edges of C, so, to be specific, let $e = e_k$. Then the walk W either contains the sequence

$$v_{k-1} e_k v_k \quad \text{or} \quad v_k e_k v_{k-1}.$$

If W contains $v_{k-1} e_k v_k$, connect v_{k-1} to v_k by taking the "counterclockwise" walk W' defined as follows (see Figure 11.5.6):

$$W': v_{k-1} e_{k-1} v_{k-2} \ldots v_0 e_n v_{n-1} \ldots e_{k+1} v_k.$$

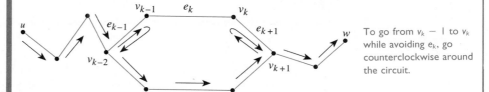

To go from $v_k - 1$ to v_k while avoiding e_k, go counterclockwise around the circuit.

FIGURE 11.5.6

If W contains $v_k e_k v_{k-1}$, connect v_k to v_{k-1} by taking the "clockwise" walk W'' defined as follows:

$$W'': v_k e_{k+1} v_{k+1} \ldots v_n e_1 v_1 e_2 \ldots e_{k-1} v_{k-1}.$$

Now patch either W' or W'' into W to form a new walk from u to w. For instance, to patch W' into W, start with the section of W from u to v_{k-1}, then take W' from v_{k-1} to v_k, and finally take the section of W from v_k to w. If this new walk still contains an occurrence of e, just repeat the process above until all occurrences are eliminated. *[This must happen eventually since the number of occurrences of e in G is finite.]* The result is a walk from u to w that does not contain e and hence is a walk in G'.

The arguments above show that both in case 1 and in case 2 there is a walk in G' from u to w. Since the choice of u and w was arbitrary, G' is connected.

The second major theorem about trees is a modified converse to Theorem 11.5.2.

THEOREM 11.5.4

For any positive integer n, if G is a connected graph with n vertices and $n - 1$ edges, then G is a tree.

Proof:

Let n be a positive integer and suppose G is a particular but arbitrarily chosen graph that is connected and has n vertices and $n - 1$ edges. *[We must show that G is a tree. Now a tree is a connected, circuit-free graph. Since we already know G is connected, it suffices to show that G is circuit-free.]* Suppose G is not circuit-free. That is, suppose G has a nontrivial circuit C. *[We must derive a contradiction.]* By Lemma 11.5.3, one edge of C can be removed from G to obtain a graph G' that is connected. If G' has a nontrivial circuit, then repeat the process above: Remove an edge of the circuit from G' to form a new connected graph. Continue repeating the process of removing edges from circuits until eventually a graph G'' is obtained that is connected and is circuit-free. By definition, G'' is a tree. Since no vertices were removed from G to form G'', G'' has n vertices just as G does. Thus, by Theorem 11.5.2, G'' has $n - 1$ edges. But the supposition that G has a nontrivial circuit implies that at least one edge of G is removed to form G''. This means that G'' has no more than $(n - 1) - 1 = n - 2$ edges! Thus we have derived a contradiction, and so we conclude that the supposition is false. Hence G is circuit-free, and therefore G is a tree *[as was to be shown]*.

Theorem 11.5.4 is not a full converse of Theorem 11.5.2. Although it is true that every *connected* graph with n vertices and $n - 1$ edges (where n is a positive integer) is a tree, it is not true that *every* graph with n vertices and $n - 1$ edges is a tree.

EXAMPLE 11.5.8   **A Graph with n Vertices and $n - 1$ Edges That Is Not a Tree**

Give an example of a graph with five vertices and four edges that is not a tree.

Solution   By Theorem 11.5.4, such a graph cannot be connected. One example of such an unconnected graph is shown below.

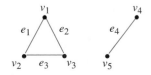

ROOTED TREES

An outdoor tree is rooted and so is the kind of family tree that shows all the descendants of one particular person. The terminology and notation of rooted trees blends the language of botanical trees and family trees. In mathematics, a rooted tree is either an empty tree or a tree in which one vertex has been distinguished from the others and is designated the *root*. Given any other vertex v in the tree, there is a unique path from the root to v. (After all, if there were two distinct paths a circuit could be constructed.) The number of edges in such a path is called the *level* of v, and the *height* of the tree is the length of the longest such path. (The height of the empty tree is defined to be 0.) It is traditional in drawing rooted trees to place the root at the top (as is done in family trees) and show the branches coming down as illustrated in Figure 11.5.7.

DEFINITION

A **rooted tree** is a tree in which one vertex is distinguished from the others and is called the **root**. The **level** of a vertex is the number of edges along the unique path between it and the root. The **height** of a rooted tree is the maximum level to any vertex of the tree. Given any internal vertex v of a rooted tree, the **children** of v are all those vertices that are adjacent to v and are one level farther away from the root than v. If w is a child of v, then v is called the **parent** of w, and two vertices that are both children of the same parent are called **siblings**. Given vertices v and w, if v lies on the unique path between w and the root, then v is an **ancestor** of w and w is a **descendant** of v.

These terms are illustrated in Figure 11.5.7.

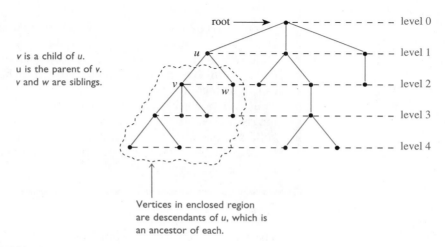

v is a child of u.
u is the parent of v.
v and w are siblings.

Vertices in enclosed region are descendants of u, which is an ancestor of each.

FIGURE 11.5.7 A Rooted Tree

EXAMPLE 11.5.9   Rooted Trees

Consider the tree with root v_0 shown in Figure 11.5.8.

a. What is the level of v_5?   b. What is the level of v_0?
c. What is the height of this rooted tree?   d. What are the children of v_3?
e. What is the parent of v_2?   f. What are the siblings of v_8?
g. What are the descendants of v_3?

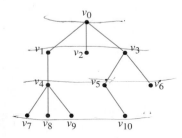

FIGURE 11.5.8

Solution   a. 2   b. 0   c. 3   d. v_5 and v_6 e. v_0 f. v_7 and v_9
g. v_5, v_6, v_{10}

BINARY TREES

When every vertex in a rooted tree has at most two children and each child is designated either the (unique) left child or the (unique) right child, the result is a *binary tree.*

> ### DEFINITION
>
> A **binary tree** is a rooted tree in which every internal vertex has at most two children. Each child in a binary tree is designated either a **left child** or a **right child** (but not both), and an internal vertex has at most one left and one right child. A **full binary tree** is a binary tree in which each internal vertex has exactly two children.
>
> Given an internal vertex v of a binary tree T, the **left subtree** of v is the binary tree whose root is the left child of v, whose vertices consist of the left child of v and all its descendants, and whose edges consist of all those edges of T that connect the vertices of the left subtree together. The **right subtree** of v is defined analogously.

These terms are illustrated in Figure 11.5.9.

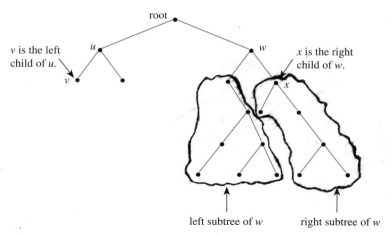

root

v is the left
child of *u*.

u

v

w

x is the right
child of *w*.

x

left subtree of *w*   right subtree of *w*

FIGURE 11.5.9

EXAMPLE 11.5.10   Representation of Algebraic Expressions

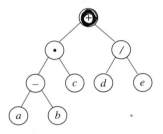

FIGURE 11.5.10

Binary trees are used in computer science to represent algebraic expressions with arbitrary nesting of balanced parentheses. For instance, the (labeled) binary tree in Figure 11.5.10 represents the expression a/b: The operator is at the root and acts on the left and right children of the root in left-right order.

More generally, the binary tree in Figure 11.5.11 represents the expression $a/(c + d)$. In such a representation, the internal vertices are arithmetic operators, the terminal vertices are variables, and the operator at each vertex acts on its left and right subtrees in left-right order.

FIGURE 11.5.11

Draw a binary tree to represent the expression $((a - b) \cdot c) + (d/e)$.

Solution

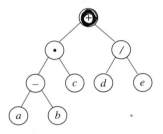

One interesting theorem about binary trees says that if you know the number of internal vertices of a full binary tree, then you can calculate the total number of vertices and also the number of terminal vertices, and conversely. More specifically, a full binary tree wih k internal vertices has a total of $2k + 1$ vertices of which $k + 1$ are terminal vertices.

THEOREM 11.5.5

If k is a positive integer and T is a full binary tree with k internal vertices, then T has a total of $2k + 1$ vertices and has $k + 1$ terminal vertices.

Proof:

Suppose T is a full binary tree with k internal vertices. Observe that the set of all vertices of T can be partitioned into two disjoint subsets: the set of all vertices that have a parent and the set of all vertices that do not have a parent. Now there is just one vertex that does not have a parent, namely the root. Also since every internal vertex of a full binary tree has exactly two children, the number of vertices that have a parent is twice the number of parents, or $2k$, since each parent is an internal vertex. Hence

$$\begin{bmatrix} \text{the total number} \\ \text{of vertices of } T \end{bmatrix} = \begin{bmatrix} \text{the number of} \\ \text{vertices that have} \\ \text{a parent} \end{bmatrix} + \begin{bmatrix} \text{the number of} \\ \text{vertices that do} \\ \text{not have a parent} \end{bmatrix}$$

$$= \qquad 2k \qquad + \qquad 1.$$

But it is also true that the total number of vertices of T equals the number of internal vertices plus the number of terminal vertices. Thus

$$\begin{bmatrix} \text{the total number} \\ \text{of vertices of } T \end{bmatrix} = \begin{bmatrix} \text{the number of} \\ \text{internal vertices} \end{bmatrix} + \begin{bmatrix} \text{the number of} \\ \text{terminal vertices} \end{bmatrix}$$

$$= \qquad k \qquad + \begin{bmatrix} \text{the number of} \\ \text{terminal vertices} \end{bmatrix}$$

Now equate the two expressions for the total number of vertices of T:

$$2k + 1 = k + \begin{bmatrix} \text{the number of} \\ \text{terminal vertices} \end{bmatrix}$$

Solving this equation gives

$$\begin{bmatrix} \text{the number of} \\ \text{terminal vertices} \end{bmatrix} = (2k + 1) - k = k + 1.$$

Thus the total number of vertices is $2k + 1$ and the number of terminal vertices is $k + 1$, as was to be shown.

EXAMPLE 11.5.11   **Determining Whether a Certain Full Binary Tree Exists**

Is there a full binary tree that has 10 internal vertices and 13 terminal vertices?

Solution   No. By Theorem 11.5.5, a full binary tree with 10 internal vertices has $10 + 1 = 11$ terminal vertices, not 13. ∎

Another interesting theorem about binary trees specifies the maximum number of terminal vertices of a binary tree of a given height. Specifically, the maximum number of terminal vertices of a binary tree of height h is 2^h. Another way to say this is that a binary tree with t terminal vertices has height of at least $\log_2 t$.

THEOREM 11.5.6

If T is a binary tree that has t terminal vertices and height h, then

$$t \leq 2^h.$$

Equivalently,

$$\log_2 t \leq h.$$

Proof:

We will use the strong form of mathematical induction on h to prove the truth of the following statement:

> For all integers $h \geq 0$, if T is any binary tree of height h then the number of terminal vertices of T is at most 2^h.

Let $P(h)$ be the property

> If T is any binary tree of height h, then the number of terminal vertices of T is at most 2^h.

To go by strong induction, we will first show that the property is true for $h = 0$. Second, we will show that if the property is true for all nonnegative integers $k < h$, then it is true for h.

The property is true for $h = 0$:   We must show that if T is any binary tree of height 0, then the number of terminal vertices of T is at most 2^0. Suppose T is a tree of height 0. Then T is the empty tree or T consists of a single vertex (the root) only. Let t be the number of terminal vertices of T. In case T is the empty tree, $t = 0$ and $h = 0$. Since $0 \leq 2^0$, then $t \leq 2^h$. In case T consists of a single vertex, $t = 1$ and $h = 0$. Since $1 = 2^0$, then in this case also $t \leq 2^h$. Hence in either case $t \leq 2^h$ *[as was to be shown]*.

If $h \geq 1$ and the property is true for all nonnegative integers $k < h$, then it is true for h: Suppose $h \geq 1$ is an integer and for all nonnegative integers $k < h$, $P(k)$ is true. In other words, suppose that for all integers $k < h$, any binary tree of height k has at most 2^k terminal vertices.

We must show that $P(h)$ is true. In other words, we must show that any binary tree of height h has at most 2^h terminal vertices.

Let T be a binary tree of height h and root v. Since $h \geq 1$, v has at least one child: v_L and/or v_R as illustrated in Figure 11.5.12.

Now v_L and v_R are the roots of the left and right subtrees of v called T_L and T_R, respectively. (One of these trees may be empty.) Now T_L and T_R are binary trees since T is a binary tree. Let h_L and h_R be the heights of T_L and T_R respectively. Then $h_L \leq h - 1$ and $h_R \leq h - 1$ since T is obtained by joining T_L

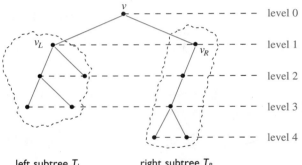

left subtree T_L right subtree T_R

FIGURE 11.5.12   A Binary Tree

and T_R and adding a level. Let t_L and t_R be the numbers of terminal vertices of T_L and T_R respectively. Then since both T_L and T_R have heights less than h, by inductive hypothesis

$$t_L \leq 2^{h_L} \quad \text{and} \quad t_R \leq 2^{h_R}.$$

But the terminal vertices of T consist exactly of the terminal vertices of T_L together with the terminal vertices of T_R. Therefore,

$$t = t_L + t_R \leq 2^{h_L} + 2^{h_R} \qquad \text{by inductive hypothesis}$$
$$\Rightarrow \quad t \leq 2^{h-1} + 2^{h-1} \qquad \text{since } h_L \leq h - 1 \text{ and } h_R \leq h - 1$$
$$\Rightarrow \quad t \leq 2 \cdot (2^{h-1}) \qquad \text{by basic algebra.}$$
$$\Rightarrow \quad t \leq 2^h$$

So the number of terminal vertices is at most 2^h [*as was to be shown*].

Since both the basis and the inductive steps have been proved, the first version of the theorem is proved.

The equivalent inequality $\log_2 t \leq h$ follows immediately from the fact that the logarithmic function with base 2 is increasing and from the definition of logarithm. For if

$$t \leq 2^h$$

then applying the logarithm with base 2 function to both sides gives

$$\log_2 t \leq \log_2 (2^h).$$

It follows from the definition of logarithm that $\log_2 (2^h) = h$ [*because $\log_2 (2^h)$ is the exponent to which 2 must be raised to obtain 2^h*]. Hence

$$\log_2 t \leq h$$

[*as was to be shown*].

EXAMPLE 11.5.12   Determining Whether a Certain Binary Tree Exists

Is there a binary tree that has height 5 and 38 terminal vertices?

Solution   No. By Theorem 11.5.6, any binary tree T with height 5 has at most $2^5 = 32$ terminal vertices. So such a tree cannot have 38 terminal vertices.   ∎

From Theorem 11.5.6 it can be deduced that any algorithm to sort a set of n data items has worst-case order of at least $n \cdot \log_2 n$. This result is obtained by analyzing a decision tree whose terminal vertices are all the $n!$ arrangements of the set to be sorted. The height of this tree is the minimum number of operations needed to sort the set. By Theorem 11.5.6 this height is at least $\log_2 (n!)$. But $\log_2 (n!) \geq M \cdot (n \cdot \log_2 n)$, where M is a positive constant (see exercises 45(a) and 46 of Section 9.4). It follows that the worst-case order for an algorithm to sort the set cannot be less than $n \cdot \log_2 n$.

EXERCISE SET 11.5

1. Read the tree in Example 11.5.2 from left to right to decide
   a. what course a student who scored 12 on part I and 4 on part II should take;
   b. what course a student who scored 8 on part I and 9 on part II should take.

2. Draw trees to show the derivations of the following sentences from the rules given in Example 11.5.3.
   a. The young ball caught the man.
   b. The man caught the young ball.

H 3. What is the total degree of a tree with n vertices? Why?

4. Let G be the graph of a hydrocarbon molecule with the maximum number of hydrogen atoms for the number of its carbon atoms.
   a. Draw the graph of G if G has three carbon and eight hydrogen atoms.
   b. Draw the graphs of three isomers of C_5H_{12}.
   c. Use Example 11.5.4 and exercise 3 to prove that if the vertices of G consist of k carbon atoms and m hydrogen atoms, then G has a total degree of $2k + 2m - 2$.
   H d. Prove that if the vertices of G consist of k carbon atoms and m hydrogen atoms, then G has a total degree of $4k + m$.
   e. Equate the results of (c) and (d) to prove Cayley's result that a saturated hydrocarbon molecule with k carbon atoms and a maximum number of hydrogen atoms has $2k + 2$ hydrogen atoms.

H 5. Extend the argument given in the proof of Lemma 11.5.1 to show that a tree with more than one vertex has at least two vertices of degree 1.

6. If graphs are allowed to have an infinite number of vertices and edges, then Lemma 11.5.1 is false. Give a counterexample that shows this. In other words, give an example of an "infinite tree" (a connected, circuit-free graph with an infinite number of vertices and edges) that has no vertex of degree 1.

7. Find all terminal vertices and all internal vertices for the following trees.

a.
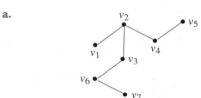

b.
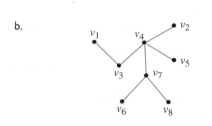

In each of 8–21, either draw a graph with the given specifications or explain why no such graph exists.

8. tree, nine vertices, nine edges

9. graph, connected, nine vertices, nine edges

10. graph, circuit-free, nine vertices, six edges

11. tree, six vertices, total degree 14

12. tree, five vertices, total degree 8

13. graph, connected, six vertices, five edges, has a nontrivial circuit

14. graph, two vertices, one edge, not a tree

15. graph, circuit-free, seven vertices, four edges

16. tree, twelve vertices, fifteen edges

17. graph, six vertices, five edges, not a tree

18. tree, five vertices, total degree 10

19. graph, connected, ten vertices, nine edges, has a nontrivial circuit

20. simple graph, connected, six vertices, six edges

21. tree, ten vertices, total degree 24

22. A connected graph has twelve vertices and eleven edges. Does it have a vertex of degree 1? Why?

23. A connected graph has nine vertices and twelve edges. Does it have a nontrivial circuit? Why?

24. Suppose that v is a vertex of degree 1 in a connected graph G and that e is the edge incident on v. Let G' be the subgraph of G obtained by removing v and e from G. Must G' be connected? Why?

25. A graph has eight vertices and six edges. Is it connected? Why?

H26. If a graph has n vertices and $n - 2$ or fewer edges, can it be connected? Why?

27. A circuit-free graph has ten vertices and nine edges. Is it connected? Why?

H28. Is a circuit-free graph with n vertices and at least $n - 1$ edges connected? Why?

29. Prove that every nonempty, nontrivial tree has at least two vertices of degree 1 by filling in the details and completing the following argument: Let T be a nontrivial tree and let S be the set of all paths from one vertex to another of T. Among all the paths in S, choose a path with the most edges. (Why is it possible to find such a P?) What can you say about the initial and final vertices of P? Why?

30. Find all nonisomorphic trees with five vertices.

31. a. Prove that the following is an invariant for graph isomorphism: a vertex of degree i is adjacent to a vertex of degree j.
    Hb. Find all nonisomorphic trees with six vertices.

32. Consider the tree shown below with root a.
    a. What is the level of n?
    b. What is the level of a?
    c. What is the height of this rooted tree?
    d. What are the children of n?

e. What is the parent of g?
f. What are the siblings of j?
g. What are the descendants of f?

33. Consider the tree shown below with root v_0.
    a. What is the level of v_8?
    b. What is the level of v_0?
    c. What is the height of this rooted tree?
    d. What are the children of v_{10}?
    e. What is the parent of v_5?
    f. What are the siblings of v_1?
    g. What are the descendants of v_{12}?

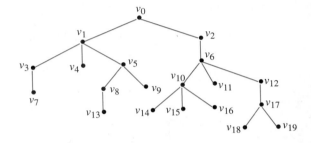

34. Draw binary trees to represent the following expressions:

    a. $a \cdot b - (c/(d + e))$ b. $a/(b - c \cdot d)$

In each of 35–50 either draw a graph with the given specifications or explain why no such graph exists.

35. full binary tree, five internal vertices

36. full binary tree, five internal vertices, seven terminal vertices

37. full binary tree, seven vertices, of which four are internal vertices

38. full binary tree, twelve vertices

39. full binary tree, nine vertices

40. binary tree, height 3, seven terminal vertices

41. full binary tree, height 3, six terminal vertices

42. binary tree, height 3, nine terminal vertices

43. full binary tree, eight internal vertices, seven terminal vertices

44. binary tree, height 4, eight terminal vertices

45. full binary tree, seven vertices

46. full binary tree, nine vertices, five internal vertices

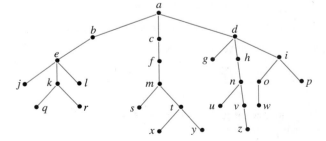

47. full binary tree, four internal vertices

48. binary tree, height 4, eighteen terminal vertices

49. full binary tree, sixteen vertices

50. full binary tree, height 3, seven terminal vertices

51. What can you deduce about the height of a binary tree if you know that it has

    a. twenty-five terminal vertices?

    b. forty terminal vertices?

    c. sixty terminal vertices?

11.6   SPANNING TREES

I contend that each science is a real science insofar as it is mathematics.
(Immanuel Kant, 1724–1804)

An East Coast airline company wants to expand service to the Midwest and has received permission from the Federal Aviation Authority to fly any of the routes shown in Figure 11.6.1.

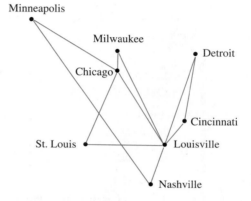

FIGURE 11.6.1

The company wishes to legitimately advertise service to all the cities shown, but, for reasons of economy, wants to use the least possible number of individual routes to connect them. One possible route system is given in Figure 11.6.2.

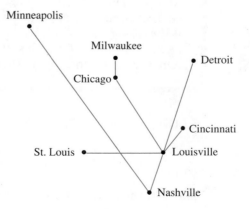

FIGURE 11.6.2

Clearly this system joins all the cities. Is the number of individual routes minimal? The answer is yes, and the reason may surprise you.

The fact is that the graph of any system of routes that satisfies the company's wishes is a tree. For if the graph were to contain a circuit, then one of the routes in the circuit could be removed without disconnecting the graph (by Lemma 11.5.3) and that would give a smaller total number of routes. But any tree with eight vertices has seven edges. Therefore any system of routes that connects all eight vertices and yet minimizes the total number of routes consists of seven routes.

DEFINITION

A **spanning tree** for a graph G is a subgraph of G that contains every vertex of G and is a tree.

The preceding discussion contains the essence of the proof of the following proposition:

PROPOSITION 11.6.1

1. Every connected graph has a spanning tree.

2. Any two spanning trees for a graph have the same number of edges.

Proof of (1):

Suppose G is a connected graph. If G is circuit-free, then G is its own spanning tree and we are done. If not, then G has at least one circuit C_1. By Lemma 11.5.3, the subgraph of G obtained by removing an edge from C_1 is connected. If this subgraph is circuit-free, then it is a spanning tree and we are done. If not, then it has at least one circuit C_2, and, as above, an edge can be removed from C_2 to obtain a connected subgraph. Continuing in this way, we can remove successive edges from circuits, until eventually we obtain a connected, circuit-free subgraph T of G. *[This must happen at some point because the number of edges of G is finite and at no stage does removal of an edge disconnect the subgraph.]* Also T contains every vertex of G because no vertices of G were removed in constructing it. Thus T is a spanning tree for G.

The proof of part (2) is left as an exercise.

EXAMPLE 11.6.1   Spanning Trees

Find all spanning trees for the graph G pictured below.

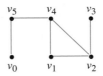

Solution   The graph G has one circuit $v_2v_1v_4v_2$ and removal of any edge of the circuit gives a tree. Thus, as shown below, there are three spanning trees for G.

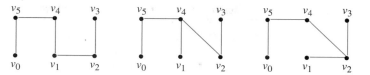

MINIMAL SPANNING TREES

The graph of the routes allowed by the Federal Aviation Authority shown in Figure 11.6.1 can be annotated by adding the distances (in miles) between each pair of cities. This is done in Figure 11.6.3.

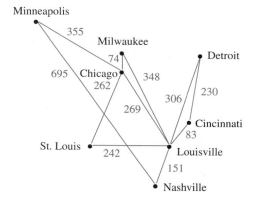

FIGURE 11.6.3

Now suppose the airline company wants to serve all the cities shown but with a route system that minimizes the total mileage. Note that such a system is a tree. For if the system contained a circuit, removal of an edge from the circuit would not affect a person's ability to reach every city in the system from every other (again, by Lemma 11.5.3), but it would reduce the total mileage of the system.

More generally, a graph whose edges are labeled with numbers (known as *weights*) is called a *weighted graph*. A *minimal spanning tree* is a spanning tree for which the sum of the weights of all the edges is as small as possible.

DEFINITION

A **weighted graph** is a graph for which each edge has an associated real number **weight**. The sum of the weights of all the edges is the **total weight** of the graph. A **minimal spanning tree** for a weighted graph is a spanning tree that has the least possible total weight compared to all other spanning trees for the graph.

If G is a weighted graph and e is an edge of G then $w(e)$ denotes the weight of e and $w(G)$ denotes the total weight of G.

The problem of finding a minimal spanning tree for a graph is certainly solvable. One solution is to list all spanning trees for the graph, compute the total weight of each, and choose one for which this total is minimal. (Note that the well-ordering principle guarantees the existence of such a minimal total.) This solution, however, is inefficient in its use of computing time because the number of distinct spanning trees is so large. For instance, a complete graph with n vertices has n^{n-2} spanning trees.

In 1956 and 1957 Joseph B. Kruskal and Robert C. Prim, working independently, described much more efficient algorithms to construct minimal spanning trees. For graphs with n vertices and m edges, Kruskal's and Prim's algorithms can be implemented so as to have worst case orders of $m \log_2 m$ and n^2 respectively.

KRUSKAL'S ALGORITHM

In Kruskal's algorithm the edges of a weighted graph are examined one by one in order of increasing weight. At each stage the edge being examined is added to what will become the minimal spanning tree provided that this addition does not create a circuit. After $n - 1$ edges have been added (where n is the number of vertices of the graph), these edges together with the vertices of the graph form a minimal spanning tree for the graph.

Joseph Kruskal

ALGORITHM 11.6.1   Kruskal

Input: G *[a weighted graph with n vertices]*

Algorithm Body:
[Build a subgraph T of G to consist of all the vertices of G with edges added in order of increasing weight. At each stage, let m be the number of edges of T.]

1. Initialize T to have all the vertices of G and no edges.

2. Let E be the set of all edges of G and let $m := 0$.
   [pre-condition: G is connected.]

3. **while** $(m < n - 1)$
   3a. Find an edge e in E of least weight.
   3b. Delete e from E.
   3c. **if** addition of e to the edge set of T does not produce a circuit
        then add e to the edge set of T and set $m := m + 1$

   end while
   [post-condition: T is a minimal spanning tree for G.]

Output T
end Algorithm 11.6.1

The following example shows how Kruskal's algorithm works for the graph of the airline route system.

EXAMPLE 11.6.2   Action of Kruskal's Algorithm

Describe the action of Kruskal's algorithm for the graph shown in Figure 11.6.4.

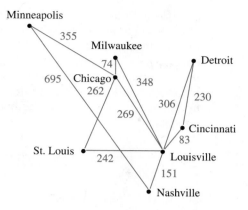

FIGURE 11.6.4

Solution

Iteration Number	Edge Considered	Weight	Action Taken
1	Chicago–Milwaukee	74	added
2	Louisville–Cincinnati	83	added
3	Louisville–Nashville	151	added
4	Cincinnati–Detroit	230	added
5	St. Louis–Louisville	242	added
6	St. Louis–Chicago	262	added
7	Chicago–Louisville	269	not added
8	Louisville–Detroit	306	not added
9	Louisville–Milwaukee	348	not added
10	Minneapolis–Chicago	355	added

The tree produced by Kruskal's algorithm is shown in Figure 11.6.5.

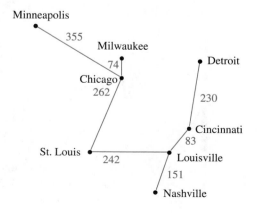

FIGURE 11.6.5

It is not obvious from the description of Kruskal's algorithm that it does what it is supposed to do. To be specific, what guarantees that it is possible at each stage to find an edge of least weight whose addition does not produce a circuit? And if such edges can be found, what guarantees that they will all eventually connect? And if they do connect, what guarantees that the resulting tree has minimal weight? Of course, the mere fact that Kruskal's algorithm is printed in this book may lead you to believe that everything works out. But the questions above are real, and they deserve serious answers.

THEOREM 11.6.2

When a connected, weighted graph is input to Kruskal's algorithm, the output is a minimal spanning tree.

Proof:

Suppose that G is a connected, weighted graph with n vertices and that T is a subgraph of G produced when G is input to Kruskal's algorithm. Clearly T is circuit-free *[since no edge that completes a circuit is ever added to T]*. Also T is connected. For as long as T has more than one connected component, the set of edges of G that can be added to T without creating a circuit is nonempty. *[The reason is that since G is connected, given any vertex v_1 in one connected component C_1 of T and any vertex v_2 in another connected component C_2, there is a path in G from v_1 to v_2. Since C_1 and C_2 are distinct, there is an edge e of this path that is not in T. Adding e to T does not create a circuit in T because deletion of an edge from a circuit does not disconnect a graph and deletion of e would.]* The preceding arguments show that T is circuit-free and connected. Since by construction T contains every vertex of G, T is a spanning tree for G.

Now we show that T has minimal weight. Let T_1 be any minimal spanning tree for G. If $T = T_1$, then T is a minimal spanning tree for G and we are done. If $T \neq T_1$, then there is an edge e in T that is not an edge of T_1. *[Since T and T_1 both have the same vertex set, if they differ at all, they must have different edge sets.]* Now adding e to T_1 produces a graph with a unique nontrivial circuit (see exercise 14 at the end of this section). Let e' be an edge of this circuit such that e' is not in T. *[Such an edge must exist because T is a tree and hence circuit-free.]* Let T_2 be the graph obtained from T_1 by removing e' and adding e. This situation is illustrated below.

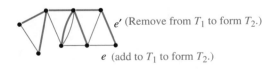

e' (Remove from T_1 to form T_2.)

e (add to T_1 to form T_2.)

Entire graph is G.
T_1 has thick edges.
e is in T but not T_1.
e' is in T_1 but not T.

Note that T_2 has $n - 1$ edges and n vertices and that T_2 is connected *[since by Lemma 11.5.3 the subgraph obtained by removing an edge from a circuit in a connected graph is connected]*. Consequently, T_2 is a spanning tree for G. In addition,

$$w(T_2) = w(T_1) - w(e') + w(e).$$

Now $w(e) \leq w(e')$ because at the stage in Kruskal's algorithm when e was added to T, e' was available to be added *[since it was not already in T, and at that stage its addition could not produce a circuit since e was not in T]*, and e' *would* have been added had its weight been smaller than that of e. Thus

$$w(T_2) = w(T_1) - \underbrace{(w(e') - w(e))}_{\geq 0}$$

$$\leq w(T_1).$$

But T_1 is a minimal spanning tree. Since T_2 is a spanning tree with weight less than or equal to the weight of T_1, T_2 is also a minimal spanning tree for G.

Finally, note that by construction, T_2 has one more edge in common with T than does T_1. If T now equals T_2, then T is a minimal spanning tree and we are done. If not, then we can repeat the process described above to find a minimal spanning tree T_3 that has one more edge in common with T than T_2. Continuing in this way produces a sequence of minimal spanning trees T_1, T_2, T_3, . . . each of which has one more edge in common with T than the preceding tree. Since T has only a finite number of edges, this sequence is finite and so there is a minimal spanning tree, T_k, which is identical to T. It follows that T is, itself, a minimal spanning tree for G.

PRIM'S ALGORITHM

Prim's algorithm works differently from Kruskal's. It builds a minimal spanning tree T by expanding outward in connected links from some vertex. One edge and one vertex are added at each stage. The edge added is the one of least weight that connects the vertices already in T with those not in T, and the vertex is the endpoint of this edge that is not already in T.

Robert Prim

ALGORITHM 11.6.2   Prim

Input: G *[a weighted graph with n vertices]*

Algorithm Body:
[Build a subgraph T of G by starting with any vertex v of G and attaching edges (with their endpoints) one by one, each time choosing an adjacent edge of least weight.]

1. Pick a vertex v of G and let T be the graph with one vertex, v, and no edges.

2. Let V be the set of all vertices of G except v.
   [pre-condition: G is connected.]

3. **for** $i := 1$ **to** $n - 1$
       3a. Find an edge e of G such that (1) e connects T to one of the vertices in V, and (2) e has the least weight of all edges connecting T to a vertex in V. Let w be the endpoint of e that is in V.
       3b. Add e and w to the edge and vertex sets of T and delete w from V.
   next i
   [post-condition: T is a minimal spanning tree for G.]

Output: T

end Algorithm 11.6.2

The following example shows how Prim's algorithm works for the graph of the airline route system.

EXAMPLE 11.6.3   Action of Prim's Algorithm

Describe the action of Prim's algorithm for the graph in Figure 11.6.6 using the Minneapolis vertex as a starting point.

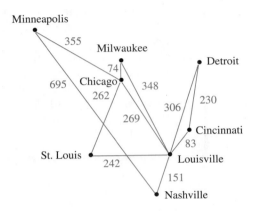

FIGURE 11.6.6

Solution

Iteration Number	Vertex Added	Edge Added	Weight
0	Minneapolis		
1	Chicago	Minneapolis–Chicago	355
2	Milwaukee	Chicago–Milwaukee	74
3	St. Louis	Chicago–St. Louis	262
4	Louisville	St. Louis–Louisville	242
5	Cincinnati	Louisville–Cincinnati	83
6	Nashville	Louisville–Nashville	151
7	Detroit	Cincinnati–Detroit	230

Note that the tree obtained is the same as that obtained by Kruskal's algorithm, but the edges are added in a different order. It is not hard to see that when a connected graph is input to Prim's algorithm, the result is a spanning tree. What is not so clear is that this spanning tree is minimal. The proof of the following theorem establishes that it is.

THEOREM 11.6.3

When a connected, weighted graph G is input to Prim's algorithm, the output is a minimal spanning tree for G.

Proof:

Let G be a connected, weighted graph and suppose G is input to Prim's algorithm. At each stage of execution of the algorithm an edge must be found that connects a vertex in a subgraph to a vertex outside the subgraph. As long as there are vertices outside the subgraph, the connectedness of G ensures that such an edge can always be found. *[For if one vertex in the subgraph and one vertex outside it are chosen, then by the connectedness of G there is a walk in G linking the two. As one travels along this walk, at some point one moves along an edge from a vertex inside the subgraph to a vertex outside the subgraph.]* Now it is clear that the output T of Prim's algorithm is a tree because the edge and vertex added to T at each stage are connected to other edges and vertices of T and because at no stage is a circuit created since each edge added connects vertices in two disconnected sets. *[Consequently, removal of a newly added edge gives a disconnected graph, whereas by Lemma 11.5.3 removal of an edge from a circuit gives a connected graph.]* Also T includes every vertex of G because T, being a tree with $n - 1$ edges, has n vertices *[and that is all G has]*. Thus T is a spanning tree for G.

Next we prove that T has minimal weight. Let T_1 be *any* minimal spanning tree for G. If $T = T_1$, then we are done. If not, then there is an edge in T that is not in T_1. Of all edges in T and not T_1, let e be the first that was added when T was constructed using Prim's algorithm. Let V be the set of vertices of T just before the addition of e. Then one endpoint, say v, of e is in T and the other, say w, is not. Since T_1 is a spanning tree for G, there is a path in T_1 joining v to w. As one travels along this path, one must encounter an edge e' joining a vertex in V to one that is not in V. Now at the stage when e was added to T, e' could also have been added and it *would* have been added instead of e had its weight been less than that of e. Since e' was not added at that stage, we conclude that

$$w(e') \geq w(e).$$

Let T_2 be the graph obtained from T_1 by removing e' and adding e. *[Thus T_2 has one more edge in common with T than T_1 does.]* Note that T_2 is a tree. The reason is that since e' is part of a path in T_1 from v to w and e connects v and w, adding e to T_1 creates a circuit. When e' is removed from this circuit, the resulting subgraph remains connected. In fact, T_2 is a spanning tree for G since no vertices were removed in forming T_2 from T_1. The argument showing that $w(T_2) \leq w(T_1)$ is left as an exercise. *[It is virtually identical to part of the proof of Theorem 11.6.2.]* It follows that T_2 is a minimal spanning tree for G with one more edge in common with T than T_1. If T now equals T_2, then we are done. If not, then, as above, we can find another minimal spanning tree T_3 having one more edge in common with T than T_2. Continuing in this way produces a sequence of minimal spanning trees $T_1, T_2, T_3, \ldots$ each of which has one more edge in common with T than the preceding tree. Since T has only a finite number of edges, this sequence is finite and so there is a tree, T_k, which is identical to T. This shows that T is itself a minimal spanning tree.

EXAMPLE 11.6.4   Finding Minimal Spanning Trees

Find all minimal spanning trees for the graph in Figure 11.6.7. Use Kruskal's algorithm and Prim's algorithm starting at vertex a. Indicate the order in which edges are added to form each tree.

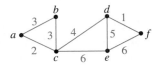

FIGURE 11.6.7

Solution   When Kruskal's algorithm is applied, edges are added in one of the following two orders:

1. $\{d, f\}, \{a, c\}, \{a, b\}, \{c, d\}, \{d, e\}$
2. $\{d, f\}, \{a, c\}, \{b, c\}, \{c, d\}, \{d, e\}$

When Prim's algorithm is applied starting at a, edges are added in one of the following two orders:

1. $\{a, c\}, \{a, b\}, \{c, d\}, \{d, f\}, \{d, e\}$
2. $\{a, c\}, \{b, c\}, \{c, d\}, \{d, f\}, \{d, e\}$

Thus, as shown below, there are two distinct minimal spanning trees for this graph.

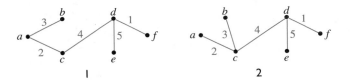

1   2

EXERCISE SET 11.6

Find all possible spanning trees for each of the graphs in 1 and 2.

1.

2.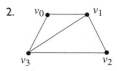

Find a spanning tree for each of the graphs in 3 and 4.

3.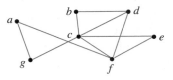

4.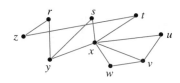

Use Kruskal's algorithm to find a minimal spanning tree for each of the graphs in 5 and 6. Indicate the order in which edges are added to form each tree.

5.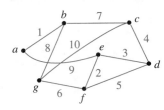

6.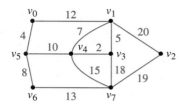

Use Prim's algorithm starting with vertex a or v_0 to find a minimal spanning tree for each of the graphs in 7 and 8. Indicate the order in which edges are added to form the tree.

7. The graph of exercise 5.

8. The graph of exercise 6.

For each of the graphs in 9 and 10, find all minimal spanning trees that can be obtained using (a) Kruskal's algorithm and (b) Prim's algorithm starting with vertex a or t. Indicate the order in which edges are added to form each tree.

9.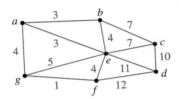

10.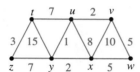

11. A pipeline is to be built that will link six cities. The cost (in millions) of constructing each potential link depends on distance and terrain and is shown in the weighted graph below. Find a system of pipelines to connect all the cities and yet minimize the total cost.

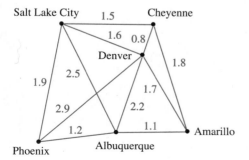

12. Prove part (2) of Proposition 11.6.1: Any two spanning trees for a graph have the same number of edges.

13. Given any two distinct vertices of a tree, there exists a unique path from one to the other.

    a. Give an informal justification for the above statement.

   ◆**b.** Write a formal proof of the above statement.

14. Prove that if G is a graph with spanning tree T and e is an edge of G that is not in T, then the graph obtained by adding e to T contains one and only one set of edges that forms a nontrivial circuit.

15. Suppose G is a connected graph and T is a circuit-free subgraph of G. Suppose also that if any edge e of G not in T is added to T, the resulting graph contains a circuit. Prove that T is a spanning tree for G.

H **16. a.** Suppose T_1 and T_2 are two different spanning trees for a graph G. Must T_1 and T_2 have an edge in common? Prove or give a counterexample.

   ◆ **b.** Suppose that the graph G in part (a) is simple. Must T_1 and T_2 have an edge in common? Prove or give a counterexample.

H **17.** Prove that an edge e is contained in every spanning tree for a connected graph G if, and only if, removal of e disconnects G.

18. Consider the spanning trees T_1 and T_2 in the proof of Theorem 11.6.3. Prove that $w(T_2) \leq w(T_1)$.

19. Suppose that T is a minimal spanning tree for a connected weighted graph G and that G contains an edge e (not a loop) that is not in T. Let v and w be the endpoints of e. By exercise 13 there is a unique path in T from v to w. Let e' be any edge of this path. Prove that $w(e') \leq w(e)$.

H **20.** Prove that if G is a connected, weighted graph and e is an edge of G (not a loop) that has smaller weight than any other edge of G, then e is in every minimal spanning tree for G.

◆ **21.** If G is a connected, weighted graph and not two edges of G have the same weight, does there exist a unique minimal spanning tree for G? Justify your answer.

◆ **22.** Prove that if G is a connected, weighted graph and e is an edge of G that (1) has larger weight than any other edge of G and (2) is in a circuit of G, then there is no minimal spanning tree T for G such that e is in T.

23. Suppose a disconnected graph is input to Kruskal's algorithm. What will be the output?

24. Suppose a disconneted graph is input to Prim's algorithm. What will be the output?

25. Prove that if a connected, weighted graph G is input to the following algorithm, the output is a minimal spanning tree for G.

Algorithm 11.6.3
Input: G *[a graph]*
Algorithm Body:
1. $T := G$.
2. $E :=$ the set of all edges of G, $m :=$ the number of edges of G
   [precondition: G is connected.]
3. **while** $(m > 0)$
   3a. Find an edge e in E that has maximal weight.
   3b. Remove e from E and set $m := m - 1$.
   3c. **if** the subgraph obtained when e is removed from the edge set of T is connected **then** remove e from the edge set of T
   end while
   [postcondition: T is a minimal spanning tree for G.]
Output: T *[a graph]*
end Algorithm 11.6.3

APPENDIX A

Properties of the Real Numbers*

In this text we take the real numbers and their basic properties as our starting point. We give a core set of properties, called axioms, which the real numbers are assumed to satisfy, and we state some useful properties that can be deduced from these axioms.

We assume that there are two binary operations defined on the set of real numbers, called **addition** and **multiplication**, such that if a and b are any two real numbers, the **sum** of a and b, denoted $a + b$, and the **product** of a and b, denoted $a \cdot b$ or ab, are also real numbers. These operations satisfy properties F1–F6, which are called the **field axioms**.

F1. *Commutative Laws* For all real numbers a and b,
$$a + b = b + a \quad \text{and} \quad a \cdot b + b \cdot a.$$

F2. *Associative Laws* For all real numbers a, b, and c,
$$(a + b) + c = a + (b + c) \quad \text{and} \quad (a \cdot b) \cdot a + a \cdot (b \cdot c).$$

F3. *Distributive Laws* For all real numbers a, b, and c,
$$a \cdot (b + c) = a \cdot b + a \cdot c \quad \text{and} \quad (b + c) \cdot a = b \cdot a + c \cdot a.$$

F4. *Existence of Identity Elements* There exist two distinct real numbers, denoted 0 and 1, such that for every real number a,
$$0 + a = a + 0 = a \quad \text{and} \quad 1 \cdot a = a \cdot 1 = a.$$

F5. *Existence of Additive Inverses* For every real number a, there is a real number, denoted $-a$ and called the **additive inverse** of a, such that
$$a + (-a) = (-a) + a = 0.$$

F6. *Existence of Reciprocals* For every real number $a \neq 0$, there is a real number, denoted $1/a$ or a^{-1}, called the **reciprocal** of a, such that
$$a \cdot \left(\frac{1}{a}\right) = \left(\frac{1}{a}\right) \cdot a = 1.$$

*Adapted from Tom M. Apostol, *Calculus, Volume I* (New York: Blaisdell, 1961), pp. 13–19.

All the usual algebraic properties of the real numbers that do not involve order can be derived from the field axioms. The most important are collected as theorems T1–T15 as follows. In all these theorems the symbols a, b, c, and d represent arbitrary real numbers.

T1. *Cancellation Law for Addition*   If $a + b = a + c$, then $b = c$. (In particular, this shows that the number 0 of Axiom F4 is unique.)

T2. *Possibility of Subtraction*   Given a and b, there is exactly one x such that $a + x = b$. This x is denoted by $b - a$. In particular, $0 - a$ is the additive inverse of a, $-a$.

T3. $b - a = b + (-a)$

T4. $-(-a) = a$

T5. $a \cdot (b - c) = a \cdot b - a \cdot c$

T6. $0 \cdot a = a \cdot 0 = 0$

T7. *Cancellation Law for Multiplication*   If $ab = ac$ and $a \neq 0$, then $b = c$. (In particular this shows that the number 1 of Axiom F4 is unique.)

T8. *Possibility of Division*   Given a and b with $a \neq 0$, there is exactly one x such that $ax = b$. This x is denoted by b/a and is called the **quotient** of b and a. In particular, $1/a$ is the reciprocal of a.

T9. If $a \neq 0$, then $b/a = b \cdot a^{-1}$.

T10. If $a \neq 0$, then $(a^{-1})^{-1} = a$.

T11. *Zero Product Property*   If $a \cdot b = 0$, then $a = 0$ or $b = 0$.

T12. *Rule for Multiplication with Negative Numbers*

$$(-a) \cdot b = a \cdot (-b) = -(a \cdot b) \quad \text{and} \quad (-a) \cdot (-b) = a \cdot b.$$

T13. *Rule for Addition of Fractions*

$$\frac{a}{b} + \frac{c}{d} = \frac{ad + bc}{bd}, \quad \text{if } b \neq 0 \text{ and } d \neq 0.$$

T14. *Rule for Multiplication of Fractions*

$$\frac{a}{b} \cdot \frac{c}{d} = \frac{a \cdot c}{b \cdot d}, \quad \text{if } b \neq 0 \text{ and } d \neq 0.$$

T15. *Rule for Division of Fractions*

$$\frac{\dfrac{a}{b}}{\dfrac{c}{d}} = \frac{ad}{bc}, \quad \text{if } b \neq 0, c \neq 0, \text{ and } d \neq 0.$$

The real numbers also satisfy the following axioms, called the **order axioms**. It is assumed that among all real numbers there are certain ones, called the **positive real numbers**, that satisfy properties Ord1–Ord3.

Ord1. For any real numbers a and b, if a and b are positive, so are $a + b$ and $a \cdot b$.

Ord2. For every real number $a \neq 0$, either a is positive or $-a$ is positive but not both.

Ord3. The number 0 is not positive.

The symbols $<$, $>$, $\leq$, and $\geq$, and negative numbers are defined in terms of positive numbers.

DEFINITION

Given real numbers a and b,

$a < b$ means $b + (-a)$ is positive; $b > a$ means $a < b$;

$a \leq b$ means $a < b$ or $a = b$; $b \geq a$ means $a \leq b$;

If $a < 0$, we say that a is **negative**;   If $a \geq 0$, we say that a is **nonnegative**.

From the order Axioms Ord1–Ord3 and the above definition, all the usual rules for calculating with inequalities can be derived. The most important are collected as theorems T16–T25 as follows. In all these theorems the symbols a, b, c, and d represent arbitrary real numbers.

T16.   *Trichotomy Law*   For arbitrary real numbers a and b, exactly one of the three relations $a < b$, $b < a$, or $a = b$ holds.

T17.   *Transitive Law*   If $a < b$ and $b < c$, then $a < c$.

T18.   If $a < b$, then $a + c < b + c$.

T19.   If $a < b$ and $c > 0$, then $ac < bc$.

T20.   If $a \neq 0$, then $a^2 > 0$.

T21. $1 > 0$.

T22.   If $a < b$ and $c < 0$, then $ac > bc$.

T23.   If $a < b$, then $-a > -b$. In particular, if $a < 0$, then $-a > 0$.

T24.   If $ab > 0$, then both a and b are positive or both are negative.

T25.   If $a < c$ and $b < d$, then $a + b < c + d$.

One final axiom distinguishes the set of real numbers from the set of rational numbers. It is called the **least upper bound axiom**.

LUB.   Any nonempty set S of real numbers that is bounded above has a least upper bound. That is, if B is the set of all real numbers x such that $x \geq s$ for all s in S and if B has at least one element, then B has a smallest element. This element is called the **least upper bound** of S.

Solutions and Hints to Selected Exercises

16. *Hint:* The following is a partial truth table.

p	q	r	$(p \vee (\sim p \vee q)) \wedge \sim (q \wedge \sim r)$
T	T	T	T
T	T	F	F
T	F	T	T
T	F	F	T
F	T	T	
F	T	F	
F	F	T	
F	F	F	

SECTION 1.1

1. common form: If p then q.

   p.

   Therefore, q.

   $(a + 2b) \cdot (a^2 - b)$ can be written in prefix notation. All algebraic expressions can be written in prefix notation.

3. Common form: $p \vee q$.

   $\sim p$.

   Therefore, q.

   My mind is shot. Logic is confusing.

5. a. It is a statement because it is a true sentence. 1,024 is a perfect square because $1,024 = 32^2$, and the next smaller perfect square is $31^2 = 961$, which has less than four digits.

6. a. $s \wedge i$ b. $\sim s \wedge \sim i$ 8. a. $(h \wedge w) \wedge \sim s$

10. a. $p \wedge q \wedge r$ c. $p \wedge (\sim q \vee \sim r)$

12.

p	q	$\sim p$	$\sim p \wedge q$
T	T	F	F
T	F	F	F
F	T	T	T
F	F	T	F

14.

p	q	r	$q \wedge r$	$p \wedge (q \wedge r)$
T	T	T	T	T
T	T	F	F	F
T	F	T	F	F
T	F	F	F	F
F	T	T	T	F
F	T	F	F	F
F	F	T	F	F
F	F	F	F	F

17.

p	q	$p \wedge q$	$p \vee (p \wedge q)$	p
T	T	T	T	T
T	F	F	T	T
F	T	F	F	F
F	F	F	F	F

same truth values so $p \vee (p \wedge q)$ and p are logically equivalent

19.

p	t	$p \vee t$
T	T	T
F	T	T

same truth values so $p \vee t$ and t are logically equivalent

21.

p	q	r	$p \wedge q$	$q \wedge r$	$(p \wedge q) \wedge r$	$p \wedge (q \wedge r)$
T	T	T ✔	T	T	T	T
T	T	F ✔	T	F	F	F
T	F	T ✔	F	F	F	F
T	F	F ✔	F	F	F	F
F	T	T ✔	F	T	F	F
F	T	F ✔	F	F	F	F
F	F	T ✔	F	F	F	F
F	F	F ✔	F	F	F	F

same truth values so $(p \wedge q) \wedge r$ and $p \wedge (q \wedge r)$ are logically equivalent

23.

p	q	r	$p \wedge q$	$q \wedge r$	$(p \wedge q) \vee r$	$p \wedge (q \vee r)$
T	T	T	T	T	T	T
T	T	F	T	T	T	T
T	F	T	F	T	T	T
T	F	F	F	F	F	F
F	T	T	F	T	T	F
F	T	F	F	T	F	F
F	F	T	F	T	T	F
F	F	F	F	F	F	F

different truth values in the fifth and seventh rows so $(p \wedge q) \vee r$ and $p \wedge (q \vee r)$ are not logically equivalent

25.

p	q	r	$\sim p$	$\sim q$	$\sim r$	$\sim p \vee q$	$p \vee \sim r$	$(\sim p \vee q) \wedge (p \vee \sim r)$	$\sim p \vee \sim q$	$((\sim p \vee q) \wedge (p \vee \sim r)) \wedge (\sim p \vee \sim q)$	$\sim (p \vee r)$
T	T	T	F	F	F	T	T	T	F	F	F
T	T	F	F	F	T	T	T	T	F	F	F
T	F	T	F	T	F	F	T	F	T	F	F
T	F	F	F	T	T	F	T	F	T	F	F
F	T	T	T	F	F	T	F	F	T	F	F
F	T	F	T	F	T	T	T	T	T	T	T
F	F	T	T	T	F	T	F	F	T	F	F
F	F	F	T	T	T	T	T	T	T	T	T

same truth values so
$((\sim p \vee q) \wedge (p \vee \sim r)) \wedge (\sim p \vee \sim q)$
and $\sim (p \vee r)$ are logically equivalent

27. Hal is not a math major or Hal's sister is not a computer science major.

29. The connector is not loose and the machine is not unplugged.

33. $-2 \geq x$ or $x \geq 7$ **35.** $1 \leq x$ or $x < 3$

37.

p	q	$\sim p$	$\sim q$	$p \wedge q$	$p \wedge \sim q$	$\sim p \vee (p \wedge \sim q)$	$(p \wedge q) \vee (\sim p \vee (p \wedge \sim q))$
T	T	F	F	T	F	F	T
T	F	F	T	F	T	T	T
F	T	T	F	F	F	T	T
F	F	T	T	F	F	T	T

↑
All T's so $(p \wedge q) \vee (\sim p \vee (p \wedge \sim q))$
is a tautology

38.

p	q	$\sim p$	$\sim q$	$p \wedge \sim q$	$\sim p \vee q$	$(p \wedge \sim q) \wedge (\sim p \vee q)$
T	T	F	F	F	T	F
T	F	F	T	T	F	F
F	T	T	F	F	T	F
F	F	T	T	F	T	F

↑
All F's so
$(p \wedge \sim q) \wedge (\sim p \vee q)$
is a contradiction

41. a. the distributive law
b. the commutative law for $\vee$
c. the negation law for $\vee$
d. the identity law for $\wedge$

43. $(p \wedge \sim q) \vee p \equiv p \vee (p \wedge \sim q)$ by the commutative law for $\vee$
$\equiv p$ by the absorption law (with $\sim q$ in place of q)

46. $\sim((\sim p \wedge q) \vee (\sim p \wedge \sim q)) \vee (p \wedge q)$
$\equiv \sim(\sim p \wedge (q \vee \sim q)) \vee (p \wedge q)$ by the distributive law
$\equiv \sim(\sim p \wedge t) \vee (p \wedge q)$ by the negation law for $\vee$
$\equiv \sim(\sim p) \vee (p \wedge q)$ by the identity law for $\wedge$
$\equiv p \vee (p \wedge q)$ by the double negative law
$\equiv p$ by the absorption law

48. a. Solution 1: Construct a truth table for $p \oplus p$ using the truth values for *exclusive or*.

p	$p \oplus p$
T	F
F	F

because an *exclusive or* statement is false when both components are true and when both components are false.

Since all its truth values are false, $p \oplus p \equiv c$, a contradiction.

Solution 2: Replace q by p in the logical equivalence $p \oplus q \equiv (p \vee q) \wedge \sim(p \wedge q)$, and simplify the result.
$p \oplus p \equiv (p \vee p) \wedge \sim(p \wedge p)$ by definition of $\oplus$
$\equiv p \wedge \sim p$ by the identity laws
$\equiv c$ by the negation law for $\wedge$

49. There is a famous story about a philosopher who once gave a talk in which he observed that whereas in English and many other languages a double negative is equivalent to a positive, there is no language in which a double positive is equivalent to a negative. To this, a person in the back row responded sarcastically, "Yeah, yeah."

  [Strictly speaking, sarcasm functions like negation. When spoken sarcastically the words "Yeah, yeah" are not a true double positive; they just mean "no."]

SECTION 1.2

1. If this loop does not contain a **stop** or a **go to**, then it will repeat exactly N times.

3. If you do not freeze, then I'll shoot.

5.

p	q	$\sim p$	$\sim q$	$\sim p \vee q$	$\sim p \vee q \to \sim q$
T	T	F	F	T	F
T	F	F	T	F	T
F	T	T	F	T	F
F	F	T	T	T	T

7.

p	q	r	$\sim q$	$p \wedge \sim q$	$p \wedge \sim q \to r$
T	T	T	F	F	T
T	T	F	F	F	T
T	F	T	T	T	T
T	F	F	T	T	F
F	T	T	F	F	T
F	T	F	F	F	T
F	F	T	T	F	T
F	F	F	T	F	T

9.

p	q	r	$\sim r$	$p \wedge \sim r$	$q \vee r$	$p \wedge \sim r \leftrightarrow q \vee r$
T	T	T	F	F	T	F
T	T	F	T	T	T	T
T	F	T	F	F	T	F
T	F	F	T	T	F	F
F	T	T	F	F	T	F
F	T	F	T	F	T	F
F	F	T	F	F	T	F
F	F	F	T	F	F	T

12. If $x > 2$ then $x^2 > 4$ and if $x < -2$ then $x^2 > 4$.

15. False. The negation of an if–then statement is not an if–then statement. It is an *and* statement.

16. a. P is a square and P is not a rectangle.
  d. n is prime and both n is not odd and n is not 2.
    Or: n is prime and n is neither odd nor 2.
  f. Tom is Ann's father and either Jim is not her uncle or Sue is not her aunt.

17. a. If $p \to q$ is false, then p is true and q is false. Hence $\sim p$ is false, and so $\sim p \to q$ is true.

18. a. If P is not a rectangle, then P is not a square.
  d. If n is not odd and n is not 2, then n is not prime.
  f. If either Jim is not Ann's uncle or Sue is not her aunt, then Tom is not her father.

19. a. *converse:* If P is a rectangle, then P is a square.
    inverse: If P is not a square, then P is not a rectangle.
  d. *converse:* If n is odd or n is 2, then n is prime.
    inverse: If n is not prime, then n is not odd and n is not 2.
  f. *converse:* If Jim is Ann's uncle and Sue is her aunt, then Tom is her father.
    inverse: If Tom is not Ann's father, then Jim is not her uncle or Sue is not her aunt.

20.

p	q	$p \to q$	$q \to p$
T	T	T	T
T	F	F	T
F	T	T	F
F	F	T	T

↑_____↑
different truth values in the second and third rows so $p \to q$ and $q \to p$ are *not* logically equivalent

22.

p	q	$\sim q$	$\sim p$	$\sim q \to \sim p$	$p \to q$
T	T	F	F	T	T
T	F	T	F	F	F
F	T	F	T	T	T
F	F	T	T	T	T

↑   ↑

same truth values so
$\sim q \to \sim p$ and $p \to q$
are logically equivalent

25. If the Cubs do not win tomorrow's game, then they will not win the pennant.

If the Cubs win the pennant, then they will have won tomorrow's game.

28. a. If a new hearing is not granted, payment will be made on the fifth.

29. a. $p \wedge \sim q \to r \equiv \sim(p \wedge \sim q) \vee r$
b. result of (a) $\equiv \sim[\sim(\sim(p \wedge \sim q)) \wedge \sim r]$
an acceptable answer
$\equiv \sim[(p \wedge \sim q) \wedge \sim r]$
by the double negative law
(another acceptable answer)

31. a. $(p \to r) \leftrightarrow (q \to r) \equiv (\sim p \vee r) \leftrightarrow (\sim q \vee r)$
$\equiv \sim(\sim p \vee r) \vee (\sim q \vee r)]$
$\wedge [\sim(\sim q \vee r) \vee (\sim p \vee r)]$
an acceptable answer
$\equiv [(p \wedge \sim r) \vee (\sim q \vee r)]$
$\wedge [(q \wedge \sim r) \vee (\sim p \vee r)]$
by De Morgan's law
(another acceptable answer)

b. result of (a) $\equiv \sim[\sim(p \wedge \sim r) \wedge \sim(\sim q \vee r)] \wedge$
$\sim[\sim(q \wedge \sim r) \wedge \sim(\sim p \vee r)]$
by De Morgan's law

$\equiv \sim[\sim(p \wedge \sim r) \wedge (q \wedge \sim r)] \wedge$
$\sim [\sim(q \wedge \sim r) \wedge (p \wedge \sim r)]$
by De Morgan's law

34. If I catch the 8:05 bus, then I am on time for work.

36. If this number is not divisible by 3, then it is not divisible by 9.

If this number is divisible by 9, then it is divisible by 3.

38. If Hal's team wins the rest of its games, then it will win the championship.

40. a. not necessarily true   **b.** must be true

SECTION 1.3

1. $\sqrt{2}$ is not rational.   **3.** Logic is not easy.

6.

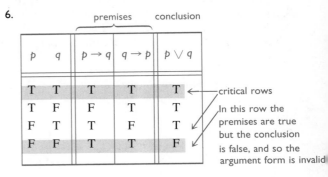

		premises		conclusion	
p	q	$p \to q$	$q \to p$	$p \vee q$	
T	T	T	T	T	←critical rows
T	F	F	T	T	
F	T	T	F	T	
F	F	T	T	F	

In this row the premises are true but the conclusion is false, and so the argument form is invalid

7.

p	q	r	p	$p \to q$	$\sim q \vee r$	r	
T	T	T	T	T	T	T	←critical r
T	T	F	T	T	F	F	
T	F	T	T	F	T	T	
T	F	F	T	F	T	F	
F	T	T	F	T	T	T	
F	T	F	F	T	F	F	
F	F	T	F	T	T	T	
F	F	F	F	T	T	F	

(premises: p, $p \to q$, $\sim q \vee r$; conclusion: r)

In the on where th premises true, the conclusio also true the argu form is v

8.

p	q	r	$p \vee q$	$p \to \sim q$	$p \to r$	r	
T	T	T	T	F	T	T	
T	T	F	T	F	F	F	
T	F	T	T	T	T	T	←critical row
T	F	F	T	T	F	F	
F	T	T	T	T	T	T	
F	T	F	T	T	T	F	
F	F	T	F	T	T	T	
F	F	F	F	T	T	F	

(premises: $p \vee q$, $p \to \sim q$, $p \to r$; conclusion: r)

In this row premises a all true but conclusion false, and s argument is invalid.

12. a.

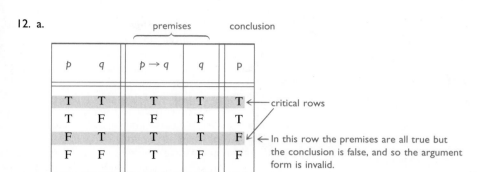

		premises		conclusion
p	q	$p \rightarrow q$	q	p
T	T	T	T	T	← critical rows
T	F	F	F	T
F	T	T	T	F	← In this row the premises are all true but the conclusion is false, and so the argument form is invalid.
F	F	T	F	F

13.

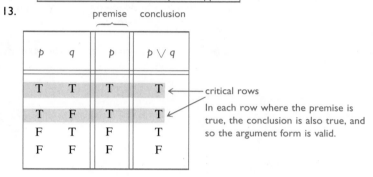

		premise	conclusion
p	q	p	$p \vee q$
T	T	T	T	← critical rows
T	F	T	T
F	T	F	T
F	F	F	F

In each row where the premise is true, the conclusion is also true, and so the argument form is valid.

17.

		premises		conclusion
p	q	$p \vee q$	$\sim q$	p
T	T	T	F	T	← critical row
T	F	T	T	T
F	T	T	F	F
F	F	F	T	F

In the only row where all the premises are true, the conclusion is also true, and so the argument form is valid.

21. logical form (one possible version):

$$\sim p \rightarrow q$$
$$\sim q \rightarrow p$$
$$\therefore \sim p \vee \sim q$$

				premises		conclusion
p	q	$\sim p$	$\sim q$	$\sim p \rightarrow q$	$\sim q \rightarrow p$	$\sim p \vee \sim q$
T	T	F	F	T	T	F	← critical rows
T	F	F	T	T	T	T
F	T	T	F	T	T	T
F	F	T	T	F	F	T

In this row the premises are all true but the conclusion is false, and so the argument form is invalid.

23. $p \rightarrow q$ invalid: converse error
      q
      $\therefore\ p$

24. $p \vee q$ valid: disjunctive syllogism
      $\sim p$
      $\therefore\ q$

26. $p \rightarrow q$ invalid: inverse error
      $\sim p$
      $\therefore\ \sim q$

35. The program contains an undeclared variable.
    One explanation:
    1. There is not a missing semicolon and there is not a misspelled variable name. *(by (c) and (d) and definition of $\wedge$)*
    2. It is not the case that there is a missing semicolon or a misspelled variable name. *(by (1) and De Morgan's laws)*
    3. There is not a syntax error in the first five lines. *(by (b) and (2) and modus tollens)*
    4. There is an undeclared variable. *(by (a) and (3) and disjunctive syllogism)*

36. The treasure is buried under the flagpole.
    One explanation:
    1. The treasure is not in the kitchen. *(by (c) and (a) and modus ponens)*
    2. The tree in the front yard is not an elm. *(by (b) and (1) and modus tollens)*
    3. The treasure is buried under the flagpole. *(by (d) and (2) and disjunctive syllogism)*

37. a. *A* is a knave and *B* is a knight.
    One explanation:
    1. Suppose *A* is a knight.
    2. $\therefore$ What *A* says is true. *(by definition of knight)*
    3. $\therefore$ *B* is a knight also. *(That's what A said.)*
    4. $\therefore$ What *B* says is true. *(by definition of knight)*
    5. $\therefore$ *A* is a knave. *(That's what B said.)*
    6. $\therefore$ We have a contradiction: *A* is a knight and a knave. *(by (1) and (5))*
    7. $\therefore$ The supposition that *A* is a knight is false. *(by contradiction rule)*
    8. $\therefore$ *A* is a knave. *(negation of supposition)*
    9. $\therefore$ What *B* says is true. *(B said A was a knave, which we now know to be true.)*
    10. $\therefore$ *B* is a knight. *(by definition of knight)*

    d. *W* and *Y* are knights; the rest are knaves.

38. The chauffeur killed Lord Hazelton.
    One explanation:
    1. Suppose the cook was in the kitchen at the time of the murder.

2. $\therefore$ The butler killed Lord Hazelton with strychnine. *(by (c) and (1) and modus ponens.)*
3. $\therefore$ We have a contradiction: Lord Hazelton was killed by strychnine and a blow on the head. *(by (2) and (a))*
4. $\therefore$ The supposition that the cook was in the kitchen is false. *(by the contradiction rule)*
5. $\therefore$ The cook was not in the kitchen at the time of the murder. *(negation of supposition)*
6. $\therefore$ Sara was not in the dining room when the murder was committed. *(by (e) and (5) and modus ponens)*
7. $\therefore$ Lady Hazelton was in the dining room when the murder was committed. *(by (b) and (6) and disjunctive syllogism)*
8. $\therefore$ The chauffeur killed Lord Hazelton. *(by (d) and (7) and modus ponens)*

40. (1) $p \rightarrow t$ premise
          $\sim t$ premise
          $\therefore\ \sim p$ by modus tollens
    (2) $\sim p$ by (1)
          $\therefore\ \sim p \vee q$ by disjunctive addition
    (3) $\sim p \vee q \rightarrow r$ premise
          $\sim p \vee q$ by (2)
          $\therefore\ r$ by modus ponens
    (4) $\sim p$ by (1)
          r by (3)
          $\therefore\ \sim p \wedge r$ by conjunctive addition
    (5) $\sim p \wedge r \rightarrow \sim s$ premise
          $\sim p \wedge r$ by (4)
          $\therefore\ \sim s$ by modus ponens
    (6) $s \vee \sim q$ premise
          $\sim s$ by (5)
          $\therefore\ \sim q$ by disjunctive syllogism

42. (1) $\sim w$ premise
          $u \vee w$ premise
          $\therefore\ u$ by disjunctive syllogism
    (2) $u \rightarrow \sim p$ premise
          u by (1)
          $\therefore\ \sim p$ by modus ponens
    (3) $\sim p \rightarrow r \wedge \sim s$ premise
          $\sim p$ by (2)
          $\therefore\ r \wedge \sim s$ by modus ponens
    (4) $r \wedge \sim s$ by (3)
          $\therefore\ \sim s$ by conjunctive simplification
    (5) $t \rightarrow s$ premise
          $\sim s$ by (4)
          $\therefore\ \sim t$ by modus tollens
    (6) $\sim t$ by (5)
          $\therefore\ \sim t \vee w$ by disjunctive addition

SECTION 1.4

1. $R = 1$ 3. $S = 1$

5.

Input		Output
P	Q	R
1	1	1
1	0	1
0	1	0
0	0	1

7.

Input			Output
P	Q	R	S
1	1	1	1
1	1	0	0
1	0	1	1
1	0	0	1
0	1	1	1
0	1	0	0
0	0	1	1
0	0	0	0

9. $P \lor \sim Q$ 11. $(P \land \sim Q) \lor R$

13.

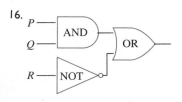

16.

18. a. $(P \land Q \land \sim R) \lor (\sim P \land Q \land R)$
b.

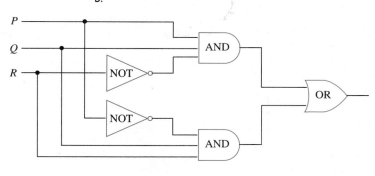

20. a. $(P \land Q \land R) \lor (P \land \sim Q \land R) \lor$
$(\sim P \land \sim Q \land \sim R)$
b.

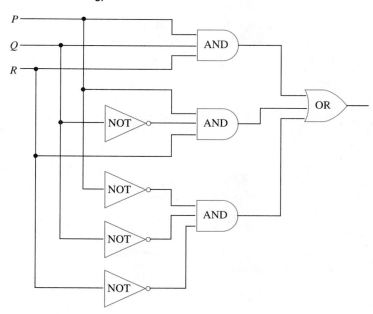

22. The input/output table is as follows.

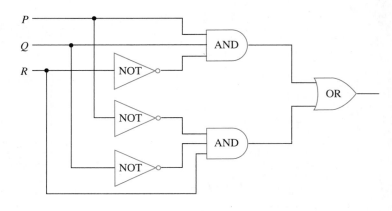

Input			Output
P	Q	R	S
1	1	1	0
1	1	0	1
1	0	1	0
1	0	0	0
0	1	1	0
0	1	0	0
0	0	1	1
0	0	0	0

One circuit (among many) having this input/output table is shown at the right.

24. If the down position is denoted 0, then the input/output table for this configuration is as follows.

Input		Output
P	Q	R
1	1	0
1	0	1
0	1	1
0	0	0

One circuit (among many) having this input/output table is the following.

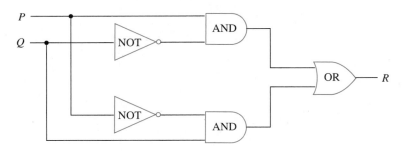

26. The Boolean expression for (a) is $(P \wedge Q) \vee Q$ and for (b) it is $(P \vee Q) \wedge Q$. We must show that if these expressions are regarded as statement forms, then they are logically equivalent. But

$(P \wedge Q) \vee Q$
$\equiv Q \vee (P \wedge Q)$ by the commutative law for $\vee$
$\equiv (Q \vee P) \wedge (Q \vee Q)$ by the distributive law
$\equiv (Q \vee P) \wedge Q$ by the idempotent law
$\equiv (P \vee Q) \wedge Q$ by the commutative law for $\wedge$

Alternatively, by the absorption laws both statement forms are logically equivalent to Q.

28. The Boolean expression for (a) is

$$(P \wedge Q) \vee (P \wedge \sim Q) \vee (\sim P \wedge \sim Q)$$

and for (b) it is $P \vee \sim Q$. We must show that if these expressions are regarded as statement forms, then they are logically equivalent. But

$(P \wedge Q) \vee (P \wedge \sim Q) \vee (\sim P \wedge \sim Q)$
$\equiv ((P \wedge Q) \vee (P \wedge \sim Q)) \vee (\sim P \wedge \sim Q)$
     by inserting parentheses (which is legal by the associative law)

$\equiv (P \wedge (Q \vee {\sim}Q)) \vee ({\sim}P \wedge {\sim}Q)$
       by the distributive law

$\equiv (P \wedge t) \vee ({\sim}P \wedge {\sim}Q)$
         by the negation law for $\vee$

$\equiv P \vee ({\sim}P \wedge {\sim}Q)$ by the identity law for $\wedge$

$\equiv (P \vee {\sim}P) \wedge (P \vee {\sim}Q)$ by the distributive law

$\equiv t \wedge (P \vee {\sim}Q)$ by the negation law for $\vee$

$\equiv (P \vee {\sim}Q) \wedge t$ by the commutative law for $\wedge$

$\equiv P \vee {\sim}Q$ by the identity law for $\wedge$

30. $(P \wedge Q) \vee ({\sim}P \wedge Q) \vee ({\sim}P \wedge {\sim}Q)$
$\equiv (P \wedge Q) \vee (({\sim}P \wedge Q) \vee ({\sim}P \wedge {\sim}Q))$
       by inserting parentheses (which is
       legal by the associative law)

$\equiv (P \wedge Q) \vee ({\sim}P \wedge (Q \vee {\sim}Q))$
       by the distributive law

$\equiv (P \wedge Q) \vee ({\sim}P \wedge t)$
        by the negation law for $\vee$

$\equiv (P \wedge Q) \vee {\sim}P$ by the identity law for $\wedge$

$\equiv {\sim}P \vee (P \wedge Q)$ by the commutative law for $\vee$

$\equiv ({\sim}P \vee P) \wedge ({\sim}P \vee Q)$ by the distributive law

$\equiv (P \vee {\sim}P) \wedge ({\sim}P \vee Q)$
       by the commutative law for $\vee$

$\equiv t \wedge ({\sim}P \vee Q)$ by the negation law for $\vee$

$\equiv ({\sim}P \vee Q) \wedge t$ by the commutative law for $\wedge$

$\equiv {\sim}P \vee Q$ by the identity law for $\wedge$

The following is, therefore, a circuit with at most two logic gates that has the same input/output table as the circuit corresponding to the given expression.

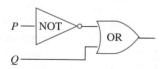

34. b. $(p \downarrow q) \downarrow (p \downarrow q) \equiv {\sim}(p \downarrow q)$ by part (a)
             $\equiv {\sim}[{\sim}(p \vee q)]$ by definition of $\downarrow$
             $\equiv p \vee q$ by the double negative law
d. *Hint:* Use the results of exercise 13 of Section 1.2, part (b) above, and part (a) above.

SECTION 1.5

1. $19_{10} = 16 + 2 + 1 = 10011_2$

4. $458_{10} = 256 + 128 + 64 + 8 + 2 = 111001010_2$

6. $1110_2 = 8 + 4 + 2 = 14_{10}$

9. $1100101_2 = 64 + 32 + 4 + 1 = 101_{10}$

11.
$$\begin{array}{r} {}^{1\ 1\ 1} \\ 1\ 0\ 1\ 1_2 \\ +\quad 1\ 0\ 1_2 \\ \hline 1\ 0\ 0\ 0\ 0_2 \end{array}$$

13.
$$\begin{array}{r} {}^{1\ 1\ 1\quad 1} \\ 1\ 0\ 1\ 1\ 0\ 1_2 \\ +\quad 1\ 1\ 1\ 0\ 1_2 \\ \hline 1\ 0\ 0\ 1\ 0\ 1\ 0_2 \end{array}$$

15.
$$\begin{array}{r} {}^{1\ 10\ \not1\ 1} \\ \not1\not0\ \not1\not0\ 0_2 \\ -\quad 1\ 1\ 0\ 1_2 \\ \hline 1\ 1\ 1_2 \end{array}$$

17.
$$\begin{array}{r} {}^{0\ 1} \\ 1\ 0\ 1\ \not1\ 0\ 1_2 \\ -\quad 1\ 0\ 0\ 1\ 1_2 \\ \hline 1\ 1\ 0\ 1\ 0_2 \end{array}$$

19. a. $S = 0, T = 1$

20. $23_{10} = (16 + 4 + 2 + 1)_{10} = 00010111_2 \rightarrow$
$11101000 \rightarrow 11101001$. So the answer is 11101001.

22. $4_{10} = 00000100_2 \rightarrow 11111011 \rightarrow 11111100$. So the answer is 11111100.

24. $11010011 \rightarrow 00101100 \rightarrow 00101101_2 =$
$(32 + 8 + 4 + 1)_{10} = 45_{10}$. So the answer is 45.

26. $11110010 \rightarrow 00001101 \rightarrow 00001110_2 =$
$(8 + 4 + 2)_{10} = 14_{10}$. So the answer is 14.

28. $57_{10} = (32 + 16 + 8 + 1)_{10} = 111001_2 \rightarrow 00111001$
$-118_{10} = -(64 + 32 + 16 + 4 + 2)_{10} =$
$-111010110_2 \rightarrow 01110110 \rightarrow 10001001 \rightarrow 10001010$.
So the 8-bit representations of 57 and -118 are
00111001 and 10001010. Adding the 8-bit representations gives

$$\begin{array}{r} \boxed{0}\ \boxed{0}\ \boxed{1}\ \boxed{1}\ \boxed{1}\ \boxed{0}\ \boxed{0}\ \boxed{1} \\ +\ \boxed{1}\ \boxed{0}\ \boxed{0}\ \boxed{0}\ \boxed{1}\ \boxed{0}\ \boxed{1}\ \boxed{0} \\ \hline \boxed{1}\ \boxed{1}\ \boxed{0}\ \boxed{0}\ \boxed{0}\ \boxed{0}\ \boxed{1}\ \boxed{1} \end{array}$$

Since the leading bit of this number is a 1, the answer is negative. Converting back to decimal form gives

$11000011 \rightarrow 00111100 \rightarrow -00111101_2$
$$= -(32 + 16 + 8 + 4 + 1)_{10} = -61_{10}.$$

So the answer is -61.

29. $62_{10} = (32 + 16 + 8 + 4 + 2)_{10}$
$$= 111110_2 \rightarrow 00111110$$

$-18_{10} = -(16 + 2)_{10}$
$$= -10010_2 \rightarrow 00010010 \rightarrow 11101101 \rightarrow 11101110$$

So the 8-bit representations of 57 and -118 are
00111110 and 11101110. Adding the 8-bit representations gives

$$\begin{array}{r} \boxed{0}\ \boxed{0}\ \boxed{1}\ \boxed{1}\ \boxed{1}\ \boxed{1}\ \boxed{1}\ \boxed{0} \\ +\ \boxed{1}\ \boxed{1}\ \boxed{1}\ \boxed{0}\ \boxed{1}\ \boxed{1}\ \boxed{1}\ \boxed{0} \\ \hline 1\ \boxed{0}\ \boxed{0}\ \boxed{1}\ \boxed{0}\ \boxed{1}\ \boxed{1}\ \boxed{0}\ \boxed{0} \end{array}$$

Truncating the 1 in the 2^8th position gives 00101100. Since the leading bit of this number is a 0, the answer is positive. Converting back to decimal form gives

$$00101100 \to 101100_2 = (32 + 8 + 4)_{10} = 44_{10}.$$

So the answer is 44.

30. $-6_{10} = -(4 + 2)_{10}$
$\qquad = -110_2 \to 00000110 \to 11111001 \to 11111010$

$-73_{10} = -(64 + 8 + 1)_{10} =$
$\qquad -1001001_2 \to 01001001 \to 10110110 \to 10110111$

So the 8-bit representations of -6 and -73 are 11111010 and 10110111. Adding the 8-bit representations gives

	1	1	1	1	1	0	1	0
+								
	1	0	1	1	0	1	1	1
1	1	0	1	1	0	0	0	1

Truncating the 1 in the 2^8th position gives 10110001. Since the leading bit of this number is a 1, the answer is negative. Converting back to decimal form gives

$$10110001 \to 01001110 \to -01001111_2$$
$$= -(64 + 8 + 4 + 2 + 1)_{10} = -79_{10}.$$

So the answer is -79.

35. $A2BC_{16} = 10 \cdot 16^3 + 2 \cdot 16^2 + 11 \cdot 16 + 12$
$\qquad = 41660_{10}$

38. $000111000000101010111110_2$

41. $2E_{16}$

SECTION 2.1

1. a. False   b. True

2. a. $\{x \in \mathbf{R} \mid x > \frac{1}{x}\}$
$\quad = \{x \in \mathbf{R} \mid x > 1 \quad$ or $\quad -1 < x < 0$.
the set of all real numbers that are greater than 1 or between -1 and 0.

3. a. True: Any real number that is greater than 2 is greater than 1.
c. False: $(-3)^2 > 4$ but $-3 \not> 2$.

4. *Counterexample: Let $x = 1$: $1 \not> \frac{1}{1}$. (This is one counterexample among many.)*

6. *Counterexample: Let $m = 1$ and $n = 1$. Then $m \cdot n = 1 \cdot 1 = 1$ and $m + n = 1 + 1 = 2$. But $1 \not\geq 2$, and so $m \cdot n \not\geq m + n$. (This is one counterexample among many.)*

8. (a), (e), (f)   9. (b), (c), (e), (f)

11. a. $\forall$ dinosaurs x, x is extinct.
c. $\forall$ irrational numbers x, x is not an integer.

12. a. $\exists$ an exercise x such that x has an answer.

13. (b), (d), (e)

15. a. $\forall x$, if x is a COBOL program, then x has at least 20 lines.
c. $\forall$ integers m and n, if m and n are even then $m + n$ is even.

16. a. $\forall n$, if n is an even integer, then n^2 is even.
$\forall$ even integers n, n^2 is even.

17. a. $\exists$ a hatter x such that x is mad.
$\exists x$ such that x is a hatter and x is mad.

19. (b) and (d) are negations.

In 20–23 there are other correct answers in addition to those shown.

20. a. Some dinosaurs are not extinct.
c. Some irrational numbers are integers.

21. a. No exercises have answers.

22. a. $\exists$ a COBOL program that does not have at least 20 lines.

Or: $\exists$ a COBOL program that has fewer than 20 lines.

c. $\exists$ integers m and n such that m and n are even and $m + n$ is not even.

Or: $\exists$ even integers m and n such that $m + n$ is not even.

23. a. There exists an even integer whose square is not even.

Or: $\exists$ an integer n such that n is even and n^2 is not even.

24. The proposed negation is not correct. Consider the given statement: "The sum of any two irrational numbers is irrational." For this to be false means that it is possible to find at least one pair of irrational numbers whose sum is rational. On the other hand, the negation proposed in the exercise ("The sum of any two irrational numbers is rational") means that given any two irrational numbers, their sum is rational. This is a much stronger statement than the actual negation: The truth of this statement implies the truth of the negation (assuming that there are at least two irrational numbers), but the negation can be true without having this statement be true.

correct negation: There are at least two irrational numbers whose sum is rational.

Or: The sum of some two irrational numbers is rational.

26. The proposed negation is not correct. There are two mistakes: The negation of a "for all" statement is not a

"for all" statement; and the negation of an if–then statement is not an if–then statement.

> *correct negation:* There exists an integer n such that n^2 is even and n is not even.

28. a. True
  c. False: $x = 16$, $x = 26$, $x = 32$, and $x = 36$ are all counterexamples.

29. $\exists$ a real number x such that $x > 3$ and $x^2 \leq 9$.

31. $\exists$ a real number x such that $x(x + 1) > 0$ and both $x \leq 0$ and $x \geq -1$.

33. $\exists$ integers a, b, and c such that $a - b$ is even and $b - c$ is even and $a - c$ is not even.

35. $\exists$ an integer n such that n is divisible by 2 and n is not even.

38. a. *One possible answer:* Let $P(x)$ be "$2x \neq 1$."

SECTION 2.2

1. a. $y = 1/2$ **b.** $y = -1$ **c.** $y = 4/3$

In 3–16 there are other correct answers in addition to those shown.

3. a. *statement:* For every color, there is an animal of that color.
  There are animals of every color.
  b. *negation:* $\exists$ a color C such that $\forall$ animals A, A is not colored C.
  For some color, there is no animal of that color.

5. a. *statement:* For every odd integer n, there is an integer k such that $n = 2k + 1$.
  Given any odd integer, there is another integer for which the given integer equals twice the other integer plus 1.
  Given any odd integer n, we can find another integer k so that $n = 2k + 1$.
  An odd integer is equal to twice some other integer plus 1.
  Every odd integer has the form $2k + 1$ for some integer k.
  b. *negation:* $\exists$ an odd integer n such that $\forall$ integers k, $n \neq 2k + 1$.
  There is an odd integer that is not equal to $2k + 1$ for any integer k.
  Some odd integer does not have the form $2k + 1$ for any integer k.

7. a. *statement:* For every real number x, there is a real number y such that $x + y = 0$.
  Given any real number x, there exists a real number y such that $x + y = 0$.
  Given any real number, we can find another real number (possibly the same) such that the sum of the given number plus the other number equals 0.

Every real number can be added to some other real number (possibly itself) to obtain 0.
  b. *negation:* $\exists$ a real number x such that $\forall$ real numbers y, $x + y \neq 0$.
  There is a real number x for which there is no real number y with $x + y = 0$.
  There is a real number x with the property that $x + y \neq 0$ for any real number y.
  Some real number has the property that its sum with any other real number is nonzero.

9. $\forall$ people x, $\exists$ a person y such that x is older than y.

10. $\exists$ a person x such that $\forall$ people y, x is older than y.

13. a. *statement:* $\forall$ even integers n, $\exists$ an integer k such that $n = 2k$.
  b. *negation:* $\exists$ an even integer n such that $\forall$ integers k, $n \neq 2k$.
  There is some even integer that is not equal to twice any other integer.

16. a. *statement:* $\exists$ a program P such that $\forall$ questions Q posed to P, P gives the correct answer to Q.
  b. *negation:* $\forall$ programs P, there is a question Q that can be posed to P such that P does not give the correct answer to Q.

18. a. *version with interchanged quantifiers:* $\exists x \in \mathbf{R}$ such that $\forall y \in \mathbf{R}$, $x < y$.
  b. The given statement says that given any real number x there is a real number y that is greater than x. This is true: For any real number x, let $y = x + 1$. Then $x < y$.
  The version with interchanged quantifiers says that there is a real number that is less than every other real number. This is false.

20. a. True. Every student chose at least one dessert: Uta chose pie, Tim chose both pie and cake, and Yuen chose pie.

21. a. The statement has the form "$\exists$ a student S in this class such that $\forall$ residence halls R at this school, S has dated at least one person from R." To determine whether this is true, you could present all the students in the class with a complete list of residence halls, asking them to check off all residence halls containing a person they have dated. Assuming all the students respond truthfully, if some student checks off every residence hall, then the statement is true. Otherwise, the statement is false.

22. *contrapositive:* $\forall x \in \mathbf{R}$, if $x^2 \leq 9$ then $x \leq 3$.
  converse: $\forall x \in \mathbf{R}$, if $x^2 > 9$ then $x > 3$.
  inverse: $\forall x \in \mathbf{R}$, if $x \leq 3$ then $x^2 \leq 9$.

24. *contrapositive:* If an integer is not divisible by 3, then it is not divisible by 6.
  converse: If an integer is divisible by 3, then it is divisible by 6.

inverse: If an integer is not divisible by 6, then it is not divisible by 3.

26. *contrapositive:* $\forall\, x \in \mathbf{R}$, if $x \leq 0$ and $x \geq -1$ then $x(x + 1) \leq 0$.
 converse: $\forall\, x \in \mathbf{R}$, if $x > 0$ or $x < -1$ then $x(x + 1) > 0$.
 inverse: $\forall\, x \in \mathbf{R}$, if $x(x + 1) \leq 0$, then $x \leq 0$ and $x \geq -1$.

28. *contrapositive:* $\forall$ integers a, b, and c, if $a - c$ is not even, then either $a - b$ is not even or $b - c$ is not even.
 converse: $\forall$ integers a, b, and c, if $a - c$ is even then both $a - b$ is even and $b - c$ is even.
 inverse: $\forall$ integers a, b, and c, if $a - b$ is not even or $b - c$ is not even, then $a - c$ is not even.

31. If a person earns a grade of C^- in this course, then the course counts toward graduation.

33. If a person is not on time each day, then the person will not keep this job.

35. It is not the case that if a number is divisible by 2 then that number is divisible by 4. In other words, there is a number that is divisible by 2 and is not divisible by 4.

37. It is not the case that if a person has a large income that person is happy. In other words, there is a person who has a large income and is not happy.

39. If a program is reasonably correct, then it does not produce error messages during translation. But some incorrect programs do not produce error messages during translation.

40. a. Yes d. $X = w_1$, $X = w_2$ f. $X = b_2$, $X = w_2$

42. a. This statement is true. The unique real number with the given property is 1. Note that

$$1 \cdot y = y \quad \text{for all real numbers } y,$$

and if x is any real number such that for instance, $x \cdot 2 = 2$, then dividing both sides by 2 gives $x = 2/2 = 1$.

44. a. No matter what the domain D or the predicates $P(x)$ and $Q(x)$ are, the given statements have the same truth value. If the statement "$\forall x$ in D, $(P(x) \wedge Q(x))$" is true, then $P(x) \wedge Q(x)$ is true for every x in D, which implies that both $P(x)$ and $Q(x)$ are true for every x in D. But then $P(x)$ is true for every x in D and also $Q(x)$ is true for every x in D. So the statement "$(\forall x$ in D, $P(x)) \wedge (\forall x$ in D, $Q(x))$" is true. Conversely, if the statement "$(\forall x$ in D, $P(x)) \wedge (\forall x$ in D, $Q(x))$" is true, then $P(x)$ is true for every x in D and also $Q(x)$ is true for every x in D. This implies that both $P(x)$ and $Q(x)$ are true for every x in D, and so $P(x) \wedge Q(x)$ is true for every x in D. Hence the statement "$\forall x$ in D, $(P(x) \wedge Q(x))$" is true.

SECTION 2.3

1. b. $(f_i + f_j)^2 = f_i^2 + 2f_i f_j + f_j^2$
 c. $(3u + 5v)^2 = (3u)^2 + 2(3u)(5v) + (5v)^2$
   $(= 9u^2 + 30uv + 25v^2)$
 d. $(g(r) + g(s))^2 = (g(r))^2 + 2g(r)g(s) + (g(s))^2$

2. 0 is even.

3. $2/3 + 4/5 = (2 \cdot 5 + 3 \cdot 4)/(3 \cdot 5) \ (= 22/15)$

5. Harry is not a healthy person.

7. Invalid by converse error

8. Valid by universal modus ponens (or universal instantiation)

9. Invalid by inverse error

10. Valid by universal modus tollens

19. $\forall x$, if x is a good car, then x is not cheap.
 a. Valid, universal modus ponens (or universal instantiation)
 b. Invalid, converse error

21. Valid (A valid argument can have false premises and a true conclusion!)

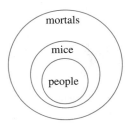

The major premise says the set of people is included in the set of mice. The minor premise says the set of mice is included in the set of mortals. Assuming both of these premises are true, it must follow that the set of people is included in the set of mortals. Since it is impossible for the conclusion to be false if the premises are true, the argument is valid.

23. Valid. The major and minor premises can be diagrammed as follows.

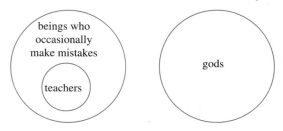

According to the diagram, the set of teachers and the set of gods can have no common elements. Hence, if the premises are true, then the conclusion must also be true; and so the argument is valid.

27. 4. If an animal is in the yard, then it is mine.
1. If an animal belongs to me, then I trust it.
5. If I trust an animal, then I admit it into my study.
3. If I admit an animal into my study, then it will beg when told to do so.
6. If an animal begs when told to do so, then that animal is a dog.
2. If an animal is a dog, then that animal gnaws bones.
∴ If an animal is in the yard, then that animal gnaws bones; that is, all the animals in the yard gnaw bones.

29. 2. If a bird is in this aviary, then it belongs to me.
4. If a bird belongs to me, then it is at least 9 feet high.
1. If a bird is 9 feet high, then it is an ostrich.
3. If a bird lives on mince pies, then it is not an ostrich.
Contrapositive: If a bird is an ostrich, then it does not live on mince pies.
∴ If a bird is in this aviary, then it does not live on mince pies; that is, no bird in this aviary lives on mince pies.

SECTION 3.1

1. a. Yes: $6m + 8n = 2 \cdot (3m + 4n)$ and $(3m + 4n)$ is an integer because 3, 4, m, and n are integers and products and sums of integers are integers.
b. Yes: $10rs + 7 = 2 \cdot (5rs + 3) + 1$ and $5rs + 3$ is an integer because 3, 5, r, and s are integers and products and sums of integers are integers.
c. Not necessarily. For instance, if $m = 3$ and $n = 2$, then $m^2 - n^2 = 9 - 4 = 5$, which is prime. (Note that because of the identity $m^2 - n^2 = (m - n) \cdot (m + n)$, $m^2 - n^2$ is composite for many values of m and n.)

3. For example, let $n = 7$. Then $2^7 - 1 = 128 - 1 = 127$, and 127 is prime.

5. For example, let $m = 2$ and $n = 2$. Then $1/m = 1/2$ and $1/n = 1/2$, and $m + n = 1$, which is an integer.

8. $2 = 1^2 + 1^2$, $4 = 2^2$, $6 = 2^2 + 1^2 + 1^2$,
$8 = 2^2 + 2^2$, $10 = 3^2 + 1^2$, $12 = 2^2 + 2^2 + 2^2$,
$14 = 3^2 + 2^2 + 1^2$, $16 = 4^2$,
$18 = 3^2 + 3^2 = 4^2 + 1^2 + 1^2$, $20 = 4^2 + 2^2$,
$22 = 3^2 + 3^2 + 2^2$, $24 = 4^2 + 2^2 + 2^2$

10. a. any even integers
Or: any particular but arbitrarily chosen integers that are even.
b. integers r and s
c. $2r + 2s$
d. $m + n$ is even

11. *Two versions of a correct proof are given below to illustrate some of the variety that is possible.*

Proof 1: Suppose n is any *[particular but arbitrarily chosen]* even integer. *[We must show that $-n$ is even.]* By definition of even, $n = 2k$ for some integer k. Multiplying both sides by -1 gives that $-n = -(2k) = 2 \cdot (-k)$. Let $r = -k$. Then r is an integer because a product of two integers is an integer, $r = -k = (-1) \cdot k$, and -1 and k are integers. Hence $-n = 2r$ for some integer r, and so $-n$ is even *[as was to be shown].*

Proof 2: Suppose n is any even integer. By definition of even, $n = 2k$ for some integer k. Then $-n = -2k = 2 \cdot (-k)$. But $2 \cdot (-k)$ is even by definition of even because $-k$ is an integer (being the product of the integers -1 and k). Hence $-n$ is even.

13. *Proof:* Suppose n is any even integer. Then $n = 2k$ for some integer k. Hence $(-1)^n = (-1)^{2k} = ((-1)^2)^k = (1)^k = 1$ *[by the laws of exponents of algebra.]* This is what was to be shown.

15. *[Note that the negation of the given statement is "∃ an integer n such that n is prime and n is even."]*
Counterexample: Let $n = 2$. Then n is prime and n is even. Hence ∃ an integer n such that n is prime and n is even, and so the given statement is false.

17. *Start of Proof:* Suppose m is an integer such that $m > 1$. *[We must show that $0 < 1/m < 1$.]*

19. *Start of Proof:* Suppose m and n are integers and $m \cdot n = 1$. *[We must show that $m = n = 1$ or $m = n = -1$.]*

21. The incorrect proof just shows the theorem to be true in one case where $k = 2$. A real proof must show it is true for *all* integers $k > 0$.

22. The mistake in the "proof" is that the same symbol, k, is used to represent two different quantities. By setting both m and n equal to $2k$, the proof specifies that $m = n$, and, therefore, it only deduces the conclusion in case $m = n$. If $m \neq n$, the conclusion is not always true. For instance, $6 + 4 = 10$ but $10 \neq 4k$ for any integer k.

23. This incorrect proof begs the question. The word *since* in the third sentence is completely unjustified. The second sentence only tells what happens *if $k^2 + 2k + 1$* is composite. But at that point in the proof it has not been established that $k^2 + 2k + 1$ *is* composite. In fact, that is exactly what is to be proved.

25. *Proof:* Suppose m and n are any odd integers. *[We must show that $m \cdot n$ is odd.]* By definition of odd, $n = 2r + 1$ and $m = 2s + 1$ for some integers r and s. Then

$$m \cdot n = (2r + 1) \cdot (2s + 1) \quad \text{by substitution}$$
$$= 4rs + 2r + 2s + 1$$
$$= 2(2rs + r + s) + 1 \quad \text{by algebra.}$$

Now $2rs + r + s$ is an integer because products and sums of integers are integers and 2, r, and s are all integers. Hence $m \cdot n = 2 \cdot$ (some integer) $+ 1$, and so by definition of odd, $m \cdot n$ is odd.

29. *Counterexample:* Let $m = 1$ and $n = 3$. Then $m + n = 4$ is even, but neither summand m nor n is even.

36. *Proof:* Suppose n is any integer. Then $4(n^2 + n + 1) - 3n^2 = 4n^2 + 4n + 4 - 3n^2 = n^2 + 4n + 4 = (n + 2)^2$ (by algebra). But $(n + 2)^2$ is a perfect square because $n + 2$ is an integer (being a sum of n and 2). Hence $4(n^2 + n + 1) - 3n^2$ is an integer, as was to be shown.

38. *Hint:* This is true.

44. *Hint:* The answer is no.

46. a. *Hint:* Note that $(x - r)(x - s) = x^2 - (r + s) + rs$. If both r and s are odd, then $r + s$ is even and rs is odd. So the coefficient of x^2 is 1 (odd), the coefficient of x is even, and the constant coefficient is odd.

SECTION 3.2

1. $-(26/7) = (-26)/7$

3. $5/6 + 3/7 = (5 \cdot 7 + 6 \cdot 3)/42 = 53/42$

4. Let $x = 0.58585858\ldots$.
   Then $100x = 58.58585858\ldots$, and so
   $100x - x = 99x = 58$. Hence
   $x = 58/99$.

6. Let $x = 20.492492492\ldots$. Then
   $1{,}000x = 20492.492492492\ldots$, and so
   $1{,}000x - x = 999x = 20{,}472$. Hence
   $x = 20{,}472/999$.

8. b. $\forall$ real numbers x and y, if $x \neq 0$ and $y \neq 0$ then $xy \neq 0$.

9. Because a and b are integers, $b - a$ and ab^2 are both integers (since differences and products of integers are integers). Also by the zero product property, $ab^2 \neq 0$ because neither a nor b is zero. Hence $(b - a)/ab^2$ is a quotient of two integers with nonzero denominator, and so it is rational.

11. *Proof:* Suppose n is any *[particular but arbitrarily chosen]* integer. Then $n = n \cdot 1$, and so $n = n/1$ by the laws of algebra. Now n and 1 are both integers and $1 \neq 0$. Hence n can be written as a quotient of integers with a nonzero denominator, and so n is rational.

12. a. any *[particular but arbitrarily chosen]* rational number
   b. integers a and b c. $(a/b)^2$ d. b^2
   e. zero product property f. r^2 is rational

13. *Proof:* Suppose r and s are rational numbers. By definition of rational, $r = a/b$ and $s = c/d$ for some integers a, b, c, and d with $b \neq 0$ and $d \neq 0$. Then

$$r \cdot s = \frac{a}{b} \cdot \frac{c}{d} \qquad \text{by substitution}$$

$$= \frac{a \cdot c}{b \cdot d} \qquad \begin{array}{l}\text{by the rules of algebra for} \\ \text{multiplying fractions}\end{array}$$

Now $a \cdot c$ and $b \cdot d$ are both integers (being products of integers) and $b \cdot d \neq 0$ (by the zero product property). Hence $r \cdot s$ is a quotient of integers with a nonzero denominator, and so by definition of rational, $r \cdot s$ is rational.

14. *Hint: Counterexample:* Let r be any rational number and $s = 0$. Then r and s are both rational, but the quotient of r divided by s is undefined and, therefore, is not a rational number.
   Revised statement to be proved: For all rational numbers r and s, if $s \neq 0$ then r/s is rational.

18. *Proof:* Suppose r is any rational number. Then $r^2 = r \cdot r$ is a product of rational numbers and is therefore rational by exercise 12.

21. *Proof:* Suppose n is any odd integer. Then $n^2 = n \cdot n$ is a product of odd integers and hence is odd by exercise 25, Section 3.1. Therefore, $n^2 + n$ is a sum of odd integers and hence is even by exercise 12, Section 3.1.

23. Let

$$x = \frac{1 - \dfrac{1}{2^{n+1}}}{1 - \dfrac{1}{2}} = \frac{1 - \dfrac{1}{2^{n+1}}}{\dfrac{1}{2}} = \frac{1 - \dfrac{1}{2^{n+1}}}{\dfrac{1}{2}} \cdot \frac{2^{n+1}}{2^{n+1}} = \frac{2^{n+1} - 1}{2^n}.$$

But $2^{n+1} - 1$ and 2^n are both integers (since n is a nonnegative integer) and $2^n \neq 0$ by the zero product property. Therefore, x is rational.

27. *Proof:* Suppose c is a real number such that

$$r_3 c^3 + r_2 c^2 + r_1 c + r_0 = 0,$$

where r_0, r_1, r_2, and r_3 are rational numbers. By definition of rational, $r_0 = a_0/b_0$, $r_1 = a_1/b_1$, $r_2 = a_2/b_2$, and $r_3 = a_3/b_3$ for some integers, a_0, a_1, a_2, a_3, and nonzero integers b_0, b_1, b_2, and b_3. By substitution,

$$r_3 c^3 + r_2 c^2 + r_1 c + r_0$$

$$= \frac{a_3}{b_3} c^3 + \frac{a_2}{b_2} c^2 + \frac{a_1}{b_1} c + \frac{a_0}{b_0}$$

$$= \frac{b_0 b_1 b_2 a_3}{b_0 b_1 b_2 b_3} c^3 + \frac{b_0 b_1 b_3 a_2}{b_0 b_1 b_2 b_3} c^2 + \frac{b_0 b_2 b_3 a_1}{b_0 b_1 b_2 b_3} c + \frac{b_1 b_2 b_3 a_0}{b_0 b_1 b_2 b_3}$$

$$= 0.$$

Multiplying both sides by $b_0 b_1 b_2 b_3$ gives

$$b_0 b_1 b_2 a_3 \cdot c^3 + b_0 b_1 b_3 a_2 \cdot c^2 + b_0 b_2 b_3 a_1 \cdot c + b_1 b_2 b_3 a_0$$
$$= 0.$$

Let $n_3 = b_0 b_1 b_2 a_3$, $n_2 = b_0 b_1 b_3 a_2$, $n_1 = b_0 b_2 b_3 a_1$, and $n_0 = b_1 b_2 b_3 a_0$. Then n_0, n_1, n_2, and n_3 are all integers (being products of integers). Hence c satisfies the equation

$$n_3 c^3 + n_2 c^2 + n_1 c + n_0 = 0.$$

where n_0, n_1, n_2, and n_3 are all integers. This is what was to be shown.

30. By setting both r and s equal to a/b, this incorrect proof violates the requirement that r and s be arbitrarily chosen rational numbers. The reason is that if both r and s equal a/b, then $r = s$.

31. The fourth sentence claims that $r + s$ is a fraction because it is a sum of two fractions. But the statement that the sum of two fractions is a fraction is a restatement of what is to be proved. Hence this proof begs the question by assuming what is to be proved.

SECTION 3.3

1. Yes, $52 = 13 \cdot 4$

3. Yes, $(3k + 1) \cdot (3k + 2) \cdot (3k + 3) =$
   $3 \cdot [(3k + 1) \cdot (3k + 2) \cdot (k + 1)]$ and
   $(3k + 1) \cdot (3k + 2) \cdot (k + 1)$ is an integer because k is an integer and sums and products of integers are integers.

5. No, $29/3 \cong 9.67$, which is not an integer.

6. Yes, $66 = (-3) \cdot (-22)$.

7. Yes, $6a(a + b) = 3a \cdot [2(a + b)]$ and $2(a + b)$ is an integer because a and b are integers and sums and products of integers are integers.

9. No, $34/7 \cong 4.86$, which is not an integer.

11. Yes, $n^2 - 1 = (4k + 1)^2 - 1 = (16k^2 + 8k + 1) - 1 = 16k^2 + 8k = 8 \cdot (2k^2 + k)$, and $2k^2 + k$ is an integer because k is an integer and sums and products of integers are integers.

13. a. $a \mid b$ b. $a \cdot k$ c. integer
    d. $-(a \cdot k)$ e. $a \mid (-b)$

14. *Proof:* Suppose a, b, and c are any integers such that $a \mid b$ and $a \mid c$. *[We must show that $a \mid (b + c)$.]* By definition of divides, $b = a \cdot r$ and $c = a \cdot s$ for some integers r and s. Then $b + c = a \cdot r + a \cdot s = a \cdot (r + s)$ (by algebra).
    Let $t = r + s$. Then t is an integer (being a sum of integers), and thus $b + c = a \cdot t$ where t is an integer. By definition of divides, then, $a \mid (b + c)$ *[as was to be shown.]*

16. *Proof:* Suppose n, $n + 1$, and $n + 2$ are any three consecutive integers. Then $n + (n + 1) + (n + 2) = 3n + 3 = 3 \cdot (n + 1)$. This is divisible by 3 because $n + 1$ is an integer (since n is an integer and a sum of integers is an integer).

18. The given statement can be rewritten formally as "∀ integers n, if n is divisible by 6 then n is divisible by 2." This statement is true.

    Proof 1: Suppose n is any integer that is divisible by 6. By definition of divisibility, $n = 6k$ for some integer k. But $6k = 2 \cdot (3k)$, and $3k$ is an integer because k is. Hence $n = 2 \cdot$ (some integer), and so n is divisible by 2.

    Proof 2: Suppose n is any integer that is divisible by 6. We know that 6 is divisible by 2 because $6 = 2 \cdot 3$. So $2 \mid 6$ and $6 \mid n$. Hence by transitivity of divisibility (Theorem 3.3.1), $2 \mid n$, or, in other words, n is divisible by 2.

21. *Hint:* This is true.

23. *Hint:* This is false.

27. No. Each of these numbers is divisible by 3, and so their sum is also divisible by 3. But 100 is not divisible by 3. Thus the sum cannot equal $100.

30. a. The sum of the digits is 45, which is divisible by 9. Therefore, 637,425,403,425 is divisible by 9 and hence also divisible by 3 (by transitivity of divisibility). The right-most digit is 5. Therefore, 637,425,403,425 is divisible by 5. The two right-most digits are 25 which is not divisible by 4. Therefore, 637,425,403,425 is not divisible by 4.

31. a. $702 = 2 \cdot 3^3 \cdot 13$

32. *Hint:* Model your answer after Example 3.3.11 and use the fact that 151 is prime.

34. $p_1^{2e_1} \cdot p_2^{2e_2} \cdot \ldots \cdot p_k^{2e_k}$

37. *Proof:* Suppose n is a nonnegative integer whose decimal representation ends in 0. Then $n = 10m + 0 = 10m$ for some integer m. By factoring out a 5, $n = 10m = 5 \cdot (2m)$, and $2m$ is an integer since m is an integer. Hence $10m$ is divisible by 5, which is what was to be shown.

40. *Hint:* You may take it as a fact that for any positive integer k, $10^k = \underbrace{99 \ldots 9}_{k \text{ of these}} + 1$, i.e.,

$$10^k = 9 \cdot 10^{k-1} + 9 \cdot 10^{k-2} + \cdot \cdot + 9 \cdot 10^1$$
$$+ 9 \cdot 10^0 + 1.$$

SECTION 3.4

1. $q = 7, r = 7$ 3. $q = 0, r = 36$

5. $q = -5, r = 10$ 7. a. 4   b. 7

11. a. When today is Saturday, 15 days from today is two weeks (which is Saturday) plus one day (which is Sunday). Hence $DayN$ should be 0. Now when today is Saturday, then $DayT = 6$, and so when $N = 15$ the formula gives

$$DayN = (DayT + N) \bmod 7$$
$$= (6 + 15) \bmod 7$$
$$= 21 \bmod 7 = 0, \text{ which agrees.}$$

13. *Solution 1:* $30 = 28 + 2$. Hence the answer is two days after Monday, or Wednesday.

Solution 2: By the formula, the answer is $(1 + 30) \bmod 7 = 31 \bmod 7 = 3$, which is Wednesday.

15. a.

A	69	19	9	
q	2			
d		1		
n			1	
p				4

16. a. 8,309

21. a. *Proof:* Suppose n, $n + 1$, and $n + 2$ are any three consecutive integers. *[We must show that $n(n + 1)(n + 2)$ is divisible by 3.]* By the result of exercise 17, n can be written in one of the three forms, $3q$, $3q + 1$, or $3q + 2$ for some integer q. We divide into cases accordingly.

Case 1 ($n = 3q$ for some integer q): In this case,

$$n(n + 1)(n + 2)$$
$$= 3q(3q + 1)(3q + 2) \qquad \text{by substitution}$$
$$= 3 \cdot [q(3q + 1)(3q + 2)] \qquad \text{by factoring out a 3.}$$

Let $m = q(3q + 1)(3q + 2)$. Then m is an integer because q is an integer and sums and products of integers are integers. By substitution,

$$n(n + 1)(n + 2) = 3 \cdot m \quad \text{where } m \text{ is an integer.}$$

And so by definition of divisible, $n(n + 1)(n + 2)$ is divisible by 3.

Case 2 ($n = 3q + 1$ for some integer q): In this case,

$$n(n + 1)(n + 2)$$
$$= (3q + 1)((3q + 1) + 1)((3q + 1) + 2) \qquad \text{by substitution}$$
$$= (3q + 1)(3q + 2)(3q + 3)$$
$$= (3q + 1)(3q + 2)3(q + 1)$$
$$= 3 \cdot [(3q + 1)(3q + 2)(q + 1)] \qquad \text{by algebra}$$

Let $m = (3q + 1)(3q + 2)(q + 1)$. Then m is an integer because q is an integer and sums and products of integers are integers. By substitution,

$$n(n + 1)(n + 2) = 3 \cdot m \quad \text{where } m \text{ is an integer.}$$

And so by definition of divisible, $n(n + 1)(n + 2)$ is divisible by 3.

Case 3 ($n = 3q + 2$) for some integer q): In this case,

$$n(n + 1)(n + 2)$$
$$= (3q + 2)((3q + 2) + 1)((3q + 2) + 2) \qquad \text{by substitution}$$
$$= (3q + 2)(3q + 3)(3q + 4)$$
$$= (3q + 2)3(q + 1)(3q + 4)$$
$$= 3 \cdot [(3q + 2)(q + 1)(3q + 4)] \qquad \text{by algebra}$$

Let $m = (3q + 2)(q + 1)(3q + 4)$. Then m is an integer because q is an integer and sums and products of integers are integers. By substitution,

$$n(n + 1)(n + 2) = 3 \cdot m \quad \text{where } m \text{ is an integer.}$$

And so by definition of divisible, $n(n + 1)(n + 2)$ is divisible by 3.

In each of the three cases, $n(n + 1)(n + 2)$ was seen to be divisible by 3. But by the quotient-remainder theorem, one of these cases must occur. Therefore, the product of *any* three consecutive integers is divisible by 3.

23. b. If $m^2 - n^2 = 56$, then $56 = (m + n)(m - n)$. Now $56 = 2^3 \cdot 7$ and by the unique factorization theorem, this factorization is unique. Hence the only representations of 56 as a product of two positive integers are $56 = 7 \cdot 8 = 14 \cdot 4 = 28 \cdot 2 = 56 \cdot 1$. By part (a), m and n must both be odd or both be even. Thus the only solutions are either $m + n = 14$ and $m - n = 4$ or $m + n = 28$ and $m - n = 2$. This gives either $m = 9$ and $n = 5$ or $m = 15$ and $n = 13$ as the only solutions.

24. *Hint:* Use Example 3.4.5 to say that $n = 4q$, $n = 4q + 1$, $n = 4q + 2$, or $n = 4q + 3$ and divide into cases accordingly.

27. *Hint:* It is not necessary to divide into cases to prove this statement.

29. *Hint:* Use the quotient-remainder theorem to say that n must have one of the forms $6q$, $6q + 1$, $6q + 2$, $6q + 3$, $6q + 4$, or $6q + 5$ for some integer q.

32. *Answer to first question:* No. Counterexample: Let $m = 1$, $n = 3$, and $d = 2$. Then $m \bmod d = 1$ and $n \bmod d = 1$ but $m \neq n$.

Answer to second question: Yes. Proof: Suppose m, n, and d are integers such that $m \bmod d = n \bmod d$. Let $r = m \bmod d = n \bmod d$. By definition of $\bmod$, $m = d \cdot p + r$ and $n = d \cdot q + r$ for some integers p and q. Then $m - n = (d \cdot p + r) - (d \cdot q + r) = d \cdot (p - q)$. But $p - q$ is an integer (being a difference of integers), and so $m - n$ is divisible by d by definition of divisible.

36. b. *Proof:* Let x be any real number. Either $x \geq 0$ or $x < 0$.

Case 1 $(x \geq 0)$: In this case, $|x| = x$, and so $-|x| \leq x \leq |x|$.

Case 2 $(x < 0)$: In this case $|x| = -x$, which implies that $-|x| = x$. So $-|x| \leq x \leq |x|$.

SECTION 3.5

1. $\lfloor 37.999 \rfloor = 37$, $\lceil 37.999 \rceil = 38$

3. $\lfloor -14.00001 \rfloor = -15$, $\lceil -14.00001 \rceil = -14$

8. $\lfloor n/7 \rfloor$. The floor notation is more appropriate. If the ceiling notation is used, two different formulas are needed, depending upon whether $n/7$ is an integer or not. (What are they?)

10. a. (i) $(2001 + \lfloor \frac{2000}{4} \rfloor - \lfloor \frac{2000}{100} \rfloor + \lfloor \frac{2000}{400} \rfloor) \bmod 7$

$= (2001 + 500 - 20 + 5) \bmod 7 = 2486 \bmod 7$

$= 1$, which corresponds to a Monday.

b. *Hint:* One day is added every four years except each century the day is not added unless the century is a multiple of 400.

12. *Proof:* Suppose n is any even integer. By definition of even, $n = 2k$ for some integer k. Then

$$\lfloor n/2 \rfloor = \lfloor 2k/2 \rfloor = \lfloor k \rfloor = k \quad \text{because } k \text{ is an integer and } k \leq k < k + 1.$$

But $\quad k = n/2 \quad$ because $n = 2k$.

Thus, on the one hand, $\lfloor n/2 \rfloor = k$, and on the other hand, $k = n/2$. It follows that $\lfloor n/2 \rfloor = n/2$ *[as was to be shown]*.

14. *Hint:* This is false.

15. *Hint:* This is true. To prove it directly from the definition of floor, suppose x is a real number. Then $\lfloor x - 1 \rfloor$ is some integer; say $\lfloor x - 1 \rfloor = n$. By definition of floor, $n \leq x - 1 < n + 1$. Adding 1 to all parts gives $n + 1 \leq x < n + 2$, and thus

$\lfloor x \rfloor = n + 1$. (Why?) Solve this equation for n and then equate the two different expressions for n to conclude that $\lfloor x - 1 \rfloor = \lfloor x \rfloor - 1$. Alternatively, the statement can be derived from Theorem 3.5.1 using $m = -1$.

17. *Proof for the case where $n \bmod 3 = 2$:*
In case $n \bmod 3 = 2$, then $n = 3q + 2$ for some integer q by definition of $\bmod$. By substitution,

$$\lfloor n/3 \rfloor = \lfloor (3q + 2)/3 \rfloor$$
$$= \lfloor 3q/3 + 2/3 \rfloor$$
$$= \lfloor q + 2/3 \rfloor = q \quad \text{because } q \text{ is an integer and } q \leq q + 2/3 < q + 1.$$

But

$$q = (n - 2)/3 \quad \text{by solving } n = 3q + 2 \text{ for } q.$$

Thus, on the one hand, $\lfloor n/3 \rfloor = q$, and on the other hand, $q = (n - 2)/3$. It follows that $\lfloor n/3 \rfloor = (n - 2)/3$.

18. *Hint:* This is false.

19. *Hint:* This is true.

23. *Proof:* Suppose x is a real number that is not an integer. Let $\lfloor x \rfloor = n$. Then by definition of floor and because n is not an integer, $n < x < n + 1$. Multiplying both sides by -1 gives $-n > -x > -n - 1$, or, equivalently, $-n - 1 < -x < -n$. Since $-n - 1$ is an integer, it follows by definition of floor that $\lfloor -x \rfloor = -n - 1$. Hence $\lfloor x \rfloor + \lfloor -x \rfloor = n + (-n - 1) = n - n - 1 = -1$, as was to be shown.

25. *Hint:* Let $n = \lfloor x/2 \rfloor$ and consider the two cases: n is even and n is odd.

26. *Proof:* Suppose x is any real number such that $x - \lfloor x \rfloor < 1/2$. Multiplying both sides by 2 gives $2x - 2 \cdot \lfloor x \rfloor < 1$, or $2x < 2 \cdot \lfloor x \rfloor + 1$. Now by definition of floor, $\lfloor x \rfloor \leq x$. Hence, $2 \cdot \lfloor x \rfloor \leq 2x$. Putting the two inequalities involving $2x$ together gives $2 \cdot \lfloor x \rfloor \leq 2x < 2 \cdot \lfloor x \rfloor + 1$. Thus, by definition of floor (and because $2 \cdot \lfloor x \rfloor$ is an integer), $\lfloor 2x \rfloor = 2 \cdot \lfloor x \rfloor$. This is what was to be shown.

30. This incorrect proof begs the question. The equality $\lfloor n/2 \rfloor = (n - 1)/2$ is what is to be shown. By substituting $2k + 1$ for n into both sides of the equality and working from the result as if it were known to be true, the proof assumes the truth of the conclusion to be proved.

SECTION 3.6

1. a. $x \leq y$ **b.** $x/2$ is a positive real number
c. multiplying both sides of the inequality $1/2 < 1$ by x, which is positive, gives $x/2 < x$
d. $x/2$ is a positive real number that is less than x

2. *Negation of statement:* There is a greatest even integer.

Proof of statement: Suppose not. Suppose there is a greatest even integer; call it N. Then N is an even integer and $N \geq n$ for every even integer n. *[We must deduce a contradiction.]* Let $M = N + 2$. Then M is an even integer since it is a sum of even integers, and $M > N$ since $M = N + 2$. This contradicts the supposition that $N \geq n$ for *every* even integer n. *[Hence the supposition is false and the statement is true.]*

5. The mistake in this proof occurs in the second sentence where the negation written by the student is incorrect: Instead of being existential, it is universal. The problem is that if the student proceeds in a logically correct manner, all that is needed to reach a contradiction is one example of a rational and an irrational number whose sum is irrational. To prove the given statement, however, it is necessary to show that there is *no* rational number and *no* irrational number whose sum is rational.

6. *Proof (by contradiction):* Suppose not. Suppose $\exists$ a rational number x and an irrational number y such that $x - y$ is rational. *[We must derive a contradiction.]* By definition of rational, $x = a/b$ and $x - y = c/d$ for some integers $a, b, c,$ and d with $b \neq 0$ and $d \neq 0$. By substitution,

$$\frac{c}{d} = x - y = \frac{a}{b} - y$$

Solving for y gives

$$y = \frac{a}{b} - \frac{c}{d} = \frac{ad - bc}{bd}.$$

But $ad - bc$ are integers (because $a, b, c,$ and d are integers and products and differences of integers are integers), and $bd \neq 0$ (by the zero product property). Thus by definition of rational, y is rational. This contradicts the supposition that y is irrational. *[Hence the supposition is false and the given statement is true.]*

7. *Proof (by contraposition):* *[To go by contraposition, we must prove that $\forall$ positive real numbers, r and s, if $r \leq 10$ and $s \leq 10$, then $r \cdot s \leq 100$.]* Suppose r and s are positive real numbers and $r \leq 10$ and $s \leq 10$. By the algebra of inequalities, $r \cdot s \leq 100$. *[To derive this fact, multiply both sides of $r \leq 10$ by s to obtain $r \cdot s \leq 10 \cdot s$. And multiply both sides of $s \leq 10$ by 10 to obtain $10 \cdot s \leq 10 \cdot 10 = 100$. By transitivity of $\leq$, then, $r \cdot s \leq 100$.]* But this is what was to be shown.

9. a. The contrapositive is the statement "$\forall$ real numbers x, if $-x$ is not irrational then x is not irrational." Equivalently (because $-(-x) = x$): "$\forall$ real numbers x, if x is rational then $-x$ is rational."

Proof by contraposition: Suppose x is any rational number. *[We must show that $-x$ is also rational.]* By definition of rational, $x = a/b$ for some integers a and b with $b \neq 0$. Then $x = -(a/b) = (-a)/b$. Since both $-a$ and b are integers and $b \neq 0$, $-x$ is rational *[as was to be shown.]*

b. *Proof:* Suppose not. *[We take the negation and suppose it to be true.]* Suppose $\exists$ an irrational number x such that $-x$ is rational. *[We must derive a contradiction.]* By definition of rational, $-x = a/b$ for some integers a and b with $b \neq 0$. Multiplying both sides by -1 gives $x = -(a/b) = -a/b$. But $-a$ and b are integers (since a and b are) and $b \neq 0$. Thus x is a ratio of the two integers $-a$ and b with $b \neq 0$. Hence x is rational (by definition of rational), which is a contradiction. *[This contradiction shows that the supposition is false, and so the given statement is true.]*

10. a. *Hint:* Begin by supposing that n is any *[particular but arbitrarily chosen]* integer such that n is not odd (i.e., n is even). Then deduce that n^2 is not odd (i.e., n^2 is even). The proof is similar to that of Proposition 3.6.3.

b. *Hint:* Begin by supposing that $\exists$ an integer n such that n^2 is odd and n is even. Then derive a contradiction. The proof is similar to that given as an alternate proof of Proposition 3.6.3.

11. *Hint:* Use the unique factorization theorem.

12. a. *Proof by contraposition:* Suppose $a, b,$ and c are any *[particular but arbitrarily chosen]* integers such that $a \mid b$. *[We must show that $a \mid bc$.]* By definition of divides, $b = a \cdot k$ for some integer k. Then $bc = (ak) \cdot c = a \cdot (kc)$. But kc is an integer (because it is a product of the integers k and c). Hence $a \mid bc$ by definition of divisibility *[as was to be shown.]*

b. *Proof by contradiction:* Suppose not. *[We take the negation and suppose it to be true.]* Suppose $\exists$ integers $a, b,$ and c such that $a \nmid bc$ and $a \mid b$. Since $a \mid b$, there exists an integer k such that $b = a \cdot k$ by definition of divides. Then $bc = (a \cdot k)c = a \cdot (kc)$ *[by the associative law of algebra]*. But kc is an integer (being a product of integers), and so $a \mid bc$ by definition of divides. Thus $a \nmid bc$ and $a \mid bc$, which is a contradiction. *[This contradiction shows that the supposition is false, and hence the given statement is true.]*

13. a. *Hint:* The contrapositive is "For all integers m and n, if m and n are not both even and m and n are not both odd, then $m + n$ is not even." Equivalently: "For all integers m and n, if one of m and n is even and the other is odd, then $m + n$ is odd."

b. *Hint:* The negation of the given statement is the following: $\exists$ integers m and n such that $m + n$ is even and either m is even and n is odd, or m is odd and n is even.

16. The negation of "Every integer is rational" is "There is at least one integer that is irrational" not "Every integer

is irrational." Deriving a contradiction from an incorrect negation of a statement does not prove the statement is true.

17. *Counterexample:* Let $x = \sqrt{2}$ and let $y = -\sqrt{2}$. Then x and y are irrational (as will be proved formally in Section 3.7), but $x + y = 0 = 0/1$. which is rational.

18. *Hint:* This is true.

19. *Hint:* This is false.

SECTION 3.7

2. *Proof by contradiction:* Suppose not. Suppose $6 - 7\sqrt{2}$ is rational. *[We must prove a contradiction.]* By definition of rational, there exists integers a and $b \neq 0$ with

$$6 - 7\sqrt{2} = \frac{a}{b}.$$

Then $\sqrt{2} = \dfrac{1}{-7}\left(\dfrac{a}{b} - 6\right)$ by subtracting 6 from both sides and dividing both sides by -7

and so $\sqrt{2} = \dfrac{a - 6b}{-7b}$ by the rules of algebra.

But $a - 6b$ and $-7b$ are both integers (since a and b are integers and products and differences of integers are integers), and $-7b \neq 0$ by the zero product property. Hence $\sqrt{2}$ is a ratio of the two integers $a - 6b$ and $-7b$ with $-7b \neq 0$, so $\sqrt{2}$ is a rational number (by definition of rational). *[This contradiction shows that the supposition is false. Hence $6 - 7\sqrt{2}$ is irrational.]*

4. *Hint:* Either use the result of exercise 11 of Section 3.6 or prove directly that for all integers n, if n^2 is divisible by 3 then n is divisible by 3. Then prove that $\sqrt{3}$ is irrational using the proof of the irrationality of $\sqrt{2}$ as a model.

5. This is false. $\sqrt{4} = 2 = 2/1$, which is rational.

7. *Hint:* This statement is true. If $a^2 - 2 = 4b$, then $a^2 = 4b + 2 = 2(2b + 1)$, and so a^2 is even. Hence a is even, and so $a^2 = (2c)^2$ for some integer c.

8. *Hint:* This is true.

11. *Hint:* Can you think of any "nice" numbers x and y so that $x^2 = y^3$?

16. *Hint:* Is it possible for all three of $n - 4$, $n - 6$, and $n - 8$ to be prime?

17. *Hint:* Prove the contrapositive: If for some integer $n > 2$ that is not a power of 2, $x^n + y^n = z^n$ has a positive integer solution, then for some prime number $p > 2$, $x^p + y^p = z^p$ has a positive integer solution. Note that if $n = kp$, then $x^n = x^{kp} = (x^k)^p$.

19. *Hint:* Every odd integer can be written as $4k + 1$ or as $4k + 3$ for some integer k. (Why?) If $p_1 \cdot p_2 \cdot \ldots \cdot p_n + 1 = 4k + 1$, then $4 \mid p_1 \cdot p_2 \cdot \ldots \cdot p_n$. Is this possible?

20. *Hint:* By Theorem 3.3.2 (divisibility by a prime) there is a prime number p such that $p \mid (n! - 1)$. Show that the supposition that $p \leq n$ leads to a contradiction. It will then follow that $n < p < n!$.

21. *Existence Proof:* When $n = 2$, then $n^2 - 1 = 3$, which is prime. Hence there exists a prime number of the form $n^2 - 1$, where n is an integer and $n \geq 2$.

Uniqueness Proof (by contradiction): Suppose m is another integer that satisfies the given conditions. That is, $m > 2$ and $m^2 - 1$ is prime. *[We must derive a contradiction.]* Factor $m^2 - 1$ to obtain $m^2 - 1 = (m - 1)(m + 1)$. But $m > 2$, and so $m - 1 > 1$ and $m + 1 > 1$. Hence $m^2 - 1$ is not prime, which is a contradiction. *[This contradiction shows that the supposition is false, and so there is no other integer $m > 2$ such that $n^2 - 1$ is prime.]*

Uniqueness Proof (direct): Suppose m is *any* integer such that $m \geq 2$ and $m^2 - 1$ is prime. *[We must show that $m = 2$.]* By factoring, $m^2 - 1 = (m - 1)(m + 1)$. Since $m^2 - 1$ is prime, either $m - 1 = 1$ or $m + 1 = 1$. But $m + 1 \geq 2 + 1 = 3$. Hence, by elimination, $m - 1 = 1$, and so $m = 2$.

23. *Proof (by contradiction):* Suppose not. Suppose a_1 and a_2 are two distinct real numbers such that for all real numbers r,

$$(1) \quad a_1 + r = r$$

and

$$(2) \quad a_2 + r = r$$

Then

$$a_1 + a_2 = a_2 \quad \text{by (1) with } r = a_2$$

and

$$a_2 + a_1 = a_1 \quad \text{by (2) with } r = a_1.$$

It follows that

$$a_2 = a_1 + a_2 = a_2 + a_1 = a_1.$$

which implies that $a_2 = a_1$. But this contradicts the supposition that a_1 and a_2 are distinct. *[Thus the supposition is false and there is at most one real number a such that $a + r = r$ for all real numbers r.]*

Proof (direct): Suppose a_1 and a_2 are real numbers such that for all real numbers r,

$$(1) \quad a_1 + r = r$$

and

$$(2) \quad a_2 + r = r$$

Then

$$a_1 + a_2 = a_2 \quad \text{by (1) with } r = a_2$$

and

$$a_2 + a_1 = a_1 \quad \text{by (2) with } r = a_1.$$

It follows that

$$a_2 = a_1 + a_2 = a_2 + a_1 = a_1.$$

Hence $a_2 = a_1$. *[Thus there is at most one real number a such that $a + r = r$ for all real numbers r.]*

SECTION 3.8

1. $z = 0$ 3. a. $z = 12$ 4. $a = 17/12$

6.   iteration number

	0	1	2	3
a	26			
d	7			
q	0	1	2	3
r	26	19	12	5

8. $\gcd(27, 72) = 9$ 9. $\gcd(5,9) = 1$

12.
$$\begin{array}{r} 3 \\ 385\overline{)1188} \\ 1155 \\ \hline 33 \end{array}$$
So $1188 = 385 \cdot 3 + 33$, and hence $\gcd(1188, 385) = \gcd(385, 33)$.

$$\begin{array}{r} 11 \\ 33\overline{)385} \\ 33 \\ \hline 55 \\ 33 \\ \hline 22 \end{array}$$
So $385 = 33 \cdot 11 + 22$, and hence $\gcd(385, 33) = \gcd(33, 22)$.

$$\begin{array}{r} 1 \\ 22\overline{)33} \\ 22 \\ \hline 11 \end{array}$$
So $33 = 22 \cdot 1 + 11$, and hence $\gcd(33, 22) = \gcd(22, 11)$.

$$\begin{array}{r} 2 \\ 11\overline{)22} \\ 22 \\ \hline 0 \end{array}$$
So $22 = 11 \cdot 2 + 0$, and hence $\gcd(22, 11) = \gcd(11, 0)$.

But $\gcd(11, 0) = 11$. So $\gcd(1188, 385) = 11$.

15.

A	1,001					
B	871					
r		130	91	39	13	0
b	871	130	91	39	13	0
a	1,001	871	130	91	39	13
$\gcd$						13

20. *Proof:* Suppose that a and d are integers with $d > 0$ and that q_1, q_2, r_1, and r_2 are integers such that

$$a = d \cdot q_1 + r_1 \quad \text{and} \quad a = d \cdot q_2 + r_2$$

where

$$0 \le r_2 < d \quad \text{and} \quad 0 \le r_1 < d.$$

Then

$$d \cdot q_1 + r_1 = d \cdot q_2 + r_2$$

and so

$$r_2 - r_1 = d \cdot (q_1 - q_2).$$

This implies that

$$d \mid (r_2 - r_1). \quad \text{[Why?]}$$

But both r_1 and r_2 lie between 0 and d. Thus the difference of $r_2 - r_1$ lies between $-d$ and d. *[For if $0 \le r_1 < d$ and $0 \le r_2 < d$, then multiplying the first inequality by -1 gives $0 \ge -r_1 > -d$, or $-d < -r_1 \le 0$. Adding $-d < -r_1 \le 0$ and $0 \le r_2 < d$ gives $-d < r_2 - r_1 < d$.]* Since $r_2 - r_1$ is a multiple of d and yet lies between $-d$ and d, the only possibility is that

$$r_2 - r_1 = 0, \quad \text{or, equivalently,} \quad r_1 = r_2.$$

Substituting back into the original expressions for a and equating gives

$$d \cdot q_1 + r_1 = d \cdot q_2 + r_1$$

[since $r_2 = r_1$]. Subtracting r_1 from both sides gives

$$d \cdot q_1 = d \cdot q_2,$$

and since $d \ne 0$,

$$q_1 = q_2.$$

Hence $r_1 = r_2$ and $q_1 = q_2$, as was to be shown.

21. a. *Hint 1:* If $a = dq - r$, then $-a = -dq - r = -dq - d + d - r = d(-q - 1) + (d - r)$.

   Hint 2: If $0 \le r < d$, then $0 \ge -r > -d$. Add d to all parts of this inequality and see what results.

24. a. $\operatorname{lcm}(12, 18) = 36$

25. *Proof:* Let a and b be positive integers and suppose $d = \gcd(a, b) = \operatorname{lcm}(a, b)$. By definition of greatest common divisor and least common multiple, $d > 0$, $d \mid a$, $d \mid b$, $a \mid d$, and $b \mid d$. Thus, in particular, $a = d \cdot m$ and $d = a \cdot n$ for some integers m and n. By substitution, $a = d \cdot m = (a \cdot n) \cdot m = a \cdot nm$. Dividing both sides by a gives $1 = nm$. But the only divisors of 1 are 1 and -1 (Example 3.3.4), and so $m = n = \pm 1$. Since both a and d are positive, $m = n = 1$, and hence $a = d$. Similar reasoning show that $b = d$ also, and so $a = b$.

SECTION 4.1

1. $\dfrac{1}{9}, \dfrac{2}{8}, \dfrac{3}{7}, \dfrac{4}{6}$

3. $1, -\dfrac{1}{3}, \dfrac{1}{9}, -\dfrac{1}{27}$

5. $0, 0, 2, 2$

8. $g_1 = \lfloor \log_2 1 \rfloor = 0$
   $g_2 = \lfloor \log_2 2 \rfloor = 1$, $\quad g_3 = \lfloor \log_2 3 \rfloor = 1$
   $g_4 = \lfloor \log_2 4 \rfloor = 2$, $\quad g_5 = \lfloor \log_2 5 \rfloor = 2$
   $g_6 = \lfloor \log_2 6 \rfloor = 2$, $\quad g_7 = \lfloor \log_2 7 \rfloor = 2$
   $g_8 = \lfloor \log_2 8 \rfloor = 3$, $\quad g_9 = \lfloor \log_2 9 \rfloor = 3$
   $g_{10} = \lfloor \log_2 10 \rfloor = 3$, $\quad g_{11} = \lfloor \log_2 11 \rfloor = 3$
   $g_{12} = \lfloor \log_2 12 \rfloor = 3$, $\quad g_{13} = \lfloor \log_2 13 \rfloor = 3$
   $g_{14} = \lfloor \log_2 14 \rfloor = 3$, $\quad g_{15} = \lfloor \log_2 15 \rfloor = 3$

   When n is an integral power of 2, g_n is the exponent of that power. For instance, $8 = 2^3$ and $g_8 = 3$. More generally, if $n = 2^k$, where k is an integer, then $g_n = k$. All terms of the sequence from g_n up to g_m, where $m = 2^{k+1}$ is the next integral power of 2, have the same value as g_n, namely k. For instance, all terms of the sequence from g_8 through g_{15} have the value 3.

Exercises 10–16 have more than one correct answer.

10. $a_n = (-1)^n$, where n is an integer and $n \ge 1$.

11. $b_n = (n - 1) \cdot (-1)^n$, where n is an integer and $n \ge 1$.

12. $c_n = \dfrac{n}{n + 2}$, where n is an integer and $n \ge 1$.

18. a. $2 + 3 + (-2) + 1 + 0 + (-1) + (-2) = 1$
    b. $a_0 = 2$
    c. $a_2 + a_4 + a_6 = -2 + 0 + (-2) = -4$
    d. $2 \cdot 3 \cdot (-2) \cdot 1 \cdot 0 \cdot (-1) \cdot (-2) = 0$

19. $2 + 3 + 4 + 5 + 6 = 20$

20. $2^2 \cdot 3^2 \cdot 4^2 = 576$

24. $\left(\dfrac{1}{1} - \dfrac{1}{2}\right) + \left(\dfrac{1}{2} - \dfrac{1}{3}\right) + \left(\dfrac{1}{3} - \dfrac{1}{4}\right) + \left(\dfrac{1}{4} - \dfrac{1}{5}\right)$

   $+ \left(\dfrac{1}{5} - \dfrac{1}{6}\right) + \left(\dfrac{1}{6} - \dfrac{1}{7}\right) + \left(\dfrac{1}{7} - \dfrac{1}{8}\right) + \left(\dfrac{1}{8} - \dfrac{1}{9}\right)$

   $+ \left(\dfrac{1}{9} - \dfrac{1}{10}\right) + \left(\dfrac{1}{10} - \dfrac{1}{11}\right) = 1 - \dfrac{1}{11} = \dfrac{10}{11}$

26. $(-2)^1 + (-2)^2 + (-2)^3 + \cdots + (-2)^n$

   $= -2 + 2^2 - 2^3 + \cdots + (-1)^n \cdot 2^n$

Exercises 29–38 have more than one correct answer.

29. $\displaystyle\sum_{k=1}^{7} (-1)^{k+1} \cdot k^2$ or $\displaystyle\sum_{k=0}^{6} (-1)^k \cdot (k + 1)^2$

32. $\displaystyle\sum_{j=2}^{6} \dfrac{j}{(j + 1) \cdot (j + 2)}$ or $\displaystyle\sum_{j=3}^{7} \dfrac{j - 1}{j \cdot (j + 1)}$

33. $\displaystyle\sum_{i=0}^{5} r^i$

35. $\displaystyle\sum_{k=1}^{n} k^3$

37. $\displaystyle\sum_{i=0}^{n-1} (n - i)$

39. When $k = 0$, then $i = 1$. When $k = 5$, then $i = 6$. Since $i = k + 1$, then $k = i - 1$. Thus,

   $$\sum_{k=0}^{5} (k - 1) \cdot k = \sum_{i=1}^{6} (i - 2) \cdot (i - 1).$$

41. When $i = 1$, then $j = 0$. When $i = n + 1$, then $j = n$. Since $j = i - 1$, then $i = j + 1$. Thus,

   $$\sum_{i=1}^{n+1} \dfrac{(i - 1)^2}{i} = \sum_{j=0}^{n} \dfrac{j^2}{j + 1}.$$

42. When $i = 3$, then $j = 2$. When $i = n + 1$, then $j = n$. Since $j = i - 1$, then $i = j + 1$. Note that n is constant as far as the sum is concerned. Thus,

   $$\sum_{i=3}^{n+1} \dfrac{i}{i + n - 1} = \sum_{j=2}^{n} \dfrac{j + 1}{(j + 1) + n - 1}$$

   $$= \sum_{j=2}^{n} \dfrac{j + 1}{j + n}.$$

45. $\displaystyle\sum_{k=1}^{n} [3(2k - 3) + (4 - 5k)]$

   $= \displaystyle\sum_{k=1}^{n} [(6k - 9) + (4 - 5k)] = \sum_{k=1}^{n} (k - 5)$

48. $\dfrac{4 \cdot \cancel{3} \cdot \cancel{2} \cdot \cancel{1}}{\cancel{3} \cdot \cancel{2} \cdot \cancel{1}} = 4$

51. $\dfrac{n \cdot \cancel{(n - 1)} \cdot \cancel{(n - 2)} \cdot \cdots \cdot \cancel{3} \cdot \cancel{2} \cdot \cancel{1}}{\cancel{(n - 1)} \cdot \cancel{(n - 2)} \cdot \cdots \cdot \cancel{3} \cdot \cancel{2} \cdot \cancel{1}} = n$

52. $$\frac{\cancel{(n-1)}\cdot\cancel{(n-2)}\cdot\ldots\cdot\cancel{3}\cdot\cancel{2}\cdot\cancel{1}}{(n+1)\cdot n\cdot\cancel{(n-1)}\cdot\cancel{(n-2)}\cdot\ldots\cdot\cancel{3}\cdot\cancel{2}\cdot\cancel{1}}$$

$$= \frac{1}{n(n+1)}$$

54. $$\frac{[(n+1)\cdot n\cdot(n-1)\cdot(n-2)\cdot\ldots\cdot 3\cdot 2\cdot 1]^2}{(n\cdot(n-1)\cdot(n-2)\cdot\ldots\cdot 3\cdot 2\cdot 1)^2}$$

$$= (n+1)^2$$

55. $$\frac{n\cdot(n-1)\cdot(n-2)\cdot\ldots\cdot(n-k+1)\cdot\cancel{(n-k)}\cdot\cancel{(n-k-1)}\cdot\ldots\cdot\cancel{2}\cdot\cancel{1}}{\cancel{(n-k)}\cdot\cancel{(n-k-1)}\cdot\ldots\cdot\cancel{2}\cdot\cancel{1}}$$

$$= n\cdot(n-1)\cdot(n-2)\cdot\ldots\cdot(n-k+1)$$

57. a. *Proof*: Let n be an integer such that $n \geq 2$. By definition of factorial,

$$n! = \begin{cases} 2\cdot 1 & \text{If } n = 2 \\ 3\cdot 2\cdot 1 & \text{if } n = 3 \\ n\cdot(n-1)\cdot\ldots\cdot 2\cdot 1 & \text{if } n > 3. \end{cases}$$

In each case, $n!$ has a factor of 2, and so

$$n! = 2\cdot k \quad \text{for some integer } k.$$

Then

$$n! + 2 = 2\cdot k + 2 \quad \text{by substitution}$$
$$= 2\cdot(k+1) \quad \text{by factoring out the 2.}$$

Since $k + 1$ is an integer, $n! + 2$ is divisible by 2 *[as was to be shown]*.

c. *Hint*: Consider the sequence $m! + 2$, $m! + 3$, $m! + 4, \ldots, m! + m$.

58. a. $m - 1$, $sum + a[i + 1]$

59.

	0	remainder $= r[6] = 1$
2	1	remainder $= r[5] = 0$
2	2	remainder $= r[4] = 1$
2	5	remainder $= r[3] = 1$
2	11	remainder $= r[2] = 0$
2	22	remainder $= r[1] = 1$
2	45	remainder $= r[0] = 0$
2	90	

Hence $90_{10} = 1011010_2$.

62.

a	23					
i	0	1	2	3	4	5
q	23	11	5	2	1	0
r[0]		1				
r[1]			1			
r[2]				1		
r[3]					0	
r[4]						1

66.

	0	remainder 1 $= r[2] = 1_{16}$
16	1	remainder 1 $= r[1] = 1_{16}$
16	17	remainder 15 $= r[0] = F_{16}$
16	287	

Hence $287_{10} = 11F_{16}$.

SECTION 4.2

1. Let $P(n)$ be the property "n cents can be obtained by using 3-cent and 8-cent coins."

The property is true for $n = 15$:

Fifteen cents can be obtained by using five 3-cent coins.

If the property is true for $n = k$ then it is true for $n = k + 1$:

Suppose k cents (where $k \geq 15$) can be obtained using 3-cent and 8-cent coins. *[Inductive hypothesis]* We must show that $k + 1$ cents can be obtained using 3-cent and 8-cent coins. If the k cents includes an 8-cent coin, replace it by three 3-cent coins to obtain a total of $k + 1$ cents. If the k cents consists of 3-cent coins exclusively, then there must be at least five 3-cent coins (since the total amount is at least 15 cents). In this case, replace five of the 3-cent coins by two 8-cent coins to obtain a total of $k + 1$ cents. Thus, in either case, $k + 1$ cents can be obtained using 3-cent and 8-cent coins. *[This is what we needed to show.]*

[Since we have proved the basis step and the inductive step, we conclude that the given statement is true for all integers $n \geq 15$.]

3. a. $P(1)$ is "$1^2 = \dfrac{1\cdot(1+1)\cdot(2\cdot 1 + 1)}{6}$." $P(1)$ is true because $1^2 = 1$ and $\dfrac{1\cdot(1+1)\cdot(2+1)}{6} = \dfrac{2\cdot 3}{6} = 1$ also.

b. $P(k)$ is "$1^2 + 2^2 + \cdots + k^2 = \dfrac{k(k+1)(2k+1)}{6}$."

c. $P(k + 1)$ is "$1^2 + 2^2 + \cdots + (k + 1)^2 =$

$$\frac{(k + 1)((k + 1) + 1)(2 \cdot (k + 1) + 1)}{6}.$$"

d. *Must show:* If for some integer $k \geq 1$,

$$1^2 + 2^2 + \cdots + k^2 = \frac{k(k + 1)(2k + 1)}{6}, \quad \text{then}$$

$$1^2 + 2^2 + \cdots + (k + 1)^2 =$$

$$\frac{(k + 1)[(k + 1) + 1][2(k + 1) + 1]}{6}.$$

5. a. 1^2 b. k^2

   c. $1 + 3 + 5 + \cdots + [2(k + 1) - 1]$
   d. $(k + 1)^2$
   e. the odd integer just before $2k + 1$ is $2k - 1$.
   f. k^2

6. The formula is true for $n = 1$:

   To prove the formula for $n = 1$, we must show that when 1 is substituted in place of n, the left-hand side equals the right-hand side. But when $n = 1$, the left-hand side is the sum of all the even integers from 2 to $2 \cdot 1$, which is just 2. The right-hand side is $1^2 + 1$, which also equals 2. So the formula is true for $n = 1$.

   If the formula is true for $n = k$, then it is true for $n = k + 1$:

   Suppose that for some integer $k \geq 1$,

   $$2 + 4 + 6 + \cdots + 2k = k(k + 1). \qquad \text{This is the inductive hypothesis.}$$

   We must show that

   $$2 + 4 + 6 + \cdots + 2(k + 1) = (k + 1)[(k + 1) + 1],$$

   or, equivalently,

   $$2 + 4 + 6 + \cdots + 2(k + 1) = (k + 1)(k + 2).$$

   But

   $$2 + 4 + 6 + \cdots + 2(k + 1)$$
   $$= 2 + 4 + 6 + \cdots + 2k + 2(k + 1)$$
   $$\qquad \text{by making the next-to-last term explicit}$$
   $$= k(k + 1) + 2(k + 1) \qquad \text{by substitution from the inductive hypothesis}$$
   $$= k^2 + k + 2k + 2$$
   $$= k^2 + 3k + 2$$
   $$= (k + 1)(k + 2) \qquad \text{by algebra.}$$

   [This is what was to be shown.]

9. *Proof:*

   The formula is true for $n = 1$:

   To prove the formula for $n = 1$, we must show that

   $$1^2 = \frac{1 \cdot (1 + 1) \cdot (2 \cdot 1 + 1)}{6}. \text{ But the left-hand side}$$

of this equation equals 1 and the right-hand side of this equation equals $\frac{2 \cdot 3}{6}$, which also equals 1. So the formula is true for $n = 1$.

If the formula is true for $n = k$ then it is true for $n = k + 1$:

[Suppose the formula $1^2 + 2^2 + \cdots + n^2$

$$= \frac{n(n + 1)(2n + 1)}{6} \text{ is true when an integer } k \geq 1 \text{ is}$$

substituted for n.]

Suppose

$$1^2 + 2^2 + \cdots + k^2 = \frac{k(k + 1)(2k + 1)}{6},$$

for some integer $k \geq 1$.   Inductive hypothesis

[We must show that the formula

$$1^2 + 2^2 + \cdots + n^2 = \frac{n(n + 1)(2n + 1)}{6}$$

is true when $k + 1$ is substituted for n.]

We must show that

$$1^2 + 2^2 + \cdots + (k + 1)^2$$
$$= \frac{(k + 1)[(k + 1) + 1][2(k + 1) + 1]}{6},$$

or, equivalently,

$$1^2 + 2^2 + \cdots + (k + 1)^2 = \frac{(k + 1)(k + 2)(2k + 3)}{6}.$$

But

$$1^2 + 2^2 + \cdots + (k + 1)^2$$
$$= 1^2 + 2^2 + \cdots + k^2 + (k + 1)^2 \qquad \text{because the terms are squares of successive integers, and so the next-to-last term is } k^2.$$

$$= \frac{k(k + 1)(2k + 1)}{6} + (k + 1)^2 \qquad \text{by inductive hypothesis}$$

$$= \frac{k(k + 1)(2k + 1)}{6} + \frac{6(k + 1)^2}{6} \qquad \text{since } \frac{6}{6} = 1$$

$$= \frac{(k + 1)[k(2k + 1) + 6(k + 1)]}{6} \qquad \text{by the rules for adding fractions and by factoring out } k + 1 \text{ from both terms of the numerator}$$

$$= \frac{(k+1)(2k^2 + k + 6k + 6)}{6} \quad \text{by multiplying out}$$

$$= \frac{(k+1)(2k^2 + 7k + 6)}{6} \quad \text{by combining like terms}$$

$$= \frac{(k+1)(k+2)(2k+3)}{6} \quad \text{since } 2k^2 + 7k + 6 = (k+2)(2k+3)$$

[This is what we needed to show.]
[Since we have proved the basis step and the inductive step, we conclude that the given formula is true for all integers $n \geq 1$.]

12. *Proof:*
The formula is true for n = 2: To prove the formula for $n = 2$, we must show that

$$\sum_{i=1}^{2-1} i(i+1) = \frac{2(2-1)(2+1)}{3}.$$

But the left-hand side of this equation equals $1 \cdot 2$, and the right-hand side equals $\frac{2 \cdot 1 \cdot 3}{3}$, which also equals 2. So the formula is true for $n = 2$.

If true for n = k then true for n = k + 1:
Suppose for some integer

$$\sum_{i=1}^{k-1} i(i+1) = \frac{k(k-1)(k+1)}{3} \quad k \geq 2. \quad \text{Inductive hypothesis}$$

We must show that

$$\sum_{i=1}^{(k+1)-1} i(i+1) = \frac{k(k+1)((k+1)-1)((k+1)+1)}{3},$$

or, equivalently,

$$\sum_{i=1}^{k} i(i+1) = \frac{(k+1)k(k+2)}{3}.$$

But

$$\sum_{i=1}^{k} i(i+1)$$

$$= \sum_{i=1}^{k-1} i(i+1) + k(k+1) \quad \text{by writing the sum of the first } k-1 \text{ terms separately from the last term}$$

$$= \frac{k(k-1)(k+1)}{3} + k(k+1) \quad \text{by inductive hypothesis}$$

$$= \frac{k(k-1)(k+1)}{3} + \frac{3k(k+1)}{3} \quad \text{since } \frac{3}{3} = 1$$

$$= \frac{k(k-1)(k+1) + 3k(k+1)}{3} \quad \text{by the rules for adding fractions}$$

$$= \frac{k(k+1)[(k-1)+3]}{3} \quad \text{by factoring out } k \cdot (k+1) \text{ from both terms of the numerator}$$

$$= \frac{(k+1)k(k+2)}{3}$$

[This is what we needed to show.]
[Since we have proved the basis step and the inductive step, we conclude that the given formula is true for all integers $n \geq 2$.]

14. *Hint:* To prove the basis step, show that $\sum_{i=1}^{1} i(i!) = (1+1)! - 1$. To prove the inductive step, suppose that $\sum_{i=1}^{k} i(i!) = (k+1)! - 1$ for some integer $k \geq 1$ and show that $\sum_{i=1}^{k+1} i(i!) = (k+2)! - 1$.
Note that $[(k+1)! - 1] + (k+1)[(k+1)!] = (k+1)![1 + (k+1)] - 1$.

19. $\frac{(1000) \cdot (1001)}{2} - (1+2) = 500,500 - 3 = 500,497$

21. $\frac{(k-1)((k-1)+1)}{2} = \frac{k(k-1)}{2}$

22. a. $\frac{2^{26} - 1}{2 - 1} = 2^{26} - 1 = 67,108,863$

b. $2 + 2^2 + 2^3 + \cdots + 2^{26}$
$= 2(1 + 2 + 2^2 + \cdots + 2^{25})$
$= 2 \cdot (67,108,863) \quad \text{by part (a)}$
$= 134,217,726$

24. $\frac{\left(\frac{1}{2}\right)^{n+1} - 1}{\frac{1}{2} - 1} = \frac{\frac{1}{2^{n+1}} - 1}{-\frac{1}{2}} = \left(\frac{1}{2^{n+1}} - 1\right) \cdot (-2)$

$= -\frac{2}{2^{n+1}} + 2 = 2 - \frac{1}{2^n}$

26. *Hint*: $a + (a+d) + (a+2d) + \cdots + (a+nd) = (n+1)a + d \cdot \frac{n(n+1)}{2}$.

29. In the inductive step both the inductive hypothesis and what is to be shown are wrong. The inductive hypothesis should be

Suppose that for some $k \geq 1$,

$$1^2 + 2^2 + \cdots + k^2 = \frac{k(k+1)(2k+1)}{6}.$$

And what is to be shown should be

$$1^2 + 2^2 + \cdots + (k+1)^2 = \frac{(k+1)((k+1)+1)(2(k+1)+1)}{6}.$$

SECTION 4.3

1. *Formula:* $\displaystyle\prod_{i=2}^{n}\left(1-\frac{1}{i}\right)=\frac{1}{n}$ for all integers $n\geq 2$.

Proof (by mathematical induction):

The formula is true for $n=2$:

The formula holds for $n=2$ because $\displaystyle\prod_{i=2}^{2}\left(1-\frac{1}{i}\right)=$

$1-\dfrac{1}{2}=\dfrac{1}{2}$.

If the formula is true for $n=k$ then it is true for $n=k+1$:

Suppose $\displaystyle\prod_{i=2}^{k}\left(1-\frac{1}{i}\right)=\frac{1}{k}$ for some integer $k\geq 2$. *[In-*

ductive hypothesis] We must show that $\displaystyle\prod_{i=2}^{k+1}\left(1-\frac{1}{i}\right)=$

$\dfrac{1}{k+1}$. But by the laws of algebra and substitution from

the inductive hypothesis, $\displaystyle\prod_{i=2}^{k+1}\left(1-\frac{1}{i}\right)$

$\displaystyle=\prod_{i=2}^{k}\left(1-\frac{1}{i}\right)\cdot\left(1-\frac{1}{k+1}\right)$

$\displaystyle=\left(\frac{1}{k}\right)\cdot\left(1-\frac{1}{k+1}\right)=\left(\frac{1}{k}\right)\cdot\left(\frac{(k+1)-1}{k+1}\right)$

$=\dfrac{1}{k+1}$ *[as was to be shown]*.

3. *General formula:*
$$\frac{1}{1\cdot 3}+\frac{1}{3\cdot 5}+\cdots+\frac{1}{(2n-1)(2n+1)}=\frac{n}{2n+1}$$
for all integers $n\geq 1$.

Proof (by mathematical induction):

The formula holds for $n=1$:

When $n=1$, the left-hand side equals $\dfrac{1}{1\cdot 3}$, and the

right-hand side equals $\dfrac{1}{2\cdot 1+1}$. But both of these

equal $\dfrac{1}{3}$, and so the formula holds for $n=1$.

If the formula holds for $n=k$, then it holds for $n=k+1$:

Suppose that for some integer $k\geq 1$,

$$\frac{1}{1\cdot 3}+\frac{1}{3\cdot 5}+\cdots+\frac{1}{(2k-1)(2k+1)}$$
$$=\frac{k}{2k+1}. \quad \text{Inductive hypothesis}$$

We must show that

$$\frac{1}{1\cdot 3}+\frac{1}{3\cdot 5}+\cdots+\frac{1}{(2(k+1)-1)(2(k+1)+1)}$$
$$=\frac{k+1}{2(k+1)+1},$$

or, equivalently,

$$\frac{1}{1\cdot 3}+\frac{1}{3\cdot 5}+\cdots+\frac{1}{(2k+1)(2k+3)}$$
$$=\frac{k+1}{2k+3}.$$

But

$\dfrac{1}{1\cdot 3}+\dfrac{1}{3\cdot 5}+\cdots+\dfrac{1}{(2k+1)(2k+3)}$

$=\dfrac{1}{1\cdot 3}+\dfrac{1}{3\cdot 5}+\cdots+\dfrac{1}{(2k-1)(2k+1)}$

$\qquad\qquad +\dfrac{1}{(2k+1)(2k+3)}$

$=\dfrac{k}{2k+1}+\dfrac{1}{(2k+1)(2k+3)}$

$\qquad\qquad\qquad$ by inductive hypothesis

$=\dfrac{k(2k+3)}{(2k+1)(2k+3)}+\dfrac{1}{(2k+1)(2k+3)}$

$=\dfrac{2k^2+3k+1}{(2k+1)(2k+3)}$

$=\dfrac{(2k+1)(k+1)}{(2k+1)(2k+3)}$

$=\dfrac{k+1}{2k+3}$ by algebra.

[This is what was to be shown.]

4. *Hint 1:* General formula is

$1-4+9-16+\cdots+(-1)^{n-1}n^2$

$=(-1)^{n-1}(1+2+3+\cdots+n)$ in expanded form

$\displaystyle\sum_{i=1}^{n}(-1)^{i-1}\cdot i^2=(-1)^{n-1}\cdot\left(\sum_{i=1}^{n}i\right)$ in closed form.

Hint 2: In the proof, use the fact that

$$1+2+3+\cdots+n=\sum_{i=1}^{n}i=\frac{n(n+1)}{2}.$$

6. a. $P(1)$ is "4^1-1 is divisible by 3." $P(1)$ is true because $4^1-1=3$, which is divisible by 3.

b. $P(k)$ is "4^k-1 is divisible by 3."

c. $P(k+1)$ is "$4^{k+1}-1$ is divisible by 3."

d. *Must show:* If for some integer $k\geq 1$, 4^k-1 is divisible by 3, then $4^{k+1}-1$ is divisible by 3.

8. *Proof (by mathematical induction):*

The property is true for $n = 1$:

To prove the divisibility property for $n = 1$, we must show that $4^1 - 1$ is divisible by 3. But $4^1 - 1 = 3$, which is certainly divisible by 3. So the property is true for $n = 1$.

If the property is true for $n = k$ then it is true for $n = k + 1$:

Suppose

$4^k - 1$ is divisible by 3 for an integer $k \geq 1$.

Inductive hypothesis

We must show that $4^{k+1} - 1$ is divisible by 3.
But

$$4^{k+1} - 1 = 4 \cdot 4^k - 1 = (3 + 1) \cdot 4^k - 1$$
$$= 3 \cdot 4^k + (4^k - 1).$$

By inductive hypothesis, $4^k - 1$ is divisible by 3, and so $3 \cdot 4^k + (4^k - 1)$ is a sum of numbers each divisible by 3. Hence $3 \cdot 4^k + (4^k - 1)$ is divisible by 3, and so $4^{k+1} - 1$ is divisible by 3 *[as was to be shown].*
[Since we have proved the basis step and the inductive step, we conclude that the given property is true for all integers $n \geq 1$.]

Alternate proof of inductive step: Suppose that for some integer $k \geq 1$, $4^k - 1$ is divisible by 3. Then $4^k - 1 = 3r$ for some integer r. Hence $4^k = 3r + 1$, and so $4^{k+1} - 1 = 4^k \cdot 4 - 1 = (3r + 1) \cdot 4 - 1 = 12r + 4 - 1 = 12r + 3 = 3(4r + 1)$. Since $4r + 1$ is an integer, it follows by definition of divisibility that $4^{k+1} - 1$ is divisible by 3.

13. *Hint:* $x^{k+1} - y^{k+1} = x^{k+1} - x \cdot y^k + x \cdot y^k - y^{k+1}$
$= x \cdot (x^k - y^k) + y^k \cdot (x - y)$

14. *Hint 1:*

$(k + 1)^3 - (k + 1) = k^3 + 3k^2 + 3k + 1 - k - 1$
$= (k^3 - k) + 3k^2 + 3k$
$= (k^3 - k) + 3k(k + 1)$

Hint 2: $k(k + 1)$ is a product of two consecutive integers. By Theorem 3.4.3 one of these must be even.

16. *Proof:*

The inequality is true for $n = 5$:

To prove the inequality for $n = 5$, we must show that $5^2 < 2^5$. But $5^2 = 25$ and $2^5 = 32$. Hence the inequality is true for $n = 5$.

If the inequality is true for $n = k$ then it is true for $n = k + 1$:

Suppose

$k^2 < 2^k$, for an integer $k \geq 5$.

Inductive hypothesis

We must show that $(k + 1)^2 < 2^{k+1}$.
But

$(k + 1)^2 = k^2 + 2k + 1 < 2^k + 2k + 1$ by inductive hypothesis

Also, by Proposition 4.3.2,

$2k + 1 < 2^k$ Prop. 4.2.4 applies since $k \geq 5 \geq 3$.

Putting these inequalities together gives

$$(k + 1)^2 < 2^k + 2k + 1 < 2^k + 2^k = 2^{k+1}.$$

[This is what was to be shown.]
[Since we have proved the basis step and the inductive step, we conclude that the given inequality is true for all integers $n \geq 5.$]

21. The formula $a_n = 3 \cdot 7^{n-1}$ holds for all integers $n \geq 1$.
Proof (by mathematical induction):

The formula holds for $n = 1$:

When $n = 1$, $a_n = a_1 = 3$ (by definition of a_1, a_2, a_3, . . .) and
$3 \cdot 7^{n-1} = 3 \cdot 7^{1-1} = 3 \cdot 7^0 = 3 \cdot 1 = 3$ also. Hence the formula holds for $n = 1$.

If the formula holds for $n = k$, then it holds for $n = k + 1$:

Suppose that for some integer $k \geq 1$, $a_k = 3 \cdot 7^{k-1}$. We must show that $a_{k+1} = 3 \cdot 7^{(k+1)-1}$, or, equivalently, $a_{k+1} = 3 \cdot 7^k$. But

$a_{k+1} = 7 \cdot a_{(k+1)-1}$ by definition of a_1, a_2, a_3, . . .
$= 7 \cdot a_k$
$= 7 \cdot (3 \cdot 7^{k-1})$ by inductive hypothesis
$= 3 \cdot 7^k$ by the laws of exponents.

[This is what was to be shown.]

27. The inductive step fails for going from $n = 1$ to $n = 2$. For when $k = 1$,

$$A = \{a_1, a_2\} \quad \text{and} \quad B = \{a_1\},$$

and no set C can be defined to have the properties claimed for the C in the proof. The reason is that if $C = \{a_1\}$, then B and C share a common element; but one element of A, a_2, is not in either B or C. On the other hand, if $C = \{a_2\}$, then B and C have no common element.

Since the inductive step fails for going from $n = 1$ to $n = 2$, the truth of the following statement is never proved: "All the numbers in a set of two numbers are equal to each other." This breaks the sequence of inductive steps, and so none of the statements for $n > 2$ is proved true either.

Here is an explanation for what happens in terms of the domino analogy. The first domino *is* tipped backward (the basis step is proved). Also, if any domino from the second onward tips backward, then it tips the one behind it backward (the inductive step works for

$n \geq 2$). However, when the first domino is tipped backward, it does *not* tip the second one backward. So only the first domino falls down; the rest remain standing.

28. The basis step is not proved, and in fact it is false because for $n = 1$, $3^n - 2 = 3^1 - 2 = 1$, which is odd.

31. *Hint:* Use proof by contradiction. If the statement is false, then there exists some ordering of the integers from 1 to 30, say $x_1, x_2, \ldots, x_{30}$, such that $x_1 + x_2 + x_3 < 45$, $x_2 + x_3 + x_4 < 45$, $\ldots$, and $x_{30} + x_1 + x_2 < 45$. Evaluate the sum of all these inequalities using the fact that $\sum_{i=1}^{30} x_i = \sum_{i=1}^{30} i$ and Theorem 4.2.2

SECTION 4.4

1. The property "a_n is odd" holds for all integers $n \geq 1$.
   Proof (by strong mathematical induction):

   The property is true for $n = 1$ and $n = 2$:

   Observe that $a_1 = 1$ and $a_2 = 3$ and both 1 and 3 are odd.

   If $k > 2$ and the property is true for all i with $1 \leq i < k$, then it is true for k:

   Let $k > 2$ be an integer and suppose a_i is odd for all integers i with $1 \leq i < k$. *[This is the inductive hypothesis.]* We must show that a_k is odd. But by definition of $a_1, a_2, a_3, \ldots, a_k = a_{k-2} + 2a_{k-1}$. Now a_{k-2} is odd by inductive hypothesis *[since $1 \leq k - 2 < k$ because $k > 2$]* and $2a_{k-1}$ is even *[by definition of even, because $a_k - 1$ is an integer]*. Thus $a_k = a_{k-2} + 2a_{k-1}$ is the sum of an odd integer and an even integer and hence is odd *[exercise 18, Section 3.1]*. This is what was to be shown. *[Since we have proved both the basis step and the inductive step, we conclude that a_n is odd for all integers $n \geq 1$.]*

4. *Proof (by strong mathematical induction):*

   The inequality holds for $n = 1$ and $n = 2$:

   Observe that $d_1 = \frac{9}{10}$ and $d_2 = \frac{10}{11}$ and both $\frac{9}{10} \leq 1$ and $\frac{10}{11} \leq 1$.

   If $k > 2$ and the inequality holds for all i with $1 \leq i < k$, then it holds for k:

   Let $k > 2$ be an integer and suppose $d_i \leq 1$ for all integers i with $1 \leq i < k$. *[This is the inductive hypothesis.]* We must show that $d_k \leq 1$. But by definition of $d_1, d_2, d_3, \ldots, d_k = d_{k-1} \cdot d_{k-2}$. Now $d_{k-1} \leq 1$ and $d_{k-2} \leq 1$ by inductive hypothesis *[since $1 \leq k - 1 < k$ and $1 \leq k - 2 < k$ because $k > 2$]*. Consequently, $d_k = d_{k-1} \cdot d_{k-2} \leq 1$ because if two positive numbers are each less than or equal to 1 then their product is less than or equal to 1. *[If $0 < a \leq 1$ and $0 < b \leq 1$, then multiplying $a \leq 1$ by b gives $ab \leq b$, and since $b \leq 1$, then*

by transitivity of order, $ab \leq 1$.] This is what was to be shown. *[Since we have proved both the basis step and the inductive step, we conclude that $d_n \leq 1$ for all integers $n \leq 1$.]*

7. *Proof (by strong mathematical induction):*

   The formula holds for $n = 0$ and $n = 1$:

   We must show that $g_0 = 5 \cdot 3^0 + 7 \cdot 2^0$ and $g_1 = 5 \cdot 3^1 + 7 \cdot 2^1$. The left-hand side of the first equation is 12 (by definition of $g_0, g_1, g_2, \ldots$), and its right-hand side is $5 \cdot 1 + 7 \cdot 1 = 12$ also. The left-hand side of the second equation is 29 (by definition of $g_0, g_1, g_2, \ldots$), and its right-hand side is $5 \cdot 3 + 7 \cdot 2 = 29$ also. So the formula holds for $n = 0$ and $n = 1$.

   If the formula holds for all i with $0 \leq i < k$, then it holds for k:

   Let $k \geq 2$ be an integer and suppose $g_i = 5 \cdot 3^i + 7 \cdot 2^i$ for all integers i with $0 \leq i < k$. *[Inductive hypothesis]* We must show that $5 \cdot 3^k + 7 \cdot 2^k$. But

   $g_k = 5g_{k-1} - 6g_{k-2}$ by definition of $g_0, g_1, g_2, \ldots$

   $\quad = 5(5 \cdot 3^{k-1} + 7 \cdot 2^{k-1})$
   $\qquad - 6(5 \cdot 3^{k-2} + 7 \cdot 2^{k-2})$ by inductive hypothesis

   $\quad = 25 \cdot 3^{k-1} + 35 \cdot 2^{k-1} - 30 \cdot 3^{k-2} - 42 \cdot 2^{k-2}$

   $\quad = 25 \cdot 3^{k-1} + 35 \cdot 2^{k-1} - 10 \cdot 3 \cdot 3^{k-2}$
   $\qquad - 21 \cdot 2 \cdot 2^{k-2}$

   $\quad = 25 \cdot 3^{k-1} + 35 \cdot 2^{k-1} - 10 \cdot 3^{k-1} - 21 \cdot 2^{k-1}$

   $\quad = (25 - 10) \cdot 3^{k-1} + (35 - 21) \cdot 2^{k-1}$

   $\quad = 15 \cdot 3^{k-1} + 14 \cdot 2^{k-1}$

   $\quad = 5 \cdot 3 \cdot 3^{k-1} + 7 \cdot 2 \cdot 2^{k-1}$

   $\quad = 5 \cdot 3^k + 7 \cdot 2^k$ by algebra.

   [This is what was to be shown.]

9. *Hint:* The idea of the proof is very similar to that of Example 4.3.3. Show that the following sequence is true for all integers $n \geq 1$: "$n - 1$ steps are needed to fit together the pieces of a jigsaw puzzle with n pieces."

11. *Proof:* Let S be the set of all integers r such that $n = 2^i \cdot r$ for some integer i. Then $n \in S$ because $n = 2^0 \cdot n$, and so $S \neq \varnothing$. Also since $n \geq 1$, each r in S is positive, and so by the well-ordering principle, S has a least element m. This means that $n = 2^k \cdot m$ (*) for some nonnegative integer k and $m \leq r$ for every r in S. We claim that m is odd. The reason is that if m were even, then $m = 2p$ for some integer p. Substituting into equation (*) gives

$$n = 2^k \cdot m = 2^k \cdot (2p) = (2^k \cdot 2)p = 2^{k+1} \cdot p.$$

It follows that $p \in S$ and $p \leq m$, which contradicts the fact that m is the *least* element of S. Hence m is odd, and so $n = m \cdot 2^k$ for some odd integer m and nonnegative integer k.

16. *Hint*: In the inductive step, divide into cases depending upon whether k can be written as $k = 3x$ or $k = 3x + 1$ or $k = 3x + 2$ for some integer x.

17. *Hint*: In the inductive step, you let an integer $k \geq 0$ be given and suppose that there exist integers q' and r' such that $k = dq' + r'$ and $0 \leq r' < d$. You must show that there exist integers q and r such that

$$k + 1 = dq + r \text{ and } 0 \leq r < d.$$

To do this, you need to consider the two cases $r' < d - 1$ and $r' = d - 1$.

18. *Proof*: Let n be any integer greater than 1. Consider the set S of all positive integers other than 1 that divide n. Since $n \mid n$ and $n > 1$, there is at least one element in S. Hence by the well-ordering principle, S has a smallest element; call it p. We claim that p is prime. For suppose p is not prime. Then there are integers a and b with $1 < a < p$, $1 < b < p$, and $p = a \cdot b$. By definition of divides, $a \mid p$. Also $p \mid n$ because p is in S and every element in S divides n. Therefore, $a \mid p$ and $p \mid n$, and so by transitivity of divisibility, $a \mid n$. Consequently, $a \in S$. But this is a contradiction because $a < p$ and p is the smallest element of S. *[This contradiction shows that the supposition that p is not prime is false.]* Hence p is prime, and we have shown the existence of a prime number that divides n.

19. *Hint*: Given a predicate $P(n)$ that satisfies conditions (1) and (2) of the principle of mathematical induction, let S be the set of all integers greater than or equal to a for which $P(n)$ is false. Suppose that S has one or more elements and use the well-ordering principle to derive a contradiction.

SECTION 4.5

1. *Proof*: Suppose the condition $m + n = 100$ is true before entry to the loop. Then

$$m_{\text{old}} + n_{\text{old}} = 100.$$

After execution of the loop,

$$m_{\text{new}} = m_{\text{old}} + 1 \quad \text{and} \quad n_{\text{new}} = n_{\text{old}} - 1.$$

So

$$m_{\text{new}} + n_{\text{new}} = (m_{\text{old}} + 1) + (n_{\text{old}} - 1)$$
$$= m_{\text{old}} + n_{\text{old}} = 100.$$

3. *Proof*: Suppose the condition $m^3 > n^2$ is true before entry to the loop. Then

$$m_{\text{old}}^3 > n_{\text{old}}^2.$$

After execution of the loop,

$$m_{\text{new}} = 3 \cdot m_{\text{old}} \quad \text{and} \quad n_{\text{new}} = 5 \cdot n_{\text{old}}.$$

So

$$m_{\text{new}}^3 = (3 \cdot m_{\text{old}})^3 = 27 \cdot m_{\text{old}}^3 > 27 \cdot n_{\text{old}}^2.$$

But since $n_{\text{new}} = 5 \cdot n_{\text{old}}$, then $n_{\text{old}} = \frac{1}{5} n_{\text{new}}$. Hence

$$m_{\text{new}}^3 > 27 \cdot n_{\text{old}}^2 = 27 \cdot \left(\frac{1}{5} n_{\text{new}}\right)^2 = 27 \cdot \frac{1}{25} n_{\text{new}}^2$$
$$= \frac{27}{25} \cdot n_{\text{new}}^2 > n_{\text{new}}^2.$$

6. *Proof*: *[The wording of this proof is almost the same as that of Example 4.5.2.]*

I. Basis Property *[I(0) is true before the first iteration of the loop.]*
$I(0)$ is "$exp = x^0$ and $i = 0$." According to the pre-condition, before the first iteration of the loop $exp = 1$ and $i = 0$. Since $x^0 = 1$, $I(0)$ is evidently true.

II. Inductive Property *[If $G \wedge I(k)$ is true before a loop iteration (where $k \geq 0$), then $I(k + 1)$ is true after the loop iteration.]*
Suppose k is a nonnegative integer such that $G \wedge I(k)$ is true before an iteration of the loop. Then as execution reaches the top of the loop, $i \neq m$, $exp = x^k$, and $i = k$. Since $i \neq m$, the guard is passed and statement 1 is executed. Now before execution of statement 1,

$$exp_{\text{old}} = x^k.$$

So execution of statement 1 has the following effect:

$$exp_{\text{new}} = exp_{\text{old}} \cdot x = x^k \cdot x = x^{k+1}.$$

Similarly, before statement 2 is executed,

$$i_{\text{old}} = k.$$

So after execution of statement 2,

$$i_{\text{new}} = i_{\text{old}} + 1 = k + 1.$$

Hence after the loop iteration, the statement $I(k + 1)$ $exp = x^{k+1}$ and $i = k + 1$ is true. This is what we needed to show.

III. Eventual Falsity of Guard *[After a finite number of iterations of the loop, G becomes false.]*
The guard G is the condition $i \neq m$, and m is a nonnegative integer. By I and II, it is known that

for *all* integers $n \geq 0$, if the loop is iterated n times, then $exp = x^n$ and $i = n$.

So after m iterations of the loop, $i = m$. Thus G becomes false after m iterations of the loop.

IV. Correctness of the Post-Condition *[If N is the least number of iterations after which G is false and $I(N)$ is true, then the value of the algorithm variables will be as specified in the post-condition of the loop.]*

According to the post-condition, the value of *exp* after execution of the loop should be x^m. But when G is false, then $i = m$. And when $I(N)$ is true, then $i = N$ and $exp = x^N$. Since *both* conditions (G false and $I(N)$ true) are satisfied, $m = i = N$ and $exp = x^m$ as required.

8. *Proof:*

I. Basis Property: $I(0)$ is "$i = 1$ and *sum* $= A[1]$." According to the pre-condition this statement is true.

II. Inductive Property: Suppose k is a nonnegative integer such that $G \wedge I(k)$ is true before an iteration of the loop. Then as execution reaches the top of the loop, $i \neq m$, $i = k + 1$, and *sum* $= A[1] + A[2] + \cdots + A[k + 1]$. Since $i \neq m$, the guard is passed and statement 1 is executed. Now before execution of statement 1, $i_{\text{old}} = k + 1$. So after execution of statement 1, $i_{\text{new}} = i_{\text{old}} + 1 = (k + 1) + 1 = k + 2$. Also before statement 2 is executed, $sum_{\text{old}} = A[1] + A[2] + \cdots + A[k + 1]$. Execution of statement 2 adds $A[k + 2]$ to this sum, and so after statement 2 is executed $sum_{\text{new}} = A[1] + A[2] + \cdots + A[k + 1] + A[k + 2]$. Thus after the loop iteration $I(k + 1)$ is true.

III. Eventual Falsity of Guard: The guard G is the condition $i \neq m$. By I and II, it is known that for all integers $n \geq 1$, after n iterations of the loop $I(n)$ is true. Hence after $m - 1$ iterations of the loop $I(m)$ is true, which implies that $i = m$ and G is false.

IV. Correctness of the Post-Condition: Suppose that N is the least number of iterations after which G is false and $I(N)$ is true. Then (since G is false) $i = m$ and (since $I(N)$ is true) $i = N + 1$ and *sum* $= A[1] + A[2] + \cdots + A[N + 1]$. Putting these together gives $m = N + 1$ and so *sum* $= A[1] + A[2] + \cdots + A[m]$, which is the post-condition.

10. *Hint:* Assume $G \wedge I(k)$ is true for a nonnegative integer k. Then $a_{\text{old}} \neq 0$ and $b_{\text{old}} \neq 0$ and

(1) a_{old} and b_{old} are nonnegative integers with $\gcd(a_{\text{old}}, b_{\text{old}}) = \gcd(A, B)$,

(2) at most one of a_{old} and b_{old} equals 0,

(3) $0 \leq a_{\text{old}} + b_{\text{old}} \leq A + B - k$.

It must be shown that $I(k + 1)$ is true after the loop iteration. That means it is necessary to show

(1) a_{new} and b_{new} are nonnegative integers with $\gcd(a_{\text{new}}, b_{\text{new}}) = \gcd(A, B)$,

(2) at most one of a_{new} and b_{new} equals 0,

(3) $0 \leq a_{\text{new}} + b_{\text{new}} \leq A + B - (k + 1)$.

To show (3) observe that

$$a_{\text{new}} + b_{\text{new}} = \begin{cases} a_{\text{old}} - b_{\text{old}} + b_{\text{old}} & \text{if } a_{\text{old}} \geq b_{\text{old}} \\ b_{\text{old}} - a_{\text{old}} + a_{\text{old}} & \text{if } a_{\text{old}} < b_{\text{old}} \end{cases}$$

[The reason for this is that when $a_{\text{old}} \geq b_{\text{old}}$, then $a_{\text{new}} = a_{\text{old}} - b_{\text{old}}$ and $b_{\text{new}} = b_{\text{old}}$ and when $a_{\text{old}} < b_{\text{old}}$, then $b_{\text{new}} = b_{\text{old}} - a_{\text{old}}$ and $a_{\text{new}} = a_{\text{old}}.]$ Thus

$$a_{\text{new}} + b_{\text{new}} = \begin{cases} a_{\text{old}} & \text{if } a_{\text{old}} \geq b_{\text{old}} \\ b_{\text{old}} & \text{if } a_{\text{old}} < b_{\text{old}} \end{cases}$$

But since $a_{\text{old}} \neq 0$ and $b_{\text{old}} \neq 0$ and a_{old} and b_{old} are nonnegative integers, then $a_{\text{old}} \geq 1$ and $b_{\text{old}} \geq 1$. Hence $a_{\text{old}} - 1 \geq 0$ and $b_{\text{old}} - 1 \geq 0$ and $a_{\text{old}} \leq a_{\text{old}} + b_{\text{old}} - 1$ and $b_{\text{old}} \leq b_{\text{old}} + a_{\text{old}} - 1$. It follows that $a_{\text{new}} + b_{\text{new}} \leq a_{\text{old}} + b_{\text{old}} - 1 \leq (A + B - k) - 1$ by the truth of (3) going into the kth iteration. Hence $a_{\text{new}} + b_{\text{new}} < A + B - (k + 1)$ by algebraic simplification.

SECTION 5.1

1. set (a) = set (c) and set (b) = set (d)

4. a. $S = \{1, -1\}$

5. a. No. $j \in B$ and $j \notin A$.
   d. Yes. Both elements of C are in A, but A contains elements (namely c and f) that are not in C.

6. a. Yes   b. No   f. No   i. Yes

7. c. $A - B = \{d, f, g\}$

8. a. $A \cup B = \{x \in \mathbf{R} \mid 0 < x < 4\}$
   b. $A \cap B = \{x \in \mathbf{R} \mid 1 \leq x \leq 2\}$
   c. $A^c = \{x \in \mathbf{R} \mid x \leq 0 \text{ or } x > 2\}$
   d. $B^c = \{x \in \mathbf{R} \mid x < 1 \text{ or } x \geq 4\}$
   e. $A^c \cap B^c = \{x \in \mathbf{R} \mid x \leq 0 \text{ or } x \geq 4\}$
   f. $A^c \cup B^c = \{x \in \mathbf{R} \mid x < 1 \text{ or } x > 2\}$
   g. $(A \cap B)^c = \{x \in \mathbf{R} \mid x < 1 \text{ or } x > 2\}$
   h. $(A \cup B)^c = \{x \in \mathbf{R} \mid x \leq 0 \text{ or } x \geq 4\}$

10. b. False. Many negative real numbers are not rational. For example, $-\sqrt{2} \in \mathbf{R}$ but $-\sqrt{2} \notin \mathbf{Q}$.
    d. False. $0 \in \mathbf{Z}$ but $0 \notin \mathbf{Z}^- \cup \mathbf{Z}^+$.

11. a. *Negation:* $\exists$ a set A such that $A \subseteq \mathbf{R}$ and $A \nsubseteq \mathbf{Z}$. The negation is true. For example, let

    $$A = \text{the open interval from 0 to 2}$$
    $$= \{x \in \mathbf{R} \mid 0 < x < 2\}.$$

    Then $A \subseteq \mathbf{R}$ but $A \nsubseteq \mathbf{Z}$ (since $\frac{1}{2} \in A$).

12. a. No. $1 \in A$ since $a = 2 \cdot 1 - 1$. But $1 \notin B$. For if 1 were an element of B, then $1 = 3j + 2$, for some integer j. Solving for j would give

    $$3j + 2 = 1$$
    $$3j = 1 - 2 = -1$$
    $$j = \frac{-1}{3}.$$

So if 1 were an element of B, then there would be an integer j such that $j = -1/3$. But $-1/3$ is not an integer, so $1 \notin B$. Since there is an element in A that is not in B, $A \ne B$.

b. Yes. $A = C$.

Proof:

$A \subseteq C$: Let $m \in A$. *[We must show that $m \in C$.]* By definition of A, $m = 2i - 1$, for some integer i. Let $r = i - 1$. Then r is an integer and $i = r + 1$. By substitution, $m = 2i - 1 = 2(r + 1) - 1 = 2r + 2 - 1 = 2r + 1$. Hence $m \in C$ by definition of C *[as was to be shown]*.

$C \subseteq A$: Let $p \in C$. *[We must show that $p \in A$.]* By definition of C, $p = 2r + 1$, for some integer r. Let $i = r + 1$. Then i is an integer and $r = i - 1$. By substitution, $p = 2r + 1 = 2(i - 1) + 1 = 2i - 2 + 1 = 2i - 1$. Hence $p \in A$ by definition of A *[as was to be shown]*. Since $A \subseteq C$ and $C \subseteq A$, then $A = C$ by definition of set equality.

Note that A and C are alternate representations for the set of odd integers.

13. a. No, $R \not\subseteq T$. $2 \in R$ but $2 \notin T$.

    b. Yes, $T \subseteq R$. Every integer y that is divisible by 6 is also divisible by 2.

14. a. $A \cup (B \cap C) = \{a, b, c\}$, $(A \cup B) \cap C = \{b, c\}$, and $(A \cup B) \cap (A \cup C) = \{a, b, c, d\} \cap \{a, b, c, e\} = \{a, b, c\}$.

Hence $A \cup (B \cap C) = (A \cup B) \cap (A \cup C)$.

15. a.

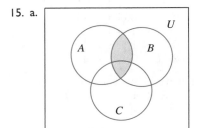

16. a.

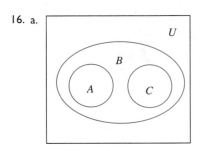

17. a. $A \times B = \{(x, a), (y, a), (z, a), (w, a), (x, b), (y, b), (z, b), (w, b)\}$

18. a. $A \times (B \times C) = \{(1, (u, m)), (2, (u, m)), (3, (u, m)), (1, (u, n)), (2, (u, n)), (3, (u, n)), (1, (v, m)), (2, (v, m)), (3, (v, m)), (1, (v, n)), (2, (v, n)), (3, (v, n))\}$

19. a. $L_1 = \{\varepsilon, x, y, xx, yy, xxx, xyx, yxy, yyy, xxxx, xyyx, yxxy, yyyy\}$

    c. $L_3 = \{\varepsilon, x, y, xx, xy, yy, xxx, xxy, xyy, yyy\}$

21.

i	1			2				3		4
j		1	2	3	1	2	3	4	1	2
found		no	yes		no			yes	no	yes
answer	$A \subseteq B$									

SECTION 5.2

1. a. (1) A (2) $B \cup C$

    b. (1) $A \cap B$ (2) C

2. a. (1) $A - B$ (2) A (3) A (4) B

    b. (1) $X \in A$ (2) A (3) B (4) A

3. a. A b. C c. B d. C e. $B \subseteq C$

5. *Proof*: Suppose A and B are sets.

$B - A \subseteq B \cap A^c$: Suppose $x \in B - A$. By definition of set difference, $x \in B$ and $x \notin A$. But then by definition of complement $x \in B$ and $x \in A^c$, and so by definition of intersection $x \in B \cap A^c$. *[Thus $B - A \subseteq B \cap A^c$ by definition of subset.]*

$B \cap A^c \subseteq B - A$: Suppose $x \in B \cap A^c$. By definition of intersection, $x \in B$ and $x \in A^c$. But then by definition of complement $x \in B$ and $x \notin A$, and so by definition of set difference $x \in B - A$. *[Thus $B \cap A^c \subseteq B - A$ by definition of subset.]*

[Since both set containments have been proved, $B - A = B \cap A^c$ by definition of set equality.]

6. (1) a. $(A \cap B) \cup (A \cap C)$ b. A c. $B \cup C$ d. $x \in C$.   e. $A \cap B$ f. by definition of intersection $x \in A \cap C$, and so by definition of union $x \in (A \cap B) \cup (A \cap C)$.

    (2) a. $A \cap (B \cup C)$ b. $A \cap B$ c. $A \cap C$ d. $x \in A$ e. $x \in B$ f. $A \cap (B \cup C)$ g. by definition of intersection $x \in A$ and $x \in C$. Since $x \in C$, by definition of union $x \in B \cup C$. Hence $x \in A$ and $x \in B \cup C$, and so by definition of intersection $x \in A \cap (B \cup C)$.

    (3) $A \cap (B \cup C) = (A \cap B) \cup (A \cap C)$

7. *Hint*: This is somewhat similar to the proof in Example 5.2.3.

8. False. One counterexample is given in exercise 14 of Section 5.1. (See the solution to 14b.) Here is another: Let $A = \{1, 3\}$, $B = \{2, 3\}$, and $C = \{4\}$. Then $(A \cap B) \cup C = \{3\} \cup \{4\} = \{3, 4\}$ whereas $A \cap (B \cup C) = \{1, 3\} \cap \{2, 3, 4\} = \{3\}$.

9. *Hint:* The statement is true. The essence of the proof is that if an element is in $A - B$ or in $A \cap B$, then it is in A and not B or it is in both A and B. In either case it is in A. Conversely, any element in A is either in B or not in B. If the former, then the element is in $A \cap B$; if the latter, it is in $A - B$.

10. *Hint:* The statement is false. Consider sets A, B, and C where A and C have common elements that are not in B.

15. This is false. *Counterexample:* Let $A = \{1, 2\}$, $B = \{2, 3\}$, and $C = \{2, 4\}$. Then $A \cap C = \{2\}$ and $B \cap C = \{2\}$, so $A \cap B = A \cap C$. But $A \neq B$.

23. (Optional) *Starting Point:* Suppose A, B, and C are arbitrarily chosen sets.

(Optional) *To Show:* $A \times (B \cup C) = (A \times B) \cup (A \times C)$ (That is, $A \times (B \cup C) \subseteq (A \times B) \cup (A \times C)$ and $(A \times B) \cup (A \times C) \subseteq A \times (B \cup C)$. That is, $\forall x$, if $x \in A \times (B \cup C)$ then $x \in (A \times B) \cup (A \times C)$ and $\forall x$, if $x \in (A \times B) \cup (A \times C)$ then $x \in A \times (B \cup C)$.)

Partial Proof: Suppose A, B, and C are arbitrarily chosen sets.

$A \times (B \cup C) \subseteq (A \times B) \cup (A \times C)$:
Suppose $(x, y) \in A \times (B \cup C)$. *[We must show that $(x, y) \in (A \times B) \cup (A \times C)$.]* Then $x \in A$ and $y \in B \cup C$. By definition of union, this means that $y \in B$ or $y \in C$.

Case 1 ($y \in B$): Then since $x \in A$, $(x, y) \in A \times B$ by definition of Cartesian product. Hence $(x, y) \in (A \times B) \cup (A \times C)$ by the inclusion in union property.

Case 2 ($y \in C$): Then since $x \in A$, $(x, y) \in A \times C$ by definition of Cartesian product. Hence $(x, y) \in (A \times B) \cup (A \times C)$ by the inclusion in union property.

Hence in either case $(x, y) \in (A \times B) \cup (A \times C)$ *[as was to be shown].*
Thus $A \times (B \cup C) \subseteq (A \times B) \cup (A \times C)$ by definition of subset.

$(A \times B) \cup (A \times C) \subseteq A \times (B \cup C)$: This half of the proof is left to the student.

25. a.

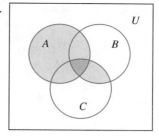

Entire shaded region is $A \cup (B \cap C)$.

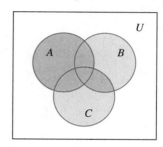

Darkly shaded region is $(A \cup B) \cap (A \cup C)$.

26. *Proof (by mathematical induction):*

The formula holds for $n = 1$:
For $n = 1$ the formula is $A_1 - B = A_1 - B$, which is clearly true.

If the formula holds for $n = k$, then it holds for $n = k + 1$:
Let k be an integer with $k \geq 1$, and suppose A_1, $A_2, \ldots, A_k, A_{k+1}$ and B are any sets such that
$$(A_1 - B) \cup (A_2 - B) \cup \cdots \cup (A_k - B) = (A_1 \cup A_2 \cup \cdots \cup A_k) - B.$$

We must show that
$$(A_1 - B) \cup (A_2 - B) \cup \cdots \cup (A_{k+1} - B) = (A_1 \cup A_2 \cup \cdots \cup A_{k+1}) - B.$$

But
$$(A_1 - B) \cup (A_2 - B) \cup \cdots \cup (A_{k+1} - B)$$
$$= [(A_1 - B) \cup (A_2 - B) \cup \cdots (A_k - B)] \cup (A_{k+1} - B) \qquad \text{by assumption}$$
$$= [(A_1 \cup A_2 \cup \cdots \cup A_k) - B] \cup (A_{k+1} - B)$$
$$\text{by the inductive hypothesis}$$
$$= [(A_1 \cup A_2 \cup \cdots \cup A_k) \cup A_{k+1}] - B$$
$$\text{by Example 5.2.5}$$
$$= (A_1 \cup A_2 \cup \cdots \cup A_k \cup A_{k+1}) - B.$$
$$\text{by assumption}$$

28. a. commutative law for $\cap$
b. distributive law
c. commutative law for $\cap$

29. a. alternate representation for set difference law
    b. alternate representation for set difference law
    c. commutative law for $\cap$ d. De Morgan's law
    e. double complement law   f. distributive law
    g. alternate representation for set difference law

30. *Hint:* First use the commutative law for $\cup$, then the distributive law, and then the commutative law for $\cup$.

32. *Proof:* Let A, B, and C be any sets. Then

$(A - B) - (B - C)$

$= (A \cap B^c) \cap (B \cap C^c)^c$ by the alternate representation for set difference law

$= (A \cap B^c) \cap (B^c \cup C)$ by De Morgan's and the double complement laws

$= [(A \cap B^c) \cap B^c] \cup [(A \cap B^c) \cap C]$ by the distributive law

$= [A \cap (B^c \cap B^c)] \cup [A \cap (B^c \cap C)]$ by the associative law for $\cap$

$= [A \cap B^c] \cup [A \cap (B^c \cap C)]$ by the idempotent law for $\cap$

$= [A \cap B^c] \cup [A \cap (C \cap B^c)]$ by the commutative law for $\cap$

$= [A \cap B^c] \cup ((A \cap C) \cap B^c]$ by the associative law for $\cap$

$= [B^c \cap A] \cup [B^c \cap (A \cap C]$ by the commutative law for $\cap$

$= B^c \cap [A \cup (A \cap C)]$ by the distributive law

$= B^c \cap A$ by the absorption law

$= A \cap B^c$ by the commutative law for $\cap$

$= A - B$ by the alternate representation for set difference law.

35. *Proof:* Let A, B, and C be any sets. Then

$((A^c \cup B^c) - A)^c = ((A^c \cup B^c) \cap A^c)^c$ by the alternate representation for set difference law

$= (A^c \cup B^c)^c \cup (A^c)^c$ by De Morgan's law

$= (A^c)^c \cap (B^c)^c \cup (A^c)^c$ by De Morgan's law

$= (A \cap B) \cup A$ by the double complement law

$= A \cup (A \cap B)$ by the commutative law for $\cup$

$= A$ by the absorption law

37. *Hint:* There is more than one error in this "proof." The most serious is the misuse of the definition of subset. To say that A is a subset of B means that for all x, **if** $x \in A$ **then** $x \in B$. It does not mean that there exists an element of A that is also an element of B. There is also a more subtle error in the internal logic of the "proof." Can you spot it?

38. The statement "since $x \notin A$ or $x \notin B$, $x \notin A \cup B$" is fallacious. Try to think of an example of sets A and B and an element x so that the statement "$x \notin A$ or $x \notin B$" is true and the statement "$x \notin A \cup B$" is false.

SECTION 5.3

1. b. No. The left-hand set is the empty set; it does not have any elements. The right-hand set is a set with one element, namely $\varnothing$.

2. *Proof:* Let A be a set. *[We must show that $A \cup \varnothing = A$.]*

    $A \cup \varnothing \subseteq A$: Suppose $x \in A \cup \varnothing$. Then $x \in A$ or $x \in \varnothing$ by definition of union. But $x \notin \varnothing$ since $\varnothing$ has no elements. Hence $x \in A$.

    $A \subseteq A \cup \varnothing$: Suppose $x \in A$. Then the statement "$x \in A$ or $x \in \varnothing$" is true. Hence $x \in A \cup \varnothing$ by definition of union.
    [Alternately. $A \subseteq A \cup \varnothing$ by the inclusion in union property.]

    Since $A \cup \varnothing \subseteq A$ and $A \subseteq A \cup \varnothing$, then $A \cup \varnothing = A$ by definition of set equality.

3. a. $(A - B) \cap (B - A)$ b. intersection   c. $B - A$
    d. B e. A f. A g. $(A - B) \cap (B - A) = \varnothing$

4. a. For any subset A of a universal set U, $A \cap A^c = \varnothing$.

    Proof: Let A be a subset of a universal set U. Suppose $A \cap A^c \neq \varnothing$, that is, suppose there is an element x such that $x \in A \cap A^c$. Then by definition of intersection, $x \in A$ and $x \in A^c$, and so by definition of complement, $x \in A$ and $x \notin A$. This is a contradiction. *[Hence the supposition is false, and we conclude that $A \cap A^c = \varnothing$.]*

    b. For any subset A of a universal set U, $A \cup A^c = U$.

    Proof: Suppose A is a subset of a universal set U.

    $A \cup A^c \subseteq U$: Let $x \in A \cup A^c$. By definition of union, $x \in A$ or $x \in A^c$. But A is a subset of U by hypothesis and A^c is also a subset of U by definition of complement. Thus by definition of subset, $x \in U$ regardless of whether $x \in A$ or $x \in A^c$.

    $U \subseteq A \cup A^c$: Let $x \in U$. It is certainly true that $x \in A$ or $x \notin A$ (this is a tautology). But by definition of complement, if $x \notin A$ then $x \in A^c$. Thus $x \in A$ or $x \in A^c$, and so by definition of union $x \in A \cup A^c$.
    [Since both set containments $U \subseteq A \cup A^c$ and $A \cup A^c \subseteq U$ have been proved, $U = A \cup A^c$ by definition of set equality.]

6.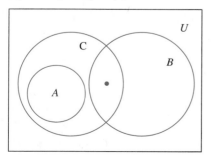

7. This is true. *Illustration:*

    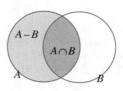

    Proof: Suppose $(A - B) \cap (A \cap B) \neq \varnothing$, that is, suppose there were an element x in $(A - B) \cap (A \cap B)$. Then by definition of intersection, $x \in A - B$ and $x \in A \cap B$. By definition of set difference, then, $x \in A$ and $x \notin B$, and by definition of intersection $x \in A$ and $x \in B$. But then $x \in B$ and $x \notin B$, which is a contradiction. *[Hence the supposition is false and so $(A - B) \cap (A \cap B) = \varnothing$.]*

9. This is true. *Illustration:*

    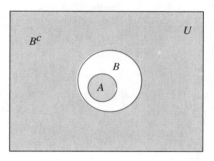

    Proof: Let A and B be sets such that $A \subseteq B$. *[We must show that $A \cap B^c = \varnothing$.]* Suppose $A \cap B^c \neq \varnothing$; that is, suppose there were an element x such that $x \in A \cap B^c$. Then $x \in A$ and $x \in B^c$ by definition of intersection. So $x \in A$ and $x \notin B$ by definition of complement. But $A \subseteq B$ by hypothesis. So since $x \in A$, $x \in B$ by definition of subset. Thus $x \notin B$ and also $x \in B$, which is a contradiction. Hence the supposition that $A \cap B^c \neq \varnothing$ is false, and so $A \cap B^c = \varnothing$.

10. This is false.
    Illustration:

    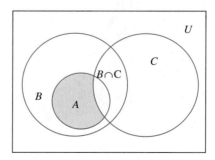

    Counterexample: Let $U = \{1, 2, 3, 4, 5, 6\}$, $A = \{1, 2, 3\}$, $B = \{1, 2, 3, 4\}$, and $C = \{3, 4, 5\}$. Then $B \cap C = \{3, 4\}$, so $(B \cap C)^c = \{1, 2, 5, 6\}$ and $A \cap (B \cap C)^c = \{1, 2, 3\} \cap \{1, 2, 5, 6\} = \{1, 2\} \neq$

∅. Find a different counterexample (simpler, if you can).

20. a. *statement:* ∀ sets S, ∃ a set T such that $S \cap T = \varnothing$.
negation: ∃ a set S such that ∀ sets T, $S \cap T \neq \varnothing$.
The statement is true. Given any set S, take $T = S^c$. Then $S \cap T = S \cap S^c = \varnothing$ by the intersection with the complement property (Theorem 5.3.3(2)(a)). Alternatively, take $T = \varnothing$.

21. *Proof:* Suppose A and B are sets. Then

$A \cup (B - A)$

$= A \cup (B \cap A^c)$ by the alternate representation for set difference law

$= (A \cup B) \cap (A \cup A^c)$ by the distributive law

$= (A \cup B) \cap U$ by the union with the complement law

$= A \cup B$ by the intersection with U law

26. *Hint:* The answer is $\varnothing$.

33. a. *Proof:* Let A and B be any sets. By definition of $\oplus$, showing that $A \oplus B = B \oplus A$ is equivalent to showing that $(A - B) \cup (B - A) = (B - A) \cup (A - B)$. But this follows immediately from the commutative law for $\cup$.

34. a. *Hint:* Yes. Proof is similar to that of Example 5.3.4(b).
b. *Hint:* No. You can find sets A, B, and C such that $(A - B) \cap (C - B) \neq \varnothing$.

35. a. No. The element d is in two of the sets.
d. No. None of the sets contains 6.

36. Yes. Every integer is either even or odd, and no integer is both even and odd.

40. a. $A \cap B = \{2\}$, so $\mathscr{P}(A \cap B) = \{\varnothing, \{2\}\}$.

b. $A = \{1, 2\}$, so $\mathscr{P}(A) = \{\varnothing, \{1\}, \{2\}, \{1, 2\}\}$.

c. $A \cup B = \{1, 2, 3\}$, so $\mathscr{P}(A \cup B) = \{\varnothing, \{1\}, \{2\}, \{3\}, \{1, 2\}, \{1, 3\}, \{2, 3\}, \{1, 2, 3\}\}$.

d. $A \times B = \{(1, 2), (1, 3), (2, 2), (2, 3)\}$, so
$\mathscr{P}(A \times B) = \{\varnothing, \{(1, 2)\}, \{(1, 3)\}, \{(2, 2)\}, \{(2, 3)\}, \{(1, 2), (1, 3)\}, \{(1, 2), (2, 2)\}, \{(1, 2), (2, 3)\}, \{(1, 3), (2, 2)\}, \{(1, 3), (2, 3)\}, \{(2, 2), (2, 3)\}, \{(1, 2), (1, 3), (2, 2)\}, \{(1, 2), (1, 3), (2, 3)\}, \{(1, 2), (2, 2), (2, 3)\}, \{(1, 3), (2, 2), (2, 3)\}, \{(1, 2), (1, 3), (2, 2), (2, 3)\}\}$.

41. a. $\mathscr{P}(A \times B) = \{\varnothing, \{(1, u)\}, \{(1, v)\}, \{(1, u), (1, v)\}\}$

42. b. $\mathscr{P}(\mathscr{P}(\varnothing)) = \mathscr{P}(\{\varnothing\}) = \{\varnothing, \{\varnothing\}\}$

43. a. *Hint:* This is false.

46. a. $S_1 = \{\varnothing, \{t\}, \{u\}, \{v\}, \{t, u\}, \{t, v\}, \{u, v\}, \{t, u, v\}\}$

47. *Hint:* Use mathematical induction. In the inductive step, you will consider the set of all nonempty subsets of $\{2, \ldots, k\}$ and the set of all nonempty subsets of $\{2, \ldots, k + 1\}$. Any subset of $\{2, \ldots, k + 1\}$ either contains $k + 1$ or does not contain $k + 1$. Thus

$$\begin{bmatrix} \text{the sum of all products} \\ \text{of elements of nonempty} \\ \text{subsets of } \{2, \ldots, k + 1\} \end{bmatrix}$$

$$= \begin{bmatrix} \text{the sum of all} \\ \text{products of elements} \\ \text{of nonempty} \\ \text{subsets of} \\ \{2, \ldots, k + 1\} \\ \text{that do not} \\ \text{contain } k + 1 \end{bmatrix} + \begin{bmatrix} \text{the sum of all} \\ \text{products of} \\ \text{elements of} \\ \text{nonempty} \\ \text{subsets of} \\ \{2, \ldots, k + 1\} \\ \text{that contain } k + 1 \end{bmatrix}$$

But any subset of $\{2, \ldots, k + 1\}$ that does not contain $k + 1$ is a subset of $\{2, \ldots, k\}$. And any subset of $\{2, \ldots, k + 1\}$ that contains $k + 1$ is the union of a subset of $\{2, \ldots, k\}$ and $\{k + 1\}$.

48. a. *Proof:* Let x be any element of B.

(i) Then

$x = x \cdot 1$ because 1 is an identity for ·

$= x \cdot (x + \bar{x})$ by the complement law for +

$= x \cdot x + x \cdot \bar{x}$ by the distributive law for · over +

$= x \cdot x + 0$ by the complement law for ·

$= x \cdot x$ because 0 is an identity for +.

(ii) Furthermore,

$x = x + 0$ because 0 is an identity for +

$= x + (x \cdot \bar{x})$ by the complement law for ·

$= (x + x) \cdot (x + \bar{x})$ by the distributive law for + over ·

$= (x + x) \cdot 1$ by the complement law for ·

$= x + x$ because 1 is an identity for ·.

b. *Proof:* Let x and y be any elements of B, and suppose that $x \cdot y = 1$.

(i) Then

$1 = x \cdot y$ by assumption

$= (x \cdot x) \cdot y$ because $x = x \cdot x$ by part (a)

$= x \cdot (x \cdot y)$ by the associative law for ·

$= x \cdot 1$ because $x \cdot y = 1$

$= x$ because 1 is an identity for ·.

(ii) Similarly,

$$1 = x \cdot y \qquad \text{by assumption}$$
$$= x \cdot (y \cdot y) \qquad \text{because } y = y \cdot y \text{ by part (a)}$$
$$= (x \cdot y) \cdot y \qquad \text{by the associative law for } \cdot$$
$$= 1 \cdot y \qquad \text{because } x \cdot y = 1$$
$$= y \cdot 1 \qquad \text{by the commutative law for } \cdot$$
$$= y \qquad \text{because 1 is an identity for } \cdot .$$

SECTION 5.4

1. The sentence is not a statement because it is neither true nor false. If the sentence were true, then because it declares itself to be false, the sentence would be false. Therefore, the sentence is not true. On the other hand, if the sentence were false, then it would be false that "This sentence is false," and so the sentence would be true. Consequently, the sentence is not false.

2. This sentence is a statement because it is true. Recall that the only way for an if–then statement to be false is for the hypothesis to be true and the conclusion false. In this case the hypothesis is not true. So regardless of what the conclusion states, the sentence is true. (This is an example of a statement that is vacuously true or true by default.)

5. This sentence is not a statement because it is neither true nor false. If the sentence were true, then either the

sentence is false or $1 + 1 = 3$. But $1 + 1 \neq 3$, and so the sentence is false. Therefore, the sentence is not true. On the other hand, if the sentence were false, then it would be true that "This sentence is false or $1 + 1 = 3$," and so the sentence would be true. Consequently, the sentence is not false.

7. *Hint:* Suppose that apart from statement (ii), all of Nixon's other assertions about Watergate are evenly split between true and false.

8. No. Suppose there were a computer program P that had as output a list of all computer programs that do not list themselves in their output. If P lists itself as output, then it would be on the output list of P, which consists of all computer programs that do not list themselves in their output. Hence P would not list itself as output. But if P does not list itself as output, then P would be a member of the list of all computer programs that do not list themselves in their output, and this list is exactly the output of P. Hence P would list itself as output. This analysis shows that the assumption of the existence of such a program P is contradictory, and so no such program exists.

12. *Hint:* Show that any algorithm that solves the printing problem can be adapted to produce an algorithm that solves the halting problem.

SECTION 6.1

2. 3/4, 1/2, 1/2

3. a.
$\{2\blacklozenge, 3\blacklozenge, 4\blacklozenge, 5\blacklozenge, 6\blacklozenge, 7\blacklozenge, 8\blacklozenge, 9\blacklozenge, 10\blacklozenge, 2\heartsuit, 3\heartsuit, 4\heartsuit, 5\heartsuit, 6\heartsuit, 7\heartsuit, 8\heartsuit, 9\heartsuit, 10\heartsuit\}$,
probability = 18/52

4. a. $\{26, 35, 44, 53, 62\}$, probability = 5/36

5. a. $\{HHH, HHT, HTH, HTT, THH, THT, TTH, TTT\}$
   b. (i) $\{HTT, THT, TTH\}$, probability = 3/8

6. a. $\{BBB, BBG, BGB, BGG, GBB, GBG, GGB, GGG\}$
   b. (i) $\{BGG, GBG, GGB\}$, probability = 3/8

7. a. $\{CCC, CCW, CWC, CWW, WCC, WCW, WWC, WWW\}$
   b. (i) $\{CWW, WCW, WWC\}$, probability = 3/8

8. a. probability = 3/8

10. a.
10 11 12 13 14 15 16 17 18 . . . 96 97 98 99
$3 \cdot 4 \quad 3 \cdot 5 \quad 3 \cdot 6 \ 3 \cdot 32 \quad 3 \cdot 33$

The above diagram shows that there are as many positive two-digit integers that are multiples of 3 as there are integers from 4 to 33 inclusive. By Theorem 6.1.1, there are $33 - 4 + 1$, or 30, such integers.

b. There are $99 - 10 + 1 = 90$ positive two-digit integers in all, and by part (a) 30 of these are multiples of 3. So the probability that a randomly chosen positive two-digit integer is a multiple of 3 is $30/90 = 1/3$.

12. c. $\dfrac{m - 3 + 1}{n} = \dfrac{m - 2}{n}$

d. $\dfrac{19 - \lfloor 19/2 \rfloor + 1}{n} = \dfrac{19 - 9 + 1}{n} = \dfrac{11}{n}$

13. b. When n is odd, $\lfloor n/2 \rfloor = (n - 1)/2$, and so the probability that a randomly chosen array element is in the subarray $A[1], A[2], \ldots, A[\lfloor n/2 \rfloor]$ is

$$\dfrac{\lfloor \frac{n}{2} \rfloor - 1 + 1}{n} = \dfrac{\frac{n-1}{2}}{n} = \dfrac{n - 1}{2n}.$$

14. a. When n is even, $\lfloor n/2 \rfloor = n/2$, and so the probability that a randomly chosen array element is in the subarray $A[\lfloor n/2 \rfloor], A[\lfloor n/2 \rfloor] + 1], \ldots, A[n]$ is

$$\dfrac{n - \lfloor \frac{n}{2} \rfloor + 1}{n} = \dfrac{n - \frac{n}{2} + 1}{n} = \dfrac{\frac{n}{2} + 1}{n} = \dfrac{n + 2}{2n} = \dfrac{1}{2} + \dfrac{1}{n}$$

15. Let k be the 27th element in the array. By Theorem 6.1.1, $k - 42 + 1 = 27$, and so $k = 42 + 27 - 1 = 68$. Thus the 27th element in the array is $A[68]$.

17. Let m be the smallest of the integers. By Theorem 6.1.1, $279 - m + 1 = 56$, and so $m = 279 - 56 + 1 = 224$. Thus the smallest of the integers is 224.

20. 1 2 3 4 5 6 7 8 9 ... 999 1000 1001
     ↕   ↕   ↕   ↕
   3 · 1   3 · 2   3 · 3 3 · 333

So there are 333 multiples of 3 between 1 and 1001.

21. a. M Tu W Th F Sa Su M Tu W Th F Sa Su ... F Sa Su M
    1 2 3 4 5 6 7 8 9 10 11 12 13 14   362 363 364 365
                 ↕   ↕   ↕
             7 · 1   7 · 2   7 · 52

Sundays occur on the seventh day of the year, the fourteenth day of the year, and in fact on all days that are multiples of 7. There are 52 multiples of 7 between 1 and 365, and so there are 52 Sundays in the year.

SECTION 6.2

1.

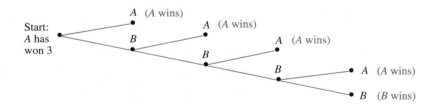

game 4   game 5   game 6   game 7

There are five ways to complete the series.

3. four ways: A–A–A–A, B–A–A–A–A, B–B–A–A–A–A, B–B–B–A–A–A–A

4. two ways: A–B–A–B–A–B–A and B–A–B–A–B–A–B

6. a.

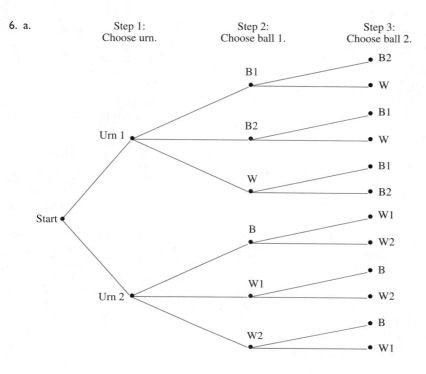

Step 1:
Choose urn.

Step 2:
Choose ball 1.

Step 3:
Choose ball 2.

b. 12   **c.** $2/12 = 1/6$ **d.** $8/12 = 2/3$

8. By the multiplication rule, the answer is $3 \cdot 2 \cdot 2 = 12$.

9. a. In going from city A to city B, any of the 3 roads may be taken. In going from city B to city C, any of the 5 roads may be taken. So by the multiplication rule, there are $3 \cdot 5 = 15$ ways to travel from city A to city C via city B.

b. A roundtrip journey can be thought of as a four-step operation:

step 1 is to go from A to B,

step 2 is to go from B to C,

step 3 is to go from C to B,

step 4 is to go from B to A.

Since there are 3 ways to perform step 1, 5 ways to perform step 2, 5 ways to perform step 3, and 3 ways to perform step 4, by the multiplication rule, there are $3 \cdot 5 \cdot 5 \cdot 3 = 225$ roundtrip routes.

c. In this case the steps for making a roundtrip journey are the same as in part (b), but since no route segment may be repeated, there are only 4 ways to perform step 3 and only 2 ways to perform step 4. So by the multiplication rule, there are $3 \cdot 5 \cdot 4 \cdot 2 = 120$ round trip routes in which no road is traversed twice.

11. a. Imagine constructing a bit string of length 8 as an eight-step process:

step 1 is to choose either a 0 or a 1 for the left-most position,

step 2 is to choose either a 0 or a 1 for the next position to the right,

step 3 is to choose either a 0 or a 1 for the next position to the right.

Since there are 2 ways to perform each step, the total number of ways to accomplish the entire operation, which is the number of different bit strings of length 8, is $2 \cdot 2 \cdot 2 \cdot 2 \cdot 2 \cdot 2 \cdot 2 \cdot 2 = 2^8 = 256$.

b. Imagine placing a 1 in the left-most position of an 8-bit string and then imagine filling in the remaining seven positions as an operation with seven steps, where step i is to fill in the $(i + 1)$st position. Since there are 2 ways to perform seven steps, there are 2^7 ways to perform the entire operation. So there are 2^7 8-bit strings that begin with a 1.

12. a. Imagine constructing a hexadecimal number from 10_{16} through FF_{16} as a two-step process, where step 1 is to fill in the left-most position and step 2 is to fill in the right-most position. There are 15 ways to perform step 1 because there are 15 hexadecimal digits from 1 through F ($= 15$), and there are 16 ways to perform step 2 because there are 16 hexadecimal digits from 0 through F. So by the multiplication rule, there are $15 \cdot 16 = 240$ hexadecimal numbers from 10_{16} through FF_{16}.

b. The steps are the same as in part (a), but there are only 11 ways to perform step 1 (because there are 11 hexadecimal digits from 5 through F). So by the multiplication rule, there are $11 \cdot 16 = 176$ hexadecimal numbers from 50_{16} through FF_{16}

13. a. In each of the four tosses there are two possible results: either a head (H) or a tail (T) is obtained. So by the multiplication rule, the number of outcomes is $2 \cdot 2 \cdot 2 \cdot 2 = 2^4 = 16$.

b. There are six outcomes with two heads:
   HHTT, HTHT, HTTH, THHT, THTH, TTHH.
   So the probability of obtaining exactly two heads in $6/16 = 3/8$.

14. a. Let each of steps 1–3 be to choose a letter of the alphabet to put in positions 1–3 and each of steps 4–6 be to choose a digit to put in positions 4–6. Since there are 26 letters and 10 digits (0–9), the number of license plates is

$$26 \cdot 26 \cdot 26 \cdot 10 \cdot 10 \cdot 10 = 17{,}576{,}000.$$

b. In this case there is only one way to perform step 1 (because the first letter must be an A) and only one way to perform step 6 (because the last digit must be a 0). So the number of license plates is $26 \cdot 26 \cdot 10 \cdot 10 = 67{,}600$.

d. In this case there are 26 ways to perform step 1, 25 ways to perform step 2, 24 ways to perform step 3, 10 ways to perform step 4, 9 ways to perform step 5, and 8 ways to perform step 6. So the number of license plates is $26 \cdot 25 \cdot 24 \cdot 10 \cdot 9 \cdot 8 = 11{,}232{,}000$.

16. a. Let step 1 be to choose either the number 2 or one of the letters corresponding to the number 2 on the keypad, let step 2 be to choose either the number 1 or one of the letters corresponding to the number 1 on the keypad, and let steps 3 and 4 be to choose either the number 3 or one of the letters corresponding to the number 3 on the keypad. There are 4 ways to perform step 1, 3 ways to perform step 2, and 4 ways to perform each of steps 3 and 4. So by the multiplication rule, there are $4 \cdot 3 \cdot 4 \cdot 4 = 192$ ways to perform the entire operation. Thus there are 192 different PINs that are keyed the same as 2133. Note that on a computer keyboard, these PINs would not be keyed the same way.

17.

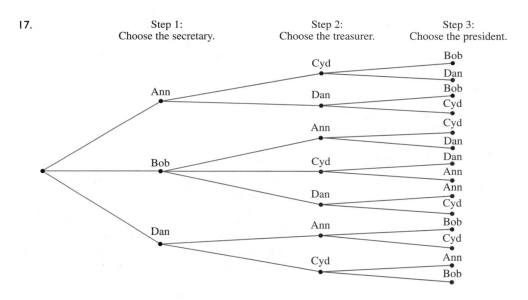

Step 1:
Choose the secretary.

Step 2:
Choose the treasurer.

Step 3:
Choose the president.

There are 14 different paths from "root" to "leaf" of this possibility tree, and so there are 14 ways the officers can be chosen. It does not appear that reordering the steps will make it possible to use the multiplication rule alone to solve this problem.

18. a. The number of ways to perform step 4 is not constant; it depends on how the previous steps were performed. For instance, if 3 digits had been chosen in steps 1–3, then there would be $10 - 3 = 7$ ways to perform step 4, but if 3 letters had been chosen in steps 1–3 then there would be 10 ways to perform step 4.

19. a. Two solutions:
    (i) number of integers

$$= \begin{bmatrix} \text{number of} \\ \text{ways to} \\ \text{pick first} \\ \text{digit} \end{bmatrix} \cdot \begin{bmatrix} \text{number of} \\ \text{ways to} \\ \text{pick second} \\ \text{digit} \end{bmatrix}$$

$$= 9 \cdot 10 = 90$$

    (ii) Using Theorem 6.1.1, number of integers = $99 - 10 + 1 = 90$.

    b. Odd integers end in 1, 3, 5, 7, or 9. So,

    number of odd integers

$$= \begin{bmatrix} \text{number of} \\ \text{ways to} \\ \text{pick first} \\ \text{digit} \end{bmatrix} \cdot \begin{bmatrix} \text{number of} \\ \text{ways to} \\ \text{pick second} \\ \text{digit} \end{bmatrix}$$

$$= 9 \cdot 5 = 45$$

    Alternate solution: Use listing method used for Example 6.1.2.

    c. $\begin{bmatrix} \text{number of integers with} \\ \text{distinct digits} \end{bmatrix}$

$$= \begin{bmatrix} \text{number of} \\ \text{ways to} \\ \text{pick first} \\ \text{digit} \end{bmatrix} \cdot \begin{bmatrix} \text{number of} \\ \text{ways to} \\ \text{pick second} \\ \text{digit} \end{bmatrix}$$

$$= 9 \cdot 9 = 81$$

    d. $\begin{bmatrix} \text{number of odd integers} \\ \text{with distinct digits} \end{bmatrix}$

$$= \begin{bmatrix} \text{number of} \\ \text{ways to} \\ \text{pick second} \\ \text{digit} \end{bmatrix} \cdot \begin{bmatrix} \text{number of} \\ \text{ways to} \\ \text{pick first} \\ \text{digit} \end{bmatrix}$$

$$= 5 \cdot 8 = 40 \qquad \text{because the first digit} \\ \text{can't equal 0, nor can it} \\ \text{equal the second digit}$$

    e. $81/90 = 9/10$, $40/90 = 4/9$

21. The outer loop is iterated 30 times, and during each iteration of the outer loop there are 15 iterations of the inner loop. Hence by the multiplication rule, the total number of iterations of the inner loop is $30 \cdot 15 = 450$.

24. The outer loop is iterated $50 - 5 + 1 = 46$ times, and during each iteration of the outer loop there are $20 - 10 + 1 = 11$ iterations of the inner loop. Hence by the multiplication rule, the total number of iterations of the inner loop is $46 \cdot 11 = 506$.

26. *Hints:* Add leading zeros as needed to make each number five digits long. For instance, write 1 as 00001. Let some of the steps be to choose positions for the given digits. The answer is 720.

28. a. There are $a + 1$ divisors: $1, p, p^2, \ldots, p^a$.
    b. A divisor is a product of any one of the $a + 1$ numbers listed in part (a) times any one of the $b + 1$ numbers $1, q, q^2, \ldots, q^b$. So by the multiplication rule there are $(a + 1)(b + 1)$ divisors in all.

29. a. Since the 9 letters of the word *ALGORITHM* are all distinct, there are as many arrangements of these letters in a row as there are permutations of a set with nine elements: $9! = 362,880$.
    b. In this case there are effectively 8 symbols to be permuted (because $\boxed{AL}$ may be regarded as a single symbol). So the number of arrangements is $8! = 40,320$.

31. The same reasoning as in Example 6.2.9 gives an answer of $4! = 24$.

32. *WX, WY, WZ, XW, XY, XZ, YW, YX, YZ, ZW, ZX, ZY*

34. a. $P(6, 4) = \dfrac{6!}{(6-4)!} = \dfrac{6 \cdot 5 \cdot 4 \cdot 3 \cdot \cancel{2 \cdot 1}}{\cancel{2 \cdot 1}} = 360$

35. a. $P(5, 3) = \dfrac{5 \cdot 4 \cdot 3 \cdot \cancel{2!}}{\cancel{2!}} = 60$

36. a. $P(9, 3) = \dfrac{9 \cdot 8 \cdot 7 \cdot \cancel{6!}}{\cancel{6!}} = 504$

    c. $P(8, 4) = \dfrac{8 \cdot 7 \cdot 6 \cdot 5 \cdot \cancel{4!}}{\cancel{4!}} = 1,680$

38. *Proof:* Let n be an integer and $n \geq 2$. Then

$$P(n + 1, 2) - P(n, 2)$$

$$= \frac{(n + 1)!}{[(n + 1) - 2]!} - \frac{n!}{(n - 2)!} = \frac{(n + 1)!}{(n - 1)!} - \frac{n!}{(n - 2)!}$$

$$= \frac{(n + 1) \cdot n \cdot \cancel{(n - 1)!}}{\cancel{(n - 1)!}} - \frac{n \cdot (n - 1) \cdot \cancel{(n - 2)!}}{\cancel{(n - 2)!}}$$

$$= n^2 + n - (n^2 - n) = 2n = 2 \cdot \frac{n \cdot (n - 1)!}{(n - 1)!}$$

$$= 2 \cdot \frac{n!}{(n - 1)!} = 2 \cdot P(n, 1).$$

This is what was to be proved.

42. *Hint:* In the inductive step, suppose there exist $k!$ permutations of a set with k elements. Let X be a set with $k + 1$ elements. The process of forming a permutation of the elements of X can be considered a two-step operation where step 1 is to choose the element to write first. Step 2 is to write the remaining elements of X in some order.

SECTION 6.3

1. a.

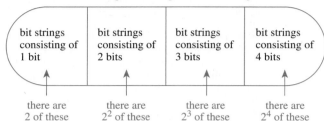

Set of bit strings consisting of from 1 through 4 bits

| bit strings consisting of 1 bit | bit strings consisting of 2 bits | bit strings consisting of 3 bits | bit strings consisting of 4 bits |

there are 2 of these — there are 2^2 of these — there are 2^3 of these — there are 2^4 of these

Applying the addition rule to the figure above shows that there are $2 + 2^2 + 2^3 + 2^4 = 30$ bit strings consisting of from one through four bits.

b. By reasoning similar to that of part (a), there are $2^5 + 2^6 + 2^7 + 2^8 = 480$ bit strings of from five through eight bits.

3. a.
$$\left[\begin{array}{l}\text{number of integers from 1 through 999}\\\text{with no repeated digits}\end{array}\right]$$

$$= \left[\begin{array}{l}\text{number of}\\\text{integers from 1}\\\text{through 9 with}\\\text{no repeated}\\\text{digits}\end{array}\right] + \left[\begin{array}{l}\text{number of}\\\text{integers from 10}\\\text{through 99 with}\\\text{no repeated}\\\text{digits}\end{array}\right]$$

$$+ \left[\begin{array}{l}\text{number of integers from}\\\text{100 through 999 with}\\\text{no repeated digits}\end{array}\right]$$

$$= 9 + 9 \cdot 9 + 9 \cdot 9 \cdot 8 = 738$$

b.
$$\left[\begin{array}{l}\text{number of integers from 1 through 999}\\\text{with at least one repeated digit}\end{array}\right]$$

$$= \left[\begin{array}{l}\text{total number}\\\text{of integers}\\\text{from}\\\text{1 through 999}\end{array}\right] - \left[\begin{array}{l}\text{number of integers}\\\text{from 1 through}\\\text{999 with no}\\\text{repeated digits}\end{array}\right]$$

$$= 999 - 738 = 261$$

The probability that an integer chosen at random has at least one repeated digit is $261/999 \cong 26.1\%$

4.

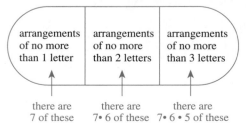

Set of arrangements (without repetition) of no more than 3 letters of NETWORK

| arrangements of no more than 1 letter | arrangements of no more than 2 letters | arrangements of no more than 3 letters |

there are 7 of these — there are 7• 6 of these — there are 7• 6 • 5 of these

Applying the addition rule to the figure above shows that there are $7 + 7 \cdot 6 + 7 \cdot 6 \cdot 5 = 259$ arrangements of three letters of the word *NETWORK* if repetition of letters is not permitted.

6. a. There are $1 + 26 + 26^2 + 26^3$ arrangements of from 0 through 3 letters of the alphabet. Any of these may be paired with an arrangement of from 0 through 4 digits, and there are $1 + 10 + 10^2 + 10^3 + 10^4$ arrangements of from 0 through 4 digits. So by the multiplication rule and the difference rule, there are

$$(1 + 26 + 26^2 + 26^3)$$
$$\cdot (1 + 10 + 10^2 + 10^3 + 10^4) - 1 = 203{,}097{,}968$$
$$\uparrow$$
the blank plate

license plates in all.

b. $(1 + 26 + 26^2 + 26^3 - 85)$
$\cdot (1 + 10 + 10^2 + 10^3 + 10^4) - 1 = 202{,}153{,}533$

8. a. Each column of the table below corresponds to a pair of values of i and j for which the inner loop will be iterated.

i	1	2→		3→			4→			
j	1	1	2	1	2	3	1	2	3	4

Since there are $1 + 2 + 3 + 4 = 10$ columns, the inner loop will be iterated ten times.

9. a. The answer is the number of permutations of the five letters in *QUICK*, which equals $5! = 120$.

b. Because *QU* (in order) is to be considered as a single unit, the answer is the number of permutations of the four symbols $\boxed{QU}$, *I*, *C*, *K*. This is $4! = 24$.

c. By part (b), there are $4!$ arrangements of $\boxed{QU}$, *I*, *C*, *K*. Similarly, there are $4!$ arrangements of $\boxed{UQ}$, *I*, *C*, *K*. So by the addition rule, there are $4! + 4! = 48$ arrangements in all.

11. a. $\begin{bmatrix} \text{number of ways to place eight people in a} \\ \text{row keeping } A \text{ and } B \text{ together} \end{bmatrix}$

$= \begin{bmatrix} \text{number of ways to arrange} \\ \boxed{AB} \; C \; D \; E \; F \; G \; H \end{bmatrix}$

$+ \begin{bmatrix} \text{number of ways to arrange} \\ \boxed{BA} \; C \; D \; E \; F \; G \; H \end{bmatrix}$

$= 7! + 7! = 5{,}040 + 5{,}040 = 10{,}080$

b. $\begin{bmatrix} \text{number of ways to arrange the eight people} \\ \text{in a row keeping } A \text{ and } B \text{ apart} \end{bmatrix}$

$= \begin{bmatrix} \text{total number of} \\ \text{ways to place} \\ \text{the eight people} \\ \text{in a row} \end{bmatrix} - \begin{bmatrix} \text{number of ways} \\ \text{to place the} \\ \text{eight people in} \\ \text{a row keeping} \\ A \text{ and } B \\ \text{together} \end{bmatrix}$

$= 8! - 10{,}080 = 40{,}320 - 10{,}080$

$= 30{,}240$

12. number of variable names

$= \begin{bmatrix} \text{number of numeric} \\ \text{variable names} \end{bmatrix} + \begin{bmatrix} \text{number of string} \\ \text{variable names} \end{bmatrix}$

$= (26 + 26 \cdot 36) + (26 + 26 \cdot 36) = 1{,}924$

13. *Hint:* In exercise 12 note that

$$26 + 26 \cdot 36 = 26 \sum_{k=0}^{1} \cdot 36^k.$$

Generalize this idea here. Use Theorem 4.2.3 to evaluate the expression you obtain.

14. a. $10 \cdot 9 \cdot 8 \cdot 7 \cdot 6 \cdot 5 \cdot 4 = 604{,}800$

b. $\begin{bmatrix} \text{number of phone numbers with} \\ \text{at least one repeated digit} \end{bmatrix}$

$= \begin{bmatrix} \text{total number of} \\ \text{phone numbers} \end{bmatrix}$

$- \begin{bmatrix} \text{number of phone numbers} \\ \text{with no repeated digits} \end{bmatrix}$

$= 10^7 - 604{,}800 = 9{,}395{,}200$

c. $9{,}395{,}200/10^7 \cong 93.95\%$.

16. a. *Proof:* Let A and B be mutually disjoint events in a sample space S. By the addition rule, $n(A \cup B) = n(A) + n(B)$. So by the equally likely probability formula,

$$P(A \cup B) = \frac{n(A \cup B)}{n(S)} = \frac{n(A) + n(B)}{n(S)}$$

$$= \frac{n(A)}{n(S)} + \frac{n(B)}{n(S)} = P(A) + P(B).$$

17. *Hint:* Count the number of combinations with two adjacent numbers the same, and subtract this number from the total.

18. a. Identify the integers from 1 to 100,000 that contain the digit 6 exactly once with strings of five digits. Thus, for example, 306 would be identified with 00306. It is not necessary to use strings of six digits because 100,000 does not contain the digit 6. Imagine the process of constructing a five-digit string that contains the digit 6 exactly once as a five-step operation that consists of filling in the five digit positions:

$$\overline{}_{1} \; \overline{}_{2} \; \overline{}_{3} \; \overline{}_{4} \; \overline{}_{5}.$$

Step 1 is to choose one of the five positions for the 6.

Step 2 is to choose a digit for the left-most remaining position.

Step 3 is to choose a digit for the next remaining position to the right.

Step 4 is to choose a digit for the next remaining position to the right.

Step 5 is to choose a digit for the right-most position.

Since there are 5 choices for step 1 (any one of the five positions) and 9 choices for each of steps 2–5 (any digit except 6), by the multiplication rule, the number of ways to perform this operation is $5 \cdot 9 \cdot 9 \cdot 9 \cdot 9 = 32{,}805$. Hence there are 32,805 integers from 1 to 100,000 that contain the digit 6 exactly once.

19. *Hint:* The answer is 2/3.

21. a. Let $A =$ the set of integers that are multiples of 4 and $B =$ the set of integers that are multiples of 7. Then $A \cap B =$ the set of integers that are multiples of 28.

But $n(A) = 250$ since 1 2 3 4 5 6 7 8 . . . 999 1,000,

$4 \cdot 1 \quad 4 \cdot 2 \quad . . . 4 \cdot 250$

or, equivalently, since $1{,}000 = 4 \cdot 250$.

Also $n(B) = 142$ since 1 2 3 4 5 6 7 . . . 14 . . . 994 995 . . . 1,000

$7 \cdot 1 \quad 7 \cdot 2 . . . 7 \cdot 142$

or, equivalently, since $1{,}000 = 7 \cdot 142 + 6$.

and $n(A \cap B) = 35$ since 1 2 3 . . . 28 . . . 56 . . . 980 . . . 1,000,

$28 \cdot 1 \quad 28 \cdot 2 . . . \quad 28 \cdot 35$

or, equivalently, since $1{,}000 = 28 \cdot 35 + 20$.

So $n(A \cup B) = 250 + 142 - 35 = 357$.

23. a. The number of students who checked at least one of the magazines is $n(T \cup N \cup U) = n(T) + n(N) + n(U) - n(T \cap N) - n(T \cap U) - n(N \cap U) + n(T \cap N \cap U) = 28 + 26 + 14 - 8 - 4 - 3 + 2 = 55$.

b. By the difference rule, the number of students who checked none of the magazines is the total number of students minus the number who checked at least one magazine. This is $100 - 55 = 45$.

d. The number of students who read *Time* and *Newsweek* but not *U.S. News* is

$$n((T \cap N) - n(T \cap N \cap U)) = 8 - 2 = 6.$$

25. Let M = the set of married people in the sample,
Y = the set of people between 20 and 30 in the sample, and
F = the set of females in the sample.

Then the number of people in the set $M \cup Y \cup F$ is less than or equal to the size of the sample. And so

$$1{,}200 \geq n(M \cup Y \cup F)$$
$$= n(M) + n(Y) + n(F) - n(M \cap Y)$$
$$\quad - n(M \cap F) - n(Y \cap F) + n(M \cap Y \cap F)$$
$$= 675 + 682 + 684 - 195 - 467 - 318 + 165$$
$$= 1{,}226.$$

This is impossible since $1{,}200 < 1{,}226$. So the polltaker's figures are inconsistent. They could not have occurred as a result of an actual sample survey.

27. Let A be the set of all positive integers less than 1,000 that are not multiples of 2, and let B be the set of all positive integers less than 1,000 that are not multiples of 5. Since the only prime factors of 1,000 are 2 and 5, the number of positive integers that have no common factors with 1,000 is $n(A \cap B)$. Let the universe U be the set of all positive integers less than 1,000. Then A^c is the set of positive integers less than 1,000 that are multiples of 2, B^c is the set of positive integers less than 1,000 that are multiples of 5, and $A^c \cap B^c$ is the set of positive integers less than 1,000 that are multiples of 10. By one of the procedures discussed in Section 6.1 or 6.2, it is easily found that $n(A^c) = 499$, $n(B^c) = 199$, and $n(A^c \cap B^c) = 99$. Thus by the inclusion/exclusion rule,

$$n(A^c \cup B^c) = n(A^c) + n(B^c) - n(A^c \cap B^c)$$
$$= 499 + 199 - 99 = 599.$$

But by De Morgan's law, $n(A^c \cup B^c) = n((A \cap B)^c)$, and so

$$n((A \cap B)^c) = 599. \quad (*)$$

Now since $(A \cap B)^c = U - (A \cap B)$, by the difference rule we have

$$n((A \cap B)^c) = n(U) - n(A \cap B). \quad (**)$$

Equating the right-hand sides of $(*)$ and $(**)$ gives $n(U) - n(A \cap B) = 599$. And because $n(U) = 999$, we conclude that $999 - n(A \cap B) = 599$, or, equivalently, $n(A \cap B) = 999 - 599 = 400$. So there are 400 positive integers less than 1,000 that have no common factor with 1,000.

SECTION 6.4

1. a. 2-combinations: $\{x_1, x_2\}$, $\{x_1, x_3\}$, $\{x_2, x_3\}$.

Hence, $\dbinom{3}{2} = 3$.

b. unordered selections: $\{a, b, c, d\}, \{a, b, c, e\}, \{a, b, d, e\}$, $\{a, c, d, e\}, \{b, c, d, e\}$.

Hence, $\dbinom{5}{4} = 5$.

3. $P(7, 2) = \dbinom{7}{2} \cdot 2!$

5. a. $\dbinom{5}{0} = \dfrac{5!}{0!(5 - 0)!} = \dfrac{\cancel{5!}}{1 \cdot \cancel{5!}} = 1$

b. $\dbinom{5}{1} = \dfrac{5!}{1!(5 - 1)!} = \dfrac{5 \cdot \cancel{4 \cdot 3 \cdot 2 \cdot 1}}{1 \cdot \cancel{4 \cdot 3 \cdot 2 \cdot 1}} = 5$

6. a. number of committees of 6

$$= \binom{15}{6} = \frac{15!}{(15 - 6)! \cdot 6!}$$

$$= \frac{\overset{7}{\cancel{15}} \cdot \cancel{14} \cdot 13 \cdot \cancel{12} \cdot 11 \cdot \overset{5}{\cancel{10}} \cdot \cancel{9!}}{\cancel{9!} \cdot \cancel{6} \cdot \cancel{5} \cdot \cancel{4} \cdot \cancel{3} \cdot \cancel{2}} = 5{,}005$$

b. $\begin{bmatrix} \text{number of committees that} \\ \text{don't contain } A \\ \text{and } B \text{ together} \end{bmatrix}$

$= \begin{bmatrix} \text{number of} \\ \text{committees with} \\ A \text{ and five} \\ \text{others—none} \\ \text{of them } B \end{bmatrix} + \begin{bmatrix} \text{number of} \\ \text{committees with} \\ B \text{ and five} \\ \text{others—none} \\ \text{of them } A \end{bmatrix}$

$+ \begin{bmatrix} \text{number of committees} \\ \text{with neither } A \text{ nor } B \end{bmatrix}$

$= \dbinom{13}{5} + \dbinom{13}{5} + \dbinom{13}{6}$

$= 1{,}287 + 1{,}287 + 1{,}716 = 4{,}290$

Alternate solution:

$$\begin{bmatrix} \text{number of committees} \\ \text{that don't contain} \\ A \text{ and } B \text{ together} \end{bmatrix}$$

$$= \begin{bmatrix} \text{total number} \\ \text{of committees} \end{bmatrix} - \begin{bmatrix} \text{number of committees} \\ \text{that contain} \\ \text{both } A \text{ and } B \end{bmatrix}$$

$$= \binom{15}{6} - \binom{13}{4}$$

$$= 5{,}005 - 715 = 4{,}290$$

c. $$\begin{bmatrix} \text{number of} \\ \text{committees with} \\ \text{both } A \text{ and } B \end{bmatrix} + \begin{bmatrix} \text{number of} \\ \text{committees with} \\ \text{neither } A \text{ and } B \end{bmatrix}$$

$$= \binom{13}{4} + \binom{13}{6} = 715 + 1{,}716 = 2{,}431$$

d. (i) $$\begin{bmatrix} \text{number of subsets} \\ \text{of three men} \\ \text{chosen from eight} \end{bmatrix} \cdot \begin{bmatrix} \text{number of subsets} \\ \text{of three women} \\ \text{chosen from seven} \end{bmatrix}$$

$$= \binom{8}{3} \cdot \binom{7}{3} = 56 \cdot 35 = 1{,}960$$

(ii) $$\begin{bmatrix} \text{number of} \\ \text{committees} \\ \text{with at} \\ \text{least one} \\ \text{woman} \end{bmatrix} = \begin{bmatrix} \text{total} \\ \text{number of} \\ \text{committees} \end{bmatrix}$$

$$- \begin{bmatrix} \text{number of} \\ \text{all-male} \\ \text{committees} \end{bmatrix}$$

$$= \binom{15}{6} - \binom{8}{6} = 5{,}005 - 28$$

$$= 4{,}977$$

e. $$\begin{bmatrix} \text{number of} \\ \text{ways to} \\ \text{choose two} \\ \text{freshmen} \end{bmatrix} \cdot \begin{bmatrix} \text{number of} \\ \text{ways to} \\ \text{choose two} \\ \text{sophomores} \end{bmatrix} \cdot \begin{bmatrix} \text{number of} \\ \text{ways to} \\ \text{choose two} \\ \text{juniors} \end{bmatrix}$$

$$\cdot \begin{bmatrix} \text{number of} \\ \text{ways to} \\ \text{choose two} \\ \text{seniors} \end{bmatrix} = \binom{3}{2} \cdot \binom{4}{2} \cdot \binom{3}{2} \cdot \binom{5}{2}$$

$$= 540$$

8. *Hint:* The answers are: a. 1,001, b. (i) 420, (ii) 1,001, (iii) 175, c. 506, d. 561.

9. b. $$\binom{19}{3} \cdot \binom{11}{3} + \binom{19}{4} \cdot \binom{11}{2} + \binom{19}{5} \cdot \binom{11}{1}$$

$$+ \binom{19}{6} \cdot \binom{11}{0} = 528{,}105$$

11. a. (i) 4 (because there are as many royal flushes as there are suits)

(ii) $$\frac{4}{\binom{52}{5}} = \frac{4}{2{,}598{,}960} \cong .0000015$$

c. (i) $13 \cdot \binom{48}{1} = 624$ (because one can first choose the denomination of the four-of-a-kind and then choose one additional card from the 48 remaining)

(ii) $$\frac{624}{\binom{52}{5}} = \frac{624}{2{,}598{,}960} \cong .00024$$

f. (i) Imagine constructing a straight (including a straight flush and a royal flush) as a six-step process: step 1 is to choose the lowest denomination of any card of the five (which can be any one of $A, 2, \ldots, 10$), step 2 is to choose a card of that denomination, step 3 is to choose a card of the next higher denomination, and so forth until all five cards have been selected. By the multiplication rule, the number of ways to perform this process is

$$10 \cdot \binom{4}{1} \cdot \binom{4}{1} \cdot \binom{4}{1} \cdot \binom{4}{1} \cdot \binom{4}{1} = 10 \cdot 4^5$$

$$= 10{,}240.$$

By parts (a) and (b), 40 of these numbers represent royal or straight flushes. So there are $10{,}240 - 40 = 10{,}200$ straights in all.

(ii) $$\frac{10{,}200}{\binom{52}{5}} = \frac{10{,}200}{2{,}598{,}960} \cong .0039$$

12. *Hint:* Under the given assumptions, the probability that no two people in a group of n people have the same birthday is

$$\frac{365 \cdot 364 \cdot \ldots \cdot (365 - n + 1)}{365^n} = \frac{P(365, n)}{365^n}.$$

Why?

13. a. $2^{10} = 1{,}024$

d. $$\begin{bmatrix} \text{number of} \\ \text{outcomes} \\ \text{with at} \\ \text{least one} \\ \text{head} \end{bmatrix} = \begin{bmatrix} \text{total} \\ \text{number of} \\ \text{outcomes} \end{bmatrix} - \begin{bmatrix} \text{number of} \\ \text{outcomes} \\ \text{with no} \\ \text{heads} \end{bmatrix}$$

$$= 1{,}024 - 1 = 1{,}023$$

15. a. 50   b. 50

c. To get an even sum, both numbers must be even or both must be odd. Hence

$$\begin{bmatrix} \text{number of subsets of two integers from} \\ \text{1 to 100 inclusive whose sum is even} \end{bmatrix}$$

$$= \begin{bmatrix} \text{number of subsets} \\ \text{of two even} \\ \text{integers chosen} \\ \text{from the 50} \\ \text{possible} \end{bmatrix} + \begin{bmatrix} \text{number of subsets} \\ \text{of two odd} \\ \text{integers chosen} \\ \text{from the 50} \\ \text{possible} \end{bmatrix}$$

$$= \binom{50}{2} + \binom{50}{2} = 2{,}450.$$

d. To obtain an odd sum, one of the numbers must be even and the other odd. Hence the answer is $\binom{50}{1} \cdot \binom{50}{1} = 2{,}500$. Alternatively, note that the answer equals the total number of subsets of two integers chosen from 1 through 100 minus the number of such subsets for which the sum of the elements is even. Thus the answer is $\binom{100}{2} - 2{,}450 = 2{,}500.$

17. a. Two points determine a line. Hence

$$\begin{bmatrix} \text{number of} \\ \text{straight lines} \\ \text{determined} \\ \text{by the nine} \\ \text{points} \end{bmatrix} = \begin{bmatrix} \text{number of} \\ \text{subsets of two} \\ \text{points chosen} \\ \text{from nine} \end{bmatrix}$$

$$= \binom{9}{2} = 36.$$

19. a. $\dfrac{10!}{2!1!1!3!2!1!} = 151{,}200$ since there are 2 A's, 1 B, 1 H, 3 L's, 2 O's, and 1 U)

b. $\dfrac{8!}{2!1!1!2!2!} = 5{,}040$ c. $\dfrac{9!}{1!2!1!3!2!} = 15{,}120$

21. a. There are two choices for each of four positions. So the answer is $2 \cdot 2 \cdot 2 \cdot 2 = 2^4 = 16.$

b. $\begin{bmatrix} \text{number of} \\ \text{strings with} \\ \text{three } a\text{'s} \\ \text{and one } b \end{bmatrix} = \begin{bmatrix} \text{number of ways} \\ \text{to pick a subset} \\ \text{of three positions} \\ \text{out of four into} \\ \text{which to place} \\ \text{the } a\text{'s} \end{bmatrix} = \binom{4}{3} = 4$

23. Rook must move seven squares to the right and seven squares up. So

$$\begin{bmatrix} \text{the number} \\ \text{of paths} \\ \text{the rook} \\ \text{can take} \end{bmatrix} = \begin{bmatrix} \text{the number} \\ \text{of orderings} \\ \text{of seven R's} \\ \text{and seven} \\ \text{U's} \end{bmatrix} \quad \begin{array}{l} \text{where R stands for} \\ \text{"right" and U} \\ \text{stands for "up"} \end{array}$$

$$= \dfrac{14!}{7!7!} = 3{,}432.$$

25. *Hint:* Use the difference rule and also the inclusion/exclusion rule.

SECTION 6.5

1. a. $\left(\dfrac{5 + 3 - 1}{5} \right) = \binom{7}{5} = \dfrac{7 \cdot 6}{2} = 21$

b. The three elements of the set are 1, 2, and 3. The 5-combinations are: [1, 1, 1, 1, 1], [1, 1, 1, 1, 2], [1, 1, 1, 1, 3], [1, 1, 1, 2, 2], [1, 1, 1, 2, 3], [1, 1, 1, 3, 3], [1, 1, 2, 2, 2], [1, 1, 2, 2, 3], [1, 1, 2, 3, 3], [1, 1, 3, 3, 3], [1, 2, 2, 2, 2], [1, 2, 2, 2, 3], [1, 2, 2, 3, 3], [1, 2, 3, 3, 3], [1, 3, 3, 3, 3], [2, 2, 2, 2, 2], [2, 2, 2, 2, 3], [2, 2, 2, 3, 3], [2, 2, 3, 3, 3], [2, 3, 3, 3, 3], [3, 3, 3, 3, 3],

2. a. $\left(\dfrac{4 + 3 - 1}{4} \right) = \binom{6}{4} = \dfrac{6 \cdot 5}{2} = 15$

3. a. $\left(\dfrac{20 + 6 - 1}{20} \right) = \binom{25}{20} = 53{,}130$

b. If at least three are eclairs, then 17 additional pastries are selected from six kinds. The number of selections is $\left(\dfrac{17 + 6 - 1}{17} \right) = \binom{22}{17} = 26{,}334.$

Note: In parts (a) and (b) it is assumed that the selections being counted are unordered.

c. By parts (a) and (b), the probability that at least three eclairs are among the pastries selected is $26{,}334/53{,}130 \cong .496 = 49.6\%$

d. If exactly three of the pastries are eclairs, then 17 additional pastries are selected from five kinds. The number of selections is

$$\left(\dfrac{17 + 5 - 1}{17} \right) = \binom{21}{17} = 5{,}985.$$

Hence the probability that a random selection includes exactly three eclairs is $5{,}985/53{,}130 \cong .113 = 11.3\%.$

5. The answer equals the number of 4-combinations with repetition allowed that can be formed from a set of n

elements. It is

$$\binom{4+n-1}{4} = \binom{n+3}{4}$$

$$= \frac{(n+3)(n+2)(n+1)n(n-1)!}{4!(n-1)!}$$

$$= \frac{n(n+1)(n+2)(n+3)}{24}.$$

8. As in Example 6.5.4, the answer is the same as the number of quadruples of integers (i, j, k, m) for which $1 \le i \le j \le k \le m \le n$. By exercise 5, this number is $\binom{n+3}{4} = \frac{n(n+1)(n+2)(n+3)}{24}$.

10. Think of the number 20 as divided into 20 individual units and the variables x_1, x_2, and x_3 as three categories into which these units are placed. The number of units in category x_i indicates the value of x_i in a solution of the equation. By Theorem 6.5.1, the number of ways to select 20 objects from the three categories is $\binom{20+3-1}{20} = \binom{22}{20} = \frac{22 \cdot 21}{2} = 231$. So there are 231 nonnegative integer solutions to the equation.

11. The analysis for this exercise is the same as for exercise 10 except that since each $x_i \ge 1$, we can imagine taking 3 of the 20 units, placing one in each category x_1, x_2, and x_3, and then distributing the remaining 17 units among the three categories. The number of ways to do this is $\binom{17+3-1}{17} = \binom{19}{17} = \frac{19 \cdot 18}{2} = 171$. So there are 171 positive integer solutions to the equation.

18. a. Because only ten eclairs are available, any selection of 20 pastries contains k eclairs, where $0 \le k \le 10$. Since such a selection includes $20 - k$ of the other five kinds of pastry, the number of such selections is $\binom{(20-k)+5-1}{20-k} = \binom{24-k}{20-k}$ (by Theorem 6.5.1). Therefore, by the addition rule, the total number of selections is $\sum_{k=0}^{10} \binom{24-k}{20-k}$. The numerical value of this expression is 51,128, which can be obtained using a calculator that automatically computes values of $\binom{n}{r}$ or using a symbolic manipulation computer program such as Derive, Maple, or Mathematica.

b. For each combination of k eclairs and m napoleon slices, choose $20 - (k + m)$ pastries from the remaining four types. By Theorem 6.5.1, the number

of ways to make such a selection is

$$\binom{(20-(k+m))+4-1}{20-(k+m)} = \binom{23-(k+m)}{20-(k+m)}.$$

Since $0 \le k \le 10$ and $0 \le m \le 8$, by the addition rule the total number of selections of 20 pastries is $\sum_{m=0}^{8} \sum_{k=0}^{10} \binom{23-(k+m)}{20-(k+m)}$. The numerical value of this expression is 46,761.

SECTION 6.6

1. $\binom{n}{0} = \frac{n!}{0!(n-0)!} = \frac{n!}{1 \cdot n!} = 1$

3. $\binom{n}{2} = \frac{n!}{(n-2)! \cdot 2!} = \frac{n \cdot (n-1) \cdot (n-2)!}{(n-2)! \cdot 2!}$

$$= \frac{n \cdot (n-1)}{2}$$

5. *Proof*: Suppose n and r are nonnegative integers and $r \le n$. Then

$$\binom{n}{r} = \frac{n!}{r! \cdot (n-r)!} \qquad \text{by Theorem 6.4.1}$$

$$= \frac{n!}{(n-(n-r))! \cdot (n-r)!} \qquad \begin{array}{l}\text{since } n-(n-r) = \\ n-n+r = r\end{array}$$

$$= \frac{n!}{(n-r)! \cdot (n-(n-r))!} \qquad \begin{array}{l}\text{by interchanging} \\ \text{the factors in the} \\ \text{denominator}\end{array}$$

$$= \binom{n}{n-r} \qquad \text{by Theorem 6.4.1.}$$

6. Apply formula (6.6.2) with $n + k$ in place of n. This is legal since $n + k \ge 1$. The result is $\binom{n+k}{n+k-1} = n + k$.

9. $\binom{6}{4} = \binom{5}{3} + \binom{5}{4} = 10 + 5 = 15$

$\binom{6}{5} = \binom{5}{4} + \binom{5}{5} = 5 + 1 = 6$

14. *Proof*:

The formula is true for n = 1: To prove the formula for $n = 1$, we must show that

$$\sum_{k=2}^{1+1} \binom{k}{2} = \binom{1+2}{3}.$$

But

$$\sum_{k=2}^{1+1} \binom{k}{2} = \sum_{k=2}^{2} \binom{k}{2} = \binom{2}{2} = 1 = \binom{3}{3} = \binom{1+2}{3}.$$

So the formula is true for $n = 1$.

If the formula is true for $n = i$, then it is true for $n = i + 1$: Suppose the formula

$$\sum_{k=2}^{n+1} \binom{k}{2} = \binom{2}{2} + \binom{3}{2} + \cdots + \binom{n+1}{2}$$
$$= \binom{n+2}{3}$$

is true when an integer $i \geq 1$ is substituted for n. That is, suppose

$$\sum_{k=2}^{i+1} \binom{k}{2} = \binom{i+2}{3} \qquad \text{for some integer } i \geq 1.$$
Inductive hypothesis

[We must show the formula $\sum_{k=2}^{n+1} \binom{k}{2} = \binom{n+2}{3}$ *is true when $i + 1$ is substituted for n.]*

We must show that

$$\sum_{k=2}^{(i+1)+1} \binom{k}{2} = \binom{(i+1)+2}{3},$$

or, equivalently,

$$\sum_{k=2}^{i+2} \binom{k}{2} = \binom{i+3}{3}.$$

But

$$\sum_{k=2}^{i+2} \binom{k}{2} = \sum_{k=1}^{i+1} \binom{k}{2} + \binom{i+2}{2} \qquad \substack{\text{by writing the}\\ \text{last term}\\ \text{separately}}$$

$$= \binom{i+2}{3} + \binom{i+2}{2} \qquad \substack{\text{by inductive}\\ \text{hypothesis}}$$

$$= \binom{(i+2)+1}{3} \qquad \substack{\text{by Pascal's}\\ \text{formula}}$$

$$= \binom{i+3}{3}.$$

[This is what we needed to show.]
[Since we have proved the basis step and the inductive step, we conclude that the formula is true for all $n \geq 2$.]

15. *Hint:* Use the results of exercises 3 and 14.

17. a. $C_1 = \dfrac{1}{2}\binom{2}{1} = \dfrac{1}{2} \cdot 2 = 1,$

$C_2 = \dfrac{1}{3}\binom{4}{2} = \dfrac{1}{3} \cdot 6 = 2,$

$C_3 = \dfrac{1}{4}\binom{6}{3} = \dfrac{1}{4} \cdot 20 = 5$

19. *Hint:* This follows by letting $m = n = r$ in exercise 18 and using the result of Example 6.6.2.

SECTION 6.7

1. $1 + 7x + \binom{7}{2} \cdot x^2 + \binom{7}{3} \cdot x^3 + \binom{7}{4} \cdot x^4$

$+ \binom{7}{5} \cdot x^5 + \binom{7}{6} \cdot x^6 + x^7$

$= 1 + 7x + 21x^2 + 35x^3 + 35x^4 + 21x^5 + 7x^6 + x^7$

2. $1 + 6(-x) + \binom{6}{2} \cdot (-x)^2 + \binom{6}{3} \cdot (-x)^3$

$+ \binom{6}{4} \cdot (-x)^4 + \binom{6}{5} \cdot (-x)^5 + (-x)^6$

$= 1 - 6x + 15x^2 - 20x^3 + 15x^4 - 6x^5 + x^6$

7. Term is $\binom{9}{3}x^6y^3 = 84x^6y^3$. Coefficient is 84.

9. Term is $\binom{12}{7}a^5(-2b)^7 = 792 \cdot a^5 \cdot (-128) \cdot b^7 = -101{,}376a^5b^7$. Coefficient is $-101{,}376$.

11. $(1.1)^{10\,000}$ is larger because

$$(1.1)^{10\,000} = 1^{10\,000} + \binom{10\,000}{1} \cdot 1^{9999} \cdot (0.1)^1$$

$$+ \text{ other positive terms}$$
$$= 1 + 1000$$
$$+ \text{ other positive terms}$$
$$= 1001 + \text{ other positive terms}$$
$$> 1000$$

13. *Proof:* Let $a = 1$, $b = -1$, and let n be a positive integer. Substitute into the binomial theorem to obtain

$$(1 + (-1))^n = \sum_{k=0}^{n} \binom{n}{k} \cdot 1^{n-k} \cdot (-1)^k$$

$$= \sum_{k=0}^{n} \binom{n}{k} \cdot (-1)^k \qquad \text{since } 1^{n-k} = 1.$$

But $(1 + (-1))^n = 0^n = 0$. So

$$0 = \sum_{k=0}^{n} \binom{n}{k} \cdot (-1)^k$$

$$= \binom{n}{0} - \binom{n}{1} + \binom{n}{2} - \binom{n}{3} + \cdots + (-1)^n \binom{n}{n}.$$

14. *Hint:* $3 = 1 + 2$

15. *Hint:* The answer is 4^n.

17. b. $n(1 + x)^{n-1} = \sum_{k=1}^{n} \binom{n}{k} kx^{k-1}.$

[The term corresponding to $k = 0$ is zero because $\dfrac{d}{dx}(x^0) = 0.$]

c. (i) Substitute $x = 1$ in part (b) above to obtain

$$n(1 + 1)^{n-1} = \sum_{k=1}^{n} \binom{n}{k} k \cdot 1^{k-1} = \sum_{k=1}^{n} \binom{n}{k} k =$$

$$\binom{n}{1} \cdot 1 + \binom{n}{2} \cdot 2 + \binom{n}{3} \cdot 3 + \cdots + \binom{n}{n} \cdot n.$$

Dividing both sides by n and simplifying gives

$$2^{n-1} = \frac{1}{n}\left[\binom{n}{1} + 2 \cdot \binom{n}{2} + 3 \cdot \binom{n}{3} + \cdots + n \cdot \binom{n}{n}\right].$$

SECTION 7.1

1. a. domain of $f = \{1, 3, 5\}$, co-domain of $f = \{s, t, u, v\}$
   b. $f(1) = v, f(3) = s, f(5) = v$
   c. range of $f = \{s, v\}$
   d. inverse image of $s = \{3\}$, inverse image of $u = \emptyset$, inverse image of $v = \{1, 5\}$
   e. $\{(1, v), (3, s), (5, v)\}$

3. a. This arrow diagram does not define a function be-cause there are two arrows coming out of the 2.
   b. This diagram does not define a function because the element 5 in the domain is not related to any element in the co-domain. (There is no arrow coming out of the 5.)

4. a. There are four functions from X to Y as shown below.

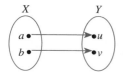

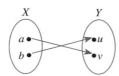

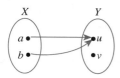

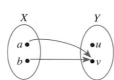

5. a. The answer is $4 \cdot 4 \cdot 4 = 4^3 = 64$. Imagine creating a function from a 3-element set to a 4-element set as a three-step process: Step 1 is to send the first ele-ment of the 3-element set to an element of the 4-element set (there are four ways to perform this step); step 2 is to send the second element of the 3-element set to an element of the 4-element set (there are also four ways to perform this step); and step 3 is to send the third element of the 3-element set to an element of the 4-element set (there are four ways to perform this step). Thus the entire process can be performed in $4 \cdot 4 \cdot 4$ different ways.

6. For all $x \in \mathbf{R}$

$$g(x) = \frac{2x^3 + 2x}{x^2 + 1} = \frac{2x(x^2 + 1)}{x^2 + 1}$$

$$= 2x = f(x) \quad \text{since } x^2 + 1 \neq 0 \text{ for any real number } x$$

Hence, $f = g$.

8. $F \cdot G$ and $G \cdot F$ are equal because for all real numbers x,

$$(F \cdot G)(x) = F(x) \cdot G(x) \quad \text{by definition of } F \cdot G$$
$$= G(x) \cdot F(x) \quad \text{by the commutative law of multiplication of real numbers}$$
$$= (G \cdot F)(x) \quad \text{by definition of } G \cdot F.$$

11. a. The sequence is given by the function $f : \mathbf{Z}^{nonneg} \to \mathbf{R}$ defined by the rule

$$f(n) = \frac{(-1)^n}{2n + 1} \quad \text{for all nonnegative integers } n.$$

12. a. 1 *[because there are an odd number of elements in $\{1, 3, 4\}$]*
    c. 0 *[because there are an even number of elements in $\{2, 3\}$]*

13. a. $f(aba) = 0$ *[because there are no b's to the left of the left-most a in aba]*
    $f(bbab) = 2$ *[because there are two b's to the left of the left-most a in bbab]*
    $f(b) = 0$ *[because the string b contains no a's]*

    range of $f = \mathbf{Z}^{nonneg}$

14. a. $2^3 = 8$ c. $4^1 = 4$

15. a. $\log_3 81 = 4$ because $3^4 = 81$
    c. $\log_3(1/27) = -3$ because $3^{-3} = 1/27$

16. Let b be any positive real number with $b \neq 1$, since $b^1 = b$, by definition of logarithm $\log_b b = 1$.

18. Suppose b and y are positive real numbers with $\log_b y = 3$. By definition of logarithm, this implies that $b^3 = y$, Then

$$y = b^3 = \frac{1}{\frac{1}{b^3}} = \frac{1}{(\frac{1}{b})^3} = \left(\frac{1}{b}\right)^{-3}.$$

So by definition of logarithm (with base $1/b$), $\log_{1/b}(y) = -3$.

20. a. $p_1(2, y) = 2, p_1(5, x) = 5$, range of $p_1 = \{2, 3, 5\}$

21. a. $mod(67, 10) = 7$ and $div(67, 10) = 6$ since $67 = 10 \cdot 6 + 7$.

22. a. $E(0110) = 000111111000$ and $D(111111000111) = 1101$

23. a. $H(10101, 00011) = 3$

24. a.

1	2	3		1	2	3		1	2	3
↓	↓	↓		↓	↓	↓		↓	↓	↓
1	2	3		2	1	3		3	2	1

   b.

1	2	3		1	2	3		1	2	3
↓	↓	↓		↓	↓	↓		↓	↓	↓
1	3	2		2	3	1		3	1	2

   c.

1	2	3		1	2	3
↓	↓	↓		↓	↓	↓
2	3	1		3	1	2

25. **a.** domain of f co-domain of f

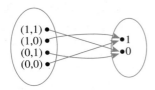

27. **a.** $f(1, 1, 1) = (3 \cdot 1 + 1 + 2 \cdot 1) \bmod 2$
$$= 6 \bmod 2 = 0$$
$f(0, 1, 1) = (3 \cdot 0 + 1 + 2 \cdot 1) \bmod 2$
$$= 3 \bmod 2 = 1$$

28. g is not well-defined. Suppose g were well-defined. Then $g(1/2) = g(2/4)$ since $1/2 = 2/4$, but also $g(1/2) \neq g(2/4)$ because $g(1/2) = 1 - 2 = -1$ and $g(2/4) = 2 - 4 = -2$. This contradiction shows that the supposition that g is well-defined is false. Hence g is not well-defined.

32. **a.** $\phi(15) = 8$ *[because 1, 2, 4, 7, 8, 11, 13, and 14 have no common factors with 15 other than ± 1]*

   b. $\phi(2) = 1$ *[because the only positive integer less than or equal to 2 having no common factors with 2 other than ± 1 is 1]*

   c. $\phi(5) = 4$ *[because 1, 2, 3, and 4 have no common factors with 5 other than ± 1]*

33. *Proof:* Let p be any prime number and n any integer with $n \geq 1$. There are p^{n-1} positive integers less than or equal to p^n that have a common factor other than ± 1 with p^n, namely, $p, 2p, 3p, \ldots, (p^{n-1}) \cdot p$. Hence, by the difference rule, there are $p^n - p^{n-1}$ positive integers less than or equal to p^n that have no common factor with p^n except ± 1.

34. *Hint:* Use the result of exercise 33 with $p = 2$.

35. *Hint:* Let A be the set of all divisors of n that are divisible by p, and let B be the set of all divisors of n that are divisible by q. Then $\phi(n) = n - (n(A \cup B))$.

37. The statement is true. *Proof:* Let f be a function from X to Y, and suppose $A \subseteq X$, $B \subseteq X$, and $A \subseteq B$. Let $y \in f(A)$. *[We must show that $y \in f(B)$.]* Then by definition of image of a set, $y = f(x)$ for some $x \in A$. Since $A \subseteq B, x \in B$, and so $y = f(x)$ for some $x \in B$. Hence $y \in f(B)$ *[as was to be shown].*

39. The statement is false. *Counterexample:* Let $X = \{1, 2, 3\}$, $Y = \{a, b\}$, and define a function $f: X \to Y$ by the arrow diagram shown below.

Let $A = \{1, 3\}$ and $B = \{2, 3\}$.
Then $f(A) = \{a, b\} = f(B)$,
and so $f(A) \cap f(B) = \{a, b\}$.
But $f(A \cap B) = f(\{3\})$
$= \{b\} \neq \{a, b\}$.

And so $f(A) \cap f(B) \neq f(A \cap B)$.

(This is just one of many possible counterexamples.)

41. The statement is true. *Proof:* Let f be a function from a set X to a set Y, and suppose $C \subseteq Y$, $D \subseteq Y$, and $C \subseteq D$. *[We must show that $f^{-1}(C) \subseteq f^{-1}(D)$.]* Suppose $x \in f^{-1}(C)$. Then $f(x) \in C$. Since $C \subseteq D, f(x) \in D$ also. Hence by definition of inverse image, $x \in f^{-1}(D)$. *[So $f^{-1}(C) \subseteq f^{-1}(D)$.]*

42. *Hint:* $x \in f^{-1}(C \cup D) \Leftrightarrow f(x) \in C \cup D \Leftrightarrow f(x) \in C$ or $f(x) \in D$

SECTION 7.2

1. **a.** 20¢ or more deposited

2. **a.** s_0, s_1, s_2 **b.** 0, 1   **c.** s_0 **d.** s_2
   e. annotated next-state table:

		input	
		0	1
$\to$	s_0	s_1	s_0
state	s_1	s_1	s_2
◎	s_2	s_2	s_2

5. **a.** A, B, C, D, E, F **b.** x, y **c.** A **d.** D, E
   e. annotated next-state table:

		input	
		x	y
$\to$	A	C	B
	B	F	D
state	C	E	F
◎	D	F	D
◎	E	E	F
	F	F	F

7. **a.** s_0, s_1, s_2, s_3 **b.** 0, 1   **c.** s_0 **d.** s_0, s_2
   e. annotated next-state table:

		input	
		0	1
$\to$ ◎	s_0	s_0	s_1
state	s_1	s_1	s_2
◎	s_2	s_2	s_3
	s_3	s_3	s_0

8. **a.** s_0, s_1, s_2 **b.** 0, 1   **c.** s_0 **d.** s_2

e.

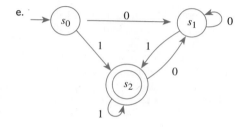

10. a. $N(s_1, 1) = s_2$, $N(s_0, 1) = s_3$
    c. $N^*(s_0, 10011) = s_2$, $N^*(s_1, 01001) = s_2$

11. a. $N(s_3, 0) = s_4$, $N(s_2, 1) = s_4$
    c. $N^*(s_0, 010011) = s_3$, $N^*(s_3, 01101) = s_4$

12. a. (i) s_2 (ii) s_2 (iii) s_1
    b. those in (i) and (ii) but not (iii)

    c. The language accepted by this automaton is the set of
       all strings of 0's and 1's that contain at least one 0
       followed (not necessarily immediately) by at least
       one 1.

14. The language accepted by this automaton is the set of
    all strings of 0's and 1's that end 00.

15. The language accepted by this automaton is the set of
    all strings of x's and y's that consist either entirely of
    x's or entirely of y's.

17. The language accepted by this automaton is the set of
    all strings of 0's and 1's with the following property: If
    n is the number of 1's in the string, then $n \bmod 4 = 0$
    or $n \bmod 4 = 2$. This is equivalent to saying that n is
    even.

18. The language accepted by this automaton is the set of
    all strings of 0's and 1's that end in 1.

20. Call the automaton being constructed A. Acceptance of
    a string by A depends on the values of three consecutive
    inputs. Thus A requires at least four states:

    s_0: initial state

    s_1: state indicating that the last input character was a 1

    s_2: state indicating that the last two input characters
    were 1's

    s_3: state indicating that the last three input characters
    were 1's, the acceptance state

    If a 0 is input to a when it is in state s_0, no progress is
    made toward achieving a string of three consecutive 1's.
    Hence A should remain in state s_0. If a 1 is input to A
    when it is in state s_0, it goes to state s_1, which indicates
    that the last input character of the string is a 1. From
    state s_1, A goes to state s_2 if a 1 is input. This indicates
    that the last two characters of the string are 1's. But if
    a 0 is input, A should return to s_0 because the wait for
    a string of three consecutive 1's must start all over

again. When A is in state s_2 and a 1 is input, then a string
of three consecutive 1's is achieved, so A should go to
state s_3. If a 0 is input when A is in state s_2, then
progress toward accumulating a sequence of three consecutive 1's is lost, so A should return to s_0. When A is
in a state s_3 and a 1 is input, then the final three symbols
of the input string are 1's, and so A should stay in state
s_3. If a 0 is input when A is in state s_3, then A should
return to state s_0 to await the input of more 1's. Thus the
transition diagram is as follows:

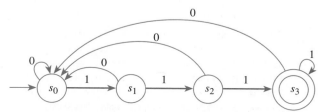

21. *Hint:* Use five states: s_0 (the initial state), s_1 (the state
    indicating that the previous input symbol was an a), s_2
    (the state indicating that the previous input symbol was
    a b), s_3 (the state indicating that the previous two input
    symbols were a's), and s_4 (the state indicating that the
    previous two symbols were b's).

23.

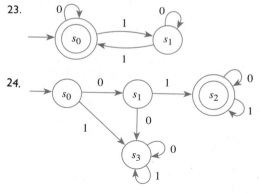

24.

26.

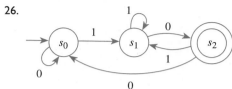

27.

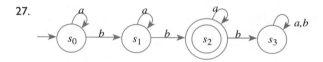

29.

SECTION 7.3

I. a. The second statement is the contrapositive of the first.

b. Consider the function defined by the arrow diagram shown below:

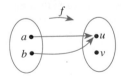

Observe that a is sent to exactly one element of Y, namely, u, and b is also sent to exactly one element of Y, namely, u also. So it is true that every element of X is sent to exactly one element of Y. But f is not one-to-one because $f(a) = f(b)$ but $a \neq b$. Note that to say "Every element of X is sent to exactly one element of Y" is just another way of saying that in the arrow diagram for the function there is only one arrow coming out of each element of X. But this statement is part of the definition of *any* function, not just a one-to-one function.

c. The statement is true.

2. a. f is not one-to-one because $f(1) = 4 = f(9)$ and $1 \neq 9$. f is not onto because $f(x) \neq 3$ for any x in X.

b. g is one-to-one because $g(1) \neq g(5)$, $g(1) \neq g(9)$, and $g(5) \neq g(9)$. g is onto because each element of Y is the image of some element of X: $3 = g(5)$, $4 = g(9)$, and $7 = g(1)$.

3. a. F is not one-to-one because $F(c) = x = F(d)$ and $c \neq d$. F is onto because each element of Y is the image of some element of X: $x = F(c) = F(d)$, $y = F(a)$, and $z = F(b)$.

5. a. One example of many is the following:

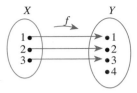

6. a. There are four choices for where to send the first element of the domain (any element of the co-domain may be chosen), three choices for where to send the second (since the function is one-to-one, the second element of the domain must go to a different element of the co-domain from the one to which the first element went), and two choices for where to send the third element (again since the function is one-to-one). Thus the answer is $4 \cdot 3 \cdot 2 = 24$.

b. none

e. *Hint:* The answer is

$$n \cdot (n-1) \cdot \ldots \cdot (n - m + 1).$$

7. a. Let the elements of the domain be called a, b, and c and the elements of the co-domain be called u and v. In order for a function from $\{a, b, c\}$ to $\{u, v\}$ to be onto, two elements of the domain must be sent to u and one to v or two elements must be sent to v and one to u. There are as many ways to send two elements of the domain to u and one to v as there are ways to choose which elements of $\{a, b, c\}$ to send to u, namely $\binom{3}{2} = 3$. Similarly there are $\binom{3}{2} = 3$ ways to send two elements of the domain to v and one to u. Therefore, there are $3 + 3 = 6$ onto functions from a set with three elements to a set with two elements.

c. *Hint:* The answer is 14.

e. *Hint:* Let X be a set with m elements and Y be a set with n elements, where $m \geq n \geq 1$, and let $x \in X$. Then

$$C_{m,n} = \begin{bmatrix} \text{the number of onto} \\ \text{functions from } X \text{ to } Y \end{bmatrix}$$

$$= \begin{bmatrix} \text{the number of onto} \\ \text{functions for which} \\ f^{-1}(f(x)) \text{ has more} \\ \text{than one element} \end{bmatrix}$$

$$+ \begin{bmatrix} \text{the number of onto} \\ \text{functions for which} \\ f^{-1}(f(x)) \text{ has one} \\ \text{element} \end{bmatrix} \quad [Why?]$$

$$= \begin{bmatrix} \text{the number of} \\ \text{choices for} \\ \text{where to send } x \end{bmatrix} \cdot \begin{bmatrix} \text{the number of} \\ \text{onto functions} \\ \text{from } X - \{x\} \text{ to } Y \end{bmatrix}$$

$$+ \begin{bmatrix} \text{the number} \\ \text{of choices} \\ \text{for where} \\ \text{to send } x \end{bmatrix} \cdot \begin{bmatrix} \text{the number of} \\ \text{onto fractions} \\ \text{from } X - \{x\} \\ \text{to } Y - \{f(x)\} \end{bmatrix} \quad [Why?]$$

8. a. (i) f *is one-to-one:* Suppose $f(n_1) = f(n_2)$ for some integers n_1 and n_2. *[We must show that $n_1 = n_2$.]* By definition of f, $2n_1 = 2n_2$, and dividing both sides by 2 gives $n_1 = n_2$, as was to be shown.

(ii) f *is not onto:* Consider $1 \in \mathbf{Z}$. We claim that $1 \neq f(n)$, for any integer n. For if there were an integer n such that $1 = f(n)$, then by definition of f, $1 = 2n$. Dividing both sides by 2 would give $n = 1/2$. But $1/2$ is not an integer. Hence $1 \neq f(n)$ for any integer n, and so f is not onto.

b. h *is onto:* Suppose $m \in 2\mathbf{Z}$. *[We must show that there exists an integer n such that $h(n) = m$.]* Since $m \in 2\mathbf{Z}$, $m = 2k$ for some integer k. Let $n = k$. Then $h(n) = 2n = 2k = m$. Hence there exists an integer (namely, n) such that $h(n) = m$. This is what was to be shown.

10. a. (i) *H is not one-to-one:* $H(1) = 1 = H(-1)$ but $1 \neq -1$.

(ii) *H is not onto:* $H(x) \neq -1$ for any real number x (since no real numbers have negative squares).

11. *Hint:* See exercise 1(b) and its answer.

12. f is one-to-one. *Proof:* Suppose $f(x_1) = f(x_2)$ where x_1 and x_2 are nonzero real numbers. *[We must show that $x_1 = x_2$.]* By definition of f,

$$\frac{x_1 + 1}{x_1} = \frac{x_2 + 1}{x_2}$$

cross-multiplying gives

$$x_1 x_2 + x_2 = x_1 x_2 + x_1,$$

and so

$$x_1 = x_2 \quad \begin{array}{l}\text{by subtracting } x_1 x_2 \\ \text{from both sides}\end{array}$$

[This is what was to be shown.]

13. f is not one-to-one. Note that

$$\frac{x_1}{x_1^2 + 1} = \frac{x_2}{x_2^2 + 1} \Rightarrow x_1 x_2^2 + x_1 = x_2 x_1^2 + x_2$$

$$\Rightarrow x_1 x_2^2 - x_2 x_1^2 = x_2 - x_1$$

$$\Rightarrow x_1 x_2 (x_2 - x_1) = x_2 - x_1$$

$$\Rightarrow x_1 = x_2 \text{ or } x_1 x_2 = 1.$$

So for a counterexample take any x_1 and x_2 with $x_1 \neq x_2$ but $x_1 x_2 = 1$. For instance, take $x_1 = 2$ and $x_2 = 1/2$. Then $f(x_1) = f(2) = 2/5$ and $f(x_2) = f(1/2) = 2/5$, but $2 \neq 1/2$.

16. a. Note that because $\dfrac{417302072}{7} \cong 59614581.7$ and

$417302072 - 7 \cdot 59614581 = 5$,

$h(417\text{-}30\text{-}2072) = 5$. But position 5 is already occupied, so the next position is checked. It is free, and thus the record is placed in position 6.

17. Recall that $\lfloor x \rfloor =$ that unique integer n such that $n \leq x < n + 1$.

a. *Floor is not one-to-one:*
Floor$(0) = 0 =$ Floor$(1/2)$ but $0 \neq 1/2$.

b. *Floor is onto:* Suppose $m \in \mathbf{Z}$. *[We must show that there exists a real number y such that Floor$(y) = m$.]* Let $y = m$. Then Floor$(y) =$ Floor$(m) = m$ since m is an integer. (Actually, Floor takes the value m for *all* real numbers in the interval $m \leq x < m + 1$.) Hence there exists a real number y such that Floor$(y) = m$. This is what was to be shown.

18. a. l is not one-to-one: $l(0) = l(1) = 1$ but $1 \neq 0$.

b. l is onto: Suppose n is a nonnegative integer. *[We must show that there exists a string s in Σ^* such that $l(s) = n$.]* Let

$$s = \begin{cases} \varepsilon \text{ (the null string)} & \text{if } n = 0 \\ \underbrace{00 \ldots 0}_{n \text{ 0's}} & \text{if } n > 0. \end{cases}$$

Then $l(s) =$ the length of $s = n$. This is what was to be shown.

19. a. *F is not one-to-one:* Let $A = \{a\}$ and $B = \{b\}$. Then $F(A) = F(B) = 1$ but $A \neq B$.

20. b. *N is not onto:* The number -1 is in $\mathbf{Z}$ but $N(s) \neq -1$ for any string s in Σ^* because no string has a negative number of a's.

23. a. Let $x = \log_8 27$ and $y = \log_2 3$. *[The question is: Is $x = y$?]* By definition of logarithm, both of these equations can be written in exponential form as

$$8^x = 27 \quad \text{and} \quad 2^y = 3.$$

Now $8 = 2^3$. So

$$8^x = (2^3)^x = 2^{3x}.$$

Also $27 = 3^3$ and $3 = 2^y$. So

$$27 = 3^3 = (2^y)^3 = 2^{3y}.$$

Hence since $8^x = 27$,

$$2^{3x} = 2^{3y}.$$

By (7.3.4), then,

$$3x = 3y,$$

and so

$$x = y.$$

But $x = \log_8 27$ and $y = \log_2 3$, and so $\log_8 27 = y = \log_2 3$ and the answer to the question is yes.

24. *Proof:* Suppose that b, x, and y are positive real numbers and $b \neq 1$. Let $u = \log_b(x)$ and $v = \log_b(y)$. By definition of logarithm, $b^u = x$ and $b^v = y$. By substitution, $\frac{x}{y} = \frac{b^u}{b^v} = b^{u-v}$ *[by (7.3.1) and the fact that $b^{-v} = \frac{1}{b^v}$]*. Translating $\frac{x}{y} = b^{u-v}$ into logarithmic form gives $\log_b(\frac{x}{y}) = u - v$, and so by substitution, $\log_b(\frac{x}{y}) = \log_b(x) - \log_b(y)$ *[as was to be shown]*.

26. *Proof:* Suppose a, b, and x are *[particular but arbitrarily chosen]* real numbers such that b and x are positive and $b \neq 1$. *[We must show that $\log_b(x^a) = a \cdot \log_b x$.]* Let

$$r = \log_b(x^a) \text{ and } s = \log_b x$$

By definition of logarithm, these equations may be written in exponential form as

$$(*) \; b^r = x^a \quad \text{and} \quad (**) \; b^s = x$$

Then

$$b^r = x^a = (b^s)^a \quad \text{by substituting } (**) \text{ into } (*)$$
$$= b^{s \cdot a} \quad \text{by (7.3.2.)}$$

It follows from property (7.3.4) that $r = s \cdot a$ which equals $a \cdot s$. By substituting the values of r and s into this equation, $\log_b(x^a) = a \cdot \log_b x$ *[as was to be shown.]*

27. No. *Counterexample:* Define $f: \mathbf{R} \to \mathbf{R}$ and $g: \mathbf{R} \to \mathbf{R}$ as follows: $f(x) = x$ and $g(x) = -x$ for all real numbers x. Then f and g are both one-to-one *[because for all real numbers x_1 and x_2, if $f(x_1) = f(x_2)$ then $x_1 = x_2$, and if $g(x_1) = g(x_2)$ then $-x_1 = -x_2$ and so $x_1 = x_2$ also]*, but $f + g$ is not one-to-one *[because $f + g$ satisfies the equa-*

tion $(f + g)(x) = x + (-x) = 0$ *for all real numbers x, and so, for instance,* $(f + g)(1) = (f + g)(2)$ *but* $1 \neq 2$].

29. Yes. *Proof:* Let b be a one-to-one function from **R** to **R**, and let c be any nonzero real number. Suppose $(c \cdot f)(x_1) = (c \cdot f)(x_2)$. *[We must show that* $x_1 = x_2$.] It follows by definition of $c \cdot f$ that $c \cdot f(x_1) = c \cdot f(x_2)$. Since $c \neq 0$, we may divide both sides of the equation by c to obtain $f(x_1) = f(x_2)$. But since f is one-to-one, this implies that $x_1 = x_2$ *[as was to be shown]*.

31.

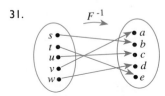

33. The function is not onto. Hence it is not a one-to-one correspondence.

34. The answer to exercise 8(b) shows that h is onto. To show that h is one-to-one, suppose $h(n_1) = h(n_2)$. By definition of h, this implies that $2n_1 = 2n_2$. Dividing both sides by 2 gives $n_1 = n_2$. Hence h is one-to-one.

Given any even integer m, if $m = h(n)$, then by definition of h, $m = 2n$, and so $n = m/2$. Thus

$$h^{-1}(m) = \frac{m}{2} \quad \text{for all } m \in 2\mathbf{Z}.$$

35. This function is not onto. Therefore it is not a one-to-one correspondence.

38. The function is not one-to-one. Hence it is not a one-to-one correspondence.

39. The function is not one-to-one. Hence it is not a one-to-one correspondence.

42. The answer to exercise 12 shows that f is one-to-one, and if the co-domain is taken to be the set of all real numbers not equal to 1, then f is also onto. *[The reason is that given any real number* $y \neq 1$, *if we take*

$x = \dfrac{1}{y - 1}$, *then*

$$f(x) = f\left(\frac{1}{y-1}\right) = \frac{\frac{1}{y-1} + 1}{\frac{1}{y-1}} = \frac{1 + (y-1)}{1} = y.]$$

$$f^{-1}(y) = \frac{1}{y - 1} \quad \text{for each real number } y \neq 1.$$

43. *Hint:* Is there a real number x so that $f(x) = 1$?

47. *Hint:* Let a function F be given and suppose the domain of F is represented as a one-dimensional array $a[1]$, $a[2], \ldots , a[n]$. Introduce a variable *answer* whose initial value is "one-to-one." The main part of the body of

the algorithm could be written as follows:

while ($i \leq n - 1$ and *answer* = "one-to-one")
    $j := i + 1$
    while ($j \leq n$ and *answer* = "one-to-one")
        if ($F(a[i]) = F(a[j])$ and $a[i] \neq a[j]$)
        then *answer* := "not one-to-one"
        $j := j + 1$
    end while
    $i := i + 1$
end while

What can you say if execution reaches this point?

48. *Hint:* Let a function F be given and suppose the domain and co-domain of F are represented by the one-dimensional arrays $a[1], a[2], \ldots , a[n]$ and $b[1], b[2], \ldots , b[m]$, respectively. Introduce a variable *answer* whose initial value is "onto." For each $b[i]$ from $i = 1$ to m, make a search through $a[1], a[2], \ldots , a[n]$ to check whether $b[i] = F(a[j])$ for some $a[j]$. Introduce a Boolean variable to indicate whether a search has been successful. (Set the variable equal to 0 before the start of each search and let it have the value 1 if the search is successful.) At the end of each search, check the value of the Boolean variable. If it is 0, then F is not onto. If all searches are successful, then F is onto.

SECTION 7.4

1. a. No. For instance, the aces of the four different suits could be selected.
   b. Yes. Consider the function S that sends each card to its suit.

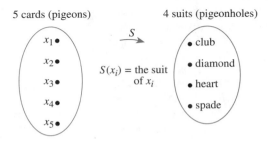

By the pigeonhole principle, S is not one-to-one: $S(x_i) = S(x_j)$ for some two cards x_i and x_j. Hence, at least two cards have the same suit.

3. Yes. Consider the function B from residents to birthdays that sends each resident to his or her birthday:

500 residents (pigeons)   366 birthdays (pigeonholes)

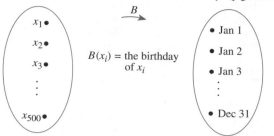

$B(x_i)$ = the birthday of x_i

By the pigeonhole principle, B is not one-to-one: $B(x_i) = B(x_j)$ for some two residents x_i and x_j. Hence, at least two residents have the same birthday.

5. a. Yes. There are only three possible remainders that can be obtained when an integer is divided by 3: 0, 1, and 2. Thus by the pigeonhole principle, if four integers are each divided by 3, then at least two of them must have the same remainder. More formally, consider the function R that sends each integer to the remainder obtained when that integer is divided by 3:

4 integers (pigeons)   3 remainders (pigeonholes)

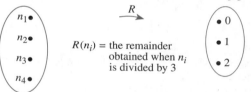

$R(n_i)$ = the remainder obtained when n_i is divided by 3

By the pigeonhole principle, R is not one-to-one: $R(x_i) = R(x_j)$ for some two integers x_i and x_j. Hence, at least two integers must have the same remainder.

  b. No. For instance, $\{0, 1, 2\}$ is a set of three integers no two of which have the same remainder when divided by 3.

7. *Hint:* Look at Example 7.4.3.

9. a. Yes.
Solution 1: Only six of the numbers from 1 to 12 are even (namely, 2, 4, 6, 8, 10, 12), so at most six even numbers can be chosen from between 1 and 12 inclusive. Hence if seven numbers are chosen, at least one must be odd.
Solution 2: Partition the set of all integers from 1 through 12 into six subsets (the pigeonholes), each consisting of an odd and an even number: $\{1, 2\}$, $\{3, 4\}$, $\{5, 6\}$, $\{7, 8\}$, $\{9, 10\}$, $\{11, 12\}$. If seven integers (the pigeons) are chosen from among 1 through 12, then by the pigeonhole principle, at least two must be from the same subset. But each subset contains one odd and one even number. Hence at least one of the seven numbers is odd.
Solution 3: Let $S = \{x_1, x_2, x_3, x_4, x_5, x_6, x_7\}$ be a set of seven numbers chosen from the set $T = \{1, 2, 3, 4, 5, 6, 7, 8, 9, 10, 11, 12\}$, and let P be the following partition of T: $\{1, 2\}$, $\{3, 4\}$, $\{5, 6\}$, $\{7, 8\}$, $\{9, 10\}$, and

$\{11, 12\}$. Since each element of S lies in exactly one subset of the partition, we can define a function F from S to P by letting $F(x_i)$ be the subset that contains x_i.

S (pigeons) P (pigeonholes)

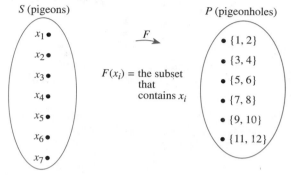

$F(x_i)$ = the subset that contains x_i

Since S has 7 elements and P has 6 elements, by the pigeonhole principle, F is not one-to-one. So two distinct numbers of the seven are sent to the same subset, which implies that these two numbers are the two distinct elements of the subset. Therefore, since each pair consists of one odd and one even integer, one of the seven numbers is odd.

  b. No. For instance, none of the 10 numbers 1, 3, 5, 7, 9, 11, 13, 15, 17, 19 is even.

10. Yes. There are n even integers in the set $\{1, 2, 3, \ldots, 2n\}$, namely $2\,(= 2 \cdot \underline{1})$, $4\,(= 2 \cdot \underline{2})$, $6\,(= 2 \cdot \underline{3})$, $\ldots$, $2n\,(= 2 \cdot \underline{n})$. So the maximum number of even integers that can be chosen is n. Thus if $n + 1$ integers are chosen, at least one of them must be odd.

12. The answer is 27. There are only 26 black cards in a standard 52-card deck, so at most 26 black cards can be chosen. Hence if 27 are taken, at least one must be red.

14. There are 61 integers from 0 to 60 inclusive. Of these, 31 are even $(0 = 2 \cdot \underline{0}, 2 = 2 \cdot \underline{1}, 4 = 2 \cdot \underline{2}, \ldots, 60 = 2 \cdot \underline{30})$ and so 30 are odd. Hence if 32 integers are chosen, at least one must be odd, and if 31 integers are chosen, at least one must be even.

17. The answer is 8. (There are only seven possible remainders for division by 7: 0, 1, 2, 3, 4, 5, 6.)

20. The answer is 20,483 [*namely* 0, 1, 2, $\ldots$, 20482].

22. *Hint:* Let A be the set of the 13 chosen numbers, and let B be the set of all prime numbers between 1 and 40. For each x in A, let $F(x)$ be the smallest prime number that divides x.

23. *Hint:* This proof is virtually identical to that of Example 7.4.5. Just take p and q in that proof so that $p > q$. From the fact that A accepts $a^p b^p$, you can deduce that A accepts $a^q b^p$. Since $p > q$, this string is not in L.

25. *Hint:* Suppose the automaton A has N states. Choose an integer m such that $(m + 1)^2 - m^2 > N$. Consider strings of a's of lengths between m^2 and $(m + 1)^2$.

Since there are more strings than states, at least two strings must send A to the same state s_i:

$$\overbrace{\underbrace{aa \ldots aaa}_{m^2} \ldots a\,aa \ldots aaa \ldots a}^{(m + 1)^2}$$

after both of these
inputs, A is in state s_i

It follows (by removing the a's shown in color) that the automaton must accept a string of the form a^k, where $m^2 < k < (m + 1)^2$.

26. Yes. This follows from the generalized pigeonhole principle with 30 pigeons, 12 pigeonholes, and $k = 2$, using the fact that $30 > 2 \cdot 12$.

27. No. For instance, the birthdays of the 30 people could be distributed as follows: three birthdays in each of the six months January through June and two birthdays in each of the six months July through December.

30. The answer is $x = 3$. There are 18 years from 17 through 34. Now $40 > 18 \cdot 2$ so by the generalized pigeonhole principle, you can be sure that there are at least $x = 3$ students of the same age. However, since $18 \cdot 3 > 40$, you cannot be sure of having more than three students with the same age. (For instance, three students could be each of the ages 17 through 20 and two could be each of the ages from 21 through 34.) So x cannot be taken to be greater than 3.

32. *Hint:* Use the same type of reasoning as in Example 7.4.7.

33. *Hints:* (1) The number of subsets of the six integers is $2^6 = 64$. (2) Since each integer is less than 13, the largest possible sum is 57. (Why? What gives this sum?)

34. *Hint:* Let X be the set consisting of the given 52 positive integers and let Y be the set containing the following elements: $\{00\}, \{50\}, \{01, 99\}, \{02, 98\}, \{03, 97\}, \ldots, \{48, 52\}, \{49, 51\}$. Define a function F from X to Y by the rule $F(x) =$ the set containing the last two digits of x. Use the pigeonhole principle to argue that F is not one-to-one and show how the desired conclusion follows.

35. *Hint:* Represent each of the 101 integers x_i as $2^{k_i} \cdot a_i$ where a_i is odd. Now $1 \le x_i \le 200$, and so $1 \le a_i \le 199$ for all i. There are only 100 odd integers from 1 to 199 inclusive.

36. b. *Hint:* For each $k = 1, 2, \ldots, n$, let $a_k = x_1 + x_2 + \cdots + x_k$. If some a_k is divisible by n, the problem is solved: the consecutive subsequence is $x_1, x_2, \ldots, x_k$. If no a_k is divisible by n, then $a_1, a_2, a_3, \ldots$ satisfies the hypothesis of part (a). Hence $a_j - a_i$ is divisible by n for some integers i and j with $j > i$. Write $a_j - a_i$ in terms of the x_i's to derive the given conclusion.

37. *Hint:* Let $a_1, a_2, \ldots, a_{n^2+1}$ be any sequence of $n^2 + 1$ distinct real numbers, and suppose that this sequence contains neither a strictly increasing subsequence of length $n + 1$ nor a strictly decreasing subsequence of length $n + 1$. Let S be the set of all ordered pairs of integers (i, d) where $1 \le i \le n$ and $1 \le d \le n$. For each term a_k in the sequence, let $F(a_k) = (i_k, d_k)$, where i_k is the length of the longest increasing sequence starting at a_k and d_k is the length of the longest decreasing sequence starting at a_k. Show that F is one-to-one and derive a contradiction.

SECTION 7.5

1. $g \circ f$ is defined as follows:

$$(g \circ f)(1) = g(f(1)) = g(5) = 1,$$
$$(g \circ f)(3) = g(f(3)) = g(3) = 5,$$
$$(g \circ f)(5) = g(f(5)) = g(1) = 3.$$

$f \circ g$ is defined as follows:

$$(f \circ g)(1) = f(g(1)) = f(3) = 3,$$
$$(f \circ g)(3) = f(g(3)) = f(5) = 1,$$
$$(f \circ g)(5) = f(g(5)) = f(1) = 5.$$

Then $g \circ f \ne f \circ g$ because, for example, $(g \circ f)(1) \ne (f \circ g)(1)$.

3. $(G \circ F)(x) = G(F(x)) = G(x^3) = x^3 - 1$ for all real numbers x.
$(F \circ G)(x) = F(G(x)) = F(x - 1) = (x - 1)^3$ for all real numbers x.
$G \circ F \ne F \circ G$ because, for instance, $(G \circ F)(2) = 2^3 - 1 = 7$ whereas $(F \circ G)(2) = (2 - 1)^3 = 1$.

6. $(f \circ l)(abaa) = f(l(abaa)) = f(4) = 4 \bmod 3 = 1$
$(f \circ l)(baaab) = f(l(baaab)) = f(5) = 5 \bmod 3 = 2$
$(f \circ l)(aaa) = f(l(aaa)) = f(3) = 3 \bmod 3 = 0$

8. $(f^{-1} \circ f)(x) = f^{-1}(f(x)) = f^{-1}(3x + 2)$

$$= \frac{(3x + 2) - 2}{3} = \frac{3x}{3} = x = i_\mathbf{R}(x)$$

for all x in $\mathbf{R}$. Hence $f^{-1} \circ f = i_\mathbf{R}$ by definition of equality of functions.

$$(f \circ f^{-1})(y) = f(f^{-1}(y)) = f\left(\frac{y - 2}{3}\right)$$

$$= 3\left(\frac{y - 2}{3}\right) + 2 = (y - 2) + 2$$

$$= y = i_\mathbf{R}(y)$$

for all y in $\mathbf{R}$. Hence $f \circ f^{-1} = i_\mathbf{R}$ by definition of equality of functions.

11. a. By definition of logarithm with base b, for each real number x, $\log_b(b^x)$ is the exponent to which b must be raised to obtain b^x. But this exponent is just x! So $\log_b(b^x) = x$.

14. a. $s_k = s_m$

15. No. *Counterexample:* Define f and g by the arrow diagrams below.

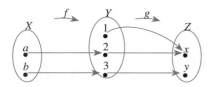

Then $g \circ f$ is one-to-one but g is not one-to-one. (So it is false that *both* f and g are one-to-one by De Morgan's law!)

17. *Hint:* Suppose $f: X \to Y$ and $g: Y \to Z$ are functions and $g \circ f$ is one-to-one. Given x_1 and x_2 in X, if $f(x_1) = f(x_2)$ then $(g \circ f)(x_1) = (g \circ f)(x_2)$. (Why?) Then use the fact that $g \circ f$ is one-to-one.

18. *Hint:* Suppose $f: X \to Y$ and $g: Y \to Z$ are functions and $g \circ f$ is onto. Given $z \in Z$, there is an element x in X such that $(g \circ f)(x) = z$. (Why?) Let $y = f(x)$. Then $g(y) = z$.

20.

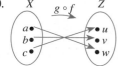
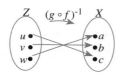

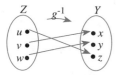
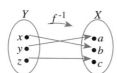

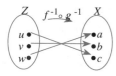

The functions $(g \circ f)^{-1}$ and $f^{-1} \circ g^{-1}$ are equal.

23. *Hints:* (1) Theorems 7.5.3 and 7.5.4 taken together ensure that $g \circ f$ is one-to-one and onto. (2) Use the inverse function identity: $F^{-1}(b) = a \Leftrightarrow F(a) = b$, for all a in the domain of F and b in the domain of F^{-1}.

24. *Hint:* Find a function f for which the given property is false.

25. *Proof:* Let $f: X \to Y$ be any function and let C be any subset of Y. Suppose $y \in f(f^{-1}(C))$. *[We must show that $y \in C$.]* Then by definition of image of a set, $y = f(x)$ for some $x \in f^{-1}(C)$, and so by definition of inverse image of a set, $f(x) \in C$. Hence $y \in C$ *[as was to be shown]*.

SECTION 7.6

1. Define a function $f: \mathbf{Z}^+ \to S$ as follows: For all positive integers k, $f(k) = k^2$.
   f is one-to-one: *[We must show that for all $k_1, k_2 \in \mathbf{Z}^+$, if $f(k_1) = f(k_2)$ then $k_1 = k_2$.]* Suppose k_1 and k_2 are positive integers and $f(k_1) = f(k_2)$. By definition of f, $k_1^2 = k_2^2$, so $k_1 = \pm k_2$. But k_1 and k_2 are *positive*. Hence $k_1 = k_2$.
   f is onto: *[We must show that for all $n \in S$, there exists $k \in \mathbf{Z}^+$ such that $n = f(k)$.]* Suppose $n \in S$. By definition of S, $n = k^2$ for some positive integer k. But then by definition of f, $n = f(k)$.
   Since there is a one-to-one, onto function (namely, f) from $\mathbf{Z}^+$ to S, the two sets have the same cardinality.

3. Define a function $f: \mathbf{Z}^+ \to \mathbf{Z}^{nonneg}$ by the rule $f(n) = n - 1$ for all positive integers n. Observe that if $n \geq 1$ then $n - 1 \geq 0$, so f is well-defined. Now f is one-to-one because for all positive integers n_1 and n_2, if $f(n_1) = f(n_2)$ then $n_1 - 1 = n_2 - 1$ and so $n_1 = n_2$. Also f is onto because if m is any nonnegative integer, then $m + 1$ is a positive integer and $f(m + 1) = (m + 1) - 1 = m$ by definition of f. So since there is a function $f: \mathbf{Z}^+ \to \mathbf{Z}^{nonneg}$ that is one-to-one and onto, $\mathbf{Z}^+$ has the same cardinality as $\mathbf{Z}^{nonneg}$. It follows that $\mathbf{Z}^{nonneg}$ is countably infinite and hence countable.

4. b. $f(n) = (-1)^n \left\lfloor \dfrac{n}{2} \right\rfloor$ for each positive integer n.

5. *Hint:* It is possible to find a formula for such a function that uses the floor notation.

7. In Example 7.6.3 is was shown that there is a one-to-one correspondence from $\mathbf{Z}^+$ to $\mathbf{Q}^+$. This implies that the positive rational numbers can be written as an infinite sequence: $r_1, r_2, r_3, r_4, \ldots$. Now the set $\mathbf{Q}$ of all rational numbers consists of the numbers in this sequence together with 0 and the negative rational numbers: $-r_1, -r_2, -r_3, -r_4, \ldots$. Let $r_0 = 0$. Then the elements of the set of all rational numbers can be "counted" as follows:

$$r_0, r_1, -r_1, r_2, -r_2, r_3, -r_3, r_4, -r_4, \ldots$$

In other words, we can define a one-to-one correspondence

$$F(n) = \begin{cases} r_{n/2} & \text{if } n \text{ is even} \\ -r_{(n-1)/2} & \text{if } n \text{ is odd} \end{cases} \quad \text{for all integers } n \geq 1.$$

Therefore, $\mathbf{Q}$ is countably infinite and hence countable.

9. *Hint:* No.

10. *Hint:* Suppose r and s are real numbers with $s > r > 0$. Let n be an integer such that $n > \dfrac{\sqrt{2}}{s - r}$. Then $s - r > \dfrac{\sqrt{2}}{n}$. Let $m = \left\lfloor \dfrac{nr}{\sqrt{2}} \right\rfloor + 1$. Then

$m > \dfrac{n \cdot r}{\sqrt{2}} \geq m - 1$. Use the fact that

$s = r + (s - r)$ to show that $r < \dfrac{m \cdot \sqrt{2}}{n} < s$.

14. a. Define a function $G: \mathbf{Z}^{nonneg} \to \mathbf{Z}^{nonneg} \times \mathbf{Z}^{nonneg}$ as follows: Let $G(0) = (0,0)$, and then follow the arrows in the diagram letting each successive ordered pair of integers be the value of G for the next successive integer. Thus, for instance, $G(1) = (1,0)$, $G(2) = (0,1)$, $G(3) = (2,0)$, $G(3) = (1,1)$, $G(4) = (0,2)$, $G(5) = (3,0)$, $G(6) = (2,1)$, $G(7) = (1,2)$, and so forth.

   b. *Hint:* Observe that if the top ordered pair of any given diagonal is $(k,0)$, the entire diagonal (moving from top to bottom) consists of $(k,0)$, $(k-1,1)$, $(k-2,2)$, . . . , $(2,k-2)$, $(1,k-1)$, $(0,k)$. Thus for all the ordered pairs (m,n) within any given diagonal, the value of $m + n$ is constant, and as you move down the ordered pairs in the diagonal, starting at the top, the value of the second element of the pair keeps increasing by 1.

16. *Hint:* There are at least two different approaches to this problem. One is to use the method discussed in Section 3.2. Another is to suppose that 1.999999. . . < 2 and derive a contradiction. (Show that the difference between 2 and 1.999999. . . can be made smaller than any given positive number.)

17. *Proof:* Define $f: S \to U$ by the rule $f(x) = 2x$ for all real numbers x in S. Then f is one-to-one by the same argument as in exercise 8a of Section 7.3 with $\mathbf{R}$ in place of $\mathbf{Z}$. Furthermore, f is onto because if y is any element in U, then $0 < y < 2$ and so $0 < y/2 < 1$. Consequently, $y/2 \in S$ and $f(y/2) = 2(y/2) = y$. Hence f is a one-to-one correspondence, and so S and U have the same cardinality.

18. *Hint:* Define $h: S \to V$ as follows: $h(x) = 3x + 2$, for all $x \in S$.

20.

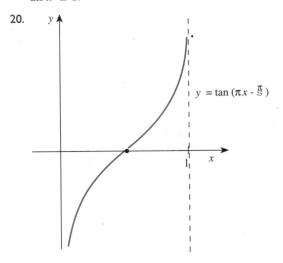

$y = \tan\left(\pi x - \frac{\pi}{5}\right)$

It is clear from the graph that f is one-to-one (since it is increasing) and that the image of f is all of $\mathbf{R}$ (since the lines $x = 0$ and $x = 1$ are vertical asymptotes). Thus S and $\mathbf{R}$ have the same cardinality.

22. Let A be an infinite set. Construct a countably infinite subset $a_1, a_2, a_3, \ldots$ of A letting a_1 be any element of A, letting a_2 be any elment of A other than a_1, letting a_3 be any element of A other than a_1 or a_2, and so forth. This process never stops (and hence $a_1, a_2, a_3, \ldots$ is an infinite sequence) because A is an infinite set. More formally,

   1. Let a_1 be any element of A.
   2. For each integer $n \geq 2$, let a_n be any element of $A - \{a_1, a_2, a_3, \ldots, a_{n-1}\}$. Such an element exists for otherwise $A - \{a_1, a_2, a_3, \ldots, a_{n-1}\}$ would be empty and A would be finite.

26. *Hint for Solution 1:* Define a function $f: \mathcal{P}(S) \to T$ as follows: For each subset A of S, let $f(A) = \chi_A$, the *characteristic function* of A, where $\chi_A: S \to \{0, 1\}$ is defined by the rule

$$\chi_A(x) = \begin{cases} 1 & \text{if } x \in A \\ 0 & \text{if } x \notin A \end{cases} \quad \text{for all } x \in S.$$

Show that f is one-to-one (for all A_1, $A_2 \subseteq S$, if $\chi_{A_1} = \chi_{A_2}$ then $A_1 = A_2$) and that f is onto (given any function $g: S \to \{0, 1\}$, there is a subset A of S such that $g = \chi_A$).

Hint for Solution 2: Define $H: T \to \mathcal{P}(S)$ by letting $H(f) = \{x \in S \mid f(x) = 1\}$. Is H a one-to-one correspondence?

27. *Partial Proof (by contradiction):* Suppose not. Suppose there is a one-to-one, onto function $f: S \to \mathcal{P}(S)$. Let

$$A = \{x \in S \mid x \notin f(x)\}.$$

Then $A \in \mathcal{P}(S)$ and since f is onto, there is a $z \in S$ such that $A = f(z)$. *[Now derive a contradiction!]*

29. *Hint:* Since A and B are countable, their elements can be listed as

$A: a_1, a_2, a_3, \ldots$ and $B: b_1, b_2, b_3, \ldots$

Represent the elements of $A \times B$ in a grid:

$$\begin{array}{llll} (a_1, b_1) & (a_1, b_2) & (a_1, b_3) & \cdots \\ (a_2, b_1) & (a_2, b_2) & (a_2, b_3) & \cdots \\ (a_3, b_1) & (a_3, b_2) & (a_3, b_3) & \cdots \\ \vdots & \vdots & \vdots & \end{array}$$

Now use a counting method similar to that of Example 7.6.3.

SECTION 8.1

1. $a_1 = 1$, $a_2 = 2 \cdot a_1 + 2 = 2 \cdot 1 + 2 = 4$,
$a_3 = 2 \cdot a_2 + 3 = 2 \cdot 4 + 3 = 11$,
$a_4 = 2 \cdot a_3 + 4 = 2 \cdot 11 + 4 = 26$

3. $c_0 = 1$, $c_1 = 1 \cdot (c_0)^2 = 1 \cdot (1)^2 = 1$,
$c_2 = 2 \cdot (c_1)^2 = 2 \cdot (1)^2 = 2$,
$c_3 = 3 \cdot (c_2)^2 = 3 \cdot (2)^2 = 12$

5. $s_0 = 1$, $s_1 = 1$, $s_2 = s_1 + 2 \cdot s_0 = 1 + 2 \cdot 1 = 3$,
$s_3 = s_2 + 2 \cdot s_1 = 3 + 2 \cdot 1 = 5$

7. $u_1 = 1$, $u_2 = 1$, $u_3 = 3 \cdot u_2 - u_1 = 3 \cdot 1 - 1 = 2$,
$u_4 = 4 \cdot u_3 - u_2 = 4 \cdot 2 - 1 = 7$

9. By definition of $a_0, a_1, a_2, \ldots$, for each integer $k \geq 1$,

(*) $a_k = 3k + 1$ and
(**) $a_{k-1} = 3(k - 1) + 1$.

Then $a_{k-1} + 3$

$= 3(k - 1) + 1 + 3$
$= 3k - 3 + 1 + 3$
$= 3k + 1$
$= a_k$

11. Call the nth term of the sequence c_n. Then, by definition, $c_n = 2^n - 1$, for each integer $n \geq 0$. Substitute k and $k - 1$ in place of n to get

(*) $c_k = 2^k - 1$ and
(**) $c_{k-1} = 2^{k-1} - 1$

for all integers $k \geq 1$. Then

$2c_{k-1} + 1$
$= 2(2^{k-1} - 1) + 1$ by substitution from (**)
$= 2^k - 2 + 1$
$= 2^k - 1$ by basic algebra
$= c_k$ by substitution from (*).

13. Call the nth term of the sequence t_n. Then, by definition, $t_n = 2 + n$, for each integer $n \geq 0$. Substitute k, $k - 1$, and $k - 2$ in place of n to get

(*) $t_k = 2 + k$,
(**) $t_{k-1} = 2 + (k - 1)$,   and
(***) $t_{k-2} = 2 + (k - 2)$

for each integer $k \geq 2$. Then

$2t_{k-1} - t_{k-2}$
$= 2(2 + (k - 1)) - (2 + (k - 2))$ by substitution from (**) and (***)
$= 2(k + 1) - k$
$= 2 + k$ by basic algebra
$= t_k$ by substitution from (*)

15. Let k be an integer and $k \geq 2$.
Case 1(k is even): Then

$$a_k = (-2)^{k/2} \quad \text{and} \quad a_{k-2} = (-2)^{(k-2)/2}.$$

So

$-2a_{k-2} = -2 \cdot (-2)^{(k-2)/2}$
$= (-2)^{1+(k-2)/2}$
$= (-2)^{k/2}$ since $1 + \dfrac{k - 2}{2} = \dfrac{2}{2} + \dfrac{k - 2}{2}$
$\qquad\qquad = \dfrac{2 + k - 2}{2}$
$\qquad\qquad = \dfrac{k}{2}$
$= a_k$.

Case 2 (k is odd): Then

$$a_k = (-2)^{(k-1)/2}$$

and

$$a_{k-2} = (-2)^{(k-3)/2} \quad \text{since} \quad \frac{(k - 2) - 1}{2} = \frac{k - 3}{2}$$

So

$-2a_{k-2} = -2 \cdot (-2)^{(k-3)/2}$
$= (-2)^{1+(k-3)/2}$
$= (-2)^{(k-1)/2}$ since $1 + \dfrac{k - 3}{2}$
$\qquad\qquad = \dfrac{2}{2} + \dfrac{k - 3}{2}$
$\qquad\qquad = \dfrac{2 + k - 3}{2}$
$\qquad\qquad = \dfrac{k - 1}{2}$
$= a_k$.

Hence in either case, $a_k = -2a_{k-2}$, as was to be shown.

18. a. $a_1 = 2$

$a_2 = 2$ (moves to move the top disk from pole A to pole C)

  $+ 1$ (move to move the bottom disk from pole A to pole B)

   $+ 2$ (moves to move the top disk from pole C to pole A)

    $+ 1$ (move to move the bottom disk from pole B to pole C)

     $+ 2$ (moves to move top disk from pole A to pole C)

$= 8$

$a_3 = 8 + 1 + 8 + 1 + 8 = 26$

c. For all integers $k \geq 2$.

$a_k = a_{k-1}$ (moves to move the top $k - 1$ disks from pole A to pole C)

  $+ 1$ (move to move the bottom disk from pole A to pole B)

   $+ a_{k-1}$ (moves to move the top disks from pole C to pole A)

    $+ 1$ (move to move the bottom disk from pole B to pole C)

     $+ a_{k-1}$ (moves to move the top disks from pole A to pole C)

$= 3a_{k-1} + 2.$

19. b. $b_4 = 40$

20. a. $s_1 = 1, s_2 = 1 + 1 + 1 = 3,$
$s_3 = s_1 + (1 + 1 + 1) + s_1 = 5$

b. $s_4 = s_2 + (1 + 1 + 1) + s_2 = 9$

21. b. $t_3 = 14$

23. a. Each term of the Fibonacci sequence beyond the second equals the sum of the previous two. For any integer $k \geq 1$, the two terms previous to F_{k+1} are F_k and F_{k-1}. Hence for all integers $k \geq 1$, $F_{k+1} = F_k + F_{k-1}$.

24. a. $F_k^2 - F_{k-1}^2$

$= (F_k - F_{k-1}) \cdot (F_k + F_{k-1})$ by basic algebra (difference of two squares)

$= (F_k - F_{k-1}) \cdot F_{k+1}$ by definition of the Fibonacci sequence

$= F_k F_{k+1} - F_{k-1} F_{k+1}$

25. Let $L = \lim_{n \to \infty} \dfrac{F_{n+1}}{F_n}$. Since each $F_{n+1} > F_n > 0$, $L > 0$. Then by definition of the Fibonacci sequence,

$$L = \lim_{n \to \infty} \left(\frac{F_{n-1} + F_n}{F_n} \right) = \lim_{n \to \infty} \left(\frac{F_{n-1}}{F_n} + \frac{F_n}{F_n} \right)$$

$$= \lim_{n \to \infty} \left(\frac{1}{\dfrac{F_n}{F_{n-1}}} + 1 \right) = \frac{1}{\lim_{n \to \infty} \left(\dfrac{F_n}{F_{n-1}} \right)} + 1$$

$$= \frac{1}{L} + 1.$$

Hence $L = \dfrac{1}{L} + 1$. Multiply both sides by L to obtain $L^2 = 1 + L$, or, equivalently, $L^2 - L - 1 = 0$. By the quadratic formula, then, $L = \dfrac{1 \pm \sqrt{5}}{2}$. But one of these numbers, $\dfrac{1 - \sqrt{5}}{2}$, is less than zero, and $L > 0$.

Hence $L = \dfrac{1 + \sqrt{5}}{2}$.

$$L = \frac{1 + \sqrt{5}}{2}.$$

27. b. $r_0 = 1, r_1 = 1, r_2 = 1 + 4 = 5, r_3 = 5 + 4 = 9,$
$r_4 = 9 + 4 \cdot 5 = 29,$ $r_5 = 29 + 4 \cdot 9 = 65,$
$r_6 = 65 + 4 \cdot 29 = 181$

28. c. There are 904 rabbit pairs or 1,808 rabbits after 12 months.

29. a. $R_k = R_{k-1} + 0.02 R_{k-1} = 1.02 R_{k-1}$

b. $R_4 = \$541.22$ (rounded to the nearest cent)

31. a. length 0: ε
length 1: 0, 1
length 2: 00, 01, 10, 11
length 3: 000, 001, 010, 011, 100, 101, 110
length 4: 0000, 0001, 0010, 0011, 0100, 0101, 0110, 1000, 1001, 1010, 1011, 1100, 1101

b. By part (a), $d_0 = 1, d_1 = 2, d_2 = 4, d_3 = 7,$ and $d_4 = 13.$

c. Let k be an integer with $k \geq 3$. Any string of length k that does not contain the bit pattern 111 either starts with a 0 or with a 1. If it starts with a 0, this can be followed by any string of $k - 1$ bits that does not contain the pattern 111. There are d_{k-1} of these. If the string starts with a 1, then the first two bits are 10 or 11. If the first two bits are 10, then these can be followed by any string of $k - 2$ bits that does not contain the pattern 111. There are d_{k-2} of these. If the string starts with a 11, then the third bit must be 0 (because the string does not contain 111), and these three bits can be followed by any string of $k - 3$ bits that does not contain the pattern 111. There are d_{k-3} of these. Therefore, for all integers $k \geq 3$, $d_k = d_{k-1} + d_{k-2} + d_{k-3}$.

d. By parts (b) and (c), $d_5 = d_4 + d_3 + d_2 = 13 + 7 + 4 = 24$. This is the number of bit strings of length five that do not contain the pattern 111.

32. c. *Hint:* $s_k = 2s_{k-1} + 2s_{k-2}$

34. When climbing a staircase consisting of n stairs, the last step taken is either a single stair or two stairs together. The number of ways to climb the staircase and have the final step be a single stair is c_{n-1}; the number of ways to climb the staircase and have the final step be two stairs is c_{n-2}. Therefore by the addition rule, $c_n = c_{n-1} + c_{n-2}$. Note also that $c_1 = 1$ and $c_2 = 2$ *[because either the two stairs can be climbed one by one or they can be climbed as a unit].*

36. a. $a_3 = 3$ (The three permutations that do not move more than one place from their "natural" positions are 213, 132, and 123.)

38. Call the set X, and suppose that $X = \{x_1, x_2, \ldots, x_n\}$. For each integer $i = 0, 1, 2, \ldots, n - 1$, we can consider the set of all partitions of X (let's call them *partitions of type i*) where one of the subsets of the partition is an $(i + 1)$-element set containing x_n along with i elements chosen from $\{x_1, \ldots, x_{n-1}\}$. The remaining subsets of the partition will be a partition of the remaining $(n - 1) - i$ elements of $\{x_1, \ldots, x_{n-1}\}$. For instance, let $X = \{x_1, x_2, x_3\}$. There are five partitions of the various types, namely,

type 0: two partitions where one set is a 1-element set containing x_3: $\{x_3\}, \{x_1\}, \{x_2\}, \{x_3\}, \{x_1, x_2\}$

type 1: two partitions where one set is a 2-element set containing x_3: $\{x_1, x_3\}, \{x_2\}, \{x_2, x_3\}, \{x_1\}$

type 2: one partition where one set is a 3-element set containing x_3: $\{x_1, x_2, x_3\}$

We can imagine constructing a partition of type i as a two-step process:

step 1 is to select out the i elements of $\{x_1, \ldots, x_{n-1}\}$ to put together with x_n,

step 2 is to choose any partition of the remaining $(n - 1) - i$ elements of $\{x_1, \ldots, x_{n-1}\}$ to put with the set formed in step 1.

There are $\binom{n-1}{i}$ ways to perform step 1 and $P_{(n-1)-i}$ ways to perform step 2. Therefore, by the multiplication rule, there are $\binom{n-1}{i} \cdot P_{(n-1)-i}$ partitions of type i. Because any partition of X is of type i for some $i = 0, 1, 2, \ldots, n - 1$, it follows from the addition rule that the total number of partitions is

$$\binom{n-1}{0} \cdot P_{n-1} + \binom{n-1}{1} \cdot P_{n-2} +$$
$$\binom{n-1}{2} \cdot P_{n-3} + \cdots + \binom{n-1}{n-1} \cdot P_0.$$

40. $S_{5,2} = S_{4,1} + 2 \cdot S_{4,2} = 1 + 2 \cdot 7 = 15$

43. *Proof (by mathematical induction):*
The formula holds for $n = 2$: For $n = 2$ the for-

mula gives $2^{2-1} - 1 = 2^1 - 1 = 1$. But by Example 8.1.11, $S_{2,2} = 1$ also. Hence the formula holds for $n = 2$.

If the formula holds for $n = k$, then it holds for $n = k + 1$: Suppose that for some integer $k \geq 2$, $S_{k,2} = 2^{k-1} - 1$. *[Inductive hypothesis.]* We must show that $S_{k+1,2} = 2^{(k+1)-1} - 1 = 2^k - 1$. But according to Example 8.1.11, $S_{k+1,2} = S_{k,1} + 2S_{k,2}$ and $S_{k,1} = 1$. So by substitution and the inductive hypothesis,

$$S_{k+1,2} = 1 + 2S_{k,2} = 1 + 2(2^{k-1} - 1)$$
$$= 1 + 2^k - 2 = 2^k - 1$$

[as was to be shown].

45. *Hint:* Observe that the number of onto functions from $X = \{x_1, x_2, x_3, x_4\}$ to $Y = \{y_1, y_2, y_3\}$ is $S_{4,3} \cdot 3!$ because the construction of an onto function can be thought of as a two-step process where step 1 is to choose a partition of X into three subsets and step 2 is to choose, for each subset of the partition, an element of Y for the elements of the subset to be sent to.

47. *Hint:* Use mathematical induction. In the inductive step, use Lemma 3.8.2 and the fact that $F_{k+2} = F_{k+1} + F_k$ to deduce that

$$\gcd(F_{k+2}, F_{k+1}) = \gcd(F_{k+1}, F_k).$$

48. c. *Hint:* If $k \geq 6$, any sequence of k games must begin with W, LW, or LLW, where L stands for "lose" and W stands for "win."

49. c. *Hint:* Divide the set of all derangements into two subsets: one subset consists of all derangements in which the number 1 changes places with another number, and the other subset consists of all derangements in which the number 1 goes to position $i \neq 1$ but i does not go to position 1. The answer is $d_k = (k - 1)d_{k-1} + (k - 1)d_{k-2}$. Can you justify it?

SECTION 8.2

1. a. $1 + 2 + 3 + \cdots + (k - 1)$
$$= \frac{(k - 1)((k - 1) + 1)}{2} = \frac{(k - 1)k}{2}$$

b. $3 + 2 + 4 + 6 + 8 + \cdots + 2n$
$$= 3 + 2(1 + 2 + 3 + \cdots + n)$$
$$= 3 + 2\frac{n(n + 1)}{2} = 3 + n(n + 1)$$
$$= n^2 + n + 3$$

2. a. $1 + 2 + 2^2 + \cdots + 2^{i-1} = \dfrac{2^{(i-1)+1} - 1}{2 - 1}$
$$= 2^i - 1$$

c. $2^n + 2^{n-2} \cdot 3 + 2^{n-3} \cdot 3 + \cdots + 2^2 \cdot 3 + 2 \cdot 3 + 3$
$$= 2^n + 3(2^{n-2} + 2^{n-3} + \cdots + 2^2 + 2 + 1)$$
$$= 2^n + 3(1 + 2 + 2^2 + \cdots + 2^{n-3} + 2^{n-2})$$

$$= 2^n + 3\left(\frac{2^{(n-2)+1} - 1}{2 - 1}\right)$$
$$= 2^n + 3(2^{n-1} - 1)$$
$$= 2 \cdot 2^{n-1} + 3 \cdot 2^{n-1} - 3$$
$$= 5 \cdot 2^{n-1} - 3$$

3. $a_0 = 1$
$a_1 = 1 \cdot a_0 = 1 \cdot 1 = 1$
$a_2 = 2 \cdot a_1 = 2 \cdot 1$
$a_3 = 3 \cdot a_2 = 3 \cdot 2 \cdot 1$
$a_4 = 4 \cdot a_3 = 4 \cdot 3 \cdot 2 \cdot 1$
$\vdots$

Guess:
$a_n = n \cdot (n - 1) \cdot \ldots \cdot 3 \cdot 2 \cdot 1 = n!$

5. $c_1 = 1$
$c_2 = 3c_1 + 1 = 3 \cdot 1 + 1 = 3 + 1$
$c_3 = 3c_2 + 1 = 3 \cdot (3 + 1) + 1 = 3^2 + 3 + 1$
$c_4 = 3c_3 + 1 = 3 \cdot (3^2 + 3 + 1) + 1$
$\quad = 3^3 + 3^2 + 3 + 1$
$\vdots$

Guess:
$c_n = 3^{n-1} + 3^{n-2} + \cdots + 3^3 + 3^2 + 3 + 1$
$\quad = \dfrac{3^n - 1}{3 - 1}$ by Theorem 4.2.3 with $r = 3$
$\quad = \dfrac{3^n - 1}{2}$

7. $e_0 = 3$
$e_1 = e_0 + 2 \cdot 1 = 3 + 2 \cdot 1$
$e_2 = e_1 + 2 \cdot 2 = [3 + 2 \cdot 1] + 2 \cdot 2$
$\quad = 3 + 2 \cdot (1 + 2)$
$e_3 = e_2 + 2 \cdot 3 = [3 + 2 \cdot (1 + 2)] + 2 \cdot 3$
$\quad = 3 + 2 \cdot (1 + 2 + 3)$
$e_4 = e_3 + 2 \cdot 4 = [3 + 2 \cdot (1 + 2 + 3)] + 2 \cdot 4$
$\quad = 3 + 2 \cdot (1 + 2 + 3 + 4)$
$\vdots$

Guess:
$e_n = 3 + 2 \cdot (1 + 2 + 3 + \cdots + (n - 1) + n)$
$\quad = 3 + 2 \cdot \dfrac{n(n + 1)}{2}$ by Theorem 4.2.2
$\quad = 3 + n(n + 1)$ by basic algebra

10. $w_0 = 1$
$w_1 = 2^1 - w_0 = 2^1 - 1$
$w_2 = 2^2 - w_1 = 2^2 - (2^1 - 1) = 2^2 - 2^1 + 1$
$w_3 = 2^3 - w_2 = 2^3 - (2^2 - 2^1 + 1)$
$\quad = 2^3 - 2^2 + 2^1 - 1$
$w_4 = 2^4 - w_3 = 2^4 - (2^3 - 2^2 + 2^1 - 1)$
$\quad = 2^4 - 2^3 + 2^2 - 2^1 + 1$
$\vdots$
$\vdots$

Guess:
$w_n = 2^n - 2^{n-1} + \cdots + (-1)^n \cdot 1$
$\quad = (-1)^n[1 - 2 + 2^2 - \cdots + (-1)^n \cdot 2^n]$
$\quad = (-1)^n[1 + (-2)$
$\qquad + (-2)^2 - \cdots + (-2)^n]$ by basic algebra
$\quad = (-1)^n\left[\dfrac{(-2)^{n+1} - 1}{(-2) - 1}\right]$ by Theorem 4.2.3
$\quad = \dfrac{(-1)^{n+1} \cdot [(-2)^{n+1} - 1]}{(-1) \cdot (-3)}$
$\quad = \dfrac{2^{n+1} - (-1)^{n+1}}{3}$ by basic algebra

11. $x_1 = 1$
$x_2 = 3x_1 + 2 = 3 + 2$
$x_3 = 3x_2 + 3 = 3(3 + 2) + 3 = 3^2 + 3 \cdot 2 + 3$
$x_4 = 3x_3 + 4 = 3(3^2 + 3 \cdot 2 + 3) + 4$
$\quad = 3^3 + 3^2 \cdot 2 + 3 \cdot 3 + 4$
$x_5 = 3x_4 + 5 = 3(3^3 + 3^2 \cdot 2 + 3 \cdot 3 + 4) + 5$
$\quad = 3^4 + 3^3 \cdot 2 + 3^2 \cdot 3 + 3 \cdot 4 + 5$
$x_6 = 3x_5 + 6$
$\quad = 3(3^4 + 3^3 \cdot 2 + 3^2 \cdot 3 + 4 \cdot 3 + 5) + 6$
$\quad = 3^5 + 3^4 \cdot 2 + 3^3 \cdot 3 + 3^2 \cdot 4 + 3 \cdot 5 + 6$
$\vdots$

Guess:
$x_n = 3^{n-1} + 3^{n-2} \cdot 2 + 3^{n-3} \cdot 3 + \cdots$
$\qquad + 3 \cdot (n - 1) + n$
$\quad = 3^{n-1} + \underbrace{3^{n-2} + 3^{n-2}}_{2 \text{ times}} + \underbrace{3^{n-3} + 3^{n-3} + 3^{n-3}}_{3 \text{ times}} + \cdots$

$\qquad + \underbrace{3 + 3 + \cdots + 3}_{(n-1) \text{ times}} + \underbrace{1 + 1 + \cdots + 1}_{n \text{ times}}$

$\quad = (3^{n-1} + 3^{n-2} + \cdots + 3^2 + 3 + 1)$
$\qquad + (3^{n-2} + 3^{n-3} + \cdots + 3^2 + 3 + 1) + \cdots$
$\qquad + (3^2 + 3 + 1) + (3 + 1) + 1$
$\quad = \dfrac{3^n - 1}{2} + \dfrac{3^{n-1} - 1}{2} + \cdots + \dfrac{3^3 - 1}{2}$
$\qquad + \dfrac{3^2 - 1}{2} + \dfrac{3 - 1}{2}$
$\quad = \tfrac{1}{2}[(3^n + 3^{n-1} + \cdots + 3^2 + 3) - n]$
$\quad = \tfrac{1}{2}[3(3^{n-1} + 3^{n-2} + \cdots + 3 + 1) - n]$
$\quad = \tfrac{1}{2}\left(3\left(\dfrac{3^n - 1}{3 - 1}\right) - n\right)$
$\quad = \tfrac{1}{4}(3^{n+1} - 3 - 2n)$

15. *Proof (by mathematical induction):*
The formula holds for $n = 0$: Since $0 = 0 \cdot d$ for any real number d, $a_0 + 0 \cdot d$. Hence the formula holds for $n = 0$.
If the formula holds for $n = k$ then it holds for $n = k + 1$: Suppose

$$a_k = a_0 + k \cdot d, \quad \text{for some integer } k \geq 0.$$

[This is the inductive hypothesis.]

We must show that $a_{k+1} = a_0 + (k + 1) \cdot d$. But

$$a_{k+1} = a_k + d \qquad \text{by definition of } a_0, a_1, a_2, \ldots$$
$$= [a_0 + k \cdot d] + d \qquad \text{by substitution from the inductive hypothesis}$$
$$= a_0 + (k + 1) \cdot d \qquad \text{by basic algebra}$$

16. Let U_n = the number of units produced on day n. Then

$$U_k = U_{k-1} + 2 \quad \text{for all integers } k \geq 1,$$
$$U_0 = 170.$$

Hence $U_0, U_1, U_2, \ldots$ is an arithmetic sequence with constant adder 2. It follows that when $n = 30$,

$$U_n = U_0 + n \cdot 2 = 170 + 2n = 170 + 2 \cdot 30$$
$$= 230 \text{ units.}$$

Thus the worker must produce 230 units on day 30.

21. $\displaystyle\sum_{k=0}^{20} 5^k = \frac{5^{21} - 1}{4} \cong 1.192 \times 10^{14} \cong$ 119,200,000,000,000 $\cong$ 119 trillion people (This is about 20,000 times the current population of the earth!)

23. b. *Hint:* Before simplification,

$$A_n = 1000(1.005)^n + 100[(1.005)^{n-1} + (1.005)^{n-2} + \cdots + (1.005)^2 + (1.005) + 1].$$

d. $A_{240} \cong \$49{,}514.29$, $A_{480} \cong \$210{,}106.53$

e. *Hint:* Use logarithms to solve the equation $A_n = 10{,}000$, where A_n is the expression found (after simplification) in part (b).

24. We must prove the following:

If $a_0, a_1, a_2, \ldots$ is a sequence defined by $a_k = k \cdot a_{k-1}$, for all integers $k \geq 1$, with initial condition $a_0 = 1$, then $a_n = n!$, for all integers $n \geq 0$.

Proof (by mathematical induction):

The formula holds for $n = 0$: Observe that $a_0 = 1$ by definition of $a_0, a_1, a_2, \ldots$. And $1! = 1$, also by definition of factorial. Hence $a_0 = 1!$, and so the formula holds for $n = 0$.

If the formula holds for $n = k$ then it holds for $n = k + 1$: Suppose

$$a_k = k!, \quad \text{for some integer } k \geq 0.$$
[This is the inductive hypothesis.]

We must show that $a_{k+1} = (k + 1)!$. But

$$a_{k+1} = (k + 1) \cdot a_k \qquad \text{by definition of } a_0, a_1, a_2, \ldots$$
$$= (k + 1) \cdot k! \qquad \text{by substitution from the inductive hypotheses}$$
$$= (k + 1)! \qquad \text{by definition of factorial.}$$

[Hence if the formula holds for $n = k$, then it holds for $n = k + 1$.]

26. We must show that if $c_1, c_2, c_3, \ldots$ is the sequence defined recursively by $c_k = 3c_{k-1} + 1$, for all integers $k \geq 2$, with initial condition $c_1 = 1$, then $c_n = \dfrac{3^n - 1}{2}$, for all integers $n \geq 1$.

Proof (by mathematical induction):

The formula holds for $n = 1$: Observe that $c_1 = 1$ by definition of $c_1, c_2, c_3, \ldots$ and that $\dfrac{3^1 - 1}{2} = \dfrac{3 - 1}{2} = 1$. Hence $c_1 = \dfrac{3^1 - 1}{2}$, and so the formula holds for $n = 1$.

If the formula holds for $n = k$ then it holds for $n = k + 1$: Suppose that

$$c_k = \frac{3^k - 1}{2}, \quad \text{for some integer } k \geq 1.$$
[This is the inductive hypothesis.]

We must show that $c_{k+1} = \dfrac{3^{k+1} - 1}{2}$. But

$$c_{k+1} = 3c_k + 1 \qquad \text{by definition of } c_1, c_2, c_3, \ldots$$
$$= 3 \cdot \left(\frac{3^k - 1}{2}\right) + 1 \qquad \text{by substitution from the inductive hypothesis}$$
$$= \frac{3^{k+1} - 3}{2} + \frac{2}{2}$$
$$= \frac{3^{k+1} - 1}{2} \qquad \text{by basic algebra.}$$

28. *Hint:*

$$[3 + k(k + 1)] + 2(k + 1)$$
$$= 3 + k^2 + k + 2k + 2 = 3 + [k^2 + 3k + 2]$$
$$= 3 + (k + 1)(k + 2)$$
$$= 3 + (k + 1)[(k + 1) + 1]$$

31. *Hint:*

$$2^{k+1} - \frac{2^{k+1} - (-1)^{k+1}}{3}$$
$$= \frac{3 \cdot 2^{k+1}}{3} - \frac{2^{k+1} - (-1)^{k+1}}{3}$$
$$= \frac{2 \cdot 2^{k+1} + (-1)^{k+1}}{3} = \frac{2^{k+2} - (-1)^{k+2}}{3}$$

32. We must show that if $x_1, x_2, x_3, \ldots$ is the sequence defined recursively by $x_k = 3x_{k-1} + k$ for all integers $k \geq 2$, with initial condition $x_1 = 1$, then $x_n = \dfrac{3^{n+1} - 2n - 3}{4}$ for all integers $n \geq 1$.

Proof (by mathematical induction):

The formula holds for $n = 1$: For $n = 1$,
$$\frac{3^{n+1} - 2n - 3}{4} = \frac{3^{1+1} - 2 \cdot 1 - 3}{4} = \frac{9 - 2 - 3}{4} = 1,$$
which is the defined value of x_1. So the formula holds for $n = 1$.

If the formula holds for $n = k$, then it holds for $n = k + 1$: Suppose that for some integer $k > 1$, $x_k = \dfrac{3^{k+1} - 2k - 3}{4}$. *[Inductive hypothesis]* We must

show that $x_{k+1} = \dfrac{3^{(k+1)+1} - 2(k + 1) - 3}{4}$, or, equiv-

alently, $x_{k+1} = \dfrac{3^{k+2} - 2k - 5}{4}$. But

$x_{k+1} = 3x_k + k$ by definition of $x_1, x_2, x_3,$

$= 3\left(\dfrac{3^{k+1} - 2k - 3}{4}\right) + k$ by inductive hypothesis

$= \dfrac{3 \cdot 3^{k+1} - 3 \cdot 2k - 3 \cdot 3}{4} + \dfrac{4(k + 1)}{4}$

$= \dfrac{3^{k+2} - 6k - 9 + 4k + 4}{4}$

$= \dfrac{3^{k+2} - 2k - 5}{4}$ by algebra.

[This is what was to be shown.]

36. $v_1 = 1$

$v_2 = v_{\lfloor 2/2 \rfloor} + v_{\lfloor 3/2 \rfloor} + 2 = v_1 + v_1 + 2$
$= 1 + 1 + 2$

$v_3 = v_{\lfloor 3/2 \rfloor} + v_{\lfloor 4/2 \rfloor} + 2 = v_1 + v_2 + 2$
$= 1 + (1 + 1 + 2) + 2 = 3 + 2 \cdot 2$

$v_4 = v_{\lfloor 4/2 \rfloor} + v_{\lfloor 5/2 \rfloor} + 2 = v_2 + v_2 + 2$
$= (1 + 1 + 2) + (1 + 1 + 2) + 2$
$= 4 + 3 \cdot 2$

$v_5 = v_{\lfloor 5/2 \rfloor} + v_{\lfloor 6/2 \rfloor} + 2 = v_3 + v_2 + 2$
$= (3 + 2 \cdot 2) + (1 + 1 + 2) + 2$
$= 5 + 4 \cdot 2$

$v_6 = v_{\lfloor 6/2 \rfloor} + v_{\lfloor 7/2 \rfloor} + 2 = v_3 + v_3 + 2$
$= (3 + 2 \cdot 2) + (3 + 2 \cdot 2) + 2$
$= 6 + 5 \cdot 2$

$\vdots$

Guess: $v_n = n + 2(n - 1) = 3n - 2$ for all integers $n \geq 1$

Proof: Let $v_1, v_2, v_3, \ldots$ be a sequence that satisfies the recurrence relation $v_k = v_{\lfloor k/2 \rfloor} + v_{\lfloor (k+1)/2 \rfloor} + 2$ for all integers $k \geq 2$ and the initial condition $v_1 = 1$. We will show by strong mathematical induction that $v_n = 3n - 2$ for all integers $n \geq 1$.
The formula holds for $n = 1$: For $n = 1$ the formula gives $3 \cdot 1 - 2 = 1$, which equals v_1.
If the formula holds for all i with $1 \leq i < k$, then it holds for $i = k$: Let k be an integer with $k > 1$ and suppose $v_i = 3i - 2$ for all integers i with $1 \leq i < k$. *[This is the inductive hypothesis.]* We must show that $v_k = 3k - 2$. But

$v_k = v_{\lfloor k/2 \rfloor} + v_{\lfloor (k+1)/2 \rfloor} + 2$ by definition of $v_1, v_2, v_3, \ldots$

$= \left[\left\lfloor \dfrac{k}{2} \right\rfloor + 2\left(\left\lfloor \dfrac{k}{2} \right\rfloor - 1\right)\right] + \left[\left\lfloor \dfrac{k + 1}{2} \right\rfloor \right.$

$\left. + 2\left(\left\lfloor \dfrac{k + 1}{2} \right\rfloor - 1\right)\right] + 2$ by substitution from the inductive hypotheses

$= \begin{cases} \dfrac{k}{2} + 2 \cdot \dfrac{k}{2} - 2 + \dfrac{k}{2} + 2 \cdot \dfrac{k}{2} - 2 + 2 & \text{if } k \text{ is even} \\ \dfrac{k - 1}{2} + 2 \cdot \dfrac{k - 1}{2} - 2 + \dfrac{k + 1}{2} \\ \qquad\qquad + 2 \cdot \dfrac{k + 1}{2} - 2 + 2 & \text{if } k \text{ is odd} \end{cases}$

$= \begin{cases} \dfrac{k}{2} + k + \dfrac{k}{2} + k - 2 & \text{if } k \text{ is even} \\ \dfrac{k - 1}{2} + k - 1 + \dfrac{k + 1}{2} + k + 1 - 2 & \text{if } k \text{ is odd} \end{cases}$

$= 3k - 2$ by the laws of algebra.
[This is what was to be shown.]

39. a. *Hint:* $w_n = \begin{cases} \left(\dfrac{n + 1}{2}\right)^2 & \text{if } n \text{ is odd} \\ \dfrac{n}{2}\left(\dfrac{n}{2} + 1\right) & \text{if } n \text{ is even} \end{cases}$

40. a. *Hint:* Express the answer using the Fibonacci sequence.

41. The sequence does not satisfy the formula. According to the formula, $a_4 = (4 - 1)^2 = 9$. But by definition of the sequence, $a_1 = 0$, $a_2 = 2 \cdot 0 + (2 + 1) = 1$, $a_3 = 2 \cdot 1 + (3 - 1) = 4$, and so $a_4 = 2 \cdot 4 + (4 - 1) = 11$. Hence the sequence does not satisfy the formula for $n = 4$.

43. a. *Hint:* The maximum number of regions is obtained when each additional line crosses all the previous lines, but not at any point that is already the intersection of two lines. When a new line is added, it divides each region through which it passes into two pieces. The number of regions a newly added line passes through is one more than the number of lines it crosses.

44. *Hint:* The answer involves the Fibonacci numbers!

SECTION 8.3

1. (a), (d), and (f)

3. a. $\left.\begin{array}{l} a_0 = C \cdot 2^0 + D = C + D = 1 \\ a_1 = C \cdot 2^1 + D = 2C + D = 3 \end{array}\right\}$

$\Leftrightarrow \begin{cases} D = 1 - C \\ 2C + (1 - C) = 3 \end{cases} \Leftrightarrow \begin{cases} C = 2 \\ D = -1 \end{cases}$

$a_2 = 2 \cdot 2^2 + (-1) = 7$

4. a. $b_0 = C \cdot 3^0 + D(-2)^0 = C + D = 0$
$b_1 = C \cdot 3^1 + D \cdot (-2)^1 = 3C - 2D = 5$

$\Leftrightarrow \begin{cases} D = -C \\ 3C - 2(-C) = 5 \end{cases} \Leftrightarrow \begin{cases} C = 1 \\ D = -1 \end{cases}$

$b_2 = 3^2 + (-1)(-2)^2 = 9 - 4 = 5$

5. *Proof:* Given that $a_n = C \cdot 2^n + D$, then for any choice of C and D and integer $k > 2$,

$$a_k = C \cdot 2^k + D,$$
$$a_{k-1} = C \cdot 2^{k-1} + D,$$
$$a_{k-2} = C \cdot 2^{k-2} + D.$$

Hence

$3a_{k-1} - 2a_{k-2}$
$= 3(C \cdot 2^{k-1} + D) - 2(C \cdot 2^{k-2} + D)$
$= 3C \cdot 2^{k-1} + 3D - 2C \cdot 2^{k-2} + 2D$
$= 3C \cdot 2^{k-1} - C \cdot 2^{k-1} + D$
$= 2C \cdot 2^{k-1} + D$
$= C \cdot 2^k + D = a_k.$

8. a. If for all $k > 2$, $t^k = 2t^{k-1} + 3t^{k-2}$ and $t \neq 0$ then $t^2 = 2t + 3$ *[by dividing by t^{k-2}]*, and so $t^2 - 2t - 3 = 0$. But $t^2 - 2t - 3 = (t - 3)(t + 1)$; hence $t = 3$ or $t = -1$.

b. It follows from (a) and the distinct roots theorem that for some constants C and D, $a_0, a_1, a_2, \ldots$ satisfies the equation

$$a_n = C \cdot 3^n + D \cdot (-1)^n \quad \text{for all integers } n \geq 0.$$

Since $a_0 = 1$ and $a_1 = 2$, then

$a_0 = C \cdot 3^0 + D(-1)^0 = C + D = 1$
$a_1 = C \cdot 3^1 + D \cdot (-1)^1 = 3C - D = 2$

$\Leftrightarrow \begin{cases} D = 1 - C \\ 3C - (1 - C) = 2 \end{cases}$

$\Leftrightarrow \begin{cases} D = 1 - C \\ 4C - 1 = 2 \end{cases}$

$\Leftrightarrow \begin{cases} C = 3/4 \\ D = 1/4 \end{cases}$

Thus

$$a_n = \frac{3}{4} \cdot 3^n + \frac{1}{4} \cdot (-1)^n \quad \text{for all integers } n \geq 0.$$

11. *Characteristic equation:* $t^2 - 4 = 0$. Since $t^2 - 4 = (t - 2)(t + 2)$, $t = 2$ and $t = -2$ are the roots. By the distinct roots theorem, for some constants C and D

$$d_n = C \cdot 2^n + D \cdot (-2)^n \quad \text{for all integers } n \geq 0.$$

Since $d_0 = 1$ and $d_1 = -1$, then

$d_0 = C \cdot 2^0 + D(-2)^0 = C + D = 1$
$d_1 = C \cdot 2^1 + D \cdot (-2)^1 = 2C - 2D = -1$

$\Leftrightarrow \begin{cases} D = 1 - C \\ 2C - 2(1 - C) = -1 \end{cases}$

$\Leftrightarrow \begin{cases} D = 1 - C \\ 4C - 2 = -1 \end{cases}$

$\Leftrightarrow \begin{cases} C = \frac{1}{4} \\ D = \frac{3}{4} \end{cases}$

Thus

$$d_n = \frac{1}{4} \cdot 2^n + \frac{3}{4} \cdot (-2)^n \quad \text{for all integers } n \geq 0.$$

13. *Characteristic equation:* $t^2 - 2t + 1 = 0$. By the quadratic formula,

$$t = \frac{2 \pm \sqrt{4 - 4 \cdot 1}}{2} = \frac{2}{2} = 1.$$

By the single root theorem, for some constants C and D

$r_n = C \cdot 1^n + D \cdot n \cdot 1^n$
$= C + nD \quad \text{for all integers } n \geq 0.$

Since $r_0 = 1$ and $r_1 = 4$, then

$r_0 = C + 0 \cdot D = C = 1$
$r_1 = C + 1 \cdot D = C + D = 4$
$\Leftrightarrow \begin{cases} C = 1 \\ 1 + D = 4 \end{cases} \Leftrightarrow \begin{cases} C = 1 \\ D = 3 \end{cases}$

Thus $r_n = 1 + 3n$ for all integers $n \geq 0$.

16. *Hint:* The answer is $s_n = \dfrac{\sqrt{3} + 2}{2\sqrt{3}} (1 + \sqrt{3})^n + \dfrac{\sqrt{3} - 2}{2\sqrt{3}} (1 - \sqrt{3})^n$ for all integers $n \geq 0$.

19. *Proof:* Suppose $r, s, a_0,$ and a_1 are numbers with $r \neq s$. Consider the system of equations

$$C + D = a_0$$
$$C \cdot r + D \cdot s = a_1.$$

By solving for D and substituting,

$$D = a_0 - C$$
$$C \cdot r + (a_0 - C) \cdot s = a_1.$$

Hence

$$C \cdot (r - s) = a_1 - a_0 s.$$

Since $r \neq s$, both sides may be divided by $r - s$. Thus the given system of equations has the unique solution

$$C = \frac{a_1 - a_0 s}{r - s}$$

and

$$D = a_0 - C = a_0 - \frac{a_1 - a_0 s}{r - s}$$

$$= \frac{a_0 r - a_0 s - a_1 + a_0 s}{r - s} = \frac{a_0 r - a_1}{r - s}.$$

Alternate Solution: Since the determinant of the system is $1 \cdot s - r \cdot 1 = s - r$ and since $r \neq s$, the given system has a nonzero determinant and therefore has a unique solution.

21. *Hint:* Use strong mathematical induction. First note that the formula holds for $n = 0$ and $n = 1$. To prove the inductive step, suppose that for some $k \geq 2$ the formula holds for all i with $0 \leq i < k$. Then show that the formula holds for k. Use the proof of Theorem 8.3.3 (the distinct roots theorem) as a model.

22. The characteristic equation is $t^2 - 2t + 2 = 0$. By the quadratic formula, its roots are

$$t = \frac{2 \pm \sqrt{4 - 8}}{2} = \frac{2 \pm 2i}{2} = \begin{cases} 1 + i \\ 1 - i \end{cases}.$$

By the distinct roots theorem, for some constants C and D

$$a_n = C \cdot (1 + i)^n + D \cdot (1 - i)^n$$

for all integers $n \geq 0$.

Since $a_0 = 1$ and $a_1 = 2$, then

$$a_0 = C \cdot (1 + i)^0 + D(1 - i)^0 = C + D = 1$$
$$a_1 = C \cdot (1 + i)^1 + D \cdot (1 - i)^1$$
$$= C \cdot (1 + i) + D \cdot (1 - i) = 2$$

$$\Leftrightarrow \begin{cases} D = 1 - C \\ C \cdot (1 + i) + (1 - C) \cdot (1 - i) = 2 \end{cases}$$

$$\Leftrightarrow \begin{cases} D = 1 - C \\ C \cdot (1 + i - 1 + i) + 1 - i = 2 \end{cases}$$

$$\Leftrightarrow \begin{cases} D = 1 - C \\ C \cdot (2i) = 1 + i \end{cases}$$

$$\Leftrightarrow \begin{cases} D = 1 - C \\ C = \dfrac{1 + i}{2i} = \dfrac{1 + i}{2i} \cdot \dfrac{i}{i} = \dfrac{i - 1}{-2} = \dfrac{1 - i}{2} \end{cases}$$

$$\Leftrightarrow \begin{cases} D = 1 - \dfrac{1 - i}{2} = \dfrac{2 - 1 + i}{2} = \dfrac{1 + i}{2} \\ C = \dfrac{1 - i}{2} \end{cases}$$

Thus

$$a_n = \left(\frac{1 - i}{2}\right) \cdot (1 + i)^n + \left(\frac{1 + i}{2}\right) \cdot (1 - i)^n$$

for all integers $n \geq 0$.

25. *Hint:* $P_{20} = \dfrac{5^{300} - 5^{20}}{5^{300} - 1} \cong 1$

26. a. *Hint:* The answer is $s_k = 2s_{k-1} + 3s_{k-2}$ for $k \geq 4$.

SECTION 8.4

1. a. (1) p, q, r, and s are Boolean expressions by I.
   (2) $\sim s$ is a Boolean expression by (1) and II(c).
   (3) $(r \vee \sim s)$ is a Boolean expression by (1), (2), and II(b).
   (4) $(q \wedge (r \vee \sim s))$ is a Boolean expression by (1), (3), and II(a).
   (5) $\sim p$ is a Boolean expression by (1) and II(c).
   (6) $(\sim p \vee (q \wedge (r \vee \sim s)))$ is a Boolean expression by (4), (5), and II(b).

2. a. (1) $\varepsilon \in \Sigma^*$ by I.
   (2) $a = \varepsilon a \in \Sigma^*$ by (1) and II(a).
   (3) $aa \in \Sigma^*$ by (2) and II(a).
   (4) $aab \in \Sigma^*$ by (1) and II(b).

3. a. (1) *MI* is in the *MIU* system by I.
   (2) *MII* is in the *MIU* system by (1) and II(b).
   (3) *MIIII* is in the *MIU* system by (3) and II(b).
   (4) *MIIIIIIII* is in the *MIU* system by (3) and II(b).
   (5) *MIUIIII* is in the *MIU* system by (4) and II(c).
   (6) *MIUUI* is in the *MIU* system by (5) and II(c).
   (7) *MIUI* is in the *MIU* system by (6) and II(d).

4. *Hint:* Can the number of *I*'s in a string in the *MIU* system be a multiple of 3? How do rules II(a)–(d) affect the number of *I*'s in a string?

5. a. (1) () is in P by I.
   (2) (()) is in P by (1) and II(a).
   (3) ()(()) is in P by (1), (2), and II(b).

6. a. This structure is not in P. Define a function $f: P \rightarrow Z$ as follows: for each parenthesis structure S in P, let

$$f(S) = \begin{bmatrix} \text{the number of left} \\ \text{parentheses in } S \end{bmatrix} - \begin{bmatrix} \text{the number of right} \\ \text{parentheses in } S \end{bmatrix}.$$

Observe that for all S in P, $f(S) = 0$. To see why, note that

1. the base element of P is sent by f to 0: $f[()] = 0$ *[because there is one left and one right parenthesis in ()]*;
2. for all $S \in P$, if $f[S] = 0$ then $f[(S)] = 0$ *[because if $k - m = 0$ then $(k + 1) - (m + 1) = 0$]*;
3. for all S and T in P, if $f[S] = 0$ and $f[T] = 0$, then $f[ST] = 0$ *[because if $k - m = 0$ and $n - p = 0$, then $(k + n) - (m + p) = 0$]*.

Items (1), (2), and (3) show that all parenthesis structures obtainable from the base structure () by repeated application of II(a) and II(b) are sent to 0 by f. But by III (the restriction condition), there are no other elements of P beside those obtainable from the base element by applying II(a) and II(b). Hence $f(S) = 0$ for all $S \in P$.

Now if ()((were in P, then it would be sent to 0 by f. But $f[()((] = 3 - 2 = 1 \neq 0$. Thus ()(($\notin P$.

7. a. (1) 2, 0.3, 4.2, and 7 are arithmetic expressions by I.

(2) $(0.3 - 4.2)$ is an arithmetic expression by (1) and II(d).

(3) $(2 \cdot (0.3 - 4.2))$ is an arithmetic expression by (1), (2), and II(e).

(4) (-7) is an arithmetic expression by (1) and II(b).

(5) $((2 \cdot (0.3 - 4.2)) + (-7))$ is an arithmetic expression by (3), (4), and II(c).

8. Let S be the set of all strings of 0's and 1's with the same number of 0's and 1's. The following is a recursive definition of S.

   I. BASE: the null string $\varepsilon \in S$.
   II. RECURSION: If $s \in S$, then
      a. $01s \in S$ b. $s01 \in S$ c. $10s \in S$
      d. $s10 \in S$ e. $0s1 \in S$ f. $1s0 \in S$
   III. RESTRICTION: There are no elements of S other than those obtained from I and II.

10. Let T be the set of all strings of a's and b's that contain an even number of a's. The following is a recursive definition of T.

   I. BASE: the null string $\varepsilon \in T$.
   II. RECURSION: If $t \in T$, then
      a. $bt \in T$ b. $tb \in T$ c. $aat \in T$
      d. $ata \in T$ e. $taa \in T$
   III. RESTRICTION: There are no elements of T other than those obtained from I and II.

12. *Proof (by mathematical induction):*

The property holds for $n = 1$: Let a_1 and c be any real numbers. By the recursive definition of sum, $\sum_{i=1}^{1}(ca_i) = ca_1$ and $\sum_{i=1}^{1} a_i = a_1$. Therefore $\sum_{i=1}^{1}(ca_i) = c \cdot \sum_{i=1}^{1} a_i$, and so the property holds for $n = 1$.

If the property holds for $n = k$, then it holds for $n = k + 1$: Let k be an integer with $k \geq 1$. Suppose that for any real numbers $a_1, a_2, a_3, \ldots, a_k$ and c, $\sum_{i=1}^{k}(ca_i) = c \cdot \sum_{i=1}^{k} a_i$. *[This is the inductive hypothesis]* *[We must show that for any real numbers $a_1, a_2, a_3, \ldots, a_{k+1}$ and c, $\sum_{i=1}^{k+1}(ca_i) = c\sum_{i=1}^{k+1} a_i$.]* Let $a_1, a_2, a_3, \ldots, a_{k+1}$ and c be any real numbers. Then

$$\sum_{i=1}^{k+1} ca_i = \sum_{i=1}^{k} ca_i + ca_{k+1} \qquad \text{by the recursive definition of } \Sigma$$

$$= c \sum_{i=1}^{k} a_i + ca_{k+1} \qquad \text{by inductive hypothesis}$$

$$= c\left(\sum_{i=1}^{k} a_i + a_{k+1}\right) \qquad \text{by the distributive law for the real numbers}$$

$$= c \sum_{i=1}^{k+1} a_i \qquad \text{by the recursive definition of } \Sigma$$

[This is what was to be shown.]
[Since the basis step and the inductive step have both been proved true, the given property of summation holds for all integers $n \geq 1$.]

16. *Proof (by mathematical induction):*

The property holds for $n = 1$: Let B_1 and A be any sets. By the recursive definition of union, $\bigcup_{i=1}^{1} B_i = B_1$ and $\bigcup_{i=1}^{1}(A \cap B_i) = A \cap B_1$. Therefore, $A \cap (\bigcup_{i=1}^{1} B_i) = \bigcup_{i=1}^{1}(A \cap B_i)$, and so the property holds for $n = 1$.

If the property holds for $n = k$, then it holds for $n = k + 1$: Let k be an integer with $k \geq 1$. Suppose that for any sets $B_1, B_2, B_3, \ldots, B_k$ and A, $A \cap (\bigcup_{i=1}^{k} B_i) = \bigcup_{i=1}^{k}(A \cap B_i)$. *[This is the inductive hypothesis.]* *[We must show that for any sets $B_1, B_2, B_3, \ldots, B_{k+1}$ and A, $A \cap (\bigcup_{i=1}^{k+1} B_i) = \bigcup_{i=1}^{k+1}(A \cap B_i)$.]* Let $B_1, B_2, B_3, \ldots, B_{k+1}$, and A be any sets. Then

$$A \cap \left(\bigcup_{i=1}^{k+1} B_i\right)$$

$$= A \cap \left[\left(\bigcup_{i=1}^{k} B_i\right) \cup B_{k+1}\right] \qquad \text{by the recursive definition of union}$$

$$= \left(A \cap \left(\bigcup_{i=1}^{k} B_i\right)\right) \cup (A \cap B_{k+1}) \qquad \text{by one of the distributive laws for sets (Theorem 5.2.2(3)).}$$

$$= \left(\bigcup_{i=1}^{k}(A \cap B_i)\right) \cup (A \cap B_{k+1}) \qquad \text{by inductive hypothesis}$$

$$= \bigcup_{i=1}^{k+1}(A \cap B_i) \qquad \text{by the recursive definition of union.}$$

[This is what was to be shown.]
[Since the basis step and the inductive step have both been proved true, the given general distributive law for sets holds for all integers $n \geq 1$.]

19. a. $M(86) = M(M(97))$ since $86 \leq 100$
$= M(M(M(108)))$ since $97 \leq 100$
$= M(M(98))$ since $108 > 100$
$= M(M(M(109)))$ since $98 < 100$
$= M(M(99))$ since $109 > 100$
$= M(91)$ by Example 8.4.7

21. a. $A(1, 1) = A(0, A(1, 0))$ by (8.4.3) with $m = 1$ and $n = 1$
$= A(1, 0) + 1$ by (8.4.1) with $n = A(1, 0)$
$= A(0, 1) + 1$ by (8.4.2) with $m = 1$
$= (1 + 1) + 1$ by (8.4.1) with $n = 1$
$= 3$

Alternate Solution:
$A(1, 1) = A(0, A(1, 0))$ by (8.1.3) with $m = 1$ and $n = 1$
$= A(0, A(0, 1))$ by (8.1.2) with $m = 1$
$= A(0, 2)$ by (8.1.1) with $n = 1$
$= 3$ by (8.1.1) with $n = 2$

22. a. *Proof (by mathematical induction):*
The formula holds for $n = 0$: When $n = 0$,

$$A(1, n) = A(1, 0) \quad \text{by substitution}$$
$$= A(0, 1) \quad \text{by (8.4.2)}$$
$$= 1 + 1 \quad \text{by (8.4.1)}$$
$$= 2.$$

On the other hand, $n + 2 = 0 + 2$ also. So $A(1, n) = n + 2$ for $n = 0$.

If the formula holds for $n = k$, then it holds for $n = k + 1$: Let k be an integer with $k \geq 1$ and suppose the formula holds for $n = k$. In other words, suppose $A(1, k) = k + 2$. *[This is the inductive hypothesis.]* We must show that the formula holds for $n = k + 1$. In other words, we must show that $A(1, k + 1) = (k + 1) + 2 = k + 3$. But

$$A(1, k + 1) = A(0, A(1, k)) \quad \text{by (8.1.3)}$$
$$= A(1, k) + 1 \quad \text{by (8.4.1)}$$
$$= (k + 2) + 1 \quad \text{by inductive hypothesis}$$
$$= k + 3.$$

[This is what was to be shown.]
[Since both the basis and the inductive steps have been proved, we conclude that the formula holds for all nonnegative integers n.]

24. Suppose F is a function. Then $F(1) = 1$, $F(2) = F(1) = 1$, $F(3) = 1 + F(6) = 1 + F(3)$. Subtracting $F(3)$ from both sides gives $1 = 0$, which is false. Hence F is not a function.

SECTION 9.1

1. a. $f(0)$ is positive.
b. $f(x) = 0$ when $x = -2$ and $x = 3$ (approximately)
c. increase
d. decrease

2.

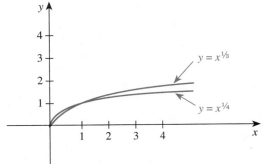

When $0 < x < 1$, $x^{1/3} < x^{1/4}$. When $x > 1$, $x^{1/3} > x^{1/4}$.

4.

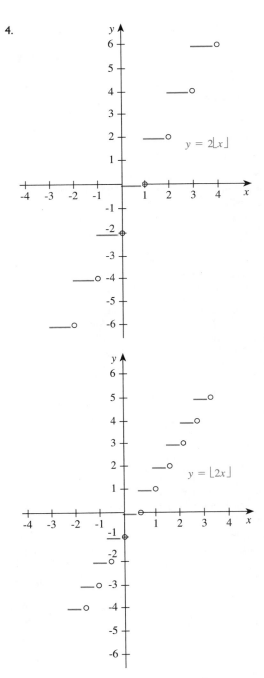

The graphs show that $2\lfloor x \rfloor \neq \lfloor 2x \rfloor$ for many values of x.

5.

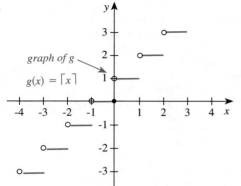

graph of g

$g(x) = \lceil x \rceil$

7.

x	$F(x) = \lfloor x^{1/2} \rfloor$
0	0
$\frac{1}{2}$	0
1	1
2	1
3	1
4	2

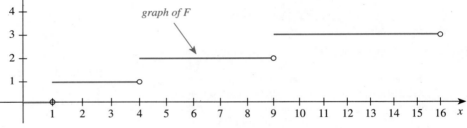

graph of F

9.

n	$f(n) = |n|$
0	0
1	1
2	2
3	3
-1	1
-2	2
-3	3

graph of f

11.

n	$h(n) = \left\lfloor \dfrac{n}{2} \right\rfloor$
0	0
1	0
2	1
3	1
4	2
5	2
-1	-1
-2	-1
-3	-2
-4	-2

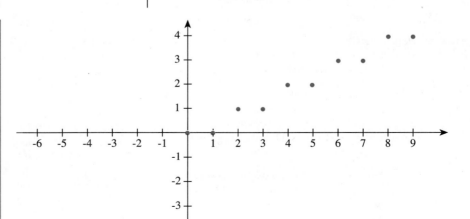

13. f is increasing on the intervals
$\{x \in \mathbf{R} \mid -3 < x < -2\}$ and
$\{x \in \mathbf{R} \mid 0 < x < 2.5\}$, and f is decreasing on
$\{x \in \mathbf{R} \mid -2 < x < 0\}$ and $\{x \in \mathbf{R} \mid 2.5 < x < 4\}$
(approximately).

14. *Proof:* Suppose x_1 and x_2 are particular but arbitrarily chosen real numbers such that $x_1 < x_2$. *[We must show that $f(x_1) < f(x_2)$.]* Since

$$x_1 < x_2$$
then
$$2x_1 < 2x_2$$
and
$$2x_1 - 3 < 2x_2 - 3$$

by basic properties of inequalities. But then by definition of f,

$$f(x_1) < f(x_2)$$

[as was to be shown]. Hence f is increasing on the set of all real numbers.

16. a. *Proof:* Suppose x_1 and x_2 are real numbers with $x_1 < x_2 < 0$. *[We must show that $h(x_1) > h(x_2)$.]* Multiply both sides of $x_1 < x_2$ by x_1 to obtain $x_1^2 > x_1x_2$ *[by T22 of Appendix A since $x_1 < 0$]*, and multiply both sides of $x_1 < x_2$ by x_2 to obtain $x_1x_2 > x_2^2$ *[by T22 of Appendix A since $x_2 < 0$]*. By transitivity of order *[Appendix A, T17]* $x_2^2 < x_1^2$, and so by definition of h, $h(x_2) < h(x_1)$.

17. a. *Preliminaries:* If both x_1 and x_2 are positive, then by the rules for working with inequalities (see Appendix A)

$$\frac{x_1 - 1}{x_1} < \frac{x_2 - 1}{x_2} \Rightarrow x_2(x_1 - 1) < x_1(x_2 - 1)$$
by multiplying both sides by x_1x_2 (which is positive)

$$\Rightarrow x_1x_2 - x_2 < x_1x_2 - x_1$$
by multiplying out

$$\Rightarrow -x_2 < -x_1$$
by subtracting x_1x_2 from both sides

$$\Rightarrow x_2 > x_1 \quad \text{by multiplying by } -1.$$

Are these steps reversible? Yes!
Proof: Suppose that x_1 and x_2 are positive real numbers and $x_1 < x_2$. *[We must show that $k(x_1) < k(x_2)$.]* Then

$$x_1 < x_2 \Rightarrow -x_2 < -x_1 \quad \text{by multiplying by } -1$$
$$\Rightarrow x_1x_2 - x_2 < x_1x_2 - x_1 \quad \text{by adding } x_1x_2 \text{ to both sides}$$
$$\Rightarrow x_2(x_1 - 1) < x_1(x_2 - 1) \quad \text{by factoring both sides}$$
$$\Rightarrow \frac{x_1 - 1}{x_1} < \frac{x_2 - 1}{x_2} \quad \text{by dividing both sides by the positive number } x_1x_2$$
$$\Rightarrow k(x_1) < k(x_2) \quad \text{by definition of } k.$$

[This is what was to be shown.]

18. *Proof:* Suppose $f: \mathbf{R} \to \mathbf{R}$ is increasing. *[We must show that f is one-to-one. In other words, we must show that for all real numbers x_1 and x_2, if $x_1 \neq x_2$ then $f(x_1) \neq f(x_2)$.]* Suppose x_1 and x_2 are real numbers and $x_1 \neq x_2$. By the trichotomy law *[Appendix A, T16]*, $x_1 < x_2$ or $x_1 > x_2$. In case $x_1 < x_2$, then since f is increasing, $f(x_1) < f(x_2)$ and so $f(x_1) \neq f(x_2)$. Similarly in case $x_1 > x_2$, then $f(x_1) > f(x_2)$ and so $f(x_1) \neq f(x_2)$. Thus in either case, $f(x_1) \neq f(x_2)$ *[as was to be shown]*.

20. a. *Proof:* Suppose x_1 and x_2 are nonnegative real numbers with $x_1 < x_2$. *[We must show that $f(x_1) < f(x_2)$.]* Then $x_2 = x_1 + h$ for some positive real number h. By substitution and the binomial theorem

$$x_2^m = (x_1 + h)^m = x_1^m + \left[\binom{m}{1}x_1^{m-1}h$$
$$+ \binom{m}{2}x_1^{m-2}h^2 + \cdots$$
$$+ \binom{m}{m-1}x_1h^{m-1} + h^m\right].$$

The bracketed sum is positive because $x_1 \geq 0$ and $h > 0$, and a sum of nonnegative terms that includes at least one positive term is positive. Hence

$$x_2^m = x_1^m + \text{a positive number},$$

and so $f(x_1) = x_1^m < x_2^m = f(x_2)$ *[as was to be shown]*.

21.

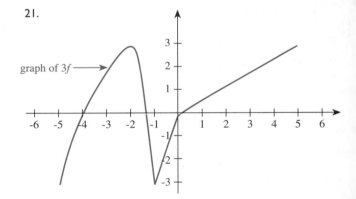

graph of $3f$

23. *Proof:* Suppose that f is a real-valued function of a real variable, f is decreasing on a set S, and M is any positive real number. *[We must show that $M \cdot f$ is decreasing on S. In other words, we must show that for all x_1 and x_2 in S, if $x_1 < x_2$ then $(M \cdot f)(x_1) > (M \cdot f)(x_2)$.]* Suppose x_1 and x_2 are in S and $x_1 < x_2$. Since f is decreasing on S, $f(x_1) > f(x_2)$ since M is positive, $M \cdot f(x_1) > M \cdot f(x_2)$. Since M is positive, $M \cdot f(x_1) > M \cdot f(x_2)$ *[because when both sides of an inequality are multiplied by a positive number, the direction of the inequality is unchanged]*. It follows by definition of $M \cdot f$ that $(M \cdot f)(x_1) > (M \cdot f)(x_2)$ *[as was to be shown]*.

26.

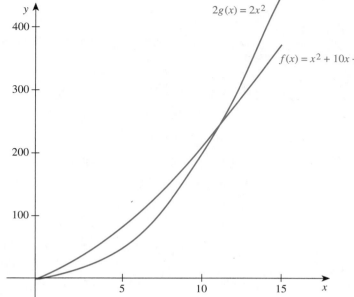

The zoom and trace features of a graphing calculator or computer indicate that when $x \geqslant 10.292$ then $f(x) \leqslant 2 \cdot g(x)$. Alternatively, to find the answer algebraically, solve the equation $2x^2 = x^2 + 10x + 3$. Subtracting $x^2 + 10x + 3$ from both sides gives $x^2 - 10x - 3 = 0$. By the quadratic formula,

$$x = \frac{10 \pm \sqrt{100 - 4(-3)}}{2} \cong 10.2915 \text{ or } -0.2915.$$

Since f and g are only defined for positive x, the only place where the two graphs cross is at approximately 10.2915. If $x_0 = 10.292$, then, for all $x > x_0$, $f(x) \leqslant 2g(x)$.

SECTION 9.2

1.

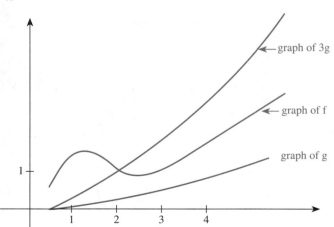

$f(x)$ is $O(g(x))$. (Note that there are many other functions that satisfy the given conditions besides those shown in the preceding figure.)

2. a. $\forall$ positive real numbers M and real numbers x_0, $\exists$ a real number x such that $x > x_0$ and $|f(x)| > M \cdot |g(x)|$.

b. *One Version:* No matter what positive number M and real number x_0 are given, you can find a real number x with the property that

$$x > x_0 \quad \text{and} \quad |f(x)| > M \cdot |g(x)|.$$

3. Let $M = 20$ and $x_0 = 1$. Then $|5x^8 - 9x^7 + 2x^5 + 3x - 1| \leq M \cdot |x^8|$ for all real numbers $x > x_0$, and so by definition of O-notation, $5x^8 - 9x^7 + 2x^5 + 3x - 1$ is $O(x^8)$.

5. a. For all real numbers $x > 1$,

$$\left| 2x^2 + 15x + 4 \right| = 2x^2 + 15x + 4$$

because $2x^2 + 15x + 4$ is positive (since $x > 1$)

$$\Rightarrow \left| 2x^2 + 15x + 4 \right| \le 2x^2 + 15x^2 + 4x^2$$

because since $x > 1$, then $x < x^2$ and $1 < x^2$

$$\Rightarrow \left| 2x^2 + 15x + 4 \right| \le 21x^2$$

because $2 + 15 + 4 = 21$

$$\Rightarrow \left| 2x^2 + 15x + 4 \right| \le 21 \left| x^2 \right|$$

because x^2 is positive.

b. $2x^2 + 15x + 4$ is $O(x^2)$.

7. a. For all real numbers $x \ge 36$, $x + 1$ is positive, and so by the rules for working with inequalities (see Appendix A),

$$\frac{8x + 45}{x + 1} \le 9 \overset{(1)}{\Rightarrow} 8x + 45 \le 9(x + 1)$$

by multiplying both sides by the positive number $x + 1$

$$\overset{(2)}{\Rightarrow} 8x + 45 \le 9x + 9$$

by multiplying out

$$\overset{(3)}{\Rightarrow} 45 - 9 \le 9x - 8x$$

by subtracting $8x + 9$ from both sides

$$\overset{(4)}{\Rightarrow} 36 \le x \quad \text{by combining like terms.}$$

Now observe that each of the above implications is reversible. For instance, to reverse (1), divide both sides of the inequality by the positive number $x + 1$, and to reverse (3), add $8x + 9$ to both sides.

Hence, if $x \ge 36$, then $\dfrac{8x + 45}{x + 1} \le 9$.

b. Let x be a real number with $x > 36$. By part (a),

$$\frac{8x + 45}{x + 1} \le 9$$

$$\Rightarrow \left| \frac{8x + 45}{x + 1} \right| \le 9 \quad \text{because } \frac{8x + 45}{+ 1} \text{ is positive}$$

$$(\text{since } x > 36)$$

$$\Rightarrow \left| 15x^4 \right| \cdot \left| \frac{8x + 45}{x + 1} \right| \le \left| 15x^4 \right| \cdot 9$$

by multiplying both sides by $\left| 15x^4 \right|$

$$\Rightarrow \left| \frac{15x^4(8x + 45)}{x + 1} \right| \le 135 \left| x^4 \right|$$

because $9 \cdot 15 = 135$.

c. $\dfrac{15x^4(8x + 45)}{x + 1}$ is $O(x^4)$

9. For all real numbers $x > 1$,

$$\left| 7x^2 + 12x \right| = 7x^2 + 12x$$

because $7x^2 + 12x$ is positive (since $x > 1$)

$$\Rightarrow \left| 7x^2 + 12x \right| \le 7x^2 + 12x^2$$

because since $x > 1$, then $x < x^2$

$$\Rightarrow \left| 7x^2 + 12x \right| \le 19x^2 \quad \text{because } 7 + 12 = 19$$

$$\Rightarrow \left| 7x^2 + 12x \right| \le 19 \left| x^2 \right| \quad \text{because } x^2 \text{ is positive.}$$

Let $M = 19$ and $x_0 = 1$. Then

$$\left| 7x^2 + 12x \right| \le M \left| x^2 \right| \quad \text{for all } x > x_0.$$

So by definition of O-notation, $7x^2 + 12x$ is $O(x^2)$.

11. For all real numbers $x > 1$,

$$\left| 100x^5 - 50x^3 - 18x^2 + 12x \right|$$
$$\le \left| 100x^5 \right| + \left| 50x^3 \right| + \left| 18x^2 \right| + \left| 12x \right|$$

by the triangle inequality

$$\Rightarrow \left| 100x^5 - 50x^3 - 18x^2 + 12x \right| \le 100x^5$$
$$+ 50x^3 + 18x^2 + 12x$$

because $x > 1$ and so each term is positive

$$\Rightarrow \left| 100x^5 - 50x^3 - 18x^2 + 12x \right|$$
$$\le 100x^5 + 50x^5 + 18x^5 + 12x^5$$

because since $x > 1$, then $x^3 < x^5$, $x^2 < x^5$, and $x < x^5$

$$\Rightarrow \left| 100x^5 - 50x^3 - 18x^2 + 12x \right| \le 180x^5$$

because $100 + 50 + 18 + 12 = 180$

$$\Rightarrow \left| 100x^5 - 50x^3 - 18x^2 + 12x \right| \le 180 \left| x^5 \right|$$

because x^5 is positive.

Let $M = 180$ and $x_0 = 1$. Then

$$\left| 100x^5 - 50x^3 - 18x^2 + 12x \right| \le M \left| x^5 \right|$$
$$\text{for all } x > x_0.$$

So by definition of O-notation, $100x^5 - 50x^3 - 18x^2 + 12x$ is $O(x^5)$.

13. By definition of ceiling, for any real number x, $\lceil x^2 \rceil$ is that integer n such that $n - 1 < x^2 \le n$. Adding 1 to all parts of this inequality gives $n < x^2 + 1 \le n + 1$. So $\lceil x^2 \rceil < x^2 + 1$. Thus if x is any real number with

$x > 1$, then

$$|\lceil x^2 \rceil| \leq \lceil x^2 \rceil \qquad \text{because } \lceil x^2 \rceil \text{ is positive}$$

$$\Rightarrow |\lceil x^2 \rceil| \leq x^2 + 1 \qquad \text{by the argument above}$$

$$\Rightarrow |\lceil x^2 \rceil| \leq x^2 + x^2 \qquad \text{because } 1 < x^2 \text{ since } x > 1$$

$$\Rightarrow |\lceil x^2 \rceil| \leq 2x^2$$

$$\Rightarrow |\lceil x^2 \rceil| \leq 2|x^2| \qquad \text{because } x^2 \text{ is positive.}$$

Let $M = 2$ and $x_0 = 1$. Then

$$|\lceil x^2 \rceil| \leq M|x^2| \quad \text{for all } x > x_0.$$

So by definition of O-notation, $\lceil x^2 \rceil$ is $O(x^2)$.

15. *Hint:* First prove by mathematical induction that for all real numbers $x > 1$ and positive integers r, $x^r > 1$. It follows that if $x > 1$ and n and m are integers with $n > m$, then $x^{n-m} > 1$.

16. a. *Hint:* Let x be a positive real number. Use mathematical induction to show that if $x \leq 1$ then $x^n \leq 1$. Then take the contrapositive, noting that $x = (x^{1/n})^n$.
   b. *Hint:* $x^{p/q} = (x^{1/q})^p$

17. a. Let f, g, and h be functions from $\mathbf{R}$ to $\mathbf{R}$, and suppose $f(x)$ is $O(h(x))$ and $g(x)$ is $O(h(x))$. Then there exist real numbers x_1, x_2, M_1, and M_2 so that $|f(x)| \leq M_1|h(x)|$ for all $x > x_1$ and $|g(x)| \leq M_2|h(x)|$ for all $x > x_2$. Let $M = M_1 + M_2$ and let x_0 be the greater of x_1 and x_2. Then for all $x > x_0$,

$$|f(x) + g(x)| < |f(x)| + |g(x)|$$

$$\text{by the triangle inequality}$$

$$\Rightarrow |f(x) + g(x)| \leq M_1 \cdot |h(x)| + M_2 \cdot |h(x)|$$

$$\text{by hypothesis}$$

$$\Rightarrow |f(x) + g(x)| \leq (M_1 + M_2) \cdot |h(x)|$$

$$\text{by algebra}$$

$$\Rightarrow |f(x) + g(x)| \leq M \cdot |h(x)|$$

$$\text{because } M = M_1 + M_2.$$

Hence by definition of O-notation, $f(x) + g(x)$ is $O(h(x))$.

   b. By exercise 15, for all $x > 1$, $x^2 < x^4$. Thus $|x^2| \leq 1 \cdot |x^4|$ for all $x > 1$. So by definition of O-notation, x^2 is $O(x^4)$. Clearly also, $|x^4| \leq 1 \cdot |x^4|$ for all x, and so x^4 is $O(x^4)$. It follows by part (a) that $x^2 + x^4$ is $O(x^4)$.

19. $\dfrac{(x+1)(x-2)}{4} = \dfrac{x^2 - x - 2}{4} = \dfrac{1}{4}x^2 - \dfrac{1}{4}x - \dfrac{1}{2}$ is $O(x^2)$ by the theorem on polynomial orders.

21. $\dfrac{n(n+1)(2n+1)}{6} = \dfrac{2n^3 + 3n^2 + n}{6}$

$$= \frac{1}{3}n^3 + \frac{1}{2}n^2 + \frac{1}{6}n \text{ is } O(n^3) \text{ by the theorem on poly-}$$
nomial orders.

23. By exercise 9 of Section 4.2, $1^2 + 2^2 + 3^2 + \cdots + n^2 = \dfrac{n(n+1)(2n+1)}{6}$, and by exercise 21 above this is $O(n^3)$. Hence $1^2 + 2^2 + 3^2 + \cdots + n^2$ is $O(n^3)$.

25. By exercise 12 of Section 4.2,

$$\sum_{i=1}^{n-1} i(i+1) = \frac{n(n+1)(n-1)}{3} \quad \text{for all integers } n \geq 2.$$

Hence by a change of variable (n in place of $n - 1$),

$$\sum_{i=1}^{n} i(i+1) = \frac{(n+1)(n+2)n}{3} = \frac{1}{3}n^3 + n^2 + \frac{2}{3}n,$$

which is $O(n^3)$ by the theorem on polynomial orders. So $\sum_{i=1}^{n} i(i+1)$ is $O(n^3)$.

Alternative Solution: Use the fact that

$$\sum_{i=1}^{n} i(i+1) = \sum_{i=1}^{n}(i^2 + i) = \sum_{i=1}^{n} i^2 + \sum_{i=1}^{n} i$$

$$= \frac{n(n+1)(2n+1)}{6} + \frac{n(n+1)}{2}$$

$$\text{by exercise 9 of Section 4.2 and Theorem 4.2.2}$$

$$= \left(\frac{1}{3}n^3 + \frac{1}{2}n^2 + \frac{1}{6}n\right) + \left(\frac{1}{2}n^2 + \frac{1}{2}n\right)$$

$$= \frac{1}{3}n^3 + n^2 + \frac{4}{6}n.$$

27. By part (b) of exercise 16, for all $x > 1$, $x \leq x^{4/3}$ and $1 = x^0 \leq x^{4/3}$. Hence by definition of O-notation (since all expressions are positive), x is $O(x^{4/3})$ and 1 is $O(x^{4/3})$. By part (c) of exercise 17, then, $-15x = (-15)x$ is $0(x^{4/3})$ and $7 = 7 \cdot 1$ is $O(x^{4/3})$. It follows by part (a) of exercise 17 (applied twice) that $4x^{4/3} - 15x + 7 = 4x^{4/3} + (-15x) + 7$ is $O(x^{4/3})$.

29. Suppose $a_0, a_1, a_2, \ldots, a_n$ are real numbers with $a_n \neq 0$, and suppose $r_0, r_1, r_2, \ldots, r_n$ and r are rational numbers with $r_0 < r_1 < \cdots < r_n \leq r$. By part (b) of exercise 16, for any real number $x > 1$,

$$x^{r_0} < x^r, x^{r_1} < x^r, \cdots, x^{r_{n-1}} < x^r, \text{ and } x^{r_n} \leq x^r.$$

So by part (c) of exercise 17,

$$a_0 x^{r_0} \text{ is } O(x^r), a_1 x^{r_1} \text{ is } O(x^r), \ldots,$$
$$a_{n-1} x^{r_{n-1}} \text{ is } O(x^r), \text{ and } a_n r^{r_n} \text{ is } O(x^r).$$

Hence by part (a) of exercise 17 (applied n times),

$$a_0 x^{r_0} + a_1 x + \cdots + a_{n-1} x^{r_{n-1}} + a_n x^{r_n} \text{ is } O(x^r).$$

31. a. *Proof:*

The inequality holds for $n = 1$: For $n = 1$ the left-hand side of the inequality is $\sqrt{1}$, and the right-hand side is $1^{3/2}$, which equals 1 also. So the inequality holds for $n = 1$.

If the inequality holds for $n = k$, then it holds for $n = k + 1$: Let k be an integer with $k \geq 1$, and suppose that

$$\sqrt{1} + \sqrt{2} + \cdots + \sqrt{k} \leq {}^{3/2}.$$

Inductive hypothesis

We must show that $\sqrt{1} + \sqrt{2} + \cdots + \sqrt{k} + 1 \leq (k + 1)^{3/2}$. But

$$\sqrt{1} + \sqrt{2} + \cdots + \sqrt{k + 1}$$
$$= (\sqrt{1} + \sqrt{2} + \cdots + \sqrt{k}) + \sqrt{k + 1}$$

by making the next-to-last term explicit

$$\Rightarrow \sqrt{1} + \sqrt{2} + \cdots + \sqrt{k + 1}$$
$$\leq k^{3/2} + \sqrt{k + 1} \quad (*)$$

by inductive hypothesis.

Note that by the algebra of inequalities (see Appendix A),

$$k^{3/2} + \sqrt{k + 1} \leq (k + 1)^{3/2}$$
$$= (k + 1)\sqrt{k + 1} = k\sqrt{k + 1} + \sqrt{k + 1}$$
$$\Rightarrow k^{3/2} \leq (k + 1)^{3/2} - \sqrt{k + 1} =$$
$$\sqrt{k + 1}(k + 1 - 1) = k\sqrt{k + 1}$$
$$\Rightarrow \frac{k^{3/2}}{k} = \sqrt{k} \leq \sqrt{k + 1}$$
$$\Rightarrow 1 \leq \sqrt{\frac{k + 1}{k}},$$

which is true because $\frac{k + 1}{k} > 1$. Are these implications reversible? Yes. For any positive integer k,

$$1 \leq \sqrt{\frac{k + 1}{k}}$$

$$\Rightarrow \sqrt{k} \leq \sqrt{k + 1}$$

because $(k + 1)/k > 1$ and the

$$\Rightarrow \frac{k^{3/2}}{k} \leq \sqrt{k + 1}$$

square root function is increasing by multiplying both sides by $\sqrt{k}$

$$\Rightarrow k^{3/2} \leq k\sqrt{k + 1} = \sqrt{k + 1}(k + 1 - 1)$$
$$= (k + 1)^{3/2} - \sqrt{k + 1}$$

$$\Rightarrow k^{3/2} + \sqrt{k + 1} \leq (k + 1)^{3/2} \quad (**)$$

by adding $\sqrt{k + 1}$ to both sides.

Putting (*) and (**) together gives

$$\sqrt{1} + \sqrt{2} + \cdots + \sqrt{k + 1} \leq (k + 1)^{3/2}$$

[as was to be shown.]

b. $\sqrt{1} + \sqrt{2} + \cdots + \sqrt{n}$ is $O(n^{3/2})$.

33. a. *Hint:* For all integers k with $1 \leq k \leq \lfloor \sqrt{n} \rfloor$, $1/k \leq 1$.

b. *Hint:* For all integers k with $\lfloor \sqrt{n} \rfloor + 1 \leq k \leq n$, $1/k \leq 1/(\lfloor \sqrt{n} \rfloor + 1)$.

34. *Hint:* The proof is similar to the solution to Example 9.2.7. (Choose a real number x so that $x > M^{1/(r-s)}$ and $x > x_0$.)

35. Note that $(x + 1)(2x^2 - 5) = 2x^3 + 2x^2 - 5x - 5$. So by property (9.2.4), $(x + 1)(2x^2 - 5)$ is $O(x^r)$ for all rational numbers $r \geq 3$ and $(x + 1)(2x^2 - 5)$ is not $O(x^s)$ for any rational number $s < 3$. Thus x^3 is the best big-oh approximation for $(x + 1)(2x^2 - 5)$ from among the set of all rational power functions.

37. Note that $\dfrac{\sqrt{x}(3x + 5)}{2x} = \dfrac{3x^{3/2} + 5x^{1/2}}{2x} = \dfrac{3}{2}x^{1/2} + \dfrac{5}{2}x^{-1/2}$. So by property (9.2.4), $\dfrac{\sqrt{x}(3x + 5)}{2x}$ is $O(x^r)$ for all rational numbers $r \geq 1/2$ and is not $O(x^s)$ for any rational number $s < 1/2$. Therefore $x^{1/2}$ is the best big-oh approximation for $\dfrac{\sqrt{x}(3x + 5)}{2x}$ from among the set of all rational power functions.

39. b. i. Note that $\lim_{x \to \infty} \dfrac{x + 4}{x + 1} = \lim_{x \to \infty} \dfrac{1 + \frac{4}{x}}{1 + \frac{1}{x}} = 1$, and by property (9.2.4.), x^2 is a best big-oh approximation for $2x^2 + 5$ from among the set of all rational power functions. Hence by part (a) *[with $g(x) = x^2$, $h(x) = x + 4$, $k(x) = x + 1$, and $f(x) = 2x^2 + 5$]*, x^2 is a best big-oh approximation for $\dfrac{(2x^2 + 5)(x + 4)}{x + 1}$ from among the set of all rational power functions.

40. *Proof:* Suppose $f(x)$ is $o(g(x))$. By definition of o-notation, $\lim_{x \to \infty} \dfrac{f(x)}{g(x)} = 0$. By definition of limit, this implies that given any real number $\epsilon > 0$, there exists a real number x_0 such that $\left| \dfrac{f(x)}{g(x)} - 0 \right| < \epsilon$ for all $x > x_0$. Then, $|f(x)| \leq \epsilon |g(x)|$ for all $x > x_0$. Choose $\epsilon = 1$, and set $M = 1$ also. Then there exists a real number x_0 so that $|f(x)| \leq M|g(x)|$ for all $x > x_0$. Hence by definition of O-notation, $f(x)$ is $O(g(x))$.

SECTION 9.3

1. a. $\log_2(200) = \dfrac{\log_{10}(200)}{\log_{10}(2)} \cong 7.6$ microseconds $= .0000076$ second

d. $200^2 = 40{,}000$ microseconds $= .04$ second

e. $200^8 = 2.56 \times 10^{18}$ microseconds $\cong$

$$\frac{2.56 \times 10^{18}}{10^6 \cdot 60 \cdot 60 \cdot 24 \cdot (365.25)} \text{ years} \cong 81{,}121.5 \text{ years}$$

[because there are 10^6 microseconds in a second, 60 seconds in a minute, 60 minutes in an hour, 24 hours in a day, and 365.25 days in a year on average].

2. a. When the input size is increased from m to $2m$, the number of operations increases from cm^2 to $c(2m)^2 = 4cm^2$.

b. By part (a), the number of operations increases by a factor of $(4cm^2)/cm^2 = 4$.

c. When the input size is increased by a factor of 10 (from m to $10m$), the number of operations increases by a factor of $(c(10m)^2)/(cm^2) = (100cm^2)/cm^2 = 100$.

4. a. Many correct answers are possible, but the simplest is to say that algorithm A has order n^2 and algorithm B has order $n^{3/2}$. The reason is that the best big-oh approximation for the efficiency of algorithm A from among the set of power functions is n^2 and for algorithm B it is $n^{3/2}$.

b. Algorithm A is more efficient than algorithm B when $2n^2 < 80n^{3/2}$. This occurs exactly when

$$n^2 < 40n^{3/2} \Leftrightarrow \frac{n^2}{n^{3/2}} < 40 \Leftrightarrow n^{1/2} < 40 \Leftrightarrow$$

$n < 40^2 = 1{,}600$. So algorithm A is more efficient than algorithm B when $n < 1600$.

c. Algorithm B is at least 100 times more efficient than algorithm A for values of n with $100(80n^{3/2}) \le 2n^2$. This occurs exactly when $8{,}000n^{3/2} \le 2n^2 \Leftrightarrow$

$$4{,}000 \le \frac{n^2}{n^{3/2}} \Leftrightarrow 4{,}000 \le \sqrt{n} \Leftrightarrow 16{,}000{,}000 \le n.$$

So algorithm B is at least 100 times more efficient than algorithm A when $n \ge 16{,}000{,}000$.

6. a. There are two multiplications, one addition, and one subtraction for each iteration of the loop, so there are four times as many operations as there are iterations of the loop. The loop is iterated $(n - 1) - 3 + 1 = n - 3$ times (since the number of iterations equals the top minus the bottom index plus 1). Thus the total number of operations is $4(n - 3) = 4n - 12$.

b. By the theorem on polynomial orders, $4n - 12$ is $O(n)$, so the algorithm segment is $O(n)$.

8. a. There is one subtraction for each iteration of the loop and there are $\lfloor n/2 \rfloor$ iterations of the loop.

b. $\lfloor n/2 \rfloor = \begin{cases} n/2 & \text{if } n \text{ is even} \\ (n-1)/2 & \text{if } n \text{ is odd} \end{cases}$
is $O(n)$ by the theorem on polynomial orders, so the algorithm segment is $O(n)$.

10. a. There is one comparison for each iteration of the inner loop. The number of iterations of the inner loop can be deduced from the table below, which shows the values of k and i for which the statements in the inner loop are executed.

k	1				$\rightarrow$	2				$\rightarrow$	3				$\rightarrow$	$\cdots$	$n-2$	$\rightarrow$	$n-1$
i	2	3	$\cdots$	n		3	4	$\cdots$	n		4	5	$\cdots$	n		$\cdots$	$n-1$	n	n

Hence the total number of iterations of the inner loop is

$$(n - 1) + (n - 2) + \cdots + 2 + 1 = \frac{(n-1)n}{2}$$

(by Theorem 4.2.2).

b. $\dfrac{(n-1)n}{2} = \dfrac{1}{2}n^2 - \dfrac{n}{2}$ is $O(n^2)$ by the theorem on polynomial orders, and so the algorithm segment is $O(n^2)$.

12. a. There are two subtractions and one multiplication for each iteration of the inner loop. If n is even, the number of iterations of the inner loop is

$$1 + 1 + 2 + 2 + 3 + 3 + \cdots + \frac{n}{2} + \frac{n}{2}$$

$$= 2 \cdot \left(1 + 2 + 3 + \cdots + \frac{n}{2}\right)$$

$$= 2 \cdot \left(\frac{\frac{n}{2}\left(\frac{n}{2} + 1\right)}{2}\right) \quad \text{by Theorem 4.2.2}$$

$$= \frac{n^2}{4} + \frac{n}{2}.$$

If n is odd, the number of iterations of the inner loop is

$$1 + 1 + 2 + 2 + \cdots + \frac{n-1}{2} + \frac{n-1}{2} + \frac{n+1}{2}$$

$$= 2 \cdot \left(1 + 2 + 3 + \cdots + \frac{n-1}{2}\right) + \frac{n+1}{2}$$

$$= 2 \cdot \frac{\frac{n-1}{2}\left(\frac{n-1}{2} + 1\right)}{2} + \frac{n+1}{2} \quad \text{by Theorem 4.2.2}$$

$$= \frac{n^2 - 2n + 1}{4} + \frac{n-1}{2} + \frac{n+1}{2}$$

$$= \frac{1}{4}n^2 + \frac{1}{2}n + \frac{1}{4}.$$

So the answer is $3\left(\dfrac{n^2}{4} + \dfrac{n}{2}\right)$ when n is even and

$3\left(\dfrac{1}{4}n^2 + \dfrac{1}{2}n + \dfrac{1}{4}\right)$ when n is odd.

b. Since $3\left(\dfrac{n^2}{4} + \dfrac{n}{2}\right)$ is $O(n^2)$ and $3\left(\dfrac{1}{4}n^2 + \dfrac{1}{2}n + \dfrac{1}{4}\right)$ is $O(n^2)$ also (by the theorem on polynomial orders), then this algorithm segment has order n^2.

14. *Hint:* See Section 6.5 for a discussion of how to count the number of iterations of the innermost loop.

15.

	a[1]	a[2]	a[3]	a[4]	a[5]
initial order	6	2	1	8	4
result of step 1	2	6	1	8	4
result of step 2	1	2	6	8	4
result of step 3	1	2	6	8	4
final order	1	2	4	6	8

17. 1 *[from step 1]* $+$ 1 *[from step 2]* $+$ 3 *[from step 3]* $+$ 2 *[from step 4]* $=$ 7

20. The top row of the table below shows the initial values of the array and the bottom row shows the final values. The result of each interchange is shown in a separate row.

a[1]	a[2]	a[3]	a[4]	a[5]
5	3	4	6	2
3	5	4	6	2
2	5	4	6	3
2	4	5	6	3
2	3	5	6	4
2	3	4	6	5
2	3	4	5	6

22. The number of interchanges is six, one less than the number of rows in the table constructed in exercise 20.

25. b. $n - 3 + 1 = n - 2$
    d. *Hint:* the answer is n^2.

29.

n	3								
a[0]	2								
a[1]	1								
a[2]	-1								
a[3]	3								
x	2								
polyval	2	4			0				24
i	1	2			3				
term	1	2	-1	-2	-4	3	6	12	24
j	1		1	2		1	2	3	

31. number of multiplications $=$ number of iterations of the inner loop
$$= 1 + 2 + 3 + \cdots + n$$
$$= \frac{n(n+1)}{2}$$
by Theorem 4.2.2

number of additions $=$ number of iterations of the outer loop
$$= n$$

Hence the total number of multiplications and additions is

$$\frac{n(n+1)}{2} + n = \frac{1}{2}n^2 + \frac{3}{2}n.$$

33.

n	3			
$a[0]$	2			
$a[1]$	1			
$a[2]$	-1			
$a[3]$	3			
x	2			
i		1	2	3
polyval	3	5	11	24

35. *Hint:* The answer is $t_n = 2n$.

SECTION 9.4

1.

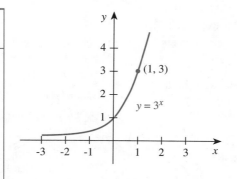

x	$f(x) = 3^x$
0	$3^0 = 1$
1	$3^1 = 3$
2	$3^2 = 9$
-1	$3^{-1} = 1/3$
-2	$3^{-2} = 1/9$
1/2	$3^{1/2} \cong 1.7$
$-(1/2)$	$3^{-(1/2)} \cong 0.6$

3.

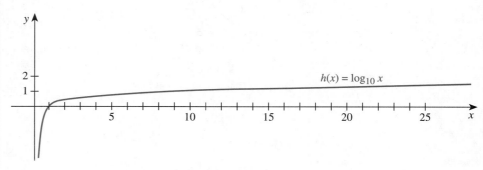

x	$h(x) = \log_{10} x$
1	0
10	1
100	2
1/10	-1
1/100	-2

5.

x	$\lfloor \log_2 x \rfloor$
$1 \le x < 2$	0
$2 \le x < 4$	1
$4 \le x < 8$	2
$8 \le x < 16$	3
$1/2 \le x < 1$	−1
$1/4 \le x < 1/2$	−2

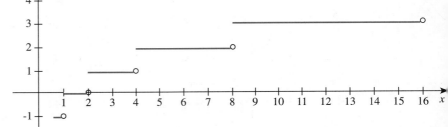

$F(x) = \lfloor \log_2 x \rfloor$

7.

x	$x \log_2 x$
1	$1 \cdot 0 = 0$
2	$2 \cdot 1 = 2$
4	$4 \cdot 2 = 8$
8	$8 \cdot 3 = 24$
1/8	$(1/8) \cdot (-3) = -3/8$
1/4	$(1/4) \cdot (-2) = -1/2$
3/8	$(3/8) \cdot (\log_2(3/8)) \cong -0.53$

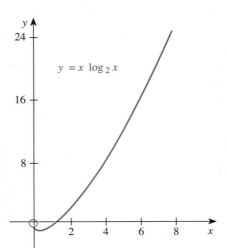

$y = x \log_2 x$

9. The distance above the axis is $(2^{64} \text{ units})(\frac{1}{4} \frac{\text{inch}}{\text{unit}}) = \frac{2^{64}}{4}$ inches $= \frac{2^{64}}{4 \cdot 12 \cdot 5280}$ miles $\cong 72{,}785{,}448{,}520{,}000$ miles. The ratio of the height of the point to the average distance of the earth to the sun is approximately $72{,}785{,}448{,}520{,}000/93{,}000{,}000 \cong 782{,}639$.

10. b. By definition of logarithm, $\log_b x$ is the exponent to which b must be raised to obtain x. So when b is actually raised to this exponent, x is obtained. That is, $b^{\log_b x} = x$.

11. b.

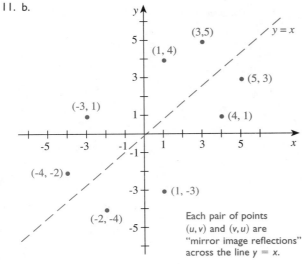

$y = x$

(3,5)
(1, 4)
(5, 3)
(-3, 1)
(4, 1)
(-4, -2)
(1, -3)
(-2, -4)

Each pair of points (u, v) and (v, u) are "mirror image reflections" across the line $y = x$.

13. *Hint:* $\lfloor \log_{10} x \rfloor = m$

16. No. *Counterexample:* Let $n = 2$. Then $\lceil \log_2(n - 1) \rceil = \lceil \log_2 1 \rceil = \lceil 0 \rceil = 0$ whereas $\lceil \log_2 n \rceil = \lceil \log_2 2 \rceil = \lceil 1 \rceil = 1$.

17. *Hint:* The statement is true.

19. a. $\lfloor \log_2 148{,}206 \rfloor + 1 = 18$

21. a.
$$a_1 = 1$$
$$a_2 = a_{\lfloor 2/2 \rfloor} + 2 = a_1 + 2 = 1 + 2$$
$$a_3 = a_{\lfloor 3/2 \rfloor} + 2 = a_1 + 2 = 1 + 2$$
$$a_4 = a_{\lfloor 4/2 \rfloor} + 2 = a_2 + 2 = (1 + 2) + 2$$
$$= 1 + 2 \cdot 2$$
$$a_5 = a_{\lfloor 5/2 \rfloor} + 2 = a_2 + 2 = (1 + 2) + 2$$
$$= 1 + 2 \cdot 2$$
$$a_6 = a_{\lfloor 6/2 \rfloor} + 2 = a_3 + 2 = (1 + 2) + 2$$
$$= 1 + 2 \cdot 2$$
$$a_7 = a_{\lfloor 7/2 \rfloor} + 2 = a_3 + 2 = (1 + 2) + 2$$
$$= 1 + 2 \cdot 2$$
$$a_8 = a_{\lfloor 8/2 \rfloor} + 2 = a_4 + 2$$
$$= (1 + 2 \cdot 2) + 2 = 1 + 3 \cdot 2$$
$$a_9 = a_{\lfloor 9/2 \rfloor} + 2 = a_4 + 2$$
$$= (1 + 2 \cdot 2) + 2 = 1 + 3 \cdot 2$$
$$\vdots$$

$$a_{15} = a_{\lfloor 15/2 \rfloor} + 2 = a_7 + 2$$
$$= (1 + 2 \cdot 2) + 2 = 1 + 3 \cdot 2$$
$$a_{16} = a_{\lfloor 16/2 \rfloor} + 2 = a_8 + 2$$
$$= (1 + 3 \cdot 2) + 2 = 1 + 4 \cdot 2$$
$$\vdots$$

Guess: $a_n \overset{.}{=} 1 + 2 \cdot \lfloor \log_2 n \rfloor$

b. *Proof:* Suppose the sequence $a_1, a_2, a_3, \ldots$ is defined recursively as follows: $a_1 = 1$ and $a_k = a_{\lfloor k/2 \rfloor} + 2$ for all integers $k \geq 2$. We will show by strong mathematical induction that for all integers $n \geq 1$, $a_n = 1 + 2 \cdot \lfloor \log_2 n \rfloor$.

The formula holds for $n = 1$: When $n = 1$, $1 + 2 \cdot \lfloor \log_2 n \rfloor = 1 + 2 \cdot \lfloor \log_2 1 \rfloor = 1 + 2 \cdot 0 = 1$, which is the value of a_1.

If $k > 1$ and the formula holds for all integers i with $1 \leq i < k$, then it holds for $n = k$: Let k be an integer with $k > 1$ and suppose $a_i = 1 + 2 \cdot \lfloor \log_2 i \rfloor$ for all integers i with $1 \leq i < k$. *[This is the inductive hypothesis.]* We must show that $a_k = 1 + 2 \cdot \lfloor \log_2 k \rfloor$.

Case 1 (k is even):

$a_k = a_{\lfloor k/2 \rfloor} + 2$	by the recursive definition of $a_1, a_2, a_3, \ldots$
$= a_{k/2} + 2$	because k is even
$= 1 + 2 \cdot \lfloor \log_2 (k/2) \rfloor + 2$	by inductive hypothesis
$= 3 + 2 \cdot \lfloor \log_2 k - \log_2 2 \rfloor$	because $\log_b (x/y) = \log_b x - \log_b y$ (exercise 24, Section 7.3)
$= 3 + 2 \cdot \lfloor \log_2 k - 1 \rfloor$	because $\log_2 2 = 1$
$= 3 + 2 \cdot (\lfloor \log_2 k \rfloor - 1)$	because for all real numbers x $\lfloor x - 1 \rfloor = \lfloor x \rfloor - 1$ (by exercise 15, Section 3.5)
$= 1 + 2 \cdot \lfloor \log_2 k \rfloor$	by algebra

Case 2 (k is odd):

$a_k = a_{\lfloor k/2 \rfloor} + 2$	by the recursive definition of $a_1, a_2, a_3, \ldots$
$= a_{\lfloor (k-1)/2 \rfloor} + 2$	because k is odd
$= 1 + 2 \cdot \lfloor \log_2((k - 1)/2) \rfloor + 2$	by inductive hypothesis
$= 3 + 2 \cdot \lfloor \log_2(k - 1) - \log_2 2 \rfloor$	because $\log_b (x/y) = \log_b x - \log_b y$
$= 3 + 2 \cdot \lfloor \log_2(k - 1) - 1 \rfloor$	because $\log_2 2 = 1$

$= 3 + 2 \cdot (\lfloor \log_2(k - 1) \rfloor - 1)$	because for all real numbers x, $\lfloor x - 1 \rfloor = \lfloor x \rfloor - 1$ (by exercise 15, Section 3.5)
$= 1 + 2 \cdot \lfloor \log_2(k - 1) \rfloor$	by algebra
$= 1 + 2 \cdot \lfloor \log_2 k \rfloor$	by property (9.4.3)

Thus in either case, $a_k = 1 + 2 \cdot \lfloor \log_2 k \rfloor$ *[as was to be shown]*.

23. *Hint:* When $k \geq 2$, then $k^2 \geq 2k$, and so $k \leq \dfrac{k^2}{2}$.

Hence $\dfrac{k^2}{2} + k \leq \dfrac{k^2}{2} + \dfrac{k^2}{2} = k^2$. Also when $k \geq 2$, then $k^2 > 1$, and so $\dfrac{1}{2} < \dfrac{k^2}{2}$. Consequently, $\dfrac{k^2}{2} + \dfrac{1}{2} < \dfrac{k^2}{2} + \dfrac{k^2}{2} = k^2$.

24. *Hint:* Here is the argument for the inductive step in case k is even.

$c_k = 2 \cdot c_{\lfloor k/2 \rfloor} + k$	by the recursive definition of $c_1, c_2, c_3, \ldots$
$\Rightarrow c_k = 2 \cdot c_{k/2} + k$	because k is even
$\Rightarrow c_k \leq 2 \cdot [(k/2) \cdot \log_2(k/2)] + k$	by inductive hypothesis
$\Rightarrow c_k \leq k(\log_2 k - \log_2 2) + k$	by algebra and the fact that $\log_b(x/y) = \log_b x - \log_b y$
$\Rightarrow c_k \leq k(\log_2 k - 1) + k$	because $\log_2 2 = 1$
$\Rightarrow c_k \leq k \cdot \log_2 k$	by algebra

25. *Solution 1:* One way to solve this problem is to compare values for $\log_2 x$ and $x^{1/10}$ for conveniently chosen, large values of x. For instance, if powers of 10 are used, the following results are obtained: $\log_2(10^{10}) = 10 \log_2 10 \cong 33.2$ and $(10^{10})^{1/10} = 10^{10 \cdot (1/10)} = 10^1 = 10$. So the value $x = 10^{10}$ does not work. However, since $\log_2(10^{20}) = 20 \log_2 10 \cong 66.4$ and $(10^{20})^{1/10} = 10^{20 \cdot (1/10)} = 10^2 = 100$ and since $66.4 < 100$, the value $x = 10^{20}$ works.

Solution 2: Another approach is to use a graphing calculator or computer to sketch graphs of $y = \log_2 x$ and $y = x^{1/10}$, taking seriously the hint to "think big" in choosing the interval size for the x's. A few tries and use of the zoom and trace features make it appear that the graph of $y = x^{1/10}$ crosses above the graph of $y = \log_2 x$ at about 4.9155×10^{17}. So for values of x larger than this, $x^{1/10} > \log_2 x$.

27. As with exercise 25, this problem can be solved either by numerical exploration or by using a graphing calculator or computer. For instance, by just raising 1.0001 to successive very large powers of 10, the solution $x = 10^6 = 1,000,000$ can be found: $(1.0001)^{1,000,000} > 2.67 \times 10^{43} > 1,000,000$. (This is the first power of 10 that works.) Alternatively, you can use a graphing calculator or computer to sketch the graphs of $y = (1.0001)^x$ and $y = x$ and then look to see where the graph of $y = (1.0001)^x$ rises above the graph of $y = x$. By experimenting with several windows, a solution like $x = 116,685$ can be discovered: $(1.0001)^{116,685} > 116,764 > 116,685$.

29. $7x^2 + 3x \log_2 x$ is $O(x^2)$.

30. *[To show that $2x + \log_2 x$ is $O(x)$. we must find real numbers M and x_0 such that $|2x + \log_2 x| \le M \cdot |x|$, for all real numbers $x > x_0$.]*

By property (9.4.9) if $x > 0$, then

$$\log_2 x \le x.$$

Adding $2x$ to both sides gives

$$2x + \log_2 x \le 3x.$$

If $x > 1$, then $\log_2 x > 0$ and so

$$|2x + \log_2 x| = 2x + \log_2 x \le 3x = 3|x|.$$

Thus let $x_0 = 1$ and $M = 3$. Then

$$|2x + \log_2 x| \le M \cdot |x| \quad \text{for all } x > x_0.$$

Hence by definition of O-notation,

$$2x + \log_2 x \text{ is } O(x).$$

32. By Theorem 4.2.3,

$$1 + 2 + 2^2 + \cdots + 2^n = \frac{2^{n+1} - 1}{2 - 1} = 2^{n+1} - 1.$$

Let $M = 1$ and $x_0 = 1$. Then by basic algebra, $|2^{n+1} - 1| \le |2^{n+1}| = M \cdot |2^{n+1}|$ for all integers $n \ge x_0$.

Hence by definition of O-notation,

$$1 + 2 + 2^2 + \cdots + 2^n \text{ is } O(2^{n+1}).$$

33. *Hint:* Factor out the n and use Theorem 4.2.3.

36. By property (9.4.10), there is a real number x_0 so that $n^2 \le 2^n$ for all $n > x_0$. Adding 2 to both sides gives $n^2 + 2^n \le 2 \cdot 2^n$. Let $M = 2$. Then $|n^2 + 2^n| \le M \cdot |2^n|$ for all $n \ge x_0$. So by definition of O-notation, $n^2 + 2^n$ is $O(2^n)$.

38. If n is any integer with $n \ge 3$, then by factoring out an n and using Example 9.4.7(c),

$$n + \frac{n}{2} + \frac{n}{3} + \cdots + \frac{n}{n}$$

$$= n\left(1 + \frac{1}{2} + \frac{1}{3} + \cdots + \frac{1}{n}\right)$$

$$\le n(2 \cdot \ln n) = 2 \cdot (n \ln n)$$

Let $M = 2$ and $x_0 = 3$. Then

$$\left|n + \frac{n}{2} + \frac{n}{3} + \cdots + \frac{n}{n}\right| \le M \cdot |n \ln n|$$

for all integers $n > x_0$. So by definition of O-notation,

$n + \dfrac{n}{2} + \dfrac{n}{3} + \cdots + \dfrac{n}{n}$ is $O(n \ln n)$.

41. a. *Hint:*

$$n! = \underbrace{n \cdot (n - 1) \cdot (n - 2) \cdot \ldots \cdot 2 \cdot 1}_{n \text{ factors}}$$

$$\le \underbrace{n \cdot n \cdot n \cdot \ldots \cdot n \cdot n}_{n \text{ factors}} = n^n.$$

43. **The inequality holds for $n = 1$:** We must show that $\log_2 1 \le 1$. But $\log_2 1 = 0$ and $0 \le 1$, so the inequality holds for $n = 1$.

If the inequality holds for $n = k$, then it holds for $n = k + 1$: Suppose the inequality holds when an integer $k \ge 1$ is substituted in place of n. That is,

suppose $\log_2 k \le k$, for an integer $k \ge 1$.

[This is the inductive hypothesis.]

We must show that the inequality holds when $k + 1$ is substituted in place of n; that is,

show $\log_2(k + 1) \le k + 1$.

But the logarithmic function with base 2 is increasing and when $k \ge 1$, then

$$k + 1 \le 2k.$$

Hence

$$\log_2(k + 1) \le \log_2(2k), \quad \text{for } k \ge 1.$$

Now

$$\log_2(2k) = \log_2 2 + \log_2 k \quad \text{since } \log_b(xy) = \log_b x + \log_b y \text{ (exercise 25 of Section 7.3)}$$

$$= 1 + \log_2 k \quad \text{since } \log_2 2 = 1$$

$$\le 1 + k \quad \text{since } \log_2 k \le k \text{ by inductive hypothesis.}$$

It follows by transitivity of $\le$ (Appendix A, T17) that

$$\log_2(k + 1) \le k + 1,$$

which is what was to be shown.

45. a. *Hint 1:*

$$(k + 1) \cdot \log_2(k + 1)$$
$$= k \cdot \log_2(k + 1) + \log_2(k + 1)$$
$$= \log_2(k + 1)^k + \log_2(k + 1)$$
$$= \log_2\left(\frac{(k + 1)^k}{k^k} k^k\right) + \log_2(k + 1)$$
$$= \log_2\left(\left(1 + \frac{1}{k}\right)^k\right) + \log_2(k^k) + \log_2(k + 1)$$
$$= \log_2\left(\left(1 + \frac{1}{k}\right)^k\right) + k \cdot \log_2(k) + \log_2(k + 1)$$

Hint 2: Let F be the function defined by the rule

$$F(x) = \left(1 + \frac{1}{x}\right)^x \text{ for all real numbers } x \geq 1.$$

Use calculus to show that F is strictly increasing for all $x \geq 1$. Combine this result with the fact that

$$\lim_{x \to \infty}\left(\left(1 + \frac{1}{x}\right)^x\right) = e \text{ to deduce that } \left(1 + \frac{1}{n}\right)^n < 3$$

for all integers $n \geq 1$.

46. *Hint:* There is more than one way to deduce this fact.
  I. **a.** Show that for all integers n and i with $1 \leq i \leq n$,
     $i \cdot (n - i + 1) \geq n$.
     b. Regroup the factors of $n!$ as indicated and use the result of part (a). When n is even,

$$n! = [n \cdot 1] \cdot [(n - 1) \cdot 2] \cdot [(n - 2) \cdot 3] \cdot \ldots \cdot \left[\left(\frac{n}{2} + 1\right)\frac{n}{2}\right]$$

And when n is odd,

$$n! = [n \cdot 1] \cdot [(n - 1) \cdot 2] \cdot$$
$$[(n - 2) \cdot 3] \cdot \ldots \cdot$$
$$\left[\left(\frac{n + 3}{2}\right) \cdot \left(\frac{n - 1}{2}\right)\right] \cdot \left(\frac{n + 1}{2}\right)$$

(When you use this method, you can show that $\log_2(n!) \geq \frac{1}{2} n \log_2 n$ for all integers $n \geq 1$.)

  2. Use the fact that $n! \geq n(n - 1)(n - 2) \cdot \ldots \cdot \lceil n/2 \rceil$ to derive the inequality $n! \geq n(n/2)^{n/2}$ for $n \geq 1$. Then take logarithms of both sides and use the assumption that $n \geq 4$ to derive the given result.

47. *Hint:* Use the binomial theorem with $a = 1$ and $b = r$ to derive the equation $(1 + r)^n = 1 + r \cdot n +$ other positive terms.

48. a. Let n be a positive integer. For any real number $x > 1$, properties of exponents and logarithms (see Section 7.3) imply that $0 \leq \log_2(x) = \log_2((x^{1/n})^n) = n \cdot \log_2(x^{1/n}) < n \cdot x^{1/n}$ (where the last inequality holds by substituting $x^{1/n}$ in place of u in $\log_2 u < u$).

  b. Let $M = n$ and $x_0 = 1$. Then if $x > x_0$, $|\log_2 x| = \log_2 x \leq M \cdot |x^{1/n}|$, and so $\log_2 x$ is $O(x^{1/n})$.

50. Let n be a positive integer, and suppose that $x > (2n)^{2n}$. By properties of logarithms,

$$\log_2 x = (2n)\left(\frac{1}{2n}\right)\log_2 x$$
$$= 2n \cdot \log_2\left(x^{1/2n}\right) < 2n \cdot x^{1/2n} \quad (*)$$

(where the last inequality holds by substituting $x^{1/2n}$ in place of u in $\log_2 u < u$). But raising both sides of $x > (2n)^{2n}$ to the $1/2$ power gives $x^{1/2} > ((2n)^{2n})^{1/2} = (2n)^n$. When both sides are multiplied by $x^{1/2}$, the result is $x = x^{1/2} \cdot x^{1/2} > x^{1/2} \cdot (2n)^n = x^{1/2} \cdot (2n)^n$, or, more compactly,

$$x^{1/2} \cdot (2n)^n < x.$$

Then since the power function defined by $x \to x^{1/n}$ is increasing for all $x > 0$ (see exercise 20 of Section 9.1), we can take the nth root of both sides of inequality and use the laws of exponents to obtain

$$(x^{1/2} \cdot (2n)^n)^{1/n} < x^{1/n}$$

or, equivalently

$$x^{1/2n} \cdot (2n) < x^{1/n} \quad (**).$$

Now use transitivity of $<$ (Appendix A, T17) to combine $(*)$ and $(**)$ and conclude that $\log_2 x < x^{1/n}$ [as was to be shown].

52. a. *Proof:*
  The formula holds for $n = 1$: By L'Hôpital's rule,
$$\lim_{x \to \infty}\frac{x^1}{b^x} = \lim_{x \to \infty}\left(\frac{0}{(\ln b)b^x}\right) = \lim_{x \to \infty} 0 = 0.$$
  So the formula holds for $n = 1$.
  If the formula holds for $n = k$, then it holds for $n = k + 1$: Let k be an integer with $k \geq 1$, and suppose that $\lim_{x \to \infty}\frac{x^k}{b^x} = 0$. *[Inductive hypothesis]*
  We must show that $\lim_{x \to \infty}\frac{x^{k+1}}{b^x} = 0$. But by L'Hôpital's rule, $\lim_{x \to \infty}\frac{x^{k+1}}{b^x} = \lim_{x \to \infty}\frac{(k + 1)x^k}{(\ln b)b^x} = \frac{(k + 1)}{(\ln b)} \lim_{x \to \infty}\frac{x^k}{b^x} = \frac{(k + 1)}{(\ln b)} \cdot 0$ *[by inductive hypothesis]* $= 0$ *[This is what was to be shown.]*

  b. By the result of part (a) and the definition of limit, given any real number $\epsilon > 0$, (in this case take $\epsilon = 1$) there exists an integer N so that $\left|\frac{x^n}{b^n} - 0\right| < 1$ for all $x > N$. It follows that for all $x > N$, $\left|\frac{x^n}{b^x}\right| = \frac{|x^n|}{|b^x|} < 1$. Multiply both sides by $|b^x|$ to obtain $|x^n| < |b^x|$. Let $M = 1$ and $x_0 = N$. Then $|x^n| < M \cdot |b^x|$ for all $x > x_0$. Hence by definition of O-notation, x^n is $O(b^x)$.

SECTION 9.5

1. $\log_2 1000 = \log_2(10^3) = 3 \log_2 10 \cong 3(3.32)$
    $\cong 9.96$
    $\log_2(1,000,000) = \log_2(10^6) = 6 \log_2 10 \cong 6(3.32)$
    $\cong 19.92$
    $\log_2(1,000,000,000,000) = \log_2(10^{12}) = 12 \log_2 10$
    $\cong 12(3.32) = 39.84$

2. a. If $m = 2^k$, where k is a positive integer, then the algorithm requires $c \lfloor \log_2(2^k) \rfloor = c \lfloor k \rfloor = ck$ operations. If the input size is increased to $m^2 = (2^k)^2 = 2^{2k}$, then the number of operations required is $c \lfloor \log_2(2^{2k}) \rfloor = c \lfloor 2k \rfloor = 2(ck)$. So the number of operations doubles.
    b. As in part (a), for an input of size $m = 2^k$, where k is a positive integer, the algorithm requires ck operations. If the input size is increased to $m^{10} = (2^k)^{10} = 2^{10k}$, then the number of operations required is $c \lfloor \log_2(2^{10k}) \rfloor = c \lfloor 10k \rfloor = 10(ck)$. So the number of operations increases by a factor of 10.
    c. When the input size is increased from 2^7 to 2^{28}, the factor by which the number of operations increases is
    $$\frac{c \lfloor \log_2 (2^{28}) \rfloor}{c \lfloor \log_2 (2^7) \rfloor} = \frac{28c}{7c} = 4.$$

3. A little numerical exploration can help find an initial window to use to draw the graphs of $y = x$ and $y = \lfloor 50 \log_2 x \rfloor$. Note that when $x = 2^8 = 256$, $\lfloor 50 \log_2 x \rfloor = \lfloor 50 \log_2 (2^8) \rfloor = \lfloor 50 \cdot 8 \rfloor = \lfloor 400 \rfloor = 400 > 256 = x$. But when $x = 2^9 = 512$, $\lfloor 50 \log_2 x \rfloor = \lfloor 50 \log_2 (2^9) \rfloor = \lfloor 50 \cdot 9 \rfloor = \lfloor 450 \rfloor = 450 < 512 = x$. So a good choice of initial window would be the interval from 256 to 512. Drawing the graphs, zooming if necessary, and using the trace feature reveals that when $n < 438$, $n < \lfloor 50 \log_2 n \rfloor$.

5. a.

index	0			1
bot	1			
top	10	4	1	
mid		5	2	1

b.

index	0			
bot	1	6		7
top	10		7		6
mid		5	8	6	7

7. a. $top - bot + 1$
    b. *Proof:* Suppose top and bot are particular but arbitrarily chosen positive integers such that $top - bot + 1$ is an odd number. Then by definition of odd, there is an integer k such that
    $$top - bot + 1 = 2k + 1$$
    Adding $2 \cdot bot - 1$ to both sides gives
    $$bot + top = 2 \cdot bot - 1 + 2k + 1$$
    $$= 2 \cdot (bot + k).$$
    But $bot + k$ is an integer. Hence by definition of even, $bot + top$ is even.

8.

n	27	13	6	3	1	0

9. For each positive integer n, n div $2 = \lfloor n/2 \rfloor$. Thus when the algorithm segment is run for a particular n and the **while** loop has iterated one time, the input to the next iteration is $\lfloor n/2 \rfloor$. It follows that the number of iterations of the loop for n is one more than the number of iterations for $\lfloor n/2 \rfloor$. That is, $a_n = 1 + a_{\lfloor n/2 \rfloor}$. Also $a_1 = 1$.

10. The recurrence relation and initial condition for a_1, a_2, a_3, . . . derived in exercise 9 are the same as those for the sequence $w_1, w_2, w_3, \ldots$ discussed in the worst-case analysis of the binary search algorithm. Thus the general formulas for the two sequences are the same. That is, $a_n = 1 + \lfloor \log_2 n \rfloor$, for all integers $n \geq 1$.

11. In the analysis of the binary search algorithm it was shown that $1 + \lfloor \log_2 n \rfloor$ is $O(\log_2 n)$. Thus the algorithm segment has order $\log_2 n$.

20.

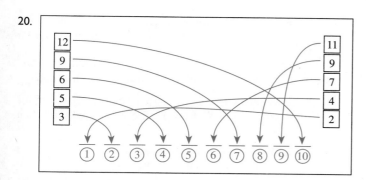

22. Initial array:

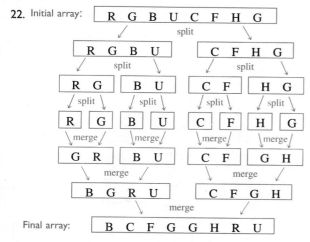

Final array: | B C F G G H R U |

24. b. Refer to Figure 9.5.3 and observe that when k is odd, the subarray $a[bot], a[bot + 1], \ldots, a[mid]$ has length $(k + 1)/2 = \lceil k/2 \rceil$, and when k is even it has length $k/2 = \lceil k/2 \rceil$ also.

25. *Hint:* Applying the inductive hypothesis in case k is odd gives

$$m_k = m_{(k-1)/2} + m_{(k+1)/2} + k - 1$$
$$\leq 2\left(\frac{k-1}{2}\right)\log_2\left(\frac{k-1}{2}\right) + 2\left(\frac{k+1}{2}\right)\log_2\left(\frac{k+1}{2}\right)$$
$$+ k - 1$$
$$\leq (k - 1)(\log_2 k - 1) + (k + 1)\log_2 k + k - 1$$
$$[Why?]$$
$$\leq 2k \log_2 k.$$

SECTION 10.1

1. a. No. Yes. No. Yes
   b. $R = \{(2, 6), (2, 8), (2, 10), (3, 6), (4, 8)\}$

3. a. $0\, E\, 0$ because $0 - 0 = 0 = 2 \cdot 0$, so $2\,|\,(0 - 0)$.
      $5\, \not{E}\, 2$ because $5 - 2 = 3$ and $3 \neq 2k$ for any integer k, so $2 \nmid (5 - 2)$.
      $(6, 6) \in E$ because $6 - 6 = 0 = 2 \cdot 0$, so $2\,|\,(6 - 6)$.
      $(8, 1) \notin E$ because $8 - 1 = 7$ and $7 \neq 2k$ for any integer k. So $2 \nmid (8 - 1)$.

4. *Hint:* To show a statement of the form $p \leftrightarrow (q \text{ or } r)$, show $p \to (q \text{ or } r)$ and $(q \text{ or } r) \to p$. To show a statement of the form $p \to (q \text{ or } r)$, you can show $(p \wedge \sim q) \to r$ (since these two statement forms are logically equivalent). To show a statement of the form $(q \text{ or } r) \to p$, you can show $(q \to p) \wedge (r \to p)$ (since these two statement forms are logically equivalent). In this case, suppose m and n are any integers, and let p be "$m - n$ is even," let q be "m and n are both even," and let r be "m and n are both odd."

5. a. $10\, T\, 1$ because $10 - 1 = 9 = 3 \cdot 3$, so $3\,|\,(10 - 1)$.
      $1\, T\, 10$ because $1 - 10 = -9 = 3 \cdot (-3)$, so $3\,|\,(1 - 10)$.
      $2\, T\, 2$ because $2 - 2 = 0 = 3 \cdot 0$, so $3\,|\,(2 - 2)$.
      $8\, \not{T}\, 1$ because $8 - 1 = 7 \neq 3 \cdot k$, for any integer k. So $3 \nmid (8 - 1)$.
   b. *One possible answer:* $3, 6, 9, -3, -6$
   e. *Hint:* All integers of the form $3k + 1$, for some integer k, are related by T to 1.

6. b.

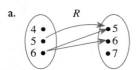

$x \geq y$ in shaded region

7. a. Yes because $4 = 2^2$.
No because $2 \neq 4^2$.
Yes because $9 = (-3)^2$.
No because $-3 \neq 9^2$.

8. a. Yes because 15 and 25 are both divisible by 5, which is prime.
  b. No because 22 and 27 have no common prime factor.

9. a. Yes because both $\{a, b\}$ and $\{b, c\}$ have two elements.

10. a. No because $\{a\} \cap \{c\} = \emptyset$.

11. a. Yes because both *abaa* and *abba* have the same first two characters *ab*.

12. *Hint:*

  a.

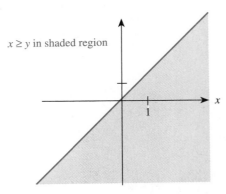

  b. R is not a function. It satisfies neither property (1) nor property (2). It fails property (1) because $(4, y) \notin R$ for any y in B. It fails property (2) because $(6, 5) \in R$ and $(6, 6) \in R$ and $5 \neq 6$.

13. a. $\emptyset, \{(0, 1)\}, \{(1, 1)\}, \{(0, 1), (1, 1)\}$
  b. $\{(0, 1), (1, 1)\}$
  c. $1/4$

15. *Hint:*
  a. The answer is 2^{mn}.
  b. The answer is n^m.

16. No because, for example, $(4, 2) \in P$ and $(4, -2) \in P$ but $2 \neq -2$.

17. $R = \{(3, 4), (3, 5), (3, 6), (4, 5), (4, 6), (5, 6)\}$
$R^{-1} = \{(4, 3), (5, 3), (6, 3), (5, 4), (6, 4), (6, 5)\}$

19. a. Yes because *aab* is the concatenation of *a* with *ab*.
  b. No because *ab* is not the concatenation of *a* with *aab*.
  d. Yes because *aba* T^{-1} *ba* $\iff$ *ba* T *aba* $\iff$ *aba* is the concatenation of *a* with *ba*, which is true.

21. a. A function $F : X \to Y$ is one-to-one if, and only if, for all $x_1, x_2 \in X$, if $(x_1, y) \in F$ and $(x_2, y) \in F$, then $x_1 = x_2$.

22. a. No. If $F : X \to Y$ is not onto, then F^{-1} is not defined on all of Y. In other words, there is an element y in Y such that $(y, x) \notin F^{-1}$ for all $x \in X$. Consequently, F^{-1} does not satisfy property (1) of the definition of function.

23.

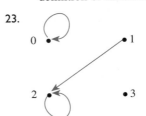

25.

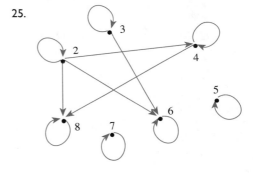

26. *Hint:* See Example 10.1.10.

28. a. 574329   Tak Kurosawa
011985   John Schmidt

29. $A \times B = \{(2, 6), (2, 8), (2, 10), (4, 6), (4, 8), (4, 10)\}$,
$R = \{(2, 6), (2, 8), (2, 10), (4, 8)\}$,
$S = \{(2, 6), (4, 8)\}$,
$R \cup S = R, R \cap S = S$

31.

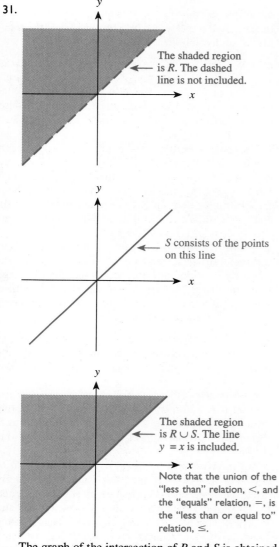

The graph of the intersection of R and S is obtained by finding the set of all points common to both graphs. But there are no points for which both $x < y$ and $x = y$. Hence $R \cap S = \varnothing$ and the graph consists of no points at all.

SECTION 10.2

1. R_1:
   a.

b. R_1 is not reflexive: $2 \not{R}_1 2$.
c. R_1 is not symmetric: $2 \ R_1 \ 3$ *but* $3 \not{R}_1 2$.
d. R_1 is not transitive: $1 \ R_1 \ 0$ and $0 \ R_1 \ 3$ but $1 \not{R}_1 3$.

3. R_3:
   a. 0 •   • 1

b. R_3 is not reflexive: $(0, 0) \notin R_3$
c. R_3 is symmetric. (If R_3 were not symmetric, there would be elements x and y in $A = \{0, 1, 2, 3\}$ such that $(x, y) \in R_3$ but $(y, x) \notin R_3$. It is clear by inspection that no such elements exist.)
d. R_3 is not transitive: $(2, 3) \in R_3$ and $(3, 2) \in R_3$ but $(2, 2) \notin R_3$

6. R_6:
   a.

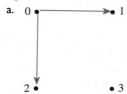

b. R_6 is not reflexive: $(0, 0) \notin R_6$
c. R_6 is not symmetric: $(0, 1) \in R_6$ but $(1, 0) \notin R_6$.
d. R_6 is transitive. (If R_6 were not transitive, there would be elements $x, y,$ and z in $\{0, 1, 2, 3\}$ such that $(x, y) \in R_6$ and $(y, z) \in R_6$ but $(x, z) \notin R_6$. It is clear by inspection that no such elements exist.)

9. $R^t = R \cup \{(0, 0), (0, 3), (1, 0), (3, 1), (3, 2), (3, 3),$
    $(0, 2), (1, 2)\}$
    $= \{(0, 0), (0, 1), (0, 2), (0, 3), (1, 0), (1, 1), (1, 2),$
    $(1, 3), (2, 2), (3, 0), (3, 1), (3, 2), (3, 3)\}$

12. *R is reflexive:* R is reflexive $\Leftrightarrow$ for all real numbers x, $x \ R \ x$. By definition of R this means that for all real numbers x, $x \geq x$. In other words, for all real numbers x, $x > x$ or $x = x$. But this is true.
    R is not symmetric: R is symmetric $\Leftrightarrow$ for all real numbers x and y, if $x \ R \ y$ then $y \ R \ x$. By definition of R, this means that for all real numbers x and y, if $x \geq y$ then $y \geq x$. But this is false. As a counterexample, take $x = 1$ and $y = 0$. Then $x \geq y$ but $y \not\geq x$ because $1 \geq 0$ but $0 \not\geq 1$.
    R is transitive: R is transitive $\Leftrightarrow$ for all real numbers x, $y,$ and z, if $x \ R \ y$ and $y \ R \ z$ then $x \ R \ z$. By definition of R this means that for all real numbers x, y and z, if $x \geq y$ and $y \geq z$ then $x \geq z$. But this is true by definition of $\geq$ and the transitive property of order for the real numbers. (See Appendix A, T17.)

14. *D is reflexive:* For *D* to be reflexive means that for all real numbers *x*, *x D x*. But by definition of *D* this means that for all real numbers *x*, $x \cdot x = x^2 \geq 0$, which is true.

D is symmetric: For *D* to be symmetric means that for all real numbers *x* and *y*, if *x D y* then *y D x*. But by definition of *D* this means that for all real numbers *x* and *y*, if $xy \geq 0$ then $yx \geq 0$, which is true by the commutative law of multiplication.

D is not transitive: For *D* to be transitive means that for all real numbers *x*, *y*, and *z*, if *x D y* and *y D z* then *x D z*. By definition of *D* this means that for all real numbers *x*, *y*, and *z*, if $xy \geq 0$ and $yz \geq 0$ then $xz \geq 0$. But this is false: there exist real numbers *x*, *y*, and *z* such that $xy \geq 0$ and $yz \geq 0$ but $xz \not\geq 0$. As a counterexample, let $x = 1$, $y = 0$, and $z = -1$. Then *x D y* and *y D z* because $1 \cdot 0 \geq 0$ and $0 \cdot (-1) \geq 0$. But $x \not\!D z$ because $1 \cdot (-1) \not\geq 0$.

15. *E is reflexive:* [*We must show that for all integers m, m E m.*] Suppose *m* is any integer. Since $m - m = 0$ and $2 \mid 0$, we have that $2 \mid (m - m)$. Consequently, *m E m* by definition of *E*.

E is symmetric: [*We must show that for all integers m and n, if m E n then n E m.*] Suppose *m* and *n* are any integers such that *m E n*. By definition of *E* this means that $2 \mid (m - n)$, and so by definition of divisibility $m - n = 2k$ for some integer *k*. Now $n - m = -(m - n)$. Hence by substitution, $n - m = -(2k) = 2 \cdot (-k)$. It follows that $2 \mid n - m$ by definition of divisibility (since $-k$ is an integer), and thus *n E m* by definition of *E*.

E is transitive: [*We must show that for all integers m, n and p, if m E n and n E p then m E p.*] Suppose *m*, *n* and *p* are any integers such that *m E n* and *n E p*. By definition of *E* this means that $2 \mid (m - n)$ and $2 \mid (n - p)$, and so by definition of divisibility $m - n = 2k$ for some integer *k* and $n - p = 2l$ for some integer *l*. Now $m - p = (m - n) + (n - p)$. Hence by substitution, $m - p = 2k + 2l = 2(k + l)$. It follows that $2 \mid m - p$ by definition of divisibility (since $k + l$ is an integer), and thus *m E p* by definition of *E*.

18. *D is reflexive:* [*We must show that for all positive integers m, m D m.*] Suppose *m* is any positive integer. Since $m = m \cdot 1$, by definition of divisibility $m \mid m$. Hence *m D m* by definition of *D*.

D is not symmetric: For *D* to be symmetric would mean that for all positive integers *m* and *n*, if *m D n* then *n D m*. By definition of divisibility this would mean that for all positive integers *m* and *n*, if $m \mid n$ then $n \mid m$. But this is false. As a counterexample, take $m = 2$ and $n = 4$. Then $m \mid n$ because $2 \mid 4$ but $n \nmid m$ because $4 \nmid 2$.

D is transitive: To prove transitivity of *D*, we must show that for all positive integers *m*, *n*, and *p*, if *m D n* and *n D p* then *m D p*. By definition of *D* this means that

for all positive integers *m*, *n*, and *p*, if $m \mid n$ and $n \mid p$ then $m \mid p$. But this is true by Theorem 3.3.1 (the transitivity of divisibility).

21. *L is not reflexive:* *L* is reflexive $\Leftrightarrow$ for all strings $s \in \Sigma^*$, *s L s*. By definition of *L* this means that for all strings *s* in Σ^*, $\ell(s) < \ell(s)$, which means that the length of *s* is less than the length of *s*. But this is false for every string in Σ^*. For instance, let $s = \varepsilon$. Then $\ell(s) = 0$ and $0 \not< 0$.

L is not symmetric: For *L* to be symmetric would mean that for all strings *s* and *t* in Σ^*, if *s L t* then *t L s*. By definition of *L* this would mean that for all strings *s* and *t* in Σ^*, if $\ell(s) < \ell(t)$ then $\ell(t) < \ell(s)$. But this is false for all strings *s* and *t* in Σ^*. For instance, take $s = 01$ and $t = 010$. Then $\ell(s) = 2$ and $\ell(t) = 3$, and so $\ell(s) < \ell(t)$ but $\ell(t) \not< \ell(s)$.

L is transitive: To prove transitivity of *L*, we must show that for all strings *s*, *t*, and *u* in Σ^*, if *s L t* and *t L u* then *s L u*. By definition of *L* this means that for all strings *s*, *t*, and *u* in Σ^*, if $\ell(s) < \ell(t)$ and $\ell(t) < \ell(u)$ then $\ell(s) < \ell(u)$. But this is true by the transitivity property of order (Appendix A, T17).

23. *# is reflexive:* *#* is reflexive $\Leftrightarrow$ for all subsets *A* of *X*, *A # A*. By definition of *#*, this means that for all subsets *A* of *X*, *A* has the same number of elements as *A*. But this is true.

 # is symmetric: *#* is symmetric $\Leftrightarrow$ for all subsets *A* and *B* of X, if *A # B* then *B # A*. By definition of *#* this means that if *A* has the same number of elements as *B* then *B* has the same number of elements as *A*. But this is true.

 # is transitive: *#* is transitive $\Leftrightarrow$ for all subsets *A*, *B*, and *C* of *X*, if *A # B* and *B # C*, then *A # C*. By definition of *#*, this means that for all subsets *A*, *B*, and *C* of *X*, if *A* has the same number of elements as *B* and *B* has the same number of elements as *C*, then *A* has the same number of elements as *B* and *B* has the same number of elements as *C*, then *A* has the same number of elements as *C*. But this is true.

26. *$\mathcal{S}$ is reflexive:* *$\mathcal{S}$* is reflexive $\Leftrightarrow$ for all subsets *X* of *A*, *X $\mathcal{S}$ X*. By definition of *$\mathcal{S}$*, this means that for all subsets *X* of *A*, $X \subseteq X$. But this is true because every set is a subset of itself.

$\mathcal{S}$ is not symmetric: *$\mathcal{S}$* is symmetric $\Leftrightarrow$ for all subsets *X* and *Y* of *A*, if *X $\mathcal{S}$ Y* then *Y $\mathcal{S}$ X*. By definition of *$\mathcal{S}$*, this means that for all subsets *X* and *Y* of *A*, if $X \subseteq Y$ then $Y \subseteq X$. But this is false because $A \neq \varnothing$ and so there is an element *a* in *A*. Take $X = \varnothing$, and $Y = \{a\}$. Then $X \subseteq Y$ but $Y \not\subseteq X$.

$\mathcal{S}$ is transitive: *$\mathcal{S}$* is transitive $\Leftrightarrow$ for all subsets *X*, *Y*, and *Z* of *A*, if *X $\mathcal{S}$ Y* and *Y $\mathcal{S}$ Z*, then *X $\mathcal{S}$ Z*. By definition of *$\mathcal{S}$*, this means that for all subsets *X*, *Y*, and *Z* of *A*, if $X \subseteq Y$ and $Y \subseteq Z$ then $X \subseteq Z$. But this is true by the transitive property of subsets (Theorem 5.2.1 (3)).

30. *I is reflexive: [We must show that for all statements p, p I p.]* Suppose p is a statement. The only way a conditional statement can be false is for its hypothesis to be true and its conclusion false. Consider the statement $p \to p$. Both the hypothesis and the conclusion have the same truth value. Thus it is impossible for $p \to p$ to be false, and so $p \to p$ must be true.

I is not symmetric: I is symmetric $\Leftrightarrow$ for all statements p and q, if $p \, I \, q$ then $q \, I \, p$. By definition of I this means that for all statements p and q, if $p \to q$ then $q \to p$. But this is false. As a counterexample, let p be the statement "10 is divisible by 4" and let q be "10 is divisible by 2". Then $p \to q$ is the statement "If 10 is divisible by 4, then 10 is divisible by 2." This is true because its hypothesis, p, is false. On the other hand, $q \to p$ is the statement "If 10 is divisible by 2, then 10 is divisible by 4." This is false because its hypothesis, q, is true and its conclusion, p, is false.

I is transitive: [We must show that for all statements p, q, and r, if p I q and q I r then p I r.] Suppose p, q, and r are statements such that $p \, I \, q$ and $q \, I \, r$. By definition of I, this means that $p \to q$ and $q \to r$ are both true. By hypothetical syllogism (Example 1.3.7 and exercise 19 of Section 1.3), we can conclude that $p \to r$ is true. Hence by definition of I, p, I r.

31. $\mathcal{R}$ *is reflexive:* $\mathcal{R}$ is reflexive $\Leftrightarrow$ for all elements (x, y) in $\mathbf{R} \times \mathbf{R}$, $(x, y)\mathcal{R}(x, y)$. By definition of $\mathcal{R}$ this means that for all elements (x, y) in $\mathbf{R} \times \mathbf{R}$, $x = x$. But this is true.

$\mathcal{R}$ *is symmetric: [We must show that for all elements (x_1, y_1) and (x_2, y_2) in $\mathbf{R} \times \mathbf{R}$, if $(x_1, y_1)\mathcal{R}(x_2, y_2)$ then $(x_2, y_2)\mathcal{R}(x_1, y_1)$].* Suppose (x_1, y_1) and (x_2, y_2) are elements of $\mathbf{R} \times \mathbf{R}$ such that $(x_1, y_1)\mathcal{R}(x_2, y_2)$. By definition of $\mathcal{R}$ this means that $x_1 = x_2$. By symmetry of equality, $x_2 = x_1$. So by definition of $\mathcal{R}$, $(x_2, y_2)\mathcal{R}(x_1, y_1)$.

$\mathcal{R}$ *is transitive: [We must show that for all elements (x_1, y_1), (x_2, y_2) and (x_3, y_3) in $\mathbf{R} \times \mathbf{R}$, if $(x_1, y_1)\mathcal{R}(x_2, y_2)$ and $(x_2, y_2)\mathcal{R}(x_3, y_3)$ then $(x_1, y_1)\mathcal{R}(x_3, y_3)$.]* Suppose (x_1, y_1), (x_2, y_2), and (x_3, y_3) are elements of $\mathbf{R} \times \mathbf{R}$ such that $(x_1, y_1)\mathcal{R}(x_2, y_2)$ and $(x_2, y_2)\mathcal{R}(x_3, y_3)$. By definition of $\mathcal{R}$ this means that $x_1 = x_2$ and $x_2 = x_3$. By transitivity of equality, $x_1 = x_3$. Hence by definition of $\mathcal{R}$, $(x_1, y_1)\mathcal{R}(x_3, y_3)$.

34. *R is reflexive:* R is reflexive $\Leftrightarrow$ for all people p in A, $p \, R \, p$. By definition of R this means that for all people p living in the world today, p lives within 100 miles of p. But this is true.

R is symmetric: [We must show that for all people p_1 and p_2 in A, if $p_1 \, R \, p_2$ then $p_2 \, R \, p_1$.] Suppose p_1 and p_2 are people in A such that $p_1 \, R \, p_2$. By definition of R this means that p_1 lives within 100 miles of p_2. But this implies that p_2 lives within 100 miles of p_1. So by definition of R, $p_2 \, R \, p_1$.

R is not transitive: R is transitive $\Leftrightarrow$ for all people p, q, and r, if $p \, R \, q$ and $q \, R \, r$ then $p \, R \, r$. But this is false. As

a counterexample, take p to be an inhabitant of Chicago, Illinois, q an inhabitant of Kankakee, Illinois, and r an inhabitant of Champaign, Illinois. Then $p \, R \, q$ because Chicago is less than 100 miles from Kankakee, and $q \, R \, r$ because Kankakee is less than 100 miles from Champaign, but $p \, \not{R} \, r$ because Chicago is not less than 100 miles from Champaign.

37. a. A binary relation is any subset of $A \times A$. So there are 2^{64} binary relations on A.

c. Form a symmetric relation by a two-step process: (1) pick a set of elements of the form (a, a) (there are eight such elements, so 2^8 sets); (2) pick a set of pairs of elements of the form (a, b) and (b, a) (there are $(64 - 8)/2 = 28$ such pairs, so 2^{28} such sets). The answer is, therefore, $2^8 \cdot 2^{28} = 2^{36}$.

38. Algorithm—Test for Reflexivity

[The input for this algorithm consists of a binary relation R defined on a set A which is represented as the one-dimensional array $a[1], a[2], \ldots, a[n]$. To test whether R is reflexive, the variable answer is initially set equal to "yes" and then each element $a[i]$ of A is examined in turn to see whether it is related by R to itself. If any element is not related to itself by R, then the answer is set equal to "no," the while loop is not repeated, and processing terminates.]

Input: *n [a positive integer], $a[1], a[2], \ldots, a[n]$ [a one-dimensional array representing a set A], R [a subset of $A \times A$]*

Algorithm Body:

```
i := 1, answer := "yes"
while (answer = "yes" and i ≤ n)
    if (a[i], a[i]) ∉ R then answer := "no"
    i := i + 1
end while
```

Output: *answer [a string]*

end Algorithm

43. a. $R \cap S$ *is reflexive:* Suppose R and S are reflexive. *[To show that $R \cap S$ is reflexive, we must show that $\forall x \in A$, $(x, x) \in R \cap S$.]* So suppose $x \in A$. Since R is reflexive, $(x, x) \in R$, and since S is reflexive, $(x, x) \in S$. Thus by definition of intersection $(x, x) \in R \cap S$ *[as was to be shown].*

b. *Hint:* The answer is yes.

44. b. Yes. To prove this we must show that for all x and y in A, if $(x, y) \in R \cup S$ then $(y, s) \in R \cup S$. So suppose (x, y) is a particular but arbitrarily chosen element in $R \cup S$. *[We must show that $(y, x) \in R \cup S$.]* By definition of union, $(x, y) \in R$ or $(x, y) \in S$. If $(x, y) \in R$ then $(y, x) \in R$ because R is symmetric. Hence $(y, x) \in R \cup S$ by definition of union. But also if $(x, y) \in S$ then $(y, x) \in S$ because S is symmetric. Hence $(y, x) \in R \cup S$ by definition of union. Thus, in either case, $(y, x) \in R \cup S$ *[as was to be shown]*.

45. R_1 is not irreflexive because $(0, 0) \in R_1$. R_1 is not asymmetric because $(0, 1) \in R_1$ and $(1, 0) \in R_1$. R_1 is not intransitive because $(0, 1) \in R_1$ and $(1, 0) \in R_1$ and $(0, 0) \in R_1$.

47. R_3 is irreflexive. R_3 is not asymmetric because $(2, 3) \in R_3$ and $(3, 2) \in R_3$. R_3 is intransitive.

50. R_6 is irreflexive. R_6 is asymmetric. R_6 is intransitive (by default).

SECTION 10.3

1. a. $R = \{(0,0), (0, 2), (1, 1), (2, 0), (2, 2), (3, 3), (3, 4), (4, 3), (4, 4)\}$

2. $\{0, 4\}, \{1, 3\}, \{2\}$

4. $\{1, 5, 9, 13, 17\}, \{2, 6, 10, 14, 18\}, \{3, 7, 11, 15, 19\}, \{4, 8, 12, 16, 20\}$

6. $\{(1, 3), (3, 9)\}, \{(2, 4), (-4, -8), (3, 6)\}, \{(1, 5)\}$

8. $\{aaaa, aaab, aaba, aabb\}, \{abaa, abab, abba, abbb\}, \{baaa, baab, baba, babb\}, \{bbaa, bbab, bbba, bbbb\}$

10. a. True. $17 - 2 = 15$ and $5 \mid 15$.

11. a. $[7] = [4] = [19]$, $[-4] = [17]$, $[-6] = [27]$

12. a. *Proof:* Suppose that m and n are integers such that $m \equiv n \pmod 3$. *[We must show that $m \bmod 3 = n \bmod 3$.]* By definition of congruence, $3 \mid (m - n)$, and so by definition of divisibility $m - n = 3k$ for some integer k. Let $m \bmod 3 = r$. Then $m = 3l + r$ for some integer l. Since $m - n = 3k$, then by substitution, $(3l + r) - n = 3k$, or, equivalently, $n = 3(l - k) + r$. Since $l - k$ is an integer, it follows by definition of *mod*, that $n \bmod 3 = r$ also, and so $m \bmod 3 = n \bmod 3$ *[as was to be shown]*.
Suppose that m and n are integers such that $m \bmod 3 = n \bmod 3$. *[We must show that $m \equiv n \pmod 3$.]* Let $r = m \bmod 3 = n \bmod 3$. Then by definition of *mod*, $m = 3p + r$ and $n = 3q + r$ for some integers p and q. By substitution, $m - n = (3p + r) - (3q + r) = 3(p - q)$. Since $p - q$ is an integer, it follows that $3 \mid (m - n)$, and so by definition of congruence, $m \equiv n \pmod 3$. *[as was to be shown]*

13. a. For example let $A = \{1, 2\}$ and $B = \{2, 3\}$. Then $A \neq B$ so A and B are distinct. But A and B are not disjoint since $2 \in A \cap B$.

14. a. There is one equivalence class for each major and double major at the college. Each class consists of all students with that major.

15. Two distinct classes: $\{x \in \mathbf{Z} \mid x = 2k$, for some integer $k\}$ and $\{x \in \mathbf{Z} \mid x = 2k + 1$, for some integer $k\}$.

19. The distinct classes are all sets of the form $\{x, -x\}$, where x is a real number.

20. There is one class for each real number x with $0 \leq x < 1$. The distinct classes are all sets of the form $\{y \in \mathbf{R} \mid y = n + x$, for some integer $n\}$, where x is a real number such that $0 \leq x < 1$.

22. There is one equivalence class for each real number. The distinct equivalence classes are all sets of ordered pairs $\{(x, y) \in \mathbf{R} \times \mathbf{R} \mid x = a\}$, for each real number a. Equivalently, the equivalence classes consist of all vertical lines in the Cartesian plane.

24. There is one equivalence class for each real number t such that $0 \leq t < \pi$. One line in each class goes through the origin and that line makes an angle of t with the positive horizontal axis.

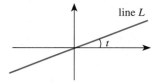

Alternatively, there is one equivalence class for every possible slope: all real numbers plus "undefined."

27. *Proof:* Suppose R is an equivalence relation on a set A and $a \in A$. Because R is an equivalence relation, R is reflexive, and because R is reflexive each element of A is related to itself by R. In particular, $a \, R \, a$. Hence by definition of equivalence class, $a \in [a]$.

29. *Proof:* Suppose R is an equivalence relation on a set A and a, b, and c are elements of A with $b \, R \, c$ and $c \in [a]$. Since $c \in [a]$, then $c \, R \, a$ by definition of equivalence class. But R is transitive since R is an equivalence relation. So since $b \, R \, c$ and $c \, R \, a$, then $b \, R \, a$. It follows that $b \in [a]$ by definition of class.

33. c. For example $(2, 6), (-2, -6), (3, 9), (-3, -9)$.

34. a. Suppose that (a, b), (a', b'), (c, d), and (c', d') are any elements of A such that $[(a, b)] = [(a', b')]$ and $[(c, d)] = [(c', d')]$. By definition of the relation, $ab' = ba'$ (*) and $cd' = dc'$ (**). We must show that $[(a, b)] + [(c, d)] = [(a', b')] + [(c', d')]$. By definition of the addition, this equation is true if, and only if,

$$[(ad + bc, bd)] = [(a'd' + b'c', b'd')].$$

And, by definition of the relation, this equation is true if, and only if,

$$(ad + bc)b'd' = bd(a'd' + b'c'),$$

which is equivalent to

$$adb'd' + bcb'd' = bda'd' + bdb'c' \quad \text{by multiplying out,}$$

which in turn is equivalent to

$$(ab')(dd') + (cd')(bb') =$$
$$(a'b)(dd') + (dc')(bb') \quad \text{by regrouping.}$$

But by substitution from (*) and (**), this last equation is true.

c. Suppose that (a, b) is any element of A. We must show that $[(a, b)] + [(0, 1)] = [(a, b)]$. By definition of the addition, this equation is true if, and only if,

$$[(a \cdot 1 + b \cdot 0, b \cdot 1)] = [(a, b)].$$

But this last equation is true because $a \cdot 1 + b \cdot 0 = a$ and $b \cdot 1 = b$.

e. Suppose that (a, b) is any element of A. We must show that $[(a, b)] + [(-a, b)] = [(0, 1)]$. By definition of the addition, this equation is true if, and only if,

$$[(ab + b(-a), bb)] = [(0, 1)],$$

or, equivalently,

$$[(0, bb)] = [(0, 1)]$$

By definition of the relation, this last equation is true if, and only if, $0 \cdot 1 = bb \cdot 0$, which is true.

35. a. Let (a, b) be any element of $\mathbf{Z}^+ \times \mathbf{Z}^+$. We must show that $(a, b)R(a, b)$. By definition of R, this relationship holds if, and only if, $a + b = b + a$. But this equation is true by the commutative law of addition for real numbers. Hence R is reflexive.

c. *Hint:* You will need to show that for any positive integers a, b, c, and d, if $a + d = c + b$ and $c + f = d + e$, then $a + f = b + e$.

d. *One possible answer:* (1, 1), (2, 2), (3, 3), (4, 4), (5, 5)

g. Observe that for any positive integers a and b, the equivalence class of (a, b) consists of all ordered pairs in $\mathbf{Z}^+ \times \mathbf{Z}^+$ for which the difference between the first and second coordinates equals $a - b$. Thus there is one equivalence class for each integer: positive, negative, and zero. Each positive integer n corresponds to the class of $(n + 1, 1)$; each negative integer $-n$ corresponds to the class of $(1, n + 1)$; and zero corresponds to the class of $(1, 1)$.

SECTION 10.4

1. a. 0-equivalence classes: $\{s_0, s_1, s_3, s_4\}, \{s_2, s_5\}$
   1-equivalence classes: $\{s_0, s_3\}, \{s_1, s_4\}, \{s_2, s_5\}$
   2-equivalence classes: $\{s_0, s_3\}, \{s_1, s_4\}, \{s_2, s_5\}$

b.

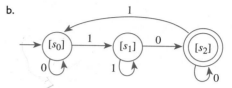

4. a. 0-equivalence classes: $\{s_0, s_1, s_2\}, \{s_3, s_4, s_5\}$
   1-equivalence classes: $\{s_0, s_1, s_2\}, \{s_3, s_5\}, \{s_4\}$
   2-equivalence classes: $\{s_0, s_2\}, \{s_1\}, \{s_3, s_5\}, \{s_4\}$
   3-equivalence classes: $\{s_0, s_2\}, \{s_1\}, \{s_3, s_5\}, \{s_4\}$

b.

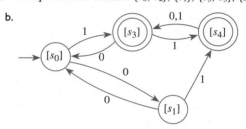

6. a. *Hint:* The 3-equivalence classes are $\{s_0\}, \{s_1\}, \{s_2\}, \{s_3\}, \{s_4\}, \{s_5\},$ and $\{s_6\}.$

7. Yes. For A:

0-equivalence classes: $\{s_0, s_2\}, \{s_1, s_3\}$
1-equivalence classes: $\{s_0\}, \{s_2\}, \{s_1, s_3\}$
2-equivalence classes: $\{s_0\}, \{s_2\}, \{s_1, s_3\}$

transition diagram for $\overline{A}$:

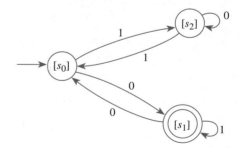

For A':

0-equivalence classes: $\{s'_0, s'_1, s'_2\}, \{s'_3\}$
1-equivalence classes: $\{s'_0, s'_2\}, \{s'_1\}, \{s'_3\}$
2-equivalence classes: $\{s'_0, s'_2\}, \{s'_1\}, \{s'_3\}$

transition diagram for $\overline{A'}$:

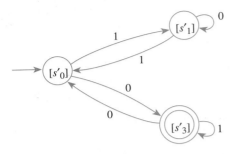

Except for the labeling of the states, the transition diagrams for $\overline{A}$ and $\overline{A'}$ are identical. Hence $\overline{A}$ and $\overline{A'}$ accept the same language, and so by Theorem 11.4.3 A and A' also accept the same language. Thus A and A' are equivalent automata.

9. For A:

  0-equivalence classes: $\{s_1, s_2, s_4, s_5\}, \{s_0, s_3\}$
  1-equivalence classes: $\{s_1, s_2\}, \{s_4, s_5\}, \{s_0, s_3\}$
  2-equivalence classes: $\{s_1\}, \{s_2\}, \{s_4, s_5\}, \{s_0, s_3\}$
  3-equivalence classes: $\{s_1\}, \{s_2\}, \{s_4, s_5\}, \{s_0, s_3\}$

Therefore, the states of $\overline{A}$ are the 3-equivalence classes of A.
For A':

  0-equivalence classes: $\{s'_2, s'_3, s'_4, s'_5\}, \{s'_0, s'_1\}$
  1-equivalence classes: $\{s'_2, s'_3, s'_4, s'_5\}, \{s'_0, s'_1\}$

Therefore, the states of $\overline{A'}$ are the 1-equivalence classes of A'.
According to the text, two automata are equivalent if, and only if, their quotient automata are isomorphic, provided inaccessible states have first been removed. Now A and A' have no inaccessible states, and $\overline{A}$ has four states whereas $\overline{A'}$ has only two states. Therefore, A and A' are not equivalent.
This result can also be obtained by noting, for example, that the string 11 is accepted by A' but not by A.

11. *Partial answer:* Suppose A is a finite-state automaton with set of states S and relation R_* of *-equivalence of states. *[To show that R_* is an equivalence relation, we must show that R is reflexive, symmetric, and transitive.]*
Proof that R_* is symmetric:
[We must show that for all states s and t, if $s\,R_*\,t$ then $t\,R_*\,s$.]
Suppose that s and t are states of A such that $s\,R_*\,t$.
[We must show that $t\,R_*\,s$] Since $s\,R_*\,t$, then for all input strings w,

$$\begin{bmatrix} N^*(s, w) \text{ is an} \\ \text{accepting state} \end{bmatrix} \Leftrightarrow \begin{bmatrix} N^*(t, w) \text{ is an} \\ \text{accepting state} \end{bmatrix}$$

where N^* is the eventual-state function on A. But then by symmetry of the $\Leftrightarrow$ relation, it is true that for all input strings w,

$$\begin{bmatrix} N^*(t, w) \text{ is an} \\ \text{accepting state} \end{bmatrix} \Leftrightarrow \begin{bmatrix} N^*(s, w) \text{ is an} \\ \text{accepting state} \end{bmatrix}$$

Hence $t\,R_*\,s$ *[as was to be shown].* So R_* is symmetric.

12. The proof is identical to the proof of property (10.4.1) given in the solution to exercise 11 provided each occurrence of "for all input strings w" is replaced by "for all input strings w of length less than or equal to k".

13. *Proof:* By property (10.4.2), for each integer $k \geq 0$, k-equivalence is an equivalence relation. But by Theorem 10.3.4, the distinct equivalence classes of an equivalence relation form a partition of the set on which the relation is defined. In this case, the relation is defined on the set of states of the automaton. So the k-equivalence classes form a partition of the set of all states of the automaton.

15. *Hint 1:* Suppose C_k is a particular but arbitrarily chosen k-equivalence class. You must show that there is a $(k - 1)$-equivalence class C_{k-1} such that $C_k \subseteq C_{k-1}$.
Hint 2: If s is any element in C_k, then s is a state of the automaton. Now the $(k - 1)$-equivalence classes partition the set of all states of the automaton into a union of mutually disjoint subsets, so $s \in C_{k-1}$ for *some* $(k - 1)$-equivalence class C_{k-1}.
Hint 3: To show that $C_k \subseteq C_{k-1}$, you must show the following: for any state t, if $t \in C_k$, then $t \in C_{k-1}$.

17. *Hint:* If $m < k$, then every input string of length less than or equal to m has length less than or equal to k.

19. *Hint:* Suppose two states s and t are equivalent. You must show that for any input symbol m, the next-states $N(s, m)$ and $N(t, m)$ are equivalent. To do this, use the definition of equivalence and the fact that for any string w', input symbol m, and state s, $N^*(N(s, m), w') = N^*(s, mw')$.

SECTION 10.5

1. a.

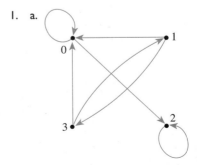

R_1 is not antisymmetric: $1\ R_1\ 3$ and $3\ R_1\ 1$ and $1 \neq 3$.

b.

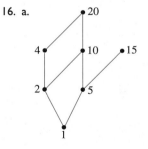

R_2 is antisymmetric: there are no cases where $a\,R\,b$ and $b\,R\,a$ and $a \neq b$.

2. R is not antisymmetric. Let x and y be any two distinct people of the same age. Then $x\,R\,y$ and $y\,R\,x$ but $x \neq y$.

5. R is a partial order relation.
Proof:
R is reflexive: Suppose $(a, b) \in \mathbf{R} \times \mathbf{R}$. Then $(a, b)\,R\,(a, b)$ because $a = a$ and $b \leq b$.
R is antisymmetric: Suppose (a, b) and (c, d) are ordered pairs of real numbers such that $(a, b)\,R\,(c, d)$ and $(c, d)\,R\,(a, b)$. Then

either $a < c$ or   both $a = c$ and $b \leq d$

and

either $c < a$ or   both $c = a$ and $d \leq b$.

Thus

$$a \leq c \quad \text{and} \quad c \leq a$$

and so

$$a = c.$$

Consequently,

$$b \leq d \quad \text{and} \quad d \leq b.$$

and so

$$b = d.$$

Hence, $(a, b) = (c, d)$.
R is transitive: Suppose (a, b), (c, d), and (e, f) are ordered pairs of real numbers such that $(a, b)\,R\,(c, d)$ and $(c, d)\,R\,(e, f)$. Then

either $a < c$ or   both $a = c$ and $b \leq d$

and

either $c < e$ or   both $c = e$ and $d \leq f$.

It follows that one of the following cases must occur.

Case 1 $(a < c$ *and* $c < e)$: Then by transitivity of $<$, $a < e$, and so $(a, b)\,R\,(e, f)$ by definition of R.
Case 2 $(a < c$ *and* $c = e)$: Then by substitution, $a < e$, and so $(a, b)\,R\,(e, f)$ by definition of R.

Case 3 $(a = c$ *and* $c < e)$: Then by substitution, $a < e$, and so $(a, b)\,R\,(e, f)$ by definition of R.
Case 4 $(a = c$ *and* $c = e)$: Then by definition of R, $b \leq d$ and $d \leq f$, and so by transitivity of $\leq$, $b \leq f$. Hence $a = e$ and $b \leq f$, and so $(a, b)\,R\,(e, f)$ by definition of R.
In each case, $(a, b)\,R\,(e, f)$. Therefore R is transitive. Since R is reflexive, antisymmetric, and transitive, R is a partial order relation.

8. R is not a partial order relation because R is not antisymmetric.
Counterexample: $1\,R\,3$ (because $1 + 3$ is even) and $3\,R\,1$ (because $3 + 1$ is even) but $1 \neq 3$.

10. No. Suppose $a\,R\,b$, $b\,S\,a$; but $b\,R\,a$ and $a \neq b$ $a\,(R \cup S)\,b$ and $b\,(R \cup S)\,a$; but $a \neq b$. Hence, $R \cup S$ is not antisymmetric.

11. a. This follows from (1).
b. False. By (1), $bba \leq bbab$.

13. $R_1 = \{(a, a), (b, b)\}$, $R_2 = \{(a, a), (b, b), (a, b)\}$, $R_3 = \{(a, a), (b, b), (b, a)\}$

14. a. $R_1 = \{(a, a), (b, b), (c, c)\}$,
$R_2 = \{(a, a), (b, b), (c, c), (b, a)\}$,
$R_3 = \{(a, a), (b, b), (c, c), (c, a)\}$,
$R_4 = \{(a, a), (b, b), (c, c), (b, a), (c, a)\}$,
$R_5 = \{(a, a), (b, b), (c, c), (c, b), (c, a)\}$,
$R_6 = \{(a, a), (b, b), (c, c), (b, c), (b, a)\}$,
$R_7 = \{(a, a), (b, b), (c, c), (c, b), (b, a), (c, a)\}$,
$R_8 = \{(a, a), (b, b), (c, c), (b, c), (b, a), (c, a)\}$,
$R_9 = \{(a, a), (b, b), (c, c), (b, c)\}$,
$R_{10} = \{(a, a), (b, b), (c, c), (c, b)\}$

15. *Hint:* R is the identity relation on A: $x\,R\,x$ for all $x \in A$ and $x\,\not\!R\,y$ if $x \neq y$.

16. a.

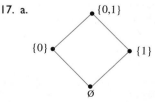

17. a.

18.

(1,1)

(1,0)

(0,1)

(0,0)

21. a. *Proof: [We must show that for all a and b in A, a | b or b | a.]* Let a and b be particular but arbitrarily chosen elements of A. By definition of A, there are nonnegative integers r and s such that $a = 2^r$ and $b = 2^s$. Now either $r \leq s$ or $s < r$. If $r \leq s$, then

$$b = 2^s = 2^r \cdot 2^{s-r} = a \cdot 2^{s-r},$$

where $s - r \geq 0$. It follows by definition of divisibility that $a \mid b$. By a similar argument, if $s < r$, then $b \mid a$. Hence either $a \mid b$ or $b \mid a$ *[as was to be shown]*.

b.

1   2 2^2 2^3 2^4

22. greatest element: none;   least element: 1;
maximal elements: 15, 20;   minimal element: 1

24. greatest element: $\{0, 1\}$;   least element: $\emptyset$;
maximal elements: $\{0, 1\}$;   minimal elements: $\emptyset$

26. greatest element: $(1, 1)$;   least element: $(0, 0)$;
maximal elements: $(1, 1)$;   minimal elements: $(0, 0)$

30. a. no greatest element, no least element
b. least element is 0, greatest element is 1

32. *Hint:* Let R' be the restriction of R to B and show that R' is reflexive, antisymmetric, and transitive. In each case, this follows almost immediately from the fact that R is reflexive, antisymmetric, and transitive.

33. $\emptyset \subseteq \{w\} \subseteq \{w, x\} \subseteq \{w, x, y\} \subseteq \{w, x, y, z\}$

36. *Proof:* Suppose A is a partially ordered set with respect to a relation $\preceq$. By definition of total order, A is totally ordered if, and only if, any two elements of A are comparable. By definition of chain, this is true if, and only if, A is a chain.

37. a. *Proof by contradiction:* Suppose not. Suppose A is a finite set that is partially ordered with respect to a relation $\preceq$ and A has no minimal element. Construct a sequence of elements $x_1, x_2, x_3, \ldots$ of A as follows.

1. Pick any element of A and call it x_1.
2. For each $i = 2, 3, 4, \ldots$, pick x_i to be an element of A for which $x_i \preceq x_{i-1}$ and $x_i \neq x_{i-1}$. *[Such an element must exist because otherwise x_{i-1} would be minimal and we are supposing that no element of A is minimal.]* Now $x_i \neq x_j$ for any $i \neq j$. *[If $x_i = x_j$ where $i < j$, then on the one hand, $x_j \preceq x_{j-1} \preceq \ldots \preceq x_{i+1} \preceq x_i$ and so $x_j \preceq x_{i+1}$,*

and on the other hand, since $x_i = x_j$ then $x_j = x_i \succeq x_{i+1}$, and so $x_j \succeq x_{i+1}$. Hence by antisymmetry, $x_j = x_{i+1}$, and so $x_i = x_{i+1}$. But this contradicts the definition of the sequence $x_1, x_2, x_3, \ldots$] Thus $x_1, x_2, x_3, \ldots$ is an infinite sequence of distinct elements and consequently $\{x_1, x_2, x_3, \ldots\}$ is an infinite subset of the finite set A. This is impossible. Hence the supposition is false and we conclude that any partially ordered subset of a finite set has a minimal element.

39.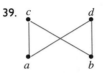

41. One such total order is 1, 5, 2, 15, 10, 4, 20.

43. One such total order is (0, 0), (1, 0), (0, 1), (1, 1).

47. b. critical path: 1, 2, 5, 8, 9.

SECTION 11.1

1. $V(G) = \{v_1, v_2, v_3, v_4\}$, $E(G) = \{e_1, e_2, e_3\}$
edge-endpoint function:

edge	endpoints
e_1	$\{v_1, v_2\}$
e_2	$\{v_1, v_3\}$
e_3	$\{v_3\}$

3.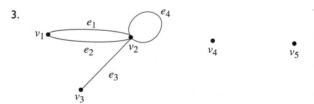

5. Imagine that the edges are strings and the vertices are knots. You can pick up the left-hand figure and lay it down again to form the right-hand figure as shown below.

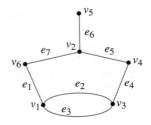

8. (i) e_1, e_2, and e_3 are incident on v_1.

(ii) v_1, v_2, and v_3 are adjacent to v_3.

(iii) e_2, e_8, e_9, and e_3 are adjacent to e_1.

(iv) Loops are e_6 and e_7.

(v) e_8 and e_9 are parallel; e_4 and e_5 are parallel.

(vi) v_6 is an isolated vertex.

(vii) degree of $v_3 = 5$

(viii) total degree $= 20$

10. a. Yes. According to the graph, *Sports Illustrated* is an instance of a sports magazine, a sports magazine is a periodical, and a periodical contains printed writing.

12. To solve this puzzle using a graph, introduce a notation in which, for example, *wc/mg* means that the wolf and cabbage are on the left bank of the river and the man and the goat are on the right bank. Then draw those arrangements of wolf, cabbage, goat, and ferryman that can be reached from the initial arrangement (*wgcf/*) and that are not arrangements to be avoided (such as (*wg/fc*)). At each stage ask yourself, "Where can I go from here?" and draw lines or arrows pointing to those arrangements. This method gives the graph shown below.

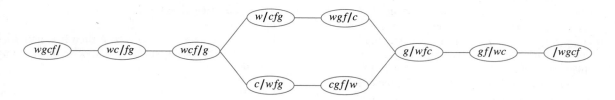

Examination of the diagram shows the solutions

$$(wgcf/) \rightarrow (wc/gf) \rightarrow (wcf/g) \rightarrow (w/gcf) \rightarrow (wgf/c) \rightarrow$$
$$(g/wcf) \rightarrow (gf/wc) \rightarrow (/wgcf)$$

and

$$(wgcf/) \rightarrow (wc/gf) \rightarrow (wcf/g) \rightarrow (c/wgf) \rightarrow (gcf/w) \rightarrow$$
$$(g/wcf) \rightarrow (gf/wc) \rightarrow (/wgcf)$$

14. *Hint:* The answer is yes. Represent possible amounts of water in jugs A and B by ordered pairs. For instance, the ordered pair $(1, 3)$ would indicate that there is one quart of water in jug A and three quarts in jug B. Starting with $(0, 0)$, draw arrows from one ordered pair to another if it is possible to go from the situation represented by one pair to that represented by the other by either filling a jug, emptying a jug, or transferring water from one jug to another. You need only draw arrows from states that have arrows pointing to them; the other states cannot be reached. Then find a directed path (sequence of directed edges) from the initial state $(0, 0)$ to a final state $(1, 0)$ or $(0, 1)$.

15. One such graph is the following.

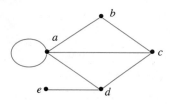

16. If there were a graph with four vertices of degrees 1, 2, 3, and 3, then its total degree would be 9, which is odd. But by Corollary 11.1.2 the total degree of the graph must be even. *[This is a contradiction.]* Hence there is no such graph. (Alternatively, if there were such a graph it would have an odd number of vertices of odd degree. But by Proposition 11.1.3 this is impossible.)

19. Suppose there were a simple graph with four vertices of degrees 1, 2, 3, and 4. Then the vertex of degree 4 would have to be connected by edges to four distinct vertices other than itself because of the assumption that the graph is simple (and hence has no loops or parallel edges). This contradicts the assumption that the graph has four vertices in total. Hence there is no simple graph with four vertices of degrees 1, 2, 3, and 4.

22.

24. a. The nonempty subgraphs are as follows.

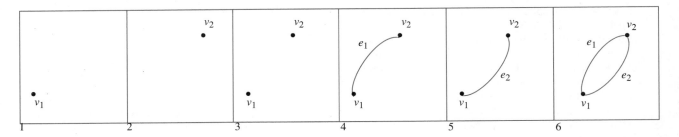

25. a. Suppose that, in a group of 15 people, each person had exactly three friends. Then you could draw a graph representing each person by a vertex and connecting two vertices by an edge if the corresponding people were friends. But such a graph would have 15 vertices, each of degree 3, for a total degree of 45. This contradicts the fact that the total degree of any graph is even. Hence the supposition is false, and in a group of 15 people it is not possible for each to have exactly three friends.

28. The total degree of the graph is $0 + 2 + 2 + 3 + 9 = 16$, so by Theorem 11.1.1 the number of edges is $16/2 = 8$.

31. We give two proofs for the following statement, one less formal and the other more formal.

For all integers $n \geq 0$, if $a_1, a_2, a_3, \ldots, a_{2n+1}$ are odd integers, then $\sum_{i=1}^{2n+1} a_i$ is odd.

Proof 1 (by mathematical induction): It is certainly true that the "sum" of one odd integer is odd. Suppose that for a certain positive odd integer r, the sum of r odd integers is odd. We must show that the sum of $r + 2$ odd integers is odd (because $r + 2$ is the next integer after r). But any sum of $r + 2$ odd integers equals a sum of r odd integers (which is odd by inductive hypothesis) plus a sum of two more odd integers (which is even). Thus the total sum is an odd integer plus an even integer, which is odd. *[This is what was to be shown.]*

Proof 2 (by mathematical induction):
The property is true for $n = 0$: Suppose a_1 is an odd integer. The $\sum_{i=1}^{2 \cdot 0 + 1} a_i = \sum_{i=1}^{1} a_i = a_1$, which is odd.
If the property is true for $n = k$ then it is true for $n = k + 1$: Let k be an integer with $k \geq 0$, and suppose that

if $a_1, a_2, a_3, \ldots, a_{2k+1}$ are odd integers,

then $\sum_{i=1}^{2k+1} a_i$ is odd.

[This is the inductive hypothesis.]
Suppose $a_1, a_2, a_3, \ldots, a_{2(k+1)+1}$ are odd integers. *[We must show that $\sum_{i=1}^{2(k+1)+1} a_i$ is odd, or, equivalently, that $\sum_{i=1}^{3k+3} a_i$ is odd.]* But

$$\sum_{i=1}^{2k+3} a_i = \sum_{i=1}^{2k+1} a_i + (a_{2k+2} + a_{2k+3}).$$

Since the sum of any two odd integers is even, $a_{2k+2} + a_{2k+3}$ is even, and by inductive hypothesis, $\sum_{i=1}^{2k+1} a_i$ is odd. Therefore, $\sum_{i=1}^{2k+3} a_i$ is the sum of an odd integer and an even integer, which is odd. *[This is what was to be shown.]*

32. *Hint:* Use proof by contradiction.

33. a. K_6:

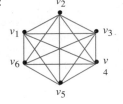

b. A proof of this fact was given in Section 8.2 using recursion. Try to find a different proof.
Hint for Proof 1: There are as many edges in K_n as there are subsets of two vertices (the endpoints) that can be chosen from a set of n vertices.
Hint for Proof 2: Use mathematical induction. A complete graph on $k + 1$ vertices can be obtained from a complete graph on k vertices by adding one vertex and connecting this vertex by k edges to each of the other vertices.
Hint for Proof 3: Use the fact that the number of edges of a graph is half the total degree. What is the degree of each vertex of K_n?

35. Suppose G is a nonempty simple graph with n vertices and $2n$ edges. By exercise 34, its number of edges cannot exceed $\dfrac{n(n-1)}{2}$. Thus $2n \le \dfrac{n(n-1)}{2}$, or $4n \le n^2 - n$. Equivalently, $n^2 - 5n \ge 0$, or $n(n-5) \ge 0$. This implies that $n \ge 5$ since G is nonempty. Hence a nonempty simple graph with twice as many edges as vertices must have at least five vertices. But a complete graph with five vertices has $\dfrac{5(5-1)}{2} = 10$ edges. Consequently, the answer to the question is yes. For example, K_5 is a graph with twice as many edges as vertices.

36. a. $K_{4,2}$:
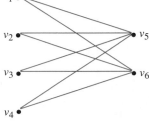

37. a.
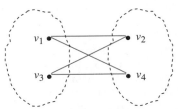

b. Suppose this graph is bipartite. Then the vertex set can be partitioned into two mutually disjoint subsets such that vertices in each subset are connected by edges only to vertices in the other subset and not to vertices in the same subset. Now v_1 is in one subset of the partition, say V_1. Since v_1 is connected by edges to v_2 and v_3, both v_2 and v_3 must be in the other subset, V_2. But v_2 and v_3 are connected by an edge to each other. This contradicts the fact that no vertices in V_2 are connected by edges to other vertices in V_2. Hence the supposition is false, and so the graph is not bipartite.

39. a.
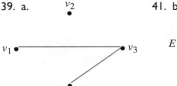

41. b.
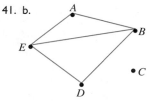

42. *Hint:* Consider the graph obtained by taking the vertices and edges of G plus all the edges of G'. Use exercise 33(b).

44. c. *Hint:* Suppose there were a simple graph with n vertices (where $n \ge 2$) each of which had a different degree. Then no vertex could have degree more than $n - 1$ (why?), so the degrees of the n vertices must be $0, 1, 2, \dots, n - 1$ (why?). This is impossible (why?).

45. *Hint:* Use the result of 44(c).

46. *Hint:* One solution is to begin by choosing a vertex of maximal degree and assigning the first time slot to it and to all other vertices that do not share an edge with it or with each other. Then choose a vertex of maximal degree from those remaining and assign the second time slot to it and to all those still unassigned that do not share an edge with it or with each other. Continue in this way until all vertices have been assigned.

SECTION 11.2

1. a. path (no repeated edge), not a simple path (repeated vertex—v_1), not a circuit
   b. walk, not a path (has repeated edge—e_9), not a circuit
   c. simple circuit (no repeated edge, no repeated vertex, starts and ends at same vertex)
   d. circuit (no repeated edge, starts and ends at same vertex), not a simple circuit (vertex v_4 is repeated)
   e. closed walk (starts and ends at the same vertex but has repeated edges)
   f. simple path

3. a. No. The notation $v_1v_2v_1$ could equally well refer to $v_1e_1v_2e_2v_1$ or to $v_1e_2v_2e_1v_1$, which are different walks.

4. a. Three (There are three ways to choose the middle edge.)
   b. $3! + 3 = 9$ (In addition to the three simple paths, there are $3!$ with vertices $v_1, v_2, v_3, v_2, v_3, v_4$. The reason is that from v_2 there are three choices of an edge to go to v_3, then two choices of different edges to go back to v_2, and then one choice of a different edge to return to v_3. This makes $3!$ paths in all.)
   c. Infinitely many (Since a walk may have repeated edges, a walk from v_1 to v_4 may contain an arbitrarily large number of repetitions of edges joining a pair of vertices along the way.)

6. a. $\{v_1, v_3\}, \{v_2, v_3\}, \{v_4, v_3\}$, and $\{v_5, v_3\}$ are all bridges.

8. a. Three connected components.

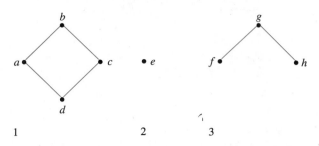

1   2   3

9. a. No. This graph has two vertices of odd degree, whereas all vertices of a graph with an Euler circuit have even degree.

12. One Euler circuit is $e_4 e_5 e_6 e_3 e_2 e_7 e_8 e_1$.

14. One Euler circuit is $iabihbchgcdgfdefi$.

19. There is an Euler path since $\deg(u)$ and $\deg(w)$ are odd, all other vertices have even degree, and the graph is connected. One Euler path is $uv_1 v_0 v_7 uv_2 v_3 v_4 v_2 v_6 v_4 w v_5 v_6 w$.

23. $v_0 v_7 v_1 v_2 v_3 v_4 v_5 v_6 v_0$

25. *Hint:* See the solution to Example 11.2.8.

26. Here is one sequence of reasoning you could use: Call the given graph G, and suppose G has a Hamiltonian circuit. Then G has a subgraph H that satisfies conditions (1)–(4) of Proposition 11.2.6. Since the degree of b in G is four and every vertex in H has degree two, two edges incident on b must be removed from G to create H. Edge $\{a, b\}$ cannot be removed because doing so would result in vertex d having degree less than two in H. Similar reasoning shows that edge $\{b, c\}$ cannot be removed either. So edges $\{b, i\}$ and $\{b, e\}$ must be removed from G to create H. Because vertex e must have degree two in H and because edge $\{b, e\}$ is not in H, both edges $\{e, d\}$ and $\{e, f\}$ must be in H. Similarly, since both vertices c and g must have degree two in H, edges $\{c, d\}$ and $\{g, d\}$ must also be in H. But then three edges incident on d, namely $\{e, d\}$, $\{c, d\}$, and $\{g, d\}$ must all be in H, which contradicts the fact that vertex d must have degree two in H.

28. *Hint:* This graph does not have a Hamiltonian circuit.

32. *Partial Answer:*

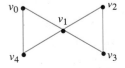

This graph has an Euler circuit $v_0 v_1 v_2 v_3 v_1 v_4 v_0$ but no Hamiltonian circuit.

33. *Partial Answer:*

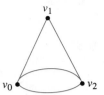

This graph has a Hamiltonian circuit $v_0 v_1 v_2 v_0$ but no Euler circuit.

34. *Partial Answer:*

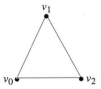

The walk $v_0 v_1 v_2 v_0$ is both an Euler circuit and a Hamiltonian circuit for this graph.

35. *Partial Answer:*

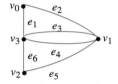

This graph has the Euler circuit $e_1 e_2 e_3 e_4 e_5 e_6$ and the Hamiltonian circuit $v_0 v_1 v_2 v_3 v_0$. These are not the same.

37. a. *Proof:* Suppose G is a graph and W is a walk in G that contains a repeated edge e. Let v and w be the endpoints of e. In case $v = w$, then v is a repeated vertex of W. In case $v \neq w$, then one of the following must occur: (1) W contains two copies of vew or of wev (for instance, W might contain a section of the form $vewe'vew$, as illustrated below); (2) W contains separate sections of the form vew and wev (for instance, W might contain a section of the form $vewe'wev$, as illustrated below); or (3) W contains a section of the form $vewev$ or of the form $wevew$ (as illustrated below). In cases (1) and (2), both vertices v and w are repeated and in case (3) one of v or w is repeated. In all cases, there is at least one vertex in W that is repeated.

(1)   (2)

(3)

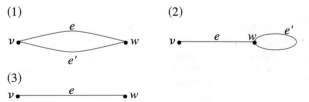

38. *Proof:* Suppose G is a connected graph. *[We must show that any two vertices of G can be connected by a simple path.]* Let v and w be any particular but arbitrarily chosen vertices of G. Since G is connected, there is a walk from v to w. If the walk contains a repeated vertex, then delete the portion of the walk from the first occurrence of the vertex to its next occurrence. (For example, in the walk $v e_1 v_2 e_5 v_7 e_6 v_2 e_3 w$, the vertex v_2 occurs twice. Deleting the portion of the walk from one occurrence to

tains a repeated vertex, do the above deletion process another time. Then check again for a repeated vertex. Continue in this way until all repeated vertices have been deleted. (This must occur eventually since the total number of vertices is finite.) The resulting walk connects v to w but has no repeated vertex. By exercise 37(b), it has no repeated edge either. Hence it is a simple path from v to w.

40. The proof of this fact is discussed in detail later in the text. See Lemma 11.5.3.

42. Yes. Suppose a graph contains a circuit that starts and ends at a vertex v. Successively delete sections of this circuit as follows. For each repeated vertex w in the circuit (excluding the first vertex if its only repetition is at the end of the circuit but including the first vertex if it is repeated in the middle of the circuit), there is a section of the circuit of the following form: $w e_1 v_1 e_2 v_2 \ldots e_{n-1} v_{n-1} e_n w$. Replace this section of the circuit by the single vertex w. Because the circuit has finite length, only a finite number of such deletions can be made, after which a simple circuit starting and ending at v will remain.

44. *Proof:* Let G be a connected graph and let C be a circuit in G. Let G' be the nonempty subgraph obtained by removing all the edges of C from G and also any vertices that become isolated when the edges of C are removed. *[We must show that there exists a vertex v such that v is in both C and G'.]* Pick any vertex v of C and any vertex w of G'. Since G is connected, there is a simple path from v to w (by Lemma 11.2.1(a)):

$$v = v_0 e_1 v_1 e_2 v_2 \ldots v_{i-1} e_i v_i e_{i+1} v_{i+1} \ldots v_{n-1} e_n v_n = w.$$

   ↑   ↑   ↑   ↑

in C   *in C   not in C*   *in G'*

Let i be the largest subscript such that v_i is in C. If $i = n$, then $v_n = w$ is in C and also in G' and we are done. If $i < n$, then v_i is in C and v_{i+1} is not in C. This implies that e_{i+1} is not in C (for if it were, both endpoints would be in C by definition of circuit). Hence when G' is formed by removing the edges and resulting isolated vertices from G, then e_{i+1} is not removed. That means that v_i does not become an isolated vertex, so v_i is not removed either. Hence v_i is in G'. Consequently, v_i is in both C and G' *[as was to be shown]*.

45. *Proof:* Suppose G is a graph with an Euler circuit. If v and w are any two vertices of G, then v and w each appear at least once in the Euler circuit (since an Euler circuit contains every vertex of the graph). The section of the circuit between the first occurrence of one of v or w and the first occurrence of the other is a walk from one of the two vertices to the other.

SECTION 11.3

1. a. By equating corresponding entries,

$$a + b = 1,$$
$$a - c = 0,$$
$$c = -1,$$
$$b - a = 3.$$

Thus $a - c = a - (-1) = 0$, and so $a = -1$. Consequently, $a + b = (-1) + b = 1$, and hence $b = 2$. The last equation should be checked to make sure the answer is consistent: $b - a = 2 - (-1) = 3$, which agrees.

2. a.

$$\begin{array}{c} \\ v_1 \\ v_2 \\ v_3 \end{array} \begin{array}{ccc} v_1 & v_2 & v_3 \\ \left[\begin{array}{ccc} 0 & 1 & 1 \\ 1 & 0 & 0 \\ 0 & 0 & 0 \end{array}\right] \end{array}$$

3. a.

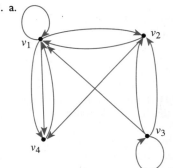

Any labels may be applied to the edges because the adjacency matrix does not determine edge labels.

4. a.

$$\begin{array}{c} \\ v_1 \\ v_2 \\ v_3 \\ v_4 \end{array} \begin{array}{cccc} v_1 & v_2 & v_3 & v_4 \\ \left[\begin{array}{cccc} 0 & 0 & 1 & 1 \\ 0 & 0 & 2 & 0 \\ 1 & 2 & 0 & 0 \\ 1 & 0 & 0 & 1 \end{array}\right] \end{array}$$

c.

$$\begin{array}{c} \\ v_1 \\ v_2 \\ v_3 \\ v_4 \end{array} \begin{array}{cccc} v_1 & v_2 & v_3 & v_4 \\ \left[\begin{array}{cccc} 0 & 1 & 1 & 1 \\ 1 & 0 & 1 & 1 \\ 1 & 1 & 0 & 1 \\ 1 & 1 & 1 & 0 \end{array}\right] \end{array}$$

5. a.

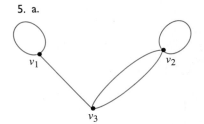

Any labels may be applied to the edges because the adjacency matrix does not determine edge labels.

6. a. The graph is connected.

8. a. $2 \cdot 1 + (-1) \cdot 3 = -1$

9. a. $\begin{bmatrix} 3 & -3 & 12 \\ 1 & -5 & 2 \end{bmatrix}$

10. a. no product (**A** has three columns and **B** has two rows.)

b. $BA = \begin{bmatrix} -2 & -2 & 2 \\ 1 & -5 & 2 \end{bmatrix}$ **f.** $B^2 = \begin{bmatrix} 4 & 0 \\ 1 & 9 \end{bmatrix}$

i. $AC = \begin{bmatrix} 2 & -1 \\ -5 & -2 \end{bmatrix}$

12. One among many possible examples is

$$A = B = \begin{bmatrix} 0 & 1 \\ 0 & 0 \end{bmatrix}.$$

14. *Hint:* If the entries of the $m \times m$ identity matrix are denoted δ_{ik}, then $\delta_{ik} = \begin{cases} 0 & \text{if } i \neq j \\ 1 & \text{if } i = j \end{cases}$. The ijth entry of **IA** is $\displaystyle\sum_{k=1}^{m} \delta_{ik} A_{kj}$.

15. *Proof:* Suppose **A** is an $m \times m$ symmetric matrix. Then for all integers i and j with $1 \leq i, j \leq m$,

$$(A^2)_{ij} = \sum_{k=1}^{m} A_{ik} A_{kj} \quad \text{and} \quad (A^2)_{ji} = \sum_{k=1}^{m} A_{jk} A_{ki}.$$

But since **A** is symmetric, $A_{ik} = A_{ki}$ and $A_{kj} = A_{jk}$ for all i, j, and k, and thus $A_{ik} A_{kj} = A_{jk} A_{ki}$ *[by the commutative law for multiplication of real numbers].* Hence $(A^2)_{ij} = (A^2)_{ji}$ for all integers i and j with $1 \leq i, j \leq m$.

17. *Proof (by mathematical induction):*
The formula is true for $n = 1$: We must show that $A^1 A = AA^1$. But this is true because $A^1 = A$ and $AA = AA$.
If the formula is true for $n = k$ then it is true for $n = k + 1$: Suppose that for some integer $k \geq 1$, $A^k A = AA^k$. We must show that $A^{k+1} A = AA^{k+1}$. But

$$A^{k+1} A = (AA^k)A \quad \text{by definition of matrix power}$$

$$= A(A^k A) \quad \text{by exercise 16}$$

$$= A(AA^k) \quad \text{by inductive hypothesis}$$

$$= AA^{k+1} \quad \text{by definition of matrix power.}$$

[This is what was to be shown.]
[Since the basis step and the inductive step have been proved true, we conclude that the given matrix formula holds for all integers $n \geq 1$.]

19. a. $A^2 = \begin{bmatrix} 1 & 1 & 2 \\ 1 & 0 & 1 \\ 2 & 1 & 0 \end{bmatrix} \begin{bmatrix} 1 & 1 & 2 \\ 1 & 0 & 1 \\ 2 & 1 & 0 \end{bmatrix} = \begin{bmatrix} 6 & 3 & 3 \\ 3 & 2 & 2 \\ 3 & 2 & 5 \end{bmatrix}$

$$A^3 = \begin{bmatrix} 1 & 1 & 2 \\ 1 & 0 & 1 \\ 2 & 1 & 0 \end{bmatrix} \begin{bmatrix} 6 & 3 & 3 \\ 3 & 2 & 2 \\ 3 & 2 & 5 \end{bmatrix} = \begin{bmatrix} 15 & 9 & 15 \\ 9 & 5 & 8 \\ 15 & 8 & 8 \end{bmatrix}$$

20. a. 2 since $(A^2)_{23} = 2$
b. 3 since $(A^2)_{34} = 3$
c. 6 since $(A^3)_{14} = 6$
d. 17 since $(A^3)_{23} = 17$

22. b. *Hint:* If G is bipartite, then its vertices can be partitioned into two sets V_1 and V_2 so that no vertices in V_1 are connected to each other by an edge and no vertices in V_2 are connected to each other by an edge. Label the vertices in V_1 as $v_1, v_2, \ldots, v_k$ and label the vertices in V_2 as $v_{k+1}, v_{k+2}, \ldots, v_n$. Now look at the matrix of G formed according to the given vertex labeling.

23. b. *Hint:* Consider the ijth entry of $A + A^2 + A^3 + \cdots + A^n$. If G is connected, then given the vertices v_i and v_j, there is a walk connecting v_i and v_j. If this walk has length k, then by Theorem 11.3.2, the ijth entry of A^k is not equal to 0. Use the facts that all entries of each power of A are nonnegative and a sum of nonnegative numbers is positive provided that at least one of the numbers is positive.

SECTION 11.4

1. The graphs are isomorphic. One way to define the isomorphism is as follows.

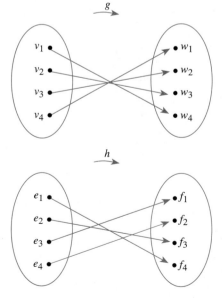

2. The graphs are not isomorphic. G has five vertices and G' has six.

6. The graphs are isomorphic. One isomorphism is the following.

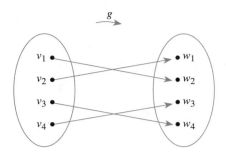

8. The graphs are not isomorphic. G has a simple circuit of length 3; G' does not.

10. The graphs are isomorphic. One way to define the isomorphism is as follows.

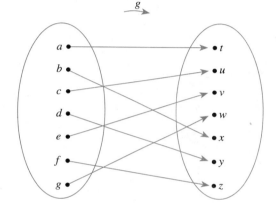

12. These graphs are isomorphic. One isomorphism is the following.

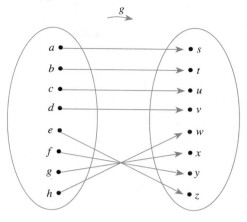

14.

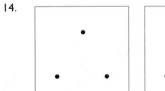 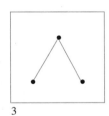 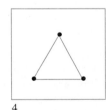 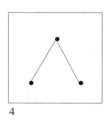

16.

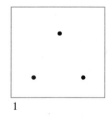 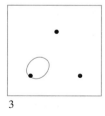 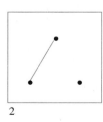

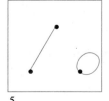 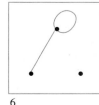 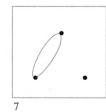 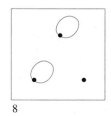 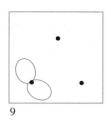

18. *Hint:* There are 19.

19.

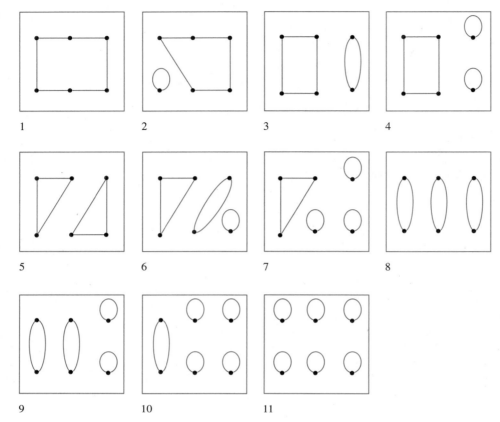

1   2   3   4

5   6   7   8

9   10   11

21. *Proof:* Suppose G and G' are isomorphic graphs and G has n vertices where n is a nonnegative integer. *[We must show that G' has n vertices.]* By definition of graph isomorphism, there is a one-to-one correspondence $g: V(G) \rightarrow V(G')$ sending vertices of G to vertices of G'. Since $V(G)$ is a finite set and g is a one-to-one correspondence, the number of vertices in $V(G')$ equals the number of vertices in $V(G)$. Hence, G' has n vertices *[as was to be shown].*

23. *Proof:* Suppose G and G' are isomorphic graphs and suppose G has a circuit C of length k, where k is a nonnegative integer. Let C be $v_0 e_1 v_1 e_2 \ldots e_k v_k (= v_0)$. By definition of graph isomorphism, there are one-to-one correspondences $g: V(G) \rightarrow V(G')$ and $h: E(G) \rightarrow E(G')$ that preserve the edge-endpoint functions in the sense that for all v in $V(G)$ and e in $E(G)$, v is an endpoint of $e \Leftrightarrow g(v)$ is an endpoint of $h(e)$. Let C' be $g(v_0)h(e_1)g(v_1)h(e_2) \ldots$. $h(e_k)g(v_k)(= g(v_0))$. Then C' is a circuit of length k in G'. The reason is that (1) because g and h preserve the

edge-endpoint functions, for all $i = 0, 1, \ldots, k - 1$ both $g(v_i)$ and $g(v_{i+1})$ are incident on $h(e_{i+1})$ so that C' is a walk from $g(v_0)$ to $g(v_0)$, and (2) since C is a circuit then $e_1, e_2, \ldots, e_k$ are distinct, and since h is a one-to-one correspondence, $h(e_1), h(e_2), \ldots, h(e_k)$ are also distinct, which implies that C' has k distinct edges. Therefore, G' has a circuit C of length k.

25. *Hint:* Suppose G and G' are isomorphic and G has m vertices of degree k; call them $v_1, v_2, \ldots, v_m$. Since G and G' are isomorphic, there are one-to-one correspondences $g: V(G) \rightarrow V(G')$ and $h: E(G) \rightarrow E(G')$. Show that $g(v_1), g(v_2), \ldots, g(v_m)$ are m distinct vertices of G' each of which has degree k.

27. *Hint:* Suppose G and G' are isomorphic and G is connected. To show G' is connected, suppose w and x are any two vertices of G'. Show that there is a walk connecting w with x by finding a walk connecting the corresponding vertices in G.

SECTION 11.5

1. a. Math 110

2. a.

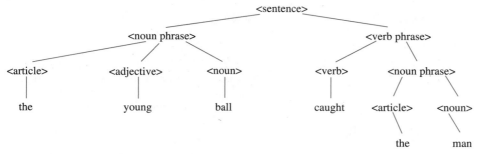

3. *Hint:* The answer is $2n - 2$. To obtain this result, use the relationship between the total degree of a graph and the number of edges of the graph.

4. a.
$$
\begin{array}{ccc}
\text{H} & \text{H} & \text{H} \\
| & | & | \\
\text{H}-\text{C}-\text{C}-\text{C}-\text{H} \\
| & | & | \\
\text{H} & \text{H} & \text{H}
\end{array}
$$

d. *Hint:* Each carbon atom in G is bonded to four other atoms in G because otherwise an additional hydrogen atom could be bonded to it and this would contradict the assumption that G has the maximum number of hydrogen atoms for its number of carbon atoms. Also each hydrogen atom is bonded to exactly one (carbon) atom in G because otherwise G would not be connected.

5. *Hint:* Revise the algorithm given in the proof of Lemma 11.5.1 to keep track of which vertex and edge were chosen in step 1 (by, say, labeling them v_0 and e_0). Then after one vertex of degree 1 is found, return to v_0 and search for another vertex of degree 1 by moving along a path outward from v_0 starting with e_0.

7. a. internal vertices: v_2, v_3, v_4, v_6
   terminal vertices: v_1, v_5, v_7

8. Any tree with nine vertices has eight edges, not nine. Thus there is no tree with nine vertices and nine edges.

9. One such graph is the following.

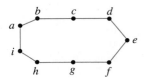

10. One such graph is the following.

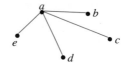

11. There is no tree with six vertices and a total degree of 14. Any tree with six vertices has five edges and hence (by Theorem 11.1.1) a total degree of 10, not 14.

12. One such tree is shown.

13. No such graph exists. By Theorem 11.5.4, a connected graph with six vertices and five edges is a tree. Hence such a graph cannot have a nontrivial circuit.

14.

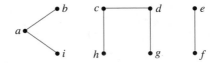

22. Yes. Since it is connected and has 12 vertices and 11 edges, by Theorem 11.5.4, it is a tree. It follows from Lemma 11.5.1 that it has a vertex of degree 1.

25. Suppose there were a connected graph with eight vertices and six edges. Either the graph itself would be a tree or edges could be eliminated from its circuits to obtain a tree. In either case, there would be a tree with eight vertices and six or fewer edges. But by Theorem 11.5.2, a tree with eight vertices has seven edges, not six or fewer. This contradiction shows that the supposition is false. So there is no connected graph with eight vertices and six edges.

26. *Hint:* See the answer to exercise 25.

27. Yes. Suppose G is a circuit-free graph with ten vertices and nine edges. Let $G_1, G_2, \ldots, G_k$ be the connected components of G. *[To show that G is connected, we will show that $k = 1$.]* Each G_i is a tree since each G_i is connected and circuit-free. For each $i = 1, 2, \ldots, k$, let G_i have n_i vertices. Note that since G has ten vertices in all,

$$n_1 + n_2 + \cdots + n_k = 10.$$

By Theorem 11.5.2,

$$G_1 \text{ has } n_1 - 1 \text{ edges,}$$
$$G_2 \text{ has } n_2 - 1 \text{ edges,}$$
$$\vdots$$
$$G_k \text{ has } n_k - 1 \text{ edges.}$$

So the number of edges of G equals

$$(n_1 - 1) + (n_2 - 1) + \cdots + (n_k - 1)$$
$$= (n_1 + n_2 + \cdots + n_k) - \underbrace{(1 + 1 + \cdots + 1)}_{k \text{ 1's}}$$
$$= 10 - k.$$

But we are given that G has nine edges. Hence $10 - k = 9$, and so $k = 1$. Thus G has just one connected component, G_1, and thus G is connected.

28. *Hint:* See the answer to exercise 27.

31. b. *Hint:* There are six.

32. a. 3   **b.** 0   **c.** 5   **d.** u, v
   e. d **f.** k, l **g.** m, s, t, x, y

34. a.

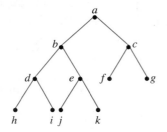

Exercises 35 and 39–41 have other answers in addition to the ones shown.

35.

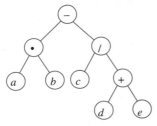

36. There is no full binary tree with the given properties because any full binary tree with five internal vertices has six terminal vertices, not seven.

37. Any full binary tree with four internal vertices has five terminal vertices for a total of nine, not seven, vertices in all. Thus there is no full binary tree with the given properties.

38. There is no full binary tree with 12 vertices because any full binary tree has $2k + 1$ vertices, where k is the number of internal vertices. But $2k + 1$ is always odd and 12 is even.

39.

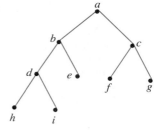

40.

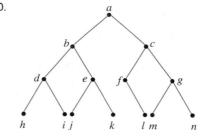

41.

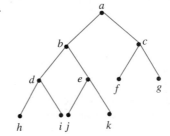

42. There is no binary tree that has height 3 and nine terminal vertices because any binary tree of height 3 has at most $2^3 = 8$ terminal vertices.

51. a. height of tree $\geq \log_2 25 \cong 4.6$. Since the height of any tree is an integer, the height must be at least 5.

SECTION 11.6

1.

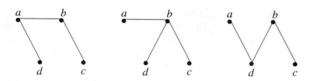

3. One of many spanning trees is as follows.

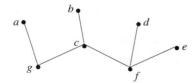

5. Minimal spanning tree:

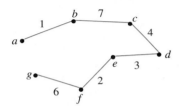

Order of adding the edges:
$\{a, b\}, \{e, f\}, \{e, d\}, \{d, c\}, \{g, f\}, \{b, c\}$

7. Minimal spanning tree: same as in exercise 5
Order of adding the edges: $\{a, b\}, \{b, c\}, \{c, d\}, \{d, e\},$
$\{e, f\}, \{f, g\}$

9. There are four minimal spanning trees:

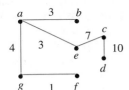 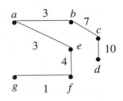 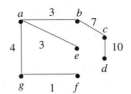 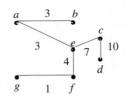

When Prim's algorithm is used, edges are added in any
of the orders obtained by following one of the eight
paths from left to right across the diagram below.

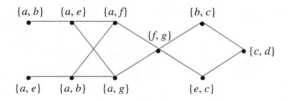

When Kruskal's algorithm is used, edges are added in
any of the orders obtained by following one of the eight
paths from left to right across the diagram below.

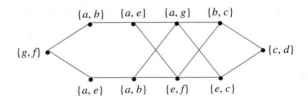

13. b. *Proof:* Suppose not. Suppose that for some tree T, u and v are distinct vertices of T, and P_1 and P_2 are two distinct paths joining u and v. *[We must deduce a contradiction. In fact, we will show that T contains a circuit.]* Let P_1 be denoted $u=v_0, v_1, v_2, \ldots, v_m=v$, and let P_2 be denoted $u=w_0, w_1, w_2, \ldots, w_n=v$. Because P_1 and P_2 are distinct, and T has no parallel edges, the sequence of vertices in P_1 must diverge from the sequence of vertices in P_2 at some point. Let i be the least integer such that $v_i \neq w_i$. Then $v_{i-1} = w_{i-1}$. Let j and k be the least integers greater than i so that $v_j = w_k$. (There must be such integers because $v_m = w_n$.) Then

$$v_{i-1}v_iv_{i+1} \ldots v_j(=w_k) \, w_{k-1} \ldots w_iw_{i-1}(=v_{i-1})$$

is a circuit in T. The existence of such a circuit contradicts the fact that T is a tree. Hence the supposition must be false. That is, given any tree with vertices u and v, there is a unique path joining u and w.

15. *Proof:* Suppose G is a connected graph, T is a circuit-free subgraph of G, and if any edge e of G not in T is added to T, the resulting graph contains a circuit. Suppose that T is not a spanning tree for G. *[We must derive a contradiction.]*

Case 1 (T is not connected): In this case, there are vertices u and v in T such that there is no walk in T from u to v. Now since G is connected, there is a walk in G from u to v, and hence by Lemma 11.2.1, there is a simple path in G from u to v. Let $e_1, e_2, \ldots, e_k$ be the edges of this path that are not in T. When these edges are added to T, the result is a graph T' in which u and v are connected by a path. In addition, by hypothesis, each of the edges e_i creates a circuit when added to T. Now remove these edges one by one from T'. By the same argument used in the proof of Lemma 11.5.3, each such removal leaves u and v connected since each e_i is an edge of a circuit when added to T. Hence, after all the e_i have been removed, u and v remain connected. But this contradicts the fact that there is no walk in T from u to v.

Case 2 (T is connected): In this case, since T is not a spanning tree and T is circuit-free, there is a vertex v in G such that v is not in T. *[For if T were connected, circuit-free, and contained every vertex in G, then T would be a spanning tree for G.]* Since G is connected, v is not isolated. Thus there is an edge e in G with v as an endpoint. Let T' be the graph obtained from T by adding e and v. *[Note that e is not already in T because if it were, its endpoint v would also be in T and it is not.]* Then

T' contains a circuit because, by hypothesis, addition of any edge to T creates a circuit. Also T' is connected because T is and because when e is added to T, e becomes part of a circuit in T'. Now deletion of an edge from a circuit does not disconnect a graph. So if e is deleted from T', the result is a connected graph. But the resulting graph contains v, which means that there is an edge in T connecting v to another vertex of T. This implies that v is in T *[because both endpoints of any edge in a graph must be part of the vertex set of the graph]*, which contradicts the fact that v is not in T.

Thus in either case, the supposition that T is not a spanning tree leads to a contradiction. Hence the supposition is false, and T is a spanning tree for G.

16. a. *Counterexample:* Let G be the following graph.

Then G has the spanning trees shown below.

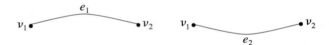

These trees have no edge in common.

17. *Hint:* Suppose e is contained in every spanning tree of G and the graph obtained by removing e from G is connected. Let G' be the subgraph of G obtained by removing e and let T' be a spanning tree for G'. How is T' related to G?

19. *Proof:* Suppose that $w(e') > w(e)$. Form a new graph T' by adding e to T and deleting e'. By exercise 15, addition of an edge to a spanning tree creates a circuit, and by Lemma 11.5.3 deletion of an edge from a circuit does not disconnect a graph. Consequently, T' is also a spanning tree for G. Furthermore, $w(T') < w(T)$ because $w(T') = w(T) - w(e') + w(e) = w(T) - (w(e') - w(e)) < w(T)$ *[since $w(e') > w(e)$, which implies that $w(e') - w(e) > 0$]*. But this contradicts the fact that T is a minimal spanning tree for G. Hence the supposition is false, and so $w(e') \leq w(e)$.

20. *Hint:* Suppose e is an edge that has smaller weight than any other edge of G and suppose T is a minimal spanning tree for G that does not contain e. Create a new spanning tree T' by adding e to T and removing another edge of T (which one?). Then $w(T') < w(T)$.

21. Yes. *Proof by contradiction:* Suppose G is a weighted graph in which all the weights of all the edges are distinct and suppose G has two distinct minimal span-

ning trees T_1 and T_2. Let e be the edge of least weight that is in one of the trees but not the other. Without loss of generality, we may say that e is in T_1. Add e to T_2 to obtain a graph G'. By exercise 14, G' contains a non-trivial circuit. At least one other edge f of this circuit is not in T_1 because otherwise T_1 would contain the complete circuit, which would contradict the fact that T_1 is a tree. Now f has weight greater than e because all edges have distinct weights, f is in T_2 and not in T_1, and e is the edge of least weight that is in one of the trees and not the other. Remove f from G' to obtain a tree T_3.

Then $w(T_3) < w(T_2)$ because T_3 is the same as T_2 except that it contains e rather than f and $w(e) < w(f)$. Consequently, T_3 is a spanning tree for G of smaller weight than T_2. This contradicts the supposition that T_2 is a minimal spanning tree for G. Thus G cannot have more than one minimal spanning tree.

23. The output will be a "spanning forest" for the graph. It will contain one spanning tree for each connected component of the input graph.

Photo Credits

The following photos were supplied courtesy of the rights holders listed:

Index

Subject	Symbol	Meaning	Page
FORMAL LANGUAGES AND FINITE-STATE AUTOMATA	Σ	an alphabet of a language	239
	ε	the null string	239
	Σ^n	the set of all strings over Σ of length n	240
	Σ^*	the set of all strings over Σ of finite length	240
	$N(s, m)$	the value of the next-state function for a state s and input symbol m	360
	$\rightarrow \enclose{circle}{s_0}$	initial state	360
	$\enclose{circle}{s_a}$	accepting state	360
	$L(A)$	language accepted by A	362
	$N^*(s, w)$	the value of the eventual-state function for a state s and input string w	363
	$s\,R_*\,t$	s and t are *-equivalent	574
	$\overline{A}$	the quotient automaton of A	578
MATRICES	$\mathbf{A}$	matrix	640
	$\mathbf{I}$	identity matrix	650
	$\mathbf{A} + \mathbf{B}$	sum of matrices $\mathbf{A}$ and $\mathbf{B}$	656
	$\mathbf{AB}$	product of matrices $\mathbf{A}$ and $\mathbf{B}$	647
	$\mathbf{A}^n$	matrix $\mathbf{A}$ to the power n	651
GRAPHS AND TREES	$V(G)$	the set of vertices of a graph G	603
	$E(G)$	the set of edges of a graph G	603
	$\{v, w\}$	the edge joining v and w in a simple graph	609
	K_n	complete graph on n vertices	610
	$K_{m,n}$	complete ⬛⬛⬛ (m, n) vertices	610
	$deg(v)$	degree ⬛	

Continued on next page.

LIST OF SYMBOLS

Subject	Symbol	Meaning	Page
GRAPHS AND TREES	$v_0 e_1 v_1 e_2 \ldots e_n v_n$	A walk from v_0 to v_n	621
	$w(e)$	the weight of edge e	685
	$w(G)$	the total weight of graph G	685

REFERENCE FORMULAS

Topic	Name	Formula	Page
LOGIC	De Morgan's law	$\sim(p \wedge q) \equiv \sim p \vee \sim q$	10
	De Morgan's law	$\sim(p \vee q) \equiv \sim p \wedge \sim q$	10
	Negation of $\rightarrow$	$\sim(p \rightarrow q) \equiv p \wedge \sim q$	20
	Equivalence of a conditional and its contrapositive	$p \rightarrow q \equiv \sim q \rightarrow \sim p$	21
	Nonequivalence of a conditional and its converse	$p \rightarrow q \not\equiv q \rightarrow p$	22
	Nonequivalence of a conditional and its inverse	$p \rightarrow q \not\equiv \sim p \rightarrow \sim q$	22
	Negation of a universal statement	$\sim(\forall x \text{ in } D, Q(x)) \equiv \exists x \text{ in } D \text{ such that } \sim Q(x)$	83
	Negation of an existential statement	$\sim(\exists x \text{ in } D \text{ such that } Q(x)) \equiv \forall x \text{ in } D, \sim Q(x)$	84
SUMS	Sum of the first n integers	$1 + 2 + \cdots + n = \dfrac{n(n+1)}{2}$	197
		$1 + r + r^2 + \cdots + r^n = \dfrac{r^{n+1} - 1}{r - 1}$	201